Y0-EHE-068

INTRODUCTION TO MECHATRONIC DESIGN

J. EDWARD CARRYER
R. MATTHEW OHLINE
THOMAS W. KENNY

Mechanical Engineering
Stanford University

Prentice Hall

Boston Columbus Indianapolis New York San Francisco Upper Saddle River
Amsterdam Cape Town Dubai London Madrid Milan Munich Paris Montreal Toronto
Delhi Mexico City Sao Paulo Sydney Hong Kong Seoul Singapore Taipei Tokyo

Vice President and Editorial Director, ECS: *Marcia J. Horton*
Senior Editor: *Tacy Quinn*
Acquisitions Editor: *Norrin Dias*
Editorial Assistant: *Coleen McDonald*
Vice President, Production: *Vince O'Brien*
Senior Managing Editor: *Scott Disanno*
Production Liaison: *Jane Bonnell*
Production Editor: *Pavithra Jayapaul, TexTech International*
Senior Operations Supervisor: *Alan Fischer*
Operations Specialist: *Lisa McDowell*
Executive Marketing Manager: *Tim Galligan*
Marketing Assistant: *Mack Patterson*
Senior Art Director and Cover Designer: *Kenny Beck*
Cover Image *Somatuscan/Shutterstock*
Art Editor: *Greg Dulles*
Media Editor: *Daniel Sandin*
Composition/Full-Service Project Management: *TexTech International*

Library of Congress Cataloging-in-Publication Data
Carryer, J. Edward.
Introduction to mechatronic design / J. Edward Carryer, R. Matthew Ohline, Thomas W. Kenny.
p. cm.
Includes bibliographical references and index.
ISBN-13: 978-0-13-143356-4 (alk. paper)
ISBN-10: 0-13-143356-3
1. Mechatronics. I. Ohline, R. Matthew. II. Kenny, Thomas William. III. Title.
TJ163.12.C37 2010
621—dc22 2010033713

Prentice Hall
is an imprint of

10

www.pearsonhighered.com

ISBN-13: 978-0-13-143356-4
ISBN-10: 0-13-143356-3

To Sheri Sheppard, for always being there with support and encouragement through the preparation of this book and everything else in life, and to Portia Carryer, for the frequent reminders of IALAC and showing me the joy in writing.

To my parents, John and Lottie Jean Carryer, for instilling in me the belief that I could do this.

J. E. C.

This book, like everything else, is for my lovely and talented wife Jill, my son Aidan, and my daughter Erin.

R. M. O.

To my first great teacher, Bruce Bauer of Stillwater High School in Minnesota.

To all my friends and family who supported this and many other extracurricular adventures.

To my Dad, who taught me everything nontechnical I've ever needed to know.

T. W. K.

Contents

Preface xix
Trademark Acknowledgments xxiii
About the Authors xxv

PART 1 **INTRODUCTION** 1

CHAPTER 1 Introduction 1

1.1 Philosophy 3
1.2 The Organization of This Book 3
1.3 Who Should Study Mechatronics? 4
1.4 How to Use This Book 4
1.5 Summary 5

PART 2 **SOFTWARE** 7

CHAPTER 2 What's a Micro? 7

2.1 Introduction 7
2.2 What's a Micro? 7
2.3 Microprocessors, Microcontrollers, Digital Signal Processors, and More 8
2.4 Microcontroller Architecture 10
2.5 The Central Processing Unit 11
2.5.1 Representing Numbers in the Digital Domain 11
2.5.2 The Arithmetic Logic Unit 12
2.6 The Data Bus and the Address Bus 13
2.7 Memory 14
2.8 Subsystems and Peripherals 15
2.9 Von Neumann Architecture 17
2.10 The Harvard Architecture 19
2.11 Real World Examples 21
2.11.1 The Freescale MC9S12C32 Microcontroller 21
2.11.2 The Microchip PIC12F609 Microcontroller 24
2.12 Where to Find More Information 26
2.13 Homework Problems 27

CHAPTER 3 Microcontroller Math and Number Manipulation 29

3.1 Introduction 29
3.2 Number Bases and Counting 29
3.3 Representing Negative Numbers 33
3.4 Data Types 35
3.5 Sizes of Common Data Types 36
3.6 Arithmetic on Fixed Size Variables 36
3.7 Modulo Arithmetic 37
3.8 Math Shortcuts 39
3.9 Boolean Algebra 40
3.10 Manipulating Individual Bits 40
3.11 Testing Individual Bits 42
3.12 Homework Problems 43

CHAPTER 4 Programming Languages 45

4.1 Introduction 45
4.2 Machine Language 46
4.3 Assembly Language 47
4.4 High-Level Languages 48
4.5 Interpreters 48
4.6 Compilers 50
4.7 Hybrid Compiler/Interpreters 51
4.8 Integrated Development Environments (IDEs) 53
4.9 Choosing a Programming Language 54
4.10 Homework Problems 55

CHAPTER 5 Program Structures for Embedded Systems 57

5.1 Background 57
5.2 Event Driven Programming 58
5.3 Event Checkers 58
5.4 Services 62
5.5 Building an Event Driven Program 62
5.6 An Example 64
5.7 Summary of Event Driven Programming 65
5.8 State Machines 65
5.9 A State Machine in Software 67
5.10 The Cockroach Example as a State Machine 69
5.11 Homework Problems 71

CHAPTER 6 Software Design 73

6.1 Introduction 73
6.2 Building as a Metaphor for Creating Software 73
6.3 Introducing Some Software Design Techniques 75
6.3.1 Decomposition 75
6.3.2 Abstraction and Information Hiding 76
6.3.3 Pseudo-Code 77
6.4 Software Design Process 78
6.4.1 Generating Requirements 78
6.4.2 Defining the Program Architecture 79
6.4.3 The Performance Specification 80
6.4.4 The Interface Specification 80
6.4.5 Detail Design 81
6.4.6 Implementation 82
6.4.7 Unit Testing 84
6.4.8 Integration 84
6.5 The Sample Problem 84
6.5.1 Requirements for the Morse Code Receiver 85
6.5.2 The Morse Code Receiver System Architecture 85
6.5.3 The Morse Code Receiver Software Architecture 86
6.5.4 The Morse Code Receiver Performance Specifications 89
6.5.5 The Morse Code Receiver Interface Specification 89
6.5.6 The Morse Code Receiver Detail Design 92
6.5.7 The Morse Code Receiver Implementation 102
6.5.8 The Morse Code Receiver Unit Testing 102
6.5.9 The Morse Code Receiver Integration 103
6.6 Homework Problems 104

CHAPTER 7 **Inter-Processor Communications** 105

7.1 Introduction 105
7.2 Without a Medium, There Is No Message 106
7.3 Bit-Parallel and Bit-Serial Communications 107
7.3.1 Bit-Serial Communications 107
7.3.2 Bit-Parallel Communications 117
7.4 Signaling Levels 118
7.4.1 TTL/CMOS Levels 118
7.4.2 RS-232 118
7.4.3 RS-485 119
7.5 Communicating over Limited Bandwidth Channels 120
7.5.1 Limited Bandwidth and Modems 120
7.6 Communicating with Light 123
7.7 Communicating over a Radio 126
7.7.1 RF Remote Controls 126
7.7.2 RF Data Links 127
7.7.3 RF Networks 127
7.8 Homework Problems 127

CHAPTER 8 **Microcontroller Peripherals** 129

8.1 Accessing the Control Registers 129
8.2 The Parallel Input/Output Subsystem 130
8.2.1 The Data Direction Register 130
8.2.2 The Input/Output Register(s) 130
8.2.3 Shared Function Pins 131
8.3 Timer Subsystems 132
8.3.1 Timer Basics 133
8.3.2 Timer Overflow 134
8.3.3 Output Compare 135
8.3.4 Input Capture 136
8.3.5 Combining Input Capture and Output Compare to Control an Engine 137
8.4 Pulse Width Modulation 138
8.5 PWM Using the Output Compare System 140
8.6 The Analog-to-Digital Converter Subsystem 141
8.6.1 The Process for Converting an Analog Input to a Digital Value 142
8.6.2 The A/D Converter Clock 142
8.6.3 Automating the A/D Conversion Process 143
8.7 Interrupts 143
8.8 Homework Problems 144

PART 3 **ELECTRONICS** 145

CHAPTER 9 **Basic Circuit Analysis and Passive Components** 145

9.1 Voltage, Current, and Power 145
9.2 Circuits and Ground 147
9.3 Laying Down the Laws 149
9.4 Resistance 150
9.4.1 Resistors in Series and Parallel 151
9.4.2 The Voltage Divider 153

9.5 Thevenin Equivalents 154
9.6 Capacitors 155
9.6.1 Capacitors in Series and Parallel 157
9.6.2 Capacitors and Time-Varying Signals 158
9.7 Inductors 159
9.7.1 Inductors and Time-Varying Signals 160
9.8 The Time and Frequency Domains 161
9.9 Circuit Analysis with Multiple Component Types 162
9.9.1 Basic RC Circuit Configurations 162
9.9.2 Low-Pass RC Filter Behavior in the Time Domain 163
9.9.3 High-Pass RC Filter Behavior in the Time Domain 166
9.9.4 RL Circuit Behavior in the Time Domain 167
9.9.5 Low-Pass RC Filter Behavior in the Frequency Domain 169
9.9.6 High-Pass RC Filter Behavior in the Frequency Domain 172
9.9.7 High-Pass RC Filter with a DC Bias 173
9.10 Simulation Tools 174
9.10.1 Limitations of Simulation Tools 174
9.11 Real Voltage Sources 175
9.12 Real Measurements 176
9.12.1 Measuring Voltage 176
9.12.2 Measuring Current 177
9.13 Real Resistors 178
9.13.1 A Model for a Real Resistor 178
9.13.2 Resistor Construction Basics 178
9.13.3 Carbon Film Resistors 179
9.13.4 Metal Film Resistors 180
9.13.5 Power Dissipation in Resistors 181
9.13.6 Potentiometers 183
9.13.7 Choosing Resistors 184
9.14 Real Capacitors 184
9.14.1 A Model for a Real Capacitor 185
9.14.2 Capacitor Construction Basics 185
9.14.3 Polar vs. Nonpolar Capacitors 186
9.14.4 Ceramic Disk Capacitors 187
9.14.5 Monolithic Ceramic Capacitors 188
9.14.6 Aluminum Electrolytic Capacitors 188
9.14.7 Tantalum Capacitors 189
9.14.8 Film Capacitors 190
9.14.9 Electric Double Layer Capacitors/Super Capacitors 190
9.14.10 Capacitor Labeling 191
9.14.11 Choosing a Capacitor 194
9.15 Homework Problems 195

CHAPTER 10 Semiconductors **198**

10.1 Doping, Holes, and Electrons 199
10.2 Diodes 199
10.2.1 The *V-I* Characteristic for Diodes 200
10.2.2 The Magnitude of V_f 201
10.2.3 Reverse Recovery 201
10.2.4 Schottky Diodes 201
10.2.5 Zener Diodes 202
10.2.6 Light Emitting Diodes 204
10.2.7 Photo-Diodes 205

10.3 Bipolar Junction Transistors 205
10.3.1 The Darlington Pair 210
10.3.2 The Photo-Transistor 211
10.4 MOSFETs 212
10.5 Choosing between BJTs and MOSFETs 216
10.5.1 When Will a BJT Be the Best (or Only) Choice? 216
10.5.2 When Will a MOSFET Be the Best (or Only) Choice? 217
10.5.3 How Do You Choose When Either a MOSFET or a BJT Could Work? 217
10.6 Multitransistor Circuits 217
10.7 Reading Transistor Data Sheets 219
10.7.1 Reading a BJT Data Sheet 219
10.7.2 Reading a MOSFET Data Sheet 221
10.7.3 A Sample Application 223
10.7.4 A Potpourri of Transistor Circuits 224
10.8 Homework Problems 225

CHAPTER 11 Operational Amplifiers 231

11.1 Op-Amp Behavior 231
11.2 Negative Feedback 232
11.3 The Ideal Op-Amp 232
11.4 Analyzing Op-Amp Circuits 233
11.4.1 The Golden Rules 233
11.4.2 The Noninverting Op-Amp Configuration 233
11.4.3 The Inverting Op-Amp Configuration 235
11.4.4 The Unity Gain Buffer 238
11.4.5 The Difference Amplifier Configuration 240
11.4.6 The Summer Configuration 241
11.4.7 The Trans-Resistive Configuration 242
11.4.8 Computation with Op-Amps 243
11.5 The Comparator 244
11.5.1 Comparator Circuits 246
11.6 Homework Problems 249

CHAPTER 12 Real Operational Amplifiers and Comparators 251

12.1 Real Op-Amp Characteristics—How the Ideal Assumptions Fail 251
12.1.1 Noninfinite Gain 251
12.1.2 Variation in Open-Loop Gain with Frequency 252
12.1.3 Input Current Is Not Zero 253
12.1.4 The Output Voltage Source Is Not Ideal 255
12.1.5 Other Nonidealities 256
12.2 Reading an Op-Amp Data Sheet 258
12.2.1 Maxima, Minima, and Typical Values 259
12.2.2 The Front Page 259
12.2.3 The Absolute Maximum Ratings Section 259
12.2.4 The Electrical Characteristics Section 259
12.2.5 The Packaging Section 263
12.2.6 The Typical Applications Section 264
12.3 Reading a Comparator Data Sheet 264
12.3.1 Comparator Packaging 265
12.4 Comparing Op-Amps 265
12.5 Homework Problems 268

CHAPTER 13 Sensors 270

13.1 Introduction 270
13.2 Sensor Output and Microcontroller Inputs 270
13.3 Sensor Design 271
13.3.1 Measuring Temperature with a Thermistor 271
13.3.2 Measuring Acceleration 272
13.3.3 Definitions of Sensor Performance Characteristics 273
13.4 Fundamental Sensors and Interface Circuits 281
13.4.1 Switches as Sensors 281
13.4.2 Interfacing to Switches 282
13.4.3 Resistive Sensors 284
13.4.4 Interfacing to Resistive Sensors 286
13.4.5 Capacitive Sensors 290
13.4.6 Interfacing to Capacitive Sensors 290
13.5 A Survey of Sensors 292
13.5.1 Light Sensors 293
13.5.2 Strain Sensors 302
13.5.3 Temperature Sensors 306
13.5.4 Magnetic Field Sensors 310
13.5.5 Proximity Sensors 314
13.5.6 Position Sensors 315
13.5.7 Acceleration Sensors 323
13.5.8 Force Sensors 326
13.5.9 Pressure Sensors 328
13.6 Homework Problems 329

CHAPTER 14 Signal Conditioning 333

14.1 Basic Operations in Signal Conditioning 333
14.2 Offset Removal 334
14.2.1 Amplification Relative to an Offset 334
14.2.2 Offset Removal by AC Coupling 336
14.3 Amplification 337
14.3.1 Multistage Amplification with DC Coupling 338
14.3.2 Multistage Amplification with AC Coupling 339
14.4 Filtering 339
14.4.1 Filter Terminology 339
14.4.2 What Is Noise and Where Does It Come From? 341
14.4.3 Passive Filters 342
14.5 Other Signal Conditioning Techniques 343
14.5.1 The Instrumentation Amplifier 344
14.5.2 Peak Detection 346
14.6 Case Studies 347
14.6.1 Information in the Amplitude 347
14.6.2 Information in the Timing 348
14.7 Homework Problems 350

CHAPTER 15 Active and Digital Filters 352

15.1 Active Filters 352
15.1.1 Phase Delay 352
15.1.2 Filter Response Characteristics 353
15.1.3 Active Filter Topologies 355

15.2 Digital Techniques 360
15.2.1 Digital Filtering 361
15.2.2 Digital Signal Processing 363
15.2.3 Synchronous Sampling 364
15.3 Homework Problems 365

CHAPTER 16 Digital Inputs and Outputs 367

16.1 Introduction 367
16.2 Representing Logical States 367
16.3 Ideal Behavior for Digital Devices 368
16.4 Real Behavior of Digital Devices 368
16.5 Reading Device Data Sheets 369
16.6 Digital Inputs 370
16.6.1 Digital Input Voltage Requirements 370
16.6.2 Digital Input Current Requirements 373
16.6.3 Pull-Ups and Pull-Downs 374
16.6.4 Digital Input Timing Requirements 378
16.7 Digital Outputs 381
16.7.1 Digital Output Voltage and Current Specifications 381
16.7.2 Digital Output Timing Specifications 383
16.8 Output Meets Input 384
16.8.1 Evaluating Compatibility 384
16.8.2 Pull-Ups and Pull-Downs for Disconnectable and Indeterminate Inputs 386
16.8.3 Interconnecting Incompatible Devices 389
16.9 Homework Problems 394

CHAPTER 17 Digital Outputs and Power Drivers 398

17.1 Totem-Pole Outputs 398
17.1.1 Totem-Pole Output Specifications 398
17.2 Open-Collector/Open-Drain Outputs 401
17.2.1 Open-Collector/Open-Drain Output Specifications 402
17.3 Three-State Outputs 403
17.3.1 Three-State Output Specifications 405
17.4 Low Side Drivers 405
17.4.1 Low Side Driver Specifications 407
17.5 High Side Drivers 407
17.5.1 High Side Driver Specifications 408
17.6 Half-Bridges and Full-Bridges 409
17.6.1 Shoot-Through Currents and Dead Time 410
17.6.2 H-Bridge Specifications 411
17.7 Thermal Design Issues 411
17.8 Homework Problems 414

CHAPTER 18 Digital Logic and Integrated Circuits 416

18.1 Basic Combinatorial Logic 416
18.1.1 Truth Tables 417
18.1.2 Describing Microcontroller Subsystems Using Combinatorial Logic 418

18.2 A Survey of Useful Functions Implemented with Combinatorial Logic 419
18.2.1 The Digital Comparator 419
18.2.2 The Digital Multiplexer 419
18.2.3 The Decoder 420
18.3 Introduction to Sequential Logic 421
18.4 A Survey of Useful Functions Implemented with Sequential Logic 422
18.4.1 The D-Type Flip-Flop 422
18.4.2 The J-K Flip-Flop 422
18.4.3 The Counter 423
18.4.4 The Shift Register 425
18.5 Logic Families 426
18.6 Using Available Logic Chips to Expand the Capabilities of a Microcontroller 427
18.6.1 Using a Multiplexer to Expand Input Capabilities 427
18.6.2 Using a Decoder to Expand Output Capabilities 428
18.6.3 Using Shift Registers to Expand Input Capabilities 429
18.6.4 Using Shift Registers to Expand Output Capabilities 431
18.6.5 Using the SPI Subsystem with Shift Registers 432
18.7 The 555 Timer 432
18.7.1 Inside the 555 Timer 432
18.7.2 Astable Operation 432
18.7.3 Mono-Stable Operation 434
18.7.4 Other Uses of the 555 Timer 435
18.8 Homework Problems 436

CHAPTER 19 A-to-D and D-to-A Converters 438

19.1 Interfacing between Digital and Analog Domains 438
19.2 Digitizing Continuous Signals 439
19.3 A/D and D/A Converter Performance 442
19.3.1 Ideal A/D Converter Performance 444
19.3.2 Sources of Error for A/D Converters 446
19.3.3 Ideal D/A Converter Performance 448
19.3.4 Sources of Error for D/A Converters 449
19.4 D/A Converter Designs 450
19.4.1 Using Pulse Width Modulation to Generate Analog Voltages 450
19.4.2 Summing Amplifier D/A Converters 451
19.4.3 String D/A Converters 455
19.4.4 R-2R Ladder D/A Converters 456
19.5 A/D Converter Designs 458
19.5.1 Single Slope and Dual Slope A/D Converters 458
19.5.2 Flash A/D Converters 460
19.5.3 Half-Flash A/D Converters 462
19.5.4 Successive Approximation Register A/D Converters 463
19.5.5 Sigma-Delta A/D Converters 464
19.6 Homework Problems 466

CHAPTER 20 Voltage Regulators, Power Supplies, and Batteries 468

20.1 Introduction 468
20.2 Power Requirements and Power Sources 468
20.3 Voltage Regulators 469
20.3.1 Voltage Regulator Terms and Definitions 470
20.3.2 Linear Voltage Regulators 473
20.3.3 Switch-Mode (Switching) Voltage Regulators 481

20.4 Power Supplies 487
20.4.1 Linear Power Supplies 487
20.4.2 Switching Power Supplies 490
20.5 Batteries and Electrochemical Cells 493
20.5.1 Battery Performance and Characteristics 495
20.5.2 Primary Batteries 499
20.5.3 Secondary Batteries 500
20.5.4 Battery Safety and Environmental Issues 506
20.6 Homework Problems 507

CHAPTER 21 Noise, Grounding, and Isolation 510

21.1 Noise Coupling Channels 510
21.2 Conductively Coupled Noise 511
21.2.1 The Origins of the Conductive Coupling Channel 511
21.2.2 Reducing Conductively Coupled Noise 512
21.2.3 Reducing Noise at the Source: Decoupling 512
21.2.4 Reducing the Coupling of Conductive Noise 514
21.2.5 Reducing Conductive Noise at the Receptor: Power Supply Filtering 515
21.2.6 Best Practices for Reducing Conductive Noise 516
21.3 Capacitively Coupled Noise 516
21.3.1 The Origins of the Capacitive Coupling Channel 516
21.3.2 Reducing Capacitively Coupled Noise 517
21.3.3 Reducing Capacitively Coupled Noise at the Source 519
21.3.4 Reducing the Coupling of Capacitively Coupled Noise 520
21.3.5 Shielding 520
21.3.6 Reducing Capacitively Coupled Noise at the Receiver 521
21.3.7 Best Practices for Reducing Capacitively Coupled Noise 522
21.4 Inductively Coupled Noise 523
21.4.1 The Origins of the Inductive Coupling Channel 523
21.4.2 Reducing Inductively Coupled Noise at the Source 523
21.4.3 Reducing the Coupling of Inductively Coupled Noise 524
21.4.4 Reducing Inductively Coupled Noise at the Receptor 524
21.4.5 Best Practices for Reducing Inductively Coupled Noise 524
21.5 Isolation 524
21.5.1 Optical Isolation 525
21.5.2 Capacitive Isolation 527
21.5.3 Inductive Isolation 527
21.5.4 Comparing Isolation Technologies 528
21.6 Homework Problems 528

PART 4 **ACTUATORS** **531**

CHAPTER 22 Permanent Magnet Brushed DC Motor Characteristics 531

22.1 Introduction 531
22.2 Subfractional Horsepower Permanent Magnet Brushed DC Motors 531
22.3 Electrical Model 535
22.4 Back-EMF and the Generator Effect 536
22.5 Characteristic Constants for Permanent Magnet Brushed DC Motors 536
22.6 Characteristic Equations for Constant Voltage 538
22.7 Power Characteristics 541

22.8 DC Motor Efficiency 543
22.9 Gearheads 547
22.10 Homework Problems 549

CHAPTER 23 Permanent Magnet Brushed DC Motor Applications 552

23.1 Introduction 552
23.2 Inductive Kickback 552
23.2.1 Inductive Kickback Summary 560
23.3 Bidirectional Control of Motors 561
23.3.1 Commercially Available H-Bridge Integrated Circuits 564
23.3.2 H-Bridges for Higher Current Applications 566
23.4 Speed Control with Pulse Width Modulation 567
23.5 Homework Problems 572

CHAPTER 24 Solenoids 575

24.1 Introduction 575
24.2 Solenoid Construction 575
24.3 Solenoid Performance 577
24.4 Driving a Solenoid 580
24.5 Mechanical Response Time 582
24.6 Applications 582
24.7 Homework Problems 583

CHAPTER 25 Brushless DC Motors 588

25.1 Introduction 588
25.2 BLDC Motor Construction 588
25.3 BLDC Motor Operation 590
25.3.1 Sensored Commutation 591
25.3.2 Sensor-less Commutation 592
25.4 Driving a BLDC Motor 593
25.5 Commutating a BLDC Motor 594
25.6 BLDC Motor Driver Integrated Circuits 596
25.7 Comparing Brushed and Brushless DC Motors 597
25.8 Homework Problems 598

CHAPTER 26 Stepper Motors 600

26.1 Introduction 600
26.2 Stepper Motor Construction 600
26.3 The Variable Reluctance Stepper Motor 603
26.4 The Hybrid Stepping Motor 604
26.5 Comparing Stepping Motor Types 605
26.6 Stepper Motor Internal Wiring 606
26.7 Driving Stepper Motors 608
26.8 Stepping Sequences for Stepper Motors 609
26.9 Generating the Drive Sequences for Stepper Motors 614
26.10 Stepper Motor Dynamics 615
26.11 Stepper Motor Performance Specifications 617
26.12 Optimizing Stepper Motor Performance with Drive Electronics 620
26.13 The Role of Snubbing in Performance 622
26.14 Homework Problems 623

CHAPTER 27 **Other Actuator Technologies** 626

27.1 Introduction 626
27.2 Pneumatic and Hydraulic Systems 626
27.2.1 Solenoid Valves 627
27.2.2 Servo Valves 631
27.2.3 Pneumatic and Hydraulic Actuators 632
27.3 RC Servos 636
27.4 Piezo Actuators 637
27.4.1 Types of Piezo Actuators 639
27.5 Shape Memory Alloy Actuators 644
27.6 Summary 648
27.7 Homework Problems 648

CHAPTER 28 **Basic Closed-Loop Control** 651

28.1 Introduction 651
28.2 Terminology 651
28.3 Open-Loop Control 652
28.4 On-Off Closed-Loop Control 653
28.5 Linear Closed-Loop Control 654
28.5.1 Getting Started 655
28.5.2 Getting Smarter 657
28.5.3 Disturbance Rejection 659
28.5.4 Improving Performance Further by Adding Derivative Control 661
28.5.5 Choosing the Right Gains 663
28.6 System Type and the Need for Integral Control 666
28.7 Selecting the Control Loop Rate 666
28.8 Ad Hoc Methods 667
28.9 Homework Problems 671

PART 5 **MECHATRONIC PROJECTS AND SYSTEMS ENGINEERING** 673

CHAPTER 29 **Rapid Prototyping** 673

29.1 Introduction 673
29.2 Why Make Prototypes? 674
29.3 Prototyping Philosophies: Build or Simulate? 674
29.4 Rapid Prototyping of Mechanical Systems 675
29.4.1 Solid Modeling Tools 675
29.4.2 Modeling System Dynamics 676
29.4.3 Foamcore, X-Acto Knives, and Hot Glue 676
29.4.4 2-D Rapid Prototyping: Laser Cutting/LaserCAMM 680
29.4.5 2-D Rapid Prototyping, Cheap! 682
29.4.6 Tab and Slot Construction 682
29.4.7 Toys 682
29.4.8 3-D Rapid Prototyping: SLA, SLS, FDM, and Soft-Mold Castings 684
29.5 Rapid Prototyping of Electrical Systems 687
29.5.1 Schematic Capture and Circuit Simulation Tools 687
29.5.2 Circuit Prototyping: Breadboards, Wire Wrap, and Perf Boards 689

29.5.3 Prototype PCBs 692
29.5.4 Soldering Technique 694
29.6 Suppliers and Resources 696
29.7 Homework Problems 698

CHAPTER 30 **Project Planning and Management** **699**

30.1 Introduction 699
30.2 Increasing Project Complexity and the Need for Managing the Process 700
30.3 Planning and Executing a Project 702
30.3.1 System Requirements 702
30.3.2 Generating Design Candidates and Alternatives 703
30.3.3 Design Concept Evaluation: Prototypes and Iteration 707
30.3.4 Specifications 707
30.4 Management Tools 708
30.4.1 Project Management 708
30.4.2 Systems Engineering 712
30.4.3 Concurrent Design 713
30.5 Communication and Documentation 714
30.6 Pitfalls and Suggestions 715
30.7 Homework Problems 715

CHAPTER 31 **Troubleshooting** **718**

31.1 Introduction 718
31.2 Staring into the Void 719
31.3 An Ounce of Prevention Is Worth an All-Nighter of Troubleshooting 721
31.4 Troubleshooting Attitude 728
31.5 Final Thoughts 729
31.6 Homework Problems 730

CHAPTER 32 **Mechatronic Synthesis** **731**

32.1 Introduction 731
32.2 The Project Description 732
32.3 System Requirements 732
32.4 Design Candidates and Alternatives 734
32.4.1 Team Zero's Concepts 734
32.4.2 Team InTheRuff's Concepts 739
32.4.3 Review of the Design Candidates 741
32.5 Morphology Charts 741
32.6 Design Concept Evaluation: Prototypes and Iteration 743
32.7 Implementation Phase 746
32.7.1 InTheRuff Drive Motor Choice 746
32.7.2 Team Zero Ball Release Motor Choice 747
32.7.3 Team Zero Beacon Sensor Circuit Evolution 747
32.7.4 Team Zero Support Stiffness Issue 748
32.7.5 Team Zero Compass Sensor Failure 749
32.8 The Completed Designs 749
32.8.1 Team InTheRuff 749
32.8.2 Team Zero 752

32.9 Performance Results 752
32.10 Gems of Wisdom from the Students 753

Appendix A Resistor Color Code and Standard Values 755

Appendix B Sample C Code 757

Appendix C Project Description for the Chapter 32 Case Study 769

Index 773

Preface

This book is an introduction to mechatronics suitable for upper level undergraduates as well as graduate students. The principal target audience is engineering students who have studied or are currently studying mechanical engineering. The material covered in this book has been the basis of an introductory mechatronics course for both undergraduate and graduate mechanical engineers that has been delivered at Stanford University 18 times over the past 14 years.

With the addition of supplemental materials in the areas of mechanics and mechanical design, this book is also suitable for an introductory mechatronics course targeted to electrical engineering or computer science students. One of the authors teaches such a course using the material and approaches found in this text.

The approach that we take in all of these courses is to introduce the students to electronics, software, sensors, and actuators in a way that emphasizes practical applications. We assume that incoming students have had an introductory course where they learned basic circuit analysis (Ohm's Law, Kirchoff's Laws, etc.) and another course in which they learned to program. After a brief review of the basics of circuit analysis and the behavior of basic electronic components, we focus on using that knowledge to make sense of the data sheets for the most common types of interface devices. The goal is to show students how to apply those devices in realistic circuits.

This book is unlike, for example, a statics book, in that mechatronics involves several independent bodies of material. As a result, we have chosen to arrange the book around four parts (Software, Electronics, Actuators, and Mechatronic Projects and Systems Engineering) devoted to the various disciplines, technologies, and processes involved in creating a mechatronic product. Within the parts the content builds on what came before, but dependencies among the sections are kept to a minimum. Some common themes, such as decomposition and incremental integration, emerge in several contexts, and therefore several sections. There is no electronics prerequisite for the Software section or the Mechatronic Projects and Systems Engineering section; however the Actuators section does assume the knowledge covered in the Electronics section. The order of these parts within the book should not be construed as an endorsement of a particular sequence in presenting the material. There is no particular reason that electronics could not be taught before software. You can make a good case for doing either first, and over the years we have taught it both ways. The Actuators, Electronics, and Software sections of the book each deal with their respective topic and also include integration strategies. The Mechatronic Projects and Systems Engineering section addresses system architecture as well as integration issues.

In the Software section, as in industrial applications today, the C programming language is the most often used. We cover the spectrum of programming language options and describe the features of those that are in use in embedded systems in Chapter 4. We feel that the programming language for this type of course should be a compiled language that provides for what are generally considered to be the "modern language" constructs. We say a "modern language" so that the students have available the range of program constructs that make expressing the programs easiest. While it is true that features like case selection [the switch() case construct in C] can be constructed from more primitive elements, the burden of learning to do that would be placed on a neophyte programmer who may already be struggling to get *any* algorithm expressed at all. We specify a compiled language so performance limitations of the microcontroller are less likely to get in the way of successfully implementing the algorithms. As an example, Chapter 5 (event driven programming and state machines) assumes that the processor executes quickly enough that all events can be processed in a timely manner. For the kinds of projects students are most likely to take on, this will be true for solutions coded in almost any modern compiled language. However, the performance limitations of, for example, interpreted BASIC languages would likely present problems when using this approach.

With regard to prior programming experience, we assume that students begin our course (and this text) having successfully completed a first course in programming. The emphasis in this book is on a few software *design* techniques and structures that are most relevant to programming for embedded systems. We introduce students to program design using a "Program Design Language" (PDL), which is not specific to any one programming language, and show them how completing a design first in a PDL will make implementation in a regular programming language easier.

Within each of the sections, there is a body of technical material that builds in complexity as the section progresses. Each chapter is designed to be read from beginning to end, with the later portion of a chapter often assuming familiarity with the material presented earlier in that chapter or in a preceding chapter within that section. Understanding and taking command of the technology is important in creating an integrated solution. In our courses, this material is reinforced by a series of semi-structured laboratory assignments that not only support the material from the text, but also incrementally introduce the students to a set of hardware and software modules designed to give them first-hand experience using the material. The semi-structured nature of the lab assignments forces them to understand and "own" the material in order to be able to successfully complete the lab assignments. The capstone experience for each of these courses is a multiweek design project in which student teams are expected to create their own solutions that meet a set of project specifications while applying the material from the lectures, textbook, and laboratory assignments. Chapter 32 follows the progress of two student teams in completing a representative project.

RESOURCES FOR INSTRUCTORS

All instructor resources are available at www.pearsonhighered.com. If you are in need of a login and password for this site, please contact your Pearson representative.

The authors have developed syllabi, designed support hardware and software, and written laboratory assignments and many suitable project descriptions for this book. All of this documentation and the figures from the text are available to instructors.

The site also includes a community wiki where adopters of the text can pose questions about content and share teaching techniques and pointers to useful resources.

ACKNOWLEDGMENTS

While there are three authors listed at the front of this book, there have been many more contributors to the content that deserve some measure of credit. First among these must be the students who have been using the drafts of the manuscript and who took the time to provide feedback to us about what worked (and didn't work) for them in understanding the material. The book also benefitted greatly from the input made by the reviewers, some of whom persevered through multiple drafts, and, in some cases, taught with early drafts and provided valuable input that helped us to improve our presentation. The reviewers were: Raj Amireddy, Penn State Hazleton; Larry Banta, West Virginia University; Daniel J. Block, University of Illinois; James E. Bobrow, University of California, Irvine; Meng-Sang Chew, Lehigh University; Gabriel Hugh Elkaim, University of California, Santa Cruz; David Fisher, Rose Hulman Institute of Technology; Sooyong Lee, Texas A&M University; Richard

B. Mindek, Jr., Western New England College; William R. Murray, California Polytechnic State University; Mark Nagurka, Marquette University; and Howard A. Smolleck, New Mexico State University.

Most of the part photos in the book were shot in a studio that was graciously made available to the authors by Dave Beach and Craig Milroy of the Product Realization Lab at Stanford. Jonathan Edelman taught us how to light and set up the studio to enable us to make the kinds of photos that we wanted for the book. Nick Streets created the solid models that were used in several figures.

Chris Kitts taught us how to teach project management (see Chapter 30).

Chris Gerdes showed us how to simply add friction to the basic motor behavior chapter (see Chapter 22).

Our thanks also go to the members of Team Zero and Team InTheRuff: Adam Bernstein, Ho Lum Cheung, Nancy Dougherty, Derianto Kusuma, Jordan LeNoach, Matthew Norcia, Kanya Siangliulue, and Wesley Zuber (Chapter 32).

Trademark Acknowledgments

Alibre is a trademark of Alibre Inc. Design Xpress is a registered trademark of Alibre Inc.

Altium, Altium Designer, Board Insight, DXP, Innovation Station, LiveDesign, NanoBoard, NanoTalk, OpenBus, P-CAD, SimCode, Situs, TASKING, and Topological Autorouting and their respective logos are trademarks or registered trademarks of Altium Limited or its subsidiaries.

Autodesk is a registered trademark of Autodesk, Inc.

AutoTRAX © Ilija Kovacevic.

BASIC Stamp is a registered trademark of Parallax Inc.

Bluetooth is a registered trademark of Bluetooth SIG, Inc.

Burleigh is a trademark and Inchworm is a registered trademark of EXFO Burleigh Products Group Inc.

Cadence, Allegro, orCAD, and PSpice are registered trademarks of Cadence Design Systems, Inc.

Canon and Canon product and services names are the trademark or registered trademark of Canon Inc. and/or other members of the Canon Group.

Delrin is a registered trademark of E.I. du Pont de Nemours and Company.

DREMEL is a registered trademark of Credo Technology Corporation, an affiliate of Dremel.

Electronics Workbench, Multisim, and Ultiboard are registered trademarks of National Instruments Corporation.

Erector is a registered trademark of Meccano SN.

expresspcb is a trademark of ExpressPCB.

FedEx is a registered trademark of Federal Express.

FlexiForce is a registered trademark of Tekscan, Inc.

Freescale is a trademark of Freescale Semiconductor, Inc.

Google and Google SketchUp are registered trademarks of Google Inc.

Intel, Pentium, and Intel Core are registered trademarks of Intel Corporation in the U.S. and/or other countries.

iPod is a registered trademark of Apple, Inc.

K'NEX is a registered trademark of K'NEX Brands, L.P.

LaserCAMM is a trademark of LaserCAMM.

LEGO is a registered trademark of the LEGO Group.

LTspice is a registered trademark of Linear Technology.

Masonite is a registered trademark of Masonite International Corporation.

MATLAB and MathWorks are registered trademarks of The MathWorks, Inc.

ON Semiconductor is a registered trademark of Semiconductor Components Industries, LLC.

PIC is a registered trademark of Microchip Technology Inc.

Pro/ENGINEER is a registered trademark of Parametric Technology Corporation (PTC).

SolidWorks is a registered trademark of DS SolidWorks Corporation.

TINA is a trademark of DesignSoft, Inc.

TINA-TI is a trademark of Texas Instruments Incorporated.

TurboCAD is a registered trademark of IMSI/Design, LLC.

UGS, Transforming the process of innovation, Teamcenter, Solid Edge, and NX are trademarks or registered trademarks of UGS Corp. or its subsidiaries in the United States and other countries.

Unigraphics is a registered trademark of UGS PLM.

X-ACTO is a registered trademark of X-ACTO.

ZigBee is a registered trademark of ZigBee Alliance.

About the Authors

Ed Carryer graduated from the Illinois Institute of Technology in 1975 with a BSE as a member of the first graduating class of the Education and Experience in Engineering Program. This innovative project-based learning program taught him that he could learn almost anything that he needed to know and set him on a path of lifelong learning. That didn't, however, keep him from going back to school.

Upon completion of his Master's Degree in Bio-Medical Engineering at the University of Wisconsin Madison in 1978, he was seduced by his love of cars, and instead of going into medical device design, he went to work for Ford on the 1979 Turbocharged Mustang. In later programs at Ford, he got to apply the background that he had gained in electronics and microcontrollers during his graduate work to the 1983 Turbocharged Mustang and Thunderbird and the 1984 SVO Mustang. After leaving Ford, Ed worked on the design and implementation of engine control software for GM and on a stillborn development program to put a turbocharged engine into the Renault Alliance at AMC before deciding to return once again to school. At Stanford University, he did research in the engine lab and earned his PhD in 1992.

While working on his PhD, Ed got involved in teaching the graduate course sequence in mechatronics that is known at Stanford as Smart Product Design. He took over teaching the courses first part time in 1989, then full time after completing his PhD. In teaching mechatronics, Ed has found his calling. The integration of mechanical, electronic, and software design with teaching others how to use all of this to make new products hits all his buttons. He is currently a Consulting Professor and the Director of the Smart Product Design Lab (SPDL). He teaches graduate courses in mechatronics in the Mechanical Engineering department and an undergraduate course in mechatronics in the Electrical Engineering department.

Since 1984, Ed has maintained a consultancy focused on helping firms apply electronics and software in the creation of integrated electromechanical solutions (in 1984, almost no one was using the term mechatronics). The projects that he has worked on include an engine controller for an outboard motor manufacturer, an automated blood gas analyzer, a turbocharger boost control system for a new type of turbocharger, and a heated glove for arctic explorers. His most recent project involved using ZigBee radios and local structural model evaluation to create a wireless network of intelligent sensors to monitor and evaluate the structural health of buildings and transportation infrastructure.

Matt Ohline was a dedicated reader of *Road & Track* and thoroughly fascinated by cars even before graduating from high school in 1987. During more than a few misspent evenings in his high school years, he had already unknowingly fallen under the influence of Ed Carryer while burning considerable rubber in a 1984 SVO Mustang. As far as either he or Ed knows, this was the first time their paths crossed.

Matt attended Stanford University as an undergraduate and began work on a degree in English. After two years, he realized that the Mechanical Engineering department offered courses with material that was more directly applicable to cars. He then shifted his focus, eventually obtaining a dual degree in Mechanical Engineering and English. (This textbook has been the best, and first, opportunity to put both degrees to use.) When Matt began working on his Master's Degree in ME in the early 1990s, most systems in cars were electronically controlled and actuated, and those that weren't yet soon would be. Stanford's Smart Product

Design program directed by Ed Carryer seemed the perfect way to acquire the knowledge and skills that would enable him to apply mechatronics in the automotive arena. Upon graduating with an MS in 1994, however, the automotive industry (like much of the U.S. economy at the time) was a less than dynamic scene.

Luckily, the Silicon Valley was abuzz with interesting companies and technical problems to solve. Matt took his first job at a small manufacturer of automotive measurement equipment and software, after which he returned to Stanford to lecture in mechatronics and began performing design consulting for a variety of Silicon Valley start-up companies.

Consulting for start-ups convinced Matt that he wanted to try his own hand at being an entrepreneur. In 2000, mechatronics was seldom applied in the field of medical devices, and this seemed to present an excellent opportunity. Along with a senior lecturer in ME and a doctor from the Stanford Medical Center, he cofounded NeoGuide Systems, a manufacturer of robotic flexible endoscopes that was acquired by Intuitive Surgical (ISRG) in 2009. Matt is currently hard at work on a new company, which, in true Silicon Valley form, he won't discuss with anybody unless they sign a nondisclosure agreement.

Throughout this time, Matt has continued to teach mechatronics at Stanford as a Consulting Associate Professor. His focus is Stanford's introductory mechatronics course, which is taught to upper-level undergraduate and graduate students in Mechanical Engineering. He considers himself incredibly fortunate to work with Ed Carryer and Tom Kenny and is honored to have collaborated on this textbook project with them.

Tom Kenny has always been interested in the properties of small structures. His PhD research was carried out in the Physics department at UC Berkeley and focused on a measurement of the heat capacity of a monolayer of helium atoms. After graduating, his research at the Jet Propulsion Laboratory focused on the development of a series of microsensors that use tunneling displacement transducers to measure small signals, with an ultimate goal of enabling smaller payloads for smaller spacecraft.

In 1994, Tom moved back to the San Francisco Bay Area to join the Design Group of the Department of Mechanical Engineering at Stanford University. There he directs MEMS-based research in a variety of areas, including resonators, wafer-scale packaging, cantilever beam force sensors, gecko-inspired adhesion, energy harvesting, microfluidics, and novel fabrication techniques for micromechanical structures.

Living in the Silicon Valley, Tom has been unable to resist the siren song of "The Startup." He is the cofounder and CTO of Cooligy, Inc., a microfluidics chip cooling component manufacturer currently owned by Emerson, and cofounder and member of the board of directors of SiTime Corporation, a developer of CMOS timing references using MEMS resonators. He is currently a Stanford Bosch Faculty Development Scholar and was the General Chairman of the 2006 Hilton Head Solid State Sensor, Actuator, and Microsystems Workshop.

While on leave from Stanford from October 2006 through September 2010, Tom served as Program Manager in the Microsystems Technology Office at the Defense Advanced Research Projects Agency, starting and managing over $200M in programs involving the development of next-generation thermal management, controlled nanomanufacturing, manipulation of Casimir forces, and the Young Faculty Award. He has authored or coauthored over 250 scientific papers and is a holder of 48 issued patents.

PART 1 **Introduction**

Introduction

If you look around at the products that you interact with on a daily basis, you will find that over the last 10 to 15 years there has been a dramatic shift in the capabilities and complexity of these devices. Many have gotten "smarter" as small computers, called microcontrollers, have become an integral part of such devices. The number of products that employ these embedded microcontrollers has grown dramatically and this growth shows no sign of slowing. Today, embedded microcontrollers are practically ubiquitous in products. They have replaced mechanical timers in dishwashers and washing machines. They have enabled the addition of sensors and complex decision making in these machines that allow them to, for example, wash until things are clean rather than running a fixed cycle. Office copiers now include multiple microcontrollers communicating over an internal network to coordinate their many tasks. Automobiles include processors communicating over in-vehicle and worldwide networks to control the engine, transmission, braking, navigation, interior climate, air-bag deployment, and more. The process of sensing the state of the world, making often complex decisions based on that data and producing actions (outputs) to effect a desired change of conditions, is at the core of all of these products. The success of these products depends on the successful integration of the core technologies that are necessary to implement these "smart" products.

The benefits of this technology are not isolated in the affluent cultures of the world. These small computers are being integrated into clean, solar powered lighting systems for parts of the world that are not on the power grid (Figure 1.1). These new lights will not only provide better indoor lighting but, by removing the smoky kerosene lamp, improve the health of their users. A more mechanical example is that of Driptech, which has created a portable manufacturing system (Figure 1.2) that uses a computer controlled laser to perforate plastic

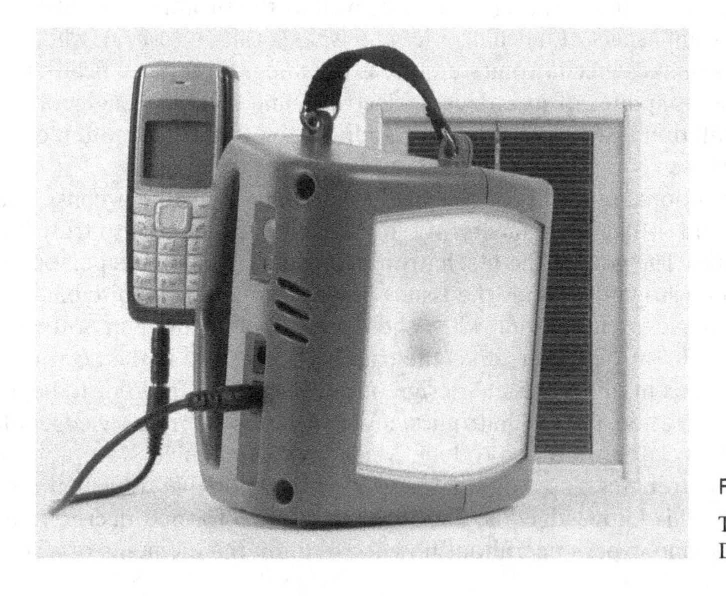

FIGURE 1.1

The Nova solar powered lamp from D.light. (Courtesy of D.light Design.)

FIGURE 1.2
The Driptech irrigation tubing manufacturing system. (Courtesy of Driptech Inc.)

tubing at regular intervals and thereby create extremely inexpensive drip irrigation systems that open up new regions to agriculture, improve the income of farmers, and increase food production in the developing world.

These new "smart" products are enabled by a body of knowledge and an approach to design known as **mechatronics**. The term originated at the Yaskawa Corporation from the combination of *mecha*nical and elec*tronics,* though its meaning has been broadened to include software and computation. While there have been many definitions of mechatronics proposed, there is no single, universally accepted definition. We believe that Dinsdale and Yamazaki have a working definition that we feel captures the essence of mechatronics [1]:

> the synergistic integration of fine mechanical engineering with electronics and intelligent computer control in the design of products and manufacturing processes.

This quote best embodies of our philosophy of what mechatronics is about. We have chosen this definition because it captures not only the breadth of applications for mechatronics, encompassing both manufacturing process as well as the products, but also what we feel is the most important aspect of mechatronics and what sets it apart from what came before. In a nutshell, what makes mechatronics unique is the integration. Specifically, it is the integration across the disciplines of mechanical, electrical, and software engineering that sets the mechatronic solution apart from the more traditional solutions that often didn't cross disciplinary boundaries.

In the traditional model, products might be designed by teams consisting of mechanical, electrical, and software engineers. Missing was the person able to truly integrate *across* all the disciplines. The goal of the mechatronics engineer is to be the person on (or leading) the design team who understands the issues involved in all of the disciplines and is able to make informed judgments about what should be implemented in software, electronics, or mechanicals. He or she understands the trade-offs involved in the decision and is able to balance the choices in the successful design of novel systems. In order to be able to achieve that level of integration, the mechatronics engineer needs to be knowledgeable in all of the technologies that will go into a product. Mechatronics engineers are able to understand the capabilities, strengths, and weaknesses of the many possible ways that a problem *could* be solved. With this knowledge, they are able to make informed decisions, based on their own knowledge and experience, about how to partition the elements of a solution into the various disciplines. This need to educate a cadre of engineers knowledgeable across the domains of electronics, software, mechanics, and project planning has driven the content of this book.

1.1 PHILOSOPHY

From the notion that the core of mechatronics is in the integration comes our assertion that the emphasis in mechatronics education should be on applications. To the authors, the study of electronics or software is only here to allow its effective integration and application to the solution of real problems. In this book, the emphasis is on the behavior of the devices that go into a mechatronic product, rather than the underlying physics, because how a device behaves is what the designer needs to understand in order to effectively *use* the device. To be sure, the designers of new devices (e.g., integrated circuits or actuators) require a deep understanding of the underlying physics, but that task is delegated to the domain specialists rather than to the integrators.

The choice of material and presentation used in this book is a reflection of our belief that the *application* of the core technologies to the solution of problems is the prime role of the mechatronics engineer. Our goal is to present this material in such a way that students will be empowered to apply these technologies wherever appropriate in their designs. Electronics, software, and mechanics should all be comfortable tools in the hands of a mechatronic engineer.

1.2 THE ORGANIZATION OF THIS BOOK

Mechatronics is generally understood to combine electronics, software, and mechanical systems. The four sections to this book cover these areas and adds a section on Project Planning and Systems Engineering that contains chapters on Rapid Prototyping techniques (Chapter 29), Project Planning and Management (Chapter 30), and Troubleshooting (Chapter 31), an often neglected topic in textbooks. The section closes with a chapter on Mechatronic Synthesis (Chapter 32) that follows the development of two mechatronic projects from the initial project description, through the brainstorming and ideation phase, into the refinement and detailed design, and all the way through implementation and prove out.

The Software section begins with an introduction to the terminology used to describe embedded computers and the peripherals that are most commonly encountered (Chapter 2—What's a Micro?). Chapter 3 (Microcontroller Math and Number Manipulation) introduces the low-level math and bit manipulation concepts that are essential in embedded systems but often omitted in beginning programming courses. The chapter on programming languages (Chapter 4) surveys the range of options available for embedded systems and provides guidelines for the selection of a language for a new project. Chapters 5 (Program Structures for Embedded Systems) and 6 (Software Design) work together to demonstrate effective paradigms for the design and implementation of the software for embedded systems. Chapter 7 (Inter-Processor Communications) covers alternate techniques (other than direct sensing) in which a program in a microcontroller can collect the input data necessary to making decisions. The chapter describes various techniques available to allow multiple embedded computers to share information and cooperate within a system. Chapter 8 finishes off the Software section with a description (from the programmer's point of view) of the typical operation of the parallel input/output (I/O), timer, pulse width modulation (PWM), and analog-to-digital (A/D) converter subsystems found in modern microcontrollers.

The Electronics section is the largest of the four sections. It provides a chapter on basic circuits (Chapter 9) that presents all of the aspects of circuit analysis that are necessary for the following chapters in the Electronics and Actuators sections. Most, if not all, of this material should be a review for our target audience but we provide it as an alternative view and as reinforcement of the material. We also sought to put the most useful and necessary background material into a single, easily-referenced text. Our treatment of semiconductors (Chapter 10) places a heavy emphasis on the use of both bipolar and MOSFET transistors as *switches*. For amplification, we direct the reader to the operational amplifier (treated as ideal devices in Chapter 11, and real devices with real nonidealities in Chapter 12). Chapters 13 (Sensors), 14 (Signal Conditioning), and 15 (Active and Digital Filters) introduce readers to several ways physical phenomena are sensed, and how to sculpt the output of a sensor into a signal suitable for presentation to either a digital input (Chapters 14 and 15) or an analog input (Chapter 19). The characteristics of digital outputs, both logic level as well as power outputs, are treated in Chapters 16 and 17. The section finishes with chapters on Digital Logic and Integrated Circuits in the context of microcontrollers (Chapter 18), Voltage Regulators, Power Supplies, and Batteries (Chapter 20), and Noise, Grounding, and Isolation (Chapter 21).

The Actuators section focuses primarily on electromagnetic actuators (Chapters 22–26), but also offers a survey of alternative actuators in Chapter 27. Because of their popularity in such a wide range of applications, the brushed DC motor gets the most attention (Chapters 22 and 23) though the behaviors and drive techniques for solenoids (Chapter 24), brushless DC motors (Chapter 25), and stepper motors (Chapter 26) are also covered. The section closes with an introduction to closed-loop control (Chapter 28). In recognition of the range of controls backgrounds likely among students, this chapter focuses on developing an intuitive understanding of why proportional, integral, derivative (PID) control works and the role that each of the components plays in the control.

1.3 WHO SHOULD STUDY MECHATRONICS?

While anyone willing to devote the time and effort *can* study mechatronics, those students with backgrounds in one of the core technologies [electrical engineering (EE), mechanical engineering (ME), or computer science (CS)] will find the task easier. Of these three, EE and ME curricula typically require more programming than a CS curriculum requires mechanics or electronics. As a result, the ME and EE students will usually find that they have less new material to learn than do CS students. The differences in background between these three majors are not enough to keep any of these students from being successful in studying mechatronics. The material from this text has been used successfully to teach mechatronics to students of all three of these majors as well as to aeronautics and civil engineering students.

1.4 HOW TO USE THIS BOOK

The technical contents of this book are the foundation of mechatronics. However, becoming a mechatronic engineer involves not only learning the theory and analysis techniques but also gaining experience in applying that information to solve problems. Building mechatronic projects is an essential part of learning to be a mechatronic engineer. The most effective use of this book will be to use it to develop an understanding of the set of tools available to the designer of mechatronic systems. To be truly successful, you will need to supplement what you learn from this text with experience in the laboratory and in solving open-ended problems. As with most tools (indeed, as with most anything), only hands-on experience will allow you to develop and hone your skills. So take what you learn here and practice, practice, practice.

1.5 SUMMARY

Our goal for courses based on this text is that they will bring the aspiring mechatronics students to the point where they have acquired the background knowledge and skill to become capable electromechanical system designers. We seek to pack the student's "design toolbox" full of electronic, software, and mechanical design tools, where before it may have contained tools from only one of these disciplines. It takes longer than one introductory course or text to get there, but this text has much of what they need, and we hope that it can get them well on their way.

REFERENCE

[1] "Mechatronics and ASICs," Dinsdale, J., and Yamazaki, K., *Annals of the CIRP,* 1989; (38): 627–634.

CHAPTER

What's a Micro?

2.1 INTRODUCTION

The term "microprocessor" (often shortened to just "micro") is, by now, familiar to almost everyone. In student or professional life, it has become very difficult—or even impossible—not to use at least one micro every day. They comprise the heart of personal computers, digital alarm clocks, electronic organizers, automobile engine controllers, cell phones, and answering machines—and countless other everyday items. But what exactly is a micro? In fact, the term "micro" is a generalized term that encompasses several different types of processors and other supporting circuitry required to create functional systems. To the designer of mechatronic systems, the micro is the brain; it is the component that imparts a measure of intelligence and enables devices to express complex behavior. Without the ability to effectively incorporate micros, designers are limited to systems that are strictly mechanical or electrical in nature, which can be a severe restriction. Designers with micros in their "tool-kits" have the ability to create vastly more powerful and capable systems.

In this chapter, you will learn:

1. what a "micro" is, with emphasis on microcontrollers,
2. the differences between microcontrollers, microprocessors, and digital signal processors (DSPs),
3. the types of microcontroller architectures,
4. the subsystems most commonly incorporated into microcontrollers,
5. how microcontrollers execute programs and perform operations,
6. what peripheral systems are commonly available and the function of these systems.

2.2 WHAT'S A MICRO?

"**Micro**" is actually a nonspecific term that covers a wide range of electronic devices, including microprocessors, microcontrollers, DSPs, and other more customizable devices. They are all closely related, but differ in several significant respects. Each type of device is tailored to perform better in specific circumstances. For example, a microcontroller may be capable of performing a task better suited to a DSP, but it will probably do so more slowly and less efficiently. Usually, it isn't possible to tell the difference simply by looking at the package; it's what's inside that differentiates the devices. It is, however, useful to have a sense of the scale of the packaging for these devices, and Figure 2.1 shows one of the types of packaging used in a wide range of microcontrollers.

FIGURE 2.1

Typical microcontroller packaging.

2.3 MICROPROCESSORS, MICROCONTROLLERS, DIGITAL SIGNAL PROCESSORS, AND MORE

The key component common to microcontrollers, microprocessors, and DSPs is the **central processing unit**, or **CPU**. The CPU is the fundamental control logic hardware that executes instructions. The device architecture and the types of other systems and subsystems that are combined on the same chip with the CPU, if any, determine whether the device is a microcontroller, a microprocessor, or a DSP. The term may also be applied to field programmable gate arrays (FPGAs), application specific integrated circuits (ASICs), and programmable systems-on-chips (PSoCs), which can be configured by the designer to incorporate a **core**, or CPU, and allow for extensive customization by the end user. These devices differ as follows:

Microprocessors are, at their core, a CPU on a chip. Such fundamental components as program and data memory, parallel input/output, and communications peripherals are all physically separate from the microprocessor and must be integrated by the designer with separate components. Microprocessors can fetch and execute instructions, perform mathematical and logical operations, store data, and request actions from peripheral systems—and nothing more. All operations and tasks require interaction with external components. Some familiar examples of microprocessors that form the basis of most modern personal computers include Intel's Pentium and Core i7, i8, i9 processors, and so on, and AMD's Athlon, Opteron, and Sempron processor families.

Microcontrollers integrate the CPU of a microprocessor with everything else necessary to create a single-chip computer. For this reason, they were originally known as **microcomputers**. (Motorola first used the term "microcontroller" in marketing literature, and this term evolved to describe the entire device family—much like "Xerox" is to "copy" or "Coke" is to "soda.") At a minimum, this requires non-volatile memory for programs (such as ROM, Read-Only Memory), memory for data (known as RAM, Random Access Memory), and digital inputs and/or outputs (usually called **I/O**). Often, other convenient features are added, including serial communications modules, analog-to-digital converters, timers, and a variety of other peripheral systems. A clock source is also required, and though several processors incorporate internal clock sources, the time-base part of the clock source (typically a crystal) is often an external component that is added by the designer. There are literally many hundreds of different microcontrollers available, each with different combinations of resources, capabilities, and speed. Each microcontroller is appropriate for different applications and markets. In general, microcontrollers target "embedded control," and are intended for use in stand-alone, single-purpose systems. Specific examples of microcontrollers that have become household names are rare (unlike the Pentium microprocessor), but devices that incorporate microcontrollers are easy to identify. One such device is a standard computer mouse, which optically tracks the position of the mouse as it moves across a flat surface, and communicates mouse displacement data via the Universal Serial Bus (USB) protocol or PS/2 protocol to a personal computer.

Digital Signal Processors (DSPs) are microprocessors that are specifically designed to perform signal processing operations very rapidly. They are distinguished by their specialized instruction sets and their memory architectures. Their instruction sets are designed so that they are capable of rapidly executing the kinds of mathematical operations that are common in signal processing. An example of this is the "multiply-accumulate" instruction, in which a number is multiplied by a factor, summed with the results of previous operations, and stored to memory—all in a single step. Their architectures are often designed to significantly increase the speed of instruction execution by simultaneously executing one instruction while the next instruction is fetched from memory. Applications for DSP technology have become increasingly commonplace as cell phones, digital voice recorders, high-definition television, and other consumer electronic devices become widely available. In addition, manufacturers of DSPs now routinely incorporate many of the subsystems previously found only in microcontrollers, such as digital I/O and analog-to-digital converters, in order to make them attractive choices in a wider variety of applications. As with microcontrollers, few DSPs have become household names, but devices that incorporate DSPs are easy to identify. For example, DSPs are the key components in digital cell phones, where they are responsible for digitizing and compressing sound from the mouthpiece microphone for the outgoing signal and decompressing and converting the incoming digital data stream into analog signals for the earpiece speaker.

Other devices: There are several types of highly configurable and flexible devices that can be configured by the designer to incorporate a CPU. One example of this is the **field programmable gate array (FPGA)**, a customizable logic chip that contains very large numbers of fundamental logical building blocks. Since these building blocks are the same as those used to design and construct microprocessors, microcontrollers, and DSPs, FPGAs may also be configured as any of these devices, with the additional benefit of being highly customizable by the end user. A **programmable system-on-chip (PSoC)** incorporates a microcontroller as well as arrays of configurable analog and digital logic elements. The most flexible of all micros, of course, would be one that is completely custom. **Application specific integrated circuits (ASICs)** can be designed to incorporate whatever circuitry is needed, including a CPU. The design or use of each of these "other" devices is beyond the scope of this text; interested readers are encouraged to consult more specific and advanced references to learn more about them.

Microprocessors, microcontrollers, and DSPs (as well as FPGAs, PSoCs, and ASICs, in some cases) are all closely related: each contains a CPU as the computational core, and interacts with memory and peripherals in order to perform operations and tasks. The location and arrangement of the memory and peripherals, and the types of instructions they execute, are what distinguish one type of device from the others. In this introductory text, we focus primarily on microcontrollers, as they are the most common choice in mechatronic systems. However, once the reader has become familiar with microcontrollers, he or she should be able to design with microprocessors and DSPs with some effort, as they all share many basic characteristics.

2.4 MICROCONTROLLER ARCHITECTURE

The functional heart of the microcontroller is the central processing unit, or CPU. The CPU's job is to fetch instructions and data from memory or peripheral systems and manipulate the data in useful ways. Logical operations, mathematical operations, or data reorganization (movement) are the most common types of manipulations. Every task performed by the CPU is fundamentally comprised of moving numbers within the memory space and performing operations with those numbers. And, very importantly, *everything* in the memory space is numbers. **Data** is the easiest to conceive of as being made up of numbers: often numbers in the data space are literally numbers (such as the unsigned integer 50), or they represent types of numbers that aren't easily expressed with straight binary digits (such as floating point numbers), or they represent other quantities, like letters (such as ASCII character code, in which the number 85 represents the character "U"). **Instructions** to the micro are also represented by numbers (e.g., the number "212" may represent the instruction "jump to address 4 and continue executing code"). All interactions with the microcontroller's hardware, such as registers and ports, are done with numbers as well. The memory, input/output hardware (I/O), and peripheral systems of a microcontroller are all interconnected and communicate directly with the CPU (Figure 2.2).

Program memory is usually implemented with specific types of memory that retain their contents during times when the microcontroller system is not powered. This type of memory is called **non-volatile memory**, because it is stable and retains its contents without power. Common examples of non-volatile memory are **Read Only Memory (ROM)**, **Electrically Erasable Programmable Read Only Memory (EEPROM)**, and **Flash EEPROM**. EEPROM and Flash EEPROM are much easier to work with, since they are user programmable and erasable, making it easy and inexpensive to use while developing new programs. Data memory, on the other hand, is not usually required to be retained when a microcontroller system is switched off and then back on. Each time the program runs, new variables are created and updated, new information is read in from sensors, and so on. **Volatile memory** is suitable for this type of application. This class of memory derives its

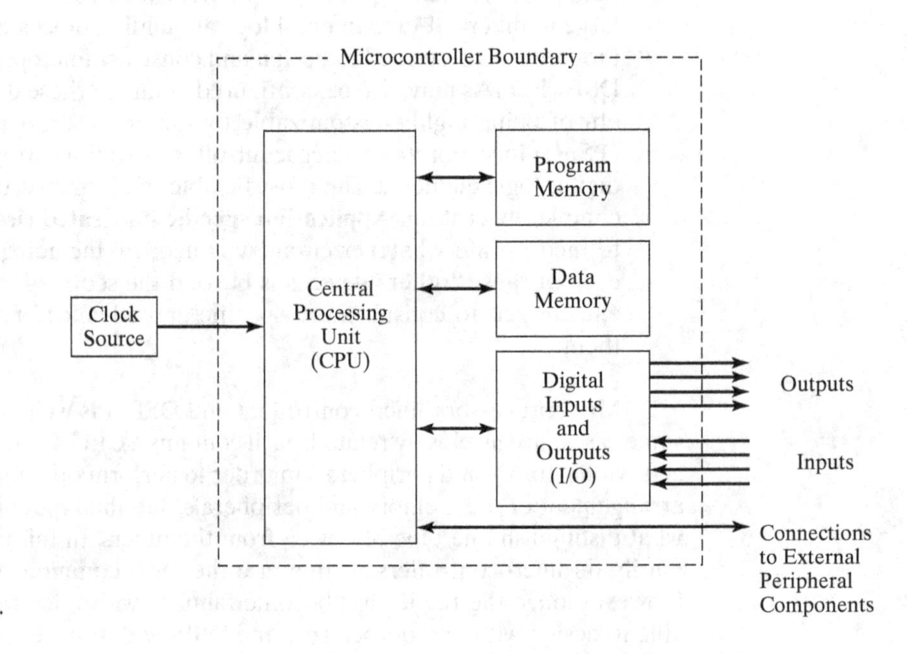

FIGURE 2.2
Typical microcontroller components, subsystems, and interconnections.

name from the fact that data is not retained when the system is not powered. **Random Access Memory (RAM)**—probably so named because the more meaningful acronym Read Write Memory (which would be "RWM") is unpronounceable—is the standard choice for implementing volatile memory.

Often, additional connection points to the CPU are brought out to the physical pins of the microcontroller. This allows for the connection of external peripheral components, such as additional memory or specialized communication subsystems. In this way, microcontroller systems can be customized so that they better fill a wide variety of system requirements.

The CPU also requires a clock source in order to function. The internals of the CPU are inherently sequential logic that require a source of regular transitions (a clock) to synchronize the operations. Most commonly, internal oscillator circuitry built into the microcontroller works with an external quartz crystal or ceramic resonator to provide the clock input to the CPU. It is usually desirable to have a very precise and stable clock source, as this is used to provide timing to the CPU and most other components in the microcontroller. An external quartz crystal provides the most accurate and stable clock source, with external ceramic resonators and internal or external RC oscillators providing less accurate (and less expensive) alternatives to the designer. The internal oscillator circuit is the component in the system that determines when the microcontroller CPU performs the next operation, whether that operation is fetching data from memory, writing data to a register, or executing a math instruction. In general, the faster the clock, the faster the CPU functions (though it is not always the case that the CPU executes an instruction every clock cycle). The clock, however, is an instrumental aspect of how quickly a CPU executes programs (instruction sets and bus architecture also play major roles, as we will discuss below).

2.5 THE CENTRAL PROCESSING UNIT

In order to understand microcontroller architecture, it's essential to have a basic knowledge of how digital electronic components, including microcontrollers, represent numbers.

2.5.1 Representing Numbers in the Digital Domain

Microcontrollers are digital electronic devices and as such represent numbers with electronic switches. A switch that is on is typically agreed to represent a logical true or 1, and a switch that is off a logical false or 0. A single switch can then count, in a very simple fashion, from 0 to 1. This single switch is referred to as a "**bit**" (derived from the phrase "**bin**ary dig**it**"), which has become synonymous with a digit in **base-2**, or **binary**, numbering. Switches, and therefore bits or binary digits, can be aggregated into groups of arbitrary length to represent numbers larger than 1 by assigning each switch a numerical place. For example, we could designate two adjacent switches and agree that together they represent a single binary number that is 2 bits in length. The first bit, which by convention is called "bit 0," is in the 1's place (which is $base^0 = 2^0$). The second bit is in the 2's place (which is $base^1$ or 2^1), and is called "bit 1." Counting in binary follows the same rules as counting in any number base, including decimal: the lowest place number increments until the unique digits are exhausted, then rolls over while the next higher place number increments. When that place reaches the largest number that can be represented in that number base, the next higher place increments, and so on. Table 2.1 shows a rudimentary example of how this works in base-2, as we count from 00 to 11 (the largest number we can represent with 2 bits in binary). This is the same as counting from 0 to 3 in decimal.

TABLE 2.1 Using two switches in a microcontroller to count from 0 to 3 (decimal) in binary.

bit 1	bit 0	Binary	Decimal
0	0	00	0
0	1	01	1
1	0	10	2
1	1	11	3

By extension, numbers much larger than 3 may be represented by designating a larger number of switches to act as additional bits. In general, the number in decimal D that can be represented with a number of bits N is given by the expression:

$$D = 2^N - 1 \tag{2.1}$$

While any number of bits could be arbitrarily chosen to represent larger numbers, it is customary to use groupings of 4, 8, 16, 32, and 64 bits. This allows for the representation of 0 to 15 with 4 bits, 0 to 255 with 8 bits, 0 to 65,535 with 16 bits, 0 to 4,294,967,295 with 32 bits, and 0 to 18,446,744,073,709,551,615 with 64 bits. [In Chapter 3, we discuss number bases in much greater detail, with special emphasis on base-2 (binary) and base-16 (hexadecimal), as well as conventions for representing negative and floating point numbers.]

2.5.2 The Arithmetic Logic Unit

The computational core of the CPU is called the **arithmetic logic unit**, or the **ALU**, and commonly consists of several highly specialized **registers** where mathematical and logical operations are performed using digital logic circuitry. In addition to the implementation of the circuits in digital logic, the size (specified as the width of the registers, in bits) of the registers in the ALU has a major influence on how quickly and efficiently operations are performed. Eq. (2.1) applies in this situation, so an 8 bit wide ALU register is capable of holding numbers ranging from 0 to 255 (if we restrict the discussion to positive integers). Operations on numbers larger than 255 can be managed, but this requires that the operation be decomposed into several steps, none of which contains numbers or parts of numbers that exceed 8 bits in length. For this reason, it is common for ALU registers to be ganged together so that they can be used to represent larger numbers (e.g., 16-bit registers capable of representing any positive integer between 0 and 65,535). In general, larger ALU registers allow for faster mathematical operations; however, microcontrollers with larger ALU registers typically cost more. Usually, the larger the registers, the higher the performance (and cost) of the microcontroller. In fact, this parameter is such a primary determinant of microcontroller performance that microcontrollers are categorized by the width of their ALU registers and the width of the path between those registers and memory. A list of the most common classifications of microcontroller, by register width, is as follows:

4-bit microcontrollers: These are the ultra-low end of the microcontroller spectrum. Early in the evolution of microcontrollers (e.g., the 1970s), these micros were in widespread use because they were less expensive than the more powerful 8-bit microcontrollers, but the price difference has eroded over time. 4-bit microcontrollers are not generally used in new designs as a result.

8-bit microcontrollers: These are the low end of the modern microcontroller spectrum, and combine low cost with reasonable capabilities. 8-bit registers are wide enough to perform relatively complex tasks, though many operations will require considerable movement of 8 bit wide operands and results in memory. Microcontrollers with 8-bit registers can be obtained at very reasonable prices (less than $1 in even small quantities), and are available in a wide variety of configurations. As a result, 8-bit microcontrollers are very popular.

16-bit microcontrollers: Since it is possible with this class of microcontroller to manipulate twice as much data at a time, 16-bit microcontrollers are a good choice when performance is a higher priority. As you would expect, these cost more than 8-bit microcontrollers (typically $1 to $15 in high quantities), but the price premium for 16-bit microcontrollers has been eroding over time, just as the difference in price between 4- and 8-bit microcontrollers became smaller in the past.

32-bit microcontrollers: Currently, this market segment is one of the most dynamic, with devices that provide superior performance *and* challenge 16- and, in some cases, even 8-bit microcontrollers on price. Many are based on architectures originally developed for desktop computers, and advances in integrated circuit technology and manufacturing techniques have made them cost competitive in microcontroller applications. The segment is dominated by processors based on the ARM architecture (the acronym was originally based on the full name of the architecture, the Acorn RISC Machine, later called the Advanced RISC Machine), though this is only one of many 32-bit architectures available. The biggest shortcoming of these devices is their lower code density, which results in requirements for more program memory compared to more traditional 8- and 16-bit microcontrollers. Given the current program memory density available, this is only a minor limitation.

2.6 THE DATA BUS AND THE ADDRESS BUS

In addition to representing numbers, a microcontroller must have a means of transferring and communicating them between the various on-chip and off-chip components in order to perform useful operations. Since microcontrollers use switches to represent the bits that make up numbers, this is accomplished by connecting the output of these switches among all the components that either need to read in numbers or write numbers for other components to use. (This is a simplification of real implementations, but illustrates the basic concept.) A separate connection, or line, is required to communicate each bit of data. Together, the grouping of connections that serve to communicate numbers between components is called a "**bus.**"

The size of each bus, that is, the number of bits, has a significant impact on the performance of a microcontroller. As in the case of the ALU registers, the number of bits in the **data bus** determines how much data can be moved from a source to a destination in a single operation. For example, if you were working with a microcontroller that had a data bus that was 8 bit wide and needed to transfer the number 210 from data memory to the CPU, this could be done in a single step, since an 8-bit data bus can represent numbers as large as 255 (as long as we are talking about unsigned numbers, at least). However, if you needed to transfer the number 31,552 instead, this would take two steps with an 8-bit data bus, since with two separate transfers of 8 bits (for a total of 16 bits) we are able to represent numbers as large as 65,535. Microcontrollers that have data buses 16 bit wide are very common and have the benefit of significantly speeding up transfers of larger numbers. Microcontrollers with bus widths of 32 bits are available as well. There is a trade-off to having a large data bus, however, it requires more silicon to implement the chip, and hence costs more. As the cost of chips falls due to improvements in manufacturing, the premium for wider bus widths continues to diminish. This drop in cost is at the root of the increased market penetration by 32-bit microcontrollers.

2.7 MEMORY

For the **address bus**, the width of the bus, or the number of bits that comprise it, determines how many unique addresses in memory the CPU can specify. Consider an imaginary but not-so-useful microcontroller that only has an 8-bit address bus. There are 256 unique locations in memory that can be specified, and these must be divided between program memory, data memory, control registers, and any peripheral devices that may be required. Though such a microcontroller would be suitable for only very simple systems (that require very short programs), it would probably be simple to use and inexpensive, and might address a large market. Nevertheless, most microcontrollers have address buses larger than 8 bits, and are able to address far more than 256 unique memory locations.

Each memory location is identified by its address, just as street numbers correspond to unique houses on a street, as shown in Figure 2.3. Each location in memory contains what is generically referred to as "data." This "data" could be interpreted as an instruction, the contents of a variable, configuration data, or something else, depending on the specific context. Most often, the data is 8 bits (1 byte) long, but this is not required. Regardless of what the data at any particular address is used for, the address serves to identify it so that it may be read, written, or otherwise put to use.

Most microcontrollers have address buses much larger than 8 bits; 16-bit address buses are very common, and much larger address buses are also available. A 16-bit address bus allows for 65,536 unique locations in memory to be specifically identified. Again, these locations are often divided into program memory, data memory, registers, and other peripherals, but certainly having thousands of addresses available affords the designer far greater flexibility.

Regardless of how many unique addresses are available in a microcontroller's memory space, they are most commonly divided up among the various types of memory, registers, and other subsystems. If a device or subsystem requires a number of addresses for its function (for instance, an EPROM device used for program memory might be capable of storing 2,048 bytes), it is usually convenient and customary to connect it so that it responds at contiguous addresses. The standard way of graphically representing these divisions and the overall organization of the memory space is with a **memory map**. A representative memory map is shown in Figure 2.4.

Address Space

Address	Data Contents
0000	
0001	
0002	
0003	
0004	
0005	
0006	
0007	
0008	
0009	
000A	
000B	
000C	
000D	
000E	
000F	
0010	
⋮	⋮

FIGURE 2.3

Address space organization scheme. Each address corresponds to a unique location in memory, which contains data. Data may be storage for numbers, configuration information, or other numerical quantities.

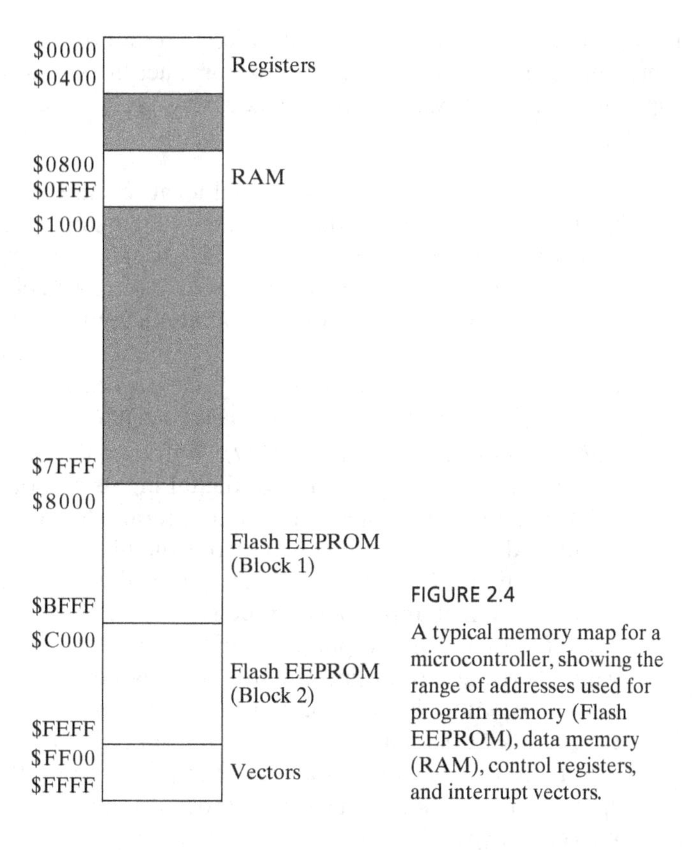

FIGURE 2.4

A typical memory map for a microcontroller, showing the range of addresses used for program memory (Flash EEPROM), data memory (RAM), control registers, and interrupt vectors.

Memory addresses are shown on the left side of the memory map and that they start with 0 at the top and increase in value towards the bottom. Addresses are given in **hexadecimal** (sometimes shortened to just "**hex**"), or base-16, which is discussed in greater detail in Chapter 3. The names of the devices and subsystems connected to the microcontroller are shown on the right of the memory map. The entire range of addresses should be shown in order for the map to be complete and as useful as possible. The control register section, where the microcontroller and many of the subsystems and peripherals are configured and controlled, starts at the lowest address, $0000 (the dollar sign indicates hexadecimal), and ends at $0400, indicating that there are 400 hex (1,024 decimal) possible registers for this microcontroller. Note that there are several regions of the memory map that are gray, indicating that they are not in use. There is no requirement that every address be in use, and these gray regions show where the unused regions of the memory space are. Also, note that there are two regions labeled "Flash EEPROM." Each of these regions refers to different blocks of non-volatile program memory. Finally (though the use of interrupts and interrupt vectors is beyond what we cover in this introductory textbook), the interrupt vectors (from $FF00–$FFFF) are special areas in the program memory that are used by the processor hardware when servicing an interrupt.

2.8 SUBSYSTEMS AND PERIPHERALS

In addition to the program memory (instructions) and data memory, most microcontrollers include a number of additional subsystems. Each type of subsystem is useful in specific types of applications. In combination with a microcontroller's architecture, the subsystems that are incorporated determine which applications are appropriate. Often, many different subsystems can be mixed and matched within product lines to afford a great variety of cost and capabilities. The subsystems summarized here are covered in more detail in Chapters 7, 8, and 19.

Control Registers: These are special locations in the memory onboard the microcontroller used to configure the chip and its subsystems. All microcontrollers have control registers, as all microcontrollers require some setup specifications from the user. Control registers are used for such things as setting a clock speed, putting a micro to sleep to save power, and configuring a communications port for a desired baud rate, among many other possibilities. Generally, where there is hardware that can be used in more than one way or turned on and off, there is a register or set of registers associated with it for configuring and exchanging data with it.

Ports & Parallel Input/Output (I/O): **Ports** are the most common peripheral available on micros. Ports are groupings of digital inputs and/or outputs, so they are one of the most direct means a micro can use to interact digitally with its environment. Ports are comprised of several individually controllable digital inputs or outputs, called "**pins**" (instead of "bits" because they are physically connected to the external pins of the microcontroller). Input pins can be read to determine the digital state of devices connected to them, and output pins can be written so that they take on a specified digital value. Ports, and often individual pins, may be input-only, output-only, or user selectable to be either input or output. A very common task performed by micros is to turn external devices on and off, or read the state of an external device's output. I/O ports are the most convenient way of accomplishing this. Ports are generally mapped to a specific memory location in the microcontroller's control registers. The state of output pins can be specified by writing data to the register associated with the output port, and the state of input pins can be determined by reading data from the input port's associated register.

Counters: These are registers in the memory of a micro that count events. For example, a counter can monitor transitions on a digital input without requiring the input pin to be continually monitored by the CPU. Counters are covered in depth in Chapter 18.

Timers: These are counters that specifically increment based on intervals of time from a clock source. Timer designs can be very simple, such as those that keeps track of transitions of a micro's system clock, or very complex, such as those that coordinate complex sequences of events at specified times. Timers are also covered in depth in Chapters 8 and 18.

Serial I/O: Serial communication is a means of communicating data between components sequentially, one bit of data at a time. With high-speed serial communication, it is straightforward to transfer large quantities of information quickly and easily between two or more points. Since links between microcontrollers, peripheral chips, and other miscellaneous external systems are very common, it is very convenient to work with micros that incorporate serial communication subsystems. There are several standard methods of performing serial communications, and microcontrollers usually incorporate serial I/O peripherals that adhere to one or more of the conventions. Examples of this are **USB** (Universal Serial Bus), **CAN** (Controller Area Network), **SPI** (Serial Peripheral Interface), and **I^{2}C** (Inter-IC) **Bus**. The two major categories of serial communication are **synchronous** (where a separate line acts as a clock, informing receivers when the current bit of data is ready to be read) and **asynchronous** (where a clock rate is agreed upon before communications begin, and each bit of data is asserted for an interval of time determined by that clock rate). Serial communications are covered in depth in Chapter 7.

Analog-to-Digital Converters: Whenever a microcontroller must interact with components that have an analog output (an output that can take on any value between certain limits, rather than just logical values of "0" or "1"), such as a sensor, the value of the output must be converted into a digital, numerical representation before the micro can make use of it. This is the task of the **analog-to-digital converter**, also called the **A/D converter** or **A-to-D converter**. Many microcontrollers have on-chip analog-to-digital converters. These on-chip converters may be of varying resolution (a measure of how finely they are able to convert the analog signal to a digital representation), with 8-bit (1 part in 256), 10-bit (1 part in 1,024), and 12-bit (1 part in 4,096) converters common. Also, it is common to have multichannel converters incorporated so that the micro can process several individual signals. Analog-to-digital converters are covered in depth in Chapter 19.

Digital-to-Analog Converters: The converse operation of the analog-to-digital converter is also a very commonly needed task for microcontrollers: the need to be able to create an output with a continuously varying analog value, rather than the digital logic levels representing "1" or "0." A **digital-to-analog converter** (also called "**D-to-A**" and "**D/A**") is required to accomplish this. This is a less commonly needed operation than the analog-to-digital conversion, and as a result, there are very few microcontrollers that incorporate true digital-to-analog converters on-chip. An alternate means of creating an analog value using only a digital output is to rapidly pulse it on and off. By varying the amount of time it is "on" compared with the amount of time it is "off" and using a means of filtering or averaging the output, an analog voltage can be created. This means of creating analog voltages using digital outputs is called **pulse width modulation**, or **PWM**, and is often incorporated as a peripheral subsystem in microcontrollers. Both of these means of converting digital values into analog values are further discussed in Chapters 8 and 19.

2.9 VON NEUMANN ARCHITECTURE

There are two common ways of organizing and arranging the connections between the CPU, the memory, and the peripheral systems. The most common type of bus architecture in use with microcontrollers is one that incorporates two buses to create the interconnections between the various on-chip microcontroller system components: one data bus and one address bus (Figure 2.5). The data bus serves to communicate data between components in a system, and the **address bus** serves to identify the locations in memory of the source and/or destination of the data. All components in the system share the same data bus and address bus as they perform tasks and communicate with each other. This arrangement of microcontroller buses is known as the **von Neumann architecture** and is named after John von Neumann, the mathematician who first proposed it. The von Neumann architecture has been widely adopted because of its flexibility and relative simplicity, both of which allow for lower-cost microcontroller and system design costs. Other approaches are capable of optimizing various aspects of operation (such as the Harvard architecture, discussed in Section 2.10), but in a great many applications, the von Neumann architecture serves very well.

In the von Neumann architecture, the data bus and the address bus always work together to fully describe the "what" and the "where" of every transfer of data within a microcontroller. The CPU coordinates the flow of information for the system as a whole. It

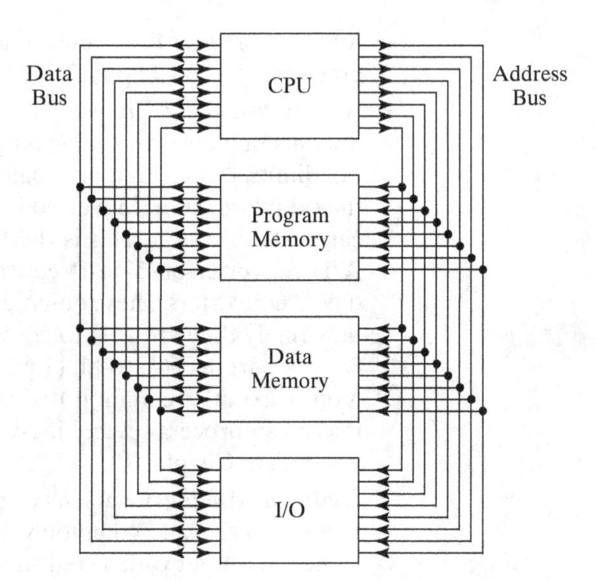

FIGURE 2.5

A microcontroller's CPU connected to digital input/output (I/O) and memory via data and address buses.

can either request to read information from another component on the bus at a specific address or request to store information into a component on the bus at a specific address. The CPU initiates all communication and data transfers for all components on the buses.

The important implication of the von Neumann architecture is that all devices connected to the bus are essentially equivalent from the perspective of the CPU. For instance, there's no difference between the program memory (which would normally be filled with instructions) and the data memory (which would normally be filled with variables). Since instructions and data look very much the same to the CPU, there is nothing to prevent the CPU from interpreting data as instructions and gamely executing it. This can be disastrous in the wrong circumstances, resulting in a system crash (or virus infection). However, there are many scenarios when this is desirable, for instance, when a programmer would like to rapidly iterate on a program under development. "Data" space can be quickly loaded with the latest version of a new program, and the CPU can immediately begin executing it. As long as the instructions are valid and there's no confusion about which addresses are to be used for instructions and which are to be used for data, this will work perfectly. This is actually a very common practice in programming microcontrollers where a small "boot-loader" program will be loaded into what is normally the data memory (RAM) and then executed to control the programming of the non-volatile program memory (typically Flash EEPROM).

When microcontrollers with von Neumann architectures execute instructions, they follow the basic pattern as shown in Figure 2.6.

Since fetching instructions, fetching data, and storing data all require the use of the shared data bus and address bus, this process is inherently sequential. Microcontroller instructions can range from the very simple (e.g., "jump to this new location in memory and continue executing instructions") to the very complex (e.g., "divide the integer X by the integer Y and place the result in memory starting at address A"). For simpler instructions, this means that a relatively large percentage of the microcontroller's time (and thus execution speed) is spent fetching the next instruction to be executed. For more complex instructions, fetching it may take far less time than executing it.

Instructions are usually several bytes long, and are not necessarily all the same length. Depending on the size of the data bus, several sequential transfers may be required to get an entire instruction from the program memory to the CPU so that it can be executed. For

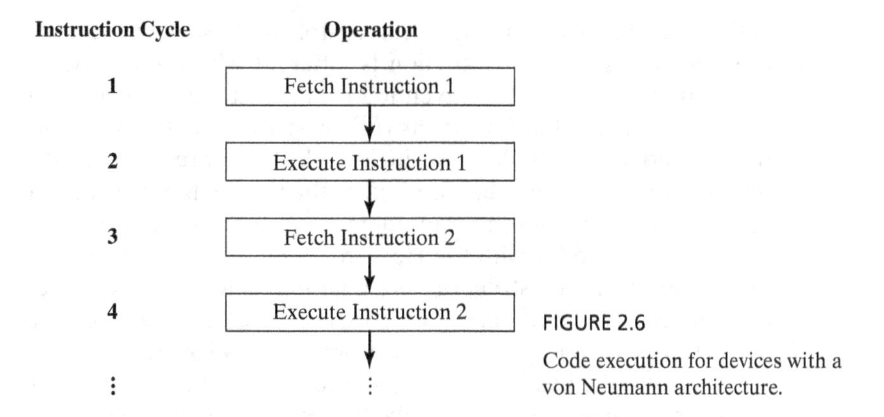

FIGURE 2.6

Code execution for devices with a von Neumann architecture.

an 8-bit microcontroller, a 4-byte instruction will require four sequential transfers for the CPU to fetch the complete instruction, for example. A 16-bit microcontroller, on the other hand, would only require two sequential transfers. Simple instructions, such as a command to "jump" to a new address and continue executing, may be only 3 bytes long. More complex instructions, such as a command to multiply two integers together and store the results, might be 4 or more bytes in length, requiring numerous sequential transfers to move it from program memory into the CPU for execution.

2.10 THE HARVARD ARCHITECTURE

A different approach to organizing the program and data memory is to separate the program memory completely from the data memory. If they no longer share a common data bus and address bus, they are then free to transfer their information independently, which eliminates the traffic jam created by the necessity to fetch an instruction first and then execute it and store the results. One arrangement that implements this concept is referred to as the **Harvard architecture**, and is incorporated into DSPs and microcontrollers that have been optimized for execution speed. In general, a micro incorporating the Harvard architecture will be arranged as shown in Figure 2.7. The width of the buses will vary depending on the architecture of the device. The bus widths illustrated in Figure 2.7 represent one particular example.

Since the Harvard architecture separates the instruction memory completely from the data memory, there is no longer a requirement for the width of the buses to be the same for

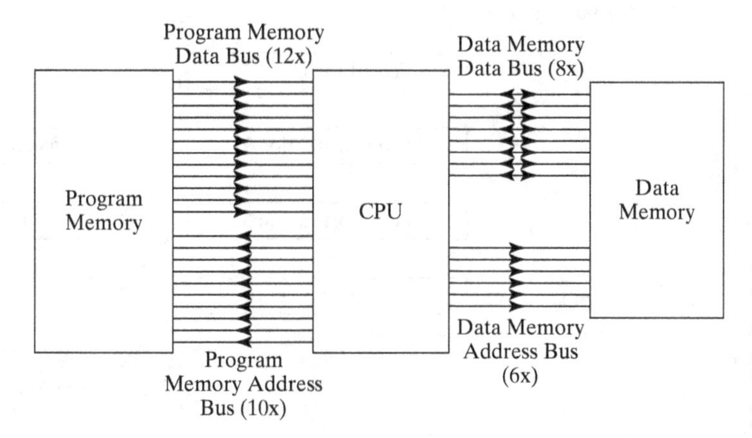

FIGURE 2.7

Typical arrangement for a microcontroller with a Harvard architecture, showing separate buses for instruction memory and data memory.

both spaces. In the data memory, it is customary to stick with the standard bus sizes (8, 16, 32, or even 64 bits), but the situation is different with the instruction space. There is no real requirement to adhere to convention for the instruction space, and in fact, it is often advantageous to create a bus for the instruction space that allows all instructions to be transferred from memory to the CPU in a single cycle. Compare this with the von Neumann architecture, where instructions that are longer than the data bus width can require multiple cycles to be transferred from program memory into the CPU before they can be executed. The wider bus allowed for with the Harvard architecture can result in a great increase in execution speed, since all instructions are guaranteed to move from the memory into the CPU in only one cycle. In addition, knowing that one cycle is always required allows designers to calculate the execution speed of their programs very easily.

Usually, the bus for the instructions will need to be larger than 8 bits, which wouldn't allow enough space to fully describe the instruction to be executed, include any references to addresses for variables or jump destinations, contain constants, and so on, so the bus width is typically 12 bits or more. Any number of bits is possible—the only requirement is that it must be sufficiently large to make it possible to fully specify every instruction understood by the CPU and incorporate all the necessary references and options.

In the case of a microcontroller that employs a Harvard architecture, the CPU can fetch an instruction from the program memory without tying up the data memory bus, so the CPU is free to perform manipulations using the data memory at the same time. So while a microcontroller that employs a von Neumann architecture is limited to the sequential "fetch, execute, fetch, execute, and so on" cycle, a microcontroller with a Harvard architecture can fetch an instruction, and then simultaneously execute it while it fetches the next instruction. Graphically, this looks like Figure 2.8.

With the Harvard architecture, after the first fetch operation has primed the system, it is able to execute one instruction per instruction cycle. Von Neumann architecture systems, by comparison, are only able to execute at most one instruction for every two instruction cycles. The execution process for devices employing a Harvard architecture is inherently parallel with respect to the fetching and execution of instructions, compared with the von Neumann architecture's sequential execution process. The additional complexity of the design of this architecture (which usually translates directly into increased expense for devices that use it) results in the substantial return of increasing the execution speed. This is the primary reason for its use in DSPs, where speed of execution is a primary requirement.

Having separate data and instruction spaces, as well as different bus widths for each space, generally results in no longer being possible to execute instructions stored in the data memory space, which was easy to do—and often convenient—with a micro using the

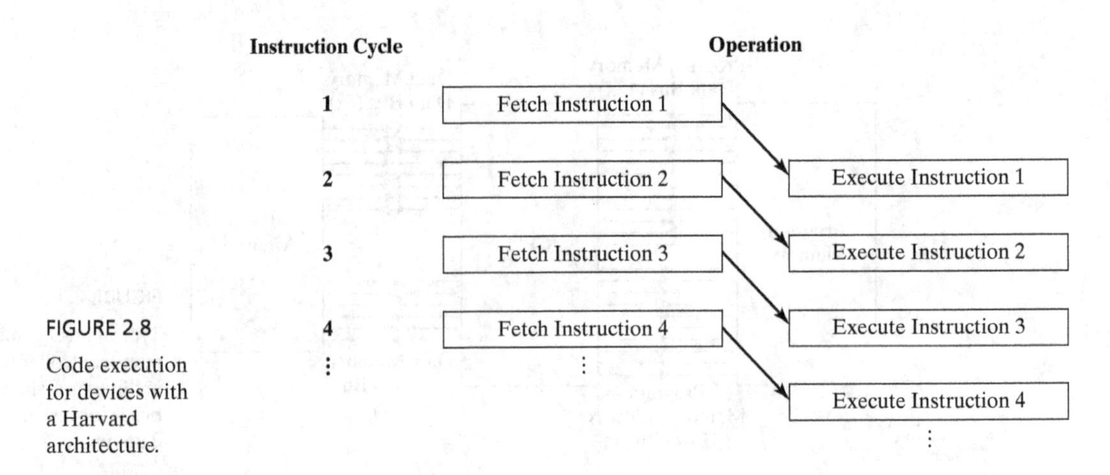

FIGURE 2.8

Code execution for devices with a Harvard architecture.

von Neumann architecture. With the Harvard architecture, the instructions are not guaranteed to be the right size to fit in the memory space provided (consider an architecture that uses 14 bit long instructions and has an 8 bit wide data bus). And, if they were broken up and distributed over adjacent data memory locations, the CPU doesn't have any mechanism for reconstructing them so that it can then execute them. For micros with Harvard architectures, the distinctions between instructions and data are significant, and it's not usually possible to interchange them.

2.11 REAL WORLD EXAMPLES

Up to this point in the chapter, the discussion has been as general as possible, in order to broadly describe how microcontrollers are designed and how they function. In this section, two specific examples of microcontrollers will be explored. One example uses a von Neumann architecture (the Freescale MC9S12C32) and one example uses a Harvard architecture (the Microchip PIC12F609). These examples are representative of current 8-bit and 16-bit microcontrollers, are readily available, and are in widespread use.

2.11.1 The Freescale MC9S12C32 Microcontroller

The Freescale MC9S12C32 16-bit microcontroller is one member of a subfamily of microcontrollers that was introduced in 2003 as an expansion of the HCS12 family of controllers. The HCS12 family devices were initially designed for use as automotive controllers but have been very popular outside the automotive market as well. This particular subfamily, the C series, all have the same set of peripherals but differ in the amounts of RAM and Flash EEPROM included in the devices. Devices are available with as little as 32K bytes of Flash and 2K bytes of RAM, or as much as 128K bytes of Flash and 4K bytes of RAM. Each of the family members is available in a choice of three different packages from 48 to 80 pins (Figure 2.9). This allows the designer the option to trade fewer pins for a smaller package size.

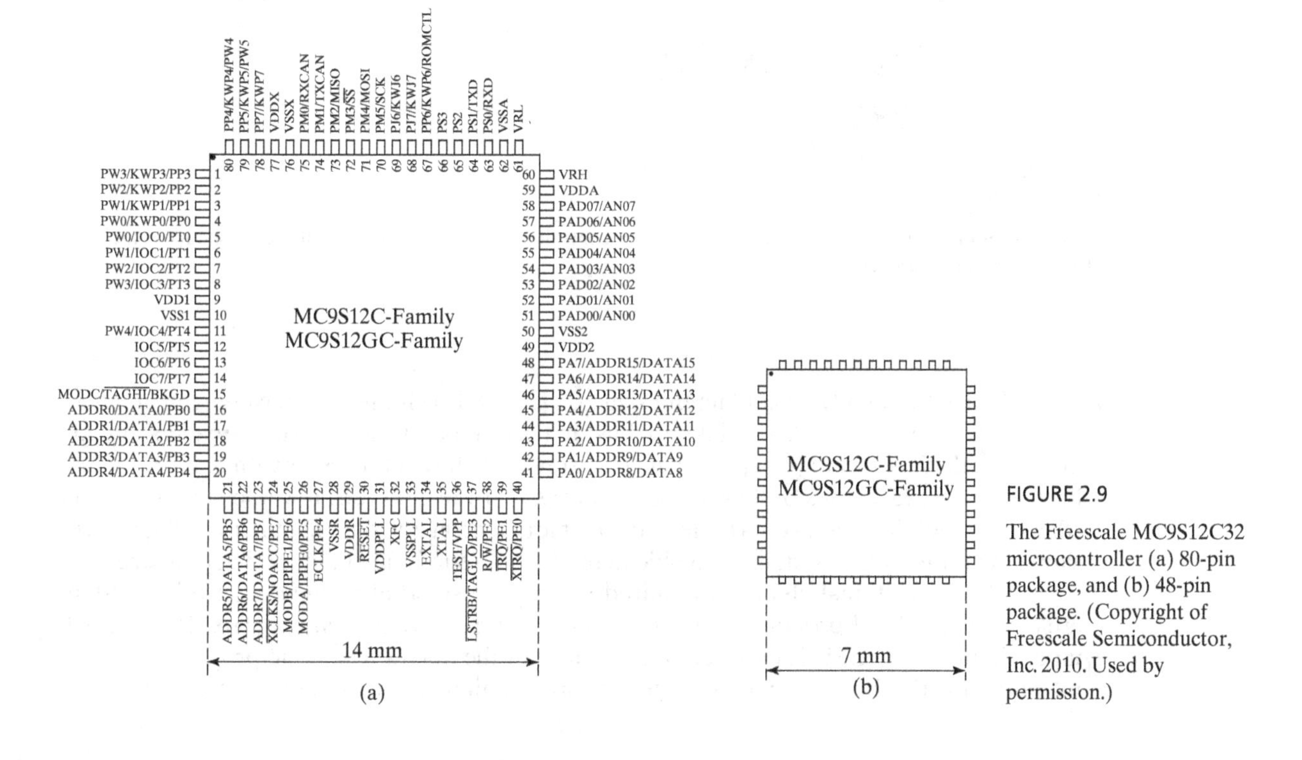

FIGURE 2.9

The Freescale MC9S12C32 microcontroller (a) 80-pin package, and (b) 48-pin package. (Copyright of Freescale Semiconductor, Inc. 2010. Used by permission.)

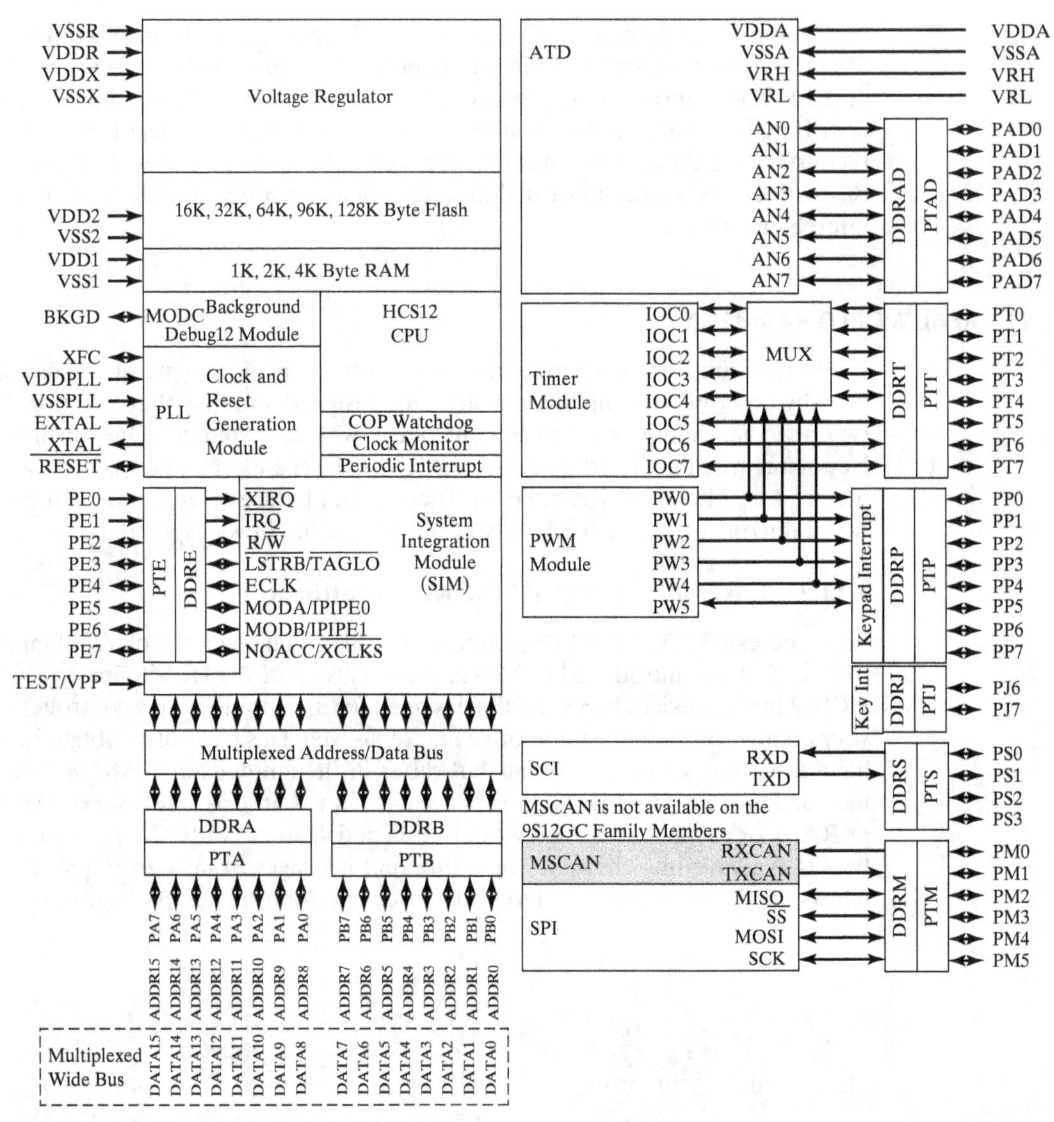

FIGURE 2.10

A block diagram of the Freescale HCS12C series microcontrollers. (Copyright of Freescale Semiconductor, Inc. 2010. Used by permission.)

The HCS12 family all contain the same CPU core which uses a von Neumann architecture and has a 16-bit ALU, a 16-bit data bus, and a 16-bit address bus. Those readers already familiar with binary math (Chapter 3) may be wondering how a processor with only 16 address lines accesses 128K bytes on Flash. The answer is that the chip uses a windowing scheme in which not all of the memory is visible in the memory map at any one point in time. Figure 2.10 shows the ports and subsystems available in the C series subfamily in block diagram form.

All the functional elements required to create a stand-alone microcontroller system are present: the CPU, program memory (Flash, in this case), data memory (RAM), and clock input circuitry. The CPU is depicted in the center of the diagram, and the program and data memory directly above it. In addition to the basic minimum components necessary to make

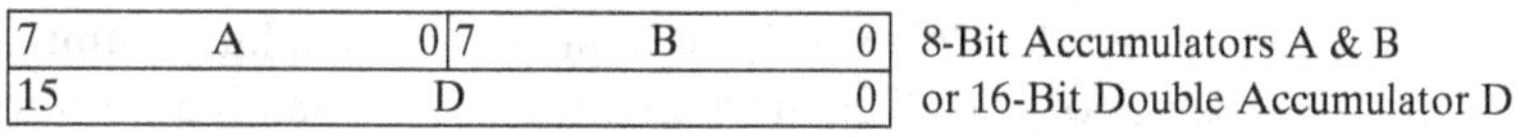

7 A 0	7 B 0	8-Bit Accumulators A & B
15 D 0		or 16-Bit Double Accumulator D

FIGURE 2.11

The Freescale HCS12 CPU's accumulator registers. (Copyright of Freescale Semiconductor, Inc. 2010. Used by permission.)

a microcontroller, the HCS12C family includes a wide range of I/O ports, analog-to-digital converter, timing, and communications subsystems.

The CPU in the HCS12 family is a direct descendant of the Motorola MC6800 introduced in the mid-1970s. It includes the same two 8-bit accumulators (Accumulator A and Accumulator B) as the original (Figure 2.11) but also allows those two 8-bit accumulators to be concatenated to form a single 16-bit accumulator register (Accumulator D). This gives the HCS12 CPU the ability to perform operations on 8- or 16-bit numbers with 8- or 16-bit results in a single operation. The other notable addition to the HCS12 CPU over its great-grandfather (the MC6800) is a second 16-bit pointer register that significantly improves the efficiency of programs written in C.

In addition to the CPU, the program memory and data memory, a wide variety of useful peripherals and subsystems are represented. The following list is not comprehensive (several functional blocks shown in Figure 2.10 are not mentioned) but covers many of the most important and useful:

Ports: Most of the ports of the 9S12C family can be configured for use as general-purpose inputs or outputs (I/O), or for special purposes. Many of the port pins may be configured as digital inputs or digital outputs, but some of them may be input only, and others may be configured to read analog or digital signals.

A/D Converter: 9S12C family devices include an 8 channel, 10 bit analog-to-digital converter. This subsystem can be configured to perform a single conversion on command, or to continually scan and convert a range of channels automatically. Port AD can be configured for use with the A/D converter, or as general-purpose digital inputs or outputs.

Timer: The timer subsystem is based on a single free-running timer but it can be configured to time external events or to trigger outputs to change at specific points in time on up to eight different port lines. Port T can be configured for use with the timer subsystem, or as general-purpose inputs or outputs.

PWM: While the timer subsystem can be used to generate pulse width modulated (PWM) outputs, the HCS12 family also includes up to six channels of hardware generated PWM that significantly lowers the overhead on the CPU necessary to maintain these outputs.

Communication: The HCS12C family devices include a rich set of communications subsystems. These include the SPI system for synchronous serial communications, the SCI (Serial Communications Interface) for asynchronous serial communications, and the CAN subsystem for implementing a networking standard that was developed for the automotive market but now sees broader usage.

External access to data and address buses: Though not used in the vast majority of applications, the 80-pin package offers a mode in which the internal address and data bus are brought out onto the lines of Port A and Port B. This gives designers the option of supplementing the internal memory of the chip with external memory or other special purpose devices that interface to the CPU in a way similar to memory chips.

The Freescale MC9S12C32 is available from a large number of distributors and retailers, and costs about $6.00 each when purchased in small quantities. The cost varies somewhat depending on the package chosen. The cost drops substantially for large quantity orders. For orders in the thousands, the price falls to under $5.00. In addition to the versatility and attractive price of the HCS12 family, there are a large number of high-quality (and often inexpensive) development tools available. Cross compilers that allow programmers to write code in high-level languages such as C are widely available at low (some even free) cost, and several companies manufacture evaluation boards using the HCS12 family devices, making it very easy to begin development.

2.11.2 The Microchip PIC12F609 Microcontroller

The Microchip PIC12F609 microcontroller takes a completely different approach to fill the market's needs for inexpensive, moderate-performance microcontrollers. The philosophy behind the PIC12F609 is to target applications where designers wish to minimize the total parts count, and hence the cost and/or space, required to implement a design. To that end, the PIC12F609 was designed to use an internal clock oscillator, though it is possible to use anexternal crystal if that extra precision is required. When using the internal oscillator, the 8-pin device has up to 6 pins available for I/O. The device includes 1,024 program words, each 14 bits wide, of internal program memory, and 64 bytes of internal data memory. The PIC12F609 (Figure 2.12) uses a Harvard architecture, so the program and data memory are completely separate. Because they are of different widths (14-bit program instructions vs. 8-bit data memory) the restrictions discussed in Section 2.10 apply with respect to executing instructions from data memory or storing data in program memory. A limited but very useful number of input pins and output pins are included, as are provisions for timing with one 8-bit and one 16-bit timer as well as an analog comparator.

The system block diagram for the PIC12F609 is shown in Figure 2.13. In this diagram, the CPU isn't shown as a single block: the ALU and a register called the "W" or "working" register are shown. The program memory is shown in the upper-left corner of the diagram as"Flash 1K X 14 Program Memory," and the data memory is shown at the upper center as "RAM 64 Bytes File Registers." Just as with the Freescale MC9S12C32, discussed in Section 2.11.1, all subsystems necessary to create a stand-alone microcontroller are included.

Many aspects of the PIC12F609 are significantly different from the MC9S12C32, however. One key difference is that the number of peripherals and subsystems included are much more limited. For instance, there is no synchronous or asynchronous serial communications

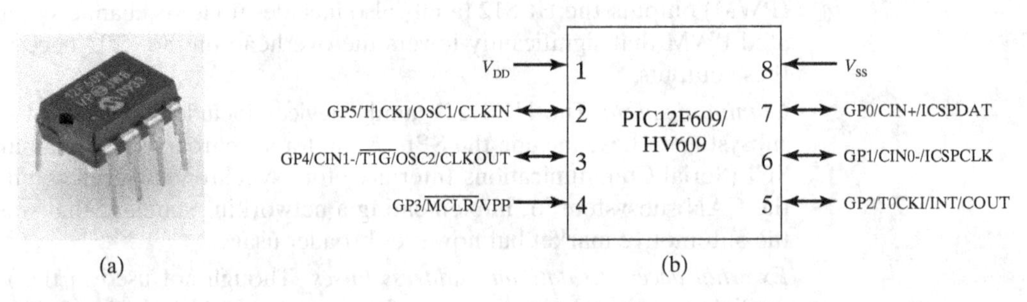

FIGURE 2.12

The Microchip PIC12F609 microcontroller (a) DIP package integrated circuit and (b) pinout. (© 2010 Microchip Technology Incorporated.)

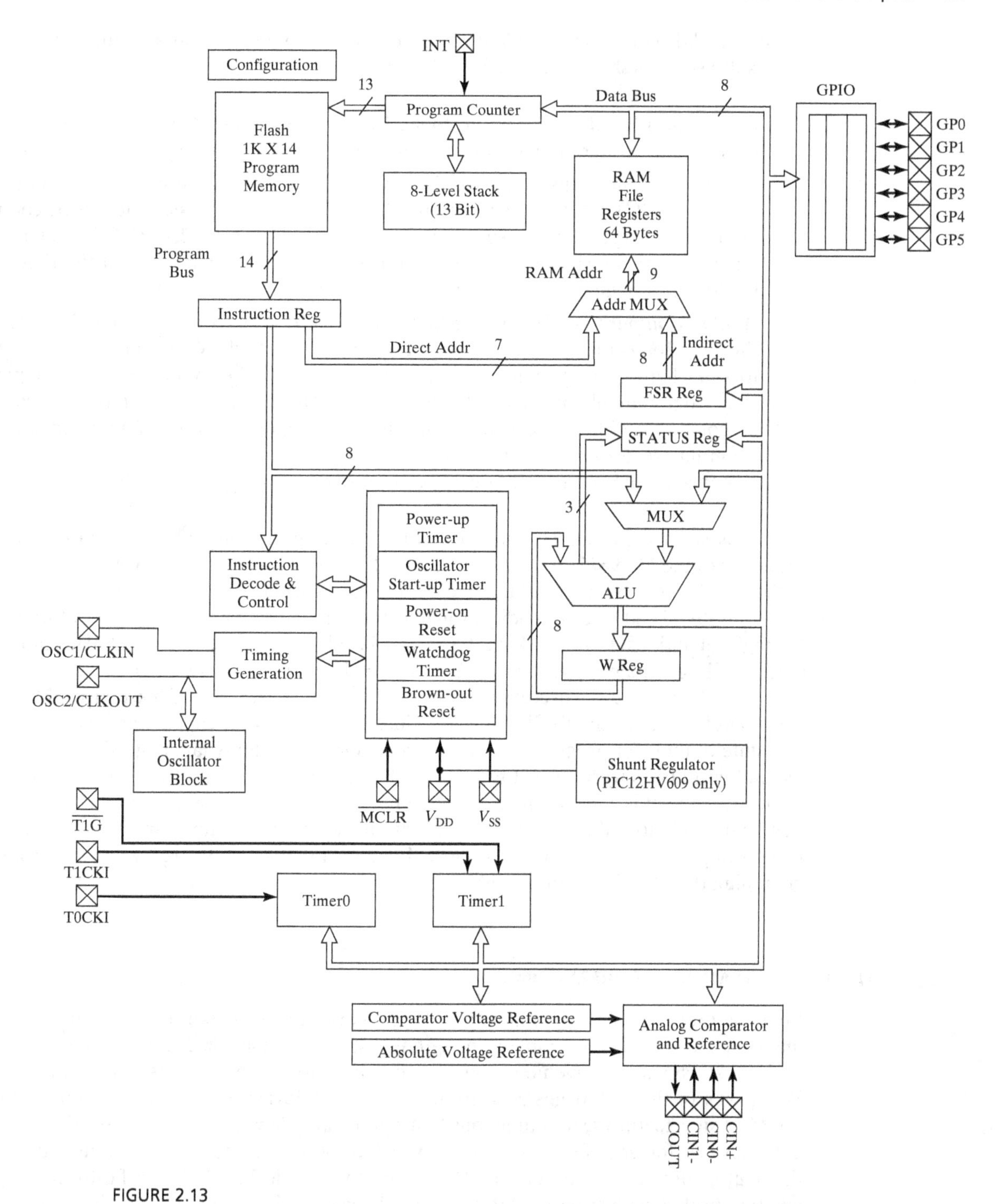

FIGURE 2.13

A block diagram of the Microchip PIC12F609 microcontroller architecture. (© 2010 Microchip Technology Incorporated.)

module, no A/D converter, no PWM generator, and the timer systems are much simpler. The list of subsystems is shorter for the PIC12F609:

Ports: The PIC12F609 offers up to six I/O pins, labeled GP0–GP5 in Figure 2.13. Of these, five are programmable as either inputs or outputs, and one is limited to input only.

Timer: The timer subsystems for the PIC12F609 are significantly simpler than that of the MC9S12C32, but still very useful. In addition to the basic function of counting internal clock cycles to measure time intervals, three pins (GP2, GP4, GP5) can be configured to create what are really counter/timer systems, since either of the timers can be driven from an external source to produce a counter of external events.

Analog Comparator: The analog comparator subsystem is a very versatile feature of the PIC12F609 that the MC9S12C32 does not offer at all. (We cover analog comparators in Chapter 11.) The inputs to the comparator may be connected to device pins or to a programmable internal voltage reference. The output of the comparator can also be connected either externally or internally and operates completely independent of program execution. The polarity of the output of the comparator is programmable and a change on the comparator output can cause an interrupt to the CPU.

This short list of subsystems may sound severely limiting when compared with the resources of the MC9S12C32, but in fact the PIC12F609 is considered a very capable device in the very low-cost segment of the microcontroller market.

The PIC12F609 is also readily available and very inexpensive (one of its real strengths). In single quantities, some package types can be obtained for under $0.71. The price falls quickly in larger quantities such as to below $0.60 in quantities of just 100.

Development tools are easy to find and in some cases free. For example, Microchip (the manufacturer of the PIC12F609) provides MPLAB, a very capable and well developed free suite of tools for programming their processors in assembly language. Despite its architectural limitations with respect to high-level languages, there are also several good C compilers available. It is a testimony to the ingenuity of the compiler writers that it is possible to write genuinely useful applications in C within the severe memory and architectural constraints imposed by this device. The most "bang for the buck" however, will come from programming the PIC12F609 in assembly language.

2.12 WHERE TO FIND MORE INFORMATION

Manufacturers of microcontrollers and DSPs work hard to make information about their many products readily available to potential customers—a group that now includes readers of this text. The most convenient way to access this information is to visit the manufacturers' Web pages, or the Web pages of distributors who sell devices from a variety of manufacturers. Most sites feature search functions that significantly facilitate the identification of candidate devices for specific applications. A list of some of the major manufacturers and distributors of microcontrollers and DSPs is given in Table 2.2. This list is far from complete, and this market evolves particularly rapidly. The list is only intended to provide a good starting point when initiating a search for a microcontroller or DSP for a specific design.

Another tool that manufacturers use to inform potential customers about their products is the "selection guide": a document that lists their products, grouped by performance and feature sets. Selection guides can be an efficient way to begin, and can be used to quickly determine which devices from a given manufacturer might be appropriate for a given design.

TABLE 2.2 Major manufacturers and distributors of microcontrollers and DSPs.

Manufacturers	Distributors
Atmel Corporation	Digikey Corporation
Freescale Semiconductor, Inc.	Mouser Electronics
Infineon Technologies, AG	Newark
Maxim Integrated Products	
Microchip Technology, Inc.	
SHARP Microelectronics	
Texas Instruments, Inc.	
Analog Devices Inc.	

Finally, manufacturers typically offer "evaluation boards" for most of their products. Evaluation boards are convenient, ready-to-use, assembled circuits (often provided with a compiler) that allow designers to begin using a specific microcontroller or DSP immediately—usually at minimal cost. Evaluation boards may be simple and bare bones (e.g., provide power to a device and access to its pins), or they may be complex and feature-rich, (e.g., provide basic power and access, and also facilitate the use of a device's features and peripherals: I/O, A/D converters, Ethernet, USB, CAN bus, and so on). Most evaluation boards include a schematic, which may then serve as a reference design for new circuits based on the device. Evaluation boards enable designers to begin using a micro immediately, and allow a device's capabilities to be explored, software to be written and tested, and custom circuitry to be designed—without having to reinvent the support circuitry needed to simply get the micro to run.

2.13 HOMEWORK PROBLEMS

2.1 Compile a list of at least 25 everyday objects that incorporate microcontrollers, microprocessors, or DSPs.

2.2 Using the Internet, locate the data sheet for the Atmel ATmega128A microcontroller, and answer the following:

(a) Does the ATmega128A have a von Neumann or a Harvard architecture?

(b) How much non-volatile Flash program memory is incorporated?

(c) How much volatile RAM data memory is incorporated?

(d) List at least five other important peripheral systems that are included (there are many more than five).

2.3 How is a microprocessor different from a microcontroller?

2.4 How big is the address space for a microcontroller whose address bus is 20 bits, wide?

2.5 What is the biggest number that can be represented with 24 bits?

2.6 List three different types of non-volatile memory.

2.7 Using the Internet, locate the data sheet for the Microchip PIC16LF727, and answer the following questions:

(a) What is the fastest clock source that can be used with this chip? What is the slowest?

(b) Is there an analog-to-digital converter peripheral included on this chip?

(c) How wide is the program memory bus? That is, how many bits are the program instructions?

(d) How many input/output (I/O) pins does the PIC16LF727 have?

2.8 What kind of microcontroller peripheral is present on the PIC12F609 that is not present on the MC9S12C32?

2.9 What prevents a microcontroller with a von Neumann architecture from attempting to execute data? What about a microcontroller with a Harvard architecture?

2.10 How many I/O pins are available in the largest MC9S12C32? How many are input only?

CHAPTER

Microcontroller Math and Number Manipulation

3

At its core, numbers are what a computer manipulates. In this chapter, we review some of the properties of binary numbers and introduce you to how numbers are represented and manipulated in a computer. When you have mastered the material in this chapter, you will be able to:

1. convert a number between its decimal, binary, and hexadecimal representations,
2. know the difference between 2's complement and sign-magnitude representations for negative numbers,
3. know how to form the 2's complement of a number,
4. know why integers are preferred to floating point numbers on microcontrollers,
5. construct a set of operations to set or clear one bit or a group of bits in a byte without affecting the other bits in the byte.

3.1 INTRODUCTION

Early human beings learned to count things using their fingers to keep track of the number of items. As a result, the mathematics that they developed and we grew up with is based on groups of 10. We call this the decimal system from the Latin word "decima," meaning a tenth part. We refer to the numbers as "digits" which comes from the Latin word for finger "digitus." Digital computers are different. Despite the name, they don't have 10 fingers and in current commercial technology, the only way that they have to keep track of things (memory cells) can only represent two states: 0 and 1. As a result, the mathematical operations that go on inside computer memory are based on binary, as opposed to decimal, mathematics. Since we'd ultimately like to program computers to do things for us, it is important that we understand how numbers are represented inside the computer.

Given that we are dealing with binary numbers, how do we manipulate these numbers to make decisions? Boolean algebra is a closely related field that describes the logical operations between elements that can take on only two possible states. We will use Boolean algebra operations to manipulate and test the binary numbers that we use in our computer programs.

3.2 NUMBER BASES AND COUNTING

The terms binary and decimal refer to the number of possible values that a digit can take on. This number of possible values is referred to as the **number base** for a representation. In general, numbers in any base are represented by a series of symbols that we generally call

numerals. The rightmost numeral is referred to as the **least significant digit**. For a given number base-N, the **place** of the least significant digit will be the number base-N raised to the 0th power (this is always 1, of course). This is then referred to as the "1's place" of the number. The next least significant digit is the numeral to the left of the 1's place, and its place is determined by taking the number base-N to the 1st power. For the case of base-10, then, that is 10^1, or the 10's place. This continues for the numerals to the left of the N^1 place, with the progression being N^2, N^3, N^4, and so on. For example, in base-10, the places are:

...	10^4	10^3	10^2	10^1	10^0
...	10,000's place	1,000's place	100's place	10's place	1's place

Counting proceeds by incrementing the least significant place until the number base minus 1 (N–1) is reached. On the next count, that place rolls over to 0 and the next most significant place is incremented by 1. Again, for the case of base-10:

Place:	10^2	10^1	10^0
			0
			1
			2
			.
			.
			.
			9
		1	0
		1	1
		1	.
		1	.
		1	.
		1	9
		2	0
		.	.
		.	.
		.	.
		9	9
	1	0	0
	1	0	1

The same rules hold for any number base. For base-2, the places are:

...	2^4	2^3	2^2	2^1	2^0
...	16's place	8's place	4's place	2's place	1's place

We refer to bits in a binary number by the exponent associated with the bit position. The 1's place is referred to as bit 0, the 2's place as bit 1, the 4's place as bit 2, and so on. Counting from 0 to 9 proceeds as:

Place:	2^3	2^2	2^1	2^0	(Base-10)
				0	(0)
				1	(1)
			1	0	(2)
			1	1	(3)
		1	0	0	(4)
		1	0	1	(5)
		1	1	0	(6)
		1	1	1	(7)
	1	0	0	0	(8)
	1	0	0	1	(9)

We sometimes need to be able to figure out what the binary representation is for a decimal number. The quick and easy way to do this is to grab a scientific calculator or one of the online equivalents. If neither of them is handy, you can use this procedure to convert a number in decimal to binary:

1. Divide the decimal number by 2.
2. If the result has a fractional part, write down a 1 to the left of the last binary digit. If the result is a whole number, write down a 0.
3. Throw away the fractional part, but keep the whole number part of the result.
4. If there is a nonzero whole number part left, repeat from step 1 replacing the original number with the result of the division, without the fractional part.

The algorithm works because dividing by 2 is the equivalent of shifting a binary number 1 bit to the right. After the division, a nonzero fractional part indicates that the original number was odd and therefore had a 1 in the least significant position. By repeatedly dividing by 2, we are effectively sliding the bits of the binary number "off the right" 1 bit at a time.

Example 3.1

Convert decimal 22 to binary.

22/2 = 11	result has no fractional part, add a 0	0
11/2 = 5.5	result has a fractional part, add a 1	10
5/2 = 2.5	result has a fractional part, add a 1	110
2/2 = 1	result has no fractional part, add a 0	0110
1/2 = .5	result has a fractional part, add a 1	10110

At this point, there is nothing left of the original number, so the task is done. To check the result, perform the inverse operation: 10110 = 16 + 4 + 2 = 22.

Given that we work in base-10 and the computer works in base-2, you might think that all you really need to master in life is base-10 and base-2. However, you will find that understanding base-16 (**hexadecimal** or simply **hex**) will also be required. Being comfortable with hex will help make sense of the data sheets and programming examples for many devices such as data acquisition and I/O cards even if you are not programming a microcontroller.

Hex has become the standard way of expressing microcontroller-related numbers in programming because it is a more compact and efficient way of expressing numbers than binary. It is also much more closely related to the numbers, as they are used by microcontrollers, than is base-10. Although the examples are probably getting tedious at this point, here is how the places are identified in hex:

...	16^4	16^3	16^2	16^1	16^0
...	65,536's place	4096's place	256's place	16's place	1's place

Counting gets a little more interesting. Since we are now using a number base that needs more symbols than we have at our disposal with the Arabic numerals used in base-10, we borrow some symbols (A–F) from the Roman alphabet. Counting from 0 to 20 in base-10 (0 to 14 in hex) proceeds as follows:

Place:	16^1	16^0	(Base-10)	(Binary)
		0	(0)	0000
		1	(1)	0001
		2	(2)	0010
		3	(3)	0011
		4	(4)	0100
		5	(5)	0101
		6	(6)	0110
		7	(7)	0111
		8	(8)	1000
		9	(9)	1001
		A	(10)	1010
		B	(11)	1011
		C	(12)	1100
		D	(13)	1101
		E	(14)	1110
		F	(15)	1111
	1	0	(16)	10000
	1	1	(17)	10001
	1	2	(18)	10010
	1	4	(20)	10100

It is an especially useful skill to be able to convert between the hex and binary representations of a number. This is made easier by the fact that, since base-16 is an even power of 2, each hex numeral must represent four digits worth of binary. For example, the binary number 1010 is equal to the hex number A. Each group of four adjacent binary digits, no matter how many total digits there are, can always be represented by a single hexadecimal numeral. This is the primary reason that hex is so useful—a number that might be 12 digits long in binary would only be three digits long in hex, and is therefore much easier for the human mind to grasp.

Example 3.2

Convert the binary number 101101010001 to hexadecimal.

In binary, it is a challenge to comprehend the number:	101101010001
But by recognizing that each grouping of four binary digits can be represented by a single hex digit, we break the number into more easily understood chunks:	1011 0101 0001
Then, each group of four binary digits can be represented in hex:	B 5 1
So the number in hex is:	B51

This becomes a straightforward exercise with practice. There are only 16 combinations of binary-to-hex conversions to remember, and once you know how to count in binary, you can quickly recreate a binary-to-hexadecimal conversion crib sheet when needed. Compared to converting to decimal and back, this is far easier.

To indicate that a number is hexadecimal and not decimal, there are a few common conventional notations used. One approach is to precede the number with a dollar sign, so for our example, it would be written $B51. Another is the way used in the C programming language: precede the number with "0x" (i.e., a zero followed by the letter x), for example, 0xB51. The last of the common notations (the method used by many versions of the BASIC programming language) is to append a lower case h to the number: B51h.

3.3 REPRESENTING NEGATIVE NUMBERS

Now that we can represent positive numbers in decimal, binary, and hexadecimal, how do we deal with negative numbers? There are two basic approaches to representing the sign of the number. One approach mimics the way we write negative numbers in base-10. The other seeks to take advantage of a property of negative numbers.

One of the first solutions that comes to mind when puzzling over how to represent negative numbers is that we should simply do it as we do decimal numbers: add a symbol to indicate the sign. In the case of binary numbers, that would be a **sign bit**. Using this approach, dubbed **sign-magnitude**, we would simply use the leftmost bit in the binary number to indicate the sign of the number. This would make it easy for a human to look at a binary number and quickly tell if it were negative, but doesn't do much to help a computer. As a result, this format sees most use when formatting numbers to be displayed for humans to read. In these situations, the sign bit can be used to turn on the negative sign in a display, for example.

The other common way of representing negative numbers makes use of a mathematical concept known as **complement notation**. Since we'll be working in binary, we'll use 2's complement notation. The power of 2's complement notation is that it fits well with the property of a negative number: $-A + A = 0$.

To form a **2's complement**, you take the 1's complement and add 1. The **1's complement** is often simply referred to as the **complement**, and in binary, it means flipping all the 1s to 0s and all the 0s to 1s.

Example 3.3

Express the decimal number −6 using 2's complement notation.

6 in binary is	0110
The 1's complement is	1001
Adding 1 to form the 2's complement gives −6	1010

Counting, with a 4-bit binary number, using 2's complement notation proceeds as:

Decimal	2's Complement Binary
7	0111
6	0110
5	0101
4	0100
3	0011
2	0010
1	0001
0	0000
−1	1111
−2	1110
−3	1101
−4	1100
−5	1011
−6	1010
−7	1001
−8	1000

The real power of 2's complement notation, and why it is the dominant format in computers, is that it simplifies the logic inside the computer's arithmetic logic unit (ALU). Using 2's complement notation, the computer does not need to know how to subtract, per se. To subtract, it forms the 2's complement and adds. In addition, the logic to add and subtract both signed and unsigned numbers is exactly the same and, as we will see in Section 3.7, the process of decrementing through zero results naturally in the 2's complement representation for −1.

Example 3.4

Perform $-5 + 7 = 2$ using 2's complement notation.

5 in binary is	0101
The 1's complement is	1010
Adding 1 to form the 2's complement gives −5	1011
Add 7	0111
The result is 2	10010

Here we have a dilemma. To fully represent the result of the add operation shown in Example 3.4 requires 5 bits, but we have been working with 4-bit numbers. What has occurred is an **overflow** beyond of the limited range of numbers that we can represent with a given number of bits. Whenever we deal with arithmetic in a fixed number of bits, there is always the possibility of overflow. In general, what happens is that the carry out of the most significant bit (in our case, the fourth) is lost. This is true, no matter how many bits that we use to represent −5 and 7. You will always have an overflow out of the most significant bit. In the case of 2's complement math, we are depending on that overflow being lost (discarded), so that the 4 bits that will remain after the add operation will be 0010, which is the correct result for $-5 + 7$.

3.4 DATA TYPES

The numbers that we have been representing so far have all been whole numbers. In our everyday life, we don't limit ourselves to whole numbers and regularly use numbers that also contain a fractional part. In fact, in normal use, we don't usually make any distinction between them. In computer math, it is necessary to make a distinction between data types that can only represent whole numbers and those that can also represent fractional numbers. The two most common forms of these data types are called **integers** for the whole number only form, and **floating point** for the form that can represent fractional numbers. When integers are stored in memory, they are stored directly in the forms that we discussed in Sections 3.2 and 3.3. When floating point numbers are stored in memory, they are stored using a form similar to scientific notation with a **mantissa** and an **exponent**. Single precision floating point numbers are stored in memory using 4 bytes, where the bits within the 32-bit field describe the various parts of the number:

Sign	Exponent	Mantissa
bit 31 (1 bit)	bits 23–30 (8 bits)	bits 0–22 (23 bits)

When you write a number as 12, you are specifying an integer number. If you write it as 12.0, you are specifying a floating point number.

Why are we making such a fuss over this distinction? In a word: performance. The designers of microcontrollers have included electronic circuits (hardware) with the ability to deal directly with integer numbers. Adding, subtracting, and, in some cases, even multiplying and dividing integers are done very quickly and efficiently by the hardware of the microcontroller. On the other hand, most microcontrollers do not have hardware to deal with floating point numbers directly. In those cases, the microcontroller must use software to execute a series of integer operations on the various parts of the floating point number in order to perform floating point arithmetic. The result of this is that it takes significantly longer, often 10–100 times longer, to perform floating point calculations than it does to perform integer calculations. To write software that executes quickly, we will want to use floating point numbers judiciously and reserve them for those situations where they are absolutely necessary.

Another thing to be careful of with floating point numbers is comparisons. Because the mantissa has a limited number of bits with which to represent a number, calculations inevitably have **round-off errors** as less significant bits are dropped to be able to fit the result into the 23 bits of mantissa. Depending on the sequence of calculations performed, this might result in two numbers that differ only by round-off errors. This can be a problem if a comparison statement is used to test these two numbers for equality. They will not be exactly equal, though they may be close enough for your purposes. If that is the case, you must use care in constructing the comparison so that you allow for some amount of difference in the test for equality.

As an example of calculation using integers vs. floating points, consider the situation where we have a number in the range of 1–1,000 and we wish to multiply it by 1.5 but are only interested in the integer part of the result of the multiplication. In C, we could write:

```
Result = Variable * 1.5;
```

Because we have used a floating point constant (1.5), the compiler will generate code to first convert `Variable` to a floating point number, then do a floating point multiplication with 1.5, and finally convert the floating point result back into an integer.

On the other hand, if we wrote the assignment as:

```
Result = Variable + (Variable / 2);
```

The entire calculation would be done with integers and run much more quickly.

3.5 SIZES OF COMMON DATA TYPES

The memory of a microcontroller is organized into groups of bits. The **byte** (8 bits, in C a **char**) is the smallest unit of storage on most microcontrollers that you will encounter. While the byte is the smallest unit of storage, we do find occasion to talk about smaller chunks. The chunk smaller than a byte is called a **nibble** and consists of a group of 4 bits. A nibble is half a byte, and holds enough bits to represent one hexadecimal digit. For data sizes larger than a byte, there is less agreement. For most microcontrollers, the next step up is a 2-byte, 16-bit, quantity. In C, on 8- and 16-bit microcontrollers, this is referred to as an **int**. Here is where the commonality ends. When programming in C on 32-bit and larger computers, an int is a bigger number, most commonly 32 bits (4 bytes). On microcontrollers, the 4-byte quantity is usually referred to as a **long integer**, or simply **long**.

So, you've got a number and you need a place to store it: you need a variable. The first decision that you have to make when creating a variable is how big the variable is going to be. Since resources on microcontrollers are limited, we want to choose the smallest size that will work for our application. To answer the question of how big a variable we need, we must decide what range of values the number will take on. If you need to store a number in the range of 0–255, that will require 8 bits (a byte) to represent. In fact, that is the largest unsigned range that will fit in a byte. If the numbers are signed numbers (using 2's complement notation), we will need to share the range between positive and negative numbers. A byte-sized variable holding a signed number can represent the values −128 to +127. For 2-byte variables (an int in C), we can store unsigned numbers in the range of 0 to 65,535 or signed numbers in the range of −32,768 to +32,767. In general, an unsigned variable of N bits can take on the range of 0 to 2^N-1. A signed variable of N bits can cover the range from -2^{N-1} to $2^{N-1}-1$. Table 3.1 summarizes the most common data sizes and the ranges of values that they can hold.

TABLE 3.1 Data sizes and their ranges of values.

Data Size	Minimum Value	Maximum Value
Unsigned byte	0	255
Signed byte	−128	127
Unsigned 2-byte	0	65,535
Signed 2-byte	−32,768	32,767
Unsigned 4-byte	0	4,294,967,295
Signed 4-byte	−2,147,483,648	2,147,483,647

3.6 ARITHMETIC ON FIXED SIZE VARIABLES

The contents of variables in a computer program behave a little bit differently than the normal numbers that we are used to using. As we saw in Sections 3.4 and 3.5, variables have a fixed size and limited range of values that they can hold. This can lead to some frustrating problems for the programmer who doesn't pay close attention to the sizes of the variables that they are using. Consider this: What is the sum of 253 + 5? We can quickly say: 258. However, if a computer is doing the math, then before we can decide what the result of adding

253 and 5 is, we must know the size of the variables holding those values and the size of the result. If the variables are 2 bytes or larger variables, then the computer will get the same answer as we do: 258. If the variables are 1-byte variables, the answer we will get is 2. The sum of 253 and 5 will not fit into a byte-sized variable, so the sum will overflow and we will be left with the part that is less than 256. In this case, the number 258 requires 9 bits to represent: 100000010. Since we have only 8 bits in our variable, we lose the most significant bit, leaving us with $253 + 5 = 2$, which probably isn't what we intended. In general, if we add two variables, each with N bits, and the values of those variables can use the full range of the variable, then we will need $N + 1$ bits to represent the sum. With multiplication, the issue is even more pronounced. When multiplying 2 numbers, each with N bits, in general, we need $N + N$ bits to hold the biggest possible result. A 1-byte value multiplied by a 1-byte value requires a 2-byte variable to store the largest possible result.

The place that this overflow issue most commonly raises its head is when the result fits in the variable that we expect, but an intermediate result of a multistep operation requires a larger data size. Consider: $(250 \times 3)/5$. You probably look at that and expect the result to be 150, a number that comfortably fits into a byte-sized variable. However, if the variables in the expression are all byte-sized variables, we will get a result of 47. To understand what is happening, take it step by step. First we multiply 250 by 3, yielding a result of 750. The number 750 requires 10 bits to represent (1011101110). We multiplied a number (3) that needed 2 bits to represent it by a number (250) that needed 8 bits to represent it. The result will need $8 + 2 = 10$ bits to represent it. Since we have only 8 bits available, we lose the most significant 2 bits leaving us with a result of 238. When we divide 238 by 5, we get 47.6. Since we are dealing with integer numbers and not floating point numbers, we lose everything to the right of the decimal, leaving us with 47.

There are a number of different ways of dealing with this issue, and the best choice will depend on the particular circumstances. In some cases, we can rearrange the sequence of operations using parenthesis to prevent the overflow of the intermediate result. If we divided 3 by 5, we would get 0 (the integer part of the result) which would yield a result of 0 for the operation. This is not right either. Instead, we could divide 250 by 5 first, then do the multiplication by 3. This gives us the correct result. We could make this choice in this case because we knew what the values of the parameters were. But consider the situation where some or all of the parameters are variables and you can't know a priori what operation order will give the proper result. In cases like this, the only option is to create an intermediate result with a sufficient number of bits. Depending on the language you are using, you may have several techniques available. In C, you can use the **cast operation** to "promote" a byte-sized variable (char) to a 2-byte variable (int) to hold the intermediate result: `((int)250 * 3)/5`. In BASIC, you would need to store one of the variables in a larger data size in order to get the correct result.

The lesson here is to not take for granted even the simplest mathematical operations when programming a microcontroller. While addition is still addition, the rules are slightly different from the ones that you learned in grade school.

3.7 MODULO ARITHMETIC

In our day-to-day use of arithmetic, if we add 1 to a number we can always write the result, even if it takes an extra digit to express it. The same is true for subtraction where we might need to add a negative sign, but we can still express the result. As we learned in Section 3.5, variables in the computer need to fit into fixed size storage locations that limit the range of values that they can take on. This means that we cannot always add 1 to a variable and get the same result that we would if we did the addition on paper.

Example 3.5

Add 1 to the unsigned 8-bit variable whose current value is 255.

255 in binary is	1111 1111
If we then add 1	1
We get a 9-bit result	1 0000 0000
Since we only have 8 bits available, we get 0	0000 0000

In Example 3.4 and Section 3.6, we encountered other situations where the result of the operations would not fit into the size of the variable available. In those cases, we said that the results "overflowed" and we ignored the bits that wouldn't fit into the variable. Overflow is a particular example of a kind of arithmetic that is in general known as **modulo arithmetic**. All integer arithmetic performed in microcontrollers is done using modulo arithmetic. In modulo arithmetic, if you add 1 to the largest possible value for a variable, the result will be 0. If you subtract 1 from the smallest possible value for a variable, the result will be the largest possible value for that variable.

Example 3.6

Subtract 1 from the unsigned 8-bit variable whose current value is 0.

0 in binary is	0000 0000
To subtract 1, we add its 2's complement	1111 1111
We get	1111 1111

Modulo arithmetic is best visualized by thinking of a number "circle" rather than the traditional number "line" (Figure 3.1).

Another useful property of modulo arithmetic is that the count of the steps between two points (from point A to point B) on the number circle is always (B − A), even if the count has crossed 0 (rolled over) between points A and B.

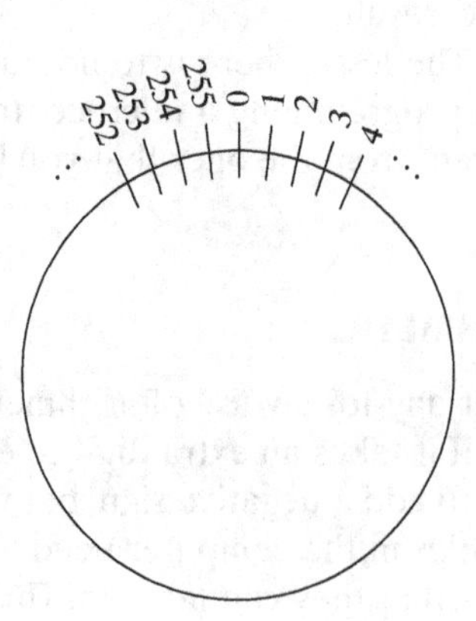

FIGURE 3.1
The number circle for an 8-bit unsigned variable.

Example 3.7

If point A is a count of 1 and point B is a count of 4, what is the number of steps between points A and B?

Point B is	4
Subtract the count at point A	1
The number of steps from A to B is	3

Example 3.8

If point A is a count of 254 and point B is a count of 4, what is the number of steps between points A and B?

Point B is 4 in binary	0000 0100
To subtract the count at point A, add its 2's complement	0000 0010
We get 6	0000 0110

3.8 MATH SHORTCUTS

If you study the form of binary numbers, you can begin to see that there are some interesting properties that drop out of a binary representation. For example, take a look at the binary numbers represented in Table 3.2.

Each successive number is twice as big as the one before, and in binary, the larger numbers each have one more zero on the right than the one before. We can move from 6 to 12 by adding a 0 to the right of the pattern for 6. This process of adding zeros to the right-hand side of a binary number is referred to as a **left shift operation** and most languages for microcontrollers have a built-in operator to do this. Every position that a number is shifted to the left represents a multiplication by 2. This is interesting because the left shift operation is usually faster than an arbitrary multiplication operation.

If a left shift is a multiplication by 2, what is a right shift? Yes: division. This works even when the number to be divided is odd, provided that we are only interested in the integer part of the result. Table 3.3 shows some examples of the **right shift operation**.

The C programming language provides two operators: << to perform left shifts and >> to perform right shifts. These are two operand expressions where the first operand is the number to be shifted and the second is the number of places to shift: `A >> 2` (bit shift right two places) or `B << 3` (bit shift left three places).

TABLE 3.2 Comparing decimal and binary representations.

Decimal Number	Binary Representation
6	110
12	1100
24	11000
48	110000

TABLE 3.3 Examples of right shift to divide by 2.

Decimal Number	Binary Representation	Right Shifted Binary	Decimal of Right Shifted Binary
6	110	11	3
12	1100	110	6
25	11001	1100	12
49	110001	11000	24

3.9 BOOLEAN ALGEBRA

Boolean algebra is a field of study that was developed to deal with the mathematical description of logic. Prior to George Boole's work in this area in 1847, logic was the domain of philosophers and not mathematicians. Today, you find Boolean algebra most often used in the design of computers and their programs. It provides us with a set of rules for manipulating logical (true or false) states that maps well to the 0 and 1 states of the binary computer.

While Boolean algebra has the typical set of mathematical laws such as the associative, commutative, and distributive laws, we can do most of what we would like to do in programming with an understanding of only the fundamental laws. These laws describe the behavior of the AND (shown as •), OR (shown as +) , NOT (shown as an over-bar), and Exclusive OR (XOR)(shown as ⊕) operators and their interactions with the constants 1 (true) and 0 (false).

OR	$A + 0 = A$	Rule 3.1
	$A + 1 = 1$	Rule 3.2
	$A + A = A$	Rule 3.3
	$A + \overline{A} = 1$	Rule 3.4
AND	$A \bullet 0 = 0$	Rule 3.5
	$A \bullet 1 = A$	Rule 3.6
	$A \bullet A = A$	Rule 3.7
	$A \bullet \overline{A} = 0$	Rule 3.8
NOT	$\overline{\overline{A}} = A$	Rule 3.9
XOR	$A \oplus 0 = A$	Rule 3.10
	$A \oplus 1 = \overline{A}$	Rule 3.11
	$A \oplus A = 0$	Rule 3.12
	$A \oplus \overline{A} = 1$	Rule 3.13

Since in most high-level programming languages, such as C and BASIC, the smallest data element that we can deal with is a byte, we will make use of these rules in manipulating and testing individual bits within bytes in the microcontroller.

3.10 MANIPULATING INDIVIDUAL BITS

In programming, the strength of the basic Boolean operators is that they can be applied to allow us to manipulate individual bits or groups of bits in a byte without disturbing the remaining bits. This is often necessary when manipulating the control registers (Chapter 8) within a microcontroller. The description of the behavior of the Boolean operators is defined on the basis of a single binary element, the bit. When those operations are performed on larger groups of bits, the operation is performed on a bit-by-bit basis, that is, bit 0 in one operand is paired with bit 0 in the second operand, bit 1 with bit 1, and so on. Because of their operation on a bit-by-bit basis, these operators are typically known as the **bit-wise operators.** In C, the symbol for bit-wise OR is |, the symbol for bit-wise AND is **&**, bit-wise XOR is indicated by ^, and the bit-wise NOT operator is a tilde, ~, that precedes the variable or constant.

Now, we can start applying the fundamental Boolean laws to setting and clearing individual bits in a byte. To **set a bit** in a variable, we want to guarantee that the result of the operation will have a 1 in that bit position. Looking through the rules, we see that we can apply Rule 3.2 and OR the variable with a number that has 1s in all the bit positions that we would like to have set to 1 in the result. We refer to this number that we create for the second operand as the **mask**.

Example 3.9

Set bits 0, 1, and 3 in an 8-bit variable whose current value is 16.

16 in binary is	0001 0000
To set bits 0, 1, and 3, we use a value of 11 (0x0B) for the mask	0000 1011
The result of the OR operation has bits 0, 1, and 3 set to 1	0001 1011

In C, we would write this as `Result = Var1 | 0x0B;`

The mask is typically, though not necessarily, a constant. Every bit in the mask that is set will cause the corresponding bit in the result to be set (Rule 3.2). Bits that are cleared (0) in the mask have no effect on the result (Rule 3.1). This allows us to leave those bits of the original value undisturbed.

To **clear a bit** in a variable (make it a 0), we want to apply Rule 3.5 and AND that variable with a mask that has 0s in all the bit positions that we would like to have cleared to 0 in the result. Every bit in the mask that is cleared will cause the corresponding bit in the result to be clear (Rule 3.5). Bits that are set (1) in the mask have no effect on the result (Rule 3.6).

Example 3.10

Clear bits 0, 1, and 3 in an 8-bit variable whose current value is 251.

251 in binary	1111 1011
To clear bits 0, 1, and 3, we use a value of 244 for the mask	1111 0100
The result of the AND operation has bits 0, 1, and 3 set to 0	1111 0000

In C, we would write this as `Result = Var1 & 244;`

For readability, it is common to set up the mask for a bit-wise AND by defining a value that has a 1 in the position that we would like to have clear in the result. In this way, if we express the constant as a hexadecimal value, it will be easy to see which bits will be cleared. The best way to achieve that readability is to use the bit-wise NOT operator on a mask defined by 1s in the positions that are to be cleared and then AND the result of the NOT operation with the variable.

Example 3.11

Clear bits 0, 1, and 3 in an 8-bit variable whose current value is 251 using a more readable mask.

A mask with bits 0, 1, and 3 set would be hexadecimal B	0000 1011
Applying the bit-wise NOT operator gives us the 1's complement	1111 0100
251 in binary	1111 1011
The result of the AND operation has bits 0, 1, and 3 set to 0	1111 0000

In C, we would write this as `Result = Var1 & ~(0x0B);`

Occasionally, the need arises to flip the state of a bit from its current state to the opposite state. If it is currently a 1, it would become a 0, if it is currently a 0, it would become a 1. This is often referred to as "toggling" a bit. The quick way to do this without making any decisions based on the current state is to make use of the XOR operator and Rule 3.11. If we XOR a value with a mask that has bits set in the positions that we want to toggle , the result of the XOR operation will have the state of those bits flipped to the opposite state.

Example 3.12

Toggle the states of bits 0 and 2 in an 8-bit variable whose current value is 251.

251 in binary	1111 1011
To toggle bits 0 and 2, we use a value of 5 for the mask	0000 0101
The result of the XOR operation has the states of bits 0 and 2 flipped	1111 1110

In C, we would write this as `Result = Var1 ^ 5;`

3.11 TESTING INDIVIDUAL BITS

Testing individual bits is really just a variation on the manipulations covered in Section 3.9. The idea in testing is to manipulate the variable to clear the bits that we are not interested in testing while preserving the state of the bit(s) that we are interested in. To do this, we will apply Rules 3.5 and 3.6 with the AND operator to isolate the bit(s) of interest while clearing the others. In this case, we want to construct the mask value so that it has 1s in all of the positions of interest and use it directly in the AND operation. Once we have isolated the bit(s) of interest, we can test if the result is all zeros. As an alternative to testing for zero, we can also compare the result with the mask value. If we are testing for multiple bits, the comparison to zero will be true if *all* of the tested bits are clear and false if *any* of the tested bits are set. This is true because, in C, if the result of an operation is 0, the result is considered logically false. If the result has any bits set (i.e., a nonzero result), the result is considered logically true. When comparing against the mask value, the comparison will be true if *all* of the tested bits are set and false if *any* of the tested bits are clear. In C, you might write the test as:

```
if ( (Variable & Mask) == 0)       true if all of the tested bits are clear
if ( (Variable & Mask) != 0)       true if any of the tested bits are set
if ( (Variable & Mask) != Mask)    true if any of the tested bits are clear
if ( (Variable & Mask) == Mask)    true if all of the tested bits are set
```

Example 3.13

Testing the state of an individual switch stored in a variable.

Consider the situation where there are four switches (S3, S2, S1, S0) connected to a microcontroller input port so that when you read a variable called Switches, the states of those four switches are stored in the four least significant bits of the value (S3 = bit 3, S2 = bit 2, S1 = bit 1, S0 = bit 0). To test if the state of switch S3 is 0, we would first form a mask (S3_Mask) with a single 1 in the position corresponding to S3 (0000 1000 = 8). We will then AND this mask with the variable to isolate the state of S3 `(Switches & S3_Mask)`. If the result of the AND operation is 0, then the switch state is 0. We could test this in C using `if((Switches & S3_Mask) == 0)`. If the result of the AND operation is 0000 1000 = 8, then the switch state is 1. We could test this in C using `if((Switches & S3_Mask) == S3_Mask)`.

Example 3.14

Testing if all of the switches in Example 3.13 are in the 1 state.

If we wanted to test to see if all of the switches were 1, we would form a mask (All_Mask) with 1s in the four least significant bits (0000 1111 = 0x0F = 15) where the switch information resides. We would then use that mask and the AND operator to isolate the bits of the variable Switches that actually carry the switch information. We could then test the result of the AND operation to see if the result was equal to the mask `if((Switches & All_Mask) == All_Mask)`. Since our mask, All_Mask, had 1s in the positions of all of the switches, if the result of the AND operation is equal to All_Mask, then all of the switches were in the 1 state.

3.12 HOMEWORK PROBLEMS

3.1 Convert the following binary patterns to hexadecimal:

1101
01010
10011
11001100
110011000001

3.2 Construct a comparison expression that will test if bits 0, 2, and 4 in a byte are all in the high state without altering the state of the byte.

3.3 Construct an expression that will clear bits 1, 3, and 5 in a byte-sized variable called Bumpers while not altering any of the other bits.

3.4 Construct an expression that will set bit 1 in a byte-sized variable called PortA without affecting any of the other bits in the variable.

3.5 Write out the 16-bit hexadecimal representation of the following signed decimal numbers (assume the representation is 2's complement):

10
17
27
−45
−128
127

3.6 Construct an expression that will test if either bit 0 or bit 3 in a byte-sized variable is in the high state without altering the state of the byte.

3.7 Construct an expression that will test if either bit 1 or bit 3 in a byte-sized variable is in the low state without altering the state of the byte.

3.8 What result would you expect from the following expressions if all values were contained in byte-sized variables?

$(240 \times 2)/4$
$(240 \times 2)/10$
$(240/10) \times 2$
$(240/4) \times 2$

3.9 Construct an expression that will clear bit 1 in a byte-sized variable called PortA without altering any of the other bits.

3.10 Construct an expression to toggle (change a 0 to a 1 or a 1 to a 0) bit 3 in a byte-sized variable called PortB that doesn't alter the value of any of the other bits.

3.11 Construct an expression that quickly divides an integer variable by 32 without using the division operator.

3.12 Complete the blank cells in the following table:

	Binary	Hexadecimal
1	1100 0111	
2	0110 1010	
3	0001 1100	
4	0101 1001	
5	1011 0110	
6		0xB4
7		0x1E
8		0x72
9		0x39
10		0xFF

3.13 What is the range of a 24-bit 2's complement number?

CHAPTER

Programming Languages

Programming a computer is all about giving a machine instructions to perform some set of operations. Since the native capabilities of computers are so limited, we are forced to express our instructions in something other than English, as English is far too nuanced and subtle to be able to effectively convey instructions to a computer. To act as a language of instruction, computer scientists have invented a large number of computer programming languages that seek to balance the capabilities of the computer with our own need to express complex ideas. In this chapter, we will look at what the native "language" of a representative microcontroller looks like and examine a number of the ways that have been developed to mediate between the programmer and the machine. Specifically, you will learn about:

1. what "machine language" is and how it relates to the execution of programs,
2. why assembly language was invented and how it differs from "machine language,"
3. what compiled languages are, with some examples,
4. what interpreted languages are and what their strengths are,
5. how to go about choosing a language to use for a project.

4.1 INTRODUCTION

Coaxing a microcontroller to do your bidding is a subtle business. The basis of the challenge is that the machines that we are giving instructions to are only capable of performing a very limited and exceedingly simple set of operations. Indeed, some examples of early microcontrollers (e.g., Intel 8048) did not have specific instructions for subtracting two numbers, and even today, many microcontrollers don't know how to divide. Our task as programmers is to translate our higher-level needs into a sequence of instructions from the very simple set of operations available in the microcontroller. We must "teach" the microcontroller to subtract and divide and do everything else that we would like for it to do for us. Ultimately, this comes down to getting the correct pattern of 1s and 0s into the microcontroller's memory that will cause it to behave as desired. In this chapter, we'll start off by looking briefly at the lowest level of instructions but fairly quickly move up the hierarchy to the so-called "higher-level" languages that are closer to how we think and therefore make the process of giving instructions easier for us.

Before we get into talking about languages, we need to introduce a few terms that are related to the software development process. These terms relate to where various parts of the processes are taking place. In the modern software development process, the writing of programs and most of the processing of those programs takes place on a "development computer." These days, that computer is usually a PC or Mac. The development computer is most often (but not always) distinct from the "target computer," where our program will actually be executed (Figure 4.1).

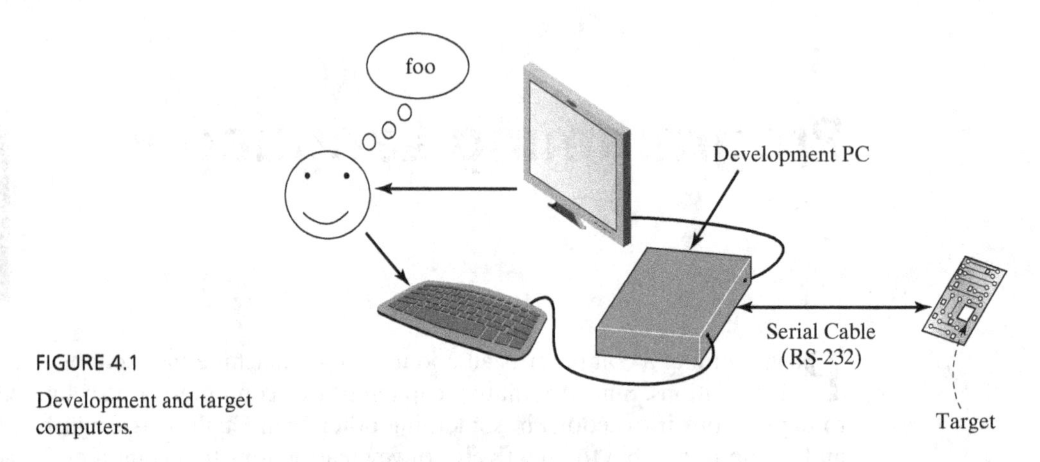

FIGURE 4.1
Development and target computers.

Along with these two different computers, we will talk about "compile time" or "assembly time" as distinct from "run time." **Compile time** and **assembly time** both refer to operations that take place on the development computer as it processes a **source program** (the instructions that we programmers write) to prepare it for the target computer. **Run time** refers to the actual execution of the program instructions on the target computer.

4.2 MACHINE LANGUAGE

The brain of any computer resides in an element known as the **central processing unit** or **CPU**. This is the part of a computer that actually executes the most primitive operations of the machine. In human terms, the capabilities of the CPU of even the most powerful computer are exceedingly limited. Its claim to fame is that it can execute these simple instructions very quickly. The CPU is capable of performing operations like adding two numbers (as long as the numbers are not too big) or determining if one number is larger than another. The operations that the CPU can perform are encoded in a pattern of 1s and 0s that are fetched from memory and presented at the input to the CPU. In response to the pattern of 1s and 0s, the CPU performs some logical or arithmetic operation. These patterns of 1s and 0s are the **machine language** of the computer, and the set of meaningful patterns defines the **instruction set** of the CPU.

The 1s and 0s of the machine language control the logic in various parts of the CPU to specify the operation. We can see this by looking at a few instructions and comparing the patterns with the functions that the instructions perform. As our example machine, we'll look at the machine language for the midrange family of PIC® microcontrollers made by Microchip. For this family of microcontrollers, the instruction to call a subroutine is encoded in binary as `10 0xxx xxxx xxxx`, where the x's represent the address of the subroutine to call. For example, to call a subroutine whose first instruction is located at address 0x100, the PIC machine language instruction (in binary) would be: 10 0001 0000 0000. To simply jump to an address, without saving a return address, the machine code is `10 1xxx xxxx xxxx`. Notice that the left two bits of both instructions are 10 and the third bit is a 0 to indicate a subroutine call and a 1 to indicate a simple jump. The instruction to clear (set to 0) a particular bit in memory is encoded as `01 00bb bfff fff`. The pattern of `01 00` in the leftmost bits specifies the bit clear operation. The three bits `bbb` specify which of the 8 bits in a byte to clear and `ff ffff` specifies the memory address of the byte in which the bit resides.

Each different kind of computer or microcontroller has its own machine language. In some cases, these languages are said to be "backward compatible," when a new processor can execute all of the exact same instructions as an older processor, but adds new capabilities. An example of this appears in the desktop PC, where the latest processors from both Intel and AMD can still execute all of the instructions for the 8088 processor used in the original IBM PC from the 1980s.

The machine language that a processor executes is a function of the arrangement of logic inside the CPU, which is, in turn, a reflection of the importance that the CPU designers placed on various instructions. In recent years, the trend has been to focus on designing CPUs that have a smaller, rather than larger, set of instructions, but execute each of these instructions much more quickly. This has led to the proliferation of a range of what are called **reduced instruction set computers (RISC)**.

Humans make only limited use of machine language. In the early days of computing, a programmer might manually enter a machine language program into memory (usually to get the systems started), but these days, dealing with machine language is largely left to our software tools during the development process and the hardware of the CPU at run time.

To give you a flavor of the differences between some of the languages that we will discuss, we'll show a very simple set of operations coded in different languages. The operations test a bit in a memory location used to control the speed of a motor. Based on the state of this bit (1 or 0), the code will set a variable to indicate whether the motor speed is 50% or 75% of the maximum speed. In midrange PIC machine code (expressed as hexadecimal), this would look like:

```
0x1921
0x2803
0x3032
0x2805
0x304B
0x00A0
```

While this code segment is completely understandable by the CPU, if we programmers were still writing programs at this level, we would not have been able to create the complex systems that are commonplace today.

4.3 ASSEMBLY LANGUAGE

In contrast to machine language, programmers regularly deal with something called **assembly language**. This is a representation of the machine language of the processor that is more understandable to human beings. While the machine language instruction that causes a CPU to jump to a subroutine located at address 0x100 might be 10 0001 0000 0000 (in binary), the corresponding assembly language instruction might be "CALL PulseEnable." These human readable forms of the machine language instructions are referred to as **mnemonics**. There is a 1:1 correspondence between the core machine language of the processor instructions (not including the address specifications) and the assembly language mnemonics. In this example, the mnemonic CALL corresponds to a bit pattern of 100 in the three most significant bits of the machine language instruction. The balance of the bits in the instruction specify where in memory the subroutine to be called is located. In our example, the location in memory of the subroutine has been given the symbolic name "PulseEnable," to further free the programmer from the tedium of tracking the actual numerical address of the code being called (not to mention eliminating a source of bugs).

The programmer wishing to write an assembly language program would prepare a text file that listed, one instruction per line, the sequence of assembly language mnemonics that correspond to the machine language instructions that are needed to describe the program. This text file would be read by a computer program called an **assembler**. The job of the assembler is to take the human readable list of mnemonics and numeric constants and produce the corresponding machine language instructions. The assembler is a relatively simple program, since it need only know how to look up the machine language (pattern of 1s and 0s) associated with an assembly language mnemonic. Continuing the example introduced at the end of the preceding section, an assembly language version (for midrange PIC microcontrollers) of the machine code shown in Section 4.2 might look like the following.

```
                btfsc       Var2Test,Bit2Test
                goto        Use75
                movlw       50
                goto        SaveSpeed
Use75           movlw       75
SaveSpeed       movwf       Speed
```

While it probably still looks pretty cryptic, you can see that the addition of variable names (Var2Test, Bit2Test), some of the instruction mnemonics (goto), and line labels (Use75, SaveSpeed) have added to the readability of the code.

Programming in assembly language is as "close to the hardware" as human programmers usually get. Very few programs today are coded entirely in assembly language because it typically takes much longer than to use a higher-level language. Assembly language is reserved for programs that must be absolutely as small and/or as fast as possible. With the falling cost of program memory and the rising performance of microcontrollers, in most cases, it is not cost-effective to develop entire programs in assembly language. It is much more common to write most of the program in a higher-level language and resort to assembly language for those routines that need to run more quickly, run in a precisely known amount of time, or when a few bytes of program memory need to be saved to fit into the memory available on a particular chip.

4.4 HIGH-LEVEL LANGUAGES

Assembly language and machine language are what are considered **low-level programming languages**. These languages are intimately tied to a particular microcontroller architecture and their designs are determined more by the hardware design of the chip than by their understandability and usability by programmers. The contrast to low-level languages is, naturally enough, **high-level languages**. The thing that all high-level languages have in common is a disconnection between the design of the language and the underlying hardware on which these programs are executed. High-level languages are designed for programmers first and the hardware second (if at all). High-level languages facilitate the expression of algorithms in a way that aids the programmer in writing the program. We have drawn the line between high- and low-level languages at the hardware. This is not universally the case. In some circles, C and Forth (to be introduced shortly) are also considered low-level languages and it is not until you get to C++ or Ada that the language is considered to be a truly high-level language.

4.5 INTERPRETERS

For many years, the first programming language that many people learned was a language called **BASIC**. BASIC, which stands for **B**eginners **A**ll-purpose **S**ymbolic **I**nstruction **C**ode, was invented in 1964 by Kemney and Kurtz at Dartmouth College as a language that they could use to teach people how to program computers. The language not only had a syntax that was very

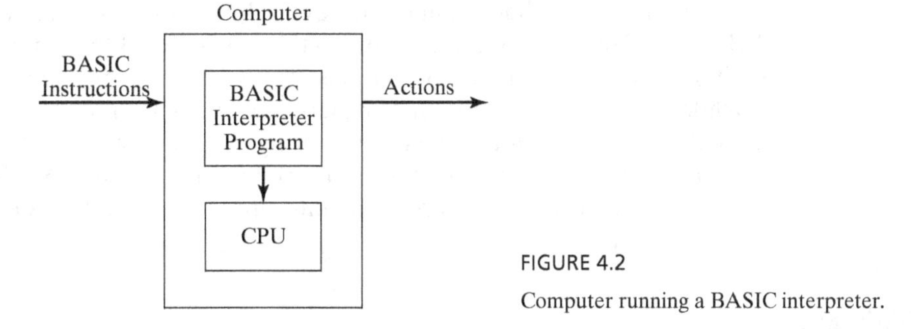

FIGURE 4.2

Computer running a BASIC interpreter.

simple to grasp, but in its most popular form it was implemented in such a way that people could sit at a keyboard and type a line of instruction to the computer and the computer would immediately execute that instruction. The computer was actually running a program, called the BASIC **interpreter**. The BASIC interpreter program was reading the instructions written by the programmer in BASIC and performing the operations indicated by those instructions. From the outside, it looked as if the computer was executing the BASIC commands. This is the essence of an interpreter: it is a computer program that takes as its input a stream of instructions in a specific language and performs the operations described by those instructions (Figure 4.2).

The language that is being interpreted need bear no resemblance to the underlying machine code or even assembly language of the computer that is executing the instructions. Interpreted languages are "high-level languages" as described above. The languages have been designed to be easier for a human to understand and manipulate. Do not confuse the interpreter with a translator. An **interpreter** extracts meaning from a command and performs the necessary actions to carry out the command. A **translator** simply translates from one language to another without performing any other actions. A language translator program might take the phrase "Close the door" as its input and produce "Schließen Sie die Tür" as its output, whereas an interpreter program would take that same phrase and actually close the door. While BASIC is not always (and, in fact, initially it was not) implemented as an interpreter, its availability as an interpreter on early personal computers exposed a whole generation to their first programming language, and so it became closely associated with interpreted languages.

The great attraction of the interpreter is the interactivity that it provides. With an interpreter, your instructions can be executed immediately. This allows you to quickly see where your thought process might have gone awry when you don't get the response to an instruction that you had expected. This interactivity can, unfortunately, also lead to sloppy programming practice as "quick fixes" are made to cure problems as they are discovered. This approach to programming led to the derogatory term "spaghetti code" to describe the kind of program that resulted from that approach to programming.

Over the years, there have been many dialects of BASIC that have been invented to address the shortcomings of the original language (i.e., variables were limited to A–Z, A0–A9, B0–B9, ..., and so on). Our sample set of operations, expressed in the original Dartmouth BASIC might look like the following.

```
10 IF ((V/4)-((V/8)*2)) = 1 THEN 40
20 LET S = 75
30 GOTO 50
40 LET S = 50
50
```

Notice the use of line numbers on every line. Note also the convoluted expression needed to test if the state of bit 2 is 1 or 0—this version of BASIC didn't provide easy ways to do bit-level logical manipulations or operations (discussed in Chapter 3).

While BASIC was not the first higher-level language, nor the first interpreter, it was certainly the language that did the most to bring programming to the masses. Modern engineers are most likely to encounter a pure interpreter in the Matlab, Mathcad, or Mathematica environments. In microcontrollers, pure interpreters have fallen out of favor and have become relatively rare.

4.6 COMPILERS

Compilers share some similarity to interpreters in that they are computer programs that extract meaning from a sequence of program instructions. Compilers differ from interpreters in that compilers do not perform any of the actions described by the source program.

In an interpreter, the extraction of meaning happens when the high-level program is being executed. This is often referred to as "run time." If a line of a program is, for example, part of a loop and is executed multiple times, then it is interpreted each time through the loop. Once the interpreter has extracted the meaning from the statement, it then performs the actions indicated by the statement.

When using a compiled language, the extraction of meaning from the program is separated from the actions necessary to execute the program (Figure 4.3).

The process is broken into two parts: *compiling* (on the development computer) and *executing* (on the target computer). The compiler's job is to perform a translation from the higher-level language into the machine language for the processor on which the program will execute. This means that the compilation and execution can actually happen on two different computers. This is, in fact, the most common scenario for embedded systems. The compilation typically happens on a desktop computer, but the program is actually executed on a microcontroller.

Since the compilation stage happens separately from the execution, it is no longer necessary to interpret a line that is part of a loop multiple times. The compiler will translate the line once and add instructions to form the loop around the generated code. At run time, the machine language associated with that line will be executed multiple times, but the translation only occurred once during compilation. This leads to a vast improvement in the speed of execution of compiled programs over interpreted programs.

A compiled language loses the interactivity of the interpreter. In return, it gains execution speed. Since the interpretation of the line does not happen while the program is running, it is possible to execute instructions much more quickly. An interpreter might execute a few thousand instructions per second while a compiled version might run at many tens to hundreds of thousands of instructions per second.

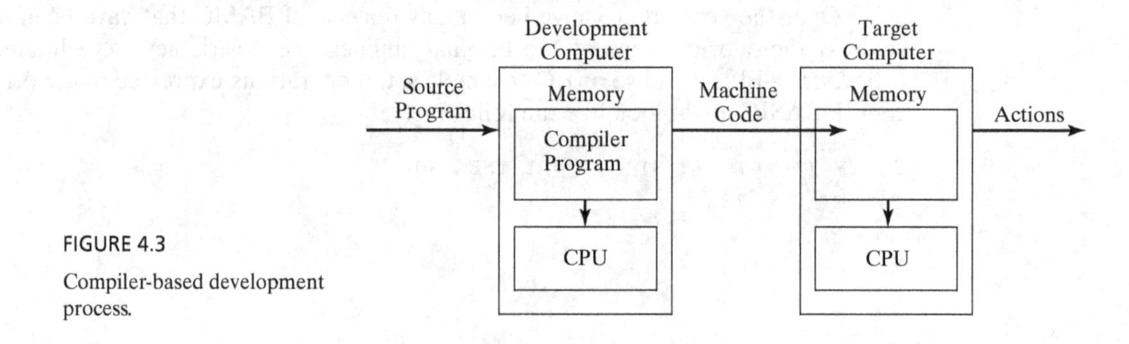

FIGURE 4.3

Compiler-based development process.

For relatively small- to medium-sized programs, the compiler will also produce less total code. Because of the need to have an interpreter program resident in the final system, there is a certain amount of overhead in even the smallest programs. The compiler avoids this. As program size grows, the higher density of meaning that can be conveyed with a higher-level language can lead to a situation where the size of the interpreter plus the program to be interpreted is smaller than the compiled version of that same program. In this case, the interpreter approach will yield a smaller, though still more slowly executing, program. The trade-off between speed and size is one that will come up time and again when you try to optimize a system. It is rare to find the solution that is both smaller and faster.

The most common compilers for embedded systems are translators for the **C** or **C++** languages. For systems employing larger microprocessors, especially those in avionics, compilers for the **Ada** language exist. C has become the de facto high-level language for use in embedded systems. It provides a good mix of low-level access with the ability to define arbitrary data types and structured programming constructs. C compilers exist for most microcontrollers. C++ is an **object-oriented programming** (**OOP**) language that is built on C. As OOP has become a more widely accepted programming paradigm, it has moved into the embedded systems world as well. The overhead of implementing some of the object-oriented features in C++ has limited its application to larger (i.e., more resource rich) embedded systems. The perceived shortcomings have also led to the development of a "stripped down" version (**EC++**) designed specifically to address the overhead issues for use in embedded applications. While C++ compilers are available for a few 8-bit microcontrollers, the language is much more popular for use in 32-bit systems.

At the level of our example code sequence, there is no difference between C and C++ in how the sequence of operations would be expressed:

```
if ((Var2Test & Bit2Test) == Bit2Test)
    Speed = 75;
else
    Speed = 50;
```

4.7 HYBRID COMPILER/INTERPRETERS

At this point, you might be wondering "What happens if we cross an interpreter with a compiler?" The answer is that you get a useful hybrid that combines features from the two approaches. Languages such as **Forth** and **Java** and products such as the **BASIC Stamp®** have taken this approach and won converts. While Forth, Java, and the BASIC Stamp share the hybrid compiler and interpreter approach, each has taken a very different road to implementing the hybrid.

Forth has been around since the early days of minicomputers, having been invented by Chuck Moore in the early 1970s. There is a relatively small but very active community that employs Forth in a wide range of embedded systems, including some high-profile devices like the handheld scanner used by FedEx® delivery agents. In the early days, memory was very precious and most programming was done in assembly language. Forth was designed to provide a significant improvement in programmer productivity over assembly language but remain very frugal with the resources of the microcomputer. To do this, Forth combined the compiler and the interpreter (called simply the interpreter) and ran both processes on the target computer. The "compiler" part of the Forth interpreter operates as you define new **words** (analogous to functions in conventional languages) to produce a very compact form of your program, but not one that is in the native machine language of the processor. When the program runs, the Forth interpreter reads this compact form to decide what actions are necessary to carry out your instructions. This approach avoids the overhead of reading each line of code at execution time, resulting in better execution speed than pure interpreters. Because the Forth interpreter is relatively simple, it also produces very compact applications. Combining the compiler and interpreter on the embedded system also produces a very interactive development system. New words are defined (compiled) on the embedded system as the definition is entered. As soon as a new word has been defined, it is available for testing, either by incorporating it into a new higher-level word, or by immediate execution from the command line.

The principal drawback of Forth lies in its early roots. In order to make the interpreter very simple and relatively fast, the structure of Forth requires the programmer to think more like a machine than more modern languages. This is one of the reasons that Forth is sometimes considered a low-level programming language. The most obvious place to see this is the near exclusive reliance on reverse polish notation for just about everything. For those of you not familiar with many of the scientific calculators made by HP, reverse polish notation places the operands before the operator (3 4 +), rather than the more common in-fix notation which places the operator between the operands (3 + 4). Reverse polish notation makes the job of the compiler and interpreter easier at the expense of making the programmer think more like the machine. Once a programmer adjusts to the "Forth Way," the language lends itself to very compact and reasonably quickly executing applications. Both attributes are sought after for embedded applications.

In Forth, our sample set of operations might look like the following.

```
Var2Test @
Bit2Test AND IF
    50 Speed !
ELSE
    75 Speed !
THEN
```

Java, on the other hand, is a much more recently developed language and one that is built completely around the OOP paradigm. OOP is a very different way of thinking about programming that traces its roots to the **Smalltalk** language invented at Xerox PARC in the mid-70s. The Java system has taken a very different approach than Forth, splitting the compilation and interpretation into two completely separate operations often running on different computers. The Java compiler runs on a "host" computer such as a desktop PC. It translates the Java source code into a series of standard **byte codes**. These byte codes are very much like the machine language for an idealized processor. The difference is that instead of the pattern of 1s and 0s being chosen to fit the hardware of a particular processor, the byte codes are designed to be easy to interpret, resulting in good performance when the Java application is being run by the interpreter. Because the byte codes that the Java compiler produces have been standardized, the output of the compiler is not tied to any specific programming hardware. It is what is called **platform independent**. What this means is that the compiled Java program can be executed on any hardware for which a Java interpreter (called a **Java virtual machine** or **JVM**) exists. JVMs exist for the common desktop computers as well as a number of embedded processors. In the case of Java, the designers chose to emphasize platform portability over interactivity, so while the JVM is an interpreter, it is not one that humans interact with directly. The program development process for Java is very similar to that of purely compiled languages such as C or C++.

At the level of the syntax required to express our sample code snippet, Java shows its heritage in C:

```
if ((Var2Test & Bit2Test) == Bit2Test)
    Speed = 75;
else
    Speed = 50;
```

Java and Forth share two other characteristics. In the interests of improved performance, both languages have inspired people to write compilers that will translate the intermediate representation of the language into the machine code for a particular processor. This produces a result that is functionally identical to a pure compiler system. The final common characteristic that both languages share is that, in the interests of even better speed of execution, processors have been designed that will directly execute the languages with no need of an interpreter. In the case of Java, this means that the hardware directly implements the JVM and uses the byte codes as its machine language.

Parallax's BASIC Stamp makes use of a custom language called **PBASIC®,** which shares characteristics of both Forth and Java. Like Forth, the variant of BASIC that the BASIC Stamp executes is focused on the efficient use of the hardware. Like Java, it uses an intermediate language based on byte codes and a separate interpreter of those byte codes that runs on the embedded system. In the case of the BASIC Stamp, the interpreter runs on a very tiny processor making the resulting system both small and inexpensive. As its name implies, the BASIC Stamp and PBASIC are based on the BASIC language. The simplicity and low cost of the BASIC Stamp system has made it easily accessible to even neophyte programmers, making it very popular and spawning a range of competitors. Because of the limited capabilities of the inexpensive microcontrollers on which they are based, the performance of a BASIC Stamp is more in line with pure interpreters, ranging from a few thousand instructions per second in the simplest BASIC Stamps to about 19,000 instructions per second in the highest performing variants. Our sample set of operations, expressed in PBASIC, might look like the following.

```
IF (Var2Test & Bit2Test) = Bit2Test THEN Speed = 75
ELSE Speed = 50
ENDIF
```

4.8 INTEGRATED DEVELOPMENT ENVIRONMENTS (IDEs)

Early in their development, compilers were stand-alone applications. In fact, it was not uncommon for the various phases in the compilation to each be a separate program. These programs typically operated on plain text files (not formatted as a word-processor does). The programmer would use a separate text editor application to write and edit the program. The first stage of the compilation would take this programmer-written file as its input and output a new text file that would be passed to the next phase in the compilation. These days, working with these individual components has almost completely disappeared. In its place, we have **integrated development environments** (IDEs). IDEs package together in one application all of the tasks necessary to prepare a file for execution in the target environment. They will include a text editor along with the compiler and the other programs necessary to compile a program into the machine language for a given target microcontroller. IDEs provide a graphical interface to configure all of the possible settings of a compiler and other tools that greatly ease the task of the programmer who wishes to write embedded code.

4.9 CHOOSING A PROGRAMMING LANGUAGE

With all the choices of language available, how do you decide which one is right for a given project? This is a question that is not often actually asked out loud. All too often the language of choice is "the one I used last time" or "the one I'm most comfortable with." This despite the fact that making a rational decision is generally not that hard to do. Pausing to think about the requirements of the project at hand and answering a few questions about the project and the programmers working on the project can lead to a language choice based in reason rather than emotion or convenience.

Probably the first question to be asked is about the scope of the project. If the project is small, then it is also likely that the time frame for completing the project is short. In the case where the project is small and the time frame is short, the choice of language will likely not revolve around the features of the language so much as it will revolve around the familiarity of the programmer and the ease of use of the development environment. If the project is small enough to be completed by a single programmer and speed of development is of the essence, then the appropriate choice of language is probably whichever one is most comfortable for the programmer. This answer needs to be tempered by the knowledge that the solution must also meet the performance requirements of the project. The simplicity of development with the BASIC Stamp and its competitors has been one of their big attractions for small projects. Most programmers have come across a dialect of BASIC at some point in their careers, and the language itself is so simple that it is very fast to relearn or refresh your memory. These types of implementations often include special purpose commands to do pulse width modulated output and serial communications, making them ideal for simple jobs where all that is necessary is to monitor a few simple sensors, do a little bit of processing and generate a few outputs. As projects become more complex, the limited program and data space, integer-only arithmetic, and relatively slow execution speed become more significant issues.

If the project is not small, and especially if it is large enough to require multiple programmers, then features of the language environment may drive the choice of language. While it is possible to do large multiperson projects in any of these languages, some languages have development features that better facilitate this kind of project. Pure interpreters of any language are probably not the best choice in a situation like this. Their ease of interactive development, that is such a strength in some settings, detracts from the kind of structure that is necessary in multiprogrammer projects. In addition, these kinds of language implementations are usually limited to a single global data space that makes such proven programming concepts as information hiding (more on this in Chapter 6) impossible to implement. In these situations, programming environments that support separate compilation, such as C, C++, and Ada compilers, are a better choice.

Performance, in terms of speed of execution, is another requirement that may drive language choice. In order of increasing levels of execution speed, we have pure interpreters, byte-code interpreters, compiled languages, and finally assembly language. Within the range of compilers, the object-oriented languages such as C++ and the Ada language, designed to produce more reliable software, carry more overhead (more machine code generated) than C. While very few applications are coded entirely in assembly language any more, there are still a few applications, such as very high volume consumer products, that may justify that approach. More common is the incorporation of sections of human-written assembly language into programs that have been largely coded in a higher-level language, often using a technique called **in-line assembly**. This is a process that is fairly easy to do with most of the compiled languages. It is more difficult, but still workable, with some of the byte-code interpreters and very difficult with pure interpreters.

So where do the hybrids like Forth and Java fit? That is not an easy question to answer. At its most basic level, Forth is considered by many programmers to be a low-level language, requiring the programmer to adopt the programming model envisioned by Chuck Moore 40+ years ago. However, Forth is also "extensible." What this means is that as a programmer adds functionality to a project by defining new words, those words have the same stature in the language as the built-in words. From a practical standpoint, this means that as a project develops, Forth looks more and more like a language customized to the application and is very fast to develop with. The issue is getting over the initial hurdle, scaling the Forth learning curve. There are Forth implementations available for many microcontrollers, including some that, in the interest of performance, discard the interpreter and compile to machine code.

Java is a language whose role in embedded development is still evolving. Java's strengths are in its platform independence and its "ground up" object-oriented nature. The existence of hardware JVM implementations on single-chip microcontrollers notwithstanding, there are still performance issues with Java in real time applications that have yet to be fully resolved. For now, the future of Java in serious embedded systems development is unclear.

TABLE 4.1 Comparing language suitability based on project size and number of programmers.

Project Size	Small	Medium	Medium	Large
# of Programmers	1	1	Multiple	Multiple
PBASIC	+	OK	–	––
Forth	+	+	+	+
Java	OK	+	+	+
C	+	+	+	–
C++	–	OK	+	++
Ada	–	OK	+	++
Assembly Language	OK	OK	–	––
– = Exposes language's weakness	–– = These weaknesses seriously impact project	OK = Neutral	+ = Plays to the language's strengths	++ = The language's forte

Table 4.1 categorizes the languages that we have discussed according to their general suitability for various sizes of projects with varying numbers of programmers. This is a complex issue, so any sort of a summary table must be taken as a general indication of trends rather than as an absolute evaluation.

4.10 HOMEWORK PROBLEMS

4.1 Describe the differences between an interpreter and a compiler.

4.2 Explain how this line of BASIC code tests the state of bit 2:

```
10 IF ((V/4)-((V/8)*2)) = 1
```

4.3 True or False: It is easy to improve the performance of a program written in an interpreted language by selectively coding in assembly language.

4.4 True or False: Assembly language is considered a high-level language.

4.5 When is the meaning of a compiled program converted into actions?

4.6 True or False: Assembly language and machine language are synonymous.

4.7 Which of the languages described in this chapter was designed to teach programming?

4.8 Which are the object-oriented languages described in this chapter?

4.9 Describe a scenario in which programming in an interpreted language might be an advantage.

4.10 Describe two similarities between Java and the PBASIC language used by the BASIC Stamp.

4.11 Name at least three other widely-used computer languages that were not mentioned in this chapter.

4.12 You are beginning to test a new circuit in which a microcontroller is connected to a large number of peripheral devices—a process commonly referred to as "bring-up." What language(s) would you choose for this task? Why?

4.13 You are writing final production code for a microcontroller used to control a servomotor. What language(s) would be reasonable choices for this application? Why?

FURTHER READING

C Programming Language, Kernighan, B.W., and Ritchie, D.M., 2nd ed., Prentice Hall, 1988.

History of Programming Languages, Wexelblat, R.L. (Ed.), Academic Press, 1981.

Threaded Interpretive Languages, Loeliger, R.G., Byte Books, 1981.

Program Structures for Embedded Systems

In your introductory programming course, you learned the basic program structures, assignments, loops, tests, and so on, associated with the language used for the course. If you go back and look at those programs, you will likely see that your program was always in control of when the input occurred. Your code looked for input at specific places in your program and waited for it to appear. While your program was looking for a particular kind of input, be it a mouse click or a key press, it wasn't doing any other work. And, when it was working on a computation, it was unable to detect inputs. If, at a particular place in your program, you were looking for a mouse click and a key was pressed, generally, the key press was ignored (at least at that instant). While this approach to programming is functional in many situations, it doesn't work very well for most mechatronics systems, where there will be multiple possible sources of input which may become active at times you cannot predict and where it is unacceptable to ignore any of the inputs. This chapter introduces two complementary programming structures that are extremely useful in designing and implementing software for mechatronic systems. By the time you have mastered these topics, you should be able to:

1. identify the events that will be relevant to a given system,
2. draw a state diagram to graphically capture the behavior of a system,
3. understand how to deal with noise on signals that indicate the occurrence of events,
4. write functions to implement event checking routines, service routines, and state machines.

5.1 BACKGROUND

As an example of a mechatronic system, consider the case of a DVD player. On a typical DVD player, there are a number of buttons on the front of the player, sensor inputs for disc present, and for disc tray closed, as well as commands that can be received from the remote control. The program for the embedded computer controlling the DVD player must be able to respond to any of those inputs at any time and in any sequence. For example, while running fast forward, you can stop, play, pause, or any number of other actions can be carried out. The program for the DVD player can't predict if the next command from the remote control will come before reaching the end of the disc, for example, so a program structure that can only manage one input at a time would fail in this application.

What we need is a structure that will allow the program to respond to a multitude of inputs, any of which may require attention at unpredictable times and in an unpredictable sequence. While this problem is very common in embedded systems, like the DVD player, it is also very common in modern desktop computers. Think about using a word processor.

While you are typing, the program is accepting your keys and entering the characters into your document. You can also click with the mouse at any place on the screen. If the mouse pointer is at another place in your document, the point of insertion moves. If it is in the menu bar, a menu will drop down. If it is over a formatting button, it may change the format of the next character entered. Your word processor, and most other programs running under Windows, MacOS, X-Windows, or any other modern graphical user interface (GUI), must also be capable of dealing with the same kind of variable input that the DVD player must handle. All of these operating systems, and many embedded applications, have adopted a common program structure to deal with the need to respond to many asynchronous inputs. This program structure is generally called **event driven programming**.

5.2 EVENT DRIVEN PROGRAMMING

Under the event driven programming model, the program structure is divided into two groups: events and services. An **event** represents the occurrence of something interesting in your application. A **service** is the set of actions that are performed in response to an event. While events most often originate from outside your program, such as a mouse click, events can also be generated by other parts of the program, such as the expiration of a timer. The program executes by constantly checking for possible events, and when an event is detected, it executes the associated service.

In order for this approach to work, the event checking routines must run continuously and often. This implies that the services must execute quickly so that the program can get back to checking for other events. To meet this requirement, a service cannot go into a state where it is waiting for some long or indeterminate time. The most common example of this would be a WHILE loop, where the condition for termination was not under program control, such as a switch closure. This kind of program structure, an indefinite loop, is referred to as **blocking code**. In order for the event driven programming model to work, you must only write **nonblocking code**.

So, how do you deal with the need to wait for the switch to close? You make the switch closure an event. Then, rather than waiting in a loop for the switch to close, you program a service that performs a desirable action when the switch closes. Until the switch closure event occurs, however, your program continually scans for all the other possible events. In this way, you could also react to another event, if necessary, while you were waiting for the switch to close.

5.3 EVENT CHECKERS

An event, by its nature, occurs at a discrete point in time, instantaneously. At one instant, the event has not yet occurred, at the next, it has. In the case of a switch, the opening or closing of the switch represents an event. The switch being open or closed does not represent an event. The event is marked by the *transition* from open to closed (Figure 5.1).

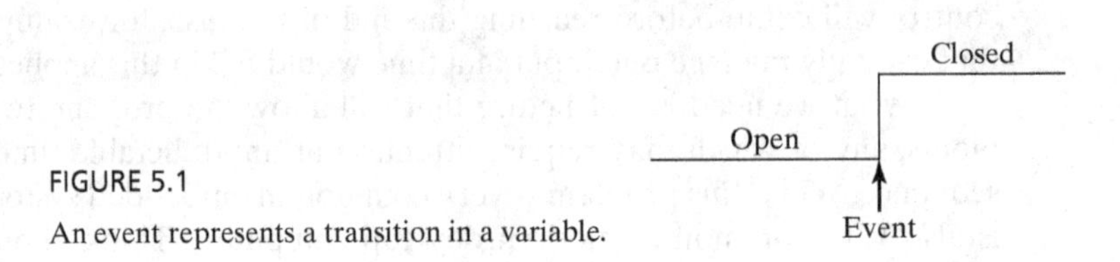

FIGURE 5.1
An event represents a transition in a variable.

The event checkers, then, are small pieces of code that test for the occurrence of an event. In order to test for an event when the input is something like a switch, whose value is continuously available, the event checker must have some record of the past. An event has only occurred when the current value is different from the past value. In pseudo-code, the test would look something like the following.

IF switch is closed now AND switch was open last time
THEN SwitchClosed event has occurred.

To complete the event checker, the code must update the value kept as the last state of the switch. The variable used for the last state of the switch must be able to retain its value between successive calls to the event checker. In C, this might look like Listing 5.1.

```
unsigned char CheckSwitchClosed( void)
{
    static unsigned char LastState = OPEN;
    unsigned char CurrentState;
    unsigned char EventStatus;

    CurrentState = GetSwitchState();
    if((CurrentState != LastState) &&
           (CurrentState == CLOSED))          /* the event test */
           EventStatus = TRUE;
    else
           EventStatus = FALSE;

    LastState = CurrentState;                 /* update state variable */

    return(EventStatus);
}
```

LISTING 5.1

An event checking routine in C.

In examining this function, notice that the LastState variable is local because it is only needed within this function. It is also declared with the `static` modifier so that it will retain its value between function calls. This is not required for the other two variables, which are also only needed within the function, but do not need to retain their values between function calls. As a reminder: when a static variable has an initializer, the initialization only takes place once (before your program starts running), not every time the function is entered (as is the case for nonstatic variable initializations).

The use of the CurrentState variable also demonstrates an important concept. By using the CurrentState local variable, we read the switch state in only one place. We will need the state of the switch in two places. We need the switch state first to test for an event and, then, later to update the LastState variable. By reading the switch state only once and assigning the result to a local variable, we avoid the possibility that the switch state might change between the time that we test it to check for an event and the time that we update the LastState variable. If this were to happen, we would miss an event since the test would fail, but the LastState variable would be updated with a new value.

This basic function outline can be used for most event checkers that test for events that can be represented as the transition between two discrete states. Events can also be triggered by variables that can take on a range of values. In that case, the event would be defined

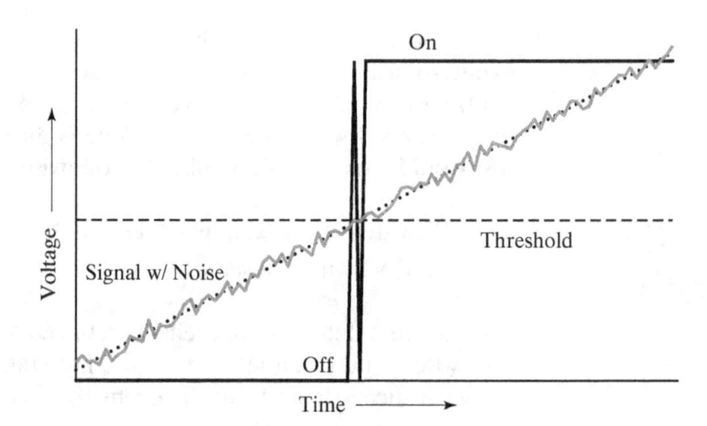

FIGURE 5.2
Noise on a signal induces chatter in the event detection.

as the transition across some threshold value. An example of this might be a light sensor that would generate an event when the light level rose above some threshold value. If you apply this same basic function outline to the light sensor [by substituting a test of above/below a threshold for the call to GetSwitchState()], you will probably find that you get a string of ON-OFF-ON-OFF events as the light level rises from below the threshold to above the threshold. To understand why that might be, consider the plot shown in Figure 5.2.

Any measured variable that represents the state of some real world (analog) signal will have a certain amount of noise associated with it. As the measured value of the variable approaches the threshold, the noise will alternately take it above, then below the threshold. This will continue until the combination of the signal plus noise stays completely above the threshold. While the combination of signal plus noise is dancing back and forth across the threshold, a sequence of spurious events is being generated. While this might be desirable in some cases, most of the time we would like to be able to respond to the signal while ignoring the noise. Ideally, we'd like such a transition, from low to high, to generate only a single event.

One approach to this problem would be to go back to the source of the signal and reduce the noise (Chapters 14 and 15). This would most likely be done with filtering to reduce the noise level to an undetectable level. This solution requires additional hardware and has the potential to adversely affect the time response of the desired signal. There is another approach that allows us to add a small amount of additional software complexity to eliminate the effect of the noise with no additional hardware.

We can use hysteresis (a different response to rising and falling signals) to overcome the chatter. If we modified our test to implement the correct amount of hysteresis at the switch point, we could eliminate the multiple events associated with the noise. If the hysteresis band was slightly larger than the amplitude of the noise, we would get a single clean transition as the signal passes through the hysteresis band from the low threshold to the high threshold (Figure 5.3).

The way to implement hysteresis in software is to change the threshold from a single fixed value to a variable that can take on one of two different values. For a signal that starts below the desired trip-point, the value is initially set at the higher of the two threshold values. As the signal plus noise rises, it will eventually exceed this threshold, generating an event. The service response to this detected event is to lower the threshold to the lower value and then do whatever else this event requires. Changing the threshold to a lower value, at this point in time, will prevent the noise superimposed on the signal from immediately triggering another event. To generate another event, the signal would need to fall by the width of the hysteresis band to get below the new threshold. Since we set the spacing between the two thresholds to be larger than the amplitude of the noise, this noise will not, by itself, generate additional

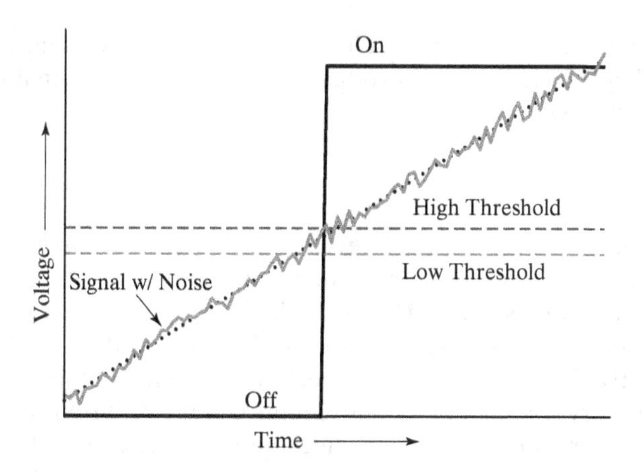

FIGURE 5.3

Adding a high and low threshold implements hysteresis and eliminates the chatter.

spurious events. When the falling signal plus noise does get below the new (lower) threshold, an event signifying that the light has gone off will be generated, and the threshold will then be raised to its higher value. The hysteresis behavior can be encapsulated into a GetLightState function that will implement the hysteresis and create a two-valued system like the switch. This will allow us to use the GetLightState function in the same basic event checking template that we used for the switch. In C, this might look like Listing 5.2.

```
unsigned char GetLightState( void)
{
    static unsigned char Threshold= HI_THRESHOLD;
    unsigned char LightState;

    if( LightValue >= Threshold)   /* above the threshold ? */
    {
          LightState = LIGHT_ON;
          Threshold = LO_THRESHOLD;
    }else                          /* must be below */
    {
          LightState = LIGHT_OFF;
          Threshold = HI_THRESHOLD;
    }
    return(LightState);
}
```

LISTING 5.2

Implementing hysteresis in software, an example in C.

Here, again, we use a static local variable to maintain internal information that needs to be preserved between successive calls to the function. In this case, it is the threshold value. Whenever the current light value is above the threshold, the light is considered on and the threshold is forced to its lower value. Changing the threshold to the lower value allows the light level to drop somewhat (due to noise) and still be considered on. Only when the light level drops below the new lower threshold will the light be declared off and the threshold raised back to its higher value. Similarly, if the light level then rises slightly (due to noise), it will still be below the new higher value and the state will remain off. It is worthwhile noting again that the LightValue is read only once per call to the function. This prevents any confusion that might be caused by a rapidly changing light value that was sampled twice within a single function call.

The solution of adding hysteresis to eliminate chatter is not limited to software implementations. In Chapter 11, we will introduce an electronic functional block that can implement hysteresis in hardware.

5.4 SERVICES

Now that we can recognize events, what are services? Services are simply the actions that you want your program to perform when an event has been detected. As an example, think about a mobile robot with bump sensors. The event detector would check for bumping into something, and the service would execute the response. In this case, the response would most likely be to stop or reverse the drive motors and begin to move in another direction.

Services should be very compact functions that initiate the required action and quickly return. This allows the program to get back to checking for other events. The core assumption in event driven programming is that you can check for events quickly enough so that none are missed. This can only happen if both the event checkers and the service routines execute quickly. Neither event checkers nor service routines should enter into an indefinite loop (i.e., blocking code).

If you find yourself wanting to code a WHILE loop for something other than stepping through an array, you probably need to add an additional event. The way to handle a situation like that is to have a service routine that starts the activity going and an event checker to detect the end condition. Using the mobile robot as an example, you might want to back up for a second after a bump in order to move away from the obstacle. You might, improperly, code that as the response to the bump event (Listing 5.3).

```
void BumpResponse( void)
{
     unsigned int StartTime = GetTime();
     DriveMotors(REVERSE);
     while((GetTime() - StartTime) < BACKUP_TIME)
          ;                    /* wait for backup time to expire */
     DriveMotors(STOP);
}
```

LISTING 5.3

A poor example: a response routine that implements blocking code.

If you did this, you would not be able to respond to any other inputs during the time that you were waiting for the backup timer to expire. This would mean that if you hit something while backing up, you would ignore it. A much better approach to this problem would be to write a response routine that started the motors in reverse and also started a timer to run for the backup time. Then you would need to write a new event checker to check for when the backup timer had expired (Listing 5.4).

The response routine for BackupTimeoutCheck() would take care of stopping the motors and initiating any other evasive maneuvers. This approach preserves the ability of the software to perform other checks while it waits for the backup time to expire.

5.5 BUILDING AN EVENT DRIVEN PROGRAM

The main body of an event driven program will consist of calls to the initialization routines followed by an endless loop to check for events and execute the service routines. While there are software frameworks available to simplify the event checking and calling of the services, they are simply optimizations of a structure that looks something like Listing 5.5.

```
Void BumpResponse( void)
{
     unsigned int StartTime = GetTime();
     DriveMotors(REVERSE);
     StartTimer(BACKUP_TIMER, BACKUP_TIME);
}
unsigned char BackupTimeoutCheck( void)
{
     static unsigned char LastTimerState = TRUE;
     unsigned char ReturnVal;
     unsigned char CurrentTimerState = IsTimerExpired(BACKUP_TIMER);
     if((CurrentTimerState == TRUE) &&
        (LastTimerState == FALSE))
           ReturnVal = TRUE;
     else
           ReturnVal = FALSE;
     LastTimerState = CurrentTimerState;
     return (ReturnVal);
}
```

LISTING 5.4

A better response with a new event checking routine.

```
void main(void)
{
    DoInitializations();
    while(1)
    {
          if(EventChecker1())
                ServiceRoutine1();
          if(EventChecker2())
                ServiceRoutine2();
          .
          .
          .
    }
}
```

LISTING 5.5

The main() function for a program using an event driven framework.

The bulk of your programming effort will be spent designing and coding the event checkers and service routines. The emphasis on short event checking and service routines has the added benefit of making the design and implementation of each of these routines simpler and less prone to errors. This implies that, with reasonable care, the major problems to be debugged will be at the design level, (i.e., choosing what the events will be and what should be done in response to those events) not at the code, or implementation, level.

By emphasizing the design phase as a place for debugging, we put ourselves in the enviable position of not needing either hardware or software to get started debugging! Usually, the debugging phase is the longest and most difficult, so getting such an early start will likely get us to a working program sooner than if we had tackled the problem without using event driven programming.

5.6 AN EXAMPLE

You are to design the software for a simple machine that, on a small scale, mimics the behavior of a cockroach. At the simplest level, this means that the cockroach runs straight ahead while in the light and stops when in the dark. If the cockroach encounters an object, it turns left, reverses direction, and runs backwards for 3 seconds. If, at the end of that time, it is not in the dark, it should begin running straight ahead again.

Identifying the basic events and services is pretty straightforward.

Light Goes On	*Event*
Drive Forward	*Service*
Light Goes Off	*Event*
Stop	*Service*
Contact Object	*Event*
Turn left, reverse for 3 sec.	*Service*

Notice the wording of the event descriptions. An event represents a transition (an instantaneous change), not a state (a continuing condition), and each of these event descriptions is written to capture the transition. The most common mistake made by programmers when beginning to use this methodology is to write event checking routines that actually detect the state of a sensor or input. The result is a continuous stream of "events" being detected and the corresponding actions being executed. In the example above, "Light Goes On" is an *event* that happens when the light transitions from off to on, not a *state* of the light being on. You need to be sure that your event checkers are testing for a transition in the variable so that they are checking for *events*.

Another interesting aspect is the service associated with contacting an object. How do you handle the "reverse for 3 sec." service? As in the earlier example, we shouldn't just sit in the service routine for 3 seconds. So we need to introduce a new, internal, event: Reverse Timer Expires. Adding this new event and a little more thought yields:

Light Goes On
    Drive Forward
Light Goes Off
    Stop
Contact Object
    Turn left, drive in reverse, set Reverse Timer for 3 sec.
Reverse Timer Expires
    Turn straight, simulate Light Goes Off event

Now, when the internal event Reverse Timer Expires occurs, the machine should straighten its wheels and simulate a Light Goes Off event. The simulation of a Light Goes Off event must modify the "last state" variable of the light.[1] In this way, if the light is still on, it will trigger a Light Goes On event and begin driving forward. This structure appears to be entirely governed by the events and services and should behave as the specifications demand. To be sure, we can start testing its behavior. No hardware is required, simply the hypothesis for a sequence of events. The designer can, in his/her head or on a sheet of paper, work through how the machine would behave.

---

[1]This will require that the scope of the state variable be expanded beyond the event checking function, an undesirable state of affairs. In the State Machines section, we will examine how to eliminate this.

As an example of debugging the design, consider what appears to be a reasonable simplification of the service for the Reverse Timer Expires event. What would happen if, rather than simulate Light Goes Off, we simply performed Drive Forward? It seems like a reasonable simplification, let's test it. What would happen if, while we were backing up, we ran into the dark? The Light Goes Off event would fire and we would stop. Then, when the Reverse Timer Expires, we would Drive Forward out of the dark, not exactly the behavior that we were looking for. Here, we tested a possible simplification and found a potential error before we even started coding.

In working through this process, it will become clear that there is a possible conflict between the design and the specifications. Consider this sequence:

Contact Object

While Reverse Timer is running, Light Goes Off

In the design described above, as soon as the light goes off, the motors will stop. A closer look at the specification shows that this may not be the correct behavior.

The specifications state: "If the cockroach encounters an object, it turns left, reverses direction, and runs backwards for 3 seconds." Running backwards for 3 seconds may, if the machine runs into the dark during that time, conflict with the specification that the machine "stops when in the dark."

The process of evaluating the proposed software design has uncovered an ambiguity in the specifications. Without the emphasis on design that a methodology like this provides, it is likely that this ambiguity would not have been uncovered until the customer actually saw the behavior and said "That's not what I wanted." One of the real strengths of an event driven programming model is that it allows the designer to begin to evaluate and test the program design very early in the process.

5.7 SUMMARY OF EVENT DRIVEN PROGRAMMING

The event driven programming model represents a way of thinking about software that lends itself very nicely to situations where there are many inputs, and the timing of events on the inputs is unpredictable. It supports the decomposition of the problem into a set of relatively simple and short event checking and service routines. The behavior of the resulting program is defined by the events that are tested for and the services carried out in response to the events. It is an approach that supports the testing and evaluation of program designs before implementation (coding).

While the event driven programming model alone is capable of dealing with many problems, it may fall short when tackling many of the more complex problems that you may encounter. However, when combined with the concept of state machines, event driven programming is capable of tackling almost any problem, independent of complexity.

5.8 STATE MACHINES

While event driven programming by itself will allow you to successfully tackle a wide variety of embedded software problems, it really comes into bloom when it is combined with a concept called **state machines**. This combination can tackle just about any problem that you can throw at it while retaining the simplicity and design focus of event driven programming.

State machines are formally referred to as **finite state machines** (**FSM**) or **finite state automata** (**FSA**). These "machines" are constructs that are used to represent the behavior of a reactive system. **Reactive systems** are those whose inputs are not all active at a single point in time. Hopefully, this sounds a lot like the situations where event driven programming would be useful because the two concepts go hand in hand. Events are the driving forces in state machines.

At its simplest, an FSM consists of a set of states and a set of defined transitions between those states. At any point in time, the state machine can be in only one of the possible states. States embody the concept of time duration. Time is passing while the state machine is in a state. For example, the cockroach we explored in Section 5.6 may be in a state of driving forward for an indefinite period of time. As a result, states are most commonly associated with gerund words (e.g., pump*ing*, driv*ing*, fill*ing*, wait*ing*). In contrast, transitions between the states are treated as instantaneous and occur in response to events. It is probably easiest to understand state machines by examining the graphical depiction, known as a **finite state diagram (FSD)** or **state transition diagram (STD)**, of a simple example (Figure 5.4).

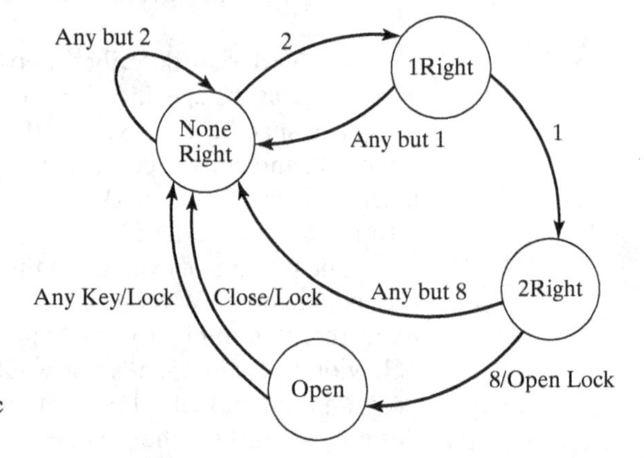

FIGURE 5.4
The state transition diagram for a simple electronic keypad lock.

The FSD shown in Figure 5.4 describes the behavior of an electronic keypad lock whose combination is 2-1-8. The FSM can be in one of the four possible states: NoneRight, 1Right, 2Right, and Open. The arrows between the state bubbles represent the transitions, labeled with the event that triggers that transition. The bubbles represent the states. While in a state, the FSM is waiting (gerund word) for an event to cause it to transition.

The lock FSM transitions from NoneRight to 1Right when the digit "2" is entered. When in the 1Right state, entering a "1" causes a transition to the 2Right state; any other entry returns the FSM to NoneRight. This pattern continues until we are ready to examine the transition from 2Right that is triggered by an entry of "8." This transition is labeled in two parts. The upper part ("numerator") calls out the event that triggers the transition and the lower part ("denominator") describes an action associated with taking the transition. In this case, the action is to open the lock. The actions are one of the ways in which state machines actually *do* something.

When the state diagram appears to be complete, the next step is to test it. In this phase, we imagine sequences of events and examine how the system described by the FSM would behave. It is pretty easy to convince yourself that for any sequence of three digits, the FSM will only unlock for 2-1-8. But what about four-digit sequences? Let's try 2-1-8-1. The FSM ends up in the NoneRight state with the lock locked. That's good! From this example sequence, it is pretty easy to extrapolate that any four-digit sequence that begins with the correct combination will leave the lock locked. What about 1-2-1-8? That leaves the lock open, as will any other four-digit sequence that ends in 2-1-8!

We are now faced with a decision. We could accept this behavior, though it is probably not exactly what we expected. Or we could modify the design in some way to eliminate the undesired response. For the purposes of this discussion, the decision about how to treat the situation is of less importance than the fact that we found a possible error in our design. Notice, we just tested our design and found a potential flaw before writing the first line of code! This is one of the real strengths of a methodical use of state machines. The *design* can be tested easily and repeatedly before any code is written. This is another example of "design time" debugging vs. "implementation time" debugging.

5.9 A STATE MACHINE IN SOFTWARE

While there are a number of ways that you can go about implementing a state machine in software, the most straightforward (though generally not the best) approach is a series of nested IF-THEN-ELSE statements. Using this approach, there is a series of IF clauses that test for the possible machine states. Within each of those state tests, there is another series of nested IF clauses that handle the events that could occur while in that state. This combination of IF statements are wrapped inside a function that maintains a static local variable to track the current state of the machine. This state machine function takes at least one parameter that passes the most recent event to the state machine. For the lock example that we were just considering, the state machine function, implemented in C, might look like Listing 5.6.

```
void LockStateMachineIF( unsigned char NewEvent)
{
    static unsigned char CurrentState = NoneRight;
    unsigned char NextState;

    if( CurrentState == NoneRight)
    {
          if( NewEvent == KeyEqual2) /* Key == '2' ? */
                NextState = OneRight;
          /* no else clause needed, we are already in NoneRight */
    }else if( CurrentState == OneRight)
    {
          if( NewEvent == KeyEq1) /* Key == '1' ? */
                NextState = TwoRight;
          else
                NextState = NoneRight; /* Bad Key go back to none */
    }else if( CurrentState == TwoRight)
    {
          if( NewEvent == KeyEq8) /* Key == '8' ? */
          {
                NextState = Open;
                OpenLock();
          }
          else
                NextState = NoneRight; /* Bad Key go back to none */
    }else if( CurrentState == Open)
    {
          NextState = NoneRight;
          LatchLock();
    }
    CurrentState = NextState; /* update the current state variable */
    return;
}
```

LISTING 5.6

C implementation of the combination lock state machine using a nested IF structure.

The main function to run this state machine would look something like Listing 5.7.

```
void main(void)
{
    unsigned int KeyReturn;
    while(1)
    {
          KeyReturn = CheckKeys(); /* check for events */
          if( KeyReturn != NO_KEYS)
                LockStateMachineIF(KeyReturn); /* run state machine */
    }
}
```

LISTING 5.7

The main() function that would be used with the state machine function of Listing 5.6.

This program would run forever ["while(1)"], calling the event checking routine CheckKeys() to check for key input. Every time that the event checker found that there had been a new key pressed, the LockStateMachine() function would be called with a parameter that indicated *which* key had been pressed. In this case, the event:service pair would be CheckKeys:LockStateMachineIF.

Although the nested IF clause approach works for problems like this, where there are only a few states and only one or two paths out of each state, as the state machine becomes more complex, it will be more robust, easier to read, and often generate more efficient code if the problem is expressed using a nested SWITCH:CASE structure, rather than the nested IF-THEN-ELSE clauses. As an example of this structure, the following function (Listing 5.8) duplicates the lock state machine function using a nested SWITCH:CASE structure in C.

```
void LockStateMachineCASE( unsigned char NewEvent)
{
    static unsigned char CurrentState = NoneRight;
    unsigned char NextState;
    switch(CurrentState)
    {
          case NoneRight :
                switch(NewEvent)
                {
                      case '2':
                            NextState = OneRight;
                            break;
                      default:     /* we are already in NoneRight */
                                   break;
                }
                break;
          case OneRight :
                switch(NewEvent)
                {
                      case '1':
                            NextState = TwoRight;
                            break;
                      default:     /* anything else sends us back */
                            NextState = NoneRight;
                            break;
```

```
                }
                break;
          case TwoRight :
                switch(NewEvent)
                {
                      case '8':
                            NextState = Open;
                            OpenLock();
                            break;
                      default:
                            NextState = NoneRight;
                            break;
                }
                break;
          case Open :
                NextState = NoneRight;
                LatchLock();
                break;
    }
    CurrentState = NextState;
    return;
}
```

LISTING 5.8

C implementation of the combination lock state machine using a SWITCH:CASE structure.

While the C code for this simple function appears noticeably longer than that for the nested IF clause version, the resulting machine code generated by the compiler is comparable. The clarity of the code is improved by the labeling of the event cases as well as the explicit default case. As the complexity of the state machine grows, both in terms of the number of states and number of active events, the clarity and efficiency of the SWITCH:CASE structure becomes more evident. For anything more complex than the simplest state machines, the SWITCH:CASE structure will be preferable.

5.10 THE COCKROACH EXAMPLE AS A STATE MACHINE

As another example, let's cast the cockroach behavior from the Event Driven Programming section as a state machine. As the events and actions remain the same, this is simply another way of representing the problem. In this case, you might come up with a state diagram that looks something like Figure 5.5.

Here, the behavior of the cockroach is described as being in one of the three states (Hiding, Driving Forward, and Backing Up), and the same set of events are used to trigger transitions between these states.

The diagram also introduces another feature of state machines: **guard conditions.** Notice that, while backing up, the Timer Expires event triggers one of the two possible transitions, depending on the current state of the light. The current state of the light is the guard condition on those two transitions. In order to take either of the transitions, the event must occur *and* the guard condition must be true.

It is important to be very careful whenever a single event will take one of two transitions based on guard conditions. In particular, the two guard conditions on the transitions

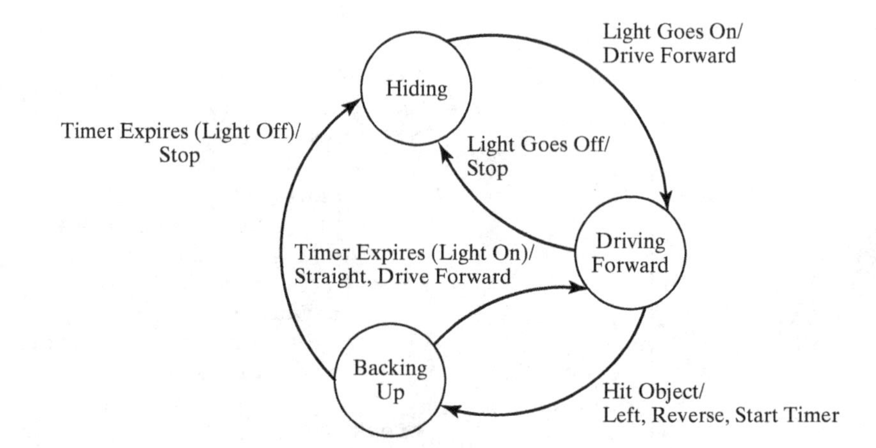

FIGURE 5.5

The state transition diagram for a simple cockroach behavior.

must not overlap one another. If this were not the case, it would be possible to have the event occur and the response would be indeterminate, since both guards had been met.

As drawn, this state machine captures the specifications without the ambiguity that we discovered in the Event Driven Programming section. Notice that the Light Goes Off event is only responded to if the state machine is in the Driving Forward state. In this way, if the light goes out while backing up, it will continue to back up for 3 seconds. Only when the timer expires will it either resume going forward or stop, based on the current state of the light.

The state machine representation has another benefit over a pure events and services approach. In the events and services solution, we needed to simulate a Light Goes Off event in order to get the desired behavior. This required that the scope of the variable holding the last state of the light be expanded beyond the Light Goes On event checker. That is not necessary in the state machine solution.

If we make the problem just a little more complex, we can really see how the state machine representation becomes useful. Let's add a response to the back bumper that will only be active while we are backing up. Much like the response to the front bumper, the cockroach should change direction (go forward) for a period of time and turn (to the right this time). When this maneuver is complete, it should return to the Backing Up State. Adding this to the state machine requires adding one new state (Evading Forward), one new event (Rear Bumper Hit), one new action (Wheels Right; Motor Forward), and two new state transitions (Figure 5.6).

In this revised state machine, there are two distinct responses to the Timer Expires event. If we are Backing Up, we should straighten the wheels and go forward or stop, depending on the guard condition. If we are Evading Forward, we should always set the wheels left and go into reverse. The response to the event depends on what we are currently doing—that is, what state we are in when the event occurs. This is a situation that would be messy to handle using only events and services. A state machine is an excellent way to capture that type of behavior.

While this state machine might not be the final design (e.g., it won't stop while Evading Forward), it does provide an easily understood representation of the behavior. That is one of the key strengths of using state diagrams: they allow you to capture the behavior of the design very early in the process, easily present it to others, and immediately begin testing it.

The thing to emphasize at this point is that we have described the software to control a system with an interesting behavior in very short order. Equally importantly, the actual code that would need to be written to implement the individual functions described is relatively simple. The combination of focusing on events and a state machine representation allows us

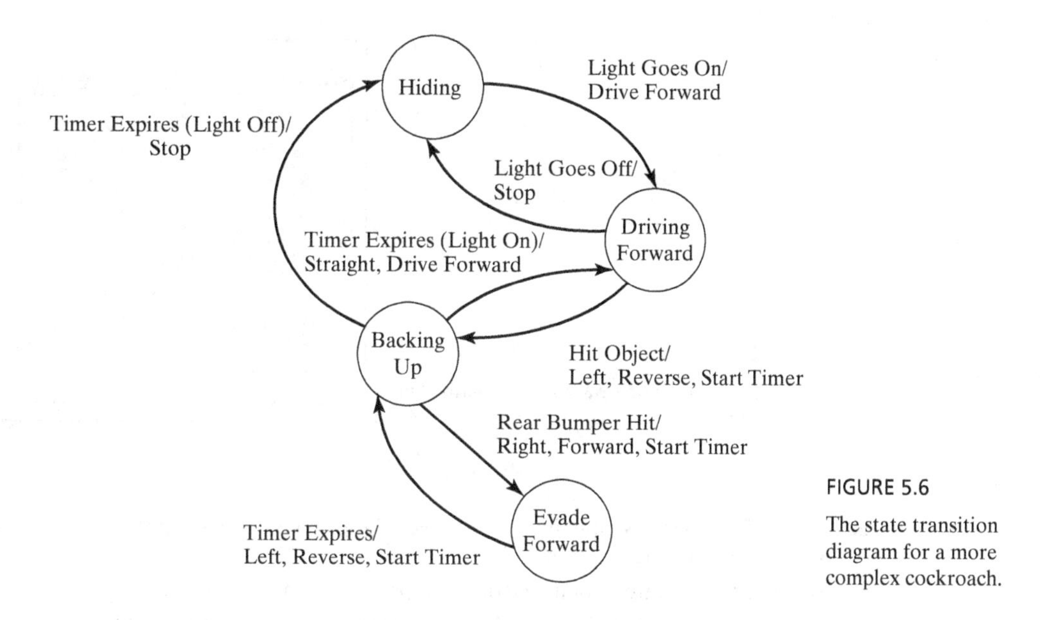

FIGURE 5.6
The state transition diagram for a more complex cockroach.

to easily decompose the problem into relatively simple subproblems (test bumper, test for light goes on, drive forward ...) that are then easier to code without errors. The state machine captures the complexity of the desired design behavior in an easily understood framework. Filling in the framework with the code from the simple subproblems will give us a working program in a much shorter time than would be possible without the event driven/state machine paradigm.

5.11 HOMEWORK PROBLEMS

5.1 Work with a DVD player and develop an exhaustive set of events that you think the software must sense and respond to.

5.2 Identify the events and draw an STD for the behavior of a DVD player. Limit yourself to the events generated by the DVD player (ignore the remote control).

5.3 Given a function GetRoomTemperature() that returns the room temperature and a variable, SetPoint, write a pair of functions in pseudo-code called IsTemperatureTooHot() and IsTemperatureTooCool() that return TRUE-FALSE values and together implement a hysteresis band around the SetPoint temperature.

5.4 Work with an answering machine to develop the list of events that the machine's software must respond to.

5.5 You are writing code to implement cruise control for a car. Given a function GetVehicleSpeed() that returns the speed of a vehicle, write a single function TestAccelDecel() in pseudo-code that returns, NeedAccelerate, NeedDecelerate, or SpeedOK based on the speed relative to a set point DesiredSpeed. To provide noise immunity, implement hysteresis around the switch point.

5.6 In an application, the state of a switch is determined by reading a variable called PortA. The state of the switch is indicated by the state value of bit 4 within the byte variable PortA. In pseudo-code, write an event checking function TestSwitch() that returns Opened, Closed, or NoChange depending on what has happened since the last time TestSwitch() was called.

5.7 Work with a microwave oven to develop the list of events that the oven's software must respond to.

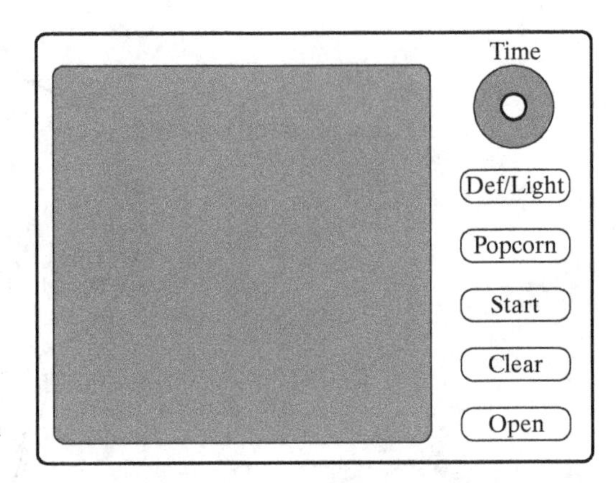

FIGURE 5.7

Front view of the microwave oven described in Problem 5.8.

5.8 Figure 5.7 shows a simple microwave oven. Draw an STD that captures the behavior described by the following description as follows.

(a) The Open button opens the door, stops cooking, and holds time.

(b) The Clear button clears the timer. If cooking was active, it is disabled.

(c) The Start button starts cooking for whatever time has been set. While cooking, the timer decrements to zero and turns off cooking when it reaches zero.

(d) The Popcorn button forces a time of 2 minutes to be set into the timer.

(e) The Def/Light button forces the power level to 50%.

(f) Time is set by twisting the dial and then pressing the button at the center of the dial. No action takes place until the button is pressed. At that point, the time on the dial is entered into the timer.

(g) There is a switch connected to the door to show whether it is open or closed.

5.9 Assume that the switches from the microwave oven in Problem 5.8 are available in a single variable (Switches), that the timer can be set with a function SetTimer(), that the timer can be read with a function IsTimerExpired(), and that the power level can be set with a function SetPower(). In pseudo-code, write routines for the required event checkers, state machine function, and main() to implement your design from Problem 5.8.

5.10 Draw the STD for a soft drink machine that accepts nickels, dimes, and quarters and sells drinks that cost $0.75 each.

5.11 Draw a state diagram for modified combination lock like that shown in Figure 5.4. Make your lock such that it will not accept unlock if four or more digits are entered that end with the combinations that end in 2-1-8 sequence.

5.12 There are many common behaviors that can be described using state machines. As an example, identify the relevant events, and draw a state diagram to describe the behavior of a "Shy Party Guest" (a guest who does not initiate conversations but participates when someone else initiates).

FURTHER READING

"Build Applications Faster with State Transition Automatons," Cline, A., *C/C++ Users Journal,* December 1992.

"State Machines in C," Fischer, P., *C Users Journal,* December 1990; 8(12).

"Who Moved My State?" Samek, M., *Dr. Dobbs Journal,* April 2003.

Software Design

CHAPTER 6

As with many of the other topics in this book, a complete discussion of software design could easily fill a book on its own. Rather than attempt to condense the topic into a chapter, we will address only a limited subset of material that we feel is likely to have the highest impact on improving the quality of the software you generate. In this chapter, we discuss software design in the context of an extended metaphor that gives a broad perspective on the importance of design in the creation of both houses and software. In the process, we explore the rationale for taking the time to do software *design*.

The chapter is structured in two parts. In the first part (Sections 6.1–6.4), we introduce a physical analogy to the software design process and describe the subset of software design techniques that we are going to focus on, which we feel can have the most impact when applied to the software design process. The second part of the chapter (Section 6.5) is organized around a sample problem that we will take through several stages of development and use as the vehicle to demonstrate the concepts and techniques that we introduced in the first part of the chapter.

When you have completed this chapter, you should have a better understanding of why software design is important and be able to effectively apply decomposition, information hiding, pseudo-code, and software interface specifications to the design of your software.

6.1 INTRODUCTION

Beginning (and even some experienced) programmers often don't spend enough time designing their software. Perhaps one of the reasons behind this has to do with the typical ways that we describe the creation of software. We usually talk about *writing* software. This comes from the mechanical similarity of the process of typing source code to the typing of prose. As it turns out, "writing" is a *terrible* description of the process of programming. No matter how eloquent we strive to make our comments and code, we cannot escape the fact that we are not telling a story, persuading or presenting facts. We are creating a very literal set of detailed instructions. Good software is more like a set of blueprints for a building. When you are writing a program, you are describing how to *build* an application, not *writing* a paper. The building metaphor is much more appropriate to describe the creation of software, even though the actions involved are mechanically similar to writing prose. As you will see, there are many parallels between building a house and building an application, which will serve as the enabling metaphor for this discussion.

6.2 BUILDING AS A METAPHOR FOR CREATING SOFTWARE

In building, simple jobs can be done with ad hoc planning and in some cases, no formal planning at all. More complex jobs require detailed planning in order to have any chance of success. Consider the differences between building a dog house and building a new single family home[1]. The scope (size) of the dog house is relatively small, and the expectations for the complexity of the structure are relatively low. It needs to keep the rain off and most of the wind out. It needs a single way in and there are generally no interior improvements. In contrast, the single family home is quite a bit larger in size and there are a lot more expectations for the structure. It too needs to keep the rain off, but there are times (like winter) when we want to be able to keep all of the wind out, yet other times (like a hot summer day) when we want to allow the breeze in. The interior is not a single room, but divided into several rooms. There is an expectation of many interior improvements: trim, paint, plumbing, electricity, lights, heat, and ventilation. All of these features need to be considered and incorporated into the structure if it is to be successful.

Someone reasonably skilled with a saw, hammer, and nails could build a dog house in an afternoon. It is not essential that there be a detailed plan as long as the builder has a reasonably good vision of what they want the dog house to look like and enough experience to achieve the vision. Detailed design and implementation can happen concurrently when the project is simple enough. However, no one would seriously consider building a single family home without a detailed design. The scope and complexity of the task are simply too great. The various parts of the structure and the infrastructure all interact with one another and need to fit together properly if the structure is to be safe, functional, and reliable as a home.

There are strong parallels to building dog houses and single family homes—and indeed skyscrapers—to be found in software engineering and mechatronics projects as a whole. This analogy works very well across a range of construction and software projects. The more complex the project, the greater the need for advance planning in order to be successful at achieving the desired result. Someone reasonably skilled in building software (the skill level here is critical) could sit down and, in a fairly short period of time, write and debug a simple program with little formal design. The design is improvised and ad hoc, its structure is based on the experience of the programmer. This approach is effective only for very simple programs that can be expressed in a page or two of code. This kind of programming is done regularly by experienced programmers, most often showing up as a simple utility to automate a process or as a test program to try out an idea. It is very different than something that is intended to be a significant application, be it a shrink-wrapped product or something written for a client. In the case of a commercial program, the problem to be solved by the software is nontrivial, otherwise why would you pay someone else to write it? Nontrivial problems require planning and attention to detail in order to create successful solutions. The solution will also need to be maintained after the first finished version is complete. Logical organization and good design facilitate maintainability. Like the house, the successful solution needs to be designed before it is implemented.

When thinking about the overall process of development, the analogy to construction is appropriate as well. When designing and building a house, you start by first deciding what kind of a house that you want to build. Here you are defining the requirements that the house must meet. Will it be a large elaborate house with many requirements, or a simple cabin with only a few needs, or something in between? Once you have the basic concept that you want to pursue, you decide on things like the size and layout of the rooms and so on. During this phase, it is not uncommon to develop several possible arrangements that might meet the requirements and then choose the best among those ideas. Once the basic layout is set, there are many design details to attend to. You must decide how the plumbing will be routed, how the walls will tie into one another and into the foundation, how the roof loads will be supported, and a myriad of other design details. With a complete design in hand, you

[1]This comparison was inspired by an example in Steve McConnell's book *Code Complete* [1].

are ready to enter the construction phase. There are several phases to construction and the entire construction process is punctuated by inspections, done mostly by the independent building inspectors, to be sure that the house is being built to the local building standards.

In software design and construction, we go through an analogous set of steps. Deciding what kind of house to build is akin to deciding what the application should do and what the feature set should be. In the parlance of classic software engineering, we talk about a **problem definition** that leads to a set of **requirements**. Analysis of the requirements leads to a set of **specifications** which, in turn, drives the architecture. Finally, the architecture is used to guide the detailed design that is followed by construction—for example, actually writing the code.

In mechatronic systems, our problem definition is often quite broad and it is not immediately clear exactly what parts of the functionality will need to be implemented in the software. These systems require a few more steps before getting to the software design. Mechatronic system design starts with a problem definition from which we derive a set of requirements. At this point, we will need to prepare a system architecture that defines how the system should be partitioned into subsystems and how the functionality in the subsystems will be partitioned between electronic hardware solutions, software solutions, and mechanical solutions. The system architecture will drive the development of a set of detailed specifications for each of these subsystems. Once we have identified the software-related specifications, the embedded software project follows a path similar to the classic software project. We'll deal with the details of these early steps in Chapter 30, where we discuss project planning.

6.3 INTRODUCING SOME SOFTWARE DESIGN TECHNIQUES

In this section, we describe the basics of several software design techniques that we feel, when implemented, have the biggest potential to improve the quality of the code generated by neophyte programmers. The following subsections introduce the terms commonly used for these techniques and briefly describe them and how they fit into the software design and construction process. In the later sections of this chapter, we will apply these techniques to a sample problem to demonstrate how they work in the context of designing a complete application.

6.3.1 Decomposition

The most basic of the high-value concepts that we will focus on is **decomposition**: breaking down the big problem into a series of smaller problems that are easier to understand and that will, when linked together, solve the larger problem. Decomposition is the essential first step in creating a design. Without it, it is all too easy to get caught up in the details before you see how they fit together to make the big picture.

Decomposition happens repeatedly on several different levels during the design process. It begins at the system level in deciding how to break down the problem into tasks that are best solved using hardware or software. It happens again as each of those subsystems is further decomposed into the hardware or software components necessary to implement their required functions. It continues to happen in software all the way down to the detailed code design stage in deciding which pieces of code ought to be broken out as separate functions as opposed to simply written as in-line code.

As we apply decomposition to our software, we will group the program functions and data necessary to work with those functions into **modules**. Some languages use slightly different terms for this grouping of functions and data. In C++, they are referred to as classes. In Ada, they are referred to as packages. Physically, modules are generally implemented as separate files. In the context of this chapter, when we talk about a module, we are referring to a group of functions and the data that those functions operate on that are grouped together to provide some higher-level functionality.

6.3.2 Abstraction and Information Hiding

Decomposition is linked to another programming concept known as **abstraction.** If you look at what you are doing as you decompose a problem, you are breaking out tasks that can be described in a few words. Generally, the details of what it will take to implement the functionality described in those few words are much more complex than the few words themselves. You have created an abstraction: an entity that does not physically exist that performs the functions described. You can use that abstraction to understand how it could fit with other abstractions to allow you to solve the problem at hand. The real power in abstraction is that you do not need to understand the details of *how* an abstraction will implement its functionality in order to be able to use that abstraction. In the language of software engineers, this separation between the function and its implementation is part of a larger concept known as **information hiding**.

One of our goals in designing the modules that will make up our program is to maximize the amount of information about the implementation that is hidden within the module. This will make it easier to use the module to solve the higher-level problem and it will also allow us to tinker with the internal implementation without disturbing the way in which the functions of the module are used. By isolating the interface (more in Section 6.4.4) from the implementation (Section 6.4.6), we help to limit the scope of changes necessary when, for instance, a bug is discovered.

Looking at a few concrete examples of information hiding will hopefully make the use of the concept clearer. Consider a module that provides the interface to an LCD display (Section 7.3.2). The kinds of things that you would want to hide inside the module would be things such as:

1. The sequence of operations necessary to initialize the display.
2. The sequence of operations necessary to write a character to the display.
3. Which port lines serve which purpose in the hardware interface.
4. What timing is necessary when transferring data or commands to the display.

If the rest of the program only used a few well defined high-level functions from the module, it would be very easy to change the details of the display without affecting the rest of the code in the program. The need to change the details of the display might arise if the display hardware were changed to accommodate a different vendor. It would also allow, for example, a data logging system to be added by modifying the Display module to log every write to the display. By hiding the details of writing to the display inside the module, any of these changes could be implemented without affecting the rest of the code in the application.

As another example, consider the interface to a keypad. On a product that one of the authors worked on, the keypad hardware for the instrument was cost-reduced during the development process. The change in the hardware resulted in the need to completely change the way in which the program interacted with the keypad. The project had been developed by a team with a strong culture of information hiding, so all of the details necessary to test for and decode key-presses were isolated in a single module. That module was rewritten to accommodate the new hardware over the course of an afternoon and no further changes were necessary outside the keypad module.

In these two examples, the concept of information hiding is applied to the interface to external hardware. While this is an area ripe for the application of information hiding, the concept is equally applicable to the details *inside* a module that has no interface to the hardware.

We will review a couple of examples of that kind of situation later in this chapter as we explore the application of software design to a sample problem.

6.3.3 Pseudo-Code

As we get deeper into the program design process, we will need a way to express the descriptions of how to carry out the details of the programming tasks. One tool that was adopted early on to help with this issue was the flow chart. Flow charts are still used in some environments, but have largely been supplanted by a more flexible and less graphically intensive technique known by any of several names, most commonly **program design language** (**PDL**) or simply **pseudo-code** [1].

Pseudo-code or PDL is most simply described as structured English. The goal is to describe an algorithm without resorting to code for a specific programming language. In this way, the pseudo-code could be used to code in any programming language. In practice, pseudo-code assumes the existence of the typical structured programming constructs available in most modern programming languages (definite and indefinite loops, if-then-else, and so on). Even in the absence of direct support for these structures, the pseudo-code can still be used as a guide to code the algorithm, albeit with a bit more effort in creating the structures.

Less experienced programmers often look on the generation of pseudo-code as waste of time, thinking "Why not just code it 'right' to start with?" Don't let yourself fall into this trap! The closer your pseudo-code is to a specific language, the more time you will be spending working out how to express your ideas in the idioms of that programming language and the less time that you will be spending on expressing the idea itself. The creation of pseudo-code is *never* wasted time. In some instances, the effort to express an idea in pseudo-code will lead you to see a flaw in the idea (and hopefully a way to eliminate it as well). Even if you get all of your pseudo-code completed without finding flaws, you still have created something useful for your code. Your pseudo-code can, and should, be the first level of comments in your code. Take the document in which you have written your pseudo-code, make every line a comment in the programming language of your choice, then proceed to fill in the actual code necessary to implement the pseudo-code. This way, you'll never be in the position of saying "It's all done, except for the comments."

This approach to using your pseudo-code as comments is also useful in refining your creation of pseudo-code. If you find yourself repeating the pseudo-code as the real code, then you are probably writing your pseudo-code too close to the target language. Consider moving up a level of abstraction. As an example, consider how to express a definite loop used to iterate through the elements of an array:

```
For every element in the array
 Do something interesting to the element.
```

This conveys much more information than a similar piece of pseudo-code that is closer to the programming language:

```
For i=1 to 100
 Do something interesting to the ith element.
```

In this case, the pseudo-code doesn't convey that the code should process *all* the elements in the array and unnecessarily specializes the pseudo-code to an implementation that the writer had in mind when they wrote the pseudo-code. Because it lacks the higher-level intent, it would be much harder to hand a piece of pseudo-code like this off to someone else to actually code. While the actual coding might go more quickly, the likelihood of it meeting the intent is diminished because that intent was not captured in the pseudo-code.

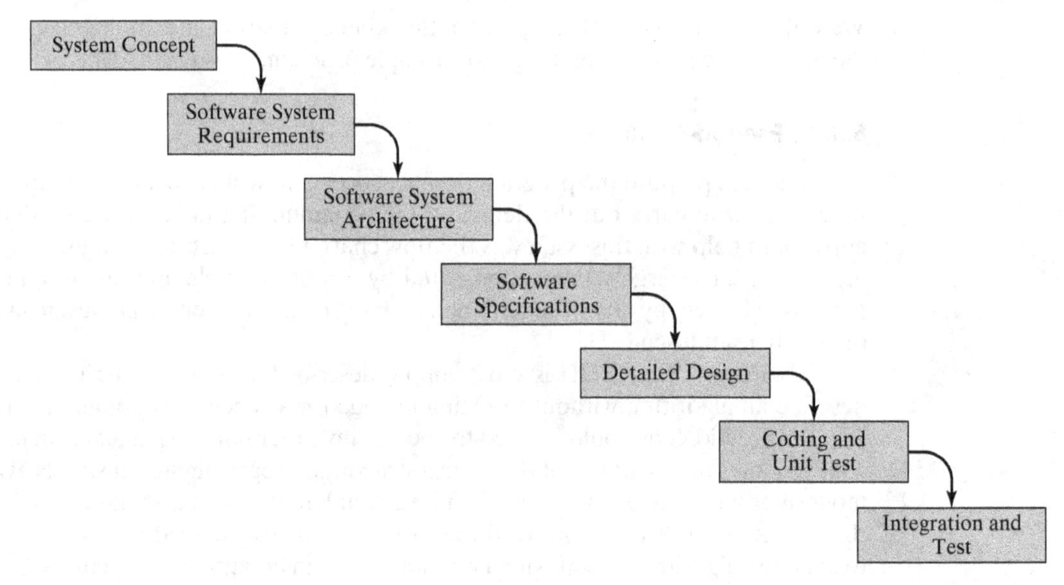

FIGURE 6.1

The classic software development model.

6.4 SOFTWARE DESIGN PROCESS

There is no single software design process. In fact, in recent years, several new software design processes have come into use. These new processes all attempt to address one or more perceived deficiencies in what is the classic software development model (Figure 6.1) that is often referred to as the **waterfall model**.

While these new development models have benefits, they also bring an added complexity. As our starting point for introducing you to a software design methodology, we are going to use a variation of the waterfall model that does away with the linear connection and acknowledges that the software design and development process is iterative in nature and often loops back on itself as the process uncovers new information that must be included in the design (Figure 6.2).

6.4.1 Generating Requirements

The first step in addressing a design challenge is to develop a set of requirements that the system must be capable of meeting. For a mechatronic system project, the system architect works with the client at this stage. The goal is to identify those things that the solution must be able to do if it is to be considered successful. Continuing the analogy with house building that we introduced at the beginning of this chapter, the architect would consult with the client in a similar way, to find out more specifics about the requirements for the home, such as preferences for sizes of rooms, traffic flow between the rooms, functionality of the spaces, and so on. A basic set of house requirements might look like the following.

1. Must house a family of four with separate bedrooms for each child.
2. Must provide for a guest bedroom.
3. Must provide a formal entertaining area.
4. Must provide for informal family space.
5. Must fit into a 1,500 sq. ft. footprint on the lot.

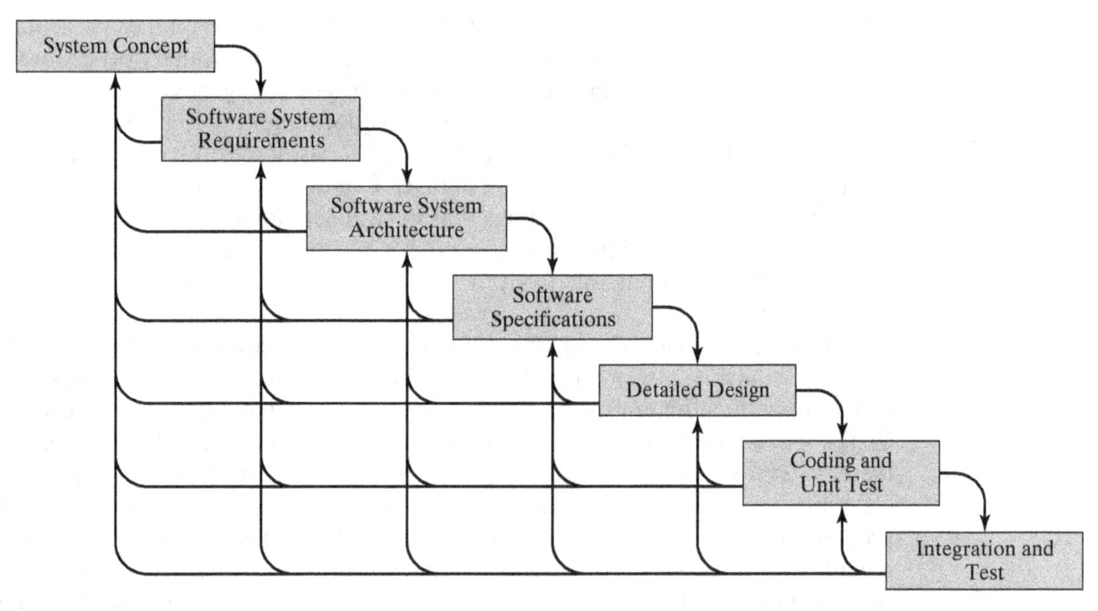

FIGURE 6.2

A more realistic software development model.

The requirements for a project are a very (some might say *the most*) important part of the project development process. As we emphasize in Chapter 30, all of the decisions made subsequently in the design and development process should be traceable to a requirement. This is true not only at the system level but at the level of software (and hardware) design as well. When developing the requirements, it is important to stick to those things that are actually *required* to meet the function of the system. Since so many later decisions will be traced to the requirements, and each feature can result in substantial time and effort, this is one of the places where including an optional requirement ("It would be nice if . . .") can unnecessarily drive the complexity of the solution.

In academic settings, it is tempting to simply declare that the lab or project assignment document is the requirements document for the assignment. It is rarely (if ever) the case that the assignment will be structured as a list of requirements that must be met. An important first step, and one that is not always done correctly (see Section 32.3), is to thoroughly understand the assignment and translate it into a concise set of requirements that the solution *must* meet.

6.4.2 Defining the Program Architecture

Once the requirements are understood and defined, the designer must decompose the problem to a reasonable level and develop a vision for how all these pieces are going to fit together to create the complete program. At this stage, it is best to seek multiple solutions that you could implement. Are you going to use an event driven paradigm (Chapter 5)? If so, what are the events going to be? Will you use a state machine structure (also in Chapter 5)? If so, what should the state transition diagram look like? These are all questions involved in deciding on the **software architecture** for the program.

In the building analogy, at this stage, you need to decide things like what kind of foundation the house will have: poured slab or footings? Will you use wood frame or masonry construction for the walls? There are many different approaches to specifying software architectures, many more than we can reasonably treat here, so we will focus on the event driven paradigm and state machines that we introduced in Chapter 5. These may be supplemented, as in the example we show later in this chapter, by ordinary procedural functions of the type that you learned in your introductory programming class.

Another task at the architectural level is deciding how to partition the program into individual **modules**. In this context, a module is considered to be a collection of local data and the routines that operate on that data. In C, a module is synonymous with a source file. It is in this partitioning into modules that we will actually implement the principle of information hiding. We would like a set of modules in which the details of the internal operations are as well hidden as possible from the "outside world"—in this case, the other modules of the program.

When partitioning the program into modules, there are several concepts that have come out of the study of software design that we should keep in mind. **Cohesiveness** is the property of a group of functions and data that makes them seem to hang together. This is a good thing. Cohesiveness allows us to build mental images of the program that we are designing, and thereby improve our understanding of it. **Coupling** is the property that connects things. Coupling is a necessary part of any program, but it is at odds with getting our arms around a program or a part of the program. The greater the level of coupling (or dependency) between modules, the more we need to understand about those modules in order to see how they will work together. With less coupling, it is easier to isolate the parts in order to understand and test them.

When we start to think about partitioning the problem into modules, it is important to keep in mind the idea of information hiding. You can help this process along by asking yourself "What information am I going to hide in this module?" To be most helpful in easing the task of understanding the overall structure of the program, you should be looking for those details that you will hide inside each module.

The last part of defining the program architecture should be to develop an **integration plan**. The integration plan should detail which modules are to be integrated in what order, building up finally to a complete application. As with most other aspects of the design, there is no one "right" integration plan. Generally, the plan should build from the lowest level (closest to the hardware) modules to the top level of functionality. In C, the integration phase may take the form of an ever increasing `main()` function as more and more modules are integrated. As the complexity of the project increases, this approach will start to have shortcomings. When debugging a problem, it will be difficult to go back to a lower level of integration to insure that a recent change has not broken some interaction between modules. This sort of **regression testing** is facilitated by the development of separate test programs that integrate the various modules of the program.

6.4.3 The Performance Specification

In software, as at the system level (Chapter 30), requirements are the key functions that the system must exhibit in order to fill a particular need. Specifications, on the other hand, are the particular and quantifiable metrics that enable a system to meet the overall requirements. In some instances, the specifications will be derived directly from the requirements. In some cases, they are elaborations or rewordings of the requirements. In other cases, the necessary specifications are things that a particular design must be able to do in order for that design to meet the requirements. In the example specifications that we present in Section 6.5.4, we will see examples of both kinds of specifications.

6.4.4 The Interface Specification

Concurrent with the partitioning of the software into modules, you also need to specify the interactions among those modules. This specification will describe which functions in other modules will be called to perform the necessary activity. The best way to be sure that the description of the interactions is complete is to prepare a set of detailed **interface specifications** for each of the modules. The interface specifications will describe the details of which functions will be provided by the module (the "**public interface**"), what the parameters will be to each of these functions, what the return values will be from each of the functions, any **side effects** (effects that the function has on the state of the program or hardware that might not be obvious from the function description) that the execution of the function will have (such as starting a motor), and the definition for any data structures necessary to use the **public functions** (those functions that are part of the public interface). Notice that the interface specifications do not deal at all with *how* the function will accomplish the activity, only the external view of the functionality. When describing the interactions between modules, the only functions available are those that are described in the interface specification. There will likely be a number of **private functions** (functions only made available for use within the module) that are necessary to implement the module but they will not be part of the public interface.

When writing the interface specification, the designer should also consider the possible outcomes from a function call. If the function could possibly fail, for example, due to an external hardware failure or bad parameters being passed, then the return value from the function should include some mechanism for indicating the failure.

Writing the interface specifications at design time serves several useful purposes. First, it allows the designers to better evaluate the design. By knowing which functions will be available, they can better study the interaction between the modules to insure that the overall intent of the design will be preserved. Second, it facilitates the distribution of labor during the implementation phase. With an accurate set of interface specifications, any member of the team may proceed to implement the module and all of its functions. The interface represents the view that the module presents to the outside world. As long as the implementation of the module meets the interface specification, it will be usable by the other members of the team to implement the finished product. Finally, a set of interface specifications will guide the detailed design of the module and its functions. Knowing in detail all of the things that the functions of the module must do will help the detail designer implement a structure internal to the module that will best meet the requirements. The specifications also provide a natural starting point for the design of a **test harness** (Section 6.4.7).

6.4.5 Detail Design

With a module breakdown and an interface specification in place, we are ready to move on to the **detailed design** of each of the modules. We can view the detailed design as a mini-design process within our larger design. In the detailed design phase, the designer decides how the functionality necessary to meet the interface specification will be implemented. In this phase, we develop an architecture *internal* to the module. What data structures will be used to represent the internal data of the module, what algorithms will be used and what private functions will be necessary to manipulate that data are all questions to be answered at this stage. As with the system level architecture, there are often multiple ways that the module level architecture could develop, and it is a good practice to seek multiple possible solutions and evaluate each of them relative to meeting the needs of the project.

During the detailed design phase, the designer will develop pseudo-code descriptions for each of the functions in the module. This includes both the public as well as all of the private functions. The pseudo-code should capture the intent of the designer as well as specify the desired algorithms to be implemented. The goal of the pseudo-code is a description that could be handed off to coders who understood the subtleties of the implementation language but did not need to concern themselves with the interactions between this module and the other modules of the program.

6.4.6 Implementation

Once the pseudo-code is complete, the process of generating code to implement the pseudo-code should be straightforward. This is not to say that it is easy or without challenge. The major benefit of pseudo-code is that the person doing the coding can focus on constructing code a few lines at a time. The pseudo-code should have captured the next higher level of intent, so the coder can focus on the implementation in the programming language of choice.

6.4.6.1 Intra-Module Organization

With a detailed design in hand, you are almost ready to start coding. However, before you get started, it will pay big dividends if you stop to think about the relatively mundane issue of module organization before you actually start coding.

Once you have decomposed your application into a set of modules, developed a detailed design and are ready to begin implementation, you are still faced with a wide range of choices that you will need to make. One of those things, layout within the module, is a detail that often "just happens" in the rush to write code. In fact, one of the simplest things that you can do to improve the understandability of your code is to adopt some standards for module organization and coding style. To realize the truth in this statement, think about how often, in the course of debugging a piece of code, you need to know the definition of a constant, the names of the members of a structure, or the size of an array. Where do you go to find the answer? A standard layout to your modules will allow you to spend the minimum amount of time answering questions like these.

While there are many possible arrangements that you could make that would fulfill the need for a standard layout, some things should naturally appear before others. In the following paragraphs, we propose one layout that seeks to provide the value that can be derived from a well thought out standard layout. The specifics details of this section are based on the C programming language, and the order of layout is not carved in stone, but the principles apply to any language. To make this all easier, it is useful to make up a module template that contains generic headers for the various sections. Then, when you need to make a new file, you can open this template file and use the "Save As" function of the editor to save it with the name of the module that you want to create.

At the very top of every file should be a `#define` for a preprocessor, typically named TESTING. This will be used with the test harness that we will discuss in the next section. By placing the definition at the very top of the file, we make it very easy to find to comment out or uncomment when we move between module level testing and integration between modules.

Next should be a block comment that identifies what this module is and generally what functionality it will provide to the rest of the program.

This should be followed by the section that contains all of the header files for the standard libraries and any other modules from your program that need to be referenced (using `#include`) from within this module.

Follow this up by a section containing any preprocessor macro definitions that are only used within this module.

Next are the data type definitions that are only used internally to the module. Any macro definitions or data types that are necessary to use the module should appear in the public header file for the module and be included along with the standard header references in the earlier section.

These should be followed by a section defining all of the private variables that are only used in this module. These are sometimes referred to as **module level variables** and, where the language allows, they should be defined in such a way as to limit their access to references from within this file only.

The last of the standard sections contains the prototypes for the private functions used in the module. These, along with the private data types and the private data, were worked out during the detailed design phase for the module. The prototypes for the public functions should appear in the public header file for the module, and the public header file for the module should be included along with the standard header references in the earlier section.

A template for a C program might look like the following.

```
#define TESTING
/****************************************************************************
Module
        Filename
 Revision
        1.0.1

 Description

 Notes

 History
 When          Who  What/Why
 ------------ ---  --------
****************************************************************************/
/*---------------------------- Include Files -----------------------------*/
#include <stdio.h>
#include <me218_c32.h> // local standard header file for 9S12C32 programming
/*---------------------------- Module Defines ----------------------------*/
/*---------------------------- Module Types ------------------------------*/
/*---------------------------- Module Variables --------------------------*/
/*---------------------------- Module Functions --------------------------*/
/*---------------------------- Module Code -------------------------------*/
/*---------------------------- Test Harness ------------------------------*/
#ifdef TESTING
void main(void)
{
}
#endif /* TESTING */
/*---------------------------- End of file -------------------------------*/
```

6.4.6.2 Writing the Code

With all of this preparation in place, writing the code should be a fairly straightforward process. Not only do you have a good handle on what the code needs to do, but you also understand how the various pieces will fit together to make up the finished application. The interface specifications describe each of the public functions. Copy those into the template and turn them into block comments in the programming language. Copy the pseudo-code from the detailed design, paste it into the Module Code area of the template, and turn the pseudo-code into comments. Now you are ready to start writing the actual code. To do this, simply work your way through the pseudo-code comments and write the actual programming language instructions necessary to implement what is described in the pseudo-code. While it is true that you have done a lot of work before you ever got to writing code, experience has shown that the time is well spent and will reliably lead to working code faster than an ad hoc approach to coding.

6.4.7 Unit Testing

Unit testing is the process of testing the functionality of an individual module. This is usually done by the person responsible for coding the module and involves the creation of a test harness specific to the module. The test harness is simply a program that will be used to verify that the module provides the required functionality and meets the interface specification. In C, this most often appears as a `main()` function that can be conditionally compiled to produce an executable program from just the single module under test. The test harness is used to validate the module before it is integrated with other modules. The test harness should exercise each of the public and private functions over the full range of possible input parameters. The testing implemented by the test harness should include not only the expected range of values for the input parameters, but also those that lie outside the expected range. When out of range or nonsense inputs are provided to a public function, it should behave in a safe manner and return an indication of failure. Every module should have a test harness associated with it.

6.4.8 Integration

When you have completed several modules and used the test harness for each module to validate the functionality of the module, you are ready to begin integration. When integrating subsystems, be they hardware or software, the approach should always be **incremental integration**. If you try to bring many modules together all at once and everything doesn't work just right, it can be very difficult to sort out just where the problem is. Incremental integration, when followed strictly, will actually get you to a finished working program faster than simply throwing all the modules together and hoping for the best.

The integration phase should take place according to the integration plan developed during the architecture definition. Integration should always be a step-wise process, with pretested individual modules being integrated with other pretested modules in as small a group as makes sense. In the ideal, this will be one module at a time. However, coupling between modules may dictate that more than one module be added at a given stage. This phase will entail the creation of increasingly more complex test programs that integrate the various modules of the complete program.

6.5 THE SAMPLE PROBLEM

These program design concepts and processes will be easier to grasp once you've seen them applied in the context of a concrete example. To that end, throughout the rest of this chapter we will develop the design for an application and point out how we are applying the software design concepts and processes that we have introduced in the preceding sections of the chapter. The problem that we will address is to interpret the input from an optical [infrared (IR) light] for which the relationship between when the light is on and when the light is off represents a message encoded in Morse code (see below text). After the Morse code characters are received and decoded, they should be presented as a scrolling, ticker-tape style message on a display. The speed of transmission will be constant within a message, but is not known beforehand and may change to any rate between 5 and 10 words per minute between messages. Since the speed of the message may change from time to time, the solution should include a button that may be pressed at any time. When pressed, the button should cause the software to immediately readapt to the current transmitting rate. While adapting, no characters should be decoded.

Morse code is an encoding scheme invented by Samuel Morse in the 19th century and used extensively for telegraph and, later, early radio communication. The elements of Morse code are dots, represented by a logical true state (light on) for a short period of time, dashes, represented by a logical true state for a longer period of time, and the spaces between dots and dashes, represented by a logical false state (light off). A character, for example the letter **A**, will be represented by a group of dots and dashes (in this case, dot-dash). The periods of time are defined such that the time of a dot, a dot time, is the basic unit of measurement. Dots are 1 dot time long. The space between dots and dashes is 1 dot time long. Dashes are 3 dot times long. The space after the dots and dashes of a character is 5 dot times long. The space at the end of a word is 7 or more dot times long. For speed specifications, a standard "word" is considered to be 50 dot times long.

With this as our problem definition, we will proceed to work through the various steps to demonstrate the high-value concepts in modular software design applied to this problem. With the problem statement in place, let's move on to looking at the activities in software design.

6.5.1 Requirements for the Morse Code Receiver

By studying the textual description above, we can come up with a set of requirements that the problem solution must meet in order to perform as described:

1. Must respond to pulses of IR light that encode any printable message in Morse code.
2. Must handle Morse code transmission rates between 5 and 10 words per minute.
3. Must be able to readapt to the current transmission speed whenever a button is pressed.
4. Must be able to scroll the received message across a display, ticker-tape style.

As with most initial requirements lists, this one is not complete because not all the information needed was given in the problem description. The system architect would need to consult with the client to get answers to such questions as "What range of rise and fall times will the light pulses exhibit?" and "What is the acceptable delay between the pressing of the button and the beginning of the speed recalibration?" At this stage, it is very common to find that this kind of important information is missing and must be solicited from the client. We cover the development of requirements further in Chapter 30. For the purposes of the software portion of this design, the only critical missing specification is the delay between the button press and the beginning of the readaptation. When asked, the client stated "50 ms or less should be fine."

6.5.2 The Morse Code Receiver System Architecture

The system architect will study the problem description and requirements, decompose the problem into a set of functional blocks, and propose a system level architecture to address the complete problem. For a system that crosses discipline boundaries, as this one does, the system level architecture will address which portions of the task will be delegated to a hardware solution and which to software. For most problems, the architect will develop multiple possible architectures and evaluate them based on how they impact the specifications necessary to meet the requirements. This problem is simple enough that it is hard to imagine a system architecture that would be substantially different than that shown in Figure 6.3.

The light pulses that represent the message are first transformed in electronic hardware into a signal voltage that is at a logic high or logic low level for various periods of time. This digital signal is applied to an input on a microcontroller that will run our software. The software interprets the digital input as a series of 1s and 0s and extracts the message, which it writes to an intelligent LCD module for display. Since this chapter is focused on software

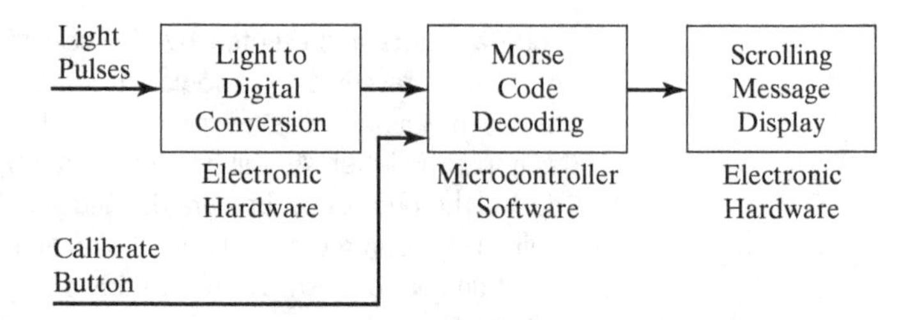

FIGURE 6.3

System level architecture for a Morse code decoder.

development, we refer the reader to Section 14.4.2 for a possible light-to-voltage pulse conversion circuit and Section 7.3.2 for possible display hardware. In the balance of this chapter, we will delve into the Morse Code Decoding block in Figure 6.3.

6.5.3 The Morse Code Receiver Software Architecture

Given the nature of the interaction with the outside world (responding to light pulses and button presses), this is a natural application for event driven programming. Similarly, since there will be different responses to a rising or falling edge based on whether the system is calibrating or decoding the signal, it is also a natural fit for state machines. This is not to say that this is the only way to approach the software for this problem. You could write blocking code that simply waited for rising and falling edges to process the incoming pulses. However, in order to meet the requirement to be able to respond to the button press at any time, you would end up needing to insert code to test for the button press into each of the loops that was waiting for rising and falling edges. Rather than follow this more convoluted approach with blocking code, we will continue the architecture development adhering to an event driven paradigm. We'll leave the question of the application of state machines until later in the individual module designs.

In developing the software architecture, we are again going to apply decomposition to get a handle on what our program will need to do. A logical decomposition of the Microcontroller Software block from Figure 6.3 might involve viewing the overall software task as being composed of five smaller tasks:

1. Calibrating to determine the time of a Morse code dot.
2. Recognizing dots, dashes, and spaces in the incoming data stream.
3. Segmenting the dots and dashes into character groups.
4. Translating the groups of dots and dashes into printable characters.
5. Displaying those translated characters on the display device.
6. Recognizing the button press to initiate recalibration.

There are certainly other decompositions possible, but all variants should capture the elements listed here.

We can study this list of tasks to determine which items are possibly related to others and therefore might be grouped together. We can note that items 1 and 2 both revolve around the rate at which the Morse code is being transmitted and only require that we know information about the timing definitions in Morse code. Item 4 requires that we know all of the possible legal groups of dots and dashes, but nothing about the timing. Item 5 is totally unrelated to any of the Morse code tasks and requires only that we know how to control the display device. Item 6 is related to item 1 in that the button press triggers the calibration process. Item 3 sits at the juncture between the items 1 and 2 and item 4. Performing the item 3 task requires that we assemble a group of dots and dashes and recognize the character and word spaces (as events, not necessarily the timing) to decide when to decode the group. Based on this analysis, we can propose a first pass at the software architecture as shown in Figure 6.4.

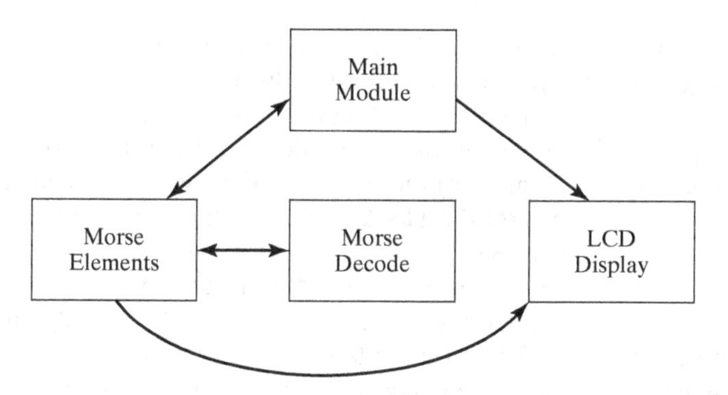

FIGURE 6.4

First proposal for a Morse code decoding architecture.

In this design, the tasks for items 1, 2, 3, and 6 are handled in the module called Morse Elements because they all revolve around the timing aspects of receiving Morse code. Item 4 is handled in the module labeled Morse Decode and item 5 is handled in LCD Display. The arrows indicate the interactions (coupling) between the modules. The Main module calls the initialization routine for the LCD display to get it ready to display the message and then calls a function in the Morse Elements module to actually receive the Morse code. Inside the Morse Element module, the dots, dashes, and spaces are recognized and grouped for decoding by a function in Morse Decode and subsequent display by LCD Display. The Morse Elements module needs to interact with the display in this design because at the end of every word, two characters need to be written to the display: the last character of the word and the space between words.

While this could be a functional architecture, it is less than ideal. While the button processing is being handled inside Morse Elements because it only interacts with functions in that module, putting the button in with the Morse code processing reduces the cohesiveness of that module. The horizontal lines between modules at the same level in the hierarchy represent inter-module coupling. As we discussed in Section 6.4.2, minimizing coupling should be one of our goals in the design. Coupling at the level of function calls is often unavoidable and not necessarily something that would disqualify a design. However, the coupling described between Morse Elements and Morse Decode goes much deeper. As described, the Morse Elements module is responsible for collecting the dots and dashes into groups for decoding. On one level this makes sense, since the decision to decode is based on the detection of a character space or a word space within the stream of dots and dashes. However, partitioning the modules in this way will require that a data structure that represents the character group of dots and dashes be passed between Morse Elements and Morse Decode. This represents a very strong level of coupling and also violates the principle of information hiding. Details of the internal representation of the group of dots and dashes must be shared between these two modules. We could eliminate this deeper coupling by moving the representation of the group of dots and dashes completely inside the Morse Decode module and provide a function-level interface to the module to allow building the group of dots and dashes. This still leaves a function-level coupling between the modules. In the process of being diligent about minimizing coupling, we should see if we can find a way to eliminate even that.

Before we begin developing an alternative architecture, it is useful to examine the breakdown that we have in this architecture with our information hiding "hat" on. When evaluating modules with respect to information hiding, it is useful to ask: "What information is being hidden in this module?" For the modified architecture that we currently have, we can see that we actually have a good level of information hiding. The Morse Elements module hides all of the details related to receiving the signal from the hardware and the timing of Morse code. The Morse Decode module hides the representation of the groups of dots and dashes and all of the knowledge about the translation of the groups of dots and dashes into printable characters. The LCD module hides the hardware interface and the details of the steps and timing necessary to write a character to the display.

If the information hiding aspect of the design is good, perhaps we don't need an all-new design but simply need to focus on the interactions between the modules to see if we can find a way to minimize (or eliminate) the coupling. Since we have decided that we want this to be an event driven design, lets take a look at the events that we have in the system and see what sort of responses are necessary.

The most obvious events are those related to the hardware: rising edge, falling edge, and button press. Perhaps less immediately obvious is that the falling edge hardware event causes different things to happen based on the length of time that the light was on (dot recognized, dash recognized). Similarly, the rising edge represents the end of a dot space, a character space, or a word space. Perhaps we could treat each of these as software generated events and deal with them in the way that we described in Sections 5.3 and 5.4: with an event checker and a service routine. We could improve the cohesiveness of the Morse Elements module by extracting the button processing into a separate module. The Main module would then become a loop that called the Button and Morse Elements event checkers and processed those events by interacting with the Morse Decode module and the LCD Display module. When dot or dash events were detected, the dot or dash would be added to the internal representation in Morse Decode. When a character-space event was detected, the current group of dots and dashes would be decoded and sent to the display. When a word-space event was detected, the current group would also be decoded and displayed but in this case an extra space would be appended to the display. The dot-space event could be ignored.

Using this approach, the diagram of the software architecture could be redrawn as shown in Figure 6.5.

This architecture preserves all of the information hiding aspects of the original design and eliminates the inter-module coupling at the second level of the hierarchy. Putting the details of the button processing into its own module improves the cohesiveness of the modules. The Morse Elements module will encapsulate all of the information about the timing aspects of Morse code. In this module, we will calibrate our timing to the incoming Morse code stream and use the knowledge of the timing to break the data stream up into dots, dashes, character spaces, and word spaces. The Morse Decode module will hide all of the information about the relationship between groups of dots and dashes and the letters that they represent. In this module, we will need a way to represent the dots and dashes as they arrive but before a complete character has arrived. The details of how we represent these groups inside the module need not be presented to the rest of the program. All that this module needs as input is notification that a new dot or dash or character space or word space has arrived. When either a character or word space arrives, the functions inside the module will translate the collected group of dots and dashes into a character to pass on to the display. The Display module abstracts the details of operating the intelligent LCD into a very simple

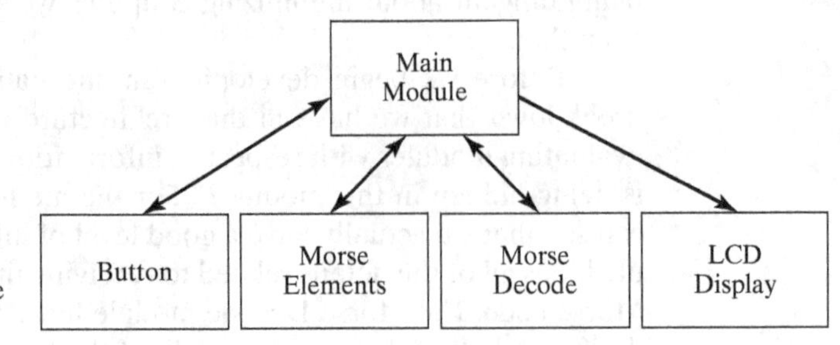

FIGURE 6.5

Improved Morse code software architecture.

paradigm. The display is a write-only device that you send characters to make them visible to the user. Outside the module, it is not necessary to understand the details of the kind of display or what manipulations of port lines or memory are necessary to get the characters to appear.

6.5.4 The Morse Code Receiver Performance Specifications

The **performance specifications** are the particular and quantifiable metrics that the software must fulfill in order to enable the system to meet the overall requirements. With so few requirements, the software for our example needs only to meet a few key specifications in order to fulfill the requirements:

1. The software must be able to decode all printable Morse code groups.
2. The software must be able to display all printable Morse code characters and space to the display.
3. The software must be able to scroll the display in a ticker-tape style as new characters arrive.
4. The calibration process must be started within 50 ms of a button press.
5. The software must be able to decode a character and write it and a space to the display in less than 120 ms.

The first three specifications are directly related to the requirements to be able to receive the Morse code message and scroll it across a display. The fourth comes from a requirement that we elicited from the client during the requirement development process. The fifth specification is one that results from the particular software architecture that we have chosen. We have designed an event driven architecture with the principal hardware events being the rising and falling edges of the Morse code signal. This implies that we will not know that we have reached the end of a character until we see the first rising edge of the first dot or dash of the next character. Only then will we be able to see that the space is long enough to indicate the end of a character or word. In order not to miss any edges on the input signal, we must be able to decode the Morse character and display it along with a space (in case it was the end of a word) before the fall of that first Morse element of the following character. That first Morse element could be as short as a dot time, so we need to be able to complete the decode-and-display operation in less than 1 dot time. At the fastest speed that we are required to receive, that amounts to

$$\frac{1}{\left(\frac{10\text{ words}}{\text{minute}} \times \frac{50\text{ dot times}}{\text{word}} \times \frac{1\text{ minute}}{60\text{ seconds}}\right)} = \frac{120\text{ ms}}{\text{dot time}}$$

6.5.5 The Morse Code Receiver Interface Specification

With a software architecture in place and a vision for how the modules are going to interact, we can start to develop the module interface specifications. Some of the necessary functions will be easy to identify based on our understanding at the beginning of the interface specification process. It is also common to find that the need for other functions becomes clear as we sharpen our vision of the interactions between the modules.

6.5.5.1 The Button Module Interface Specification

Given the event driven nature of our design, we should design the Button module to provide for debouncing the button hardware and reporting on the events related to the button:

ButtonDown and ButtonUp. Since we will need to initialize that port line(s) for the button interface and the static variable that we will need for event checking, we should also include a module initialization function that would be called as part of the start-up sequence before we start processing the incoming signal. In general, it is a good practice to allow for an initialization function for every module.

> The `InitializeButton()` function does not need a parameter and does not need a return value. It is responsible for doing all of the initialization of port lines and data structures internal to the module that are necessary to prepare the module to begin monitoring the calibrate button. When execution of the function is complete, the module should be ready to begin accepting calls to the other function of the module. There are no external side effects.
>
> The `CheckButtonEvents()` function does not need a parameter and should return one of the events related to button presses: ButtonDown, ButtonUp, or NoEvent. This function tests for the button being pressed and released and implements a software debounce on the input. Neither this function, nor any of the functions that it calls, may contain blocking code.

6.5.5.2 The Morse Elements Module Interface Specification

Based on the description so far, there are only three public functions that we have identified for the Morse Elements module. The first function would be a checker for the software generated events related to the arrival of dots and dashes. The second would be a function to initiate the calibration process. This function would be called by the Main module when a button press was detected and hides whatever details are necessary internal to the module to start the calibration sequence. In addition to these two functions, we should also include a module initialization function. At this time, we do not know the details of the internal design of the module, so it is safest to assume that there will be at least some module level variables that may need to be initialized and provide for a function that does this.

With that addition, we have three public functions in the Morse Elements module: `InitMorseElements()`, `StartCalibration()`, and `CheckMorseEvents()`. The interface specifications that we developed for these functions are as follows.

> The `InitMorseElements()` function does not need a parameter and does not need a return value. It is responsible for doing all of the initialization of port lines and data structures internal to the module that are necessary to prepare the module to begin capturing rising and falling edges from the IR light sensor circuit, and categorizing the intervals according to the type of Morse element represented. When execution of the function is complete, the module should be ready to begin accepting calls to the other function of the module. There are no external side effects.
>
> The `StartCalibration()` function does not need a parameter and does not need a return value. It is responsible for initiating the calibration process internal to the Morse Elements module. After this function is called, no further characters will be printed to the display until the calibration process is complete.
>
> The `CheckMorseEvents()` function does not need a parameter and should return one of the events related to the reception of Morse code or calibration: DotDetected, DashDetected, CharSpaceDetected, WordSpaceDetected. This function tests for rising and falling edges, times the intervals between the edges, and categorizes these intervals to generate the Morse events. Neither this function, nor any of the functions that it calls, may contain blocking code.

6.5.5.3 The Morse Decode Module Interface Specification

The Morse Decode module builds and maintains an internal representation of the group of Morse elements (dots and dashes) and decodes the group into the appropriate ASCII[2] character. To achieve this, it will need functions that add a dot to its internal representation, add a dash to its internal representation, and decode the current state of the internal representation of the group of Morse elements. As with most modules, it should also have an initialization function, which in this case can be used to clear out the data from the last character and prepare to accumulate a new group of Morse elements. The interface specification that we developed looks like the following.

> The `AddDot()` function does not need a parameter and should not need a return value. This function is responsible for adding the representation of a dot to the internal group of Morse elements. Neither this function, nor any of the functions that it calls, may contain blocking code.
>
> The `AddDash()` function does not need a parameter and should not need a return value. This function is responsible for adding the representation of a dash to the internal group of Morse elements. Neither this function, nor any of the functions that it calls, may contain blocking code.
>
> The `ClearMorseChar()` function does not need a parameter and should not need a return value. This function is responsible for clearing the data structure used to represent the internal group of Morse elements. Neither this function, nor any of the functions that it calls, may contain blocking code.
>
> The `DecodeMorseChar()` function does not need a parameter and should return an ASCII character. This function is responsible for using the current state of the internal data structure of Morse elements to determine which ASCII character corresponds to that group. In the event that no valid translation exists for the group of Morse elements, then a "#" character should be returned. Neither this function, nor any of the functions that it calls, may contain blocking code.

Notice that in the description of `DecodeMorseChar()` a new behavior has appeared. In thinking about what the return value should be for this function, we realized that there is the possibility that the group of Morse elements might not represent a valid character. There is nothing written in the problem description to specify how this condition should be handled, so we decided to handle it by giving an error indication (a reasonable choice, since the "#" character is not defined by Morse code). This is an example of the kind of insight that can develop as we create the interface specifications. If we were preparing this design for a client, we would go back to the client and ask how they would like this condition to be handled and add their response to the project requirements.

6.5.5.4 The LCD Display Module Interface Specification

The LCD Display module interface is the simplest of the modules. It requires only a function to initialize the display subsystem and the associated hardware and a second function to write a character to the display.

> The `InitDisplay()function` does not need a parameter and should not need a return value. This function is responsible for initializing the hardware associated with

[2]ASCII stands for the American Standard Code for Information Interchange and is the standard that is used to encode printable characters (and a few control characters as well) into a 7-bit numeric code. This code is used by display and printing devices to translate between numeric values and the characters to be printed or displayed.

the display as well as the initialization of any internal data structures necessary to communicate with the display. When execution of the function is complete, the module should be ready to begin accepting calls to the `WriteChar()` function to display characters to the LCD display in a ticker-tape style presentation. The external side effects relate to the state of the display hardware and will be specific to that hardware.

The `WriteChar()` function takes a single parameter: the new character to be written to the display. Execution of this function will cause the external display to scroll one position to the left to create the required "ticker-tape" scrolling and the new character will be added to the rightmost position of the display. Neither this function, nor any of the functions that it calls may contain blocking code.

6.5.6 The Morse Code Receiver Detail Design

In the detailed designs that we present here, we have assumed the existence of a function that will read a clock to determine the current time and that we will implement a fully event driven system so that the occurrence of events is being checked rapidly enough that the interval between the actual occurrence of the event, the detection of that event, and the reading of the time will be small relative to the intervals being timed.

6.5.6.1 Button Module Detail Design

The detailed design of the Button module must provide for debouncing the switch while also meeting the specification for the maximum delay between the press of the button and the start of a new calibration. While there are several different approaches to debouncing in software, we have chosen what is arguably the simplest: we won't look at the switch often enough that we will see the bouncing. What this will involve is timing inside the event checker so that no matter how quickly the event checker is called, the sampling of the switch hardware will only happen at the debounce rate. For this to work and meet the specifications for response time, the sampling interval must be shorter than the 50 ms response time specification. Since debounce intervals almost never need to be even 50 ms long, this should present no problem.

```
Pseudo-code for the Button module
Data private to the module: LastButtonState, TimeOfLastSample

InitializeButton
Takes no parameters, returns no value.
Initialize the port line to monitor the button
Sample the current button state and use it to initialize
LastButtonState,
Set TimeOfLastSample to the current time
End of InitializeButton

CheckButtonEvents
Takes no parameters, returns one of the following events:
     NoEvent, ButtonDown, ButtonUp
Local Variables: ReturnValue
Set ReturnValue to NoEvent
If more than the debounce interval has elapsed since the last sampling
     If the current button state is different from the LastButtonState
          If the current button state is up
               Set ReturnValue to ButtonUp
          Else
```

```
                Set ReturnValue to ButtonDown
            Endif
            Set LastButtonState to the current button state
        Endif
        Record current time as TimeOfLastSample
    Endif
    Return ReturnValue
    End of CheckButtonEvents
```

6.5.6.2 Morse Elements Detail Design

In approaching the detailed design of the Morse Elements module, we need to keep in mind that we need a fully event driven solution, so no blocking code will be allowed. If `CheckMorseEvents()` is internally going to be monitoring and responding to hardware events, then we would expect that its structure would mimic that of the `main()` for an event driven program that we described in Section 5.5.

We can apply decomposition to the Morse Elements functionality to determine that the task of recognizing dots and dashes involves a number of subordinate tasks, some of which are shared with the calibration task:

1. Detecting rising and falling edges in the incoming signal.
2. Determining the time between edges of the incoming signal.
3. Classifying and sorting the incoming data stream into dots, dashes, dot spaces, character spaces, and word spaces.

There are two hardware events that need to be responded to: rising edges and falling edges. In addition, we can see that the response to the rising and falling edges will be different depending on whether we are calibrating or decoding the incoming pulse stream. The fact that we need to respond differently to the rising and falling edges depending on the mode tells us that the history of the events will be important in determining how we respond. This, in turn, suggests that a state machine might be a good way to capture the necessary behavior for the module. Taking this approach, we developed a state machine (Figure 6.6) to describe the response to the hardware events.

There are three general regions to the state diagram: Calibrating, Waiting for End of Character, and Decoding. The state machine is initialized into the calibration region, waiting for a rising edge. The Waiting for End of Character functionality is necessary to insure that the first character decoded after the calibration completes is a legitimate character and not the ending portion of a character that was coming in when the calibration completed. This was a need that was discovered as we worked out the state diagram, using it to play "what if" games with the input stream. Since the time associated with every rising and falling edge is always needed, we have factored the logging of that time out of the state machine response and delegated it to the event checker, which has all of the necessary information available at the time that one of these events is detected. In working out the design, we also discovered the need for two more software generated events: the detection of an end-of-character or end-of-word condition and the detection of the completion of the calibration. The end-of-calibration event will trigger the entry into the wait for end-of-character region. The end-of-character or word event will trigger the entry into the decoding region at the trailing edge of a character. The characterize pulse length and characterize space length functions will categorize the pulse or space length according to the rules of Morse code. The TestCalibration function will determine if the latest edge has given us enough data to produce a valid calibration for the dot time being used in this transmission.

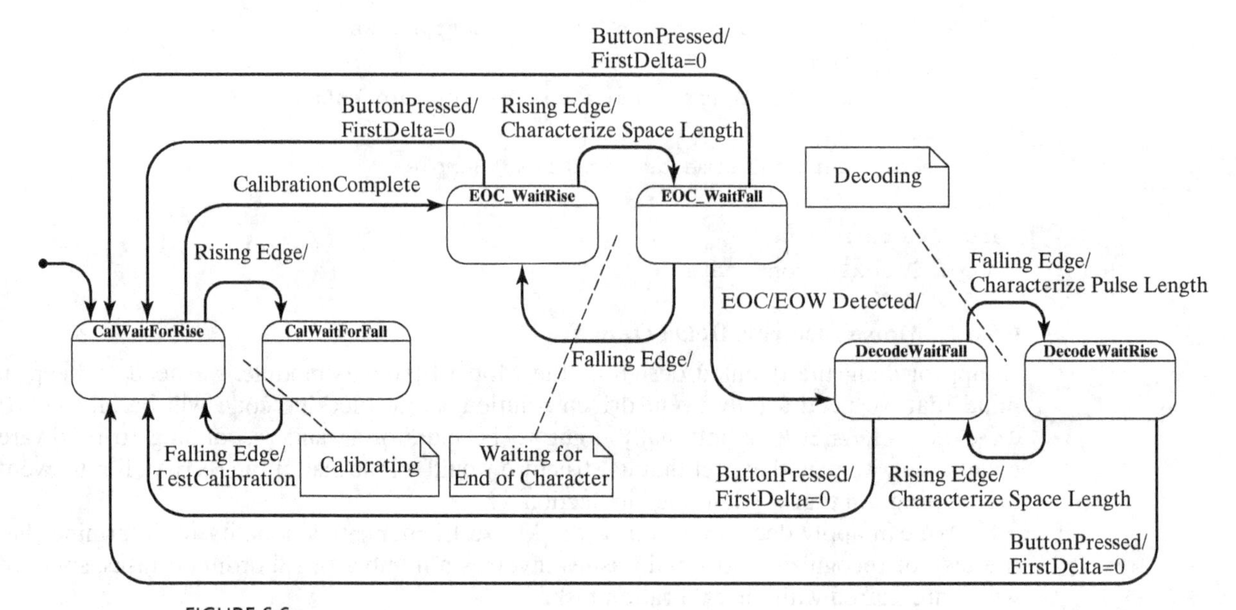

FIGURE 6.6

State diagram for the Morse Elements module.

One thing that the state diagram does not capture is the algorithm that we will be using to calibrate to the Morse code transmission speed. One possibility is a simple, but slow, algorithm:

```
Capture a large number of time intervals between rising and falling edges
Prepare a histogram of those time intervals
Find the most common time interval. This is the dot time.
```

This algorithm makes use of the fact that both the time of a dot and the time between dots and dashes within a character is 1 dot time long. For any meaningful English message, this will make the dot time the most common interval that occurs in the message.

The algorithm that we have chosen to implement will likely find the calibration much quicker. This algorithm makes use of the fact that the ratio between dots and dashes should be 1:3. We will collect the times of two adjacent pulses and look to see if those times are in a 1:3 or 3:1 ratio. Because of the uncertainty in the time measurement, we can only know the time of a measured interval to ±1 timer count. If we wait until we have a pair of measurements that are in a 1:3 or 3:1 ratio, we can be confident that the shorter of them is actually the dot time and not the dot time −1 or dot time +1.

With the interface specification, state diagram and calibration algorithm in place, we are in a position to start developing the pseudo-code to describe the public and private functions necessary for the module.

```
Pseudo-code for the Morse Elements module
Data private to the module: LengthOfDot, TimeOfLastRise, TimeOfLastFall,
    FirstDelta, CurrentState
```

InitializeMorseElements

```
Takes no parameters, returns no value.
Initialize the port line to receive Morse code
Set CurrentState to be CalWaitForRise
```

```
Set FirstDelta to 0
End of InitializeMorseElements
```

StartCalibration

```
Takes no parameters, returns no value.
Call MorseElementsSM with the event ButtonPressed and CurrentTime
End of StartCalibration
```

CheckMorseEvents

```
Takes no parameters, returns one of the following events:
      NoEvent, DotDetected, DashDetected, CharSpaceDetected
      WordSpaceDetected
Local Variables: ReturnValue
Set ReturnValue to NoEvent
If the state of the Morse input line has changed
      If the current state of the input line is high
            Record current time as TimeOfLastRise
            Call MorseElementsSM with the event RisingEdge and CurrentTime
            If the return from MorseElementsSM is EndOfCharacterFlag
                  Call MorseElementsSM with EOCDetected and CurrentTime
                  Set ReturnValue to CharSpaceDetected
            ElseIf the return from MorseElementsSM is EndOfWordFlag
                  Call MorseElementsSM with EOWDetected and CurrentTime
                  Set ReturnValue to WordSpaceDetected
            EndIf
      Else (current input state is low)
            Record current time as TimeOfLastFall
            Call MorseElementsSM with the event FallingEdge and Current-
            Time
                  If the return from MorseElementsSM is CalCompleteFlag
                        Call MorseElementsSM with CalibrationCompleted
                        and CurrentTime
                        Set ReturnValue to NoEvent
                  ElseIf the return from MorseElementsSM is
                        DotDetectedFlag
                        Set ReturnValue to DotDetected
                  ElseIf the return from MorseElementsSM is
                        DashDetectedFlag
                        Set ReturnValue to DashDetected
                  EndIf
Else (Morse input line is unchanged)
      Set ReturnValue to NoEvent
EndIf
Return ReturnValue
```

MorseElementsSM (implements the state machine for Morse Elements)

```
Takes two parameters
      (ThisEvent), one of the following events:
      RisingEdge, FallingEdge, CalibrationCompleted, EOCDetected
```

```
      EOWDetected, ButtonPressed
      (ThisTime), the time that the event was detected
Returns EndOfCharacterFlag, EndOfWordFlag, DotDetectedFlag,
    DashDetectedFlag, CalCompleteFlag, NoFlag, BadPulseFlag,
    BadSpaceFlag
Local Variables: ReturnValue, NextState
Set ReturnValue to NoFlag
Set NextState to CurrentState
Based on the state of the CurrentState variable choose one of the follow-
ing blocks of code:
      CurrentState is CalWaitForRise
            If ThisEvent is RisingEdge
                  Set TimeOfLastRise to ThisTime
                  Set NextState to CalWaitForFall
            Endif
            If ThisEvent is CalibrationComplete
                  Set NextState to EOC_WaitRise
            Endif
      End CalWaitForRise block
      CurrentState is CalWaitForFall
            If ThisEvent is FallingEdge
                  Set TimeOfLastFall to ThisTime
                  Set NextState to CalWaitForRise
                  Call TestCalibration
                  If the return from TestCalibration is CalComplete
                        Set ReturnValue to CalCompleteFlag
                  EndIf
            Endif
      End CalWaitForFall block
      CurrentState is EOC_WaitRise
            If ThisEvent is RisingEdge
                  Set TimeOfLastRise to ThisTime
                  Set NextState to EOC_WaitFall
                  Call CharacterizeSpace function
                  If return from CharacterizeSpace shows EOC or EOW
                        Set ReturnValue to EndOfCharacterFlag
                  EndIf
            Endif
            If ThisEvent is ButtonPressed
                  Set NextState to CalWaitForRise
                  Set FirstDelta to 0
            Endif
      End EOC_WaitRise block
      CurrentState is EOC_WaitFall
            If ThisEvent is FallingEdge
                  Set TimeOfLastFall to ThisTime
                  Set NextState to EOC_WaitRise
                  EndIf
            Endif
```

```
        If ThisEvent is ButtonPressed
                Set NextState to CalWaitForRise
                Set FirstDelta to 0
        Endif
        If ThisEvent is EOCDetected
                Set NextState to DecodeWaitFall
        Endif
    End EOC_WaitFall block
    CurrentState is DecodeWaitRise
        If ThisEvent is RisingEdge
                Set TimeOfLastRise to ThisTime
                Set NextState to DecodeWaitFall
                Call CharacterizeSpace function
                Set ReturnValue to the return value from
                CharacterizeSpace
                EndIf
        Endif
        If ThisEvent is ButtonPressed
                Set NextState to CalWaitForRise
                Set FirstDelta to 0
        Endif
    End DecodeWaitRise block

    CurrentState is DecodeWaitFall
        If ThisEvent is FallingEdge
                Set TimeOfLastFall to ThisTime
                Set NextState to DecodeWaitRise
                Call CharacterizePulse function
                Set ReturnValue to the return value from
                CharacterizePulse
                EndIf
        Endif
        If ThisEvent is ButtonPressed
                Set NextState to CalWaitForRise
                Set FirstDelta to 0
        Endif
    End DecodeWaitFall block
Return ReturnValue
End of MorseElementsSM
```

When preparing the pseudo-code for a state machine, it is very helpful to make a hard copy of the state diagram and use a highlighter to mark the transitions on the diagram as you add the code for that transition. This way, you can be sure that you have included the code to deal with all of the defined transitions.

In the TestCalibration function (below), we need to determine if the latest edge has given us enough data to produce a valid calibration for the dot time being used in this transmission.

```
TestCalibration
Takes no parameters, returns either CalInProgress or CalCompleted.
Local variable SecondDelta, ReturnValue
```

```
Set ReturnValue to CalInProgress
If calibration is just starting (FirstDelta is 0)
      Set FirstDelta to most recent pulse width
Else
      Set SecondDelta to most recent pulse width
      If (100 * FirstDelta / SecondDelta) less than or equal to 33
            Save FirstDelta as LengthOfDot
            Set ReturnValue to CalCompleted
      ElseIf (100 * FirstDelta / Second Delta) greater than or equal to
             300
            Save SecondDelta as LengthOfDot
            Set ReturnValue to CalCompleted
      Else (prepare for next pulse)
            Set FirstDelta to SecondDelta
      EndIf
EndIf
Return ReturnValue
End of TestCalibration
```

To follow the pseudo-code for CharacterizeSpace and CharacterizePulse, it is helpful to remember from the state diagram (Figure 6.6) that CharacterizePulse will be called immediately after a falling edge (the situation shown in Figure 6.7a) and CharacterizeSpace will be called immediately after a rising edge (the situation shown in Figure 6.7b).

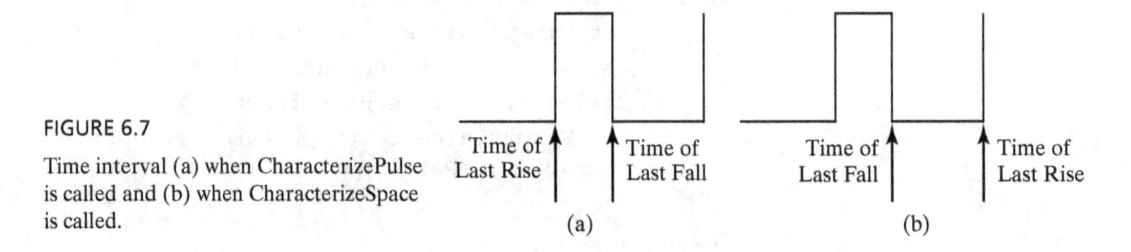

FIGURE 6.7
Time interval (a) when CharacterizePulse is called and (b) when CharacterizeSpace is called.

```
CharacterizeSpace
Takes no parameters, returns one of EndOfCharacterFlag, EndOfWordFlag,
BadSpaceFlag
Local variable ReturnValue, LastInterval
Set ReturnValue to BadSpaceFlag
Calculate LastInterval as TimeOfLastRise - TimeOfLastFall
If LastInterval OK for a Character Space
      Set ReturnValue to EndOfCharacterFlag
Else
      If LastInterval OK for Word Space
            Set ReturnValue to EndOfWordFlag
      EndIf
EndIf
Return ReturnValue
End of CharacterizeSpace
```

```
CharacterizePulse
Takes no parameters, returns one of DotDetectedFlag, DashDetectedFlag,
BadPulseFlag,
Local variable ReturnValue, LastPulseWidth
Set ReturnValue to BadPulseFlag
Calculate LastPulseWidth as TimeOfLastFall - TimeOfLastRise
If LastPulseWidth OK for a dot
      Set ReturnValue to DotDetectedFlag
Else
      If LastPulseWidth OK for dash
            Set ReturnValue to DashDetectedFlag
      EndIf
EndIf
Return ReturnValue
End of CharacterizePulse
```

6.5.6.3 Morse Decode Detail Design

The internal design details of the Morse Decode module will revolve around the internal data structure(s) that you choose to represent the correspondence between the group of Morse elements and printable characters. For example, you might choose to represent all of the decoding knowledge in a linked tree structure. Each node on the tree would have three parts: a reference to the next node if the current Morse element is a dot, a reference to the next node if the current Morse element is a dash, the printable character if this node is at the end of a group of Morse elements. This data structure would make for very fast decoding but requires that the programmer understand programming of linked list structures.

A data structure that is simpler to understand involves using a character string. Using this approach, as dots and dashes are detected, characters are added to a string. The decoding process would then involve searching for a match between the string that you have dynamically built of the incoming Morse elements and a predefined set of strings that represent the legal printable Morse code characters. In C, you might use the standard library function strcmp() to do the string comparison and a set of arrays like those shown below to capture all the legal Morse code groups.

```
char LegalChars[] = "ABCDEFGHIJKLMNOPQRSTUVWXYZ1234567890?.,:'-/()\"=
!$&+;@_";
char MorseCode[][8] ={".-","-...","-.-.","-..",".","..-.","--.",
                      "....","..",".---","-.-",".-..","--","-.","---",
                      ".--.","--.-",".-.","...","-","..-","...-",
                      ".--","-..-","-.--","--..",".----","..---",
                      "...--","....-",".....","-....","--...","---..",
                      "----.","-----","..--..",".-.-.-","--..--",
                      "---...",".----.","-....-","-..-.","-.--.-",
                      "-.--.-",".-..-.","-...-","-.-.--","...-..-",
                      ".-...",".-.-.","-.-.-.",".--.-.","..--.-"
                     };
```

In these arrays, the first element in the `LegalChars` array corresponds to the group of dots and dashes captured in the first element of the `MorseCode` array. The second element in the `LegalChars` array corresponds to the group of dots and dashes captured in the second element of the `MorseCode` array, and so on. In general, the array index for each character in

LegalChars corresponds to the Morse code representation string in the MorseCode array at that same index.

Since the character array data structures are simpler and the decoding algorithm is easier to understand and implement, we will present the pseudo-code for this module based on this design.

```
Pseudo-code for the Decode Morse module
Data private to the module: MorseString, the arrays LegalChars and
MorseCode shown above

ClearMorseChar
Takes no parameters, returns nothing
Clear (empty) the MorseString variable
End of ClearMorseChar

AddDot
Takes no parameters, returns a symbolic value indicating success or failure
      If there is room for another Morse element in the internal
         representation
            Add a Dot to the internal representation
            Return Success
      Else
            Return Failure
End of AddDot

AddDash
Takes no parameters, returns a symbolic value indicating success or failure
      If there is room for another Morse element in the internal
         representation
            Add a Dash to the internal representation
            Return Success
      Else
            Return Failure

DecodeMorse
Takes no parameters, returns either a character or a symbolic value indi-
cating failure
      For every entry in the array MorseCode
            If MorseString is the same as current position in MorseCode
                   return contents of current position in LegalChars
            EndIf
      EndFor
return ERROR, since we didn't find a matching string in MorseCode
```

6.5.6.4 Display Detail Design

For the display device, we will assume an intelligent LCD display like that shown in Section 7.3.2. The pseudo-code for most of these devices would be the same. The differences would be in the initialization sequences necessary to place the display into the proper mode. Rather than going into the details of the initialization sequence for a particular display, we will write higher level pseudo-code for the initialization function.

```
Display module

 returns nothing
f commands necessary to place the display into a
h auto-scrolling to the left and the insertion point
 point on the display.


eter, the character to write, returns nothing
ine to choose the data register in the display
e of the character to write onto the port lines
le line high, then low
```

enefits of implementing information hiding in a design is that the over into the other modules. The designs for the other three mod- hange at all if we replaced the LCD display that we have assumed fferent display. While the internal design and pseudo-code for this hange, as long as we maintained the same interface specification, rogram would need to change.

ign

ponsible for coordinating all of the other modules. It is an event button presses and the high-level events associated with the recep- elements.

```
Main module

 returns nothing
n module
Elements module
Decode module
ay module

on events
d event is ButtonDown
e calibration process

eturn from CheckMorseEvents do one of the following

ted
all AddDot function
ectedBlock
ted
all AddDash function
tectedBlock
Detected
all DecodeMorse function
 the return values from DecodeMorse was not ERROR
```

```
                        Print returned value to the display
                  Else
                        Print a '?' to the display
                  EndIf
            EndCharSpaceDetectedBlock
            WordSpaceDetected
                  Call DecodeMorse function
                  If the return values from DecodeMorse was not ERROR
                        Print returned value to the display
                  Else
                        Print a '?' to the display
                  EndIf
                  Print a space to the display
            EndWordSpaceDetectedBlock
Forever
End of Main Function
```

6.5.7 The Morse Code Receiver Implementation

The pseudo-code in the previous section could be implemented in any of the languages available to an embedded programmer. The only caveat is that the event driven paradigm on which the design is based assumes that the event checking is happening fast enough that no hardware events are missed and the timestamps for the rising and falling edges are reasonably accurate. This places some performance constraints on the program execution. All compiled language implementations should be able to meet the performance requirements on even modest embedded processors (e.g., a Freescale MC68HC11 with a 2 MHz clock). Interpreted languages may face performance issues with this design and require a significantly faster microcontroller in order to be successful.

6.5.8 The Morse Code Receiver Unit Testing

Unit testing would be handled using the test harness in each module. When writing the test harness, as with many other tasks, it is best to take an incremental approach. Start with a test harness for one of the private functions that doesn't depend on any other functions in the module. Work your way up by adding tests to the harness for each of the other functions that don't have lower-level dependencies before starting to write the tests for the functions that will need to call these lowest level functions. The last iteration of the test harness should test the public functions that the module provides. When constructing your test harnesses, don't forget to test how the functions behave when bad parameters or parameters which might cause internal overflow (e.g., AddDot and AddDash in MorseDecode) are passed as input. The body of a fairly complete test harness for the Morse Decode module might look like:

```
Call ClearMorseChar
Print MorseString
Call AddDot and make sure that it reports success
Print MorseString
Call AddDot 6 more times, this is the maximum size of a character group
Make sure that each call reports success
Print MorseString
Call AddDot make sure that it reports failure
```

```
Print MorseString

Call ClearMorseChar
Print MorseString
Call AddDash and make sure that it reports success
Print MorseString
Call AddDash 6 more times, this is the maximum size of a character group
Make sure that each call reports success
Print MorseString
Call AddDash and make sure that it reports failure
Print MorseString

Call ClearMorseChar
Construct some legal strings, especially at the beginning and end of the
array limits
For each of these strings
      ClearMorseChar
      Use AddDot & AddDash to build a representation
      Call DecodeMorse
      If no error reported
            Print the returned character
      Else
            Print a '?' to the display
      EndIf
EndFor

Construct some illegal strings
For each of these strings
      ClearMorseChar
      Use AddDot & AddDash to build a representation
      Call DecodeMorse
      If no error reported
            Print message to alert to the failed test
      Else
            Print message to alert to the successful test
      EndIf
EndFor
```

6.5.9 The Morse Code Receiver Integration

After each of the modules passes its unit tests, you can start integration. For our sample problem, it probably makes the most sense to begin the integration by adding the Morse Elements module to the Main module. In place of making the calls to the Morse Decode modules in the Main module, print the dots and dashes that are received to the terminal with spaces between the character groups. When you have a line or two of dot-dash groups, manually decode the groups to make sure that they make sense based on the input. If you are finding problems here, then you probably missed something in the testing of the Morse Elements module, since what we have done in this step of integration is really nothing more than what should have been done in the test harness for Morse Elements. Once you have convinced yourself that the Morse Element module is returning reasonable groups of dots and dashes, go ahead and put in the calls to the AddDot, AddDash, and DecodeMorse functions in the Morse Decode module and print the decoded characters. If things aren't working here, it's time to go back to the test harness for Morse Decode to figure out why. If the printing of the characters to the terminal is working OK, then it is time to add in the Display module. If your unit testing on the Display module was thorough, then this step should go smoothly as well. Congratulations! You now have a working Morse code decoder that meets all the system requirements.

6.6 HOMEWORK PROBLEMS

6.1 Describe three ways in which programming is different than writing prose.

6.2 Describe three ways in which programming is similar to building a house.

6.3 Define decomposition in the context of software design.

6.4 True or False: Coupling is a desirable characteristic between modules.

6.5 True or False: All functions in a module should be represented in the public interface to the module.

6.6 Write a pseudo-code test harness for the Morse Elements module described in Section 6.5.6.2.

6.7 Describe the function of a test harness in the context of an individual module.

6.8 True or False: Pseudo-code is a part of the design process, not the implementation process.

6.9 Using the pseudo-code from Section 6.5.6.2, write code in a real programming language to implement the `InitializeMorseElements` function.

6.10 Write code in a real programming language to implement the following snippet of pseudo-code:

```
For every element in the array
multiply the element value by Gain and add the result to the running
sum
```

REFERENCE

[1] *Code Complete,* Steve McConnell, 2nd ed., Microsoft Press, 2004.

CHAPTER 7

Inter-Processor Communications

7.1 INTRODUCTION

Any time you design a circuit or a system that incorporates more than one active component and these components interact, they communicate with each other in some way. This communication can take on simple forms, such as a single analog voltage or a few time-varying digital signals, but there may also be a requirement to transfer a large quantity of data quickly from one component to another. For example, any time a computer sends a file to a printer, it is generally a good thing to have this occur quickly and without error. Discussions of inter-processor communications include a wide range of methods for moving information from one processor to another. Here, we cover the more complex, higher-level cases where applications require the movement of larger quantities of data at higher speeds. We cover the subject of two digital logic components exchanging information more completely in Chapter 16, Digital Inputs and Outputs and Chapter 17, Digital Outputs and Power Drivers. Analog interfaces are discussed in a great number of chapters, but most specifically in Chapter 19, A-to-D and D-to-A Converters.

As with personal computers and printers, inter-processor communications among microcontrollers ideally occur quickly and without errors. In addition, methods that are simple to implement are also of great value. To the great benefit of mechatronics designers, several techniques have evolved for accomplishing these goals and numerous standards have been established. The first challenge for designers is to become familiar with a wide variety of these common solutions and conventions. The challenges that follow involve mastering the details of specific implementations best suited to a given application. This chapter seeks to familiarize readers with the most common methods of communications and the most widely accepted standards. It does not attempt to catalog all the details of these methods and standards, which are well beyond the scope of an introductory discussion of inter-processor communications. These details are best sought out by the designer for specific applications.

In this chapter, you will learn:

1. the two fundamental approaches to digital communications: bit-parallel and bit-serial,
2. the distinguishing characteristics of bit-parallel communications,
3. the distinguishing characteristics of bit-serial communications,
4. inter-processor communications depend on timing and may be accomplished synchronously (i.e., with a common clock signal) or asynchronously (without a clock signal),
5. several approaches to synchronous serial communications,
6. the most common approach to asynchronous serial communications,
7. what the common signaling level conventions are,
8. how communications are accomplished over media that don't allow for the interpretation of DC signals, such as radio frequency (RF) links and standard telephone lines.

FIGURE 7.1

An early method of wireless communication.

7.2 WITHOUT A MEDIUM, THERE IS NO MESSAGE

The first decision about communications the designer must make is how the data will be transmitted from source to destination. In many cases, this will be via direct electrical connection, with a wire or group of wires. However, a great many other media exist that can carry data from one point to another. Figure 7.1 illustrates one early medium used to relay information. Remarkably, the same general principles that governed smoke signal communications apply today, namely:

- The source and the destination must have access to the medium.
- They must agree on the meaning of the symbols used.
- They must follow the rules that govern who gets to speak, and how.

There are a wide variety of media choices available today, each with its own features, strengths, and weaknesses. A representative list with a few examples is given in Table 7.1.

TABLE 7.1 Examples of communications media and standards.

Medium	Examples
Direct, wired, and physical connections	RS-232, RS-485 Ethernet Universal Serial Bus (USB)
Light	Fiber optics Infrared (IR) emitters and detectors
Radio frequency	WiFi (802.11a/b/g/n) Bluetooth AM radio FM radio

7.3 BIT-PARALLEL AND BIT-SERIAL COMMUNICATIONS

At the simplest level, digital communications involve connecting the output of one digital device to the input of another, and establishing the meanings of logic levels and timing. In this way, the output of one digital device transmits data to the inputs of a second digital device, where it can be read and interpreted. This is the basis of all digital inter-processor communications. Complexity is heaped on these foundations in the interest of making the exchange of data fast, efficient, and reliable, but the basis of it all is actually straightforward.

Recall that digital data is represented with base-2, or "binary," numbers. Each binary digit (a "bit") can take on a value of either 0 or 1. Numbers of arbitrary length can be generated by grouping the requisite number of bits together. For example, the number 65,000 can be represented using 16 bits as

$$65{,}000 = 1111\ 1101\ 1110\ 1000$$

Data of all sorts (such as characters, signed and unsigned integers, strings, floating point numbers, and so on) can be represented by defining the meaning of each of the bits in a grouping or with streams of groupings. These issues are more fully discussed in Chapter 2, What's a Micro? and Chapter 3, Microcontroller Math and Number Manipulation.

When bits are relayed from a source to a destination one at a time, this is commonly called **serial** or **bit-serial communications**. When more than 1 bit is relayed at a time, this is called **parallel** or **bit-parallel communications**. Both means of communication are popular today. Examples of serial communications links with which many readers are familiar are computer serial COM ports and USB. Familiar examples of parallel communications include personal computer parallel printer ports and peripheral component interconnect (PCI) buses. Microcontroller data and address buses are also examples of parallel communications (see Chapter 2, What's a Micro?).

7.3.1 Bit-Serial Communications

Serial communications links relay 1 bit of data at a time. Regardless of whether the data is 3 bits in length or 100 MB, this means of communicating can get the job done and (depending on the speed of transfer, or **bit rate**) often surprisingly quickly. The transmitting component must break data streams down to individual bits and send them in a prescribed order and with prescribed timing, while the receiving component must reassemble the bits from the data stream in order to understand and use them.

For now, we will assume that we are using standard logic-level voltages to represent bits: 0 V will represent logic-level 0 and 5 V will represent logic-level 1 (however, we will see later in this chapter that there are good reasons to deviate from this). Using this convention for now, we may represent a data stream as a digital signal that is transitioning between 0 and 5 V, as shown in Figure 7.2.

Something fundamental is missing from Figure 7.2, however—the communications link will not work as depicted unless more information is shared between sender and

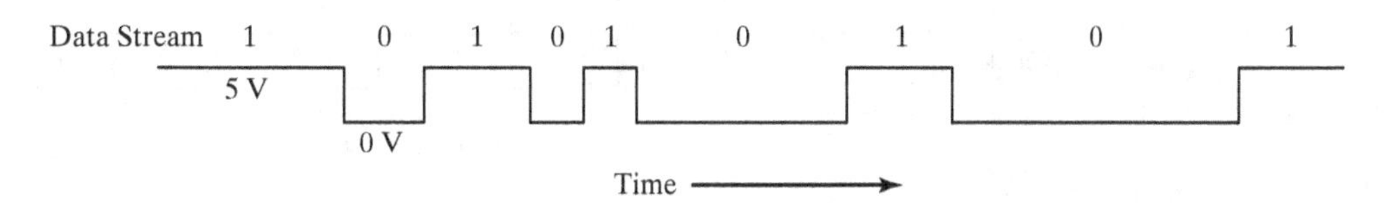

FIGURE 7.2

Basic bit-serial communications: a digital connection with logic-level transitions.

receiver. The missing element is *timing*. When should the sender transition between bits in the data stream? When should the receiver read a bit and prepare to read the next one? Timing is so critical to the success of serial communications that the major categories are classified by how they use time to organize the transfer of data. The first we will discuss in this chapter is **synchronous serial communication**. Literally, *syn* = *with* and *chronos* = *time* (in Greek), indicating that this means of serial communications incorporates a shared "clock" signal between the sender and the receiver as well as the data signal(s). Next, we will consider **asynchronous serial communications**. For this case, *asyn* = *without* and *chronos* = *time*, indicating that there is no explicit clock signal shared between the sender and the receiver.

7.3.1.1 Synchronous Serial Communications

Synchronous serial communications incorporate at least one connection for serial data, one connection for a clock used to indicate when a bit of data is valid on the data line, and finally a shared ground between the components communicating. Figure 7.3 depicts a typical example of synchronous serial data transmission. For Figure 7.3, the processor sending the data is also controlling the clock line. This is the most straightforward implementation of synchronous serial communications, where the sender is called the system **master**, and is able to send data to a receiving processor or peripheral (the **slave**) and simply tell it when to read the data line.

Taking a closer look at Figure 7.3, several important details emerge. Note that the clock signal has arrows calling attention to the rising edges (the GND-to-5 V transitions). These arrows indicate that the clock is **rising-edge active**—that is, the sender ensures that the data placed on the data line is valid at the time that it causes a rising edge on the clock line. In practice, the active edge of the clock may either be rising- or **falling-edge active**; both are common. The clock line's **idle state** is also specified and controlled to be either idle high or low. This example shows an idle high clock: when no data is transmitted, the clock line is set to 5 V indefinitely. All four combinations of clock active edge and idle state may be used; however, some are more popular than others. Note also in Figure 7.3 that the clock and data transitions may occur at any time—for synchronous serial communications links, there is no implication of any particular data rate or any consistency to its timing. Whenever the receiver detects a rising edge on the clock line, it knows that it is time to read in a bit from the data line. However, there are usually constraints on the maximum data transfer rate, dictated by the capabilities of the sender and the receiver.

The scenario shown in Figure 7.3 is one of the simpler uses of synchronous serial communications. More complex uses are also common. A few implementations of synchronous communications have become standardized and commonplace, and many microcontrollers and peripheral integrated circuits include hardware modules that adhere to one or more of the standards. For this reason, it is uncommon to create a synchronous serial interface from

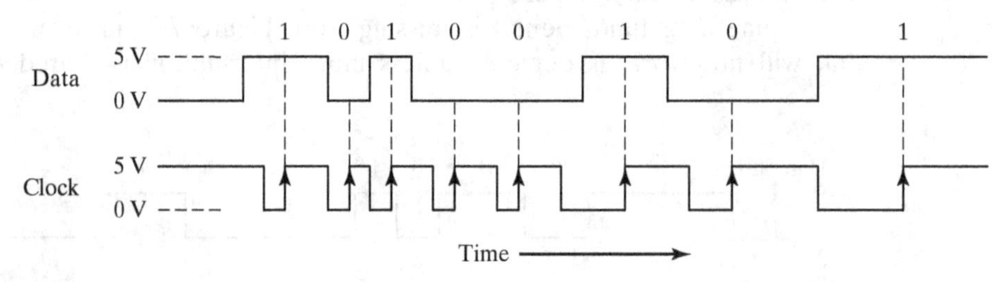

FIGURE 7.3

A representative synchronous serial communications data stream.

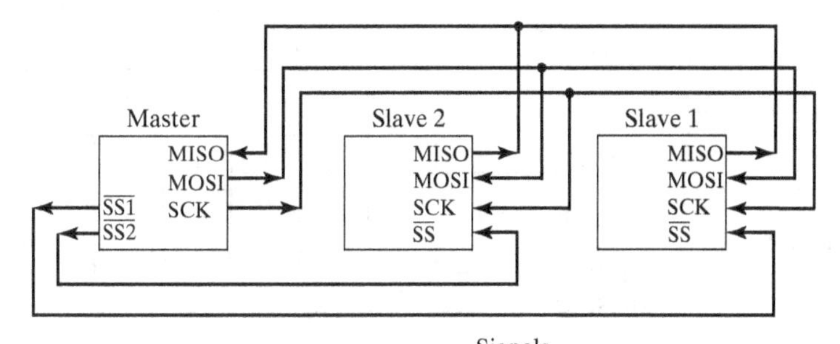

Device	Signals			
	MISO	MOSI	SCK	$\overline{SSx}$
Master	Input	Output	Output	Output
Slave	Output	Input	Input	Input

FIGURE 7.4

An implementation of an SPI synchronous serial communications system with a single master and several slaves.

scratch in hardware and software—rather, it is more common to select components that already incorporate these modules. This greatly simplifies the design task. Examples of these standardized interfaces are Freescale's **Serial Peripheral Interface** (**SPI**), Philips' **I^{2}C** (**Inter-Integrated Circuit**) **Bus**, and National Semiconductor's **Microwire**. Each is slightly different, but all share the basic characteristics described above. As an example, we will cover Freescale's Serial Peripheral Interface, commonly called SPI, in greater detail.

Freescale's SPI requires that the system has a **master** and at least one **slave** for data transfers to occur. The system master does not always need to be the same node in the network. It is possible to implement a **multimaster system** in which different members of the network may take on the role of the master at different times. For a transfer of data, the system master controls a unidirectional data line and a shared clock line. This data line is called **Master Out Slave In**, or its more common acronym, **MOSI**. The slave or slaves share another unidirectional data line, called **Master In Slave Out** or **MISO**. The clock is commonly called the **serial clock** or **SCK**. Figure 7.4 shows a system with a single master and several slaves.

As shown in Figure 7.4, MOSI is a unidirectional output line from the master to the slave(s). Only the master uses this line as an output, and all slaves treat it as an input. The master must indicate which slave is to be addressed by asserting a separate **slave select** line (denoted as $\overline{SS}$ or **SS***, with the overbar or "*" indicating that the line is active low, enabling the chip when it is in the logical 0 state). The slave select line enables a specific slave to receive and transmit data as required. When a slave is selected, or enabled, it will treat the MISO line as an output that it must control. The MISO line is shared by all the slaves in a system. All slaves that have not been selected must keep their MISO output in a high-impedance state (to see how, see Section 17.3), so that there is no contention for control of the data line. In some designs, there is no need for slaves to send data back to the master during a transfer, and it is often desirable and expedient to select several slaves at the same time.

The SPI interface has four modes defined for transferring data. All four modes require the slave select line (SS*) to be asserted before the transfer of data. As indicated by the "*" and shown in Figure 7.5, the slave select line is active low. The four SPI modes are enumerated in Table 7.2 and shown graphically in Figure 7.6.

The most common SPI configuration is Mode 3, which always requires the SCK line to make a high-to-low transition in advance of any valid data or request. This gives the receiver an indication that it should anticipate data in the near future. In Modes 0 and 2, slaves must

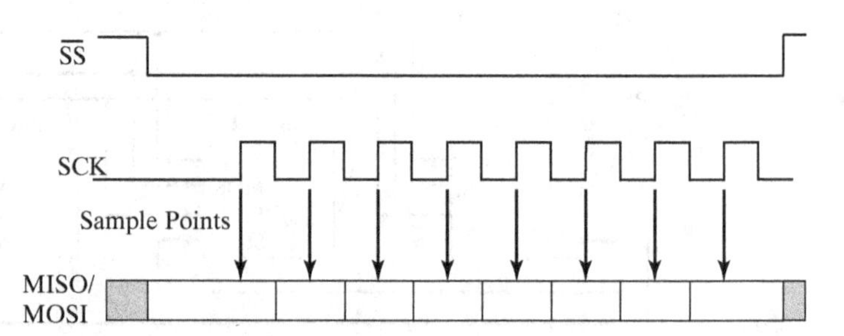

FIGURE 7.5
The SPI connections (MOSI, MISO, SCK, and SS) and their timing relationships.

be ready at any time to read data once their SS* line is asserted and the SCK line makes its first transition.

Other standardized implementations of synchronous serial communications, such as I²C and Microwire, function in much the same way. Their implementations, features, strengths, and weaknesses differ, but the common characteristic between them is that they incorporate a separate line that serves the function of a clock, alerting receivers that data is valid and should be read within a specified period of time. SPI and Microwire are very similar in nature, requiring two unidirectional data lines (such as MOSI and MISO) and a clock line, while I²C specifies only a single bidirectional data line and a clock line. I²C is a more elaborate protocol that includes individual device addressing, transfer acknowledgment, and speed and multimaster arbitration all within the two-wire transfer. It was developed in the early 1980s by Philips Semiconductors (now NXP) as an internal bus for consumer products.

TABLE 7.2 The four SPI modes of operation.

SPI Mode	SCK Idle State	SCK Active Edge
0	Low	Rising
1	Low	Falling
2	High	Falling
3	High	Rising

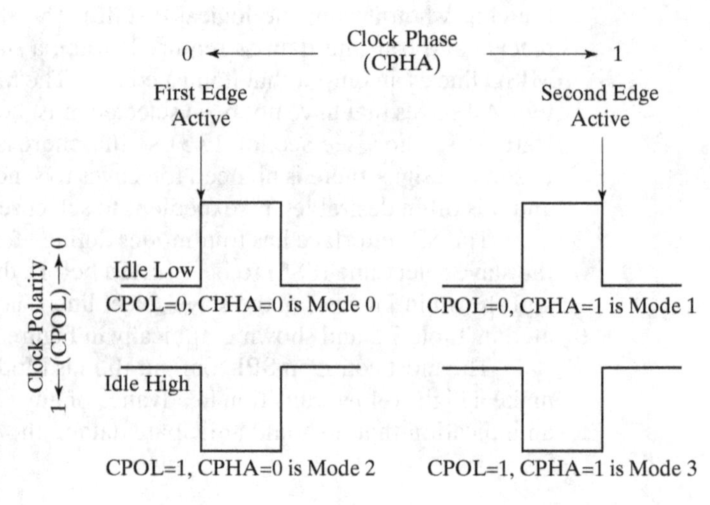

FIGURE 7.6
Illustration of the idle line state and polarity of the SCK line for each of the four SPI modes.

There are a wide range of slave ICs with I^2C interfaces available that implement functionality such as TV tuning, TV decoding, and audio gain control as well as the more common mechatronics functions, such as A/D and D/A converters, I/O line expansion, sensors, and clock/calendars. Microwire and SPI tend to offer higher overall data communication rates, while I^2C offers better support for multimaster systems and a minimal hardware interface. Each application will have differing requirements, and designers should consider the options before making a choice.

7.3.1.2 Asynchronous Serial Communications

In many cases, it is either inconvenient or impossible to include a separate connection to serve as a clock signal. One example of this is when communication is occurring over an RF link: it is not convenient or cost-effective to add a separate connection to indicate when data is ready, which might require a separate signal at another frequency. And even though it usually isn't too difficult to add another wire in a wired implementation, these serial communication links are commonly implemented without separate clock lines. When a separate clock line is not implemented, this is known as **asynchronous serial communication**. To say that they are literally without clocks is not exactly correct, though, as each node in the network must maintain its own clock and its own measure of when data is valid, when to read, and when to write. This approach does away with the necessity to add additional connections to share a clock, but it adds the responsibility to each network node to keep track of time on its own. Asynchronous serial communication is very common, and is implemented in PC serial COM ports, for example, as well as in USB ports.

A wired asynchronous serial connection typically consists of at least one data line and a common ground. If the data line is bidirectional (not the most common situation), then two or more nodes on a network can communicate with each other, with any given node acting as the sender or a receiver, as needed. If the data line is not bidirectional, then communications may occur only from one processor to another (sender to receiver) per data line. In order to allow bidirectional communications using unidirectional data connections, additional connections are required. Central to the function of an asynchronous serial link is the concept of a **bit time**, which is the duration that a single bit of data is present on a data line. The sender and the receiver of data must agree on this length of time, as well as indications of when bit times are to begin and end, in order to communicate effectively. The overall **data transfer rate**, usually stated in **bits per second (bps)** or **kilobits per second (kbps)**, is related to the bit time as:

$$\text{Data Rate [bps]} = \frac{1}{\text{Bit time [s]}} \tag{7.1}$$

A typical asynchronous serial data transfer waveform is shown in Figure 7.7. For this example, we will continue to assume that we are using standard logic-level voltages to represent bits, 0 V will represent logic-level 0 and 5 V will represent logic-level 1. Figure 7.7 demonstrates the transfer of 8 bits of data. Prior to the execution of the transfer, the data line is idle and awaiting data. This state is referred to as the **marking state**, and unless otherwise specified is at logic-level 1. The first indication that a transfer of data is beginning is a transition from the marking state to the **spacing state**. This begins the **start bit**, which is 1 bit time of logic-level 0. This transition is required, even though this initial bit time after the marking state ends does not convey data. To understand why we need a start bit, consider the case where the first bit to be transmitted is in the same logic state as the marking state. Without a start bit, no transition would occur and the receiver would have no way of distinguishing the first bit of data from the marking state. The start bit is the sender's way of alerting the receiver that data will follow starting exactly 1 bit time after the transition between the marking state and the spacing state of the start bit.

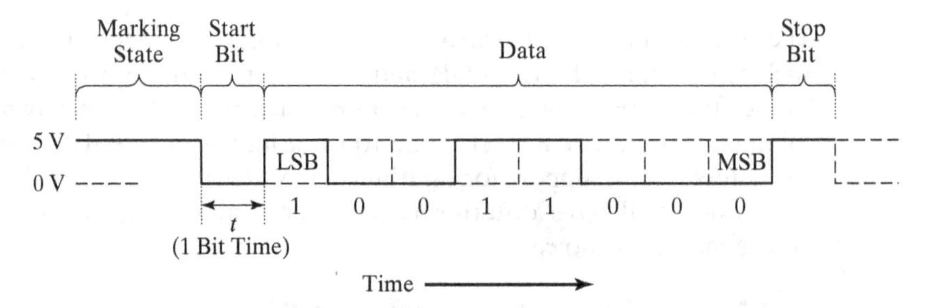

FIGURE 7.7

A representative asynchronous serial data waveform, showing an 8-bit data field, 1 start bit, and 1 stop bit.

After the start bit (i.e., after 1 bit time), the first bit of data is transmitted on the data line. For asynchronous serial communications, data is transferred **least significant bit** (**LSB**) first (i.e., the 1's place in binary, or base-2), followed by the next LSB (in binary, the 2's place), and so on, until the **most significant bit** (**MSB**) is transferred. Figure 7.7 shows a data field of 8 bits, which is very common; however, it is not unusual to see data fields that are 7 bits. Some configurations allow for as few as 5 or as many as 9 bits of data. As with the timing, it is critical that the sender and the receiver agree in advance on the number of bits in the data field.

After the MSB, one final bit, the **stop bit**, is transferred to signal the end of the data transfer. After the stop bit has been transferred (again, a length of 1 bit time), a new transfer may begin. Some specifications require that there be more than 1 stop bit, which is equivalent to returning to the marking state for a specified minimum length of time. Systems that require extra stop bits usually require a little extra time to read and process data. Requirements for multiple stop bits have become increasingly uncommon as computing power and processing speeds have increased. In many cases, there is no requirement for any additional time after the end of the stop bit, and the next start bit may occur immediately.

Together, the start and the stop bits are referred to as the **framing bits** because they frame, or surround, the data. Although the framing bits don't serve to transfer data from the sender to the receiver, they are necessary to initiate a data transfer and indicate the end of the transfer. In this mode of serial communications, at least 2 bit times are consumed in every transfer that do not serve to relay data.

In addition to the framing bits and the data field, a **parity bit** is sometimes added as an error-checking measure. This bit is added after the last data bit, the MSB. As the name implies, the parity bit's purpose is to balance the number of 1s in each of the data transfers. If **even parity** is specified, then a data field with an even number of 1s present will receive a parity bit equal to 0, while a data field with an odd number of 1s present will receive a parity bit equal to 1. If **odd parity** is specified, then a data field with an even number of 1s present will receive a parity bit equal to 1, while a data field with an odd number of 1s present will receive a parity bit equal to 0. The use of a parity bit will alert the receiver if a single error or an odd number of errors has occurred during data transfer, but it will not indicate when two errors or an even number of errors have occurred. Since it isn't a terribly robust error checking scheme, the use of parity bits has become less common. More elaborate error checking schemes, such as the use of **cyclic redundancy checks** (**CRCs**), have become more commonplace. These schemes employ much more sophisticated and robust algorithms to detect when errors have occurred. A discussion of CRCs is beyond the scope of this chapter, and interested readers are encouraged to investigate the subject further [1]. Figure 7.8 shows a data frame that incorporates a parity bit.

Asynchronous serial communications links are described by the bit rate, the number of data bits, the type of parity used (if any), and the number of stop bits. There are a great many

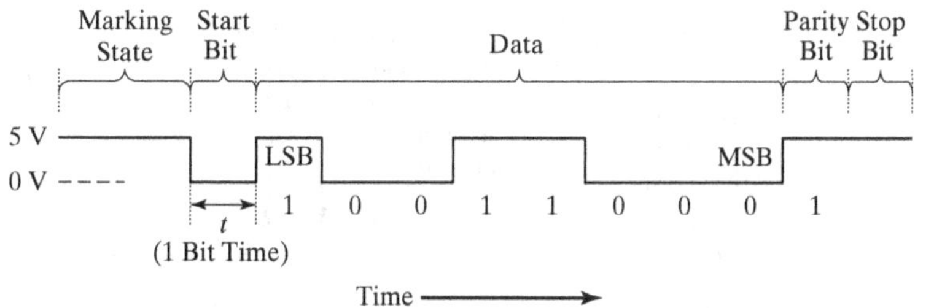

FIGURE 7.8

An asynchronous serial data transfer with a parity bit. Data format shown is 8 data bits, even parity, 1 stop bit (8E1).

possibilities and combinations of configurations. For example, the bit rate is usually one of the standard values (e.g., 300, 1,200, 2,400, 4,800, 9,600, 19,200, 38,400, 56,700, or 115,200 bps) but could theoretically be any value, the number of data bits can typically range from 5 to 9 bits (but is almost always either 8 or 7 bits), the parity can be even (E), odd (O), or no parity (N), and the number of stop bits can be 1, 1.5, or 2. The most common configuration is 8 data bits, no parity, and 1 stop bit. For a data rate of 9,600 bps, this is written in an abbreviated format as 9600 8N1. In this shorthand, a 7-bit transfer with even parity and 2 stop bits and the same bit rate would be noted as 9600 7E2.

Though the mechanics of an asynchronous serial communications link are straightforward, building a robust implementation can be quite challenging. However, designers are not often required to build their own from scratch. Most microcontrollers and many peripheral devices include specialized hardware that manages all low-level implementation requirements and allows the designer to send and receive data without having to keep track of the details. This type of hardware module is known as a **universal asynchronous receiver transmitter**, and is usually shortened to its acronym **UART**. In some cases, the UART hardware is combined with additional hardware that allows users to configure it to perform synchronous serial data communications, as described in the previous section. When the hardware has this additional capability, it is referred to as a **universal synchronous/asynchronous receiver transmitter** or **USART** for short. Some manufacturers have created standardized implementations of UARTs across their product offerings. This standardization allows designers to learn the subtleties of an implementation once and apply the experience across several designs. Freescale's **Serial Communications Interface** (**SCI**) is an example of such a standardized UART module.

When creating systems that incorporate a UART or USART, designers need only perform a number of setup operations, such as specifying the bit rate, the number of data bits in a frame, parity (if any), the number of stop bits, and so on. Once setup is complete, the designer will be in control of enabling and disabling the receiver and transmitter, checking for errors, and managing the data that flows in and out of the receive and transmit registers. This too can get complex, but the overall task of dealing with asynchronous serial data transfers is greatly simplified.

In order to implement a robust asynchronous serial receiver, a UART typically monitors incoming data 16 times during each bit time. The UART takes a series of samples of the state of the receive line at regular intervals and uses the results to determine if a frame is beginning, when data is being transferred, if there is noise on the line, and if all of the rules of asynchronous serial communications, as described above, are being met or if errors have occurred.

When the receive line is in the marking state, the UART checks for the beginning of a start bit at a rate of 16 times the specified bit rate. The conditions that must be met for a valid start bit to be detected are that the receive line must be read in the marking state for three consecutive reads prior to the transition to a logic-level 0. Once this transition has

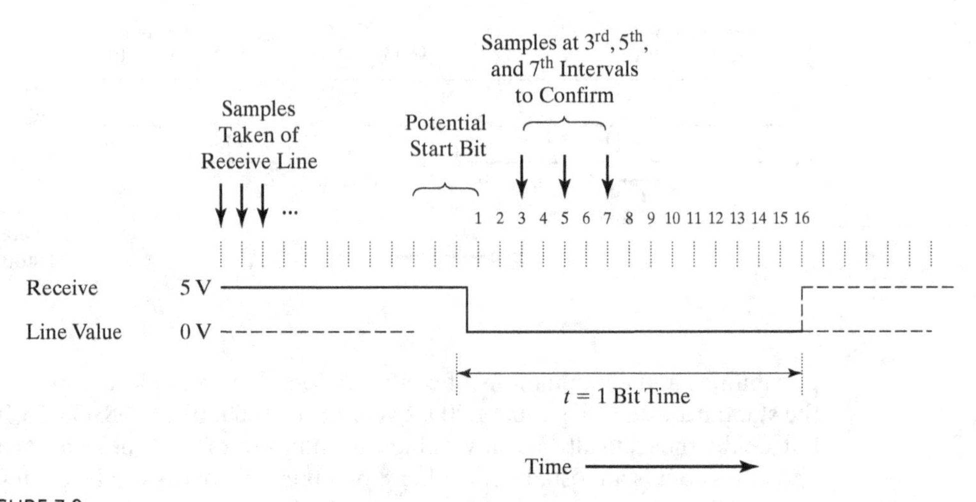

FIGURE 7.9

Standard scheme for a UART monitoring the receive line for the beginning of a start bit.

occurred, the UART will check the state of the receive line three more times, as shown in Figure 7.9.

Figure 7.9 depicts the ideal case where no unexpected samples are obtained: there is a clean transition from the marking state to the start bit, and the receive line behaves perfectly and stays at logic-level 0 for the entire bit time. But what about the cases where there is noise on the line or the case where there is disagreement about the bit rate? In these cases, the UART may report that a start bit has been detected or not, depending on the nature of the errors it detects. In any event, when there is doubt about the validity of the data, the UART will indicate this through the use of error flags.

The basic strategy employed by the UART is to use the best two out of three samples it obtains. Recall that a potential start bit is identified when three consecutive samples indicate that the receive line is in the marking state followed by a single sample at logic-level 0. Since this first sample at logic-level 0 may be the first sample at the beginning of a start bit, three additional samples are obtained at the 3rd, 5th, and 7th sampling intervals. The expectation is that all three samples will show that a valid start bit has begun, but there are three other possibilities:

1. two out of three samples indicate that the start bit has begun, but one sample is in the marking state,
2. one out of three samples indicates that the start bit has begun, but two samples are in the marking state,
3. all three subsequent samples indicate that the receive line has returned to the marking state.

In the first case, where two out of three samples are in the proper state to indicate the beginning of a start bit, but one sample disagrees, the UART will decide that a start bit has indeed been detected, but that noise has also been detected on the line. It will continue to collect data bits in the bit intervals that follow, but it will also indicate that noise has been detected by setting a noise status flag. In the second case, where only one out of the three samples was in the proper state to indicate a start bit, the UART will decide that this is not the beginning of a valid start bit. The UART will not attempt to read data in this case. For the

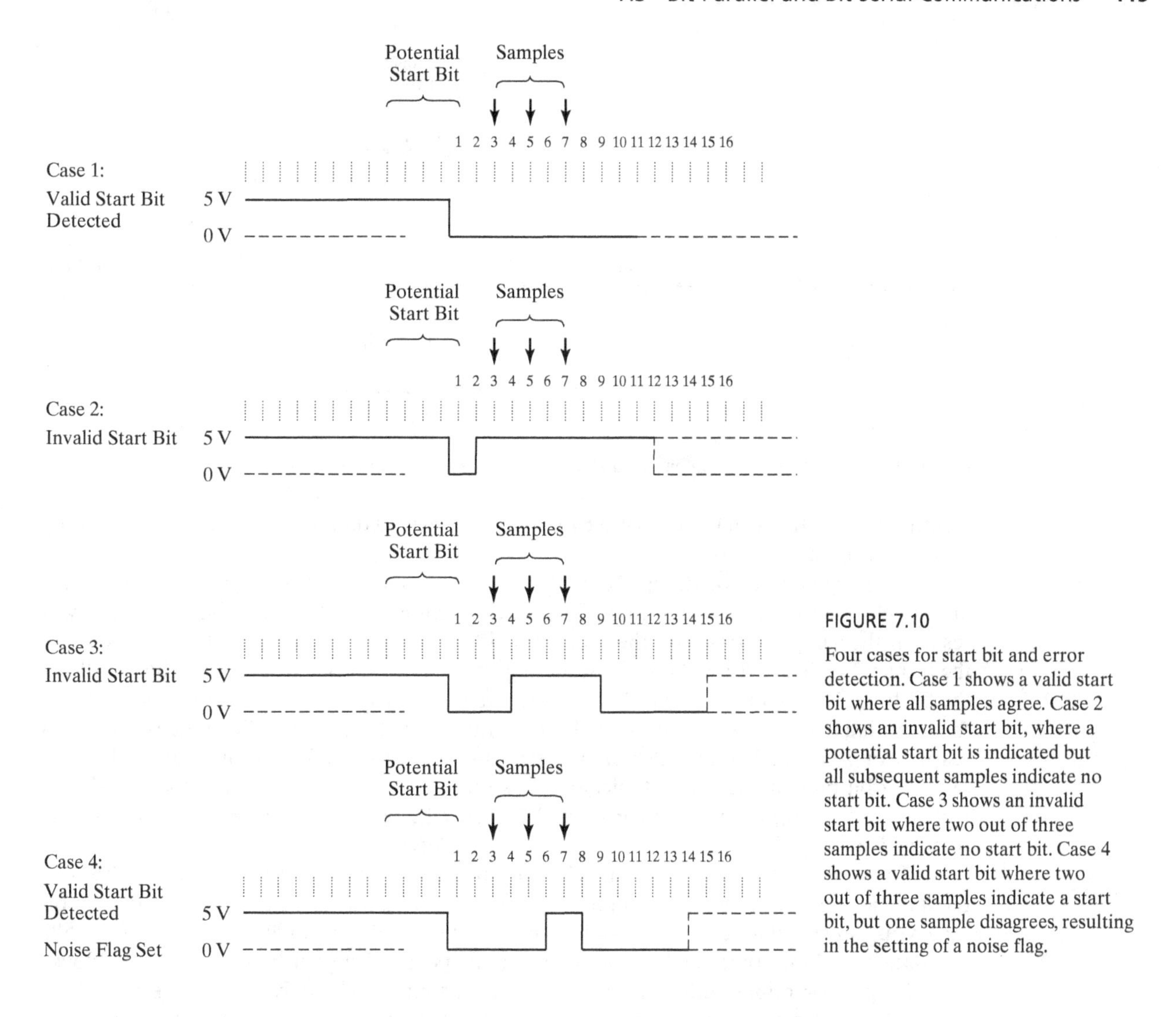

FIGURE 7.10

Four cases for start bit and error detection. Case 1 shows a valid start bit where all samples agree. Case 2 shows an invalid start bit, where a potential start bit is indicated but all subsequent samples indicate no start bit. Case 3 shows an invalid start bit where two out of three samples indicate no start bit. Case 4 shows a valid start bit where two out of three samples indicate a start bit, but one sample disagrees, resulting in the setting of a noise flag.

third case, the most logical conclusion is that a spurious noise signal has caused the receive line to transition from logic level high to low briefly, but that this is not the beginning of a start bit. This is how the UART interprets the information. It will not begin to accumulate bits of data after the spurious transition. In general, the rule is that two out of three samples determine the decision made by the UART, but that any disagreement during a transfer results in the setting of a noise status flag. Figure 7.10 graphically depicts the various scenarios.

Once a valid start bit has been detected, data bits are read in one at a time according to a similar strategy (Figure 7.11). Three samples are obtained during each bit time, though for these bits the samples are taken at the 8th, 9th, and 10th intervals to center the samples in the middle of the bit time. Once again, the state of two out of the three samples determines whether the UART decides the bit is a 1 or a 0. In the event that there is disagreement between the samples, a status flag indicating the presence of noise on the line is set.

After each of the data bits has been received, a final cluster of three samples is taken during the middle of the stop bit time. The purpose of these samples is to insure that a final stop bit has been found. If the samples do not indicate a valid stop bit, a **framing error** is

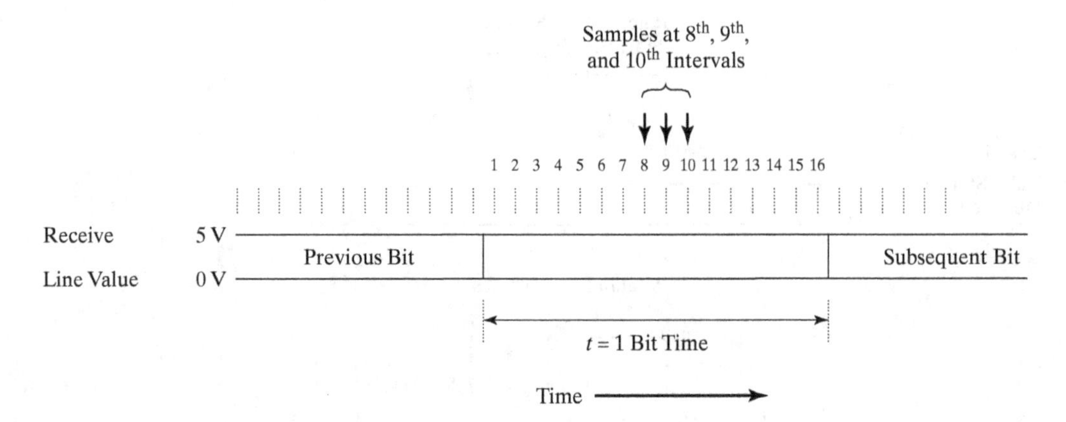

FIGURE 7.11

Determination of the value of data bits in an asynchronous serial communications frame.

identified, and a flag will be set to indicate that the received data was not properly framed by both a start bit and a stop bit.

Sampling at a rate of 16 times the bit rate allows for some mismatch between the transmitting clock and the receiving clock. The inevitable mismatch between the two clocks will result in the sampling times on the receiver drifting relative to the actual bit times sent by the transmitter. The drift will put the sampling times somewhere other than in the middle of the bit time. This is acceptable as long as the mismatch in speeds is not so great as to cause the sampling to drift outside the actual bit time for the transmitted bit. Since the two systems resynchronize at the leading edge of each start bit, the drift is only significant over the time of a start bit plus the data bits, typically 9 bits total. The question is "Given the speed mismatch, will the stop bit be read correctly?" Figure 7.12 shows the two possible cases of the transmitter being either faster or slower than the receiver.

If you account for the inherent 1/16th bit time uncertainty in the detection of the start bit and the fact that none of the three stop bit samples should fall outside the true stop bit time, you can calculate that this system can tolerate speed mismatches up to about ±4.5% without error. Note that 4.5% is the actual total speed mismatch between the two clocks with the plus or minus indicating that either transmitter or receiver could be the "fast" device. In most microcontrollers, the bit-rate clock is derived from the system clock. In many cases where asynchronous communications is necessary at standard bit rates, it is

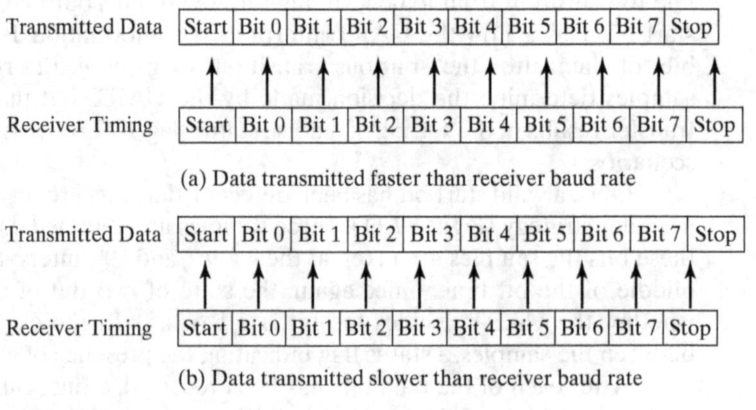

FIGURE 7.12

Effects of baud rate mismatch.

necessary to choose the system clock frequency to allow the bit-rate clock to fall within these tolerances for the most common bit rates (300, 1,200, 2,400, 4,800, 9,600, 19,200, 38,400, 56,700, 115,200 bps).

7.3.2 Bit-Parallel Communications

In modern designs, bit-parallel communication is most often conducted between devices that are part of the same instrument. For example, the parallel ports on personal computers that were used in the past to communicate with printers and other peripheral devices are rapidly being replaced by serial (USB) ports, but the PCI bus remains a very common means of connecting a PC's motherboard to expansion cards. Another classic example of such parallel communications is the device that we will use as the example in this section, the **intelligent liquid crystal display (LCD) module**. Intelligent LCDs are based on controllers that communicate with a computer and manage the generation of the characters on the LCD. These are typified by the Hitachi HD44780 and the many functionally compatible chips that are available.

HD44780 compatible displays all communicate with the microcontroller over a group of four or eight data lines and three handshake lines. Figure 7.13 shows the connections between a PIC16C84 microcontroller and a typical intelligent LCD module.

The data, on lines D_0–D_7 (or D_0–D_3 depending on whether you are using the module in 8-bit or 4-bit mode), are transferred between the microcontroller and the display via a process mediated by the E, RS, and R/$\overline{\text{W}}$ lines. The E (enable) line is used to synchronize the transfer. The RS (register select) line is used to determine whether command data or display data is being transmitted and the R/$\overline{\text{W}}$ (read/not write) line controls the direction of the data transfer (i.e., data going from the microcontroller to the display or vice versa). A description of the complete operation of the intelligent LCD module is beyond the scope of this discussion, so we will concentrate on describing how information is transferred between the two devices.

Internally, the display can be thought of as having two locations that can be written to or read from. Which of these locations will be accessed is determined by the state of the RS line. When the RS line is high, display data is being transferred, when it is low, commands are being transferred. The direction of the data transfer is controlled by the R/$\overline{\text{W}}$ line. When this line is low, the transfer is from the microcontroller to the display.

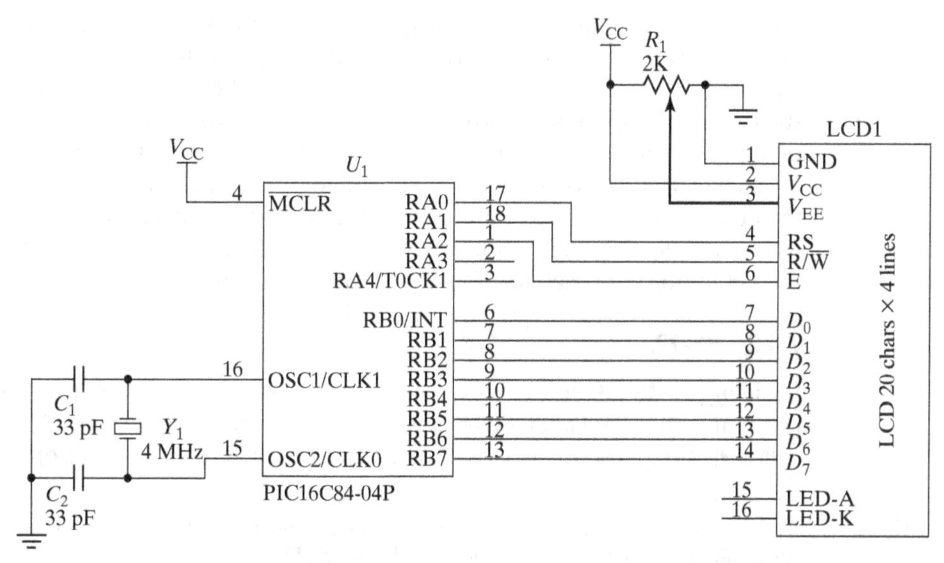

FIGURE 7.13

Interface between a PIC microcontroller and an intelligent LCD module.

The bit-parallel transfer of data between the devices occurs in the sequence shown below.

1. The microcontroller places data (be it command or data) to be transferred to the display onto the port lines connected to D_0–D_7.
2. The microcontroller chooses the destination for the data (LCD command or LCD display) by placing the corresponding level on the RS line.
3. The microcontroller chooses whether this is going to be a write or read operation and places the appropriate level on the $R/\overline{W}$ line.
4. The microcontroller raises the E line. If this is a transfer from the microcontroller to the display, the microcontroller holds the E line high for at least 500 ns, then lowers it to complete the transfer. If this is a transfer from the display to the microcontroller, after the microcontroller raises the E line, it should read the data from the port lines connected to the display before lowering the E line to complete the transfer.

This sequence of operations is representative of the process necessary to complete a bit-parallel exchange between any two devices, not just the LCD module that we are using as our example. In some cases, the $R/\overline{W}$ and E lines are replaced by separate RD and WR lines, where the RD line is pulsed to perform a read transfer and the WR line is pulsed to perform a write transfer. In either case, the data to be transferred must be presented in parallel to the receiver by the transmitter, and then the handshake lines must be manipulated to effect the transfer.

7.4 SIGNALING LEVELS

In the discussion, up to this point, we have talked about the signal states in terms of logical 1s and 0s or marking and spacing states. We have not focused on the actual voltage levels that are used to represent those logical states. The reason for sticking with logical states for the earlier discussion is that there are a number of different standards for the signaling levels used to represent the logical 1s and 0s. In this section, we will examine and compare a number of the most common signaling level standards.

7.4.1 TTL/CMOS Levels

The simplest signaling standard is to simply use the so-called "logic-level" signals from the pins of microcontrollers and peripheral chips. As you will see, when we examine digital inputs and outputs more closely in Chapters 16 and 17, these are not really governed by a single standard, but a set of standards that vary according to the characteristics of the devices and their power supply. In general, logic-level signaling is typically accomplished by representing logic-level 0 with a voltage at or near ground (0 V), and logic-level 1 with a voltage at or near the power supply voltage (often, but not always, 5 V). Signaling at logic levels is most often used when the communication is between devices that all lie within a single product. It is the most common approach for bit-parallel and synchronous serial communications.

7.4.2 RS-232

Signaling at **RS-232** levels is one of the two most common standards used for asynchronous communications between devices. This is the signaling standard used by the **serial ports**, also called **COM ports**, on personal computers[1] and is a very common way to communicate

[1] As hardware COM ports disappear, first from laptop computers and eventually from desktops as well, the need to communicate with RS-232-type devices will be handled by USB-to-serial adapters. As their name implies, these adapters plug into a USB port on a PC and reproduce a serial port that can be used to communicate with devices that adhere to the RS-232 standard.

between a PC and an embedded system. Using RS-232 in this context usually involves two data lines on each device (one for transmit and one for receive) plus a common ground. The transmit line of one device is hooked to the receive line of the other. The term RS-232 is the name of an official standard published by the Electronic Industries Alliance (EIA). In its most recent form, the standard has been renamed to EIA-232, though most people continue to refer to it as RS-232. The standard specifies more than just the signaling. It also includes naming conventions for signals, pin assignments for standard connectors, and protocols (conversation rules) for data transfer. We are going to focus on the voltage levels used to indicate logical 1s and 0s.

The EIA-232 standard is based on what is referred to as a **single-ended signaling standard**. In this type of system, the voltage levels on the pins are defined relative to a common ground. In this respect, it is not unlike the TTL/CMOS signaling that we discussed in Section 7.4.1. The voltage levels that are used are, however, very different. EIA-232 is a bipolar signaling standard in which the signal lines are required to take on both positive and negative voltages with respect to ground. The idle state for an EIA-232 signal is the marking state (logical 1), and it is defined as having a voltage level of between −3 and −25 V relative to ground. This means that a transmitter must produce an output of at most −3 V and no less than −25 V, while the receiver must tolerate signals of as low as −25 V without damage. The spacing state (logical 0) is defined as having a voltage level of between 3 and 25 V relative to ground. Since most microcontroller systems only require a single 3–5 V supply, manufacturers have developed interface chips that operate on 3–5 V and produce the required negative and positive voltages as well as translate between EIA-232 and logic levels.

7.4.3 RS-485

The **RS-485** standard improves noise immunity and adds capabilities relative to the EIA-232 standard. The RS-485 standard improves on the noise immunity of the single-ended EIA-232 by going to a **balanced differential signaling standard**, where the logical state is indicated by the difference in voltage between two signal lines and not by the absolute voltage on those lines relative to a shared ground. This allows for the inevitable difference in ground potential that occurs between widely separated devices to be tolerated without loss of signal integrity.

The EIA-232 standard defines a point-to-point system with a single transmitter connected to a single receiver. The RS-485 standard provides for multiple devices sharing a single pair of signal lines (Figure 7.14).

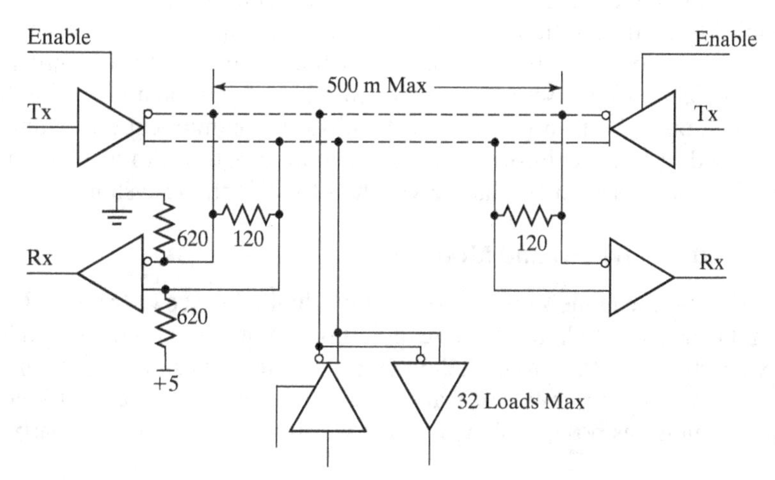

FIGURE 7.14
A typical RS-485 network.

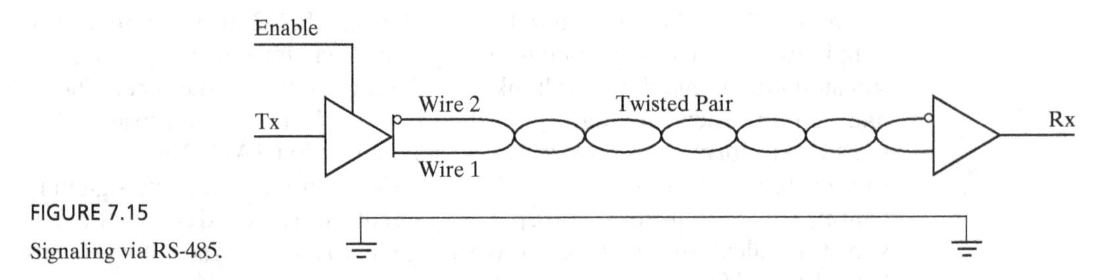

FIGURE 7.15
Signaling via RS-485.

This approach is referred to as **multidrop** and allows for a network of up to 32 standard nodes to be connected by a single pair of signal lines. Each of the devices on the network must share a common ground (to allow for current flow) but the voltage level of the signal lines relative to the ground pin does not enter into the determination of the logical state on the signal lines. The logical state on the signal lines is determined by looking at the voltage difference between the two signal lines.

If wire #1 is at a potential greater than wire #2 by at least 200 mV, that indicates a logical 1, while if the relative potentials are reversed (i.e., wire #1 is at least 200 mV *less* than wire #2), it represents a logical 0. Notice that we are only looking at the voltage difference, not the voltage relative to ground. As an example, if wire #1 were at 3 V relative to ground, a logical 1 would be indicated if the potential relative to ground at wire #2 were less than 2.8 V, while a logical 0 would be indicated if the potential relative to ground at wire #2 were greater than 3.2 V.

As drawn in Figures 7.14 and 7.15, the RS-485 network is **half-duplex**. This indicates that information can only be moving from one device to another at any given instant in time. This is different from the most common implementations of EIA-232, where there are separate signal lines for transmitting and receiving. Two signal lines allow for information to flow in both directions simultaneously, providing for **full-duplex** communications. There is nothing inherent in the RS-485 standard or the chips that implement it that prevents a designer from doubling the number of signal wires and creating a full-duplex channel between two or more devices.

7.5 COMMUNICATING OVER LIMITED BANDWIDTH CHANNELS

So far, we have been talking about communicating between two or more devices using wires. The advantage of directly wired communications channels is that the response of the channel extends from DC (0 Hz) to some fairly high frequency (~10 MHz for RS-485), typically limited by the drive electronics and not by the channel itself. However, other important scenarios don't fit this description. For example, when communicating over the telephone, or light beams, or radio waves, the capabilities of the channel will prove much more limiting than wires. In Section 7.5.1, we will discuss a number of generic modulation techniques that allow information to be fit into a limited bandwidth communications channel. These techniques are used to transfer information in the limited bandwidth available in RF channels (Section 7.7) as well as acoustic channels such as the telephone system.

7.5.1 Limited Bandwidth and Modems

All communications channels operate over only a limited range of frequencies. For nondirect contact, that is, not wired, channels, these limitations exist both at the low and high ends of the frequency spectrum. At the low end, their response often does not extend all the way to DC and at the high end the number of transitions per second must be kept below some upper limit. These limitations pose challenges when we start trying to transmit arbitrary data over

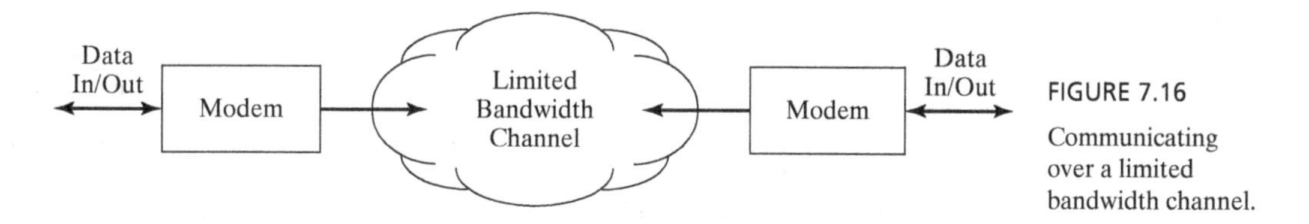

FIGURE 7.16
Communicating over a limited bandwidth channel.

these channels. Consider the case of a long idle line that might exist between transmissions on two devices using asynchronous communications. That long idle time will represent a very low frequency signal, in the limit approaching DC. This will not pass over any communications channel whose response does not extend down to DC. In order to communicate over one of these limited bandwidth channels, we will need to modify our signal so that it always fits within the bandwidth provided by the communications channel.

The way that we modify our signal is to interpose a pair of new devices between the two devices that wish to communicate (Figure 7.16).

The **modem** (MOdulator DE-Modulator) is responsible for translating our logical signal into something that fits in the frequency range provided by the channel. This translation can take place using any of a number of different modulation techniques, which we discuss in the following sections.

7.5.1.1 Modulation Techniques

The modulation techniques used over limited bandwidth channels, including both RF and the telephone system, fall mostly into a group of techniques involving amplitude modulation (AM), frequency modulation (FM), phase modulation (PM), or some combination of these. All of these techniques involve modulating (modifying) a carrier or base-band signal in order to encode digital information onto that signal. The basic operation of RF modems mimics that of the telephone modem except that the carrier frequencies are much higher.

7.5.1.2 Amplitude Modulation

In **amplitude modulation**, or **AM**, the carrier is turned on and off to indicate the changes in logical state (Figure 7.17).

As a result, this type of modulation is also referred to as **on-off-shift keying (OOSK)**, **amplitude shift keying (ASK)**, or **on-off keying (OOK)**. This simple approach to modulation is not actually employed by modems for use over the telephone system, but finds application in simple RF communications links. It serves as an excellent and easy-to-grasp starting point for a discussion on signal modulation. In an ASK system, if the carrier signal is present, it represents a “1” in the data stream. The absence of the carrier represents a “0.” Asynchronous

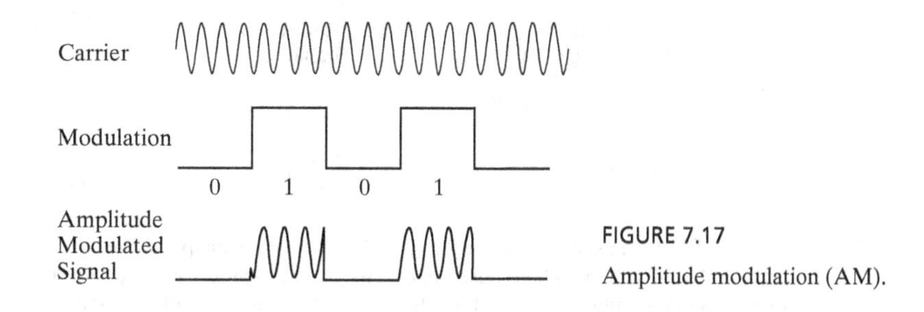

FIGURE 7.17
Amplitude modulation (AM).

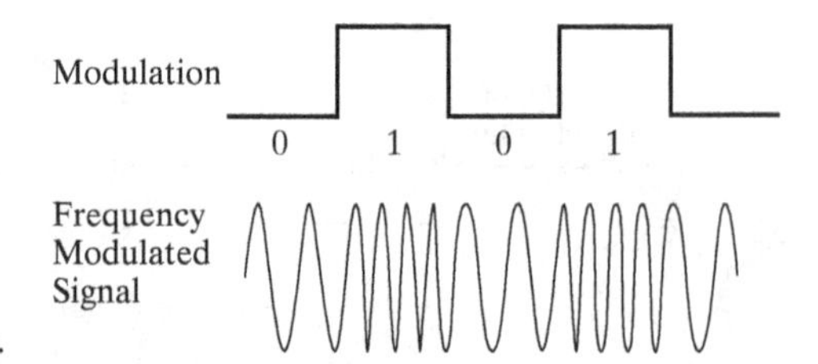

FIGURE 7.18
Frequency modulation (FM), also known as frequency shift keying (FSK).

communications is used with this and all of the other modulation techniques that we discuss, so both sides of the conversation must have agreed on what a bit time will be.

7.5.1.3 Frequency Modulation

When we modify the frequency of the carrier signal rather than its amplitude, it is called **frequency modulation** or **FM**[2]. This is also referred to as **frequency shift keying (FSK)**, and is shown in Figure 7.18.

FSK is the modulation technique used in the simplest of telephone modems. One early standard governing the function of such systems is the Bell 103 standard (released by AT&T in 1962). Under this standard, one device represents logical 1s with a 2,225 Hz signal and logical 0s with a 2,025 Hz signal. The other device in the pair represents logical 1s with a 1,270 Hz signal and logical 0s with a 1,070 Hz signal. In this way, both devices can transmit at the same time (full-duplex) and filters can be used to separate out the different frequency bands corresponding to the two devices. However, the maximum data rates allowed with this approach are quite low, since a few cycles are needed to determine the frequency present at any given time. In practice, devices employing the Bell 103 standard achieved data rates of 300 bps.

7.5.1.4 Phase Modulation

If we modulate the carrier by creating sudden phase shifts, it is referred to as **phase modulation (PM)** or **phase shift keying (PSK)**, and is depicted in Figure 7.19.

Because it is relatively easy (though outside the scope of this textbook) to detect phase shifts, even over a telephone system, it is possible to have more than two possible phases involved in the modulation. Figure 7.20 shows a system in which a pair of bits (a **dibit**) is represented by the relative shift in phase between two adjacent cells.

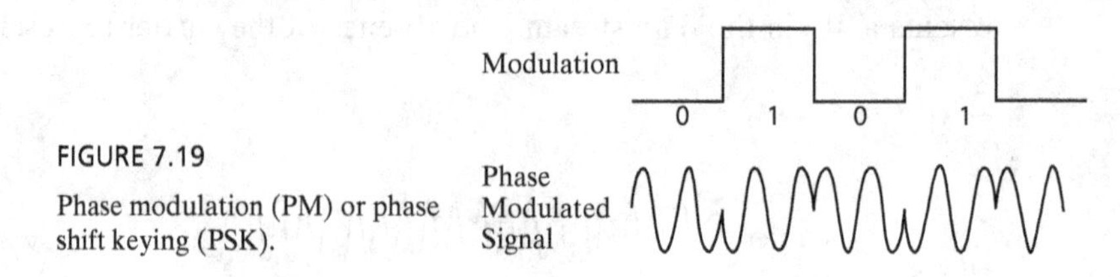

FIGURE 7.19
Phase modulation (PM) or phase shift keying (PSK).

[2]This use of FM is the same as that used in FM radio. The principal difference being that in FM radio an analog, rather than digital, voltage is being used to control the modulation. FM modems for use with phone lines operate on audio signals in the range of a few kHz, while FM radio operates with RF signals in the range of 88–108 MHz.

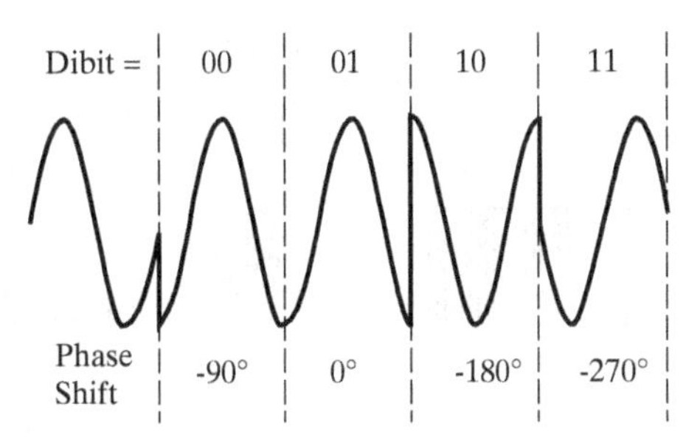

FIGURE 7.20

PM encoding 2 bits per cell.

7.5.1.5 Quadrature Amplitude Modulation

We can combine PSK with a variation on AM to encode even more information into a signal. While this approach is mathematically equivalent to modulating both phase and amplitude, the scheme is implemented by modulating only the amplitude of two carriers that are 90° out of phase (in quadrature) and then summing the two carriers together. It goes by the name of **quadrature amplitude modulation** or **QAM**. We represent this system with a plot that shows amplitude of the two carriers (I and Q) on the *x*- and *y*-axes (Figure 7.21). Another way to view this presentation is as a polar plot with the length of the vector representing its amplitude and the angle with respect to the positive *x*-axis, the phase.

For the increased complexity that this approach requires, it returns benefits in increased encoding efficiency (more bits transferred in a given period of time). The system of Figure 7.21 uses two amplitude levels and eight phases to give 16 distinct phase-amplitude combinations. This allows us to encode 4 bits of information into a single tone. This approach can be extended to larger numbers of phase-amplitude combinations. Telephone modems have been built with up to 512 combinations of amplitude and phase allowing 9 bits of information to be transmitted by a single tone with a unique amplitude and phase.

7.6 COMMUNICATING WITH LIGHT

Every time you use the remote control on your TV or DVD player, you are using a form of communications that employs IR light. The range of light wavelengths in the IR spectrum is

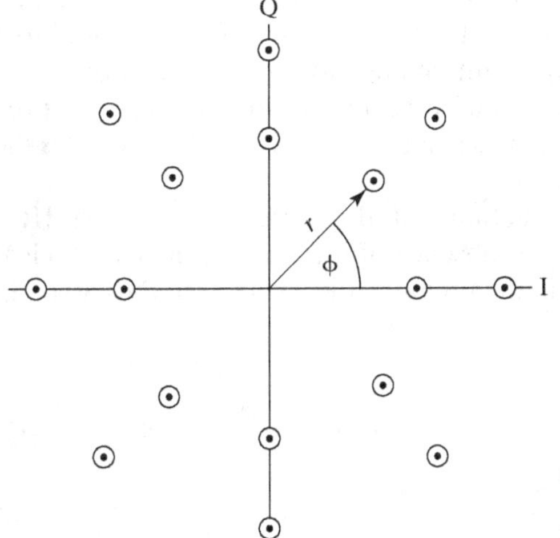

FIGURE 7.21

An example of quadrature amplitude modulation (QAM).

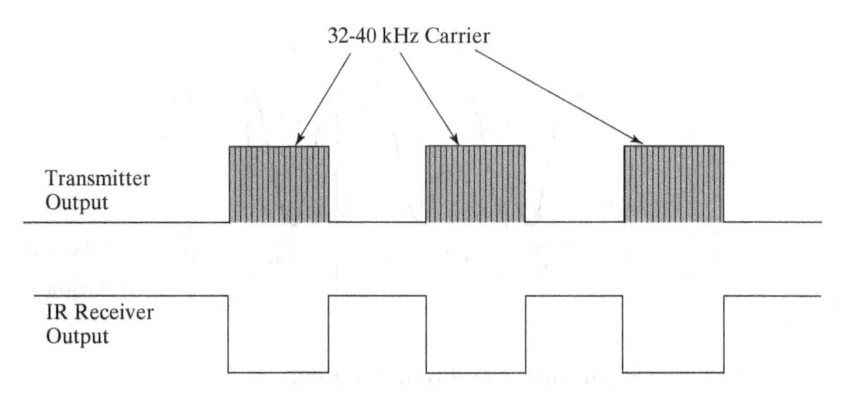

FIGURE 7.22
IR remote control modulation of the carrier.

very large, spanning from 750 to 1 mm, though most IR communications devices operate in the region between 850 and 950 nm. The remote control encodes your command as variations of the IR light level produced by an IR light emitting diode (LED).

It is not easy to modulate the frequency of the light directly, so IR remote controls generate a carrier by turning the IR LED on and off at the carrier frequency and then modulate the carrier. Most remote controls use a carrier frequency in the range of 32–40 kHz. The IR receivers used by the appliances are designed to interpret the presence of the carrier as a logical 0 and the absence of the carrier as a logical 1, see Figure 7.22.

While it might be possible to transmit an asynchronous data stream using this approach, the realities of the response of the IR receivers make this difficult. There are time shifts between the appearance of the IR signal and when the presence of the carrier is detected. These delays vary to some degree depending on ambient light levels, making it difficult to maintain framing between the transmitter and receiver. As a result, most remote controls use OOSK to modulate the carrier, but use a different encoding technique to encode the 1s and 0s. There are a number of different approaches that have been developed but all rely on incorporating timing information into every message and sending discrete messages rather than arbitrary strings of data.

The timing information is included by adding a **preamble** to every message. This preamble will consist of a predefined pattern of pulses of the carrier. The receiver uses the preamble and its knowledge of what the pattern should be to determine the bit time associated with this particular message. In this way, the receiver adapts to the transmitting rate for every message and the transmit timing only needs to be stable for a very short period of time (usually just a small fraction of a second). Figure 7.23 shows a preamble consisting of three equal pulses followed by a pulse that is twice as long that indicates the end of the preamble and the beginning of data.

To increase the reliability of the transmission, most IR remote controls do not use the simple presence or absence of a carrier to indicate logical 1s and 0s. Instead, they encode the logical data in the timing of the period that the carrier is present or absent. There

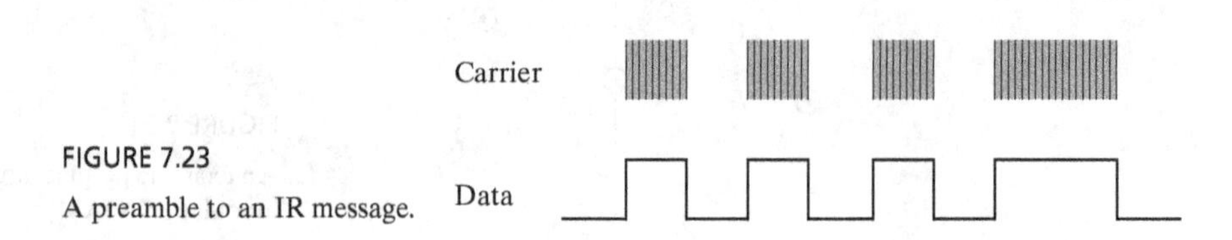

FIGURE 7.23
A preamble to an IR message.

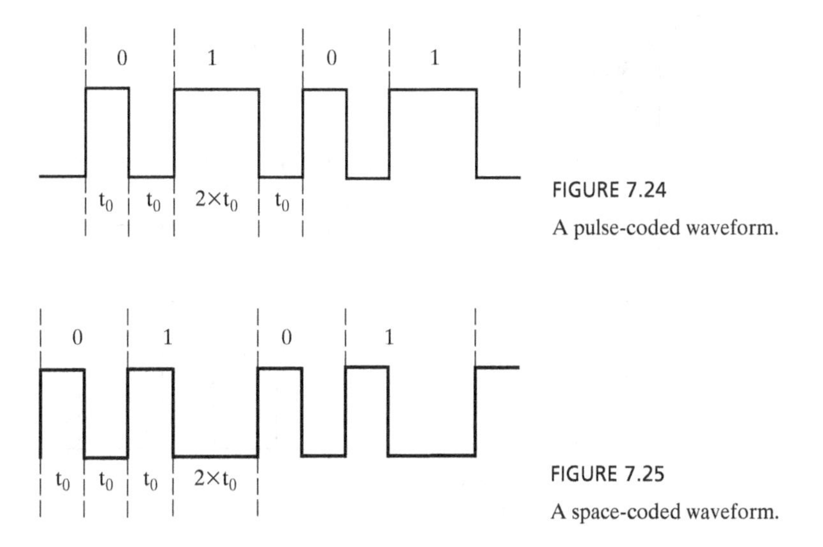

FIGURE 7.24

A pulse-coded waveform.

FIGURE 7.25

A space-coded waveform.

are many possible ways to do this. We show three possible approaches in Figures 7.24, 7.25, and 7.26.

In a **pulse-coded waveform**, the logical information is contained in the timing of the high time of the data. A logic 0 is represented by a short time high, and a logic 1 is represented by a high that is twice as long as a zero. The logical 1s and 0s are separated by a low signal equal in width to a logic 0 time.

The **space-coded waveform** trades the roles between low and high signals. In space coding, the information is contained in the length of time the signal spends low, and the data is separated by constant intervals of a logical high level.

The **shift-coded waveform** takes a very different approach to encoding. Rather than encoding the information directly in the time spent high or low, it focuses on the direction of the transitions at the center of the bit times. In the implementation shown in Figure 7.26, a high-low transition at the center of the bit time encodes a 0, while a low-high transition encodes a 1.

Each of these approaches to encoding has its adherents. Different consumer electronics companies employ their own encoding standards as well as defining the meanings of the message contents (e.g., volume up/down, channel up/down). There is no single clearly superior solution to the problem of encoding, so manufacturers have chosen to develop their own solutions to the problem.

The last form of IR communications that we will discuss is one in which there *is* an official standard. This is the **IrDA** standard promulgated by the **Infrared Data Association**. This standard is used in communications between PCs, Personal Digital Assistants (PDAs), and some other peripherals like printers. It differs markedly from the IR remote control approaches in that it does not use a modulated carrier to transfer the information. Instead, it uses the presence or absence of a pulse of IR light to encode 0s and 1s. The data are transferred as 8-bit bytes using the standard asynchronous format (Figure 7.27).

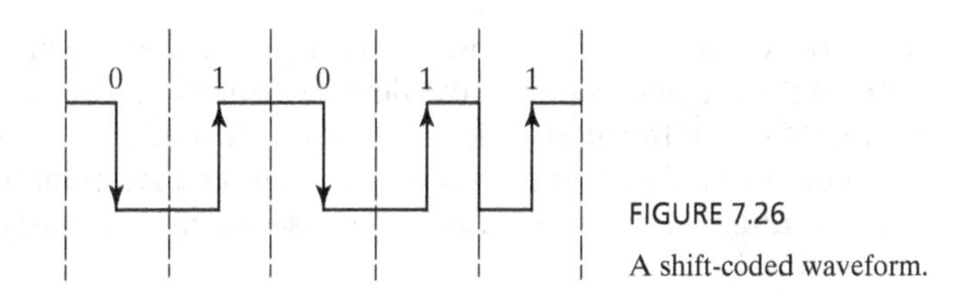

FIGURE 7.26

A shift-coded waveform.

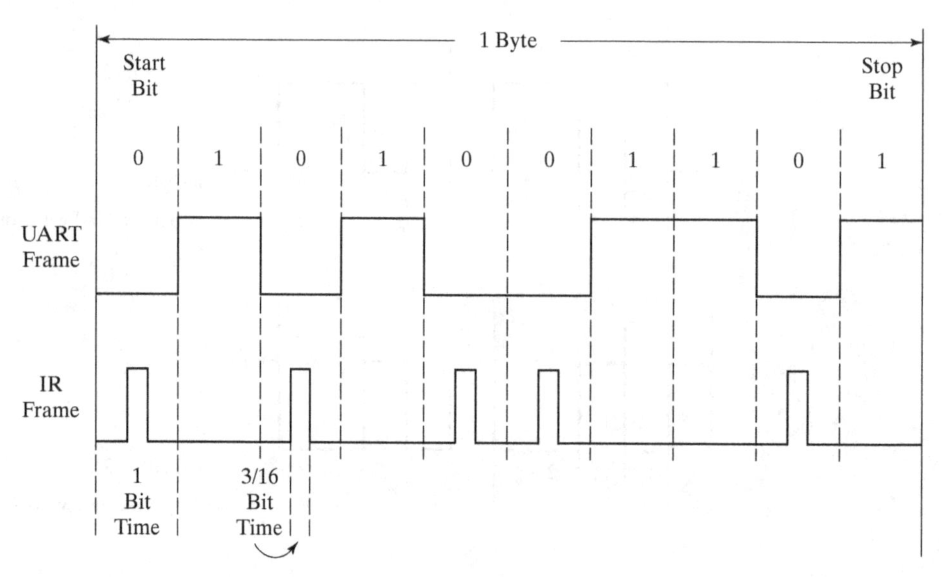

FIGURE 7.27
Infrared Data Association encoding.

Without a carrier, it is possible to achieve much higher data rates (up to 16 Mbps), but at the expense of much tighter timing requirements on both the transmitter and receiver. IrDA also does not have the long range of IR remote controls (usually several meters). IrDA communications are typically conducted over distances of under 1 m. To facilitate implementing the IrDA standard, manufacturers have developed chips that take a standard asynchronous data stream and drive an LED to the IrDA standard. Some of these chips also include the circuitry to receive an IrDA data stream and produce a data stream suitable for input to a standard UART.

7.7 COMMUNICATING OVER A RADIO

Communications over a radio link typically employs some of the same techniques that we have seen in telephone modems and IR links. The complexity of these systems varies dramatically depending on the application. We will treat three classes of **RF communications**: RF remotes, single channel data links, and RF networks. Each of these application areas has different goals and therefore different approaches to addressing the challenges associated with RF communications.

In general, the issues encountered when attempting to implement a radio link are very similar to those encountered in telephony and IR communications. These revolve around the limited bandwidth of the communications medium and the need to deal with varying signal levels and the presence of interference.

7.7.1 RF Remote Controls

RF remote controls, such as those used for garage door openers, are examples of the simplest of RF link implementations. These devices typically use a single RF frequency as the carrier and use OOSK as the modulation technique. Beyond that, there can be the same wide choice of encoding techniques as are used in IR remotes. RF remotes are designed to be inexpensive, so limiting the requirements for precise timing standards is a design goal in these applications as well.

Unlike IR remotes, RF remotes often have to provide security as well as robustness in the communications channel. For example, you would probably like *your* garage door opener to be the only one that can open *your* garage door. To achieve this, messages from RF remotes include several additional bits with the goal of achieving a degree of security. This, despite the fact that there are really only two possible commands: open and close. The extra bits are used in two ways. Some of the bits are there to be sure that a valid message has been received. This makes the link more robust in the face of interference from other RF sources in the area. The other bits are part of the security system to insure that only your remote opens your garage door.

Modern garage door openers use a rolling key security system. Rather than transmit a fixed message to open or close the garage door, the remote transmits a new message from one of billions of possible messages every time you open or close the door. In order for this to work, the transmitter and the receiver need to know what the next code in the sequence should be. This is accomplished by a synchronization process. For garage door openers, this typically involves pressing a button on the garage door controller (in your garage) while simultaneously pressing the open button on the remote.

There are dedicated ICs available from manufacturers such as Holtek that are designed to encode and decode the messages necessary to build a simple RF remote control system. These encoders and decoders can be combined with simple AM transmitters and receivers to implement a fixed message system. If a rolling key system is necessary, a small microcontroller can be programmed to generate the messages and also to vary them according to an algorithm that will allow it to work with a receiver that also knows the algorithm.

7.7.2 RF Data Links

A number of manufacturers manufacture a wide range of RF transmitters and receivers that are designed for use in RF data links [2]. The majority of devices intended for these uses also employ a single RF frequency. The simplest of these devices use the same OOSK modulation technique used by the RF remotes described above. The more sophisticated devices achieve higher performance by using FM or FSK.

There is a range of solutions available in the marketplace. On the low end are simple devices that emphasize low cost and require that the user provide encoding and link management. On the high end, so-called "smart" devices supply all of the encoding and link management so that they look like a "virtual wire" to the user's application.

7.7.3 RF Networks

The most sophisticated RF communications devices are those participating in RF networks. These involve communications standards such as **WiFi** (IEEE 802.11a/b/g/n), **Bluetooth** (IEEE 802.15.1), and **Zigbee** (IEEE 802.15.4). The RF communications in these devices tend to be much more sophisticated. They employ more elaborate modulation techniques such a QAM to improve the data throughput rate. They also employ more sophisticated RF techniques rather than operating on a single frequency. This latter approach improves their ability to operate successfully in the presence of interfering noise sources.

RF networks also define much more than just the mechanism to get bits from one place to the next. All of these standards also specify a set of protocols (rules for holding conversations) that allow multiple devices to share the same RF spectrum and interoperate reliably.

7.8 HOMEWORK PROBLEMS

7.1 True or False: Asynchronous communications involve a shared clock line.

7.2 Describe the fundamental difference between bit-serial and bit-parallel communications.

7.3 Under what conditions will the noise flag in a typical UART be set?

7.4 What is the role of the SS* line in synchronous serial communications?

7.5 True or False: All standard synchronous serial communications protocols involve two data lines and one clock line.

7.6 What is the purpose of the parity bit?

7.7 True or False: IR remote controls typically use FSK modulation.

7.8 True or False: RS-485 is an example of a differential signaling standard.

7.9 Explain the basic difference between the RS-232 and RS-485 signaling standards.

7.10 What UART error bit (if any) would be set if the center of the 10th bit time after the falling edge of the start bit in an 8N1 asynchronous serial bit stream was found to be at the spacing level?

7.11 Identify the type of modulation shown in Figure 7.28. Which types of devices use this scheme?

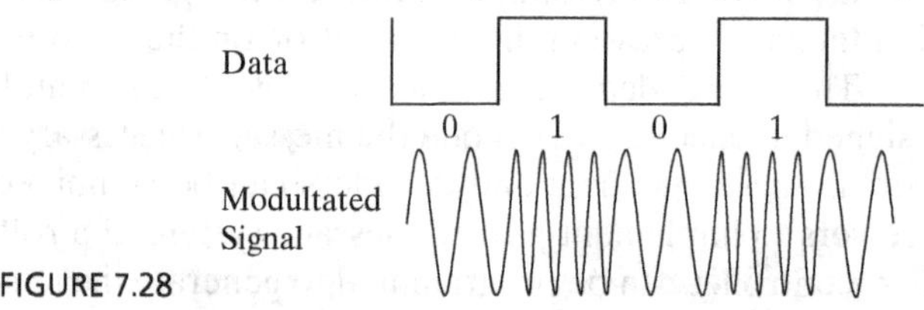

FIGURE 7.28

7.12 Identify the type of waveform shown in Figure 7.29, which is commonly implemented with OOSK. Which types of devices commonly use this scheme?

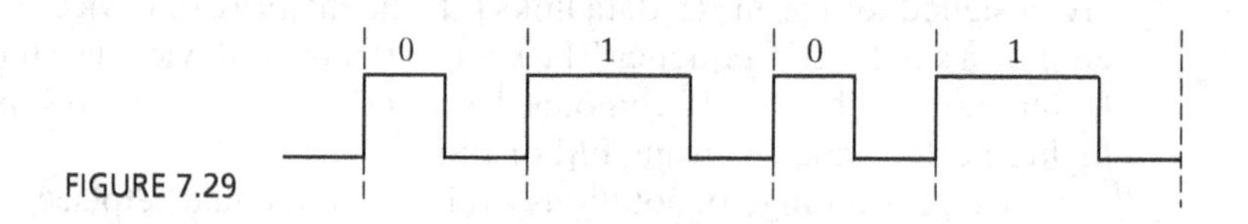

FIGURE 7.29

REFERENCES

[1] *Understanding Data Communications,* Friend, G.E., Fike, J.L., Baker, H.C., and Ballamy, J.C., Howard W. Sams & Company, 1988.

[2] RF link sources:

Linx: http://www.linxtechnologies.com/
Digi International: http://www.digi.com/
Ember: http://www.ember.com/
Helicomm: http://www.helicomm.com/
Ewave: http://www.electrowave.com/
RFM: http://www.rfm.com/

FURTHER READING

Microprocessor Based Design, Slater, M., Mayfield Publishing Co., 1987.

CHAPTER 8

Microcontroller Peripherals

One of the defining characteristics of a given microcontroller, and something most commonly associated with more "complex" microcontrollers, is the range of hardware subsystems that are added alongside of the CPU. Peripherals augment the capabilities of the main processor, and relieve it of performing tasks that would otherwise require a great deal of instruction execution time and software complexity. A study of the types of subsystems that most commonly appear in modern microcontrollers shows that there are many similarities among microcontroller subsystems across a wide range of manufacturers. Today's students of mechatronics benefit from several generations of evolution and refinements, which have led to a convergence of the most commonly needed set of functionality in microcontroller peripheral subsystems. In this chapter, we look at the features that you will most likely encounter in:

1. parallel input/output subsystems,
2. timer subsystems,
3. pulse width modulation (PWM) subsystems,
4. analog-to-digital (A/D) converter subsystems,
5. interrupts.

The subsystems that you will encounter in a given microcontroller are likely to be implemented slightly differently than the ones we describe here; however, it is also highly likely that they are similar enough that you will be able to very quickly understand those specific subsystems and be able to program them to perform the most common operations. Though what you will find in this chapter is a description of hardware subsystems, it is in the Software section of this text because the way that we interact with these hardware subsystems is through programming on the microcontroller. While we are discussing hardware, the purpose is to prepare you to configure that hardware through programming.

The descriptions of the subsystems that we use in this chapter are typical of what you find in manufacturers' data sheets. As such, it will be useful to study Chapter 18 before tackling this material so that you will be familiar with the functional schematics used to describe these subsystems.

8.1 ACCESSING THE CONTROL REGISTERS

The peripheral subsystems on microcontrollers are configured and manipulated through a set of **hardware status** and **control registers**. These are special locations in the memory map that appear to the programmer like ordinary memory, but in fact are connected to hardware used to query the state of (for status registers) or manipulate the behavior of (for control registers) peripheral subsystems. Depending on the particular microcontroller, these status and control registers may be mapped into a combined memory map that also includes the program and scratchpad memory (see Chapter 2), exist in a separate I/O memory space with its own memory map, or be part of a combined I/O plus scratchpad memory space. Interacting with registers that are in a separate I/O memory space often involves the use of special machine language instructions in order to access the I/O space. Fortunately, most high-level languages that target embedded systems use one of several available compiler-specific mechanisms to hide this complexity from the programmer and simplify the task. These compilers (or interpreters) provide a mechanism that provides the programmer with symbolic names with which to access the status and control registers. To the programmer, these status/control registers appear as predefined variables. This allows them to be read from and written to exactly as if they were ordinary variables in the language. As we discuss manipulating these registers throughout this chapter, remember that the manipulation is no more complex than reading a variable (put its name on the right-hand side of an assignment statement) or writing to a variable (put its name on the left-hand side of an assignment statement).

8.2 THE PARALLEL INPUT/OUTPUT SUBSYSTEM

The most basic needs of a microcontroller are to sense the state of the physical world and to "reach out and touch" it. In the digital domain, this is done by reading the states on digital input pins and by manipulating the states on digital output pins. Depending on the particular microcontroller and the specific pins involved, these **digital input/output (I/O)** pins may be dedicated to either the input or output functionality or, more commonly, selectable as either input or output under software control. Your software will need to configure those multipurpose pins so that they are configured properly for your application.

8.2.1 The Data Direction Register

Ports (groups of eight I/O lines) with programmable direction (input or output) are most often associated with at least two registers. One of these registers will be a **data direction register (DDR)**. Different manufacturers use different names for this register. For example, Freescale, Atmel, and others refer to this as the DDR, while Texas Instruments refers to this simply as the Direction Register, and Microchip refers to it as a TRIS register. Despite the differing names, these registers all function the same way: each bit in the register corresponds to the direction of one of the pins associated with that port. The specific bit state (1 or 0) that it takes to make a pin an input or an output varies from device to device, so you will need to consult the data sheet for your particular microcontroller. Programming the DDR to choose the direction for each of the pins on the port requires nothing more than writing the appropriate bit pattern to the DDR. For example, for a Freescale MC9S12C32, to program the lower 4 bits on port T to be outputs and the upper 4 bits to be inputs requires assigning the value 0x0F (00001111 in binary) to the DDR for port T (which is conveniently named "DDRT"). Since most C compilers for embedded systems provide definitions for the control registers that allow the programmer to treat them as variables, programming the DDR for port T is as simple as:

```
DDRT = 0x0F;      /* bits 4-7 = 0 (inputs), bits 0-3 = 1 (outputs) */
```

8.2.2 The Input/Output Register(s)

With the direction of the pins defined, the next task is to read from the input pins and/or write to the output pins. Here, there are basically two approaches that you will find across the range of available microcontrollers: separate input and output registers or a single register that serves for both reading and writing.

With separate registers, to read the state of the pins, you read the register which specifically corresponds to the inputs; to write the state of the output pins, you write to the separate register corresponding to the outputs. Typically, you can also read from the output register, which returns the last value that was written to the output port (which may or may not be what you want, so exercise caution when using this approach).

With a single register that serves for both inputs and outputs, the hardware of the port automatically processes the read and write operations to that register differently. Writes to the register cause the new data to be written to a latch that drives the pins that are configured as outputs. Reads from the register for pins that are configured as inputs will read the state of the associated input pins. The question then becomes: what state do I read for the pins that are configured as outputs? The answer to this question varies from microcontroller to microcontroller. On some devices, a read of the port always returns the actual state on the pins. On others, the read returns the last value written for those pins that are configured as outputs. In this latter case, the read of output pins returns the state of the internal output latch, not the actual state on the pins at the time of the read operation.

On those microcontrollers with a combined I/O port register and reads that always report the state of the pin (e.g., the Microchip PIC family), you must be careful if an external load drags the pin state outside the legal logic state for the desired output. For example, this occurs if we write a 1 to a port pin configured as an output and the external load requires so much current that the output voltage level on the pin drops below the minimum input high voltage. If we were only to write to the port, then this would not constitute any problem, so long as the maximum output current requirements for the pin are not being exceeded. The most common problem occurs when we want to be able to manipulate other pins on that port while keeping the original port pin at the proper level. Operations that modify only some of the bits in the data stored in a memory location are referred to as **read-modify-write** operations. The bits in the location are manipulated by first reading the contents of the location, modifying only the bits that we want to change, then writing the modified result back to the memory location. This works fine if the read portion of the operation returns the same state that was written to the output pin. However, when that is not the case, then the read of the data register returns the opposite state of the output's intended value. When the modified result is written back to the port, it will unintentionally change the state of the heavily loaded output pin. This is obviously undesirable, and a rather insidious bug to identify. The easiest way to work around this issue is to keep a "shadow copy" of the values that are written to the port. Any manipulation of bits, with the necessary read-modify-write operations, takes place on this internal shadow copy of the port contents. After the manipulation is finished, the resulting value is then written to the physical port register. In this way, the only operations performed on the actual port register are write operations.

Some microcontrollers, such as the Freescale MC9S12 family, have both a register that always reads the state of the pins and a combined I/O register. This provides great flexibility as well as the ability to easily determine if an external load is dragging the state of the output pin up or down. To determine if heavy loading exists on an output pin, compare the value in the port I/O register (which will report the state of the output latch) with the value in the port input register (which will report the actual pin state). If the two states don't agree, then the external load connected to the pin is heavy enough that the output is no longer able to maintain the requested output state.

8.2.3 Shared Function Pins

Microcontrollers are never designed with enough pins for all the ports and all of the other subsystems to be in use at the same time—that is, with a separate pin dedicated to each function. To save real estate, pins generally share many functions and the mechatronic system designer is required to commit pins to either be parallel I/O port pins or dedicate those pins to one of the other microcontroller subsystems, such as the timers, PWMs, and A/D converters covered

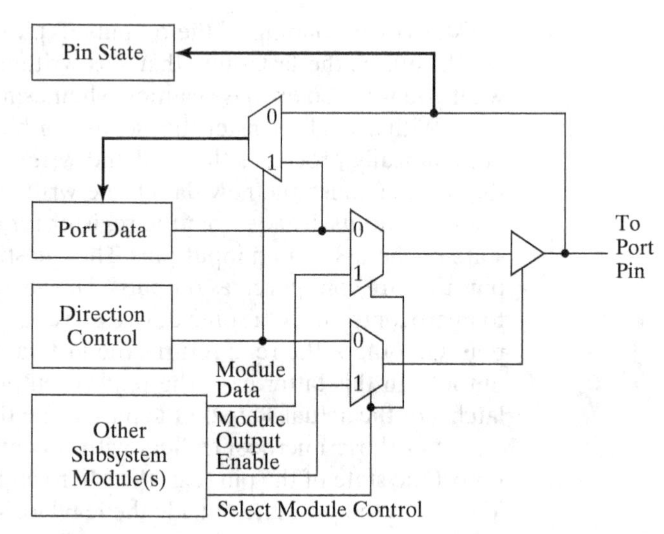

FIGURE 8.1

Functional schematic of internal connections to a port pin.

later in this chapter. In some recently introduced microcontrollers, it is possible to map any subsystem function to any port pin. More commonly though, a subsystem is associated with a particular set of port pins and the choice will be to use that pin for the subsystem or for parallel I/O. This can be particularly problematic when multiple subsystems share a single pin, since the designer will be forced to choose only one of them. For example, while a device may have both serial communications interface (SCI) and serial peripheral interface (SPI) communications subsystems, if they share any common port pins, the designer may only be able to use one of them.

The interaction between the parallel I/O, the subsystem modules, and the pins is often described using functional schematics like that shown in Figure 8.1.

Here we see a parallel I/O port pin that is shared with the parallel I/O port and some "other subsystem." The *Select Module Control* line from the other subsystem determines whether the pin will be controlled by the I/O port or by the subsystem module. If the subsystem module is enabled, then the module data and module output enable are routed to the port pin via a pair of multiplexers (shown as trapezoidal blocks). If the subsystem module is not controlling the pin, then the Direction Control line controls the direction of the pin and the Port Data and Pin State registers are used to read and write data to the pin.

Something to look for in the documentation is the default assignment of pins to subsystems. In most cases, the default assignment is to the parallel I/O subsystem associated with a pin, and the pin is typically configured as an input. However, this is not always the case. For example, for midrange PIC family microcontrollers, the pins associated with the analog-to-digital converter are configured as analog inputs at reset. In order to use them for digital I/O, the programmer must not only correctly configure the TRIS register (the PIC version of a DDR) but also change the value in the ANSEL(H) register, which provides the PIC equivalent of the *Select Module Control* line shown in Figure 8.1.

8.3 TIMER SUBSYSTEMS

Measuring the passage of time is one of the most fundamental tasks that a microcontroller can perform. All modern microcontrollers include dedicated hardware for measuring the passage of time, generally referred to as **timers**. In this chapter, we will examine a number of basic timer architectures that are so useful that they have been implemented in a wide range of microcontrollers. Specifically, we will examine the operation of:

1. timer overflow systems,
2. output compare systems,
3. input capture systems.

For each of these systems, we will describe the basic operation and the typical set of registers and bits provided to control them.

8.3.1 Timer Basics

All timers are architected in essentially the same way: they're comprised of a set of flip-flops chained together so that they count the number of clock cycles applied to the input (see Chapter 18). It is intentional and significant that there is no mention of timing in this description. This is because the essential timing information does not come from the timer per se, but from the clock source applied at the input. The timing information is deduced from knowing that the count is incremented at some well-known and stable interval. Many microcontrollers actually implement the timer as a **counter/timer subsystem**. In systems like these, the programmer can choose to drive the counter from an internal time base, creating a timer, or drive it from an external source, simply using the counting ability. If the external signal is a regular clock, then it too can be used as a timer. If not, it can be used to count events that happen at irregular intervals.

The counters used to implement counter/timer systems are most commonly implemented as either a group of 8 or 16 flip-flops—that form 8-bit or 16-bit timer registers, respectively. This allows them to count from 0 to 255 (in the case of the 8-bit timer register) or 0 to 65,535 (in the case of the 16-bit register). The counter is driven by a clock source that may have an optional divider inserted between the clock source and the counter. This allows a single clock source to increment the counter at different rates depending on the divisor (the **divide-by ratio**) chosen.

Figure 8.2 shows such an arrangement. In this figure, M is the root clock source that feeds a string of dividers. The divider chain produces outputs at fractional multiples of the root clock source, in this case 1/2, 1/4, 1/8, 1/16, 1/32, 1/64, and 1/128 of the root clock frequency. A multiplexer (MUX) is controlled by three bits (PR_0, PR_1, PR_2) in a control register to

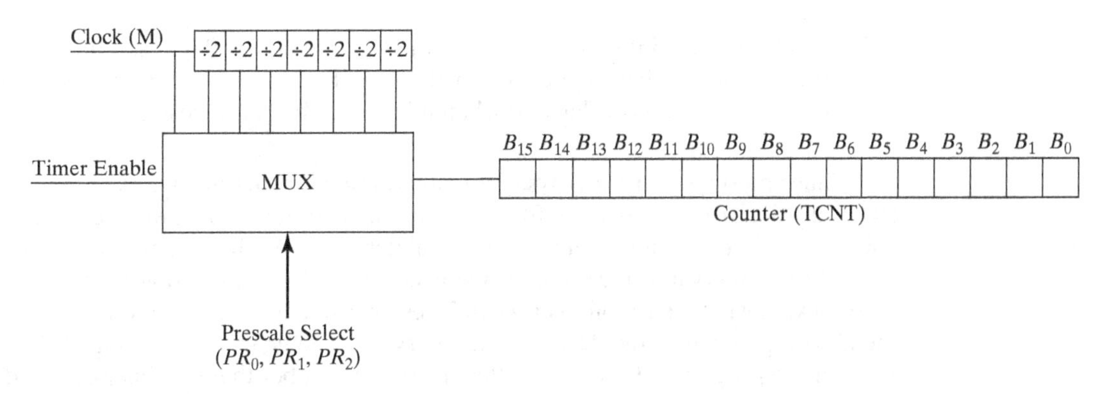

FIGURE 8.2

Clocking the counter at multiple rates.

choose which of the eight possible clock rates will be used to drive the actual counter, TCNT. The entire timer subsystem can be disabled to save power by clearing the Timer Enable bit.

There are two major categories of counters: **resettable counters** and **free-running counters**. Resettable counters can be written under program control and the count will then proceed from the value that was written. Most (but not all) resettable counters count up. Free-running counters simply increment or decrement on every active transition of the clock. A write to the counter is usually ignored. In some implementations, a write to the free-running counter will force the counter to some specified count independent of the value that is written, called a **reload value**. For the most part, the more advanced timer features (such as output compare and input capture, described in detail in Sections 8.3.3 and 8.3.4) are implemented with free-running counters.

8.3.2 Timer Overflow

No matter how many bits the hardware includes in the timer, eventually it will reach the maximum (for counters that count up) or minimum count (for counters that count down) and then **roll over** to start counting from 0 again. The ability to detect the roll-over, or **overflow**, condition is one of the basic capabilities that counter/timer systems provide. This is typically implemented as a bit in a status register, which indicates that a roll-over has happened. It will be up to the programmer to write code to periodically monitor the status of that bit and take whatever action is necessary when the hardware sets the bit. The most basic action is to clear the bit to prepare for the detection of the next counter overflow. Other responses to the overflow will vary from application to application, but one of the common uses of the overflow is to use it to extend the effective length of the timer. If, under software control, a variable is incremented each time an overflow is detected, then that variable can be combined with the value from the timer to effectively extend the time interval that can be measured. In effect, you have manually (in your software, that is) increased the number of bits in the timer register to enable the timing of events that have longer duration that could otherwise be measured.

When using a resettable timer, the timer overflow is often used to indicate the expiration of some time period. The process works like this:

1. First, an appropriate divide-by ratio is chosen to allow the desired time interval to be measured using the number of bits available in the counter.
2. Next, the value to be written to the timer is calculated and written to the timer register.
3. Finally, the time interval expires and the timer overflow flag is set. If the software needs to make something happen when this event occurs, the code needs to continuously monitor the overflow flag and when it is set, perform the desired action.

Since most counter/timer systems count up (increment), and we want to use the overflow (the transition from full-scale to 0) to indicate the end of the time, we need to figure out what count needs to be written to give the interval desired. We cannot simply write the value of the desired time interval to the counter. We need to reset the timer register with the number of timer ticks it will take to roll over when it resumes counting from that new value. This will be the 2's complement of the desired time interval. Another way of arriving at the number that you want to program is to start with the maximum number that the timer can hold, add 1, then subtract the number of timer ticks that you want your timer to have. As an example, to time an interval of three ticks on a 16-bit counter, the value that we would write would be:

0000	0000	0000	0011	Desired count = 3
1111	1111	1111	1100	1's complement
			0001	+1 to form 2's complement
1111	1111	1111	1101	Value to write = 65,533

or

65,535	Maximum 16-bit count value
65,536	+1
−3	−desired count = 3
65,533	Value to write = 65,533

The counter would then count: 65,534, 65,535, 0, and the overflow flag would be set.

This would seem to guarantee an interval of three clock periods. In reality, the interval may be as long as four clock periods less some small epsilon. The reason for the uncertainty lies in the fact that the write to counter will most likely not be synchronized to the incrementing of the counter. If the increment had happened just before the write, then the interval would be a little under four periods. If the counter were just about to increment at the instant of the write, the interval would be three periods plus the time difference between when the counter was written and when the next increment occurred. This uncertainty is inherent in timing with a clock that is not synchronized to the events (which is almost always the case).

8.3.3 Output Compare

Timers that are more complex than the basic timer described in the preceding section typically add input capture (IC) and output compare (OC) functions. We'll cover output compare here and input capture in the next section. The **output compare (OC)** function is implemented by adding a digital comparator and a register to hold the desired compare value to the basic timer/counter.

As shown in Figure 8.3, the value in the timer/counter register (TCNT) is continuously compared to the value in the output compare register (TCx). The comparator detects when two multibit digital numbers are identical (for more on this function, see Section 18.2.1). When

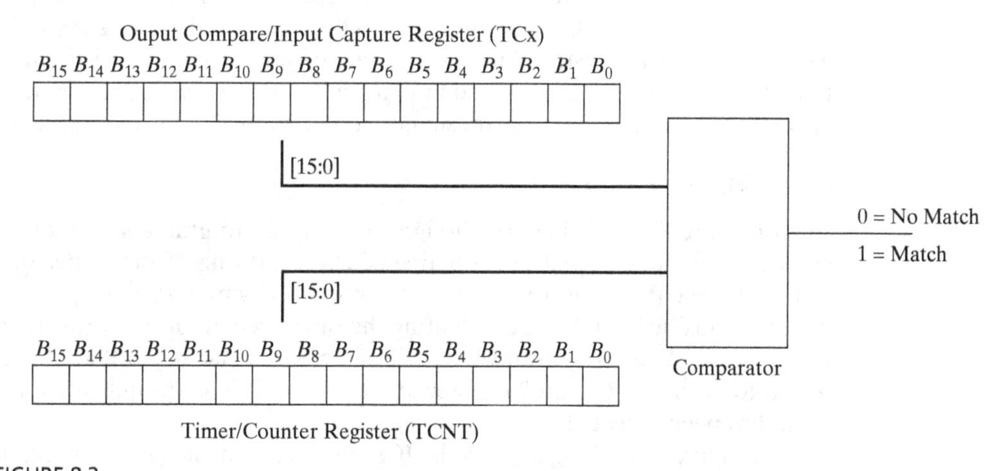

FIGURE 8.3

An output compare system.

this match occurs, the hardware of the output compare system sets a bit in a status register and optionally performs an action that has been selected based on settings in the associated control registers. Typical actions are to raise, lower, or toggle (change states) an associated output pin. In some implementations, it is possible to perform actions on multiple output pins when a successful compare occurs. A less commonly seen, yet very useful, capability is to cause the counter/timer to reset to 0 when the output compare occurs. This can allow a system with multiple output compare channels to dedicate one of those channels to generating an arbitrary counting modulus for the other channels (recall from Chapter 3 that a modulus is the number of different numbers that can be represented in a given system).

One of the simple ways in which you can use an output compare is to detect when some specific period of time has expired. Suppose that you had a situation where you were looking to see if the interval between incoming events ever exceeded some maximum value. If you coded the event detector so that every time it detected one of the events, it reprogrammed an output compare to happen at a time in the future that was at the limit of the allowable interval between events, as long as the next event happened before the output compare had expired, the output compare event would never happen. If the output compare event ever *did* happen, you would know that the maximum time limit between events had occurred and take whatever action you would need to under those circumstances. The code for the incoming event checker might look like:

```
IncomingEventChecker
  If CurrentPinState not equal LastPinState
        Event = IncomingEventDetected
        Program output compare to CurrentTime + MaxInterEventInterval
  Else
        Event = NoEvent
Save CurrentPinState as LastPinState
Return Event
```

The current time is available by reading the value in TCNT (Figure 8.3). A separate event check would then be written to test if the output compare flag ever got set. If it did, then the time was exceeded.

Using the hardware actions capability of the output compare systems, like those described above, provides the means to generate very precisely timed outputs without dedicating the computing power of the microcontroller to the timing task. That is, once the system is set up with a few software instructions, no further software interaction is required for their function. As an example, they are used in many automotive electronic engine control systems to generate the spark timing control signals and to control the width of fuel injector pulses.

8.3.4 Input Capture

Input capture (**IC**) can best be thought of as a way to grab a snapshot of the count in the counter when an external event happens. The capturing of the snapshot happens in hardware, relieving the software of the task of continually monitoring a pin to detect the time at which it transitions, and also eliminating the time needed for the software to detect and react to the event. When the capture event is detected, the current value of the counter is transferred to an "input capture" register and a status bit is set to indicate that an input capture event has been detected.

In Figure 8.4, a rising edge on the ICx pin will trigger an input capture, and the value of the TCNT will be transferred to the input capture register (TCx). Trigger events can be chosen to be a rising edge on the corresponding input pin, a falling edge, or either edge (any change of state).

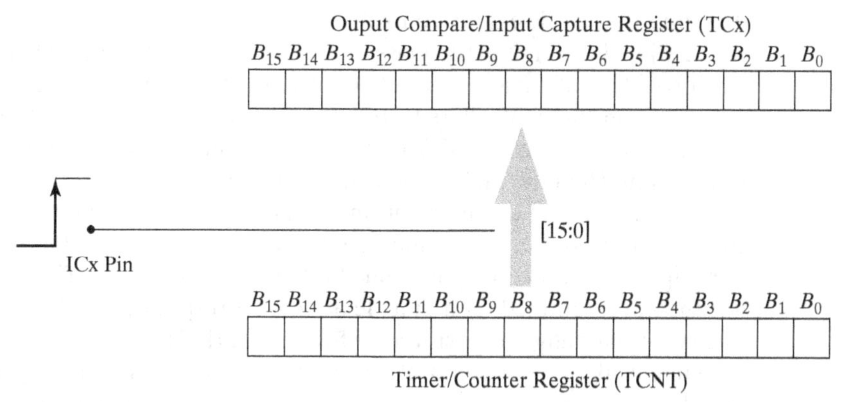

FIGURE 8.4

The input capture process.

Because the hardware captures the count of the timer, the software need only guarantee that the input capture flag is checked often enough to make sure that no events are missed. Reading the TCx can happen at any time after the event was captured, with the register holding the exact time (plus or minus the timing uncertainty) of the event. For input captures, the timing resolution is governed by how fast the clock ticks (the period) and therefore the rate at which the counter increments. It is *not* determined by how often the software checks for transitions on the pin, nor how often the software checks the input capture flag. Input captures allow for very accurate timing of external events, and the work to perform this timing all takes place in hardware. This allows the CPU to do other useful work (e.g., running your software) while a timing operation takes place.

For example, consider that case in which a motor is equipped with an encoder (Chapter 13) that produces pulses as the motor rotates. If the input capture system input were connected to the encoder output and programmed to capture the time at every rising edge, then the following snippet of pseudo-code could be used to calculate the motor's speed:

```
InputCaptureEventChecker
  If InputCapture flag is set (new edge has occurred)
        Save the contents of the input capture register as CurrentTime
        Calculate motor speed (in Revolutions/time) as
              1/(EncoderPulsesPerRevolution*((CurrentTime - LastTime)*
              TimePerTimerTick))
  Endif
Save CurrentTime as LastTime
Return
```

Input capture systems often find use in motor speed control systems, where they can be used with incremental encoders to measure the shaft speed of the motor. In automotive electronic engine control systems, input captures are used to measure the rotational speed of the engine and synchronize the generation of spark plug ignition pulses to the rotation of the crankshaft.

8.3.5 Combining Input Capture and Output Compare to Control an Engine

Input capture and output compare subsystem first appeared in the late 1970s, in microcontrollers specifically designed for use as electronic engine controllers in cars and trucks. At that time, electronic engine control systems had as inputs a mass air flow sensor (or an alternative way of estimating the mass of air flowing into the engine) and a crank position sensor that provided information about the instantaneous position of the engine crankshaft. The outputs from the controllers were connected to power drivers to control the fuel injectors and ignition coil. The signal from the crank position sensor was connected to the input capture system. The interval between edges on the crank position sensor signal corresponded to a known amount of engine rotation. By capturing the time interval between two successive crank position pulses, the engine controller could calculate the engine speed and the relationship between degrees of crankshaft rotation and time. The output compare subsystem was then programmed for a time delay that corresponded to the correct moment (and therefore crankshaft angular position) to fire the spark. The mass flow rate and engine speed were used to calculate the amount of air that was inducted into each cylinder. The chemically balanced ratio between fuel and air is 14.7:1 for complete combustion, which, in combination with the known mass of the inducted air, allows us to calculate the amount of fuel that should be delivered. If the fuel flow rate for the fuel injectors is known, the required fuel mass flow may be translated into an opening time for the injectors. A second output compare channel was used to precisely control the on time of the fuel injector to deliver just the right amount of fuel for the amount of air in the cylinder. In practice, things were a bit more complex than this, but the essence of electronic engine control can be performed with an input capture and two output compare channels.

8.4 PULSE WIDTH MODULATION

An inherent limitation of a digital output is that it can only take on one of two possible states and corresponding voltage levels: on or off, high or low. There often arises the need to produce an output that has an analog value somewhere between the voltage levels defined as the "high" and "low" states. One way to do this is to add a digital-to-analog converter (discussed in Chapter 19). A simpler way to do this is to switch the output very quickly between the high and low states. If this happens much faster than the device being controlled can respond, the effect is generally that the device effectively averages the output. For example, if an output swings between 0 and 5 and spends 50% of the time at 5 V, the effective output voltage is 50% of 5 or 2.5 V. It is possible to produce an analog output voltage by applying a low-pass filter to the digital output using a simple RC filter. If the frequency of the output switching is much faster (10x–100x or more) than the corner frequency of the low-pass filter, then the output will be an analog voltage whose value is the duty cycle (the percentage of the period the signal spends in the "high" state) × 5 V with only a small amount of voltage ripple (Figure 8.5).

While it is possible to produce low-frequency PWM signals that are generated purely in software, producing high-frequency PWM signals requires hardware support. Because the need for high-speed PWM signals is such a common requirement, microcontrollers often

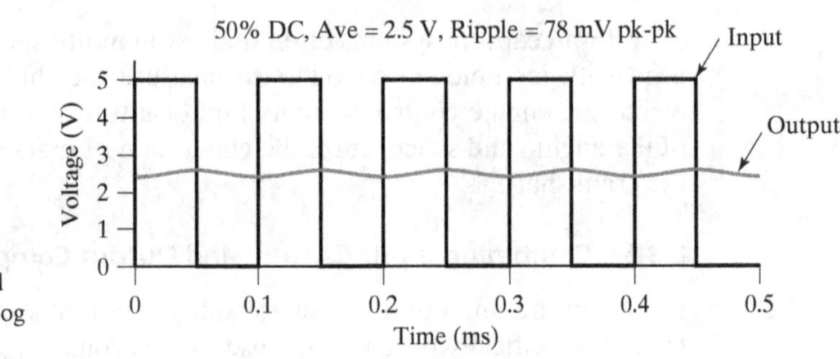

FIGURE 8.5
Filtered PWM used to generate an analog voltage.

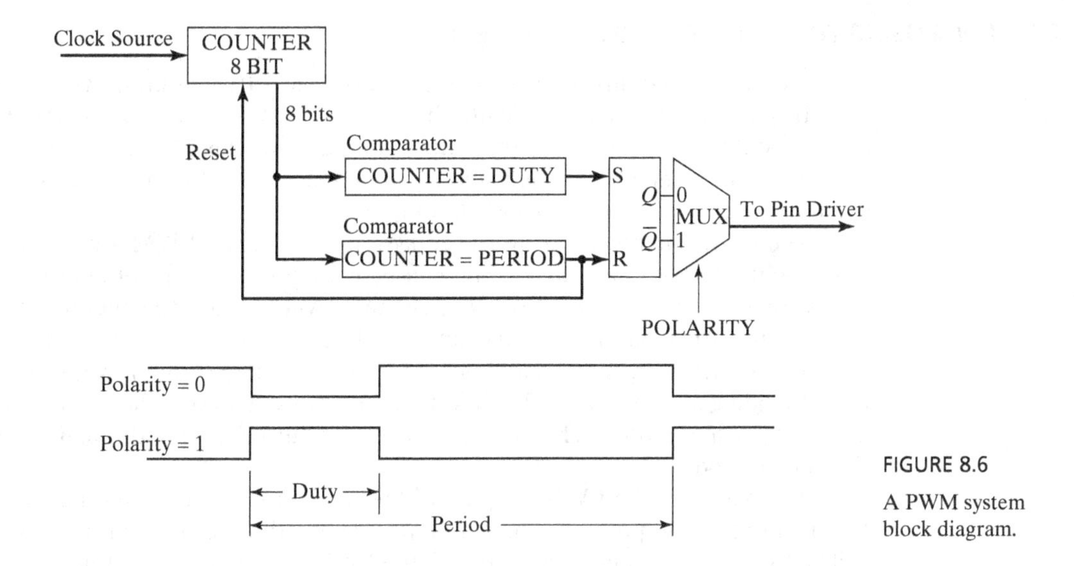

FIGURE 8.6

A PWM system block diagram.

include hardware dedicated to producing PWM output(s). To understand how to program these PWM subsystems requires that you have an understanding of how the hardware features are being used to generate the PWM output. In this section, we will examine how these systems typically work, and in a later section, we'll examine how to produce PWM using only an output compare system.

The essential aspects of producing a PWM output are to set the period of the output waveform and set either the high or the low time portion of that period. In this way, we fully determine the PWM signal characteristics by setting the period and the duty cycle. Figure 8.6 shows a block diagram for a typical PWM system drawn from the Freescale MC9S12 series.

In this system, COUNTER is the counter used to time the operations for generating the PWM output. This is typically a different counter/timer register than that used for input capture and output compare, which allows the PWM and IC/OC subsystems to function independently. COUNTER is driven by a clock that, together with the value written to the PERIOD register, will control the period of the PWM output. The clock will drive COUNTER to increasing values until the value in COUNTER matches the value in the PERIOD register. At this point, two things happen simultaneously: a flip-flop controlling the PWM output state is reset causing the Q output to go low and the value in COUNTER is reset to 0, which restarts the cycle.

In the system shown, the value in the DUTY register is also being compared against the value in the counter (COUNTER). When a match occurs between these two registers, the flip-flop controlling the state of the PWM output is set, causing the Q output to go high. This creates the second of the two events necessary to create a PWM output, and determines the duty cycle. In typical operation, COUNTER will start at 0 with the flip-flop output low. This is called the "reset state," which is the state of the hardware when it is first used after power-up. The next event happens when a match occurs between COUNTER and DUTY. At this point, the flip-flop output is set high. COUNTER continues counting until it matches the value in PERIOD. At this point, the flip-flop output is set low and COUNTER is reset to 0 and another cycle begins.

The actual PWM output in the example shown comes from a multiplexer connected to the Q and $\overline{Q}$ outputs from the flip-flop. The state of the POLARITY bit determines which of the flip-flop outputs is chosen. If Q is chosen, the value in DUTY represents the low time of the duty cycle output. If $\overline{Q}$ is chosen, the value in DUTY represents the high time of the duty cycle. The ability to control the polarity adds flexibility in controlling devices that may be active high or active low, though not all microcontrollers incorporate this particular feature.

8.5 PWM USING THE OUTPUT COMPARE SYSTEM

Hardware PWM systems offer the ability to produce a PWM output with no software overhead beyond the initial setup. While these kinds of systems are becoming more common, they are still not as common as the output compare systems. However, it is possible to use a pair of output compare channels and a minimal amount of periodic software overhead to maintain an arbitrary frequency PWM output.

The requirements for using an OC system to generate PWM are at least two channels of OC with at least one channel that can be configured to control multiple output pins in hardware. One of the output compare channels will be used to generate the period of the output waveform, and the second channel will be used to generate the active portion of the output. The "active portion" could be either the high time or the low time of the PWM signal, depending on how you define it. If there are more than two channels of output compare available, each additional channel offers the possibility for an additional PWM output with the same period.

The period of the PWM outputs will be set by the OC channel that is capable of controlling multiple output pins. The action programmed to occur when this compare happens will be to set the outputs associated with the PWM channels to either high or low, depending on the polarity of the output being generated. The ongoing software overhead is associated with responding to the occurrence of this output compare. What needs to happen in response to the period output compare event is to prepare for the next PWM cycle by reprogramming the period output compare to happen one PWM period into the future, clear the flag associated with the period output compare, and then to program each of the other output compares to generate the proper transition at a point determined by the duty cycle desired. Figure 8.7 shows a diagram of the timeline from three output compares (A, B, and C) being used to generate two channels of PWM (1 and 2).

At time A, the period output compare happens, raising both output channels in this example to the high state. In response to the period output compare flag, the software clears the period output compare flag and programs the output compares that control the duty cycle channels 1 and 2. It does this by adding the high time for each of the channels to the time at point A (the output compare time) and writing those values to the output compare registers so that a high-to-low transition occurs on channel 1 at time B and on channel 2 at time C. Before completing, the routine reprograms the period output compare to happen at point D. These are the only software actions required to maintain the PWM output. The output compare actions for each of the channels are programmed to set the output pin low, so the transitions at points B and C happen in hardware without further software intervention.

This approach to generating PWM does have some limitations. Since the time for points B and C are programmed from within the response to the period OC at point A, there will be a minimum output duty cycle that can be generated. It is possible in software to account for this in the case of a 0% duty cycle, but the transition from 0% to the minimum nonzero PWM is governed by the time it takes to reprogram the OC system, not by the resolution of the OC timers. The higher the desired PWM frequency, the larger (in percent duty cycle terms) this period will be. In most applications, this step can be made small enough as to be acceptable.

If setting the period of the PWM to a specific frequency is not a requirement, it is possible to use a similar approach with the OC system to generate **zero-software-overhead**

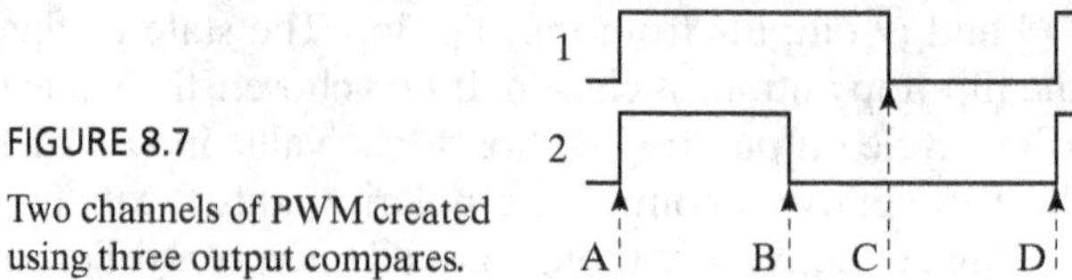

FIGURE 8.7

Two channels of PWM created using three output compares.

PWM. Consider the situation in which the time for point A in Figure 8.7 is programmed to be zero. There will be a match with the value in the free-running counter every time the counter rolls over. The period of the PWM will be set by the period of the free-running counter's overflow, removing the need to reprogram the period OC as each cycle occurs. Since the time at the rise of the output waveform is now fixed, the values in the OC registers that program for points B and C only need to be changed when the requested duty cycle is changed. There is no longer any need for software intervention to maintain the PWM output. One potential drawback to this approach is that PWM frequency may be relatively low. In an application with a 2 MHz clock rate and a 16-bit free-running counter, the overflow rate is once every 32 ms or about 31 Hz. Raising the clock rate to 10 or 20 MHz, which is within the capabilities of many current microcontrollers, brings the PWM frequency up to 165 or 330 Hz. While not ideal for all situations, these are workable rates for providing PWM control of brushed DC motors in some applications.

8.6 THE ANALOG-TO-DIGITAL CONVERTER SUBSYSTEM

In all but a few special purpose devices, the analog-to-digital (A/D) converter subsystem (Figure 8.8) on microcontrollers is implemented using a single successive approximation A/D converter (see Chapter 19). As of 2009, most recently introduced general purpose microcontrollers have A/D converters with 10–12 bits of resolution. A few devices, targeted primarily at electric power monitoring and instrumentation markets, use sigma-delta (see Chapter 19) converters with up to 16 bits of resolution, though with lower output data rates (samples per second). In order to provide for the ability to convert more than one analog input, the single converter is generally mated to an on-chip analog multiplexer that allows the single converter to be connected to one of a number of input pins. Configuring the A/D

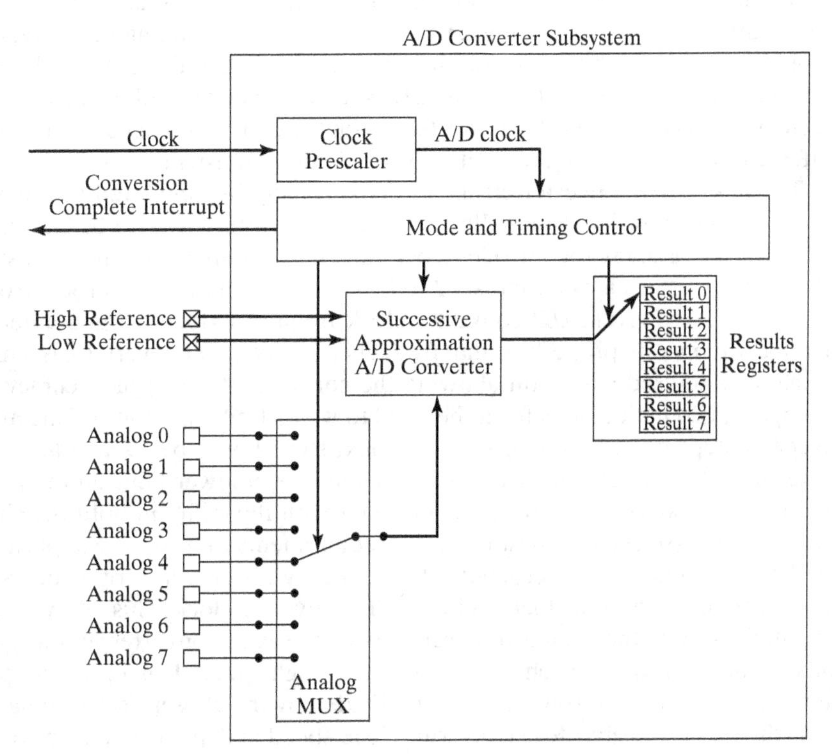

FIGURE 8.8

Typical A/D converter block diagram.

converter subsystem involves making choices about the operation of both the actual A/D converter as well as the associated multiplexer.

8.6.1 The Process for Converting an Analog Input to a Digital Value

In most cases, getting a microcontroller to convert an analog input voltage into a digital value will consist of several steps:

1. selecting the channel to be converted,
2. waiting for the voltage at the input to stabilize after the selection of a new channel,
3. starting the A/D converter,
4. allowing sufficient time for the converter to finish the conversion. This is usually indicated by the setting of a flag in the A/D converter status register,
5. reading the digital result from the A/D converter's result register.

As we discuss in Section 8.6.3, some of these steps may be automated, but they must all occur in sequence to convert an analog input voltage to a digital value.

8.6.2 The A/D Converter Clock

As described in Section 19.5.4 of Chapter 19, the successive approximation conversion process involves multiple steps. The timing of these steps is determined by an A/D converter clock that is typically derived from the central clock for the microcontroller. The choice of the speed of the A/D converter clock is part of the configuration of the A/D converter, and it involves trade-offs between the speed of conversion, power consumed by the converter, and the resolution of the conversion. Since power is consumed at every transition of the clock, the faster the conversion clock, the greater the power consumption. For a successive approximation converter, the number of steps involved in making the conversion is a function of the number of bits of resolution in the result, thus the greater the resolution, the longer the conversion time at a given A/D converter clock speed. If the conversion of multiple channels is automated (Section 8.6.3), then the choice of A/D converter clock may also be affected by the timing of the multiplexer switching transients.

The basic A/D converter clock system is usually implemented as a digital frequency divider (chain of flip-flops) that allows the A/D converter clock to be chosen from a set of fractions of the base microcontroller clock rate. For example, it might be possible to choose an A/D converter clock rate that is 1/2, 1/4, 1/8, and so on of the microcontroller clock. The ability to choose a given A/D converter clock frequency does not guarantee that the A/D converter will operate properly at that frequency. Every A/D converter has a maximum clock rate that can be used while still allowing the converter to meet its accuracy specifications. This upper speed limit results from the need to wait a finite amount of time after each of the conversion steps before judging what the next step should be. Depending on the internal architecture of the A/D converter, there may also be a lower limit on the permissible A/D converter clock speed. To provide the designer with flexibility in both the choice of microcontroller clock (selected to best meet the requirements of a specific application) as well as the A/D converter clock, microcontrollers typically provide a fairly wide range of divider ratios between the system clock and the A/D converter clock. This allows the system to be configured to meet the timing requirements of the A/D converter and any other requirements of the application, such as low power or high speed. It is up to the programmer to insure that the microcontroller clock rate divided by the chosen factor remains at or below the maximum permissible A/D converter clock speed and, if specified, above the minimum.

8.6.3 Automating the A/D Conversion Process

Some of the more elaborate A/D converter subsystems provide for a level of automation of the conversion process on a single channel or multiple A/D channels. Automating the A/D conversions reduces the programming overhead involved in monitoring the analog inputs. In this section, we introduce several of the more common schemes for automating the process and discuss the implications for programming.

The first level of automation that we will discuss is the elimination of the need to manually initiate every conversion. Some A/D converter subsystems provide for a **continuous conversion** mode in which a new conversion will be started as soon as the last conversion has been completed. When used to convert a single channel, this greatly eases the programming task. Once the A/D conversion process has begun, the programmer needs only to read the A/D result register in order to retrieve the most recent conversion result.

Often implemented in conjunction with the continuous conversion mode is a method to automatically select the channel to be converted. **Multiple channel conversions** are typically begun by specifying a list of channels to be converted. When the conversion process is started, the first channel in the list is converted and its result stored into a dedicated result register for that channel. As soon as that first conversion completes, a conversion is begun on the next channel in the list and the process repeats. When all of the channels in the list have been processed, these automated converter subsystems usually provide for the option to either stop after one pass through the list or repeat the processing of the list. The latter option provides for continuous conversions on multiple channels that can result in significant simplification of the software necessary to handle multiple analog inputs.

8.7 INTERRUPTS

Most of the peripheral subsystems on microcontrollers have status bits that are set by the hardware when certain conditions are met. In this chapter, we have seen examples of these bits in places like the timer subsystem where status bits indicate (among other things) that the timer has overflowed, an input capture has occurred or an output compare has occurred. In our discussions of these subsystems, we pointed out that the programmer can write software to test the state of these bits to determine when these events have occurred. This process of manually testing bits is referred to as **polling**, and it is considered the simplest way to monitor the status of the subsystems. It is not, however, the only way to monitor the status. Most processors also include a hardware announcement mechanism that allows the program to be notified when an event has occurred without the need to constantly monitor the state of a bit. This announcement mechanism is generally referred to as an **interrupt**.

The term interrupt is very descriptive of what goes on when one of these events occurs. Just as you might be interrupted by someone tapping you on your shoulder while you are intent on a task, the processor can be interrupted in its execution of your program. When an interrupt occurs, the stream of instruction execution (your program) is interrupted (temporarily stopped) and instruction execution is diverted to a section of the code that you have designated to run when the interrupt occurs. When this new section of code completes its task, the flow of instruction execution returns to the point where it was interrupted and continues as if the interrupt had never happened. The piece of code to which execution is diverted is typically referred to as the **Interrupt Service Routine** or **ISR**. Depending on the processor, there may be only a single ISR or one for each possible subsystem that can generate an interrupt.

There are a number of issues that need to be attended to in order to be able to return to the point of interruption without affecting the program flow and the details of these operations vary from processor to processor. When programming ISRs from within a high-level language, the compiler, in conjunction with the processor hardware, takes care of these details so that the high-level language programmer only needs to focus on what must happen in response to the interrupt, not the process of getting into and returning from the ISR. The ISR is typically written as an ordinary procedure with a compiler-specific keyword applied to let the compiler know that the procedure is to be an ISR. The details of what needs to happen inside an ISR are mostly specific to the particular application. The one thing that commonly needs to be done in any interrupt response is to acknowledge to the hardware that the interrupt has been handled. This is often described as **clearing the source of the interrupt** and generally involves setting or clearing a bit in a status or control register. If the programmer forgets to clear the source of the interrupt in the ISR, the situation can be much like being interrupted by a child who is not satisfied with your response: you will be immediately interrupted again. And again. And again...

8.8 HOMEWORK PROBLEMS

8.1 Based on Figure 8.1, what value should you write into the DDR to make the pin an input?

8.2 Given a microcontroller that implements separate input and output registers (call them INPUT and OUTPUT), write C code to sample input bit 3 and transfer its value to output bit 7 while leaving all other bits undisturbed.

8.3 In terms of the registers introduced in this chapter, what architecture feature is necessary to be able to tell if an output pin is being loaded heavily enough that its output voltage is outside the input voltage range for that state?

8.4 A particular application requires both high-resolution timing and timing of long enough intervals that the total time cannot be counted using a single 16-bit register. Write pseudo-code using other features of the timer systems described in this chapter to implement a virtual 32-bit input capture timer. Write the pseudo-code using an events and services paradigm (a set of event checkers and associated services).

8.5 Write pseudo-code for an initialization routine and an input capture event checker and service pair that would calculate the speed of a motor assuming that the output of a 100 pulse per revolution motor was connected to the input capture pin of a timer system being clocked by a 100 kHz clock.

8.6 Given a PWM subsystem like that shown in Figure 8.6, what duty cycle would be generated if DUTY contained 128 and PERIOD contained 240 and POLARITY = 1?

8.7 Given a PWM subsystem like that shown in Figure 8.6, with DUTY = 128 and PERIOD = 129, what sequence of values would you expect to see in COUNTER?

8.8 Given a microcontroller with a clock rate of 24 MHz, an A/D converter clock prescaler with divide-by ratios of 2, 4, 6, 8, 10, 12, ..., 64, an A/D converter with a minimum clock speed of 500 kHz, and a maximum clock speed of 2 MHz, what range of prescale dividers result in A/D clock speeds that fall within the A/D converter limits?

8.9 How are the Freescale DDR registers and the Microchip TRIS registers alike?

8.10 Given a PWM subsystem like that shown in Figure 8.6 with a Clock Source of 24 MHz, what is the maximum PWM output frequency that can still achieve 1% resolution on the duty cycle?

FURTHER READING

Embedded C Programming and the Atmel AVR, Barnett, R.H., Cox, S., and O'Cull, L., 2nd ed., Thomson Delmar Learning, 2006.

Embedded Microcontrollers, Morton, T.D., Prentice Hall, 2001.

MSP430 Microcontroller Basics, Davies, J.H., Newnes, 2008.

PART 3 **Electronics**

CHAPTER 9

Basic Circuit Analysis and Passive Components

This chapter offers an introduction to the most important concepts in basic electronics and linear circuits that are necessary to analyze and design a broad range of electromechanical systems. Our goal in this chapter is to help students develop both the analytic skills and physical intuition necessary to be able to effectively use the most common electronic components and integrated circuits. We complement the analytic information with descriptions of the kinds of commercial products that are available and how to navigate among them to choose an appropriate set of components for a particular application. Topics covered in this chapter include:

1. basic rules of circuit analysis and methods,
2. the behaviors of the most commonly used passive components,
3. practical and physical aspects of resistors, capacitors, and measuring instruments.

We feel that it is important for the student to develop an intuitive feel for the behavior of circuits, as well as the ability to approach them analytically. To that end, in our descriptions, we will demonstrate techniques to quickly develop a sense of how a circuit might behave and use those techniques as a litmus test for our analytic solutions. To help bring some physicality to things and foster an intuitive sense for the behavior of electrical phenomena and components that are inconveniently invisible to us, we will be introducing a fluid flow analogy in conjunction with many of the electrical descriptions. For convenience, we'll assume that the working fluid is water and refer to this as the "hydraulic analogy" throughout the chapter.

After reading this chapter, you should be able to look at simple circuits of linear elements (voltage sources, resistors, and capacitors) and be able to figure out the basic functionality of stages of the entire circuit without any formal analysis, as well as be able to complete the analysis to develop analytic expressions quantifying circuit behavior. You should understand that there are many different kinds of resistors and capacitors, why they exist, and the important features to keep in mind when making selections and when using them.

9.1 VOLTAGE, CURRENT, AND POWER

Voltage and current are the two fundamental quantities that we use to describe the behavior of circuits and electrical components. They are intimately related and the relationship between them (the *V-I* characteristic) is often the defining aspect of a component's behavior. We will start by examining current flow, since all useful electrical circuits involve the flow of current.

While physicists correctly describe electrical effects in terms of the flow of electrons, we will follow the electrical engineering tradition (originating with Benjamin Franklin) of treating current as the flow of positively charged particles. The two descriptions are equivalent from an external standpoint. Much of the foundation of electrical engineering was developed before we understood that it was electrons (with a negative charge) and not the hypothetical positively charged particles that were actually moving. All of that development resulted in a functional description of current as the flow of these positively charged particles, and electrical engineering upholds that tradition.

Current flow, then, is described in terms of the flow of these (hypothetical) positively charged particles. We measure the quantity of charged particles using a unit of **charge** (Q) called the **coulomb** (C). A coulomb is the charge associated with a specific number (6.241506×10^{18}) of charged particles. **Current**, as a flow, is measured in terms of the number of charged particles that move past any given point in an increment of time (dQ/dt). This flow rate is measured in coulombs of charge per second and is given the unit of **amperes** (1 A = 1 C/s). The symbol I or i is most commonly used to denote the flow of current. In the kinds of mechatronic systems that we'll deal with in this book, we'll encounter currents in the range from femtoamps (fA, 10^{-15} A) through tens of amps.

In the first of the hydraulic analogies that we'll introduce, we note that current flow is analogous to the flow of water in a pipe. This is not a perfect analogy for a number of reasons, but it is nonetheless useful in that it helps us to develop intuition about the behavior of these invisible electrical flows. While it is technically feasible to draw strict analogies between many hydraulic and electrical systems and their governing equations, we will not be going that far. We present analytical descriptions of the electrical components, circuits and phenomena, and introduce the water analogies as a means to help you picture what is going on in the invisible world of voltages and currents.

Current flows in response to an externally applied electric field. We speak about the **potential** of the electric field in a way that is analogous to potential energy in a gravitational field. The electric field does work on the charged particles as it moves them from place to place in the circuit. We measure the strength of the electric field, which is given the name **voltage**, in terms of the amount of work done per unit charge (dW/dQ). The unit that we give to voltage is the **volt** (V or, in some contexts, E), and it represents the amount of work (**joules**) done in moving a coulomb of charge, 1 V = 1 J/C. In the context of this book, we will have occasion to deal with voltages in the range from microvolts (μV, 10^{-6}) to a few tens of volts. This definition of a volt is particularly useful in that we will often be interested in the rate at which work is being done, which is **power**. If the work is being done at a constant rate, we can express the power as:

$$\text{Power} = \frac{\text{Work}}{\text{Time}} = \frac{\text{Joules}}{\text{Second}} = \frac{\text{Joules}}{\text{Coulomb}} \times \frac{\text{Coulombs}}{\text{Second}} = VI \tag{9.1}$$

Power, in electrical terms, is the product of voltage and current. If you know the voltage applied to a device and the amount of current flowing into that device, the power consumed by the device is simply the product of the voltage and current.

In electrical circuits, current flows in response to differences in voltage. If there is no difference in voltage, there will be no flow of current. In a water analogy, water flows in a pipe in response to differences in pressure. A conductor is analogous to the pipe, voltage is analogous to pressure, and current flow is analogous to the flow of water.

An important aspect of voltage is that it is not an absolute quantity, it is relative. Voltage is always measured as a difference between two points in a circuit and the behavior of a device is determined by the difference(s) in voltage across or between the pins of the device. This does not mean that you will not hear someone speaking about "the voltage at the base of the transistor"—you will. The important thing in this case is that if only a single point is mentioned in regards to voltage, the implied reference point is ground. We'll learn more about ground in Section 9.2.

The differential nature of voltage measurements has a parallel in the hydraulic analogy. The most common type of pressure measurement, so-called **gauge pressure**, is a pressure measurement relative to atmospheric pressure. The hydraulic analogy does break down somewhat with respect to pressure and voltage. Pressure, being the result of the collisions between particles, can have an absolute zero reference point in which there are no collisions (a vacuum). No such absolute zero reference exists for voltage.

In electrical circuits, voltage is produced by **voltage sources**. Examples of voltage sources include the **electro-chemical cell** or **battery**, **power supplies**, **signal generators**, as well as some kinds of **sensors**. An ideal voltage source would deliver a constant voltage independent of the amount of current drawn from the source. In the water analogy, pressure is produced by a pump or, in keeping with the potential term sometimes used to describe voltage, as a result of the potential energy present in a column of water that results in a pressure at the base of the column of water. Voltage sources come in two basic varieties. The first, referred to as a **direct current** (**DC**) source, produces a constant voltage of a single polarity. The second, called an **alternating current** (**AC**) source, produces an output voltage that varies with time. An AC voltage source is sometimes described (inaccurately) as having a changing polarity. Don't be misled. In an AC source, the voltage may vary in time but always remains in a constant polarity relative to ground *or* its polarity may also change with time. Either would be considered an AC source.

The dual of the voltage source is a **current source**. An ideal current source would generate whatever voltage was necessary to produce a constant flow of current. If you think about what would be required to do this in reality, it would truly be an amazing device. It would essentially require the device to instantaneously provide voltages anywhere in the range of $-\infty$ to $+\infty$. Clearly, you could never actually construct such a device. There are no real current sources, but they remain a useful concept in circuit analysis. It is possible to construct circuits which behave like a current source over a limited range of conditions.

For most circuits, current flows will be in the range of milliamps (mA, 10^{-3}) or microamps (μA, 10^{-6}). The most common exceptions to these ranges are those parts of a circuit associated with actuators, which often require currents of hundreds of milliamps to several amps to flow, and circuits containing certain kinds of integrated circuits, where very tiny currents (as low as a few femtoamps) flow.

Figure 9.1 shows the schematic symbols used for some example voltage and current sources.

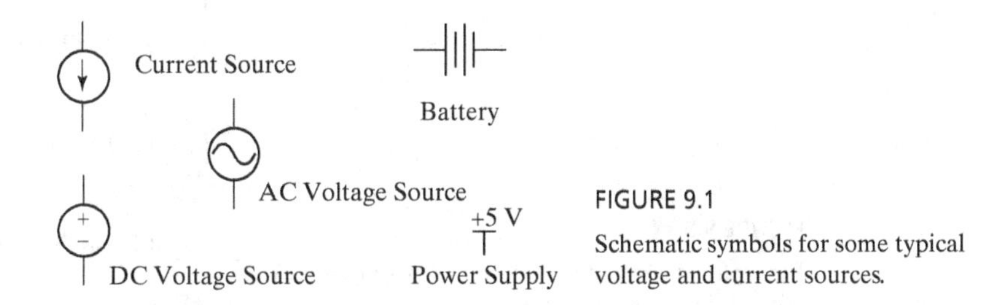

FIGURE 9.1

Schematic symbols for some typical voltage and current sources.

9.2 CIRCUITS AND GROUND

For current to flow, there must be what is referred to as a **complete circuit**. A complete circuit is one in which there is a path from one terminal of the source to the other terminal of that same source (Figure 9.2).

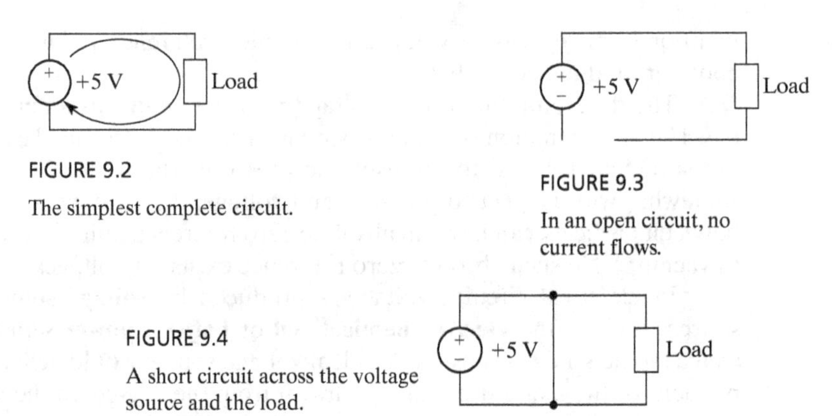

FIGURE 9.2
The simplest complete circuit.

FIGURE 9.3
In an open circuit, no current flows.

FIGURE 9.4
A short circuit across the voltage source and the load.

If the path from one terminal of the source to the other is broken or not continuous (Figure 9.3), this is referred to as an **open circuit** and no current flows.

If a path between two terminals of the same device exists that does not go through any other devices (Figure 9.4), it is referred to as a **short circuit**.

Shorting (the process of creating a short circuit) across passive devices results in at most a transient effect. Shorting across voltage sources will cause very high current to flow and should be avoided. It often results in damage to, or the destruction of, the voltage source.

In Section 9.1, we learned that when a voltage is specified by a single point, there was an implied reference point at ground. So, what and where is "ground"? In a circuit that is connected to the building power system, ground can be physically traced back to a stake or conductive pipe that is literally buried in the earth. In the case of a device powered by batteries or some power source not directly connected to building power, ground is a somewhat arbitrary point, often connected to the negative terminal of a battery or other power source. In either case, ground is a common reference point throughout a circuit and represents the point that will be assigned a value of 0 V.

There are a number of different symbols (Figure 9.5) that are used to indicate the different types of grounds that you will encounter in circuits and schematics.

It is unfortunate, but there is no real consistency in the use of the different ground symbols. The most commonly used symbol is the one labeled "Earth Ground" in Figure 9.5. This is despite the fact that in many (perhaps even most) cases, the point is not actually connected to earth.

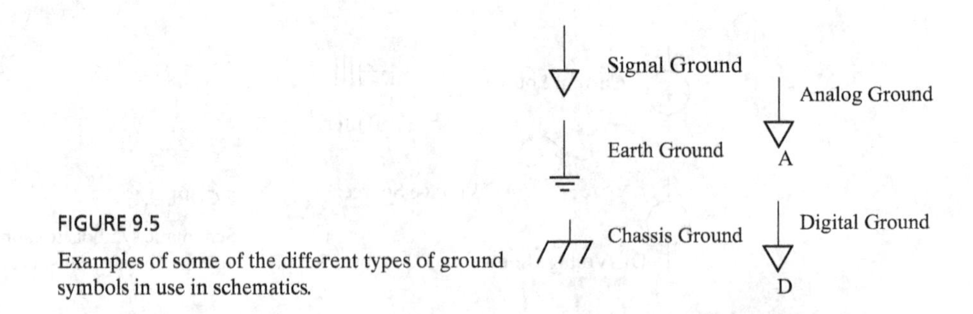

FIGURE 9.5
Examples of some of the different types of ground symbols in use in schematics.

In many cases, a ground symbol is used in a schematic to stand in for the return path for current to flow back to the negative terminal of the voltage source to complete the circuit. Thus, the simple, complete circuit might also be drawn as shown in Figure 9.6a or 9.6b. This convention allows us to considerably simplify most circuit schematics, as there will usually be

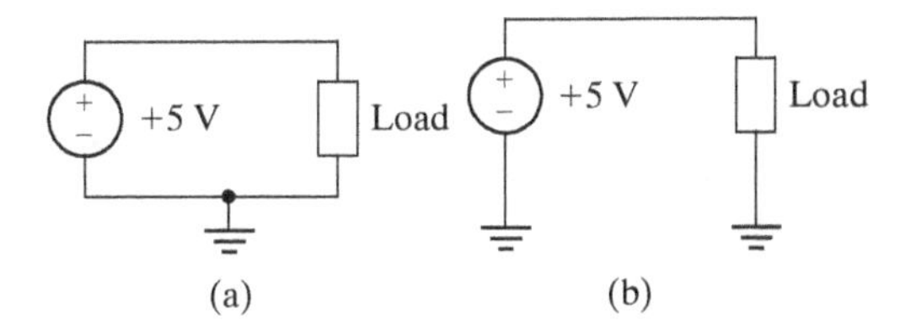

FIGURE 9.6

Alternative indications of ground in a simple circuit.

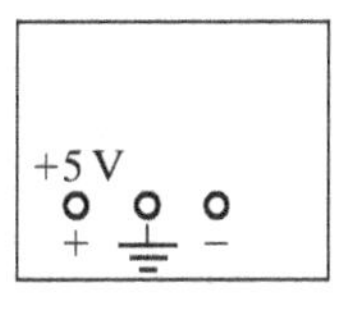

FIGURE 9.7

Power supply outputs from a supply with floating outputs.

a large number of connections to ground, and using the approach shown in Figure 9.6b allows us to avoid drawing every line back to the power supply.

Something to be aware of is that many laboratory or bench-top power supplies, despite being connected to the building power, produce outputs that are referred to as **floating**. A **floating power supply** is one in which there is no direct connection back to the building ground. These types of supplies often have three output terminals (or 2N+1 for N output voltages) similar to that shown in Figure 9.7.

In the power supply shown in Figure 9.7, the 5 V output is produced between the terminals labeled + and −. If you were powering a circuit with this power supply you would connect the terminal labeled − to the ground node of your circuit. You could also connect a jumper between the power supply's − terminal and its earth ground terminal (the one in the center in Figure 9.7) in order to actually reference your circuit to earth ground, but this is not strictly necessary for your circuit to function.

9.3 LAYING DOWN THE LAWS

The behavior of most electrical components is described in terms of their responses to voltages and currents. Before we get into analyzing circuits though, we need to lay down a few fundamental "laws" that govern the flow of electrical current, specifically, Kirchhoff's two laws governing voltages and current flow:

Kirchhoff's Current Law (also known as **Kirchhoff's First Law**):

> The sum of all the current entering a node of a circuit is the same as the sum of all the current leaving that same node.

If we agree on the convention that current entering a node has a positive sign and current leaving the node has a negative sign (or vice versa), then we can state Kirchhoff's Current Law mathematically as

$$\sum_{\mathrm{k}=1}^{n} I_{\mathrm{k}} = 0 \tag{9.2}$$

Kirchhoff's Voltage Law (also known as **Kirchhoff's Second Law**):

> The sum of the voltage differences around any closed loop in a circuit is zero.

This can be stated mathematically as

$$\sum_{\mathrm{k}=1}^{n} V_{\mathrm{k}} = 0 \tag{9.3}$$

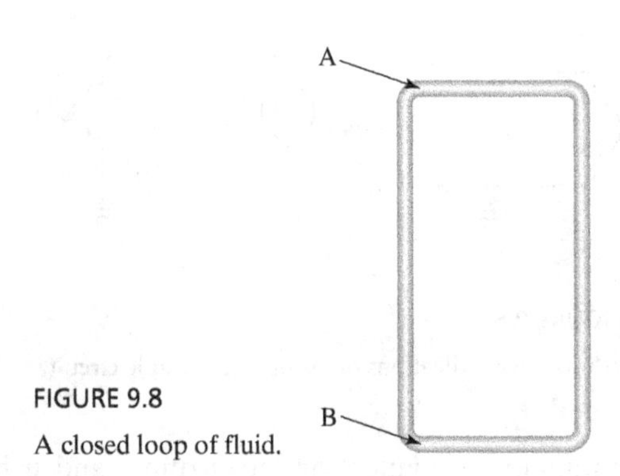

FIGURE 9.8

A closed loop of fluid.

Kirchhoff's Current Law (KCL) is really nothing more than the conservation principle applied to charge. Since the flow of charge is at its root a flow of electrons and we are not creating or destroying electrons, it stands to reason that the sum of all currents entering and leaving a node has to be zero. If it were not zero, we would either be destroying electrons or they would be accumulating on a circuit node. For our water analogy, KCL tells us that all the water entering a pipe must leave that pipe and no water will be created or destroyed inside the pipe.

Kirchhoff's Voltage Law (KVL) is probably most easily understood by thinking about a water analogy to a closed electrical circuit, as illustrated in Figure 9.8.

If we measure the pressure at the very top of either the left or right vertical columns (labeled "A" in Figure 9.8), we would find zero gauge pressure, since there is no mass of water above that point. As we move down a vertical column, the pressure would increase as the height of the water column above the measurement point increases. Eventually, we would get to the very bottom of the vertical pipe (labeled "B") where the pressure would be the greatest. If we continued to follow the pipe, we would start to rise up through the other vertical column and the pressure would start to fall. By the time that we got to the top again, the pressure would once again be 0. We could repeat this experiment starting at any point in the hydraulic circuit. After traveling around the loop and returning to the place that we started, the pressure will have the same value we initially measured.

9.4 RESISTANCE

When water flows through a pipe, the flow encounters resistance due to phenomena such as a restriction (a reduction of the cross-sectional area of the pipe) and the viscous drag of the moving fluid against the fixed wall of the pipe. In order to get water to flow through the resistance of the pipe or a restriction, we need to apply pressure. Resistance to flow results in the pressure in a pipe falling as the flow moves, for example, further away from a pump. Though the reasons are very different, electrical current also encounters resistance as it flows through devices in a circuit. This electrical resistance results in a drop in the voltage as the flow moves through the resistance.

The relationship between voltage and current in a resistive element is known as **Ohm's law** and is described by the equation:

$$V = I \times R \tag{9.4}$$

where V is a voltage in volts, I is a current in amps, and R is a resistance measured in a unit known as **ohms** (Ω). Voltage and resistance are often somewhat circularly defined by using Ohm's law and the definition of an ampere: 1 volt is the potential necessary to force 1 ampere

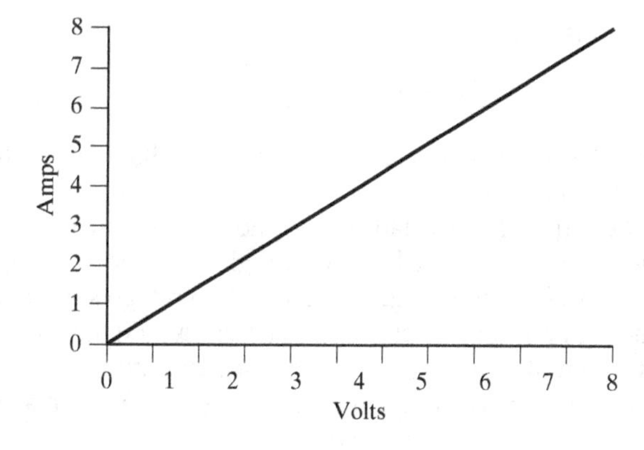

FIGURE 9.9
The *V-I* characteristic for a 1 Ω resistor.

of current to flow through a resistance of 1 ohm. While there is also a more fundamental, physical definition for the ohm[1], from a practical standpoint, simply accepting $V = IR$ in volts, amps, and ohms as a given law will prove more useful.

Graphically, Ohm's law describes the *V-I* characteristic shown in Figure 9.9. For a **resistor**, the value of the slope of the *V-I* characteristic is called the **resistance**. It is very common, even for devices that have a nonlinear *V-I* relationship, to describe their behavior in terms of the *V-I* characteristic. In many of these cases, we will speak of the instantaneous slope of the *V-I* curve as the **effective resistance** of the device. Figure 9.9 also makes it clear that the relationship between V and I for the resistor is linear, making it a linear circuit element.

9.4.1 Resistors in Series and Parallel

We can use these three laws (KVL, KCL, and Ohm's law) to describe the voltages and currents in any circuit consisting of an arbitrary combination of the three elements that we have introduced thus far: voltage sources, current sources, and resistances. For example, consider the circuit shown in Figure 9.10.

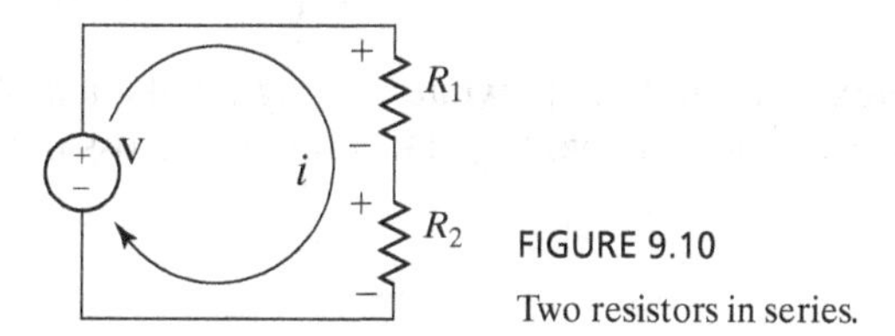

FIGURE 9.10
Two resistors in series.

This figure shows a circuit in which two resistors are in **series**. By definition, when two elements are in series, the same current flows in both elements. We can use KVL to write an equation relating voltage, current, and resistance in this circuit by simply traversing the circuit and summing the voltage drops as we go:

$$V - iR_1 - iR_2 = 0 \tag{9.5}$$

[1]The resistance offered to an unvarying electric current by a column of mercury at the temperature of melting ice 14.4521 grams in mass, of a constant cross-sectional area, and of the length of 106.3 centimeters (International Electrical Congress, Chicago, 1893).

We can rearrange this to give us:

$$V = i(R_1 + R_2) \tag{9.6}$$

from which we can see that the resistance that results from having resistors in series is simply the sum of their individual values.

Note that to write Eq. (9.5), we had to assume a direction for the current flow. The assumption of direction is completely arbitrary, though it will be more intuitive if you chose the direction to imply a rise in potential going across the voltage source. If you had assumed the other direction, the sign on the expression for current would have been negative, indicating that the current was actually flowing in the opposite direction.

Figure 9.11 shows two resistors in a circuit configuration known as **parallel.**

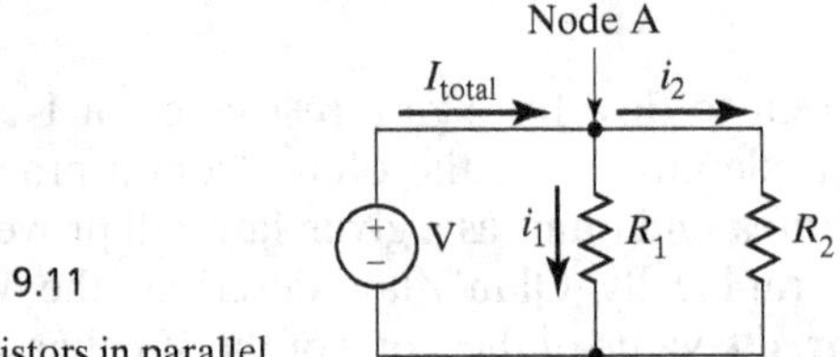

FIGURE 9.11

Two resistors in parallel.

Two circuit elements are considered to be in parallel when the same potential appears across both elements.

Since we are trying to understand how these two resistances combine, what we are looking for is an expression in the form of Ohm's law that will describe the relationships between voltage, current, and the combination of the two resistances:

$$V = I_{total}R_{parallel} \tag{9.7}$$

We can start the analysis of how these resistances combine by applying KCL at Node A to write Eq. (9.8):

$$I_{total} = i_1 + i_2 \tag{9.8}$$

Next, we can look at what is going on at each of the resistors separately, making use of the fact that the same voltage (V) appears across both resistors:

$$i_1 = \frac{V}{R_1} \tag{9.9}$$

$$i_2 = \frac{V}{R_2} \tag{9.10}$$

We can then combine Eqs. (9.8), (9.9), and (9.10) and substitute into Eq. (9.7) to get:

$$V = \left(\frac{V}{R_1} + \frac{V}{R_2}\right)R_{parallel} \tag{9.11}$$

Rearranging Eq. (9.11) gives an expression for $R_{parallel}$:

$$R_{parallel} = \frac{V}{\left(\frac{V}{R_1} + \frac{V}{R_2}\right)} \tag{9.12}$$

which we can simplify to:

$$R_{\text{parallel}} = \frac{R_1 R_2}{R_1 + R_2} \tag{9.13}$$

We can generalize this for cases with any number of resistors in parallel as:

$$\frac{1}{R_{\text{parallel}}} = \frac{1}{R_1} + \frac{1}{R_2} + \cdots \frac{1}{R_n} \tag{9.14}$$

9.4.2 The Voltage Divider

The circuit shown in Figure 9.12, though very simple, actually appears so often in circuits, that it has earned a name of its own: it is a **voltage divider**.

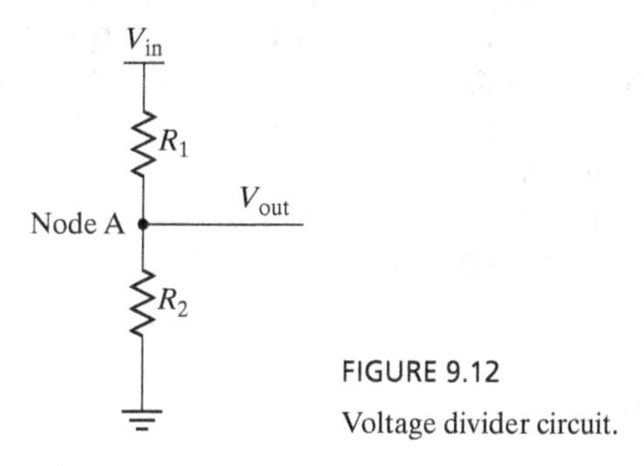

FIGURE 9.12
Voltage divider circuit.

For this analysis, we will assume that there is no current flowing to V_{out}. According to KCL, all the current entering the node labeled A must also be leaving. Therefore, the current through R_1, moving from V_{in} to Node A must be equal to the current through R_2, moving from Node A to ground. Using Ohm's law, we can write down a pair of equations for the voltages across R_1 and R_2. In this pair of equations, we have already assumed that the current passing from R_1 is the same as the current through R_2, labeled simply as I. Also note that the lower terminal of R_2 is connected to ground, which is our reference point and therefore at zero potential.

$$V_{in} - V_{out} = IR_1 \tag{9.15}$$

$$V_{out} - 0 = IR_2 \tag{9.16}$$

In writing these equations, we've assumed that the current is flowing from V_{in} to ground. This assumption may not be accurate—if V_{in} is negative, it would flow in the other direction. However, the main thing that is necessary in the analysis is that we make a *consistent* assumption throughout the analysis. When we're done, if in fact the assumed current direction was wrong, the resulting equations for the current would all end up with negative signs. Consistency in your assumptions about the direction of current flow is critical to your success when setting up the equations throughout the analysis.

This pair of two equations with two unknowns (I and V_{out}) can be solved using a number of approaches. We can add these equations together to get

$$V_{in} = I(R_1 + R_2) \tag{9.17}$$

which we can rearrange into an expression for I:

$$I = \frac{V_{in}}{(R_1 + R_2)} \tag{9.18}$$

We may substitute this value for I into Eq. (9.16) to get

$$V_{out} = V_{in}\frac{R_2}{(R_1 + R_2)} \tag{9.19}$$

We will refer to this result as the **voltage divider equation** and return to it many times throughout this chapter and in other chapters as well. Before we go any further though, it is worthwhile to look at the structure of this equation to verify that the result makes sense (this is always a good practice when doing analysis!). We see that the output voltage is proportional to the input voltage, with a constant scaling factor, which we should expect for any linear circuit. This is all reassuring. Let's also test this relationship for extreme cases. For example, if R_2 were "infinite" in value, there would be no current flow and therefore, no voltage drop across R_1, so we'd expect that V_{out} would be equal to V_{in}. Eq. (9.19) gives the correct result. Finally, if R_2 were a short (i.e., $R_2 = 0\ \Omega$), then V_{out} would be connected directly to ground and we would expect $V_{out} = 0$ V. Eq. (9.19) also gives us this result. This may seem tedious, but it is a good habit to get into at the end of calculations as a method for verifying that you haven't made any algebra mistakes. Mistakes happen, of course, and it's good practice to be vigilant.

Now that we can combine resistances in parallel and series, we can move on to using these results to simplify circuit analysis.

9.5 THEVENIN EQUIVALENTS

For complicated networks, it is often useful to be able to simplify parts of the circuit to facilitate analysis. One powerful simplification technique is the concept of the **Thevenin equivalent circuit**. **Thevenin's theorem** states that any two-terminal network of resistors (actually any linear elements) and voltage and/or current sources is equivalent to a single resistor in series with a single voltage source.

The voltage source (called the **Thevenin equivalent voltage**, or V_{th}) is equal to the open-circuit voltage between the two nodes. The series resistance (called the **Thevenin equivalent resistance**, or R_{th}) can be found using a couple of different approaches. One method is to assert Ohm's law in the equivalent circuit and describe R_{th} as V_{th} divided by the current that would flow if Nodes A and B were shorted together. Another approach is to analyze the equivalent resistance of the network without the voltage or current sources. To carry out this analysis, the voltage sources in the circuit need to be removed and replaced by short circuits. Current sources should simply be removed, leaving the resulting open circuits. When this is done, you will be left with a collection of resistances that you can simplify using the rules for parallel and series combinations.

In Figure 9.13, the open-circuit voltage would be $V_A - V_B$:

$$\left(\frac{V_{in}}{R_1 + R_2} \times R_2\right) - 0 = V_{in}\frac{R_2}{R_1 + R_2} = V_{th} \tag{9.20}$$

We could find R_{th} by removing the V_{in} voltage source and then shorting between the nodes labeled V_{in} and ground. This would leave R_1 in parallel with R_2, so this combination would be the value of the Thevenin equivalent resistance.

$$R_{th} = \frac{R_1 R_2}{R_1 + R_2} \tag{9.21}$$

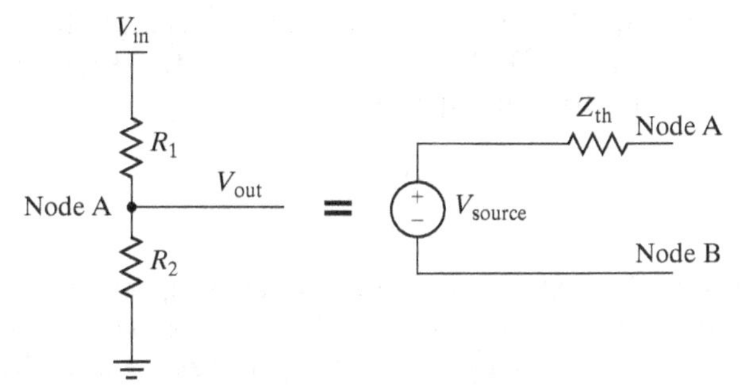

FIGURE 9.13

The Thevenin equivalent circuit for a voltage divider.

This process can be repeated to carve off parts of a circuit and replace them with simpler equivalent circuits that contain only a single voltage source (V_{th}) and resistor (R_{th}). In an example later in this chapter, we'll show an example using Thevenin equivalents to determine the response of a network of resistances and a capacitor.

9.6 CAPACITORS

Capacitors are the next important passive circuit element to understand. A **capacitor** is a two-terminal device which cannot pass an unvarying, steady current, but appears to pass oscillating currents. The water analogy for a capacitor is illustrated in Figure 9.14.

The flexible membrane prevents any real flow between the terminals. However, a sudden pressure appearing on one side (A) can cause the membrane to distort, displacing fluid on the other side of the membrane (B), and creating a transient current flow. Eventually, the stretching membrane comes into equilibrium with the applied pressure and stops distorting. At this point, the transient current flow will cease. When equilibrium is reached, one side of the membrane will hold back a store of higher pressure fluid. Just as in this hydraulic analogy, capacitors are electrical energy storage elements.

Given that the capacitor is an energy storage element, it shouldn't come as a big surprise that its *V-I* characteristic is an integral equation [Eq. (9.22)].

$$V(t) = \frac{1}{C}\int_{0}^{t} I(\tau)\mathrm{d}\tau = \frac{Q(t)}{C} \tag{9.22}$$

This equation introduces *C*, the measure of capacitance. The unit of capacitance is the **farad**, which represents 1 coulomb/volt. Typical capacitors that we will deal with range in value from a few tens of picofarads (pF, 10^{-12}) to a few hundreds of microfarads (μF, 10^{-6}). From Eq. (9.22), we can also see that there is a linear relationship (with slope $= 1/C$) between the voltage across the capacitor and the charge stored on the capacitor. This makes the capacitor another linear circuit element.

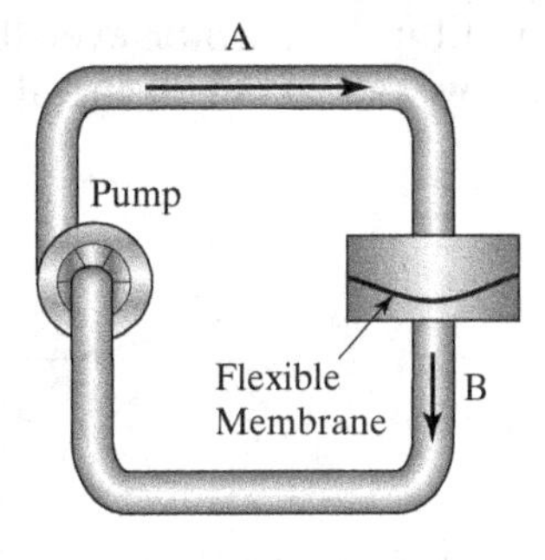

FIGURE 9.14

The water analogy for a capacitor.

As useful as the integral equation is in understanding the energy storage nature of the capacitor, if we differentiate Eq. (9.22), the results can provide insight into the dynamic behavior of the capacitor. This is shown in Eq. (9.23).

$$I(t) = C\frac{dV}{dt} \tag{9.23}$$

Eq. (9.23) shows us that the apparent current through a capacitor is a function of the size of the capacitor and the rate of change of the voltage across the capacitor. As a sanity check, let's compare this behavior to what we would expect based on the hydraulic model.

1. In the presence of a steady-state pressure (voltage), there should be no fluid (current) flow. Since at steady state dV/dt is 0, Eq. (9.23) would predict no current flow under those conditions. We see consistency between the hydraulic model and Eq. (9.23).
2. A given pressure (voltage) difference will cause a consistent displacement of the membrane, resulting in a consistent displacement of fluid (charge) on the other side of the membrane. The shorter the period of time over which that displacement takes place, the higher the flow rate (volume/time) that will be produced on the other side. Eq. (9.23) predicts that the faster the voltage is changed (higher dV/dt), the greater the current (charge/time). This too is consistent between the models.
3. In the hydraulic model, if the area of the diaphragm is increased, the volume of stored fluid will increase. If the displacement of the diaphragm happens over the same time interval, the larger diaphragm would displace a greater volume and produce a higher volumetric flow rate. The electrical analogy to a bigger diaphragm is a bigger capacitor, and Eq. (9.23) predicts that increasing the size of the capacitor (i.e., its capacitance) with the same dV/dt will result in a higher current flow. This is also consistent from the behavior we would expect in our hydraulic analogy.

Using Eq. (9.23), our general expression for power as VI and the definition of power as dE/dt, we can develop an expression for the amount of energy (E, joules) stored in a charged capacitor, Eq. (9.24):

$$E_C = \int P dt = \int VI dt = \int VC\frac{dV}{dt} = \int CV dV = \frac{1}{2}CV^2 \tag{9.24}$$

Inside the capacitor, the terminals are connected to a pair of plates that are separated by a thin insulator (dielectric), Figure 9.15. The value of the capacitance is determined by the area of these two plates, the spacing between them (d), and the characteristics of the insulating layer. The details of capacitor construction and the impact of the various parameters on the performance of real capacitors are covered in Section 9.14.

In the physical realm, it is important to be aware of the fact that *any* two conductors separated by an insulating gap constitute a capacitor, even if we didn't intend it. This fact will come up again in Chapter 21 when we talk about electrical noise.

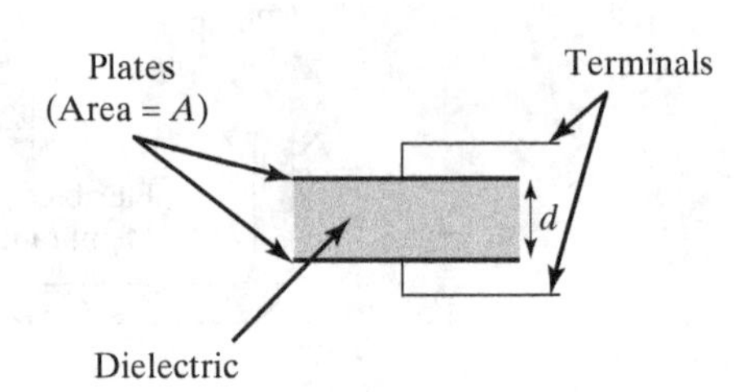

FIGURE 9.15
A simple parallel plate capacitor.

The basic idea behind a capacitor is that charge can build up on the plates of the capacitor, and oscillations in current and voltage on one side can be transmitted to the other side by borrowing from the stored charge. This only works for a limited time, as there can only be a finite charge stored, but if the time scales of the oscillations are well matched to the amount of charge stored, useful levels of voltage and current result.

We can think about current oscillations passing *through* the capacitor just as static currents pass through resistors. In the case of capacitors, the key thing that allows this to take place is that the current leaving the left electrode of the capacitor does not need to involve the same electrons as the current entering the right side of the capacitor. From the outside, it looks like current is "passing through" when there is current entering one side and leaving the other, and we'll analyze the behavior of these systems as if this is exactly what is happening.

Another way of looking at Eq. (9.22) is to note that whenever the voltage across the capacitor is changing, there must be some charge being added to or subtracted from the plates of the capacitor. Since changes in the amount of charge over time constitute a current, changes in voltage across the capacitor require the presence of a current. When there is an oscillating voltage applied to the plates of the capacitor, there is an oscillation in the current as well.

9.6.1 Capacitors in Series and Parallel

We can analyze circuits with multiple capacitors in much the same way that we did for resistors: using KCL and KVL. Figure 9.16 shows a circuit in which we have several **capacitors in parallel**.

Since the same voltage appears across all elements in parallel, we can write

$$I_{\text{total}} = I_1 + I_2 + I_3 = C_1\frac{dV}{dt} + C_2\frac{dV}{dt} + C_3\frac{dV}{dt} = (C_1 + C_2 + C_3)\frac{dV}{dt} \tag{9.25}$$

from which we see that the total capacitance of a number of capacitors in parallel is simply the sum of the individual capacitances.

If we take that group of capacitors and put them in series (Figure 9.17), we can use KVL and the fact that the current through devices in series must be the same in all devices to write

$$V_{\text{total}} = V_1 + V_2 + V_3 = \frac{1}{C_1}\int I dt + \frac{1}{C_2}\int I dt + \frac{1}{C_3}\int I dt = \left(\frac{1}{C_1} + \frac{1}{C_2} + \frac{1}{C_3}\right)\int I dt \tag{9.26}$$

From this we can see that **capacitors in series** add in much the same way the resistors in parallel add:

$$\frac{1}{C_{\text{total}}} + \frac{1}{C_1} + \frac{1}{C_2} + \frac{1}{C_3} + \ldots \frac{1}{C_n} \tag{9.27}$$

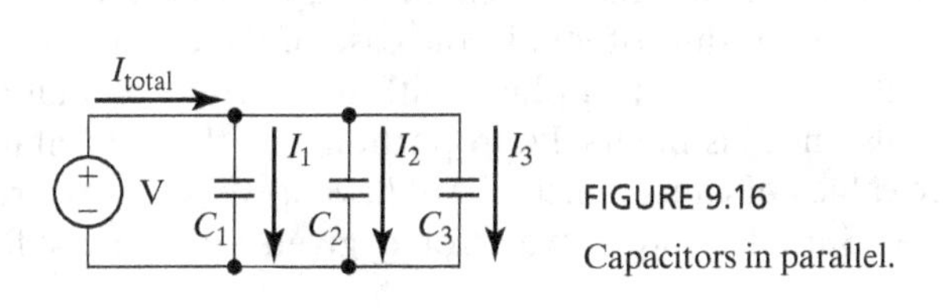

FIGURE 9.16
Capacitors in parallel.

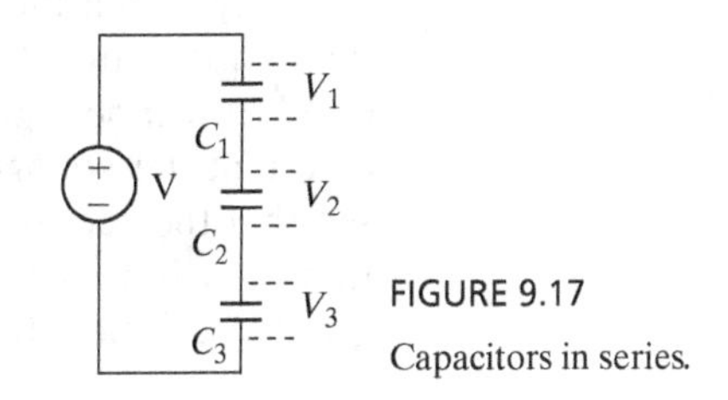

FIGURE 9.17
Capacitors in series.

9.6.2 Capacitors and Time-Varying Signals

As we go about describing the behavior of circuit elements in the presence of different kinds of time-varying signals, it will be useful to be able to apply the principle of **superposition** and the **Fourier theorem**. The superposition principle states that a circuit with multiple sources can be analyzed by treating each of the sources independently and then summing the results from each of the sources. The superposition principle applies to all circuits made up of linear circuit elements. The Fourier theorem (Jean Baptiste Joseph Fourier, 1768–1830) states that any periodic waveform of frequency f can be represented by a (possibly infinite) sum (i.e., the superposition) of sine and cosine terms of frequencies: f, $2f$, $3f$, ..., nf. Because of Fourier's theorem, we can focus our efforts at understanding circuit behavior by studying the responses of those circuits to sinusoids, confident that with that understanding and Fourier's theorem, we can describe any circuit's response to any periodic waveform.

If the current and voltage are both sinusoidal oscillations, they will have the form:

$$I = I_o e^{j\omega t} \tag{9.28}$$

$$V = V_o e^{j\omega t} \tag{9.29}$$

where these exponential functions of an imaginary argument ($j = \sqrt{-1}$, and "j" is chosen rather than "i" to avoid confusion with the symbol for current in these equations) are one convenient way to represent sinusoidal voltages with an angular frequency ω ($\omega = 2\pi f$). If we take these two functions and combine them with the relationship between current and voltage [Eq. (9.23)] for a capacitor, we get

$$I_o e^{j\omega t} = Cj\omega V_o e^{j\omega t} \tag{9.30}$$

$$I_o = Cj\omega V_o \tag{9.31}$$

Solving for V_o gives an expression for the amplitude of the voltage oscillation as a function of the amplitude of the current oscillation and the frequency and the capacitance value.

$$V_o = I_o \frac{1}{j\omega C} \tag{9.32}$$

In this resulting expression, there is a strong similarity to Ohm's law for resistors. In the original Ohm's law, $V = IR$, and in this relationship, we also have a linear relationship between the amplitudes of the voltage and current oscillations. In the relationship for a capacitor, the amplitude of the voltage oscillation is proportional to the amplitude of the current oscillation with a proportionality constant of $1/j\omega C$.

We can complete the analogy to Ohm's law by identifying the quantity, $1/j\omega C$, as the **equivalent resistance for a capacitor**, also called its **reactance**. This "resistance" has some unusual properties. First, it is frequency dependent—in this case, "resistance" increases as the frequency of the oscillation is reduced, and actually becomes infinite as the frequency goes to 0. This should seem intuitively correct. We know that the plates of the capacitor are separated by an insulator, so there can be no steady ($f = 0$ Hz, or DC) current flow. Another property is that the "equivalent resistance" is imaginary—meaning that there is a $\sqrt{-1}$ in the denominator. In the analysis of circuits with capacitors, we will find that these factors of $\sqrt{-1}$ can be regarded as a phase shift of 90°. In the case of the capacitor, it is indeed the case that the voltage oscillation is 90° out of phase with the current oscillation. Finally, we see that the "equivalent resistance" is inversely proportional to the capacitance, indicating that capacitors with larger values of capacitance C will have lower effective resistance for a given oscillation frequency. Stated another way: for a given voltage oscillation, a large capacitor will pass more current than a small capacitor. For comparison, with the same voltage oscillation, a large resistor would pass less current than a small resistor. In this sense, the "resistance" of capacitors is inversely proportional to the capacitance value.

In electrical engineering, the term **impedance** is used to denote a resistance with a complex component. In the case of a resistor, the impedance is all real (i.e., no phase difference between voltage and current) and for a capacitor, the impedance is all imaginary (i.e., a 90° phase difference). In common usage, the term impedance is used to denote an expression for resistance that may have either or both complex and real components. In a few pages, we'll show some examples of how this equivalent resistance or impedance can be used in circuit analysis, leading to a simple understanding of how capacitors behave in real circuits. The most important things to remember about capacitors are:

1. For DC voltages, no current flows through the capacitor.
2. For oscillating currents and voltages, capacitors behave a lot like resistors—there is a simple linear relationship between the amplitudes of the current and voltages, and the relationship is similar in many ways to Ohm's law.

9.7 INDUCTORS

The final passive circuit component that we will discuss in this chapter is the **inductor**. Inductors generally consist of coils of wire wound around metal or magnetically permeable cores (see Figure 9.18). The basic property of inductors is the tendency to resist changes in current flow. Like a capacitor, the inductor is an energy storage device, this time with the energy stored in a magnetic field. The unit used to measure inductance is the **henry** [1 H = 1 V/(1 A/s)].

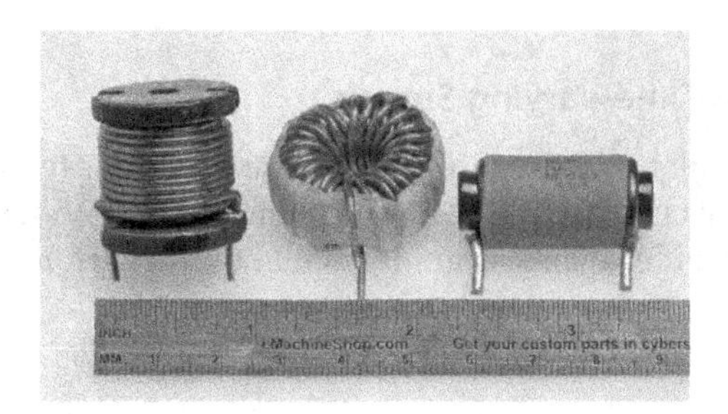

FIGURE 9.18
Photograph of several inductors.

The integral and differential forms of the equation that describe inductor behavior are

$$I(t) = \frac{1}{L}\int_0^t V(\tau)\mathrm{d}\tau \tag{9.33}$$

$$V = L\frac{\mathrm{d}I}{\mathrm{d}t} \tag{9.34}$$

From this, we can observe that a voltage must be applied to change the current in an inductor, or that changes in current will cause voltages to appear across the inductor. The hydraulic analogy to an inductor is shown in Figure 9.19.

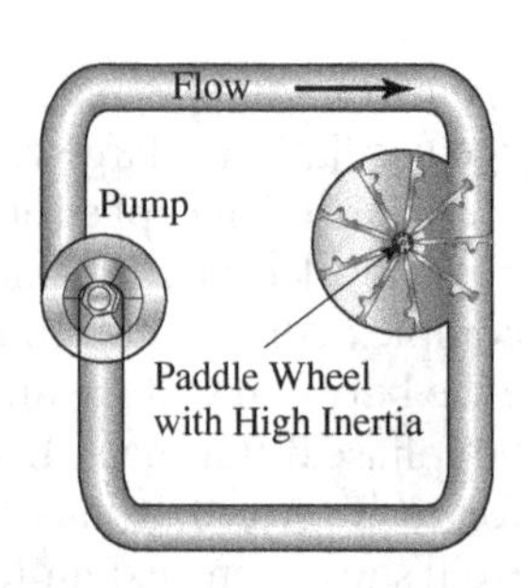

FIGURE 9.19
The hydraulic analogy for an inductor.

Here, the inertia of the paddle wheel initially limits the flow of water past the blades of the wheel, though with time the wheel will spin up and present no further resistance to the flow. If the pump were to stop pumping (equivalent to opening the circuit), the inertia of the wheel would keep it spinning, turning the paddle wheel into a pump, and creating a pressure differential across its inlet and outlet. Analogously, if we suddenly stop the flow of current in a circuit that includes an inductor, the inductor's collapsing magnetic field will generate a voltage difference across the terminals of the inductor.

The amount of energy stored in an inductor can also be derived using an approach similar to that used for the capacitor, using the generalized expression for power $P = VI$, the definition of power as $\mathrm{d}W/\mathrm{d}t$, and equating work with energy E (joules):

$$\mathrm{E_L} = \int P\mathrm{d}t = \int VI\mathrm{d}t = \int L\frac{\mathrm{d}I}{\mathrm{d}t}I\mathrm{d}t = \int LI\mathrm{d}I = \frac{1}{2}LI^2 \tag{9.35}$$

The amount of energy stored in the magnetic field of an inductor will be important when we deal with snubbing in Chapter 23.

9.7.1 Inductors and Time-Varying Signals

Using an approach similar to the analysis that we carried out for capacitors, we can analyze the effects of sinusoidal currents and voltages on the inductor to see the relationship between the amplitudes. We begin by again assuming oscillations in current and voltage are sinusoidal:

$$I = I_o e^{j\omega t} \tag{9.36}$$

$$V = V_o e^{j\omega t} \tag{9.37}$$

Substituting Eqs. (9.36) and (9.37) into Eq. (9.34), we get:

$$V_o e^{j\omega t} = j\omega L I_o e^{j\omega t} \tag{9.38}$$

which we can simplify to

$$V_o = I_o j\omega L \tag{9.39}$$

Once again, we can see that this looks like Ohm's law with an **equivalent resistance for an inductor** being given by $R_I = j\omega L$. In contrast to the case for capacitors, this "resistance" increases with inductance value and with frequency. There are also phase shifts associated with inductors as there were for capacitors, as evidenced by the j term in the equivalent resistance.

Real inductors have some very undesirable properties, such as hysteresis, large size and weight, high cost, and the tendency to generate magnetic fields that can cause electrical noise issues throughout a circuit. With a few exceptions, such as radio frequency (RF) and switching power supply circuits, modern circuit designs use very few inductors. Neither of these topics are emphasized in this text: we will not be covering RF design, and we explore inductors used in switching power supplies only enough to understand their theory of operation. However, we will encounter and need to deal with inductance in real circuits in a number of contexts. Every wire in every circuit exhibits some amount of inductance, and even the small amount that is present can play an important role in some instances. The mechatronic engineer will most likely run into this issue when dealing with the noise associated with switching high-current loads (Chapter 21). Similarly, the leads and construction of capacitors (Section 9.14.1) contribute to a parasitic inductance in capacitors, which affects their suitability for various applications (Section 9.14). Finally, the electro-magnetic actuators that make use of coils to generate forces, such as motors and solenoids (Chapters 22–26) exhibit the same electrical behavior as inductors. So, while we will not be putting any explicit inductors into our circuits, we will not be able to avoid encountering inductance, and a basic understanding of inductor behavior will help us anticipate how it will manifest itself in our circuits.

9.8 THE TIME AND FREQUENCY DOMAINS

Before we move on to analyzing circuits with combinations of resistors and capacitors, it will be useful to pause and describe the two most common ways of analyzing and representing the dynamic behavior of circuits and systems. These two approaches, the time domain and frequency domain, are equally useful and popular for electronic systems and mechanical systems as well.

Examining the behavior of a system in the **time domain** is probably the most intuitive because it most closely follows our everyday experiences. When looking at the time domain behavior, we are simply observing how the system behaves as a function of time. This is the view of the world that we get when we display a signal on an oscilloscope. Figure 9.20 shows examples of a number of common and useful signals in the time domain.

The **frequency domain** is really no more complex than the time domain, but it is probably not as intuitive to most people, and it's not the view of the world that we get with typical measurement instruments. The frequency domain view looks at how a system's

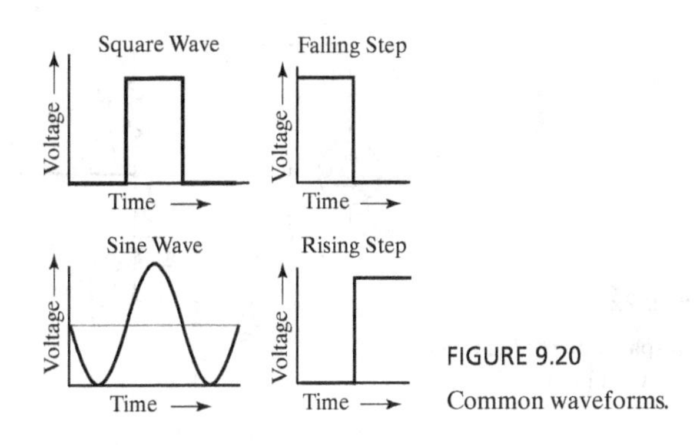

FIGURE 9.20
Common waveforms.

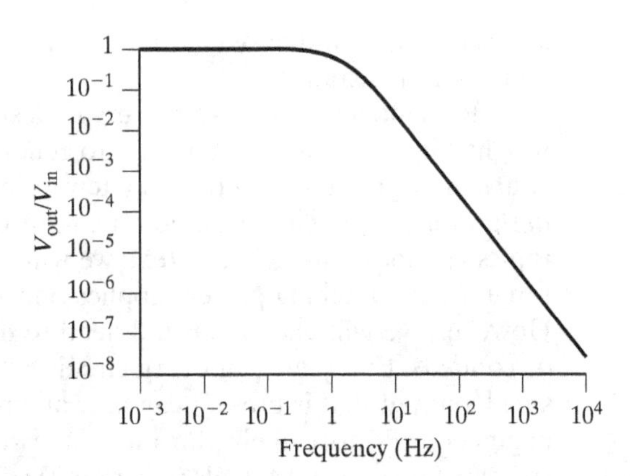

FIGURE 9.21
A Bode magnitude plot.

behavior varies as a function of frequency. There are multiple ways in which the frequency domain is represented but the one that we will encounter most often is known as a **Bode magnitude plot**, and it shows gain (or amplitude ratio) as a function of frequency, as shown in Figure 9.21.

We will use Bode magnitude plots (often shortened to simply **Bode plots**) to describe the behavior of circuits as a function of frequency. It is important to remember when looking at a Bode plot such as this that the response is being plotted as a function of *sine waves* of the frequencies shown on the x-axis. Signals made up of other waveforms (square, triangular, and so on) would need to be analyzed by summing the series of sine waves (the **Fourier series**) that make up these waveforms.

9.9 CIRCUIT ANALYSIS WITH MULTIPLE COMPONENT TYPES

As discussed above, Kirchhoff's laws allow us to find the current and voltage at any node in a circuit by writing down equations for the voltage and current across each component, and for the net currents into and out of the junctions, and then solving these sets of equations. In this section, we will apply those principles to studying the behavior of circuits composed of multiple component types, with a focus on **RC circuits** (i.e., circuits that contain resistors and capacitors).

9.9.1 Basic RC Circuit Configurations

When studying RC circuits, there are basically two configurations that have the most importance for us. These are the two series configurations shown in Figure 9.22.

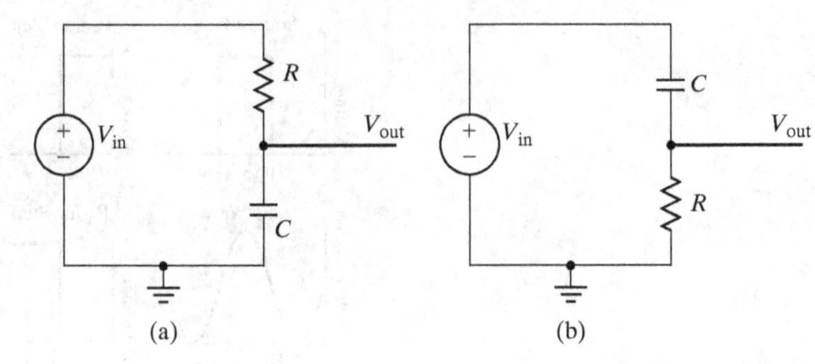

FIGURE 9.22
(a) Low-pass and (b) high-pass RC filter configurations.

For reasons that we will make clear in Section 9.9.5, configuration ***a*** is called a **low-pass filter** and configuration ***b*** is called a **high-pass filter**, in reference to their frequency domain behavior. The voltage source shown as V_{in} may be any voltage or signal source. The high-pass filter configuration is also sometimes referred to as **AC-coupled** because the current that creates a voltage across the resistor R has passed "through" the capacitor C.

9.9.2 Low-Pass RC Filter Behavior in the Time Domain

We'll begin our examination of RC circuits by studying the time domain behavior of the low-pass RC circuit shown in Figure 9.23, when the switch is moved from position A to B. This is equivalent to the rising step function from Figure 9.20.

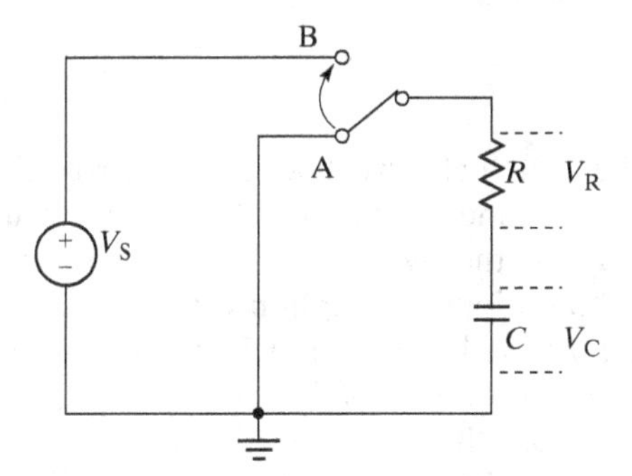

FIGURE 9.23

RC circuit beginning to charge the capacitor.

If the switch has been in position A for a sufficiently long time (to be defined shortly), then capacitor C will be fully discharged and V_C and V_R will both be 0 V. After the switch moves to position B, we can apply KVL to write:

$$V_S = RI + \frac{1}{C}\int I dt \tag{9.40}$$

By differentiating this equation and applying the initial condition that at the time of switch closure $V_C = 0$ V (the voltage across a capacitor cannot change instantaneously), we can develop an expression for I as a function of time [Eq. (9.41)].

$$I = \frac{V_S}{R}e^{\frac{-t}{RC}} \tag{9.41}$$

We can then use this expression for I to calculate expressions for the voltage across the resistor and capacitor as a function of time:

$$V_R = V_S e^{\frac{-t}{RC}} \tag{9.42}$$

$$V_C = V_S\left(1 - e^{\frac{-t}{RC}}\right) \tag{9.43}$$

The RC product appears in each of the expressions and is given a special name, the **time constant**, and symbol τ. Plotting V_R and V_C as a function of time (Figure 9.24) shows the behavior graphically for a circuit with a 5 V driving voltage (V_S) and a time constant of 0.1 second.

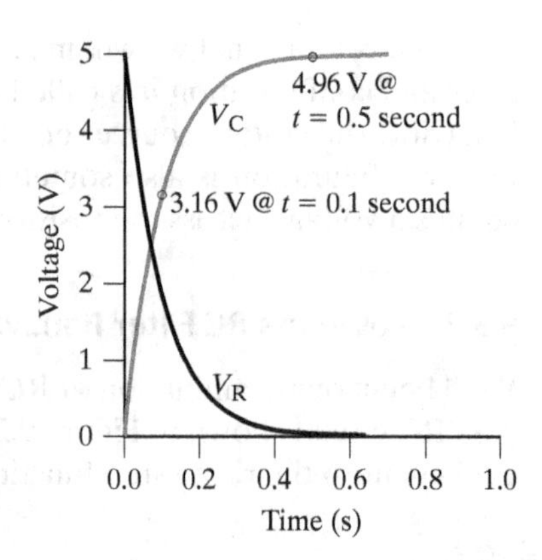

FIGURE 9.24
V_R and V_C from Figure 9.23 during charging transient.

At a time $t = \tau$ after the switch closure, the voltage across the capacitor has risen to 63.2% of the driving voltage. At a time $t = 5\tau$ after the switch closure, the voltage has risen to 99.24% of the driving voltage. If the switch had been closed for at least this long before the transition, then our initial conditions assumption would have been satisfied.

If the switch remains in position B for at least 5τ, the capacitor will be essentially fully charged. If the switch is then returned to position A, the charged capacitor will discharge through the resistance R. We can use the same sort of analysis that we used for the charging case, with the exception that in the charged case, the initial condition is that the voltage on the capacitor is 5 V. After completing that analysis, we would find that Eqs. (9.41) and (9.42) are the same for the charging and discharging case. In the case of the voltage across the capacitor during discharge, we find that the expression is

$$V_C = V_S e^{\frac{-t}{RC}} \tag{9.44}$$

This should make intuitive sense. We would expect the voltage on the capacitor to decay as it was discharged. Eq. (9.44) predicts that the voltage on the capacitor starts at V_S at $t = 0$ and decays (due to the negative exponent for e) as time increases. Figure 9.25 shows this graphically.

Square waves, which are just a regular succession of rising edges and falling edges, are such a common occurrence in digital systems that it is useful to examine how this circuit

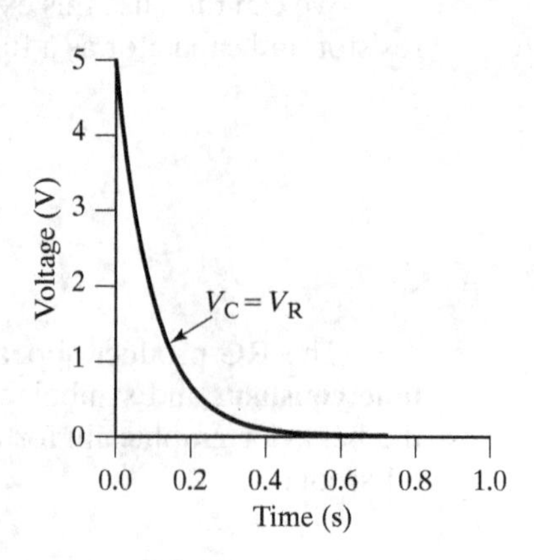

FIGURE 9.25
V_R and V_C during the discharge transient.

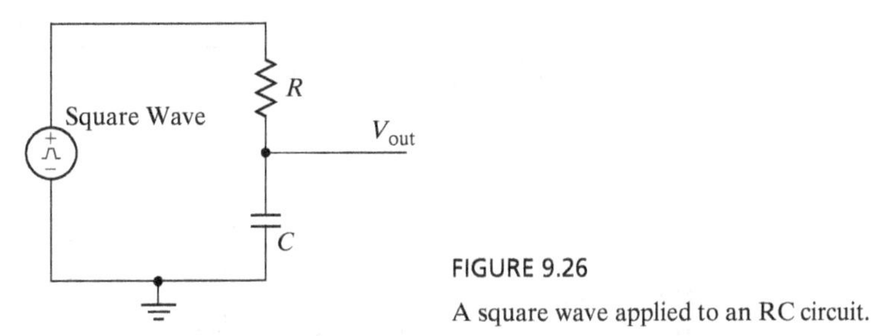

FIGURE 9.26

A square wave applied to an RC circuit.

(Figure 9.26) responds to a square wave input so that you can learn to recognize this behavior on an oscilloscope trace.

In Figure 9.26, we replace the circuit elements V_S and the switch with a square wave generator, labeled "Sq Wave." If the period of the square wave is long relative to the time constant of the RC (at least 10 times as long), then we can treat the rising and falling edges as distinct events and predict the responses using Eqs. (9.41)–(9.44). The resulting waveforms appear as Figure 9.27. The rounded edges of the waveform are characteristic of a low-pass filter response to a square wave.

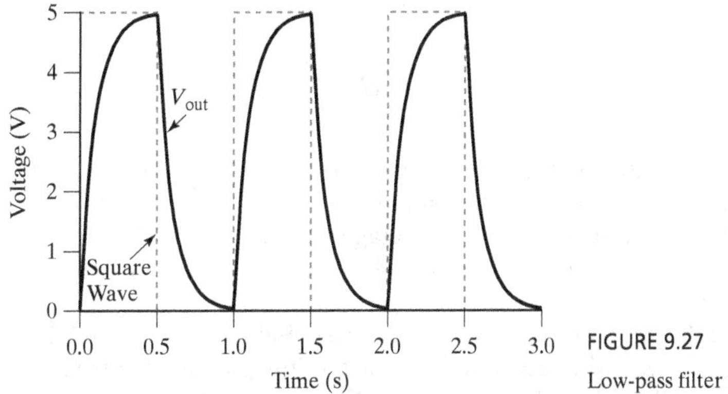

FIGURE 9.27

Low-pass filter response to a square wave.

If the period of the square wave is much shorter than the RC time constant, the output does not have time to rise to the full amplitude of the drive, nor fall all the way to zero (Figure 9.28) resulting in a very distorted waveform of a smaller amplitude than the drive.

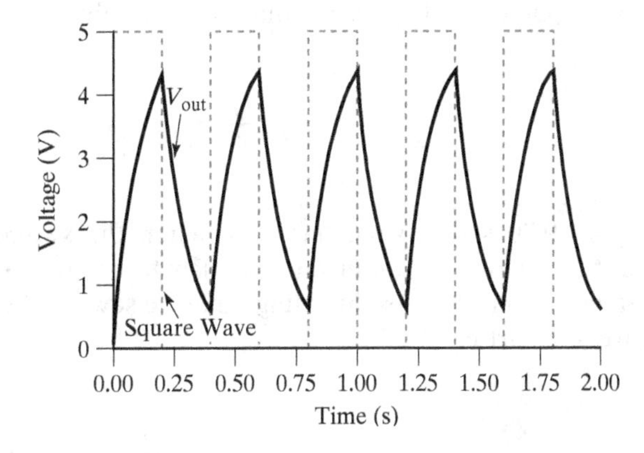

FIGURE 9.28

Low-pass filter response to higher frequency square wave.

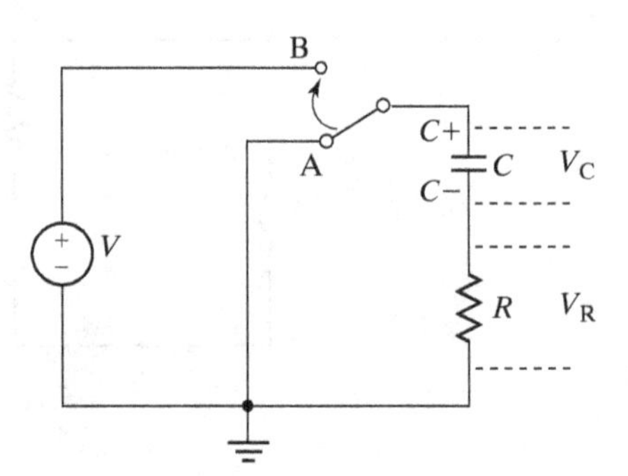

FIGURE 9.29
High-pass filter circuit for charge and discharge transients.

9.9.3 High-Pass RC Filter Behavior in the Time Domain

The circuit that we will use to examine the charge and discharge behavior of a **high-pass filter** is shown in Figure 9.29. We have simply exchanged the positions of R and C from Figure 9.23.

If the switch has been in position A for a sufficiently long time (once again, something in excess of 5τ), then the capacitor C will be fully discharged, and V_C and V_R will both be 0 V. After the switch moves to position B, we can again apply KVL to write:

$$V_S = RI + \frac{1}{C}\int I\mathrm{d}t \tag{9.45}$$

A comparison of Eq. (9.45) with Eq. (9.40) will show that they are identical. This should make sense. We have the same three circuit elements in series, so the same current behavior should exist and the same voltage differences should appear across the devices. As a result, the same expressions [Eqs. (9.42) and (9.43)] also describe the voltages across the resistor and capacitor in this case.

Once the capacitor is fully charged, we could move the switch from position B to position A, just as we did in Section 9.9.2 for the low-pass filter analysis. This case, however, is different than that of the low-pass filter. Notice that after the capacitor is fully charged, the voltage at the terminal of the capacitor labeled C+ will be at the source voltage V *above* the C− terminal of the capacitor. When the switch is moved to position A, the C+ terminal of the capacitor will be instantaneously placed at ground potential. Since the voltage across the capacitor cannot change instantaneously, the voltage at the C− terminal must remain V volts *below* the V+ terminal, causing it to be at a potential of −V. This yields a slightly different expression for V_C and V_R [Eq. (9.46)] than what we obtained in the low-pass filter analysis.

$$V_C = V_R = -V_S e^{\frac{-t}{RC}} \tag{9.46}$$

Graphically, Figure 9.30 shows what happens. When the switch closes, the voltage across both the resistor and capacitor are instantaneously forced to −V (in this case −5 V). What follows is the same exponential rise in voltage that we saw for the case of the low-pass filter, only this time we are starting at −V.

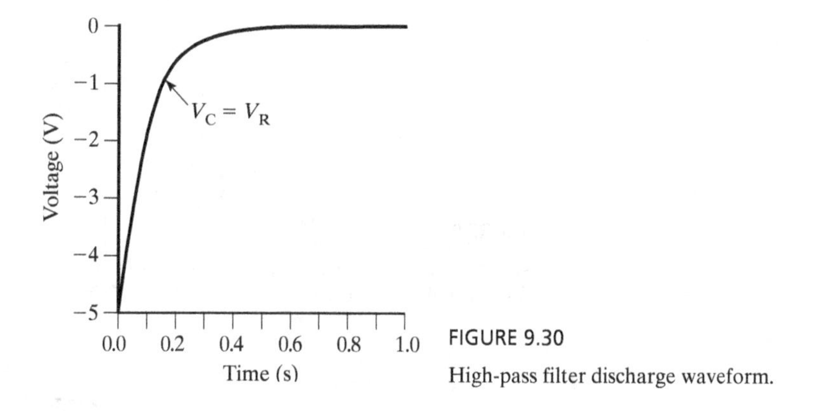

FIGURE 9.30

High-pass filter discharge waveform.

When we apply a square wave as the driving voltage to the high-pass filter (as we did above for the low-pass filter), we get the waveform shown in Figure 9.31. The positive-going spikes on rising edges and negative-going spikes on falling edges are the classic signs associated with a high-pass filter driven by a square wave (or an AC-coupled square wave).

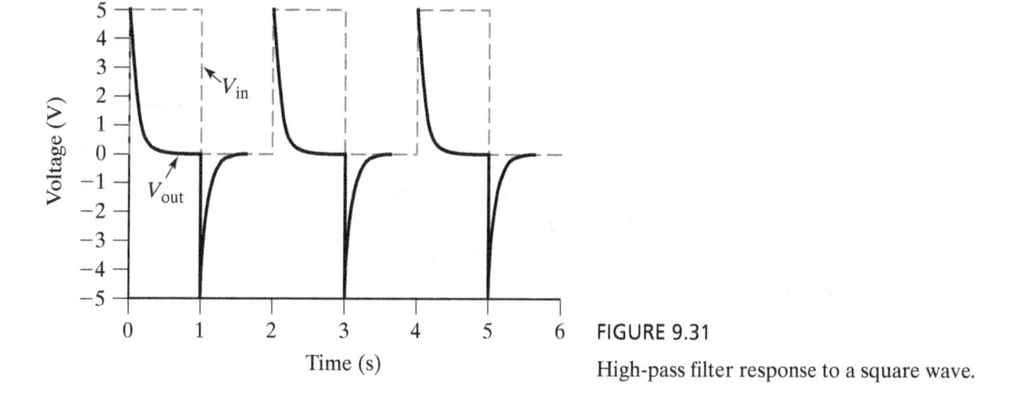

FIGURE 9.31

High-pass filter response to a square wave.

9.9.4 RL Circuit Behavior in the Time Domain

Figure 9.32 shows a circuit very similar to that of Figure 9.23, with an inductor replacing the capacitor. While, as we stated earlier, we will not be putting explicit inductors into our circuits, the situation shown in Figure 9.32 as the switch swings from A to B will occur in your circuits every time that you activate a solenoid or turn on a motor.

In this case, the equation describing the rise of current in the inductor is:

$$I = \frac{V_S}{R}\left(1 - e^{\frac{-t}{(L/R)}}\right) \tag{9.47}$$

This is shown graphically in Figure 9.33 for a circuit with $R = 10\,\Omega$, $L = 1$ H, and $V_S = 5$ V. In comparing Eq. (9.47) to Eq. (9.43) (describing voltage across the capacitor during the charge time in the RC low-pass filter), we can see a number of similarities. Perhaps what is most relevant is that the *current* rise in the inductor mimics the *voltage* rise on the

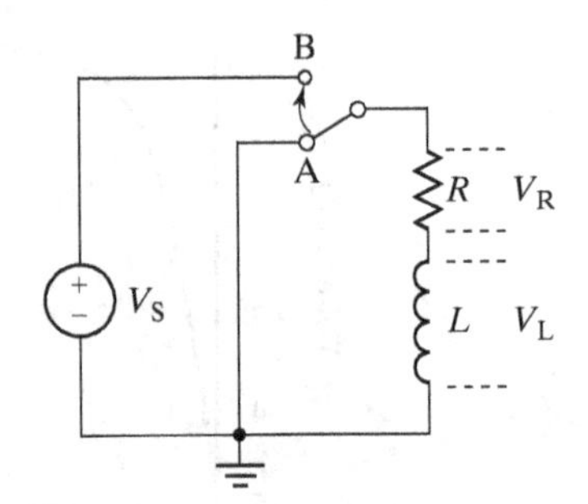

FIGURE 9.32

An RL circuit initiating current flow through the inductor.

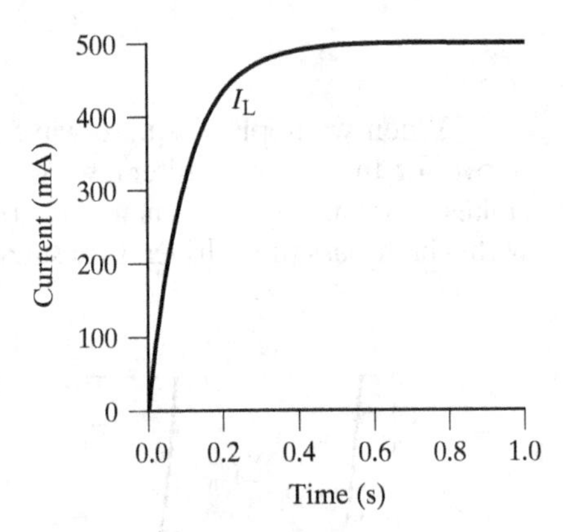

FIGURE 9.33

Current rise in RL circuit.

capacitor, with a similar exponential term. We refer to the L/R term in this case as the time constant τ for an RL circuit.

We can use Eq. (9.47) along with Ohm's law to write an expression for the voltage across the resistor Eq. (9.48).

$$V_R = V_S\left[1 - e^{\frac{-t}{(L/R)}}\right] \tag{9.48}$$

We can then use Eq. (9.48) in combination with the knowledge that $V_L = V_S - V_R$ to write an expression for V_L [Eq. (9.49)].

$$V_L = V_S\left(e^{\frac{-t}{(L/R)}}\right) \tag{9.49}$$

Note the similarity between Eq. (9.48) and Eq. (9.43) and between Eq. (9.49) and Eq. (9.42). We see that in the RL circuit, V_R mimics the behavior of V_C in the RC circuit, and V_L in the RL circuit mimics the behavior of V_R in the RC circuit. This is shown graphically in Figure 9.34.

If, after allowing the current in the inductor to build to its maximum value, we moved the switch back to position A, we could use an initial condition that $I_L = V_S/R$ to develop expressions for the I [Eq. (9.50)], V_R [Eq. (9.51)], and V_L [Eq. (9.52)] during the de-energizing process.

$$I = \frac{V_S}{R}\left(e^{\frac{-t}{(L/R)}}\right) \tag{9.50}$$

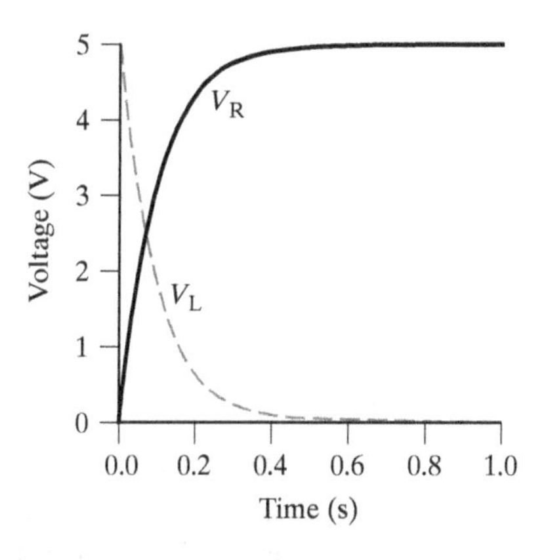

FIGURE 9.34

V_R and V_L during the energizing of the RL circuit in Figure 9.32.

$$V_R = V_S\left(e^{\frac{-t}{(L/R)}}\right) \tag{9.51}$$

$$V_L = -V_S\left(e^{\frac{-t}{(L/R)}}\right) \tag{9.52}$$

As useful as these expressions describing the de-energizing process are, the situation presented in Figure 9.32 as the switch moves from B to A is not the most common way in which an inductor will be de-energized. The circuit shown in Figure 9.35, where the circuit is opened to de-energize the inductor, is more common.

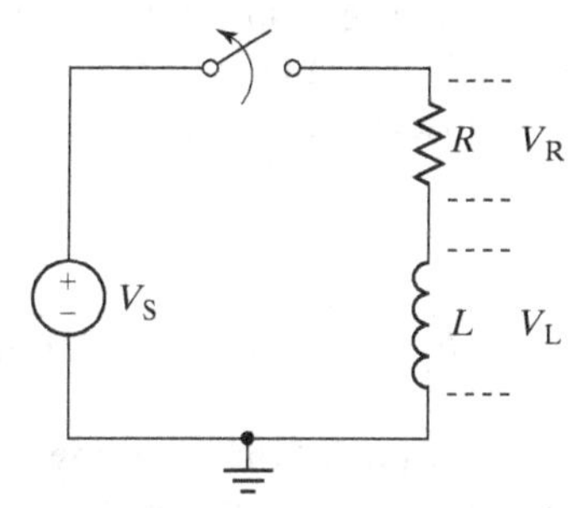

FIGURE 9.35

Opening the circuit to de-energize the inductor.

Here, we have an entirely different situation. When the switch is opened, there is no longer any path for the current to flow. Theoretically, this will produce an instantaneous halt to the current flow (infinite dI/dt). The instantaneous collapse of the magnetic field in the inductor would then produce an infinitely large voltage difference across the inductor. Practically, the collapsing field induces a voltage that quickly reaches the point of being able to ionize the gas in the gap at the switch. An arc is then generated by the high voltage, current flows through the arc and the energy stored in the inductor is dissipated. This voltage spike (referred to as **inductive kickback** or **flyback**) is damaging to the switch, resulting in localized heating and pitting of the contacts. When the switch is a transistor (Chapter 10), this situation can very easily exceed the voltage rating for the device. In Chapter 23, we show how to use a diode to avoid this voltage spike (a process known as **snubbing**), as well as several other possible methods of snubbing that offer a range of trade-offs.

9.9.5 Low-Pass RC Filter Behavior in the Frequency Domain

The discussion of resistors in Sections 9.4–9.5 can be directly applied to circuits with capacitors and inductors as well. The difference being that for capacitors and inductors we need to

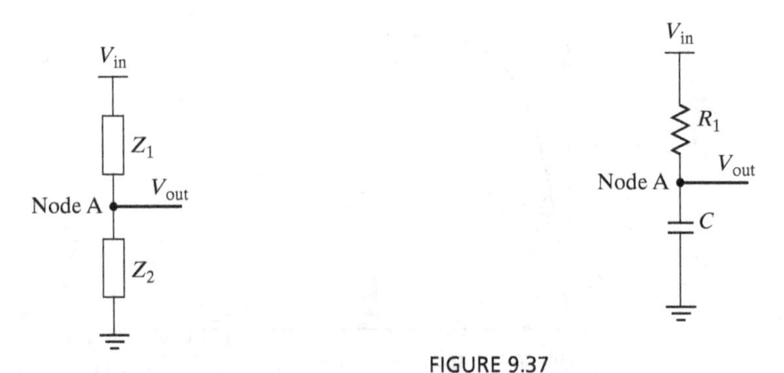

FIGURE 9.36
Generalized voltage divider.

FIGURE 9.37
The voltage divider from Figure 9.12 (or, more generally Figure 9.36) with a capacitor C in place of R_2 (Z_2).

utilize the equivalent resistances, that is, the impedances, of those elements in the analysis. For example, the voltage divider of Figure 9.12 could be redrawn as shown in Figure 9.36, where the resistors have been replaced by generalized impedances.

If we had a resistor in place of Z_1 and a capacitor in place of Z_2, as in Figure 9.37, the analysis might at first seem to be very different.

However, by relying on the "equivalent resistance" of the capacitor discussed earlier in this chapter, the analysis is actually no more difficult than for a resistor network. There are many ways to approach this problem, but the easiest is to treat the capacitor just like a resistor, except that the resistance is the "equivalent resistance" or impedance that we found in Eq. (9.32). We could simply insert this expression for R_2 at the beginning of the voltage divider analysis, and then proceed through the same algebra as before. It is even simpler to take the result from Eq. (9.19), and then insert the impedance of the capacitor in place of R_2. In this case, we quickly arrive at

$$V_{out} = V_{in}\frac{R_2}{(R_1 + R_2)} = V_{in}\frac{(1/j\omega C)}{(R_1 + (1/j\omega C))} = \frac{V_{in}}{(j\omega CR_1 + 1)} \tag{9.53}$$

If we are only interested in the magnitude of the resulting voltage, we can get the magnitude of the complex V_{out} by multiplying the complex conjugate and taking the square root of the product.

$$|V_{out}| = (V_{out} \times V^*_{out})^{1/2} = \frac{V_{in}}{(1 + R_1^2C^2\omega^2)^{1/2}} \tag{9.54}$$

We see that, for small values of ω, the denominator approaches a value of 1, and so $V_{out} = V_{in}$ for small values of ω. For large values of ω, the denominator becomes very large, and so V_{out} becomes very small. This behavior is the source of the name that we give to this configuration: the "low-pass filter," because low-frequency signals propagate through the circuit without much **attenuation** (reduction in the magnitude of V_{out}), while high-frequency signals are attenuated.

Intuitively, we might have expected this. When we replace R_2 with a capacitor, we should think of this as a "resistive" element whose resistance which gets smaller at high frequencies. In the extreme case of very high frequency, the "resistance" of this capacitor goes

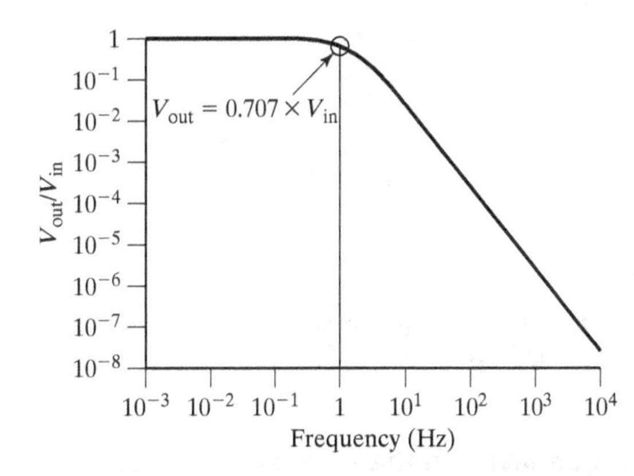

FIGURE 9.38

Bode magnitude plot for the low-pass filter circuit in Figure 9.37.

very nearly to zero, and we expect the output of the voltage divider in Figure 9.12 to be near zero when R_2 is small.

Whenever confronted with a circuit containing a mixture of resistors and capacitors, it can be useful to quickly consider what happens for the case when the capacitors are all replaced by large resistors (as would be the case at some low-frequency limit), and then again when they are all replaced by small resistors (as would be the case at some high-frequency limit). Usually, thinking through these boundary condition cases provides a very good prediction of what to expect from the circuit. Also, these extreme cases are a very handy way to check the results of the analysis.

Since we now have an expression that is not only a function of the values of the circuit components but also frequency, another way to look at this analytical result for the low-pass filter is to make a graph of the log of the ratio of V_{out}/V_{in} as a function of frequency: a Bode magnitude plot. In the graph in Figure 9.38, we used a capacitor value of 10 μF and a resistor value of 10 kΩ to make the filter.

At high frequency, the –1 slope of the curve indicates that V_{out}/V_{in} scales as $1/\omega$. This relationship between slopes and the exponent of the frequency dependence is a helpful result of the log-log plot format, and makes it easy to see the dependencies visually.

When specifying a filter, the principal characteristic is the frequency at which the filter transitions from "passing" the incoming signal [the flat portion of the plot where $\log(V_{out}/V_{in})$ is essentially equal to 1] to attenuating that signal (the sloped portion of the plot). As the Bode plot of Figure 9.38 shows, the changeover is not particularly crisp in the transition region—that is, there is a range of frequencies over which the slope is curved as it changes from horizontal to a diagonal straight line. The accepted practice in specifying filters is to refer to the **corner frequency** as that point at which the output power has fallen to 1/2 that of the input. This point occurs when the output amplitude is smaller than the input amplitude by a factor of $1/\sqrt{2} = 0.707$. This reference point is also often referenced using a decibel (dB) scale, where $\text{dB} = 20\log_{10}(V_1/V_2)$ with V_1 as the output voltage and V_2 as the input voltage. Using the dB scale, the filter corner is the **−3 dB point**. For the simple low-pass filter that we have been discussing, the −3 dB point (the corner frequency, f_c, that is) is defined as

$$f_c = \frac{1}{2\pi RC} \tag{9.55}$$

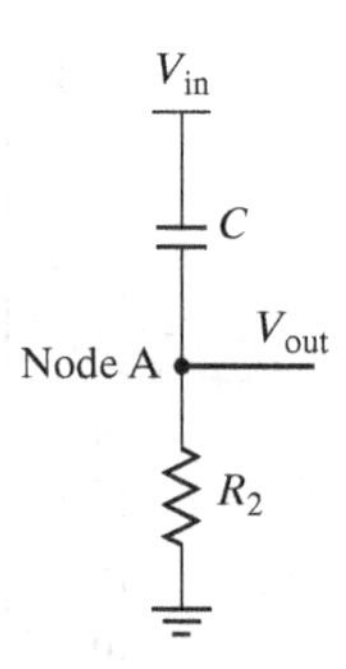

FIGURE 9.39

The voltage divider from Figure 9.12 (or, more generally Figure 9.36) with a capacitor C in place of R_1.

9.9.6 High-Pass RC Filter Behavior in the Frequency Domain

If we go back to the voltage divider circuit of Figure 9.12 [or its more generalized representation (Figure 9.36)] and this time we replace R_1 with a capacitor, as shown in Figure 9.39, the analysis is similar, but the response is different.

Using the intuitive tools we've developed in the discussion above for approximating the behavior at very low and very high frequencies, we can see that at very low frequencies (DC), the capacitor will pass no current, so the output voltage will be zero. At very high frequencies, the effective resistance of the capacitor will be very small and most or all of the amplitude of V_{in} will appear at V_{out}. This circuit then passes high frequencies and blocks low frequencies. This is known as a **high-pass filter**. We can develop an analytical expression for the behavior of this circuit by following the same approach that we used for the low-pass filter in Section 9.9.5: substitute the expression for the impedance of a capacitor for that of R_1 into Eq. (9.19) to get

$$V_{out} = V_{in}\frac{R_2}{(R_1 + R_2)} = V_{in}\frac{R_2}{[(1/j\omega C) + R_2]} = V_{in}\frac{j\omega CR_2}{(j\omega CR_2 + 1)} \tag{9.56}$$

As with the low-pass filter, we will most commonly be interested in the magnitude of the output which we can arrive at as

$$|V_{out}| = (V_{out} \times V^*_{out})^{1/2} = V_{in}\frac{\omega R_2 C}{(1 + R_2^2C^2\omega^2)^{1/2}} \tag{9.57}$$

For the frequency response graph (Bode plot) in Figure 9.40, we once again used a capacitor value of 10 μF and a resistor value of 10 kΩ to make the filter. For this figure, we have used the more common units of dB for the y-axis.

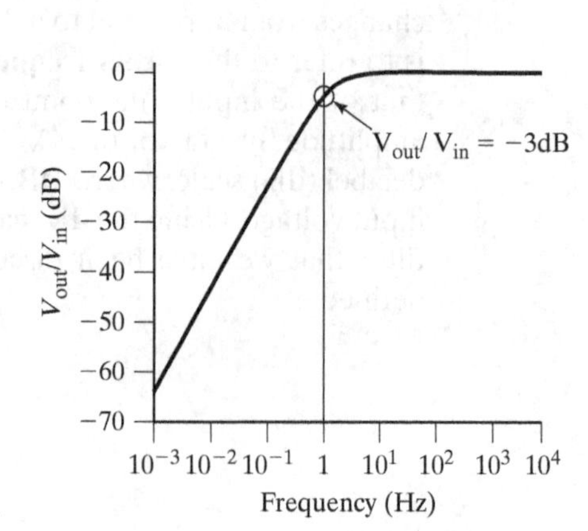

FIGURE 9.40

Bode plot of high-pass filter frequency response.

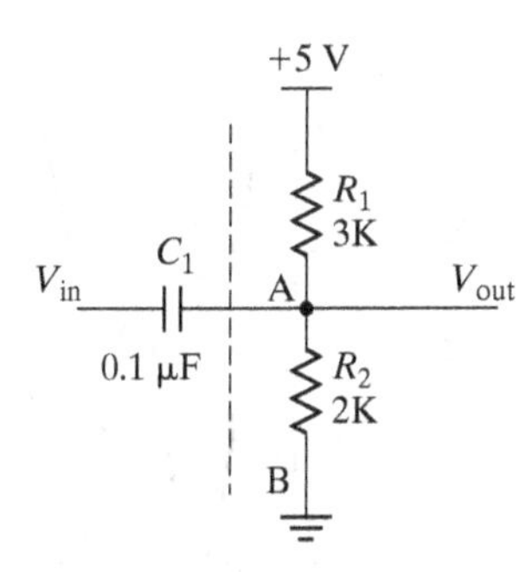

FIGURE 9.41

A slightly more complex circuit example using resistors and capacitors.

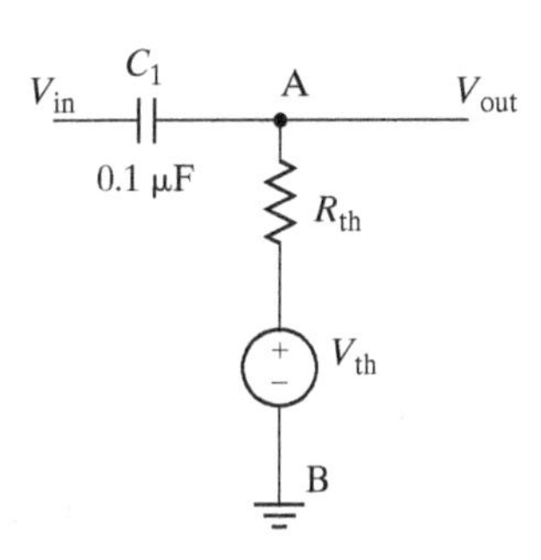

FIGURE 9.42

Thevenin equivalent for the circuit in Figure 9.41.

9.9.7 High-Pass RC Filter with a DC Bias

As a slightly more complicated example, consider the circuit in Figure 9.41. In this circuit, an oscillating input signal is applied to the V_{in} node, and the output is measured at the V_{out} node. For a 1 V amplitude (2 V peak-to-peak) oscillation with a frequency of 1 kHz, centered at 0 V what is the output?

We can use Thevenin's theorem on the portion of the circuit to the right of the dotted line and simplify Figure 9.41 into the form shown in Figure 9.42. The open-circuit voltage of the right-hand part of the network, measured between A and B is 2 V, as this is determined by the 3–2 kΩ voltage divider and the +5 V supply. So, V_{th} is 2 V in the simplified circuit, as shown in Figure 9.42. R_{th} is given by V_{th} divided by the short-circuit current, which is $2\text{ V}/(5\text{ V}/3\text{ k}\Omega) = 2\text{ V}/1.66\text{ mA} = 1.2\text{ k}\Omega$.

So, in Figure 9.42, V_{th} is 2 V and R_{th} is 1.2 kΩ, and we can see that this circuit is a high-pass filter that has the additional feature that the output oscillation is riding on a 2 V offset. The output voltage, V_{out}, includes an AC term due to V_{in} and a DC term due to the Thevenin equivalent voltage contributed by the voltage divider to the right of the dotted line in our original circuit (Figure 9.41).

An alternative way of finding the Thevenin resistance is to short all the voltage sources (and open all the current sources, if there are any) and evaluate the resulting resistive network. In this approach, we see that shorting the voltage sources amounts to connecting the +5 V and the ground together, and this places the 3 and 2 kΩ resistors in parallel. A 3 kΩ resistor in parallel with a 2 kΩ resistor gives a Thevenin equivalent resistance of 1.2 kΩ, the same result we obtained in the analysis above.

Finally, using Eq. (9.57) for this high-pass filter, operating at 1 kHz, we have

$$\frac{V_{out}}{V_{in}} = \frac{\omega RC}{(\omega^2 R^2 C^2 + 1)^{1/2}}$$

$$\frac{V_{out}}{V_{in}} = \frac{(2\pi)(1{,}000)(1{,}200)(0.1e^{-6})}{\{[(2\pi)(1{,}000)]^2(1{,}200)^2(0.1e^{-6})^2 + 1\}^{1/2}}$$

$$\frac{V_{out}}{V_{in}} = 0.6$$

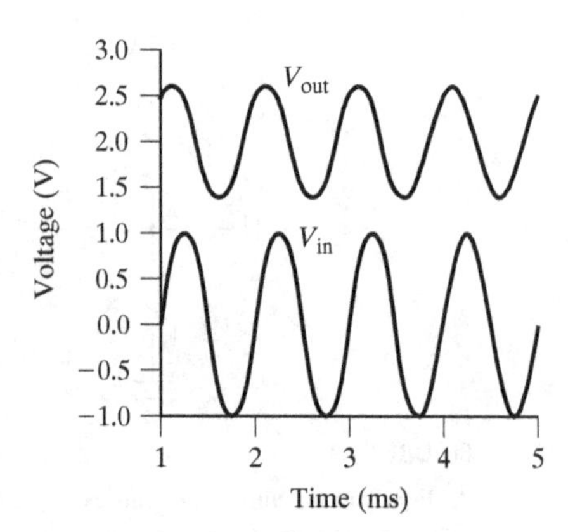

FIGURE 9.43

Input and output waveforms for the circuit of Figure 9.41.

Therefore, the output oscillation has an amplitude of 0.6 V (1.2 V peak-to-peak), and it is riding on a 2 V offset (Figure 9.43).

With the completion of this somewhat more complex example, you should be starting to feel comfortable applying Ohm's law, KCL, KVL, Thevenin equivalents, and the concept of impedance to analyze arbitrary circuits that incorporate voltage sources, resistors, capacitors, and inductors.

9.10 SIMULATION TOOLS

While the manual analysis techniques that we have described so far can be used to analyze much more complex circuits than those that we have shown, in engineering practice that is only rarely done. For circuits of more than 1–2 current loops, some sort of computer aided tool is typically used. For the linear circuits that we have been discussing in this chapter, matrix tools such as Matlab, MathCad, and Mathematica can be used to solve the equations resulting from KVL and/or KCL. However, most circuit analysis is done using more specialized tools. These tools handle the simulation of not only the linear elements that we have been discussing here, but also the nonlinear responses of transistors and diodes, which we cover in Chapter 10.

Most of these circuit simulation tools rely, at least in part, on an underlying analysis engine based on a program called **SPICE** (**Simulation Program with Integrated Circuit Emphasis**). It is very common to find these simulation tools integrated with schematic capture so that the designer simply draws a schematic and the tool extracts the resulting equations and performs the analysis. The basic analyses include DC operating point, transient response (time domain behavior), and AC response (Bode plots). A simulation tool (Protel DXP) was used to generate the transient and Bode plots that we have shown in this chapter. More advanced analyses available include Monte Carlo analysis (which analyzes how the behavior of a circuit changes as the tolerances on each of the components are varied) and sweeps in temperature and voltage that allow the designer to determine the sensitivity of a design to variations in the operating conditions.

These schematic capture and simulation tools are available as commercial products (see Further Reading) and also in free and limited student versions. It is worthwhile to acquire at least one of the free tools for use in your explorations of circuit behavior.

9.10.1 Limitations of Simulation Tools

Circuit simulation, like all simulation, is only as good as the models used by the simulation. In addition, the results need to be interpreted in the context of what was expected. Without an idea about how the circuit is expected to behave, you will be unable to recognize when the simulation results are due to a simulation issue (such as wrong initial conditions) as opposed to actual problems with the circuit design. Finally, as we will discuss in the following sections of this chapter, the behavior of real voltage and current sources, resistors, capacitors, and inductors is more complex than the ideal components that we have described thus far. For most components, the typical simulation models capture the major behaviors but not all of the subtle behaviors. In addition, the more complex models involved in the simulation of such things as operational-amplifiers (Chapter 11) can be sensitive to getting the initial conditions correct. In general, simulation tools are excellent at allowing a designer to do initial explorations of a circuit design and probe the sensitivity of the design to component tolerances. This makes them excellent tools for learning. Using them will allow the designer to root out the basic behavior of the circuit early in the development process. However, because the simulation models do not capture all of the behaviors of the real parts, they are not a substitute for building and testing a physical prototype.

9.11 REAL VOLTAGE SOURCES

In Section 9.1, we described the behavior of an ideal voltage source as delivering a constant voltage into a load independent of the current drawn by the load. Some real voltage sources come very close to that ideal while others miss the mark by a wider margin. We will explore voltage sources by creating a Thevenin equivalent circuit to represent the behavior of a real voltage source, as shown in Figure 9.44.

In this case, V_{source} is the nominal output voltage from the voltage source. A Thevenin impedance, known as the **output impedance**, represents a simple approach to modeling the fact that there may be dynamic as well as static effects in play. What the model shows clearly is that the amount of current drawn by a representative load placed across the terminals labeled A and B will affect the voltage that appears across those terminals. The amount of current being drawn from the source causes this deviation in the voltage measured between the terminals. The load current may consist of two components: a relatively static component and a dynamic component associated with activity, such as the switching of transistors, within the load.

For most power supplies (which, along with batteries, are the two most common voltage sources that you will encounter), the value of Z_{th} at low frequencies is very small, on the order of milliohms or less. This impedance increases with frequency so that typical bench-top power supplies have output impedances on the order of 1 to 10 mΩ at frequencies below 1 kHz rising to 1–2 Ω at 1 MHz. Fortunately, the magnitude of the current variations at the higher frequencies is typically much smaller than the static portion of the current load. For the static component of the load current, the power supply behaves very close to an ideal voltage source, with one caveat: there are current limits associated with any real source. While a power supply might behave very much like an ideal voltage source for low to moderate currents relative to its maximum specifications, above some current limit the power supply will no longer behave as even a close approximation of an ideal voltage source. As long as we operate within the limitations of the power supply, we can usually treat the supply as an ideal voltage source.

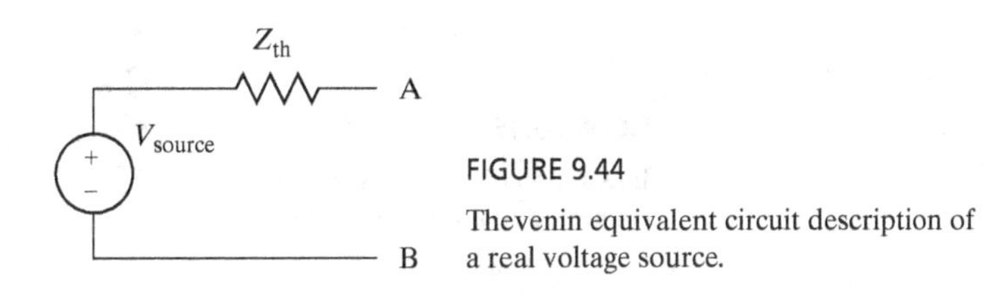

FIGURE 9.44

Thevenin equivalent circuit description of a real voltage source.

Power supplies are not the only types of voltage sources that we will need to deal with. Another major category of voltage source that you are likely to deal with frequently is that of sensors. Many types of sensors (Chapter 13) appear as a voltage source whose output voltage is related to the parameter being measured. We will need to design circuits to interface with these sensors and need to consider the output impedance of the sensor in our designs. The output impedance of sensors varies widely, from a few ohms for sensors with built-in signal conditioning up to hundreds of megaohms for sensors, such as pH probes.

9.12 REAL MEASUREMENTS

In addition to calculating voltages and currents in the circuits you design, you will also frequently perform actual measurements. These measurements will generally be regarded as a way to confirm that the calculations have been done correctly or that the circuit is functioning as designed. However, it is important to realize that the measurement will have an effect on the circuit being measured, and so it is necessary to think about the characteristics of the measurement equipment in order to obtain credible and useful data.

9.12.1 Measuring Voltage

Consider the situation in Figure 9.45, where a **digital voltmeter (DVM)** is being used to measure the output voltage from a simple voltage divider.

The DVM does not make a perfect measurement (e.g., a measurement that doesn't disturb the system it is measuring), as represented by the 1 MΩ internal resistance. The issue here is that the simple analysis that we used to calculate the output of this circuit assumed that there was no current flowing to the output. However, when a real instrument is connected to this node of a real voltage divider, a finite amount of current is actually diverted from the circuit to allow the voltage measurement. This diverted current results in a change in the output voltage. How large is the change? The best way to tell is to carry out a new analysis that includes the resistance of the measuring instrument added to the circuit.

In general, DVMs and oscilloscopes have internal resistances from 1–10 MΩ. The actual value of the resistance can depend on the specific instrument and on the settings of the instrument.

In the example discussed here, the figure shows the instrument as having a 1 MΩ internal resistor. This resistor diverts a fraction of the current away from R_2, with the fraction being related to the ratio of the instrument resistance to the R_2 value. If the instrument resistance is 10 times bigger than R_2, $1/11^{\text{th}}$ of the total current is diverted, and this causes a corresponding error in the voltage measured. This effect of instruments drawing current

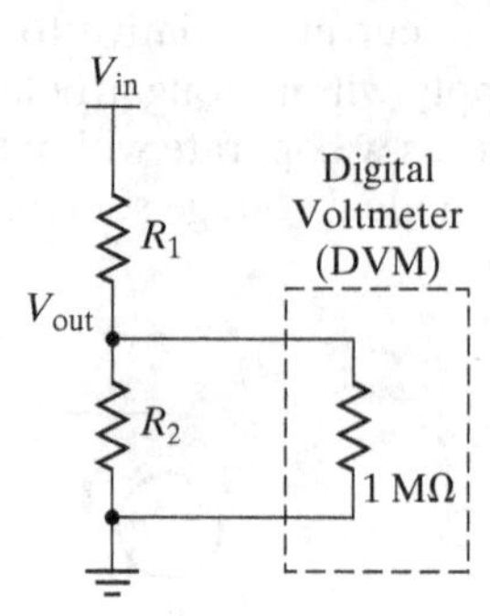

FIGURE 9.45
Illustration of a digital voltmeter connected to a voltage divider.

away from circuits in order to take measurements is often called "**loading**," and can be thought of as the "load" of the instrument "pulling down" on the voltage of the circuit. The magnitude of the error in the voltage reading will depend on the value of R_1. For example, if $R_1 = R_2$, the error in the voltage measurement would be 4.7% lower than for an ideal meter. If $R_1 = 10 \times R_2$, the error would be 9.2% lower.

As a general rule of thumb, this effect of current being diverted is only important if the resistors (or, more generally, the impedances) in the measurement instrument are comparable to that of the resistors (impedances) in the circuit. If the measurement resistors are much larger (orders of magnitude) than the circuit resistors, then their effect on the measurement can usually be neglected. This principle guides the design of measurement instruments. For example, if in Figure 9.45 $R_1 = R_2$, and the measurement resistor is the same size as R_2 (all resistors have a value of 1 MΩ, that is), there will be as much current going through the instrument as there is going through R_2. This causes a significant error, indeed. The combined resistance of R_2 in parallel with the instrument's resistance is $R_2/2$, so the voltage observed at the measurement node will be $V_{in}/3$ instead of $V_{in}/2$. This type of significant error can occur whenever DVMs are used to measure voltages in circuits with resistors on the order of 1 MΩ or higher.

On the other hand, if $R_1 = R_2 = 10$ kΩ, the current drawn through the instrument is only about 1% of the current through R_2. This would result in a 1% reduction in the current in R_2 and a 1% error in the voltage measurement. If the measurement resistance were exactly 1 MΩ, the effective parallel resistance of R_2 and the instrument would be 9.9 kΩ, and node voltage would be reduced from $V_{in}/2$ to $V_{in} \times (99/199)$, which is about 0.5% smaller. Of course, we don't usually know the instrument resistance very accurately, and the actual value of the resistors may have 5% errors, so in many applications we would safely neglect this error.

In sum, accurate voltage measurements rely on the internal resistances of the measuring instruments being much larger than the resistances in the circuit being measured. For circuits with 1 and 10 kΩ resistances, even inexpensive DVMs will serve just fine—their typical 1 MΩ resistances are certainly large enough to avoid problems. However, if the resistances in the circuit are in the megaohm range, you will need a higher quality DVM with an internal resistance in the range of tens to hundreds of megaohms to avoid large errors.

9.12.2 Measuring Current

In DC circuits, current measurements are performed by inserting a small-value resistor in series with the circuit and measuring the voltage drop across the resistor (Figure 9.46).

Resistors used in this way are called **shunt resistors**, and typically have resistances ranging from a few milliohms to a few ohms, with tolerances in the 0.1% to 1% range. This is how digital multimeters (DMMs) measure current for both DC and AC.

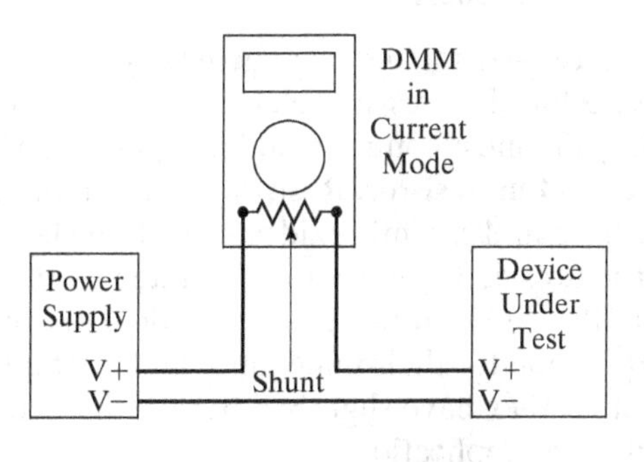

FIGURE 9.46

Measuring current with a digital multimeter in current mode.

This approach has the drawback of reducing the voltage in the circuit downstream of the meter's shunt resistor, possibly changing the response of the circuit. There is really nothing that we can do to eliminate the drop in voltage, since it is essential to measuring the current, but we may need to be aware of it and account for it in our analysis. The voltage drop is referred to as the **burden** that the meter imposes during a measurement. For high-quality multimeters measuring currents under a few hundred milliamps, the burden is in the range of 2 mV/mA. At higher current levels (> 500 mA), the burden typically drops to the range of 30 mV/A.

9.13 REAL RESISTORS

The behavior of an ideal resistor is completely described by Ohm's law. For the ideal resistor, the resistance is constant across all environmental and circuit conditions. For the most part, we shall assume that our resistors are ideal, and indeed for most practical purposes, they are. Depending on our application, however, there are a few important ways in which real resistors differ from their ideal counterparts that we may need to consider, as well as some limitations on their performance that you must keep in mind when you design and build circuits.

9.13.1 A Model for a Real Resistor

Figure 9.47 shows a model that captures much of the behavior of real resistors. In addition to the actual resistance, there is a series inductance (L_S) and a parallel capacitance (C_p). For mechatronic systems, the C_p term is completely negligible, though for very high-frequency (> 1 GHz) RF circuits it can contribute significantly. The amount of series inductance varies dramatically with resistor construction techniques. For the most common types of resistors that we will likely encounter (carbon film, described below), the term is small enough to neglect. If your design requires the stability, accuracy, and power dissipation capability of wire-wound resistors (which closely resemble in several important ways the inductors we described in Section 9.6), be sure to investigate the series inductance. Wire-wound resistors have significant series inductance.

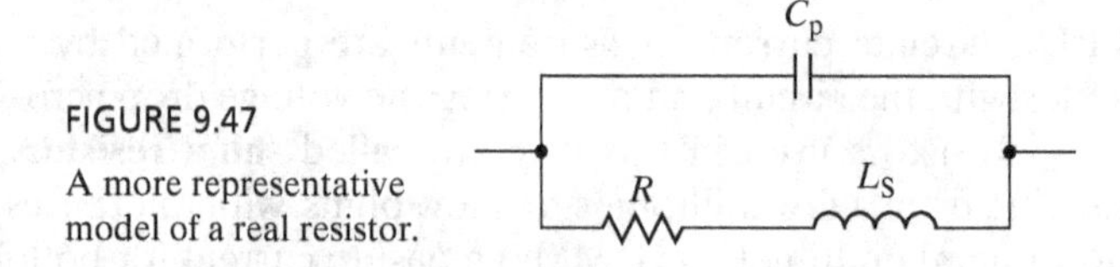

FIGURE 9.47
A more representative model of a real resistor.

9.13.2 Resistor Construction Basics

Physically, the resistor is a two-terminal device, generally consisting of a pair of metal wires attached to the (usually) cylindrical resistor element. The resistive element can be made of several possible materials, the most common and inexpensive of which is a thin film of carbon, hence the name **carbon film resistor**. Resistors can also be made from metal films, thin wires, doped semiconductors, and ceramic oxides, though the later two are relatively rare for discrete devices. The structures and materials are dictated by the value of the resistance and other characteristics that the different materials provide. They are all alike from an analytic perspective (e.g., the expression for their impedance and their adherence to Ohm's law), but from a practical perspective, they have slightly different characteristics that make each type particularly suited to different applications.

The resistance of a real piece of material is related to the resistivity of the material and the dimensions of the resistive element. Generally, the resistance is given by:

$$R = \frac{\rho l}{A} \tag{9.58}$$

where ρ is the resistivity with units of ohm-cm, l is the length of the resistive element, and A is the cross-sectional area. The resistance of a wire can be calculated using this formula, the dimensions, and the resistivity of the material. Metals have the lowest resistivities, and so bulk metal resistors are only commonly used to make resistors with very low values of resistance. It is possible to create specific small-value resistors (mΩ level) from a length of fine wire inserted into a circuit path.

9.13.3 Carbon Film Resistors

Carbon film resistors are the most common type of resistor because carbon is inexpensive and has high resistivity, and the thin film construction keeps cross-sectional area small so that a wide range of resistance values can be fabricated in a small package. The most important feature of carbon film resistors is their low cost. Typically, carbon film resistors can be purchased in quantities of 100 for 1–2¢ each. This is significantly less than any of the other resistor technologies we'll describe and is the main reason for their popularity.

Carbon film resistors are most commonly labeled with color bands painted on the body of the resistor. A resistor color code (which we include in Appendix A) helps decode a resistor's value. This code provides an easily recognizable scheme allowing visual identification of the resistance value while keeping the manufacturing costs low. A typical carbon film resistor is shown in Figure 9.48.

The exact resistance of each individual carbon film resistor will deviate slightly from its specified value, within some tolerance (typically 1%, 2%, 5%, or 10%). Some level of variation (i.e., tolerances) exists for all types of resistors, and indeed all types of electronic components. It is important to remember to allow for this variation in value when designing with these resistors. Accounting for these tolerances in electronic design is very much the same as allowing for tolerances in mechanical design: no single component is ever *exactly* as specified, and your designs should work as long as each of the components is within the specified tolerance range. When selecting resistors for use in a circuit, it is critical to ensure that actual resistors that deviate from the nominal value within the tolerance would still result in a circuit that functions within the requirements. Furthermore, it does not make sense to add small resistors to large resistors in hopes of achieving higher precision. For example, adding a 1 kΩ 5% resistor to a 100 kΩ 5% resistor might, on the surface, appear to result in a 101 kΩ resistor. However, because of the 5% errors in the value of the 100 kΩ resistor—it could be anywhere between 95 and 105 kΩ—this uncertainty swamps the value of the added 1 kΩ resistor. If you actually need 101 kΩ resistor, and 100 kΩ would not be acceptable, you need to start with a family of resistors that is more precisely specified than carbon film resistors, since 1% tolerances are the best they can offer, and 2% and 5% are the most commonly available.

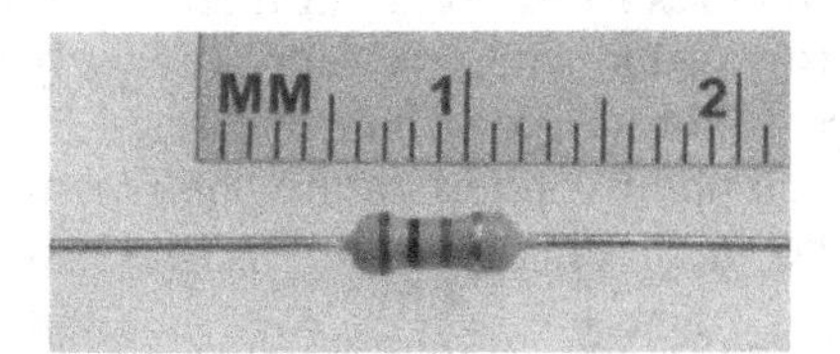

FIGURE 9.48

Photograph of a typical carbon film resistor.

Carbon film resistors are widely available in resistance values ranging from 1 Ω to more than 20 MΩ. For extremely cost-sensitive applications they are available specified to 10% accuracy, though 5% and 2% resistors have become the most common tolerances. Standard values for 2% and 5% accuracy resistors are chosen such that adjacent values have a relative ratio of about 10%. Appendix A lists the standard values for 5% tolerance resistors. You should get into the habit of consulting the table to find the appropriate standard value resistor to use once you have calculated an ideal value. You should also exercise care when choosing a standard value resistor based on the ideal calculated value. If, for example, the ideal resistor size was chosen to limit current to some maximum value (i.e., a minimum resistor size), then you would not want to choose simply the standard value closest to the value you calculated, but the standard value whose value, less the tolerance, is still greater than your calculated value.

Carbon resistors have another important characteristic (or flaw, really): the resistance value is dependent on temperature and can vary as much as 15%–20% over the temperature range from 0 to 100°C. This temperature dependence is specified in the form of a temperature coefficient stated as some number of **parts per million (PPM)** per °C. PPM is simply another way of specifying a fractional relationship, much as percent is an expression of parts per hundred. For example,

$$100 \text{ ppm} = \frac{100}{1{,}000{,}000} = 0.01\% \tag{9.59}$$

Expressing a number in PPM is a convenience because it allows you to avoid having to enumerate a large number of leading zeros for small numbers.

This temperature dependency even allows the use of carbon resistors as very low cost thermometers in applications where accurate temperature measurements are not required, but some indication of temperature variation is useful. For example, the temperature dependence of Allen-Bradley carbon resistors is very repeatable and can be calibrated with 5°C accuracy for measurements near room temperature. Their resistance increases very rapidly at low temperatures, allowing sensitive measurements of temperature in cryogenic applications. However, if you do not want your circuit parameters to be a function of operating temperature, this "feature" can pose a significant challenge and may force the use of a different, and more costly, resistor technology.

9.13.4 Metal Film Resistors

The next most common type of resistors is the **metal film resistor**. Physically, these resistors are similar in size and shape to carbon film resistors.

Metal film resistors are available with tolerances ranging from 0.01% to 5%, with 0.1% and 1% being the most commonly used. As you might expect, specifying resistors with very tight tolerances (such as 0.01% or 0.1%) costs much more, so this should only be done in cases where the exact value of the resistor is highly important to the function of your design. Metal film resistors are also available in a wider range of resistance values than carbon film (0.01 Ω up to about 33 MΩ). In addition to offering higher precision and a wider range of available resistances, metal film resistors also feature much smaller temperature coefficients of resistance than the carbon film resistors. For all of these reasons, metal film resistors are favored in circuits where resistor values must be more precisely controlled. As you might expect, metal film resistors cost more than carbon film resistors. Prices range between 3¢ each (for 5% tolerances and 350 ppm/°C temperature coefficients) to over $5 each (0.01% tolerance and 5 ppm/°C). The main functional differences with respect to carbon film

FIGURE 9.49

Photograph of a typical metal film resistor.

resistors are the improved precision (0.01%–1% vs. 1%–10%), reduced temperature coefficient of resistance (5–100 ppm/°C vs. 500–1,500 ppm/°C), and increased cost (3¢ to several dollars each vs. 1–2¢ each).

A typical metal film resistor might be labeled with the standard color code, or it might have a somewhat cryptic numeric label, such as "1102," as shown in Figure 9.49. When labeled with colored bands, they follow the same pattern as that for carbon film resistors with addition of a fourth band to allow for the increased precision. When labeled numerically, there should be (for 1% resistors) a four-digit number. For example, if the resistor is marked "1102," as in the Figure 9.49, that represents three digits of mantissa and a single-digit exponent. This example resistor has a value of 110 (the first three digits of **110**2) times 10^2 (the last digit of 110**2** is the exponent, or number of 0s, to append to the first three digits). The resistor marked as 1102 is a 11 kΩ resistor. For values less than 100 Ω, a letter will be inserted to indicate the position of the decimal point.

Identifying the value on resistors with numeric labels can be confusing at times. This is because in addition to the code indicating the resistor value, there are often other clusters of numbers and letters that indicate the resistor's temperature coefficient, the manufacturing date, and the lot number. There are some common sense guidelines that you can keep in mind when trying to decipher these codes. For example, it takes four numbers to specify the resistor value to 1%, with the last digit reserved for the exponent, so number clusters that are related to the resistance value should be exactly four numbers long. This convention is common among manufacturers. It is useful to also remember that metal film resistors are available from 0.01 Ω to a few megaohms. If the value resulting from converting a four-digit code is outside of this range, it is probably incorrect. So, "1002" is probably a code for the value of the resistor and "5137" would not be, because the value implied (51.3 MΩ) would be far outside the range available for this resistor technology.

9.13.5 Power Dissipation in Resistors

Another way in which real resistors differ from ideal resistors is in their limited ability to dissipate power. Any real resistor with current passing through it dissipates power.

$$P = VI \tag{9.60}$$

There are cases where there is enough power dissipated to cause significant heating of the resistor. Carbon and metal film resistors have maximum operating temperatures that can easily be exceeded with only modest amounts of current, so it is important to keep track of the allowed power for each resistor and to work out the power that each will dissipate. Most carbon and metal film resistors are designed to tolerate 1/8 or 1/4 W of power dissipation. When using such a resistor in a circuit, it is a good idea to perform rough calculations of the current and voltage present, and if the product exceeds 0.1 W, more careful calculations should be carried out or design changes should be made.

For example, if a 1 kΩ, 1/8 W resistor has 20 V across it, the power dissipation is found from Eq. (9.60), in combination with Ohm's law to get:

$$P = VI = V\frac{V}{R} = \frac{V^2}{R} = \frac{20^2}{1e^3} = 0.4 \text{ W}$$

In this case, the power dissipated is 0.4 W, which is much higher than the 0.125 W that the resistor is rated for. We would expect this resistor to heat up beyond its rated temperature very quickly. The speed of this heating is governed by the laws of heat transfer and depends on many factors. Some of the heat generated in the resistor is conducted away by the leads, some is carried off of the heated surface to the surrounding air by convection, and some is passed by infrared radiation (the amount passed by radiation is very small—unless the temperature of the resistor is very high indeed, at which point it is probably too late to save that particular resistor!). If a resistor operates at temperatures outside its specifications for an extended time, it is likely to be permanently damaged. If your calculations show that typical 1/8 or 1/4 W resistors are inadequate, other resistor types are available that will dissipate 0.5 W, 1 W, 2 W, and more. These types of resistors are referred to as **power resistors**.

This raises an important point about manufacturer's specifications and how we interpret them. The specifications usually describe safe operating conditions and provide details describing the behavior of the devices when operated within the safe ranges. They don't guarantee failure if you exceed the specified safe limits. For a typical 1/8 W resistor, the manufacturer guarantees that the resistance value will be within the specified accuracy (1% for a typical metal film resistor) if the power dissipation is less than 1/8 W and the ambient temperature is below 70°C. If there is more power dissipation in this device, all guarantees are null and void, and the present and future behavior of this resistor is no longer guaranteed to fall within specifications. It is not guaranteed to fail either—all that happens is that all bets (and legally binding agreements with the manufacturer) are off.

From a thermal perspective, the heat generated in the resistor must eventually escape to the environment in order to prevent very large temperature increases within the resistor. The dominant mechanisms for heat to escape are by conduction and convection. This explains why 1/4 W carbon resistors are larger than 1/8 W carbon resistors—the additional surface area is the means by which the 1/4 W resistors exhaust the additional heat that the 1/8 W resistor can't tolerate. So, the first clue that a resistor may be intended for higher power applications is its size—they tend to be bigger than the 1/8 W resistors found in most applications. They are also more expensive, so they are only used when necessary.

Power resistors capable of dissipating large amounts of heat and withstanding high temperatures are usually constructed from and packaged in materials such as ceramics or metals, and the package may incorporate mounting holes so that they can be firmly attached to external elements that can help transfer heat away from the resistor, such as heat sinks. Figure 9.50 shows a photograph of several different kinds of such power resistors.

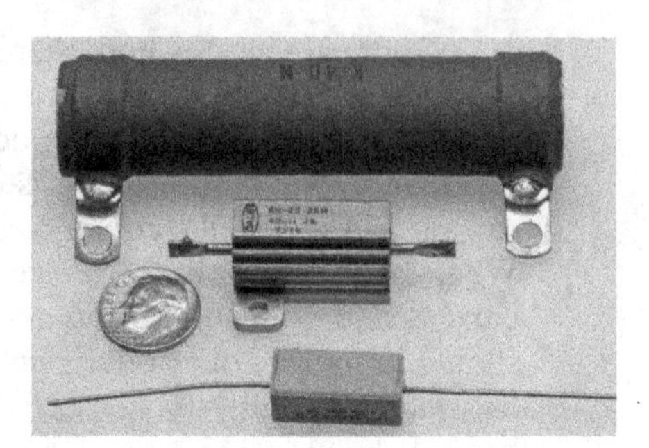

FIGURE 9.50

Photograph of several power resistors.

Power resistors are more readily available in smaller resistance values. Since $P = IV = V^2/R$, it is hard to have high power dissipation in a very large value resistor. Consider, as an example, the following: what voltage across a 1 MΩ resistor would be required to cause it to dissipate 1 W?

$$\text{Answer: } P = \frac{V^2}{R};\ PR = V^2;\ V = \sqrt{PR} = \sqrt{1 \times 1e^6} = 1000\ \text{V}$$

Since such high voltages are not common in mechatronic applications, power resistors with large resistances are also uncommon. However, there are often power resistors with resistance values well below 1 Ω. If a 0.1 Ω resistor has 1 V applied across its terminals, the power dissipated will be 10 W. Therefore, any resistor with resistance value around 1 Ω or less probably needs to be a power resistor in order to be appropriate for commonly used voltages.

One convenient and helpful rule of thumb to remember is that resistor values above 1 kΩ in circuits with voltages up to 5 V always dissipate less than 0.1 W. In the kinds of circuits that we will deal with in mechatronics, there is rarely a need to worry about power dissipation for resistors with values of 1 kΩ or greater. However, when dealing with resistor values of 100 Ω or less, the power dissipation can reach levels that might be a concern. A 100 Ω resistor with 5 V across it dissipates 1/4 W. So, any time you're thinking about using a small-value resistor in some application (e.g., smaller than 1 kΩ), it is important to consider the power dissipation.

9.13.6 Potentiometers

In addition to fixed value resistors, there are also **variable resistors** in which the resistance between the two terminals is a function of the orientation of a knob, or screw, or the position of a slider. More common than the variable resistor is a three-terminal device called a **potentiometer**. As indicated by their schematic symbols, shown in Figure 9.51, two terminals (the top and bottom) of the potentiometer have a fixed resistance between them. The third terminal (on the left in Figure 9.51b) is connected to a moving contact, called a **wiper**, that contacts the resistive element somewhere between the two extremes. In effect, the wiper creates a voltage divider, splitting the overall resistance of the device between the two fixed terminals, with one resistor (call it R_1) between the wiper and one end terminal, and another resistor (call it R_2) between the wiper and the other end terminal. The potentiometer is more common than a simple variable resistor (which simply has a single resistance that is adjustable) because it is more versatile. If you don't need to take advantage of the variable voltage divider that a potentiometer offers, you can always create a variable resistor using a potentiometer by tying the wiper to one of the other two terminals (Figure 9.51c). Variable resistors and potentiometers are very useful in applications where the optimal resistance value is not known prior to testing or when repeated adjustments or calibrations are expected. Figure 9.52 shows a photograph of some typical potentiometers.

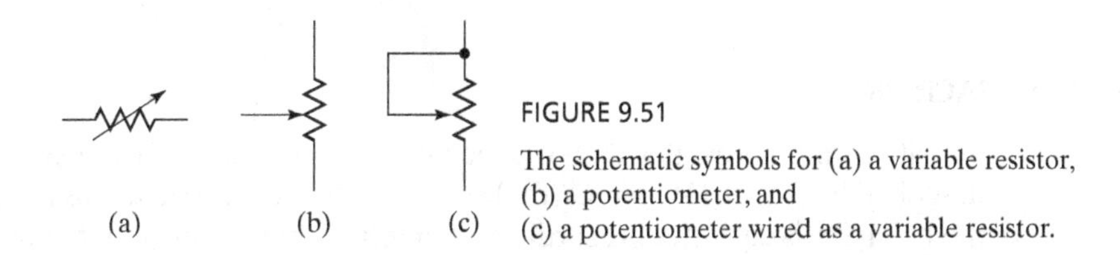

FIGURE 9.51

The schematic symbols for (a) a variable resistor, (b) a potentiometer, and (c) a potentiometer wired as a variable resistor.

FIGURE 9.52

Photograph of several types of potentiometers.

Relative to fixed value resistors, potentiometers suffer from poor stability and temperature sensitivity. In addition, they are much more expensive than ordinary resistors, so they are only used when the adjustability is very important in development or use.

9.13.7 Choosing Resistors

Choosing resistors for an application comes down to answering a few questions about the application.

1. Do I know from analysis the precise value that I need, or will the precise value depend on other components that are determined at the time the circuit is built? If we can't calculate a value because we don't have the information at the design stage, then a potentiometer will be necessary. If we can limit the range of values needed, we can improve the adjustability by combining a fixed value resistor that represents the minimum value that will be necessary with a potentiometer whose maximum value, in series with the fixed resistor, will take us up to the maximum value that we may need.
2. If I know the necessary value, how close do I need to be to that value in order to insure that the function of my circuit meets the requirements? The answer to this question will determine the required accuracy of the resistor and therefore the choice between carbon film and metal film resistor.
3. How much power will be dissipated in the resistor? This will determine the power rating that the resistor must meet.
4. How stable does the value of the resistance need to be over the range of operating temperatures? This will determine the necessary temperature coefficient of the resistor, which may, in turn, drive the choice between carbon film and metal film resistor.
5. Is my application more sensitive to resistor ratios over temperature than it is to the absolute value of the resistors? If this is true, using a SIP or DIP package may be more economical than buying precision resistors to maintain the ratios over temperature.
6. Do I need a large number of the same resistor values in my circuit? If so, how many of my devices will I be making? Do I want to make them easier to assemble correctly? If so, consider SIP and DIP resistor packs.

9.14 REAL CAPACITORS

Recall from Section 9.6 that the behavior of a theoretical, ideal capacitor is completely described by Eq. (9.23): $I = C/(dV)/(dt)$. For the ideal capacitor, as it was for the ideal resistor, the behavior is consistent regardless of environmental and circuit conditions. When doing our analyses, we assumed that all of our capacitors were ideal. But, when it comes time to choose real capacitors to put into our real circuits, we will need to consider many characteristics of real capacitors that deviate from the ideal. In addition to the parasitic effects that we cover in Section 9.14.1, all capacitors have a rated **working voltage (WV)** and a temperature coefficient to describe the variation in capacitance with temperature. We will refer to these characteristics in our discussions of the different types of capacitors in the following sections.

In general, applications for capacitors can be broken down into four broad categories: timing, filtering, AC coupling, and bypassing/decoupling. We have already discussed timing (Sections 9.9.2 and 9.9.3) and filtering (Sections 9.9.5 and 9.9.6) in our exploration of RC circuit behavior. Coupling is related to filtering, but focuses on blocking DC while passing AC signals. Bypassing/decoupling is an application that we haven't discussed in this introductory chapter—we'll devote significant attention to the topic in Chapter 21. In bypassing/decoupling applications, capacitors are used to provide a local store of charge that can provide current to a device during transient events. Capacitors used for bypassing/decoupling have dual (and often conflicting) requirements to provide a large capacitance and also be capable of delivering their stored charge quickly. The need for precision and stability in the value of the capacitor varies among these applications. Timing and filtering require the most precision, with coupling requiring somewhat less, and bypassing/decoupling being the least sensitive of all.

9.14.1 A Model for a Real Capacitor

In addition to the desired capacitance (C), a real capacitor has an **equivalent series resistance (ESR)**, labeled in Figure 9.53 as R_S, an **equivalent series inductance (ESL)**, labeled L_S, a parallel resistor (R_L) that serves to leak charge off the capacitor and a small capacitor and resistor in parallel with the main capacitance that models a phenomenon known as **dielectric absorption (DA)**. Dielectric absorption is the result of a migration of charge away from the electrodes and into the dielectric material. The effect that you see from dielectric absorption is that if you were to discharge a charged capacitor (by quickly shorting the leads, for example) and then measure the voltage on the capacitor, you would observe a voltage develop across the leads with no external source of charge applied. The charges that had been absorbed into the dielectric have migrated back to the electrodes. One of the major effects of these parasitic components of the behavior is that effective value of capacitance changes as a function of frequency. The different construction techniques used to make capacitors result in different values of the parasitic components and make some types of capacitors more suitable than others for particular applications. For example, if you are using a capacitor in a timing application, where the ability to hold a charge is critical, it will be important to choose a capacitor with a very large value for R_L and a low value for C_{DA}.

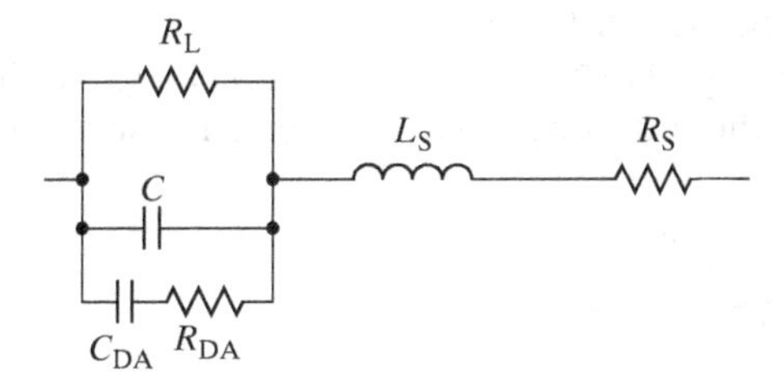

FIGURE 9.53

A more representative model for a real capacitor.

9.14.2 Capacitor Construction Basics

Physically, capacitors can be constructed in a great variety of ways, and the following discussion reviews the issues related to the size, shape, and materials, as well as the cost and performance of some of the most common types of commercially available capacitors. The simplest geometry to analyze for a capacitor is a device consisting of a pair of large area parallel plates. This is illustrated in Figure 9.15. For a capacitor made from a pair of parallel plates, the capacitance is related to the dimensions of the plates, the area between them, the spacing between the plates, and the dielectric material that separates them:

$$C = \frac{\varepsilon\varepsilon_0 A}{d} \tag{9.61}$$

In this expression, ε_0 is a physical constant called the "permittivity of free space" and has value of 8.85×10^{-12} F/m. ε is the "relative permittivity," and is equal to 1 for air, but can be as high as 1,200 for other materials which might be placed between the electrodes of the capacitor.

Consider a scenario where we construct a capacitor by placing a pair of 1 cm $\times$ 1 cm metal plates as close as possible without touching, say an air gap of 25 microns (0.001 in.). This would lead to a capacitance of

$$C = \frac{\varepsilon\varepsilon_0 A}{d} = \frac{(1)(8.85 \times 10^{-12})(1 \times 10^{-4})}{25 \times 10^{-6}} = 35 \text{ pF}$$

This is a small but useful value of capacitance for many circuits. It is easy to imagine making capacitors this way with smaller capacitance values. However, it would be very hard to make one with a significantly larger value. How large a capacitance is 35 pF? In a typical circuit, where oscillation frequencies might be around 10 kHz, the "equivalent resistance" of this capacitor (see the discussion in Section 9.6.2) would be

$$\text{“}R_C\text{”} = \frac{1}{j\omega C} = \frac{1}{j(2\pi 10{,}000)(35 \times 10^{-12})} = 5 \text{ M}\Omega$$

This is a very large value for the effective resistance—resistance values in the 1–10 kΩ range might be more useful in applications like RC filters. In order for a capacitor to have an equivalent resistance of 10 kΩ at 10 kHz, its capacitance needs to be larger than 35,000 pF. Unfortunately, it is hard to construct a capacitor with a much larger value of capacitance out of parallel plates because of limitations on area, gap, and machining tolerances.

There are many important applications for larger value capacitors in small packages, and as a result there have been strong motivations to develop good, practical solutions to this challenge. The prevailing technologies all work to optimize one or more of the variables in the equation for the capacitance—minimizing the gap between the conductors, maximizing the electrode area, and/or increasing the value of the relative permittivity of the material between the conductors. There are a large number of different possible solutions, and this has resulted in a large variety of capacitors types that are available. These different kinds of capacitors have different properties, different strengths and weaknesses, come in different sizes, and have different price tags. For practical applications (e.g., designing a circuit), it is necessary to be familiar with the different capacitor types, and in the next few pages, we'll cover the most common ones.

9.14.3 Polar vs. Nonpolar Capacitors

We open the discussion of the different types of capacitors by dividing them first into one of two camps: **polar** and **nonpolar**. For **nonpolar capacitors**, there is no difference between the two terminals, and they can be connected to a circuit in either orientation. **Polar capacitors**, on the other hand, are made using materials that require a specific electrical polarity be

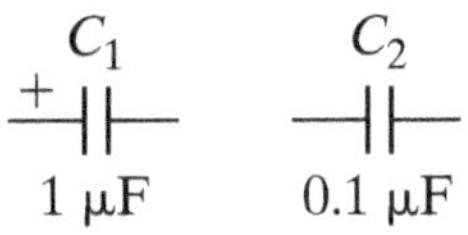

C_1 C_2

$+$

1 μF 0.1 μF

FIGURE 9.54

Polar (C_1) and nonpolar (C_2) capacitor schematic symbols.

maintained between the two terminals of the capacitor. As such, they are noted in schematics by adding a plus symbol next to the lead that must remain at the higher potential (Figure 9.54). This is clearly less than ideal as now it matters how we insert the capacitor into a circuit, and what voltages are applied to them when the circuit is operating. We accept this limitation because the materials used to make polar capacitors allow for a much higher capacitive density (greater values of capacitance for a given volume of capacitor) than can be achieved with nonpolar materials.

9.14.4 Ceramic Disk Capacitors

Ceramic disk capacitors are the least expensive type of capacitor available. They take advantage of the high value of the permittivity of certain ceramic materials [lead zirconate titanate (PZT), for example] to make capacitors that consist of thin ceramic disks with electrodes deposited (by screen printing) on opposite sides. These familiar capacitors are usually encased in a brown insulating material and labeled with the component value, which usually ranges from a few picofarads up to 0.22 μF. A photograph of several disk capacitors is shown in Figure 9.55. One useful property of ceramic disk capacitors is that the value of the capacitance is maintained for signals from low frequencies up to very high frequencies (several MHz or more).

The least expensive ceramic capacitors have the primary disadvantages that the capacitance value is highly dependent on temperature, and part-to-part variations that occur during manufacturing that lead to component tolerances of ±25% for parts from the same batch with the same nominal capacitance value. It is common to find examples of less expensive ceramic capacitors specified to +80%/−20% tolerance. For higher prices, the variations are reduced, mostly because the capacitors are tested and sorted by value. There are many applications where the exact value of the capacitor is not critical and standard ceramic capacitors are very inexpensive, so they are widely used. If you have an application that does not depend on having a specific value of the capacitance, or having a particularly stable value of that capacitance, ceramic capacitors might be an excellent choice. If you are replacing a capacitor in a product or device, and the previous capacitor was not a ceramic disk capacitor, it usually means that there is some accuracy requirement in this circuit that a ceramic capacitor cannot meet, so you should be careful. As we'll show, as the discussion continues, capacitors are not all alike—far from it!—even if they have the same nominal capacitance value.

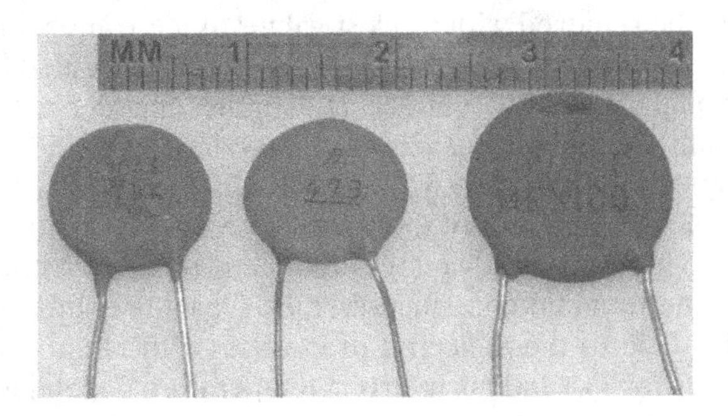

FIGURE 9.55

Photograph of ceramic disc capacitors.

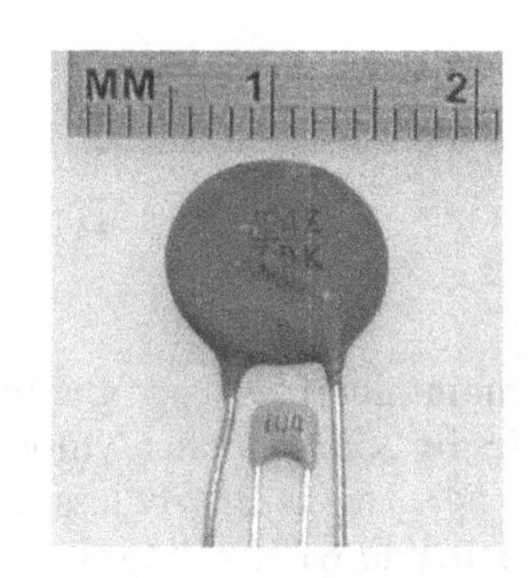

FIGURE 9.56

Comparing ceramic disk (top) and monolithic ceramic (bottom) capacitors, both 0.1 μF.

9.14.5 Monolithic Ceramic Capacitors

Monolithic ceramic capacitors are also made with a ceramic dielectric. However, their construction technique is different, being composed of a number of layers of ceramic that are fired together into a monolithic block. This leads to the alternate term for these capacitors: **multilayer ceramic** (**MLC**). Monolithic ceramic capacitors are a big step up from ceramic disks in terms of their capacitive density (capacitance per unit volume) and also offer much lower ESL. As a result of their smaller size (Figure 9.56), monolithic ceramic capacitors are favored over ceramic disks wherever size is an issue, a common requirement in many applications today. In other respects, especially their ability to maintain much of their capacitance to high frequencies, they are similar to ceramic disk capacitors. Something to be aware of in both types of ceramic capacitors is the dependence of the capacitance on voltage. Some of the most inexpensive ceramic dielectrics can lose up to 80% of their capacitance when operated at only 50% of their rated working voltage, and it is very common to begin losing a significant amount of capacitance at only 20% of the rated working voltage. For a higher price, it is possible to buy monolithic ceramic capacitors using a different dielectric formulation (often referred to as C0G or NP0) that have no variation in capacitance with applied voltage and a very low (30 ppm/°C) temperature coefficient. In modern designs, monolithic ceramic capacitors are the most common choice for local bypassing of both digital and analog integrated circuits (Chapter 21).

9.14.6 Aluminum Electrolytic Capacitors

Aluminum electrolytic capacitors take advantage of the property of aluminum that leads to the formation of a very thin (2–5 nm) insulating oxide on the metal's surface. When immersed in a conducting solution, this forms a *very* thin insulating layer between a pair of conducting elements (the foil and the solution). Remember that the capacitance is inversely proportional to the thickness of the insulating layer [d in Eq. (9.61)]—therefore large values of capacitance can be achieved with modest values of area, and without any need for precision manufacturing. Furthermore, a large area can be easily formed by taking a large sheet of aluminum foil, rolling it into a cylinder, and immersing the entire roll in a conducting solution. Aluminum electrolytic capacitors usually have a cylindrical shape reflecting this low-cost assembly method (Figure 9.57) and are available in values ranging from a few hundred nanofarads (nF) to hundreds of thousands of microfarads (μF). As great as this sounds, there are several significant disadvantages of electrolytic capacitors that must be enumerated. Because of the terminal construction, they typically exhibit relatively large value of inductance (ESL). Due to the dielectric properties of the fluid, they exhibit high-leakage currents (low values for R_L) and fairly strong temperature dependence, typically resulting in an overall change in the capacitance value over the operating temperature range of ±20%

FIGURE 9.57

Photograph of several typical aluminum electrolytic capacitors.

from the value specified at 25°C. With the typical initial component value accuracy of ±20%, parts can vary as much as ±44%! This is not quite as bad as some of the ceramic dielectrics, but it's still not impressive. Also, for applications with frequencies above 20 kHz, these aluminum electrolytic capacitors do not perform well. It is not uncommon for an aluminum electrolytic capacitor to have lost 30% of the capacitance that it exhibited at 120 Hz when operated with signals at 50 kHz.

Most importantly from an operational standpoint is that the oxide insulating layer in aluminum electrolytic capacitors is stable for voltages of one polarity only, but decomposes for the opposite polarity. As a result, aluminum electrolytic capacitors are polar capacitors. Therefore, it is important to make sure that the electrodes of the electrolytic capacitor are connected in the right orientation relative to the voltages in a circuit. If connected backwards, electrolytic capacitors can quickly self-destruct.

In many circuits, a capacitor is connected across the power supply terminals to help eliminate noise—we'll discuss more completely in Chapter 21 on Noise, Grounding and Isolation. Aluminum electrolytic capacitors are a common choice for this application, but care must be exercised to be sure that the capacitor is inserted with the proper polarity. An incorrectly inserted electrolytic capacitor exhibits a very low value for R_L (as labeled in Figure 9.53). In this situation, there is considerable current leaking through the capacitor, and there is some power dissipation (due to current through a resistance). This causes heating, further degrading the capacitor, which leads to further reductions in the value of the leakage resistance. If the power supply is capable of providing a sufficient amount of current (and they usually are), this situation can proceed until there is a lot of current, a lot of heat, and the potential for a catastrophic failure—even an explosion. Conductive electrolytic fluid can potentially be sprayed over your entire circuit. Generally speaking, it is very bad to shower your circuits with conducting fluids! It should be avoided. The reason aluminum electrolytic capacitors are in use is that they are the least expensive capacitors with values in the microfarad range, and the only types available above the hundreds of microfarads. When used properly and carefully, they're a great choice in many applications. However, the "+" and "−" labels for the terminals are very important.

9.14.7 Tantalum Capacitors

Tantalum capacitors are similar to aluminum electrolytic capacitors in that they also utilize a thin oxide layer to form the insulator. However, for tantalum, this insulating layer is thinner,

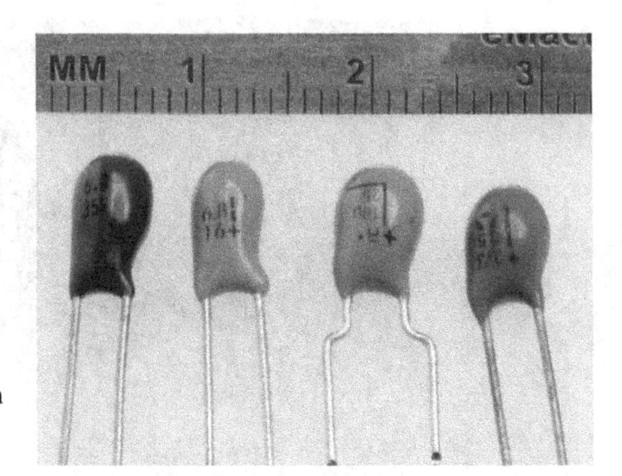

FIGURE 9.58

Photograph of several tantalum capacitors.

and has a larger value of relative permittivity, so a smaller package can contain a larger value of capacitance. Figure 9.58 shows a picture of several tantalum capacitors. In addition, the second electrode is usually an evaporated metal film on top of the insulating oxide, so there is no messy electrolytic fluid in these capacitors. Like aluminum electrolytic capacitors, tantalum capacitors are also polarized—so that applying the wrong bias voltage can lead to a failure of the device, just as for aluminum electrolytic capacitors. In contrast to aluminum electrolytic capacitors, tantalum capacitors get very hot when reverse-biased, and eventually, there is a good possibility of a heated pellet of tantalum exploding out of the package. Like electrolytic capacitors, tantalum capacitors also exhibit a relatively low initial accuracy (typically ±20%), though parts specified as tightly as ±5% are available at a price premium. Tantalum capacitors are more expensive than aluminum capacitors, but are often found in applications which need high capacitance in a small physical size. Finally, tantalum capacitors have good performance at frequencies up to about 100 kHz, but begin to degrade above that frequency. Their frequency performance is thus better than aluminum electrolytics but inferior to ceramic capacitors. Tantalum capacitors see widespread use in decoupling applications (Chapter 21).

9.14.8 Film Capacitors

Film capacitors are based on other thin insulating films, such as mica, mylar, polyester, polypropylene, and polystyrene. To achieve higher densities, they are sometimes made by vapor depositing a metal layer onto the plastic films, yielding what are referred to as **metalized film capacitors**. Even then, these capacitors are much larger than monolithic ceramic capacitors (Figure 9.59). Film capacitors have much better properties [accuracy of capacitance value (available at 1% tolerance), low temperature coefficient (2% total capacitance change over a temperature range from −40 to 80°C), low dielectric absorption] than the capacitor types described in the sections above, but are generally larger and higher in cost. However, if your application calls for an accurate, stable capacitor, these are usually the best choice.

9.14.9 Electric Double Layer Capacitors/Super Capacitors

These capacitors (sometimes generically referred to as **supercapacitors**) depend on the use of relatively recent advancements in dielectric materials and electrodes to produce capacitors

FIGURE 9.59

Photograph comparing 0.1 μF polyester film capacitor to a 0.1 μF monolithic ceramic capacitor.

with values that were previously unavailable in a single device (up to 5,000 F). They have much more limited operating voltage ranges than other types of capacitors, with most devices exhibiting a working voltage below 5.5 V and the largest values only achieved with working voltages in the 2.5 V range. They are typically used in energy storage applications where they deliver very large surge currents or maintain system power for short periods of time during battery replacement.

9.14.10 Capacitor Labeling

Capacitors are labeled according to a number of characteristics that are important to the designer and often with manufacturing information from the vendor. All capacitors will be labeled with at least the capacitance value, and if the capacitor is polarized, the polarity of the leads. In addition to this, many capacitors are marked with codes to indicate the tolerance on the capacitance value (see Table 9.1), the rated working voltage (see Table 9.2), and the temperature ratings; often for both working temperature range (see the left 2 columns of Table 9.3) and the variation in capacitance within that range (see the right column of

TABLE 9.1 EIA capacitor tolerance codes.

Code	Tolerance	Code	Tolerance
A	±0.05 pF	M	±20%
B	±0.1 pF	N	±30%
C	±0.25 pF	P	−0% ~ +100%
D	±0.5 pF	Q	−10% ~ +30%
E	±0.5%	S	±22%
F	±1%	T	−10% ~ +50%
G	±2%	U	−10% ~ +75%
H	±2.5%	Y	−20% ~ +50%
J	±5%	Z	−20% ~ +80%
K	±10%		
L	±15%		

TABLE 9.2 EIA capacitor working voltage codes.

Code	VDC
0G	4.0
0J	6.3
1A	10
1C	16
1E	25
1V	35
1H	50
1J	63
2A	100
2D	200

TABLE 9.3 Ceramic dielectric temperature ratings.

Low Temp	High Temp	Max. Cap. Change
X = −55°C	5 = +85°C	F = ±7.5%
Y = −30°C	6 = +105°C	P = ±10%
Z = +10°C	7 = +125°C	R = ±15%
		S = ±22%
		T = + 22% ~ −33%
		U = +22% ~ −56%
		V = +22% ~ −82%

Table 9.3). The conventions used to mark this information on capacitors vary somewhat based on capacitor type and, to a lesser degree, on the vendor, though there are industry-wide standards for many of the codes used. With an understanding of the standards and the ability to recognize the type of capacitor based on its appearance, it is relatively straightforward to interpret the markings.

When interpreting the markings on a capacitor, it is often helpful to have an idea about the range of values available for a particular type of construction. Figure 9.60 shows the ranges of values available using the types of capacitors that we've introduced.

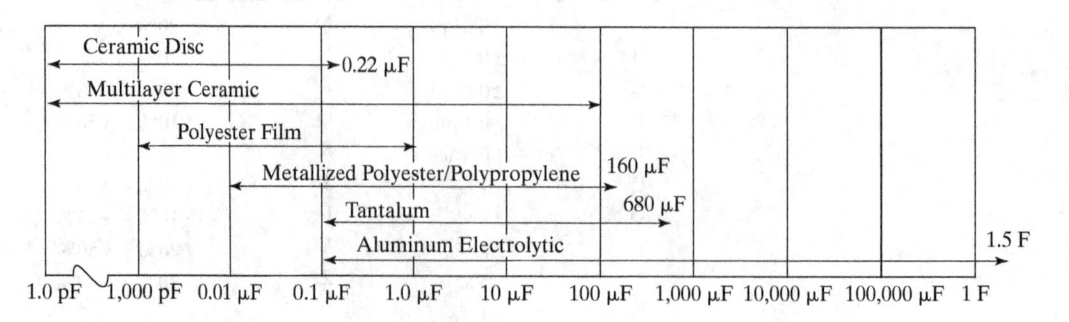

FIGURE 9.60

Capacitor value ranges by construction technology.

In the following sections, we'll present guidelines for determining the value of a capacitor based on the type of capacitor and its markings. Before we get to that, however, we need to introduce some of the cryptic standard codes that are used in these markings. These codes are based on standards promulgated by the Electronic Industries Alliance (EIA).

Tolerance on the capacitance value is either indicated directly in percent, or with a letter code, as shown in Table 9.1.

The working voltage rating will either be indicated directly with a numeric value followed by V or VDC or WVDC (Working Voltage DC), a one digit manufacturer-specific code, or by an EIA standard two-digit numeric code. The most common EIA codes are shown in Table 9.2.

9.14.10.1 Ceramic Capacitor (Disc and MLC) Labeling

Depending on their size, ceramic capacitors will be labeled directly in the number of pF or μF, or with a three-digit numeric code following an EIA standard that is similar to the scheme used in resistor markings. For the three-digit code, the first two digits represent the mantissa, with the decimal point to the right of the second digit, and the third digit is an exponent. This three-digit code indicates the number of pF. The most common standard values for the two-digit mantissa are: 10, 12, 15, 18, 22, 27, 33, 39, 47, 56, 68, and 82. There are values available in between these most common values, but they are generally special order parts. So, a capacitor labeled as "104" would be a 10×10^4 pF = 0.1 μF. As Figure 9.60 indicates, the range of values for ceramic capacitors only extends into the fractions of μF, so when labeled directly in μF, there will be a decimal point in the label.

Ceramic capacitors are most commonly available in tolerance codes J (±5%), K (±10%) and M (±20%), though special purpose parts are available with tighter tolerance. The variation in capacitance with temperature for ceramic capacitors is labeled in one of two ways. The temperature stable parts are labeled as C0G or NP0 (that's a zero, not an "O") and typically exhibit temperature coefficients of about ± 30 ppm/°C. The less stable parts are marked with a three-digit code to indicate the low-temperature operation limit, the high-temperature operation limit, and the maximum change in capacitance from the 25°C value. The most common values for these codes are shown in Table 9.3.

A very common low-cost ceramic dielectric is labeled as Z5U indicating a capacitor with a working temperature range from +10°C to +85°C and a maximum capacitance change between +22% and −56% from the nominal 25°C value.

9.14.10.2 Aluminum Electrolytic Capacitor Labeling

Aluminum electrolytic capacitors tend to be large, both in their physical size and their capacitance. This results in a lot of real estate on the capacitor body for marking, so electrolytic capacitors are usually marked in a highly readable fashion. The capacitance is most often simply printed as the value, most often followed by μF or MFD. The working voltage is indicated by a voltage value followed by V, VDC, or WVDC. Since these are polar devices, some indication of the correct polarity is required. On modern aluminum electrolytic capacitors, the polarity is indicated by a negative symbol (−), typically as part of an arrow, which indicates the negative lead.

9.14.10.3 Tantalum Capacitor Labeling

Because of their smaller size, tantalum capacitors may be marked either directly in μF, like aluminum electrolytics, or using a three-digit EIA numeric code. This can result in confusion that can be resolved by referring to Figure 9.60. If we find what appears to be a tantalum capacitor labeled as "220," what is its value? The number 220 could reasonably be interpreted as an indication of the number of μF (220 μF), or as a three-digit EIA code (22×10^0 pF = 22 pF). Figure 9.60 shows that the range of values for tantalum capacitors extends from 0.1 to 470 μF. Since 22 pF would not be a reasonable value for a tantalum capacitor, we can deduce that the actual value is 220 μF. Tantalum capacitors are also polar and require an indication of the correct polarity of the leads. The convention for tantalum capacitors is to place a plus sign (+) or a dot near the positive lead.

9.14.10.4 Film Capacitor Labeling

Film capacitors are usually marked either directly in μF (with a decimal) or using the three-digit EIA code. They exhibit better tolerance control and are typically available and marked with tolerance codes F, G, and J.

9.14.11 Choosing a Capacitor

Given the great variety of capacitors available, the process of choosing a particular type for a given application may seem daunting. To get you started, we will provide some guidance that will get you to a choice that will work well, though it may not be the optimum choice. To simplify the process of choosing a capacitor, we'll look at the four broad categories of capacitor usage that we introduced in Section 9.14: timing, filtering, AC coupling, and bypassing/decoupling.

Timing is most influenced by the accurate charging and discharging of the capacitor. Performing this task well requires the capacitor to have at least relatively high initial accuracy (tight tolerances on the value of C), good stability over temperature and working voltage, and very low leakage and dielectric absorption. As a group, the plastic film capacitors are the clear choice here.

Filtering relies on the behavior of the capacitor when exposed to a range of frequencies, since the purpose of filtering is to pass one range of frequencies while rejecting others. For this type of application, we need a capacitor that is ideally initially accurate, stable over the range of operating temperatures, and maintains its capacitance over the range of operating frequencies. For relatively low-frequency applications (< 100 kHz), the plastic film capacitor is again an excellent choice. Monolithic ceramic capacitors (especially in the C0G dielectric) can also be a good choice and at higher frequencies (> 1 MHz) become the best choice.

AC coupling is most concerned with blocking a DC voltage while passing the AC component(s). In applications like this, the leakage characteristics are most important, and initial accuracy and temperature dependency are typically less important. Some of the plastic film capacitors, especially Teflon, polystyrene, and polypropylene have the lowest leakage values and are therefore excellent choices for these applications. Monolithic ceramic capacitors, though showing somewhat higher leakage, are also a reasonable choice and lower in cost.

Bypassing/decoupling is mostly concerned with providing a local store of charge that can be delivered quickly to provide for transient current demands. It is almost insensitive to capacitance value on the high side, requiring only that some minimum value be maintained. Leakage is not usually an issue since most bypassing is done across the power and ground rails of a system or a specific device. Bypassing/decoupling is applied in two complementary cases: bulk bypassing (for circuits or systems as a whole), and local bypassing, (for individual devices within a circuit). For bulk bypassing, relatively large amounts of capacitance are required—typically tens to hundreds of microfarads. This is the domain of the aluminum electrolytic and tantalum capacitor. If size is of significant importance or the speed of the transients high, then tantalums are a better, though more expensive, choice. It is common to see several types of capacitors used in parallel in this case: a printed circuit board might have one large electrolytic capacitor near the power connection, with a few tantalums distributed around the board. Distributing capacitance around the board minimizes the effects of the naturally occurring inductance in the printed circuit board traces. For local bypassing, it is common to place a small ceramic capacitor (0.01–0.1 μF) at each integrated circuit. In order to minimize the effects of lead and wiring inductance and provide the fastest possible response to transients, these local bypass capacitors should be placed as close as practical to the power and ground pins of each device. It is advisable to add such decoupling capacitors to all integrated circuits (this will be covered in more detail in Chapter 21, when we discuss electrical noise and grounding).

9.15 HOMEWORK PROBLEMS

9.1 If, for a particular two-terminal device the voltage and current entering one terminal is 7.5 V and 1 A, and the voltage and current leaving the other terminal is 5 V and 1 A, how much power is being dissipated in the device?

9.2 What is the resistance seen across terminals A and B in Figure 9.61?

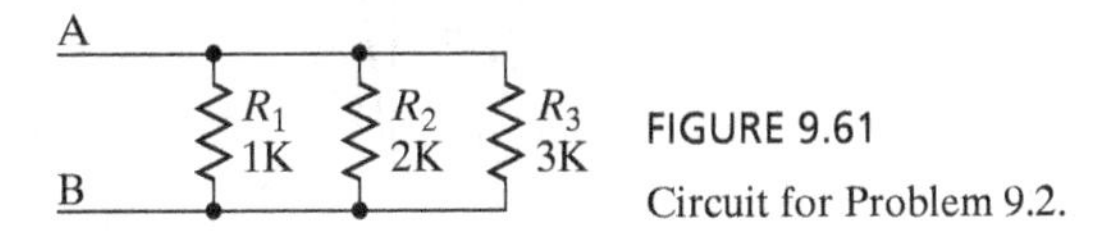

FIGURE 9.61

Circuit for Problem 9.2.

9.3 Determine the Thevenin equivalent circuit for the subcircuit block labeled A in Figure 9.62.

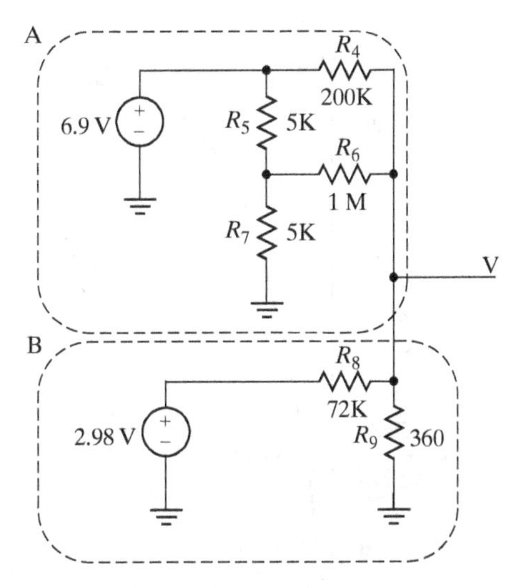

FIGURE 9.62

Circuit for Problems 9.3–9.6.

9.4 Determine the Thevenin equivalent circuit for the subcircuit block labeled B in Figure 9.62.

9.5 What is the voltage at the point labeled V in Figure 9.62?

9.6 If the voltage at point V in Figure 9.62 were measured with a voltmeter with a 100 kΩ internal resistance, what voltage would be measured?

9.7 The graph in Figure 9.63 represents the output of an RC circuit excited by a 0–5 V rising edge. Draw the RC circuit configuration. Based on the data shown, what is the time constant?

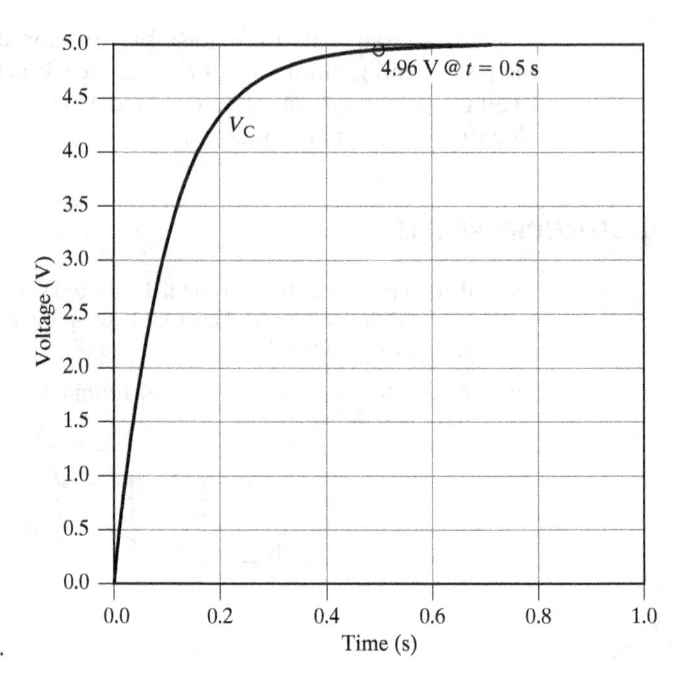

FIGURE 9.63

Waveform for Problem 9.7.

9.8 Figure 9.64 shows the current waveform for a series RL circuit excited by a 0–5 V rising edge.

(a) Estimate the total resistance seen by the exciting voltage source.

(b) Estimate the inductance in the circuit.

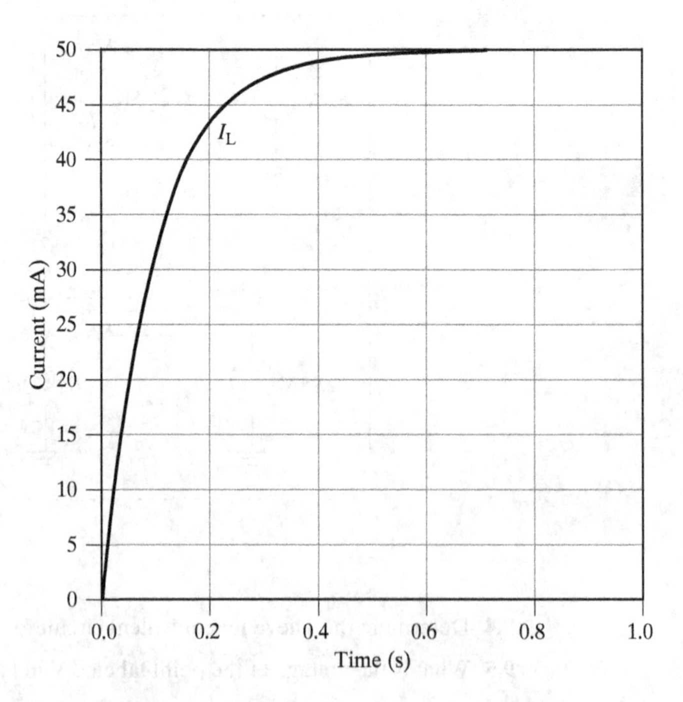

FIGURE 9.64

Waveform for Problem 9.8.

9.9 Design a circuit that takes a sinusoidal input of varying frequency at 1 V amplitude and produces an output amplitude of approximately 1 V when the frequency is 1 Hz and approximately 0.5 V when the frequency is 1,000 Hz. Choose real component values to get as close as possible to the specified amplitudes.

9.10 For the circuit that you designed in Problem 9.9, at what frequency is the output approximately 0.707 V?

9.11 What is the minimum voltage that you would need to apply across a 10 kΩ, 1/8 W resistor in order to cause its power rating to be exceeded?

9.12 If a resistor with a temperature coefficient of 1,000 ppm that measures 997 Ω at 25°C is cooled to −40°C (a cold night in Minnesota), what range of resistances would you expect to see? How about if it were heated to 65°C (inside a car on a sunny summer day in Arizona)?

9.13 Which capacitor would you expect to be physically larger: a 0.1 μF ceramic disk or a 0.1 μF monolithic ceramic capacitor?

9.14 Which capacitor would you expect to be physically larger: a 10 μF electrolytic or a 10 μF tantalum capacitor?

9.15 If a monolithic ceramic capacitor were labeled 103X5F, what would you expect its capacitance value to be?

9.16 You have been presented with two monolithic ceramic capacitors. One is labeled 104C0G, the other 104Z5U. How are these capacitors alike? How are they different?

9.17 If you were designing an RC low-pass filter to operate with a corner frequency of 10 kHz for a signal of a few volts amplitude at less than 0.1 mA, what types of resistors and capacitors (i.e., which values and technologies) would be good choices?

9.18 Design a circuit that takes a sinusoidal input of varying frequency at 1 V amplitude and produces an output amplitude of approximately 1 V when the frequency is 1,000 Hz and approximately 0.5 V when the frequency is 1 Hz. Choose real component values to get as close as possible to the specified amplitudes.

9.19 Some monolithic ceramic capacitor materials exhibit a voltage dependency. What happens as the applied voltage is increased?

9.20 List two examples each of polar and nonpolar capacitor types.

FURTHER READING

Practical Electronics for Inventors, Scherz, P., 2nd ed., McGraw-Hill, 2006.

Simulation tools with free versions:

- PSpice 9.1 Student Version, Orcad/Cadence
- LTSpice, Linear Technology
- TopSPICE, Penzar
- TinaTI, Texas Instruments

Commercial products with low-cost student versions:

- Tina, DesignSoft
- MultiSim Student Edition, National Instruments

CHAPTER 10

Semiconductors

In this chapter, we continue our introduction to electronic components with material on the broad class of devices known as **semiconductors**. As the name implies, the materials that make up these devices exhibit neither the low resistance typical of metals and other **conductors** nor the very high resistance associated with **insulators**. In the pure state, these materials, the most common of which is silicon (Si), are generally unremarkable. Their usefulness increases dramatically when very small amounts of specific materials are added to the pure semiconductor, the conductivity of the resulting material can be modified by applied voltages. By combining regions or layers that are seeded (in the parlance of physics: **doped**) with different materials, we can make many different kinds of devices with very useful properties. Semiconductors are the basic building blocks for all of the more complex electronic devices that we will deal with in later chapters. When many semiconductors and resistors or capacitors are combined into a single device, the device is referred to as an **integrated circuit** (**IC**). In this chapter, we will focus on the semiconductor devices that are most useful as discrete (stand-alone) devices and cover the behavior of ICs in later chapters in this section of the book. In keeping with the overall philosophy for the book, we focus our attention on the external behavior of semiconductors (rather than the internal device physics) and how we can use those behaviors to build useful circuits. When dealing with transistors, we will further limit ourselves to dealing with the devices as switches. Topics covered in this chapter include:

1. the types of diodes and their basic behaviors,
2. the behavior of the bipolar junction transistor (BJT) and how to use it as a switch,
3. the behavior of the metal oxide semiconductor field effect transistor (MOSFET) and how to use it as a switch,
4. how to extract the most pertinent information from a device data sheet in order to be able to apply a device in a particular application.

To help foster the development of intuition about the behavior of these devices, we will also continue using the fluid flow analogy that we introduced in Chapter 9 to describe semiconductor behavior.

After reading this chapter, you should be able to look at simple circuits of both linear elements (voltage sources, resistors, and capacitors) and semiconductors and be able to qualitatively describe the basic functionality of the stages of the overall circuit. You should be able to expand upon that qualitative description with the analysis necessary to determine that each transistor is being operated in a regime that will cause it to behave as a switch that is either off or on. You should understand the characteristics of the different types of diodes and transistors and be able to select devices that are appropriate for a particular application.

10.1 DOPING, HOLES, AND ELECTRONS

Silicon (Si) is by far the most common semiconductor material used to make electronic devices. While there are other semiconductors that are used, most notably germanium and gallium arsenide, these materials are used for special purpose devices where their different properties yield some specific advantage. While an understanding of semiconductor device physics is not necessary to make use of these devices in design, a bit of background information about the properties of the materials will go a long way in understanding the terminology that we use to describe these devices.

Pure, crystalline silicon is not of much use to a device designer. Its four valence electrons are tied up in covalent bonds with the adjacent silicon atoms in the crystal. Silicon becomes a useful electronic material only when it is **doped** by adding small amounts of other atoms to the crystal structure. The most common of these **dopants** (the atoms added to the silicon crystal) are boron (B) and phosphorus (P).

Silicon has four valence electrons, while phosphorus has five. When phosphorus is placed into a crystal with silicon, four of the valence electrons from the phosphorus will participate in covalent bonds with the adjacent silicon atoms, while one of the electrons will be more loosely bound to the crystal (since it is not participating in a covalent bond). If an external voltage (electric field) is applied to this material, the loosely bound electron can migrate through the crystal lattice, resulting in a current flow. Because it has a loosely bound electron (with a corresponding negative charge), this type of material is referred to as an **n-type** semiconductor. Don't be misled by the name. The material does not have a net negative charge; the extra valence electron in the phosphorus is balanced by an extra proton in the nucleus of the atom.

Boron, unlike silicon or phosphorus, has only three valence electrons. When boron is placed into a crystal with silicon, the three valence electrons from the boron will participate in covalent bonds with three of the adjacent silicon atoms, leaving a "**hole**" with the fourth adjacent silicon atom. If a voltage is applied to this material, one of the electrons from an adjacent silicon atom can be induced to move and to fall into this hole. In so doing, this electron has left another hole where it used to reside. In this way, the hole traverses the crystal lattice under the influence of the applied voltage. Even though the hole has no charge of its own, it is considered to be a positive charge carrier and this material is referred to as a **p-type** semiconductor. As with the n-type material, there is no net charge in p-type material, since the boron atom, in addition to having one fewer valence electrons, has one fewer protons in its nucleus.

With this bit of background about semiconductor materials and the related terminology, we can move on to the most interesting parts: what kinds of devices are made with these materials and how we can use them.

10.2 DIODES

Diodes are the simplest semiconductor devices. They allow current to flow in one direction while preventing flow in the reverse direction. They are most commonly formed when a region of p-type material is placed so that it is in contact with a region of n-type material. Their electrical behavior is analogous to the hydraulic check valve. In Figure 10.1a, the pressure on the right is higher than the left, but the check valve is forced closed and prevents flow. In Figure 10.1b, the pressure on the left is higher than the pressure on the right, which forces the valve open to allow flow.

The analogous situations for diodes are shown in Figure 10.2. When the cathode is at a higher potential than the anode, there is essentially no current flow (Figure 10.2a). When the anode is at a higher potential than the cathode, current flows (Figure 10.2b) but with some loss (a voltage drop). To remember the direction of current flow, it is helpful to note that current flows in the direction of the arrowhead in the symbol. To remember the naming of the

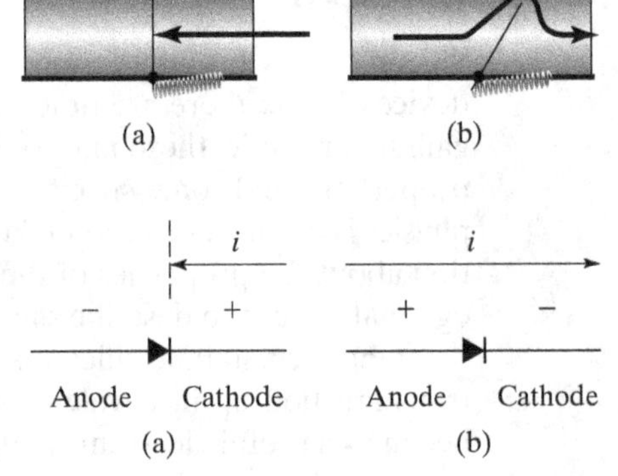

FIGURE 10.1

A check valve: the hydraulic analogy to a diode. In (a) flow is being blocked, while in (b) flow is being allowed.

FIGURE 10.2

Diode behavior. In (a) the diode is blocking flow, while in (b) it is allowing flow.

terminals, note that when current is flowing, the flow is from **a**node to **c**athode—that is, the current flows alphabetically.

10.2.1 The *V-I* Characteristic for Diodes

The basic voltage vs. current (or *V-I*) characteristic for a diode is shown in Figure 10.3. The right-hand side of the graph, where the anode is at a higher potential than the cathode, is referred to as the **forward bias** region. Once the voltage exceeds a given threshold, known as the **forward voltage** (V_f) for a diode, the current flow through the diode increases dramatically. Below this voltage, there is very little current flow. The left-hand side of the graph shows the opposite condition, when the anode is at a lower potential than the cathode. This is referred to as the **reverse bias** region. Under these conditions, only a tiny **leakage current** (I_l) flows in the diode from the cathode to the anode, until the reverse voltage exceeds a threshold known as the **reverse-breakdown voltage** (V_r). When the reverse-breakdown voltage is exceeded, significant currents begin to flow in the direction opposite that for forward bias. For all but Zener diodes (Section 10.2.5), exceeding the reverse-breakdown voltage is a destructive process that should be avoided.

We will make use of this behavior to propose a simplified behavioral model for the diode:

1. When reverse-biased, no current flows through the diode.
2. When forward-biased by less than V_f, no current flows.
3. When forward-biased by a source capable of more than V_f, the voltage drop across the diode will be V_f and the diode will not limit current flow.

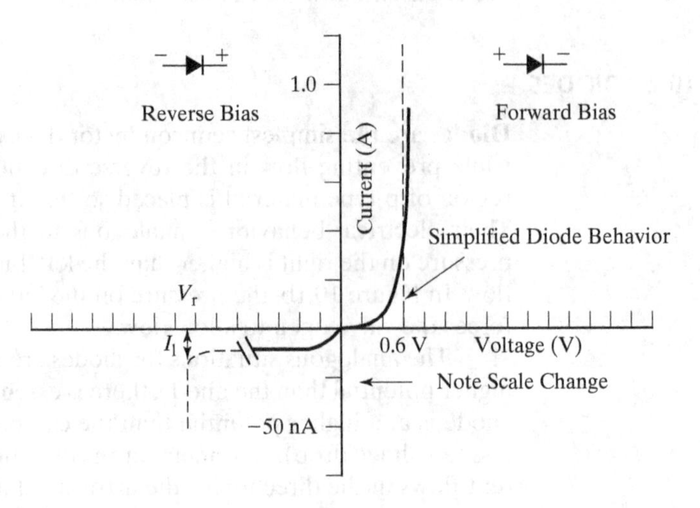

FIGURE 10.3

The *V-I* characteristic for a diode.

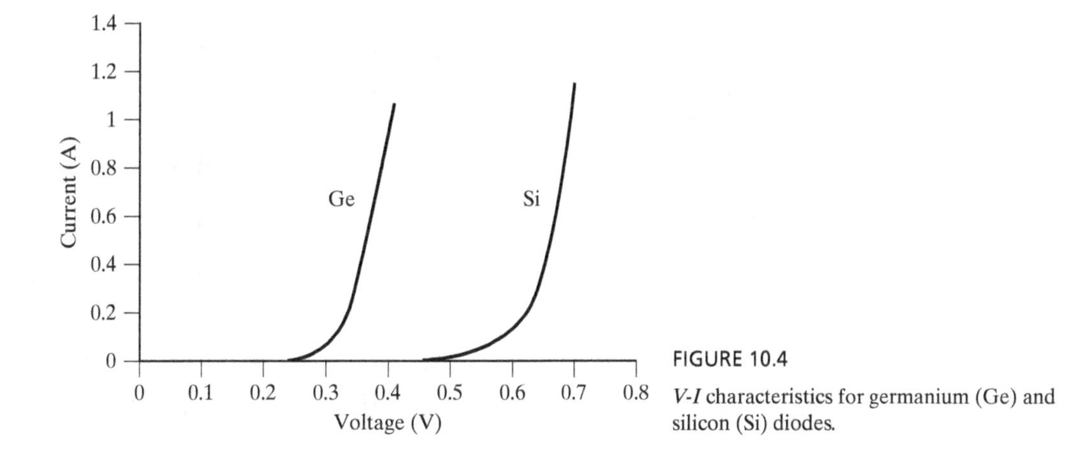

FIGURE 10.4

V-I characteristics for germanium (Ge) and silicon (Si) diodes.

10.2.2 The Magnitude of V_f

The magnitude of V_f depends on the materials used in the construction of the diode and the specific dopants used to make the p- and n-type materials.

Figure 10.4 shows a comparison of the *V-I* characteristics for diodes whose base semiconductor material is germanium vs. those based on silicon. When current on the order of a few tens of milliamps flows through these diodes (the most common conditions for signal level circuits), the forward voltage (V_f) for a germanium diode (e.g., 1N34A) is on the order of 0.3 V, while that of a silicon diode (e.g., 1N4001) is on the order of 0.6 V. While the germanium diode does exhibit significantly lower V_f compared to the silicon diode, it also exhibits leakage currents that are 100–1,000 times larger than those found in silicon diodes when it is reverse-biased.

We will use these values of V_f to augment our simplified behavioral model of the diode to create a rule of thumb: in the absence of specific information about a particular diode, we will treat silicon diodes as having a V_f of 0.6 V and germanium diodes as having a V_f of 0.3 V.

10.2.3 Reverse Recovery

When a diode transitions from being forward-biased to being reverse-biased, it isn't capable of instantaneously eliminating the flow of current. Rather, a small but finite amount of time is required, which is called the **reverse recovery period**. This is shown graphically in Figure 10.5.

In the short period of time after the bias voltage is reversed, the forward current first falls and then turns into significant reverse conduction for a short period of time (hundreds of nanoseconds). For most ordinary diode applications, the current levels are low enough and the reverse recovery period is short enough that this does not cause significant problems. The reverse recovery period of the diodes is important in some applications, for example, when substantial reverse currents can flow. We will encounter such conditions in Chapter 23, when using diodes in DC motor drive circuits.

10.2.4 Schottky Diodes

There are many more diode types than just the germanium and silicon diodes that we have introduced so far. One of the more common diodes that we encounter in mechatronic systems is the Schottky diode (Figure 10.6).

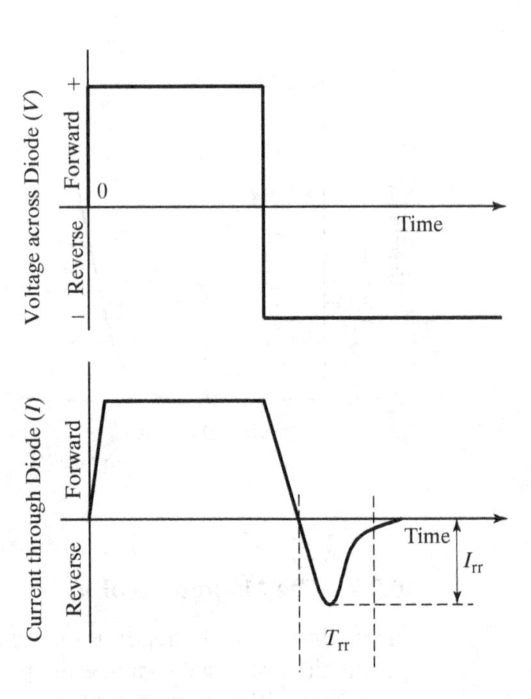

FIGURE 10.5
Reverse recovery time for a diode.

FIGURE 10.6
Schematic symbol for a Schottky diode.

Where silicon and germanium diodes are made by creating a junction between two differently doped semiconductor materials (a p-n junction), the **Schottky diode** is created at the junction between a metal and a semiconductor. This gives the Schottky diode several useful characteristics as compared with p-n junction diodes.

The forward voltage for a Schottky diode (e.g., 1N5817) is significantly less than that for a silicon diode. Typically between 0.3 and 0.4 V, the V_f of a Schottky diode approaches that of a germanium diode. Schottky diodes also exhibit faster switching between conducting and blocking, having essentially 0 reverse recovery time. Leakage current levels for Schottky diodes fall between those of germanium and silicon diodes. The lowest leakage Schottky diodes have leakage currents comparable with ordinary silicon diodes (as opposed to "low-leakage" silicon diodes). However, some Schottky diodes exhibit reverse leakage levels on the order of 10 times that of ordinary silicon diodes.

10.2.5 Zener Diodes

Zener diodes (Figure 10.7) are the exception to the rule about avoiding reverse breakdown that we introduced in Section 10.2.1. Zener diodes are designed in such a way that the reverse breakdown process is not destructive.

When forward-biased or reverse-biased at voltages less than the **Zener voltage** (V_Z), Zener diodes behave much like ordinary silicon diodes (Figure 10.8).

FIGURE 10.7
Schematic symbol for a Zener diode.

As the reverse bias voltage is increased to the point approaching the Zener voltage, the diode begins conducting in the reverse direction. For an ideal Zener diode, the *V-I* characteristic goes vertical at the Zener voltage, and the voltage drop across the diode from cathode to anode is clamped at the Zener voltage, V_Z. For real Zener diodes, the *V-I* characteristic at the Zener voltage has a small slope that results in small changes in the voltage drop across the diode as a function of the current flowing through the diode.

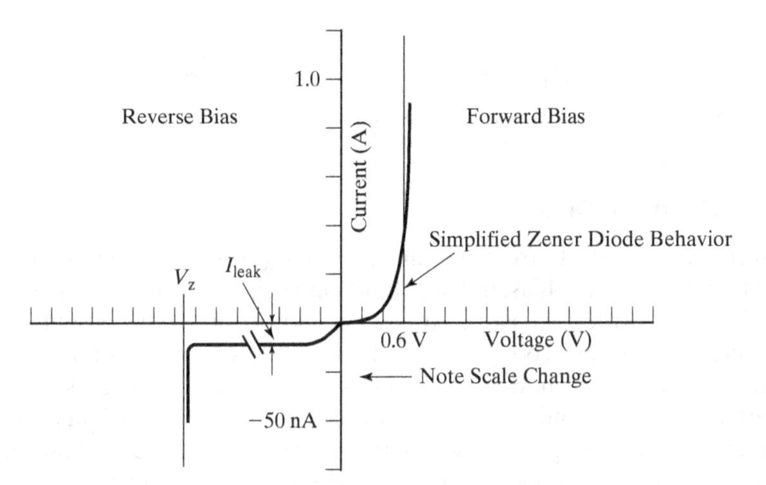

FIGURE 10.8

V-I characteristic for a Zener diode.

If we combine a Zener diode with a resistor in a circuit, such as that shown in Figure 10.9, we can make a circuit that serves as a voltage regulator. To understand how it works, first consider the situation when no current is drawn at Node A toward V_o ($i_l = 0$). If V_i exceeds V_z for the Zener diode D_1, then the voltage at Node A will be held at V_z and a current $I_z = (V_i - V_z)/R_1$ will flow in D_1. Now think about what would happen if some current, i_l, were drawn at V_o. If all other conditions were held constant, the added current going to the output would cause an increased voltage drop across R_1 and the voltage V_o would fall. However, if the voltage across the diode were to fall, then the current through the diode would also fall, compensating for the increase in current i_l. When the current flowing out (i_l) changes, the current through the Zener diode also changes so that i_t is held constant, and the voltage drop across the diode is fixed at V_z. As long as the current i_1 does not exceed V_z/R_1, the current through R_1 will be constant, and the Zener diode will modulate the current i_z to maintain point A at V_z.

This does produce a functional voltage regulator circuit, but it suffers from several drawbacks: inefficiency, limited power dissipation, and lack of regulation. For this approach, the current drawn from V_i is constant, even when there is no current delivered to the output, i_l. In this case, all of the input power is being turned into heat. Notice that if no current is drawn from i_l, then all of the current necessary to create the voltage drop ($V_i - V_z$) across R_1 must flow through D_1, where the power $i_z \times V_z$ is converted to heat. Any real Zener diode will have a limited power dissipation capability, placing a limit on the power that could be delivered to the load. If the *V-I* characteristic were vertical at V_z, then this would at least be a very stable regulator with respect to voltage. Unfortunately, in real devices, there is a slope to the *V-I* characteristic beyond V_z, so the output voltage does vary somewhat as a function of the current drawn (i_l). For some applications, this variation is acceptable. For those where it is not, or where more efficiency or high levels of output current are necessary, we cover IC voltage regulators in Chapter 20.

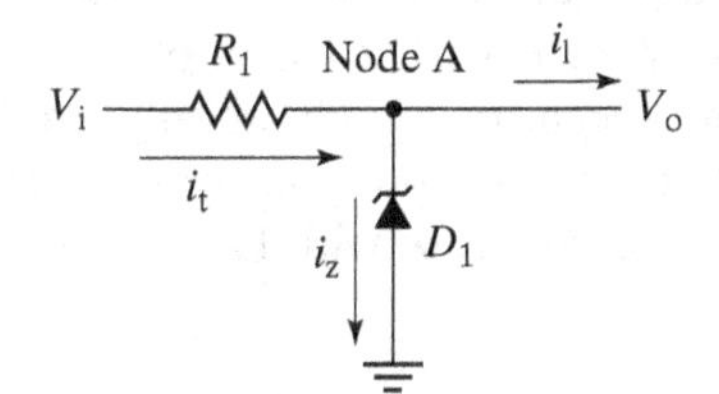

FIGURE 10.9

A simple Zener voltage regulator.

FIGURE 10.10
Schematic symbol for an LED.

10.2.6 Light Emitting Diodes

Semiconductor diodes made from certain materials (most often containing gallium) emit light when they are forward-biased. These **light emitting diodes** (or **LEDs**) were first used as indicators. Recently, with increases in the maximum possible light output, they are finding use in lighting applications. Today, we find them used not only as indicators on electronic equipment but also in automotive brake lights, traffic lights, flashlights, bicycle headlights, and they are the newest trend in home and office lighting because of their high efficiency in converting electricity into light. The schematic symbol for an LED is shown in Figure 10.10.

From an electrical standpoint, their behavior differs from silicon diodes in only a few substantive ways. The first is that the forward voltage drop, V_f, is much larger than that of a standard silicon diode. In addition, the forward voltage depends on the color of light emitted, with higher frequency corresponding to higher values of V_f (Figure 10.11). The second, more subtle difference, is that in the forward conducting region, there is more of a distinct slope to the *V-I* characteristic, rather than approaching a vertical line. This indicates that there is a more substantial change in forward voltage with current than we see in silicon diodes.

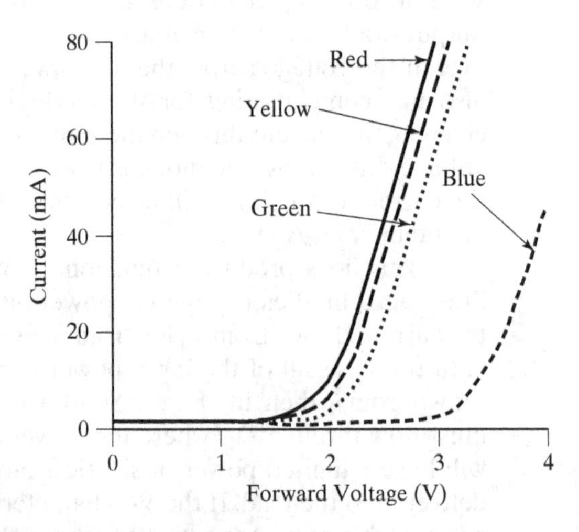

FIGURE 10.11
V-I characteristics for LEDs.

In its simplest application, using an LED requires nothing more than a voltage source and a series resistor to limit the current flow through the device (Figure 10.12). The resistor, R_1, is chosen such that $(V_S - V_f)/R_1$ meets the forward current limitations for the LED. If the voltage source (V_S) is an output pin from a microcontroller, then the designer must be careful to also select R_1 so that excessive current is not drawn from the pin. In either case, the resistor R_1 is essential to avoid damage to either the LED or the driving source. Typical indicator LEDs are designed to operate with 20 mA of forward current, though low power LEDs are available that produce significant light at only 2 mA. These latter devices are especially well matched to microcontroller output pins, which typically can only source a few milliamps at most. Illumination LEDs operate with DC currents in the range of a few hundred milliamps.

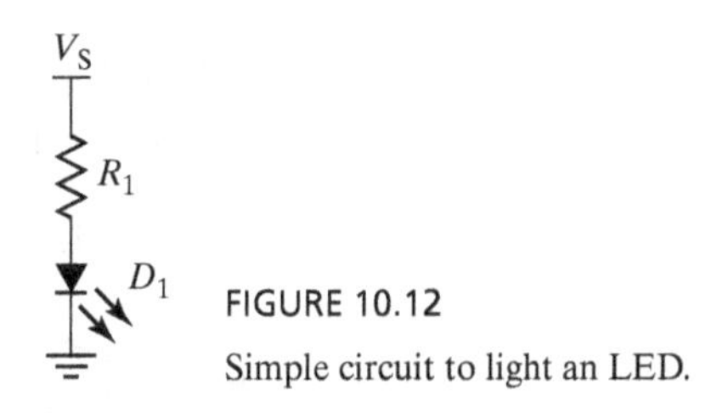

FIGURE 10.12
Simple circuit to light an LED.

10.2.7 Photo-Diodes

One final type of diode that we'll introduce briefly here and describe in far greater detail in Chapter 13 on Sensors, is the photo-diode. Photo-diodes are, in a sense, the complement to LEDs. Rather than emitting light (as LEDs do), **photo-diodes** exhibit a sensitivity to light. The schematic symbol for a photo-diode is shown in Figure 10.13. Note that the photo-diode symbol is very similar to the symbol for the LED, differing only in the direction of the arrows indicating the trajectory of photons.

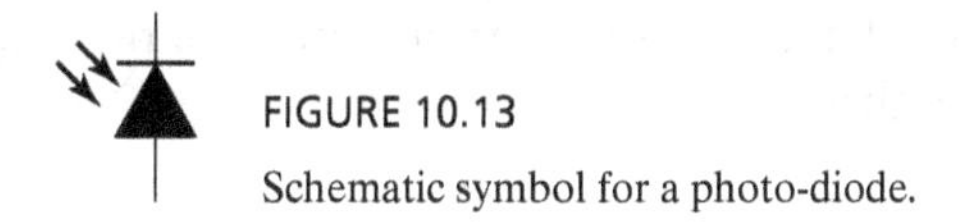

FIGURE 10.13
Schematic symbol for a photo-diode.

The amount of light falling on the junction varies the amount of leakage current that flows through the photo-diode (Figure 10.14).

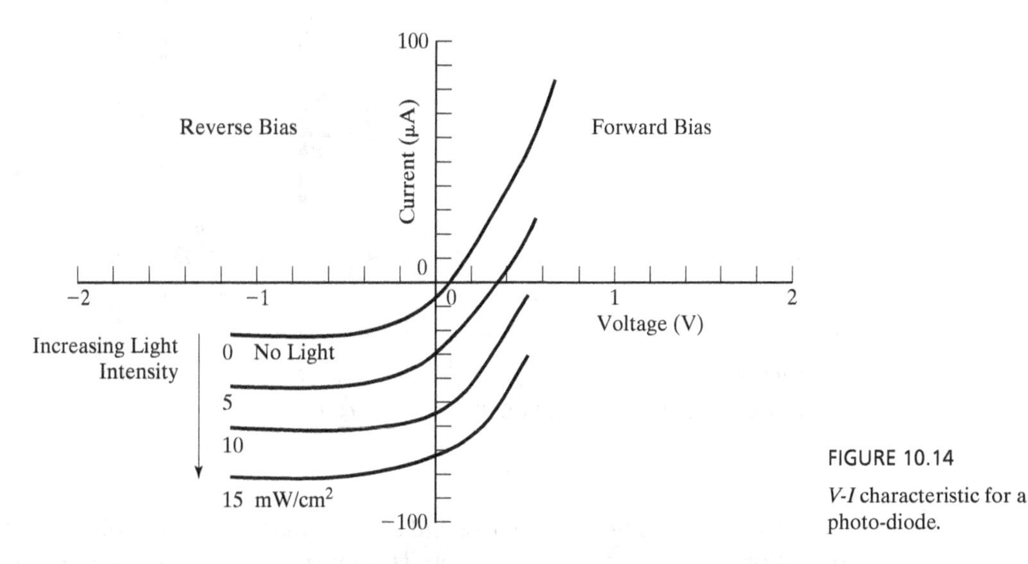

FIGURE 10.14
V-I characteristic for a photo-diode.

We'll revisit photo-diodes and discuss circuits to use them in more detail in Chapter 13 on Sensors.

10.3 BIPOLAR JUNCTION TRANSISTORS

A p-n junction is made by joining layers of p-type and n-type material. If we add a third layer to the mix, we create a **bipolar junction transistor** (**BJT**). Given the two types of materials, there are two possible ways to add that layer, one creates a P-N-P sequence and the other

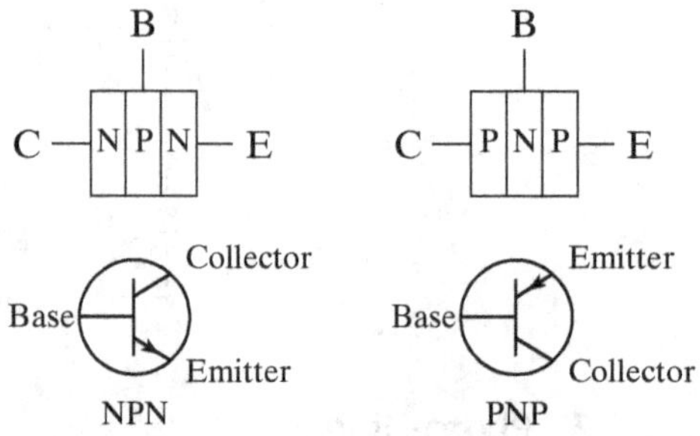

FIGURE 10.15
PNP and NPN BJTs.

creates an N-P-N sequence. These are the two basic types of BJT (Figure 10.15). The surrounding circle is sometimes omitted from the symbols.

The BJT acts as a current controlled valve, with a small current at the **base** controlling a larger current flowing between the **collector** and **emitter** (Figure 10.16). Using a BJT transistor, then, devices that require a substantial amount of current can be controlled using a much smaller amount of current. For example, a microcontroller can switch a motor on and off through a BJT.

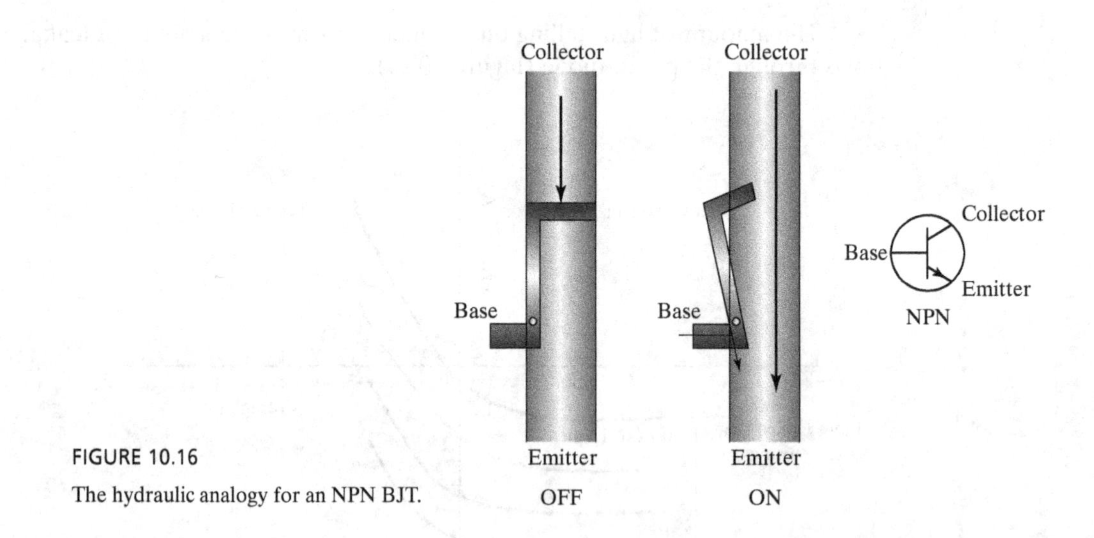

FIGURE 10.16
The hydraulic analogy for an NPN BJT.

The diode formed by the base and emitter must be forward-biased and conducting current in order to turn the transistor on. This means, for example, that for an NPN transistor, the base voltage must be approximately 0.6 V *above* the voltage at the emitter. For a PNP transistor, the base must be approximately 0.6 V *below* the emitter voltage and conducting in order to turn the device on. Figure 10.17 summarizes the conditions necessary to turn BJTs on and off.

Figure 10.18 shows the preferred circuit for creating a switch using an NPN transistor. In this circuit, the control voltage, V_{in}, causes a current, i_B, to flow into the base of the transistor. The base current controls a larger current, I_C, that flows from collector to emitter. Because the switching element is between the load and ground, this configuration is often referred to as **low-side drive**.

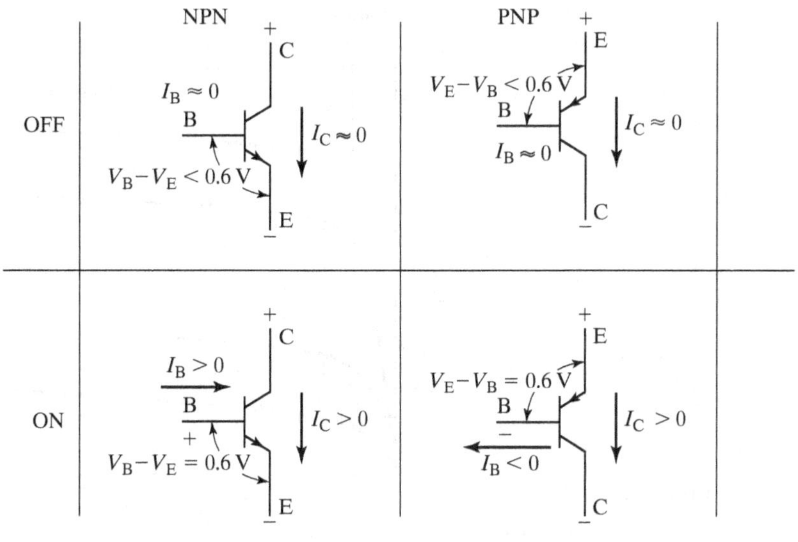

FIGURE 10.17
Controlling BJTs.

To characterize the behavior of the transistor in this configuration, we plot collector current, I_C, as a function of the collector-to-emitter voltage, V_{CE}, for varying amounts of base current, I_B (Figure 10.19).

The region to the right of the dotted line is known as the **linear region**. This region is characterized by relatively large collector-to-emitter voltage drops (V_{CE}) and a strong correlation between base current (I_B) and collector current (I_C). The region to the left of the dotted line is called the **saturation region**, and it is characterized by smaller collector-to-emitter voltage drops and a distinct lack of a simple or obvious relationship between base current and collector current. Note that there is a region, where the curves for the different levels of base current merge into a single line. In this region, the points along the three curves overlap, and it is impossible to tell how much base current was present based on measurements of V_{CE} and I_C. In this region, the transistor is said to be in **saturation**. the transistor's V_{CE} isn't significantly affected by increasing the base current. For example, at the middle level of base current, adding more base current does not change either the collector-to-emitter voltage or the amount of collector current flowing.

When we are trying to use a transistor to make a switch, we would like the collector-to-emitter voltage drop, V_{CE}, to be as small as possible. In this way, we maximize the amount of the supply voltage that appears across our load. This means that to make a switch, we want to operate the BJT in the saturation region, not in the linear region. Operating in the saturation region will deliver the greatest voltage to the load and result in the minimum power dissipation ($V_{CE} \times I_C$), or waste heat, in the transistor. Operating in the linear region, with its larger V_{CE}, will result in much greater power dissipation in the transistor.

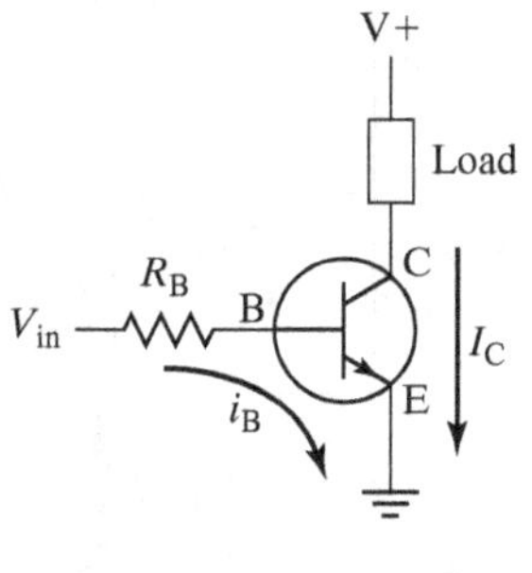

FIGURE 10.18
The common emitter configuration.

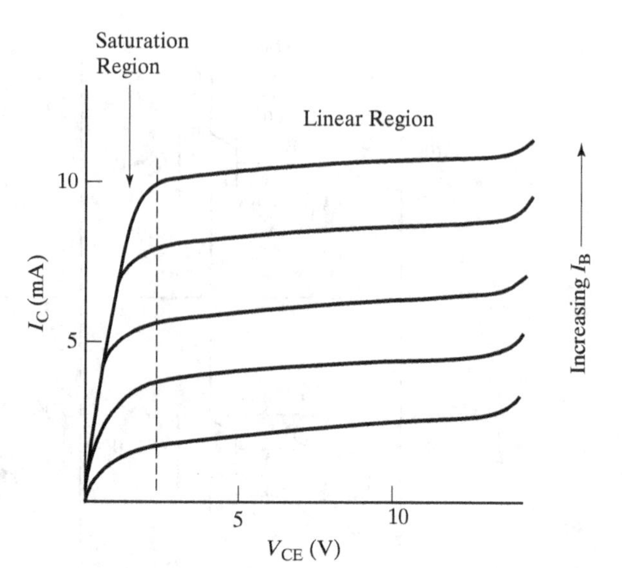

FIGURE 10.19

Transistor behavior curves.

Since we want to operate the transistor in the saturation region, the next question that we need to answer is "How can we be sure that we have enough base current to insure that we are in the saturation region?" In the linear region, the correlation factor that relates base current to collector current (the effective **current gain**) is known as h_{fe} or β. This factor, which is tabulated in the data sheet for the device, can range from 50 to several hundred and depends not only on the type of transistor but also varies among samples of a particular type. When using a BJT as a switch, this current gain value is not relevant because it is based on the transistor's behavior in the linear region. For any BJT, as the operating point moves from the linear region into the saturation region, the effective current gain drops as we get more deeply into saturation and achieve lower and lower V_{CE}. This is shown graphically in Figure 10.20, which plots current gain (β or h_{fe}) as a function of V_{CE}.

Notice that there is not much improvement (lowering) in V_{CE} for current gains (I_C/I_B) below 10–20. This behavior is consistent across a wide range of BJTs, from small low-current switching devices up to large high-current power devices. The lowest value of V_{CE} that is achieved will vary across devices and the absolute value of I_C, but the fact that there is little

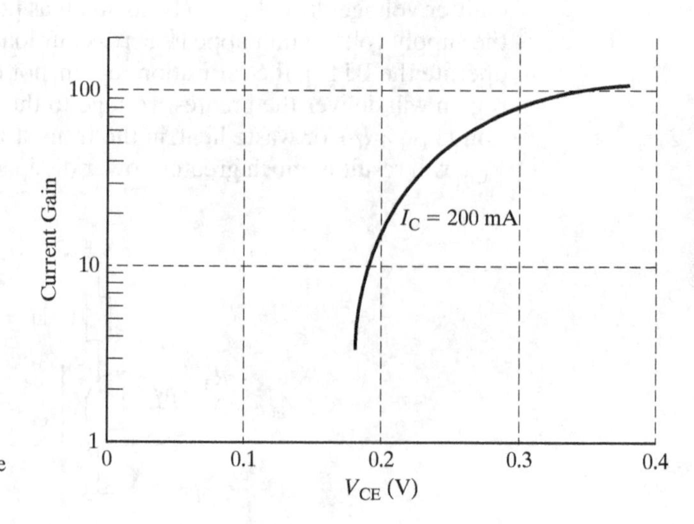

FIGURE 10.20

Transistor β as operation enters the saturation region.

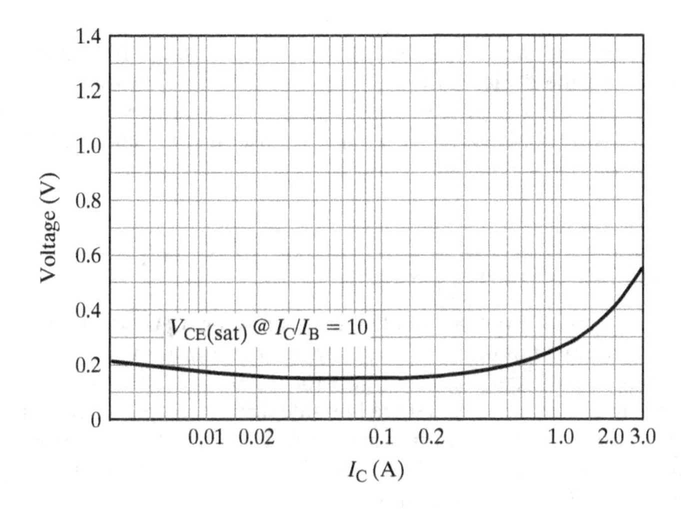

FIGURE 10.21

On voltage for a TIP32A BJT (Used with permission from SCI LLC, DBA ON Semiconductor.)

improvement in V_{CE} by using a current gain below 10–20 is very consistent. This leads to using a collector:base current ratio of 10:1 as the target for insuring that a BJT is in saturation. This standard is enshrined in many data sheets, which provide graphs like that shown in Figure 10.21 to describe the "on voltage" of a saturated BJT.

This graph also demonstrates why, in the absence of more specific data, it is common to assume that the collector-emitter voltage drop, or V_{CE}, in the saturated state ($V_{CE(sat)}$) is approximately 0.2 V.

Example 10.1 BJT low-side drive problem.

In the circuit of Figure 10.22, if V_{in} = 5 V, is Q_1 off, on, and in saturation, or operating in the linear region? If in the linear region, how would you change the circuit to operate the transistor in saturation?

To answer this question, we need to know I_C and I_B so that we can calculate the ratio and compare the result to our standards for saturation. We can calculate I_B using the given voltage and component values along with our 0.6 V rule of thumb for the forward voltage drop of the base-emitter diode: $I_B = (5\text{ V} - 0.6\text{ V})/2.2\text{ K} = 2\text{ mA}$. To calculate I_C, we need to know the state of Q_1. Since we have already seen that we do have some current flowing into the base of Q_1, we will assume that it is on and saturated, yielding a rule of thumb V_{CE} drop of 0.2 V. In that case, we can calculate the collector current as: $I_C = (10\text{ V} - 0.2\text{ V})/100\ \Omega = 98\text{ mA}$. This yields an I_C:I_B ratio of 98/2 = 49:1, which is well above the 10–20 that we would require for a transistor in saturation. Because we have *some* base current, but not enough to drive the transistor into saturation, we conclude that the transistor is operating in the linear region. Since we are trying to turn the lamp

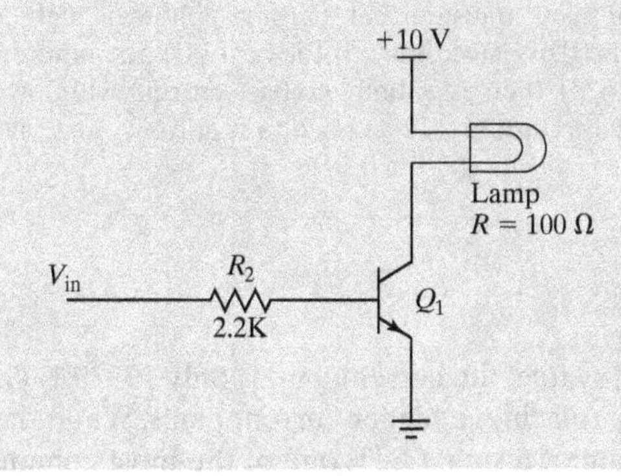

FIGURE 10.22

Circuit for BJT low-side drive example problem.

on with the available voltage, we will assume that the desired I_C is fixed at 98 mA, as we have already calculated. To get the device into saturation, we need to increase the base current to get it up to about 9.8 mA (1/10th of I_C). This will require that we lower the value of R_2 to 470 Ω (using standard resistor values). This change would increase the base current to 9.4 mA yielding a 98/9.4 = 10.5:1 current ratio.

Example 10.2 BJT high-side drive problem.

The schematic of Figure 10.23 shows a lamp being driven using **high-side drive**. High-side drive is characterized by the placement of the switching element (the transistor) between the power supply and the load (the lamp in this example). The transistor is said to be sourcing current into the load, which is typically grounded on one side. Where low-side drive is almost universally associated with NPN transistors, high-side drive is most often implemented using PNP transistors. To turn on transistor Q_2 in Figure 10.23, we need to pull the base of the transistor down to 0.6 V below the emitter and provide a path for current to flow out of the base.

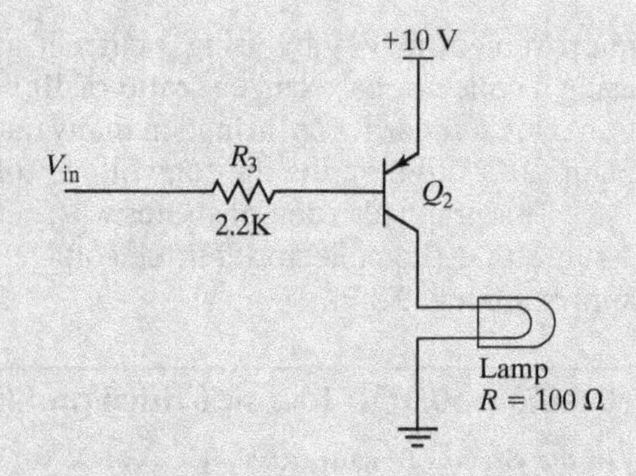

FIGURE 10.23

Schematic diagram for BJT high-side drive example.

The questions for this problem are: 1) If V_{in} is at 0.25 V, is Q_2 on and saturated? 2) What voltage at V_{in} is necessary to turn Q_2 off?

As with the last example, to judge whether or not Q_2 is in saturation, we need to know both I_C and I_B. With the same power supply voltage and same lamp resistance, the collector current should also be the same: $I_C = (10\text{ V} - 0.2\text{ V})/100\ \Omega = 98$ mA. The base current is calculated in a similar way to the prior example, only this time there are more voltage drops to consider. We have a V_{BE} drop at the transistor and V_{in} is at 0.25 V, so we would expect the base current to be $I_B = (10\text{ V} - 0.6\text{ V} - 0.25\text{ V})/2.2\text{K} = 9.4$ mA. This yields an I_C:I_B ratio of 98/9.4 = 10.5:1, which is not only in the 10–20 range that we would expect for a transistor in saturation but almost exactly the 10:1 value that we would shoot for in design. We therefore conclude that when V_{in} is at 0.25 V, Q_2 will be on and saturated.

To get Q_2 to turn off, we will need to insure that the base voltage is high enough to prevent the V_{BE} potential from reaching the 0.6 V necessary to forward bias the emitter-base junction. If we keep V_B above 9.4 V (10 V – 0.6 V), then essentially no base current will flow and there will be no voltage drop across R_3. Therefore, to turn Q_2 off, we need to supply a voltage at V_{in} that is greater than 9.4 V.

10.3.1 The Darlington Pair

When using a BJT as a saturated switch, the current gain is only 10–20:1. There will be many instances in which it is necessary to achieve higher current gains. While there are a number of ways to address the limited gain of a single BJT, one of the most common is to employ a

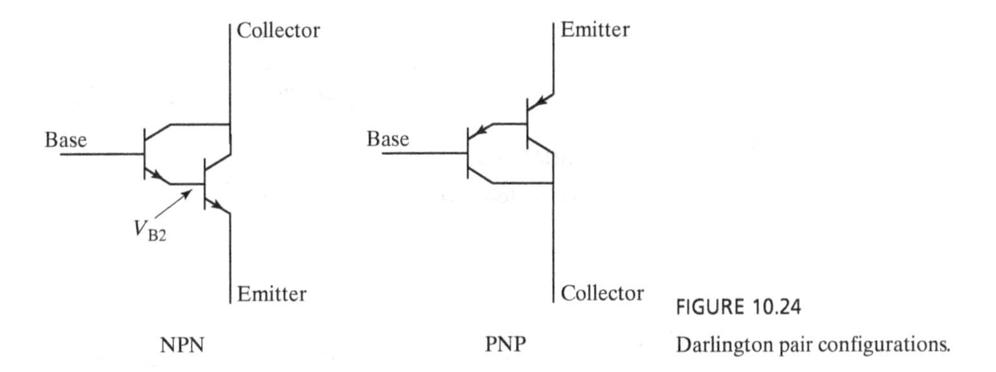

FIGURE 10.24
Darlington pair configurations.

configuration known as Darlington pair (Figure 10.24). This configuration is so common that it is available prepackaged in the same kind of three-terminal packages used for single transistors.

The behavior of the Darlington pair differs from that of a single BJT in a number of ways. We will start with an analysis for the NPN transistor. The first thing to notice is that if you trace the current path from the base to the emitter, you will see that it traverses two base-emitter junctions. As a result, V_{BE} for a Darlington is twice that of a single BJT: 2 V × 0.6 V = 1.2 V. Figuring out what the V_{CE} is requires a little more study. Let's start by assuming that we can get the left transistor into saturation. By our rule of thumb, that would place the base of the right-hand transistor (V_{B2}) 0.2 V below the collector voltage. For the right-hand transistor to be on, this voltage (V_{B2}) will need to be 0.6 V above the emitter voltage. This means that the V_{CE} for the right-hand device (and the package as a whole) will be 0.2 V + 0.6 V = 0.8 V. Using our 10:1 rule of thumb for current gain in each transistor would yield an expected current gain of 100 for the Darlington pair. However, since the right-hand transistor, with V_{CE} = 0.8 V, is not as deeply into saturation as a single BJT, it will exhibit a slightly higher current gain. In practice, Darlington pairs are generally considered to be saturated at current gains in the range of 200–500:1.

Darlington transistors are available as discrete devices, such as the TIP100 series of transistors. These devices, in the common TO-220 package, are suitable for collector currents up to a few amps. Darlington transistors are also available in multitransistor packages such as the ULN2003. This device has seven Darlington pairs, complete with base resistors suitable for logic-level drive, and is covered in more detail in Chapter 17.

The analysis for a PNP Darlington follows the same logic as for an NPN Darlington. The only differences are due to the fact that the required polarities are reversed for a PNP transistor. For a PNP Darlington, the base voltage needs to be approximately 1.2 V *below* the emitter voltage in order to turn the device on. When the PNP Darlington is turned on and saturated, the collector voltage will be approximately 0.8 V *below* the emitter voltage.

In summary, the Darlington pair provides substantially increased current gain at the expense of higher V_{BE} and V_{CE}.

10.3.2 The Photo-Transistor

If the area of the base terminal of a BJT is made large enough and exposed to light, the energy of the photons falling on the base region will act as base current, and can be used to control collector-to-emitter current flow. This is the basis of the photo-transistor (Figure 10.25).

The behavior of the photo-transistor is similar to that of an NPN BJT, except that the base current has been replaced by the light falling on the device (Figure 10.26).

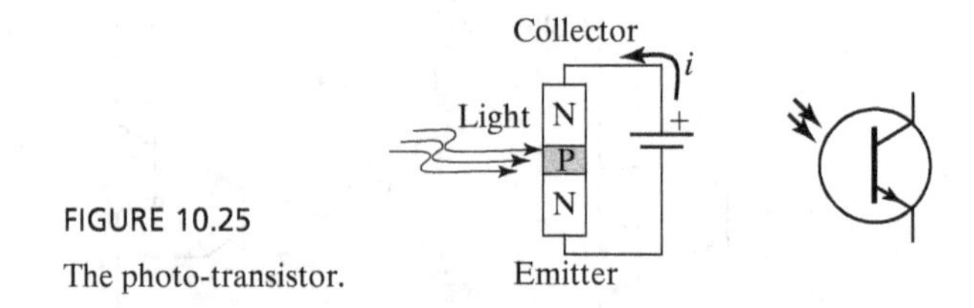

FIGURE 10.25

The photo-transistor.

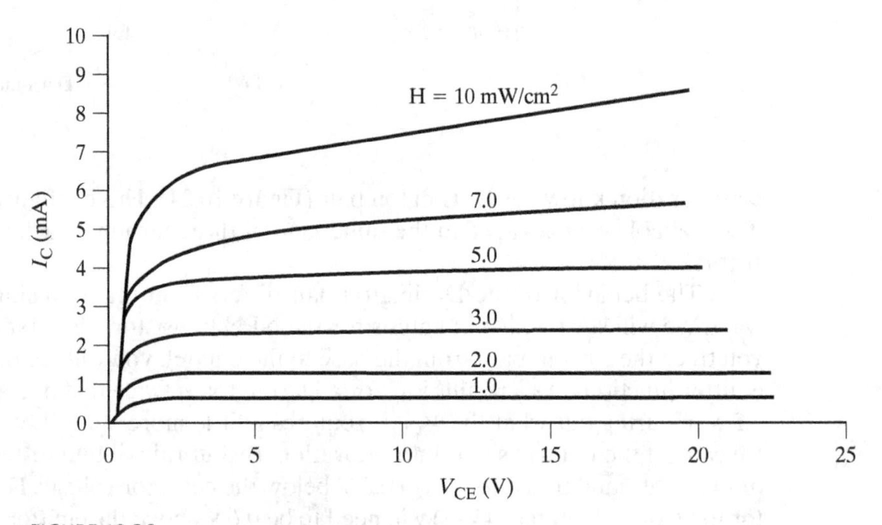

FIGURE 10.26

Transistor curves for a photo-transistor.

The photo-transistor is the one type of transistor in this text that we will use in both the saturation and the linear region. However, we will defer the discussion of how to use the photo-transistor until Chapter 13 on Sensors. This will allow us to cover op-amps in Chapters 11 and 12, which are essential elements of many photo-transistor interface and signal conditioning circuits.

10.4 MOSFETs

The **metal oxide semiconductor field effect transistor (MOSFET)** is the second major type of transistor that we will cover in this chapter. It differs from the BJT in two principal ways:

1. It is controlled by a voltage, not a current.
2. When fully on, the conduction path behaves like a small value resistor, not a fixed voltage drop.

Like BJTs, MOSFETs come in two different, complementary forms: N-channel devices and P-channel devices (Figure 10.27). There is actually another distinction that is made with MOSFETs (enhancement mode vs. depletion mode), but we will deal exclusively with enhancement mode devices in this text, as they are the most common in mechatronic applications.

MOSFETs are four-terminal devices, although in commercially produced discrete devices, the body terminal is always connected to the source, resulting in a three-terminal device. The **gate** is the control terminal, and the controlled current flows from **drain** to **source**

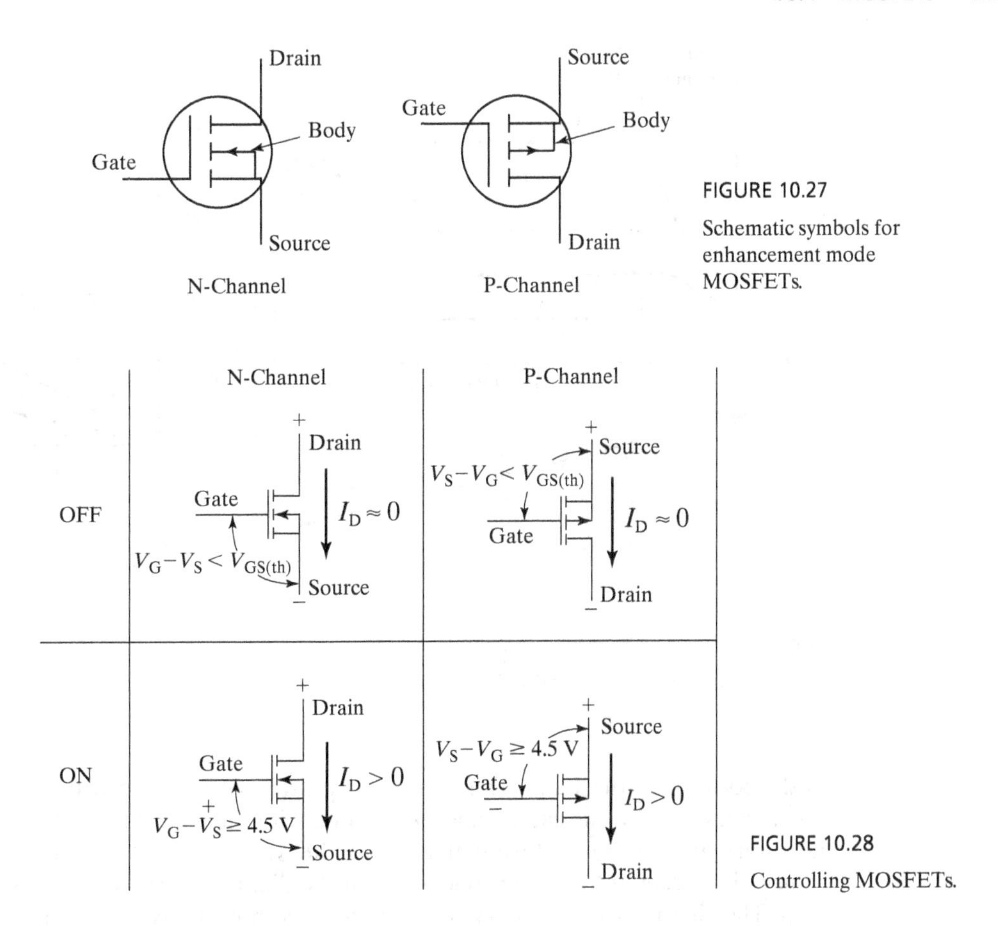

FIGURE 10.27
Schematic symbols for enhancement mode MOSFETs.

FIGURE 10.28
Controlling MOSFETs.

in an N-channel device and from source to drain in a P-channel device. The N-channel device is turned on when the gate voltage is more positive than the source voltage by a device specific amount (Section 10.7.2). The P-channel device is turned on when the gate is more negative than the source voltage by a similar amount. The simplest of these devices to use are the so-called **logic-level MOSFETs**. These devices are fully turned on when the gate-to-source voltage is only 4.5 V, as opposed to the 10 V or more typical of nonlogic-level MOSFETs. Figure 10.28 summarizes the conditions necessary to turn logic-level MOSFETs on and off.

The hydraulic analogy for an N-channel MOSFET is shown in Figure 10.29. Pressure at the gate terminal (analogous to voltage) causes the diaphragm to displace the attached

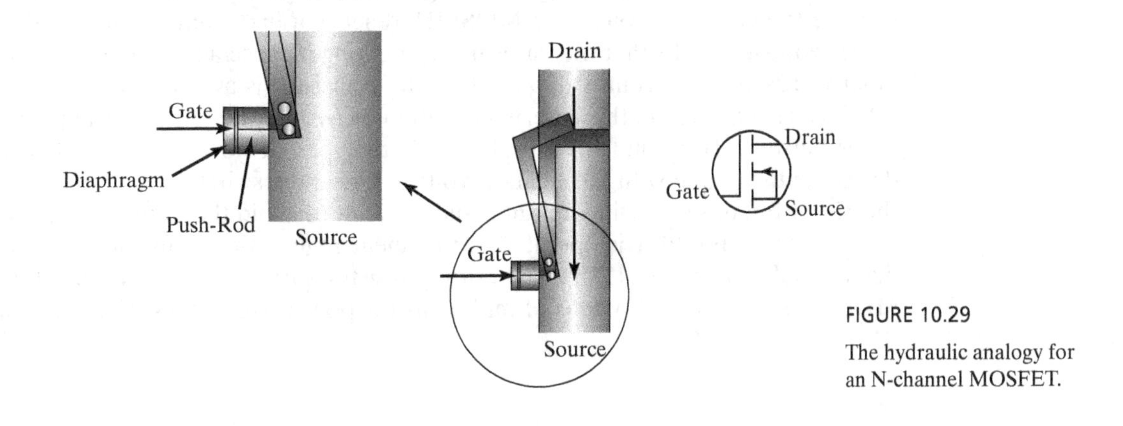

FIGURE 10.29
The hydraulic analogy for an N-channel MOSFET.

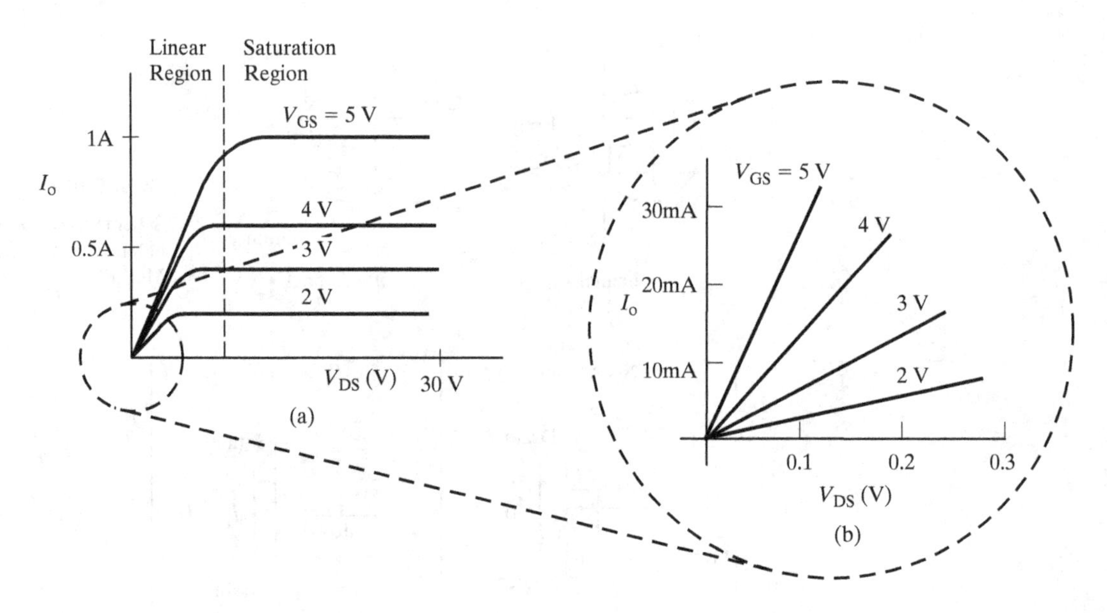

FIGURE 10.30
Characteristic curves for a MOSFET.

push-rod to open the valve. This analogy captures not only the voltage control behavior of the devices but also the capacitance associated with the gate. At steady state, there is no flow necessary into the gate terminal to hold the valve open. However, a transient flow is required to account for the displacement of the diaphragm as the valve is opened.

The characteristic *V-I* curves for MOSFETs show many similarities to those of the BJT (Figure 10.30). For the MOSFET, we plot curves for varying levels of gate-source voltage (V_{GS}) rather than I_B as for a BJT.

While it may be confusing when compared to descriptions of BJT characteristics, the left-hand region in a MOSFET is called the linear region and the saturation region is to the right. To understand why the left-hand region is called the linear region, look at Figure 10.30b. This shows at an expanded scale the behavior of the MOSFET in the region near the origin in Figure 10.30a. In Figure 10.30b, we can see that, unlike the BJT, the lines do not converge into a single line near the origin. In fact, for low values of V_{DS} with a constant V_{GS}, the conduction channel in the MOSFET exhibits a *linear V-I* characteristic. In this region, it is effectively a voltage controlled resistor, with the gate-to-source voltage (V_{GS}) determining the drain-to-source resistance (R_{DS}). The right-hand region is called the saturation region because the current through the MOSFET reaches a maximum value based on the gate-source voltage, and further increases in V_{DS} no longer cause an increase in drain current. In other words, past a certain value for V_{GS}, the transistor is as "on" as it can get—it is saturated. The saturation, in this case, is saturation of I_D. Despite the confusing switching of the terms between BJTs and MOSFETs, for both types of devices, we want to operate the devices in the region with the smallest voltage loss in the switch (i.e., nearest the *y*-axis). For the MOSFET, this is the linear region and we want to avoid the saturation region.

When a MOSFET is operated in the linear region, the drain-source channel behaves like a small value resistor. The value of the resistor varies from a few ohms for signal level MOSFETs down to a few tens of milliohms for power MOSFETs. This can result in much

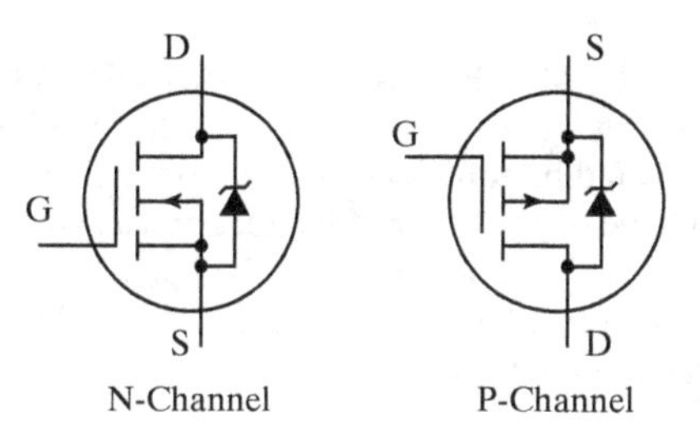

FIGURE 10.31

N-channel and P-channel MOSFET symbols showing the body diode.

smaller voltage drops from drain to source than can be achieved across the collector and emitter of a BJT. For example, the BJT, whose on voltage is shown in Figure 10.21, exhibits a $V_{CE(sat)}$ of about 0.15 V when conducting 100 mA. A power MOSFET of similar size might have an on resistance ($R_{DS(on)}$) of only 100 mΩ (and for some devices, less) yielding a V_{DS} of only 10 mV. Power dissipation is also dramatically reduced by the low resistance characteristics of the MOSFET. While the BJT of Figure 10.21 would be dissipating 10 mW when conducting 100 mA (0.1 V × 0.1 A), a MOSFET with a 100 mΩ $R_{DS(on)}$ would only dissipate 1 mW [$(0.1\text{ A})^2 \times 0.1\ \Omega$].

While not an intrinsic component of a MOSFET, all modern manufacturing techniques result in the creation of a parasitic body diode connected between the drain and source terminals (Figure 10.31).

For signal level MOSFETS (I_{DS} on the order of a few hundred mA or less), this is an ordinary diode. For power MOSFETS, the reverse breakdown characteristic is controlled to make the diode behave as a Zener diode. In either case, it is the breakdown voltage of this diode that limits the voltage that can be switched with a given MOSFET. The breakdown voltage range for conventional MOSFETs extends from about 50 V up to a few hundred volts; significantly less than the over 1 kV possible with BJT devices. The forward characteristics (V_f) for this diode are similar to ordinary silicon diodes with V_f on the order of 0.6 V for forward currents of a few milliamps.

Example 10.3 MOSFET low-side drive problem.

In the circuit of Figure 10.32, if $V_{in} = 5$ V, is Q_1 (a logic-level MOSFET) off, on, and operating in the linear region, or operating in the saturation region? If in the saturation region, how would you change the circuit to operate the transistor in the linear region?

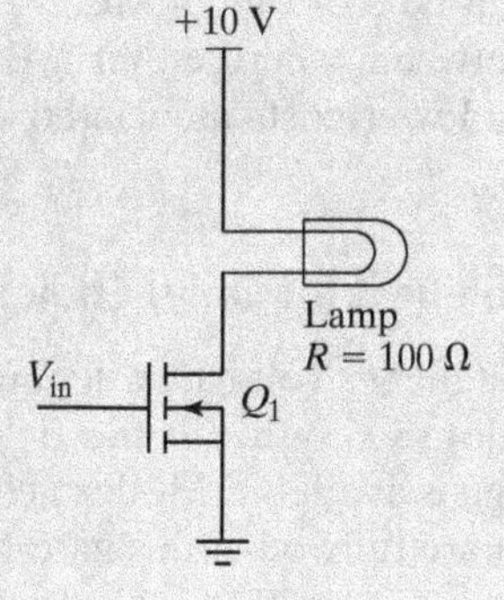

FIGURE 10.32

N-channel MOSFET in low-side drive circuit.

Since the linear region for a MOSFET is defined by having a low V_{DS}, we need to figure out what value of V_{DS} we expect in this circuit. Since Q_1 is a logic-level MOSFET [one specified with a low $R_{DS(on)}$ at 5 V V_{GS}], then 5 V would be sufficient to turn it on. When a logic-level power MOSFET is on, then its on resistance would be on the order of 100 mΩ or less. We can use this number to calculate estimates of the drain current and V_{DS} for the device. If $R_{DS(on)}$ is 100 mΩ, then we would expect a drain current of 10 V/100.1 Ω = 99.9 mA to flow, creating a V_{DS} of 99.9 mA × 100 mΩ = 9.99 mV. This is so close to 0 V as to be clearly in the linear region.

Example 10.4 MOSFET high-side drive problem.

In the circuit of Figure 10.33, if V_{in} swings between 0 and 10 V, in which of the two states will the lamp be lit?

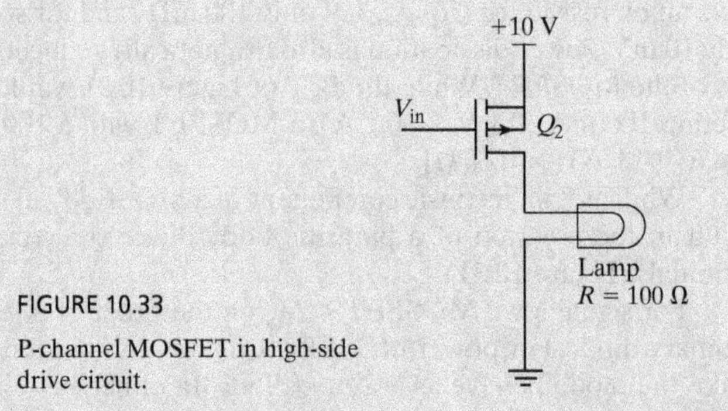

FIGURE 10.33

P-channel MOSFET in high-side drive circuit.

Note that when a P-channel MOSFET is used in a sourcing circuit, the source lead is tied to the positive supply. In this example, this places it at 10 V. If V_{in} is at 0 V, then there will be a 10 V differential across V_{GS} (−10 V in this case) and that is more than sufficient to turn on almost any P-channel MOSFET, therefore the lamp will be lit. When the gate is at 10 V, then there is no V_{GS} potential so the device will be off. When the device is turned on, the currents that would flow would be the same as the currents that we calculated for the circuit of Figure 10.32 for the power N-channel MOSFET.

10.5 CHOOSING BETWEEN BJTs AND MOSFETs

From a purely technical point of view, the choice between BJT and MOSFET will largely boil down to how much control voltage and current is available. As you might expect, there will be applications where only a BJT or only a MOSFET will meet the requirements and there will also be a region in between, where either a BJT or MOSFET would be a viable choice. In that latter region, the lower costs associated with BJTs may make them a more attractive solution.

10.5.1 When Will a BJT Be the Best (or Only) Choice?

In the realm of mechatronics, the higher voltage switching capability of BJTs will only rarely be the deciding factor. Much more likely to tilt the field in favor of BJTs is the situation when only a small control voltage is available. Modern construction techniques make it possible to produce MOSFETs that are fully on with a gate-to-source voltage, V_{GS}, of only 4.5 V and that exhibit reasonably low $R_{DS(on)}$ with a V_{GS} of only about 3 V. However, if you only have 1 V available for control, the only viable choice is a BJT. For those situations, where only a small control voltage *and* only a small amount of control current are available, you will need to resort to using multiple transistors (see Section 10.6) to meet the required current gain.

10.5.2 When Will a MOSFET Be the Best (or Only) Choice?

A MOSFET will be the best solution, when sufficient control voltage is available, and it is important to have the lowest possible voltage drop across the switch and/or the lowest possible power dissipation within the switch. If the requirements call for voltage drop across the switching element of less than 0.1 V, then a MOSFET will be the only viable solution. A MOSFET will also likely be the best solution when there is sufficient voltage available, but not sufficient current. In these situations, using a single MOSFET would be an alternative to a multitransistor solution using BJTs to achieve the required current gain.

10.5.3 How Do You Choose When Either a MOSFET or a BJT Could Work?

In situations where both sufficient control voltage and control current are available, then either a MOSFET or a BJT could be chosen to do the job. In situations like this, there is not a clear technical answer and we are forced to fall back on secondary considerations, such as cost and complexity. Using a MOSFET will be simpler (requiring no base resistor, as a BJT would), but it will be more expensive than a comparable BJT (even when the extra resistor is included). Although the BJT will be the less expensive solution, it will usually result in greater losses in the switching element, which translates into lower efficiency and greater heat dissipation. The relative importance of cost, complexity, and efficiency in a particular application should guide the designer in choosing the type of transistor used.

10.6 MULTITRANSISTOR CIRCUITS

Many situations arise where a single transistor simply won't do the job. It might be that in addition to having only a small control voltage available, there is also only a small amount of current available. In another situation, it might be that sufficient control current exists but the voltage to be controlled exceeds the control voltage swing. In both of these situations, you will need to combine several transistors in order to meet the requirements.

Figure 10.34 shows a circuit that addresses the situation when current must be sourced into the load and only a small amount of control voltage and current are available. With only up to 6 mA of input current available at V_{in}, a single BJT does not have enough current gain to supply the current to the 10 Ω load. In addition, with only a 2 V input swing, it is not possible to

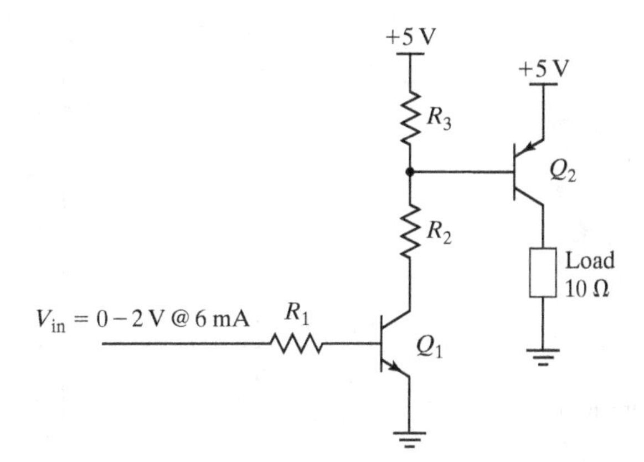

FIGURE 10.34

A two-transistor circuit with a bipolar input transistor.

control (specifically, to turn off) the PNP transistor Q_2 in a high-side drive configuration. This circuit addresses both issues. With V_{in} at 2 V, there is sufficient voltage available to turn on Q_1. If R_1 is sized appropriately, we can drive it into saturation. With Q_1 on, the base of Q_2 will be pulled low, and base current will flow out of the base of Q_2. If R_2 is sized properly, Q_2 will be driven into saturation.

With ideal transistors, which allow no current to flow when they are off, R_3 would not be necessary. In reality, all transistors leak a small amount of current, even when they are turned off. If R_3 were not present, then the leakage through Q_1 would create base current through Q_2, which would be amplified and deliver current into the load. If R_3 is sized so that the voltage drop across from the leakage current through Q_1 is less than 0.6 V, then all of the leakage current will come through R_3 and not from the base of Q_2. Similarly, because Q_2 is a real transistor, it will still leak a little bit (on the order of a few hundred μA for typical power transistors) of current into the load even when R_3 is present. However, the amount of current that flows into the load will be decreased by an order of magnitude or more relative to what would flow if R_3 were not present.

We can use our rules of thumb about transistor behavior to calculate values for R_1 and R_2 in this circuit. If Q_2 is turned on and saturated ($V_{CE} = 0.2$ V), then the collector current will be (5 V − 0.2 V)/10 Ω = 480 mA. To insure saturation, we will need to sink at least 48 mA of base current from Q_2 through Q_1 (I_C:I_B = 10:1), and we will ignore the current from R_3 because its value is typically much larger than that of R_2 and will therefore contribute only a small amount of current, which we can safely neglect. We can calculate the required value of R_2 as (4.6 V − 0.2 V)/48 mA = 91.7 Ω. We would choose a standard resistor value of 82 Ω because choosing 91 Ω would mean that a 5% resistor could be as much as 95 Ω, which is too large.

With 48 mA of collector current flowing in Q_1, we want to provide at least 4.8 mA of base current to insure that Q_1 is saturated (again, applying the I_C:I_B = 10:1 rule of thumb). This means that we need to size R_1 as no more than (2 V − 0.6 V)/4.8 mA = 291.7 Ω. Sticking with standard 5% resistor values, we would choose a 270 Ω resistor for R_1.

Figure 10.35 shows a circuit to address the situation where current must be sourced into the load and there is sufficient current available at the input, but there is not sufficient control voltage swing to control the output voltage. While V_{in} is capable of providing enough input current to use a single BJT to drive the load, the maximum input voltage is not enough to turn off Q_2. At 5 V, V_{in} is high enough to turn on a logic level MOSFET at Q_1, which allows us to eliminate the base resistor (R_1) that was necessary in Figure 10.34. Note that because we have chosen to use a MOSFET at the input stage, V_{in}, even though it is capable of sourcing up to 50 mA, will only source a few microamps of leakage current (Section 10.7.2) into the gate of Q_1.

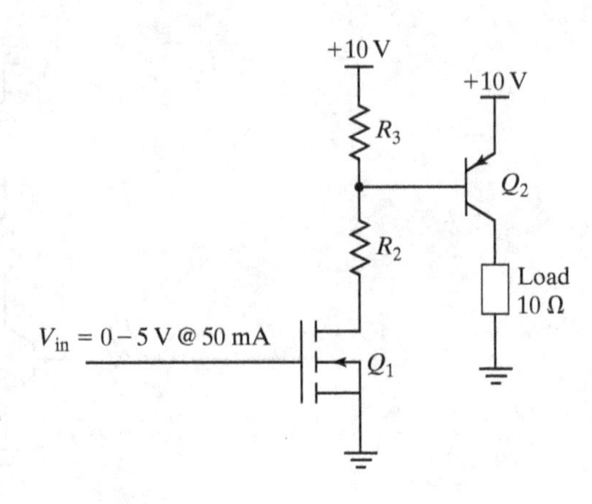

FIGURE 10.35

Two-transistor circuit with a MOSFET input stage.

10.7 READING TRANSISTOR DATA SHEETS

There are thousands of different varieties of transistors available. From this fact, you may correctly deduce that there are many parameters that are important when designing with these devices over the full range of their application possibilities. Fortunately for us, if we are willing to limit ourselves to using transistors as switches and operating those switches well within the limitations of the devices, the number of parameters that we need to understand in order to effectively choose and use transistors is manageable. We will take that approach and focus on the most important parameters when using transistors as switches. With these criteria in place, we will be interested in the specifications that describe how well the device performs as a switch. We will need to know:

1. How much voltage can the device switch?
2. How much current can the device switch?
3. How much effort does it take to switch the device?
4. How does it behave when on?
5. How does it behave when off?

The answers to these questions will guide us in selecting devices and in predicting their behavior in our circuit designs.

When studying the data sheets for devices, it is just as important to note the listed test conditions as it is to pay attention to the actual parameters. Often there will be multiple specifications under differing test conditions. When you find this situation, you should choose the parameters for the conditions that most closely match those of your application.

10.7.1 Reading a BJT Data Sheet

For this discussion, we refer to the TIP31/32 BJT data sheet in Figure 10.36. The data sheet parameter that describes how much voltage a BJT can switch is called $V_{CEO(sus)}$. This stands for the **V**oltage from **C**ollector to **E**mitter in the **O**ff state that the device can **sus**tain. In some data sheets, the "sus" portion of the label is omitted. For small signal-level transistors, this will likely be 40 V or more. For higher power BJTs, this parameter is often 60 V or more.

When trying to determine how much current a device can switch, there are multiple factors and specifications that come into play. Each device will have a specified maximum power dissipation that must not be exceeded. When this limit is exceeded, the device overheats, sometimes destructively. This specification is often called out for conditions when the ambient temperature is 25°C (T_A = 25°C) and when the case temperature (T_C) is held at 25°C. The specification for T_A = 25°C is appropriate when operating at room temperature and the device has no heat sink installed. This represents what the part is capable of under "average" conditions: in an environment with a typical ambient temperature, and no heat sink. On the other hand, the specification for the case where the *case* temperature is maintained at 25°C (T_C = 25°C) is really a theoretical upper bound on power dissipation. Unless you have an active cooler capable of cooling below ambient (or are only operating at a very low ambient temperature), you will not realistically be able to maintain the case temperature at 25°C.

Because of the relatively high power dissipation in a power BJT, it is not uncommon for the operation to be limited by the power dissipation specifications. When calculating the power dissipation in a BJT, don't forget that there will be two sources of power dissipation: the collector-emitter path ($I_C \times V_{CE(sat)}$) as well as the base-emitter path ($I_B \times V_{BE(sat)}$).

Power BJTs will also typically have a maximum specification for both continuous and peak collector current. The peak current limit derives from the capabilities of the bond wires that carry the current into the semiconductor and is independent of the average power dissipation. The continuous specification is given because, in extreme cases, active cooling of the

TIP31A, TIP31B, TIP31C, (NPN), TIP32A, TIP32B, TIP32C, (PNP)

THERMAL CHARACTERISTICS

Characteristic	Symbol	Max	Unit
Thermal Resistance, Junction to Ambient	$R_{\theta JA}$	62.5	°C/W
Thermal Resistance, Junction to Case	$R_{\theta JC}$	3.125	°C/W

ELECTRICAL CHARACTERISTICS (T_C = 25°C unless otherwise noted)

Characteristic		Symbol	Min	Max	Unit
OFF CHARACTERISTICS					
Collector–Emitter Sustaining Voltage (Note 2) (I_C = 30 mAdc, I_B = 0)	TIP31A, TIP32A TIP31B, TIP32B TIP31C, TIP32C	$V_{CEO(sus)}$	60 80 100	– – –	Vdc
Collector Cutoff Current (V_{CE} = 30 Vdc, I_B = 0) (V_{CE} = 60 Vdc, I_B = 0)	TIP31A, TIP32A TIP31B, TIP31C TIP32B, TIP32C	I_{CEO}	– – –	0.3 0.3 0.3	mAdc
Collector Cutoff Current (V_{CE} = 60 Vdc, V_{EB} = 0) (V_{CE} = 80 Vdc, V_{EB} = 0) (V_{CE} = 100 Vdc, V_{EB} = 0)	TIP31A, TIP32A TIP31B, TIP32B TIP31C, TIP32C	I_{CES}	– – –	200 200 200	μAdc
Emitter Cutoff Current (V_{BE} = 5.0 Vdc, I_C = 0)		I_{EBO}	–	1.0	mAdc
ON CHARACTERISTICS (Note 2)					
DC Current Gain (I_C = 1.0 Adc, V_{CE} = 4.0 Vdc) (I_C = 3.0 Adc, V_{CE} = 4.0 Vdc)		h_{FE}	25 10	– 50	–
Collector–Emitter Saturation Voltage (I_C = 3.0 Adc, I_B = 375 mAdc)		$V_{CE(sat)}$	–	1.2	Vdc
Base–Emitter On Voltage (I_C = 3.0 Adc, V_{CE} = 4.0 Vdc)		$V_{BE(on)}$	–	1.8	Vdc
DYNAMIC CHARACTERISTICS					
Current–Gain – Bandwidth Product (I_C = 500 mAdc, V_{CE} = 10 Vdc, f_{test} = 1.0 MHz)		f_T	3.0	–	MHz
Small–Signal Current Gain (I_C = 0.5 Adc, V_{CE} = 10 Vdc, f = 1.0 kHz)		h_{fe}	20	–	–

2. Pulse Test: Pulse Width ≤ 300 μs, Duty Cycle ≤ 2.0%.

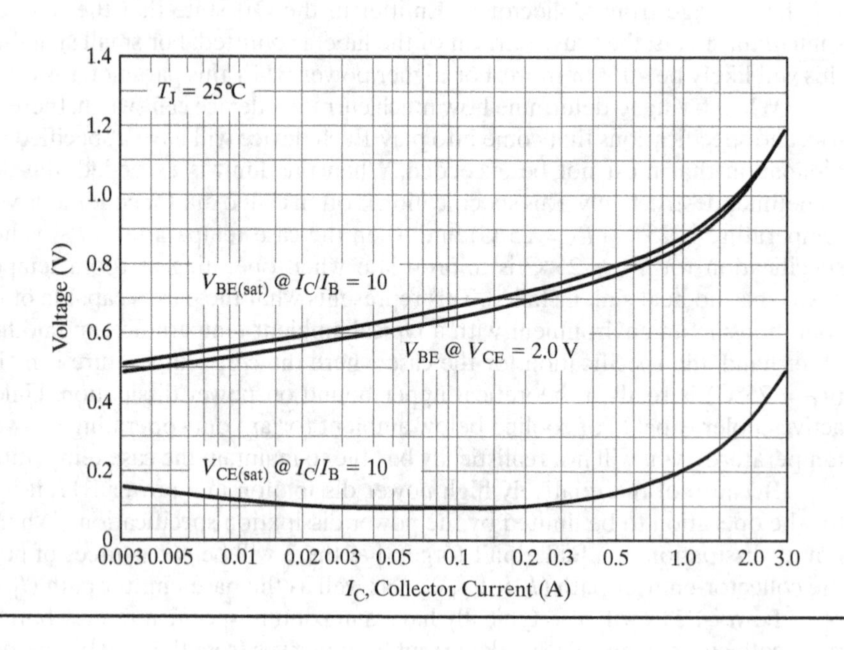

FIGURE 10.36

Excerpt from the data sheet for a TIP31/32 power BJT. (Used with permission from SCI LLC, DBA ON Semiconductor.)

case is used to try and extract a bit more power from a device. Even under these circumstances, there will always be a temperature differential between the junction and the case. So, even if you keep the case cool, the junction will be at a higher temperature.

The amount of control effort required for a BJT is directly related to the amount of current that we are trying to switch. For single BJTs, we will design for a current ratio (I_C:I_B) of 10–20:1. This is typical across a wide range of BJTs. For Darlington pairs, the ratio is on the order of 200–500:1. To find the value that a particular device manufacturer is using, look at the test conditions for the parameter $V_{CE(sat)}$. The test conditions for this parameter will include values for I_C and I_B that you can use to choose an appropriate ratio of I_C:I_B. The other aspect of control effort for the BJT is the amount of base voltage necessary to turn the device on and drive it into saturation. In the tabular part of the data sheet (often labeled "On Characteristics"), you will find a maximum value given for a parameter called $V_{BE(sat)}$, which stands for **V**oltage, **B**ase to **E**mitter in **sat**uration. This parameter also usually appears in a graphical form as a plot of $V_{BE(sat)}$ as a function of collector current (I_C).

For a BJT, the critical parameter in describing the on behavior is $V_{CE(sat)}$. This stands for **V**oltage, **C**ollector to **E**mitter in **sat**uration. It is a parameter that will typically be specified by a maximum value in a table of values as well as with a graph of typical values as a function of collector current.

The behavior of a BJT in the off state is characterized by some amount of leakage current that flows from collector to emitter even though the device is off. This parameter is usually specified for two different conditions. When the base current is 0, the specification is called I_{CEO}, which stands for current (***I***) from **C**ollector to **E**mitter, **O**pen base. In this case, "open base" implies no base current flow. This characteristic describes how the device behaves when the connection to the base is removed. The more typically useful parameter is I_{CES}, which stands for current (***I***) from **C**ollector to **E**mitter, **S**horted base. This is the condition that exists when the base lead is actively driven to (or very close to) the voltage at the emitter lead.

10.7.2 Reading a MOSFET Data Sheet

For this discussion, we refer to the 2N7000 N-channel MOSFET data sheet in Figure 10.37. The maximum voltage that can be switched by a MOSFET is limited by the breakdown voltage on the body diode. This parameter is usually known as BV_{DSS}, for **B**reakdown **V**oltage **D**rain to **S**ource **S**ustaining. This specification is sometimes given by the symbol $V_{(BR)DSS}$. Typical devices have breakdown voltages in the range of 50–100 V, though they are available with higher breakdown voltages at a price premium.

Like BJTs, the maximum current that a device can handle will be limited both by maximum allowable power dissipation (P_D) as well as by maximum continuous and peak drain currents (I_D).

The voltage (V_{GS}) necessary to turn a MOSFET fully on will be listed as a test condition for the parameter $R_{DS(on)}$. You should be careful here and not be deceived into thinking that the relevant parameter is the threshold voltage ($V_{GS(th)}$). A careful examination of the test conditions for the threshold voltage will show that it is measured with a relatively large V_{DS} (voltage between the drain and the source) and with an I_D of only a few hundred μA. Since in the on state the drain-source channel behaves like a resistor, we can combine these test conditions to see that the drain-to-source on resistance at the threshold is on the order of 2–8 kΩ. This is not a very good value for a switch and much higher than the few ohms to milliohms that we would expect if the device were fully on. The threshold voltage ($V_{GS(th)}$) is, however, still a useful parameter: it tells you how low you need to keep the gate-source voltage to insure that the device remains off.

On a real (as opposed to ideal) MOSFET, when the drain-to-source voltage is high enough to turn the device on, a small amount of leakage current will flow into the gate. This

2N7000

Electrical Characteristics T_A = 25°C unless otherwise noted

Symbol	Parameter	Conditions		Type	Min	Typ	Max	Units
OFF CHARACTERISTICS								
BV_{DSS}	Drain-Source Breakdown Voltage	V_{GS} = 0 V, I_D = 10 μA		All	60			V
I_{DSS}	Zero Gate Voltage Drain Current	V_{DS} = 48 V, V_{GS} = 0 V		2N7000			1	μA
			T_J=125°C				1	mA
		V_{DS} = 60 V, V_{GS} = 0 V		2N7002 NDS7002A			1	μA
			T_J=125°C				0.5	mA
I_{GSSF}	Gate - Body Leakage, Forward	V_{GS} = 15 V, V_{DS} = 0 V		2N7000			10	nA
		V_{GS} = 20 V, V_{DS} = 0 V		2N7002 NDS7002A			100	nA
I_{GSSR}	Gate - Body Leakage, Reverse	V_{GS} = -15 V, V_{DS} = 0 V		2N7000			-10	nA
		V_{GS} = -20 V, V_{DS} = 0 V		2N7002 NDS7002A			-100	nA
ON CHARACTERISTICS (Note 1)								
$V_{GS(th)}$	Gate Threshold Voltage	V_{DS} = V_{GS}, I_D = 1 mA		2N7000	0.8	2.1	3	V
		V_{DS} = V_{GS}, I_D = 250 μA		2N7002 NDS7002A	1	2.1	2.5	
$R_{DS(ON)}$	Static Drain-Source On-Resistance	V_{GS} = 10 V, I_D = 500 mA		2N7000		1.2	5	Ω
			T_J =125°C			1.9	9	
		V_{GS} = 4.5 V, I_D = 75 mA				1.8	5.3	
		V_{GS} = 10 V, I_D = 500 mA		2N7002		1.2	7.5	
			T_J =100°C			1.7	13.5	
		V_{GS} = 5.0 V, I_D = 50 mA				1.7	7.5	
			T_J =100C			2.4	13.5	
		V_{GS} = 10 V, I_D = 500 mA		NDS7002A		1.2	2	
			T_J =125°C			2	3.5	
		V_{GS} = 5.0 V, I_D = 50 mA				1.7	3	
			T_J =125°C			2.8	5	
$V_{DS(ON)}$	Drain-Source On-Voltage	V_{GS} = 10 V, I_D = 500 mA		2N7000		0.6	2.5	V
		V_{GS} = 4.5 V, I_D = 75 mA				0.14	0.4	
		V_{GS} = 10 V, I_D = 500mA		2N7002		0.6	3.75	
		V_{GS} = 5.0 V, I_D = 50 mA				0.09	1.5	
		V_{GS} = 10 V, I_D = 500mA		NDS7002A		0.6	1	
		V_{GS} = 5.0 V, I_D = 50 mA				0.09	0.15	

FIGURE 10.37

Excerpt from the data sheet for a 2N7000 signal level MOSFET. (Courtesy of Fairchild Semiconductor.)

current is referred to as Gate-Body leakage or Gate-Source leakage (I_{GSS}). This current ranges from a few nA up to a few hundred μA at elevated temperatures. This is dramatically less than the levels of base current required with a BJT.

When a MOSFET is on, the drain-source channel behaves like a low-value resistor. The parameter used to describe that resistance is $R_{DS(on)}$, which stands for **R**esistance, **D**rain to **S**ource in the **on** state. This will be listed in the tabular part of the data sheet under "On Characteristics." The typical $R_{DS(on)}$ as a function of V_{GS} can be determined from the plots of I_D vs. V_{DS} that are included as the "On Characteristics" in the graphical portion of the data sheet.

When a MOSFET is off, the leakage current through the drain-source channel is almost independent of the drain-to-source voltage (V_{DS}). This leakage current is typically specified with $V_{GS} = 0$ V as a parameter called Zero Gate Voltage Drain Current or Drain-to-Source leakage current and given the symbol I_{DSS}.

10.7.3 A Sample Application

To give you a chance to explore reading a data sheet, we will close this chapter with an example of a typical circuit that uses a 2N7000 signal level MOSFET and a TIP32A power BJT to control power to a 15 Ω grounded load resistance. A load like this might represent, for example, the windshield washer motor on an automobile. Please follow along by finding the quoted specifications from the data sheet excerpts in Sections 10.7.1 and 10.7.2.

The circuit of Figure 10.38 is a version of Figure 10.35 to which we have added explicit specifications for the transistors and the values of the resistors. Let's use the specifications for these real parts from their respective data sheets to evaluate the performance of this circuit when the input voltage (V_{in}) is set to both 0 and 5 V.

Based on our earlier analysis, when the input is at 5 V, we expect the devices to both be on and current to be flowing through the load. We can start the process of pulling data sheet numbers by assuming, as we did before, that the collector-emitter voltage drop for the TIP32 is 0.2 V. This will allow us to calculate the collector current under those conditions as (10 V − 0.2 V)/15 Ω = 653 mA. Now, we can use this current value and refer to the TIP32 data sheet excerpt (Figure 10.36) which includes the manufacturer's "Fig. 10 'On' Voltages." For a collector current of 653 mA, we find that we should expect the actual device to have a $V_{CE(sat)}$ of about 0.17 V. This is slightly lower than our 0.2 V rule of thumb assumption, so we can recalculate the expected collector current as (10 V − 0.17 V)/15 Ω = 655 mA. From that same graph, we determine that the expected $V_{BE(sat)}$ is about 0.85 V. If we were to get Q_2 into saturation, the base current would need to be about 65 mA to use this value of $V_{BE(sat)}$. From the 2N7000 data sheet (Figure 10.37), we can see that $R_{DS(on)}$ with a 4.5 V gate drive and 75 mA of I_D (the specification that best approximates our operating conditions) is listed as a maximum of 5.3 Ω. If we use this maximum value, we will calculate the minimum base current that we would achieve for Q_2, which is the most conservative position to take. Using these numbers, we calculate I_B as (10 V − 0.85 V)/120 Ω + 5.3 Ω = 73 mA. This is actually $I_B + I_{R3}$, but I_{R3} is only (0.85 V)/100 K = 8.5 μA, making it insignificant by comparison. With an I_B of 73 mA, we have more than enough base current to justify our assumption of saturation and the use of "Fig. 10" from the TIP32A data sheet excerpt in Figure 10.36.

For the case when $V_{in} = 0$ V, the 2N7000 will be off, because we do not have sufficient gate voltage to turn it on. In the off state, it will leak a maximum of 1 μA (I_{DSS}) from drain to

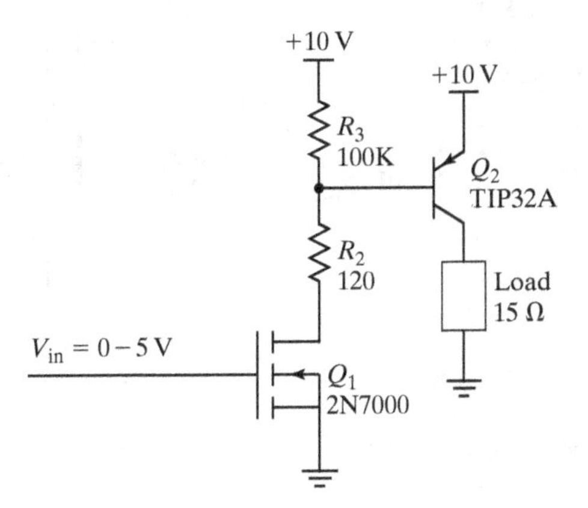

FIGURE 10.38

Circuit for Section 10.7.3 sample application.

source. This current will be supplied through R_3, causing a voltage drop of 1 μA × 100 kΩ = 0.1 V. This will put the base of Q_2 at 9.9 V, which is more than adequate to force it into the off state. With Q_2 off, it will leak as much as 200 to 300 μA into the load. The 200 μA value comes from the TIP32 I_{CES} specification. We don't quite meet the test conditions for this parameter (V_{EB} = 0.1 V rather than the 0 V test condition given), so we've estimated the actual value as between that specification and I_{CEO} (300 μA). The nature of the load will determine if this amount of leakage current is acceptable, but since it is less than 1/2,000th of the on current, it is likely to work for most real loads.

10.7.4 A Potpourri of Transistor Circuits

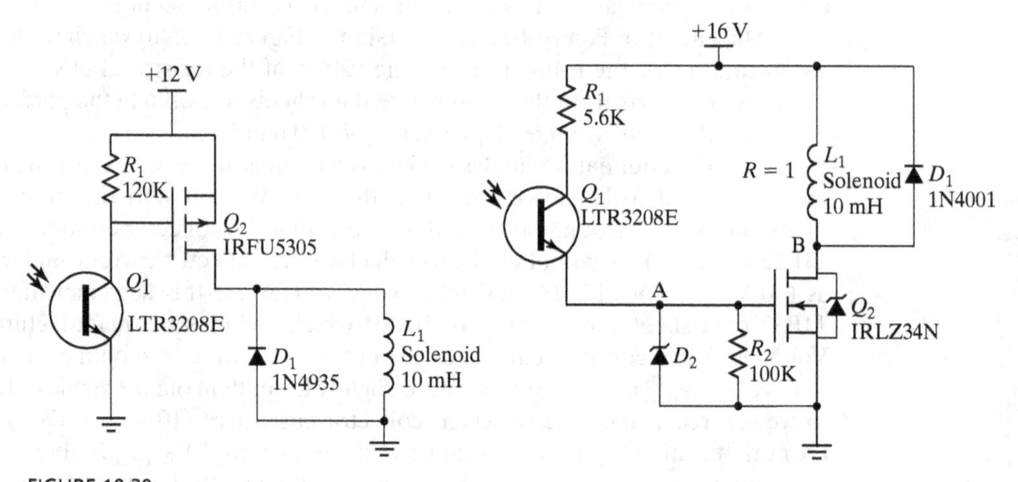

FIGURE 10.39

Light controlled solenoid drive circuits.

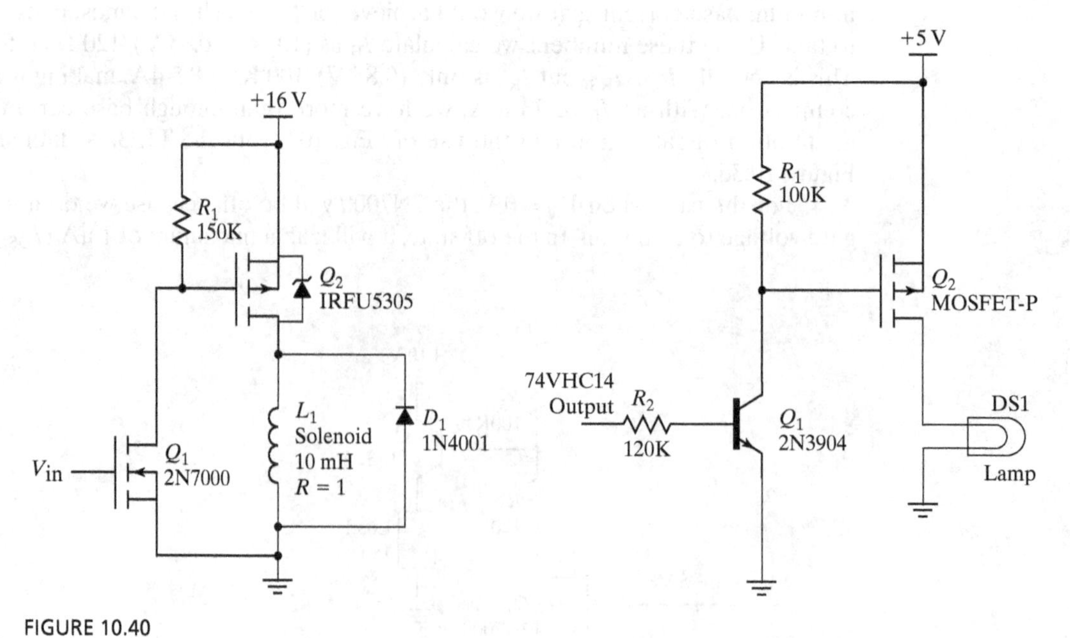

FIGURE 10.40

Circuits to activate a grounded load with a logic input.

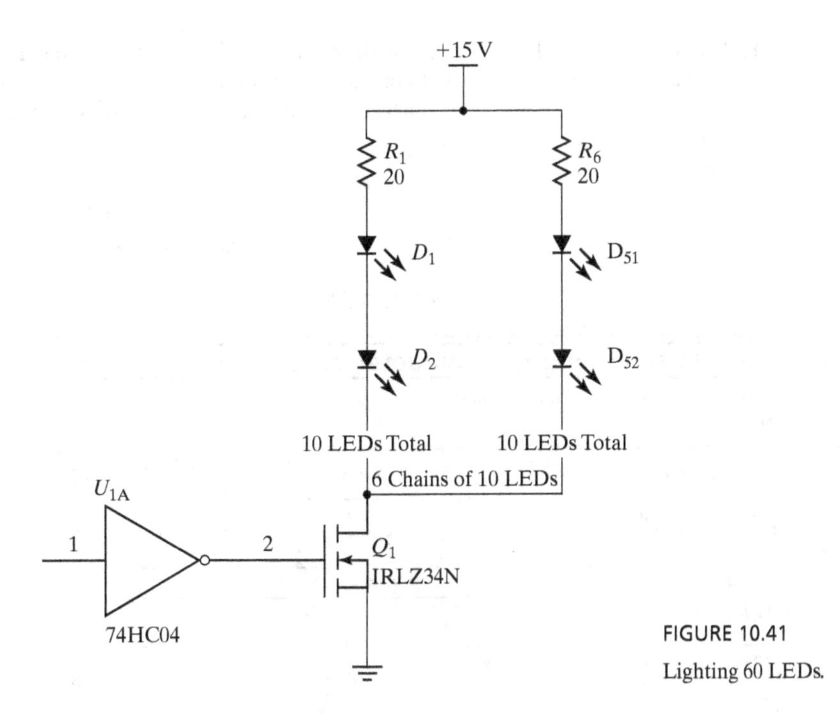

FIGURE 10.41
Lighting 60 LEDs.

10.8 HOMEWORK PROBLEMS

10.1 If the forward voltage drop for D_1 is 0.6 V, what is the current flowing in the circuit shown in Figure 10.42? Assume that the diode behaves as an ideal diode.

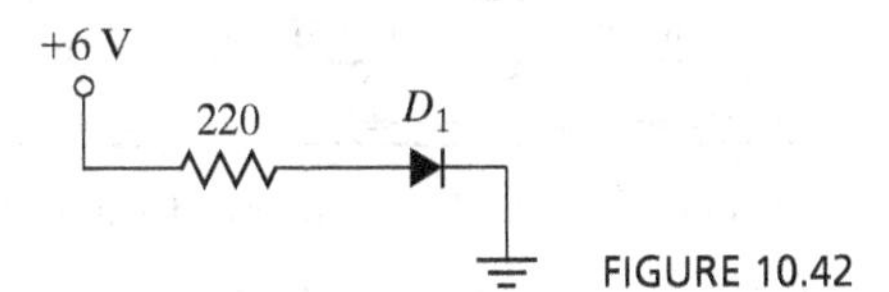

FIGURE 10.42

10.2 If the forward voltage drop for D_1 is 0.6 V, what is the current flowing in the circuit in Figure 10.43? Assume that the diode behaves as an ideal diode.

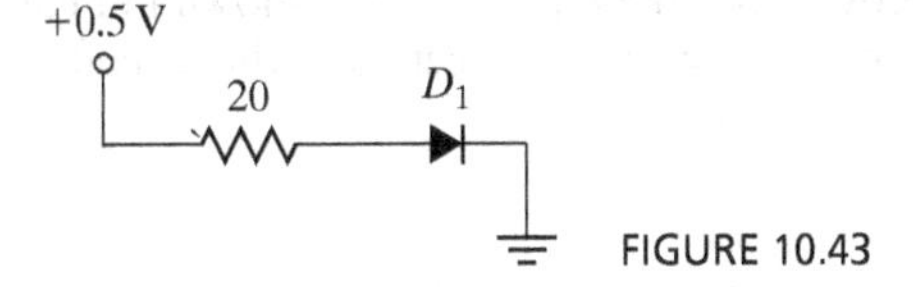

FIGURE 10.43

10.3 The arrangement of diodes shown in Figure 10.44 is known as a "diode bridge" or a "full-wave rectifier." If the output of the AC voltage source V_S is 24 V peak-to-peak (±12 V) with a frequency of 100 Hz, draw a plot of the resulting voltage drop across the resistor (assume ideal diode characteristics).

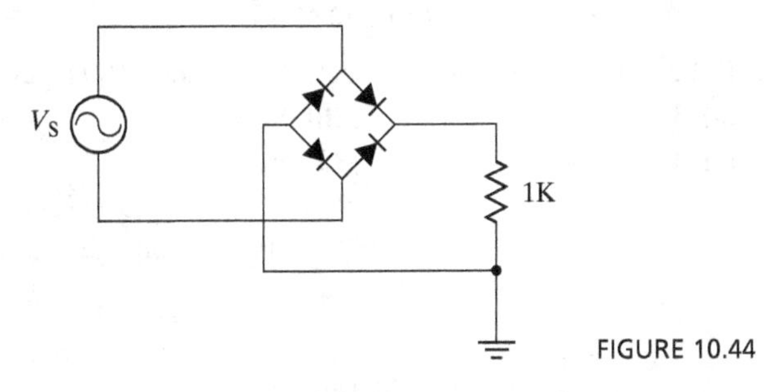

FIGURE 10.44

10.4 For the circuit in Figure 10.45, use the diode specifications provided in Figure 10.46 to determine the maximum amount of current you can expect to flow when $T_A = 25°C$. If you were to increase the power supply voltage, how much could you increase it before exceeding the diode's reverse-breakdown voltage?

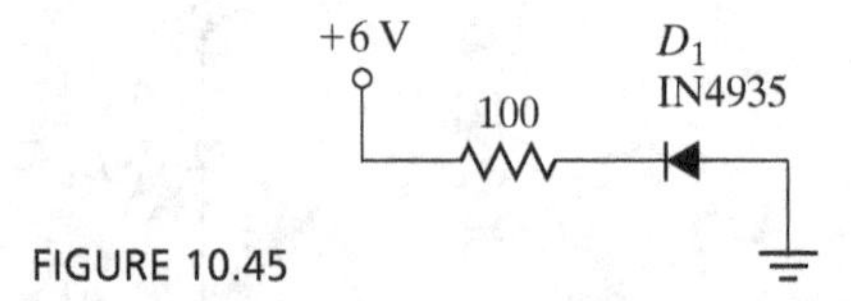

FIGURE 10.45

Maximum Ratings and Electrical Characteristics @ T_A = 25°C unless otherwise specified

Single phase, half wave, 60Hz, resistive or inductive load.
For capacitive load, derate current by 20%.

Characteristic		Symbol	1N4933 G/GL	1N4934 G/GL	1N4935 G/GL	1N4936 G/GL	1N4937 G/GL	Unit
Peak Repetitive Reverse Voltage Working Peak Reverse Voltage DC Blocking Voltage		V_{RRM} V_{RWM} V_R	50	100	200	400	600	V
RMS Reverse Voltage		$V_{R(RMS)}$	35	70	140	280	420	V
Average Rectified Output Current (Note 1)	@ T_A = 75°C	I_O			1.0			A
Non-Repetitive Peak Forward Surge Current 8.3ms single half sine-wave superimposed on rated load (JEDEC Method)		I_{FSM}			30			A
Forward Voltage	@ I_F = 1.0A	V_{FM}			1.2			V
Peak Reverse Current at Rated DC Blocking Voltage	@T_A = 25°C @ T_A = 100°C	I_{RM}			5.0 100			μA
Reverse Recovery Time (Note 3)		t_{rr}			200			ns
Typical Junction Capacitance (Note 2)		C_j			15			pF
Typical Thermal Resistance Junction to Ambient		$R_{\theta JA}$			100			°C/W
Operating and Storage Temperature Range		T_j, T_{STG}			-65 to +150			°C

FIGURE 10.46 Data sheet excerpt. (Courtesy of Diodes Incorporated.)

10.5 If D_1 and D_2 are treated as ideal diodes:

(a) Draw a graph of V_{out} when the input voltage V_{in} ranges from −10 V to +10 V.

(b) Draw a graph of the current flowing through the resistor over the same range of values for V_{in}.

(c) Given that the type of circuit shown in Figure 10.47 is commonly called a "voltage clamp," briefly describe the function of this circuit.

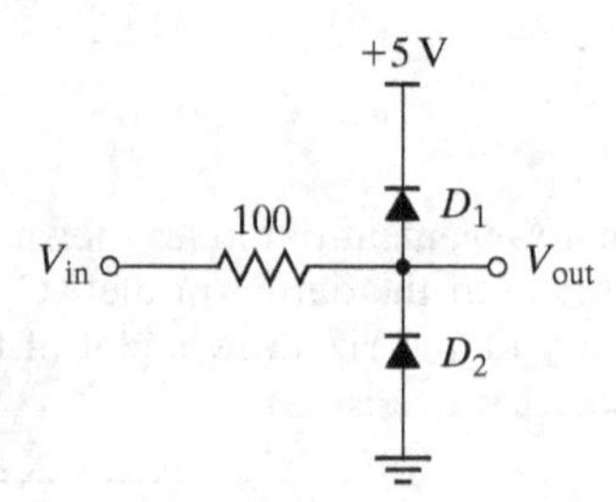

FIGURE 10.47

10.6 If the Zener voltage of the (ideal) Zener diode D_1 (Figure 10.48) is 4.3 V, what is V_{out} when:

(a) $V_S = 3.3$ V **(b)** $V_S = 5$ V

(c) $V_S = 12$ V **(d)** $V_S = -5$ V

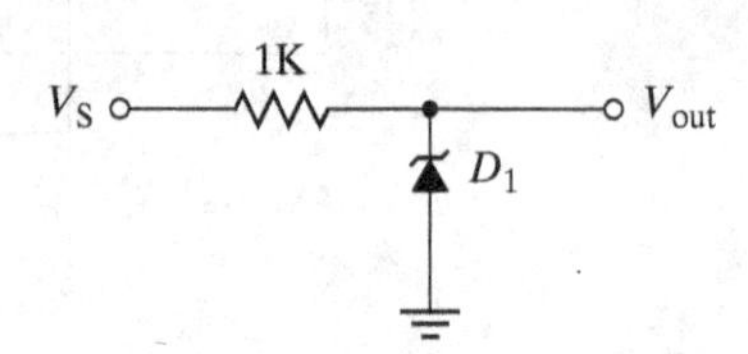

FIGURE 10.48

10.7 Design a circuit that uses a 5 V power supply to illuminate an LED that has a forward voltage drop of $V_f = 2.2$ V. Use only standard 5% resistors in your design, and insure that the current flowing through the LED is between 17 and 20 mA.

10.8 If Q_2 in Figure 10.23 were replaced by an NPN transistor, what voltage would be necessary at V_{in} in order to insure that Q_2 is on and saturated?

10.9 What is the minimum value of V_{in} required to insure that at least 2.125 A flow through the load resistor in the circuit from Figure 10.49? Use the specifications included for the IRLZ34N N-channel MOSFET (Figure 10.50). What is the power dissipated in the resistor under these conditions? In the MOSFET?

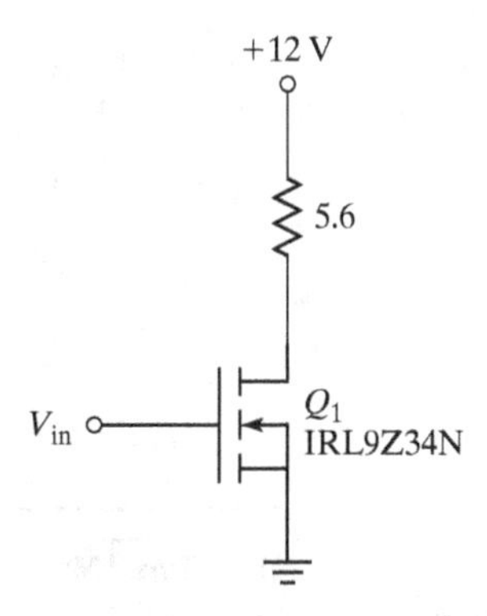

FIGURE 10.49

	Parameter	Min.	Typ.	Max.	Units	Conditions
$V_{(BR)DSS}$	Drain-to-Source Breakdown Voltage	55	—	—	V	$V_{GS} = 0V$, $I_D = 250\mu A$
$\Delta V_{(BR)DSS}/\Delta T_J$	Breakdown Voltage Temp. Coefficient	—	0.065	—	V/°C	Reference to 25°C, $I_D = 1mA$
$R_{DS(on)}$	Static Drain-to-Source On-Resistance	—	—	0.035	Ω	$V_{GS} = 10V$, $I_D = 16A$
		—	—	0.046		$V_{GS} = 5.0V$, $I_D = 16A$
		—	—	0.060		$V_{GS} = 4.0V$, $I_D = 14A$
$V_{GS(th)}$	Gate Threshold Voltage	1.0	—	2.0	V	$V_{DS} = V_{GS}$, $I_D = 250\mu A$

FIGURE 10.50 IRL9Z34N data sheet excerpt. (Courtesy of International Rectifier © 1998.)

10.10 For each of the circuits from Figures 10.51 and 10.53, state whether the transistor is off, on but not operating in the linear region, or on and operating in the linear region. Where possible, calculate the minimum current that will flow through the load resistor using the specifications given in Figures 10.52 and 10.54.

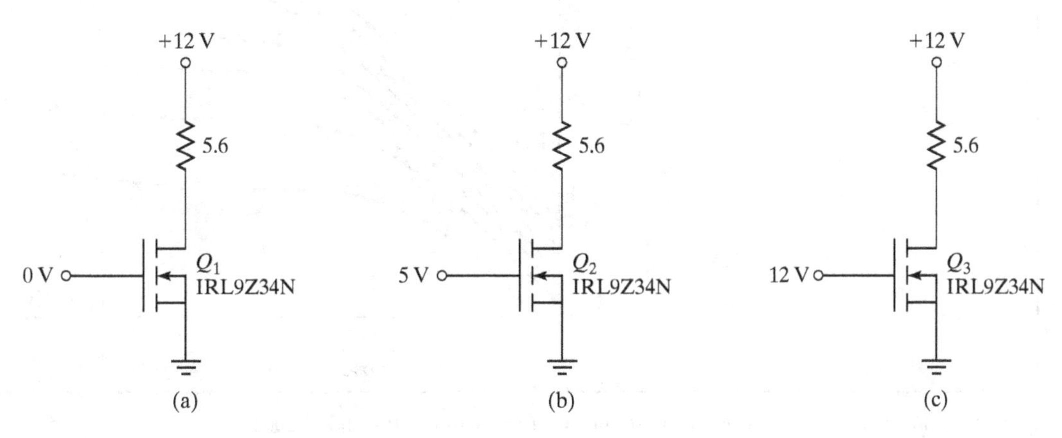

FIGURE 10.51

	Parameter	Min.	Typ.	Max.	Units	Conditions
$V_{(BR)DSS}$	Drain-to-Source Breakdown Voltage	55	—	—	V	V_{GS} = 0V, I_D = 250μA
$\Delta V_{(BR)DSS}/\Delta T_J$	Breakdown Voltage Temp. Coefficient	—	0.065	—	V/°C	Reference to 25°C, I_D = 1mA
$R_{DS(on)}$	Static Drain-to-Source On-Resistance	—	—	0.035	Ω	V_{GS} = 10V, I_D = 16A
		—	—	0.046		V_{GS} = 5.0V, I_D = 16A
		—	—	0.060		V_{GS} = 4.0V, I_D = 14A
$V_{GS(th)}$	Gate Threshold Voltage	1.0	—	2.0	V	$V_{DS} = V_{GS}$, I_D = 250μA

FIGURE 10.52 IRL9Z34N data sheet excerpt. (Courtesy of International Rectifier © 1998.)

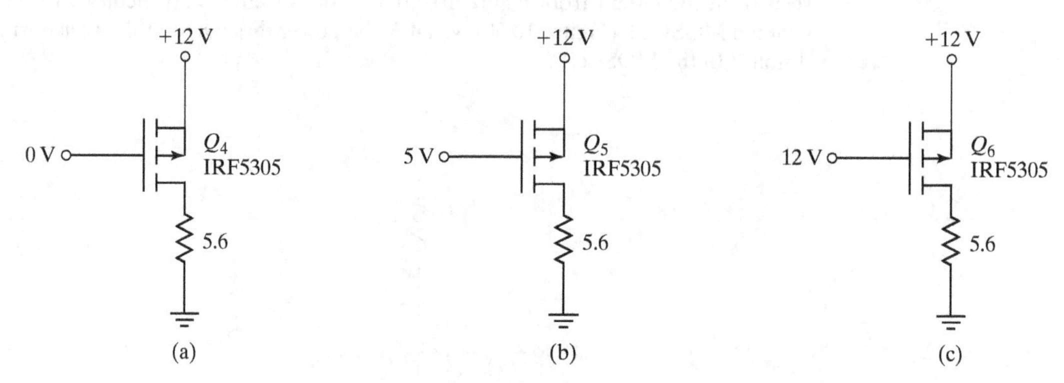

FIGURE 10.53

	Parameter	Min.	Typ.	Max.	Units	Conditions
$V_{(BR)DSS}$	Drain-to-Source Breakdown Voltage	-55	—	—	V	V_{GS} = 0V, I_D = -250μA
$\Delta V_{(BR)DSS}/\Delta T_J$	Breakdown Voltage Temp. Coefficient	—	-0.034	—	V/°C	Reference to 25°C, I_D = -1mA
$R_{DS(on)}$	Static Drain-to-Source On-Resistance	—	—	0.065	Ω	V_{GS} = -10V, I_D = -16A
$V_{GS(th)}$	Gate Threshold Voltage	-2.0	—	-4.0	V	$V_{DS} = V_{GS}$, I_D = -250μA

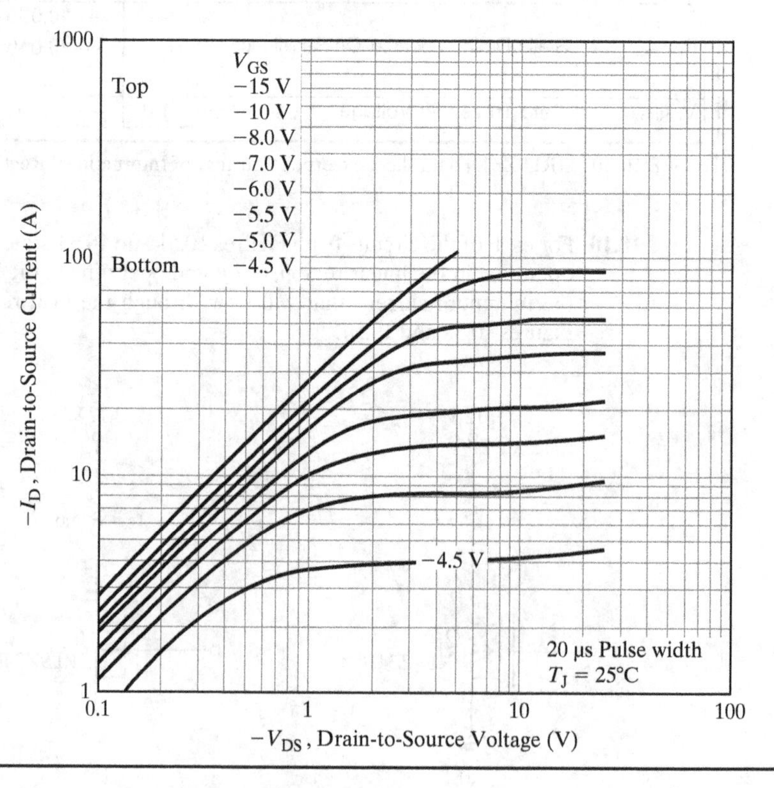

FIGURE 10.54 IRF5305 data sheet excerpt (Courtesy of International Rectifier © 2000.)

10.11 Design a circuit that satisfies all of the following conditions:

(a) Use a single BJT (use "rule of thumb" characteristics for the BJT, and insure operation in the saturated region).

(b) Use the BJT to switch the current flowing through an LED ($V_f = 2.2$ V).

(c) Use a 5 V power supply.

(d) The base of the BJT is driven by a signal that is 0 V when the LED is to be off and 3 V when the LED is to be on.

(e) The current through the LED is approximately 0 mA in the off state and 100 mA (± 10%) in the on state.

(f) Use standard 5% resistor values.

(g) Answer the following:

i. Is this a low-side drive or a high-side drive configuration?

ii. How much steady-state current is required to drive the base of the BJT?

10.12 Design a circuit that satisfies all of the following conditions:

(a) Use a single IRL9Z34N N-channel MOSFET (the data sheet for this component may be found on the Internet). Operate the MOSFET in the linear region.

(b) Use the MOSFET to switch the current flowing through an LED ($V_f = 2.2$ V).

(c) Use a 5 V power supply.

(d) The gate of the IRL9Z34N is driven by a signal that is 0 V when the LED is to be off and 5 V when the LED is to be on.

(e) The current through the LED is approximately 0 mA in the off state and 100 mA (± 10%) in the on state.

(f) Use standard 5% resistor values.

(g) Answer the following:

i. Is this a low-side drive or a high-side drive configuration?

ii. How much steady-state current is required to drive the gate of the IRL9Z34N?

10.13 Design a circuit that satisfies all of the following conditions:

(a) Use a single BJT (assume ideal characteristics, and operate the BJT in the saturated region).

(b) Use the BJT to switch the current flowing through a 50 Ω load.

(c) Use a 5 V power supply.

(d) The base of the BJT is driven by a signal that is 5 V when the load is to be off and 0 V when the load is to be on.

(e) The current through the load is approximately 0 mA in the off state and 35 mA (± 10%) in the on state.

(f) Use standard 5% resistor values.

(g) Answer the following:

i. Is this a low-side drive or a high-side drive configuration?

ii. How much steady-state current is required to drive the base of the BJT?

10.14 Design a circuit that satisfies all of the following conditions:

(a) Use a single IRFU5305 P-channel MOSFET (the data sheet for this component may be found on the Internet). Operate the MOSFET in the linear region.

(b) Use the MOSFET to switch the current flowing through a 50 Ω load.

(c) Use a 12 V power supply.

(d) The base of the BJT is driven by a signal that is 12 V when the load is to be off and 0 V when the load is to be on.

(e) The current through the load is approximately 0 mA in the off state and 200 mA (± 10%) in the on state.

(f) Use standard 5% resistor values.

(g) Answer the following:
i. Is this a low-side drive or a high-side drive configuration?
ii. How much steady-state current is required to drive the gate of the IRFU5305?

10.15 For the circuit shown in Figure 10.55, what is the voltage at V_{out} when V_{in} = 0 V? What is V_{out} when V_{in} = 3.3 V? How much current will V_{in} be required to source or sink in each state?

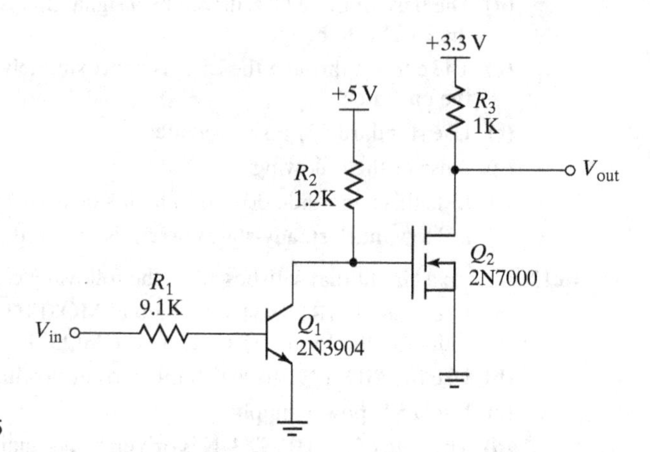

FIGURE 10.55

10.16 What is the current that flows through R_L when V_{in} = 0 V? What is the current through R_L when V_{in} = 5 V? How much current will V_{in} be required to source or sink in each state? Refer to Figure 10.56.

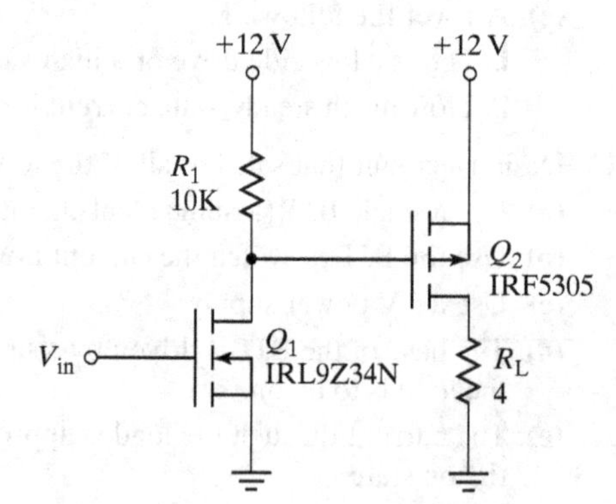

FIGURE 10.56

FURTHER READING

The Art of Electronics, Horowitz, P., and Hill, W., 2nd ed., Cambridge University Press, 1989.
Practical Electronics for Inventors, Scherz, P., 2nd ed., McGraw-Hill, 2006.

Operational Amplifiers

Operational amplifiers (usually shortened to simply "op-amps") are electronic devices that are the basic building blocks used for conditioning, manipulating, and preparing analog signals for presentation to a microcontroller. While the uses of op-amps can range from very simple to quite sophisticated circuits, the basic behavior of these devices can be described with only a few simple rules. With these rules, and the knowledge of a few basic circuit configurations, it is possible to construct many common circuits used to condition signals from sensors. This chapter will present the basic operation of the ideal op-amp, demonstrate how to analyze circuits that employ op-amps, and discuss the most important aspects of how real op-amps differ from ideal op-amps, including how to extract the necessary information from data sheets. It will also include a brief discussion of comparators, which are a subspecies of op-amps that are optimized for certain specific uses. After reading this chapter, the student should be able to:

1. recognize the basic op-amp circuit configurations,
2. use the Golden Rules of Op-Amps to analyze op-amp circuits and develop transfer functions to describe the input-to-output relationship,
3. explain the behavior of a comparator and how it differs from an op-amp,
4. explain the manifestations of the most common deviations from ideal op-amp behavior,
5. extract the relevant parameters from a data sheet and use them to choose among op-amps for a given application.

11.1 OP-AMP BEHAVIOR

Op-amps are the most common circuit elements used to manipulate analog signals. They are active electronic devices, requiring connection to a power supply in order to operate. Functionally, op-amps are 2-input, 1-output devices whose schematic representation is shown in Figure 11.1. One input is called the **inverting input** (labeled "–") and the other is called the **noninverting input** (labeled "+"). Analytically, the output voltage is equal to the difference between the noninverting input and the inverting input multiplied by the **gain** of the op-amp, ***G***,

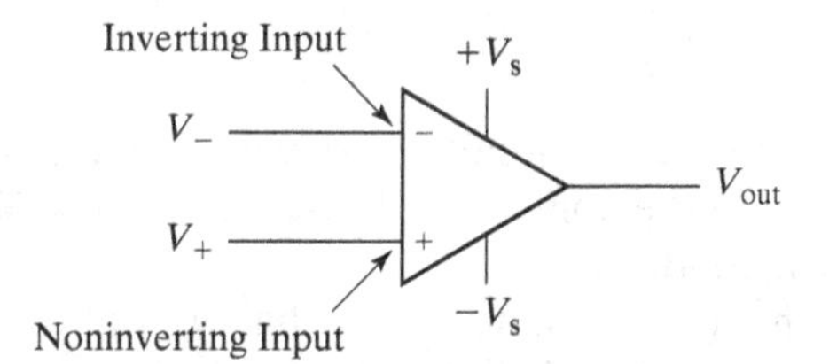

FIGURE 11.1
Schematic symbol for an op-amp.

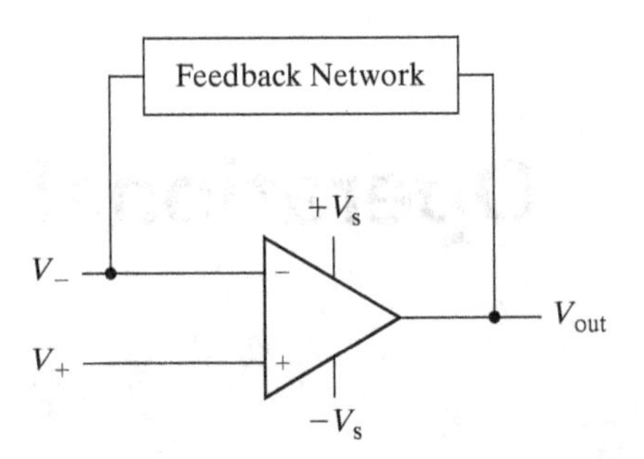

FIGURE 11.2
The closed-loop negative feedback configuration.

as shown in Eq. (11.1). For real devices, this gain is an extremely large value (more than 10^5 for typical real devices), and the output voltage is bounded by the power supply voltages (no matter what is happening at the inputs, the output voltage is always somewhere between $-V_s$ and $+V_s$).

$$V_{out} = G(V_+ - V_-) \quad (11.1)$$

When drawing generic op-amp circuits, it is common to omit the power supply connections. However, when preparing a schematic diagram for a design that you want to build, it is essential that they be included; otherwise it is very likely that power will not actually be connected, and the op-amp will not operate without it.

11.2 NEGATIVE FEEDBACK

The use of op-amps as simple differential amplifiers, as shown in Figure 11.1, is very unusual. This configuration, known as **open loop** (i.e., no feedback), is almost never used. The magnitude of amplification, which is the gain G, is so large that almost any finite voltage difference at the inputs would be amplified so much that the output voltage would be driven to one of the two power supply voltages. Instead, op-amps are most often used in a configuration known as **closed-loop negative feedback**, shown generically in Figure 11.2. The closed-loop portion of the term comes from the fact that there is a connection from the output back to the input, hence "closing a loop." The negative feedback portion of the term refers to the fact that the connection from the output is made back to the inverting input (as opposed to the noninverting input).

When operated in closed-loop negative feedback, the characteristics of the overall circuit are determined almost exclusively by the details of the feedback network and other components connected to the op-amp. For typical circuits, the exact open-loop gain of the op-amp has little impact on the actual gain of the resulting op-amp circuit. This is a very good thing, since it allows us to pick the circuit's gain so that it has a value that's useful in a particular application, with minimal dependence on the op-amp characteristics.

11.3 THE IDEAL OP-AMP

We will begin our study of op-amp circuits by developing a model for an ideal op-amp. In Chapter 12, we will introduce the nonideal behaviors and show that, under many conditions, modern op-amps are close to ideal.

Figure 11.3 shows the ideal op-amp model that we will use. The output voltage is created by an ideal voltage source that is linked to an ideal measurement of the difference in

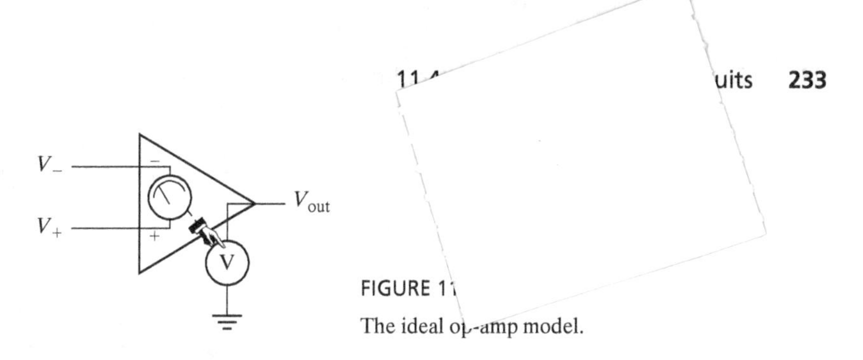

FIGURE 11

The ideal op-amp model.

the voltages present at the two inputs. This model requires three assumptions that we will make whenever we use the ideal op-amp model:

1. The gain $\left[V_{out}/(V_+ - V_-)\right]$ is infinite.
2. The inputs draw no current. This is necessary for an ideal voltage measurement.
3. The output impedance is zero. This is necessary for an ideal voltage source.

The use of this model for an ideal op-amp and acceptance of these three assumptions will greatly simplify the analysis of op-amp circuits. Even when we know that the nonideal aspects of an op-amp will impact a design, it is common to begin with the assumption of an ideal op-amp and then calculate the impact of the important nonidealities after the basic analysis.

11.4 ANALYZING OP-AMP CIRCUITS

11.4.1 The Golden Rules[1]

We can use the ideal op-amp assumptions to develop a pair of rules that have become known as **The Golden Rules** that will allow us to reduce the complexity of the equations we will develop during our analysis of op-amp circuits.

Rule #1: The inputs draw no current.

Rule #2: When operated with negative feedback, the output will do whatever is necessary to make the two input voltages the same.

Rule #1 comes directly from our assumptions about the behavior of an ideal op-amp. Rule #2 is a result of assuming that the op-amp has infinite gain. If the output is connected to the inverting input (even if there are other circuit elements in this path, such as resistors) and the noninverting input is at a higher voltage than the inverting input, then the output voltage will be driven upwards until the voltage at the inverting and noninverting inputs match. In cases where the inverting input is initially at a higher voltage than the noninverting input, the output voltage will be driven lower—also resulting in the voltages at the inputs achieving the same value. Since the gain is infinite, they will match exactly.

11.4.2 The Noninverting Op-Amp Configuration

Figure 11.4 shows what is referred to as the **noninverting amplifier** configuration. The reason for that name will become clear as we analyze this circuit. Before we get into the analysis of the circuit though, it is helpful to walk through the circuit on an imagined, very small time

[1]This usage comes from *The Art of Electronics*, Horowitz, P., and Hill, W., 2nd ed., Cambridge University Press, 1989.

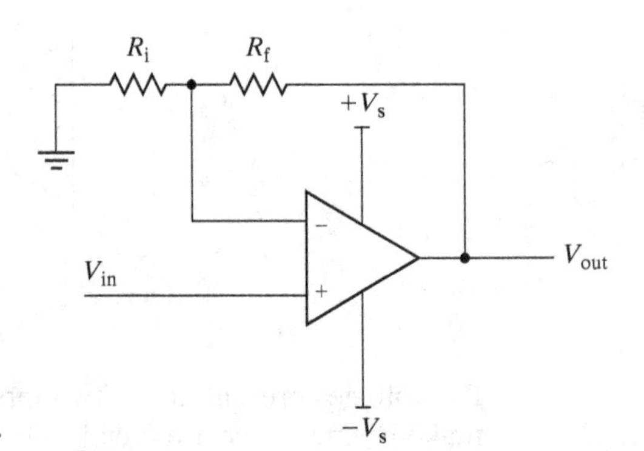

FIGURE 11.4
The noninverting amplifier configuration.

scale. Let's begin the walkthrough by imagining that V_{in} is initially at 1 V and V_{out} is at 0 V. With V_{out} at 0 V, the connection through the voltage divider formed by R_f and R_i will place the inverting input also at 0 V. We therefore have an initial condition in which the noninverting input is at a voltage greater than that at the inverting input. The model of the ideal op-amp tells us that this will result in the output voltage rising (since $V+$ is greater than $V-$). The rising output voltage, acting through the voltage divider, will cause the voltage at the inverting input to also rise. The output voltage will continue to rise until the voltage at the inverting input matches the 1 V at the noninverting input. The resistor values, R_f and R_i, determine the current flowing in the feedback loop and the voltage required at the output to make the input voltages match.

We can analyze the behavior of the circuit symbolically through the use of the Golden Rules. Applying Golden Rule #2 allows us to state that the voltage at Node A (Figure 11.5) is V_{in}. With this information, we can calculate the current i_1 as

$$i_1 = \frac{V_{out} - V_{in}}{R_f} \tag{11.2}$$

And the current i_2 as

$$i_2 = \frac{V_{in} - 0}{R_i} \tag{11.3}$$

Golden Rule #1 tells us that there is no current going into the inverting input of the op-amp, therefore $i_1 = i_2$.

$$\frac{V_{out} - V_{in}}{R_f} = \frac{V_{in} - 0}{R_i} \tag{11.4}$$

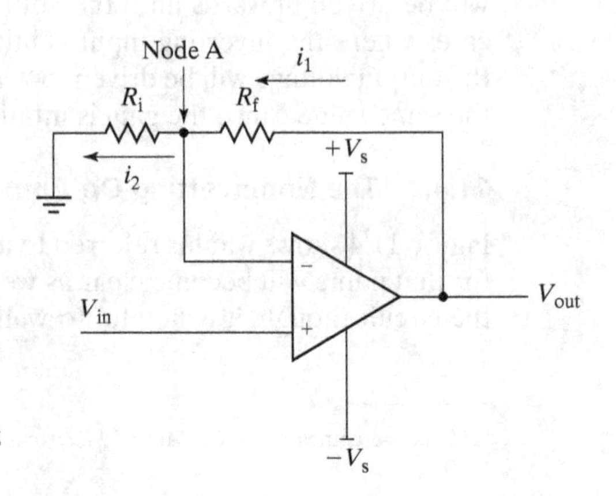

FIGURE 11.5
Noninverting op-amp circuit showing current flows for analysis.

which we can rearrange to relate V_{out} to V_{in} as

$$V_{out} = V_{in}\left(1 + \frac{R_f}{R_i}\right) \tag{11.5}$$

From this expression, we can see that the output voltage is the same polarity as the input voltage (hence, this is a noninverting amplifier) and that the input and output voltages are related by the gain for the noninverting amplifier: $(1 + R_f/R_i)$.

11.4.3 The Inverting Op-Amp Configuration

Figure 11.6 shows a variation on the noninverting amplifier configuration called the **inverting amplifier**. As you may have guessed by its name, this circuit gives a very different result. To create the inverting amplifier circuit, we swapped the connections to the noninverting input and the left-hand side of R_i. The analysis for this circuit follows much the same sequence as before: we start by invoking Golden Rule #2 to assert that the voltage at Node A is 0 V. With that helpful simplification, we can write expressions for i_1 and i_2:

$$i_1 = \frac{V_{out} - 0}{R_f} \tag{11.6}$$

$$i_2 = \frac{0 - V_{in}}{R_i} \tag{11.7}$$

Now we can invoke Golden Rule #1 to assert that since there is no current flowing into the inverting input of the op-amp, again $i_1 = i_2$.

$$\frac{V_{out} - 0}{R_f} = \frac{0 - V_{in}}{R_i} \tag{11.8}$$

which we can rearrange to relate V_{out} to V_{in} as

$$V_{out} = -V_{in}\left(\frac{R_f}{R_i}\right) \tag{11.9}$$

For the inverting configuration, we can see that the output voltage is the opposite polarity as the input voltage (hence the name "inverting") and that the input and output voltage amplitudes are related by the factor R_f/R_i. We combine the change in polarity with the amplitude factor to say that the gain for the inverting configuration is $-R_f/R_i$.

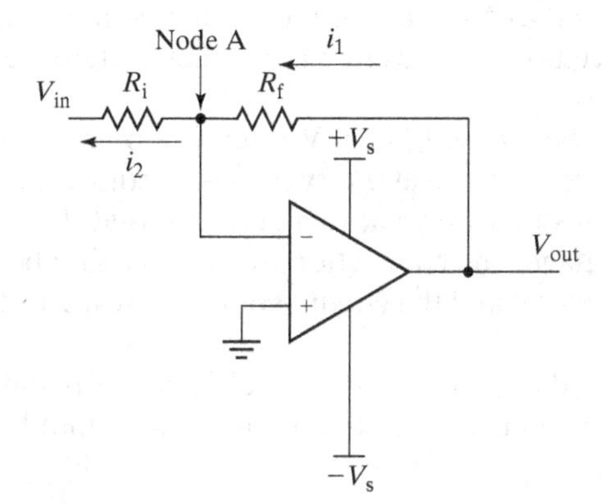

FIGURE 11.6

The inverting amplifier configuration.

As we saw before, with the noninverting configuration, the gain of closed-loop op-amp circuits is governed by the ratio of the feedback resistor to the input resistor. In the ideal case, it is only the *ratio* that matters and not the absolute values of these resistors. For practical circuits, the values of these resistors should be kept in the range of a few kilohms to a few hundreds of kilohms. These guidelines are rooted in the nonideal behavior of real op-amps. We cover op-amp nonidealities and point out which of them contribute to the guidelines in Section 12.1.

Here, it is helpful to take a look at a concrete example. Suppose that in the circuit of Figure 11.6, the input voltage, V_{in}, is 1 V and that $R_f = R_i = 10\ k\Omega$. Using Eq. (11.7), we can calculate the current i_2 as

$$i_2 = \frac{0 - V_{in}}{R_i} = \frac{-1\ V}{10\ k\Omega} = -0.1\ mA$$

The fact that the calculated current is negative tells us that the direction of current flows that we had assumed in writing the equations is opposite to the actual current flow in this situation.

With a value for i_2, we can use Golden Rule #1 to assert that $i_1 = i_2$, then rearrange Eq. (11.6) to write an expression for V_{out}:

$$V_{out} = R_f i_1 = 10\ k\Omega \times -0.1\ mA = -1\ V$$

This value for V_{out} agrees with the result that we would get from applying Eq. (11.9) directly.

11.4.3.1 The Virtual Ground

It is interesting to note that Node A in Figure 11.6 is held at ground potential by the action of the op-amp. It is not actually connected to ground, but is always at ground potential. This means that, for example, when analyzing the current through resistor R_i, we can treat the right-hand side of the resistor as if it were connected to ground. However, the current flowing through R_i is not going directly to ground. This can be a useful characteristic and this has been given the name **virtual ground.** Node A would be said to be "at a virtual ground."

11.4.3.2 There Is Nothing Magic about Ground

The role of ground in electronic circuits is often somewhat of a mystery to beginning circuit designers. Ground in electronic circuits serves two purposes: 1) it is the common voltage reference point across the circuit and 2) it provides a return path for current to flow back to the power source. It is important to learn to tell which of these purposes is being served by any given connection to ground. The return path function is of greatest importance when dealing with the power being supplied to devices. This will be primarily the power connections to device pins that are explicitly labeled as ground. For most other connections to ground, the purpose is to supply a 0 V reference point to the circuit. It is these reference voltage connections that we want to deal with next.

In instances where ground is being used as a 0 V reference point, you should think of this as a voltage source that just happens to be at 0 V (which is, of course, the same voltage as ground potential). It is in these cases that we want to assure you that *there is nothing magic about ground.* It is just another voltage. The connection to ground could be substituted with a voltage source at some other potential and the circuit would still work, just with a different transfer function.

To explore this, let's look at the circuit of Figure 11.7, which is the same circuit as Figure 11.6 except that the ground connection to the noninverting input has been replaced by a connection to a voltage source, V_{ref}.

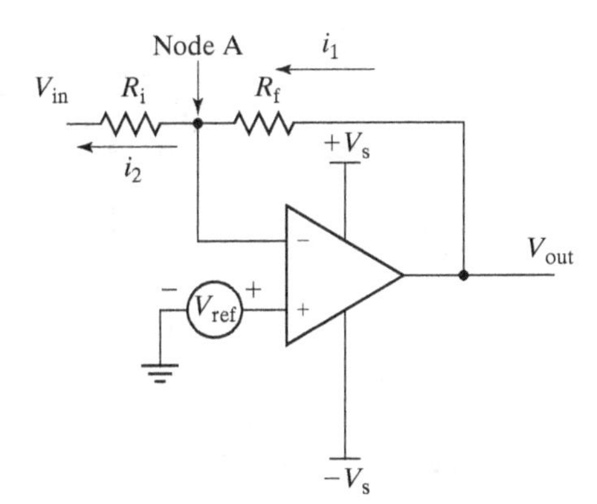

FIGURE 11.7

Inverting op-amp with offset voltage.

We can proceed just as we did in developing Eqs. (11.6)–(11.9) to get an expression relating V_{out} to V_{in}. This time, Node A will be at V_{ref}:

$$i_1 = \frac{V_{out} - V_{ref}}{R_f} \tag{11.10}$$

$$i_2 = \frac{V_{ref} - V_{in}}{R_i} \tag{11.11}$$

$$\frac{V_{out} - V_{ref}}{R_f} = \frac{V_{ref} - V_{in}}{R_i} \tag{11.12}$$

$$V_{out} = (V_{ref} - V_{in})\left(\frac{R_f}{R_i}\right) + V_{ref} \tag{11.13}$$

In Eq. (11.13), we can see the manifestation of the basic differential amplifier behavior of the op-amp in the fact that the gain is now applied to the difference between V_{ref} and V_{in}. We can see that the output has been shifted (offset) by the magnitude of the reference voltage, V_{ref}. V_{in} still appears in the equation with a negative sign, indicating that increases in V_{in} will result in reductions in V_{out}, which is the basic inverting configuration behavior. Whether or not here is an actual polarity reversal in the output will depend on the relative values of V_{in} and V_{ref}. Finally, we see that if $V_{ref} = 0$, then we get back the transfer function for the basic inverting op-amp configuration.

Example 11.1

As a concrete example of what we have seen so far, consider the design of the amplification circuit for an LM34 temperature sensor to be used to measure room temperatures. The LM34 is a precision temperature sensor that outputs a voltage proportional to temperature in Fahrenheit, with a scale factor of 10 mV/°F. It is to be used in an application that measures room temperature, where the range of interest is 50°F to 90°F and the goal is to make this temperature range correspond to an output voltage range from 0.5 to 4.5 V.

Solution: From these specifications, we can determine that we need a gain of 10 [(10 mV/°F × 40°F)/(4.5 V–0.5 V) = 10]. The specifications also tell us that we need to use a noninverting amplifier since 0.5 V is to correspond to 50°F and 4.5 V to 90°F. With that information, we can draw a first-pass schematic (Figure 11.8).

To achieve a gain of 10, we need to make the ratio of R_f to $R_i = 9{:}1$. What remains is to determine what value we need to supply for V_{ref} in order to meet the system requirements. To do that, it is easiest to look at the situation when the temperature is at 50°F and the LM34's output will be

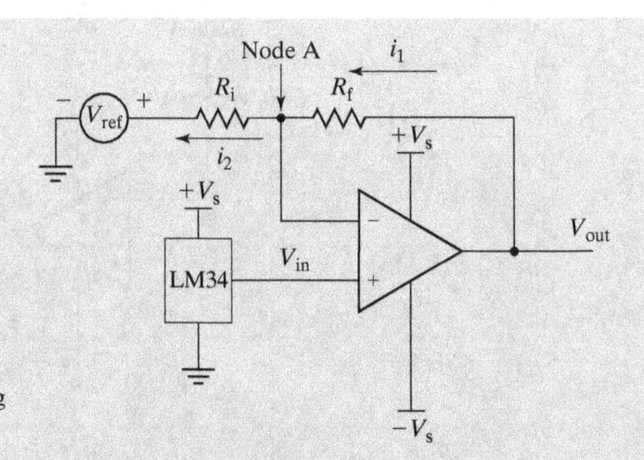

FIGURE 11.8

First-pass signal conditioning for LM34.

50°F × 10 mV/°F = 500 mV. The second Golden Rule tells us that Node A will also be at 0.5 V. The specifications require that at this temperature the output voltage also be 0.5 V. If both Node A and V_{out} are at 0.5 V, then i_1 will be 0 and therefore, i_2 will be 0, so V_{ref} must also be at 0.5 V.

We can test this solution at the upper end of the temperature range, when the output of the LM34 will be 90°F × 10 mV/°F = 0.9 V. Node A will also be at 0.9 V, and the output should (according to the specifications) be at 4.5 V, yielding 4.5 V − 0.9 V = 3.6 V across R_f and forcing i_1 to be 3.6 V/R_f. Because of the first Golden Rule, i_2 will also be 3.6 V/R_f. This current will flow across R_i (which is 1/9R_f) to create a voltage drop of $3.6\ \text{V}/R_f \times R_f/9 = 3.6\ \text{V}/9 = 0.4$ V. Since Node A was at 0.9 V, this places V_{ref} at 0.5 V, which agrees with our original calculation. Therefore, the solution is the circuit of Figure 11.8 with V_{ref} = 0.5 V. We cover how to create the 0.5 V source for V_{ref} in the next section.

11.4.4 The Unity Gain Buffer

There are many situations in which you would like the characteristics of an amplifier with a gain of exactly 1. For example, suppose that you would like to use a voltage divider to generate a reference voltage, but need the reference voltage to be able to source or sink significant amounts of current (a few milliamps). When we analyzed the voltage divider in Chapter 9, we assumed that no current was flowing into or out of the V_{ref} node in Figure 11.9. Currents flowing into or out of V_{ref} would disturb the reference voltage by changing the voltage divider expression. In fact, it would be very difficult to keep the disturbances down to only a few percent. This would require that the normal current flowing in R_1 and R_2 be approximately 50–100 times larger than the current flowing into or out of V_{ref}. If the V_{ref} current were 2 mA, that would mean 100–200 mA would be required to flow in the voltage divider. This is a relatively large amount of current that represents wasted power that is dissipated as heat in the resistors of the voltage divider. And in any case, this would be an inelegant approach: the resistors R_1 and R_2 would be dissipating somewhere between 0.5 and 1.0 W, which would require the use of power resistors.

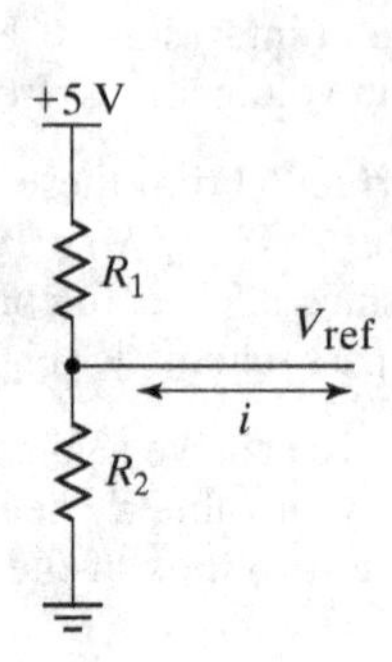

FIGURE 11.9

A voltage divider used to provide a reference voltage.

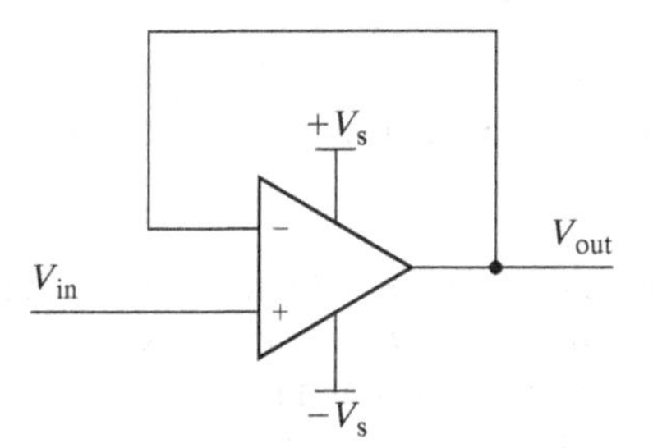

FIGURE 11.10

The unity gain buffer.

This is a situation in which the input characteristics of the op-amp can be very useful. If V_{ref} were connected to an input of an op-amp, Golden Rule #1 tells us that no current will be drawn and therefore, there would be no disturbance in V_{ref}. The current provided to the load would be delivered by the output of the op-amp. This isolation of currents between the input and the output is referred to as **buffering**.

Of the circuits that we have discussed so far, our needs would seem to be best met by the noninverting configuration, since in that circuit the input voltage is tied directly to the op-amp input and there is no inversion of the input voltage. However, at first glance, it appears that the gain expression for the noninverting amplifier, $(1 + R_f/R_i)$, indicates that it is only capable of gains greater than 1. A bit of outside-the-box thinking would bring us to the conclusion that if we were to make $R_f = 0$ and $R_i = \infty$, the resulting gain would be exactly 1. Actually, from a theoretical standpoint, we would achieve a gain of exactly 1 if either $R_f = 0$ or $R_i = \infty$, but from a practical standpoint, it's easiest to make both modifications. Modifying the noninverting amplifier circuit in this way yields a circuit known as a **unity gain buffer** (Figure 11.10).

By combining the circuits of Figures 11.9 and 11.10 (with one additional embellishment), we arrive at a practical reference voltage source (Figure 11.11).

The capacitor, C_1, has been added to combat fluctuations in the power supply. Without C_1, any disturbance in $+V_s$ will be immediately reflected in V_{ref}. Adding C_1 to the voltage divider creates a low-pass filter that will reduce the magnitude of disturbances on $+V_s$ that are seen at the input to the op-amp. Since $+V_s$ should not be changing at all (a desirable characteristic for a power supply), the value of C_1 is usually chosen to give a relatively long time constant (tens to hundreds of milliseconds, for example).

The circuit of Figure 11.11 works well when we can achieve the desired voltage using standard resistor values and are willing to tolerate the uncertainty associated with standard resistor tolerances. For those situations when this does not produce an acceptable result, it is possible to replace either R_1 or R_2 with an adjustable resistance. In order to have fine control over the adjustable resistance, it is often a good idea to implement this as a combination of a fixed value resistor and a potentiometer wired to act as a variable resistor (Figure 11.12). A single potentiometer could be used, but the resolution of the adjustment would be much

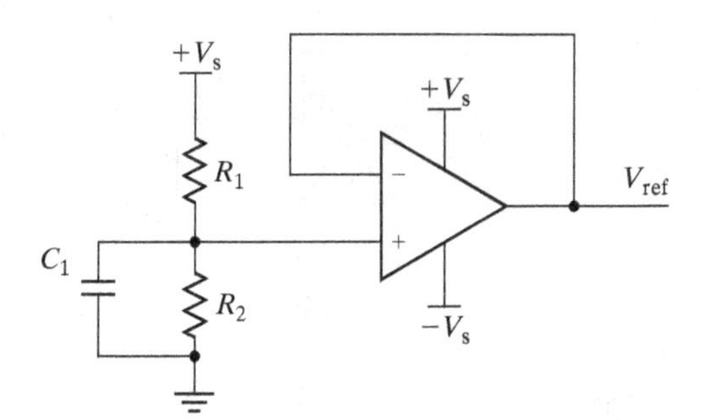

FIGURE 11.11

A practical voltage reference.

FIGURE 11.12

Potentiometer used to fine tune a resistor value.

more coarse. Additionally, it may not be desirable to decrease the resistance all the way to 0 Ω, which a single potentiometer would allow. A combination of a fixed resistor (R_{1a}) and a potentiometer (R_{1b}) solves both problems.

The fixed resistor, R_{1a}, should be chosen such that at the maximum range of its tolerance it is less than the minimum required total resistance. The value of the variable resistor, R_{1b}, should be chosen so that when R_{1a} is at its minimum possible value, based on tolerance, the sum of that minimum value plus the minimum value of R_{1b}, when the wiper is positioned for maximum resistance (i.e., labeled resistance of the potentiometer minus its worst case tolerance), slightly exceeds the maximum required value of the combination. By choosing the values of the resistors in this way, we maximize the resolution with which we can adjust the combined resistance and never have a situation when the combination has zero resistance (which often creates problems). This approach to achieving an exact resistance can be applied in place of any resistor in any circuit. Having said that, it is exceedingly rare to need this kind of adjustability for gain resistors outside the realms of prototype circuits and measurement instruments. Most mechatronics systems can be designed to meet requirements despite the tolerance in resistors. In some instances, it will be necessary (and far preferable) to specify resistors with tighter tolerances (2%, 1%, or better) to meet requirements.

11.4.5 The Difference Amplifier Configuration

Figure 11.13 shows a circuit in which there are two input voltages. The caption claims that this is a **difference amplifier**. Let's work out the details of what the transfer function is for this circuit to see if the caption writer knows what he or she is talking about.

To start the analysis, it is helpful to look for parts of the circuit with minimal dependencies on other parts of the circuit. The subcircuit composed of V_2, R_i, Node B, and R_f forms a voltage divider with the only other connection being to the noninverting input to the op-amp. Because of Golden Rule #1 (the inputs draw no current), we can treat this as an ideal voltage divider and write the expression for the voltage at Node B.

$$V_B = V_2\left(\frac{R_f}{R_i + R_f}\right) \tag{11.14}$$

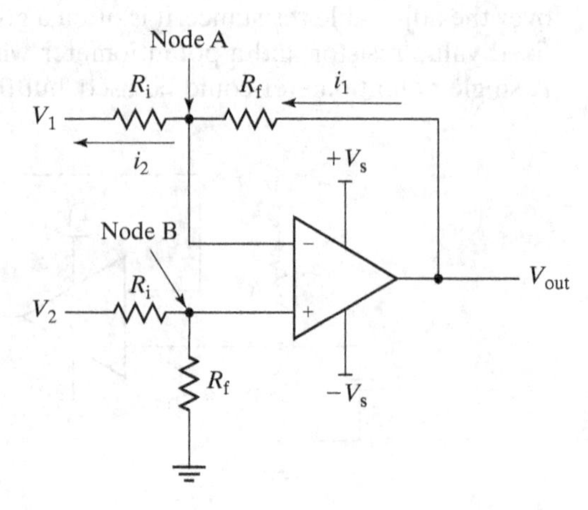

FIGURE 11.13

The difference amplifier.

We can then invoke Golden Rule #2 to assert that the voltage at Node A is the same as that at Node B, since the op-amp's output is fed back to the inverting input (negative feedback). This will allow us to write the expressions for i_1 and i_2 as

$$i_1 = \frac{V_{out} - V_2\left(\frac{R_f}{R_i + R_f}\right)}{R_f} \tag{11.15}$$

$$i_2 = \frac{V_2\left(\frac{R_f}{R_i + R_f}\right) - V_1}{R_i} \tag{11.16}$$

Applying Golden Rule #1 to the inverting input allows us to say that $i_1 = i_2$ and hence,

$$\frac{V_{out} - V_2\left(\frac{R_f}{R_i + R_f}\right)}{R_f} = \frac{V_2\left(\frac{R_f}{R_i + R_f}\right) - V_1}{R_i} \tag{11.17}$$

Now a bit of algebra allows us to get to an expression relating the output voltage to the two input voltages:

$$V_{out} = (V_2 - V_1)\frac{R_f}{R_i} \tag{11.18}$$

Eq. (11.18) shows that the caption was indeed correct. This circuit produces an output voltage that is proportional to the difference between the two input voltages with an overall gain of R_f/R_i.

From a practical standpoint, some care is necessary in using the circuit of Figure 11.13 to produce a signal that is the difference between two other signals. The first issue relates to the accuracy of the subtraction. The derivation of Eq. (11.18) assumed that both resistors labeled R_i were *exactly* the same value (that's how we were able to make many of the algebraic simplifications) and that both resistors labeled R_f also had *exactly* the same value. Any differences in the actual values of these resistors in a real circuit will change the transfer function, resulting in a more complex expression that will only be an approximation of the difference. The second issue relates to the absolute values of the resistors used for R_f and R_i. To minimize the current that must be supplied by V_1 and V_2, we would like to make the resistors as large as possible. As we shall see in Chapter 12, this is at odds with minimizing errors due to the nonideal behavior of real op-amps. As a result, the circuit of Figure 11.13 is typically only used for low precision subtraction when the voltage sources are capable of delivering significant currents (i.e., they have a low output impedance). For more accurate difference amplification, you should use what is known as an **instrumentation amplifier**. A discussion of the instrumentation amplifier appears in Chapter 14 on Signal Conditioning.

11.4.6 The Summer Configuration

Figure 11.14 shows another circuit with two input voltages and a single output voltage. Whereas the circuit in Figure 11.13 produced the difference between the two input voltages, the caption for this circuit indicates that it will produce an output that is the sum of the two input voltages: that is, this circuit acts as a **summer**. Again, let's double-check the work of the caption writer.

When looking for where to begin the analysis of this circuit, we find that it doesn't offer an obvious clean separation of the subcircuits for the two input voltages that we enjoyed in

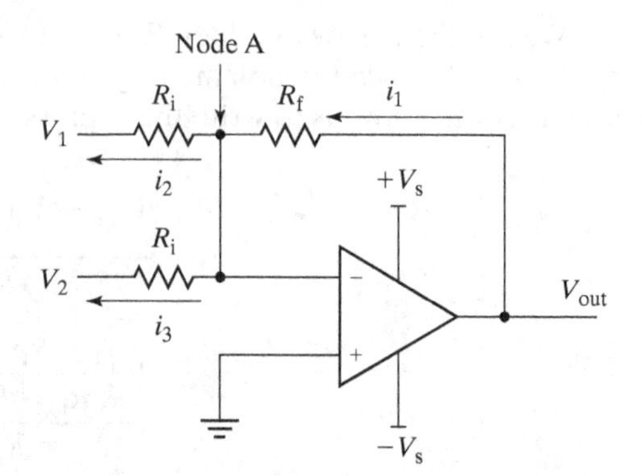

FIGURE 11.14
The summer.

our analysis of the circuit in Figure 11.13. However, the fact that we have two independent input voltages suggests that this might be a good place to apply the principle of superposition. In Chapter 9, we saw that we could analyze circuits with multiple sources by treating each source separately and then summing the results from each of the sources. That's what we'll do here.

We begin the analysis by removing one of the sources. As with Thevenin equivalent circuits and analysis, we remove voltage sources by replacing them with a short circuit, and we remove current sources by replacing them with an open circuit. Let's start by removing V_2. This is the same as setting this voltage to 0 V. This circuit shares a characteristic with the inverting amplifier configuration in that the noninverting input is tied to ground, producing a virtual ground at Node A. With Node A at 0 V and V_2 at 0 V, there is no voltage difference across the lower input resistor R_i and so $i_3 = 0$. With $i_3 = 0$, this circuit is identical to the inverting amplifier that we analyzed in Section 11.4.3. The output that results from V_1 is therefore,

$$V_{out} = -V_1 \frac{R_f}{R_i} \tag{11.19}$$

Now we can go back and reinstate V_2 in the circuit, and remove V_1 by setting it equal to 0 V. Having done that, we find that there is no longer a voltage difference across the upper input resistor R_i and therefore, $i_2 = 0$. With $i_2 = 0$, we have a circuit that is very similar to the inverting amplifier configuration shown in Figure 11.6, with the only difference being that i_2 has been replaced with i_3. That will give us a similar output description, with the term V_2 replacing the term V_1 from Eq. (11.9):

$$V_{out} = -V_2 \frac{R_f}{R_i} \tag{11.20}$$

Having calculated the results from each of the input voltages independently, we obtain the total response by adding the two results together:

$$V_{out} = -V_2 \frac{R_f}{R_i} + \left(-V_1 \frac{R_f}{R_i}\right) = -\frac{R_f}{R_i}(V_2 + V_1) \tag{11.21}$$

Eq. (11.21) shows that the caption writer was not being entirely truthful. The circuit of Figure 11.14 is actually an **inverting summer**—the sign of the resulting sum is changed by this circuit.

11.4.7 The Trans-Resistive Configuration

The circuits that we've looked at so far have all had voltage inputs and voltage outputs. While a voltage output is the most useful type of output when interfacing to microcontrollers, not all interesting inputs appear as voltages. Prime examples of devices that produce

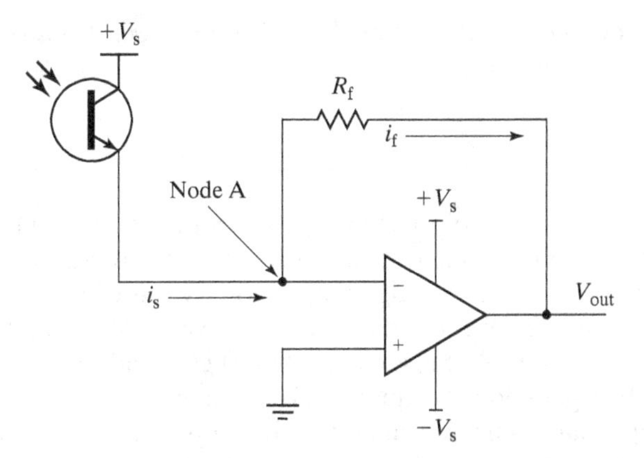

FIGURE 11.15

Trans-resistive amplifier for a photo-transistor.

an output as a current rather than a voltage are the photo-diode and photo-transistor introduced in Chapter 10. Both of these devices produce current flows that are proportional to the amount of light incident on the sensor. Both of these types of devices also produce the most linear transfer function from light to current when a constant voltage is maintained across their terminals. Using an op-amp, we can build a simple circuit that achieves both of these goals: the **trans-resistive amplifier** circuit.

In the circuit of Figure 11.15, Node A is once again a virtual ground node (held by the op-amp at 0 V). This fixes the voltage across the photo-transistor to $+V_s$, independent of how much light is falling on the photo-transistor. This will yield the most linear light-to-current transfer function from the device. The photo-current, i_s, flows into Node A. Because of Golden Rule #1, $i_f = i_s$ and therefore, $V_{out} = -i_f R_f$. This amplifier circuit's output voltage is the product of the amount of current flowing through the sensor, i_s, times a gain term which is the value of the feedback resistor, R_f. Note too, that the resulting sign is inverted.

The circuit is called trans-resistive because it performs a function similar to that of a resistor: it converts a current into a voltage. The circuit is superior to a simple resistor in a number of respects. Since Node A is held at a constant voltage, the current is being delivered into a constant voltage, independent of the amount of current flowing. Also (and this is a subtler point), the fact that Node A is held at a constant voltage means that any stray capacitance associated with the circuit is not being charged or discharged by changes in the photo-current. This, in turn, means that the rise and fall times of the photo-current will not be limited by the RC charge and discharge times that would result from using a simple resistor.

11.4.8 Computation with Op-Amps

So far we have discussed op-amp circuits that will amplify (multiply), sum (add), and take the difference (subtract). If we add in the voltage divider (divide), we have the basic mathematical operations. Indeed, in the past, analog computers were built to quickly perform mathematical transforms on analog signals because they could do the operations much more quickly than the digital computers available at that time. As part of analog computation field, op-amp circuits were also developed to perform the calculus operations of integration and differentiation. In this text, we do not deal directly with op-amp circuits to create integrators and differentiators, though in Chapter 14, we cover low-pass filters (integrators) and high-pass filters (differentiators) in the context of their frequency domain behavior. Today the integration and differentiation computations are most commonly performed in software executed by microcontrollers. While it is true that addition and subtraction of signals are also done digitally, implementing these processes digitally requires more inputs to the microcontroller, a very limited resource. Understanding the summer and difference amplifier circuits is also important in the context of signal conditioning operations (Chapter 14) and digital to analog converters (Chapter 19).

11.5 THE COMPARATOR

In designing the electronics for mechatronic systems, it is common to need to know whether a particular analog voltage is above or below some reference voltage. This would typically involve, for example, producing a high output voltage if the input voltage were above the reference and a low output voltage otherwise. While we could attempt to do this with an amplifier circuit that has a suitably chosen (very large) gain and reference voltage (Figure 11.16), that would not be a good idea because of a phenomenon known as saturation in the op-amp.

Saturation in an op-amp occurs when the output of the op-amp is not able to force the two inputs to the same voltage. The output goes to the highest (or lowest) voltage that it can produce, and the output transistors enter the saturated state as they try to drive the output voltage further. While this does not harm the op-amp, driving the transistors into saturation takes them outside of their normal operating regime. If, in the circuit of Figure 11.16, the reference voltage were 1 V and the input voltage was 0.999 V (1 mV below the reference), Eq. (11.13) predicts that the output voltage would be

$$V_{out} = (V_{ref} - V_{in})\left(\frac{R_f}{R_i}\right) + V_{ref} = (1\text{ V} - 0.999\text{ V})(10{,}000) + 1\text{ V} = 11\text{ V}$$

However, since the circuit has only a 5 V positive supply, the output would actually be driven into a state of saturation as it attempted to drive the output to a positive voltage that it cannot achieve.

If the input voltage then rose from 0.999 to 1.001 V (from 1 mV below to 1 mV above the reference), Eq. (11.13) predicts that the output voltage should go to

$$V_{out} = (V_{ref} - V_{in})\left(\frac{R_f}{R_i}\right) + V_{ref} = (1\text{ V} - 1.001\text{ V})\ (10{,}000) + 1\text{ V} = -9\text{ V}$$

Ideally, the output of the op-amp would immediately switch to the low output voltage. However, the saturated transistors in the op-amp take considerable time (often microseconds) to come out of saturation, and then more time to drive the output voltage to the new voltage level (which is again in saturation at the voltage limit imposed by the negative power supply). This produces an undesirable delay in the response to the change in the input voltage. To get around this problem, integrated circuit designers have produced another type of device

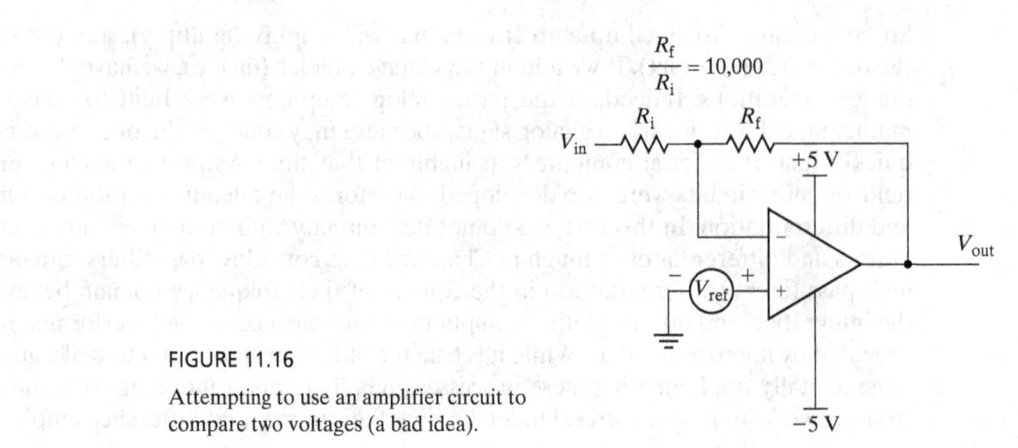

FIGURE 11.16

Attempting to use an amplifier circuit to compare two voltages (a bad idea).

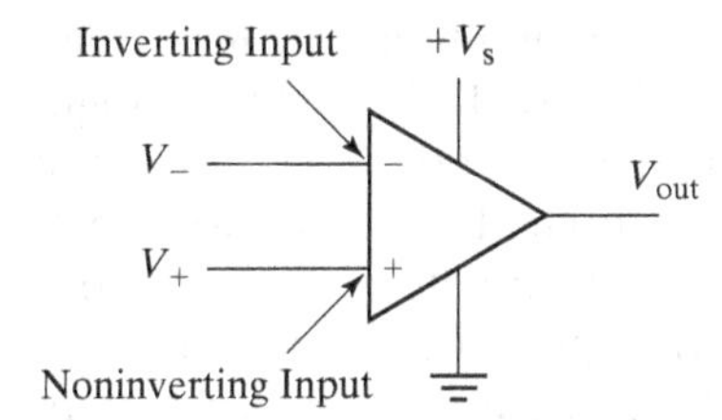

FIGURE 11.17
The schematic symbol for a comparator.

that is very similar to an op-amp, but whose internal design is optimized for a high speed of response, rather than for a linear response (as is the goal for op-amps). The comparator is designed to switch its output as quickly as possible whenever the voltage at the noninverting input exceeds the voltage at the inverting input, and vice versa. This response is largely independent of the magnitude of the difference between these two voltages. Figure 11.17 shows the schematic symbol for a comparator.

Comparing the schematic symbol for the comparator (Figure 11.17) with the schematic symbol for the op-amp (Figure 11.1) shows that they are identical except for the change of the lower supply voltage from $-V_s$ in the op-amp to ground in the comparator. This reflects the fact that the vast majority of comparators are designed to operate on a single supply voltage.

So, how do you tell the difference between an op-amp and a comparator in a schematic diagram? There are several ways. First and foremost, each device in a schematic should be labeled with a part number. With experience, you will begin to recognize the most common part numbers for op-amps and comparators. When faced with a part number that you don't recognize, you could look up the part in any number of online references or you could look at how the device is used in the circuit. Op-amps are almost always used in one of the closed-loop negative feedback configurations. Comparators are *never* used in negative feedback (they make lousy amplifiers) and are most commonly used with positive feedback (e.g., the output is connected to the noninverting or "positive" input) or in open-loop mode (e.g., no connection between the output and either of the inputs).

Most comparators are designed with a different type of output than op-amps. Op-amps are designed with what is known as a **totem-pole output** (sometimes called a **push-pull output**) as shown in a simplified form in Figure 11.18a. This type of output is capable of driving the output voltage high and sourcing current to the output by turning on Q_1 and turning Q_2 off. It is also capable of driving the output low and sinking current by turning on Q_2 and turning Q_1 off. Comparators are sometimes designed with totem-pole style outputs but much

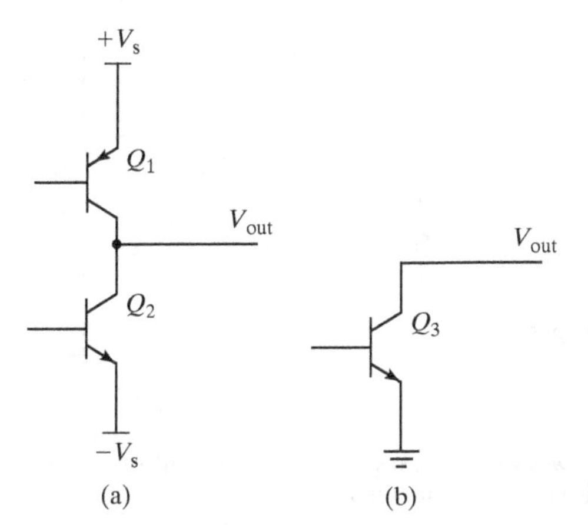

FIGURE 11.18
Totem-pole (a) vs. open-collector (b) outputs.

more often[2] (including the most commonly used comparators) they are designed with an output stage like that shown in Figure 11.18b. This configuration is known as an **open-collector output** when Q_3 is a bipolar junction transistor, (or an **open-drain output** when Q_3 is a MOSFET). With this style of output, the comparator can drive the output low by turning on Q_3, but it cannot drive V_{out} to a high voltage and must depend on external devices that the circuit designer adds in order to produce a high output voltage at the output when Q_3 is turned off. The open-collector style of output provides the circuit designer with a great deal of flexibility in what the high output voltage will be (e.g., it could come from a voltage source other than the power supply for the comparator) as well as in how this output is connected with other outputs. We will cover totem-pole and open-collector outputs in more detail in Chapter 17 on Digital Outputs and Power Drivers.

11.5.1 Comparator Circuits

The simplest practical comparator application circuit is shown in Figure 11.19. We have assumed that this comparator has an open-collector output, as evidenced by the **pull-up resistor**, R_{PU}, connected between the output and the power supply. The output, V_{out}, will be at approximately 5 V whenever V_{in} is greater than the voltage, V_{ref}, set by the voltage divider formed by R_1 and R_2. Otherwise, the output voltage will be at approximately 0 V. This circuit behaves as a **noninverting comparator**. In order to construct an **inverting comparator**, we can simply reverse the connections to the two inputs. When we do this, the circuit output will be low when V_{in} is above V_{ref}, and otherwise, the output will be high.

The circuit of Figure 11.19 works well as long as V_{in} transitions quickly across the reference voltage level. If the input voltage transitions relatively slowly, this circuit will be subject to the same chatter issue that we saw with the software comparator discussed in Chapter 5, Section 5.3. A slowly changing signal with even a little bit of electrical noise will cross back and forth across the threshold voltage multiple times as it passes completely over the threshold, as shown in Figure 11.20, and this will cause multiple output transitions for what should, in fact, be treated as a single input transition.

The cure for this problem is to take the same approach as for the software comparator: add some hysteresis to the system. For the hardware comparator, this means feeding the output signal back to the reference level input so that the reference level is shifted depending on the output level. This is most simply accomplished by modifying the inverting comparator circuit to produce the circuit shown in Figure 11.21, which gives us an **inverting comparator**

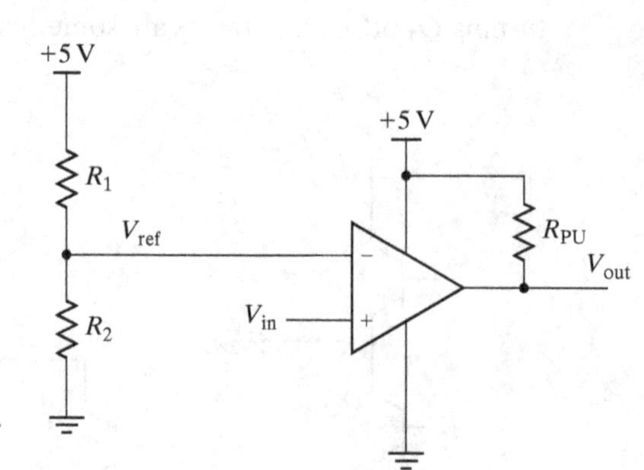

FIGURE 11.19

Simple noninverting comparator circuit.

[2]For example, at the time that this was written, Texas Instruments offered 104 comparators with open-collector or open-drain outputs but only 26 devices with push-pull outputs.

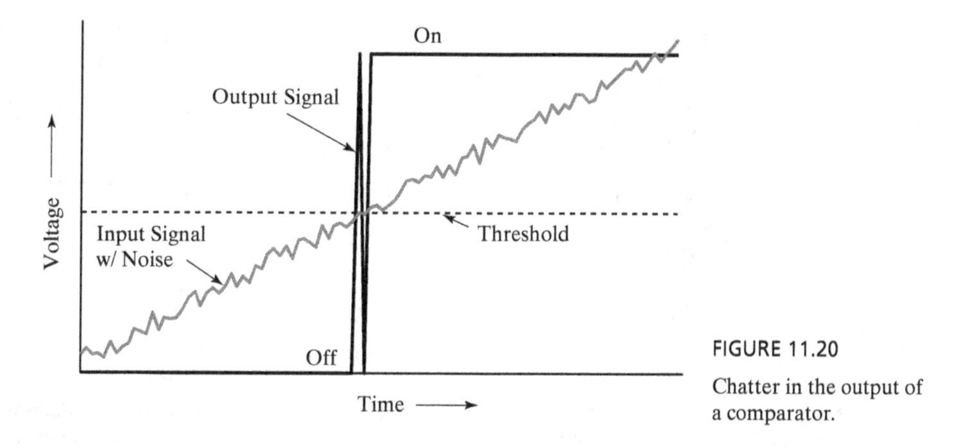

FIGURE 11.20

Chatter in the output of a comparator.

with hysteresis. A comparator with hysteresis (inverting or noninverting) also goes by the name **Schmitt trigger**.

The value of R_3, for reasonable levels of hysteresis, is usually over 100 times larger than R_{PU}, so we can simplify the analysis here by ignoring R_{PU}. For our circuit, when V_{in} is below V_{ref}, the output will be high and V_{ref} will be given by Eq. (11.22), which was derived from the simplified schematic form in Figure 11.22.

$$V_{ref(hi)} = V_{CC}\frac{R_2}{R_2 + (R_1 \| R_3)} \tag{11.22}$$

When V_{in} rises above this value of $V_{ref(hi)}$, the output will go low. This will ground the right-hand end of R_3 putting it in parallel with R_2 (Figure 11.23) and giving an expression for $V_{ref(low)}$ as

$$V_{ref(low)} = V_{CC}\frac{(R_2 \| R_3)}{R_1 + (R_2 \| R_3)} \tag{11.23}$$

The magnitude of the hysteresis is the difference between $V_{ref(hi)}$ and $V_{ref(low)}$. Once the input voltage V_{in} rises above $V_{ref(hi)}$, it must fall below $V_{ref(low)}$ before it can trigger another change in the output state. If the amplitude of the noise on the signal is less than the magnitude of the hysteresis, then there will be only one transition in the output as the input signal rises through the hysteresis band.

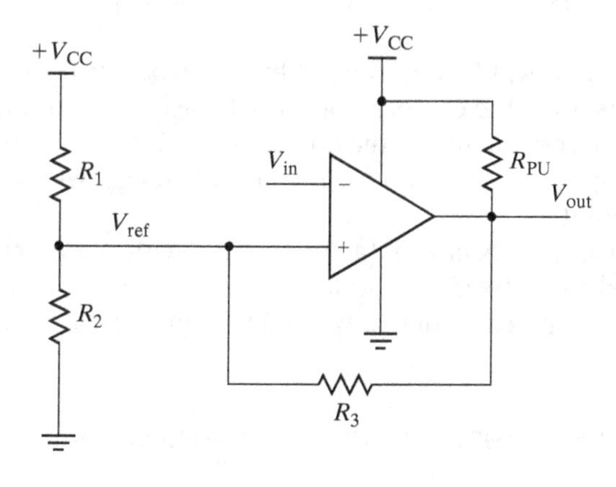

FIGURE 11.21

Inverting comparator with hysteresis.

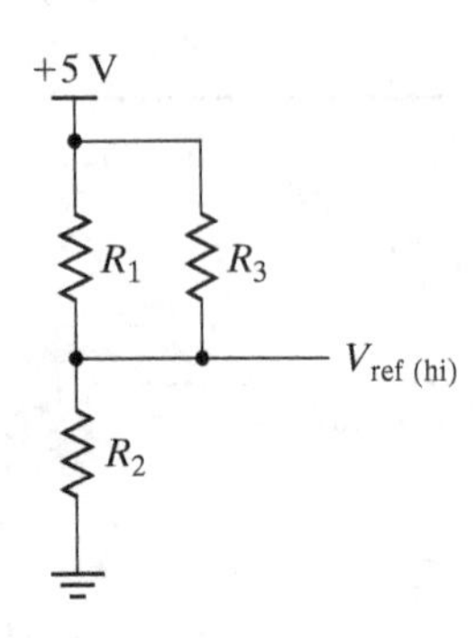

FIGURE 11.22

Schematic for V_{ref} when comparator output is high.

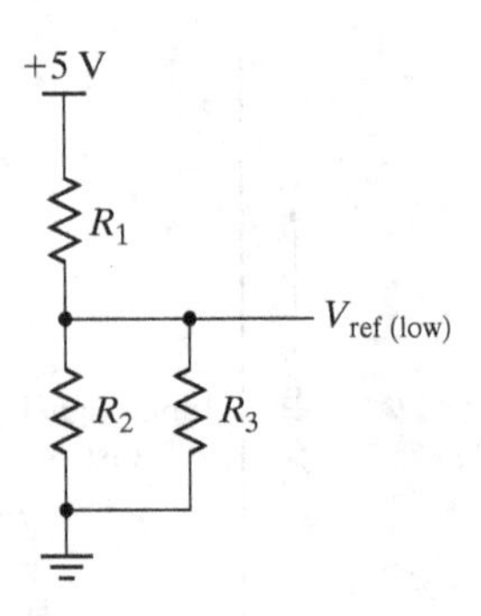

FIGURE 11.23

Schematic for V_{ref} when comparator output is low.

From a circuit analysis standpoint, we can use Eqs. (11.22) and (11.23) to calculate the upper and lower threshold voltages for an existing circuit. From a circuit design standpoint, though, we are in a bit of a quandary. We have two thresholds that we are trying to set and three resistor values (i.e., three variables) to determine. The system is under-constrained. To address this issue, we offer the following design procedure[3] for determining values for R_1, R_2, and R_3.

1. Let the upper threshold be VA_1.
2. Let the lower threshold be VA_2.
3. Calculate ΔVA as $VA_1 - VA_2$.
4. Calculate n as $n = \dfrac{\Delta VA}{VA_2}$.
5. Choose $R_3 = 1\ \text{M}\Omega$.
6. Calculate R_1 as $R_1 = nR_3$.
7. Calculate R_2 as $R_2 = \dfrac{R_1 \| R_3}{\dfrac{V_{CC}}{VA_1} - 1}$.
8. If the values for R_1 and R_2 are deemed to be too large (more than a few hundred kilohms), go back to step 5 and choose a smaller value for R_3. If the values for R_1 and R_2 are too small (less than a few kilohms), go back to step 5 and choose a larger value for R_3.

The value of R_{PU} will be determined by the needs of the device(s) connected to V_{out} and the characteristics of the comparator output stage. A good rule of thumb, however, is to use a 3.3 kΩ pull-up resistor, unless the application calls for something else. We deal with the determination of appropriate values for R_{PU} in Chapter 17, where we go into more detail on open-collector outputs.

The Schmitt trigger behavior (a comparator with hysteresis) is so useful that it has been incorporated (with fixed thresholds) into some of the inputs on microcontrollers as well as other digital logic chips that may need to deal with slowly changing inputs.

[3]Based on AN-74 "LM139/LM239/LM339 A Quad of Independently Functioning Comparators," National Semiconductor Corp., 1973.

11.6 HOMEWORK PROBLEMS

11.1 Show how you could use the voltage divider expression to develop the transfer function (input-output relationship) for the noninverting op-amp configuration.

11.2 Which Golden Rule enables you to use the voltage divider expression in Problem 11.1?

11.3 Design a circuit that will function as a noninverting summer with a gain of 2 on the sum of the input voltages. Assume an ideal op-amp and choose standard 5% resistors.

11.4 Design a trans-resistive amplifier for a photo-transistor such that the circuit will have an output of 2.5 V with no light falling on the sensor and an output voltage that rises with increasing light levels. You have only a +5 V supply available, but may assume an ideal op-amp. Write the gain expression relating photo-current to output voltage for the circuit that you design.

11.5 If the voltage divider of Figure 11.9 was to produce a voltage of 2.5 V and deliver 0–2 mA into the external circuit, what values for R_1 and R_2 would keep the variation in V_{ref} below 1% of its no-load value? What would the power rating of the resistors need to be?

11.6 Describe how the output stage of most comparators is different from the output stage of op-amps.

11.7 In the circuit shown in Figure 11.24, as the input voltage, V_{in}, is ramped from 0 to 5 V:

(a) Graph the state of the LED (D_1) over the specified range of input voltages.

(b) Graph the output voltage over the specified range of input voltages and include the corresponding state of the LED.

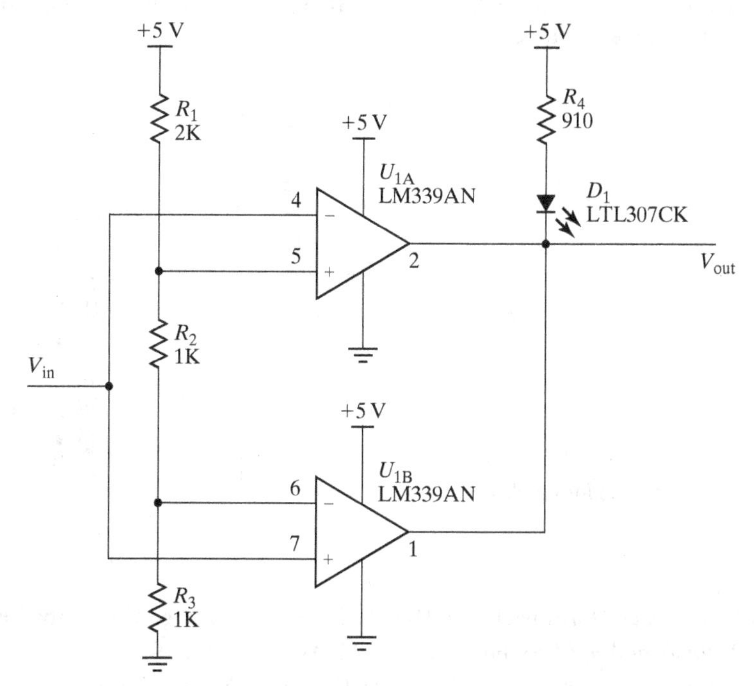

FIGURE 11.24
Circuit for Problem 11.7.

11.8 Design a circuit to implement an inverting comparator with hysteresis having an approximately 200 mV hysteresis band centered at approximately 1 V. Choose standard 5% resistors and report the expected thresholds if the values of those resistors were exact.

11.9 What range of thresholds (min to max) would you expect to see in the circuit of Problem 11.8 if the 5% resistors were at the outer limits of the tolerance?

11.10 For the circuit of Figure 11.25, answer the following questions:

(a) What is V_{out} for the circuit as drawn?

(b) What would V_{out} be if R_2 were changed to 2 kΩ?

(c) How does the current through R_1 change between the conditions of part (a) and part (b)?

(d) Write a general expression for V_{out} as a function of the value of R_2.

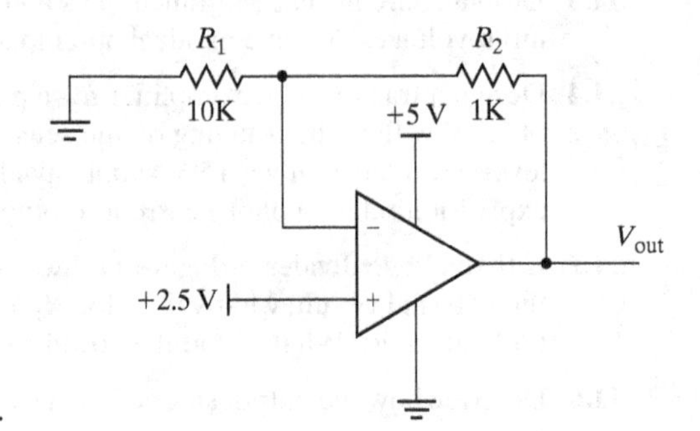

FIGURE 11.25

Circuit for Problem 11.10.

11.11 For the circuit shown in Figure 11.26, answer the following questions:

(a) If $V_1 = V_2 = 1$ V, what are the voltages at points A and B?

(b) If $V_1 = 1.1$ V and $V_2 = 1$ V, what are the voltages at points A and B?

(c) If $V_1 = 1$ V and $V_2 = 1.1$ V, what are the voltages at points A and B?

(d) Write an expression for the voltage at B minus the voltage at A in terms of V_1, V_2, and the values of the resistors.

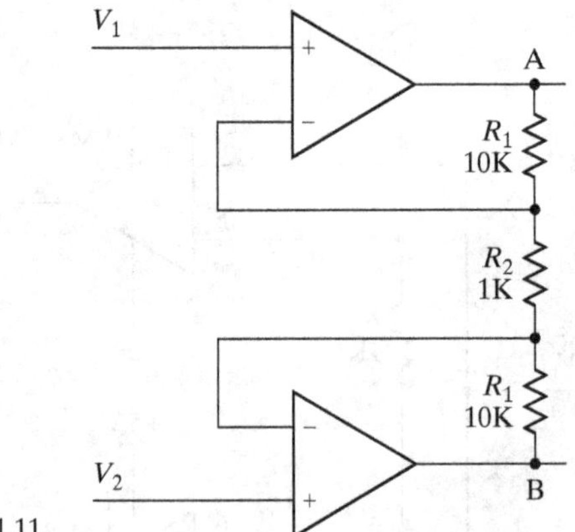

FIGURE 11.26

Circuit for Problem 11.11.

FURTHER READING

The Art of Electronics, Horowitz, P., and Hill, W., 2nd ed., Cambridge University Press, 1989.

Intuitive Operational Amplifiers, Frederiksen, T.M., McGraw-Hill, 1988.

Op Amps for Everyone, Carter, B., and Mancini, R., 3rd ed., Newnes, 2009.

CHAPTER 12

Real Operational Amplifiers and Comparators

Although the characteristics of ideal op-amps that we used in Chapter 11 are not truly representative of real op-amps, modern op-amps come amazingly close to ideal behavior for a fairly wide range of applications; so much so that you can put together many working op-amp circuits using nothing more than the Golden Rules and analyses based on ideal assumptions. However, if your requirements are somewhat demanding, or the signals that you are dealing with have important frequency content above a few kilohertz, then you need to be aware of these nonideal op-amp behaviors and, in some cases, pay very close attention to them. In this chapter, we will first detail the nonidealities that are most likely to come into play, and then we will show you how to deal with them in your designs. We'll follow that up with a run-down of some of the other nonidealities that might raise their heads, but are somewhat less likely to be a limiting factor. We'll finish this chapter on real op-amp characteristics with a guided tour of a typical op-amp data sheet to show you how to extract the important information that you'll need for your designs. After reading this chapter, the student should be able to:

1. understand how real op-amps differ from theoretical, "ideal" op-amps,
2. explain the manifestations of the most common deviations from ideal op-amp behavior,
3. extract the relevant parameters from a data sheet and use them to choose among op-amps for a given application.

12.1 REAL OP-AMP CHARACTERISTICS—HOW THE IDEAL ASSUMPTIONS FAIL

12.1.1 Noninfinite Gain

Real op-amps do not actually have infinite gain. One of the impacts of finite gain is that the simple expressions we developed for the gain of our op-amp circuits do not tell the whole story—they're a simplification. For the case of the inverting op-amp, recall that the gain expression (assuming infinite gain in the op-amp) is:

$$V_{out} = -V_{in}\frac{R_f}{R_i} \tag{12.1}$$

But, without this assumption, the transfer function that accounts for the finite gain of the op-amp is actually:

$$V_{out} = -V_{in}\frac{A\left(\dfrac{R_f}{R_f + R_i}\right)}{1 + A\left(\dfrac{R_i}{R_f + R_i}\right)} \tag{12.2}$$

where A is the finite open-loop gain of the op-amp. While the development of this equation requires no more circuit analysis than we have dealt with so far, we won't go into the process because the development itself is not really terribly instructive. The result, Eq. (12.2), *is* useful to us, however. Using Eq. (12.2) and information about a particular op-amp's open-loop gain (A), we can see how much error results from assuming that we have an ideal op-amp. For a very inexpensive op-amp, the open-loop gain at DC (0 Hz) is typically at least 15,000 (84 dB), and often 25,000 (88 dB). If we built an inverting amplifier using $R_f = 20\ \text{k}\Omega$ and $R_i = 2\ \text{k}\Omega$, we would expect the circuit to have a gain of −10 for a DC input. From Eq. (12.2), we see that the actual gain that we should get, using an op-amp with an open-loop gain of 15,000, would be:

$$-\frac{A\left(\dfrac{R_f}{R_f + R_i}\right)}{1 + A\left(\dfrac{R_i}{R_f + R_i}\right)} = -\frac{15{,}000\left(\dfrac{20\text{k}}{20\text{k} + 2\text{k}}\right)}{1 + 15{,}000\left(\dfrac{2\text{k}}{20\text{k} + 2\text{k}}\right)} = -\frac{15{,}000(0.909)}{1 + 15{,}000(0.0909)} = -9.993$$

This represents a 0.07% error in the gain. Increasing the open-loop gain with a better op-amp will reduce the gain error. If the open-loop gain of an op-amp was constant across all input frequencies, that is, if they had the same gain at 100 kHz that they did at 0 Hz, we would only need to consider it in the most critical applications. Unfortunately, the gain for real op-amps is not constant with frequency.

12.1.2 Variation in Open-Loop Gain with Frequency

The open-loop gain for op-amps is greatest at DC. The gain at DC is maintained out to a frequency of a few hertz for inexpensive op-amps and up to a few hundred hertz for better op-amps. All op-amps that have been compensated to be stable at unity gain (the vast majority of op-amps on the market) exhibit this behavior, shown graphically in Figure 12.1.

The slope of the gain curve is common across most op-amps, since it is also the result of the compensation necessary for stability at unity gain. Because of this, this graph is sometimes omitted and replaced by a pair of numbers indicating the open-loop gain at DC and the frequency at which the open-loop gain has fallen to a value of 1. The frequency at which the gain has fallen to 1 is called the **unity gain bandwidth**. This will have the same value as a typically reported data sheet parameter known as the **gain bandwidth product (GBW)**. This reflects the fact that, due to the slope of the line, the gain times the bandwidth (frequency) is a constant over a wide range of frequencies. The op-amp whose characteristics are shown in Figure 12.1 has a unity gain bandwidth and a GBW of 1 MHz.

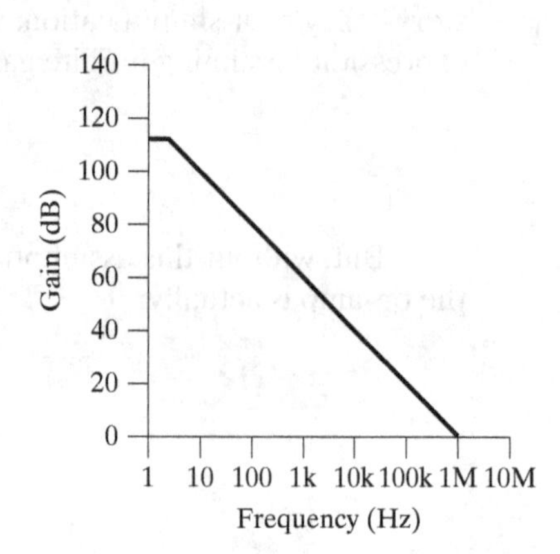

FIGURE 12.1

Open-loop gain vs. frequency for a real (nonideal) op-amp.

So, how can we use this information, and what are its implications for our designs? If we can decide on how much error in the gain we are willing to accept, we can use the GBW to predict what range of frequencies we can amplify without exceeding that error. For example, if we are willing to tolerate up to a 10% gain error as compared to ideal, we can use Eq. (12.2) to determine that we need to have an open-loop gain of at least 99 (~40 dB). With the op-amp whose frequency response is shown in Figure 12.1 and a circuit requiring a gain of −10, we could amplify signals with a frequency up to 10 kHz. If we were looking for better performance and wanted to limit the error to only 1%, we would need an open-loop gain of 10,990 (~81 dB), and with our op-amp we could only amplify signals out to a frequency of about 90 Hz or less. Another way of looking at these results is to note that to limit the error to 10% or less, we need the open-loop gain to be a factor of approximately 10 (20 dB) greater than the desired closed-loop gain. To limit the error to 1% or less, we need the open-loop gain to be greater than the desired closed-loop gain by a factor of approximately 1,000 (60 dB). Using these guidelines, we can then calculate the GBW required for a particular closed-loop gain at a given frequency.

Example 12.1

What GBW is required to amplify a 40 kHz signal by a factor of 10 with less than 10% gain error?

Solution For a closed-loop gain of 10 with less than 10% gain error, the guidelines tell us that we need an open-loop gain of 100. Using the desired frequency of 40 kHz and the required open-loop gain of 100, we calculate the GBW as 100×40 kHz = 4 MHz.

Devices are available with DC open-loop gains as high as 120 dB (1,000,000) but are more typically in the range of 90–100 dB. There is one unfortunate trade-off to be aware of: the GBW performance of an op-amp design affects the input bias currents (see Section 12.1.3.1) for most op-amp designs. The result is that the devices with the largest GBW typically do not have the lowest input bias currents. Op-amps are available with GBW specifications from approximately 100 kHz up to a few GHz.

12.1.3 Input Current Is Not Zero

One of the realities of real transistors is that current flow is required to make them work. This is especially true for those transistors being operated as linear amplifiers, as they are in an op-amp. Current flows into and out of the inputs of op-amps due to two phenomena: input bias current and input impedance.

12.1.3.1 Input Bias Current and Input Offset Current

For real op-amps, **input bias currents** flow into or out of the input pins (in most cases, out of) more or less independent of the input voltage applied. Recall that Golden Rule #1 states expressly that this doesn't happen, but for real devices, a small amount of current flow is necessary to place the internal transistors into a regime in which they can act as linear amplifiers. The magnitude of these currents is often used as one of the primary selection criteria for differentiating among op-amps, since these currents will interact with the components placed around the op-amp.

As these bias currents flow across the Thevenin equivalent resistance of the feedback network, they create voltage drops. These voltage drops appear at the output of the op-amp amplified by the **noise gain**, Eq. (12.3).

$$G_{\text{noise}} = 1 + \frac{R_{\text{f}}}{R_{\text{i}}} \tag{12.3}$$

The noise gain is the same for both inverting and noninverting configurations. The output voltage due to the bias currents represents an offset in the output that is independent of the signal.

The bias currents interact with the resistors chosen for the feedback network to create a voltage. The larger the value of the resistors in the feedback network, the greater the offset voltages induced by the bias currents. This nonideality is at the root of the practical upper limit for the absolute value of resistors that we stated in Section 11.4.3: for practical circuits, the values of these resistors should be kept in the range of a few kilohms to a few hundreds of kilohms. The voltage developed by the bias currents flowing across the feedback network creates a problem because it results in a differential voltage between the two op-amp inputs. In our original inverting op-amp circuit configuration (Figure 11.6), the voltage developed by the bias current flowing from the inverting input may be very different than the voltage developed by the bias current flowing from the noninverting input, due to the differing impedances seen by the bias currents from the two inputs. However, it is possible to compensate for the effects of input bias current by equalizing the impedances seen by the two inputs.

The circuit shown in Figure 12.2 can compensate for the effects of bias currents for the inverting amplifier, if resistor R_B is chosen such that $R_B = R_f \| R_i$. In this case, the bias currents flowing from the two inputs will see the same impedances and therefore both produce the same voltage. There will be no differential voltage resulting at the inputs of the op-amp that would be amplified and appear at the output. If the value of R_B was an exact match and the bias currents at each of the input pins were exactly equal, this circuit would compensate perfectly and completely for the effects of bias currents. Unfortunately, the input bias current at the inverting input pin is usually not the same as the input bias current at the noninverting pin. The difference in the bias currents at the input pins is called the **input offset current** and this cannot be easily corrected. Even if you were to trim R_B to exactly match the Thevenin equivalent resistance of the feedback network, the voltages at the inputs resulting from the two bias currents will not be the same, because the current flowing from (or into) the pins is not equal. Fortunately, the offset current is typically about 3–10 times smaller than the bias currents, so the circuit of Figure 12.2 is still a reasonable way to compensate for most of the effects of bias current.

Input bias currents range from a few hundred nanoamps (10^{-9} A) for inexpensive bipolar op-amps and those optimized for high-frequency performance all the way down to a few tens of femtoamps (10^{-15} A) for devices intended for electrometer, photodetector, and other ultra-low input bias current applications. These are typically trans-resistive circuits that require large feedback resistors due to the small sensor currents involved.

12.1.3.2 Input Impedance

While the input bias current is more or less independent of the input voltage, there is another effect, the **input impedance**, which causes currents that vary as a function of the input voltage. Actually, there are two input impedances: **common mode input impedance** and **differential**

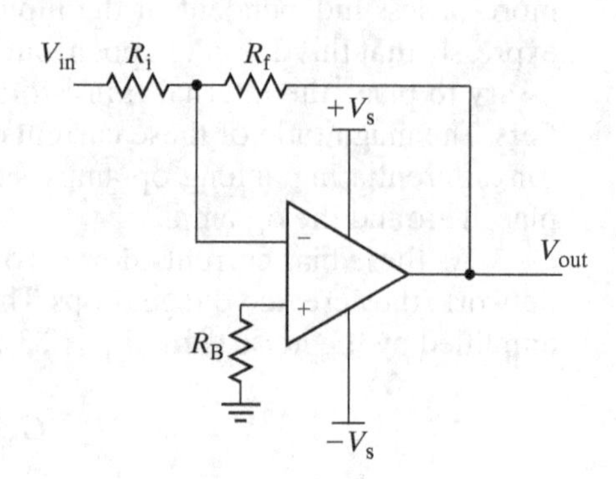

FIGURE 12.2

Inverting op-amp with bias current compensation.

input impedance. Common mode input impedance appears between either of the inputs and ground. The differential input impedance appears between the two input pins. Both impedances typically range from about 2 MΩ for low-cost op-amps to the teraohm (10^{12} Ω) region and beyond for very high input impedance devices. The input impedance of an op-amp is relevant for designs using the noninverting configuration, since the input voltage is connected directly to an op-amp input. In those applications that use the inverting configuration, the input impedance for the circuit is set by the designer's choice of R_i. For most applications, the input impedance of the op-amp does not place limits on the circuit's performance. It is only in situations in which the signal source connected to an input presents a very large output impedance that the input impedance of the op-amp may become important. This might occur, for example, in the case of pH probes which commonly exhibit output impedances in excess of 250 MΩ. The output impedance of the sensor will form a voltage divider with the input impedance of the amplifier. In this case, getting the error due to the voltage divider effect down to 1% or less requires that the input impedance be greater than 25 GΩ. Fortunately, output impedances like those of the pH probe are relatively rare, with most sensor output impedances falling in the range of tens of kilohms or less.

12.1.4 The Output Voltage Source Is Not Ideal

Our model of the op-amp treats the output of the op-amp as a voltage source. Real, as opposed to ideal, voltage sources have a nonzero output impedance. Op-amps are no different, and they too exhibit a small output impedance (typically a few ohms to a few hundred ohms). Of our original ideal assumptions about op-amps, the finite output impedance of the op-amp has the least impact on our circuits' performance. This is because negative feedback tends to compensate for the output impedance.

In Figure 12.3, we see an op-amp with some output impedance (shown as a resistor inside the op-amp symbol, at the output) in a circuit with negative feedback. Notice that, because the output impedance is "inside the loop," any voltage drop that is generated across the impedance will reduce the voltage at the inverting input, which will, in turn, cause the voltage source to create a slightly higher voltage to compensate for the voltage drop across the output impedance. With an infinite open-loop gain, the effective output impedance is zero. With a finite open-loop gain, the effective output impedance is given by

$$Z_{\text{effective}} = \frac{Z_{\text{open loop}}}{1 + A\left(\dfrac{R_i}{R_f + R_i}\right)} \tag{12.4}$$

For the op-amp with an open-loop gain of 15,000, configured with a 10:1 ratio of R_f:R_i and an open-loop output impedance of 100 Ω, the effective output impedance is given by

$$Z_{\text{effective}} = \frac{Z_{\text{open loop}}}{1 + A\left(\dfrac{R_i}{R_f + R_i}\right)} = \frac{100}{1 + 15{,}000\left(\dfrac{1}{10 + 1}\right)} = 0.073\ \Omega$$

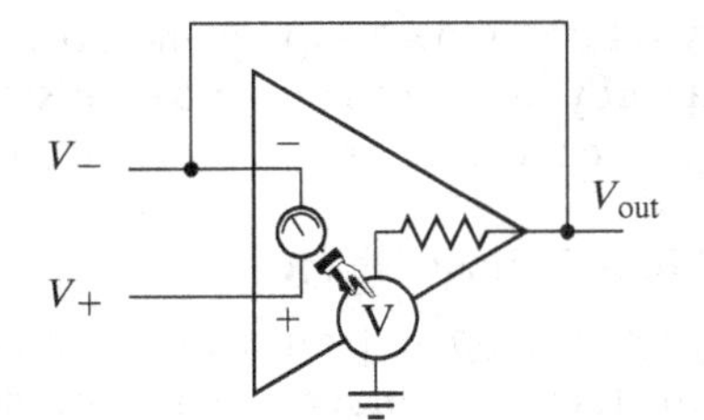

FIGURE 12.3

Op-amp with output impedance in negative feedback.

This output impedance is so small that it is negligible for a vast majority of applications. This would likely only be an issue in instrumentation applications where an output voltage error on the order of a few hundred microvolts might be important. When the output of the amplifier will be sent to a microcontroller, where the voltage resolution is on the order of a millivolt, this error is too small to be measured.

12.1.5 Other Nonidealities

In the following sections, we introduce other nonidealities that are likely to be relevant in op-amp design for mechatronic systems. This is not intended as an exhaustive list, nor will we attempt to demonstrate how to compensate for these nonidealities in this text. That's beyond the scope of an introductory chapter on the use of op-amps. Our intent for the material we cover in the following sections is to aid in the selection of devices when evaluating real op-amps against the needs of a particular design.

12.1.5.1 Input Offset Voltage

If the two inputs to an ideal op-amp operated open-loop were shorted, then the output voltage would be zero. Due to inevitable variations in the manufacturing process, this is never true for real op-amps. The **input offset voltage** is the differential voltage that you would need to apply to an op-amp's inputs to force the open-loop output to exactly 0 V. The magnitude of the input offset voltage is one of the primary specifications that differentiates commercial op-amps. The magnitude of input offset voltage varies from a few millivolts for inexpensive op-amps down to 1 μV or less for very high-precision op-amps. In closed-loop circuits, this offset voltage is amplified by the noise gain [Eq. (12.3)] and appears as an offset in the output voltage.

There are two ways to deal with the effects of input offset voltage: 1) accept it as it is or 2) build into the circuit the ability to remove (zero out or null) the effect. Option 1 is clearly the easier route and in many microcontroller-based systems represents a perfectly acceptable choice. The presence of the microcontroller often opens up the possibility to use software to read the value of the offset and then subtract that offset from the data collected after the amplifier has been characterized. For those situations where this approach isn't acceptable, there are circuit designs that can be used to compensate for the effects. Some op-amps that are specifically designed for high-precision applications include dedicated pins and circuitry within the op-amp IC that allow the input offset voltage to be zeroed during a calibration process conducted for each circuit built using the op-amp.

12.1.5.2 Power Supplies

While not technically a nonideality of the op-amp itself, the choice of power supply for the op-amp is an important issue and one that is often used as a selection criterion when evaluating op-amps. If you were watching for it, you might have noticed that the op-amps we have shown so far were all equipped with two symmetric power supplies of opposite polarity (e.g., +15 and −15 V). With a few rare exceptions, any op-amp can be operated with such a "dual power supply." However, dual supplies are increasingly rare in microcontroller-based applications, and in any case they cost more to include in a design than a single supply (e.g., +5 V only). This has led to the development of a category of op-amps whose internal design allows them to operate effectively on one power supply and ground. These op-amps are typically called **single supply op-amps**. If you want to operate your signal conditioning circuits from a single power supply, be sure to choose an op-amp specified for single supply operation.

12.1.5.3 Input Common Mode Voltage Range

The voltages seen at the inputs to an op-amp must almost always be limited to values between the power supply rails. Taking the inputs more than a few tenths of a volt above or below the power supply voltages will often cause current flows that are large enough to damage the op-amp.

But even when the input voltages are kept well within the power supply rails, many op-amps further limit the allowable voltages that can be applied to the inputs while maintaining the expected operation of the op-amp. The specification describing this limitation is called the **input common mode voltage range**, and it describes the acceptable range of voltages that may appear at the op-amp inputs during normal operation. Some op-amps exhibit a phenomenon known as phase reversal when this range is exceeded. Phase reversal occurs when the output voltage suddenly swings in the opposite direction from that expected by the normal operation of the op-amp. This can impose a serious limitation on the range of signal voltages that can be used. For example, one of the most widely used op-amps, LM324A, has a common mode input voltage range that extends from ground (making it suitable for single supply operation) up to $V_{CC} - 1.5$ V. When operated on a 5 V supply, this only allows for signals in the range of 0–3.5 V. If voltages above 3.5 V are applied to the inputs, the op-amp is subject to possible phase reversal.

Op-amp manufacturers have responded to this shortcoming by designing a range of op-amps that will operate properly with input voltages anywhere within the power supply rails. These devices are typically referred to as **rail-to-rail input** devices, often abbreviated as **RRI**.

12.1.5.4 Output Voltage Swing

As we pointed out in Section 11.1, the output voltages produced by real op-amp are always bounded by the power supply rails. The output voltage cannot exceed the power supply voltage, and for some op-amps, it can't even get particularly close. For example, the LM324A, when operated from a single +30 V supply, can produce a maximum output voltage of only 26 V. It is very common for the output voltage swing to be limited to 1–2 V (or more) below the supply voltage.

Once again, the op-amp manufacturers have responded by designing and supplying op-amps that can produce output voltage swings that come much closer to the power supply rails. These so-called **rail-to-rail output**, abbreviated **RRO**, op-amps produce maximum output voltages that are within a few tens of millivolts of the power supply voltage. When combined in a device that also exhibits RRI characteristics, we get what is perhaps the ultimate op-amp for use with microcontrollers: the single supply, **rail-to-rail input and output** (**RRIO**) op-amp.

One of the valuable specifications for an op-amp is often buried in the test conditions for the output voltage swing. The test conditions will specify some resistive load under which the output voltage swing will be guaranteed. This turns out to be something important in choosing the absolute value of resistors to be used with the op-amp. The load resistance used for this specification represents the parallel combination of the load resistance that the next stage of your design represents combined with the Thevenin equivalent resistance of the feedback network. This test load resistance is used to arrive at the lower recommended value for resistance that was put forth in the guideline mentioned in Section 11.4.4.

12.1.5.5 Input Common Mode Rejection Ratio

If we were to apply the same voltage to both inputs of an op-amp, the fact that they are fundamentally differential amplifiers would lead us to believe that the output should be 0 V. We're sure you can see this coming by now: for real devices, this is not exactly the case. For any voltage applied to both op-amp inputs (hence the name **common mode voltage**), there will be some response at the output. The parameter used to describe the magnitude of that response is called the **common mode rejection ratio** (**CMRR**) and it is defined as

$$\mathrm{CMRR} = \frac{A_{\text{differential}}}{A_{\text{common mode}}} = \frac{A_{\text{open loop}}}{A_{\text{common mode}}} \tag{12.5}$$

$A_{\text{open loop}}$ is defined as

$$A_{\text{open loop}} = \frac{V_o}{\Delta V_{\text{in}}} \quad (12.6)$$

$A_{\text{common mode}}$ is defined as

$$A_{\text{common mode}} = \frac{V_o}{V_{\text{common mode}}} \quad (12.7)$$

We can combine Eqs. (12.5) through (12.7) and rearrange the results to give an expression for the equivalent differential input voltage that would result from the common mode voltage:

$$\Delta V_{\text{in}} = \frac{V_{\text{common mode}}}{\text{CMRR}} \quad (12.8)$$

This equivalent differential input voltage would then be amplified by the closed-loop gain of the circuit.

For example, let's use the same inexpensive op-amp that we have been using so far. The data sheet for our example device (the LM324A) lists the CMRR as being a minimum of 65 dB (1,778). If the common mode input voltage were 2.5 V, the equivalent differential input voltage would be

$$\Delta V_{\text{in}} = \frac{2.5}{1{,}778} = 1.4 \text{ mV}$$

This differential voltage would then be subject to the same closed-loop gain as our signal. If we are dealing with millivolt-level signals riding on a few volts of common mode voltage, it is easy to see how the response to the common mode voltage could be comparable to the signal of interest. If we find ourselves in that situation, we want to seek out an op-amp with much better CMRR specifications. Other op-amps are available at reasonable cost with CMRR specifications on the order of 100 dB. This would lower the equivalent input voltage to 25 μV.

12.1.5.6 Temperature Effects

All of the offset voltage and bias current effects that we have described are also subject to variation with temperature, often referred to as **drift**. Each parameter will have a corresponding temperature coefficient which is called out in the data sheet. For modern op-amps and applications that are operating indoors, these variations are seldom a concern. For products that may need to operate over the full range of temperatures possible with these devices (often −40°C to +125°C), these temperature drifts will need to be considered. It is very difficult to compensate for temperature drift in hardware, so if the drift over temperature is a problem for an application, the best solution is often to buy a better op-amp with better drift specifications.

12.2 READING AN OP-AMP DATA SHEET

In order to make use of the material in Section 12.1, you will need to extract the specifications from op-amp data sheets. For most parameters, this extraction is pretty straightforward. However, there is a substantial amount of useful information in the data sheet in addition to just the values for the various nonidealities that we've introduced. In the next few sections, we'll take you on a guided tour of a typical op-amp data sheet and show you where to find the information of greatest interest to you as a circuit designer.

12.2.1 Maxima, Minima, and Typical Values

As we take our tour through a representative op-amp data sheet, you will notice that information is commonly presented in three columns, labeled Max, Typical, and Min. As designers, we are going to focus on the maxima and minima wherever they are provided. Why would we want to do that if the Typical column tells us how the typical device will behave? Because not all our devices will be typical! We would like our designs to work with any op-amp that we might obtain from the manufacturer, not just the ones that were hand picked to match the "typical" specifications. To achieve this, we need to make sure that our designs will accommodate devices whose actual performance falls anywhere within the range of performance specifications that the manufacturer will guarantee. The Max and Min columns represent performance guarantees from the manufacturer. They go to great trouble to insure that every device that they ship falls within those bounds. If we design our circuits to work within those bounds, the designs should work with any individual device. Just like the resistors and capacitors we discussed in Chapter 9 on Basic Circuit Analysis and Passive Components, all op-amps (indeed, all components) have tolerances. Good designs accommodate this and meet all requirements when using devices that fall within the specified tolerance range.

12.2.2 The Front Page

We don't reproduce an example of a front page of a data sheet here because they typically don't contain any of the critical numerical information that we are interested in. The front page is often mostly a marketing tool: the purpose is to grab your attention and get you to read the rest of the data sheet and then hopefully specify the part for your design. Even when viewed (skeptically) as a marketing tool, there is still usually a great deal of useful information here, but you need to take it with a grain of salt. The front page of the data sheet may claim that this op-amp has a "10 MHz GBW," while close examination of the data reveals that 10 MHz is a typical GBW, while the guaranteed minimum GBW is really only 6 MHz. There can also be some very important information here. For example, this is probably the place that you will find an op-amp explicitly referred to as a "single supply RRIO" op-amp. You could extract that same information from a detailed examination of the electrical characteristics described in the following pages, but it is very helpful to have the marketer put it right out there in big letters. The front page is useful as a tool to make a first-pass evaluation of the op-amp. If the front page says "1 MHZ GBW" and you know that your design will require at least a 4 MHz GBW, there is not much point in reading further.

12.2.3 The Absolute Maximum Ratings Section

The Absolute Maximum Ratings section of the data sheet (Figure 12.4) tells you about the device's physical limits. The key word for this section is "absolute": exceeding these specifications will likely cause permanent damage to the device. It is not a good idea to plan to operate the device out to the edges of the absolute maximum ratings. This section also contains the specifications for the operating and storage temperature ranges for the different devices covered in the data sheet. Operating a device out to the limits of the operating temperature range is acceptable.

12.2.4 The Electrical Characteristics Section

The section describing the electrical characteristics of an op-amp (Figures 12.4 and 12.5) is the heart of the data sheet. Here, you will find not only the specifications for all the parameters that we have described above, but also information on the conditions under which those specifications are valid. This section also contains descriptions of the test conditions that yielded the specifications, and this information can also relay important information to the

LM124/LM224/LM324/LM2902

Absolute Maximum Ratings (Note 12)

If Military/Aerospace specified devices are required, please contact the National Semiconductor Sales Office/ Distributors for availability and specifications.

	LM124/LM224/LM324 LM124A/LM224A/LM324A	LM2902
Supply Voltage, V^+	32V	26V
Differential Input Voltage	32V	26V
Input Voltage	−0.3V to +32V	−0.3V to +26V
Input Current		
($V_{IN} < -0.3V$) (Note 6)	50 mA	50 mA
Power Dissipation (Note 4)		
Molded DIP	1130 mW	1130 mW
Cavity DIP	1260 mW	1260 mW
Small Outline Package	800 mW	800 mW
Output Short-Circuit to GND		
(One Amplifier) (Note 5)		
$V^+ \le 15V$ and $T_A = 25°C$	Continuous	Continuous
Operating Temperature Range		−40°C to +85°C
LM324/LM324A	0°C to +70°C	
LM224/LM224A	−25°C to +85°C	
LM124/LM124A	−55°C to +125°C	
Storage Temperature Range	−65°C to +150°C	−65°C to +150°C
Lead Temperature (Soldering, 10 seconds)	260°C	260°C
Soldering Information		
Dual-In-Line Package		
Soldering (10 seconds)	260°C	260°C
Small Outline Package		
Vapor Phase (60 seconds)	215°C	215°C
Infrared (15 seconds)	220°C	220°C

See AN-450 "Surface Mounting Methods and Their Effect on Product Reliability" for other methods of soldering surface mount devices.

ESD Tolerance (Note 13)	250V	250V

Electrical Characteristics

$V^+ = +5.0V$, (Note 7), unless otherwise stated

	Parameter	Conditions	LM124A			LM224A			LM324A			Units
			Min	Typ	Max	Min	Typ	Max	Min	Typ	Max	
1	Input Offset Voltage	(Note 8) $T_A = 25°C$		1	2		1	3		2	3	mV
2	Input Bias Current (Note 9)	$I_{IN(+)}$ or $I_{IN(-)}$, $V_{CM} = 0V$, $T_A = 25°C$		20	50		40	80		45	100	nA
3	Input Offset Current	$I_{IN(+)}$ or $I_{IN(-)}$, $V_{CM} = 0V$, $T_A = 25°C$		2	10		2	15		5	30	nA
4	Input Common-Mode Voltage Range (Note 10)	$V^+ = 30V$, (LM2902, $V^+ = 26V$), $T_A = 25°C$	0		$V^+-1.5$	0		$V^+-1.5$	0		$V^+-1.5$	V
5	Supply Current	Over Full Temperature Range $R_L = \infty$ On All Op Amps $V^+ = 30V$ (LM2902 $V^+ = 26V$) $V^+ = 5V$		1.5 0.7	3 1.2		1.5 0.7	3 1.2		1.5 0.7	3 1.2	mA
6	Large Signal Voltage Gain	$V^+ = 15V$, $R_L \ge 2k\Omega$, (V_O = 1V to 11V), $T_A = 25°C$	50	100		50	100		25	100		V/mV
7	Common-Mode Rejection Ratio	DC, $V_{CM} = 0V$ to $V^+ - 1.5V$, $T_A = 25°C$	70	85		70	85		65	85		dB

FIGURE 12.4

LM324 op-amp Absolute Maximum Ratings and the first of the Electrical Characteristics pages. (Courtesy of National Semiconductor Corporation.)

Electrical Characteristics (Continued)

$V^+ = +5.0V$, (Note 7), unless otherwise stated

Parameter		Conditions		LM124A			LM224A			LM324A			Units
				Min	Typ	Max	Min	Typ	Max	Min	Typ	Max	
Power Supply Rejection Ratio		$V^+ = 5V$ to 30V (LM2902, $V^+ = 5V$ to 26V), $T_A = 25°C$		65	100		65	100		65	100		dB
Amplifier-to-Amplifier Coupling (Note 11)		f = 1 kHz to 20 kHz, $T_A = 25°C$ (Input Referred)			−120			−120			−120		dB
Output Current	Source	$V_{IN}^+ = 1V$, $V_{IN}^- = 0V$, $V^+ = 15V$, $V_O = 2V$, $T_A = 25°C$		20	40		20	40		20	40		mA
	Sink	$V_{IN}^- = 1V$, $V_{IN}^+ = 0V$, $V^+ = 15V$, $V_O = 2V$, $T_A = 25°C$		10	20		10	20		10	20		
		$V_{IN}^- = 1V$, $V_{IN}^+ = 0V$, $V^+ = 15V$, $V_O = 200$ mV, $T_A = 25°C$		12	50		12	50		12	50		μA
Short Circuit to Ground		(Note 5) $V^+ = 15V$, $T_A = 25°C$			40	60		40	60		40	60	mA
Input Offset Voltage		(Note 8)				4			4			5	mV
V_{OS} Drift		$R_S = 0Ω$			7	20		7	20		7	30	μV/°C
Input Offset Current		$I_{IN(+)} - I_{IN(-)}$, $V_{CM} = 0V$				30			30			75	nA
I_{OS} Drift		$R_S = 0Ω$			10	200		10	200		10	300	pA/°C
Input Bias Current		$I_{IN(+)}$ or $I_{IN(-)}$			40	100		40	100		40	200	nA
Input Common-Mode Voltage Range (Note 10)		$V^+ = +30V$ (LM2902, $V^+ = 26V$)		0		V^+-2	0		V^+-2	0		V^+-2	V
Large Signal Voltage Gain		$V^+ = +15V$ (V_OSwing = 1V to 11V) $R_L \geq 2$ kΩ		25			25			15			V/mV
Output Voltage Swing	V_{OH}	$V^+ = 30V$ (LM2902, $V^+ = 26V$)	$R_L = 2$ kΩ	26			26			26			V
			$R_L = 10$ kΩ	27	28		27	28		27	28		
	V_{OL}	$V^+ = 5V$, $R_L = 10$ kΩ			5	20		5	20		5	20	mV
Output Current	Source	$V_O = 2V$	$V_{IN}^+ = +1V$, $V_{IN}^- = 0V$, $V^+ = 15V$	10	20		10	20		10	20		mA
	Sink		$V_{IN}^- = +1V$, $V_{IN}^+ = 0V$, $V^+ = 15V$	10	15		5	8		5	8		

Note 4: For operating at high temperatures, the LM324/LM324A/LM2902 must be derated based on a +125°C maximum junction temperature and a thermal resistance of 88°C/W which applies for the device soldered in a printed circuit board, operating in a still air ambient. The LM224/LM224A and LM124/LM124A can be derated based on a +150°C maximum junction temperature. The dissipation is the total of all four amplifiers—use external resistors, where possible, to allow the amplifier to saturate of to reduce the power which is dissipated in the integrated circuit.

FIGURE 12.5

LM324 op-amp Electrical Characteristics, continued. (Courtesy of National Semiconductor Corporation.)

Note 5: Short circuits from the output to V^+ can cause excessive heating and eventual destruction. When considering short circuits to ground, the maximum output current is approximately 40 mA independent of the magnitude of V^+. At values of supply voltage in excess of +15V, continuous short-circuits can exceed the power dissipation ratings and cause eventual destruction. Destructive dissipation can result from simultaneous shorts on all amplifiers.

Note 6: This input current will only exist when the voltage at any of the input leads is driven negative. It is due to the collector-base junction of the input PNP transistors becoming forward biased and thereby acting as input diode clamps. In addition to this diode action, there is also lateral NPN parasitic transistor action on the IC chip. This transistor action can cause the output voltages of the op amps to go to the V^+voltage level (or to ground for a large overdrive) for the time duration that an input is driven negative. This is not destructive and normal output states will re-establish when the input voltage, which was negative, again returns to a value greater than −0.3V (at 25°C).

Note 7: These specifications are limited to $-55°C \le T_A \le +125°C$ for the LM124/LM124A. With the LM224/LM224A, all temperature specifications are limited to $-25°C \le T_A \le +85°C$, the LM324/LM324A temperature specifications are limited to $0°C \le T_A \le +70°C$, and the LM2902 specifications are limited to $-40°C \le T_A \le +85°C$.

Note 8: V_O . 1.4V, R_S = 0Ω with V^+ from 5V to 30V; and over the full input common-mode range (0V to V^+ − 1.5V) for LM2902, V^+ from 5V to 26V.

Note 9: The direction of the input current is out of the IC due to the PNP input stage. This current is essentially constant, independent of the state of the output so no loading change exists on the input lines.

Note 10: The input common-mode voltage of either input signal voltage should not be allowed to go negative by more than 0.3V (at 25°C). The upper end of the common-mode voltage range is V^+ − 1.5V (at 25°C), but either or both inputs can go to +32V without damage (+26V for LM2902), independent of the magnitude of V^+.

Note 11: Due to proximity of external components, insure that coupling is not originating via stray capacitance between these external parts. This typically can be detected as this type of capacitance increases at higher frequencies.

Note 12: Refer to RETS124AX for LM124A military specifications and refer to RETS124X for LM124 military specifications.

Note 13: Human body model, 1.5 kΩ in series with 100 pF.

FIGURE 12.5 continued (Courtesy of National Semiconductor Corporation.)

designer. It is especially important to be on the lookout for footnotes to the specifications. Critically important and useful information is often discovered in the footnotes (just like financial reports of public companies!). We'll walk through the specifications and index them with our own numbering scheme to help you connect the discussion to the appropriate lines in the LM324 data sheet. The bold numbers that open the following paragraphs refer to labeled portions of the op-amp data sheet shown in Figures 12.4 and 12.5.

1. This is the input offset voltage specification for an ambient temperature of 25°C. We discussed this in Section 12.1.5.1. There is a second entry on the next page for the offset over the full range of operating temperatures. The LM124A, LM224A, and LM324A are all made from the same silicon die. LM124A and LM224A are the parts with better performance, and these are identified by testing after all the parts come from the same manufacturing process.

2. This is the input bias current specification for an ambient temperature of 25°C. We discussed this specification in Section 12.1.3.1. Again, there is a second entry on the next page for the input bias current over the full range of operating temperatures. This number is given as a positive number that would normally indicate that the current was flowing into the device pin. However, here we find our first instance of a footnote containing very important information for the designer: Note 9 on the next page of the data sheet indicates that the current is actually flowing *out* of the pin.

3. This is the input offset current specification for an ambient temperature of 25°C. This is also covered in Section 12.1.3.1. Note that the maximum offset current is only one-third of the maximum bias current and the typical offset current is only 11% of the typical bias current. Again, the next page of the data sheet gives the specifications for the input offset current over the full range of operating temperatures.

4. This is the input common mode voltage range for an ambient temperature of 25°C. We introduced this parameter in Section 12.1.5.3. The next page of the data sheet has a second entry for the input common mode voltage range over the full range of temperatures. Note 10 tells us that it is important never to expose the inputs to voltages more than 0.3 V below ground when operated on a single supply. It also notes that the inputs may be taken as high as 32 V without damage, independent of the power supply voltage. In this case, the footnotes have revealed an important "gotcha" for the designer as well as a surprising feature that the designer may be able to leverage elegantly.

5. This is the internal current drawn by the op-amp over the full range of temperatures. It does not include any current that goes into the feedback network or the external load. This current is required by the device to perform its function. This will be an important parameter in battery-powered applications or other situations in which minimizing power consumption is important.

6. The large signal voltage gain is a stand-in for the open-loop gain of the op-amp (which we covered in Sections 12.1.1 and 12.1.2). Here, it is given for operation at 25°C. There is a second entry on the next page for the large signal voltage gain over the full range of temperatures. The test conditions here provide additional useful information. The load under which this gain is specified is given as $\geq 2\ k\Omega$. This load value will come up again in other specifications, and is a guide to us as to the total load that the op-amp is capable of driving. It is important to realize that the load is made up of a combination of the load presented by the feedback network and any external load that must be driven by the op-amp.

7. This is the common mode rejection ratio that we presented in Section 12.1.5.5 for an ambient temperature of 25°C.

8. This specification for output current is one that causes quite a bit of confusion among neophyte designers. You might be tempted to interpret this specification as providing guidance about the sizing of the resistors in the feedback network, or the size of the load that the op-amp can be expected to drive. That is not an appropriate use of the specification. If you look closely at the Conditions column, you will see that the specification is given for a differential input voltage of 1 V. When the op-amp is operated in closed-loop negative feedback without exceeding the output voltage swing limitations of the op-amp (i.e., not saturated), the differential input voltage that appears at the two inputs will be essentially 0 V, and certainly nowhere near 1 V. The current specification here is useful to know how much current the op-amp can provide when driven into saturation, but it is not useful in describing the behavior of the device in linear operation, which is most likely the regime you will be designing for.

9, **10**, **11**, and **12.** These are the specifications for the input offset voltage, input bias current, input offset current, and input common mode voltage range that are valid over the full range of operating temperatures.

12.2.5 The Packaging Section

The data sheet will also include information and diagrams describing the packaging of the op-amps. Many op-amps are available in a range of packaging options with differing form factors and differing numbers of op-amps per package. Op-amps are available as single op-amps (one amplifier per package), dual op-amps (two amplifiers per package), and quad op-amps (four amplifiers per package). They come in packages ranging most commonly from 8–14 pins. The pin-outs of the most popular arrangements for dual inline packages (DIP) are shown in Figure 12.6.

Surface-mount packaging has also become very common, mainly because the pin-pin spacing is at least 2x smaller, and this allows for a much smaller footprint on the circuit board as well as reducing packaging costs. In an increasing number of applications, such as cell phones, reducing the size of the overall device, and hence all the components, is critical. Op-amps in surface-mount packages often deviate from the DIP layouts shown in Figure 12.6. This is especially true for the single op-amp packages, where a 5-pin package eliminates the unused pins on the DIP package.

One convenient aspect of op-amp packaging is that all manufacturers use the same pin arrangement for the DIP versions of most single, dual, and quad op-amps. The arrangement

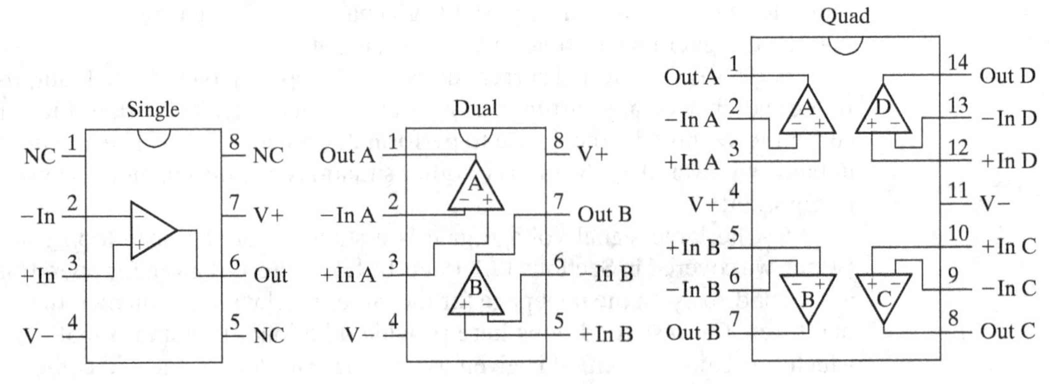

FIGURE 12.6

Common pin-outs for dual inline package op-amps.

of the quad packages is especially convenient and easy to remember—note that the arrangement makes it easy to make buffers by simply making connections from the pins on the corners to the next inboard pin. The third pin in from either corner is the noninverting input. All that is tricky is remembering which of the two middle pins is the V+ (the positive power supply) and which is V− (the negative power supply). This is very important, because mistakenly wiring the power supply backwards will destroy the entire chip immediately when the power is turned on. One way to get this right for the quad op-amp package is to remember that "if the notch is on the right, the V+ is on the top."

12.2.6 The Typical Applications Section

After the numeric portion of the data sheet, there is often a "Typical Applications" section. At its simplest, this section often consists of a series of sample application circuits that the data sheet writers feel show off the capabilities of the particular op-amp. This section of the data sheet can be a gold mine of design ideas to be added to your tool kit. Perusal of the applications section of the data sheet (and the *Analog Applications Manuals* that some manufacturers publish) is a great way to glean ideas about how you might use op-amps in practical circuits.

For many modern op-amps, especially where the designers have made specific trade-offs to tailor the behavior of the op-amp to a specific application domain, the applications section will detail how to apply the op-amp. This will likely include suggestions for such things as by-pass capacitor requirements, how to achieve stability when driving a capacitive load (if this is an issue for a particular op-amp), how to achieve unity gain stability (when the op-amp is not internally compensated), and how to lay out the circuit board to maintain the very low bias currents that a particular device is capable of. Reading this section of the data sheet should be considered just as stringently required as the Electrical Characteristics section.

12.3 READING A COMPARATOR DATA SHEET

Reading a comparator data sheet is in many respects very similar to reading an op-amp data sheet. You will find similar specifications for input offset voltage, input bias current, input offset current, input common mode voltage range, and voltage gain (analogous to open-loop gain in the op-amp). What is new are specifications for the speed of response and specifications related to the open-collector (or open-drain) output. We'll defer dealing with the specifications of open-collector outputs to Chapter 17, where we deal with open-collector outputs in detail.

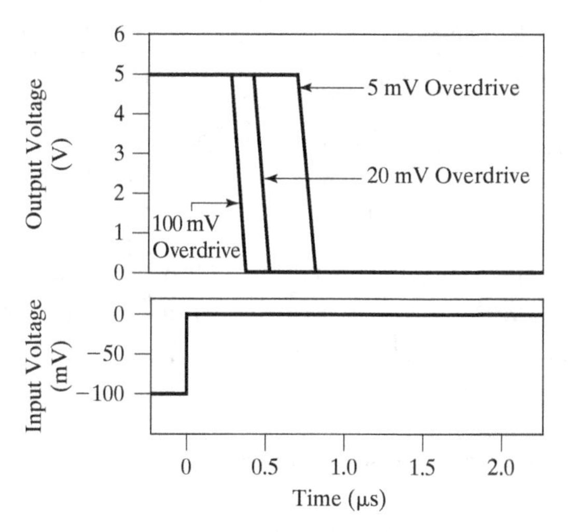

FIGURE 12.7

Comparator response times with varying overdrive.

The speed of response parameter, called simply "response time," specifies the delay from the time the input transitions across the threshold voltage until the output changes state. While a single number is given in a data sheet's Electrical Characteristics table, there is typically a graph like that of Figure 12.7 that details how the device responds with varying levels of overdrive.

Overdrive is the difference in voltage between the input signal and the threshold voltage at the time of the input transition.

Stated another way, the overdrive is the amount that an input exceeds the threshold at the time of a transition. Figure 12.8 illustrates two cases that result in different levels of overdrive. As Figure 12.7 shows, increasing amounts of overdrive result in shorter delay times to the output transition.

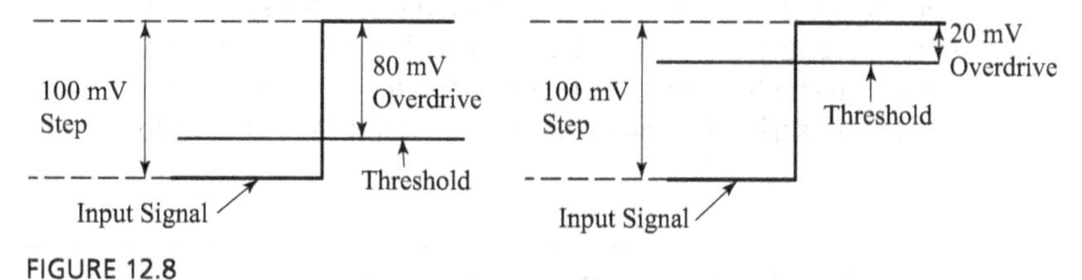

FIGURE 12.8

A 100 mV step producing two different levels of overdrive.

12.3.1 Comparator Packaging

Comparators are also available in single, dual, and quad packages, but be cautious—the standard pin-out for quad comparators differs dramatically from that used for op-amps (Figure 12.9). Care is required here, as not all manufacturers adhere to this pin-out. Always check the data sheet!

12.4 COMPARING OP-AMPS

Having introduced a number of ways in which commercial op-amps differ from one another, it is useful to make some side-by-side comparisons as a means of illustrating the process you might go through when you select an op-amp for a given application. Table 12.1 shows a selection of commercial op-amps that we chose deliberately to point out how the various device

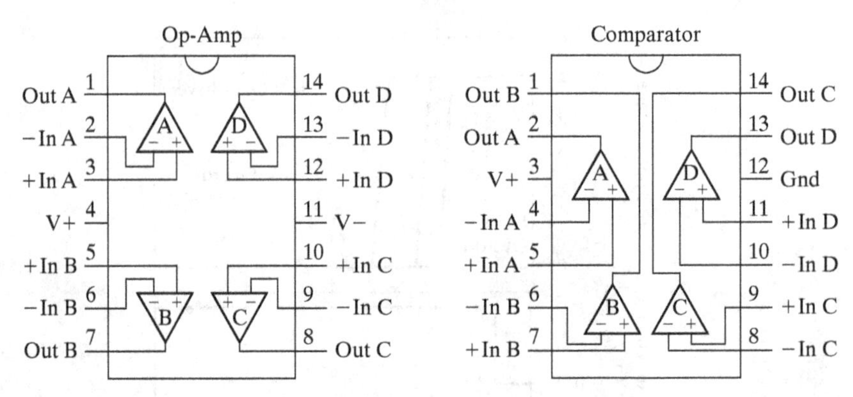

FIGURE 12.9

Comparing quad op-amp and quad comparator IC pin-outs.

designers have chosen to mix and match performance specifications to target different audiences, and how evolution in the manufacturing process is moving us closer to the ideal op-amp. These are all general purpose op-amp designs that seek to achieve a balance among the trade-offs. By focusing on a particular aspect of the performance, you can find devices that better one or more of the specifications of the devices presented here. All of the devices in the table are quad package op-amps and the prices are 2008 prices, quoted in quantities of 1,000 for DIP packages (except for the LTC6088, which is only available in surface-mount packages).

Based on sales volume, the LM324 is currently the most popular op-amp being made. It is a relatively old design, having been introduced over 30 years ago (1974), and is available from a number of manufacturers under similar part numbers and all with similar specifications. The magnitude of the input bias current reflects the use of bipolar junction transistors (BJTs) in the input stage. The limited input common mode voltage range and limited output voltage swing are also the result of the BJTs used in this device. The 1 MHz GBW means that a circuit with a gain of 10 will be limited to amplifying signals below 10 kHz if the gain error is to be held to 10% or less. By modern standards, it offers performance that might be politely described as "adequate." But don't take that in the wrong way: this means that it will

TABLE 12.1 Comparison of specifications for several op-amps.

Parameter	LM324	LM6144	LMC6484I	LTC6088
Input Offset Voltage	7 mV	3.3 mV	3.7 mV	1.4 mV
Input Bias Current	250 nA	526 nA	4 pA	40 pA
Input Offset Current	50 nA	80 nA	2 pA	30 pA
Output Voltage Hi @10 kΩ	3.5 V	4.87 V		4.96 V
Output Voltage Lo @10 kΩ	0.02 V	0.05 V		0.04 V
Output Voltage Hi @2 kΩ	3 V	4.8 V	4.7 V	4.88 V
Output Voltage Lo @2 kΩ		0.13 V	0.24 V	0.12 V
Output Voltage Hi @600 Ω			4.24 V	
Output Voltage Lo @600 Ω			0.65 V	
Unity Gain Bandwidth	1 MHz	6 MHz	1.5 MHz	9 MHz
Slew Rate	0.42 V/μs	11 V/μs	0.63 V/μs	7.2 V/μs
Input Common Mode Range	0–3.5 V	0–5 V	0–5 V	0–5 V
Quiescent Current (Chip)	1.2 mA	3.5 mA	3.6 mV	5 mA
Cost @ 1k	$0.144	$2.94	$1.25	$1.72
Temperature Range	0° to 70°C	−40° to 85°C	−40° to 85°C	−40° to 85°C

(Data courtesy of National Semiconductor Corporation and Linear Technology Corporation.)

work in a wide range of applications. The cost shown in the table is not a mistake. That is, 14.4¢ for a package of four op-amps (3.6¢ per amplifier). Prices drop below 10¢ per package in very high volumes. Performance that is "good enough" coupled with a very low price is the reason that this op-amp is so popular.

The LM6144 is a much more modern design, dating from the early 1990s. This device is representative of a class of amplifiers that use bipolar transistors in the input stage and MOSFETs in the output stage. New input circuit designs allowed the BJT input to offer RRI common mode voltage range, while the CMOS output stage achieved a RRO voltage swing. The LM6144 was one of the early RRIO op-amp offerings. With BJT input stages, there has always been a trade-off between input bias current and speed. The higher the speed of the device, the greater the input bias current required. The designers of the LM6144 opted to emphasize higher frequency performance and accepted input bias currents that are twice those of the humble LM324 in exchange. The higher bandwidth of the LM6144 allows for the creation of circuits with a closed-loop gain of 10 capable of amplifying signals of up to 60 kHz with less than 10% gain error.

The LMC6484 was introduced a few years after the LM6144 and shows what became possible as the manufacturing processes for CMOS transistors became more refined and better controlled. The LMC6484 demonstrated that it was possible to make an all-CMOS op-amp with transistors well matched enough to result in input offset voltage specifications that can compete effectively. Notice that the input offset voltage is comparable to those of the BJT-input LM6144. The use of a CMOS input stage endows the LMC6484 with extremely low input bias currents. At 4 pA (4×10^{-12} A), the input bias current is, for most purposes, essentially zero. The measurement of currents this low requires extreme care so that the leakage currents across the surface of a test apparatus circuit board do not swamp the bias currents [for reference: 1 pA represents the current through a resistance of 1 GΩ (a very very large resistance) with only a 1 mV ΔV applied]. The output stage of the LMC6484 has been optimized to deliver a wide voltage swing even into the relatively low load resistance of 600 Ω. The GBW of only 1.5 MHz is one of the drawbacks of this class of RRIO op-amp.

The LTC6088 is the newest of the op-amps in Table 12.1, having been introduced in late 2007. This design gives us a view to where general purpose op-amp performance is headed and just how close we already are to our ideal assumptions for op-amps. The input offset voltage, at 1.4 mV, is the best of those in the table and, for general purpose amplification at reasonable closed-loop gains, is small enough to be ignored. While the input bias current (40 pA) is 10 times the stunning specification for the LMC6484, it is still so low that it would be difficult to measure. The output voltage swing is the best of the table, although the drop of the output voltage with a 2 kΩ load shows that we are still quite a ways away from the ideal voltage source output. The 9 MHz guaranteed GBW allows this op-amp to be used for signals up to 90 kHz in the gain of 10 configuration with no more than 10% gain error that we have used as a benchmark throughout our discussion. If there is a drawback to the LTC6088, it is that the demand for physically smaller devices has led the manufacturer to offer this op-amp only in surface-mount packages. That complicates the task of prototyping, but is the almost inevitable result of the push for more functionality in tiny packages.

The op-amps in Table 12.1 give an indication of both the range of op-amps that are available and a hint about where the future lies. In the future, we can expect to have a range of competitors to the LTC6088 that will get even closer to our ideal assumptions, in more categories. While we will never achieve absolute perfection, we are already close enough that, for general purpose applications, the most difficult task for the circuit designer today is not creating a design that achieves the performance goals but optimizing the choice of op-amp characteristics to meet those goals at minimum cost. To this latter end, many manufacturers of op-amps offer Web-based selection tools that will help you use the set of specifications that are most important in your particular design to quickly narrow the huge array of available parts.

12.5 HOMEWORK PROBLEMS

12.1 Use the Internet to find the data sheet for the Microchip MCP6294 amplifier and use the specifications for the device when it is operated with a single 5 V supply to answer the following questions.

(a) For a circuit with a closed-loop gain of 10 and desired maximum gain error of 10%, use the guidelines provided in this chapter to determine the maximum signal frequency that can be amplified.

(b) What would the actual gain be at that frequency?

(c) How would these answers change if the power supply voltage were reduced to 3.3 V?

12.2 Use the Internet to find the data sheet for the Microchip MCP6294 amplifier. If the MCP6294 amplifier operates with a single 5 V supply across the full range of temperatures, what is the possible range of output voltages for the circuit shown in Figure 12.10? You may neglect the effects of temperature on all other circuit elements.

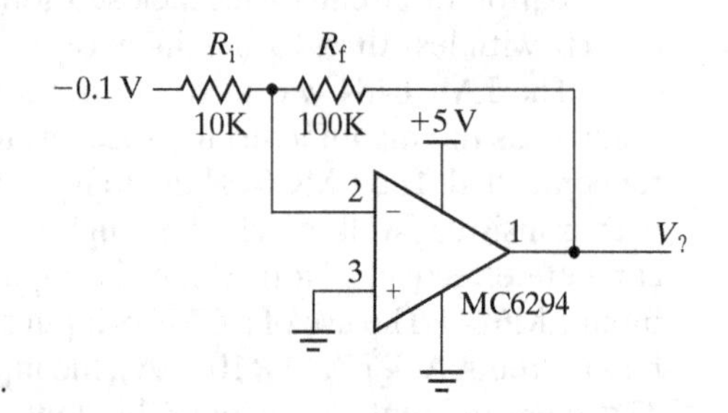

FIGURE 12.10

Circuit for Problem 12.2.

12.3 How much GBW is required of an op-amp in order for it to amplify a 10 kHz sine wave by a factor of 10 with a gain error of less than 1%?

12.4 Design a circuit to implement an inverting comparator having an approximately 200 mV hysteresis band centered at approximately 1 V. Choose standard 5% resistors and report the expected thresholds if the values of those resistors were exact.

12.5 What range of thresholds (min to max) would you expect to see in the circuit of Problem 12.4 if the 5% resistors were at the outer limits of the tolerance?

12.6 For the circuit shown in Figure 12.11, what is V_{out}? Write an expression for V_{out} in terms of the relevant device parameters.

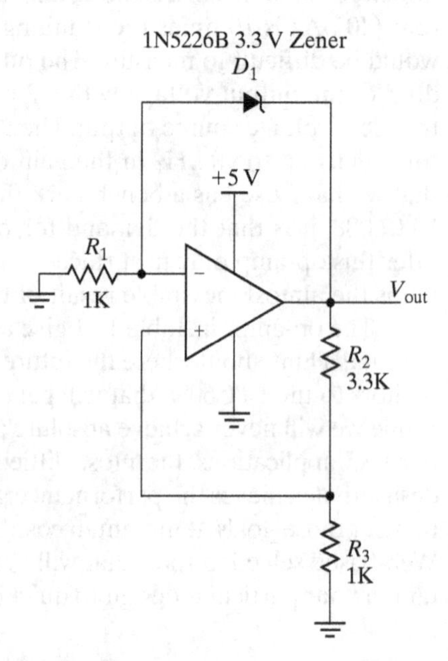

FIGURE 12.11

Circuit for Problem 12.6.

12.7 In the circuit of Figure 12.11, what purpose does resistor R_1 serve?

12.8 Using the Web, classify the following op-amps according to their input and output characteristics (i.e., RRI, RRO, or RRIO): LM139A, MCP6031, OPA337, MC34072A, ADA4051-2, and LT1012.

12.9 For the MC6294, across the range of temperatures from −40°C to 125°C, what is the maximum expected input offset voltage and input bias current? How do these numbers change if the temperature range is reduced to −40°C to 85°C?

12.10 For the circuit of Figure 12.12, what is V_{out}? Write an expression for V_{out} in terms of the relevant device parameters.

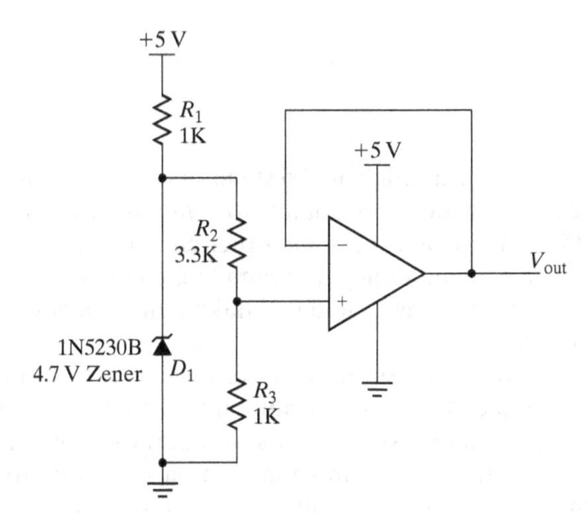

FIGURE 12.12

Circuit for Problem 12.10.

12.11 Locate a data sheet for a MAX9024 and answer the following questions.

(a) How does its pin-out compare with that of the comparator shown in Figure 12.9?

(b) How does its output stage compare with that described in Section 11.5?

(c) How does its propagation delay compare with that of the specifications for the LM339A?

FURTHER READING

Intuitive Operational Amplifiers, Frederiksen, T.M., McGraw-Hill, 1988.

Op Amps for Everyone, Carter, B., and Mancini, R., 3rd ed., Newnes, 2009.

CHAPTER 13

Sensors

13.1 INTRODUCTION

Sensors serve to inform our mechatronic systems about their environment, and enable them to make decisions and respond appropriately to stimulus. Sensors provide information about the physical world to microcontrollers and other circuit elements—parameters and characteristics such as position, light, sound, force, and human interactions, to name just a few. Sensors comprise one of mechatronics' fundamental building blocks, and as such are essential to almost all mechatronic tasks.

There are entire books on sensors, so this chapter cannot and does not attempt to be comprehensive. Many such texts concentrate on how to design sensors, and there are many other texts whose focus is on using sensors in laboratory instrumentation settings. Philosophically, we feel that the field of instrumentation is distinct from mechatronics, though they bear many similarities and share much material. We will introduce the types of sensors that are most commonly used in mechatronic systems. We'll briefly describe how they work. We'll also describe how to successfully integrate a wide variety of sensors into your designs. Rather than attempting to prepare you to design new sensors from scratch, our primary goal for this chapter is to enable you to effectively use commercially available sensors in your designs. After reading this chapter, you should:

1. be familiar with typical ways sensors translate physical quantities into an electrical output that can be read and interpreted by other circuit elements,
2. be familiar with the terminology used to characterize and specify sensor performance,
3. be familiar with the most common types of sensors used in mechatronics,
4. be able to design and build sensor interface circuitry,
5. be prepared to select and use sensors for mechatronic applications.

13.2 SENSOR OUTPUT AND MICROCONTROLLER INPUTS

As the interface between physical signals (e.g., light, pressure, temperature, force) and a microcontroller, the primary purpose of a **sensor** and its associated circuitry is to produce a measurable electrical signal correlated to the quantity it is intended to measure, also referred to as the **measurand** (Figure 13.1). Sensors are a category of **transducer**, a generic term which applies to devices that transform energy from one form to another. For example, a force transducer translates a mechanical force into an electrical signal. Actuators (which we discuss in several subsequent chapters) are also transducers. Many actuators serve to translate electrical energy into mechanical energy (e.g., a solenoid). To avoid imprecision and confusion, we will use the term "sensor" throughout the discussion in this chapter.

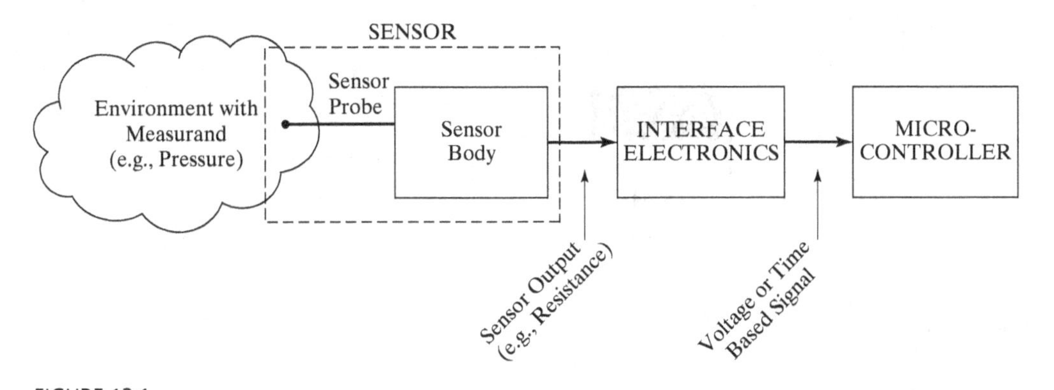

FIGURE 13.1

Translating physical phenomenon into a signal that a microcontroller can read.

In order to successfully connect a sensor to a microcontroller, we must first know what microcontrollers accept as inputs, and then make sure that the sensor translates the physical quantity of interest into one of the appropriate input types. It turns out that microcontrollers are very limited with respect to the type of inputs they can accept. Fundamentally, microcontrollers are only capable of measuring two things: voltage and time. Practically, all microcontrollers accept digital inputs, so they can read voltages that are above and below threshold levels that indicate logic levels on (1) and off (0). Most microcontrollers also include an **analog-to-digital converter** (Chapter 19) on the chip itself, so that they may also read and interpret analog voltages. And all microcontrollers make use of a clock source in order to function. Most microcontrollers provide a means of using this clock source (or optionally, an alternate clock source) to time external events.

Fundamentally, then, the job of a sensor is to translate a *single* physical parameter into a digital voltage, an analog voltage, or a voltage signal whose variations in time encode its output. We emphasize the word "single" because a key characteristic of good sensors is that they are insensitive to physical parameters other than the one you are interested in. For example, a great temperature sensor is not affected by changes in pressure, acceleration, humidity, or any other environmental factor.

13.3 SENSOR DESIGN

A sensor's fundamental purpose is to perform a translation from the physical phenomenon you wish to measure into a voltage or time-varying signal. In order for a sensor to do an effective job with this translation, the physical quantity to be measured must cause a change in some property of the sensor. Sometimes, the translation process requires multiple steps. The sensor must then be connected electrically in such a way that the final result is a change in the sensor's interface electronics output voltage. The best way to illustrate this is with a few concrete examples using real sensors.

13.3.1 Measuring Temperature with a Thermistor

A **thermistor** is a popular sensor whose resistance changes as a function of temperature. There are several types of thermistors, and we will discuss them as a class of sensors in more depth in Section 13.5.3.2, but for now, it is sufficient to say that their resistance is a function of their temperature.

Figure 13.2 shows a thermistor exposed to an environment at temperature T_{amb}. When the thermistor is in equilibrium with the environment, its resistance will reflect that temperature.

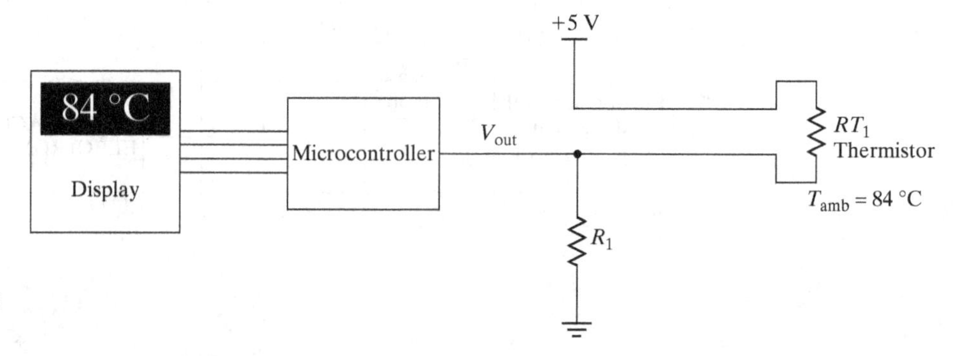

FIGURE 13.2
Using a thermistor to measure temperature.

Since it is the *resistance* of the thermistor that reflects the measurand, you still have some work to do before you can make use of this information. You can't hook the sensor directly to a microcontroller and measure resistance directly; you need to translate this into a voltage or time-varying signal in order for the microcontroller to accept it as an input. This is achieved by connecting the thermistor to **interface electronics**—a circuit that will perform the necessary translation. In this example, we've chosen one of the simplest ways of doing this: by creating a voltage divider circuit with the thermistor on the high side (connected to the 5 V power supply). With this arrangement, the output of the voltage divider, V_{out}, is an analog voltage that varies as the resistance of the thermistor varies. This is something that a microcontroller with a built-in ADC can easily read and interpret. This is only one of many possible ways to interface to a sensor whose resistance is affected by a measurand. We'll discuss several other approaches in Section 13.4.4.

13.3.2 Measuring Acceleration

Measurements of acceleration are typically more challenging than measuring temperature with a thermistor. Of course, measuring acceleration would be easy if we knew of a material whose resistance changes as a function of acceleration—then we could use the same approach we described in the section above. Sadly, no such wonder material has yet been identified, so we have to be more creative in our quest to measure acceleration. We will combine our knowledge of the laws of physics and available sensing technologies to arrive at a means of measuring acceleration. Accelerometers will serve as an excellent second example for us then, as we explore ways of translating a measurand into a useful output.

One way we might go about this is to use a simple mass attached to a spring (Figure 13.3). We know from introductory physics that in order to displace the mass and cause the spring to elongate, we must apply a force. We also know that this force can be applied to the mass by subjecting it to an acceleration. Finally, we can determine the acceleration by measuring the displacement of the mass.

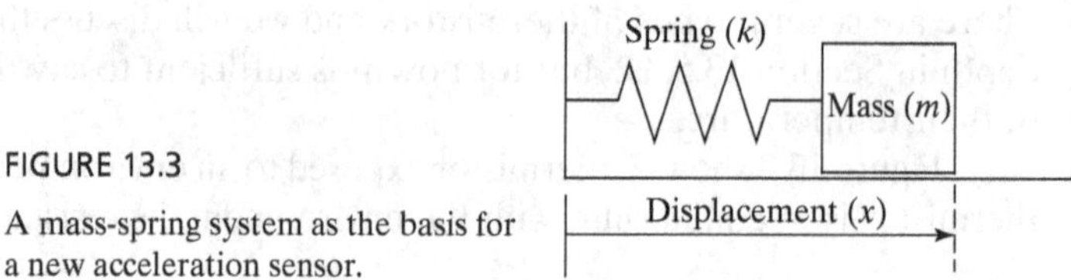

FIGURE 13.3
A mass-spring system as the basis for a new acceleration sensor.

We can start by recalling from introductory physics Newton's Second Law of Motion:

$$F = ma \tag{13.1}$$

Also, recall that the following relationship holds for a linear spring:

$$F = kx \tag{13.2}$$

If we combine these two equations, and rearrange the results slightly, we get

$$\frac{kx}{m} = a \tag{13.3}$$

The result relates the measurand (acceleration a) to the displacement x of a known mass m attached to a spring with a known spring constant k. Therefore, if we are able to determine the displacement x, we will have successfully created an accelerometer. Then the question becomes: how should we measure the displacement x?

This is an excellent example of how the translation of a measurand into an electrical output may be performed in multiple steps (the "SENSOR" block in Figure 13.1). We were initially interested in acceleration, but the first thing we did was to translate acceleration into displacement. Fortunately, there are a number of ways to measure displacement. This has been a measurand of great interest to engineers and scientists for a very long time, and as such there are many well-developed options. They range from simple and inexpensive to highly complex and costly; from very low accuracy to extremely high accuracy. We need to know a little more about the requirements for our accelerometer in order to pick wisely. A short list of specifications that may be important would likely include things like accuracy, speed of response, cost, and size. Working to optimize the sensor to fill the requirements will guide all our design choices, including how we sense the displacement.

One simple, inexpensive method of sensing the displacement of the mass would be to attach the mass directly to the wiper of a potentiometer. This would result in a device whose resistance changes with acceleration. With this approach, we would achieve low cost and simplicity, however, potentiometers used as displacement sensors have a few drawbacks. They don't offer particularly good accuracy or repeatability, and friction adds several challenges at every phase: when the mass starts to move, while it's moving, and when it stops moving. In order to achieve higher performance, it would be advantageous to seek other means of measuring the displacement, perhaps something that doesn't involve direct contact or friction. As we discuss below, reflective infrared (IR) emitter–detector pairs (Section 13.5.1.3) or capacitive sensors (Section 13.5.6.5) could measure the displacement of the mass without making contact. Finally, another variation on the approach of measuring the displacement of a mass on a spring is to flip the strategy around: build a device that actively works to hold a moveable mass at a fixed position. If we measure the amount of control effort required to maintain the position of the mass, we could correlate this to an external acceleration applied to the mass. This type of device is commercially available, and is called a **force-balance accelerometer,** or sometimes a **servo accelerometer**.

13.3.3 Definitions of Sensor Performance Characteristics

In the sensor design examples above, we were faced with design choices where the requirements for the sensors' performance determined the outcome. There are a great many terms used to describe sensor performance, and the goal of this section is to provide a list of such terms and their definitions. When specifying the performance characteristics of any component, especially sensors, it's helpful to define terms precisely, and to ensure that those definitions are shared by everyone involved. That is *almost,* but sadly not always, achieved in the sensors field, as the different communities that deal with sensors have occasionally developed their own, slightly different definitions.

Transfer Function: The functional relationship between the physical input signal (the measurand) and the electrical output signal. This may be expressed as a formula, or plotted as a graph.

Sensitivity: The relationship between input physical signal and output electrical signal. The sensitivity is often expressed as the ratio between a change in output electrical signal to a given change in the input physical signal. As such, it may be expressed as the derivative of the transfer function with respect to the input physical signal. For example, a thermistor (which measures temperature) may state its sensitivity in units of Ω/°C.

Span or Dynamic Range: The range of input physical signals which may be converted to electrical signals by the sensor. Input signals outside of this range are not guaranteed to produce output signals that meet the device's specifications. Span or dynamic range is usually specified by the sensor supplier as the range over which other performance characteristics described in the data sheets apply. For input signals outside of this range, it is possible that there will be large errors from nonlinearity, saturation, clipping, or other distortions. In some cases, signals outside the specified range may permanently damage the sensor. For example, a temperature sensor may state the span or dynamic range in units of °C. Alternatively, when the dynamic range of a sensor is particularly large, it may be stated in decibels (dB), which is:

$$\mathrm{dB} = 20\log_{10}\left(\frac{\text{Max Measurable Signal}}{\text{Min Measurable Signal}}\right) \tag{13.4}$$

For example, if the ratio between the largest and smallest measurable signal is 10,000, the dynamic range would be 80 dB. Decibel units are common in sensors like microphones and photo-sensors, where the dynamic range is large. Sensors with narrower dynamic range, such as thermometers, typically specify only the range and the accuracy.

Accuracy: The largest expected error between actual and ideal output signal. For example, a temperature sensor might state the accuracy as better than 1°C. Alternatively, the accuracy might be stated as a percentage of full-scale output (F.S.O.). For example, a temperature sensor with a span of 0 to 100°C that has an accuracy of ±1°C can also be said to be accurate to 1% F.S.O. It is usually possible to substantially improve a sensor's accuracy by calibrating it against a known standard. In cases where a sensor's offset error is significant but not its gain error, a simple single-point calibration technique may suffice, where the value of the measurand reported by the sensor is compared with the known value of the measurand at a single point somewhere within the sensor's dynamic range. When both the offset error and gain error are significant, a two-point calibration technique may be required. If the output of a sensor is highly nonlinear (see below), a multipoint calibration is likely to be necessary. Specific calibration techniques are described in greater depth in references that focus more on instrumentation applications (see Further Reading).

Nonlinearity (often optimistically called **linearity**): The maximum deviation from a linear transfer function over the specified dynamic range. This is a case of a specification that has several definitions: there are multiple ways that nonlinearity is commonly described. The most common compares the actual transfer function with the "best straight line," which lies midway between the two parallel lines which encompass the entire transfer function over the specified dynamic range of the device. This choice of comparison method is popular because it produces the smallest value for this parameter and makes most sensors look their best. Another possible definition compares the possible range of errors with the least squares fit line. Nonlinearity usually has the units of the measured signal, but you are also likely to see it expressed as a percentage of the F.S.O.

Hysteresis: For a given value of the input physical signal, this is the difference between a sensor's output reading when approached from a previous reading below the new value

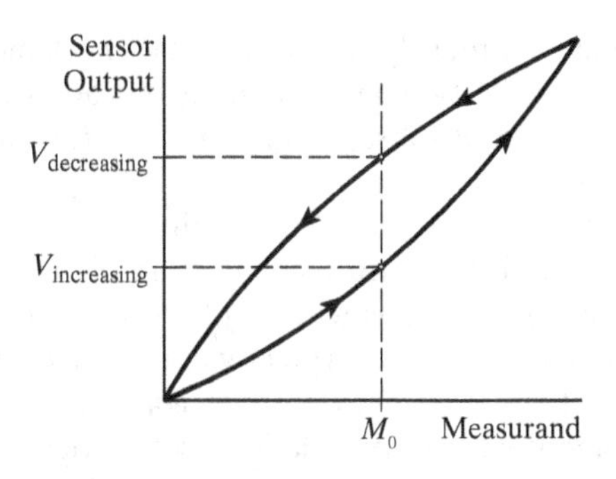

FIGURE 13.4
Hysteresis in a sensor's output.

and when approached from a previous reading above (Figure 13.4). Hysteresis is expressed in the units of the measurement quantity. For example, a temperature sensor might have a hysteresis of 2°C, indicating that its output will differ by up to 2°C, depending on whether the sensor was heated to the new conditions or cooled to the new conditions.

Noise: A constituent of a sensor's output that does not contain information about the input physical signal being measured. Noise may be thought of as "corrupting" the signal of interest, and is generally an undesirable characteristic. All sensors produce some output noise in addition to their output signal. In some cases, the noise of the sensor is less than the noise of the other elements in the electronics, or less than the fluctuations in the physical signal, in which case it is not significant. Many other cases exist in which the noise of the sensor limits the performance of the system based on the sensor. Figure 13.5a shows an oscillating signal which represents the output of a sensor with low noise, and Figure 13.5b illustrates a case where the noise levels are very large.

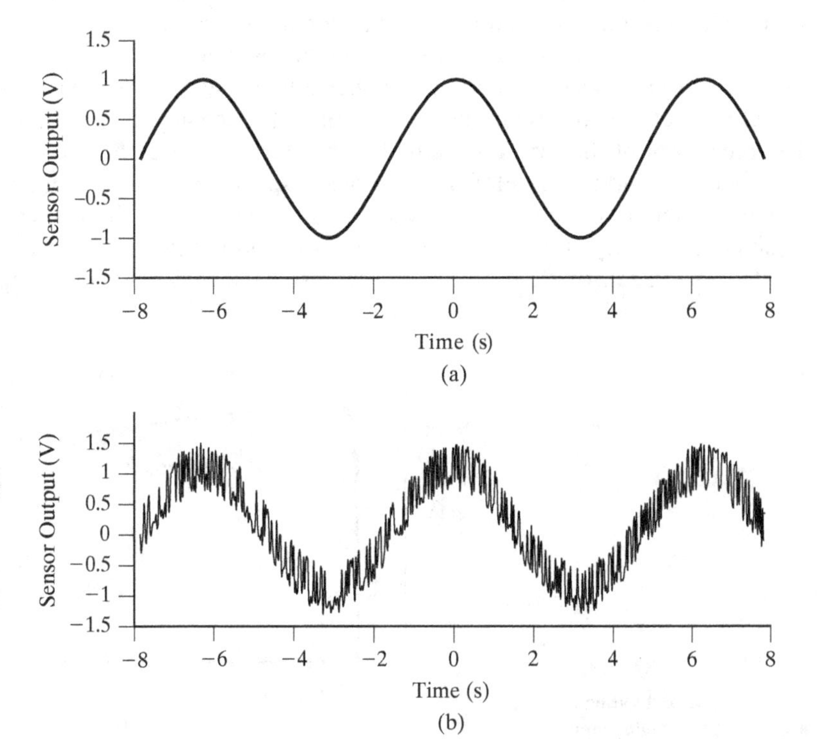

FIGURE 13.5
The output waveform of a sensor exhibiting (a) low noise levels and (b) high noise levels.

Most sensor data sheets merely quote the rms (root mean square) noise amplitude. Some will also include plots of typical noise distributions with respect to frequency. It is often left to the user to interpret these sparse data in order to estimate the noise that will appear in their application.

Many common noise sources produce a white noise distribution, where the spectral noise density is the same at all frequencies. One example of this is **Johnson noise**, which is present in conductive materials such as wires and resistors. Johnson noise is caused by thermal agitation of the charge carriers, and typically appears as white noise. For white noise, the noise spectral density is characterized in units of $volts/\sqrt{Hz}$. A distribution of this nature adds noise to a measurement with amplitude inversely proportional to the square root of the measurement bandwidth. Since there is an inverse relationship between bandwidth and measurement time, the noise decreases with the square root of the measurement time. This is why signal averaging, which is effectively a low-pass filter that reduces the bandwidth of the signal by increasing the averaging time of the measurement, usually improves the **signal-to-noise ratio** (a measure of how big the desired output signal is relative to the noise present). In order to design effective signal conditioning circuits (which we cover in the next two chapters), the spectral distribution of the noise is an important parameter to understand.

Resolution: The minimum detectable change in the measured quantity. Since fluctuations are temporal phenomena, there is a relationship between the timescale for the fluctuation and the minimum detectable amplitude. Therefore, a definition of resolution must include some information about the nature of the measurement being carried out. Many sensors are limited by noise with a white spectral distribution (see the discussion above). In these cases, the resolution may be specified in units of the *physical signal* $/\sqrt{Hz}$. Then, the actual resolution for a particular measurement may be obtained by multiplying this quantity by the square root of the measurement bandwidth. Sensor data sheets generally quote resolution in units of *signal*$/\sqrt{Hz}$, or they give a minimum detectable signal for a specific measurement. For sensors that are limited by other error sources (other than noise with a white spectral distribution), the resolution is often not even mentioned in data sheets.

Bandwidth: The range of frequencies of an input physical signal that a sensor can detect. All sensors have finite response times to an instantaneous change in the physical signal. In addition, many sensors exhibit a behavior in which they follow a step change reasonably quickly, but have an output that then decays in the absence of further changes in the input (Figure 13.6).

The reciprocal of the time constants for the step response and the decay time correspond to the upper and lower cutoff frequencies, respectively. The bandwidth of a sensor is the frequency range between these two cutoff frequencies. Bandwidth is specified in units of Hz. If a single frequency is specified, the underlying assumption is that the range starts at 0 Hz. In the case of sensors that do not produce signals when their input signals are at or

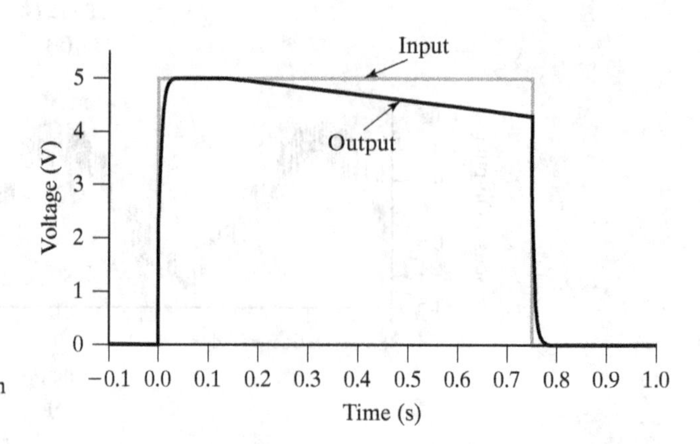

FIGURE 13.6

A sensor response exhibiting decay in the absence of a changing input.

near DC, or 0 Hz (such as a microphone), the bandwidth will be stated as a range (e.g., from 20 Hz to 20 kHz).

In order to help make these specifications and their definitions more concrete, an example of a specific off-the-shelf sensor will be helpful. Since we devoted considerable space above to considering a couple of approaches for measuring acceleration, we select a commercially available accelerometer as our example: the Freescale MMA1250 (Figure 13.7).

In contrast to the macroscale designs we considered above, this sensor is a MEMS (Micro Electro-Mechanical Systems) accelerometer, fitting entirely in a compact 16-pin SOIC (small outline integrated circuit) package. It incorporates all the mechanical elements we discussed (a proof mass, a spring element, and a displacement sensor), as well as the circuitry necessary to translate the output to a varying voltage. For this design, a very small mass is created from the base silicon wafer, and capacitive displacement sensors on either side of the proof mass determine how far from equilibrium it has moved due to externally applied accelerations (Figure 13.7). MMA1250 even incorporates filter circuits to perform signal conditioning and optimize the sensor's final output. For this example, we'll examine the MMA1250 accelerometer's data sheet with respect to each of the specifications.

Transfer Function: In the case of the MMA1250, the data sheet neither gives a formula for the transfer function nor does it supply a graph of the output vs. the input. Rather, the transfer function is implied from the description of the device's output, and from the specifications for "Zero g" output, sensitivity, and nonlinearity. From this, we can write the equation for the straight line describing its output for the range of allowable inputs [Eq. (13.5)].

$$V_{out} = 2.5 \text{ V} + 400 \text{ mV/g} \tag{13.5}$$

This device can measure accelerations in the positive and negative directions, so zero g operation represents the midspan operating point—in this case, the y-intercept. The data sheet tells us that, for $V_{DD} = 5.0$ V (where V_{DD} is the power supply voltage), the output of

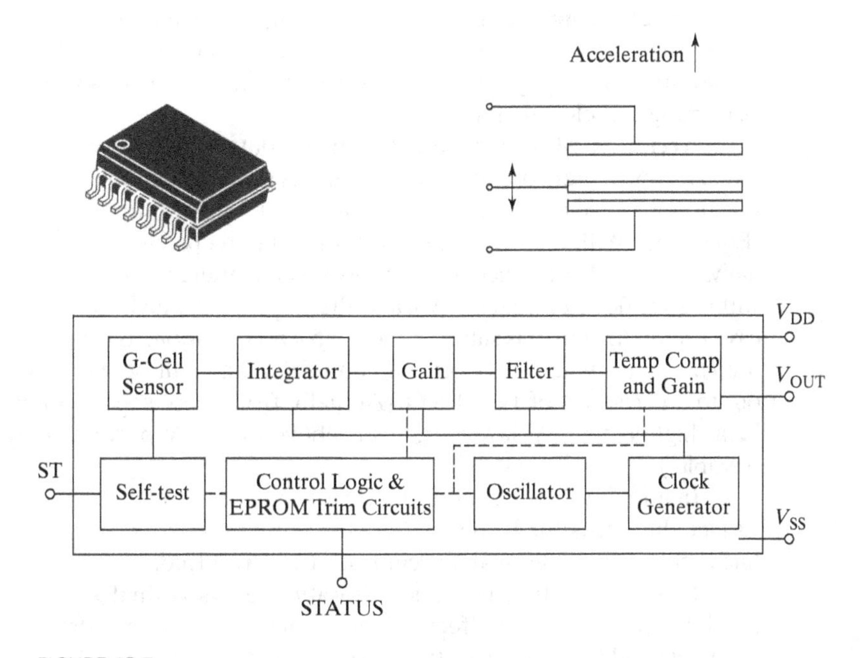

FIGURE 13.7

Freescale MMA1250 MEMS accelerometer. (Copyright Freescale Semiconductor, Inc. 2010. Used by permission.)

the sensor will be 2.5 V at 0 g. There are tolerances around this value: for the case where the ambient temperature is 25°C, the output voltage at 0 g may be anywhere between 2.25 and 2.75 V; and over the full range of allowable operating temperatures (−40 to 105°C), the output may be as low as 2.0 V or as high as 3.0 V. In order to complete the expression for the transfer function, we need the slope of the line, which is the sensitivity. The typical value for this specification is 400 mV/g for V_{DD} = 5.0 V, and the data sheet also reveals that the sensitivity for any given device may be as low as 370 mV/g or as high as 430.1 mV/g over the full temperature range. Note that all of these specifications are given for the special case where the power supply voltage is 5.0 V. In real systems, the supply voltage will usually vary by a small amount, and the first specification given shows that the device functions within specifications for power supplies with 5% of 5.0 V: 4.75–5.25 V. Further descriptions of the device's performance over the full range of power supply values are provided below the Operating Characteristics table in notes 3 and 4 (Figure 13.8). For the purposes of this discussion, we will stick to the manufacturer's convention of specifying the MMA1250's performance for V_{DD} = 5.0 V, in which case the transfer function is given in Eq. (13.5) and plotted in Figure 13.9.

The manufacturer didn't state the transfer function explicitly, but did supply enough information to deduce it. The manufacturer also did an admirable job translating the input physical signal (acceleration) into a varying voltage (V_{out}) with a very convenient range. When the sensor is powered by a 5.0 V supply voltage, the output has a value of 2.5 V for 0 g, 4.5 V for +5 g, and 0.5 V for −5 g. (Keep in mind that there are tolerances on both the 0 g value and the sensitivity, so calibration will be required to achieve the best accuracy with this sensor.) Little, if any, additional interface circuitry or signal conditioning would be necessary to connect this sensor to a microcontroller's ADC input.

Sensitivity: The data sheet gives us the sensitivity (400 mV/g) for the case where V_{DD} = 5.0 V, which is how most customers are likely to use the sensor. This can be thought of as the slope of the transfer function. The tolerances around this value for sensitivity are large: ±7.5% over the full range of allowable operating temperatures for the device.

Span/Dynamic Range: In the Operating Characteristics for the MMA1250 (Figure 13.8), we see that the performance specifications apply (given the symbol g_{FS}; FS for "full scale") for accelerations up to 5 g's. This corresponds to the data sheet's front page, which describes this as a "±5 g" accelerometer.

Accuracy: MMA1250 accelerometer data sheet provides no specification for accuracy. However, there are other specifications that we can look to for an indication. The transfer function we developed for the sensor for a 5.0 V power supply was V_{out} = 2.5 V + 400 mV/g [Eq. (13.5)]. Without calibration, both the *y*-intercept (nominally, 2.5 V) and the slope (nominally, 400 mV/g) are subject to tolerances. The *y*-intercept may vary as much as 20% and still be within specifications (2.0–3.0 V), and the slope by as much as 7.5% (370–430.1 mV/g). For the case where the sensor is subjected to a +5 g acceleration, its output voltage *should* be 4.5 V, but it could be anywhere between 3.85 and 5.0 V and still be within specifications, which works out to an accuracy of 14% F.S.O. (Actually, from these specifications, V_{out} could theoretically be as high as 5.151 V; however, this is above the 5.0 V power supply specified, and so it isn't possible.)

In addition to offset and gain errors, another issue that will affect accuracy is nonlinearity (see the discussion below). Where the sensor's output doesn't match the transfer function's linear equation, additional inaccuracies are introduced.

If an application permits, calibration of individual sensors will greatly reduce—or possibly eliminate—the effects of these tolerances for the zero-g response (the *y*-intercept, or offset) and the sensitivity (slope, or gain). To reduce or eliminate the effects of nonlinearity, a multipoint calibration is required.

Table 2. Operating Characteristics
(Unless otherwise noted: $-40°C \le T_A \le +105°C$, $4.75 \le V_{DD} \le 5.25$, Acceleration = 0g, Loaded output.(1))

Characteristic	Symbol	Min	Typ	Max	Unit
Operating Range(2)					
Supply Voltage(3)	V_{DD}	4.75	5.00	5.25	V
Supply Current	I_{DD}	1.1	2.1	3.0	mA
Operating Temperature Range	T_A	−40	—	+105	°C
Acceleration Range	g_{FS}	—	5	—	g
Output Signal					
Zero g (T_A = 25°C, V_{DD} = 5.0 V)(4)	V_{OFF}	2.25	2.5	2.75	V
Zero g (V_{DD} = 5.0 V)	V_{OFF}	2.0	2.5	3.0	V
Sensitivity (T_A = 25°C, V_{DD} = 5.0 V)(5)	S	380	400	420	mV/g
Sensitivity (V_{DD} = 5.0 V)	S	370	400	430.1	mV/g
Bandwidth Response	f_{-3dB}	42.5	50	57.5	Hz
Nonlinearity	NL_{OUT}	−1.0	—	+1.0	% FSO
Noise					
RMS (0.1 Hz – 1.0 kHz)	n_{RMS}	—	2.0	4.0	mVrms
Spectral Density (RMS, 0.1 Hz – 1.0 KHz)(6)	n_{SD}	—	700	—	$\mu g/\sqrt{Hz}$
Self-Test					
Output Response (V_{DD} = 5.0 V)	ΔV_{ST}	1.0	1.25	1.5	V
Input Low	V_{IL}	V_{SS}	—	0.3 V_{DD}	V
Input High	V_{IH}	0.7 V_{DD}	—	V_{DD}	V
Input Loading(7)	I_{IN}	−300	−125	−50	µA
Response Time(8)	t_{ST}	—	2.0	25	ms
Status(9), (10)					
Output Low (I_{load} = 100 µA)	V_{OL}	—	—	0.4	V
Output High (I_{load} = 100 µA)	V_{OH}	V_{DD} – 0.8	—	—	V
Output Stage Performance					
Electrical Saturation Recovery Time(11)	t_{DELAY}	—	—	2.0	ms
Full Scale Output Range (I_{OUT} = 200 µA)	V_{FSO}	V_{SS} + 0.25	—	V_{DD} – 0.25	V
Capacitive Load Drive(12)	C_L	—	—	100	pF
Output Impedance	Z_O	—	50	—	Ω
Mechanical Characteristics					
Transverse Sensitivity(13)	$V_{XZ,YZ}$	—	—	5.0	% FSO

1. For a loaded output the measurements are observed after an RC filter consisting of a 1 kΩ resistor and a 0.1 µF capacitor to ground.
2. These limits define the range of operation for which the part will meet specification.
3. Within the supply range of 4.75 and 5.25 volts, the device operates as a fully calibrated linear accelerometer. Beyond these supply limits the device may operate as a linear device but is not guaranteed to be in calibration.
4. The device can measure both + and – acceleration. With no input acceleration the output is at midsupply. For positive acceleration the output will increase above $V_{DD}/2$ and for negative acceleration the output will decrease below $V_{DD}/2$.
5. The device is calibrated at 3g. Sensitivity limits apply to 0Hz acceleration.
6. At clock frequency ≅70 kHz.
7. The digital input pin has an internal pull-down current source to prevent inadvertent self test initiation due to external board level leakages.
8. Time for the output to reach 90% of its final value after a self-test is initiated.
9. The Status pin output is not valid following power-up until at least one rising edge has been applied to the self-test pin. The Status pin is high whenever the self-test input is high, as a means to check the connectivity of the self-test and Status pins in the application.
10. The Status pin output latches high if a Low Voltage Detection or Clock Frequency failure occurs, or the EPROM parity changes to odd. The Status pin can be reset low if the self-test pin is pulsed with a high input for at least 100 µs, unless a fault condition continues to exist.
11. Time for amplifiers to recover after an acceleration signal causes them to saturate.
12. Preserves phase margin (60°) to guarantee output amplifier stability.
13. A measure of the device's ability to reject an acceleration applied 90° from the true axis of sensitivity.

FIGURE 13.8

MMA1250 accelerometer data sheet excerpt. (Copyright Freescale Semiconductor, Inc. 2010. Used by permission.)

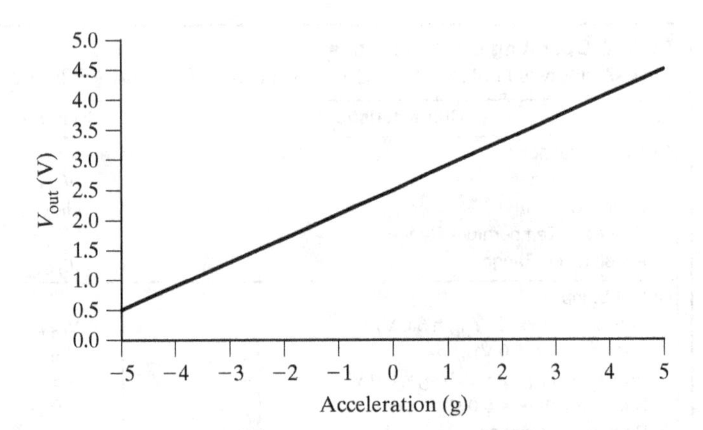

FIGURE 13.9

MMA1250 accelerometer response for V_{DD} = 5.0 V.

Hysteresis: No specification for hysteresis is provided for the MMA1250 accelerometer. It is probably reasonable to assume that the amount of hysteresis exhibited by the sensor is low or negligible, but no specification is provided, so this is far from certain. To get a definitive answer, contacting the manufacturer's applications engineers would be the best course of action. Testing a representative sampling of sensors might give you insights about their performance, but assuming your results apply (and will continue to apply in the future) to all such sensors carries some risks.

Nonlinearity: From Figure 13.8, we find that nonlinearity for the MMA1250 is specified as ±1% F.S.O. In the section above, we note that nonlinearities impact the perceived accuracy of the device, since we assume the transfer function is linear, when in fact it can deviate from a straight line by up to 1% F.S.O. Performing a multipoint calibration of the sensor and fitting a polynomial curve (or other fit, as appropriate) may help you mitigate the effects of nonlinearity for an individual sensor.

Noise: The noise specifications for the MMA1250 are particularly thorough and precise. Noise is specified both in terms of maximum magnitude (2.0 mV rms typical for frequencies between 0.1 Hz and 1 kHz) and spectral density (700 $\mu g/\sqrt{Hz}$).

Resolution: No resolution specification is provided in the data sheet. However, the noise characteristics (< 4 mV rms) and the sensitivity of the sensor (~400 mV/g) help us to generate some reasonable expectations about resolution. The resolution is defined as the minimum detectable signal, and has the form:

$$\text{Resolution} = \frac{\text{Noise}}{\text{Sensitivity}} \tag{13.6}$$

With this in mind, we should be able to expect the sensor to achieve the following resolution:

$$\text{MMA1250 Resolution} = \frac{4 \text{ mVrms}}{400 \text{ mV/g}} = 0.01 \text{ g} \tag{13.7}$$

We can also get an indication of the resolution we can expect if we average multiple samples with a given bandwidth from the noise spectral density specification, which is 700 $\mu g/\sqrt{Hz}$. For a 0.1 s averaging interval (a measurement bandwidth of 10 Hz), we can expect a resolution of $700\ \mu g/\sqrt{Hz} \times \sqrt{10\ Hz} = 2210\ \mu g$. Increasing the averaging interval by a factor of 100, to 10 s (0.1 Hz bandwidth), improves resolution by an order of magnitude, to $700\ \mu g/\sqrt{Hz} \times \sqrt{0.1\ Hz} = 221\ \mu g$.

Bandwidth: The data sheet specifies the sensor's bandwidth as 50 Hz typical (±7.5 Hz). Measuring accelerations with frequency content above the *minimum* guaranteed bandwidth (42.5 Hz) is not recommended, as the sensor's output is not guaranteed to accurately reflect them.

13.4 FUNDAMENTAL SENSORS AND INTERFACE CIRCUITS

Now that we've covered the basics and speak a common language with regards to sensor performance characteristics and specifications, we'll group sensors into fundamental categories based on the form their output takes, and show some typical ways of interfacing to each sensor type.

13.4.1 Switches as Sensors

One of the simplest and most common of all sensors is the humble **switch**—a device with two or more electrical terminals that form an open circuit in one configuration (i.e., the terminals are not in electrical contact), and a short circuit in another configuration (i.e., they make electrical contact). There are a great many different types of switch constructions and technologies available. A partial list includes: pushbutton, slide, toggle, rotary, knife, Hall, tactile, micro, lever, membrane, momentary, keyed, mercury, and reed. A sampling of switch types is depicted in Figure 13.10. Though there are many switch designs, they all share a common purpose: to sense and indicate when an object or force has interacted with them in such a way that it causes them to take on a given state or to change states. Some typical applications for switches include enabling and disabling power to a device, providing a human interface (e.g., a keypad), sensing limits of travel, detecting proximity, and sensing angles. Just for fun, the next time you get into a car, count the number of switches on the dashboard. Easy examples include the ignition switch, the headlight switch, and the turn signal stalk built into the steering column, but there are at least 20 more in most modern vehicles.

(a) Toggle

(b) Slide

(c) Rocker

(d) Pushbutton

(e) DIP

(f) Micro Switch, or Snap-Action Limit Switch

FIGURE 13.10

A sampling of switch types.

There is also a great variety in the electrical configurations of switches. Fortunately, the terminology for sorting through the variations is limited: switches are categorized according to the number of poles and throws. The number of **poles** a switch has is the number of independent circuits it can switch. A single pole switch is represented schematically in Figure 13.11a, and switches a single circuit. A double pole switch is shown in Figure 13.11b, and—as you probably guessed—it switches two separate circuits. Comparing the diagrams for the single pole and the double pole switches shows that the double pole switch is two single pole switches actuated with the same lever. Single, double, and triple pole switches are the most common, but higher numbers are also available in the marketplace when the need arises.

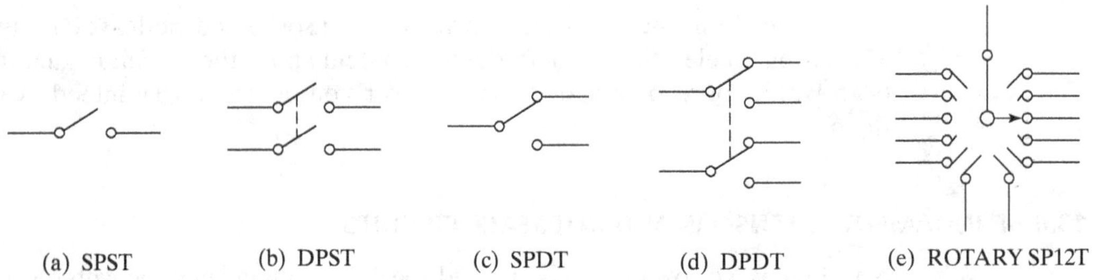

(a) SPST (b) DPST (c) SPDT (d) DPDT (e) ROTARY SP12T

FIGURE 13.11

A representative selection of switch configurations.

The number of **throws** of a switch is the number of discrete active positions the switch contacts may assume. Now that we understand the basic meaning of the two important terms, we can also introduce the shorthand notation for describing switches, which concisely relays a switch's electrical configuration. The general scheme is to describe a switch as "xPyT," where "x" is the number of poles (S = single, D = double, T or 3 = triple, followed by numbers 4, 5, 6, and so on to indicate cases of more than 3 poles) and "y" is the number of throws (again, S = single, D = double, and so on). A simple ON/OFF switch, as shown in Figure 13.11a, is "single throw": it features two discrete positions, one with the contacts open and one with the contacts closed. Based on what we learned about poles in the paragraph above, we can also say that this is a SPST (single pole, single throw) switch. Double throw switches are also common, such as the single pole, double throw (SPDT) switch in Figure 13.11c, and the double pole, double throw (DPDT) switch in Figure 13.11d. Switches are available from single throw to many 10s of throws. Rotary switches (Figure 13.11e), for example, are available with a very large number of throws (>30).

13.4.2 Interfacing to Switches

The general goal when interfacing with a switch is to create a circuit that will allow you to determine its state. Ideally, the circuit's output voltage is a digital logic level low in one state, and in the opposite state, the circuit's output voltage is a digital logic level high. The exact voltages required to achieve this vary depending on the device that is reading this voltage, and this is described in more detail in Chapter 16, Digital Inputs and Outputs. In general, however, a logic level low is very close to 0 V, and a logic level high approaches the supply voltage, often 5 V. In this section, we'll assume a 5 V power supply, and in Chapter 16, we cover more general cases.

The simplest switch interface circuit involves the use of a single resistor, and achieves basic functionality for many applications. This circuit is shown in Figure 13.12.

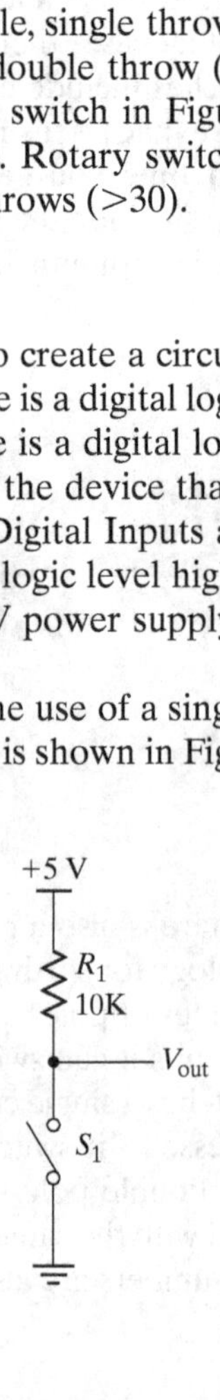

+5 V

R_1 10K

V_{out}

S_1

FIGURE 13.12

A simple switch interface circuit.

You can think of this circuit as a voltage divider, with the switch serving as a resistor that has a very small resistance when it is closed ($< 1\ \Omega$) and a very large resistance when it is open ($> 10\ \text{M}\Omega$). We developed the general expression for a voltage divider in Chapter 9, and for this circuit, we can use it to write the following equation for our switch interface circuit:

$$V_{out} = 5\ \text{V}\frac{R_{S1}}{R_1 + R_{S1}} \tag{13.8}$$

In the case where the switch is open, the effective resistance of the switch, which we write as R_{S1}, is very large relative to R_1 ($> 10\ \text{M}\Omega$ vs. $10\ \text{k}\Omega$), so the numerator and the denominator are essentially equal. This results in a V_{out} very close to the supply voltage, 5 V. In the case where the switch is closed, R_{S1} is very small relative to R_1 ($< 1\ \Omega$ vs. $10\ \text{k}\Omega$), so the fraction approaches zero and V_{out} approaches 0 V. For this analysis to hold true, we assumed that the current flowing through the node V_{out} is very small. If the current through that node becomes significant, then V_{out} will be affected, and may no longer exhibit valid digital logic voltage levels. Put another way: the input impedance of the device connected to our circuit must be significantly higher than $10\ \text{k}\Omega$ for our analysis to hold. This is generally not a problem for the digital and analog inputs of microcontrollers and other logic devices, but you should always be aware that this could be an issue any time you connect a voltage divider because of its high output impedance ($10\ \text{k}\Omega$, in this case).

Interfacing to mechanical switches also requires one additional consideration, since many switch types exhibit a phenomenon called **switch bounce**. Switches that incorporate moving electrical contacts which are forced into physical contact by spring elements form a classic friction-damped mass-spring dynamic system. When the switch changes state, the electrical contacts literally bounce against each other, making and breaking the circuit several times (~10, typically) within a few milliseconds, even though the switch is only really "actuated" once (Figure 13.13). In many applications, this is undesirable: consider how frustrating it would be to type on a keyboard if it registered each key press multiple times—anywhere between 1 and 20. Eliminating switch bounce is also referred to as **switch debouncing**.

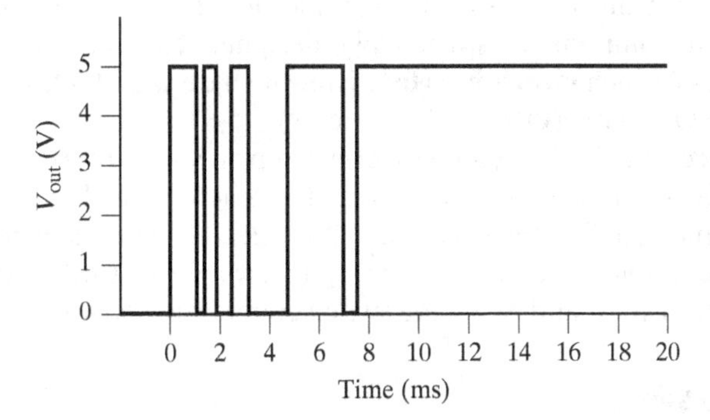

FIGURE 13.13

Output of a switch undergoing switch bounce, where it is actuated only once but its output voltage makes multiple transitions.

There are a number of ways to effectively debounce a switch. The simplest hardware approach involves adding an RC low-pass filter between the switch and its pull-up resistor and the rest of the circuit (Figure 13.14). The effect of the low-pass filter is to slow down the switch's transitions from low to high. If the component values are selected so that the rise times are longer than the period between the switch bounces (again, usually just a few milliseconds), then only a single switch transition will occur at the output of the circuit.

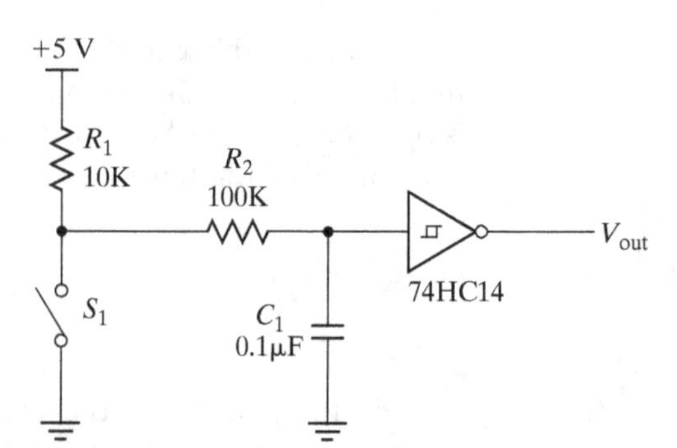

FIGURE 13.14

Debouncing a switch by adding a low-pass filter and a Schmitt trigger.

In this circuit, a Schmitt trigger (Section 11.5.1) inverter (the 74HC14) is used to translate the relatively slowly changing voltage at the output of the low-pass filter into a sharp, clean transition that is appropriate for the digital logic input that we will connect at V_{out}.

The low-pass filter's resistor and capacitor should be chosen so that the resulting rise time doesn't reach the Schmitt trigger's upper voltage threshold when the switch is bouncing. Rather, it should reach the threshold only when the switch has finished bouncing and has a stable output, as shown in Figure 13.15. Here, we've chosen a 100 kΩ resistor and a 0.1 μF capacitor, so the RC rise time is

$$\tau = (100\ \text{k}\Omega)(0.1\ \mu\text{F}) = 10\ \text{ms} \tag{13.9}$$

Depending on the switch you have selected, the bounce time may be longer or shorter than this representative value (10 ms), so you should look at each switch's data sheet for guidance. Switch manufacturers usually specify the **bounce time** as the amount of time it takes for the contacts to stop bouncing after the switch changes state, rather than the interval between bounces.

There are several other hardware-based techniques for dealing with switch bounce, however, we will limit our discussion to one additional, very common method. Several IC manufacturers sell components specifically designed for the sole task of debouncing switches. Examples of such **switch interface chips** include the MAX6816 (Maxim Integrated Products) and the ON14490 (ON Semiconductor).

Switch bounce can also be dealt with effectively in software. This is a standard approach in systems where a switch is interfaced directly to a microcontroller (e.g., a microcontroller's digital input is connected directly to V_{out} in Figure 13.12). Since it's impossible for a human to actuate a switch multiple times within 10 ms, it's straightforward to distinguish between a legitimate switch transition and bouncing switch contacts in software.

13.4.3 Resistive Sensors

Potentiometers, thermistors, photo-cells, strain gages, and flex sensors—what do these sensors have in common? Their resistance is a function of the measurand. A great many sensors behave essentially as variable resistors, and the task of determining the value of the measurand boils down to determining the resistance of these sensors. As we discussed in the example in Section 13.3.1 above, where we explored measuring temperature with a thermistor, we can't measure resistance directly with a microcontroller. First, we must translate the varying resistance into a varying voltage, which may then be connected to an analog-to-digital converter (ADC) and measured.

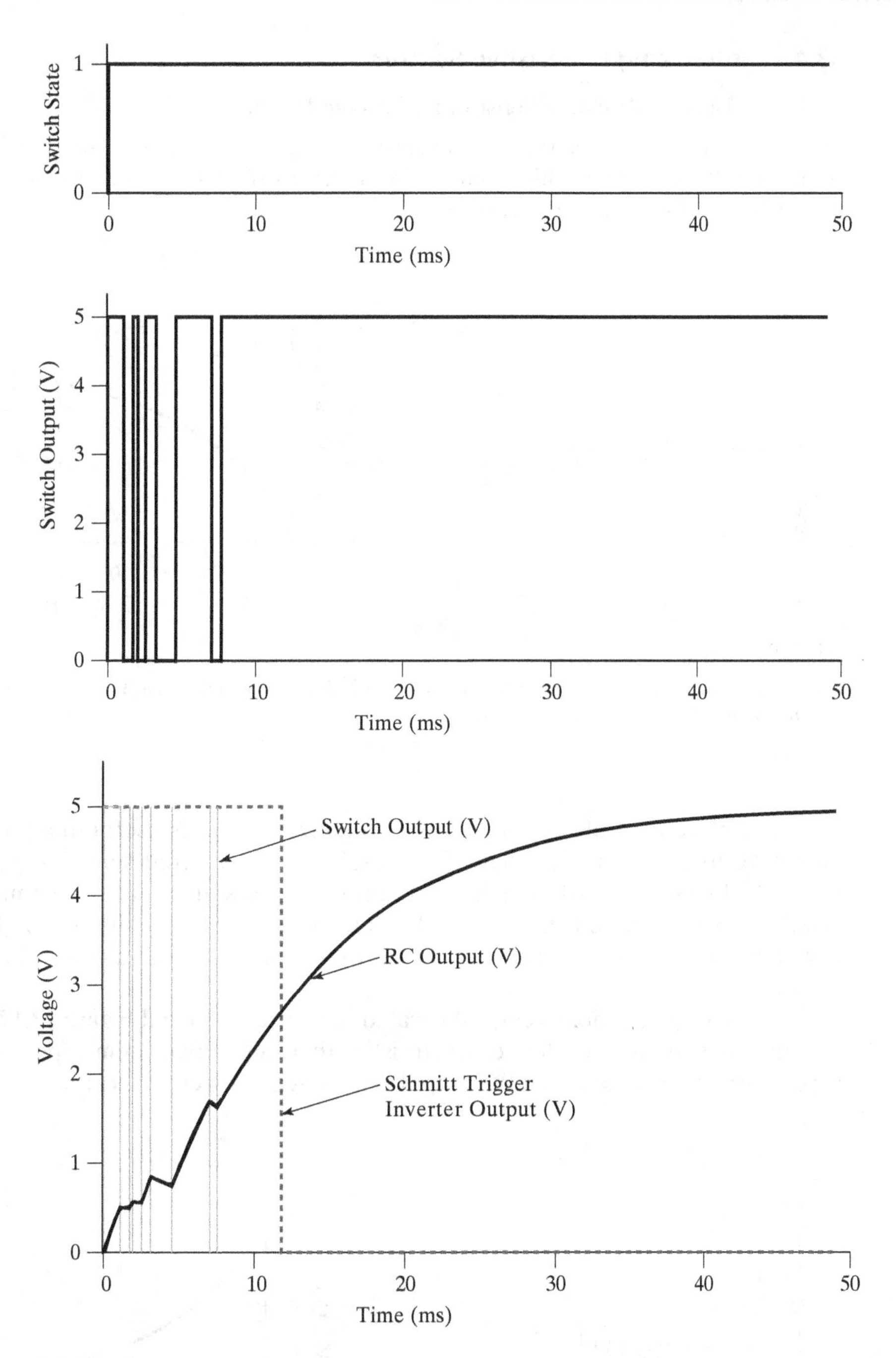

FIGURE 13.15

Response of the switch debouncing circuit shown in Figure 13.14 when the switch is turned on at $t = 0$ (upper plot), and the voltage at the bouncing switch contacts (middle plot). The bottom plot shows the bouncing switch output (gray trace), the output of the RC filter (black trace), and the output of the Schmitt trigger inverter (dashed gray trace).

13.4.4 Interfacing to Resistive Sensors

13.4.4.1 Using a Resistive Sensor in a Voltage Divider

One of the simplest ways to create a varying voltage from a varying resistor is to incorporate it into a voltage divider. This is shown in Figure 13.16, and it's how we connected the thermistor in our discussion in Section 13.3.1.

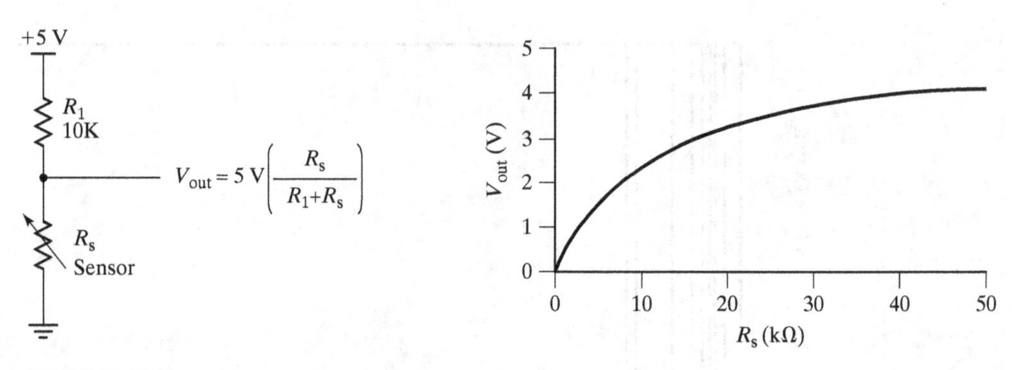

FIGURE 13.16

A resistive sensor as the lower resistor in a voltage divider (left) and the ideal V_{out} as a function of R_s (right).

The placement of the sensor can be changed if this is useful in a given application. Swapping its position with that of R_1 results in an inverted sense at V_{out}, as shown in Figure 13.17. For cases where it is useful for lower resistances to correspond to lower voltages, connecting the sensor between V_{out} and ground will give that result. If you prefer to have low values of resistance correspond to higher V_{out}, swap the position of R_1 and the sensor, R_s.

The voltage divider circuits illustrated in Figure 13.16 and Figure 13.17 are simple and effective, but there are a few characteristics that may make them a poor choice in some applications. One issue is that the output is not linear with changes in the sensor's resistance,

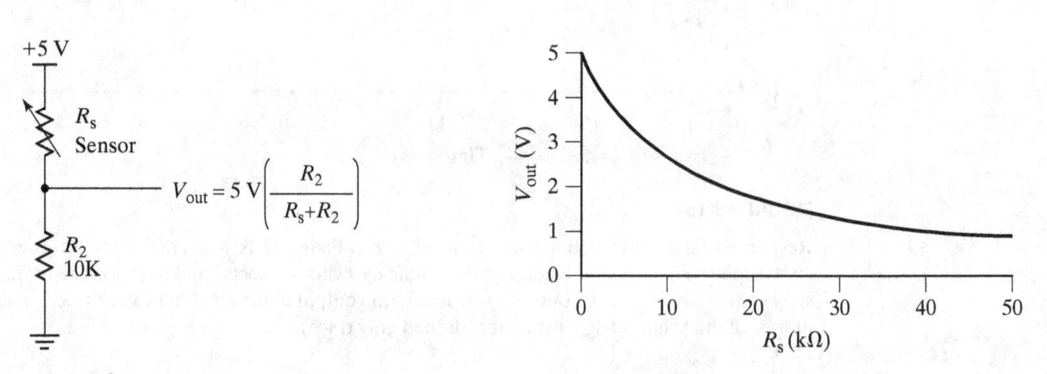

FIGURE 13.17

A resistive sensor as the upper resistor in a voltage divider (left) and the ideal V_{out} as a function of R_s (right).

R_s. For many circuits, this is fine. In cases where high accuracy is needed, the sensor and interface circuit can be calibrated and a curve can be fit to the results to determine the transfer function.

The other issue with the voltage divider approach is its high output impedance. In the circuit shown in Figure 13.16, it's 10 kΩ in parallel with R_s. Connecting V_{out} to any circuit element that has a low input impedance will draw appreciable current through the V_{out} node, and this will effectively "load" the sensor interface circuit and affect the voltage at V_{out}. Regardless of what we connect to V_{out}, some current, I_{out}, will flow. Our goal is to minimize this current flow, and thus minimize its effect on V_{out}, since that's what we hope to accurately measure.

If the performance characteristics of a voltage divider interface circuit eliminate it as an option, there are many other choices, though few are as simple.

13.4.4.2 Measuring Resistance Using a Current Source

If linear circuit response is a requirement, we can think back to one of electrical engineering's most popular linear relationships: Ohm's law ($V = IR$). We can rearrange to get an expression for R,

$$R = \frac{V}{I} \tag{13.10}$$

If we cause a known and constant current I to flow across the resistor, then we can measure the resulting voltage drop across the resistor, V, and calculate the resistance. This requires that we build or buy a **current source**, which is a circuit that causes a constant current to flow regardless of what the other elements in a circuit are doing. Figure 13.18 shows the most general case of using a current source to measure V_{out}, which can be used to solve for R_s since I is known.

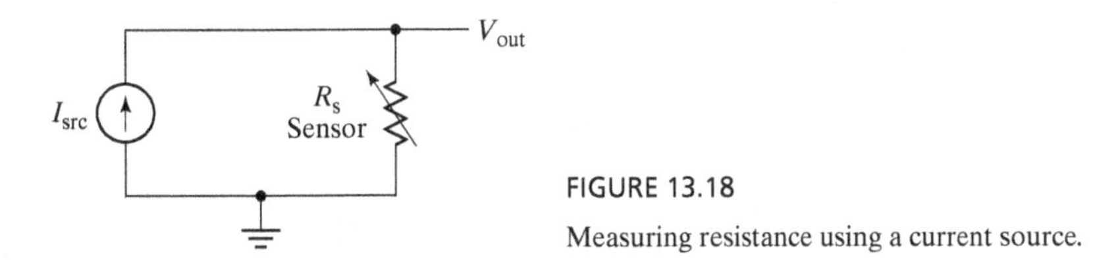

FIGURE 13.18

Measuring resistance using a current source.

Sadly, ideal general-purpose current sources don't exist, since the range of operating voltages would have to be enormous (essentially, $-\infty$ to $+\infty$) to cause a constant flow of current under all conditions, including open circuits and short circuits. Real current sources can be constructed to perform well over limited ranges, however, and this is the technique used to measure resistance in modern ohm meters.

One common method for implementing this approach in practice is the **constant current circuit**, shown in Figure 13.19.

With this circuit, we use an op-amp to fix the current flow through R_1 as a constant. In the ideal case, the Golden Rules of Op-Amps allow us to assume that the inputs draw no current, and that the output does whatever is necessary to hold the voltages at the inputs equal (when connected in a negative feedback configuration, as is the case here). With these assumptions, the voltage at the noninverting input will be very close to 1 V (to be precise, our choices for R_2 and R_3 from standard 5% resistor values result in a value of 1.02 V at the noninverting input when we don't take the resistors' tolerances into account), and

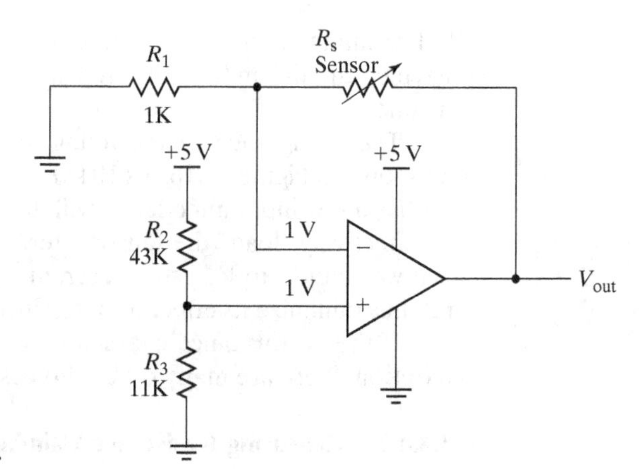

FIGURE 13.19

Constant current circuit configuration.

thus the voltage at the inverting input will also be held at 1 V. To achieve this, the op-amp's output will source whatever current is necessary through R_1 and R_s to hold the node between them at 1 V. The current flowing through R_1 is fixed since the resistance is constant, and the voltage drop is fixed (1 V – 0 V = 1 V). A current of 1 mA flows through R_1 (1 V/1 kΩ = 1 mA), and since the (ideal) inverting input draws no current, all of that current must also flow through the sensor, R_s. Depending on the resistance of R_s, the op-amp's output will range from 1 to 5 V, and source 1 mA. For values of R_s above 3.9 kΩ, the op-amp's output will saturate at 5 V, since that's as high as V_{out} can go (we powered the op-amp with a single supply: +5 V and ground). The output characteristics of this circuit are shown in Figure 13.20. Note the linear relationship between R_s and V_{out}: this makes it very easy to deduce the circuit's transfer function and determine the value of the measurand we're trying to sense in the first place.

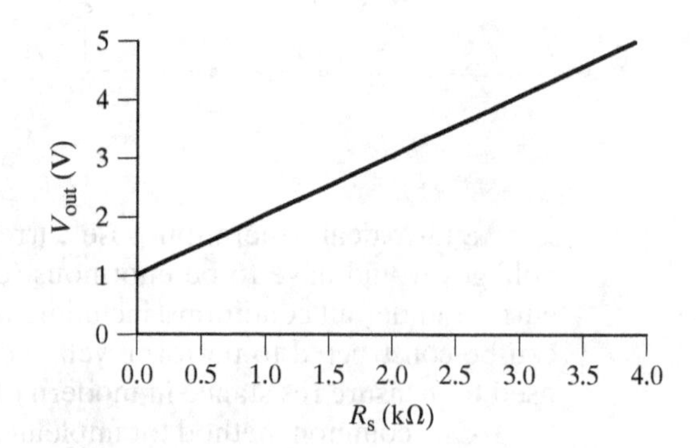

FIGURE 13.20

Response of constant current circuit shown in Figure 13.19.

13.4.4.3 The Wheatstone Bridge

The approaches discussed above work well for sensors whose resistance takes on a wide range of values. For the voltage divider example, we investigated the circuit's response as our sensor's resistance ranged from 0 to 50 kΩ; and for the constant current circuit, the range was 0 to 4 kΩ. However, there are a number of sensors, such as strain gages, whose resistance has some nominal, nonzero baseline value (e.g., 100 Ω), with a range that deviates from this

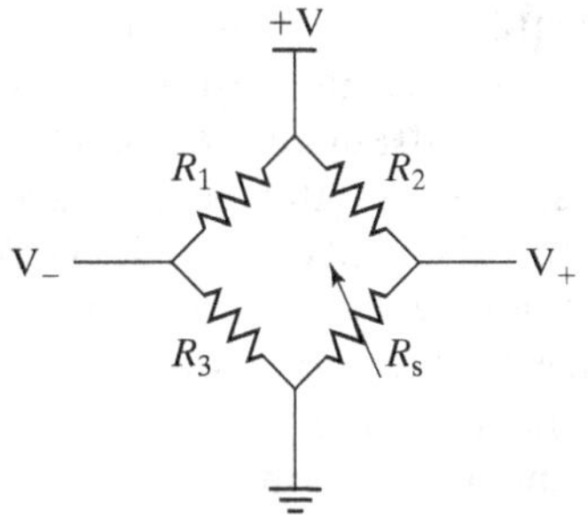

FIGURE 13.21
Wheatstone bridge.

value by a very small amount (only a few ohms). In these cases, a very good approach is to use a Wheatstone bridge circuit, as illustrated in Figure 13.21.

Closer inspection of the Wheatstone bridge circuit shows that it is comprised of two voltage dividers: one formed by R_1 and R_3 (on the left side), and the other by R_2 and R_s (on the right side). The output voltage of the bridge circuit is the difference between the midpoints of the two voltage dividers. Eq. (13.11) gives the general expression for V_{out} based on the input voltages and resistances.

$$V_{out} = \mathrm{V}_+ - \mathrm{V}_- = +V\left[\frac{R_s}{R_s + R_2} - \frac{R_3}{R_3 + R_1}\right] \tag{13.11}$$

It is common (but not required) to choose all the resistors so that they have the same nominal value. In this case, the bridge is said to be "balanced" when R_s has its nominal value and $V_{out} = 0$ V. This is also true any time the resistors are chosen so that the output of each side of the bridge is equal under nominal conditions, that is, when $R_1/R_3 = R_2/R_s$. Choosing other resistor values has the effect of altering the effective offset and gain of the bridge's differential output voltage. As R_s deviates from its nominal value, the difference between V_+ and V_- indicates how much it has changed. For example, if the nominal value of all the resistors, including the sensor, R_s, is 100 Ω and R_s varies by ±2 Ω, the differential voltage ($V_{out} = \mathrm{V}_+ - \mathrm{V}_-$) has the characteristic shown in Figure 13.22.

In this example, the range of V_{out} is small (±25 mV), which is typical of Wheatstone bridge circuits. Because of this, it is common to amplify V_{out} before connecting it to other circuit elements, such as an A/D converter pin on a microcontroller. Instrumentation amplifiers (which we discuss in Section 14.5.1) are particularly well-suited for amplifying differential signals like the output of Wheatstone bridges.

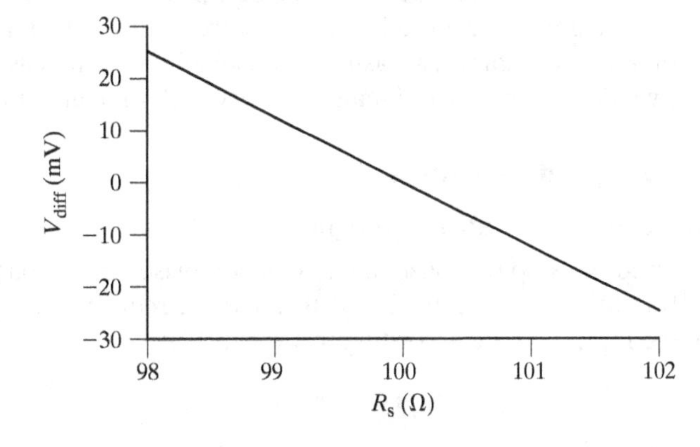

FIGURE 13.22
The differential output voltage of a Wheatstone bridge for $R_s = 100\ \Omega \pm 2\ \Omega$ and $R_1, R_2, R_3 = 100\ \Omega$ (supply voltage = 5 V).

Some sensors have strong temperature dependencies, and dealing with this can often pose serious challenges for the designer. Strain gages are an excellent example of this: their resistance is a function of both strain and temperature. A very useful feature of the Wheatstone bridge is that the resistors can often be arranged so that they're all at the same temperature. For example, mounting them all on a single block of aluminum, or in close proximity on the same printed circuit board, will ensure that the temperature of all the elements is essentially the same at all times. If all the resistors have the same or similar temperature coefficients and they're all at the same temperature, then this greatly reduces or eliminates the effects of temperature on the output of the circuit, since the values of all the resistors will drift together with temperature, in lock step.

The effective gain of a Wheatstone bridge can also be increased by using sensing elements in more than one position within the circuit, and arranging them so that they result in larger differential voltages (the difference between V_- and V_+). For example, you might start out by putting a strain gage in the position marked R_s in Figure 13.21, and this would result in a baseline response. If another strain gage is subjected to the same strain as the gage connected at R_s and is connected to the Wheatstone bridge in place of R_1, this will effectively double the differential voltage for a given strain level. In the case where the resistance of the strain gages is increasing, the output of the voltage divider on the left side of the bridge (V_-) will decrease as R_1 increases (recall that in the baseline case, R_1 just had a constant resistance). The output of the voltage divider on the right side of the bridge (V_+) will increase when R_s increases (as it did in the baseline case). This can be extended farther by replacing all the elements of the bridge with sensors, and arranging them so that the output voltages are driven further in opposite directions. Many types of load cells, which we introduce below in the discussion about strain sensors (Section 13.5.2.2), take full advantage of the Wheatstone bridge's many features.

Consider using a Wheatstone bridge in the following cases:

1. The ability to initially balance (or zero) the bridge output enables you to use high gains with a differential measurement.
2. You are using resistive sensors that have a nonzero nominal resistance and a range that deviates only a small amount from the nominal value.
3. You can mitigate the impact of temperature effects by ensuring that the temperature of all bridge elements is held uniform.

13.4.5 Capacitive Sensors

Another important category of sensors is capacitive sensors, whose capacitance varies as a function of the physical input signal. Examples of capacitive sensors include proximity sensors, displacement sensors, computer touch pads, wall stud sensors, liquid level sensors, and chemical composition sensors. Determining the value of a measurand sensed with a capacitive sensor requires us to be able to accurately measure capacitance, which is more challenging than measuring voltage or resistance. Measuring capacitance requires us to take dynamic measurements, where we need to manage signals that vary with time and amplitude.

13.4.6 Interfacing to Capacitive Sensors

13.4.6.1 Measuring Capacitance with a Step Input

In Chapter 9, we explored the step response of a circuit comprised of a capacitor and a resistor: an RC circuit. If a step input is introduced to an RC circuit, the output voltage will achieve the value of the input voltage according to Eq. (13.12):

$$V_{out} = V_{in}\,(1 - e^{(\frac{-t}{RC})}) \tag{13.12}$$

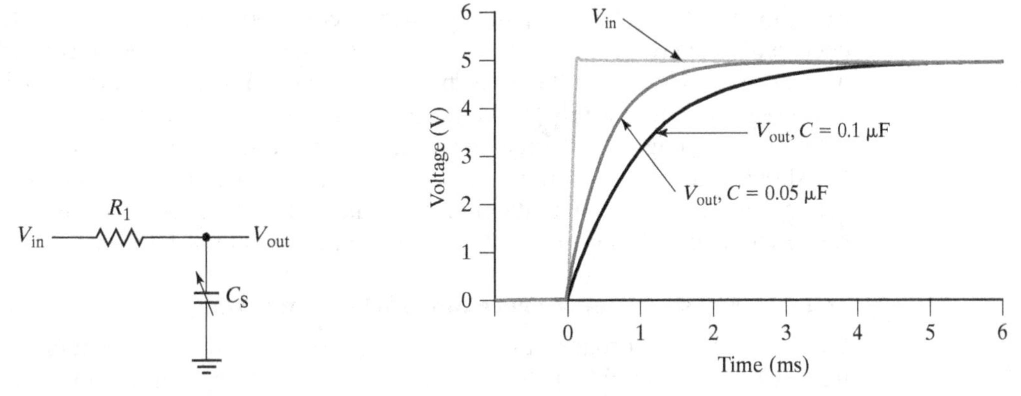

FIGURE 13.23
An RC circuit's step response.

Figure 13.23 shows an archetypal RC circuit, and for $V_{in} = 5$ V and $R = 10$ kΩ, V_{out} is shown for two values of C_s: 0.1 and 0.05 μF.

The circuit's time constant is $\tau = RC$, which is the amount of time required for the output voltage to rise to 63.2% of V_{in} after a step input. If the values of V_{in} and R are known, then C_s is the only variable. Determining C_s involves introducing a step response and determining how much time is required for V_{out} to reach a given value (τ, 2τ, or any voltage that is convenient for you to measure), and then solving Eq. (13.12) for C_s. This approach requires either the use of a comparator to indicate when the voltage threshold is crossed (see Section 11.5), or else the ability to perform rapid analog-to-digital conversions. You must also be able to accurately time the interval between the step input and the crossing of the threshold voltage. Microcontrollers are very good at performing the timing and the A-to-D conversions, but for some values of C_s and R, very short timing intervals may be required, making this approach potentially problematic.

13.4.6.2 Measuring Capacitance with an Oscillator

Another common method for measuring capacitance is with an **oscillator circuit** that relies on the combination of resistors and capacitors to determine the frequency of oscillation. Figure 13.24 shows a simple square-wave oscillator circuit setup to measure an unknown capacitance C_s.

The baseline frequency of the oscillator circuit (without C_s connected) is determined by the values of R_2, and C_1, and will be approximately $f \approx 1/(2.2R_2C_1)$ (R_1 should be $\approx 10 \times R_2$). When C_s is added in parallel with C_1, the total capacitance value increases (since the equivalent capacitance of multiple capacitors in parallel is the sum of their individual

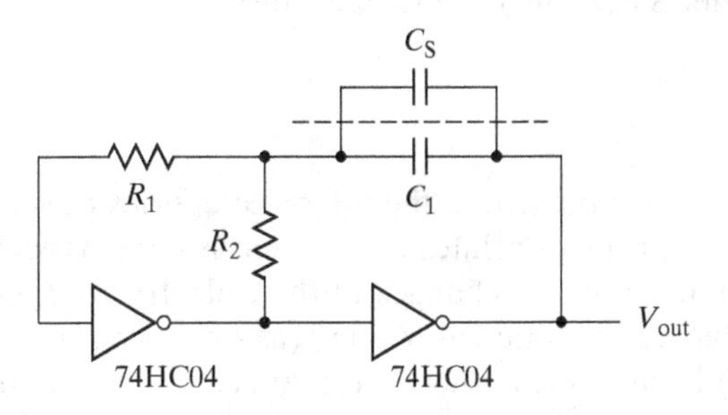

FIGURE 13.24
A simple square-wave oscillator circuit used to measure unknown capacitance C_s.

capacitances: $C_1 + C_s$), and the frequency correspondingly decreases. The amount that decreases depends on the value of the added capacitance C_s, and thus provides a means of measuring C_s. This method of determining the value of the unknown capacitance C_s has the advantage of being entirely frequency based: only the timing of the output square wave's rising and falling edges is affected by changes to the capacitance ($C_1 + C_s$), not the amplitude. We should also note that there are a large number of ways to construct oscillator circuits with an output frequency that depends on the values of resistors and capacitors. The circuit shown in Figure 13.24 is one of the simplest, and a great place to start.

13.4.6.3 Measuring Capacitance with a Wheatstone Bridge

We can take the approach of treating capacitors as frequency-dependent resistors one step further by using the Wheatstone bridge circuits and replacing the resistors with capacitors. This is shown in Figure 13.25.

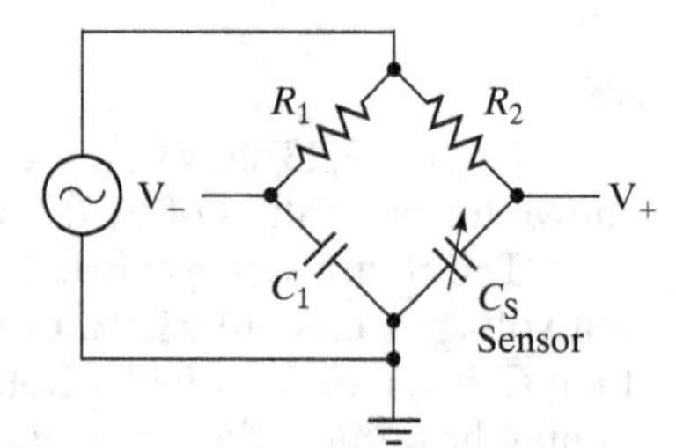

FIGURE 13.25
A capacitive Wheatstone bridge.

For the capacitive Wheatstone bridge, the left- and right-side voltage dividers that make up the circuit are both frequency dependent. Generally, R_1, C_1, and R_2 are all known prior to measuring the unknown capacitance C_s. The behavior of this circuit is similar to that of the resistive Wheatstone bridge (discussed in Section 13.4.4.3), but the excitation voltage must be time varying (e.g., a sine wave) so that the dynamic response of the RC networks on both sides of the bridge can be differentially compared. This approach shares the same advantages for measuring capacitance as for resistance:

1. The bridge's output may be initially balanced (or "zeroed"), removing any offset. Changes in the variable capacitance can then be measured as deviations away from the zero point, rather than having to account for an initial offset.
2. Capacitive Wheatstone bridges allow for a direct measurement of a capacitor's deviation from a nominal value.
3. If all the elements comprising a capacitive Wheatstone bridge are at the same temperature and have the same characteristics with respect to temperature, this mitigates or eliminates the effects of changes in temperature.

13.5 A SURVEY OF SENSORS

In the preceding sections, we presented the major categories of sensors, and grouped them based on which electrical property relates to the measurand. For each category, we proposed several methods for translating the changes in their electrical properties into varying voltages that can then be further modified as needed (as we discuss in Chapters 14 and 15 on signal conditioning and filtering techniques), or connected directly to a microcontroller. The balance of this chapter is devoted to a discussion of specific sensors that are commonly used in mechatronic applications.

13.5.1 Light Sensors

Light sensors are among the most versatile and commonly used sensors in mechatronics. In addition to enabling us to detect the amount of incident light, they can be used as building blocks and configured in a number of ways that allow for the determination of many other measurands. We'll discuss several in the sections that follow. Our goal in this section is to discuss the most common types of light sensors, which often become the basis of other sensors.

13.5.1.1 Photo-diodes

The most fundamental light sensor is the photo-diode (Figure 13.26). A **photo-diode** is a semiconductor device that is composed of a semiconductor junction housed in an optically clear package that allows incident light to fall directly on the junction. This is usually done by encapsulating the semiconductor in a material that is optically clear at the desired wavelength(s) of light, such as polycarbonate. The semiconductor junction is sensitive to light within a relatively narrow range of wavelengths. When exposed to light within its range of sensitivity, each photon liberates an electron, causing a small amount of electrical current to flow. The resulting current, called the **photo-current**, corresponds to the **irradiance** (the amount of radiant flux per unit area) of the incident light.

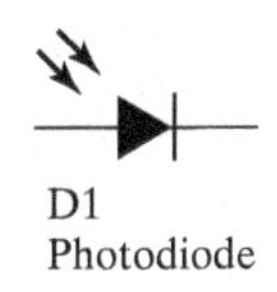

FIGURE 13.26

Schematic symbol for a photo-diode (top); typical photo-diode packages (bottom).

To obtain the desired response from a photo-diode, it must either be either zero biased (where a voltage potential of 0 V is held across the anode and cathode) or reverse-biased (where the cathode is held at a higher potential than the anode, usually only 1 V or so). A common means of achieving either of these schemes is to use a trans-resistive circuit to maintain a constant voltage across a reversed photo-diode, and a gain resistor R_g to translate the photo-current (the output of the photo-diode, which is a function of the incident light) into a voltage. A typical interface circuit is shown in Figure 13.27. In this example, R_1 and R_2 form a voltage divider with an output of 1 V, which is connected to the op-amp's noninverting input. The trans-resistive circuit connects the output to the inverting input across R_g, establishing negative feedback, so we can apply Golden Rule #2 to say that the voltage at the inverting input is held at the same potential as the noninverting input (1 V). This maintains a voltage difference of −1 V across the photo-diode, which is reverse-biased (the anode is connected to the lower voltage and the cathode is connected to the higher voltage). Under these conditions (and invoking Golden Rule #1), all of the photo-current flowing into the photo-diode comes from the op-amp's output, through the gain resistor, R_g.

The wavelength of light that a photo-diode can detect depends on the semiconductor used to make the diode and the manufacturing process. Silicon photo-diodes (the most

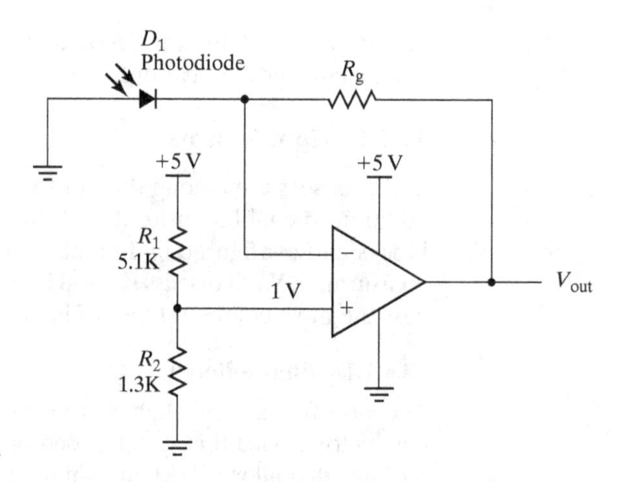

FIGURE 13.27

Photo-diode trans-resistive interface circuit.

common type) are sensitive to electromagnetic radiation at the low end of the near IR spectrum, with a peak in the region near $\lambda = 1$ μm. For comparison, the visible spectrum for the human eye extends from about 380 nm $< \lambda <$ 750 nm. A plot showing the sensitivity of a representative Si photo-diode (Lite-On LTR-516AD) is shown in Figure 13.28.

Several characteristics of photo-diodes deserve special attention. First, they have a *very* large dynamic range. In Figure 13.28, we can see that our representative photo-diode is most sensitive to light at a wavelength of 900 nm (but sensitive from ~800 to ~1050 nm), and that the response is very linear for irradiance between 0.01 and 10 mW/cm^2. Thus, the LTR-516AD has a dynamic range for irradiance of 1,000:1 (10 mW/cm^2:10^{-2} mW/cm^2) or 60 dB, which is impressively wide. Secondly, Figure 13.28 also shows that the output signal (the photo-current, I_p) is very small and also has a very wide range—from 0.4 μA at 10^{-2} mW/cm^2 to 400 μA at 10 mW/cm^2. With such small signal levels, considerable care is necessary to condition the output signal into a voltage with a usable range of values. Thirdly, even in very dark environments that have essentially no IR light within the sensor's range of sensitivity, a small amount

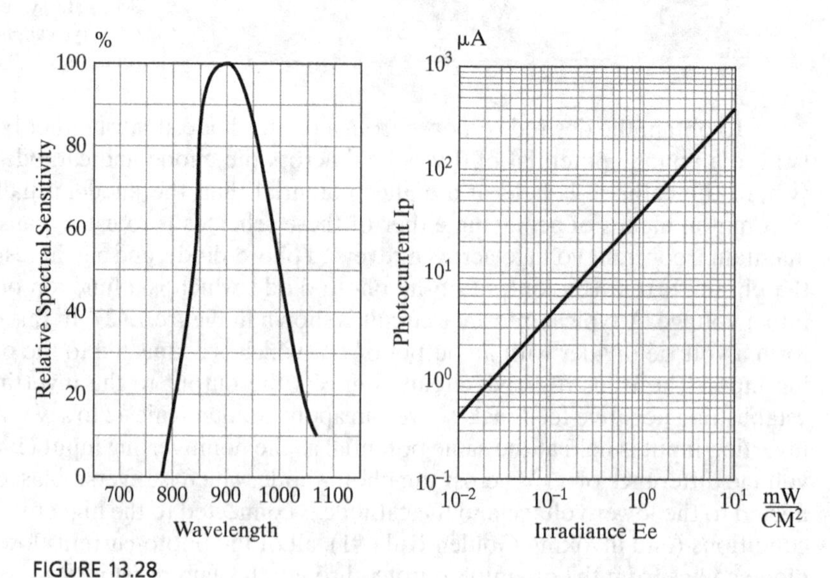

FIGURE 13.28

Representative photo-diode (Lite-On LTR-516AD) spectral response and photo-current output. (Courtesy of Lite-On, Inc.)

of current will still flow. For photo-diodes, this is referred to as the **dark current**, and it is the result of thermal excitation of the diode's semiconductor charge carriers. Because it is a thermal effect, the dark current scales with temperature. LTR-516AD has a typical dark current of 1 nA when reverse-biased with 5 V at 25°C, and a maximum guaranteed value of 30 nA (with a 10 V reverse bias at 25°C). The dark current is usually small compared with the photo-current at 25°C, however, as the temperature of the device rises, the dark current increases substantially: at 80°C, it reaches 0.4 μA (the same as the photo-current for light levels of 10^{-2} mW/cm^2). This contributes error to a measurement without some form of compensation.

Finally, the response time of this sensor is impressively quick: the rise and fall times of the output specified in the data sheet are 50 ns (when reverse-biased by $V_R = 10$ V with $R_g = 1$ kΩ). Because of this combination of characteristics, photo-diodes are commonly used in high-speed applications (such as fiber optic data communications) and where wide dynamic range is needed. Photo-diodes are the sensing elements used to detect IR remote control signals for most consumer electronics products. In this application, they must reliably produce usable signal levels (and not saturate) when the environment is drenched in direct sunlight during the day (which has very high levels of IR light), as well as in completely dark rooms (where there is essentially no ambient IR light). Also, because this is a consumer electronics application, extremely low cost is a strict requirement as well. In large quantities, photo-diodes can be purchased for as little as 10 cents. The photo-diode is a good fit for all of these requirements.

13.5.1.2 Photo-transistors

In cases where the low sensitivity characteristic of photo-diodes is problematic or eliminates them from consideration, photo-transistors offer another choice. **Photo-transistors** function similarly to photo-diodes in that they allow a current to flow (a collector–emitter current, in this case) related to the amount of incident light. However, photo-transistors provide much higher output currents than photo-diodes, albeit over a much narrower dynamic range with much slower response times. Photo-transistors are bipolar junction transistors (BJTs) whose semiconductor base region is optically exposed. Rather than connecting a wire to the base and externally causing base current to flow and control the flow of collector current, the base current in a photo-transistor results from photons striking the base region with sufficient energy to liberate electrons. This base current is then multiplied by the gain of the BJT (typically a factor of about 100), resulting in a collector current in the milliamp range, rather than the microamps characteristic of photo-diodes. Figure 13.29 shows the schematic symbol

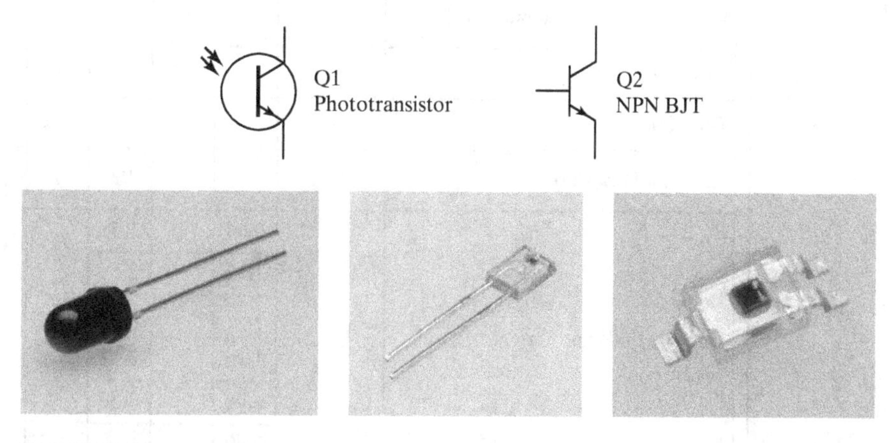

FIGURE 13.29

Top row: the schematic symbol for a photo-transistor (left), and, for comparison, the schematic symbol for a standard NPN BJT (right). Note the photo-transistor symbol indicating that photons provide the base current. Bottom row: a selection of common photo-transistor packaging options.

for an NPN photo-transistor next to a standard NPN transistor for comparison (note that the standard NPN's base terminal is replaced by a symbol indicating that photons supply the base current), and a few commonly available packaging options for photo-transistors.

For photo-transistors, the output current is referred to as the **collector current**. For the purposes of this discussion, we'll use the Lite-On LTR-3208E as a representative example. Figure 13.30 shows the irradiance vs. collector current output for a sensor, normalized to its response at 1 mW/cm^2. The specification is presented in this normalized fashion because the LTR-3208E's sensitivity can vary widely on a unit-to-unit basis, as the excerpt from the data sheet below the plot shows. The LTR-3208E is available in several grades (which the manufacturer calls Bin A through Bin F), which are sorted according to the amount of collector current that is flowing at 1 mW/cm^2 (essentially, the sensor's gain). But even within bins, the

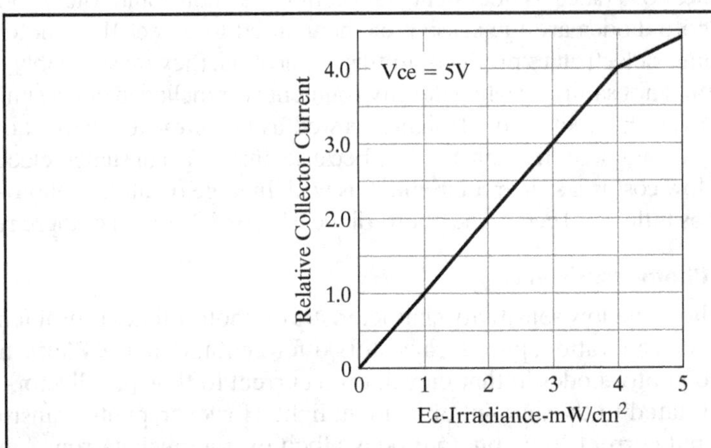

ELECTRICAL/OPTICAL CHARACTERISTICS AT TA=25°C

PARAMETER	SYMBOL	MIN.	TYP.	MAX	UNIT	TEST CONDITION	BIN NO.
Collector-Emitter Breakdown Voltage	$V_{(BR)CEO}$	30			V	I_C = 1mA Ee = 0mW/cm²	
Emitter-Collector Breakdown Voltage	$V_{(BR)ECO}$	5			V	I_E = 100 μA Ee = 0mW/cm²	
Collector Emitter Saturation Voltage	$V_{CE(SAT)}$		0.1	0.4	V	I_C = 100 μA Ee = 1mW/cm²	
Rise Time	T_r		10		μs	V_{CC} = 5V I_C = 1mA R_L = 1KΩ	
Fall Time	T_f		15		μs		
Collector Dark Current	I_{CEO}			100	nA	V_{CE} = 10V Ee = 0mW/cm²	
On State Collector Current	$I_{C(ON)}$	0.64		1.68	mA	V_{CE} = 5V Ee = 1mW/cm² λ =940nm	BIN A
		1.12		2.16			BIN B
		1.44		2.64			BIN C
		1.76		3.12			BIN D
		2.08		3.60			BIN E
		2.40					BIN F

FIGURE 13.30

Representative photo-transistor data (Lite-On LTR-3208E). (Courtesy of Lite-On, Inc.)

gains vary widely, especially Bin F which has no upper bound. The important conclusions to draw from this are that the output is linear for each individual part (regardless of what its gain is), and that there can be wide part-to-part variation.

A key difference between photo-transistors and photo-diodes is their dynamic range. Note that the dynamic range of the linear region for the LTR-3208E is 0–4 mW/cm^2, or 0–5 mW/cm^2 if the abrupt change in slope at 4 mW/cm^2 isn't problematic for a given application. This is tiny and typical, compared with the 1,000:1 range of the LTR-516DA photo-diode. Another very important detail to note in the figure is the "V_{CE} = 5 V" label, which indicates that the response is given for the case where the voltage across the photo-transistor's collector and emitter (V_{CE}) is held at a constant 5 V. If we wish to rely on the photo-current vs. irradiance specifications in the data sheet, our interface circuit must maintain V_{CE} = 5 V. More generally, if we wish to make use of a photo-transistor's linear output characteristics, we must maintain a constant voltage across the collector and emitter of about 2.5 V or more. Figure 13.31 shows that as long as V_{CE} is held constant, the collector current that flows for a given irradiance varies only slightly or not at all over a wide range of V_{CE} values. This behavior doesn't hold for collector–emitter voltages below about 2 V, where all the curves bend and join together.

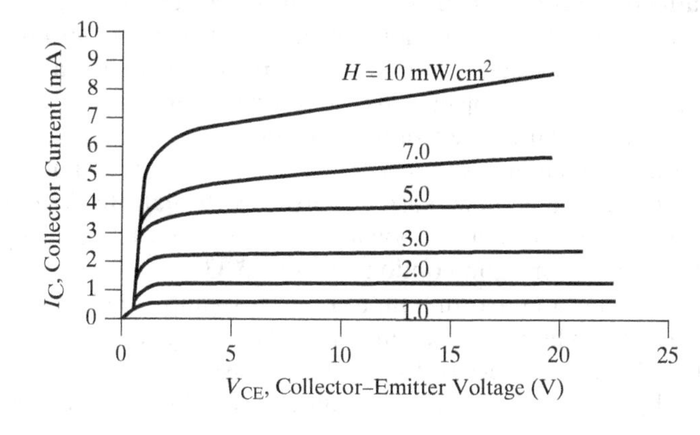

FIGURE 13.31

Typical photo-transistor collector current (I_C) characteristics over a range of V_{CE}.

For photo-transistors, there are two approaches to interface circuits that are most commonly used. In cases where linearity of the output is not a requirement, a simple pull-up or pull-down resistor will suffice. Each case is shown in Figure 13.32.

This approach is quick and simple, and yields an output that varies with incident light, however, it is nonlinear. It also links response time to gain, since the value of the resistor not only sets the gain but also determines the time required to charge and discharge any stray capacitance (which is often significant). For the sourcing configuration (Figure 13.32, left), V_{out} has its minimum value (close to 0 V) when no light is incident on the photo-transistor Q_1, and increasing light causes V_{out} to increase. The opposite is true for the sinking configuration (Figure 13.32, right). When no light is present, V_{out} has its maximum value (near 5 V), and increasing light levels cause V_{out} to decrease.

The nonlinearity is a result of not holding a constant voltage drop across the photo-transistor. For both configurations, V_{out} is essentially the output of a voltage divider, and its value is determined by the voltage drop across the gain resistor R_1, which Ohm's law tells us is $V = IR$. In this circuit, I is the collector current from the photo-transistor and R is the resistor value. In a dizzyingly circular fashion, the amount of current flowing is determined by the light level incident on the sensor, which determines the voltage drop across the resistor, which in turn determines the voltage drop across the photo-transistor's collector and

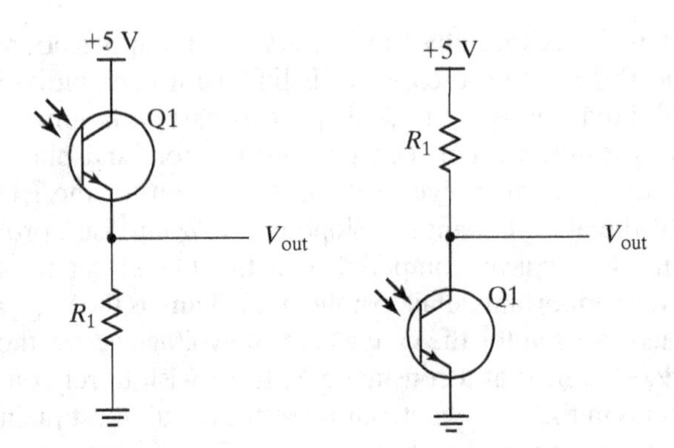

FIGURE 13.32
Left: Photo-transistor sourcing configuration. Right: Photo-transistor sinking configuration.

emitter, which affects the collector current. This all works out quite transparently in practice, but the output is far from linear, and not particularly straightforward to model. Nonlinear output is perfectly adequate and useful in many applications, such as detecting whether a light is on or off, or whether an object is in close proximity to a sensor, or whether a beam of IR light is bouncing off a white object or a black object.

For applications which involve measurements of irradiance, an interface circuit with a linear output characteristic is much easier to work with than the nonlinear response of a sourcing or sinking configuration as shown in Figure 13.32. In such applications, a trans-resistive circuit is once again a good option (Figure 13.33).

For this circuit, the noninverting input of the op-amp is held at 2.5 V with a voltage divider. Golden Rule #2 allows us to assume that the voltage at the inverting input will also be 2.5 V, and this fixes the V_{CE} across the photo-transistor at 5 V − 2.5 V = 2.5 V. We saw in Figure 13.31 that V_{CE} = 2.5 V should be adequate to result in linear collector current output

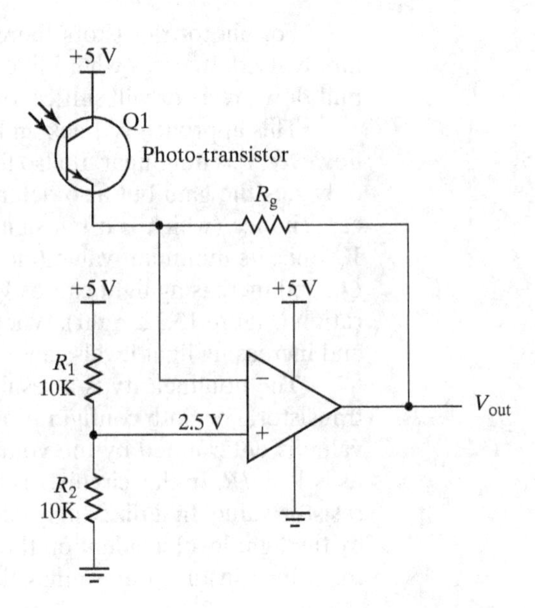

FIGURE 13.33
Trans-resistive photo-transistor interface circuit.

from the photo-transistor. In this configuration, the trans-resistive interface circuit will translate the collector current into a voltage with a gain determined by R_g, since Golden Rule #1 allows us to assume that all the photo-current flows across R_g to the op-amp's output, V_{out}. The trans-resistive circuit performs a linear current-to-voltage transform by multiplying the collector current by the gain resistor R_g. Additionally, the trans-resistive circuit maintains a constant voltage across the photo-transistor's collector and emitter, so it eliminates the response time effects of charging and discharging any stray capacitance. Since there are no changes in voltage, there are no contributions to the sensor output rise and fall times due to stray capacitance, which can be significant. This improves the photo-transistor's dynamic response.

Figure 13.34 shows the output voltage for the trans-resistive interface circuit shown in Figure 13.33 for $R_g = 560\ \Omega$. For this example, we assumed that the collector current flowing through the photo-transistor is 1 mA when illuminated at 1 mW/cm^2.

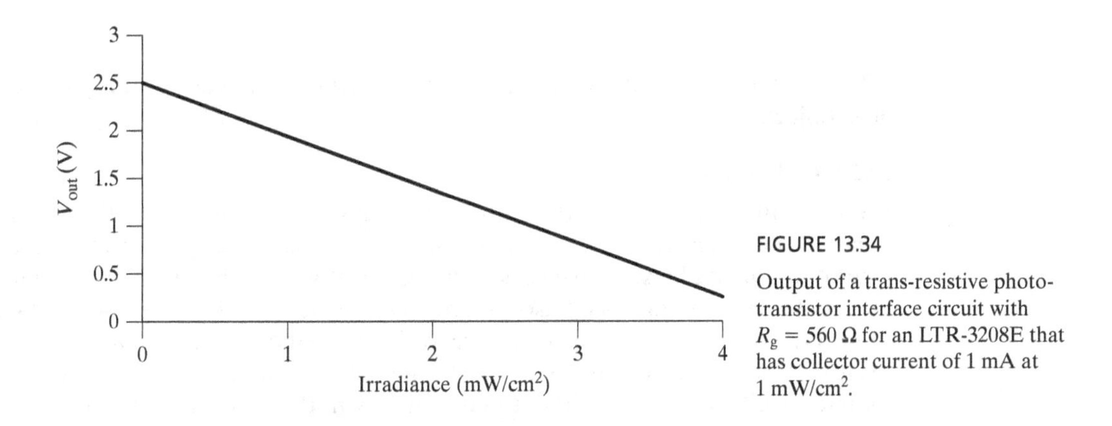

FIGURE 13.34

Output of a trans-resistive photo-transistor interface circuit with $R_g = 560\ \Omega$ for an LTR-3208E that has collector current of 1 mA at 1 mW/cm^2.

The output voltage of the trans-resistive interface circuit shown will be 2.5 V when there is no light incident on the sensor, with a linearly decreasing trend as the light level increases. This is a result of the op-amp's output sinking current from the photo-transistor. This characteristic could be flipped by connecting the photo-transistor's emitter to ground and the collector to the op-amp's inverting input. V_{CE} is also 2.5 V for this new configuration, however, the photo-transistor will now be sinking current from the trans-resistive circuit, rather than sourcing. If we keep the same value of R_g, the resulting output voltage will be 2.5 V when there's no incident light, and the output voltage will increase with increasing light levels.

13.5.1.3 Emitter–Detector Pair Modules

Photo-transistors are often paired in a single package with IR LEDs that emit light at wavelengths matched to the photo-transistor's sensitivity, and packaged so that they can be applied very conveniently for a few specific applications. One very common arrangement of emitter and detector pairs are **opto-interrupters** (Figure 13.35a) in which an emitter (LED) shines directly toward a detector (photo-transistor). When an object passes in front of the detector and interrupts the beam of light, the output of the detector changes. Another common configuration is the **reflective optical sensor** (Figure 13.35b) in which the emitter and detector are arranged side-by-side. This configuration relies on light from the emitter reflecting off a nearby object and reaching the detector. If the reflectance characteristics of the object are known, the amount of reflected light can be used to measure relatively short distances accurately (a few

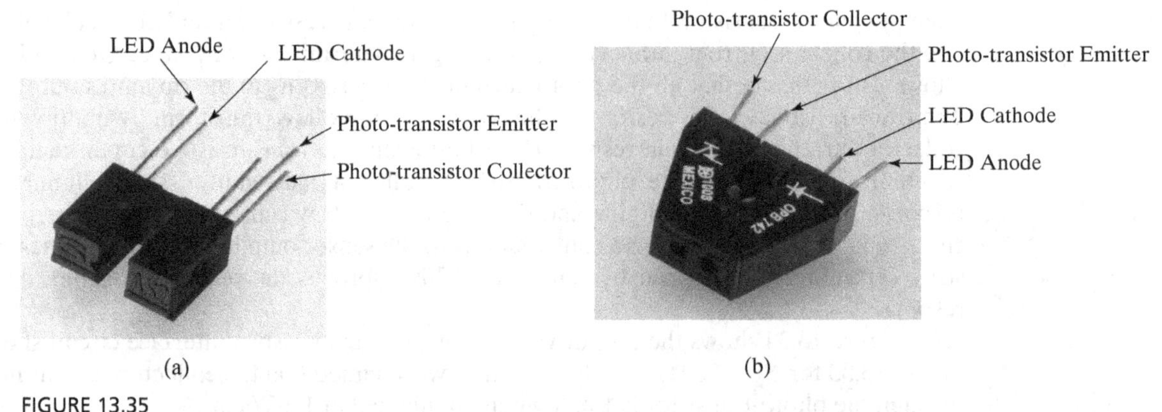

FIGURE 13.35

Emitter–detector pair optical sensor modules: (a) slot-type opto-interrupter; (b) reflective optical sensor.

millimeters to a few centimeters), or it can be used to simply detect the presence or absence of an object.

13.5.1.4 Photo-cells

For applications that require detection of light at wavelengths outside the near IR spectrum, standard photo-diodes or photo-transistors may not have the appropriate spectral response. One common and inexpensive sensor that is sensitive to light within the visible spectrum is the photo-cell. **Photo-cells** (also commonly referred to as **photo-detectors** and **photo-resistors**) are sensors whose resistance varies with the amount of incident light across a range of wavelengths similar to those detectable by the human eye. The mechanism of action for photo-cells is similar to that of photo-diodes, in that photons with sufficient energy that strike the sensing surface liberate electrons. For photo-cells, changes in the electrons' energy alter the material's ability to conduct electricity—that is, its resistance. Figure 13.36 (left) shows the schematic symbol for a photo-cell. In the schematic symbol, the arrow indicates that the value of the device's resistance is influenced by photons. The right side of the figure shows three typical photo-cell component packages.

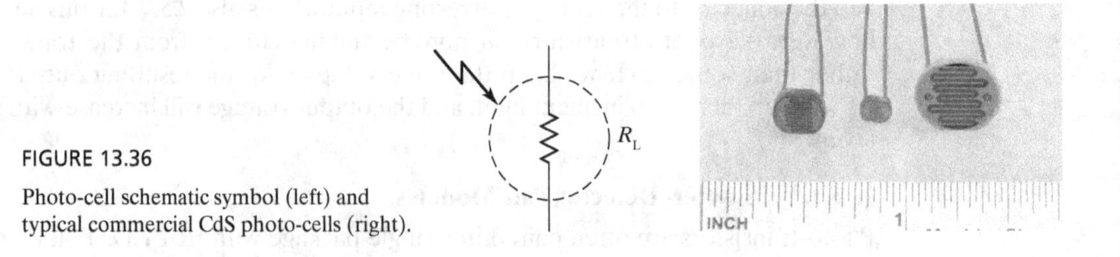

FIGURE 13.36

Photo-cell schematic symbol (left) and typical commercial CdS photo-cells (right).

The **CdS photo-cell** (cadmium sulfide) is among the most common and inexpensive types of photo-cell sensors. Figure 13.37 shows a plot of the normalized spectral response of a typical CdS photo-cell and that of the human eye for comparison. The photo-cell shown is more sensitive in the near IR and extends beyond the human limit of about 750 nm, but is otherwise a close match.

Depending on the amount of light striking the cadmium sulfide sensing material, the resistance may vary from a few kilohms in bright light to several megaohms in complete

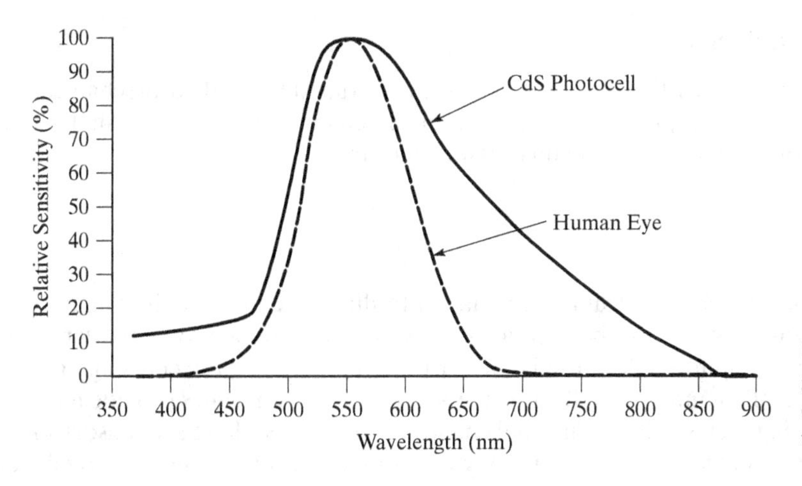

FIGURE 13.37
Spectral response of a typical CdS photo-cell compared with that of the human eye.

darkness. Part-to-part variation is wide for photo-cells, and data sheets usually specify absolute resistance vs. light intensity with wide tolerances. The relative response of each individual sensor, however, is well characterized with respect to its response at reference light levels, usually 10 lux (1 lumen/m^2) and 100 lux. Figure 13.38 shows the response for a typical CdS photo-cell.

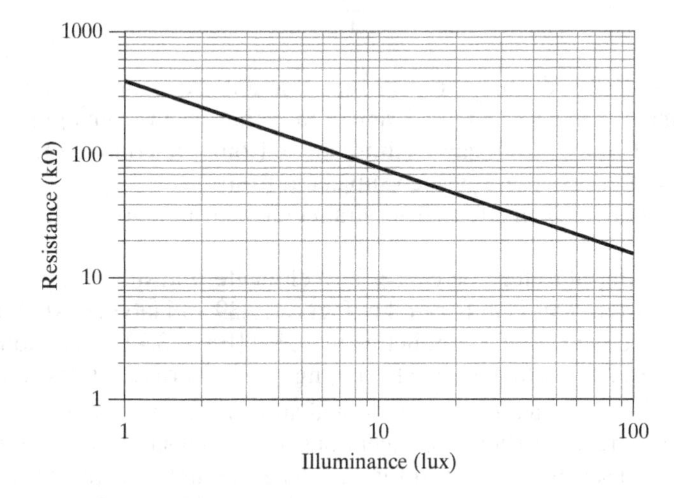

FIGURE 13.38
The change in illuminance vs. resistance for a typical CdS photo-cell.

Interfacing to a photo-cell is straightforward, given that its resistance varies with light levels. Section 13.4.4 introduces a number of methods for interfacing with resistive sensors, any of which is applicable to photo-cells.

Given all the strengths of photo-cells, you are probably starting to suspect that they must have at least one serious weakness, and you'd be right. Photo-cell response times are very slow: a rise time of 50 ms, and a fall time of 25 ms is typical. This compares poorly with photodiodes (~10 ns) and photo-transistors (~10 μs). This makes them unsuitable for medium- to high-speed applications. For lower speed applications, however, they can be a great choice. Examples of devices that use photo-cells to sense light levels include toys (e.g., the Furby), street lights that automatically switch on at dusk/off at dawn, alarm systems, and light meters for photography.

13.5.2 Strain Sensors

Strain is defined as the deformation of a material subjected to mechanical stresses from loading. Strain is expressed as the ratio of the change of length of a material under stress over the original length of the unstressed material:

$$\epsilon = \frac{\ell - \ell_0}{\ell_0} \tag{13.13}$$

Measurements of strain can be taken to directly determine how materials respond to loading, and they can also be used as the basis of measurement of secondary quantities, such as deflection and force. Accurately measuring strain can be quite challenging: in most cases, it is a very small quantity (usually much less than 1%). **Strain gages** provide an excellent means of measuring such small dimensional changes of materials. These are sensors whose resistance changes as a function of the strain they undergo. When strain gages are rigidly attached to a base material (typically with an adhesive such as cyanoacrylate), the strain experienced by the base material is imparted to the strain gage. As long as the coupling between the gage and the base material is secure, the deformation (e.g., strain) of the base material and the gage is the same. Strain gages are conceptually very simple, and consist of little more than sensing elements whose resistance changes slightly when deformed along the sensing axis.

For a material with resistivity ρ, length L, and area A, its resistance is given by Eq. (13.14):

$$R = \frac{\rho L}{A} \tag{13.14}$$

All strain gages take advantage of two fundamental properties of materials that affect the overall resistance: changes to length L and area A that result from applying loads to and deforming a material. Some strain gages (those made from semiconductors) also exhibit significant changes to their resistivity ρ when subjected to strain.

There are two-dimensional effects from strain that contribute to the changes in resistance:

1. Subjecting a material to tension causes it to elongate, and subjecting it to compression causes it to shorten. This is illustrated in Figure 13.39 and labeled ΔL. We see from our expression for resistance of a conducting material [Eq. (13.14)] that elongating a material will increase its resistance, and shortening it will decrease its resistance.
2. When a material elongates, its cross-sectional area decreases. When it shortens, its cross-sectional area increases. (There are exceptions to this, but such materials are exceedingly rare and aren't used to make strain gages.) The amount that the cross-sectional area changes for a given change of length is a quantity called Poisson's ratio [Eq. (13.15)].

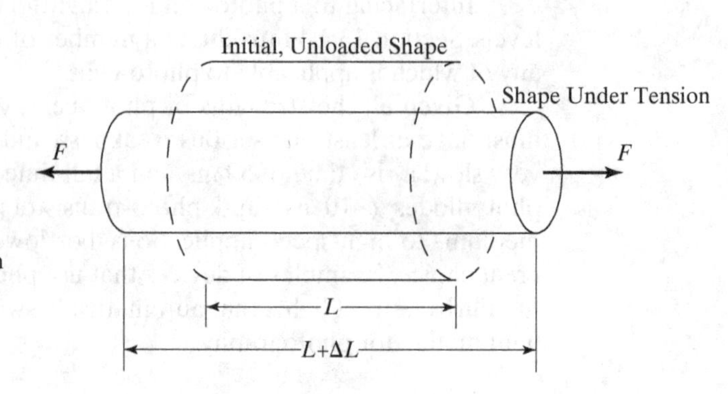

FIGURE 13.39
A material under load, resulting in changes to length (ΔL) and cross-sectional area (dashed vs. solid outlines).

$$\nu = -\frac{\epsilon_{\text{transverse}}}{\epsilon_{\text{axial}}} \tag{13.15}$$

For Poisson's ratio, $\epsilon_{\text{transverse}}$ is the strain in the transverse direction (normal to the direction of loading), and ϵ_{axial} is the strain in the axial direction (the direction of loading). Almost all materials have positive Poisson's ratios, and range from near 0 (cork) to 0.5 (rubber). Figure 13.39 illustrates this effect with dashed and solid lines. Eq. (13.14) shows that decreasing the cross-sectional area A increases the resistance R, while increasing A decreases R.

For all materials with a positive Poisson's ratio, the resistance R of a given length under tension increases because L increases and A decreases, and under compression, R decreases because L decreases and A increases.

The change of the strain gage's resistance relates to the strain of the base material as:

$$GF = \frac{\mathrm{d}R/R}{\epsilon} \tag{13.16}$$

In Eq. (13.16), the term GF is the strain gage's **gage factor**, which is determined by the properties of the materials used to make the strain gage. Metal foil strain gages have constant gage factors that range from 1 to 2, while semiconductor strain gages (e.g., piezoresistive strain sensors) have much higher gage factors, such as 50–100, but the value can vary depending on the strain.

13.5.2.1 Metal Foil Strain Gages

The left side of Figure 13.40 shows the details of how a metal foil strain gage is laid out: the dark regions are a conductive strain-sensing material and the transparent regions are an insulating substrate. On the right, a single-element metal foil strain gage is mounted on a tube in an orientation that allows it to measure tensile and compressive strain.

In practice, strain gages can be finicky. First and foremost, they must be securely bonded to the base material, or else the strain that they undergo and report will not accurately reflect the strain of the base material. Secondly, temperature effects can dominate those of strain. These result from thermal expansion or contraction of the base material relative to the thermal expansion of the strain gage material. Many metal foil strain gages are

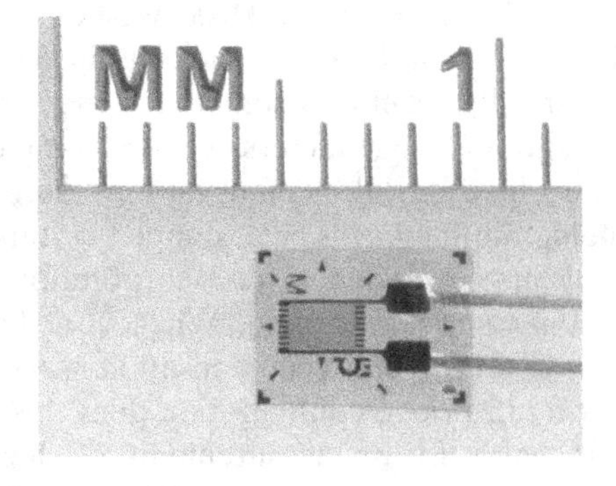

FIGURE 13.40

Single-element metal foil strain gage layout (left), and a metal foil strain gage mounted on a tube to measure strain in tension and compression (right).

made with thermal expansion coefficients that match a particular base material and mostly cancel. For example, a strain gage designed for use with aluminum will compensate for the dimensional changes that occur as the temperature of the base aluminum material changes. If the same gage were mounted instead to a sample of stainless steel, the temperature effects would be pronounced.

In Section 13.4.4.3, we explored the Wheatstone bridge, which is a very good choice as an interface circuit to use with strain gages. We showed that the Wheatstone bridge is best at measuring very small deviations in resistance (a few milliohms) from some relatively large nominal value (a few hundred ohms or kilohms), and allows for the compensation of temperature effects. Both of these characteristics are matched with the challenge of interfacing to strain gages. We also showed that the output of a Wheatstone bridge is the difference between the midpoints of two parallel voltage dividers, so interpreting the output of the bridge involves amplifying small differential voltages with large common mode DC offsets. Instrumentation amplifiers are a very good choice for achieving this, and we cover this in Chapter 14 on signal conditioning.

13.5.2.2 Load Cells

Load cells are sensors that measure force or, stated another way, an applied load. To make a load cell, a material that deforms under load (e.g., anything that exhibits strain) is fitted with strain gages. When the material's properties, geometry, and responses to loading are known, the force that caused the strain may be determined. Load cells most accurately belong in the category of instrumentation, and are rarely seen integrated into mechatronic devices outside of laboratory settings. Since this doesn't fit well with our philosophy for this chapter, we'll describe them only briefly.

Load cells are available in a very wide variety of configurations. Some are capable of measuring forces as low as a few grams, while others can measure millions of kilograms. Load cells are often highly precise sensors, and the normal course of action is to carefully calibrate each one. The resulting cost of each load cell can be substantial: the least expensive examples cost about $50, and the most expensive cost several thousand dollars. The left side of Figure 13.41 shows several common load cell configurations. The right side of the figure shows typical strain gage mounting locations for each type of load cell.

Designers of load cells strive for designs that accurately measure forces in a desired direction (or axis), while optimizing linearity, sensitivity, temperature stability, and rejecting loads in other directions ("off-axis loading"). The shape of the strained element; the way forces are applied to it; and the location, number, and type of strain gages used are all critical to achieving the design goals. The most straightforward load cell designs are those that couple strain gages to a simple strain element such as a disk, cylinder, or tube to measure axial compressive or tensile loads (Figure 13.41, top row). **Disk**, **washer**, and **canister load cells** are examples of this approach. Larger disk and washer load cells have room for, and make use of, metal foil strain gages, while smaller examples sometimes use semiconductor strain gages because of their compact size and high gage factors. A more complex approach is to apply a bending force to the load cell's mechanical element and measure the resulting deflection. **Beam**, **cantilever**, and **parallelogram load cells** are examples of this approach (Figure 13.41, bottom row, middle right). This approach allows for incorporation of several strain gages—some under tension and others in compression. A full Wheatstone bridge can be built into the load cell, bringing the benefits of temperature compensation as well as increasing the sensitivity. For a discussion of how a full-bridge configuration does this, refer to Section 13.4.4.3. The most complex load cells make use of measurements of shear strain to measure the applied force. **Shear beam load cells** are examples of this type (Figure 13.41, bottom row, bottom right). The strain gages are mounted at 45° angles on the neutral axis of a beam, with half the gage above and half below the neutral axis. In this arrangement, strain resulting from bending is rejected, and only the shear strain registers. The advantage of this is that shear strain is not dependent on the location of the loading, so how the load cell is mounted becomes less critical. Load cells using this approach are capable of achieving high load capacities in compact packages with low compliance—all highly desirable characteristics.

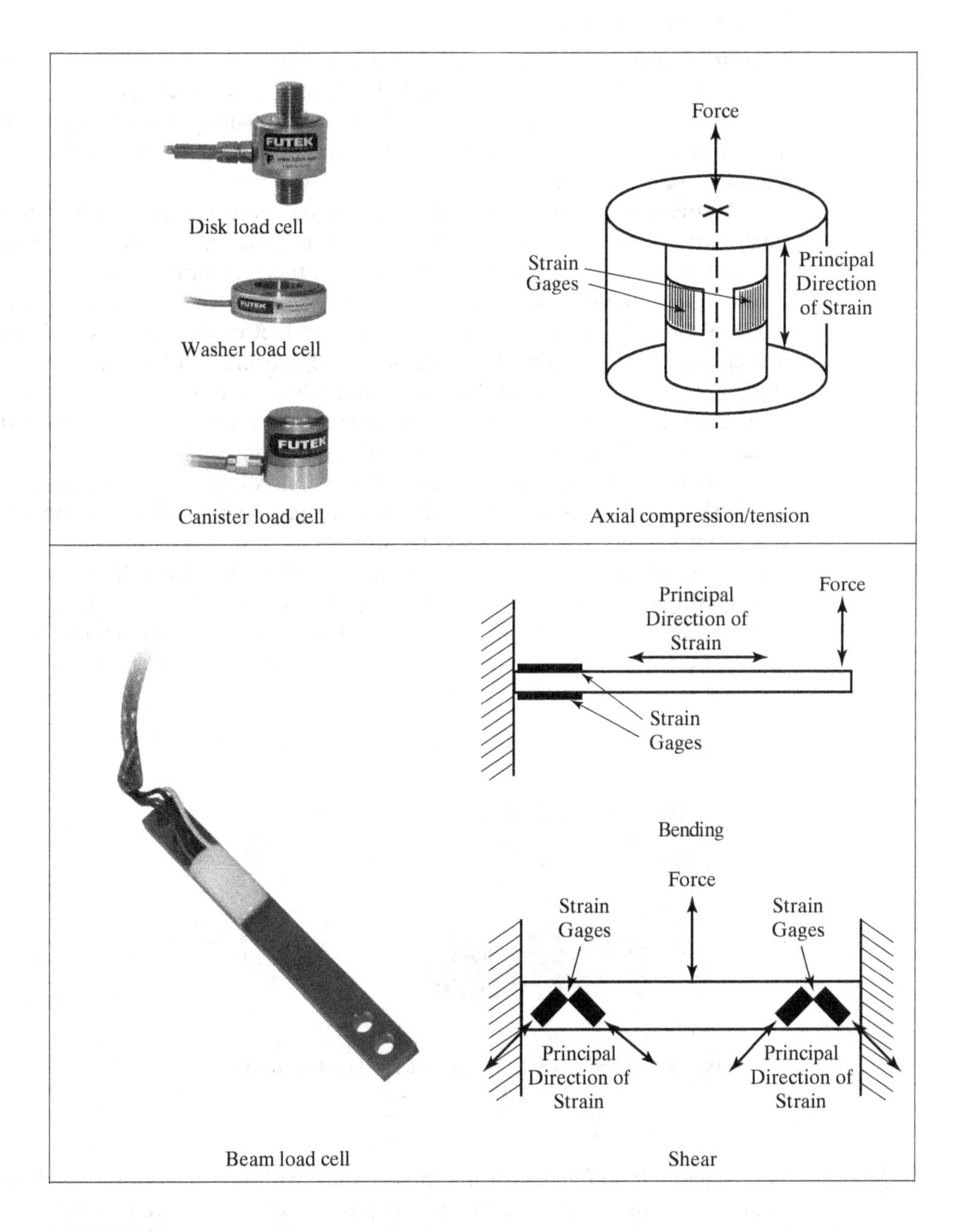

FIGURE 13.41

A variety of commercially available load cells (left) and corresponding strain gage mounting schemes (right). (Courtesy of FUTEK Advanced Sensor Technology, Inc.)

13.5.3 Temperature Sensors

Measurements of temperature are critical to the function of a large number of devices, and temperature transducers are among most commonly used sensors as a result. There are several fundamentally different physical phenomena that allow for accurate and convenient measurements of temperature, and we will discuss three of the most common in this section.

13.5.3.1 Thermocouples

Thermocouples are temperature sensors that make use of the **thermoelectric effect** (discovered by Thomas Seebeck in 1821, and also known as the **Seebeck effect**). The thermoelectric effect occurs when a thermal gradient is established along a conductive material. This results in the development of a small voltage potential between any two points which are at different temperatures along the conductor.

Rather than directly measuring the voltage between a hot point and a cold point on a single conductor, it is much more common in practice to join two dissimilar conductors together (usually by welding them) to form a thermocouple. The junction is sometimes left exposed (as with the bead-type thermocouple), or sheathed in a housing to create a thermocouple probe (Figure 13.42). Selecting two conductors that have very different thermoelectric characteristics results in more sensitive thermocouples. There are several standard combinations of conductors that are commonly used to make thermocouples, and each has been assigned a letter to indicate its construction. Types K (made with chromel and alumel conductors) and J (iron and constantan) are the most often used types. Types E ant T are also common, and there are a great variety of less commonly used types as well (e.g., B, C, N, R, S). The conductors used have different material properties (inertness, magnetic properties, temperature ranges, and so on) and result in different thermocouple sensitivities (in mV/°C), so special applications and requirements may call for the use of a particular type of thermocouple. Uncalibrated thermocouples are typically accurate to within a relatively unimpressive ±2°C, but they can be calibrated (by the manufacturer or by you) to be far more accurate, if needed. Figure 13.43 shows the characteristics for the most commonly used thermocouples: type K, J, E, and T.

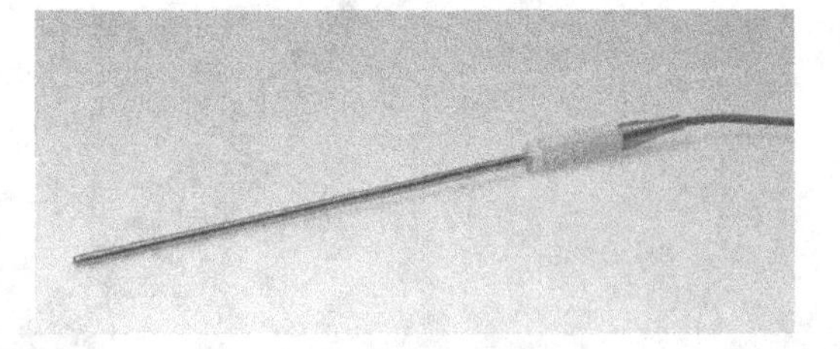

FIGURE 13.42
Type K thermocouples: bead type (left) and probes (right).

One method for measuring temperature with thermocouples involves the use of a reference temperature. Two thermocouple junctions are connected together to form a loop: one junction is held at a reference temperature, such as an ice bath at 0°C and the other junction is exposed to an unknown temperature to be measured. This setup is illustrated in Figure 13.44. If a voltmeter is introduced at any point in between junctions, the voltage measured corresponds to the difference in temperature between the two junctions (see Figure 13.43).

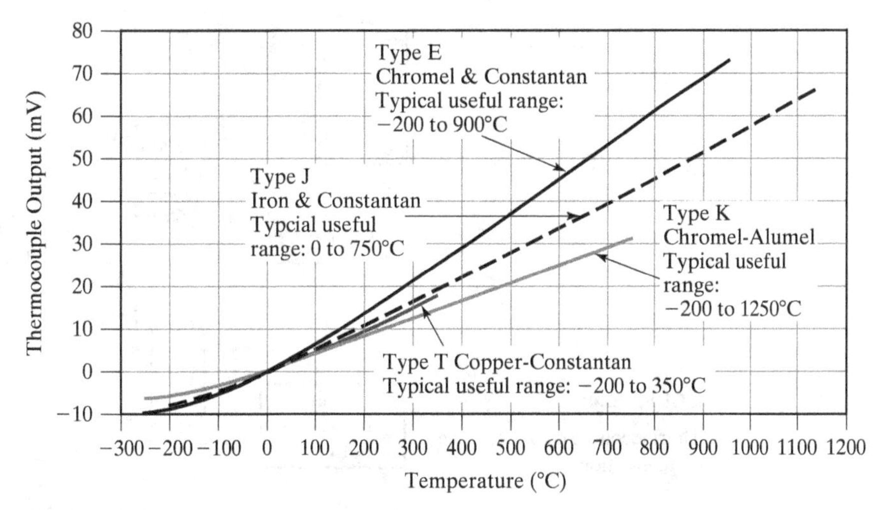

FIGURE 13.43

Selected thermocouple types, construction, and output characteristics [1].

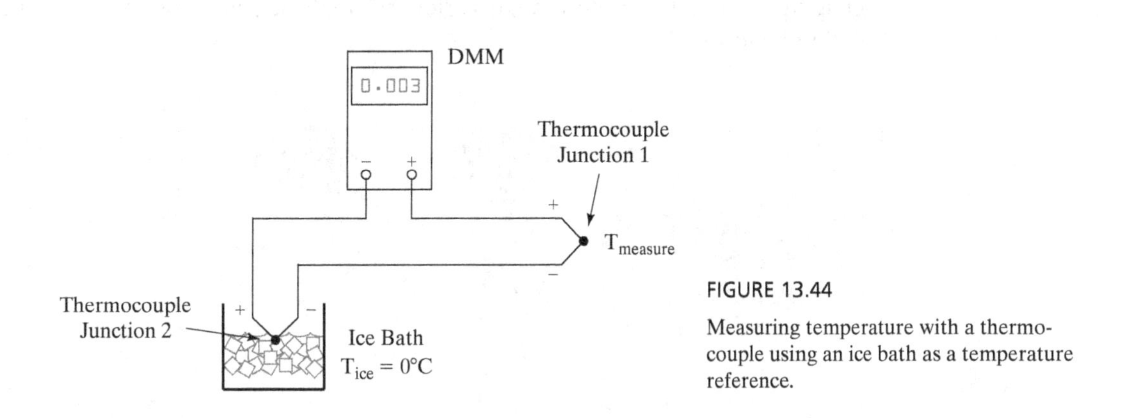

FIGURE 13.44

Measuring temperature with a thermocouple using an ice bath as a temperature reference.

Using the reference temperature approach is straightforward, but in many circumstances it's inconvenient. Another method is to connect the leads of a single thermocouple at a point whose temperature can be measured using another type of sensor (such as a thermistor, discussed below), and then compensate for the difference in temperature of that point to a 0°C reference. This is called **cold reference compensation** and is the scheme used by thermocouple meters, readouts, and data loggers. Figure 13.45 illustrates an application of this approach, where the reference temperature T_{ref} is measured by a thermistor (R_T) shown at the base of the isothermal block that incorporates connection terminals.

13.5.3.2 Thermistors

Another major type of temperature sensor that is distinct from the thermoelectric effect and thermocouples is the thermistor. A **thermistor** is a device whose resistance changes as a function of temperature. In many ways, this is an example of "making lemonade out of lemons." Nearly all devices (including resistors) have some temperature dependency, and this is usually an undesirable characteristic. Here, the temperature dependency is exploited to measure the temperature itself, so it all turns out rather conveniently indeed. Thermistors are made

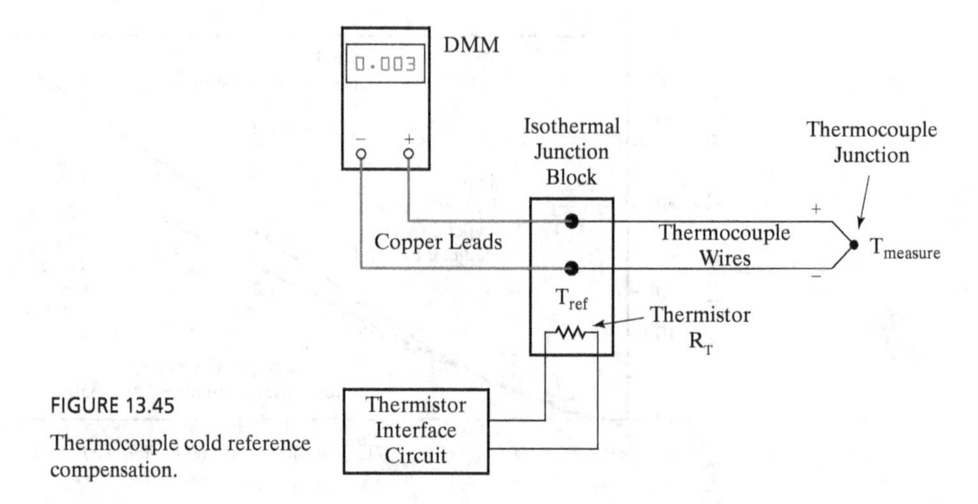

FIGURE 13.45
Thermocouple cold reference compensation.

from ceramic and polymer materials, and they are different from **resistive temperature detectors (RTDs)**, which are made from conductive metals such as platinum and typically have lower sensitivity and more linear response. Thermistors are available in a wide range of packaging options targeting a huge variety of applications. A sampling of options is illustrated in Figure 13.46.

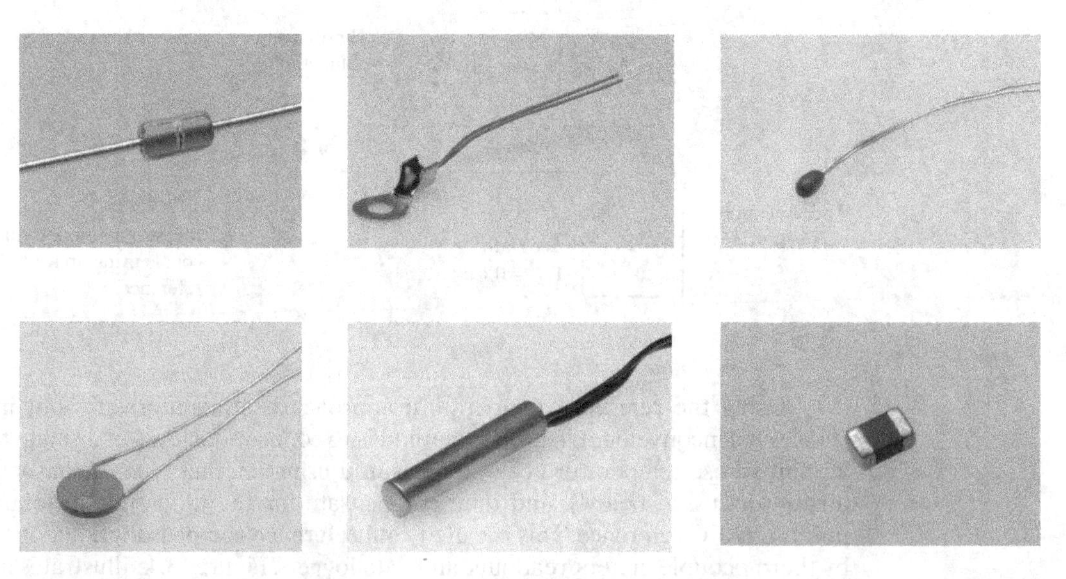

FIGURE 13.46
A variety of thermistor packaging options.

Functionally, thermistors are divided into two major categories depending on the slope of the relationship between temperature and resistance. The first category is **negative temperature coefficient** (or **NTC**) thermistors, which have a resistance that decreases with increasing temperature. A representative response curve for an NTC thermistor is shown in Figure 13.47a (in linear coordinates) and Figure 13.47b (semi log). A high degree of nonlinearity is readily apparent, and conversions between resistance and temperature require a

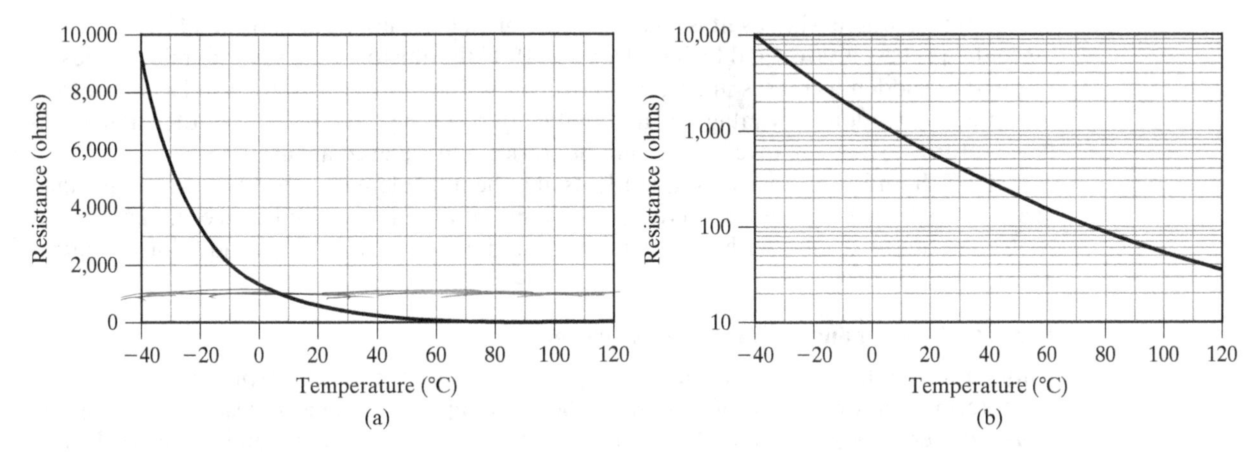

FIGURE 13.47

Temperature vs. resistance characteristics for a typical NTC thermistor.

calculation from a transfer function or lookup table supplied by the manufacturer, or a calibration curve generated by the user.

While thermistors with negative temperature coefficients are commonly used to measure temperature, thermistors with a **positive temperature coefficient** (or **PTC**) are not. Their resistance is characterized by a relatively flat, stable region at lower temperatures, a transition point, and then a region of high resistance at higher temperatures (see Figure 13.48). This behavior lends itself well to applications in current limiting and overload protection.

PTC thermistors are very useful as resettable fuses. In this type of application, they are placed in series with an electrical load, such as a power supply or a motor. When the current flowing through the load is within allowable limits, the internal heating from power dissipation within the thermistor (according to $P = I^2R$) is low, and the resistance of the PTC thermistor is low (usually between a few ohms and a few hundred ohms). However, when the current exceeds the allowable range, the PTC thermistor is internally heated to a much higher temperature, and the resistance increases significantly. When the current increases due to a fault condition, this causes increased internal heating within the PTC thermistor, which in turn causes the resistance of the device to increase greatly—usually by several

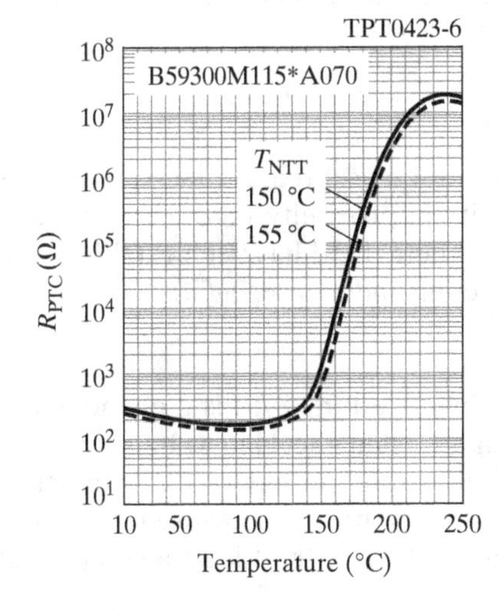

FIGURE 13.48

Resistance characteristics of a family of PTC thermistors (the EPCOS B59300M115*A070 family). (Courtesy of EPCOS CC TD.)

orders of magnitude. The result is a large reduction in the amount of current allowed to flow. Since the process is governed by heat transfer, PTC thermistors take several milliseconds to respond to sudden increases in current above allowable limits. But when used as fuses, they have the tremendous advantage of automatically resetting: after they cool off, their resistance returns to a much lower value and the device may be used again.

PTC thermistors are also frequently used as heating elements. Because their resistance increases as their temperature increases, the amount of current—and hence the amount of heat generated—is inherently self-regulating, and requires little or no additional control effort.

13.5.3.3 Semiconductor Temperature Sensors

Another example where a device's temperature characteristic is exploited to create a temperature sensor is that of a **semiconductor temperature sensor**, also called a **diode thermometer**. A popular example of this type of device, the National Semiconductor LM35, uses changes in V_{BE} (the voltage drop between a BJT's base and emitter terminals), which is a function of temperature. The LM35 (shown in Figure 13.49) costs less than $2 in low quantities, has a temperature range of −55°C to 150°C, is accurate to ±0.5°C at 25°C, and operates on any power supply from 4 to 30 V. Its transfer function couldn't be more convenient: V_{out} = 10 mV/°C. Using the LM35 is exceptionally straightforward: power the device, and read temperature in °C as a voltage at the output pin (scaled by a factor of 0.01).

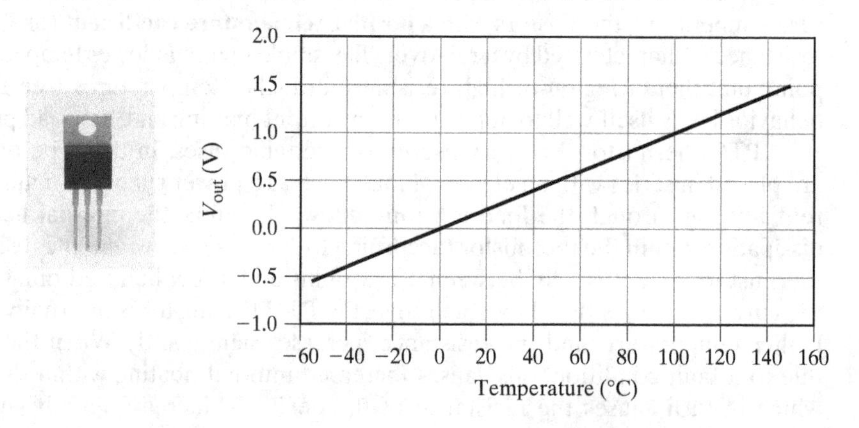

FIGURE 13.49

The LM35 from National Semiconductor, a semiconductor thermometer-based temperature sensor. (Courtesy of National Semiconductor Corporation.)

13.5.4 Magnetic Field Sensors

Measurements of magnetic fields can be useful in a wide variety of applications. Examples include the determination of position, proximity or orientation, and the use of magnetic media for data storage. Once a magnetic field has been established or identified, there are several common methods for detecting and measuring it.

13.5.4.1 Hall Effect Sensors

The Hall effect (discovered by Edwin Hall in 1879) is a phenomenon in which a magnetic field induces a measurable voltage (called the Hall voltage) across a conductor through which a current is flowing. This is illustrated in Figure 13.50: the current flowing across the conductor (into the page, as shown) is warped and caused to deviate from a straight path across the conductor by the magnetic field with at least a component perpendicular to the

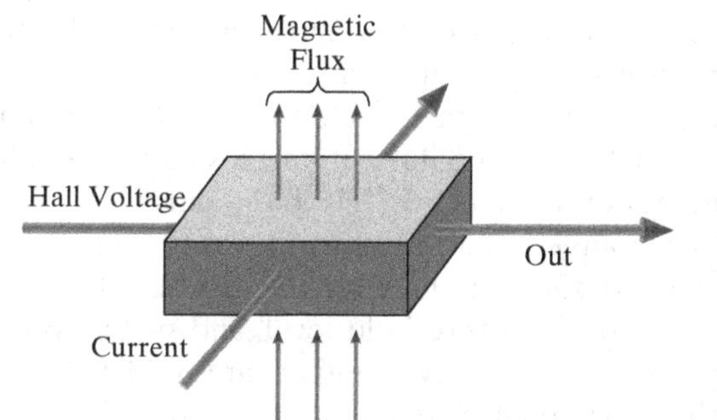

FIGURE 13.50

The Hall effect, in which a magnetic field creates a voltage across a conductor carrying a current.

current flow (from bottom to top, as shown), and this displacement results in the development of the Hall voltage (shown left to right). The Hall effect occurs for currents flowing across metallic conductors and semiconductors, and the magnitude of the Hall voltage depends on the magnetic flux and the properties of the conductive material. Hall effect sensors typically have a range of hundreds to thousands of gauss, so they are most appropriate for measuring strong magnetic fields.

Hall effect sensors are available with analog or digital outputs. Since the Hall effect results in small voltages, it is common to integrate a Hall sensor and amplifier together into a single package, and the result is known as a **linear output Hall effect transducer**, sometimes called a **LOHET**. A representative linear Hall sensor, the Allegro A1386LLHLT-T, and its characteristic output are shown in Figure 13.51. Linear Hall sensors are bipolar, meaning that they are capable of sensing both north and south magnetic polarities, and their output is proportional to the magnetic flux detected. The integrated amplifier helps minimize the effects of nearby noise sources by buffering and amplifying the small signals from the Hall element immediately adjacent to the sensing material.

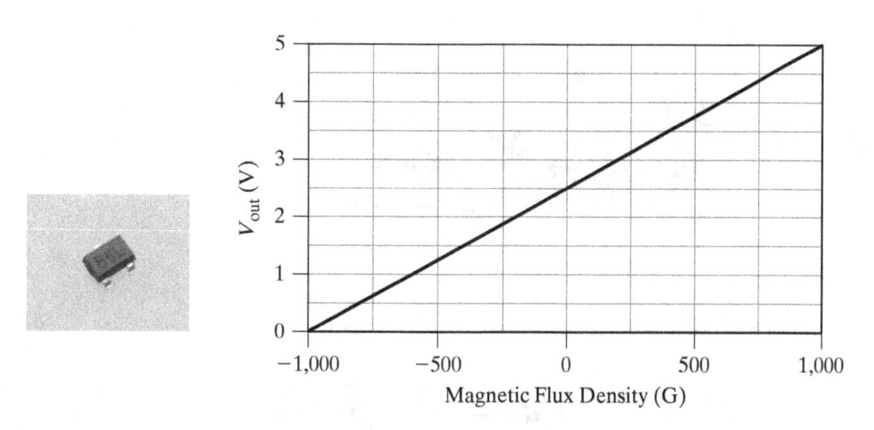

FIGURE 13.51

The Allegro A1386LLHLT-T, a linear output Hall effect sensor with programmable sensitivity.

In practice, magnetic field strength is highly nonlinear with distance, since it varies as $1/d^3$ (where d is the distance from a magnetic pole). One implication of this is that it isn't practical to use the simple arrangement between a sensor and a magnet illustrated in Figure 13.52a to measure distances greater than a few millimeters: the magnetic field strength very quickly decreases to immeasurable levels. Instead, there are other arrangements that allow for measurements of larger displacements, such as those shown in Figure 13.52b and c. Linear Hall sensors are also useful for measuring angles and continuous rotation. Ultimately, applications for these sensors are limited only by the magnetic field strength and your imagination.

Hall sensors are also available with digital outputs. Instead of coupling a Hall sensing element with a linear amplifier, as is done in a LOHET, the sensing element in a digital Hall "switch" is amplified and then coupled to a Schmitt trigger to create the digital output and add hysteresis. Hall switches are designed to take on one digital state (usually off) when the applied magnetic flux density is below a threshold level, and switch to the opposite digital state (usually on) when the flux density is above another threshold. Because the Schmitt trigger implements hysteresis, the transition back to the original state will occur at a lower threshold. The hysteresis serves its usual purpose in this application: it prevents multiple crossings of a single threshold level from causing multiple transitions of the output state.

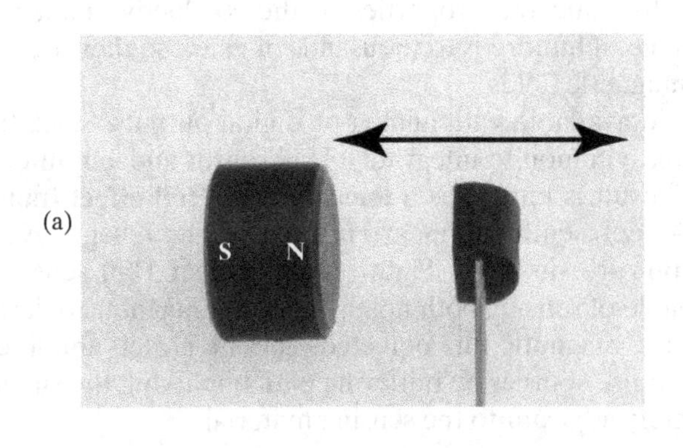

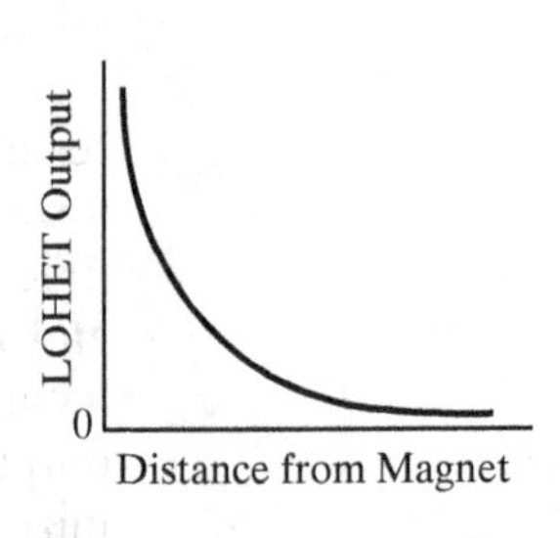

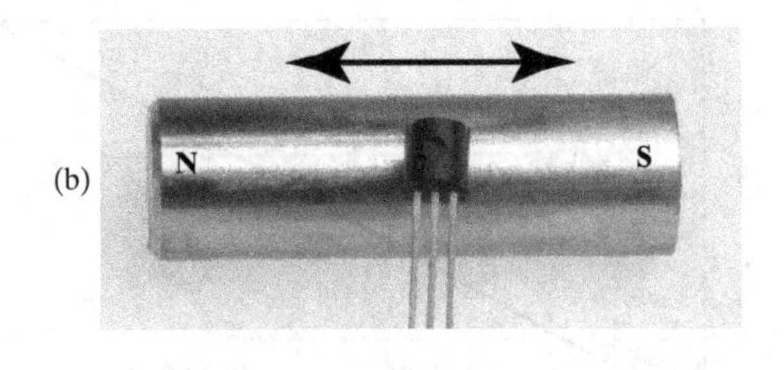

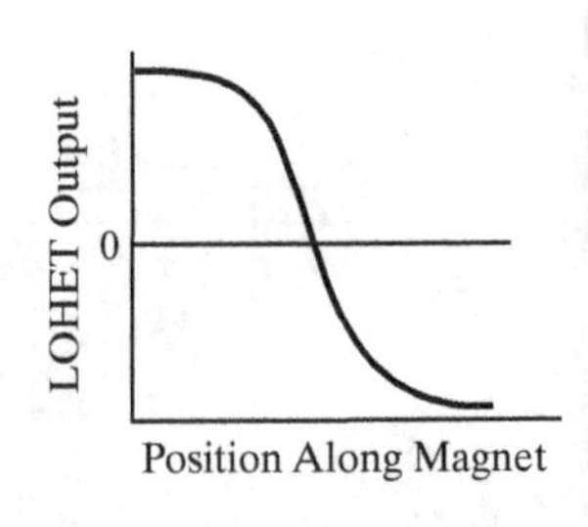

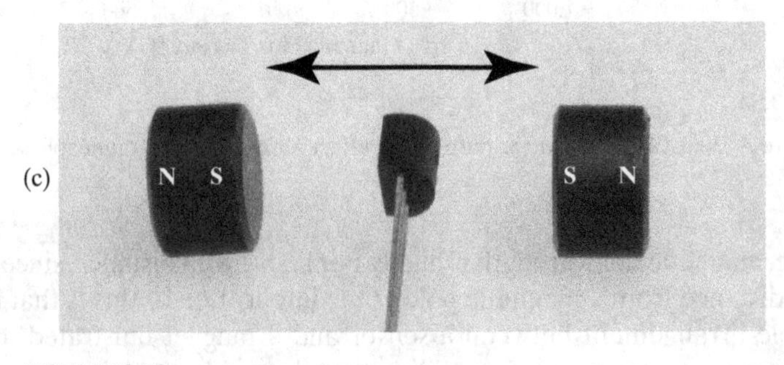

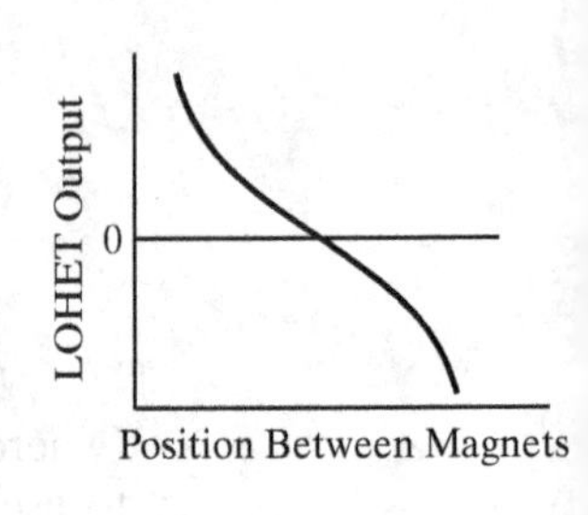

FIGURE 13.52
Methods for measuring distance with a linear Hall effect sensor.

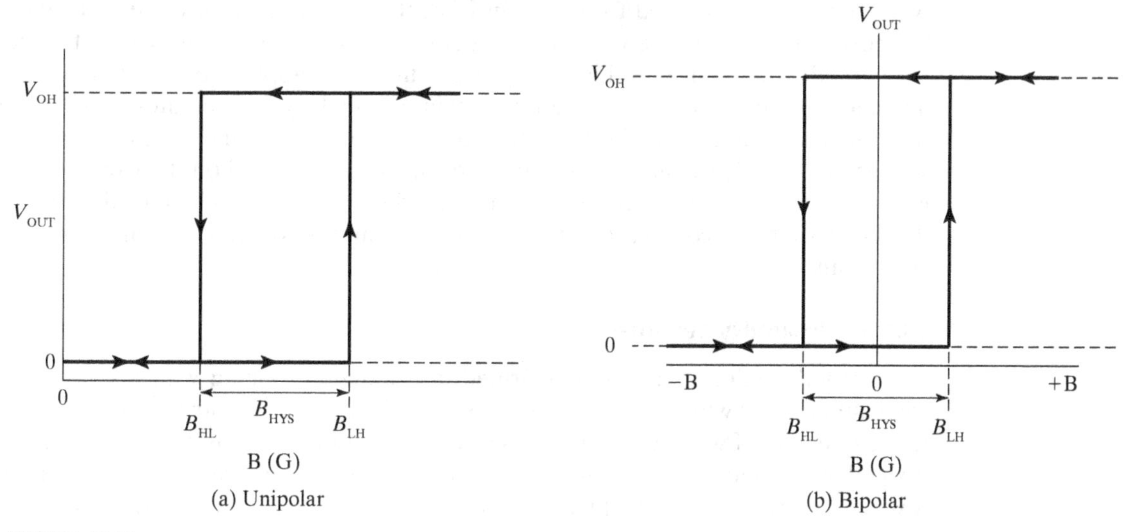

FIGURE 13.53

Unipolar and bipolar digital Hall sensor trigger levels and hysteresis.

Digital Hall sensors can be configured to be unipolar, so that a single polarity and magnitude of magnetic flux density is required to trigger a change in the output state (Figure 13.53a). Alternatively, they can be configured to be bipolar, where one magnetic polarity and magnitude of magnetic flux density is required to cause a change of state, and the other magnetic polarity at another magnitude of flux density is required to change it back (Figure 13.53b). Bipolar digital Hall sensors are also referred to as "latched" since they require the opposite magnetic pole to change states. The field that caused them to take on the current state may go all the way to zero and the switch will retain the state.

13.5.4.2 Reed Switches

Reed switches are perhaps the simplest of magnetic sensors. They consist of a set of contacts, at least one of which is ferromagnetic. In the presence of a strong magnetic field (which can be supplied by either a permanent magnet or an electromagnet), the contacts are either drawn together to close a normally open switch, or pulled apart to open a switch that is normally closed. When the magnetic field is removed, the contacts revert to their normal unactivated state. As the name suggests, the shape of the contacts is reed like (long and thin), which imparts the spring tension that allows the contacts to assume their normal, inactive state (Figure 13.54). Of course, switching tasks can be, and often are, performed with Hall sensors and appropriate interface circuitry; however, such solutions are more complex and correspondingly more costly. A reed switch is a simple way to detect the presence or absence of a strong magnetic field with minimal expense. As of mid-2008, a simple SPST NO reed

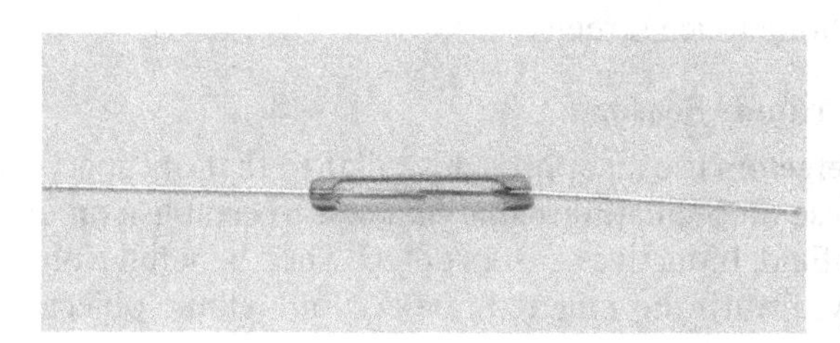

FIGURE 13.54

A reed switch, showing a set of contacts in a hermetically sealed glass capsule.

switch could be purchased for less than 25 cents in high quantities (1,000s or more). When low cost and simplicity are strict requirements, reed switches are an excellent choice.

Reed switches are used in a very large number of applications, including automobiles, home appliances, and toys. One familiar example is a bicycle computer, which senses wheel rotation and sometimes pedal rotation with reed switches. A small magnet is mounted to the spokes of a bicycle's front wheel, and a reed switch is mounted on the front fork so that the magnet passes very near the switch as the wheel rotates. Each time the magnet passes the reed switch, the contacts close and the cycle computer calculates quantities such as speed and distance.

13.5.5 Proximity Sensors

Proximity detection is usually performed in the digital domain with a sensor that detects when an object is within the range of the sensor. Proximity sensors give a "yes or no" answer to the question of whether an object is present or absent. Usually, the range of these sensors is small: rarely more than a few centimeters. A common application of proximity sensing is when a motorized stage in a piece of automated equipment reaches the home, or zero, position. Several of the sensors we've already described above are used for proximity detection, and are commonly used to detect the home position for motion stages. These include contact switches, microswitches, opto-interrupter emitter–detector pairs, reflective emitter–detector pairs, Hall switches, magnetoresistive switches, and reed switches. To round out the discussion, we present three other commonly used proximity sensors.

13.5.5.1 Capacitive Proximity Sensors

Capacitive proximity sensors detect changes in the capacitance between two plates of a capacitor when an object moves close enough to measurably disturb the electric field. An example of a **capacitive proximity detector** is the Omron E2K-X, which is shown in Figure 13.55. This noncontact sensor can detect a wide variety of metallic and nonmetallic objects like glass, wood, water, and plastic at distances ranging from 3 to 30 mm. These characteristics make it a good choice for use in automated production lines.

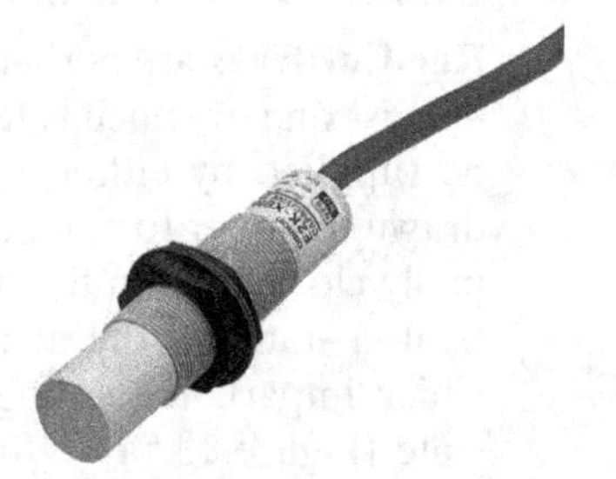

FIGURE 13.55

A capacitive proximity sensor (Omron E2K-X). (Courtesy of Omron Electronics, LLC.)

Capacitive proximity sensors have also become very popular in user interfaces. Capacitive touch sensors capable of detecting the presence of one or more fingers are now commonplace in a wide variety of applications, such as computer touch pads, mp3/media player controls, and mobile phone touch screens.

13.5.5.2 Inductive Proximity Sensors

Inductive proximity detectors use an approach similar to that of capacitive proximity detectors, except that they use an oscillating signal in a coil to establish an electromagnetic field rather than an electric field. Inductive sensors detect when a metallic object has moved close enough to measurably disturb the magnetic field. Nonmetallic objects do not disturb the field, and are not detected by inductive sensors. Figure 13.56 shows a typical inductive sensor, the Omron E2F, which closely resembles a capacitive proximity sensor. This sensor can detect metallic objects that are up to 8 mm away.

FIGURE 13.56
An inductive proximity sensor (Omron E2F). (Courtesy of Omron Electronics, LLC.)

13.5.5.3 Ultrasonic Proximity Sensors

Ultrasonic proximity detectors emit high-frequency sound waves and measure the amount of time required for reflections to return. Shorter return times indicate that an object has moved into close proximity. Figure 13.57 shows the Omron E4C-DS100 ultrasonic proximity sensor, which has a range of 9 to 100 cm. Any object that reflects at least a portion of the ultrasonic pulse will be detected, regardless of the material. Note that this sensor also closely resembles the capacitive and inductive sensors discussed above, but that the piezo element at the distal tip hints that this is an ultrasonic sensor.

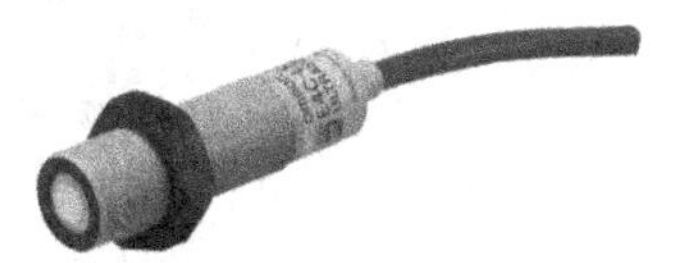

FIGURE 13.57
An ultrasonic proximity sensor (Omron E4C-DS100). (Courtesy of Omron Electronics, LLC.)

13.5.6 Position Sensors

Distinct from the task of reporting whether or not an object is within range of a sensor is the task of determining the distance from a sensor to a location or the distance that something has moved. These are the jobs of **position sensors**, also called **displacement sensors**. In this section, we present a selection of the most commonly used linear and angular position sensors.

13.5.6.1 Potentiometers

One of the simplest position sensors is the humble potentiometer. A **potentiometer**, often referred to in the diminutive as a **pot**, is a three-terminal device with a fixed resistance between two terminals, and the third terminal is connected to a movable **wiper** that contacts the resistance somewhere midspan. This creates two adjustable resistances that depend on the relative position of the wiper between the two ends of the resistance. The position of the wiper may be determined by measuring either of the variable resistances. The schematic symbol for a potentiometer is particularly effective at illustrating its construction and function (Figure 13.58).

Recall from the discussion in Section 13.3.2 that we chose a potentiometer as a simple means of determining the displacement of a mass on a spring to infer acceleration. In general,

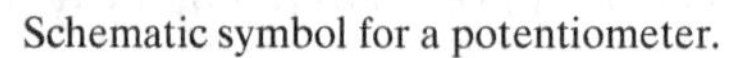

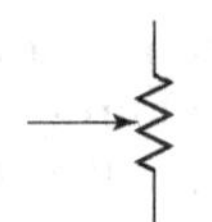

FIGURE 13.58

Schematic symbol for a potentiometer.

potentiometers are not particularly high-performance position measurement sensors, as they add frictional loads to the system being measured, and are subject to temperature effects, vibration, and long-term drift resulting from wear and tear. However, they are very inexpensive and readily available in a wide variety of packaging options and resistance values. They are often the first choice to consider for position measurements where requirements aren't highly demanding. Potentiometers are currently used to determine position in a great many applications. For example, rotary and linear pots (Figure 13.59a and b) are used as user input devices on stereo equipment. Potentiometer-based joysticks are used in video game console user interfaces (Figure 13.59c). Another interesting and useful variant is the **cable-driven linear displacement transducer** (more commonly called a **string pot**), which is comprised of a precision multiturn rotary potentiometer whose input shaft is attached to a constant tension torsional spring and a spool holding a long string. This device shown in Figure 13.59d is made by UniMeasure, Inc. The resistance of the string pot output is proportional to the length of string that has been pulled from the spool. This is a simple and effective means of measuring relatively large distances (up to about 3 m/10 ft is practical) at low cost with impressive accuracy (around 0.5%), repeatability (0.05%), and linearity (0.5%).

(a) Rotary potentiometer

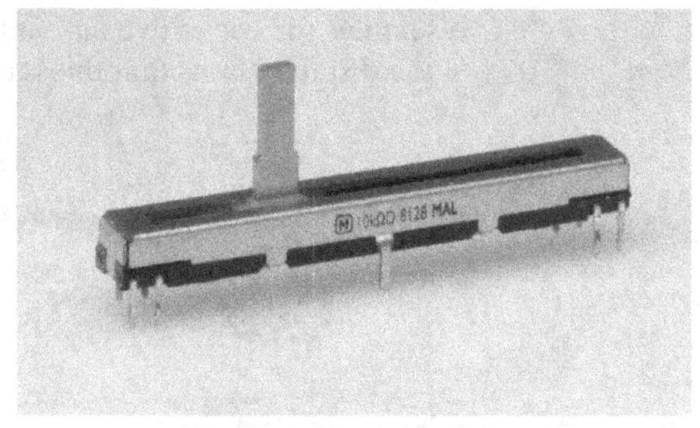

(b) Linear potentiometer, or slide pot

(c) Joystick

(d) Cable-driven linear displacement transducer, or string pot

FIGURE 13.59

A selection of potentiometers.

When used to measure position, potentiometers are in effect **absolute position sensors**: the output corresponds to the location of the wiper and does not require a separate position measurement to be used as a relative reference or serve as the origin. This contrasts with **relative position sensors**, which measure incremental motion away from an initial position.

13.5.6.2 Optical Encoders

A very popular type of position sensor that is available in both relative and absolute versions is the **optical encoder**. The most common approach used in commercially available devices is illustrated in Figure 13.60. Light from an LED is directed toward an optically opaque material such as a thin stainless steel disk that has a series of holes in it. When the disk rotates relative to the light source, a photo-diode on the opposite side detects transitions in the levels of light that occur as the light is alternately blocked or allowed to pass through a hole. Counting the transitions enables the user to track any rotation that occurs. Because of the incremental nature of these measurements, these sensors are often called **incremental encoders**. Very high resolutions are possible with this approach, as the holes can be made to be very tiny and closely packed through the use of manufacturing techniques such as chemical etching or laser cutting. Resolutions of better than 500 counts per turn are readily available.

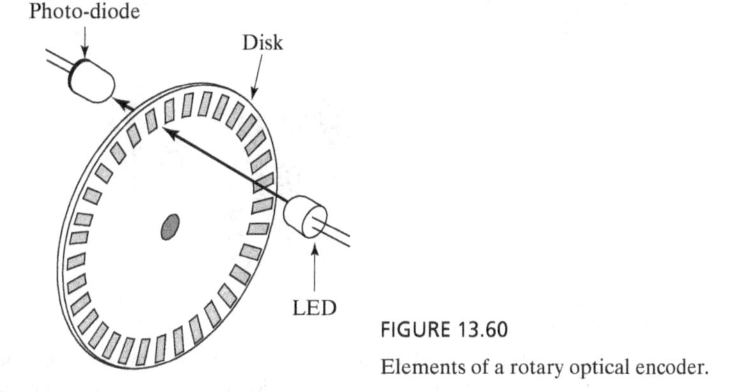

FIGURE 13.60

Elements of a rotary optical encoder.

Quadrature encoding can be used to determine the direction of motion or rotation. By using two photo-diodes (labeled A and B in Figure 13.61) to detect light transitions and offsetting them by half the spacing between holes so that one sensor is blocked when the other is illuminated, the relative phase of the resulting signals indicates the direction of motion.

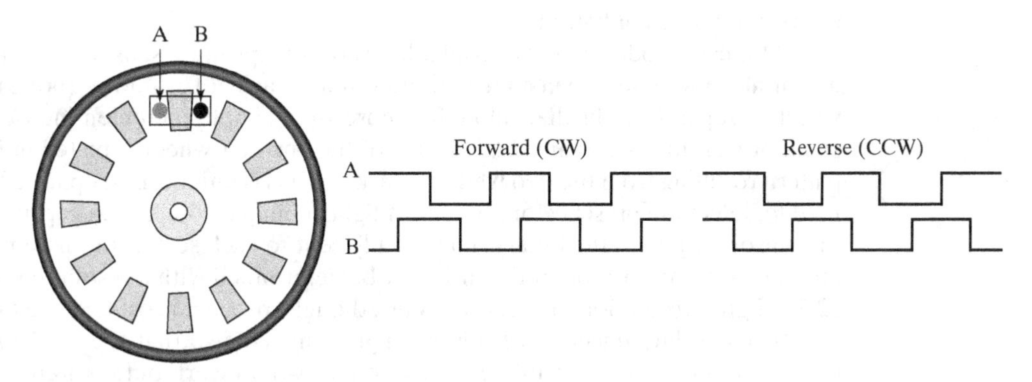

FIGURE 13.61

Quadrature encoding scheme. Left: An encoder wheel with two sensors offset by half the spacing between holes. Right: The output waveforms from each sensor.

The right side of Figure 13.61 shows the wave trains that result from clockwise rotation and counterclockwise rotation. For clockwise rotation, the rising edge of the Channel A precedes the rising edge from Channel B. For counterclockwise, the opposite is true: Channel B's falling edge precedes Channel A's. As a bonus, quadrature encoding increases the resolution of the sensor by a factor of four. For each hole in the encoder disk, a total of four transitions (rising or falling edges) occur for Channels A and B. An optical encoder with quadrature encoding that has 500 holes in the encoder wheel has a resolution of 2000 counts per rotation.

Figure 13.62 shows a typical commercially available optical encoder mounted to a brushed DC motor in a typical assembly (left) and disassembled to show the internal components (right, and below).

FIGURE 13.62

Top row: A DC motor with encoder (left) and a view of the mounted encoder with the dust cover removed (right). Bottom: DC motor, encoder wheel, sensor assembly and dust cover.

Reflective optical encoders are also available. Reflective encoders bounce the light from an LED off a surface that has alternating reflecting and nonreflecting regions, and use a photo-diode to sense the level of reflected light. Both varieties of encoders are available in rotary and linear configurations.

Optical encoders are also available as absolute position sensors. One method for creating an **absolute optical encoder** is to add more tracks of windows (holes) to the encoder wheel at separate radii that allow for more than simple incrementing or decrementing a count of transitions. For any 22.5° sector of the encoder wheel depicted in Figure 13.63, the pattern resulting from the two white vs. dark regions is unique. If a separate light sensor were used to detect the presence or absence of light shining through each gap, the combination of the sensor outputs is the binary number adjacent to each sector. In this simple example, the absolute angular position of the disk may be determined with a resolution of 1 part in 16, or 22.5°. Higher resolutions are also possible, though space and cost become limiting factors.

For absolute encoder wheels, arranging the slots so that the tracks simply count up or down in binary is an intuitive approach that would work but is rarely used because two or more bits of data would transition simultaneously at several locations. Very nearly, but not quite, simultaneous transitions are about the best we could hope for, and additional logic and error checking would be required to correctly interpret this. To avoid this, the

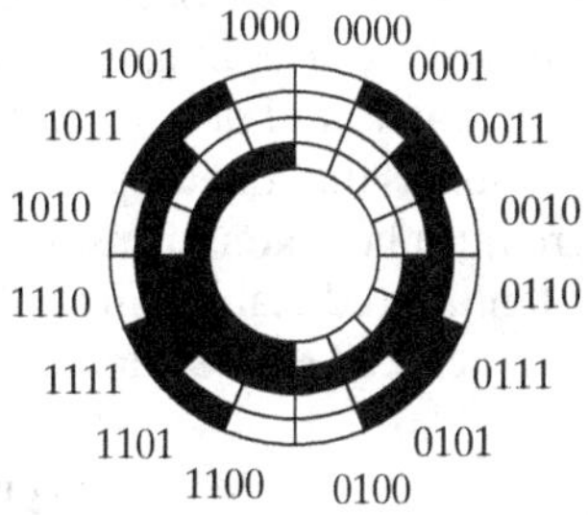

Position #	Gray Code
0	0000
1	0001
2	0011
3	0010
4	0110
5	0111
6	0101
7	0100
8	1100
9	1101
10	1111
11	1110
12	1010
13	1011
14	1001
15	1000

FIGURE 13.63

A Gray code encoder wheel with 22.5° resolution for absolute position measurements.

code is rearranged so that only one bit changes at a time. This approach is called **Gray code encoding**.

13.5.6.3 Inductive Pickups/Gear Tooth Sensors

Some applications call for measuring the rotation of a rotating gear or shaft, and this can be done with a sensor called an **inductive pickup**, also known as a **gear tooth sensor**. This scheme is popular in the automotive industry to determine crankshaft position and rotational speed (Figure 13.64). Because these sensors tend to be inexpensive and robust, as well as noncontact and aren't optical (they don't have to stay clean to work), they are used in a wide variety of applications across many industries, even though interfacing effectively to inductive pickups can pose some challenges, as we describe below.

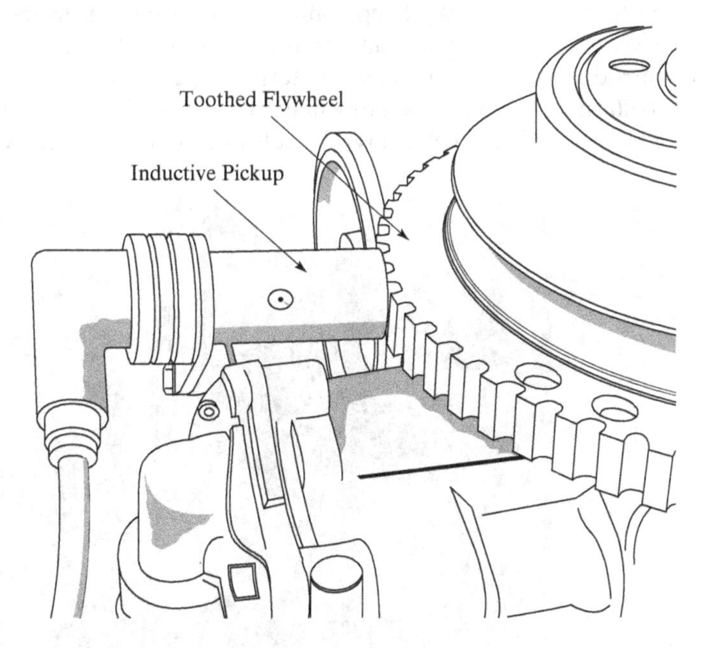

FIGURE 13.64

An inductive pickup used as a crankshaft position sensor in an automobile engine.

Inductive pickups consist of a magnet whose magnetic field is directed through the center of a coil by a flux concentrator (Figure 13.65). When placed in proximity to a ferrous material with an air gap that varies as the ferrous material moves, such as a rotating gear or a pulley with a groove in it, voltage spikes occur across the leads of the coil as shown in Figure 13.65 because of the changes induced in the magnetic field. The magnitude of the output voltage increases as the speed of the ferrous material increases, and also when the air gap is reduced. One of the challenges of interfacing to these sensors is that the output voltages can vary greatly—variation in the 10s of volts is typical—so, care must be taken to ensure that the interface electronics function properly across a wide and often unpredictable range of input voltages, and that the electronics aren't damaged by input voltages well above and below the power supply rails.

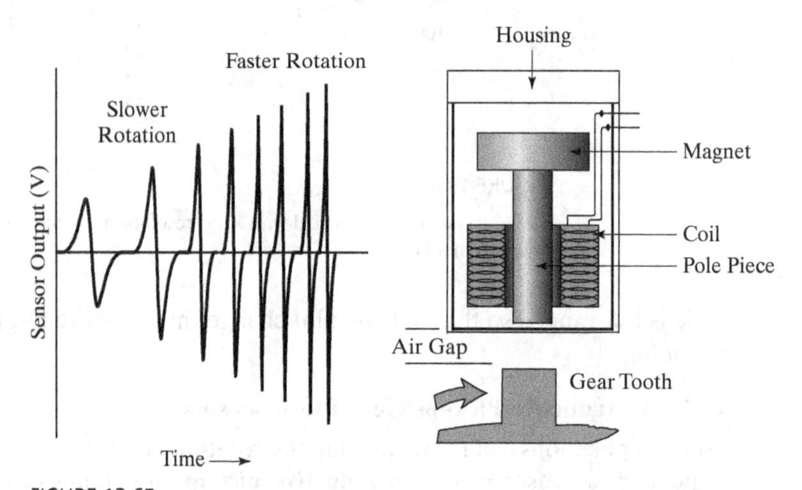

FIGURE 13.65

Construction of an inductive pickup sensor and its characteristic output signal.

13.5.6.4 Reflective Infrared Sensors

We introduced reflective IR sensors in Section 13.5.1.3 and also described their use as proximity detectors in Section 13.5.5. We keep coming back to these sensors because they are compact, inexpensive, easy to integrate, and useful in a very wide variety of applications. In this section, we introduce them as a means of determining the distance to a target.

Figure 13.66 illustrates a typical experimental setup: the emitter and detector are side-by-side in the sensor housing. When an object is relatively close to the emitter (within a centimeter

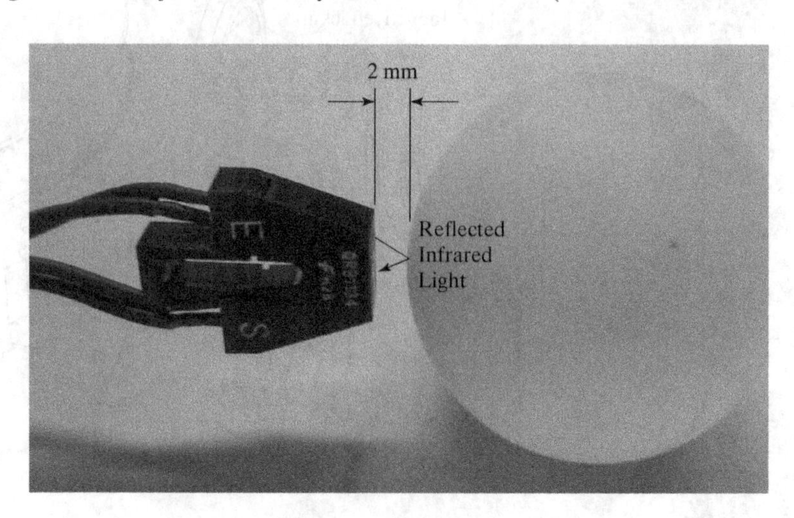

FIGURE 13.66

Using a reflective IR emitter–detector pair sensor to determine distance to a target.

or so), some of the light from the emitter is reflected by the object and sensed by the detector. The amount of light reflected, and hence the amount of collector current that flows, depends on the optical properties of the object and the distance. If the optical properties are known and consistent, then the output depends simply on the distance.

In cases where the system is well characterized and the distances are small, the phototransistor's collector current will exhibit characteristics similar to that shown in Figure 13.67, which is for the OPTEK® OPB700. When an object (in this case, a specific type of paper, label, or reflective material) is at a distance of 0.1–0.13 in. or less from the sensor, the collector current increases from zero to its maximum value (the plot is normalized). For distances beyond about 0.11 in. (depending on the reflective properties of the material) and up to 1 in. the output drops from the maximum value to about 5% of that value. The response in the region beyond 0.11 in. isn't linear, but the curvature is limited and a linear assumption would be reasonable for small variations in distance.

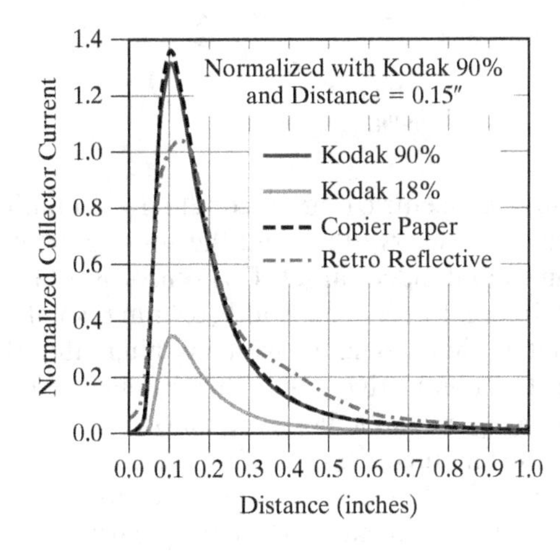

FIGURE 13.67

Normalized collector current vs. distance to various targets for the OPTEK OPB700. (Courtesy of TT Electronics OPTEK Technology.)

13.5.6.5 Capacitive Displacement Sensors

Capacitive displacement sensors provide another means of measuring the distance to an object when that distance is small. In this case, *very* small: capacitive sensors are capable of achieving subnanometer resolution, with a typical maximum range of less than a centimeter. A capacitive displacement sensor is created whenever two conductive objects are brought into close proximity, since this is all that's required to create a capacitor. Recall from Chapter 9 that a simple plate capacitor is made by placing two conductive plates in close proximity. The equation for the resulting capacitance is:

$$C = \frac{\epsilon\epsilon_0 A}{d} \tag{13.17}$$

If we hold the contributions from a dielectric material (the $\epsilon\epsilon_0$ term) and the shared area of the conductors (A) constant, then the capacitance depends only on the distance between the conductors, d. In a simplistic sense, a capacitive displacement sensor consists of one half of a simple plate capacitor, the target makes up the other plate, and the resulting capacitance depends on the spacing between them. Measuring the capacitance is all that's required to determine the distance to the target, and we've discussed several methods to measure capacitance in Section 13.4.6. Subnanometer resolution measurements require an extremely sensitive and low-noise capacitance measurement, which is a difficult challenge. As a result, higher precision measurement requirements are often best met with equipment purchased from companies focused on capacitive sensing equipment. Lower precision measurements can be made with circuits that we've introduced above. Figure 13.68 shows how a typical capacitive displacement sensor is arranged.

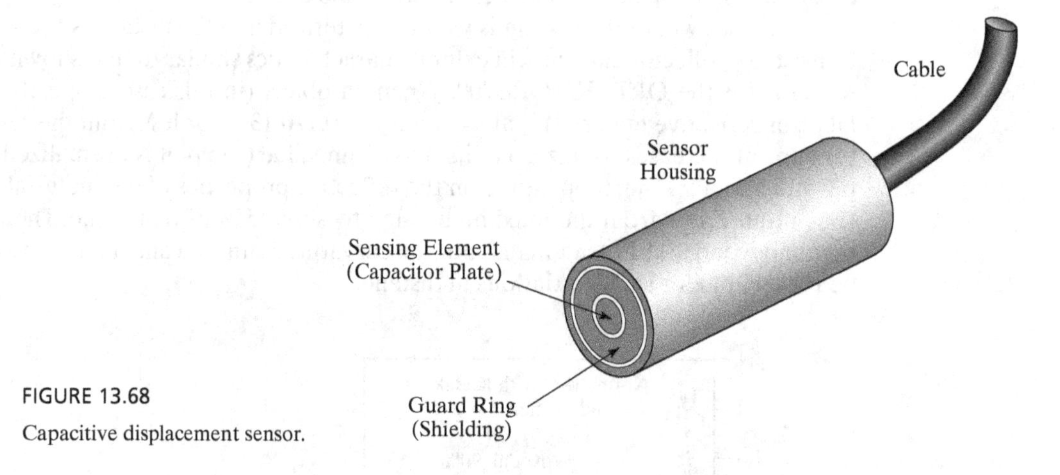

FIGURE 13.68
Capacitive displacement sensor.

In the figure, there are three concentric rings of conductive metal. The sensing element is the inner circle, and the other rings serve to shield the sensing element from capacitive effects with objects other than the intended target. This sensor is a little less than 5 mm in diameter and has a range of 0–10 μm, 15 kHz bandwidth, 0.4 nm resolution, and 0.2% linearity. Those are specifications that are hard to match with anything other than interferometry! In addition to sensing distance to a conductive target, they can be used to sense the distance to nonconductive targets by placing a conductive material on the opposite side. Capacitive sensors are also capable of measuring the thickness of coatings on conductive surfaces, such as adhesives or surface finishes, as well as the thickness of conductive materials. They are, however, sensitive to target size, alignment, and surface irregularities.

13.5.6.6 Ultrasonic Displacement Sensors

We've looked at a few options for measuring short distances with high resolution and accuracy, but what about longer distances? What if an application requires measuring the distance to a target that is greater than a few centimeters away, perhaps even several meters away? This is an application in which ultrasonic range finders excel. These sensors employ the same operating principles as the ultrasonic proximity sensors we discussed above: a high-frequency pulse of sound is emitted by a piezo or electrostatic element, and the time between the emission of the sound and the receipt of its reflection is measured (Figure 13.69). This sensor reports the sound wave's round trip time-of-flight, which is proportional to the distance between the sensor and the object that reflected the sound wave.

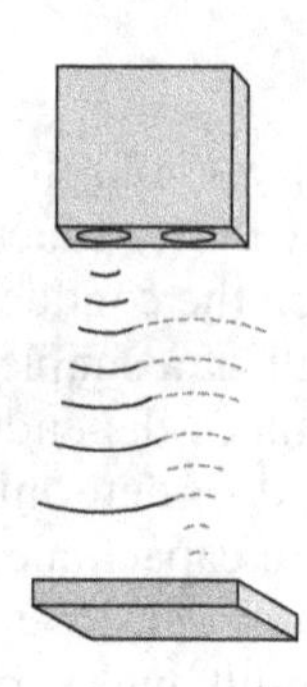

FIGURE 13.69
Ultrasonic waves emitted, reflected, and detected by an ultrasonic range finder.

Figure 13.70 shows two ultrasonic range finder modules. The Parallax 28015 (Figure 13.70a) is a fully integrated module that communicates with a host microcontroller via a single digital I/O pin. It has a range of 2 cm to 3 m. Bare ultrasonic transducers, without any interface electronics of software, such as the Maxbotix-UT, are also available (Figure 13.70b). Fully integrated versions using the Maxbotix transducer, with all necessary interface electronics, are also available from the manufacturer. Figure 13.70c shows an example, the Maxbotix LV-EZ1, which has a range of 0–254 in. (6.45 m), and communicates using a digital I/O pin, an analog output pin, or a serial communications link.

(a) Parallax 28015 module

(b) Maxbotix UT transducer

(c) Maxbotix LV-EZ1 module

FIGURE 13.70

Ultrasonic range finders.

13.5.6.7 Flex Sensors

Some applications call for measuring radius of curvature rather than straight-line distance. In some cases, curvature can be related to a quantity you'd like to measure, but isn't the primary measurand, such as arc length or spring deflection. In these cases, a compact, inexpensive sensor called a **flex sensor** or **bend sensor** may be a good choice (Figure 13.71). Flex sensors are made by printing a thin coating of a material that forms a resistive element onto a flexible substrate such as mylar or polyimide. The entire sensor can be as little as 0.004 in. thick. When the sensor is bent or flexed, microscopic cracks form in the printed sensing material, and the bulk resistance increases as the radius of curvature decreases. That is, the more they are flexed, the higher their resistance. A typical flex sensor has a nominal unflexed resistance of around 10 kΩ, and the resistance may rise well to hundreds of kilohm at high curvatures.

Flex sensors have strong temperature and humidity dependencies, so they are not typically used for high-precision measurements or applications that require stability. They are best for measurements that involve relatively fast changes and for those where relative changes in radius of curvature are more important than quantifying the actual radius. Examples of typical applications include interface devices that measure body motion for computer games and virtual reality simulations, automotive horn buttons, and sensing if a passenger is seated in a car for air bag activation/deactivation.

13.5.7 Acceleration Sensors

As of 2008 (when this chapter was written), measurements of acceleration could not be performed directly with a material whose properties change solely as a function of an applied acceleration. Instead, acceleration is inferred by measuring another quantity that is influenced by acceleration, such as the motion of a mass attached to a spring element. Earlier in this chapter, we introduced the most common way this is sensed in commercially available

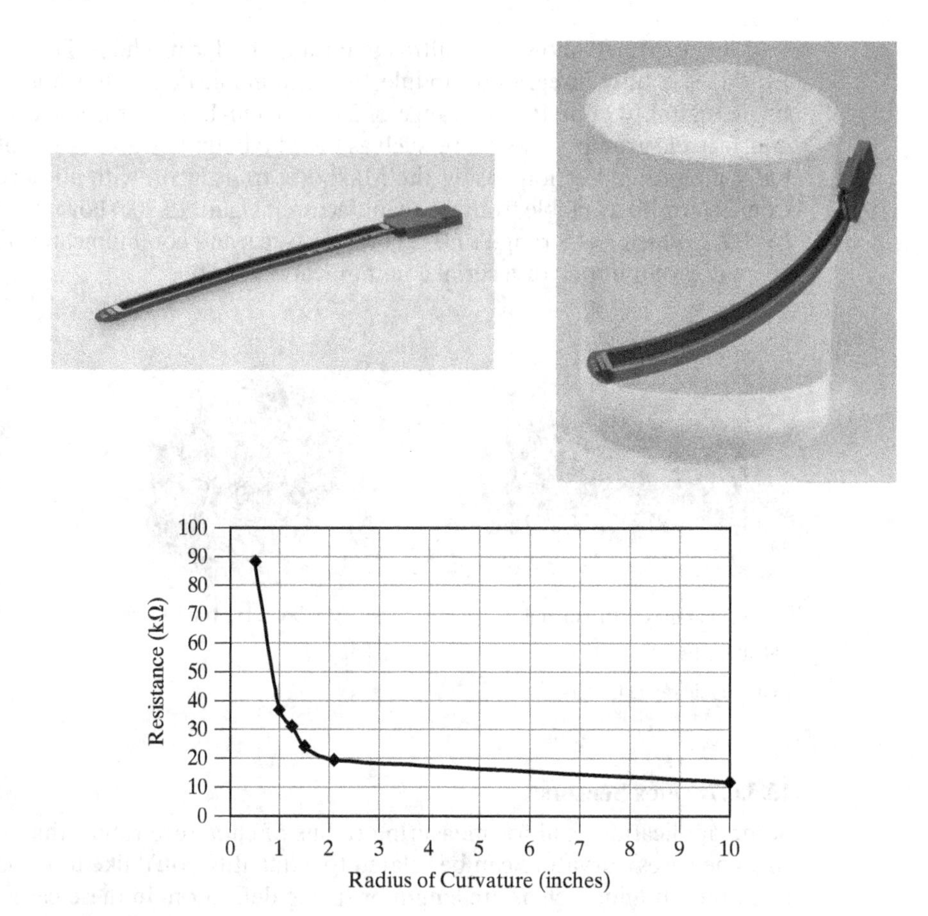

FIGURE 13.71

Flex sensor and representative output characteristic.

devices: **MEMS accelerometers** (Section 13.3.3). We used a MEMS accelerometer (the Freescale MMA1250) to explore definitions of general sensor specifications such as accuracy, resolution, and linearity. Rather than repeat this material, we refer the reader to Section 13.3.3.

13.5.7.1 Tilt Sensors

One aspect of acceleration we have not yet covered is tilt. **Tilt sensors** compare the direction of acceleration due to earth's gravity to the orientation of the sensor. This measurement can be performed with traditional accelerometers, or it can be achieved with sensors intended specifically for this purpose.

A very simple type of tilt sensors only determines if the orientation has changed in large increments, such as 90°. This is actually only a **tilt direction sensor**, and it provides a go/no-go indication of tilt. This would be useful, for example, in cases where you wish to determine the orientation of a camera as a picture is taken so the image can be stored automatically in either landscape or portrait format. Gross determination of tilt can be achieved in a great number of ways, such as with an element that moves with gravity and blocks the light between an opto-interrupter pair. The NKK® DSBA1P is an example of such a sensor (Figure 13.72). It indicates when a tilt of greater than 30° is experienced in two axes with a small sphere that interrupts light between an LED and a photo-transistor.

FIGURE 13.72

A tilt sensor that detects tilt of 30° or more in two axes (NKK DSBA1P).

In cases where a measurement of the degree of tilt is required with higher resolution, inclinometers [which consist of a rotating disk that is heavier at a point away from the center and a protractor (Figure 13.73)] and MEMS accelerometer-based devices are often employed.

FIGURE 13.73

Standard inclinometer.

Incremental inclinometers (Figure 13.74) are a clever means of sensing tilt angle with very high resolution and linearity, though there are some inherent dynamic limitations. In this simple but effective approach, the weighted disk from a standard inclinometer is coupled to the input shaft of an optical encoder (either incremental or absolute encoders—see Section 13.5.6.2), and the force of gravity draws the heavy portion of the weighted disk to

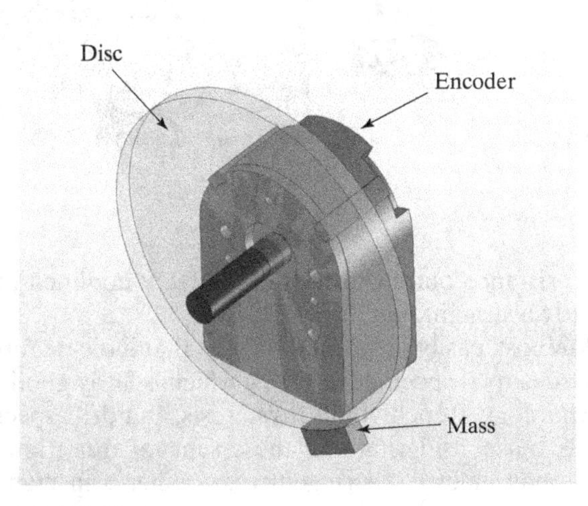

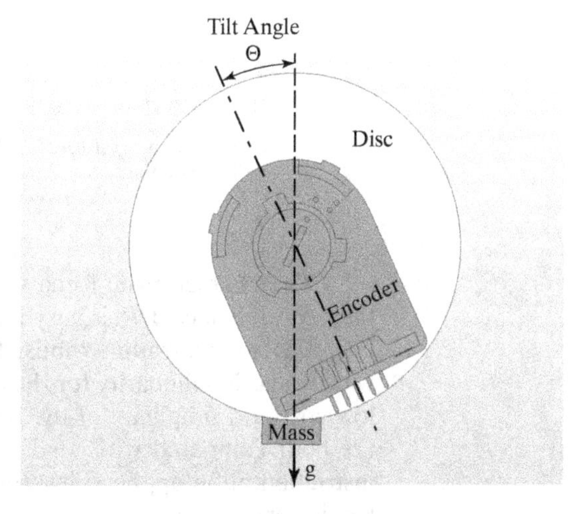

FIGURE 13.74

Incremental inclinometer.

the position where it has the lowest potential energy. The position of the weight, relative to the orientation of the encoder body, gives an indication of the tilt of the device.

One potentially undesirable aspect of this approach is that a weight pivoting around a point creates a pendulum, and oscillations may be induced if inputs to the sensor have frequency content at or near the natural frequency of the system. To reduce these effects, versions of these sensors that incorporate damping of the rotor's motion are available. This reduces the potential for oscillations and decreases overshoot and settling time, but it also decreases the sensor's response time.

13.5.8 Force Sensors

Measurements of force can be achieved using a variety of methods. The most common method involves determining the force indirectly, for example, by measuring the strain in a known material and solving for the force that must have created it. Other methods involve measuring fundamental properties of specific materials that change as the material is subjected to forces.

13.5.8.1 Force Sensitive Resistors

The heart of a **force sensitive resistor**, or **FSR**, is a pressure sensitive ink sandwiched between flexible conductive layers. The pressure sensitive ink is comprised of conductive and non-conductive materials, suspended in a polymer matrix. When the material is compressed, the density of conductive material increases, resulting in a reduction of the bulk resistance. The sensor assembly is made by first applying conductive ink layers to two separate pieces of flexible substrate, such as polyimide or polyester, and then applying the pressure sensitive ink on the conductive ink of one side. The two pieces of substrate are then bonded together so that the printed surfaces are sandwiched together and sealed inside. The resulting sensor is less than 0.01 in. thick. Figure 13.75 shows a variety of FlexiForce™ Sensors (Tekscan®, Inc.).

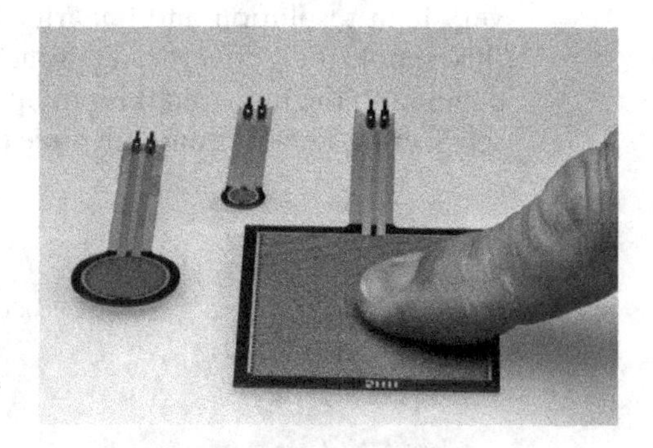

FIGURE 13.75

Force sensitive resistors (FSRs), made by Tekscan, Inc.

The characteristic force vs. resistance output of FSRs are highly nonlinear, however, their conductance, $1/R$, is very nearly a straight line (Figure 13.76).

FSRs are ultrathin, robust, low cost, easily integrated sensors that indicate force over a wide range. Nonlinearity for these sensors is specified as $\pm 3\%$, which is fairly good given the low cost and simplicity of the technology. Repeatability, hysteresis, and drift specifications are poor compared with load cells, but considering that these sensors aren't intended for instrumentation applications, they may meet less stringent requirements in many mechatronic designs quite well.

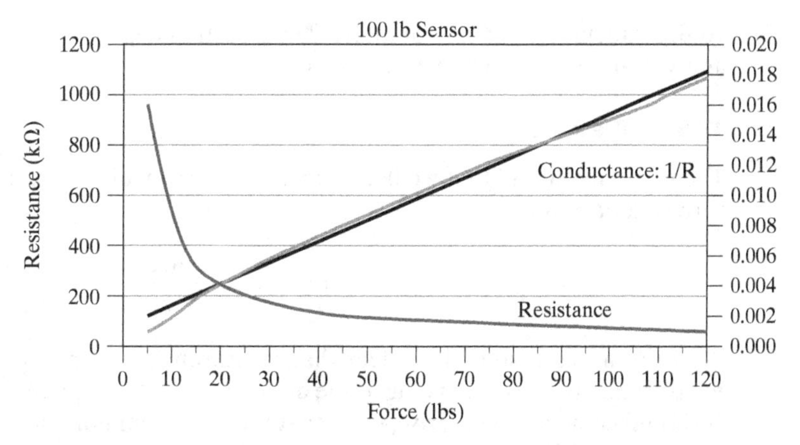

FIGURE 13.76

Characteristic output of FlexiForce™ Sensor FSRs. (Courtesy of Tekscan, Inc.)

13.5.8.2 Piezo Force Sensors

In a **piezoelectric force sensor**, a load is applied to a piezo element, which results in the development of electrical charge proportional to the applied load. Figure 13.77 shows the construction of a typical commercially available piezoelectric force sensor. The sensing component is the quartz element. A preload stud clamps the mounting pads together around the piezo element so that the element is in compression when no external load is applied. Because of the preload, tensile forces can be measured, since they serve to counteract the preload.

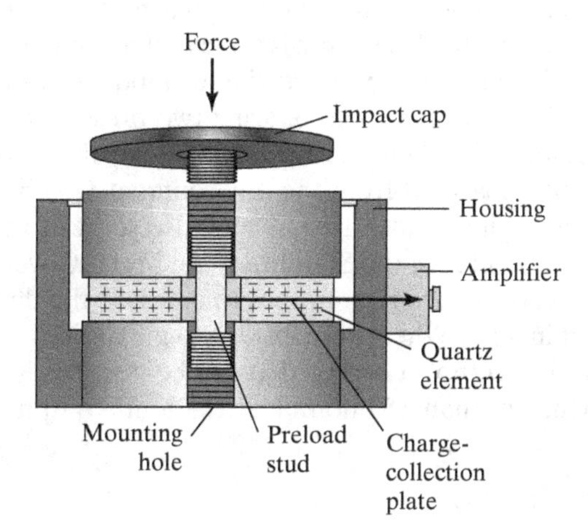

FIGURE 13.77

Sectional view of a typical piezoelectric force sensor.

Piezoelectric force sensors are only useful for dynamic load measurements. When a static load is quickly applied and then held constant, the sensor's output responds to the initial transient but then quickly decays back to zero. This happens because the charge that develops "leaks" through any adjacent material, even highly insulative materials. There are a great many applications where this isn't problematic, for instance, in automated manufacturing equipment, measurements of internal combustion engine parameters like cylinder pressure, and vibration or impact measurements.

13.5.9 Pressure Sensors

The final measurand that we'll address in this survey of sensors is pressure. Recall that pressure is defined as:

$$\text{Pressure} = \frac{\text{Force}}{\text{Area}} \tag{13.18}$$

From this, it should be clear that measurements of pressure can be directly based on measurements of force, as long as the area over which the force is applied is well understood. It should come as no surprise, then, that the most common methods for measuring pressure closely resemble the most common methods for measuring force.

13.5.9.1 Strain Gage-Based Pressure Sensors

For **strain gage-based pressure transducers**, sensing pressure involves establishing a pressure differential across a material and measuring the resulting deflection with strain gages. The strained material is typically a membrane or a diaphragm, so that even very small pressure differences result in measurable displacements. Sensors that incorporate metal foil strain gages have long been available and popular. These may be separate metal foil strain gages that are bonded to a diaphragm, or the strain gages may be formed in place on the surface of the diaphragm itself using a sputtering process. In either case, this type of sensor is more difficult and expensive than those that make exclusive use of semiconductor manufacturing techniques, which we discuss next. As a result, their popularity is waning; however, they still find use in instrumentation settings and harsh environments.

13.5.9.2 Semiconductor Pressure Transducers

Figure 13.78 shows the internal construction of a representative semiconductor pressure transducer, the Freescale MPX2010. This sensor is typical of a class of pressure sensors constructed using standard semiconductor manufacturing techniques, resulting in a highly consistent and inexpensive device. It works by comparing two pressures, labeled P_1 and P_2, by applying them to opposite sides of a silicon sensing element, labeled here as the Die. The Die is kept sealed and physically separated from the environment (P_1) by a silicone gel which fills the space inside the packaging above the Die and also serves to transmit the pressure from the inlet port (P_1) at the center of the Stainless Steel Metal Cover.

The MPX2010 uses piezoresistive strain gages to measure the deflection of the silicon die. The gages are arranged in a full Wheatstone bridge to maximize gain and minimize temperature effects. The result is a compact sensor that can be very easily integrated into a circuit, and has a 10 kPa span, less than 1% nonlinearity, 25 mV output span, 1 ms response

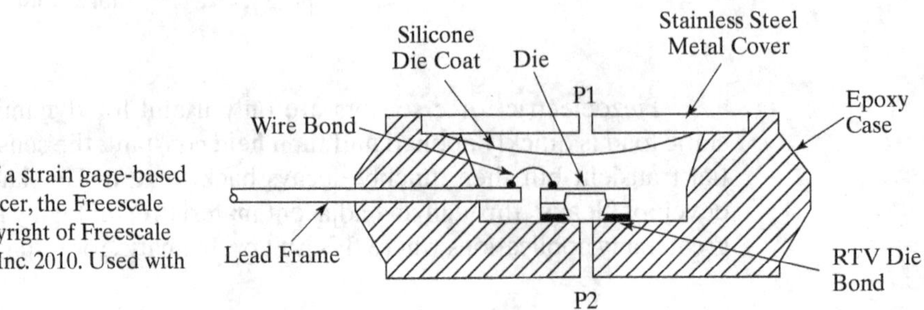

FIGURE 13.78
Cutaway view of a strain gage-based pressure transducer, the Freescale MPX2010. (Copyright of Freescale Semiconductor, Inc. 2010. Used with permission.)

time, very low hysteresis (0.1% of full scale), and low temperature sensitivity (1% over a temperature range of 0 to 80°C)— it costs less than $10 in low quantities.

Of course, there are a great many ways to sense the deflection of a diaphragm that don't involve strain gages. Capacitive position sensors are also commonly used in this application, since they feature very high resolution measurements over very small ranges. Figure 13.79 shows the internal construction of this type of pressure sensor. The bottom plate of the capacitor is fixed, and the other plate comprising the capacitor is labeled the Diaphragm. As the diaphragm deflects from the applied pressure, the capacitance between the two plates changes, and this becomes the sensor's output.

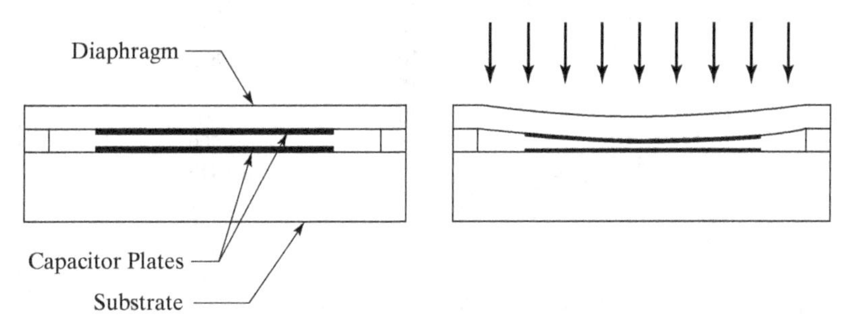

FIGURE 13.79

Construction of a pressure sensor that measures diaphragm deflection capacitively.

13.6 HOMEWORK PROBLEMS

13.1 A microphone has a dynamic range of 88 dB and a maximum output of 2.5 V rms. What is the output signal (in μV rms) that corresponds to the minimum sound level it is capable of detecting?

13.2 From the list below, pick two sensor technologies and succinctly describe their theory of operation:

(a) Geiger counter

(b) Torque sensor

(c) LVDT

(d) pH sensor

(e) Vortex flow meter

13.3 For one of the sensors you chose to describe in your answer to Problem 13.2, identify a commercially available example, locate the data sheet for the device that describes its performance, and answer the following questions:

(a) What is the range of the sensor?

(b) In what form is the output (e.g., voltage, resistance, capacitance, and so on)?

(c) What is the transfer function for the sensor?

(d) What are the categories of error for the sensor? What is the overall accuracy?

(e) Does the sensor exhibit hysteresis? If so, how much?

13.4 Using PDL (also called "pseudo-code," see Chapter 6, Software Design), describe an algorithm for performing switch debouncing in software. Does your routine report when the switch is initially activated or when it is released?

13.5 An LM35DT temperature sensor is used to monitor an environment where the temperature ranges from 0 to 50°C. Design an interface circuit that has an output ranging from 0.5 to 4.5 V that is linear with temperature over the range to be measured. For your circuit, you may make use of "ideal" op-amps, but select only standard 5% tolerance resistor values.

13.6 One method of creating a pseudo-linearized output from a standard NTC thermistor is called "voltage mode linearization" in which the thermistor is placed in series with a standard resistor (where $R_1 = R_{25C}$, the thermistor's resistance at 25°C) to create a voltage divider, as shown in Figure 13.80.

Temperature (°C)	R (Ω)
–25	4,240
–20	3,311
–15	2,607
–10	2,070
–5	1,656
0	1,335
5	1,083
10	885
15	727
20	601
25	500
30	418
35	352
40	298
45	253
50	216

FIGURE 13.80

For the thermistor characteristics shown, plot the resulting V_{out}. What is the maximum linearity error that results (the difference between the best-fit straight line through the data and the data point furthest from the line), stated both in volts and in % F.S.O.?

13.7 Design a strain gage interface circuit based on a Wheatstone bridge. The strain gage has a nominal (unstrained) resistance of 120 Ω and a gage factor of 2. The interface circuit's output should be 0 V when the gage is not subjected to strain, and sensitivity of 0.5 mV/με.

13.8 Design an interface circuit for the photo-diode whose specifications are given in Figure 13.28 that uses a 74HC14 (a logical inverter with Schmitt trigger inputs) at the output stage. The 74HC14 is to be powered by a 4.5 V power supply and its output should be guaranteed to have made a transition from logical high to logical low when the irradiance increases above 3 mW/cm^2, and guaranteed to have made the logical low to logical high transition when the irradiance then falls to 0.1 mW/cm^2.

13.9 For the photo-transistor interface circuit (Figure 13.81), use the specifications given for the LTR-3208E in Figure 13.30 to plot V_{out} for values of irradiance between 0 and 3 mW/cm^2, assuming that the LTR-3208E is selected from Bin A. If all resistors used in the circuit have 5% tolerance, what is the range of possible values of V_{out} when the irradiance is 3 mW/cm^2?

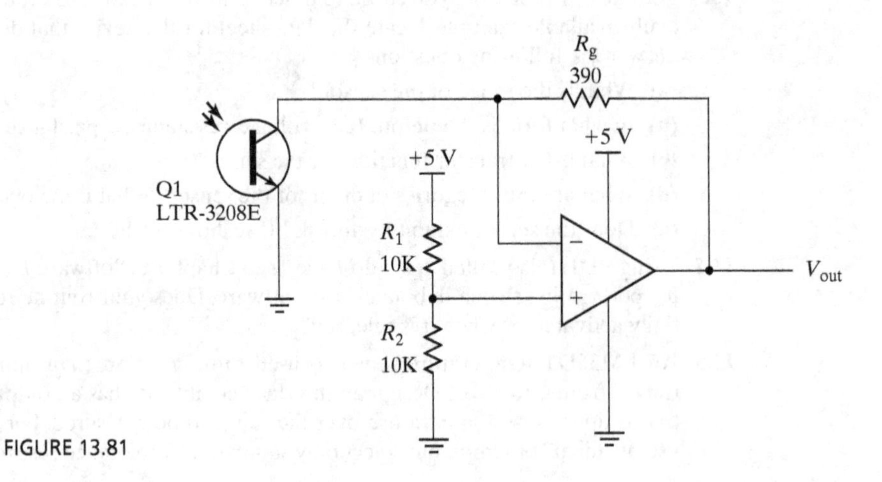

FIGURE 13.81

13.10 If R_1 in Figure 13.82 is a sensor whose resistance varies from 8 to 11 kΩ,

(a) what is the range of output voltages for the op-amp on the left (U_{1A})?

(b) what are the output voltages for the op-amp on the right (U_{1B})?

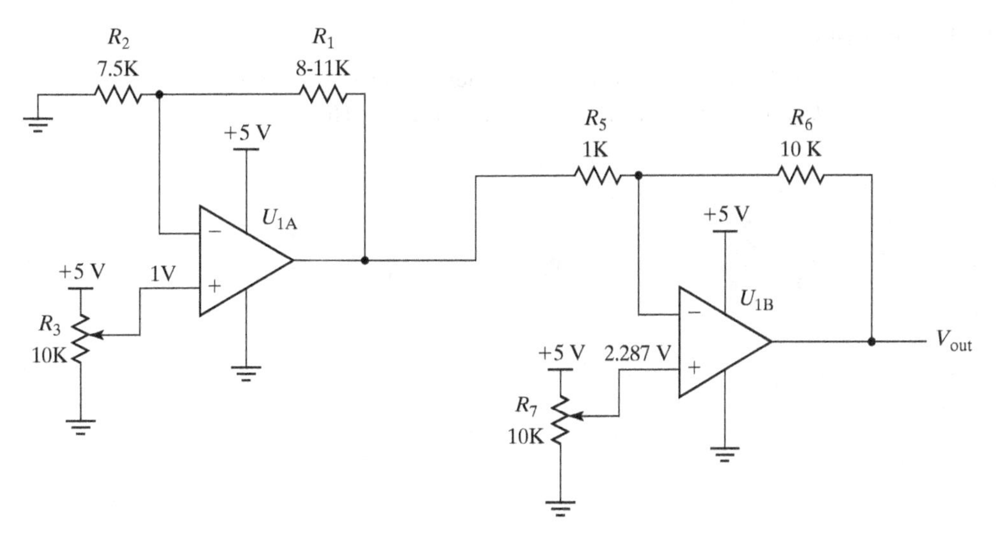

FIGURE 13.82

13.11 For the MPX2010 pressure sensor, use the specifications provided in Figure 13.83 to answer the following questions:

(a) What is the range of pressures you can measure with this device?

(b) What is the equation for the transfer function?

(c) What is the output voltage when the sensor measures 7.5 kPa?

(d) What is the total accuracy of the device at 25°C without calibration (in % F.S.O., referred to here as V_{FSS}, or "voltage at full scale span")?

(e) What is the (typical) power dissipated by the device in normal operation when it is powered at 5 V?

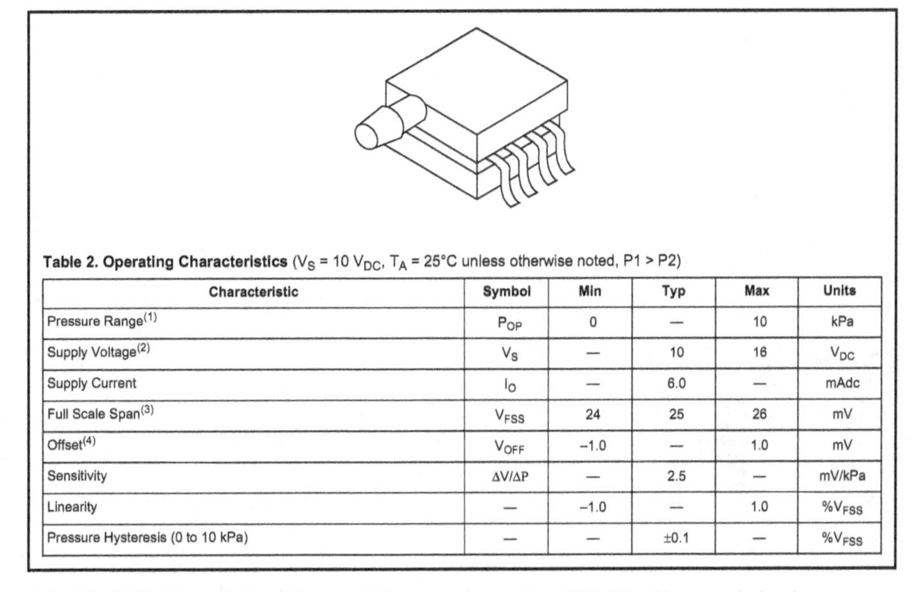

Table 2. Operating Characteristics (V_S = 10 V_{DC}, T_A = 25°C unless otherwise noted, P1 > P2)

Characteristic	Symbol	Min	Typ	Max	Units
Pressure Range(1)	P_{OP}	0	—	10	kPa
Supply Voltage(2)	V_S	—	10	16	V_{DC}
Supply Current	I_O	—	6.0	—	mAdc
Full Scale Span(3)	V_{FSS}	24	25	26	mV
Offset(4)	V_{OFF}	−1.0	—	1.0	mV
Sensitivity	ΔV/ΔP	—	2.5	—	mV/kPa
Linearity	—	−1.0	—	1.0	%V_{FSS}
Pressure Hysteresis (0 to 10 kPa)	—	—	±0.1	—	%V_{FSS}

FIGURE 13.83 (Copyright of Freescale Semiconductor, Inc. 2010. Used by permission.)

REFERENCE

[1] NIST ITS-90 Thermocouple Database, Version 2.0, http://srdata.nist.gov/its90/main/its90_main_page.html, July 27, 2010.

FURTHER READING

Handbook of Modern Sensors, Fraden, J., 2nd ed., AIP Press, 1996.

Handbook of Transducers, Norton, H., Prentice Hall, 1989.

CHAPTER

Signal Conditioning

14

The process of signal conditioning takes place in the path of a signal as it flows between a source (most frequently, a sensor) and a microcontroller (Figure 14.1).

For the purposes of this chapter, we will treat the output of the sensor block as a voltage. But for those sensors whose output is not a voltage (i.e., current, resistance), Chapter 13 presents circuits to translate these other outputs into a voltage. Even with the output of the sensor block in the form of a voltage, it is often not optimal for presentation to a microcontroller input and/or it is adulterated by signals other than the one of interest (i.e., by electrical noise). The purpose of the circuit in the signal conditioning block is to take the output of the sensor block and prepare it to be presented to a microcontroller input. The goal is to maximize the amplitude of the signal of interest while reducing the amplitude of any interfering signal. The underlying components of the signal conditioning block are those that we have just introduced in Chapters 9, 11, and 12.

After studying the material of this chapter, you should be able to:

1. understand how to remove a DC offset from a signal,
2. identify some of the reasons that you might need filtering,
3. recognize some of the situations in which more specialized signal conditioning might be useful.

14.1 BASIC OPERATIONS IN SIGNAL CONDITIONING

We categorize signal conditioning operations into one of two classes: basic and advanced. Basic signal conditioning operations are the most common and consist of **offset adjustment**, **amplification**, and **filtering**. We will treat each of these basic operations with circuits that implement the operation, as well as examples that illustrate when each is needed. Advanced signal conditioning encompasses a much broader range of operations, many of which have their origins in the era before microcontrollers existed or when they were far less capable than they are today. Many of the operations that might have been done with analog signal conditioning in the past can now be done in software inside the microcontroller. However, there are still many instances in which it will be better to process the signal first in the analog domain before presentation to the microcontroller. Later in the chapter, we will cover two of the more advanced techniques (peak detection and precision rectification) and explain why it might be preferable to perform the operations in circuitry rather than in software.

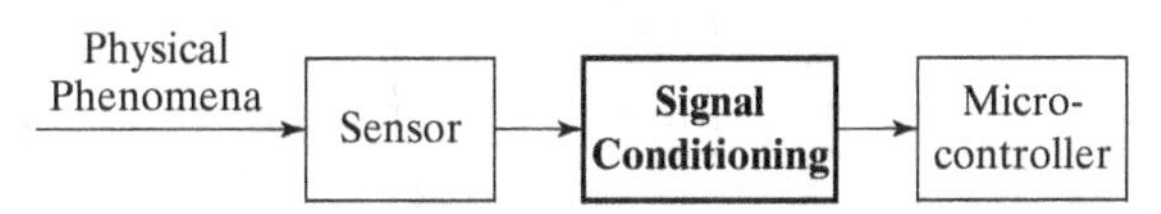

FIGURE 14.1

Signal flow path from physical phenomena to microcontroller.

14.2 OFFSET REMOVAL

It is very common for the signal from a sensor (and from numerous other sources) to consist of a time-varying signal of interest that is superimposed onto a DC level of little interest. This DC level is referred to as an **offset** (Figure 14.2).

If we attempted to amplify this signal without modifying it first, we would amplify not only the signal but also the offset. When the time-varying signal is small compared to the offset, then amplifying them both would seriously limit the maximum amount of amplification (i.e., the gain) that we can apply to the time-varying signal of interest. For the signal shown in Figure 14.2, if a rail-to-rail output op-amp powered by a single +5 V supply (a very common situation in microcontroller-based systems) were used for the amplification, the maximum amplification (gain) would be approximately 1.8. This would leave us with a 1.8 V peak-to-peak (V_{pp}) amplitude on the signal of interest. This would be much less (2.7 times less) than the approximately 5 V_{pp} signal that would be possible if we amplified after removing the offset. What we need is a way to avoid applying the gain to the offset voltage. We will probably also need to create some offset after the amplification in order to make the best use of our single supply amplifier's output range.

14.2.1 Amplification Relative to an Offset

We could take the idea of *removal* literally and use the difference-amplifier configuration from Section 11.4.5 to implement a mathematical *subtraction* of the offset from the signal. However, that would be more complex than is actually necessary. In Section 11.4.3.2, we also learned that there is "nothing magic about ground," and that we need not do our amplification relative to ground, but instead can do it relative to some other voltage. If we built a circuit like that of Figure 14.3, we could apply the gain only to the difference between the input signal and the reference voltage, rather than the full input voltage.

Figure 14.3 is a circuit that combines elements from Figure 11.7 (Inverting op-amp with offset voltage) and Figure 11.11 (A practical voltage reference). The V_{ref} from Figure 11.7 (created by the elements inside the dashed oval in Figure 14.3) is a simplification of the voltage reference shown in Figure 11.11. Since the voltage at the central node of the voltage divider (V_{ref}) is only connected to an op-amp input, there is no need for the buffer of Figure 11.11. In this case, we chose the inverting configuration for the amplifier specifically because V_{ref} would be connected directly to an op-amp input and so would not require buffering. If the amplification needs to be done with a noninverting amplifier, then a well-buffered source for V_{ref} must be used (see Figure 14.6).

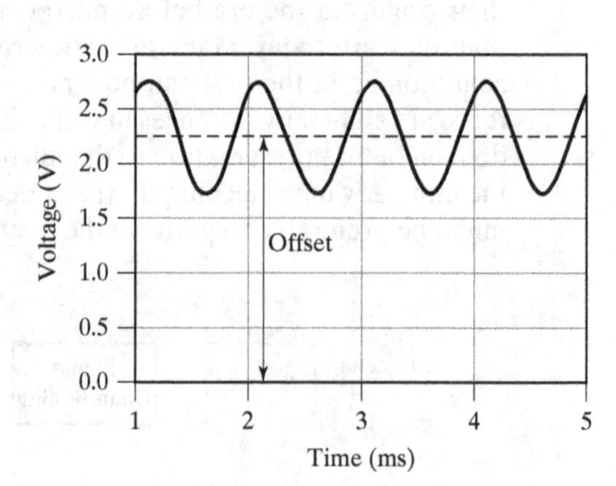

FIGURE 14.2
Time-varying signal with offset.

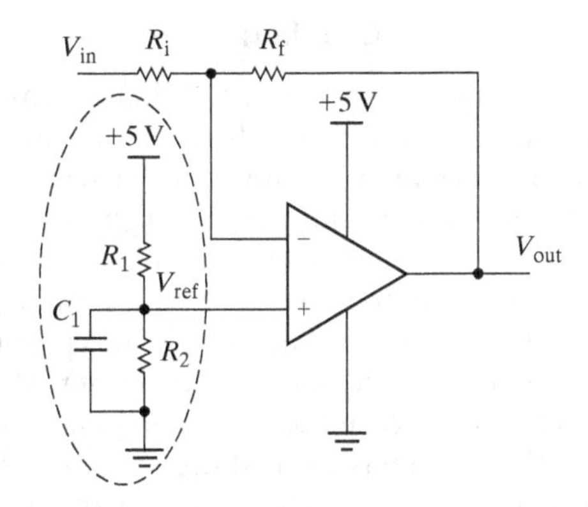

FIGURE 14.3

Amplifier with offset.

The transfer function of the circuit of Figure 14.3 is:

$$V_{out} = (V_{ref} - V_{in})\left(\frac{R_f}{R_i}\right) + V_{ref} \tag{14.1}$$

where V_{ref} is (using the voltage divider equation):

$$V_{ref} = 5\text{ V}\frac{R_2}{(R_1 + R_2)} \tag{14.2}$$

The function of capacitor, C_1, is (as explained in Section 11.4.4) to combat fluctuations in the 5 V power supply.

As we see from Eq. (14.1), the gain of the amplifier is applied to the *difference* between the signal and V_{ref}. And V_{ref} is added to the amplified difference to produce the output. If the signal of Figure 14.2 was amplified with the circuit of Figure 14.3 with V_{ref} set to 2.25 V, then the gain (R_f/R_i) could be set as high as 4.5 while still keeping the output voltage between 0 and 5 V. This would yield an output voltage in which the signal of interest was 4.5 V_{pp}, as compared to only 1.8 V_{pp} if the amplification was applied to the entire input voltage. The smaller the amplitude of the signal of interest is, the greater the improvement with offset adjustment. If the signal of interest in Figure 14.2 were only 0.5 V_{pp}, the maximum gain without offset removal would be 2, yielding a 1 V_{pp} output swing for the signal of interest. With offset removal, the gain could be as high as 9, yielding a 4.5 V_{pp} output swing for the signal of interest.

The circuit of Figure 14.3 works well when the offset is a true DC offset. As you might guess, once we start considering real, as opposed to ideal, offsets practically none of them are true DC. The offset may vary due to several possible reasons, many of which are temperature dependent, resulting in drift in the offset. For example, as the ambient temperature of the sensor changes, the offset will likely also change. Inside our signal conditioning circuit itself, the resistors used to make V_{ref} will also change with temperature, causing V_{ref} to drift. The net result is that, even if we have adjusted V_{ref} to perfectly match the sensor offset at some environmental condition, the match will not remain perfect as the environmental conditions change. By choosing low temperature coefficient components, we can minimize the effects of these temperature-induced drifts. In many applications, we can keep the difference between the actual offset and our V_{ref} small enough that we can enjoy most of the improved gain benefits of this circuit. When the frequency of the signal of interest is very low (approaching DC), a circuit like that of Figure 14.3 will be the best (and perhaps only) choice for adjusting the offset to allow for greater amplification.

14.2.2 Offset Removal by AC Coupling

In some cases, what at first appears to be a DC offset turns out to have a distinct frequency component. Consider the circuit of Figure 14.4, which is a combination of a variation on the trans-resistive amplifier of Figure 11.15 with the voltage reference of Figure 11.11 (without the buffer). It was designed to sense the infrared (IR) light from a beacon modulated at 1 kHz.

The value of R_f sets the gain of the circuit and determines the amplitude of the response to the incident IR light on the photo-transistor. If, as is often the case during student projects, the values for V_{ref} and R_f are adjusted late at night, it is possible to obtain a large signal (over 1 V) in response to the IR from the modulated emitter. Upon returning the next morning, the student discovers that the circuit appears to no longer respond to the modulated IR signal and the output of the op-amp is stuck at the upper rail. What has happened? Take a few minutes to study the circuit and think about what might be going on.

The difference between the conditions when the circuit was adjusted and the following day is the amount of background IR light due to solar radiation. It is often the case that the magnitude of the background IR due to solar radiation is substantially greater than the magnitude of the IR signal of interest. If the circuit is to work in a situation like this, there is little choice except to reduce the gain to prevent the op-amp from saturating because of the ambient IR light during the day. This will result in a much smaller amplitude for the signal of interest (from the beacon) and that signal will be riding on what appears, at any given instant, to be a significant DC offset (from the sun). Unfortunately, that DC offset varies dramatically over the course of the diurnal cycle, making it very difficult to use a circuit like that of Figure 14.4 to remove the offset and allow significant gain to be applied to the signal of interest.

To work around the problem of the large and variable offset, we can take advantage of the differences in frequency between the offset (1.16×10^{-5} Hz) and our signal of interest at 1 kHz. The circuits of Figure 14.5 use AC coupling (effectively, a high-pass filter) to block the DC (and near-DC) signals while passing the AC signals. The circuits are variations on the basic inverting and noninverting amplifiers that make use of the fact that a capacitor will not pass a DC current and that there is nothing magic about ground. Both of the circuits of Figure 14.5 provide an output voltage of 2.5 V in the absence of a varying signal at V_{in}. The capacitor labeled C_{in} in Figure 14.5 is known alternatively as an AC **coupling capacitor** or a DC **blocking capacitor**.

In Figure 14.5a, C_{in} and R_i form a high-pass filter with a corner frequency at $1/2\pi R_i C$. A circuit that has a gain of 10, with $R_f = 100\ \text{k}\Omega$, $R_i = 10\ \text{k}\Omega$, and $C_{in} = 0.1\ \mu\text{F}$, yields a corner frequency of 159 Hz. For our sample photo-transistor application, a 1 V swing due to solar radiation would result in an output voltage swing of only 80 μV (an attenuation of 80×10^{-7} or −101 dB), while the 1 kHz signal of interest would be amplified by a factor of 9.9 (19.9 dB).

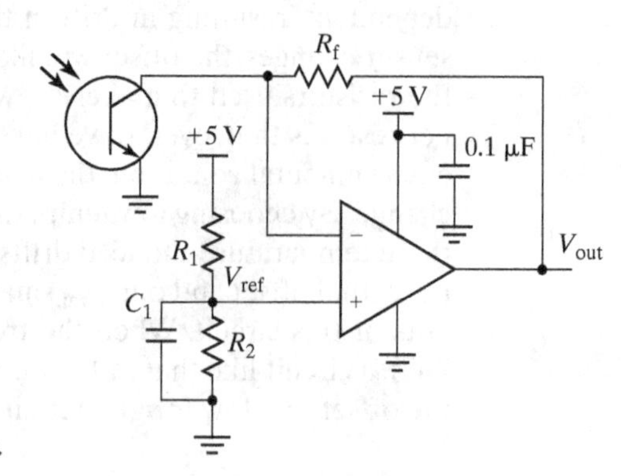

FIGURE 14.4

IR photo-transistor signal conditioning circuit.

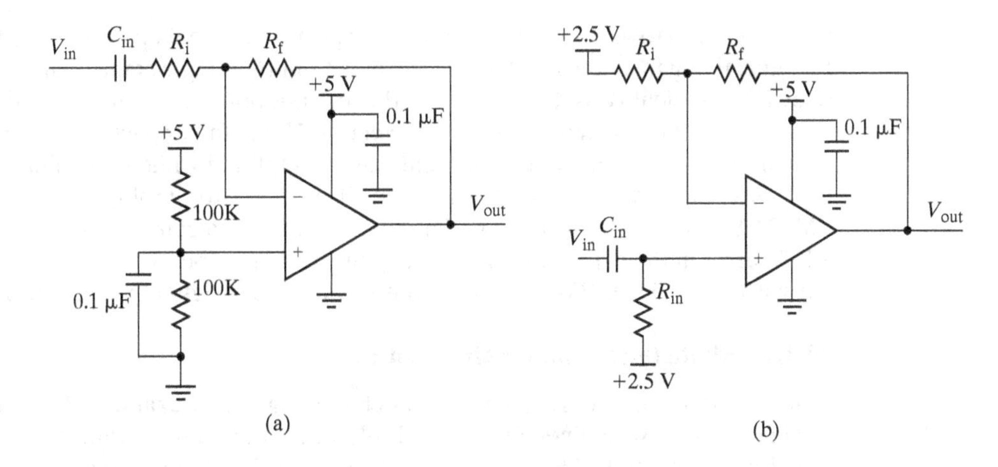

FIGURE 14.5
AC-coupled (a) inverting and (b) noninverting amplifiers.

The 2.5 V voltage source necessary for the circuit of Figure 14.5b can be created with a circuit like that of Figure 14.6. The voltage divider serves to set the output voltage halfway between ground and the positive supply, hence the name given to this circuit: a **rail-splitter**. The output is also often referred to as a **virtual ground** because in this application, the amplification takes place relative to this voltage. This is a slightly different use of the term "virtual ground" than that introduced in Chapter 11. The aspect that they share in common is being at some fixed voltage that is used as a reference point while not being physically connected to ground.

In AC coupling applications, the accuracy of the input capacitor's value is not generally of great importance, since the corner frequency is typically set far away from the signal of interest and the operation is therefore very tolerant of variations in capacitance. The characteristic of greatest importance for C_{in} in these applications is its leakage current. The leakage current represents a DC component that is being passed through the capacitor and is therefore undesirable. As we saw in Chapter 9, some of the plastic film capacitors, especially Teflon, polystyrene, and polypropylene, have the lowest leakage values and are therefore excellent choices in these applications. Monolithic ceramic capacitors, despite their somewhat higher leakage, are also a reasonable choice and lower in cost.

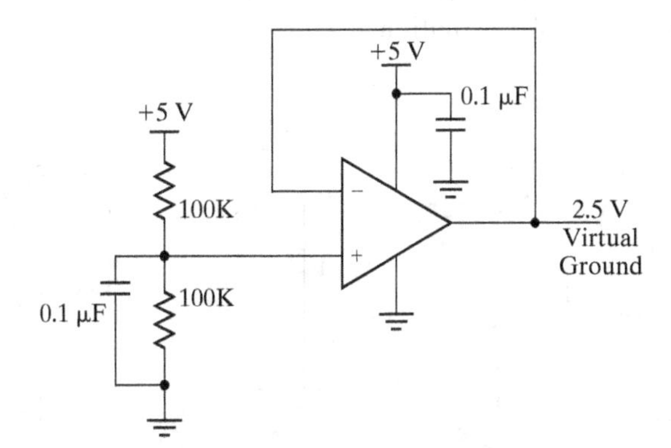

FIGURE 14.6
A 2.5 V voltage source (a.k.a. rail-splitter).

14.3 AMPLIFICATION

If our op-amps were ideal, we would never need more than a single stage of amplification for our signal conditioning. We could just choose the gain that we need to optimize the output voltage swing. However, we live in the real world, where real op-amps have real nonidealities. In determining just how much gain we can get out of a single stage of amplification, the most relevant nonideality is the fact that the open-loop gain is not only finite, but actually decreases with frequency. As we saw in Section 12.1.2, the impact of the decreasing loop gain is to limit the range of frequencies that can be amplified while maintaining some acceptable level of error in the gain that is applied. If we want to amplify a 10 kHz signal using an LM324A (which has a 1 MHz GBW), we are limited to a gain of 10—and even that results in a 10% error in the expected gain. To keep the gain error below 1%, we would need an amplifier with a 10 MHz GBW in order to amplify a 10 kHz signal by a gain of 10.

14.3.1 Multistage Amplification with DC Coupling

What are we to do if we really need a much larger gain, for example, 200 or more? To achieve a gain of 200, we would need to use multiple stages of amplification. If we were to implement this circuit using an LM324A, then each stage would be limited to a gain of no more than 10. While we could implement the necessary gain as three stages of gain (10:10:2), we will realize a lower level of gain error if we spread the total gain out evenly among the three stages, as gains of approximately 6 ($6^3 = 216$). By implementing the gain in this way, each of the stages has the greatest possible difference between the available open-loop gain and the desired closed-loop gain. Maximizing this difference will provide the best approximation of ideal behavior (see Section 12.1.2). We might initially construct a circuit like that of Figure 14.7. This figure shows the connection between these stages as a wire (whose frequency response extends all the way down to DC), so these stages are referred to as **DC coupled.**

If the ratio of R_f to R_i were kept at 5:1 (yielding a gain of 6) for each stage and the input signal, V_{in}, were already referenced to 2.5 V, then the circuit would appear to meet our requirements. However, here another nonideality raises its head.

Recall that each of the op-amps exhibits an input offset voltage, which we introduced in Section 12.1.5.1. If these are LM324A op-amps, then it could be as much as 3 mV. That doesn't sound like much, but the input offset voltage is amplified by the noise gain ($1 + R_f/R_i$) at each stage. At the input to the second stage, we could have an offset of up to (3 mV × 6) + 3 mV = 21 mV. At the input to the third stage, the offset could be as large as (21 mV × 6) + 3 mV = 129 mV. And, at the output of the third stage, it could be up to 129 mV × 6 = 774 mV. The initial 3 mV input offset voltage propagated through the three amplifiers could result in over 0.75 V of offset. We have consumed a significant fraction (22%) of the available output voltage swing of the LM324A with just the offset voltage. In the next section, we propose a solution for dealing with this undesirable situation.

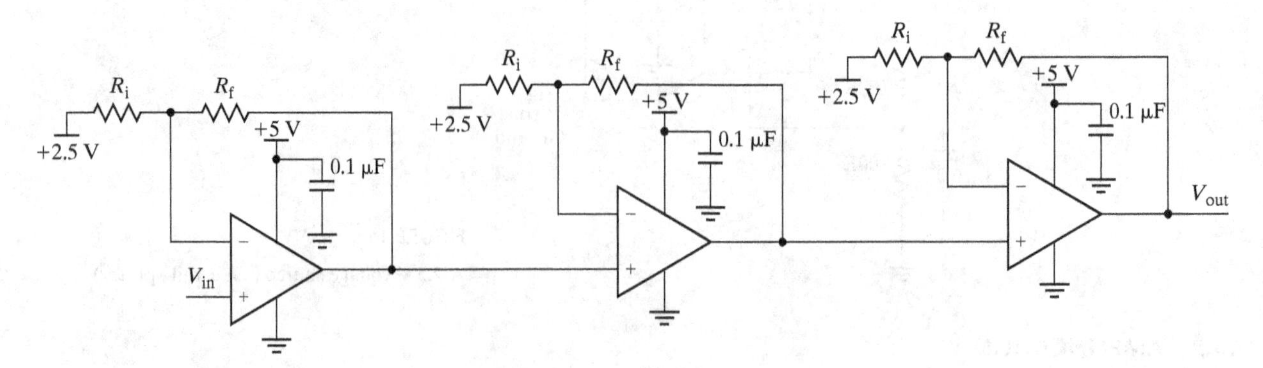

FIGURE 14.7
Multistage amplifier with DC coupling between stages.

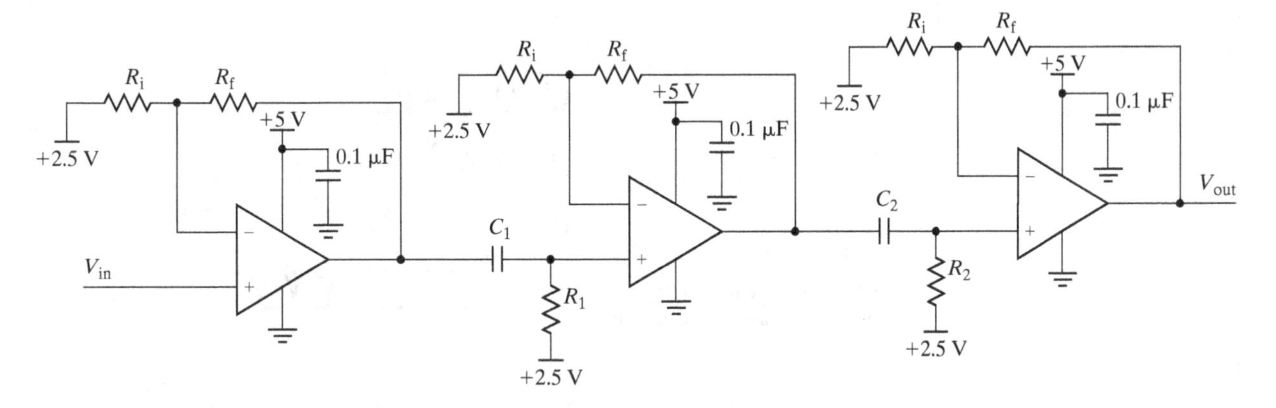

FIGURE 14.8
Multistage amplification with AC coupling.

14.3.2 Multistage Amplification with AC Coupling

To minimize the effect of input offset voltages, we could AC couple each stage, as shown in Figure 14.8.

Using this approach, the DC offset from each prior stage is blocked before the next amplification process. In this way, the output contains an offset voltage due to the input offset voltage of only the last stage (multiplied by the noise gain of that stage). When using AC coupling, it turns out that there is an advantage in spreading the gain unequally among the stages. Since the gain of the *last* stage will determine the magnitude of the output contribution due to the op-amp's input offset voltage, it is advantageous to design this stage to have the lowest practical gain.

Using AC coupling, because of the presence of the high-pass filters formed by R_1:C_1 and R_2:C_2, sets a lower limit on the frequencies that can be amplified. For many applications, this is not a problem, as the signal of interest has a frequency well above DC. When that is not the case (i.e., when the information in the signal is very near to DC), then DC coupling will be the only choice. Since DC coupling is so strongly affected by the input offset voltage, these are applications that require op-amps with very low input offset voltage specifications.

14.4 FILTERING

In Chapter 9, we encountered low-pass and high-pass filters made from resistors and capacitors. In this section, we cover some of the most accessible filters available to beginning mechatronic engineers.

14.4.1 Filter Terminology

Before we get into our discussion of filters, we need to introduce some of the most common terms used to describe their behavior. Figure 14.9 illustrates these terms in the context of the frequency response of a low-pass filter.

The **pass-band** is the range of frequencies that are passed with very little or essentially no attenuation. Some filter designs exhibit a **flat** response in the pass-band (constant amplitude independent of frequency), while other designs exhibit some variation in amplitude known as **ripple**. The filter **corner frequency** (F_C) is typically the point where the amplitude has fallen by −3 dB from the amplitude in the pass-band. An exception to this definition is for filters that exhibit some ripple in the pass-band, where the corner frequency is defined as

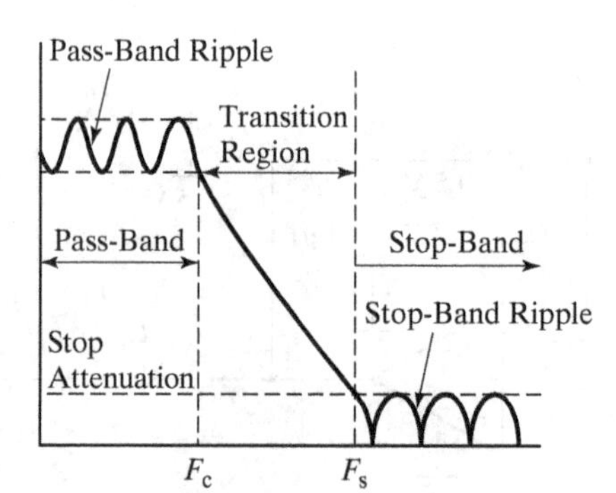

FIGURE 14.9
Filter response terminology (using low-pass filter example).

the point at which the output falls below the maximum allowable ripple. In either case, the corner frequency marks the end of the pass-band.

The **transition region** is the initial region of attenuation and extends from the end of the pass-band (F_C) to the beginning of the stop-band (F_S). In the transition band, the attenuation is increasing as the frequency gets further away from the pass-band. The **stop-band** is comprised of the range of frequencies, where there is some minimum defined level of attenuation. In the stop-band, the attenuation may continue to increase or it may reach some limiting level with some amount of ripple. The value for the **stop attenuation** is defined by the application needs but is often taken to be −40 dB or a factor of 100 below the pass-band amplitude.

The terms **poles** and **zeros** are also commonly associated with descriptions of filters. These terms refer to aspects of the mathematical structure of the transfer functions of the filters. While we will not be going into the mathematics of filters to this depth, the terms (especially poles) are ubiquitous when dealing with filters and it is useful to have a sense of their role as descriptors. Outside of the mathematics, you will often hear of filters referred to as, for example, a "four-pole filter." In this context, the number of poles refers to the number of complex impedance elements in the circuit (always capacitors in our designs). An alternative way of describing a four-pole filter would be to call it a "fourth-order filter." Here, the **order** of the filter is a reference to the exponent on frequency in the complex form of the transfer function. Since each power of frequency results from a complex impedance element, these are equivalent descriptions.

In addition to the low-pass and high-pass filter responses that we have encountered so far, there are two other common filter responses that deserve mention. The **band-pass filter** has a response that attenuates frequencies both above and below the center frequency of the filter (Figure 14.10).

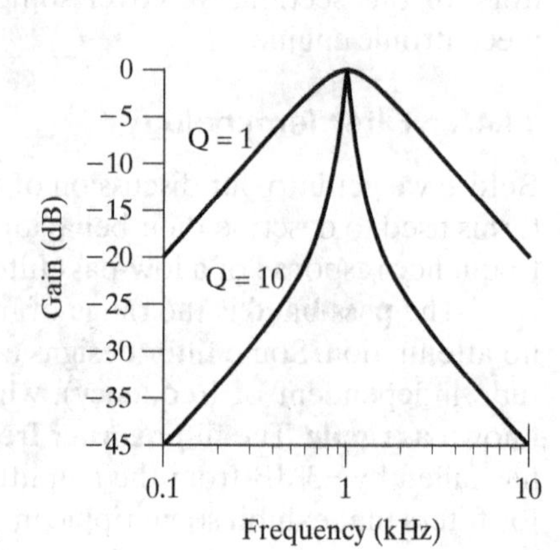

FIGURE 14.10
Band-pass filter response.

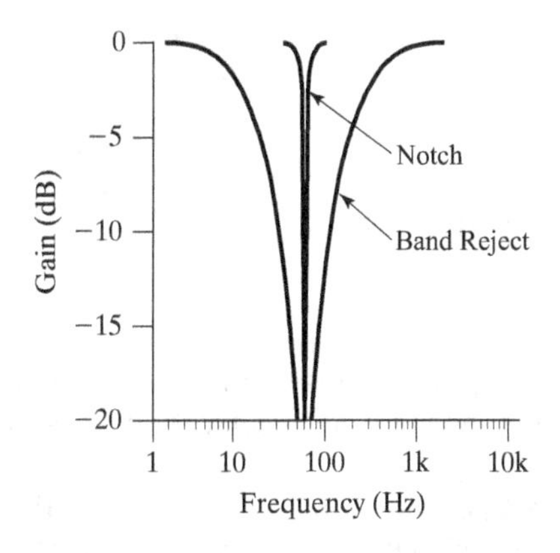

FIGURE 14.11
Band-reject and notch filter responses.

A band-pass filter response can be obtained through cascading a high-pass filter followed by a low-pass filter or through dedicated band-pass filter designs. A dedicated band-pass filter design allows for the creation of a narrower pass-band. The measure of the narrowness of the pass-band is called the **quality factor**, or **Q**, of the band-pass filter. Q is defined as $f_c/\Delta f$, where f_c is the center frequency and Δf is the width (in Hz) of the pass-band at the point where the gain has fallen by 3 dB from the peak.

The **band-reject filter**, as its name implies, attempts to pass everything *except* a range of frequencies (Figure 14.11). In this sense, it is the opposite of the band-pass filter.

Band-reject filters can only be constructed using circuits designed specifically to produce this response (by summing high- and low-pass responses). The measure of the narrowness of the rejection band is also referred to as the quality factor Q for the band-reject filter. When the rejection band becomes very narrow, it is often referred to as a **notch filter**.

14.4.2 What Is Noise and Where Does It Come From?

Noise, like beauty, is often in the eye of the beholder. While in the example of Figure 14.4, we considered the "signal" to be the 1 kHz signal and the diurnal variation in background radiation as "noise," in another application (one more focused on the overall light level), we might consider the 1 kHz variation as the noise. Generally speaking, noise is everything except the signal in which we are interested.

Noise enters our circuits through three dominant paths:

1. Through the sensor channel.
   In the IR photo-detector example above, both the response to the background IR and to the IR beacon are legitimate sensor outputs that indicate no shortcoming in the sensor. In a particular environment, a sensor might respond to more than just the sensed parameter (e.g., an air temperature sensor that also responds to solar radiation).
2. Coupled into our circuits from other circuits.
   In Chapter 21, we deal with how noise can be coupled into our circuits through several possible channels. Once combined with a signal in a circuit, the noise is indistinguishable from the signal that originated in a sensor.
3. Noise inherent in the devices.
   All devices are subject to small-scale random variations, most commonly due to temperature effects. These variations represent a stochastic noise source (with its energy distributed across the whole spectrum, rather than at a distinct frequency).

In most mechatronic devices, noise (item 3) will not often be a significant issue. The exception is the situation in which a design pushes the limits of a sensor or device. This might

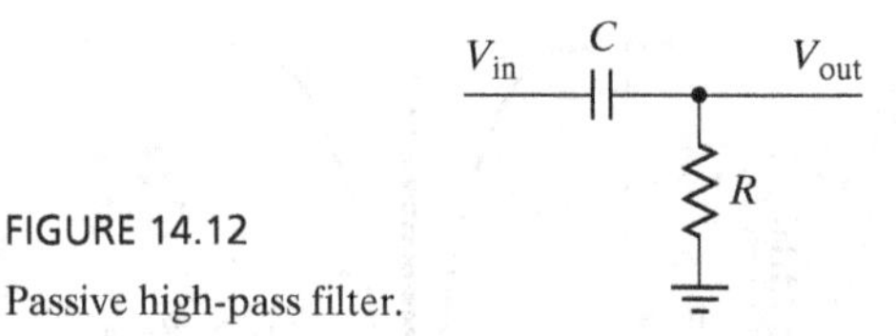

FIGURE 14.12
Passive high-pass filter.

occur, for example, when attempting to make measurements of very small accelerations where the magnitude of the response to the accelerations is comparable to the noise inherent in the accelerometer.

Of the remaining two paths, there is little we can do with sensor electronics to eliminate what we think of as "noise" but is actually a legitimate sensor response (item 1). Coupled noise (item 2) poses many significant challenges. After we have applied all the techniques presented in Chapter 21 (Noise, Grounding, and Isolation) to reduce the noise that is coupled into our circuits, we will have to employ signal conditioning techniques to deal with the coupled-in noise that remains.

In signal conditioning, the metric used to quantify the amount of noise in a system is the ratio of the amplitude of the signal of interest to the amplitude of everything else (the noise). This parameter is referred to as the **signal to noise ratio (SNR)**. When expressed in dB (the most common way), a positive SNR indicates that the signal amplitude is larger than the noise amplitude. Our goal in filtering is to optimize the SNR. If the frequency content of our signal is significantly different from the frequency content of the noise, we can use filters to selectively amplify the frequencies of interest while not amplifying (or ideally, attenuating) everything else. To do this, we must first identify the range of frequencies that are of interest to us. Only then can we effectively design circuits that will preserve the signal of interest while reducing the noise.

14.4.3 Passive Filters

The RC filters that were introduced in Chapter 9 are referred to as **passive filters** because they contain no active elements (transistors, op-amps, and so on). The transfer function for the high-pass filter (Figure 14.12) is given by Eq. (14.3).

$$|V_{out}| = V_{in}\frac{\omega RC}{(1 + R^2C^2\omega^2)^{1/2}} \tag{14.3}$$

For the low-pass filter (Figure 14.13), the transfer function is given by Eq. (14.4).

$$|V_{out}| = V_{in}\frac{1}{(1 + R^2C^2\omega^2)^{1/2}} \tag{14.4}$$

We can also express these transfer functions by reference to the frequency of the signal (in Hz) and the corner frequency of the filter (also in Hz) as shown in Eq. (14.5) for the high-pass filter and Eq. (14.6) for the low-pass filter.

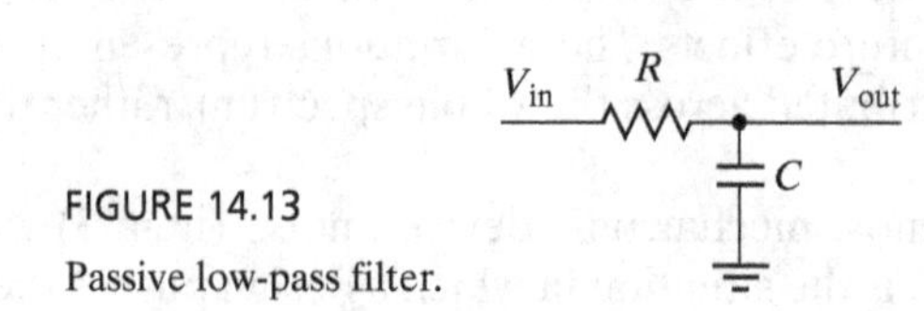

FIGURE 14.13
Passive low-pass filter.

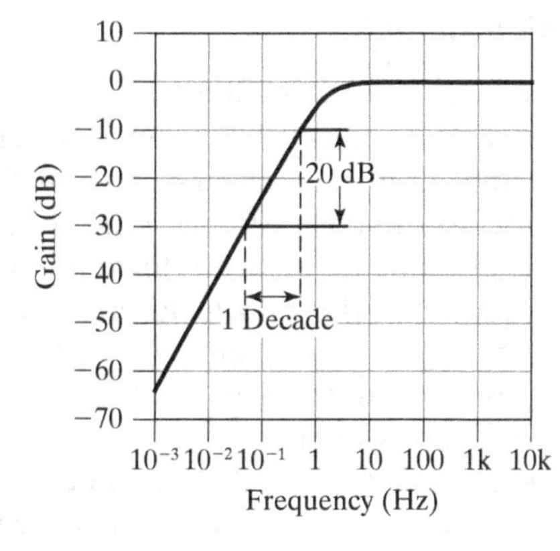

FIGURE 14.14
High-pass filter response showing −20 dB per decade slope.

$$|V_{\text{out}}| = V_{\text{in}} \frac{\left(\frac{f}{f_c}\right)}{\left[1 + \left(\frac{f}{f_c}\right)^2\right]^{\frac{1}{2}}} \tag{14.5}$$

$$|V_{\text{out}}| = V_{\text{in}} \frac{1}{\left[1 + \left(\frac{f}{f_c}\right)^2\right]^{\frac{1}{2}}} \tag{14.6}$$

These filters all exhibit a common behavior in the transition region. For every RC pair in the filter, starting at the −3 dB point, the attenuation increases by 20 dB per decade (factor of 10 increase in frequency) or 6 dB per octave (doubling in frequency). This is shown for a high-pass filter example in Figure 14.14.

In the photo-transistor example that we introduced above, there is another noise source detected by the sensor in many environments. Incandescent lights emit strongly in the IR spectrum. Even fluorescent lamps emit some IR energy, and also the photo-transistor has some sensitivity to the visible spectrum. The net result of these effects is that the desired signal from the photo-transistor will also be mixed in with 120 Hz noise.

This 120 Hz noise is much closer in frequency to our signal of interest (1 kHz) than was the diurnal variation due to the sun (1.16×10^{-5} Hz). If we were to set the corner frequency of a passive high-pass filter at 1 kHz, the 120 Hz "noise" from the room lighting would be reduced by a factor of 8.4. That doesn't sound too bad, until you realize that, in some instances, the noise may have started out as 10 times the amplitude of our signal. While this certainly improves the SNR (from −20 to −1.5 dB), the noise amplitude is still greater than the signal of interest.

Cascading passive RC filters sections to achieve greater than −20 dB per decade attenuation (in multiples of −20 dB per decade), though, is not without pitfalls. The issue is that the effective output impedance of the first RC stage adds to the *R* of the RC in the second stage, shifting its corner frequency. While it is possible to avoid this effect by using a unity gain buffer between stages, that approach is neither optimum nor elegant. Adding an op-amp to the design opens up a wide range of **active filters** that offer many more possibilities than a simple pair of RC filters. We cover active filters in Chapter 15.

14.5 OTHER SIGNAL CONDITIONING TECHNIQUES

Although the basic signal conditioning operations cover a wide range of signal conditioning applications, there are many other circuits that deal with more specialized situations. In this section, we cover two of what we feel are the most relevant circuits that fall outside the realm of basic signal conditioning.

14.5.1 The Instrumentation Amplifier

In many applications, the signal is differential in nature and, as you saw in the last chapter with the Wheatstone bridge, is riding an often substantial level of DC offset. A differential signal is fundamentally different than the typical signals that we have been dealing with. Most common signals are based on a single wire with the voltage of interest being the voltage on that wire relative to ground. These are referred to as **single-ended signals**. With **differential signals** what is important is the voltage difference between the two wires and not the voltage at either of the wires relative to ground. In the case of a fully active Wheatstone bridge made up of strain gages, when the sensors are loaded, the voltage on one side of the bridge (relative to ground) goes up, while the voltage on the other side of the bridge (relative to ground) goes down. The information that we are interested in is contained in the voltage level of one wire relative to the other wire. This sounds like the perfect situation in which to apply a difference amplifier.

While it is definitely true that what we need is a measurement of the voltage difference between the two wires, the difference amplifier that we encountered in Chapter 11 is not the best circuit to use. The difference amplifier of Chapter 11, shown again in Figure 14.15, has a number of shortcomings when put into practice with real components.

The input impedance seen at each of the inputs is only $R_f + R_i$. Since, for real op-amps, we want to keep the values of these resistors to a maximum of a few hundred kilohms, this does not provide a very high input impedance. This circuit is also very sensitive to the matching of the ratios of R_f to R_i on the two inputs. If the ratios do not match exactly (and they never will), then the circuit will amplify some part of the common mode voltage as well as the differential voltage, degrading the common mode rejection ratio, or CMRR (Chapter 12, Section 12.1.5.5).

The problem of insufficient input impedance could be tackled by adding a unity gain buffer ahead of each of the voltages. This would indeed greatly increase the input impedance. However, the problem of matching the resistors to achieve a high CMRR would still remain. The solution to both problems that has emerged is to build a **monolithic** (built on a single IC) version of what is known as a three-op-amp **instrumentation amplifier** (Figure 14.16).

To the right of the dotted line, the instrumentation amplifier is a simple difference amplifier. We'll see momentarily how the resistor matching is handled. To the left of the dotted line, the same two op-amps that would have been used to provide simple unity gain buffers are combined with three resistors (labeled R_1 and R_2) to produce a circuit that improves the CMRR. To understand how the circuit works, first consider the situation when the same voltage appears at both V_1 and V_2, call it V_{bias}. Using the second Golden Rule of op-amps, we can assert that the voltage at the inverting inputs to both op-amps is also V_{bias}. With both inverting inputs at V_{bias}, the voltage at each end of R_2 is also V_{bias}, meaning that no current flows in R_2. If no current flows in R_2, then there can be no current flowing in either of the resistors labeled R_1, and the voltage at both Node A and Node B must also be V_{bias}.

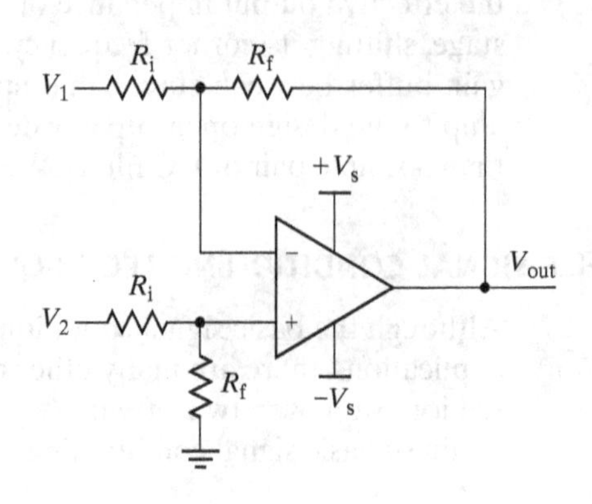

FIGURE 14.15
The basic difference amplifier.

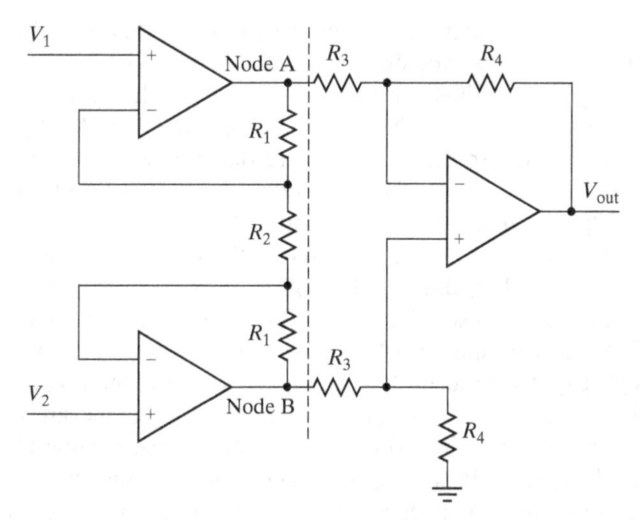

FIGURE 14.16
The three-op-amp instrumentation amplifier.

The same common mode voltage that appeared at the two inputs, V_{bias}, also appears at the inputs to the difference amplifier. Another way of describing this is to say that the circuit to the left of the dotted line has a **common mode gain** of 1.

Now, let's consider what happens when a small difference signal, ϵ, is applied to the inputs. The voltage at V_1 goes to V_{bias} $+1/2$ ϵ, while the voltage at V_2 goes to V_{bias} $-1/2$ ϵ. The second Golden Rule then tells us that the voltage across R_2 is

$$\left(V_{bias} + \frac{\epsilon}{2}\right) - \left(V_{bias} - \frac{\epsilon}{2}\right) = \epsilon \tag{14.7}$$

And Ohm's law then tells us that the current flowing in R_2 (from Node A toward Node B) is

$$\frac{\epsilon}{R_2} \tag{14.8}$$

That current must have been driven by a voltage difference across R_1 forcing V_A to be

$$V_A = \epsilon\frac{R_1}{R_2} + \left(V_{bias} + \frac{\epsilon}{2}\right) \tag{14.9}$$

A similar analysis at Node B yields

$$V_B = -\epsilon\frac{R_1}{R_2} + \left(V_{bias} - \frac{\epsilon}{2}\right) \tag{14.10}$$

Therefore, the difference in voltage, $V_A - V_B$, that appears at the inputs to the difference amplifier is

$$V_A - V_B = \left[\epsilon\frac{R_1}{R_2} + \left(V_{bias} + \frac{\epsilon}{2}\right)\right] - \left[-\epsilon\frac{R_1}{R_2} + \left(V_{bias} - \frac{\epsilon}{2}\right)\right] = \epsilon\left(1 + \frac{2R_1}{R_2}\right) \tag{14.11}$$

The differential input voltage has been amplified by a factor of $1 + 2R_1/R_2$. Since the common mode gain is 1, the CMRR has been improved by the amount of the differential gain. The output voltage is

$$V_{out} = (V_2 - V_1)\left(1 + \frac{2R_1}{R_2}\right)\left(\frac{R_4}{R_3}\right) \tag{14.12}$$

The circuit has an additional advantage if $R_4 = R_3$. In this case, the overall differential gain of the circuit can be set by choosing a single resistor value, R_2.

The circuit of Figure 14.16 has not done anything about the need to match the resistors R_3 and R_4 in order to maintain a high CMRR. That problem is handled by the monolithic construction process that we referred to above. Note that for setting the gain of the difference amplifier, the absolute values of R_3 and R_4 are not especially important (assuming that they are in the right ballpark); only their ratios really matter. This fact accommodates a manufacturing process that uses a computer controlled laser to trim the values of R_3 and R_4 after manufacture of the IC, but before it is packaged. The values of the resistors are measured, and finally a laser cuts into all but the largest resistor. The cut reduces the cross-sectional area of the resistor, increasing its resistance. A similar laser trimming process is used to match the values of R_1. In this way, it is possible to get very accurately matched resistors to maintain a high CMRR and, by having all the resistors (except for R_2) on the IC, the effects of temperature drift are minimized.

Numerous manufacturers, including Texas Instruments, Analog Devices, Linear Technology, and Maxim Integrated Products, manufacture monolithic instrumentation amplifiers. Because of the resistor trimming process during manufacturing, these products offer performance superior to what can be achieved when building an instrumentation amplifier from discrete components.

14.5.2 Peak Detection

With the advent of the microcontroller, it became possible to perform peak detection in software rather than in hardware, as it had been done in the past. For relatively low-frequency signals (less than a few hundred hertz), this is a very functional solution. The process requires that the microcontroller sample at a significantly higher frequency than the frequency content of the signal being sampled (typically 10 times faster or more), in order to be reasonably sure of sampling at, or at least very near, the peaks. As the frequency content of the signal increases, this high-frequency sampling can represent a significant execution time burden on the microcontroller, leaving less time for the other application tasks. At higher frequencies (starting about 1 kHz), it may be desirable to allocate the peak detection task to analog hardware in order to free up processor time. We recommend a circuit such as that shown in Figure 14.17.

In this circuit, the first section (which includes U_{1A}) is the actual peak detector. When the input voltage at the noninverting input exceeds the voltage on the capacitor (C_1), the output of the op-amp will swing upwards. Since the feedback point is downstream of the diode, the voltage on the capacitor (V_{cap}) is the point that will be driven to match the input voltage. The output voltage from the op-amp will rise to the input voltage plus the forward

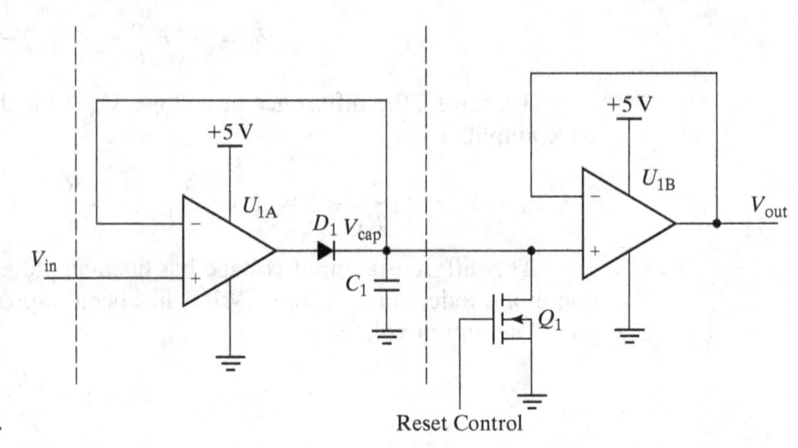

FIGURE 14.17

Peak detect circuit for use with a single power supply.

voltage drop of the diode, effectively compensating for V_f. When the input voltage falls below its most recent peak, the diode will become reverse-biased and prevent the charge stored on the capacitor from draining off and reducing its voltage. The second op-amp (U_{1B}) serves to buffer the voltage on the capacitor so that it is not discharged by a lower impedance load at the output. The MOSFET (Q_1), when controlled by a microcontroller output, allows the software to discharge the capacitor and thus reset the peak detector after a peak measurement has been made.

There are a number of factors that need to be considered when implementing a circuit such as that shown in Figure 14.17. Any charge that leaks off (or is deposited onto) the capacitor C_1 from the input bias current of the op-amp causes errors, so choosing an op-amp with low input bias currents is necessary. An unfortunate aspect of this circuit is that when the diode is blocking current from leaking off the capacitor, the output of the op-amp will be driven into saturation. When the input voltage later rises above the capacitor voltage, the op-amp will take a substantial amount of time to come out of saturation and resume charging the capacitor. This will require a very fast op-amp (> 10 MHz GBW) to be able to follow signals of even a few kilohertz with high precision. The capacitance value of the storage capacitor (C_1) is affected by the output current drive of the op-amp, which limits the rate at which it can follow input voltage changes to I_{output}/C_1. The size is also affected by the bias currents of the op-amps, and the leakage of the diode and the MOSFET causing the voltage to droop at a rate of $(I_b + I_L)/C_1$. Choosing a relatively large value for the capacitance (> 0.5 μF) minimizes the effects of the leakage and bias currents but also limits the frequency of the input signal that can be followed. Capacitors on the order of 0.01 μF are a good choice when op-amps with a few tens of picoamps of bias current and a few tens of milliamps of drive current are used. The last issue concerns the choice of op-amp. While CMOS op-amps (i.e., op-amps based on complementary MOSFET transistors; see Chapter 10, Section 10.4) offer very low input bias currents, making them desirable for this type of application, you should be aware that many RRIO CMOS op-amps (see Chapter 12, Section 12.1.5.4) become unstable when driving even moderate amounts of capacitance (more than a few hundred picofarads) at unity gain. When choosing an op-amp for use in a peak detector circuit, be sure to investigate the behavior of the op-amp when driving capacitive loads. These issues are generally dealt with in the applications section of the data sheet as well as with graphs that show the region of stability as a function of load capacitance.

14.6 CASE STUDIES

14.6.1 Information in the Amplitude

In this example, we have a signal from a photo-transistor that is being used as part of an aiming or tracking system. The amplitude of the signal contains information about how well aligned the sensor is with the emitting source. The emitter is modulated by a square wave at 1,200 Hz. One possible signal conditioning solution for this situation is shown in Figure 14.18.

The light from the emitter is converted into a voltage in the trans-resistive stage using a photo-transistor as the light sensitive element. Using a photo-transistor, rather than a photo-diode, results in a higher sensitivity for the system. The gain for this stage must be chosen so that the combination of the background radiation plus the maximum signal amplitude does not saturate amplifier U_{1A}. The voltage is then AC coupled to remove the low frequency and near-DC background component and bias the signal to 2.5 V for amplification by single supply op-amps. With the signal frequency at 1,200 Hz, setting the corner frequency of this RC to 800 Hz will minimize the loss to the signal of interest while providing good blocking of the near-DC noise. The next box provides for two stages of gain. This will provide for gains of up to about 80 per stage with op-amps having a 1 MHz GBW. If more gain is

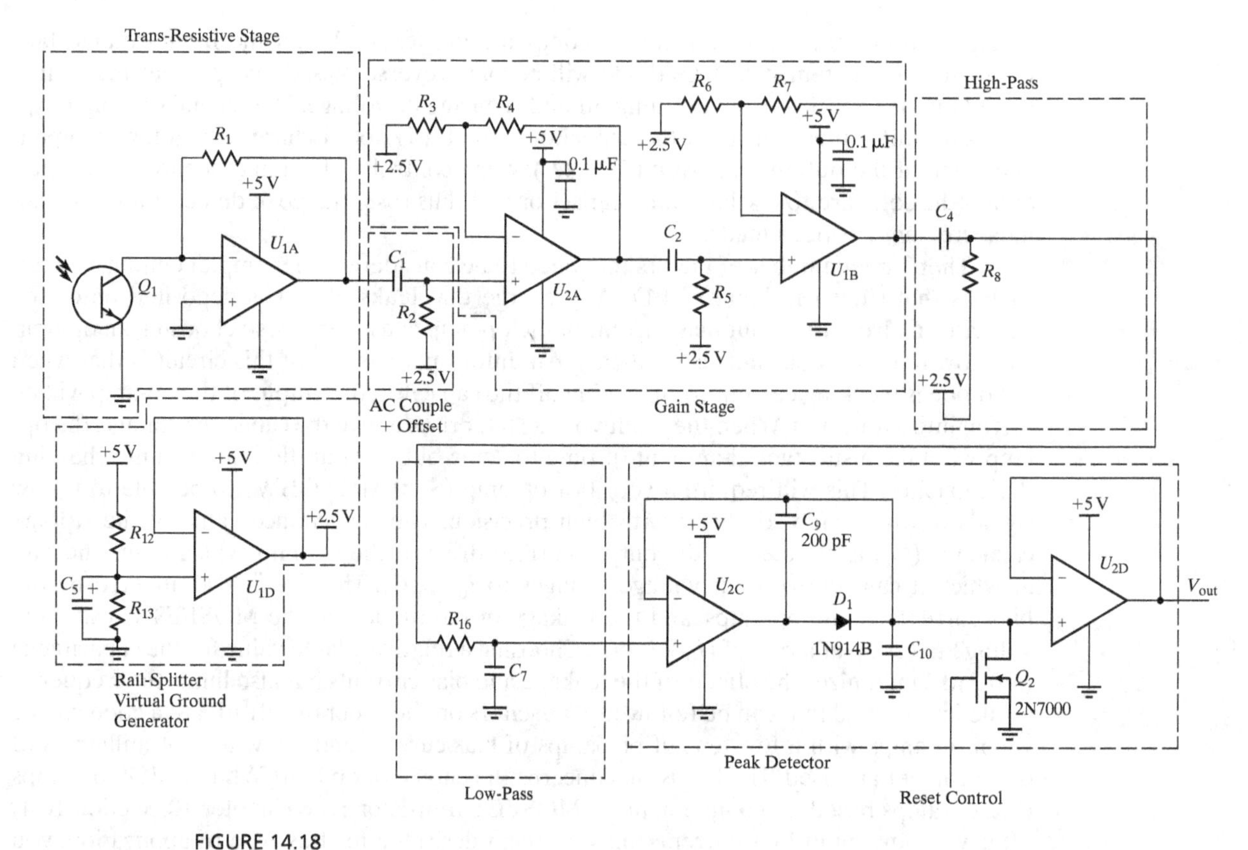

FIGURE 14.18

Signal conditioning when the information is in the amplitude.

required, then either an op-amp with a higher GBW must be used or another stage of amplification added. The R5-C2 AC coupling between stages is again set at 800 Hz to minimize signal loss. The high-pass stage serves to minimize the interference signal that will likely come at 120 Hz from the room lighting. The high-pass filter is followed by a low-pass filter to form a wide band-pass centered at 1,200 Hz. The final signal conditioning stage is a peak detector. The output of this circuit would be connected to an A/D converter input on the microcontroller. Periodically (no faster than every 1/1,200 Hz), the software should convert the analog value, which will represent the peak of the amplitude seen since the last reading, and then reset the peak detector using an output line.

14.6.2 Information in the Timing

In this second example, we once again have an optical input signal, but this time the information is in the timing of the edges of the signal. This might represent an optical signal that was being frequency-modulated (Chapter 7) to encode the information or it might be something as simple as Morse code, where dots and dashes are represented by pulses of varying duration. In any case, when the information is in the timing, the amplitude is not important. In fact, it would be best if the circuit were as insensitive to amplitude as possible. Figure 14.19 shows one possible solution to this problem.

For this situation, we have specified the use of a photo-diode as the sensing element. This is because the speed of response for the photo-diode is much better than that of a photo-transistor, and the rest of the design is predicated on having a sharp edge when light

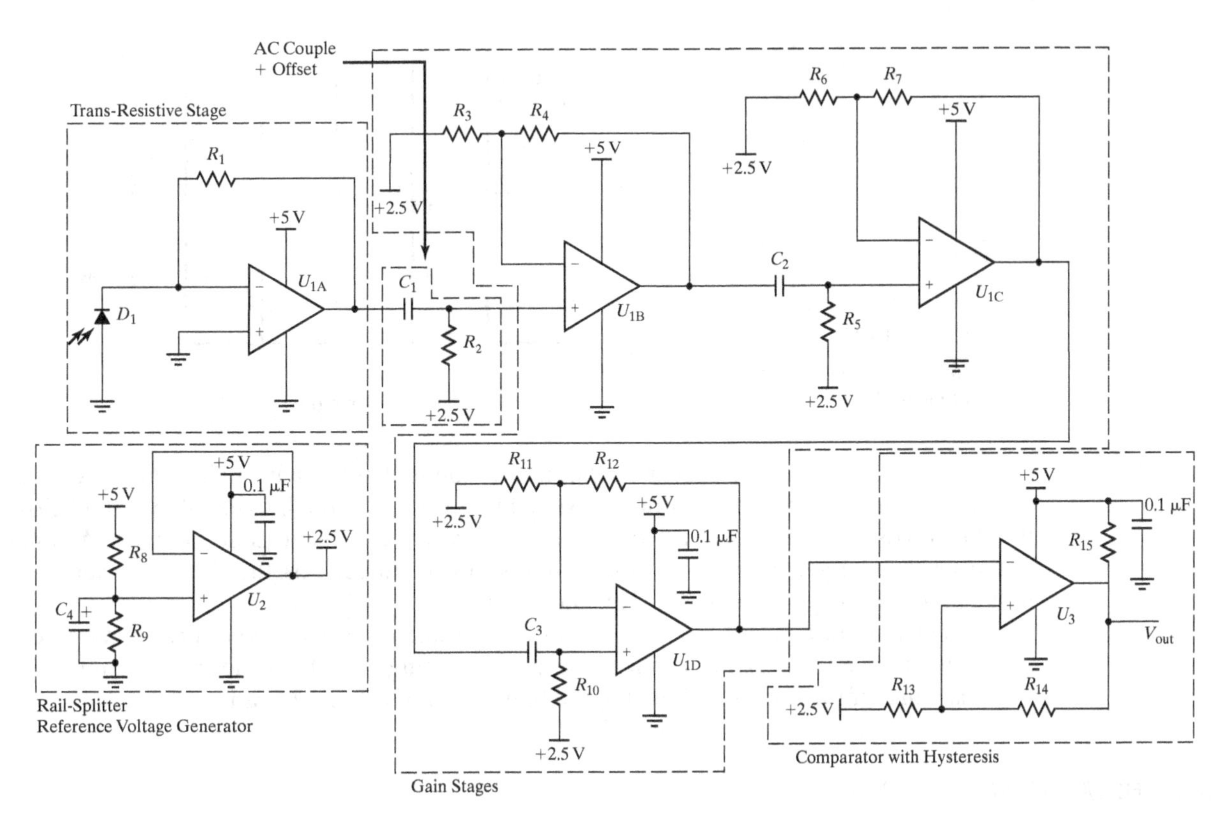

FIGURE 14.19

Signal conditioning when the information is in the timing.

transitions occur. The voltage from the trans-resistive stage is AC coupled and biased to 2.5 V next. For the AC coupling, we want to choose a corner frequency that is well above the fundamental frequency of the signal. This is because we are most interested in the edges of the signal and those are the places in the signal where the higher-frequency components are most prevalent. Since the Fourier series for a square wave ($f + 1/3 \times 3f + 1/5 \times 5f + 1/7 \times 7f + \cdots$) consists of the fundamental frequency and the odd harmonics above it, putting the corner at the frequency of the third or fifth harmonic would make the most sense. This is a trade-off between the emphasis on the higher-frequency components, which will give a sharper edge, and the signal amplitude, which decreases with each higher harmonic. A possible strategy would be to set the first coupling corner at $3f$, the second at $5f$, and the last at $7f$, taking advantage of the fact that the signal amplitude is increasing as we move through the gain stages. Because of the lower sensitivity of the photo-diode compared to a photo-transistor, this solution allows for three gain stages. The amplifiers used in these stages should be operated within a GBW based not on the fundamental frequency, but on the fifth or seventh harmonic, which is where the interesting parts of the waveform are located in the frequency domain. To maintain the sharp edges that the photo-diode is capable of generating will require op-amps with an output slew rate of 5 V/μs (or better) in order to maintain the signal rise time to 1 μs or less. The last stage of the circuit is a comparator with hysteresis with the thresholds centered around 2.5 V, which is also the offset used for amplifying the signal. The behavior of this last stage is shown graphically in Figure 14.20.

The AC-coupled square wave signal creates the spikes above and below the reference voltage at every rising and falling edge of the input signal. The comparator hysteresis band is

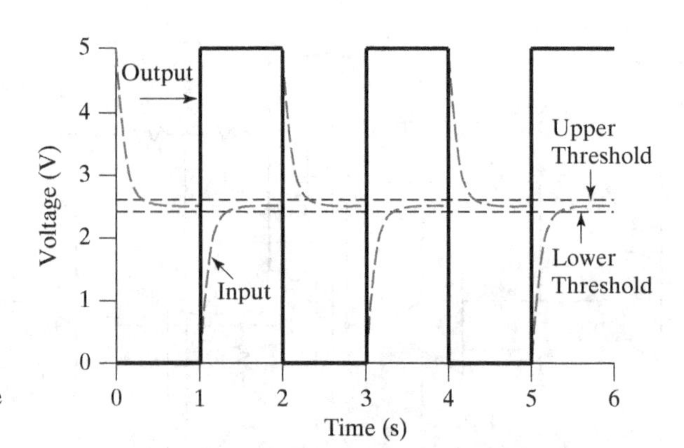

FIGURE 14.20

Behavior of the comparator stage of Figure 14.19.

set wide enough so as not to trigger on the small amount of noise that will inevitably be present on the signal out of the amplifiers. A signal that is large enough to get above and below the thresholds will recreate in the output the square wave transitions in the input. As long as the signal is large enough to get above and below the thresholds, the circuit is insensitive to the amplitude of the signal out of the amplifier.

The circuit of Figure 14.17 then serves to translate a sharp edged optical signal into a similarly sharp edged digital signal while compensating to a large degree for the finite response time of the sensor and the limited bandwidth of the amplification stages.

14.7 HOMEWORK PROBLEMS

14.1 What kind of capacitors would make a good choice for a DC blocking capacitor?

14.2 What is the minimum number of resistors that need to be changed to alter the differential gain of the circuit of Figure 14.16, while maintaining the insensitivity to common mode voltage?

14.3 You have been asked to design the signal conditioning circuit to make the signal from a new temperature sensor suitable for use with a microcontroller. The sensor has a linear response over the range from 0 to 100°C with a temperature coefficient of −100 Ω/°C. At 0°C, it has a resistance of 11 kΩ. Your circuit should produce an output that is linearly proportional to temperature with a scale factor of −20 mV/°C over a range from 0 to 100°C. The actual voltages that the circuit produces are up to you (i.e., the offset is not specified), provided no device limits are exceeded and the scale factor requested is delivered. You may assume ideal components limited only by the 5 V supply available.

14.4 Given a slowly changing (treat it as DC) signal with an amplitude that varies between 0 and 0.2 V that is superimposed on a 2 V true DC offset voltage, design a circuit that will amplify the slowly changing signal by a factor of 10 while not amplifying the DC offset. The actual range of output voltage that you choose is up to you (it may include an offset), but you must stay within the 0 and 5 V supplies that are available to power your circuit.

14.5 Given a sinusoidal signal at 6 kHz with an amplitude of between 0 and 0.2 V superimposed on a 60 Hz signal that varies between 0 and 1 V, design a circuit that will amplify the 6 kHz signal by a factor of approximately 10 and attenuate the 60 Hz signal by a factor of at least 10. You should look for a design that will use the simplest possible circuit to achieve your goals.

14.6 Some sensors produce an output that is not a single voltage but a pair of voltages in which the actual signal of interest is the difference between the two voltages. This is referred to as a differential output. In these situations, it is common for there to be some DC offset that exists on both signals. For example, consider the case of an accelerometer that produces a differential output with a nominal output voltage on each of outputs of 2.5 V. The exact value of this offset is not well controlled (it may vary from device to device). Design a circuit that will take a 0–0.2 V differential signal with frequency content from DC to 100 Hz from such an accelerometer and amplify it by a factor of approximately 10 while minimizing the effects of the offset voltage common to both inputs. You may assume ideal op-amps and exact value resistors are available for your design.

14.7 Given a 200 mV peak-peak 1 kHz sinusoidal signal offset by a 200 mV peak-peak 2.7×10^{-4} Hz (1 cycle per hr) signal, design a circuit to amplify the 1 kHz sinusoid by a factor of 10 while reducing the amplitude of the low-frequency signal below 1 μV. Use only passive filters. You may assume ideal op-amps and exact value resistors are available for your design. How would you modify your design if the only fixed resistors available were 1% resistors?

14.8 A DC motor tachometer outputs −5 to 5 V proportional to the speed and direction of the motor. Design a circuit that takes this signal and maps it onto the range of 0–5 V so that it can be presented the input of an analog-to-digital converter.

14.9 Show the design for a circuit using the Texas Instruments INA156 to amplify by a factor of 10 the output of a sensor with a Wheatstone bridge sensing element powered by 5 V.

14.10 You have been asked to design the signal conditioning circuitry for an inclinometer based on a pendulum and a potentiometer. The range of measurement should be ±45° around vertical. The potentiometer has a full-scale resistance of 25 kΩ and a 270° angular sweep. The potentiometer is at midscale when vertical and the output of your circuit should exhibit a linear transfer function of 10 mV/°.

FURTHER READING

The Art of Electronics, Horowitz, P., and Hill, W., 2nd ed., Cambridge University Press, 1989.

Op Amps for Everyone, Carter, B., and Mancini, R., 3rd ed., Newnes, 2009.

CHAPTER 15

Active and Digital Filters

While the passive filters that we introduced in Chapters 9 and 14 are useful, there will be many situations where they simply do not meet the requirements for the application. This most commonly occurs when the frequency separation between the signal of interest and the interfering signal is not very great. In that situation, it will be difficult to get significant attenuation of the unwanted signal while preserving the amplitude of the desired signal. What we need is a filter whose roll-off characteristic is steeper than the −20 dB per decade that we can achieve with a single stage passive filter. While it is possible to cascade passive RC filter sections to achieve greater than −20 dB per decade attenuation (in multiples of −20 dB per decade), as we saw in the last chapter, this approach is not without pitfalls. Adding an op-amp to the design opens up the wide range of active filters that offer many more possibilities than a simple chain of RC filters and buffers. The higher-performance alternatives, **active filters**, are a focus of this chapter. In addition to the active filtering that takes place in hardware, we will also discuss a few simple **digital signal processing** (**DSP**) techniques that can be implemented once the signal has reached the microcontroller.

After studying the material of this chapter, you should be able to:

1. choose between active and passive filters for an application,
2. understand the basics of active filters and how to implement them,
3. understand several of the more significant side effects of filtering,
4. recognize some of the opportunities available when performing signal conditioning in software.

15.1 ACTIVE FILTERS

An **active filter** is one that contains an active (i.e., powered) component. Universally, the active component of choice is the op-amp. While it is possible to construct a first-order filter (1 complex impedance element) using an op-amp, having the op-amp allows for a second-order filter with no loss of output impedance. This section will deal with one of the most popular circuit topologies used to build second-order filter blocks using op-amps. To achieve greater than second-order filters, multiple second-order stages can be cascaded. These cascaded stages are rarely all identical. We will defer more details on cascaded filters to Section 15.1.3.1, where we detail a design procedure.

15.1.1 Phase Delay

In Section 9.6.2, when we introduced the expression for the impedance of a capacitor, we noted that the presence of j ($\sqrt{-1}$) indicated a 90° phase shift in the output. While we are mostly interested in the amplitude response of filters, as the complexity of the filters increases, the phase response of the filter plays an increasingly important role. The phase response of a filter is typically described by the companion to the Bode amplitude plot, the **Bode phase plot** (Figure 15.1).

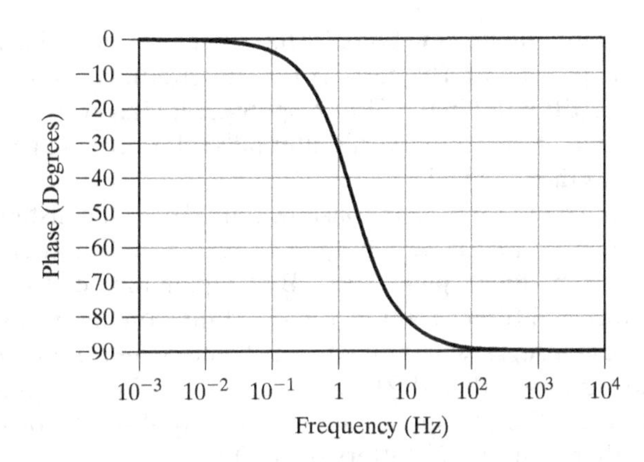

FIGURE 15.1
Bode phase plot for an RC low-pass filter.

Here, we see that in this low-pass filter response at very low frequencies, the impedance of the resistor dominates and there is no phase shift. At high frequencies, the capacitor (with its 90° phase shift) dominates. At the corner frequency, the phase shift is 45°.

If all we were interested in were the amplitude of individual frequency components, then the phase response would be of little consequence to us. However, the *shape* of a signal's waveform (time domain behavior) is often important. The shape of any waveform (other than a single-frequency sine wave) is made up of several frequency components (discussed in more depth in Section 9.6.2). As these different frequencies pass through a filter with a response like that of Figure 15.1, they experience different delays. The result is that the shape of the waveform at the filter output is distorted by the differing delays. As an example, Figure 15.2 shows the time domain output response to a square-edged negative going pulse for three different types of second-order low-pass filters with differing delay properties. We introduce each of these filters in the next section.

As we can see, some filters exhibit more accuracy in the reproduction of the pulse (time domain behavior) than others. These differences are due to the different phase characteristics of the filters.

15.1.2 Filter Response Characteristics

Over the years, many different approaches to choosing the values of the resistances and capacitances have been developed that led to, in some cases, dramatically different filter responses for what are all nominally the same order of filter. These different filters are most commonly known by the names of the people who developed either the filter circuits or the mathematics underlying them. Rather than create a litany of the possible filter types, we will limit our discussion to the three most common "named responses."

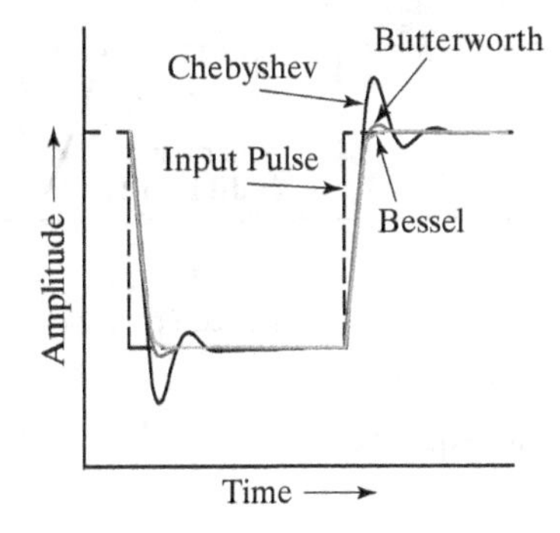

FIGURE 15.2
Pulse response for several types of second-order filter.

Butterworth filters, also known as **maximally flat filters**, are optimized for constancy of amplitude response in the pass-band. The cutoff, or corner, frequency is defined, as with the passive RC filters, as the −3 dB point in the filter response. Moving away from the corner frequency, the output amplitude of the Butterworth filter falls off (decreases) at the same −20 dB per decade per pole rate as the passive RC filter.

Bessel filters are optimized for a near-constant time delay. A constant time delay corresponds to a linear change in phase with frequency, leading to another descriptive term applied to the Bessel filter: the **linear phase filter**. By keeping the delay constant across the range of frequencies, the Bessel filter produces the best time domain behavior, which is also referred to as **transient response**, as seen in Figure 15.2. The corner frequency for the Bessel filter is defined, like the Butterworth and RC filters, as the −3 dB point in the filter's output response. Moving away from the corner frequency, the output of the Bessel filter initially falls off at a slower rate than that of the Butterworth filter.

Chebyshev filters trade both amplitude flatness in the pass-band and linearity of phase delay for improved (faster) roll-off in the transition region. Chebyshev filters intentionally allow for ripple (varying gain) in the pass-band, and the variations on the Chebyshev filter designs are principally differentiated by the peak-to-peak magnitude of ripple that is allowed in the pass-band (e.g., 0.1 dB ripple, 1 dB ripple, 3 dB ripple, and so on). Unlike Butterworth and Bessel filters, the corner frequency of a Chebyshev filter is determined by that point at which the output falls outside the allowable pass-band ripple. For a 1 dB ripple Chebyshev, for example, that would be when the output fell below −0.5 dB from the mean pass-band amplitude. The time domain behavior of the Chebyshev filter, as seen in Figure 15.2, is the worst of the three filter types exhibiting overshoot and ringing in the pulse response. In exchange for the pass-band ripple and poor time domain response, the Chebyshev filter delivers the steepest roll-off in the transition region. Increasing amounts of pass-band ripple result in steeper roll-off.

Figure 15.3 shows a comparison of the responses for sixth-order implementations of these three filter types.

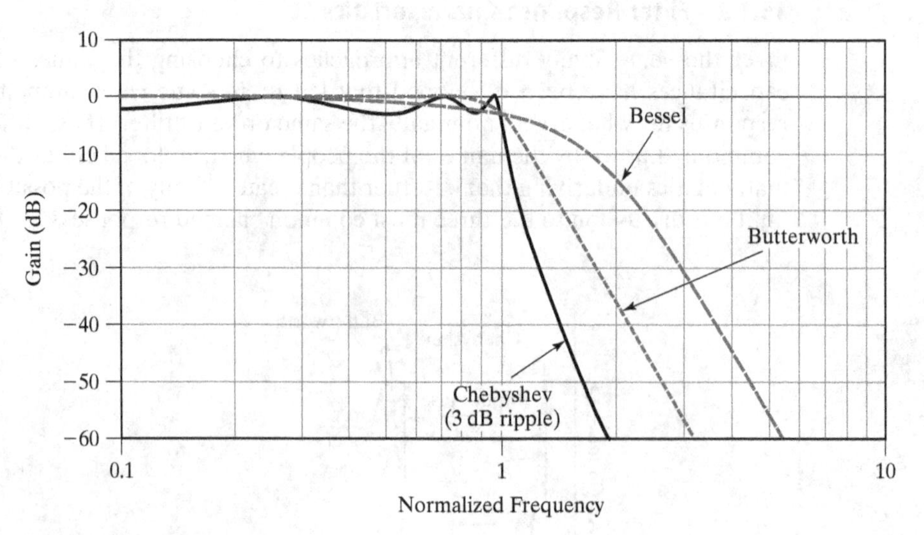

FIGURE 15.3

Comparison of sixth-order filter responses.

15.1.3 Active Filter Topologies

For each of the named responses, there are multiple arrangements of the components (circuit topologies) that can be used to build the filter. To prevent this chapter from resulting in a whole book unto itself, we limit our discussion to one of the most common topologies and present a simplified design procedure to help designers implement filters using this approach.

The circuit topology that we focus on is called a **voltage controlled voltage source filter** (or **VCVS**), also known as a **Sallen-Key filter** after its inventors. This "building block" filter produces a second-order filter that can be configured to produce a high-pass, low-pass, or band-pass filter response (Figure 15.4). These versions of the classic circuits have been adapted for single supply operation. To produce higher-order filters, second-order stages can be cascaded, though C_i and R_i are only strictly required on the first of a cascade of low-pass filters. For later stages, including C_i and R_i prevents any DC offset from the prior stage from propagating. Wherever they are used, R_i should be chosen to be large (~100 kΩ) and C_i should be chosen to be 100–1,000 times larger than C_1. This will put the corner of the high-pass filter that is inherent in an AC coupled circuit well below the filter corner frequency.

For the Sallen-Key filter configuration, there are many more variables (resistors and capacitors) than there are parameters to adjust (corner frequency, gain, response type, Q). This has led to the creation of many software packages (see the References for several freely available examples) and circuit simplifications designed to help with the selection of

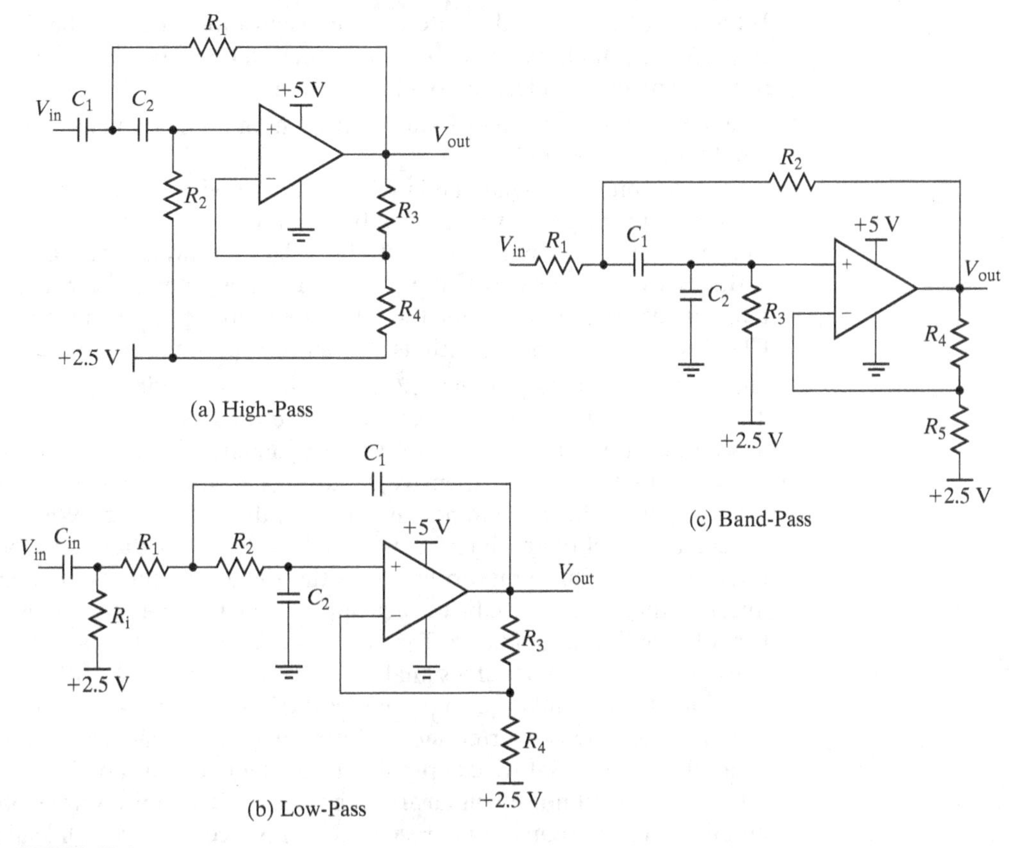

FIGURE 15.4

Sallen-Key filter circuit topologies.

components to achieve a desired response. The simplified design procedure that we will present is based on setting the frequency determining resistors and capacitors to be equal ($R_1 = R_2, C_1 = C_2$) and choosing the ratio of R_3 to R_4 to produce the gain necessary to achieve the desired filter response. The procedure only applies to the low-pass and high-pass variants of the filters. Wide band-pass filters can be constructed by cascading a high-pass stage followed by a low-pass stage, but the creation of narrow band-pass filters involves too many interactions to lend itself to reasonable simplification. Component selection for narrow band-pass versions of the Sallen-Key filter (and other useful topologies) is best handled by using one of the software packages listed in the References section of this chapter.

15.1.3.1 Simplified Sallen-Key Filter Design Procedure

Any filter design procedure must begin with identifying the frequency ranges of both the signals of interest and the most prominent noise sources. The frequency range of the signal(s) of interest tells us where we need to pass the signals, either unmolested or amplified. Knowledge of the amplitude and frequency of the interfering signals (noise) gives us the information necessary to decide how aggressive we need to be with filtering to attenuate these components. Armed with this knowledge, we can proceed to ask and answer a series of questions that will guide the selection of the type and order of filter required to achieve our filtering goals.

1. **Is the signal of interest a sine wave (or approximately sinusoidal)? If not, is the shape of the waveform important?**
   If the shape of the signal of interest is nonsinusoidal and maintaining the approximate shape is important, then a filter with a linear phase response will be necessary. This will drive the response selection to a Bessel filter.
2. **Are there multiple signals of interest present in the pass-band? If so, are their relative amplitudes important?**
   If there are multiple signals and their relative amplitude carries important information, then the filter response will need to be as uniform as possible across the frequencies of the pass-band (i.e., a maximally flat filter). This requirement argues for the selection of a Butterworth or very low ripple (0.1 dB) Chebyshev filter response. If the relative magnitudes are not important, then this opens up the opportunity to use a larger ripple Chebyshev filter response with its steeper roll-off in the transition region.
3. **How much attenuation of the interfering signals is needed?**
   This is probably the most subjective (and difficult) of the questions that need to be answered. In an ideal world, we would eliminate all trace of the interfering signals. However, that would be expensive, requiring multiple op-amps to achieve a high-order filter to squash the interference, and is generally unnecessary. What we need to do is to reduce the level of the interfering signal to the point that it no longer substantially interferes with our ability to interpret the signals of interest. Unfortunately, this is a question that is very specific to each application and does not lend itself to detailed yet broadly applicable guidance. The best specific advice that we can give is that if the amplitude of the interfering signal is reduced to below 1/2 LSB on the A/D converter (see Chapter 19), then the amplitude of the interference will be smaller than the resolution of the A/D converter and therefore will not impact the measurement. The best general guidance that we can provide is to examine your application to determine the effect that added noise will have on the system. The noise that is present adds to the amplitude of the signal of interest, so the presence of noise will lead to measurements that overstate the amplitude of the signal of interest. Since the interfering signal is often varying and its amplitude not specifically known, its presence introduces an additional level of uncertainty in the measurement. For some systems, it will be possible to

model the overall system accurately enough that analysis can lead to useful estimates of the effect of the added noise. In other cases, it will be necessary to experiment to determine the necessary attenuation. In those cases, where experimentation is necessary, start with a low-order filter (i.e., an easier solution) and add stages (i.e., complicate the design) only after it becomes evident that the low-order filter is not adequate.

Once you know the type of filter necessary and the amount of attenuation required, you can use diagrams like those of Figures 15.5–15.8 to determine the order of the filter type necessary to meet the attenuation goals.

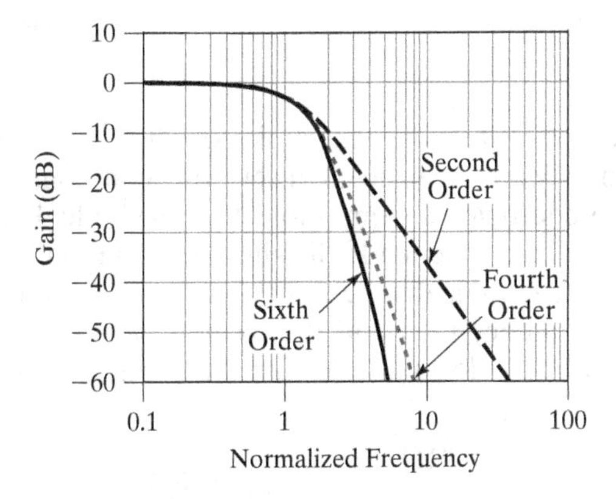

FIGURE 15.5
Bessel filter responses.

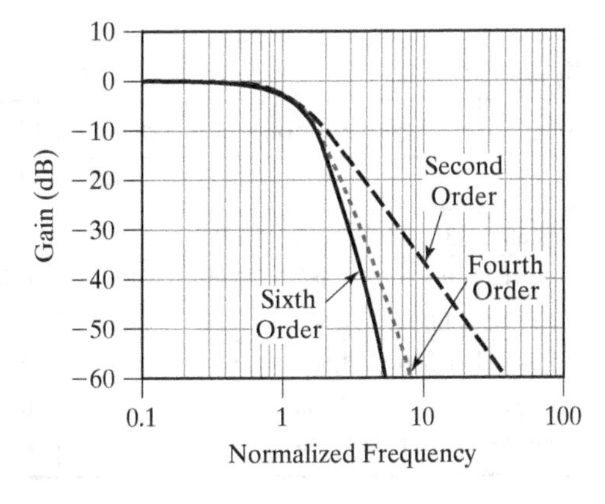

FIGURE 15.6
Butterworth filter responses.

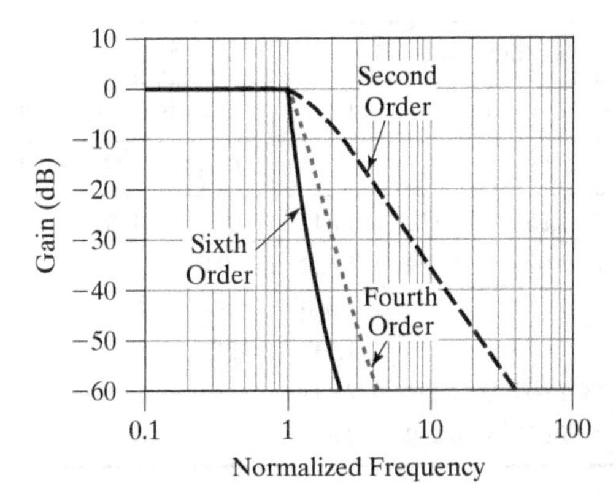

FIGURE 15.7
Chebyshev (0.5 dB ripple) responses.

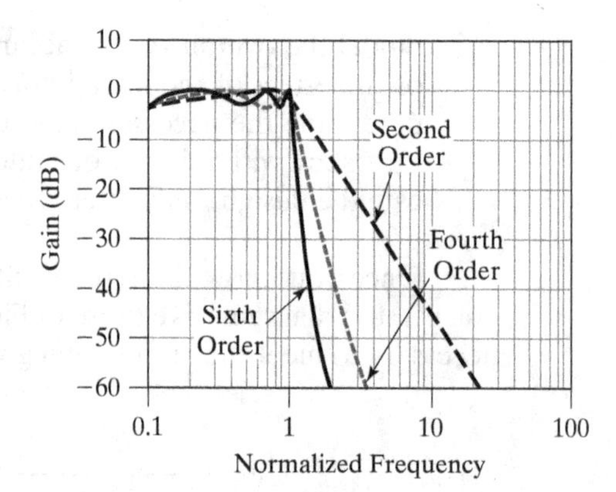

FIGURE 15.8
Chebyshev (3 dB ripple) responses.

Once you have determined the best choice of filter (response type and order) and corner frequency, you may begin the process of selecting components.

We will use the coefficients from Table 15.1 or Table 15.2 to calculate the required component values. Since we have made the simplification that $R_1 = R_2$ and $C_1 = C_2$, there is only one R value and one C value that determines the corner frequency for a stage of the filter. The desired corner frequency (F_C) and the frequency modifier (F_{mul}) from Table 15.1 or Table 15.2 are used with Eq. (15.1) to calculate the necessary RC product.

$$RC = \frac{1}{2\pi\,(F_{mul} \times F_C)} \tag{15.1}$$

TABLE 15.1 Coefficients for low-pass Sallen-Key filter design.

Order	Section	Butterworth		Bessel		Chebyshev 0.5 dB		Chebyshev 1 dB		Chebyshev 2 dB		Chebyshev 3 dB	
		F_{mul}	G	F_{mul}	G	F_{mul}	G	F_{mul}	G	F_{mul}	G	F_{mul}	G
2		1.000	1.586	1.274	1.268	1.231	1.842	1.050	1.268	0.907	2.114	0.841	2.234
4	1st	1.000	1.152	1.419	1.084	0.565	1.582	0.993	1.725	0.471	1.924	0.950	2.071
	2nd	1.000	2.235	1.591	1.759	1.031	2.660	0.529	2.719	0.964	2.782	0.443	2.821
6	1st	1.000	1.068	1.606	1.040	0.396	1.537	0.353	1.686	0.316	1.891	0.298	2.042
	2nd	1.000	1.586	1.691	1.364	0.768	2.448	0.747	2.545	0.730	2.648	0.722	2.711
	3rd	1.000	2.482	1.907	2.023	1.011	2.846	0.995	2.875	0.983	2.904	0.977	2.922

TABLE 15.2 Coefficients for high-pass Sallen-Key filter design.

Order	Section	Butterworth		Bessel		Chebyshev 0.5 dB		Chebyshev 1 dB		Chebyshev 2 dB		Chebyshev 3 dB	
		F_{mul}	G	F_{mul}	G	F_{mul}	G	F_{mul}	G	F_{mul}	G	F_{mul}	G
2		1.000	1.586	0.785	1.268	0.812	1.842	0.952	1.268	1.103	2.114	1.188	2.234
4	1st	1.000	1.152	0.705	1.084	1.769	1.582	1.007	1.725	2.123	1.924	1.052	2.071
	2nd	1.000	2.235	0.628	1.759	0.970	2.660	1.892	2.719	1.037	2.782	2.259	2.821
6	1st	1.000	1.068	0.623	1.040	2.525	1.537	2.831	1.686	3.165	1.891	3.356	2.042
	2nd	1.000	1.586	0.591	1.364	1.302	2.448	1.339	2.545	1.370	2.648	1.384	2.711
	3rd	1.000	2.482	0.524	2.023	0.989	2.846	1.005	2.875	1.017	2.904	1.023	2.922

Since capacitors are available in a much more limited range of values than are 1% resistors, start by choosing a standard value capacitor to use for C_1 and C_2 and then calculate the required value for R to make the RC product. If the calculated value for R is outside the desirable range (a few tens of kilohms to a few hundred kilohms), choose a different standard capacitor value and try again. Choose the nearest 1% resistor value to use for R_1 and R_2.

The values for R_3 and R_4 are calculated using the G entries from Tables 15.1 and 15.2 with Eq. (15.2). Make sure that both values fall into the usual desirable range for resistors.

$$R_3 = (G - 1)R_4 \tag{15.2}$$

That's all there is to the simplified design procedure. Once you've completed this last step, you have determined the value of all the components and you are ready to build the filter circuit.

As an example, suppose that we want to make a sixth-order 3 dB ripple Chebyshev low-pass filter with a corner frequency at 1 kHz. For a sixth-order filter, we can see from Table 15.1 that we will require three stages. This is because each stage provides a second-order response, so by cascading three sections we can achieve a sixth-order response. For the first section, we calculate the required RC product by substituting F_{mul} from the Table 15.1 and our desired corner frequency into Eq. (15.1).

$$\mathrm{RC} = \frac{1}{2\pi(0.298 \times 1\ \mathrm{kHz})} = 5.34 \times 10^{-4}$$

Since we would like the R to be in the range of tens to hundreds of kilohms, we will need to choose a C in the range of hundreds of picofarads to a few nanofarads. We are free to choose the R and C arbitrarily to achieve the required RC product, but the values of available capacitances are much more limited than the range of available resistances, so let's start by choosing C as 2 nF (0.002 μF). That requires that we choose an R value of 267 kΩ. It turns out that 267 kΩ is a standard 1% resistor value, so we will choose that value. To distinguish the resistors and capacitors in this section from those occupying the same positions in the later section of the filter, we will add a suffix "A" to these designators to denote those components in the first section of the multistage filter. So, we now have $R_{1A} = R_{2A} = 267$ kΩ and $C_{1A} = C_{2A} = 0.002$ μF.

Choosing the best pair of values for R_3 and R_4 is a matter of determining which combination of 1% resistor values will get us closest to our desired gain. This is probably most easily done empirically using a spreadsheet. For this stage's required gain of 2.042, the best match is using 1% resistors with multipliers of 169 and 162. Choosing $R_{3A} = 16.9$ kΩ and $R_{4A} = 16.2$ kΩ will yield a gain of 2.043, a 0.1% error.

We can repeat this process for the second and third stages determining the required values as:

	Second Stage	Third Stage
RC Product	$\mathrm{RC} = \frac{1}{2\pi\ (0.722 \times 1\ \mathrm{kHz})} = 2.2 \times 10^{-4}$	$\mathrm{RC} = \frac{1}{2\pi(0.977 \times 1\ \mathrm{kHz})} = 1.63 \times 10^{-4}$
R	147 kΩ	60.4 kΩ
C	1.5 nF	2.7 nF
G	2.711	2.922
R_3	17.4 kΩ	22.1 kΩ
R_4	10.2 kΩ	11.5 kΩ

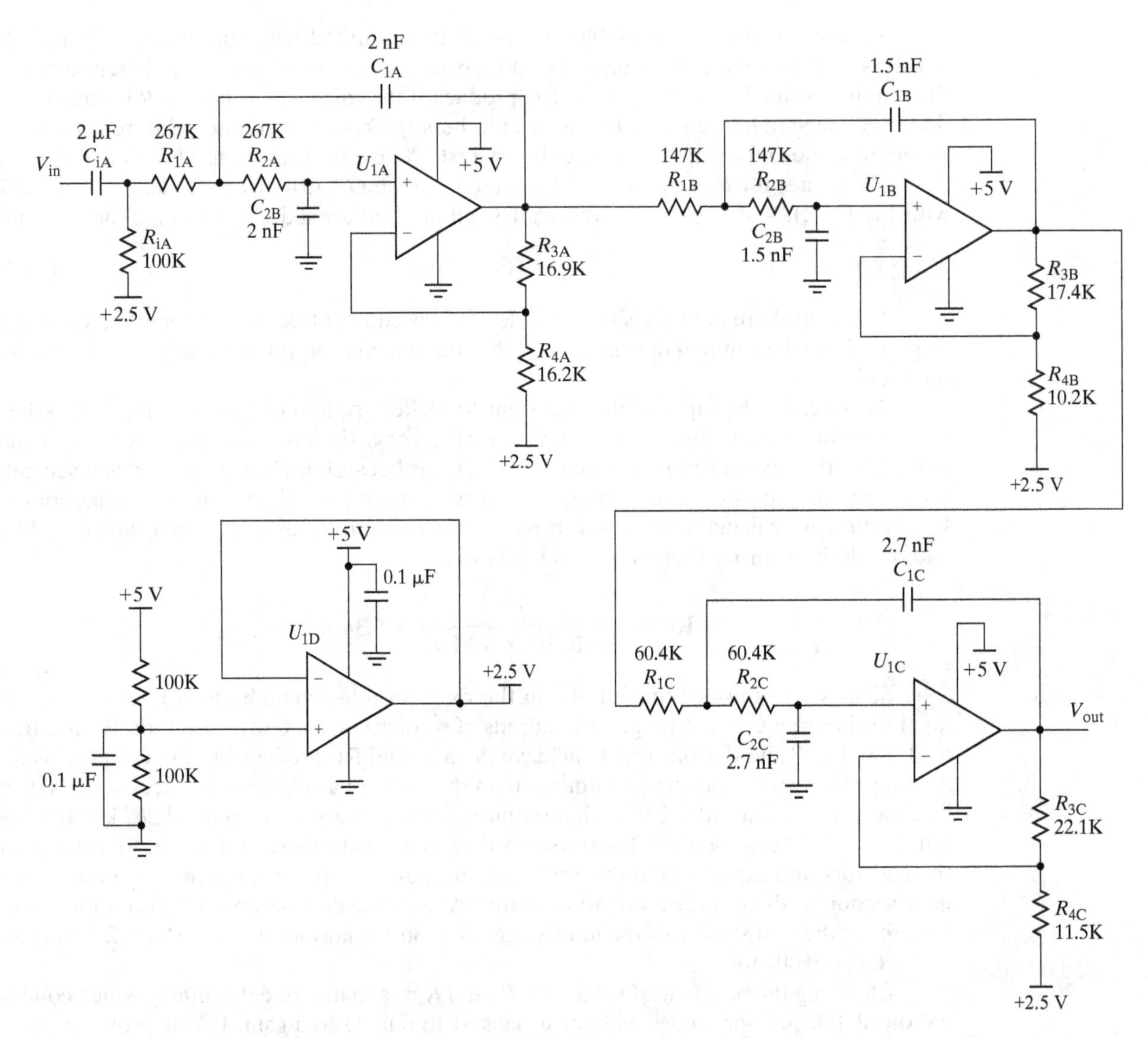

FIGURE 15.9
A complete sixth-order Chebyshev (3 dB ripple) low-pass filter.

Pulling this all together in a schematic results in the circuit shown in Figure 15.9. This design does not assume that the input signal is referenced to 2.5 V and includes C_{iA} and R_{iA} to insure that the signal is referenced to 2.5 V. The output, V_{out}, is centered at 2.5 V.

15.2 DIGITAL TECHNIQUES

Not all signal processing takes place before the signal reaches the microcontroller. It is also possible to use DSP techniques to "condition" the signal after it has been acquired, but before it is used in an algorithm in software. In this section, we introduce two (of the many) such possible DSP techniques: digital filtering and synchronous sampling.

15.2.1 Digital Filtering

Digital filtering, much like the analog filtering that we have discussed at length above, is used to modify the frequency domain characteristics of signals. With digital filters, it is possible to construct filters with much sharper transition regions and much more tightly controlled behavior than is possible with analog filters. The "cost" for these filters is in the development of software to implement the filters (which can be very low) and in the processing time necessary to execute the filtering algorithms (significant, but getting cheaper all the time). DSPs are a category of processors featuring hardware specifically designed to perform the most common digital filtering and processing operations very quickly and efficiently (see also Section 2.3). The DSP field is quite vast and understanding many of the algorithms involves mathematics that are well beyond the scope of this book. We will focus on one specific, simple, and more easily understood algorithm, and then provide a general overview of how more complex digital filters are implemented and what sets a DSP apart from an ordinary microcontroller.

15.2.1.1 The Moving Average Filter

Perhaps the most intuitive approach to digital filtering is to simply average a few samples. For random noise, this turns out to be a very effective approach. In fact, the moving average filter that we will discuss is considered to be the optimal filter for reducing random noise while best preserving the step response (time domain behavior) of an incoming signal.

We might reasonably start building an averaging filter algorithm by stating that every new value produced by the algorithm will be calculated by taking the average of the newest input data with the reading that preceded it. This would be a 2-point moving average that is expressed mathematically as

$$Y_i = (X_i + X_{i-1})/2 \tag{15.3}$$

Where Y_i is the current output from the filter, X_i is the current input to the filter, and X_{i-1} is the value of the previous input to the filter.

We could extend this to include more points (N) in the average:

$$Y_i = \left(\sum_{j=0}^{j=N-1} X_{i-j}\right)/N \tag{15.4}$$

From a computational standpoint, the moving average filter is very simple and inexpensive to implement. Addition and subtraction are, computationally, very "inexpensive"—that is, they don't require a large number of software instructions or microcontroller clock cycles to implement. The filter only requires one more "expensive" operation (division) per output value generated. If we choose the number of samples in the average to be a power of 2, then even the relatively expensive division operation can be replaced by a simple (and fast) right-shift operation (we introduced this "shortcut" for microcontrollers in Chapter 3 on microcontroller math, in Section 3.8).

For implementations of more than two samples being averaged, we can simplify the algorithm even further.

Notice in this calculation sequence (Figure 15.10) for an 8-point moving average that two successive outputs from the filter share a significant amount of commonality in the summation.

$$Y_{10} = \quad (X_{10}+X_9+X_8+X_7+X_6+X_5+X_4+X_3)/8$$
$$Y_{11} = (X_{11}+X_{10}+X_9+X_8+X_7+X_6+X_5+X_4)/8$$

FIGURE 15.10

Calculation of two successive moving average outputs.

$$\mathrm{SUM}_{11} = \mathrm{SUM}_{10} - X_3 + X_{11}$$
$$Y_{11} = \mathrm{SUM}_{11}/8$$

FIGURE 15.11
Simplified calculation for moving average.

If we were to keep the summation from the prior output calculation (we'll call it SUM), we could recast the calculation of Y_{11} as shown in Figure 15.11.

We create the new summation by removing (subtracting) the oldest value in the summation and adding in the newest value. This reduces the per point computation to 1 subtraction + 1 addition + 1 division (or right-shift operation). This creates an algorithm that executes extremely quickly. Implementing the algorithm will require an internal buffer to hold the last N samples and a variable to hold the running summation. Appendix B shows a C function to implement an 8-point moving average filter.

The moving average filter has a very different response than the RC and active filters that we discussed in the prior sections. Mathematically, the response is described by Eq. (15.5).

$$H(F_{fs}) = \frac{\sin(\pi F_{fs} M)}{M \sin(\pi F_{fs})} \tag{15.5}$$

To understand the notation of Eq. (15.5), you need to know that all digital filters are characterized relative to the sampling frequency (*fs*) of the input data. The shape of the response curve for a digital filter is independent of the sampling frequency, so the response is often evaluated in terms of the fraction of the sampling frequency (F_{fs}). The parameter F_{fs} runs from zero to one-half (due to the Nyquist limit, which we cover in Chapter 19). To make this response concrete, Figure 15.12 shows the response of several moving average filters when sampled at 1 kHz. The response of a second-order RC filter is also shown for comparison.

This figure makes it clear that the moving average filter is a low-pass filter, in fact a *very* low-pass filter. When sampling at 1 kHz, the −3 dB point is at only 115 Hz for the 4-point average and moves down to about 14 Hz for the 32-point average. In comparing its response to that of a second-order RC filter (with a −3 dB point at 115 Hz), we see that while the roll-off of the 4-point average is slightly better, adding points to the average does not improve the rate of roll-off; it simply lowers the corner frequency.

This behavior reinforces a phenomenon that we saw earlier with analog filters: filters that are good in the time domain (as the moving average filter is) are not very good in the frequency domain.

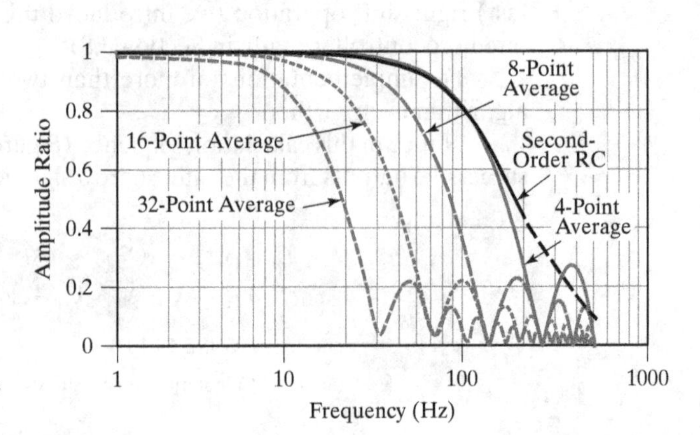

FIGURE 15.12
Bode plot for moving average filters sampled at 1 kHz.

15.2.2 Digital Signal Processing

A large fraction of DSP algorithms boil down to a single underlying mathematical operation: convolution. **Convolution** is an operation that takes two signal sets as its inputs and combines them to produce a third output signal set. Mathematically, it is described by

$$Y_i = \sum_{j=0}^{j=N-1} X_{i-j} \times H_{i-j} \tag{15.6}$$

In Eq. (15.6), Y_i is, once again, the current output value from the algorithm, X is the array of input values, and H is an array of coefficients. The H array is known by several names. Depending on the context, it may be referred to as the **impulse response**, the **filter function**, or the **filter kernel**. The impulse response is also the set of output values from the filter when the input is a single sample of 1 followed by 0s. If you study Eq. (15.6), you should be able to convince yourself that an impulse input to the filter will produce as its output the set of values in the H array. The number of entries in the H array is referred to as the number of **taps** in the filter. The characteristics of the transform performed from input to output are completely determined by the number and the values of the entries in the H array.

When a filter is made using the form of Eq. (15.6), it is referred to as a **finite impulse response** (**FIR**) filter. This descriptive term is in reference to the fact that a change in the input will contribute to the output for only a finite number of sample times (the number of taps) after the change. This is in contrast to the **infinite impulse response** (**IIR**) filter. This type of filter has a mathematical form shown in Eq. (15.7).

$$Y_i = \sum_{j=0}^{j=N-1} X_{i-j} \times H_{i-j} + \sum_{j=1}^{j=M-1} Y_{i-j} \times G_{i-j} \tag{15.7}$$

The IIR filter forms its output by using not only the recent input samples, X, but also the recently produced output values, Y, multiplied by the second filter kernel G. The fact that future outputs depend (in part) on outputs from the past is the reason that it is sometimes referred to as a **recursive filter**. The fact that a change to the input persists in the output for longer than the number of taps in the filter kernel (theoretically forever) is the root of the term *infinite* impulse response filter. Unlike the FIR filter, the IIR filter response to an impulse input does not produce a set of outputs that correspond directly to the values in either H or G.

In addition to producing all manner of elaborate filters, convolution can also be used to produce the simple moving average filter from the prior section. Take a look at Eq. (15.6) and convince yourself that if the H function were {¼, ¼, ¼, ¼}, convolving H with the input X would produce the same result as a 4-point moving average. This is not to say that convolution is a good way to implement the moving average (it is definitely not), but demonstrates the power of this general mathematical operation. Appendix B includes example C code that implements an 8-point moving average using convolution.

To introduce what is special about DSP processors, it is useful to describe, in words, what is going on in a convolution based algorithm:

> Loop through the values in the H array and multiply each by the corresponding value from the input array, X, while accumulating a running sum of the results of the multiplication.

The essential operations of convolution are looping and what is known as the **multiply-accumulate** (**MAC**) operation. What differentiates a DSP from an ordinary microcontroller is that the DSP offers an architecture designed to support **zero-overhead loops** and the

MAC operation in hardware. While a conventional processor would need to use general purpose registers to keep track of the input values, the values in the H array, the loop count, and the accumulated sum, DSPs incorporate dedicated hardware to support the core of convolution. Where a general purpose processor would need to execute many instructions to perform the data and coefficient fetches one at a time, increment pointer registers, multiply, accumulate, and manage the loop counting, the DSP processor needs only some setup and then the processor performs the fetching, index register increments, MAC operations, and loop counting all in dedicated hardware. To calculate a filter output using a DSP, the programmer would set up the calculation by setting special pointer registers to point to the array of input values and the array of filter kernel values. These pointer registers point into separate address spaces for the data and coefficients (making use of a Harvard architecture, see Section 2.10) so that reads of both data and coefficients can happen simultaneously. They would load a loop counter register with the number of elements in the filter kernel (the number of taps in the filter) and clear the accumulator register. Then a single instruction would initiate the filter calculation: fetching data and coefficients, incrementing the pointer registers, decrementing the loop counter, multiplying the data and coefficients, and accumulating the result. The calculation would proceed at a rate of typically one tap per clock cycle, with clock rates into the hundreds of MHz. This kind of processing speed makes it possible to execute filters with hundreds of taps at rates fast enough to process 50+ kHz input signals in real time.

15.2.3 Synchronous Sampling

Another powerful digital technique that is actually quite straightforward is that of **synchronous sampling**. As the name implies, this is the process of synchronizing the sampling to some signal of interest. In synchronous sampling, the synchronization can be either to only the frequency of a signal of interest or to the frequency and the phase. Synchronizing to the frequency only is most commonly used with an approximately sinusoidal noise source of known frequency in order to remove (reject) that noise from the measurements. In this case, what we want to do is sample at a rate that is some integral multiple of the frequency of the noise source and then average those samples using a running average. Since the average value of a sine wave is zero, this effectively removes the noise. To see this in action, consider the situation when we have a signal of interest at 50 Hz that is corrupted by noise from fluorescent lights at 120 Hz. As is not unusual, the polluting signal is actually bigger (in this case, two times bigger) than the signal of interest. If we sample the combined signal at 1,200 Hz and use a 10-point moving average filter, we can effectively remove the polluting signal, even though it is twice as large as our signal of interest (Figure 15.13).

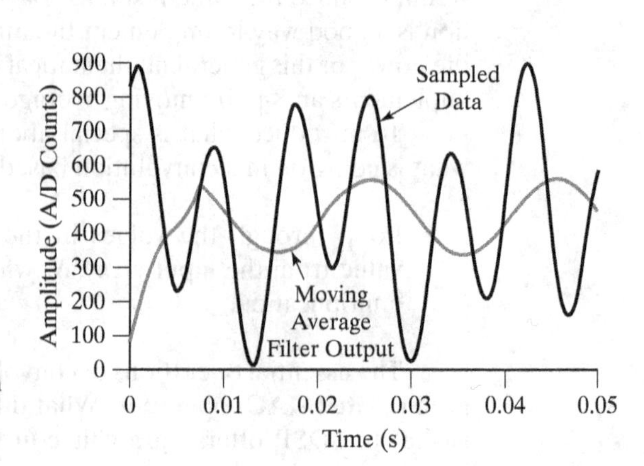

FIGURE 15.13

Signal from 120 Hz noise combined with a 50 Hz signal and output from moving average over one full cycle of the noise.

In this example, we used a simple moving average filter which resulted in the elimination of the 120 Hz signal but also created some attenuation of the 50 Hz signal (it was 300 counts peak-peak in the input). By using a more elaborate filter (such as a sinc^3 filter), it is possible to achieve similar rejection of the 120 Hz noise while only minimally impacting the 50 Hz signal. No matter what kind of filter you use to average, the key to making synchronous sampling effective is the knowledge of the exact frequency of the interfering signal. This knowledge is what allows you to average over an entire cycle of the interfering signal in order to eliminate it.

In some situations, it can be advantageous to synchronize the sampling to both the frequency and the phase of a signal, guaranteeing that the sampling always takes place at the same point in a cycle. This can be a challenging task when the signal originates completely outside the microcontroller, but becomes very straightforward when the signal is one that we have some role in generating. As an example of that kind of situation, consider a reflective object sensor (Chapter 13) used to detect the presence of a nearby object. In the simplest application, the LED emitter would be powered continuously. Unfortunately, this would expose the detector to all of the variations in background light level as well as changes caused by nearby objects. If we take control of when the LED is energized, we can use synchronous sampling of the detector to eliminate the background effects. To achieve this, we operate the LED in a pulsed fashion. The LED is energized and a short time later (a few microseconds to allow for the photo-transistor to respond), the detector is sampled. Then the LED is de-energized and a short time after that, another sample is taken. Because the LED was not on during the second sample, the measurement reflects the background light level not due to the emitter. If we subtract this result from the first reading that was obtained when the LED was on (which consisted of the background light level plus the emissions of the LED), we can remove the effect of the background light level.

15.3 HOMEWORK PROBLEMS

15.1 How does an active filter differ from a passive filter?

15.2 Which of the named filter responses provides the best preservation of the time domain shape of a waveform?

15.3 If the signal of interest is at 1 kHz and the interfering signal to be suppressed is at 4 kHz, what order filter is required to achieve at least a 40 dB reduction (–40 dB gain) in the interfering signal if a) the filter type is Bessel; b) the filter type is Butterworth?

15.4 If the filter corner frequency is 1 kHz, how much attenuation of a 4 kHz signal would you expect from a fourth-order filter if a) the filter type was Bessel; b) the filter type was Butterworth; c) the filter type was 3 dB Chebyshev?

15.5 Design a sixth-order Sallen-Key filter using the simplified design procedure laid out in this chapter. Choose from available 1% resistors.

15.6 Write a C function to implement a 16-point moving average filter using the shortcut outlined in Figure 15.11. The prototype for the function should be:

```
int MoveAvg16Pt(int);
```

15.7 Write the pseudo-code for a four-function module to implement a peak-to-peak amplitude measurement in software. The prototypes for the functions should be:

```
void InitPeakRead(void);
void AddNewValue(int);
unsigned char IsNewPeakDetected(void);
int GetPeak2PeakAmp(void);
```

15.8 Given a sinusoidal signal at 6 kHz with an amplitude of between 0 and 0.2 V superimposed on a 60 Hz signal that varies between 0 and 1 V, design a circuit that will amplify the 6 kHz signal by a factor of approximately 10 and attenuate the 60 Hz signal by a factor of at least 10. You should look for a design that will use the simplest possible circuit to achieve your goals.

15.9 Given the same input signal as in Problem 15.8, design a circuit that will amplify the 6 kHz signal by a factor of approximately 200 and attenuate the 60 Hz signal by a factor of at least 100.

15.10 Write a C function to implement the synchronous sampling algorithm described in the last paragraph of Section 15.2.3. The function should be nonblocking, assuming that it would be called repeatedly, and on each call step through the next phase of the sampling process. Each call should return the most recently completed measurement.

REFERENCES

Active Filter Synthesis Tool from Analog Devices (http://www.analog.com/Analog_Root/static/techSupport/designTools/interactiveTools/filter/filter.html), 12/03/09.

FilterCAD: PC software from Linear Technology (http://www.linear.com/designtools/software/filtercad.jsp), 12/03/09.

Filter Designer: Web-based design tool from Texas Instruments (http://focus.ti.com/docs/toolsw/folders/print/filter-designer.html), 08/30/08.

Filter Lab: PC software from Microchip (http://www.microchip.com/stellent/idcplg?IdcService=SS_GET_PAGE&nodeId=1406&dDocName=en010007), 12/03/09.

Filter-Pro: PC software from Texas Instruments (http://focus.ti.com/docs/toolsw/folders/print/filterpro. html), 12/03/09.

Resistor/Capacitor Calculator (http://www-k.ext.ti.com/SRVS/CGI-BIN/WEBCGI.EXE/,/?St=32,E=0000000000018296581,K=3113,Sxi=1,Case=obj%2832622%29), 08/25/10.

FURTHER READING

The Active Filter Cookbook, Lancaster, D., Newnes, 2nd ed., 1996.

"Analysis of the Sallen-Key Architecture," Texas Instruments Application Report SLOA024B, 2002.

The Art of Electronics, Horowitz, P., and Hill, W., 2nd ed., Cambridge University Press, 1989.

"More Filter Design on a Budget," Carter, B., Texas Instruments Application Report SLOA096, 2001.

Op Amps for Everyone, Carter, B., and Mancini, R., 3rd ed., Newnes, 2009.

CHAPTER 16

Digital Inputs and Outputs

16.1 INTRODUCTION

In this chapter, we leave the analog domain for a while and shift our focus to the digital domain. Over the course of the last several chapters, we explored circuit elements like resistors, capacitors, and op-amps, and looked at their voltage, current, and frequency responses within circuits. Here, we limit our concerns to the single issue of how digital devices represent logical states (e.g., 1/0, true/false, on/off, high/low). Since a solitary digital device representing a logical state in isolation isn't a particularly useful or interesting thing, we devote a great deal of this chapter to a discussion of how to successfully interface among digital devices. With a solid understanding of what digital devices are, what they are useful for, what their capabilities and limitations are, and how to combine them with other circuit elements, you'll be ready to include them in your own designs. After reading and understanding this chapter, you should be able to:

1. understand how digital devices represent logical states,
2. understand how current flows between digital devices,
3. understand the timing requirements for digital device inputs and outputs,
4. read and understand digital device data sheets,
5. select appropriate digital devices to perform a desired function,
6. connect digital devices to other digital devices.

16.2 REPRESENTING LOGICAL STATES

The fundamental function of digital devices is to represent information with logical states. There are two logical states (true and false), and a digital device has two states (on and off). This convenience forms the basis of a great deal of modern electronic design. We refer to the logical **true** state for digital devices with several terms that are often used interchangeably: **high**, **1**, and **on** being the most common. Similarly, we refer to the **false** state as **low**, **0**, and **off**. Digital devices represent each logical state by assigning a voltage to represent it. The exact voltages—or, more accurately, range of voltages—used to represent 1s and 0s can differ. Most commonly, the convention is to represent logical 0/off/false with 0 V (plus a small tolerance band) and to represent logical 1/on/true with a voltage near that used to power a given digital device (again, give or take a little for tolerances). For example, many digital devices are designed to be powered with 5 V, so in these cases, approximately 5 V represents 1/on/true and 0 V represents 0/off/false. Other commonly encountered standard digital device power supply voltages include 3.3, 2.5, and 1.8 V—and more options become available frequently, with a trend toward lower voltages.

16.3 IDEAL BEHAVIOR FOR DIGITAL DEVICES

Just as with op-amps, it is useful to first consider how an ideal digital device might behave. For digital devices, we find that ideal behavior is refreshingly simple compared with op-amps. Unlike op-amps, however, the behavior of real digital devices isn't close enough to the ideal for us to design as if they were—rather, it serves as a useful way to understand how they ought to function. With this as our reference point, we look at several device data sheets which describe the real, nonideal function of the devices and show how they deviate from these ideals. Once we know how far from the ideal a device's performance is, we can design circuits to accommodate it.

Digital devices typically have two types of functional digital pins: inputs and outputs. Such devices read input data from their input pins, perform an operation dictated by their specific function and the input data, and assert a result on one or more output pins. There are a great many types of digital devices, including **logic devices** [e.g., integrated circuits (ICs) that perform a specific logical function: NOT, AND, XOR, and so on], **microcontrollers**, and **programmable logic**. In this chapter, we will make use of a specific microcontroller, the Freescale MC9S12C32, as an example of a typical digital device that is representative of devices that mechatronic engineers are likely to encounter and include in designs. Though this particular device includes a little of everything (such as an analog-to-digital converter, timers, pulse width modulation, and serial communications), we are primarily interested in the characteristics of its digital inputs and outputs, which will serve as the focus of this discussion.

Feeding information to the input of a digital device requires that we introduce a voltage corresponding to one of the two logical states, 0 or 1. In an idealized scenario, we feed it exactly 0 V to represent logical 0. To represent a logical 1 at the input of a device powered at 5 V (a common power supply voltage), we would connect the input to exactly 5 V. For an ideal part, no inflow or outflow of current would be required to correctly interpret the state of the input pin. The rate of transitions between logical 0 and 1 at the input would have no limitations and could range from DC (steady over time at a single logical state, with no transitions) to the highest frequencies imaginable. We could change the state of the input at any time, at any rate, and the device would dutifully respond.

The outputs of an ideal digital device would represent logical states with a set of corresponding voltages. To represent logical 0, the pin would be held at 0 V, and for a logical 1, the output pin would be held at 5 V (again, for the case of a device powered by 5 V). Since the output is intended to be connected to other devices in order to relay information, it would be highly desirable (ideal, you might say) if there were no constraints on how much current we could cause to flow into or out of an output pin. Based on this argument, an ideal output could source or sink an infinite amount of current—whatever was necessary to accommodate the device(s) we connect to it. Finally, ideal digital outputs would be able to instantaneously respond to changes at the input.

16.4 REAL BEHAVIOR OF DIGITAL DEVICES

As we widen the discussion to real devices, we explore how they deviate from the ideals described in the section above. The primary function of a device's data sheet is to reveal just how far from ideal a part strays. Digital outputs, for example, never provide exactly 0 or 5 V (for devices powered at 5 V). Instead, they get reasonably close to 0 V, but not quite all the way, and they can't quite reach 5 V either. Since digital outputs are often connected to inputs of other devices, real inputs must be able to correctly interpret the range of voltages that real outputs provide. So rather than uncompromisingly requiring exactly 0 and 5 V at the inputs, a range of voltages around 0 and 5 V are specified that will be interpreted as valid logical states.

We asserted above that ideal inputs would require no current to flow, either into or out of a pin. In reality, some current must always flow in order for the input to detect the voltages present. Some devices come admirably close to ideal behavior in this regard and require very low levels of current—well below a microamp in some cases. However, other devices require more than 1,000 times that. Ideal digital device outputs would source or sink whatever current is asked of them. In reality, many digital devices are intended only to interface with other digital devices, and their outputs are limited to a few milliamps of current. There is a category of device called **power drivers** that are specifically designed to control large current flows (from a few hundred milliamps to several amps), which we discuss in Chapter 17; however, these are a subcategory of digital device, and the current they can provide, though higher, is still subject to limitations.

Finally, real devices have timing limitations, both for their inputs and their outputs.

- Input timing: There is an upper frequency limit to how often an input signal may change states (often a few MHz or more), and transitions between states must happen quickly, with maximum **rise times** and **fall times**.
- Output timing: Output voltage transitions between states do not occur instantaneously; real output signals exhibit a finite rise time and fall time.
- Delays: Input data isn't instantaneously reflected at the output. A short **propagation delay** occurs as the information propagates through the device and is finally reflected at the output.
- Preparation: Information at an input is often required to be present and stable for a minimum amount of time, called the **setup time**, before a device is able to reliably read it and process it. While a device is reading an input, the data must remain stable for a minimum amount of time, called the **hold time**.

16.5 READING DEVICE DATA SHEETS

Data sheets are, in effect, a promise from a manufacturer that all the devices it delivers will perform within the specifications listed. When you think about this, semiconductor manufacturers face a daunting task. They routinely produce millions of units of a device, and they guarantee that *all* of them will meet specifications. Statistically, this is an extremely tall order: it requires the utmost care in the design, manufacturing, and testing of these parts. That they almost never fail to live up to their promise is an astonishing feat.

Data sheets provide designers with a firm set of specifications defining the performance of a device. All specifications are stated as ranges. In some cases, both endpoints of the range will be given. For example, the allowable input voltages used to represent a logical 1 for a given part might be specified as having a minimum of 3.5 V and a maximum of 5.1 V. In many cases, only one endpoint is provided, the maximum or minimum value. For example, the maximum frequency for an input signal may be 100 kHz, meaning you may introduce a signal at this input that transitions at rates from 0 Hz (DC) to 100 kHz. Manufacturers strive to produce data sheets that are as sparse and succinct—terse, even—as possible while still providing the essential information. As a result, teasing out the exact values relevant to your design can be challenging. Our goal for this discussion is to enable you to be successful in this endeavor.

When deciding which specifications to use to guide your designs, we strongly advise that you always **design for the worst case**. Though this sounds as if it might be a bad thing, what we really mean is that you should design to the most conservative specifications provided in the data sheet. If a data sheet lists a typical value (e.g., an output voltage that is typically 4.5 V) and a minimum value (e.g., that same output voltage could be as low as 3.8 V), design for the minimum value. This means that your design will work with *any* part that the manufacturer delivers, not just the "typical" parts. Manufacturers don't guarantee that they'll ship you only typical parts—the performance of the parts you get are sure to vary. Manufacturers guarantee that they won't ship you parts that perform outside the minimum and maximum specifications, so your designs should accommodate any part within the specifications. Many times, the specification you need won't be explicitly listed in a data sheet, since there are far too many unique applications for manufacturers to anticipate and address. In these cases, **worst-case design** principles dictate that you select the most conservative specification that *is* explicitly stated in the data sheet. This sounds simple, but it can often require some careful thought. We give several examples of this in the sections that follow.

One very important distinction that we would like to make early in the discussion is the difference between the **recommended operating conditions** and **absolute maximum ratings** sections within data sheets. Specifications in the recommended operating conditions section describe how a device is intended to be used, and how to ensure its proper function and a long, productive life. The specifications listed in the absolute maximum ratings section (with the exception of the temperature ranges) are *not* useful for deciding how a part should be used in typical operation. The Absolute Maximum Ratings can be thought of as a list of ways to kill a device, and as such have very little to do with designing a circuit that will put the device to good use. The device is not guaranteed to operate properly under these extreme conditions, but only to survive. Useful design criteria are found in the Recommended Operating Conditions section, or, more generally, *any* of the other sections of a data sheet.

16.6 DIGITAL INPUTS

Digital device inputs interpret a specific range of voltages as logical 0 and another range of input voltages as logical 1. Input voltages outside the specified ranges are not guaranteed to be interpreted correctly, and, for egregiously out-of-spec input voltages, the devices are not guaranteed to function at all (sometimes permanently). Assuming, however, that the inputs have been provided with valid voltages and the device is properly powered and otherwise operating within specifications, it should dutifully make use of the input data and perform its intended function, whatever that may be.

To illustrate the concepts covered in this discussion, we'll refer to the data sheet of the microcontroller we've chosen as our representative example, the Freescale MC9S12C32 (which we shorten to "9S12" for convenience). The input and output characteristics of this device are typical of modern microcontrollers and the issues that arise are common among digital devices in general, so this is a great place to get started. We'll assume throughout this discussion that the power supply for the device is 5 V. From the 9S12 data sheet, we find that the sections of the data sheet that provide specifications for how the device should be used are found in the Electrical Characteristics section, in two tables labeled Operating Conditions and 5 V I/O Characteristics. Another table is included for 3.3 V I/O characteristics, which we would consult if we chose to power the 9S12 at 3.3 V (another common power supply voltage).

16.6.1 Digital Input Voltage Requirements

The valid input voltage ranges are specified by the **input high voltage** (V_{IH}), which is the *voltage* range that will be interpreted as a logic level *high* at an *input*, and the **input low voltage** (V_{IL}), the *voltage* range that will be interpreted as a logic level *low* at an *input*. The 9S12 data sheet (Figure 16.1) gives the upper and lower limits of each range relative to the positive power supply voltage (which it calls V_{DD5}), and ground (V_{SS5}). Very commonly, digital device data sheets specify only one end of the acceptable range (the maximum or minimum value),

Table A-4 Operating Conditions

Rating	Symbol	Min	Typ	Max	Unit
I/O, Regulator and Analog Supply Voltage	V_{DD5}	2.97	5	5.5	V
Digital Logic Supply Voltage[1]	V_{DD}	2.25	2.5	2.75	V
PLL Supply Voltage [1]	V_{DDPLL}	2.25	2.5	2.75	V
Voltage Difference VDDX to VDDA	Δ_{VDDX}	-0.1	0	0.1	V
Voltage Difference VSSX to VSSR and VSSA	Δ_{VSSX}	-0.1	0	0.1	V
Oscillator	f_{osc}	0.5	-	16	MHz
Bus Frequency	f_{bus}	0.5	-	25	MHz
Operating Junction Temperature Range	T_J	-40	-	140	°C

NOTES:
1. The device contains an internal voltage regulator to generate the logic and PLL supply out of the I/O supply. .

Table A-6 5V I/O Characteristics

Conditions are 4.5< VDDX <5.5V Termperature from -40˚C to +140˚C, unless otherwise noted

Num	C	Rating	Symbol	Min	Typ	Max	Unit
1	P	Input High Voltage	V_{IH}	0.65*V_{DD5}	-	-	V
	T	Input High Voltage	V_{IH}	-	-	VDD5 + 0.3	V
2	P	Input Low Voltage	V_{IL}	-	-	0.35*V_{DD5}	V
	T	Input Low Voltage	V_{IL}	VSS5 - 0.3	-	-	V
3	C	Input Hysteresis	V_{HYS}		250		mV
4	P	Input Leakage Current (pins in high ohmic input mode)[1] $V_{in} = V_{DD5}$ or V_{SS5}	I_{in}	–2.5	-	2.5	μA
5	C	Output High Voltage (pins in output mode) Partial Drive I_{OH} = –2mA	V_{OH}	V_{DD5} – 0.8	-	-	V
6	P	Output High Voltage (pins in output mode) Full Drive IOH = –10mA	V_{OH}	V_{DD5} – 0.8	-	-	V
7	C	Output Low Voltage (pins in output mode) Partial Drive IOL = +2mA	V_{OL}	-	-	0.8	V
8	P	Output Low Voltage (pins in output mode) Full Drive I_{OL} = +10mA	V_{OL}	-	-	0.8	V

FIGURE 16.1

Excerpts from the Operating Conditions and 5 V I/O characteristics tables from the MC9S12C32 data sheet. (Copyright of Freescale Semiconductor, Inc. 2010. Used by permission.)

rather than both endpoints of both ranges. In these cases, the power supply defines the upper and lower values. This particular data sheet further qualifies each specification by identifying the method used to determine it in the column labeled "C," indicating whether the specification is determined by testing each device (P), by testing a statistically significant sampling of devices (C), or by testing only a small sampling (T). These details are not usually included in digital device data sheets.

The 9S12 accepts a wide range of power supply voltages: anything between 2.97 and 5.5 V is acceptable. Other common symbols used in the industry to represent the power supply voltage are $\boldsymbol{V_{CC}}$ (for bipolar devices) and $\boldsymbol{V_{DD}}$ (for CMOS devices)[1]. The input and output

[1]Interesting aside/background: The CC in V_{CC} comes from *collector*, the bipolar transistor terminal commonly connected to the positive power supply rail, while DD in V_{DD} comes from *drain*, the corresponding terminal for a MOSFET. Similarly, V_{EE} (E for emitter) and V_{SS} (S for source) are commonly used to denote the pin that is connected to ground.

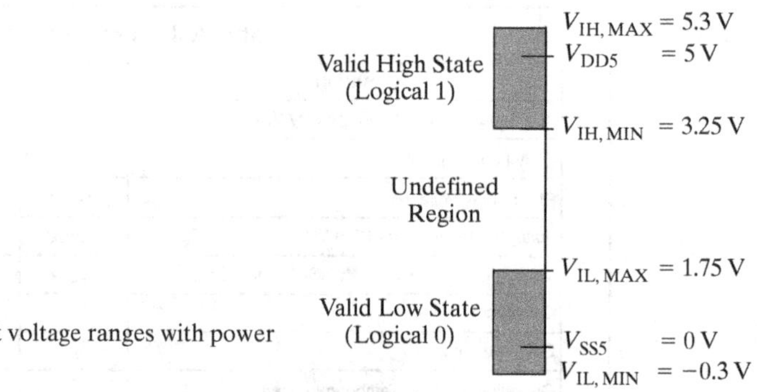

FIGURE 16.2

MC9S12C32 digital input voltage ranges with power supply $V_{DD5} = 5$ V.

voltages that represent the logic levels for the 9S12 depend on the supply voltage used. Since we've assumed that the device is powered with $V_{DD5} = 5$ V, the input voltages that the 9S12 will interpret as a valid logical 0 range from −0.3 (0.3 V below the power supply's ground, which the manufacturer calls V_{SS5}) to 1.75 V (0.35 × 5 V). Voltages ranging from 3.25 (0.65 × 5 V) to 5.3 V (0.3 V above 5 V) will be interpreted as a valid logical 1. Voltages between V_{IL} and V_{IH} are said to be in the **undefined region**, and the manufacturer doesn't specify what the device will do with an input voltage in this range. These ranges are represented graphically in Figure 16.2.

Immediately below the lines for V_{OL} and V_{OH} in Figure 16.1 is a specification called the **input hysteresis (V_{HYS})**. This relates to a special feature of the digital inputs in the 9S12: they incorporate a 250 mV (typical) hysteresis band around their voltage transition points. In Chapter 11, we discussed hysteresis in detail and showed how to create circuits with hysteresis using comparators. When the inputs of digital devices incorporate hysteresis, they are called Schmitt trigger inputs. This is a special feature and results in inputs that are much more tolerant of slowly changing input signals, which might spend an inordinate amount of time in the undefined region, where the behavior of the digital device is unspecified. This is the kind of digital input the 9S12 has. (For our discussion of hysteresis and its benefits, see Chapter 11, Section 11.5.1.) The amount of hysteresis for the 9S12 inputs is fairly loosely specified: only the "typical" value is shown, with no firm specification of how large the hysteresis band might be for any specific device.

For many devices, voltages above the positive power supply voltage and below ground are outside the valid input voltage range. The 9S12 is relatively forgiving on this point, allowing input voltages to go as much as 0.3 V beyond these values. As a general rule, introducing input voltages above V_{CC} or below ground is ill-advised. If the input voltages are allowed to exceed the recommended limits and reach the absolute limits of the part, its health and well-being are jeopardized, and its continued function is no longer guaranteed.

Locating the pertinent values of V_{IL} and V_{IH} corresponding to the power supply voltage you plan to use is a straightforward exercise for the 9S12, as long as your power supply is 3.3 or 5 V. When we power the 9S12 with a 5 V power supply, we saw above that the range of input voltages guaranteed to be interpreted as logic level low (V_{IL}) is −0.3 to 1.75 V, and the range guaranteed to be interpreted as logic level high (V_{IH}) is 3.25 to 5.3 V. In practice, however, power supply voltages are rarely exactly 3.3 or 5 V. Real power supplies have some nominal output voltage and a tolerance around that value within which the output may vary. For instance, a typical 5 V power supply has a ±10% tolerance, so the actual voltage may be anywhere between 4.5 and 5.5 V. In some circumstances, power supply voltages other than 3.3 or 5 V are required by other circuit elements and may be a better choice for the circuit as a whole. In both of these cases, we must apply worst-case design principles (i.e., the design should work under all foreseeable circumstances) to determine the acceptable input voltage ranges for an arbitrary power supply voltage within the allowable range of 2.97 to 5.5 V. This adds a new level of complexity to interpreting the data sheet.

If we expect our power supply voltage to vary 10% from a nominal 5 V value and we design for the worst case, we need to reexamine the input voltage range specifications. Both V_{IL} and V_{IH} scale proportionately with the power supply voltage V_{DD5}, and we wish to pick the most conservative specifications for our design. For V_{IL}, the worst case would be when our power supply is operating at the lowest allowable voltage, 4.5 V (5 V − 10%). Examining the top of the 5 V I/O Characteristics table in Figure 16.1 reveals that the specifications are valid for $4.5\text{ V} \le V_{DDX} \le 5.5\text{ V}$. From this note, we learn that the manufacturer has anticipated the use of a standard power supply with 10% tolerances and supplied us with specifications for this general case. The odds are reasonably high that most readers failed to notice this note the first several times they read the table. This is typical of data sheets: critical specifications are often contained (hidden, really) in subtle "notes" or "test conditions." While they are easy to overlook, these sections contain critical information and must be read as carefully as the other parts of the document. The 5 V I/O Characteristics indicate that the range of input voltages that will be interpreted as a logical 0 will be from −0.3 to $0.35\text{ V} \times V_{DD5}$, which gives a result of $-0.3\text{ V} \le V_{IL} \le 1.575\text{ V}$ for the worst case, when $V_{DD5} = 4.5\text{ V}$. This is indeed a more conservative result, considering that the maximum V_{IL} for a power supply with an output of exactly 5 V is $V_{IL} = 1.75\text{ V}$. For the worst-case range of input voltages guaranteed to be interpreted as a logical 1 (V_{IH}), we need to consider the high end of the range of possible power supply voltages, 5.5 V (5 V + 10%). Since the minimum $V_{IH} = 0.65 \times V_{DD5}$, the more conservative minimum V_{IH} will be 3.575 V (0.65×5.5 V). Comparing this with the minimum $V_{IH} = 3.25\text{ V}$ that results when using a power supply with an output of exactly 5 V, this is a narrower, and therefore more conservative, range. The high end of the range for V_{IH} is given in Figure 16.1 as $V_{DD5} + 0.3\text{ V}$, since for worst-case design, the power supply voltage might be anywhere within the tolerance band. The resulting input voltage ranges for a 5 V ± 10% power supply are shown in Figure 16.3.

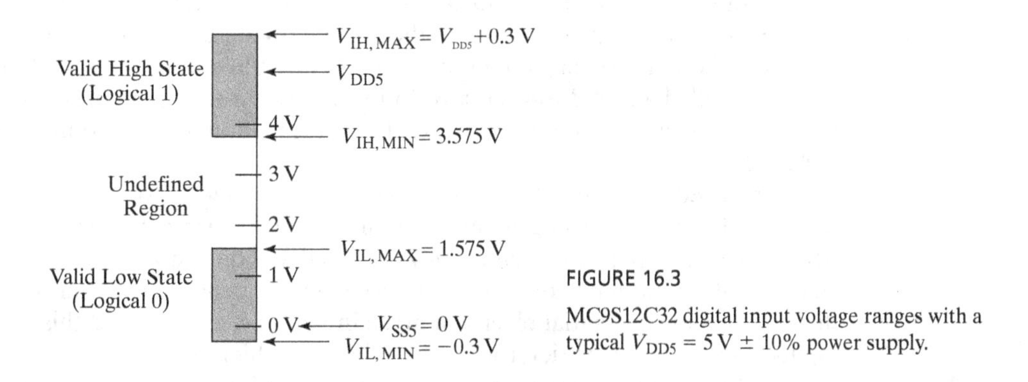

FIGURE 16.3

MC9S12C32 digital input voltage ranges with a typical $V_{DD5} = 5\text{ V} \pm 10\%$ power supply.

16.6.2 Digital Input Current Requirements

We've shown how to determine the range of input voltages that a device will accept as valid logic level inputs, so now we need to turn our attention to the range of currents the device may require to flow into or out of an input pin. This specification is referred to as the **input current** or, in some cases, the **input leakage current**, and for our example MC9S12C32 microcontroller, it is given the symbol I_{in}. For some devices, there are different values for the input current depending on whether the input is a logical low (in which case it is called the **input low**

current and given the symbol I_{IL}) or a logical high (**input high current,** I_{IH}). For our representative microcontroller, the maximum input current is specified (Figure 16.1) to be ± 2.5 μA for 4.5 V ≤ V_{DD5} ≤ 5.5 V. This specifies the maximum amount of current we can expect to flow into or out of a pin, and we can safely conclude that the minimum amount of current that might flow is 0. This is a safe (though technically impossible) assumption, because our designs will need to accommodate the maximum input current, rather than the minimum. Devices connected to the 9S12's inputs must be able to source or sink at least 2.5 μA. This level of input current is quite low and unlikely to pose much of a burden in the design. Be aware, however, that the input current requirements for digital devices can vary widely. The low input current of the 9S12 (a few microamps) is typical of CMOS devices. This category includes most modern microcontrollers and digital logic from the "HC" (or "high-speed CMOS") family. Devices with inputs that use bipolar transistors, such as the "LS" digital logic family (or "low-power Schottky"), have much higher input current requirements. Their maximum input currents can be well into the hundreds of microamps.

The **sign convention for current** used in most data sheets for digital devices is to indicate current flowing into a device with a positive value and current flowing out of a device with a negative value. This applies to any pin, regardless of whether it is an input or output. Since this is the same convention used in most engineering control volume analyses, it's relatively easy to keep straight.

16.6.3 Pull-Ups and Pull-Downs

When a digital input is not actively driven by the output of another device (which is the topic of Section 16.8), the circuit should include provisions to establish the desired state. The simplest case occurs when the state of an input is to be constant and fixed at either logical 0 or 1. When this occurs, all that's required is to introduce the appropriate constant voltage in a way that meets the device's input voltage and current requirements. This is usually done by connecting the input with a resistor to either power or ground. When the voltage at an input is fixed in the logical high state by connecting it through a resistor to the power supply voltage, the input is said to be **pulled up** or **pulled high**, and the resistor is said to be a **pull-up resistor.** When the voltage is fixed in the logical low state by connecting it to ground (either with or without a resistor), the input is said to be **pulled down** or **pulled low**. If a resistor is used, it is called a **pull-down resistor**. In those cases, where an input is to be held at ground and never driven high, the input is usually connected directly to ground and a resistor isn't included.

The need to pull inputs up and down arises frequently. For example, when a device has multiple elements and some are not used in a circuit, the state of the unused inputs should be determined. If left unconnected, they're called **floating inputs**. Unconnected inputs may change states randomly, sometimes at very high frequencies. This can result in extra power dissipation and substantial electrical noise in your circuit. To avoid this, the state of unused inputs should always be determined by tying them high or low. A similar situation involves devices that can operate in more than one mode, and the mode is determined by the state of one or more input pins (e.g., an output enable pin). If the operating mode is never intended to change, pulling the mode input pins to the appropriate states is an efficient way of configuring the device. We consider another application of pull-up and pull-down resistors in Section 16.8.2, for devices whose inputs may be connected to other devices some of the time and disconnected at other times.

First, we will examine the situation in which we want to permanently provide an input with a single constant logic level. We'll look at both pulling the input high (to logical 1) and pulling it low (to logical 0). Figure 16.4 illustrates both scenarios, using our now-familiar

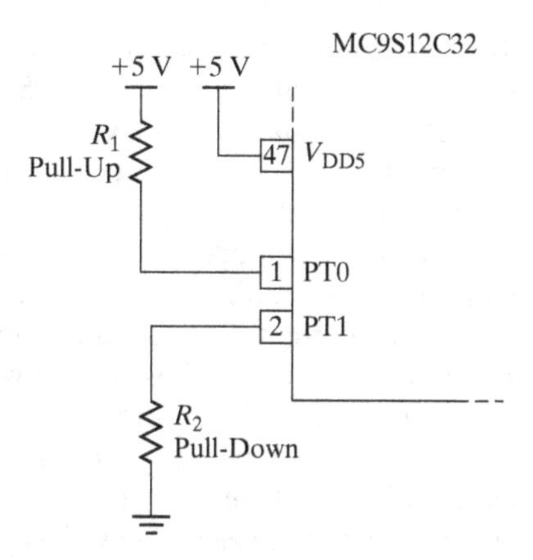

FIGURE 16.4

Pulling digital inputs up and down.

9S12 and two representative input pins, Port T pin 0 ("PT0," which we pull high with a pull-up resistor) and Port T pin 1 ("PT1," which we pull low with a pull-down resistor). We must determine the range of allowable resistor values for each case.

For PT0, we would like the input to always read a logical 1, so we need to ensure that our choice for the pull-up resistor, R_1, results in a voltage at PT0 (V_{PT0}) that meets the input high voltage specifications for the 9S12 while delivering the maximum input current that might be required. From Figure 16.1 and the discussion in Section 16.6.1, $(0.65 \times V_{DD5}) \leq V_{PT0} \leq (V_{DD5} + 0.3\text{ V})$. When we power the device with a typical 5 V ± 10% supply, we need to determine values for the pull-up resistor that meet the input high voltage specifications at the extremes of the possible power supply voltages while sourcing the maximum input current. Figure 16.1 states that the input leakage current (I_{in}) has a maximum value of ± 2.5 μA, and that voltages greater than $0.65 \times V_{DD5}$ (the minimum value of V_{IH}) will be interpreted as a valid logic level 1. We can therefore allow a voltage drop across our pull-up resistor as large as the difference between V_{DD5} and V_{IH}. Using Ohm's law, we can use these specifications to determine the maximum value of R_1 that give the desired result:

$$R_1 \leq \frac{\Delta V}{I} \tag{16.1}$$

At the low end of the power supply range (5 V – 10% = 4.5 V):

$$\Delta V = 4.5\text{ V} - (0.65 \times 4.5\text{ V}) = 1.575\text{ V} \tag{16.2}$$

If the maximum possible leakage current flows (2.5 μA), then the maximum allowable value for R_1 is

$$R_1 \leq \frac{1.575\text{ V}}{2.5\ \mu\text{A}} = 630{,}000\ \Omega = 630\text{ k}\Omega \tag{16.3}$$

Repeating the exercise for the highest possible power supply voltage, which is 5 V + 10% = 5.5 V, gives

$$R_1 \leq \frac{5.5\text{ V} - (0.65 \times 5.5\text{ V})}{2.5\ \mu\text{A}} = 770{,}000\ \Omega = 770\text{ k}\Omega \tag{16.4}$$

Of these two choices, 630 kΩ is the appropriate choice, since the drop across a 770 kΩ resistor would result in violating the input voltage requirements for any voltage less than 5.5 V.

We can also consider the impact that differing levels of input leakage current would have on our results. We know from the specifications for the 9S12 that the maximum amount of leakage current (I_{in}) is $\pm$ 2.5 μA. And, from Ohm's law ($V = IR$), we know that for a given resistance, lower levels of current flow will result in lower voltage drops. For our pull-up resistor circuit, this means that lower input leakage currents will result in lower voltage drops across the resistor, and hence input voltages that are closer to the power supply voltage (V_{DD5}) are higher than the required minimum input voltage. With this reasoning, we conclude that leakage current levels less than the guaranteed maximum result in input voltages that are higher than the minimum required. This is what we need to guarantee that the input will get what it needs and work under all conditions.

There's one additional issue to consider when specifying any real resistor: resistors are only available in certain discrete values and with certain tolerances. In this application, the exact value of the resistor isn't critical, as long as it always has a resistance less than 630 kΩ. The most common and inexpensive resistors available today are carbon film resistors with 5% tolerances. We list the available resistance values for these resistors in Appendix A. We should select the closest standard resistor value that, considered together with its tolerance, results in a resistance lower than the maximum value. From the appendix, we see that 620 kΩ is the closest value, however, because of its 5% tolerance range, this could be as high as 661.5 kΩ. This is well above our limit. The next lower value is 560 kΩ, which might be as high as 588 kΩ. Since 560 kΩ satisfies all our criteria, this is the correct choice.

We've determined the maximum pull-up resistor value, but what's the minimum allowable value? Just as we saw when we considered the effects of lower input leakage current on the input voltage through a pull-up resistor, we can use Ohm's law to see that lower values of resistance result in lower voltage drops across the pull-up resistor and input voltages that are closer to the power supply voltage. This is a favorable trend: pull-up resistors with resistances less than the maximum result in voltages higher than the minimum value of V_{IH}. Continuing this trend out to the limit, it would seem that zero resistance—no pull-up resistor at all—would be a great choice, since this guarantees the highest possible input voltage of all (the power supply voltage itself) and reduces the number of components. This is seldom done in practice, however. The practical issue is that the instantaneous power supply voltage is not the same everywhere in the circuit. Surges in current associated with outputs changing state flowing through the finite impedances within the circuit itself result in instantaneous voltage differences appearing at different points on what is nominally the same power supply. In this way, the voltage used to power the pull up at an input may instantaneously exceed the power supply voltage at the chip. Depending on how the power supply pin and the input pin for the device are connected to the power supply, this could cause the voltage at an input to exceed the device's power supply voltage by an amount that damages the part. The exact limit is device dependent. For the 9S12, this amount is 0.3 V, which is a common value. Because of this, pull-up resistors are generally used to tie an input high. While the details of calculating the minimum value that the pull-up could be are device and application specific, values of 5–10 kΩ are common as minimum pull-up values.

To sum up: for the pull-up resistor on PT0 of the 9S12 in Figure 16.4, we can use any standard resistor values (Appendix A) in the range 5 kΩ $< R_1 \leq$ 630 kΩ.

We look next at the pull-down resistor on PT1 in Figure 16.4. We wish this input to always read a logical low and need to solve for the range of values for the pull-down resistor, R_2, that satisfies the input low voltage (V_{IL}) and input current (I_{in}) requirements of the 9S12. The same approach that we used to determine the allowable pull-up resistor values applies here as well.

From Figure 16.1, the range of input voltages guaranteed to be interpreted as a logical 0 is $(V_{DD5} - 0.3\text{ V}) \leq V_{PT0} \leq (0.35 \times V_{DD5})$. The largest amount of input leakage current in this case is $-2.5\ \mu A$ (the sign is negative because whatever we connect to the input pin will hold it low by sinking this leakage current). To determine the maximum allowable pull-down resistor value (R_2), we can use equations similar to the ones we developed above for the pull-up resistor. Once again, we could allow a voltage drop across our pull-down resistor as large as the difference between ground and the maximum permissible V_{IL}. In this case, the fact that the input current specification for the 9S12 is negative tells us that current will flow from the pin to ground, with the voltage at the input pin above ground.

$$R_2 \leq \frac{\Delta V}{I} \tag{16.5}$$

At the low end of the power supply range (5 V – 10% = 4.5 V):

$$\Delta V = 0.35 \times 4.5\ \text{V} = 1.575\ \text{V} \tag{16.6}$$

If the highest possible level of leakage current flows ($-2.5\ \mu A$), then the maximum allowable value for R_2 is

$$R_2 \leq \frac{1.575\ \text{V}}{2.5\ \mu\text{A}} = 630{,}000\ \Omega = 630\ \text{k}\Omega \tag{16.7}$$

Repeating the exercise for the highest possible power supply voltage (5 V + 10% = 5.5 V), gives

$$R_2 \leq \frac{0.35 \times 5.5\ \text{V}}{2.5\ \mu\text{A}} = 770{,}000\ \Omega = 770\ \text{k}\Omega \tag{16.8}$$

These are the same results we calculated for the pull-up resistor. We could have predicted this result for the 9S12, since V_{IL} and V_{IH} are symmetrical relative to the power supply voltage (V_{DD5}) for the device. The reasoning for picking 630 kΩ as the maximum allowable pull-down resistor is the same as for the pull-up resistor as well: this meets the requirements for any power supply voltage, while 770 kΩ only meets the requirements when the power supply voltage is 5.5 V. Again, our maximum allowable pull-down resistor value is 630 kΩ, and the maximum standard 5% resistor value we could specify is 560 kΩ.

As we found for the pull-up resistor, the minimum allowable value for a pull-down resistor based solely on voltage considerations is 0 Ω. Ohm's law again shows us that, for lower values of R_2, smaller voltage drops occur across the pull-down resistor, with the resulting V_{in} closer to ground (called V_{SS5} in the 9S12 data sheet). As we mentioned in the case of the pull-up resistor, it is still a good idea to include a resistor in the pull-down circuit. That way, if the voltage at the local ground for the chip is instantaneously above the voltage at the input pin, there is a resistor in the path to limit the resulting current flow. For some logic inputs (notably those based on bipolar transistors), the input current in the low state is substantially higher than that in the high state, often approaching a milliamp. With such devices, the maximum value of the pull-down resistor is so small that it does not provide much protection against excess current flow in the case when the chip ground rises above the input voltage. In this situation, it is common to omit the pull-down resistor and be sure that the pulled down input pin is connected as directly as possible to the ground pin on the chip, which minimizes the possibility of a voltage difference developing, even instantaneously, between them.

To sum up for pull-up and pull-down resistors that fix the logical state of a digital input:

1. The maximum allowable pull-up resistor value is determined by the input high voltage (V_{IH}) and input current (I_{in}) specifications across the range of possible power supply voltages.
2. The maximum allowable pull-down resistor value is determined by the input low voltage (V_{IL}) and input current (I_{in}) specifications across the range of possible power supply voltages.
3. The minimum recommended pull-up resistor value *approaches* 0, but is not 0. Because of requirements that the voltage at the input pin of a device not exceed the voltage at the power pin by more than a certain amount (often around 0.3 V), a nonzero pull-up resistor is commonly used and recommended. 10 to 100 kΩ pull-up resistors are typical.
4. The minimum recommended pull-down resistor value is 0. Because of requirements that the voltage at the input pin of a device is not less than the voltage at the ground pin by more than specified amount (often −0.3 V), care must be taken to insure that the voltage difference between the input and ground pins is minimal.
5. When specifying pull-up or pull-down resistors, select from resistor values that are available (see Appendix A) and ensure that any resistor value within the stated tolerance meets the design criteria.

16.6.4 Digital Input Timing Requirements

In addition to meeting voltage and current requirements, input signals must also meet timing specifications. The data sheet for our representative 9S12 microcontroller only includes input timing specifications that apply to certain specialized subsystems, so we will introduce another digital device to illustrate the most common timing requirements: the MC3479 (available from ON Semiconductor). The MC3479 is a bipolar stepper motor driver chip (Figure 16.5) which has several digital inputs that determine how a stepper motor is to be controlled (e.g., when to take a step, which direction to rotate, and whether to take full steps or half steps) and outputs that control the flow of up to 350 mA of current through the phases of a bipolar stepper motor. Stepper motors are specially designed so that their rotors turn in small increments (or steps), usually a few degrees per step. To do this, current flow is

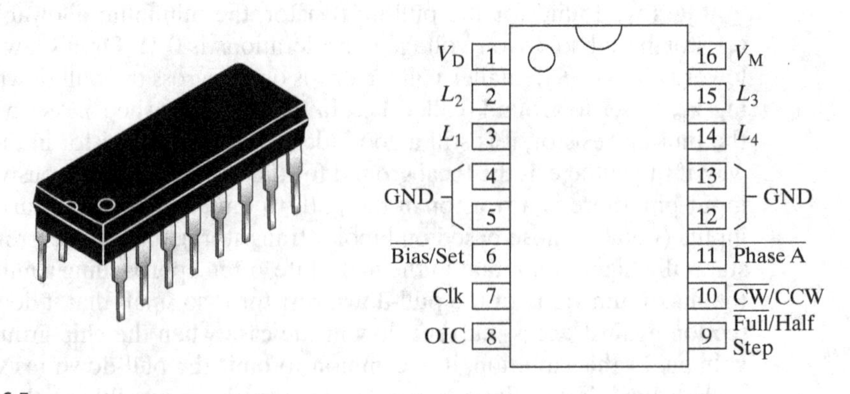

FIGURE 16.5

The MC3479 bipolar stepper motor driver device. (Used with permission from SCI LLC, DBA ON Semiconductor.)

controlled sequentially through independent motor coils. The MC3479 simplifies the task of getting the sequence of current flows that are correct to rotate the motor shaft: it has high-level inputs such as "take a step" (called the clock) and "direction" (called $\overline{\text{CW}}$/CCW). We cover stepper motors in detail in Chapter 26.

Inputs for many digital devices are required to make transitions between logical states quickly so that the length of time the input voltage might spend in the undefined region (where the behavior of the device is not specified) is limited. The length of time that an input signal takes to make a transition from a logical low state to a logical high state (a **rising edge**) is called the **input rise time** (**t_r**), and the time it takes to make a high-to-low transition (a **falling edge**) is called the **input fall time** (**t_f**). The maximum input rise time or fall time is defined as the longest allowable time for the input voltage to transition from 10% of its final voltage to 90% of its final voltage (which is the standard definition of the **10–90 rise time**).

Figure 16.6a shows what would happen in an ideal case. If this scenario was possible, the input would start in the logical 0 state (ideally, 0 V), then at the point of transition, the input would instantaneously change to its final value, the logical 1 state (ideally, V_{CC}). This isn't possible: all real signals require finite time to change states. In some cases, the transitions are very fast indeed, but never truly instantaneous. Figure 16.6b represents the real case, where the transition occurs over a length of time: the input rise time, t_r. For the case of a high-to-low transition, Figure 16.6c illustrates the ideal case, where the transition occurs instantaneously. Figure 16.6d shows the real case, where the input fall time, t_f, has a value greater than zero. Similar to the 9S12 microcontroller, our example MC3479 features Schmitt trigger inputs, with a hysteresis band guaranteed to be at least 0.4 V. The inputs for the MC3479 are much more tolerant of slowly changing input voltages (i.e., long rise and fall times) and far less likely to behave unpredictably when voltages tarry in the undefined region. (See Chapter 11, Section 11.5.1 on hysteresis.) Because it features Schmitt trigger inputs, the MC3479 does not specify a maximum t_r or t_f, however, values of 10 and 25 ns are given as test conditions for other specifications.

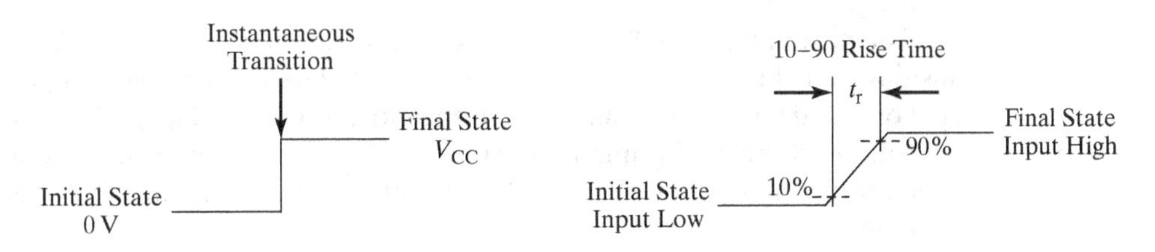

(a) Idealized low-to-high transition (rise time $t_r = 0$) (b) Actual low-to-high transition (rise time $t_r > 0$)

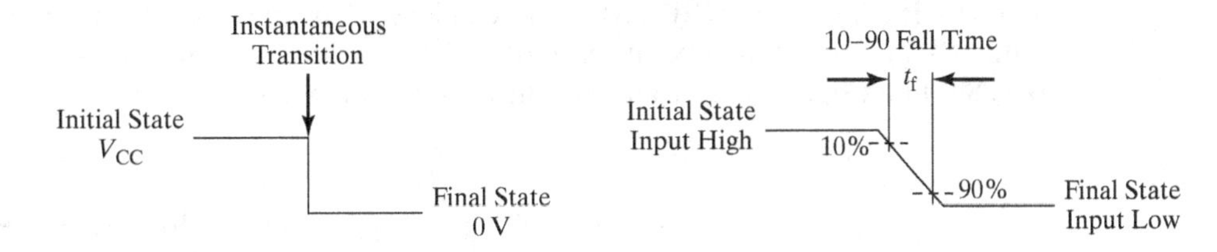

(c) Idealized high-to-low transition (fall time $t_f = 0$) (d) Actual high-to-low transition (fall time $t_f > 0$)

FIGURE 16.6

Input rise time (t_r) and fall time (t_f).

AC SWITCHING CHARACTERISTICS (T_A = + 25°C, V_M = 12 V) (See Figures 2, 3, 4) (Notes 5, 6)

Characteristic	Pins	Symbol	Min	Typ	Max	Unit
Clock Frequency	7	f_{CK}	0	–	50	kHz
Clock Pulse Width (High)	7	PW_{CKH}	10	–	–	μs
Clock Pulse Width (Low)	7	PW_{CKL}	10	–	–	μs
$\overline{\text{Bias}}$/Set Pulse Width	6	PW_{BS}	10	–	–	μs
Setup Time ($\overline{\text{CW}}$/CCW and $\overline{\text{F}}$/HS)	10–7 9–7	t_{su}	5.0	–	–	μs
Hold Time ($\overline{\text{CW}}$/CCW and $\overline{\text{F}}$/HS)	10–7 9–7	t_h	10	–	–	μs
Propagation Delay (Clk–to–Driver Output)		t_{PCD}	–	8.0	–	μs
Propagation Delay ($\overline{\text{Bias}}$/Set–to–Driver Output)		t_{PBSD}	–	1.0	–	μs
Propagation Delay (Clk–to–$\overline{\text{Phase A}}$ Low)	7–11	t_{PHLA}	–	12	–	μs
Propagation Delay (Clk–to–$\overline{\text{Phase A}}$ High)	7–11	t_{PLHA}	–	5.0	–	μs

5. Algebraic convention rather than absolute values is used to designate limit values.
6. Current into a pin is designated as positive. Current out of a pin is designated as negative.

FIGURE 16.7

The AC switching characteristics section of the MC3479 bipolar stepper motor driver data sheet. (Used with permission from SCI LLC, DBA ON Semiconductor.)

The section in the MC3479's data sheet that enumerates the device's timing requirements is labeled AC Switching Characteristics and is shown in Figure 16.7. Several of the specifications relate to the digital inputs: the maximum clock frequency, the minimum pulse widths (the high and low clock pulse width and the $\overline{\text{Bias}}$/Set pulse width), the setup time, and the hold time for the $\overline{\text{CW}}$/CCW and $\overline{\text{F}}$/HS inputs. We consider each of these specifications separately.

The maximum **clock frequency (f_{CK})** specification defines the highest allowable rate of transitions for the input that instructs the MC3479 to advance the stepper motor rotor one step. For this device, this input is called the clock ("Clk" in Figure 16.5), and the maximum clock rate is 50 kHz. The minimum step rate is 0 Hz, meaning that we are free to hold a motor's rotor stationary indefinitely. Any value of clock frequency between 0 and 50 kHz is allowable.

There are further restrictions on the clock input waveform, however, and these are the **clock pulse width (high), PW_{CKH}**, and **clock pulse width (low), PW_{CKL}**. both of which have a maximum value of 10 μs. Though the clock input need not be a square wave, the time spent in the logical high state (PW_{CKH}) must be at least 10 μs, and the time spent in the logical low state (PW_{CKL}) must also be at least 10 μs. This is illustrated in Figure 16.8. Note that the $\overline{\text{Bias}}$/Set input signal pulse width also has a minimum value of 10 μs.

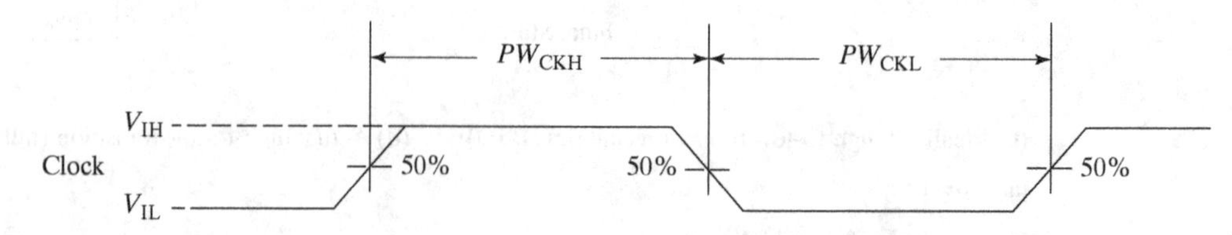

FIGURE 16.8

Graphical representation of minimum clock high and low pulse width timing specifications.

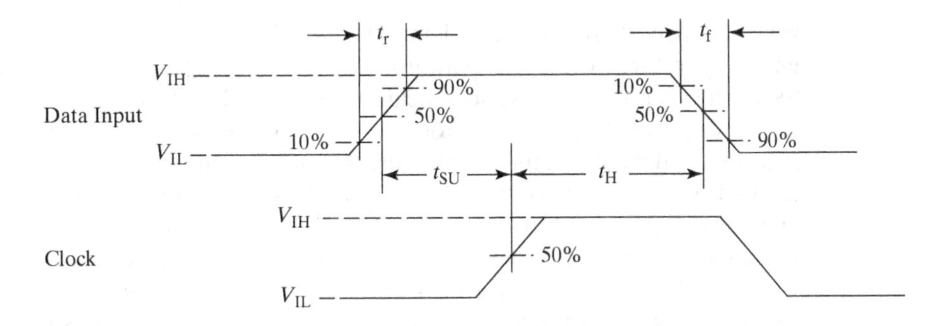

FIGURE 16.9
Graphical representation of setup and hold times.

The last input timing requirements for the MC3479 are the **setup time (t_{SU})**, which is the amount of time that an input must be applied and stable prior to the time when a device reads the input, and the **hold time (t_H)**, the amount of time that the input must be held stable while the device is reading it (Figure 16.9).

Whenever the MC3479 receives a command to advance a stepper motor's rotor by one step (a low-to-high transition, or "rising edge," on the clock input), the device needs to figure out how the user would like the step to be executed, and it does this by reading other inputs. It needs to know whether to rotate clockwise or counterclockwise (which it determines by reading the $\overline{\text{CW}}$/CCW input) and how big the step should be, that is, whether to take a full step or a half step (which it determines from the state of the $\overline{\text{F}}$/HS input). Prior to the rising edge on the clock input, the direction ($\overline{\text{CW}}$/CCW) and step mode ($\overline{\text{F}}$/HS) inputs must have been in the desired state and stable for at least the setup time t_{SU}, which is 5 μs. They must then remain stable in the same state for at least the hold time t_H (10 μs) after the clock's rising edge.

16.7 DIGITAL OUTPUTS

The specifications of a device's digital outputs describe the range of output voltages to expect for the logical states, the device's current drive capabilities, and the timing performance. The subtleties involved in interpreting the specifications for the outputs are similar to those we found for inputs: don't be surprised if you don't find the specifications you seek for your exact conditions. In that case, read all the notes and test conditions, keep your wits about you, and when confronted with a puzzling choice, use worst-case design principles to guide your decisions.

16.7.1 Digital Output Voltage and Current Specifications

In digital device data sheets, the maximum *voltage* that will represent a logic level *low* at an *output* is specified by the **output low voltage (V_{OL})**, and the **output high voltage (V_{OH})** specifies the minimum *voltage* at the *output* that will represent a logic level *high*. As we found above when considering the inputs, these values set the boundaries for the ranges of voltages we can expect to represent the logical states. For logical 0, the output will be somewhere between ground and V_{OL}, and for logical 1, the output will be somewhere between V_{OH} and the power supply voltage (called V_{DD5} for the MC9S12C32, and simply V_{DD} or V_{CC} for many other devices). Just as we found for the inputs, the output voltage specifications depend on the power supply voltage. There are also additional dependencies to consider for the output pins: the amount of current that the pin sources or sinks affects the range of output voltages we can expect (higher current demands result in less ideal performance), as does the ambient temperature (wider temperature ranges also degrade performance). For the example 9S12 microcontroller, these specifications are found in the data sheet excerpt in Figure 16.1, where V_{OL} and V_{OH} are given for two different levels of current flowing into or out of the pins. In the data sheet, the output currents are identified separately depending on whether the logical state of the output is low ($\boldsymbol{I_{OL}}$, the **output low current**) or high ($\boldsymbol{I_{OH}}$, the **output high current**). The specifications for V_{OL} and V_{OH} are given for output currents of 2 mA (described as "partial drive" in the data sheet) and 10 mA ("full drive") for T_A between −40°C and +140°C (Figure 16.10). Data sheets for other devices may present these specifications in slightly different ways. For example, the output current specifications may be combined into a single specification and called $\boldsymbol{I_{out}}$ and no distinction made for the logical states. Often, the output current is listed as a test condition on the output voltage specification.

Our challenge is to determine the range of guaranteed output voltages for the device under a particular set of circumstances. For a circuit powered at 5 V, we are again confronted with the fact that a typical 5 V power supply has a 10% voltage tolerance. For the outputs, a power supply at the lower end of the range, with V_{DD5} = 4.5 V (5 V – 10%), results in the worst-case values for both V_{OL} and V_{OH}. Designing for the worst case gives us $0\text{ V} \leq V_{OL} \leq 0.8\text{ V}$ and $3.7\text{ V} \leq V_{OH} \leq 4.5\text{ V}$. Next, we need to determine which output current conditions most closely match our application. In the case of the 9S12, this is as straightforward as it gets, since the specifications are the same for the current flow conditions cited, ± 2 and ± 10 mA. In data sheets for other devices, it is common to state specifications for very low values of output current, such as a few microamps (when the output is said to be **lightly loaded**). It is also customary to provide specifications for the device when greater amounts of current are required, which is what we have in the case of the 9S12 (± 2 and ± 10 mA). Under these conditions, the outputs are said to be more **heavily loaded.** Finally, we must identify the range of ambient temperatures that most accurately describes the conditions under which the device will operate. For many devices, separate specifications are given for room temperature operation (T_A = 25°C), the industrial temperature range (−40°C to 85°C), or the military temperature range (−55°C to 125°C). For the 9S12, specifications are given only for the full range of recommended operating temperature (−40°C to 140°C).

You may have noticed that a specification for the recommended maximum output current is conspicuously absent. In fact, recommended values for output current are only subtly mentioned in the rating column of the 5 V I/O Characteristics section (Figure 16.1). These are the operating conditions for which the output voltage specifications are valid, and as such, they are by inference the recommended maximum output current specifications. This is

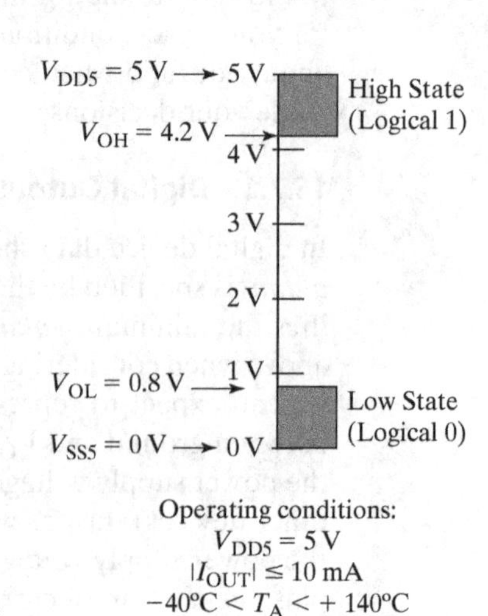

FIGURE 16.10

Output voltage ranges for the MC9S12C32 powered at V_{DD5} = 5 V for output current (I_{out}) up to 10 mA and ambient temperature between −40°C and +140°C.

another example of how it's critically important to read all the notes and test conditions in a data sheet and keep your wits about you.

16.7.2 Digital Output Timing Specifications

The final set of specifications describing the behavior of real digital outputs relates to timing. Recall from Section 16.6 that the rise times and fall times of input signals are usually required to be quick (e.g., less than 1 μs) in order to minimize the amount of time the input voltage is in the undefined region, where the device's performance is unspecified. Since all real signals require finite amounts of time to transition low-to-high or high-to-low, the output voltages have rise times and fall times specifications as well. Most digital devices will specify the **output rise time** (often given the symbol t_{TLH}, which stands for "transition low-to-high") and the **output fall time** (often t_{THL}, "transition high-to-low") in a section called the AC Electrical Characteristics. These times are stated in order to allow designers to match these outputs with inputs of other devices and their input timing requirements. As with the inputs, these are defined as the 10–90 rise times and fall times (Figure 16.11).

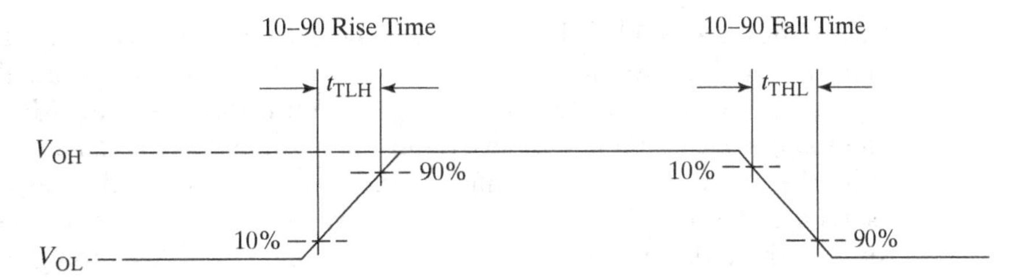

FIGURE 16.11

Output voltage rise time (t_{TLH}) and fall time (t_{THL}).

For most modern digital devices, the guaranteed rise times and fall times of the outputs are much faster than what is required of the rise times and fall times at the inputs, often at least one order of magnitude faster. If an output is going to be connected to the input of a similar digital device (as they very commonly are), it is almost certainly capable of meeting the input rise time and fall time requirements.

Another timing specification often found in the AC Electrical Characteristics section of a data sheet is the propagation delay. The MC3479 data sheet includes propagation delay specifications in the AC Switching Characteristics section, part of which is reproduced in Figure 16.12. We mentioned the concept briefly earlier, in our introduction to the behavior

AC SWITCHING CHARACTERISTICS (T_A = + 25°C, V_M = 12 V) (See Figures 2, 3, 4) (Notes 5, 6)

Characteristic	Pins	Symbol	Min	Typ	Max	Unit
Propagation Delay (Clk–to–Driver Output)		t_{PCD}	–	8.0	–	μs
Propagation Delay ($\overline{\text{Bias}}$/Set–to–Driver Output)		t_{PBSD}	–	1.0	–	μs
Propagation Delay (Clk–to–$\overline{\text{Phase A}}$ Low)	7–11	t_{PHLA}	–	12	–	μs
Propagation Delay (Clk–to–$\overline{\text{Phase A}}$ High)	7–11	t_{PLHA}	–	5.0	–	μs

FIGURE 16.12

Excerpt from the AC Switching Characteristics section of the MC3479 bipolar stepper motor driver data sheet. (Used with permission from SCI LLC, DBA ON Semiconductor.)

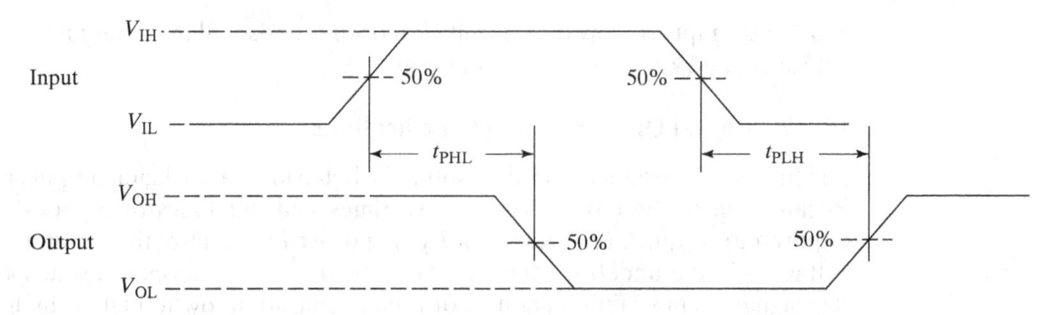

FIGURE 16.13

Propagation delay: the time required for a change at an input to be reflected at an output. Here, t_{PHL} is the propagation delay for an output that makes a high-to-low transition, and t_{PLH} is for an output that makes a low-to-high transition.

of real devices vs. ideal behavior. A **propagation delay** occurs because a finite amount of time is required for information to propagate from the input, though the circuitry of the device, and be reflected at the output (Figure 16.13). For the MC3479, the delay between the rising edge of the clock input and the response of the driver output (t_{PCD}) is typically 8 μs, the delay between a change of the $\overline{\text{Bias}}$/Set input and a response at a driver output is typically 1 μs. Finally, the delay between a rising edge of the clock and a change of the $\overline{\text{Phase A}}$ output is typically 12 μs when $\overline{\text{Phase A}}$ is going low, and 5 μs when it is going high.

16.8 OUTPUT MEETS INPUT

In order for outputs to do something useful, they must be connected to something else in a circuit. Very often, outputs are connected to inputs of other digital devices. Frequently this is very straightforward, and an output from one device can be connected directly to the input of another device. In other situations, the performance of an output on one device will not satisfy the requirements of an input on the other device, and interface circuitry must be added in order for them to share information. In the sections below, we demonstrate how compatibility can be evaluated and show how integration can be achieved when output characteristics don't match input requirements.

16.8.1 Evaluating Compatibility

We can determine if an output and an input are compatible by comparing their voltage, current, and timing specifications. If the output can provide everything the input requires, they're compatible. If not, we will have to decide what additional steps are needed to connect them. Any time two (or more) devices are connected together, you must verify that the characteristics of the outputs satisfy the requirements of the inputs. It is a very bad assumption that any output can drive any input—there are an enormous number of devices available, each satisfying a particular need in the marketplace and tailored to the requirements of that market. Some of these requirements are mutually exclusive, making it impossible to design them to all work together. While manufacturers try to make their products as flexible as possible, there are many cases where interface circuitry is required. As a result, we must always check. We illustrate the process with several examples, the simplest of which is the

DC ELECTRICAL CHARACTERISTICS (Specifications apply over the recommended supply voltage and temperature range (Notes 2, 3) unless otherwise noted.)

Characteristic	Pins	Symbol	Min	Typ	Max	Unit
INPUT LOGIC LEVELS						
Threshold Voltage (Low-to-High)	7, 8, 9, 10	V_{TLH}	–	–	2.0	Vdc
Threshold Voltage (High-to-Low)		V_{THL}	0.8	–	–	Vdc
Hysteresis		V_{HYS}	0.4	–	–	Vdc
Current: (V_I = 0.4 V) (V_I = 5.5 V) (V_I = 2.7 V)		I_{IL}	−100 – –	– – –	– +100 +20	µA

FIGURE 16.14

Excerpt from the DC Electrical Characteristics section from the MC3479 data sheet. (Used with permission from SCI LLC, DBA ON Semiconductor.)

case where two devices are interconnected and no additional interface circuitry is required. For this, we'll draw again from the two devices we've discussed throughout this chapter: outputs from a MC9S12C32 microcontroller powered at $V_{DD5} = 5$ V (± 10%) will be connected directly to the inputs of the MC3479. The MC3479 can be powered with any supply voltage between 7.2 and 16.5 V. So, as long as we stay within those constraints, the specifications in the data sheet are valid. This is one case where applying worst-case design principles is particularly straightforward.

Though we've discussed the MC3479's timing specifications, we have not as yet turned our attention to the required input voltages and currents. These are contained in the data sheet for the device in the DC Electrical Characteristics section, a portion of which is shown in Figure 16.14.

For the purposes of our comparison of output characteristics and input requirements, it is often helpful to create a table and fill it in with all the relevant specifications. For the sake of this example, we'll use the worst-case specifications throughout, including the temperature range.

MC9S12C32 Output		**MC3479 Input**		**OK?**
V_{OL} $V_{DD5} = 4.5$ V $0\text{ mA} \le I_{OL} \le 10\text{ mA}$	0.8 V	V_{THL}	0.8 V	YES $(V_{OL} = 0.8\text{ V}) \le (V_{THL} = 0.8\text{ V})$
V_{OH} $V_{DD5} = 4.5$ V $-10\text{ mA} \le I_{OH} \le 0\text{ mA}$	3.7 V	V_{TLH}	2 V	YES $(V_{TLH} = 2\text{ V}) \le (V_{OH} = 3.7\text{ V})$
I_{OL}	$0\text{ mA} \le I_{OL} \le 10\text{ mA}$	I_{IL}	−100 µA	YES $(I_{OL} = 10\text{ mA}) > (I_{IL} = 100\ \mu\text{A})$
I_{OH}	$-10\text{ mA} \le I_{OH} \le 0\text{ mA}$	I_{IH}	+100 µA	YES $(I_{OH} = 10\text{ mA}) > (I_{IH} = 100\ \mu\text{A})$

All the input requirements of the MC3479 are indeed satisfied by the outputs of the 9S12, though it was a very close call for logical low state voltage. The maximum output low voltage for the 9S12 is exactly the same as the maximum input low voltage threshold (described in the MC3479 data sheet as the threshold voltage, high-to-low, with the symbol V_{THL}). Regardless, the requirements have all been satisfied, so the devices may be connected as proposed without any additional circuitry. Note that the designer must still ensure that all timing requirements are met (delays, rise times, fall times, setup times, and hold times).

16.8.2 Pull-Ups and Pull-Downs for Disconnectable and Indeterminate Inputs

In Section 16.6.3, we showed how to configure inputs with pull-ups and pull-downs in circuits, where we want the input state to be fixed and unvarying. Here, we consider a related but more complex use for pull-up and pull-down resistors, for inputs that could be connected to the output of another device some of the time and undriven (and thus indeterminate) at other times. This can happen, for example, when digital devices interface with an external element, such as a keypad, which could be plugged in sometimes and unplugged at other times. In this case, the digital inputs should have a fixed and known default state when the keypad is disconnected, so that the inputs don't float and report fallacious keypad "input" that's indistinguishable from valid input. However, when the keypad is connected, *it* should determine the state of the inputs. This can be achieved with appropriately sized pull-up and pull-down resistors. Another common example involves circuits that incorporate a microcontroller. Most digital I/O (input/output) pins on a microcontroller can be configured in software as either inputs or outputs. When the microcontroller is initially powered on, these pins are usually configured as inputs by default. When the inputs of other digital devices are connected to microcontroller I/O pins expecting them to behave as outputs, there is some length of time after power-up when the microcontroller's pins are all inputs, and the state of these pins is indeterminate until the microcontroller's software configures them as outputs. This time may be brief or lengthy, depending on how the software is written. Pulling the inputs of these other devices into the desired default state prevents them from misbehaving during this period of uncertainty.

Our discussion from Section 16.6.3 is closely related: in that section, we dealt with how to size pull-up and pull-down resistors to meet the requirements of digital input pins that are not connected to the outputs of any other devices. In this section, we consider digital inputs that may be disconnected from other devices (equivalent to the discussion in Section 16.6.3), but they may be connected to a digital output. Sizing resistors for this situation requires that we determine the range of values that satisfy the requirements for the digital inputs when nothing else is connected and also when another digital output is connected. Figure 16.15 shows a circuit in which an output of a 9S12 (Port T pin 0, "PT0") is connected to the Clk input of an MC3479 through a connector. When the 9S12 is disconnected, the MC3479's input is pulled high through R_1, and when the 9S12 is connected, the state of PT0 determines the state at Node A.

To ensure a working result, we need to determine the range of values for the pull-up resistor, R_1, that satisfies the input requirements of the MC3479 for the two possible cases:

1. When the 9S12 is disconnected, the input of the MC3479 is pulled high.
2. When the 9S12 is connected at J_1, the output of the 9S12 determines the state of the MC3479's input.

The first case is straightforward, and closely parallels the discussion in Section 16.6.3 for the 9S12. The input high voltage and current requirements for the MC3479 are different than the examples we worked above, but the process is the same. For the MC3479, the input high-voltage threshold is called V_{TLH}, and the allowable range for the voltage at the Clk pin

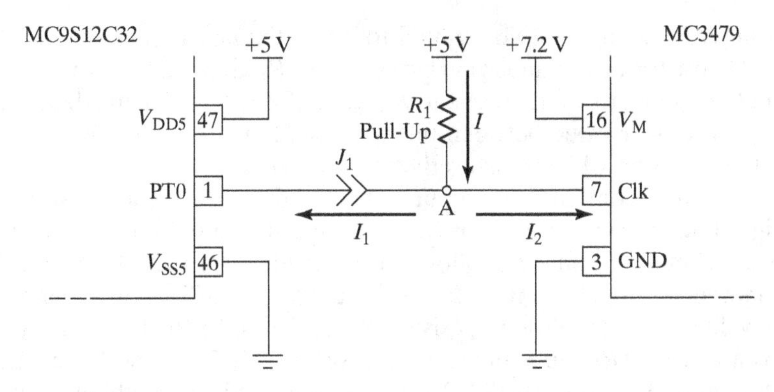

FIGURE 16.15

A pull-up resistor determines the default state of the Clk input on an MC3479 when an MC9S12C32 is disconnected and allows PT0 to control the state when the MC9S12C32 is connected.

is $2\text{ V} \leq V_{\text{Clk}} \leq 5.5\text{ V}$ (Figure 16.14). The maximum input current is 100 μA, when $V_I = 5.5$ V. Substituting these specifications into Eq. (16.1) through Eq. (16.3), we find that the maximum pull-up resistance, $R_{1(\text{max})}$, is 25 kΩ when we pull the MC3479's Clk input (pin 7) up to 4.5 V (the low end of a standard 5 V ± 10% power supply).

For the second case, we also need to insure that the 9S12's output, PT0, is capable of establishing conditions at Node A that meet the MC3479's input voltage and current requirements for the logical high and logical low states. For the logical high state, we assume that all the input current required by the MC3479's Clk input (up to 100 μA) comes through R_1, or $I = I_2$, and that no current flows from Node A into PT0, $I_1 = 0$. PT0 will do its best to output a voltage as close to V_{DD5} as possible. The pull-up resistor to V_{DD5} reduces the amount of current that PT0 might otherwise have to supply, which gives it a boost and makes its task very easy. The assumption that $I_1 = 0$ gives the same result as when the switch is open and PT0 is disconnected: $R_{1(\text{max})} = 25\text{ k}\Omega$.

The logical low state is more involved, since PT0 is required to sink both the current through the pull-up resistor (I) and the input current from the MC3479's Clk input (I_2, which can be as much as −100 μA), while maintaining a voltage at Node A of at most $V_{\text{THL}} = 0.8$ V (Figure 16.14). From Kirchhoff's current law and the maximum output current specifications for PT0 and the maximum input current for Clk:

$$I = I_{1(\max)} + I_2 = 10\text{ mA} + (-100\text{ μA}) = 9.9\text{ mA} \tag{16.9}$$

The worst case occurs when the power supply voltage is as high as possible (5 V + 10% = 5.5 V), and the voltage drop across the pull-up resistor R_1 has the maximum possible value. In the logical low state, PT0 is guaranteed to have a voltage of at most 0.8 V, but no minimum output voltage is specified. The lowest it could possibly be is 0 V, and this would result in the largest possible voltage drop across R_1. Then the minimum value for the pull-up resistor R_1 is determined by

$$R_{1(\text{min})} \geq \frac{V}{I} = \frac{5.5\text{ V}}{9.9\text{ mA}} = 556\ \Omega \tag{16.10}$$

The range of workable values for R_1 is then: $556\ \Omega \leq R_1 \leq 25\text{ k}\Omega$.

For any real circuit that you intend to build, the last step is to select from the list of available 5% resistors. The standard value closest to $R_{1(min)}$ is 620 Ω (which might be as low as 589 Ω), and the standard value closest to $R_{1(max)}$ is 22 kΩ (which could be as high as 23.1 kΩ). Any standard resistor value between these limits will work. As before, a typical choice is a 10 kΩ pull-up resistor, which satisfies the requirements.

If you wish to configure the input to have a default logical low state rather than the logical high state, the pull-up resistor (R_1) is replaced by a pull-down resistor (R_2), shown in Figure 16.16. Though the analysis follows the same logic, for the sake of completeness and to provide another example, we include a brief discussion of how to determine the range of allowable values for the pull-down resistor, R_2. In Figure 16.16, the circuit is modified so that a pull-down resistor (R_2) pulls the Clk input on the MC3479 low. In this case, a resistor is absolutely required because the 9S12's PT0 output needs to be able to drive the Clk input pin high when it is connected. It would not be able to do this if the Clk input were simply grounded.

First, we determine the maximum value of R_2 by assuming that the 9S12 is not connected, and that the pull-down resistor serves the purpose of fixing the input voltage as a logical low. In Figure 16.14, we find that the input low-voltage threshold for the MC3479, V_{THL} (the threshold voltage, high-to-low), is 0.8 V, and that the largest amount of current that will flow (I_{IL}) is −100 μA. From Ohm's law and these input specifications, we can determine $R_{2(max)}$.

$$R_{2(max)} \leq \frac{V}{I} = \frac{0.8\text{ V}}{100\ \mu\text{A}} = 8{,}000\ \Omega = 8\text{ k}\Omega \tag{16.11}$$

If we then connect the 9S12's PT0 to the Clk input at J_1, we need to determine which values of R_2 allow PT0 to assert voltages and source or sink currents that the MC3479 will interpret as valid logical states. As described for the pull-up resistor circuit above, when the 9S12 asserts a logical low, we can assume that no current flows from or into PT0 ($I = 0$), either from the Clk input or the pull-down resistor. The result of this assumption is that the value of $R_{2(max)}$ we calculated when PT0 wasn't connected also applies when PT0 is connected, and asserting a logical low. Lastly, we determine the value of $R_{2(min)}$ that allows the PT0 pin to establish conditions at Node A that the Clk input will interpret as a valid logical high. From Figure 16.1, we know that for the 9S12, V_{OH} is at least V_{DD5} − 0.8 V with a maximum output high current (I_{OH}) of −10 mA. Looking at Figure 16.16 and applying Kirchhoff's

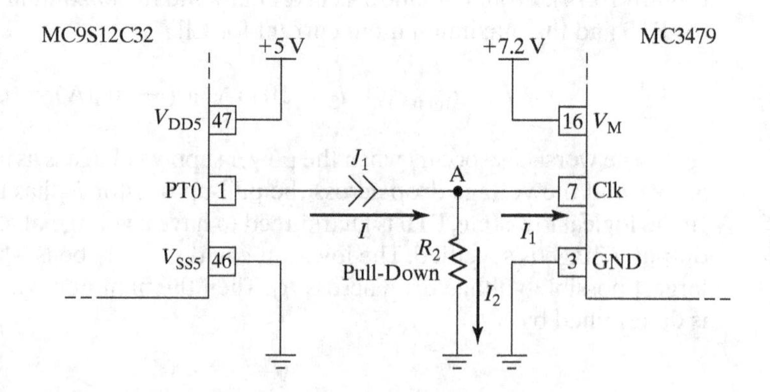

FIGURE 16.16

A pull-down resistor determines the default state of the Clk input on an MC3479 when PT0 of an MC9S12C32 is disconnected, and allows PT0 to control the state when the MC9S12C32 is connected.

current law, we can write an expression for the current through the pull-down resistor R_2, which we will use to determine the minimum allowable resistance.

$$I_2 = I - I_1 = 10\text{ mA} - 100\text{ }\mu\text{A} = 9.9\text{ mA} \quad (16.12)$$

As in the example with the pull-up resistor, the worst case occurs when the power supply voltage is as high as possible (5 V + 10% = 5.5 V), and the voltage drop across the pull-down resistor R_2 has the maximum possible value. From the specifications for the MC3479 (Figure 16.14), the input threshold voltage, low-to-high (V_{TLH}), is 2 V, and the maximum input current is 100 μA. In the logical high state, PT0 is guaranteed to have a voltage of at least 0.8 V below the power supply voltage, but no maximum output voltage is specified. The highest it could possibly be is the power supply voltage (5.5 V), and this would result in the largest possible voltage drop across R_2. For the MC3479, the maximum input current is 100 μA (Figure 16.14). Solving Ohm's law for these conditions gives $R_{2(\text{min})}$:

$$R_{2(\text{min})} \geq \frac{V}{I} = \frac{5.5\text{ V}}{9.9\text{ mA}} = 556\ \Omega \quad (16.13)$$

This is the same result we got for the pull-up resistor (which makes sense, given that the conditions are really identical except for the direction of the "pull"). Finally, selecting among the standard values available for 5% carbon film resistors (Appendix A), we determine that we are free to pick any standard value between 620 Ω and 7.5 kΩ.

16.8.3 Interconnecting Incompatible Devices

Most of the time, you won't be able to connect two digital devices together directly when they are powered by different supply voltages. This is happening more and more over time, as recent trends in design have been toward lower power supply voltages as a means of reducing power consumption. Since $P = VI$, reducing a device's supply voltage reduces its power consumption. This is an important issue for applications where efficiency is critical, such as battery powered devices like mobile phones and media players. While it's sometimes possible to design a circuit that only makes use of devices powered by the same low voltage, this often turns out to be difficult or impractical, and devices that require different power supplies are often combined together in a single circuit. These devices will need to exchange information with each other, but aren't interoperable without interface circuitry.

In Section 16.8.1, we found that we could directly drive the inputs of an MC3479 with the outputs of a MC9S12C32 powered at $V_{DD5} = 5$ V. There are many cases where it won't be that easy, and the input requirements won't be met by the outputs you'd like to connect. To illustrate a case like this, we propose a circuit in which an MC9S12C32 powered at $V_{DD5} = 5$ V (± 10%) is connected to an XBee OEM RF Module (Digi International Inc.), which is shown in Figure 16.17. The XBee module is an inexpensive, low-power, compact radio transmitter/receiver that conforms to the Zigbee radio communications standard (IEEE 802.15.4). A 9S12 connected to an XBee radio could communicate with any number of other similarly equipped devices within range, leading to many useful and exciting possibilities.

The XBee module doesn't have a 5 V power supply like the one we chose for the 9S12. Instead, it requires a power supply between 2.8 and 3.4 V. For this discussion, we will power the XBee with a 3.3 V power supply, which is a standard power supply voltage. Note that if we selected a standard power supply with ± 10% tolerances, we would exceed the XBee's maximum power supply specifications. We need a power supply with much tighter tolerances as a result. We'll select one that guarantees ± 2.4% tolerance, which results in range of supply voltages between 3.22 and 3.38 V. This will satisfy the XBee's requirements.

FIGURE 16.17
The XBee OEM RF Module from Digi International Inc.

One option that might allow the XBee and the 9S12 to connect directly would be to redesign the 9S12 circuit so that the microcontroller runs at the same power supply voltage as the XBee. This is allowable for the 9S12, since it accepts power supply voltages as low as 2.97 V. However, there are many instances when this is not a realistic option. We assume for the sake of this discussion that this is one of those times.

For the 9S12 to exchange information with the XBee, signals must go both directions. We will be connecting outputs on the 9S12 to inputs on the XBee module and outputs on the XBee module to inputs on the 9S12. As before, we need to check all the input and output specifications for both devices to see if this will work without additional circuitry. We don't expect this to work since the devices have different power supply voltages, but it will be instructive to see where the problems lie. In order to perform the comparison, we need to consult the Electrical Characteristics section of the XBee data sheet, which contains the specifications related to its inputs and outputs (Figure 16.18). We compare these specifications with the 5 V I/O Characteristics for the 9S12 shown in Figure 16.1 to construct the tables with the pertinent specifications that follow Figure 16.18.

There are two problems with connecting these two devices together directly. The first is that the 9S12's output voltage, V_{OH}, far exceeds the maximum power supply voltage for the

Table 1-03. DC Characteristics (VCC = 2.8 - 3.4 VDC)

Symbol	Characteristic	Condition	Min	Typical		Max	Unit
V_{IL}	Input Low Voltage	All Digital Inputs	-	-		0.35 * VCC	V
V_{IH}	Input High Voltage	All Digital Inputs	0.7 * VCC	-		-	V
V_{OL}	Output Low Voltage	I_{OL} = 2 mA, VCC >= 2.7 V	-	-		0.5	V
V_{OH}	Output High Voltage	I_{OH} = -2 mA, VCC >= 2.7 V	VCC - 0.5	-		-	V
II_{IN}	Input Leakage Current	V_{IN} = VCC or GND, all inputs, per pin	-	0.025		1	µA
II_{OZ}	High Impedance Leakage Current	V_{IN} = VCC or GND, all I/O High-Z, per pin	-	0.025		1	µA
TX	Transmit Current	VCC = 3.3 V	-	45 (XBee)	215, 140 (PRO, Int)	-	mA
RX	Receive Current	VCC = 3.3 V	-	50 (XBee)	55 (PRO)	-	mA
PWR-DWN	Power-down Current	SM parameter = 1	-	< 10		-	µA

FIGURE 16.18
The Electrical Characteristics section from the XBee OEM RF Module data sheet. (Courtesy of Digi International Inc.).

MC9S12C32 Output (V_{DD5} = 5 V ± 10%)		XBee RF Module Input (V_{CC} = 3.3 V ± 2.4%)		OK?
V_{OL} V_{DD5} = 4.5 V 0 mA ≤ I_{OL} ≤ 10 mA	0.8 V	V_{IL} V_{CC} = 3.22 V	1.13 V	YES (V_{OL} = 0.8 V) ≤ (V_{IL} = 1.13 V)
V_{OH} V_{DD5} = 4.5 V −10 mA ≤ I_{OH} ≤ 0 mA	3.7 V	V_{IH} V_{CC} = 3.38 V	2.366 V	NO (V_{OH} = 3.7 V) ≫ (V_{CC} = 3.38 V)
I_{OL}	0 mA ≤ I_{OL} ≤ 10 mA	II_{IN}	± 1 μA	YES (I_{OL} = 10 mA) > (II_{IN} = 1 μA)
I_{OH}	−10 mA ≤ I_{OH} ≤ 0 mA			YES (I_{OH} = −10 mA) ≤ (II_{IN} = −1 μA)

XBee RF Module Output (V_{CC} = 3.3 V ± 2.4%)		MC9S12C32 Input (V_{DD5} = 5 V ± 10%)		OK?
V_{OL} V_{CC} = 3.22 V I_{OL} = 2 mA	0.5 V	V_{IL} V_{DD5} = 4.5 V	1.58 V	YES (V_{OL} = 0.5 V) ≤ (V_{IL} = 1.58 V)
V_{OH} V_{CC} = 3.22 V I_{OH} = −2 mA	2.72 V	V_{IH} V_{DD5} = 5.5 V	3.58 V	NO (V_{OH} = 2.72 V) < (V_{IH} = 3.58 V)
I_{OL}	0 mA ≤ I_{OL} ≤ 2 mA	I_{in}	± 2.5 μA	YES (I_{OL} = 2 mA) > (I_{in} = 2.5 μA)
I_{OH}	−2 mA ≤ I_{OH} ≤ 0 mA			YES (I_{OH} = −2 mA) ≤ (I_{in} = −2.5 μA)

XBee. We mentioned in our discussion of allowable input voltage ranges for the 9S12 (Section 16.6) that introducing voltages above a device's power supply voltage (the XBee's V_{CC} in this case) or below ground was generally a bad idea and likely to result in the destruction of the device. For this reason, we should not directly connect a 9S12 powered at 5 V to an XBee powered at 3.3 V. The second problem is that the XBee's outputs cannot directly drive the 9S12's inputs, since they don't satisfy the requirement for the voltages that will be interpreted as a logical high (V_{IH}). The XBee promises that it will represent logical 1 with voltages between 2.88 V and V_{CC}, but the 9S12 requires input voltages to be at least 3.58 V. Our conclusion is that we need to take additional measures to connect these devices together.

Semiconductor manufacturers have also perceived this need (i.e., market) and address the problem by offering a category of devices called **voltage level translators**. These types of components translate the signal voltages from a device powered by a first supply voltage (e.g., 3.3 V) so that they meet the specifications for devices powered by a second power supply voltage (e.g., 5 V). Many manufacturers provide a wide variety of these devices, such as the Texas Instruments TXB0104 and the Maxim Integrated Products MAX3000E.

As a representative example, let's examine the TI TXB0104 (Figure 16.19). Rather than reproduce large sections of the data sheet in this discussion, we will populate the tables

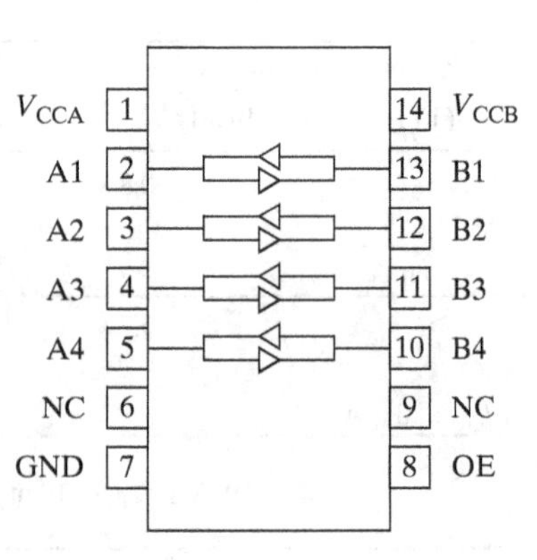

FIGURE 16.19

Pin-out diagram for the TXB0104.
(Courtesy of Texas Instruments Incorporated.)

you've come to expect and adhere to the data sheet's naming conventions so that the curious reader can obtain the data sheet from the manufacturer and work through the example.

At the highest conceptual level, this device has two ports (Port A on the left side of the diagram in Figure 16.19 and Port B on the right side). Each port requires its own separate power supply, called V_{CCA} and V_{CCB}. These are the same power supplies that power the devices you would like to interconnect. From the specifications, we see that $1.2\text{ V} \leq V_{CCA} \leq 3.6\text{ V}$ and $1.65\text{ V} \leq V_{CCB} \leq 5.5\text{ V}$, with the additional constraint that $V_{CCA} \leq V_{CCB}$. The TXB0104 takes signals from one port and relays them to the other port (in either direction) and translates the signal voltage. For example, if $V_{CCA} = 3.3\text{ V}$ and $V_{CCB} = 5\text{ V}$, and a 3.3 V input signal is connected to Port A on channel A1 (pin 2), the signal will be translated to 5 V as output on Port B channel B1 (pin 13). Conversely, if a 5 V output is connected at Port B channel B2 (pin 12), it would be output as 3.3 V on Port A channel A2 (pin 3). The device conveniently determines the signal direction automatically on a pin-by-pin basis.

Figure 16.20 shows a circuit that uses channel 1 (A1 to B1) on the TXB0104 to connect the D_{OUT} ("data out") output pin on the XBee RF Module to the RXD ("receive") input

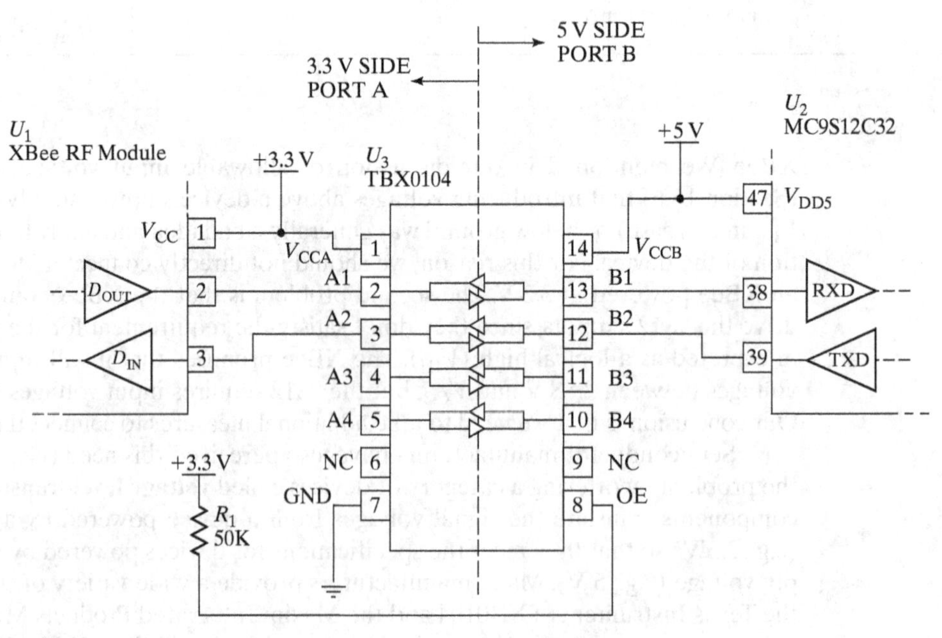

FIGURE 16.20

Interfacing devices powered with different power supply voltages.

pin on 9S12. Since the TXB0104's channels can relay data in either direction and automatically determine the direction, we can use channel 2 in the other direction (B2 to A2) to connect the 9S12's TXD ("transmit") output pin to the XBee's D_{IN} ("data in") input pin. The TXB0104's OE (the output enable pin) is pulled to V_{CCA} = 3.3 V to enable the device. When the OE pin is pulled to a logical low, all of the input/output pins (A1–A4 and B1–B4) are disabled, shutting down the chip and placing it in a power saving mode.

To be certain that the TXB0104 will work properly with the XBee and the 9S12, we need to examine the Recommended Operating Conditions and Electrical Characteristics sections in the data sheet for the inputs and outputs and compare them with the specifications of our trusty 9S12 (with V_{DD5} = 5 V) and the XBee (with V_{CC} = 3.3 V). For any circuit, including this example, we must check every connection between any two devices: the inputs and outputs on both the 3.3 and 5 V sides. We start where D_{OUT} on U_1 (the data output of the XBee) connects to the TXB0104 at A1 (pin 2).

XBee Output (V_{CC} = 3.3 V ± 2.4%)		**TBX0104 Input (V_{CCA} = 3.3 V ± 2.4%)**		**OK?**
V_{OL} V_{CC} = 3.22 V I_{OL} = 2 mA	0.5 V	V_{IL} V_{CCA} = 3.22 V	1.13 V	YES (V_{OL} = 0.5 V) ≤ (V_{IL} = 1.13 V)
V_{OH} V_{CC} = 3.22 V I_{OH} = −2 mA	2.72 V	V_{IH} V_{CCA} = 3.38 V	2.2 V	YES (V_{OH} = 2.72 V) ≥ (V_{IH} = 2.2 V)
I_{OL}	0 mA ≤ I_{OL} ≤ 2 mA	I_{in}	±5 μA	YES (I_{OL} = 2 mA) ≥ (I_{in} = 5 μA)
I_{OH}	−2 mA ≤ I_{OH} ≤ 0 mA			YES (I_{OH} = −2 mA) ≤ (I_{in} = −5 μA)

All specifications for the XBee and TBX0104 are satisfied, so we will track the signal through to the TBX0104's output at B1 (pin 13) and verify that this satisfies the input requirements of the 9S12 powered at 5 V (U_2).

TBX0104 Output (V_{CCB} = 5 V ± 10%)		**MC9S12C32 Input (V_{DD5} = 5 V ± 10%)**		**OK?**
V_{OL} V_{CCB} = 4.5 V I_{OL} = 20 μA	0.4 V	V_{IL} V_{DD5} = 4.5 V	1.58 V	YES (V_{OL} = 0.4 V) ≤ (V_{IL} = 1.58 V)
V_{OH} V_{CCB} = 4.5 V I_{OH} = −20 μA	4.1 V	V_{IH} V_{DD5} = 5.5 V	3.58 V	YES (V_{OH} = 4.1 V) ≥ (V_{IH} = 3.58 V)
I_{OL}	0 μA ≤ I_{OL} ≤ 20 μA	I_{in}	± 2.5 μA	YES (I_{OL} = 20 μA) ≥ (I_{in} = 2.5 μA)
I_{OH}	−20 μA ≤ I_{OH} ≤ 0 μA			YES (I_{OH} = −20 μA) ≤ (I_{in} = −2.5 μA)

This works too, so let's look at the signal that begins at U_2 (the TXD output of the 9S12 on the 5 V side) and track it through the TBX0104's B2 input (pin 12), out the A1 output (pin 3), to the D_{IN} input of U_1 (the XBee on the 3.3 V side).

MC9S12C32 Output (V_{DD5} = 5 V ± 10%)		TBX0104 Input (V_{CCB} = 5 V ± 10%)		OK?
V_{OL} $V_{DD5} = 4.5$ V $0\text{ mA} \le I_{OL} \le 10\text{ mA}$	0.8 V	V_{IL} $V_{CCB} = 4.5$ V	1.58 V	YES $(V_{OL} = 0.8\text{ V}) \le (V_{IL} = 1.58\text{ V})$
V_{OH} $V_{DD5} = 4.5$ V $-10\text{ mA} \le I_{OH} \le 0\text{ mA}$	3.7 V	V_{IH} $V_{CCB} = 5.5$ V	3.58 V	YES $(V_{OH} = 3.7\text{ V}) \ge (V_{IH} = 3.58\text{ V})$
I_{OL}	$0\text{ mA} \le I_{OL} \le 10\text{ mA}$	I_{in}	±5 μA	YES $(I_{OL} = 10\text{ mA}) \ge (I_{in} = 5\ \mu\text{A})$
I_{OH}	$-10\text{ mA} \le I_{OH} \le 0\text{ mA}$			YES $(I_{OH} = -10\text{ mA}) \le (I_{in} = -5\ \mu\text{A})$

TBX0104 Output (V_{CCA} = 3.3 V ± 2.4%)		XBee Input (V_{CC} = 3.3 V ± 2.4%)		OK?
V_{OL} $V_{CCA} = 3.22$ V $I_{OL} = 20\ \mu\text{A}$	0.4 V	V_{IL} $V_{CC} = 3.22$ V	1.13 V	YES $(V_{OL} = 0.4\text{ V}) \le (V_{IL} = 1.13\text{ V})$
V_{OH} $V_{CCA} = 3.22$ V $I_{OH} = -20\ \mu\text{A}$	2.82 V	V_{IH} $V_{CC} = 3.38$ V	2.36 V	YES $(V_{OH} = 2.82\text{ V}) \ge (V_{IH} = 2.36\text{ V})$
I_{OL}	$0\ \mu\text{A} \le I_{OL} \le 20\ \mu\text{A}$	II_{IN}	1 μA	YES $(I_{OL} = 20\ \mu\text{A}) \ge (II_{IN} = 1\ \mu\text{A})$
I_{OH}	$-20\ \mu\text{A} \le I_{OH} \le 0\ \mu\text{A}$			YES $(I_{OH} = -20\ \mu\text{A}) \le (II_{IN} = -1\ \mu\text{A})$

Success! Though this can get tedious, it is the only way to ensure that all the devices will work smoothly together. In general, however, voltage level translators are designed to be very flexible so that they meet the input and output specifications for the widest possible range of devices (nearly all).

16.9 HOMEWORK PROBLEMS

16.1 If an MC9S12C32 is operated on a nominal 3.3 V supply that is 10% higher than the nominal value, how would that affect the input high-voltage requirement? To answer, compare the input high voltage when operating with a supply at exactly 3.3 V to the input high voltage required when the supply voltage is 10% higher than 3.3 V.

Table A-7 3.3V I/O Characteristics

Conditions are VDDX=3.3V +/-10%, Termperature from -40˚C to +140˚C, unless otherwise noted

Num	C	Rating	Symbol	Min	Typ	Max	Unit
1	P	Input High Voltage	V_{IH}	0.65*V_{DD5}	-	-	V
	T	Input High Voltage	V_{IH}	-	-	VDD5 + 0.3	V
2	P	Input Low Voltage	V_{IL}	-	-	0.35*V_{DD5}	V
	T	Input Low Voltage	V_{IL}	VSS5 - 0.3	-	-	V
3	C	Input Hysteresis	V_{HYS}		250		mV
4	P	Input Leakage Current (pins in high ohmic input mode)[1] $V_{in} = V_{DD5}$ or V_{SS5}	I_{in}	–2.5	-	2.5	µA
5	C	Output High Voltage (pins in output mode) Partial Drive I_{OH} = –0.75mA	V_{OH}	V_{DD5} – 0.4	-	-	V
6	P	Output High Voltage (pins in output mode) Full Drive I_{OH} = –4.5mA	V_{OH}	V_{DD5} – 0.4	-	-	V
7	C	Output Low Voltage (pins in output mode) Partial Drive I_{OL} = +0.9mA	V_{OL}	-	-	0.4	V
8	P	Output Low Voltage (pins in output mode) Full Drive I_{OL} = +5.5mA	V_{OL}	-	-	0.4	V

FIGURE 16.21

3.3 V I/O Characteristics for the Freescale MC9S12C32. (Copyright of Freescale Semiconductor, Inc. 2010. Used by permission.)

16.2 For an MC9S12C32 operated on a nominal 3.3 V supply, what is the maximum expected output low voltage when the output sinks 4 mA from an external source? (Use Figure 16.21 and assume full drive strength.)

16.3 What is the maximum input low voltage for a TTL type input on the PIC16F690 when operated on a nominal 5 V supply? (Use the specifications provided in Figure 16.22.)

16.4 What is the maximum input low voltage for a Schmitt trigger type input on the PIC16F690 when operated on a nominal 3.3 V supply? (Use the specifications provided in Figure 16.22.)

16.5 If an output of PIC16F690 operating with a 5 V (± 10%) supply is connected to an input of an MC3479 operating with a 7.5 V supply, will the devices be operating within their specifications? Use Figure 16.22 and quote the relevant specifications to support your answer.

16.6 Is it possible to have a PIC16F690 operating at 3.3 V exchange data (both inputs and outputs) with an MC9S12C32 operating at 3.3 V with both devices operating within their specifications? Use Figures 16.21 and 16.22 and quote the relevant specifications to support your answer.

16.7 What is the maximum value for a pull-up resistor that is to be used on the $\overline{F}$/HS input on an MC3479 that is operated with a 5 V supply? Use the specifications given in Figure 16.14. Quote a standard 5% resistor value and don't forget to allow for resistor tolerances.

16.8 Which of the inputs of a PIC16F690 (operated at 5 V ± 10%) can an output from an XBee-Pro module (operated at 3.3 V) properly drive? (Use the specifications given in Figures 16.18 and 16.22.) If only a subset of the inputs is compatible with the XBee-Pro outputs, identify which inputs. Quote specifications to support your answer.

16.9 Design a circuit using only passive components that will allow an output from a PIC16F690 operated at 5 V (± 10%) to properly drive the inputs of an XBee-Pro module operated at 3.3 V. You may assume that the absolute maximum input high voltage for the XBee-Pro module is V_{CC} + 0.4 V. Use the specifications given in Figures 16.18 and 16.22.

16.10 Can an output from an XBee-Pro module powered by a 3.3 V supply (Figure 16.18) properly drive an input of a MC9S12C32 powered at 5 V ± 10% (Figure 16.1)? Quote specifications to support your answer.

16.11 What is the maximum value for a pull-down resistor to be used on an input to an XBee-Pro module? Use Figure 16.18 and quote specifications to support your answer.

16.12 Given the circuit in Figure 16.23, what is the range of acceptable values for the resistor *R*? Consider that the connector may be disconnected and under those conditions, the input to the PIC16F690 should be a valid logic level high. The power supplies are nominal 5 V ± 10% supplies (remember to consider the full range of output voltages possible from the full range of power supply voltages when determining the upper and lower limits for the resistor). Report the largest and smallest standard 5% resistor value that will result in all devices operating within their specifications. Use the specifications for the PIC16F690 provided in Figure 16.22. Locate the data sheet for the LM339 on the Internet. Quote the specifications you used to determine your answers.

DC CHARACTERISTICS			**Standard Operating Conditions (unless otherwise stated)** Operating temperature -40°C ≤ TA ≤ +85°C for industrial -40°C ≤ TA ≤ +125°C for extended				
Param No.	**Sym**	**Characteristic**	**Min**	**Typ†**	**Max**	**Units**	**Conditions**
	VIL	**Input Low Voltage**					
		I/O Port:					
D030		with TTL buffer	Vss	—	0.8	V	4.5V ≤ VDD ≤ 5.5V
D030A			Vss	—	0.15 VDD	V	2.0V ≤ VDD ≤ 4.5V
D031		with Schmitt Trigger buffer	Vss	—	0.2 VDD	V	2.0V ≤ VDD ≤ 5.5V
D032		$\overline{\text{MCLR}}$, OSC1 (RC mode)(1)	Vss	—	0.2 VDD	V	
D033		OSC1 (XT and LP modes)	Vss	—	0.3	V	
D033A		OSC1 (HS mode)	Vss	—	0.3 VDD	V	
	VIH	**Input High Voltage**					
		I/O Ports:		—			
D040		with TTL buffer	2.0	—	VDD	V	4.5V ≤ VDD ≤ 5.5V
D040A			0.25 VDD + 0.8	—	VDD	V	2.0V ≤ VDD ≤ 4.5V
D041		with Schmitt Trigger buffer	0.8 VDD	—	VDD	V	2.0V ≤ VDD ≤ 5.5V
D042		$\overline{\text{MCLR}}$	0.8 VDD	—	VDD	V	
D043		OSC1 (XT and LP modes)	1.6	—	VDD	V	
D043A		OSC1 (HS mode)	0.7 VDD	—	VDD	V	
D043B		OSC1 (RC mode)	0.9 VDD	—	VDD	V	**(Note 1)**
	IIL	**Input Leakage Current(2)**					
D060		I/O ports	—	± 0.1	± 1	μA	VSS ≤ VPIN ≤ VDD, Pin at high-impedance
D061		$\overline{\text{MCLR}}$(3)	—	± 0.1	± 5	μA	VSS ≤ VPIN ≤ VDD
D063		OSC1	—	± 0.1	± 5	μA	VSS ≤ VPIN ≤ VDD, XT, HS and LP oscillator configuration
D070*	IPUR	**PORTA Weak Pull-up Current**	50	250	400	μA	VDD = 5.0V, VPIN = VSS
	VOL	**Output Low Voltage(5)**					
D080		I/O ports	—	—	0.6	V	IOL = 8.5 mA, VDD = 4.5V (Ind.)
	VOH	**Output High Voltage(5)**					
D090		I/O ports	VDD – 0.7	—	—	V	IOH = -3.0 mA, VDD = 4.5V (Ind.)

FIGURE 16.22
I/O characteristics for the Microchip PIC16F690 microcontroller. (© 2008 Microchip Technology Incorporated.)

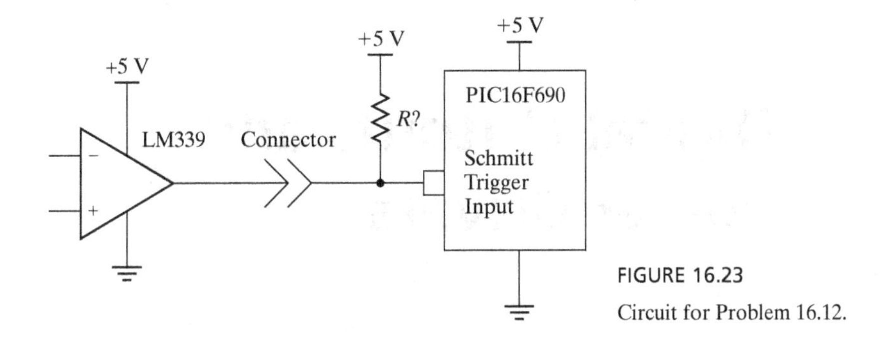

FIGURE 16.23

Circuit for Problem 16.12.

16.13 Locate the data sheet for the TI TXB0104 voltage level translator on the Internet, and use it to answer the following questions:

(a) What is the range of recommended power supply voltages for the A Port? For the B Port?

(b) If $V_{CCA} = 3.3$ V (exactly), what are the values for:

- **i.** Minimum V_{ih} and Maximum I_{ih}
- **ii.** Maximum V_{il} and Maximum I_{il}
- **iii.** Minimum V_{oh} and Maximum I_{oh}
- **iv.** Maximum V_{ol} and Maximum I_{ol}
- **v.** The allowable range of power supply voltages for the B Port

(c) What is the function of the OE input?

(d) What is the minimum value of V_{ih} for the OE input?

(e) What is the maximum value of V_{il} for the OR input?

CHAPTER 17

Digital Outputs and Power Drivers

The output characteristics that were described in Chapter 16 are not the only ones that you are likely to encounter. In this chapter, we look in more detail at both the outputs described in Chapter 16 and a range of other types of outputs that are commonly used in mechatronic systems. When you have completed this chapter, you should be able to describe the basic differences between the following output types and select the appropriate one(s) for specific applications:

- logic-level outputs,
- power-driver outputs,
- totem-pole outputs,
- open-collector outputs,
- three-state outputs,
- high and low side drivers,
- half- and full-bridges.

17.1 TOTEM-POLE OUTPUTS

The digital outputs that we described in Chapter 16 are capable of both sourcing current into an external device and sinking current from an external device. The output structure of this class of devices is usually described as being of the **totem-pole** style. The name derives from the stacking of circuit elements used to make the output stage, as shown in Figure 17.1a.

For either the bipolar or CMOS construction, the output transistors are controlled in such a way that only Q_1 or Q_2 is on at any point in time. When the control input is low, Q_1 is on and the output is forced into the high state, in which it is capable of sourcing current into an external load. When the control input is high, Q_2 is on and the output is forced into the low state, in which it is capable of sinking current from an external load.

In the bipolar version (Figure 17.1a), the diode D_1 is necessary to insure that Q_1 is off when Q_2 is turned on. The resistor R_1 is necessary to limit the transient current when changing states. Because bipolar transistors turn on more quickly than they turn off, when the output changes from a low level to a high level, there is a short period of time when both Q_1 and Q_2 are conducting significant current. During this time, the resistance of R_1 serves to limit the peak magnitude of this **shoot-through current**.

17.1.1 Totem-Pole Output Specifications

The presence of D_1 and R_1 in the bipolar totem-pole output creates an asymmetry in the output drive capability that can be seen in the specifications for typical devices, such as the one

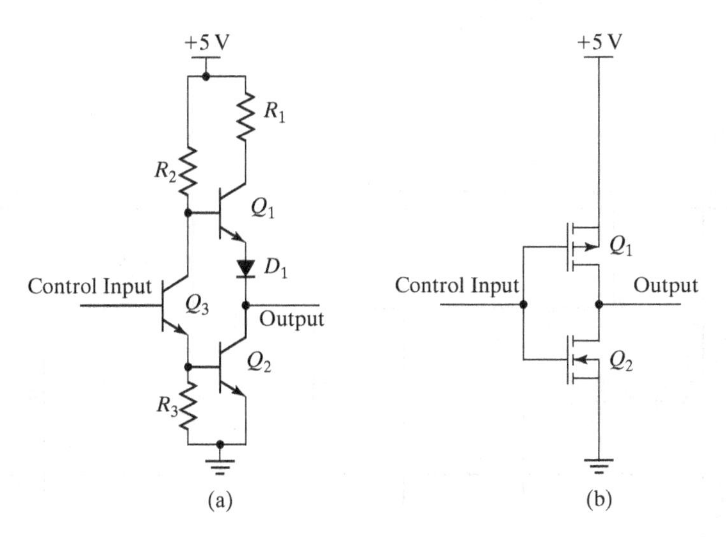

FIGURE 17.1

Totem-pole style outputs using (a) bipolar transistors and (b) CMOS transistors.

shown in Figure 17.2. While this particular device (the SN7400, a quad 2-input NAND logic chip) is capable of sinking up to 16 mA, its sourcing capability is much (40x) smaller: only 0.4 mA. It is not uncommon to see factors of 10–20 times difference between the sourcing and sinking current drive ability for bipolar devices with totem-pole style outputs.

When a totem-pole output stage is built using CMOS technology (Figure 17.1b), the symmetric nature of the structure (with complementary N-channel and P-channel devices) and the resistive nature of MOSFETs in the on-state results in an output with symmetric current sourcing and sinking capabilities. This can be seen in Figure 17.3 for the 74HC04 (a hex inverter chip), where the output source and sink capabilities are listed as test conditions on the output voltages in the high and low states.

As we discussed in Chapter 16, it is typical of CMOS device specifications to list V_{OH} and V_{OL} specifications at two different output current levels. One of these levels will be with

recommended operating conditions (see Note 3)

		SN5400			SN7400			UNIT
		MIN	NOM	MAX	MIN	NOM	MAX	
V_{CC}	Supply voltage	4.5	5	5.5	4.75	5	5.25	V
V_{IH}	High-level input voltage	2			2			V
V_{IL}	Low-level input voltage			0.8			0.8	V
I_{OH}	High-level output current			−0.4			−0.4	mA
I_{OL}	Low-level output current			16			16	mA
T_A	Operating free-air temperature	−55		125	0		70	°C

NOTE 3: All unused inputs of the device must be held at V_{CC} or GND to ensure proper device operation. Refer to the TI application report, *Implications of Slow or Floating CMOS Inputs*, literature number SCBA004.

FIGURE 17.2

Output drive specifications for the SN7400, a device with a bipolar totem-pole style output. (Courtesy of Texas Instruments Incorporated.)

DC Electrical Characteristics (Note 4)

Symbol	Parameter	Conditions	V_{CC}	$T_A = 25°C$		$T_A = -40$ to $85°C$	$T_A = -55$ to $125°C$	Units
				Typ	Guaranteed Limits			
V_{IH}	Minimum HIGH Level Input Voltage		2.0V		1.5	1.5	1.5	V
			4.5V		3.15	3.15	3.15	V
			6.0V		4.2	4.2	4.2	V
V_{IL}	Maximum LOW Level Input Voltage		2.0V		0.5	0.5	0.5	V
			4.5V		1.35	1.35	1.35	V
			6.0V		1.8	1.8	1.8	V
V_{OH}	Minimum HIGH Level Output Voltage	$V_{IN} = V_{IL}$, $\|I_{OUT}\| \leq 20\ \mu A$	2.0V	2.0	1.9	1.9	1.9	V
			4.5V	4.5	4.4	4.4	4.4	V
			6.0V	6.0	5.9	5.9	5.9	V
		$V_{IN} = V_{IL}$, $\|I_{OUT}\| \leq 4.0$ mA	4.5V	4.2	3.98	3.84	3.7	V
		$\|I_{OUT}\| \leq 5.2$ mA	6.0V	5.7	5.48	5.34	5.2	V
V_{OL}	Maximum LOW Level Output Voltage	$V_{IN} = V_{IH}$, $\|I_{OUT}\| \leq 20\ \mu A$	2.0V	0	0.1	0.1	0.1	V
			4.5V	0	0.1	0.1	0.1	V
			6.0V	0	0.1	0.1	0.1	V
		$V_{IN} = V_{IH}$, $\|I_{OUT}\| \leq 4.0$ mA	4.5V	0.2	0.26	0.33	0.4	V
		$\|I_{OUT}\| \leq 5.2$ mA	6.0V	0.2	0.26	0.33	0.4	V
I_{IN}	Maximum Input Current	$V_{IN} = V_{CC}$ or GND	6.0V		±0.1	±1.0	±1.0	μA
I_{CC}	Maximum Quiescent Supply Current	$V_{IN} = V_{CC}$ or GND, $I_{OUT} = 0\ \mu A$	6.0V		2.0	20	40	μA

Note 4: For a power supply of 5V ±10% the worst case output voltages (V_{OH}, and V_{OL}) occur for HC at 4.5V. Thus the 4.5V values should be used when designing with this supply. Worst case V_{IH} and V_{IL} occur at V_{CC}=5.5V and 4.5V respectively. (The V_{IH} value at 5.5V is 3.85V.) The worst case leakage current (I_{IN}, I_{CC}, and I_{OZ}) occur for CMOS at the higher voltage and so the 6.0V values should be used.

FIGURE 17.3

Output drive capability for the 74HC04 with a CMOS totem-pole style output. (Courtesy of Fairchild Semiconductor.)

relatively low output current (≤ 20 μA in this example) and another specification at a higher output current level (≤ 4 mA in this example).

17.2 OPEN-COLLECTOR/OPEN-DRAIN OUTPUTS

When the sourcing transistors (Q_1 in Figure 17.1) are omitted from the output stage, the result is a drive circuit like those in Figure 17.4. Because the collector or the drain of the output transistor is directly brought out of the chip, these configurations are known as **open collector** (for bipolar transistors) and **open drain** (for MOSFETs). In the case shown in Figure 17.4a where a bipolar transistor is used, the diode and current limiting resistor are no longer required and are also omitted.

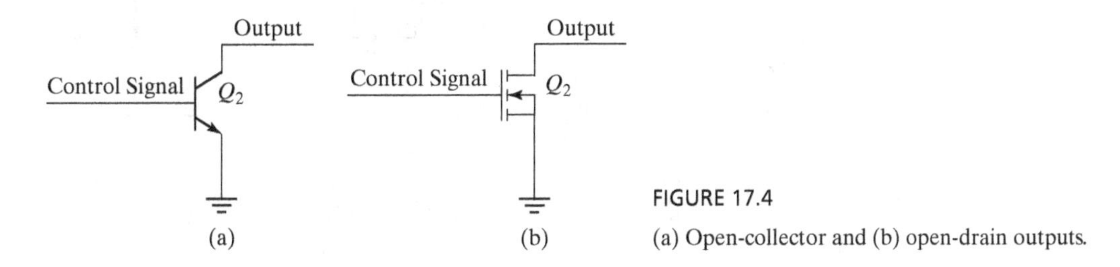

FIGURE 17.4

(a) Open-collector and (b) open-drain outputs.

Without a transistor connected to the positive power supply, these output stages cannot force the output voltage into a logical high state, and they cannot source current. An external component (typically a pull-up resistor) is required (Figure 17.5) to force the output to a high state and set the voltage to which the output will rise.

By removing the sourcing transistor from the output stage, these open-collector and open-drain outputs provide added flexibility at the expense of some performance. Since the output high voltage is no longer determined by the power supply of the chip that contains the output stage, it is possible to pull the output up to a different voltage and create an output swing with an upper voltage that is different than the chip's power supply. This would allow, for example, a chip powered by 5 V to produce an output voltage that swings between ground and 3.3 V, simplifying the interface between devices operating on different supply voltages. Many open-collector and open-drain outputs allow for the pull-up voltage to be higher than the chip power supply. This capability makes it possible with these devices to produce a larger output voltage swing than would be available using the chip power supplies.

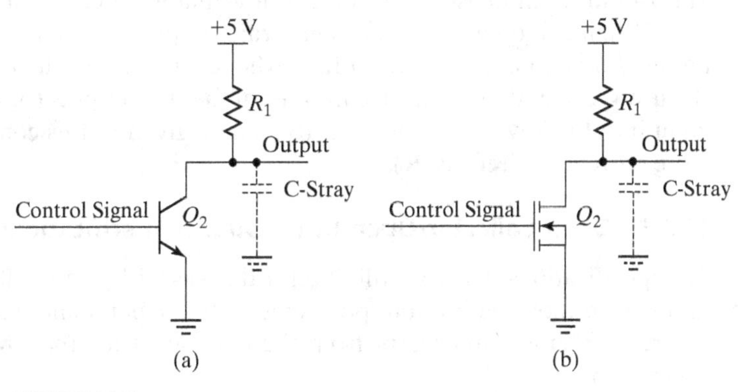

FIGURE 17.5

Open-collector and open-drain outputs with pull-up resistors.

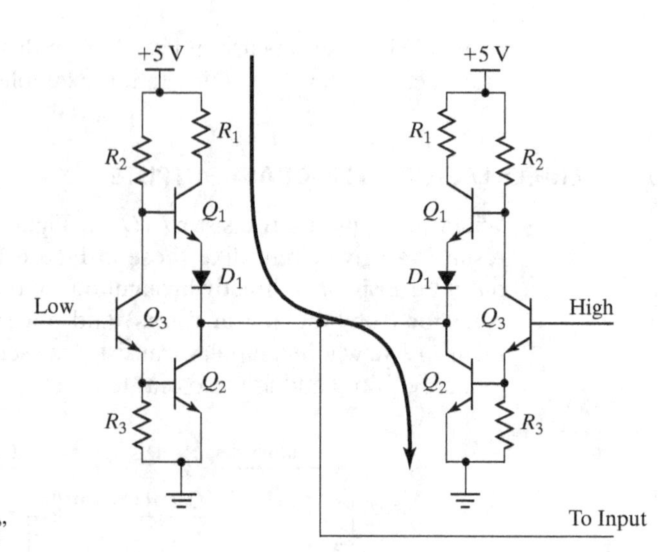

FIGURE 17.6

Two totem-pole outputs "fighting."

The major drawback to the open-collector/open-drain style output relates to the speed of the low-high transition. Since the high state voltage is supplied through pull-up resistor with a relatively high value (as compared to the typical R_1 in a totem-pole style output), the time to charge the stray capacitance (due to the wiring) and the capacitance of the input being driven (shown as C-stray in Figure 17.5) could be slow compared with typical digital rise and fall times. In sizing the pull-up resistor, the designer faces a trade-off between the speed of the low-to-high transition and power dissipation when the output is in the low state. A low value pull-up resistor will yield a shorter RC time constant and therefore faster rise time, but this comes at the expense of a larger current that will flow through the pull-up resistor when the output is in the low state.

Another configuration that is made possible by an open-collector/open-drain output is to connect two or more outputs together to drive a single input. With totem-pole style outputs, connecting two outputs together is strictly forbidden. If one of the outputs tried to go high while the other output was in the low state, there would be a very low impedance path from the power supply, through the sourcing transistor of the chip in the high state, then through the sinking transistor in the low state to ground. This is illustrated in Figure 17.6. This would result in excessive current flow that would likely damage both outputs.

When two open-collector/open-drain outputs are tied together, then the external pull-up resistor will limit the current flow whenever *either* of the outputs goes into the low state (Figure 17.7). With this approach, either of the two outputs (or both) can force the connected input into the low state. This leads to a name given to this configuration: the **wired OR** (or, more correctly, **wired NOR**).

17.2.1 Open-Collector/Open-Drain Output Specifications

The specifications that you will find for devices with open-collector/open-drain outputs will differ from those with totem-pole style outputs, but sometimes the differences are subtle. Figure 17.8 shows an excerpt from the data sheet for the LM339, a device with an open-collector output.

Close inspection of Figure 17.8 reveals that the specifications are missing a value for V_{OH}. This is because the output high voltage is determined by the voltage source that is

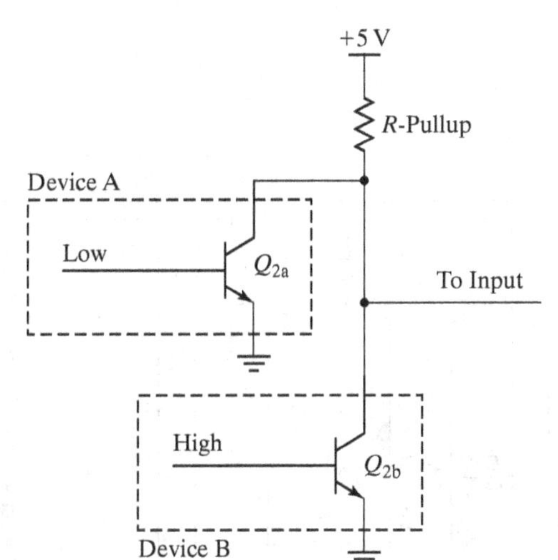

FIGURE 17.7

Two open-collector outputs in a wired OR configuration.

connected to an external pull-up resistor, not by the device itself. A specification is given for V_{OL} with the corresponding I_{OL} listed as a test condition. There is an explicit value listed for I_{OH}, but here again, more subtlety is involved. Notice that the sign of the value listed for I_{OH} is positive. From our discussions in Chapter 16, recall that a positive value for a current is current that flows *into* the pin. For a totem-pole style output, I_{OH} would have a negative sign, indicating an ability to source current out of the pin.

Not all data sheets for open-collector/open-drain output devices follow the conventions of V_{OL} and I_{OH} for describing the output low voltage and output high current. For some devices, the output low voltage will be called the **output saturation voltage** rather than V_{OL}. You may also see I_{OH} described as the **output leakage current**.

17.3 THREE-STATE OUTPUTS

To address the loss of speed associated with the open-collector style output while still allowing multiple outputs to be connected to a single input, manufacturers developed devices with three output states: low, high, and high-impedance. The first of these devices was created by National Semiconductor who trademarked the term "Tri-State" to describe these devices. In common usage today, **tri-state** has come to be generally applied to any device whose outputs can be placed into one of the three states.

An output that can take on one of the three states requires at least two input control lines. Most devices with three-state outputs use a single common control line to determine whether to place all of the device outputs into the third, high-impedance state. This is done with a circuit like that shown in Figure 17.9. Here, we see a circuit with many similarities to the totem-pole output of Figure 17.1a (three-state outputs can also be made using CMOS transistors with a similar layout).

What is different about Figure 17.9 is that a new transistor, Q_4, has been added that is controlled by the new control line, $\overline{\text{Output Enable}}$ (the over-bar indicates that the signal is active when the input is in the low state). When the $\overline{\text{Output Enable}}$ line is in the low state, the new transistor (Q_4) is off and the output behaves just as the totem-pole output did with either transistor Q_1 or Q_2 driving the output based on the state of the line labeled **Control Input**. When the $\overline{\text{Output Enable}}$ line is raised to a high level, Q_4 turns on and clamps the signals at the bases of Q_1 and Q_2, holding both of these devices in the off state independent of

Electrical Characteristics

at specified free-air temperature, V_{CC} = 5 V (unless otherwise noted)

	PARAMETER	TEST CONDITIONS(1)		T_A(2)	LM239 LM339 MIN	LM239 LM339 TYP	LM239 LM339 MAX	LM239A LM339A MIN	LM239A LM339A TYP	LM239A LM339A MAX	UNIT
V_{IO}	Input offset voltage	V_{CC} = 5 V to 30 V, V_{IC} = V_{ICR} min, V_O = 1.4 V		25°C		2	5		1	3	mV
				Full range			9			4	
I_{IO}	Input offset current	V_O = 1.4 V		25°C		5	50		5	50	nA
				Full range			150			150	
I_{IB}	Input bias current	V_O = 1.4 V		25°C		–25	–250		–25	–250	nA
				Full range			–400			–400	
V_{ICR}	Common-mode input-voltage range			25°C	0 to V_{CC} – 1.5			0 to V_{CC} – 1.5			V
				Full range	0 to V_{CC} – 2			0 to V_{CC} – 2			
A_{VD}	Large-signal differential-voltage amplification	V_{CC} = 15 V, V_O = 1.4 V to 11.4 V, $R_L \geq$ 15 kΩ to V_{CC}		25°C	50	200		50	200		V/mV
I_{OH}	High-level output current	V_{ID} = 1 V	V_{OH} = 5 V	25°C		0.1	50		0.1	50	nA
			V_{OH} = 30 V	Full range			1			1	μA
V_{OL}	Low-level output voltage	V_{ID} = –1 V,	I_{OL} = 4 mA	25°C		150	400		150	400	mV
				Full range			700			700	
I_{OL}	Low-level output current	V_{ID} = –1 V,	V_{OL} = 1.5 V	25°C	6	16		6	16		mA
I_{CC}	Supply current (four comparators)	V_O = 2.5 V,	No load	25°C		0.8	2		0.8	2	mA

(1) All characteristics are measured with zero common-mode input voltage, unless otherwise specified.
(2) Full range (MIN to MAX) for LM239/LM239A is –25°C to 85°C, and for LM339/LM339A is 0°C to 70°C. All characteristics are measured with zero common-mode input voltage, unless otherwise specified.

FIGURE 17.8

Output specifications for the LM339, a device with an open-collector output. (Courtesy of Texas Instruments Incorporated.)

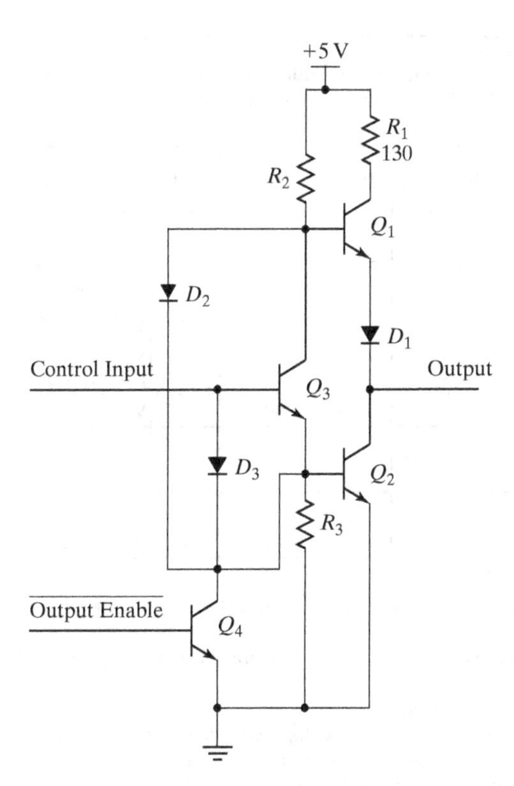

FIGURE 17.9
An example of a bipolar three-state output circuit.

the state of the control input. In this state, neither output transistor is actively driving the output line and it is said to be in a "high impedance" state.

The ability to turn off both output transistors allows for multiple outputs to be safely connected together, as long as you insure that only one of the output enable lines is active at any instant in time. When that one device's output is active, it has both sourcing and sinking transistors available to cause rapid high-low and low-high transitions, and all other devices must have their outputs in a high-impedance state.

The output pins on most microcontrollers have three-state drivers. This feature is what allows these pins to be used as either outputs or inputs under software control.

17.3.1 Three-State Output Specifications

In the specifications for a three-state output pin, you will find not only the output specifications typical of a totem-pole output (V_{OL}, V_{OH}, I_{OL}, I_{OH}), but also an added specification for the amount of leakage current that will flow into or out of a pin when it is in the high-impedance state. Figure 17.10 is an extract from the data sheet for a PIC16F690, a microcontroller whose outputs have three states.

Here, we see that parameter number D060 shows the leakage current that is to be expected when the I/O ports (the input/output pins) are in the high-impedance state. This specification is given in addition to parameters D080 (V_{OL}) and D090 (V_{OH}), with their respective test conditions specifying I_{OL} and I_{OH}.

17.4 LOW SIDE DRIVERS

The output stages that we have been discussing in Sections 17.1 through 17.3 are what are referred to as **logic-level** outputs. This term is used to describe outputs that swing between ground and something near the chip power supply voltage and are capable of sourcing or

DC CHARACTERISTICS			Standard Operating Conditions (unless otherwise stated) Operating temperature -40°C ≤ TA ≤ +85°C for industrial -40°C ≤ TA ≤ +125°C for extended				
Param No.	**Sym**	**Characteristic**	**Min**	**Typ†**	**Max**	**Units**	**Conditions**
	IIL	**Input Leakage Current(2)**					
D060		I/O ports	—	± 0.1	± 1	μA	VSS ≤ VPIN ≤ VDD, Pin at high-impedance
D061		MCLR(3)	—	± 0.1	± 5	μA	VSS ≤ VPIN ≤ VDD
D063		OSC1	—	± 0.1	± 5	μA	VSS ≤ VPIN ≤ VDD, XT, HS and LP oscillator configuration
D070*	IPUR	**PORTA Weak Pull-up Current**	50	250	400	μA	VDD = 5.0V, VPIN = VSS
	VOL	**Output Low Voltage(5)**					
D080		I/O ports	—	—	0.6	V	IOL = 8.5 mA, VDD = 4.5V (Ind.)
	VOH	**Output High Voltage(5)**					
D090		I/O ports	VDD – 0.7	—	—	V	IOH = -3.0 mA, VDD = 4.5V (Ind.)

* These parameters are characterized but not tested.

† Data in "Typ" column is at 5.0V, 25°C unless otherwise stated. These parameters are for design guidance only and are not tested.

Note 1: In RC oscillator configuration, the OSC1/CLKIN pin is a Schmitt Trigger input. It is not recommended to use an external clock in RC mode.

2: Negative current is defined as current sourced by the pin.

3: The leakage current on the MCLR pin is strongly dependent on the applied voltage level. The specified levels represent normal operating conditions. Higher leakage current may be measured at different input voltages.

4: See **Section 10.2.1 "Using the Data EEPROM"** for additional information.

5: Including OSC2 in CLKOUT mode.

FIGURE 17.10

Specifications for a device with three-state outputs. (© 2008 Microchip Technology Incorporated.)

sinking up to a few milliamps of current to or from an external load. While these outputs can be used to drive something other than the inputs to other digital devices (e.g., a low current LED), many loads require current levels beyond what logic-level outputs are capable of providing. For controlling these higher current loads, we use the logic-level outputs to drive transistors, or as the inputs to a class of integrated circuits (ICs) known as **peripheral drivers** or **power drivers**.

The most common circuit topology for driving higher current loads is shown in Figure 17.11. It takes its name, **low-side drive**, from the placement of the switching element in the current flow path between the load and ground.

In Chapter 10, we encountered both the term and the configuration when we talked about simple switching configurations using NPN transistors and N-channel MOSFETs. Indeed, the simplest low side drivers are simply transistors. When driven from a 5 V logic-level

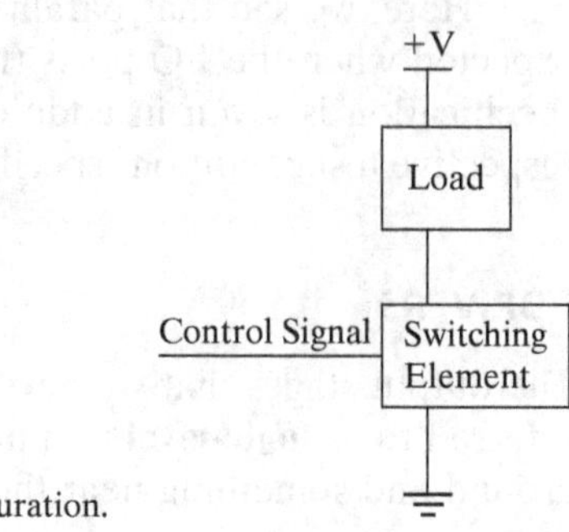

FIGURE 17.11

The generic low-side drive configuration.

output, N-channel MOSFETs (such as the 2N7000) are capable of sinking up to 200 mA while exhibiting $R_{DS(on)}$ levels on the order of 5 Ω. For larger current loads, a range of logic-level MOSFETs (e.g., IRLZ34N or MTP30N06) are available for switching voltages up to 60–80 V at currents well into the tens of amps with $R_{DS(on)}$ levels on the order of a few tens of milliohms.

While simple transistors can be used as low side drivers, IC manufacturers have produced a wide range of devices to simplify the implementation of low-side drive circuits. One of the simplest of these is a device known as a ULN2003. This device integrates an array of seven independent NPN Darlington transistors capable of switching up to 500 mA, resistors to limit base current, resistors to speed up switching, and a diode suitable for use in snubbing inductive loads into a single 16-pin DIP package. Near the high end of the integration spectrum are devices like the Fairchild Semiconductor FDMS2380. This device integrates a pair of low side switches capable of controlling 5 A and also incorporates active snubbing, over- and under-voltage lockout (which prevents the device from functioning if a specification—voltage in this case—is outside specified limits), over-current lockout, over-temperature lockout, and a diagnostic output to report these abnormal situations back to the controller.

17.4.1 Low Side Driver Specifications

Since the low side driver can only sink current, the most relevant output specifications refer to the output voltages and currents that the device can switch. While V_{OL} and I_{OL} are the common terms in use for logic-level devices, power/peripheral drivers often use slightly different terms that sometimes reflect the technology used to manufacture the device. For example, the ULN2003 specifications (Figure 17.12) use $V_{CE(sat)}$ to describe the output low voltage, which in this case is the same as the collector-emitter saturation voltage. This also serves as a clue to the fact that this device uses bipolar transistors.

The output current for the ULN2003 is referred to as I_C, a reference to this being the collector current in the Darlington transistor. In a manner similar to logic-level output specifications, it is common to find this parameter listed as a test condition under the specifications for the output voltage. For a low side driver, the high state is the off state and for bipolar devices, this is often referred to as the collector leakage current or the collector cut-off current (I_{CEX} in Figure 17.12). The output voltage in the high state will be determined by the external circuit with the low-side drive device simply specifying a maximum allowable switching voltage, typically in the Absolute Maximum Ratings section of the data sheet.

17.5 HIGH SIDE DRIVERS

The complement to the low side driver is the high side driver, shown in Figure 17.13. Here, we see that the switching element has been placed between the power supply and the load and that the other side of load is connected directly to ground. This configuration is especially common in automotive applications where loads (lamps, motors, and so on) are typically grounded to the chassis at a mounting point and a single wire from a switch or controller is used to source current into the grounded load to activate it.

Just as the simplest low side drivers are transistors, so are high side drivers. This time, however, the transistors are PNP bipolar transistors or P-channel MOSFETs. While PNP transistors achieve $V_{CE(sat)}$ levels on par with their NPN counterparts, P-channel MOSFETs typically do not achieve $R_{DS(on)}$ values as low as similar N-channel parts. Because of differences in the designs necessary to make P-channel MOSFETs, they typically require larger amounts of silicon (making them more expensive) and exhibit $R_{DS(on)}$ values that are two to five times larger than otherwise similarly specified N-channel MOSFETs.

electrical characteristics, T_A = 25°C (unless otherwise noted)

PARAMETER		TEST FIGURE	TEST CONDITIONS		ULN2003A			ULN2004A			UNIT
					MIN	TYP	MAX	MIN	TYP	MAX	
$V_{I(on)}$	On-state input voltage	6	V_{CE} = 2 V	I_C = 125 mA						5	V
				I_C = 200 mA			2.4			6	
				I_C = 250 mA			2.7				
				I_C = 275 mA						7	
				I_C = 300 mA			3				
				I_C = 350 mA						8	
$V_{CE(sat)}$	Collector-emitter saturation voltage	5	I_I = 250 μA,	I_C = 100 mA		0.9	1.1		0.9	1.1	V
			I_I = 350 μA,	I_C = 200 mA		1	1.3		1	1.3	
			I_I = 500 μA,	I_C = 350 mA		1.2	1.6		1.2	1.6	
I_{CEX}	Collector cutoff current	1	V_{CE} = 50 V,	I_I = 0			50			50	μA
		1	V_{CE} = 50 V, T_A = 70°C	I_I = 0			100			100	
		2		V_I = 1 V						500	
V_F	Clamp forward voltage	8	I_F = 350 mA			1.7	2		1.7	2	V
$I_{I(off)}$	Off-state input current	3	V_{CE} = 50 V, T_A = 70°C	I_C = 500 μA,	50	65		50	65		μA
I_I	Input current	4	V_I = 3.85 V			0.93	1.35				mA
			V_I = 5 V						0.35	0.5	
			V_I = 12 V						1	1.45	
I_R	Clamp reverse current	7	V_R = 50 V				50			50	μA
			V_R = 50 V,	T_A = 70°C			100			100	
C_i	Input capacitance		V_I = 0,	f = 1 MHz		15	25		15	25	pF

FIGURE 17.12

Specifications for the ULN2003 low side driver. (Courtesy of Texas Instruments Incorporated.)

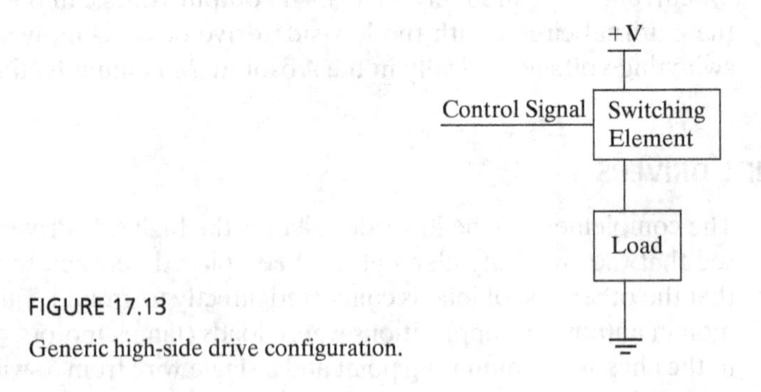

FIGURE 17.13

Generic high-side drive configuration.

17.5.1 High Side Driver Specifications

Just as IC manufacturers did with low side drivers, they supply a range of high side switching ICs that simplify implementation in a design. For example, the STMicroelectronics VN808CM integrates eight high side switches capable of sourcing 700 mA with an $R_{DS(on)}$ of 160 mΩ with circuitry to provide output current limiting, protection against short circuits to ground and over-temperature conditions, under-voltage lockout, and a status output, all in a

Table 3. Power section

Symbol	Parameter	Test conditions	Min	Typ	Max	Unit
V_{CC}	Operating supply voltage		10.5		45	V
V_{USD}	Undervoltage shutdown		7		10.5	V
R_{ON}	On state resistance	$I_{OUT} = 0.5$ A; $T_J = 25$ °C $I_{OUT} = 0.5$ A;			160 280	mΩ mΩ
I_S	Supply current	OFF state; $V_{CC} = 24$ V; $T_{CASE} = 25$ °C ON state (all channels ON); $V_{CC} = 24$ V, $T_{CASE} = 100$ °C			150 12	µA mA
I_{LGND}	Output current at turn-off	$V_{CC} = V_{STAT} = V_{IN} = V_{GND} = 24$ V $V_{OUT} = 0$ V			1	mA
$I_{L(off)}$	OFF state output current	$V_{IN} = V_{OUT} = 0$ V;	0		5	µA
$V_{OUT(off)}$	OFF state output voltage	$V_{IN} = 0$ V, $I_{OUT} = 0$ A			3	V

FIGURE 17.14

Specifications for the VN808CM high side driver IC. (Copyright STMicroelectronics. Used with permission.)

36-pin small outline package (SOP). The VN808CM, like many other high side driver ICs, actually uses N-channel MOSFETs as the switching elements and includes an on-chip voltage boosting circuit to allow them to be used in a sourcing configuration. These types of ICs require a minimum input voltage in order to create the high gate voltage necessary to insure that the N-channel MOSFETs turn on fully. Many devices that employ this technique include an under-voltage lockout that prevents the device from functioning if the minimum voltage requirement isn't met.

Figure 17.14 shows an excerpt from the data sheet for the VN808CM, which includes the kinds of specifications typical of this class of device. The presence of the under-voltage lockout is reflected in the minimum required value for V_{CC}. Note that this minimum exactly matches the maximum specification for the under-voltage shutdown feature. The on-state resistance, R_{ON}, along with the amount of current being sourced, allows us to calculate the voltage drop seen between V_{CC} and the load that is being powered by the output. The supply current specification (I_S) listed is the amount of current consumed inside the driver circuit, as opposed to the current delivered to the load. If a short circuit is detected between the output and ground, the amount of current that will flow to the output is given as I_{LGND}. When the device is in the off state, a small amount of leakage current, $I_{L(off)}$ will still flow into the load. Similarly, when the device is off and not delivering current to the load, there will still be a voltage present at the output terminal, shown as $V_{OUT(off)}$.

17.6 HALF-BRIDGES AND FULL-BRIDGES

When you need an output that is capable of both sourcing and sinking current at levels that exceed those possible with a logic-level output, it is common to use a device with an output structure like that of Figure 17.15. This example is shown with bipolar transistors but similar structures are also made using MOSFETs. This structure is referred to as a **push-pull** output or a **half-bridge**.

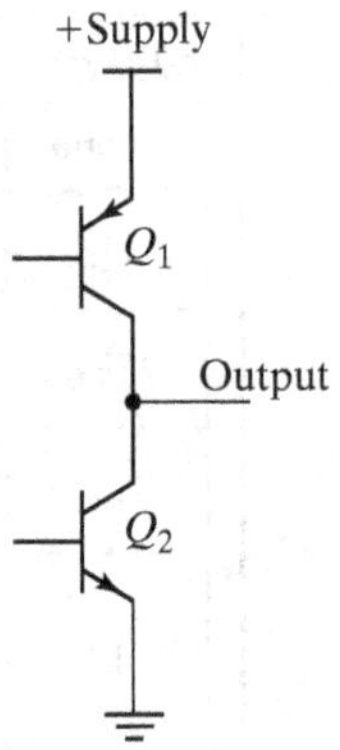

FIGURE 17.15

The basic configuration for a half-bridge or push-pull output.

Note that there is a distinct similarity to the totem-pole output, with a pair of transistors "stacked" between power and ground. Unlike the totem-pole, however, the half-bridge eliminates the resistor and diode in the sourcing path and substitutes a PNP transistor for the transistor in the sourcing position (the high side, that is). These changes are targeted toward lowering the power dissipation within the device. This is more important in a power driver which will be used to source and sink significantly higher currents than a logic-level output. In this circuit, Q_1 will be on when its base is pulled low, and Q_2 will be on when its base is pulled high. From this description, it might appear that you could control these two transistors simply by tying the base leads together. However, this is never done for the reasons that we discuss in Section 17.6.1 on shoot-through current.

Commercial implementations of half-bridge outputs are commonly done in pairs on a single IC. This makes it convenient to combine the two half-bridges to create a **full-bridge**, sometimes called an **H-bridge** (Figure 17.16).

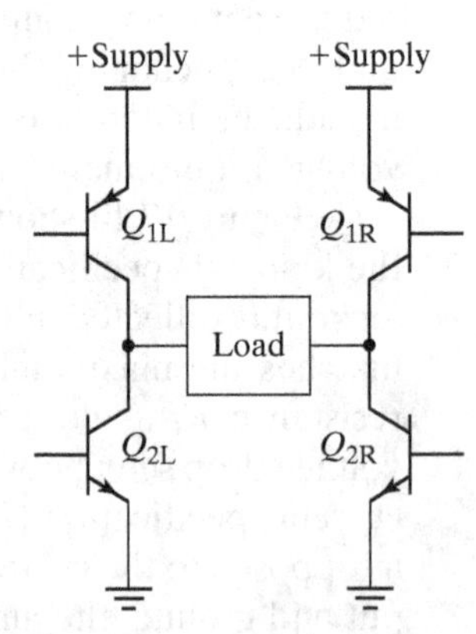

FIGURE 17.16

A full-bridge configured from two half-bridges.

By controlling the transistors in a way that turns on diagonally opposite pairs (Q_{1L} and Q_{2R} or Q_{1R} and Q_{2L}), we can control the direction of current flow through the load. This is a particularly important capability when it comes to driving motors, as we show in Chapters 23 and 26.

17.6.1 Shoot-Through Currents and Dead Time

One of the realities of transistor behavior that circuit designers are forced to deal with is the issue of transistor switching speeds. In general, transistors turn on more quickly than they turn off. When transistors are stacked and controlled from a single input, as they are in the totem-pole output, this means that there is a short period of time during the transitions in the output state when the upper and lower transistors are both somewhat on. This produces a low impedance path between power and ground and leads to what is known as **shoot-through current**. Recall that in bipolar implementations of a totem-pole output stage, a resistor was placed between the power supply and the sourcing (upper) transistor to limit the shoot-through current. This approach is not realistic for a power driver because of the high currents involved. The presence of a resistor in a power driver would introduce unacceptably large drops in the voltage supplied to the load at the output and dissipate power inside the driver whenever it was sourcing current.

The designers of bridge drivers handle the problem of shoot-through currents by independently controlling the upper and lower transistors and intentionally staggering the time when one is switched off and the other is switched on—that is, a **dead time** is introduced during each transition. This dead time, when neither the upper nor lower transistor is being activated, allows time for the transistor that is turning off to fully reach the off state before the other transistor is commanded to turn on. The introduction of a dead time at transitions allows for the elimination of shoot-through current at the expense of switching time.

17.6.2 H-Bridge Specifications

For bridge drivers, the most relevant output specifications are for the amount of voltage and current that can be controlled, the voltage losses to be expected across the switching elements, and the timing of the switching operations. The input specifications are important as well, of course, but this discussion centers on the performance of the outputs. The electrical specifications (both output and input) for the L293B, a device with four 1 A half-bridges, are shown in Figure 17.17.

This device uses separate power supply pins for the logic supply (V_{SS}) and the motor supply (V_S). Notice that the motor supply voltage can be as high as 36 V, however, it can be no lower than the logic supply voltage. This sets a lower limit on the motor voltage of 4.5 V.

The output voltage to the load is specified in terms of the voltage drop that will occur across each switching element (V_{CEsatH} for the upper, or sourcing, transistor and V_{CEsatL} for the lower, or sinking, transistor). It is important to recognize that when the L293B is operated as an H-bridge, there will be one sourcing transistor **and** one sinking transistor in the switched current path.

Near the bottom of the data sheet excerpt, we find the timing specifications for the device. As is common with power drivers, the timing specifications are given in two parts. The **delay** time is the time between a change in the input and the corresponding change in the output. In addition, there is a specification for the 10–90% **rise time** and **fall time**. Notice that the turn-on delay time is longer than the turn-off delay time plus the fall time, revealing the use of dead time to control the shoot-through current.

17.7 THERMAL DESIGN ISSUES

When switching substantial currents, the voltage losses that occur across the switching elements represent not only a loss in voltage to the load but also power ($P = V \times I$) that must be dissipated within the power driver. This power is dissipated as heat in the chip, and it must be removed to keep the temperature from rising to a level that would cause damage to the driver. To facilitate the thermal analysis, manufacturers provide data to describe the thermal performance of the chips. The data provided are generally based on a simple heat flow model that has a direct analog in circuits (Figure 17.18).

In this model, the temperatures correspond to voltages and heat flows correspond to current flow. The resistors represent the resistance to heat flow (°C/W). The capacitors represent the thermal capacity of the case (and possibly a heat sink) and, when included, allow for the analysis of thermal transients. Basic thermal analysis is generally done at steady state, equivalent to DC, in which case the thermal capacitance terms are ignored.

ELECTRICAL CHARACTERISTCS

Symbol	Parameter	Test Condition	Min.	Typ.	Max.	Unit
V_s	Supply Voltage		V_{SS}		36	V
V_{SS}	Logic Supply Voltage		4.5		36	V
I_s	Total Quiescent Supply Current	V_i = L; I_o = 0; V_{inh} = H		2	6	mA
		V_i = h; I_o = 0; V_{inh} = H		16	24	mA
		V_{inh} = L			4	mA
I_{ss}	Total Quiescent Logic Supply Current	V_i = L; I_o = 0; V_{inh} = H		44	60	mA
		V_i = h; I_o = 0; V_{inh} = H		16	22	mA
		V_{inh} = L		16	24	mA
V_{iL}	Input Low Voltage		-0.3		1.5	V
V_{iH}	Input High Voltage	$V_{SS} \leq 7V$	2.3		V_{ss}	V
		$V_{SS} > 7V$	2.3		7	V
I_{iL}	Low Voltage Input Current	V_{il} = 1.5V			-10	μA
I_{iH}	High Voltage Input Current	$2.3V \leq V_{IH} \leq V_{SS} - 0.6V$		30	100	μA
V_{inhL}	Inhibit Low Voltage		-0.3		1.5	V
V_{inhH}	Inhibit High Voltage	$V_{SS} \leq 7V$	2.3		V_{SS}	V
		$V_{SS} > 7V$	2.3		7	V
I_{inhL}	Low Voltage Inhibit Current	V_{inhL} = 1.5V		-30	-100	μA
I_{inhH}	High Voltage Inhibit Current	$2.3V \leq V_{inhH} \leq V_{ss} - 0.6V$			±10	μA
V_{CEsatH}	Source Output Saturation Voltage	I_o = -1A		1.4	1.8	V
V_{CEsatL}	Sink Output Saturation Voltage	I_o = 1A		1.2	1.8	V
V_{SENS}	Sensing Voltage (pins 4, 7, 14, 17) (**)				2	V
t_r	Rise Time	0.1 to 0.9 V_o (*)		250		ns
t_f	Fall Time	0.9 to 0.1 V_o (*)		250		ns
t_{on}	Turn-on Delay	0.5 V_i to 0.5 V_o (*)		750		ns
t_{off}	Turn-off Delay	0.5 V_i to 0.5 V_o (*)		200		ns

* See figure 1
** Referred to L293E

FIGURE 17.17

The Electrical Specifications section for the L293B, a chip with four half-bridge drivers that can be combined to create two full H-bridges. (Copyright STMicroelectronics. Used with permission.)

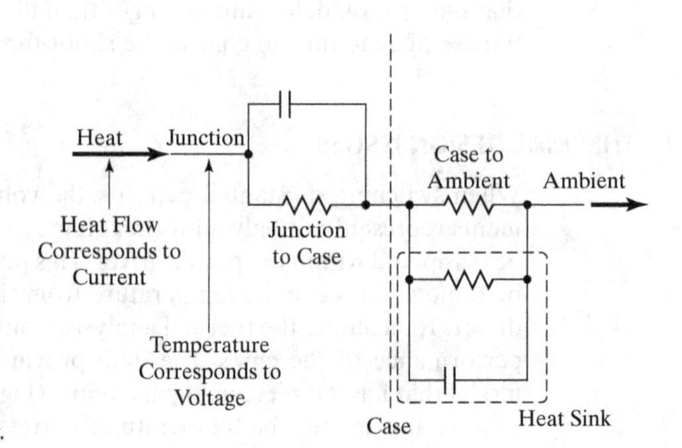

FIGURE 17.18

Electrical analogy to heat flow.

THERMAL DATA

Symbol	Parameter		Value	Unit
$R_{th\ j\text{-}case}$	Thermal Resistance Junction-case	Max.	14	°C/W
$R_{th\ j\text{-}amb}$	Thermal Resistance Junction-ambient	Max.	80	°C/W

FIGURE 17.19

Thermal specifications for an L293B. (Copyright STMicroelectronics. Used with permission.)

To make this model more concrete, let's look at an example using the L293B. The thermal specifications for the L293B are shown in Figure 17.19.

We can use the junction-ambient specification to evaluate the thermal performance without an external heat sink or combine it with the specification for the junction-case to determine the case-ambient specifications for use in evaluating the system performance with a heat sink. Let's start our example with an application that does not use a heat sink.

The Absolute Maximum section of the data sheet states that the maximum allowable junction temperature is 150°C. If we want to analyze the worst-case power dissipation in steady state, we would use the maximum output current (1 A) and the maximum voltage drops across the switching elements (1.8 V for sourcing and 1.8 V for sinking, for a total of 3.6 V). This gives us a total power dissipation of approximately 3.6 W (we have ignored the power dissipation due to the internal operation of the chip since it is much smaller than the power dissipated in the switching elements). Using the junction-ambient thermal resistance, we calculate the temperature rise across the case to be 3.6 W × 80°C/W = 288°C. If we limit the junction temperature to 150°C, as the data sheet tells us we must, we could not operate at an ambient temperature above −138°C. Clearly, this won't work! We have two choices as to how to proceed: we could ask "What current could we use and still keep the junction temperature low enough?" or we could ask "What kind of a heat sink do we need in order to keep the temperature low enough?"

If we wanted to operate at room temperature (25°C), we would be limited to a 125°C rise over ambient (ROA) temperature to meet the junction temperature limit. With a junction-ambient thermal resistance of 80°C/W, we would need to limit the power dissipation to 125°C/(80°C/W) = 1.56 W. Assuming worst-case voltage drops across the transistors that would limit the current to 1.56 W/3.6 V = 434 mA.

If 434 mA isn't enough for our application, we can consider adding a heat sink as an alternative. Given the specifications for the L293B, we can calculate the minimum thermal resistance for a heat sink that would allow us to dissipate the required power and meet the junction temperature limit. To determine this, we calculate the total thermal resistance needed to meet our maximum ROA: 125°C/3.6 W = 34.7°C/W. This total thermal resistance will be made up of the combination of the junction-case thermal resistance in series with the parallel combination of the case-ambient and sink-ambient thermal resistances (Figure 17.20). When using a heat sink, it is not uncommon (and produces a more conservative result) to ignore the case to ambient path.

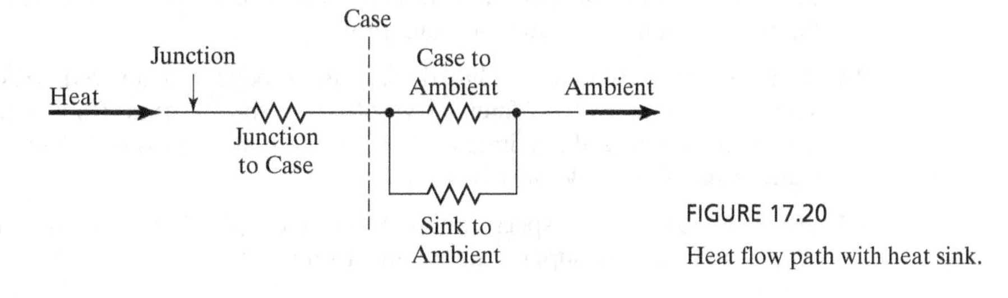

FIGURE 17.20

Heat flow path with heat sink.

FIGURE 17.21
DIP heatsink.

To find the maximum thermal resistance of the heat sink in parallel with the case-ambient resistance for the IC package, we can simply subtract the junction-case resistance (14°C/W) from the total allowable resistance (34.7°C/W) to get the result, 34.7°C/W − 14°C/W = 20.7°C/W. The case-ambient thermal resistance for the IC package is the difference between the supplied values for the junction-case and junction-ambient (80°C/W − 14°C/W = 66°C/W). Finally, by using the formulas for parallel resistors (a big benefit of using the electrical analogy for thermal systems), we can develop an expression for the required thermal resistance of the heat sink:

$$R_{\text{total}} = \frac{R_{\text{case}} R_{\text{sink}}}{R_{\text{case}} + R_{\text{sink}}} \tag{17.1}$$

$$R_{\text{sink}} = \frac{R_{\text{case}} R_{\text{total}}}{R_{\text{case}} - R_{\text{total}}} \tag{17.2}$$

Inserting the numbers for this example, we calculate a required heat sink resistance of at most 30.2°C/W. With this value in hand, we can search for heat sinks that mount to a 16-pin DIP package and provide a thermal resistance of less than 30.2°C/W. Figure 17.21 shows an example of such a heat sink which is specified for 20°C/W.

17.8 HOMEWORK PROBLEMS

17.1 If you wanted to drive an LED to the maximum possible brightness with the output of a 7400, would you arrange it so that the LED was on when the 7400's output was in the low state or the high state? Use worst-case design principles and quote specifications from Figure 17.2 to support your answer.

17.2 If you wanted to drive an LED to the maximum possible brightness with the output of a 74HC04, would you arrange it so that the LED was on when the 74HC04's output was in the low state or the high state? Use worst-case design principles and quote specifications from Figure 17.3 to support your answer.

17.3 If an LED is to be driven from either a 7400 or a 74HC04, which would you expect to produce the greater LED light output? Use worst-case design principles and quote specifications from Figures 17.2 and 17.3 to support your answer.

17.4 If the outputs from two LM339A devices are to be used in a wired OR configuration to drive a single input of an MC9S12 family device (use the specifications from Figures 17.8 and 16.1), what are the minimum and maximum values for the pull-up resistor? Choose from among commercially available 5% tolerance resistors.

17.5 Using the ULN2003A specifications from Figure 17.12, what is the minimum required input voltage to be able to support an output current of 200 mA?

17.6 For the conditions specified in Problem 17.5, what is the expected output voltage at the collector of the ULN2003A?

17.7 If an L293B powered by 5 V is used to drive a motor whose stall current is 1 A, what is the minimum voltage that you would expect to see across the terminals of the motor under stall? Use the specifications from Figure 17.17, and include a schematic of the circuit with your answer.

17.8 If an L293B is used without a heat sink to drive a motor whose stall current is 400 mA, what is the expected maximum junction temperature if the motor is stalled indefinitely? Assume the maximum specified voltage drop across the L293B and an ambient temperature of 25°C.

17.9 If, under the same conditions as Problem 17.8, the heat sink shown in Figure 17.21 were added to the L293B, what will the expected maximum junction temperature be?

17.10 In a particularly bright smart product, the designer needs to turn on a set of 40 IR LEDs (V_f = 1.4 V), each carrying 50 mA. Design a circuit that performs this task using the output from a single MC9S12 family device (use the specifications in Figure 16.1) as the drive signal and a minimal number of other components (the number of active elements should not exceed 3). You have +5, +12, and +15 V power supplies available. Show, by calculations and reference to the relevant specifications, that your design does not exceed the capabilities of any of the devices involved.

FURTHER READING

Electronic Circuits, Schilling, D.L., and Belove, C., McGraw-Hill, 1989.

Practical Electronics for Inventors, Scherz, P., 2nd ed., McGraw-Hill, 2006.

CHAPTER 18

Digital Logic and Integrated Circuits

The increases in the execution speed of microcontrollers have largely eliminated the need for the mechatronic engineer to engage in logic design at the gate level. There are only a few domains, such as interfacing to high resolution encoders on high speed motors, where software is simply not fast enough to do the job and hardware—in the form of logic gates—is required. However, the ability for software to perform so many of the required tasks has not eliminated the need for the mechatronic engineer to understand the fundamentals of digital logic. Logic concepts and symbols are used to communicate the function of many types of devices. Designers need to be familiar with this material so that they can understand the documentation for devices they are likely to use and to create documentation that communicates how their designs work. It is often the case that the documentation for microcontroller peripherals is provided in the form of a functional schematic that is used to document the behavior of a peripheral subsystem. In order to be able to understand the operation of the subsystem, you will need to be able to read and decipher these schematic diagrams. In addition, there are a number of digital logic chips that were originally designed to implement discrete logic designs, and turn out to be very useful as peripheral chips for microcontrollers. For the most part, these external functions allow the designer to make better use of the limited number of pins available on small microcontrollers.

To acquaint you with digital logic at a level that will enable you to understand functional schematics and incorporate some of the most useful digital logic chips into your designs with microcontrollers, we will introduce you to:

1. the basic building blocks of combinatorial logic,
2. how these combinatorial logic blocks are used to describe the function of microcontroller peripherals,
3. the difference between combinatorial and sequential logic,
4. the basic building blocks of sequential logic,
5. a review of some of the digital logic and other integrated circuits available that are of the most use in conjunction with a microcontroller.

18.1 BASIC COMBINATORIAL LOGIC

There are four combinatorial logic gates that implement the logical operations that are the basis of Boolean algebra (Chapter 3). **Combinatorial logic** is logic whose output is determined solely by the current states applied to the inputs. When used as schematic symbols, the four functions are drawn as shown in Figure 18.1.

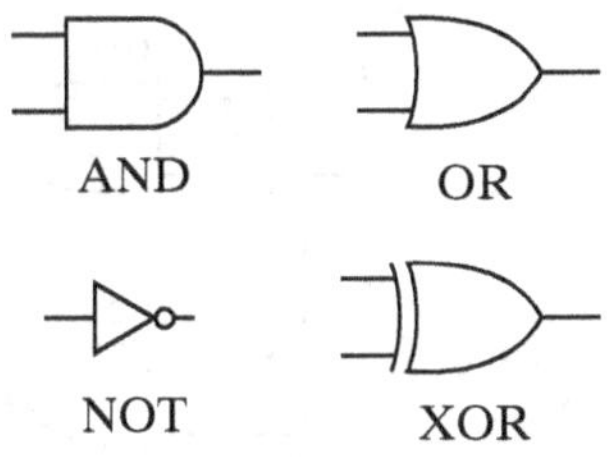

AND OR

NOT XOR

FIGURE 18.1

The schematic symbols for the four basic combinatorial gates.

The AND (expressed mathematically as •) and OR (mathematically +) gates are shown in Figure 18.1 with two inputs, although there may be more than two inputs when implementing these functions. The XOR gate always has two inputs and a single output, while the inverter that implements the NOT operation (indicated with a bar over the expression to be negated) is always a single-input single-output gate. Technically, the XOR operator is not a fundamental operation, since the function that it implements can be described using the other three operations: $[(A + B) \bullet \overline{(A \bullet B)}]$. However, the operation is so useful, and the expression is so cumbersome that it is often treated as a basic operation.

The schematic symbol for the NOT operation is referred to as an inverter and is actually made up of two elements, the triangle (indicating a buffer) and the bubble, which is actually the symbol for the NOT (or inversion) operation. In the symbol for the inverter, the bubble appears at the output, though in other instances it may also appear at the input of any of the logic symbols. When the bubble appears at the output, we often speak of that output as **active low**. This indicates that when the logical condition represented by the logic gate is true, the output line will be in the low state. In a similar way, when the bubble appears at the input, we speak of that input as active low. In this case, it indicates that when the input line is in the low state, the actual input to the logic gate will be true. It is also common to find a variation on the NOT gate (and also the buffer) in which a vertical line is added to either the top or the bottom of the gate (Figure 18.2).

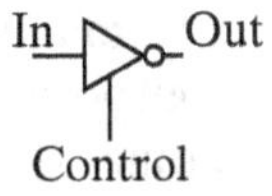

In Out

Control

FIGURE 18.2

Inverting buffer with output control.

This extra line is an output control line. When it is inactive, it disables the output of the gate, putting the output into a "tri-state" mode (see Section 17.3 in Chapter 17).

18.1.1 Truth Tables

The behavior of a gate (or collection of gates) is often described graphically in a format referred to as a **truth table**. A truth table is nothing more elaborate than an exhaustive listing of all of the possible input combinations with the corresponding output state. Figure 18.3 shows the truth tables for the basic logic operations.

I_1	I_2	O
0	0	0
0	1	0
1	0	0
1	1	1

AND

I_1	I_2	O
0	0	0
0	1	1
1	0	1
1	1	1

OR

I	O
0	1
1	0

NOT

I_1	I_2	O
0	0	0
0	1	1
1	0	1
1	1	0

XOR

FIGURE 18.3

Truth tables for the basic logic functions.

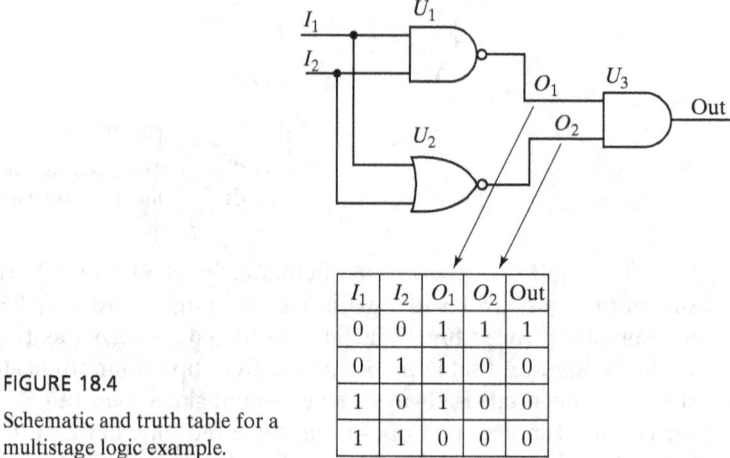

I_1	I_2	O_1	O_2	Out
0	0	1	1	1
0	1	1	0	0
1	0	1	0	0
1	1	0	0	0

FIGURE 18.4
Schematic and truth table for a multistage logic example.

When developing the truth table for a more complex set of gates, it is often useful to prepare a multistage truth table that captures the behavior of subparts of the circuit as well as the overall behavior. An example of this is shown in Figure 18.4.

18.1.2 Describing Microcontroller Subsystems Using Combinatorial Logic

A very common place you are likely to find digital logic symbols being used to describe the behavior of a microcontroller subsystem is in the timer subsystem. The logic shown in Figure 18.5 is used to describe when the Gated Clock signal will be applied to a timer/counter register.

The signals labeled TMR1ON, TMR1GE, and T1GINV are the states of bits in the microcontroller's control registers that can be manipulated under program control, and $\overline{\text{T1G}}$ is the state of an external input pin. The purpose of the logic is to decide when the clock (Clock Input signal) will be applied to the timer (Gated Clock output). The rightmost AND gate operates to either pass or block the Clock Input signal based on the result of the upper logic. Careful study of the logic reveals that the bit labeled TMR1ON must be a true in order for the Clock Input to be passed as the Gated Clock.

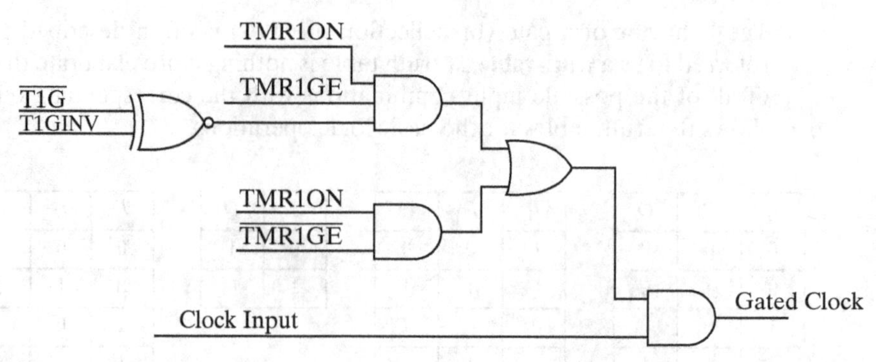

FIGURE 18.5
Gating logic from a microcontroller timer.

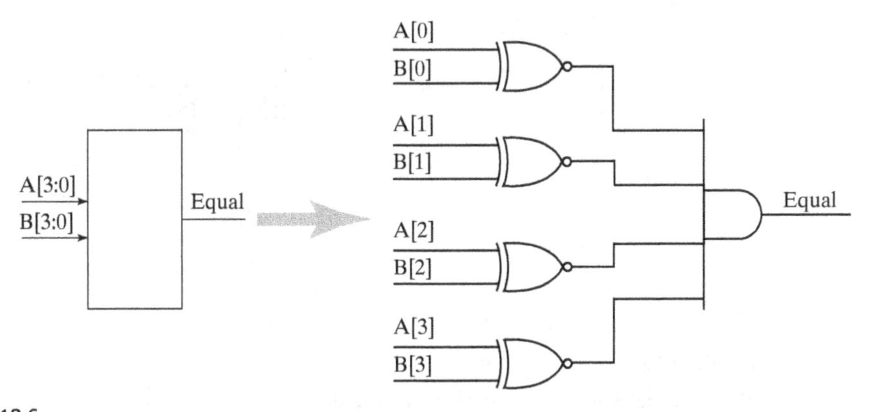

FIGURE 18.6

The digital comparator and its implementation.

18.2 A SURVEY OF USEFUL FUNCTIONS IMPLEMENTED WITH COMBINATORIAL LOGIC

In this section, we introduce several larger-scale logic functions that can be useful to the mechatronic system designer. In Section 18.5, we will present some integrated implementations of these functions that you can actually buy and use in your designs. When presenting these functions, we show the gate level implementation and truth table to demonstrate that these larger-scale functions can indeed be built from the simple set of gates that we have introduced. Our goal is to acquaint you with these larger-scale functions, rather than to try to teach you how to design them from gates.

18.2.1 The Digital Comparator

The **digital comparator** is a block of combinatorial logic that determines when a pair of bit patterns match exactly. It is a common building block in microcontroller timer systems, covered in Chapter 8. Specifically, the digital comparator is at the heart of the output compare system that is common among many microcontrollers.

In Figure 18.6, the comparator functions by comparing the 4-bit binary patterns, A[3:0] and B[3:0], and producing a true output when the patterns present on A and B match exactly. The notation A[3:0] is a shorthand way of denoting that A consists of four lines (A[3], A[2], A[1], A[0]). The XNOR gates are an example of a device with an active low output, and they are nothing more than an XOR followed by an inverter. When the 4-bit groups are thought of as representing binary numbers, then the output will be true when the two numbers are equal.

18.2.2 The Digital Multiplexer

A **multiplexer** is a circuit that allows one of many inputs to be connected to a single output under the control of a set of select lines. This allows for the time multiplexing of, for example, a single microcontroller input, to be used to monitor multiple physical inputs. The digital multiplexer differs from the analog multiplexer described in Chapter 8, Section 8.6, in that the digital multiplexer is implemented using logic gates rather than the analog switches used to make the analog multiplexer.

Figure 18.7 shows the implementation of a 2:1 digital multiplexer. In this example, a single select line is used to choose which of the inputs, D_0 or D_1, will be presented at the output. Each additional select line that is added allows for a doubling of the number of inputs from which the selection is being made. With two select lines, you can select from four possible inputs; with three select lines that goes up to eight possible inputs (Figure 18.8).

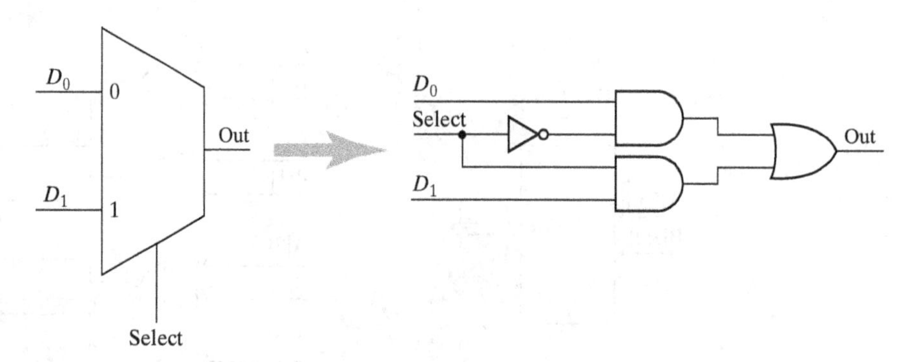

FIGURE 18.7

The 2:1 digital multiplexer and its implementation.

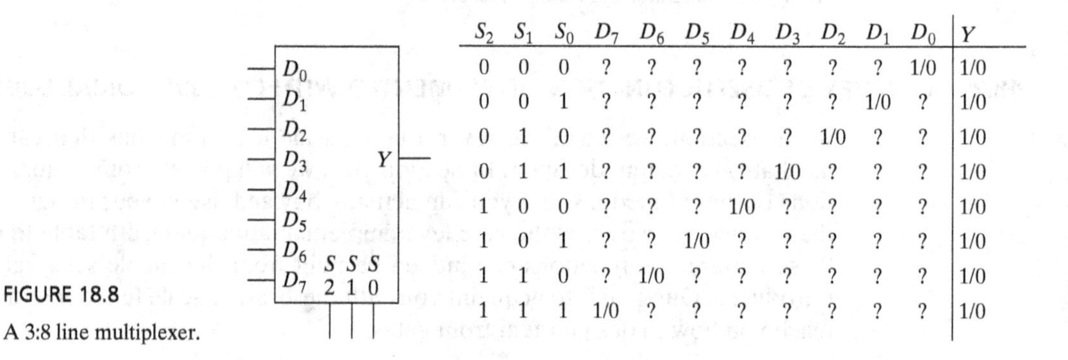

S_2	S_1	S_0	D_7	D_6	D_5	D_4	D_3	D_2	D_1	D_0	Y
0	0	0	?	?	?	?	?	?	?	1/0	1/0
0	0	1	?	?	?	?	?	?	1/0	?	1/0
0	1	0	?	?	?	?	?	1/0	?	?	1/0
0	1	1	?	?	?	?	1/0	?	?	?	1/0
1	0	0	?	?	?	1/0	?	?	?	?	1/0
1	0	1	?	?	1/0	?	?	?	?	?	1/0
1	1	0	?	1/0	?	?	?	?	?	?	1/0
1	1	1	1/0	?	?	?	?	?	?	?	1/0

FIGURE 18.8

A 3:8 line multiplexer.

18.2.3 The Decoder

A decoder is, in some sense, the inverse of a multiplexer. A **decoder** features multiple output lines, only one of which is active at any particular point in time. The purpose of the select lines on a decoder is to choose which of the outputs will be active at that particular instant in time.

Figure 18.9 shows the implementation of a 2:4 decoder. In this example, the outputs are active low (note the bubbles on the output lines in the schematic symbol). Based on the

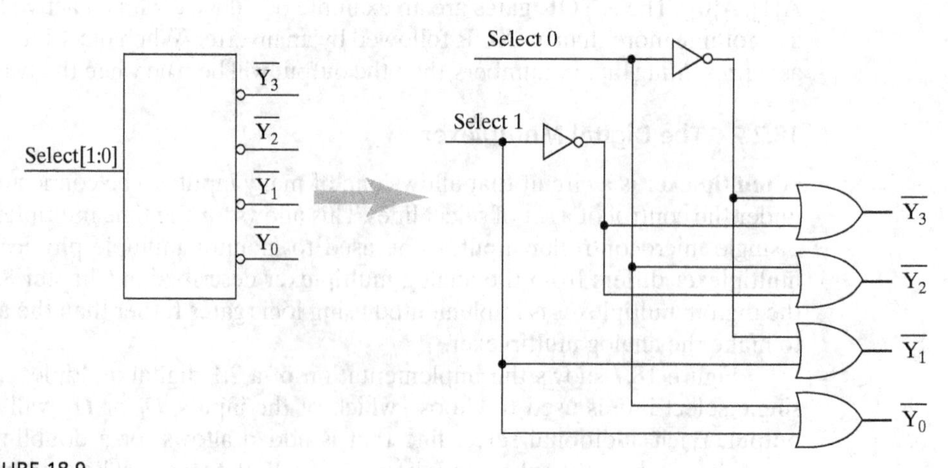

FIGURE 18.9

The 2:4 decoder and its implementation.

pattern presented at the select inputs (Select 0 and Select 1), one of the Y output lines will go into the active (low) state. The other three lines will remain in the inactive (high) state. Notice that in this implementation, one of the outputs is always active. Most commercial implementations of decoders provide an additional control input that allows for all of the outputs to be forced into the inactive state.

18.3 INTRODUCTION TO SEQUENTIAL LOGIC

All of the logic presented so far has been what is referred to as combinatorial logic. As a reminder, this means that at any given instant in time, the outputs are solely dependent on some combination of the input states at that instant. There is no notion of time or past history of the inputs reflected in the output state. **Sequential logic** introduces the notion of a logic device with a memory of past events. The simplest implementation of such a device is shown in Figure 18.10. This device goes by several names: **RS latch**, **SR latch**, **SR flip-flop**, **RS flip-flop**. In general, the terms **latch** and **flip-flop** are used interchangeably, indicating a device with two stable states. The R and the S in the names refer to the Reset and Set inputs of the device. As we shall see, naming the type of flip-flop after the labels of its inputs is very common.

To understand the operation of the RS latch, begin by assuming that the Set input is high (1) and the Reset input is low (0). The $\overline{Q}$ output of the NOR (Not OR) gate B is guaranteed to be low (see Boolean algebra identities in Chapter 3). Since the Reset input is also low, then both inputs to the NOR gate A are low, forcing its output (Q) high. Next, consider what happens if the Set input is subsequently brought low. Since the Q output is high, the output from the NOR gate B ($\overline{Q}$) will remain low. If $\overline{Q}$ remains low, then there has been no change at the inputs to the NOR gate A and its output will remain high. Notice that the inputs changed state, but the outputs did not change state. This is described in the first line of the truth table. The notation $Q_{(n)}$ indicates the previous value of the Q output. If we then raise the Reset input high, we will see that the outputs Q and $\overline{Q}$ both change state: Q goes low and $\overline{Q}$ high. If we subsequently lower the Reset input, returning to the same $R = 0$, $S = 0$ input that we had earlier, the outputs will remain with Q low and $\overline{Q}$ high. We have the same input state that we had earlier, but a different output state because of the history of how we got here. The circuit "remembers" which of the inputs (Set or Reset) was most recently raised high.

If both the Set and Rest inputs are brought high, then both outputs will be forced low. If we tried to move from the "both high" state to the "both low" state (where we see from the truth table that the outputs should remain in their prior state), we would find that we could not successfully hold the $Q = 0, \overline{Q} = 0$ output state. The output would stabilize at either $Q = 1, \overline{Q} = 0$ or $Q = 0, \overline{Q} = 1$. No matter what we would do, one of the inputs would change state a small increment of time before the other or one of the gates would switch a little faster than the other and we would go through the 1,0 or 0,1 state. This phantom transitional "state" is the state that would be reflected when both inputs are brought low to 0.

While the RS flip-flop occasionally appears in designs, it is much more common to find one of the other flip-flop types that we discuss in Section 18.4.

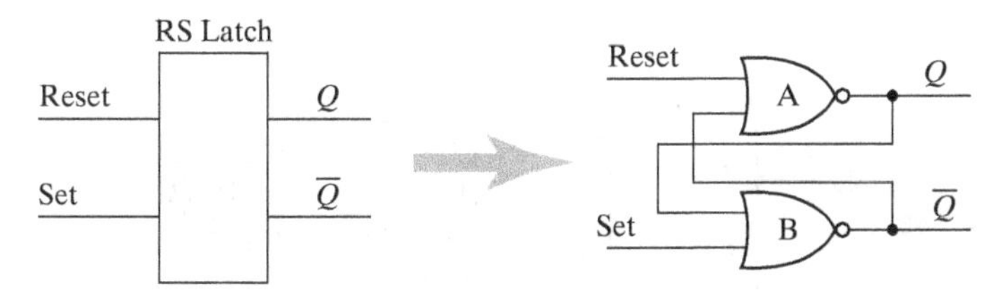

Reset	Set	Q	$\overline{Q}$
0	0	$Q_{(n)}$	$\overline{Q}_{(n)}$
0	1	1	0
1	0	0	1
1	1	0*	0*

* = state can not be held by transitioning to Set = 0, Reset = 0

FIGURE 18.10

The RS latch (flip-flop) and its implementation.

18.4 A SURVEY OF USEFUL FUNCTIONS IMPLEMENTED WITH SEQUENTIAL LOGIC

Individual flip-flops can be used to implement a number of different kinds of latches that do not have the unstable behavior of the RS latch. In addition, flip-flops can be grouped together into chains to make devices with especially useful sequential behavior.

18.4.1 The D-Type Flip-Flop

Probably the most common of the flip-flops is the D-type flip-flop. This device has a data (D) input and a clock pulse (CP) input (Figure 18.11). The right-hand portion of Figure 18.11 is a truth table that describes the dynamic behavior of the device.

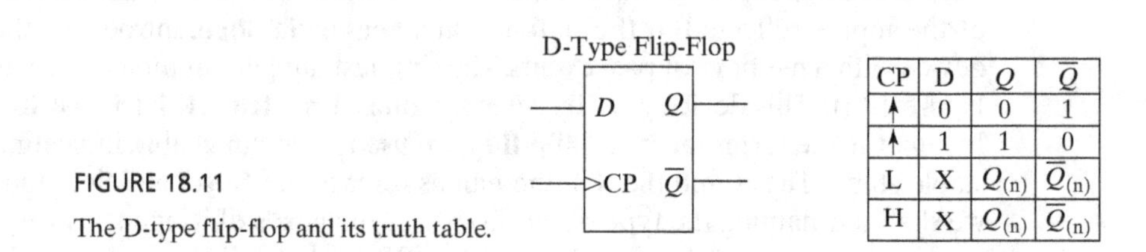

CP	D	Q	$\overline{Q}$
↑	0	0	1
↑	1	1	0
L	X	$Q_{(n)}$	$\overline{Q}_{(n)}$
H	X	$Q_{(n)}$	$\overline{Q}_{(n)}$

FIGURE 18.11

The D-type flip-flop and its truth table.

The first two lines of the truth table work together to show that a rising edge of the clock pulse (CP) causes the logical state present at the D input to be transferred to the Q output and the complement of D to the $\overline{Q}$ output. The last two lines indicate that when the CP line is in the low or high state, the outputs retain their last value, independent of the current value of D. The angled symbol at the CP input of the schematic symbol and the arrow in the truth table both indicate that the transfer of the D input to the Q output (and its complement to the $\overline{Q}$ output) happens on the rising *edge* of the CP signal. The CP line is said to be **edge triggered.** While the CP line is low or high, the Q outputs "remember" their last state, giving them the characteristic of memory. Another useful way of looking at this behavior is to think of the rising edge of the CP line as "sampling" the D input and capturing an image of its value at that instant in time.

18.4.2 The J-K Flip-Flop

If, rather than having a single data input, we build a flip-flop with two inputs, we create a device with more flexibility. When there are two inputs, there are four (2^2) possible actions that we can define when the active clock edge occurs. The J-K flip-flop uses this approach to cause the Q output to be: 1) set to 1, 2) cleared to 0, 3) toggled (change states), or 4) held at its last value. In all cases, the $\overline{Q}$ output reflects the complement of the Q output. The schematic symbol and truth table for the J-K flip-flop is shown in Figure 18.12.

The toggle operation is a new capability over those provided by the D-type flip-flop. (Here, the verb "to toggle" means to change states.) In the context of the J-K flip-flop, when the J and K inputs are both high, every active clock edge causes the state of both the Q and $\overline{Q}$ outputs to change states.

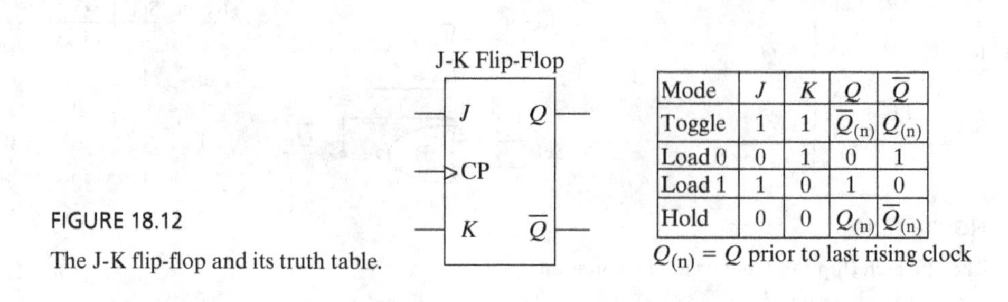

Mode	J	K	Q	$\overline{Q}$
Toggle	1	1	$\overline{Q}_{(n)}$	$Q_{(n)}$
Load 0	0	1	0	1
Load 1	1	0	1	0
Hold	0	0	$Q_{(n)}$	$\overline{Q}_{(n)}$

$Q_{(n)}$ = Q prior to last rising clock

FIGURE 18.12

The J-K flip-flop and its truth table.

18.4.3 The Counter

By putting together a group of flip-flops it is possible to build a circuit that will count how many clock pulses have occurred. The clock pulses may have been created by external physical events with no particular time relationships to one another or they may originate from a source with a regular, known frequency. In the later case, the count represented on the counter will be an indication of the amount of time that has elapsed. In fact, the counter is the basis of all timer and PWM subsystems that appear in microcontrollers (Chapter 8).

Figure 18.13 is an example of how you can connect together a group of flip-flops to create a counter. All of the J-K flip-flops have their J and K inputs tied high, causing their outputs to toggle on every falling edge of their clock lines. The Q output of the flip-flop is connected both to an output (Q_i) and to the clock input of the flip-flop to its right. In this configuration, every time the Q output undergoes a high-to-low transition (a falling edge), an active clock edge will be generated for the flip-flop to its right.

The sequence diagram below the schematic describes how the counter counts. Starting from a state when all Q outputs are low, the first active (falling) edge at the clock input causes the Q output of FF_0 (Q_0) to toggle, going high. Since the clock input to which Q_0 is connected (FF_1) is falling edge triggered, no action is triggered by this rising edge. On the next falling edge of the input clock, the output Q_0 once again toggles, this time going low. This creates a falling edge at the clock input to FF_1 which causes its Q output to toggle from a 0 to a 1.

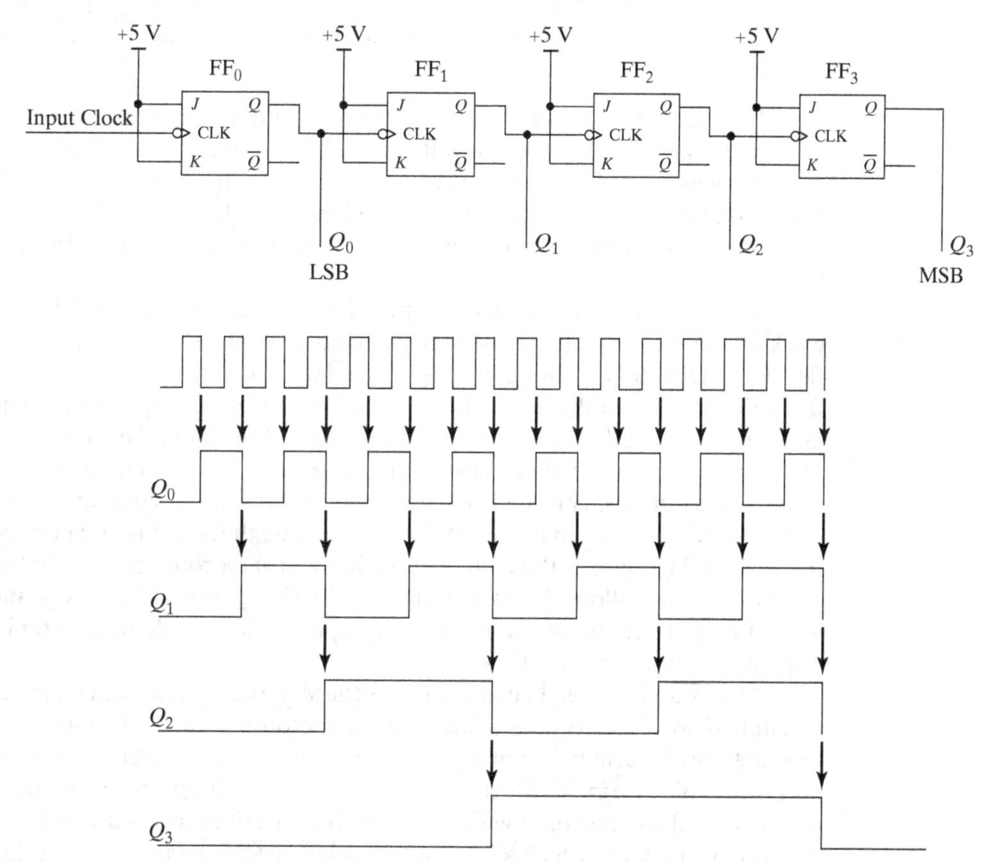

FIGURE 18.13

A group of flip-flops connected to implement a counter.

	Q_3	Q_2	Q_1	Q_0
0	0	0	0	0
1	0	0	0	1
2	0	0	1	0
3	0	0	1	1
4	0	1	0	0
5	0	1	0	1
6	0	1	1	0
7	0	1	1	1
8	1	0	0	0
9	1	0	0	1
10	1	0	1	0
11	1	0	1	1
12	1	1	0	0
13	1	1	0	1
14	1	1	1	0
15	1	1	1	1
0	0	0	0	0
1	0	0	0	1

FIGURE 18.14

Output pattern for subsequent clock pulses.

If we examine the states present on the Q outputs as the counter is triggered by subsequent clock pulses, we can see how it counts. Figure 18.14 shows states on the four Q outputs for a sequence of clock pulses starting from the initial state when all Q outputs were 0. Notice that if we arrange the states of the bits with Q_0 in the rightmost position and subsequent Qs to its left, they form a binary pattern that counts from 0 to 15, then starts over again from 0.

The arrows in the diagram (Figure 18.13) indicate that the falling edge of the input clock triggers a sequence of further transitions as the effects propagate through the later flip-flops. Due to the finite propagation delays (Chapter 16, Sections 16.4 and 16.7.2) through each of the flip-flops, the effect triggered by the input clock happens at later and later times as the effect moves through the chain of flip-flops. The upshot of all this is that the Q outputs do not all change state at the same time, but in fact, the effects "ripple" through the counter outputs. This leads to the name of this kind of counter, the **ripple counter**. The fact that the effects from a clock pulse ripple through the counter can be problematic. Imagine what will happen as the counter transitions from a count of 1111 to 0000. The Q_0 output will first go low, followed a short time later by the Q_1 output, then Q_2, and finally Q_3. If we looked very closely in time at the Q outputs, we would see them very rapidly go through the sequence shown in Figure 18.15.

This is undesirable because each of these patterns represents a plausible count on the counter. If we were to take a snapshot of the outputs at an instant in time, we would be unable to distinguish between a pattern that was a legitimate count and one that was simply an artifact of the ripple effect. This can cause undesirable behavior in a system that isn't designed to deal gracefully with the issue. It is possible to get around the ripple effect and produce a counter in which all the outputs change state simultaneously. In fact, these are the most common kind of counter. However, the logic used to implement these **parallel output counters** is beyond the scope of this discussion.

Q_3	Q_2	Q_1	Q_0
1	1	1	1
1	1	1	0
1	1	0	0
1	0	0	0
0	0	0	0

FIGURE 18.15

The Q outputs as the counter changes from 1111 to 0000.

18.4.4 The Shift Register

If we make a string of D-type flip-flops and wire them together as shown in Figure 18.16, we get another useful device: the **shift register**.

To get a sense for how the device operates, consider the case when all the Q outputs and both inputs are initially at 0. If the clock signal is brought high, nothing will happen because the bubble symbol, plus the angle symbol, on the clock input indicates that it is *falling edge triggered*. When the clock input is brought low, it will trigger all of the flip-flops, since they all are connected to the single clock input. The behavior of the D-type flip-flop is to transfer the state of its D input to its Q output whenever it is triggered. In this design, that means that the state on the input at the left of the figure, labeled Data, will be transferred to Q_0 and the state that was on Q_0 will be transferred to Q_1, the state that was on Q_1 will be transferred to Q_2, and so on. If the Data input remains low, we won't see any change in any of the outputs.

Now, think about what will happen if we raise the Data line high, toggle the clock line up, then down, lower the Data line, and toggle the clock line up and down seven more times. On the first falling edge of the clock, the high state on the Data line will be transferred to Q_0, forcing it high. Since the input states for all of the other D inputs were low, none of the other outputs will change from the low state. When the second falling edge of the clock occurs, the Data input will have been brought low, so a low will be transferred to Q_0. On that same clock edge, the next flip-flop will transfer the old state of Q_0 (a high) to its output (Q_1). All the other outputs will remain low. On the third falling edge of the clock, a low state will once again be transferred from the Data input to Q_0. The old state of Q_0 (a low) will be transferred to Q_1 and the old state of Q_1 (a high) will be transferred to Q_2. With each subsequent falling edge of the clock input, the high state will move over by one flip-flop until it finally disappears after eight falling edges. This behavior is presented in tabular form in Figure 18.17.

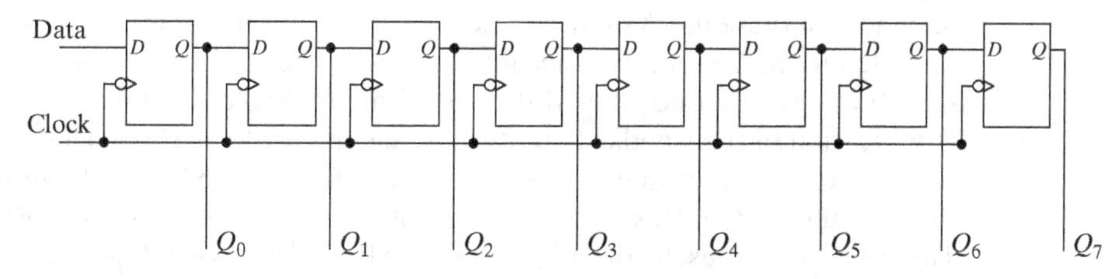

FIGURE 18.16

D-type flip-flops arranged to create a shift register.

Clock	D	Q_0	Q_1	Q_2	Q_3	Q_4	Q_5	Q_6	Q_7
0	0	0	0	0	0	0	0	0	0
1	1	1	0	0	0	0	0	0	0
2	0	0	1	0	0	0	0	0	0
3	0	0	0	1	0	0	0	0	0
4	0	0	0	0	1	0	0	0	0
5	0	0	0	0	0	1	0	0	0
6	0	0	0	0	0	0	1	0	0
7	0	0	0	0	0	0	0	1	0
8	0	0	0	0	0	0	0	0	1
9	0	0	0	0	0	0	0	0	0

FIGURE 18.17
Shift register outputs as a 1 is shifted through

There are many ways to think about (and use) this behavior. The simplest way of thinking about this is to consider that the output Q_0 shows the state of the Data input at the most recent clock pulse (one clock ago), while the output at Q_1 shows the state of the Data input two clocks ago, Q_2 shows the state of the Data input three clocks ago, and so on. Looking at a single output, you can look back in time to the state that the Data input was in some number of clock pulses ago.

If we get a bit more complex and put both the Data and clock inputs under the control of the microcontroller, we can manipulate the Data and clock lines in such a way as to place any arbitrary 8-bit pattern on the Q outputs by appropriately choosing the sequence of states to place on the Data input as we pulse the clock line eight times. This gives us a means of creating some greater number of digital outputs (eight in this example) using only two microcontroller outputs. We'll go into more detail on this capability in Sections 18.6.3 and 18.6.4.

18.5 LOGIC FAMILIES

Relatively early in the development of discrete logic chips, manufacturers settled on a set of numeric identifiers that all have adopted to indicate the functionality of a logic chip. For example, the number 00 is used to denote a chip that has four 2-input NAND gates, and 02 is used to denote a chip that has four 2-input NOR gates. One of the earliest of these families of standard logic chips was the family known as **transistor-transistor-logic** (**TTL**). The full part number for a TTL device with four 2-input NAND gates is the 7400 (or 5400 if the chip is qualified to military specifications, or "mil-spec").

As the technology used to make these chips evolved, new chips were developed that implemented the same logical functions but with different electrical characteristics. The designers of these new families of devices made differing choices about the kinds of transistors used to implement the logical functions or the trade-off between speed and power consumption. To denote the differences in technology, the manufacturers adopted a modification of the standard numbering scheme for logic chips. The modification consists of a 2–4 letter identifier that is placed between the 74 (or 54) and the number used to indicate the logical function. This has led to part numbers such as 74HC00, 74VHC00, and 74AHC00. These letters represent the high-speed CMOS (HC), very high-speed CMOS (VHC), and advanced high-speed CMOS (AHC) families and are only a few of the many families available. Most of the high-speed CMOS families are appropriate for use in interfacing to microcontrollers. To be sure though, you should get the data sheet for the chip in question and go through the procedures outlined in Section 16.8 to evaluate compatibility.

When describing commercial chips in subsequent sections of this chapter, we will adopt the common convention of replacing the family identifier with xx (e.g., 74xx00). This indicates that what is important about the chip being discussed is its logical function and that the particular family of the chip is not as important. It is worth noting here that not all logical functions are available in all of the different logic families. The functions that we will describe are all available in the high-speed CMOS family which is usually a good choice when interfacing to microcontrollers.

18.6 USING AVAILABLE LOGIC CHIPS TO EXPAND THE CAPABILITIES OF A MICROCONTROLLER

The most relevant use of discrete logic chips in the context of microcontroller-based systems is to use these logic devices to expand the capabilities of the microcontroller. For example, the number of input and output lines on a microcontroller is fixed for a particular device. The step up to the next larger microcontroller package may provide for many more I/O lines, but also entail a significant increase in cost. What can you do if you only need a few more lines? In this section, we show how you can use several different types of very inexpensive logic chips to expand the I/O capabilities of a microcontroller.

18.6.1 Using a Multiplexer to Expand Input Capabilities

As we described earlier (Section 18.2.2), you can use a multiplexer to choose which of a selection of possible external inputs will be connected to a single microcontroller input. For the example shown in Figure 18.18, we have chosen to use the 74xx151, an 8-input multiplexer. To expand the number of input lines by eight requires the use of a total of four lines from the microcontroller. Three lines (Select bits 0–2) are outputs from the microcontroller that are used to select which of the external inputs will be applied to a single microcontroller input (labeled "Data Input"). In a situation where fewer than eight input lines are needed, it is possible to tie one of the select lines permanently high or low, reducing the number of control outputs necessary to select which input will be read.

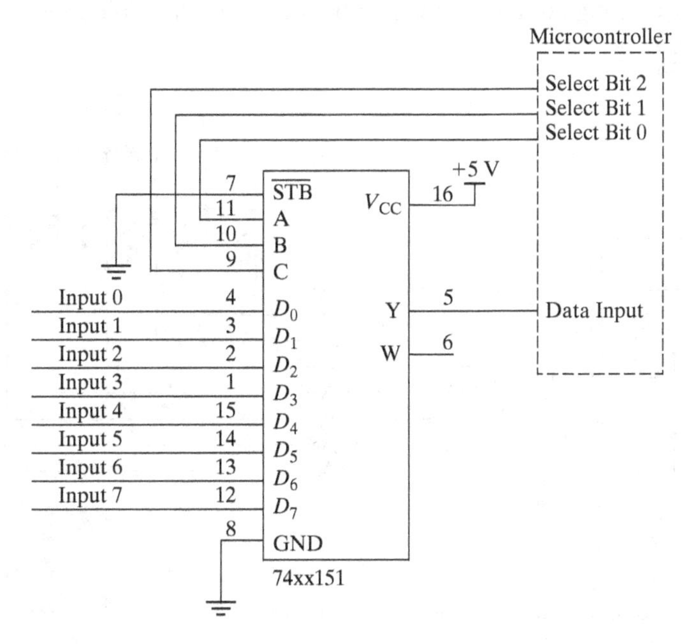

FIGURE 18.18

A 74xx151 used to expand the number of input lines.

To go along with this hardware, it will be necessary to write a function in software that will output the correct pattern to the select lines and then read and return the state of the selected input. In pseudo-code, a function that would take a parameter called WhichBit and return TRUE if the selected input line were high and FALSE otherwise, might look something like:

```
Write WhichBit to the output port connected to the multiplexer
Read the input port associated with the DataInput bit
Mask the result of the read to isolate the DataInput bit
If the result of the masking operation is TRUE
     return TRUE
else
     return FALSE
```

18.6.2 Using a Decoder to Expand Output Capabilities

Under the right circumstances, we can use the decoder that we described in Section 18.2.3 to allow us to effectively expand the number of output lines from a microcontroller. The key condition that is necessary to be able to use the decoder for this purpose is that only one of the outputs can be active at any point in time. This will commonly be the case when the outputs are being used to indicate, for example, a machine's state where it can only be in one state at a time (think about a washing machine that is filling, or washing, or spinning but never filling and spinning at the same time).

In Figure 18.19, we show a sample application where three output bits (Select Bit 0-2) are used to choose which of the LEDs will be lit. It is very common for decoder chips to include an extra "enable" input that can be used to put all of the outputs into an inactive state. In the 74xx138, there are actually three such enable inputs, two of which are active low and one which is active high. For this example, we have tied the active low enable lines to ground and shown the active high enable tied to a fourth microcontroller output line. When the Output Enable line from the microcontroller is brought low, all of the LEDs will turn off. As with the multiplexer example, if four or fewer outputs are necessary, one of the select lines can be tied low or high to reduce the number of control lines required.

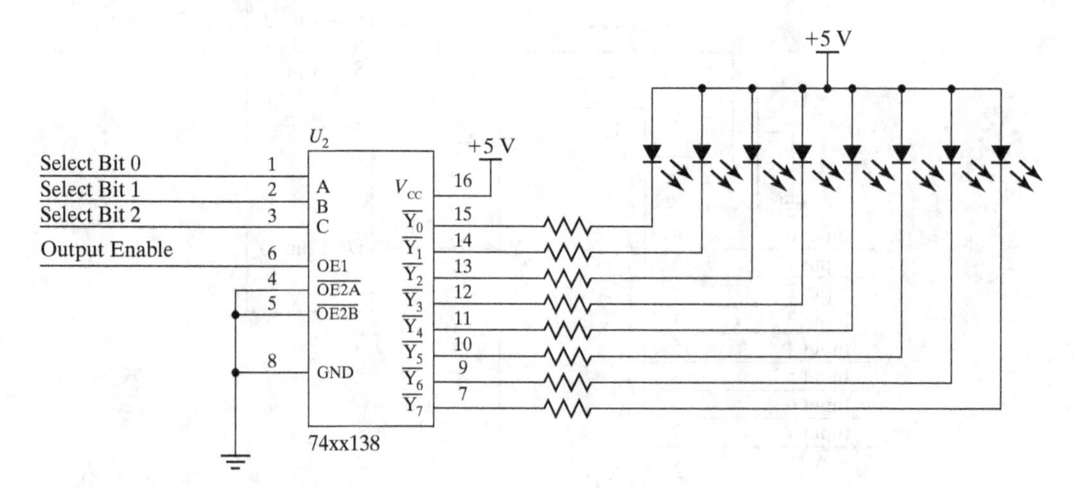

FIGURE 18.19

A 74xx138 being used to control eight LEDs from four output lines.

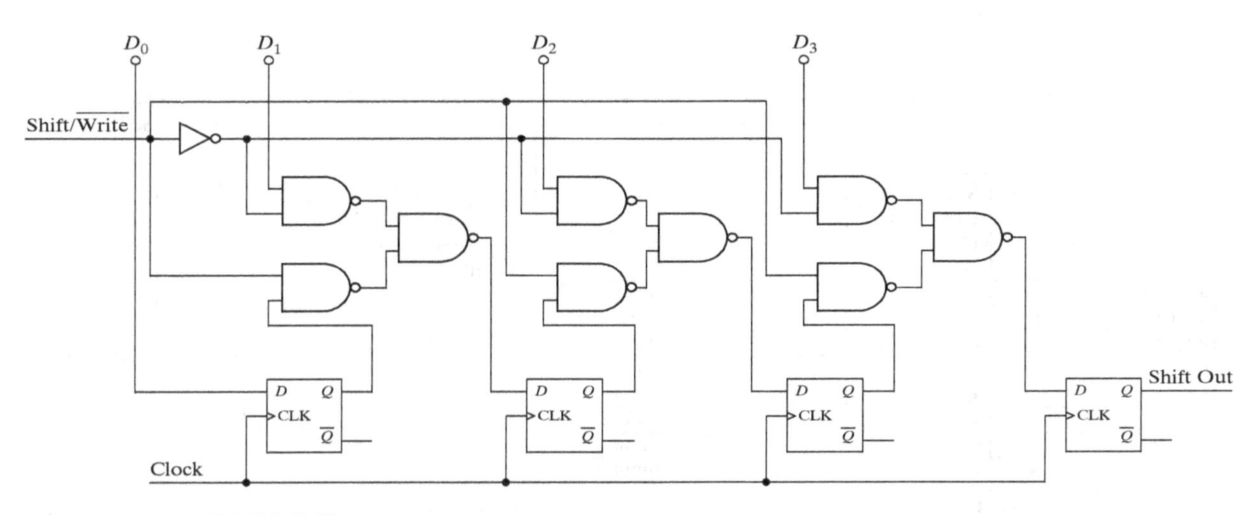

FIGURE 18.20

A 4-bit parallel input serial output shift register.

18.6.3 Using Shift Registers to Expand Input Capabilities

If the logic of the shift register of Figure 18.16 is modified to that of Figure 18.20, a **parallel input serial output** shift register is created. When the Shift/$\overline{\text{Write}}$ line is low, a pulse on the clock line will force the states of the Q outputs of the flip-flops to match the states on the input pins (D_1–D_4). If the Shift/$\overline{\text{Write}}$ line is then brought high, subsequent pulses on the clock line will sequentially move those captured states out of the rightmost Q.

This will allow three lines from the microcontroller to be used to sample an arbitrarily large number of input lines.

Figure 18.21 shows a design that uses a 74xx165 shift register to allow eight digital input lines to be sampled using three microcontroller lines: two outputs (to control the shift register chip) and one input (to read in the digital data). The 74xx165 is also capable of being cascaded (multiple 74xx165s connected together in a chain) to form longer shift registers with still more digital input lines.

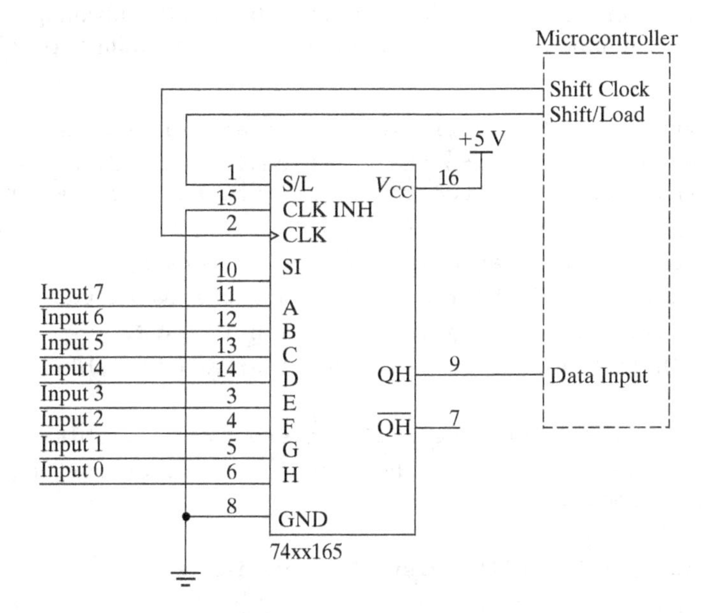

FIGURE 18.21

A 74xx165 used to expand the number of input lines available.

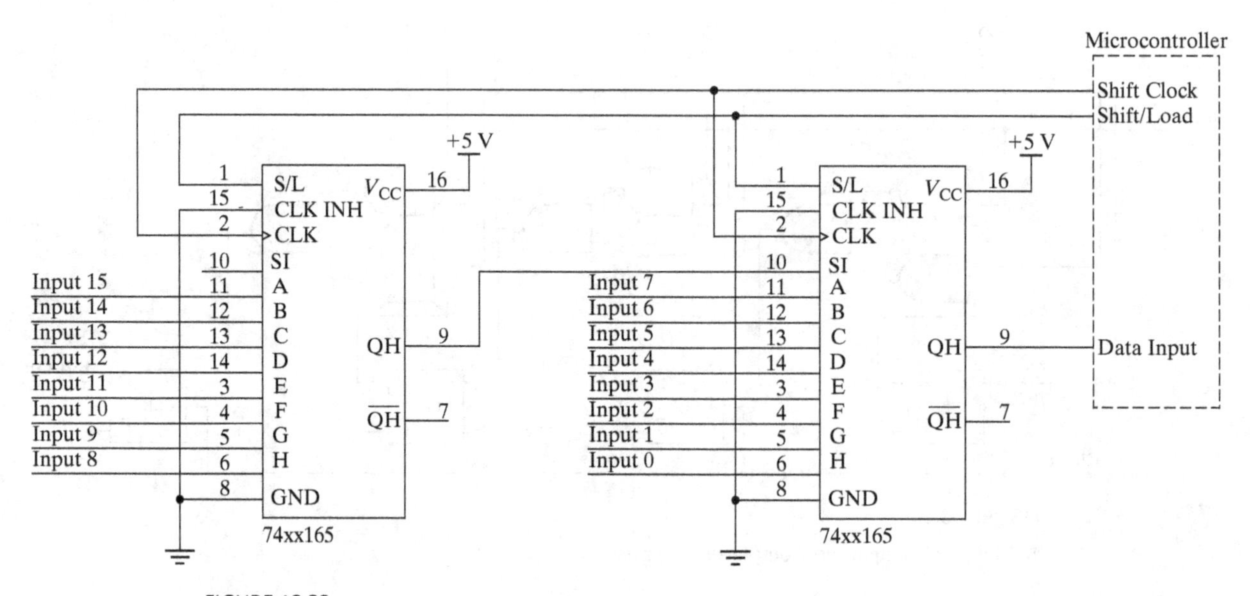

FIGURE 18.22

Cascading two 74xx165s to form a 16 input shift register.

For example, in Figure 18.22, the QH output of the left 74xx165 is tied to the SI (Serial Input) pin of the right 74xx165 (with the other two control pins, S/L and CLK, tied to both chips), forming an effective 16-bit shift register. This allows 16 inputs to be sampled using the same three microcontroller lines. However, this comes at the expense of a longer time to shift in the data from the larger number of inputs.

The software required by the shift register approach to expanding inputs is a bit more complex than that required by the multiplexer. If the microcontroller that you are using does not include an SPI subsystem (Chapter 7, Section 7.3.1.1), then you must write software to manipulate the control lines and read the input line in order to read the shift register's inputs. When I/O lines are manipulated like this in software, it is often referred to as **bit banging**. As you can see from the sample pseudo-code below, this process requires several more steps than were required by the multiplexer, but has the advantage that it is easily expandable to larger numbers of inputs with no increase in the number of microcontroller lines required.

```
Clear the ShiftRegisterData variable used to accumulate the incoming data
Latch the input pin values into the shift register by first setting the
pin of the output port connected to the Shift/Load line low then high.
Repeat 8 times:
       Read the input port associated with the DataInput bit
       Mask the result of the read to isolate the DataInput bit
       If the result of the masking operation is TRUE
              Shift a 1 into the ShiftRegisterData variable
       else
              Shift a 0 into the ShiftRegisterData variable
       pulse (raise, then lower) the pin of the output port connected to
       the ShiftClock line
End Repeat
Return the value of the ShiftRegisterData variable
```

18.6.4 Using Shift Registers to Expand Output Capabilities

Commercially available shift registers are rarely as simple as the example described in Section 18.4.4. Since it is generally undesirable to have the outputs changing state as the desired bit states are shifted through the shift register, the most useful commercial shift registers don't connect the Q outputs of the shift register directly to the chip's output pins. Instead, these chips include a second set of latches that drive the output pins. In this case, as the new data is being shifted into the shift register flip-flops, the levels on the output pins are held at their last value by the latch outputs. After the complete 8-bit pattern has been shifted in, then another control line (labeled Register Clock) is toggled to transfer all eight new bits to the outputs at one time (Figure 18.23).

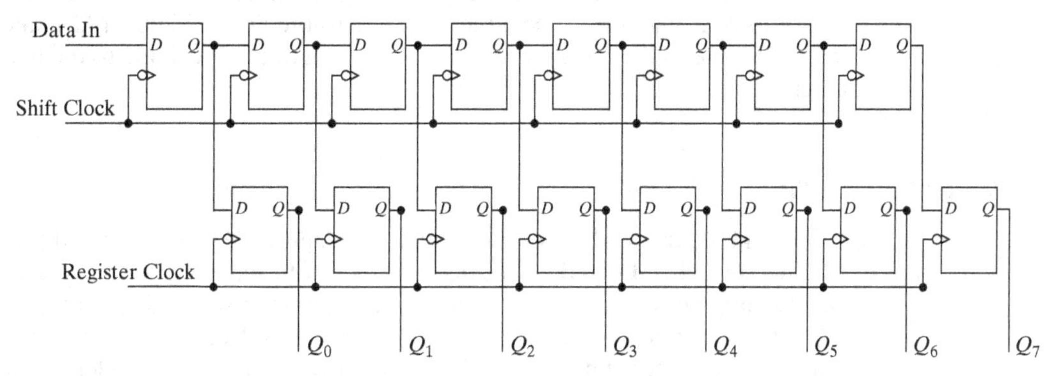

FIGURE 18.23

A registered output shift register.

Using a registered output shift register is very straightforward from a hardware standpoint. Figure 18.24 shows a circuit with similar LED drive capabilities to that of Figure 18.19. Using the shift register requires only three port lines in order to be able to control the eight LEDs. This solution is also more flexible than that of Figure 18.19 in that it can more flexibly control all of the LEDs, allowing multiple LEDs to be active at the same time. The design is capable of being expanded in a way similar to that of the input shift register described in Section 18.6.3, without requiring additional pins on the microcontroller. As with the input

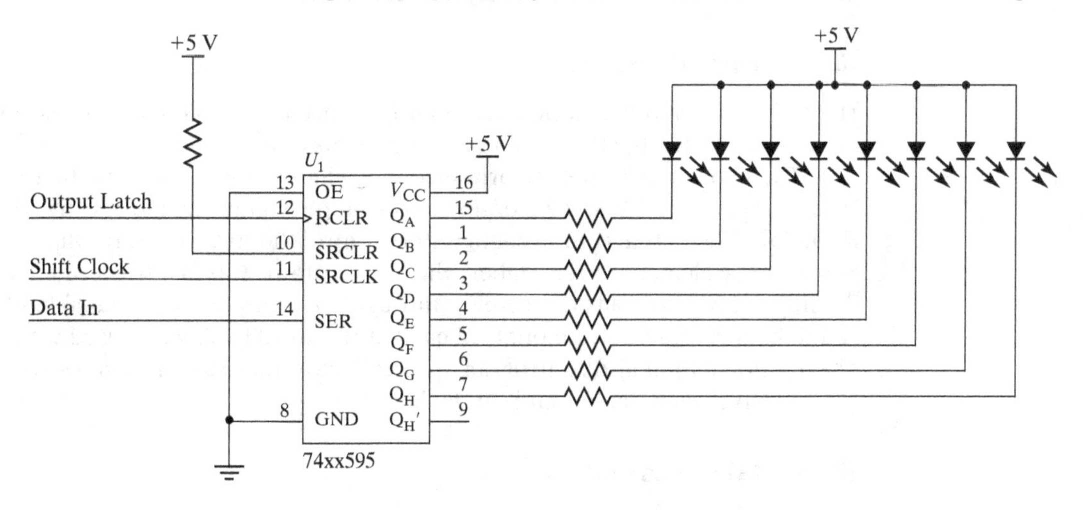

FIGURE 18.24

Using a 74xx595 to control eight LEDs using only three microcontroller output pins.

shift register, the drawback to using the output shift register is the increased complexity in the software necessary to control it, and the increased time required to update the states on all of the output lines.

18.6.5 Using the SPI Subsystem with Shift Registers

If the microcontroller you are using includes a hardware SPI (or equivalent) subsystem, then the software necessary to control the shift registers is greatly reduced. As was discussed in Section 7.3.1.1 of Chapter 7, the SPI system provides a shift clock and the hardware to automatically shift data out of and into the microcontroller, obviating the need for writing your own "bit banging" software. The SPI subsystem effectively implements in hardware the "repeat 8 times" loop in the previous pseudo-code. To implement the circuit from Figure 18.24 using SPI, connect the SCK line to the line in Figure 18.24 labeled Shift Clock, the MOSI line to the line labeled Data In, and an output port line to the line labeled Output Latch.

18.7 THE 555 TIMER

If we combine the analog comparators from Chapter 11 with the RS flip-flop from Section 18.3, we get a **mixed signal** device that is capable of using the timing associated with charging and discharging RC circuits to produce an analog timer with a digital output. While the timer subsystems in microcontrollers are very flexible and offer a great range of possibilities, sometimes the problem at hand does not warrant a microcontroller, or the hardware and software necessary to program the microcontroller are not easily available. For such cases, this type of analog timer chip is capable of providing a variety of functions. Only a few external resistors and capacitors are required to configure the devices.

Far and away the most popular of the analog timer chips that are available are those based on the original NE555 introduced in the early 1970s. This was originally a bipolar transistor design, which has since been translated into CMOS and produced in variants with two timers per chip (the NE556). It continues to be available from several manufacturers. We will introduce the basic operation of the chip and show how it can be used to implement the two most common timing tasks: producing a one time (one-shot) pulse or producing a regular waveform of controllable frequency and duty cycle.

18.7.1 Inside the 555 Timer

The 555 timer chip is composed primarily of a pair of comparators and an RS flip-flop. Figure 18.25 shows the basic internal components of a 555 timer.

The two comparators are arranged in a window comparator configuration with reference voltages set at 1/3 and 2/3 of the power supply voltage by the resistive divider chain R_1, R_2, and R_3. The outputs of the comparators control the state of the RS flip-flop which is used as a memory element to remember which comparator was last triggered. The output of the flip-flop is provided outside the chip through a totem-pole drive capable of both sourcing and sinking considerable amounts of current (up to 200 mA when used with a 15 V supply). The flip-flop output also controls an open collector transistor arranged so that the transistor (Q_1) is active when the output is in the low state.

18.7.2 Astable Operation

To configure the 555 to produce a continuous stream of pulses, two resistors and two capacitors are added external to the chip (Figure 18.26).

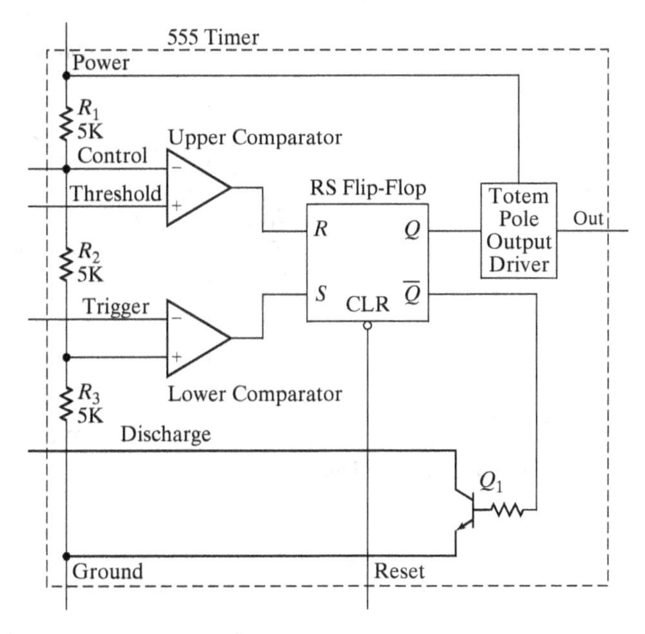

FIGURE 18.25

The internals of a 555 timer chip.

This configuration is known as the **astable** configuration because the output is not stable in either state. When the output is in the high state, transistor Q_1 is turned off. This allows the timing capacitor, C, to be charged through the combined resistance of R_a and R_b. When the voltage on the capacitor (also connected to the Threshold and Trigger pins on the 555) reaches 2/3 of the power supply voltage, the upper comparator is tripped which in turn asserts the R(eset) input on the flip-flop forcing its output and the 555's Out(put) pin low.

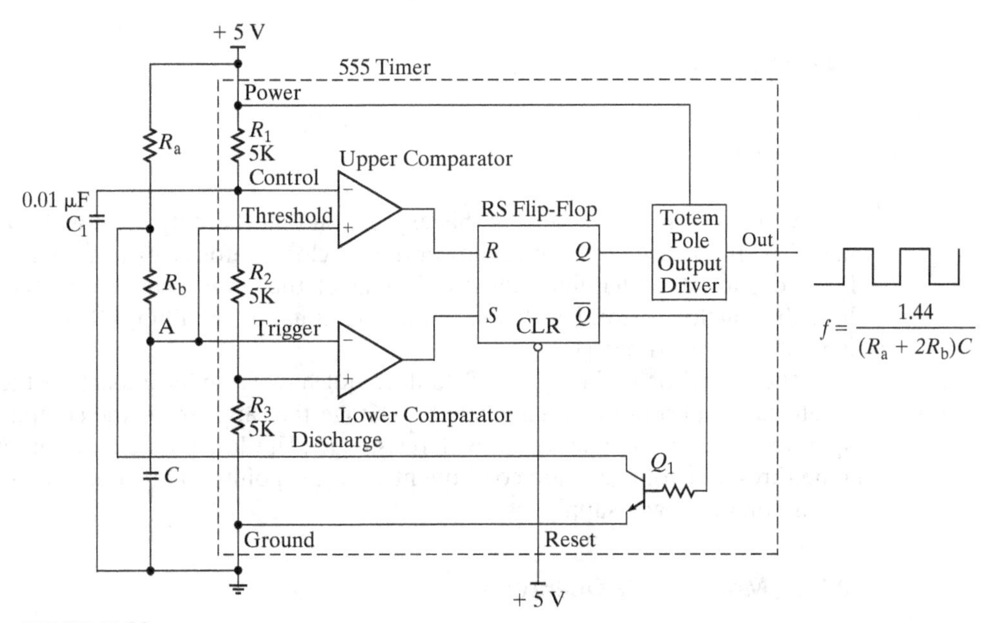

FIGURE 18.26

The 555 configured for astable operation, producing a continuous series of pulses.

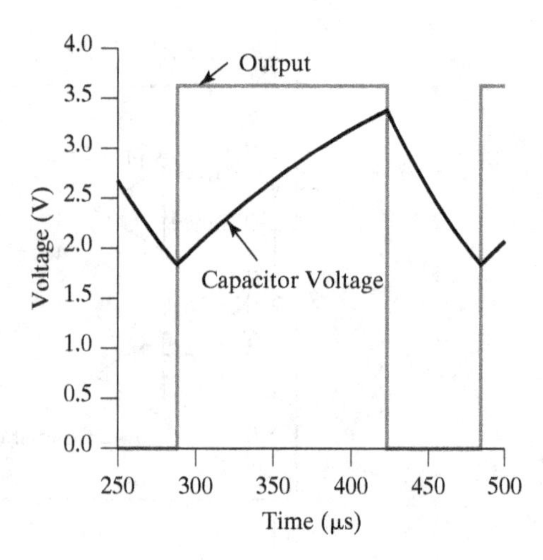

FIGURE 18.27

Voltages on the 555 pins during astable operation.

Since the base of Q_1 is driven by the inverse of the flip-flop Q output, it becomes active and begins discharging the capacitor C through the resistance R_b. The discharge continues until the voltage on the capacitor falls to 1/3 the power supply voltage, at which time the lower comparator is tripped. The lower comparator output asserts the S(et) input to the flip-flop which causes the output to return to the high state, Q_1 to be turned off and the charging of the timing capacitor C, to begin again. Figure 18.27 shows the output voltage along with the voltage that appears on the capacitor.

This process repeats indefinitely yielding a pulse train output with a frequency

$$f = \frac{1.44}{(R_a + 2R_b)\,C} \tag{18.1}$$

and a duty cycle

$$\mathrm{DC} = \frac{R_a + R_b}{R_a + 2R_b} \tag{18.2}$$

You may have noticed from this expression that the duty cycle will always be greater than 50%. It is possible to achieve lower duty cycle waveforms by adding additional components to allow the charging and discharging of the timing capacitor each to take place through a single resistor. For details of these circuits see the Philips Semiconductor application note for the NE555 [1].

The capacitor C_1 in Figure 18.26 does not have a direct impact on the timing of the waveform. Its purpose is to stabilize the voltage that appears at the comparator reference inputs by creating a low-pass filter with resistor R_1. This low-pass filter reduces the variations in the threshold voltages (and consequently trigger points) that would otherwise occur due to noise on the power supply.

18.7.3 Mono-Stable Operation

The second common configuration of the 555 timer, shown in Figure 18.28, is the **mono-stable** (also known as **one-shot**) configuration.

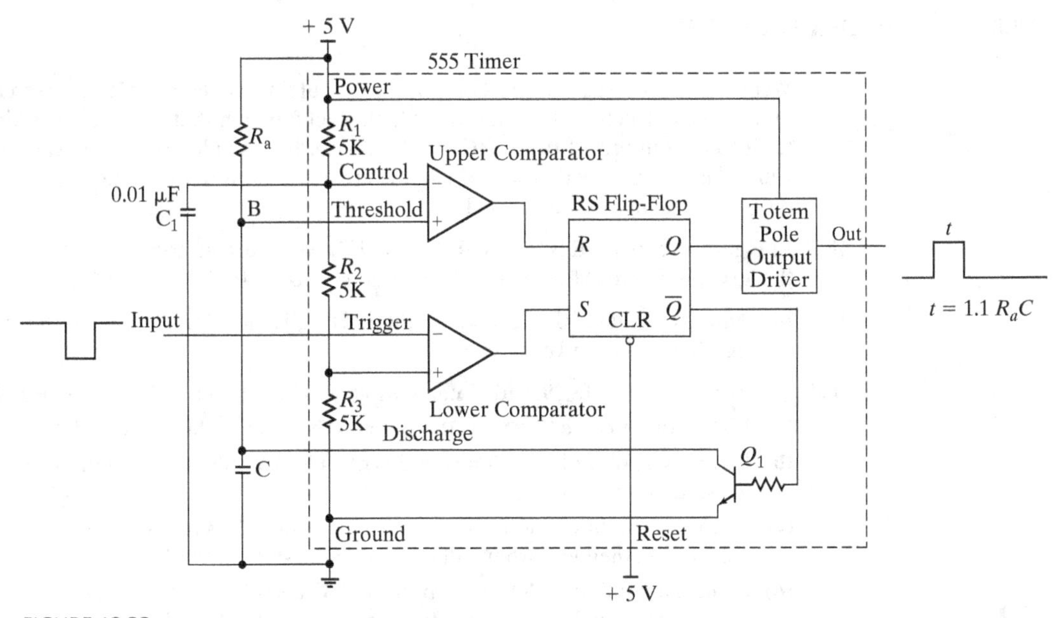

FIGURE 18.28

The monostable configuration for a 555.

In this configuration, a low-going input pulse applied to the Trigger input initiates an output pulse whose duration t is determined only by the values of the external components R_a and C:

$$t = 1.1\,R_aC \tag{18.3}$$

In the circuit's stable mode, the output is low and transistor Q_1 is turned on, holding the voltage on C at approximately 0 V. When the Trigger input is taken below 1/3 of the power supply voltage, the lower comparator is tripped, which triggers the S(et) input of the flip-flop. The output of the flip-flop goes high and the transistor Q_1 is turned off, allowing capacitor C to begin charging through resistor R_a. The charging continues until the voltage on the capacitor reaches 2/3 of the supply voltage when it will trigger the upper comparator, resetting the flip-flop, driving the output low, turning Q_1 on, and discharging the voltage on capacitor C. At this point, if the trigger input has risen above 1/3 of the supply voltage, the output will return to its stable low value until the next trigger pulse. If the input trigger is not returned above 1/3 of the supply voltage before the pulse times out, the output of the 555 will remain high. Therefore, it is important to guarantee that the trigger pulse width is shorter than the width of the one-shot pulse.

18.7.4 Other Uses of the 555 Timer

While these are the most common configurations in which the 555 is used, they are by no means the only possible ways to make use of it. The chip can be configured to produce **pulse position modulation** (**PPM**), where the high time is constant and the low time varies as a function of a control voltage and a linear voltage ramp. A pair of 555s (or a single 556) can be configured to generate pulse width modulation (PWM), where the high time and low time vary based on a control input but the total period remains constant. These are but two of the many other circuits that use the 555 and its variants. A Web search for "555 timer" will provide a wealth of information and application circuits. The device is so versatile that whole books [2, 3] have been devoted entirely to different ways to use the chip.

18.8 HOMEWORK PROBLEMS

18.1 Write the pseudo-code to "bit-bang" three output lines to use a 74HC595 to create an extra eight output lines. Structure the code with a high level function that will take an 8-bit value and transfer it to the outputs of the 74HC595. Also write the pseudo-code for lower level functions that will isolate the interaction with the hardware and therefore make it easier to change which output ports and bits are being used.

18.2 In Figure 18.5, if TMR1ON, TMR1GE, and T1GINV are all true, what state must be present on the $\overline{\text{T1G}}$ pin in order for the Clock Input to be passed to the Gated Clock?

18.3 In Figure 18.5, what is the function of the TMR1GE bit? (Describe the difference in behavior of the logic based on its two states.)

18.4 Referring to Figure 18.29, with Analog Input Mode = 0, Read TRISA = 0, Read PORTA = 0:

(a) If all Q outputs are 0, what is the state on the I/O pin (high, low, indeterminate)?

(b) If Data = 1, Write PORTA = 0, and Write TRISA pulses high, then low, what will happen to the state of the I/O pin?

(c) From the conditions left after part b), if Data = 1, Write TRISA = 0, and Write PORTA pulses high, then low, what will happen to the state of the I/O pin?

(d) From the conditions left after part c), if Data = 0, Write PORTA = 0, and Write TRISA pulses high, then low, what will happen to the state of the I/O pin?

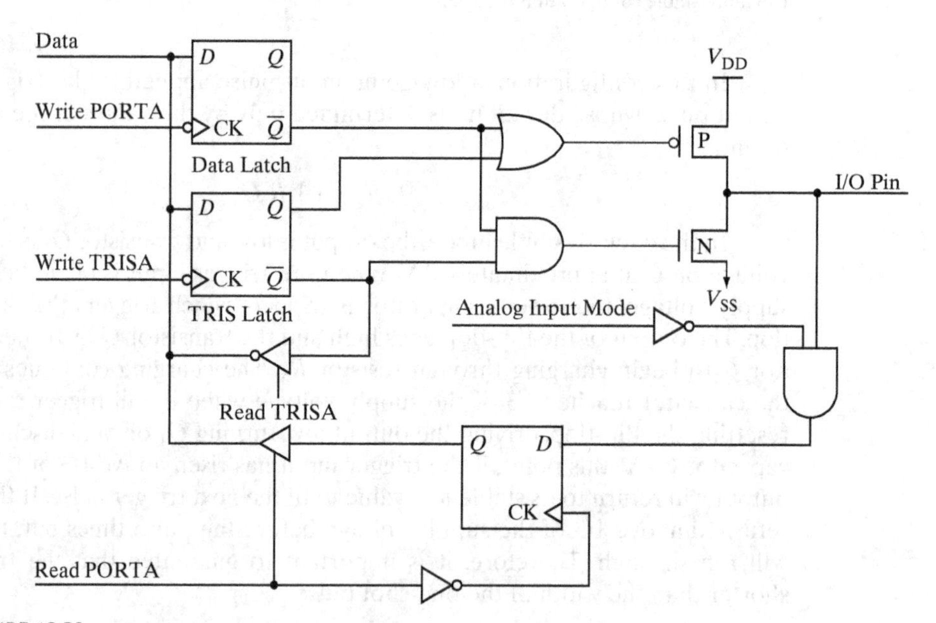

FIGURE 18.29

Schematic for circuit in Problem 18.4.

18.5 As is all too often the case, the microcontroller for your project has come up short of input and output lines. You have four programmable direction I/O lines available and you need four inputs and four outputs. Design a circuit that will expand the four available lines into the eight required lines. Changes to the expansion ports should mimic the behavior of ordinary ports in that reads of the input port should return the state of all four input bits at the same instant in time and updates of the four output bits should all appear simultaneously on those 4 bits. Show the directions for each of the four port lines that you use.

18.6 What named function does the circuit in Figure 18.30 implement?

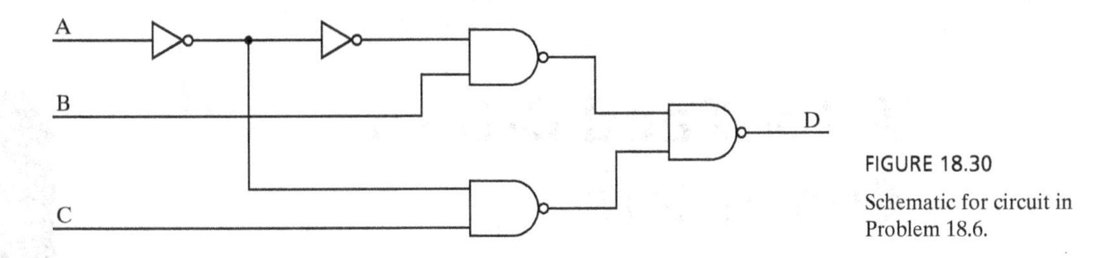

FIGURE 18.30

Schematic for circuit in Problem 18.6.

18.7 Write the pseudo-code to use the circuit from Problem 18.5 to read in the four inputs and write the four outputs. Your code should provide any required initialization function and a function to read the lines and a function to write the lines.

18.8 Design a circuit using a 555 that will light an LED with 20 mA of forward current for a period of 0.5 second for every 1 ms long low-going pulse from a 74HC04.

18.9 Using a 555, design an astable multivibrator. Shoot for a frequency of about 10 kHz (± 10%) and a duty cycle (ratio of high time to total time) of about 50% (± 10%). Do not choose R_a too small (less than 1 kΩ) or it won't work in practice. Include the calculations that you did to set the frequency and duty cycle. In your answer, include a schematic for the circuit, including component values and types.

18.10 Write pseudo-code to control the circuit of Figure 18.19. The top level function should take a symbolic constant (Light0-Light7, All_OFF) to determine which of the LEDs will be lit.

REFERENCES

[1] "AN170 NE555 and NE556 Applications," Philips Semiconductors, 1988.

[2] *555 Timer Applications Source Book,* Berlin, Howard M., Sams Publishing, 1979.

[3] *555 Timer IC Circuits,* Mims III, Forrest M., Radio Shack, 1992.

CHAPTER 19

A-to-D and D-to-A Converters

In many mechatronic applications, a microcontroller senses conditions in an environment and uses this information to control a process or a physical quantity. In recent chapters, we've discussed scenarios where all the information to and from the microcontroller is strictly in binary form (1 or 0, on or off, true or false), and a microcontroller can perform its functions using only digital inputs and outputs. Microcontrollers are fundamentally digital devices, and can only deal with digital data. For a microcontroller to make sense of information in the analog domain and produce analog outputs, a translation between the domains is required. This is the function of analog-to-digital and digital-to-analog converters. These serve as the links between the digital data that a microcontroller understands and the analog information that characterizes much of the world around them. In this chapter, you will learn:

1. the effects of converting between analog and digital data,
2. the terms used to describe the performance of analog-to-digital and digital-to-analog converters,
3. error sources and classifications for both types of converters,
4. how several types of digital-to-analog converters are designed, how they work, and what their strengths and weaknesses are,
5. how several types of analog-to-digital converters are designed, how they work, and what their strengths and weaknesses are,
6. how to use this information to select appropriate converters for your specific applications.

19.1 INTERFACING BETWEEN DIGITAL AND ANALOG DOMAINS

In Chapter 2, we described the central processing unit, or CPU, as the heart of a microcontroller. The CPU moves digital data between memory and its arithmetic logic unit (ALU), where mathematical and logical operations are performed. The CPU does not have the native ability to deal with analog values, and as a result, analog inputs and outputs must be translated into a digital format on their way to or from a microcontroller and memory. This is represented in block diagram form in Figure 19.1.

For a microcontroller to read in the state of an analog quantity (i.e., the brightness of a light source), the quantity must first be sensed (the Sensor block, see Chapter 13). A sensor's output often requires some degree of signal conditioning to optimize the information content, such as scaling, offset adjustment, and filtering (Chapters 14 and 15). If the sensor's output isn't in the form of a voltage, the Signal Conditioning block will also need to perform that translation. The final step before the microcontroller can read the analog output of the

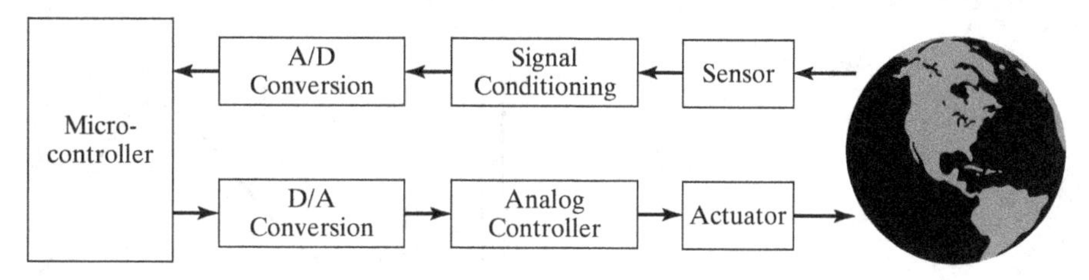

FIGURE 19.1

Interfacing a digital microcontroller to an analog world.

sensor is to convert it to a digital signal. This is done by the aptly named **analog-to-digital converter** (also called an **A-to-D** or **A/D converter**, or simply an **ADC**). The A/D converter translates the analog voltage into a digital number that represents the analog value, and the microcontroller may then manipulate that number and perform operations with it, as it would with any other element of digital data.

For a microcontroller to control the state of an analog output (to control an audio speaker, for example), a similar but reverse process is required. This is depicted in the bottom portion of Figure 19.1. Once a microcontroller determines the desired state of an analog output, it sends a digital representation of the desired value to a **digital-to-analog converter** (also called a **D-to-A** or **D/A converter**, or a **DAC**). The job of the D/A converter is to translate the digital value into an analog value. The analog signal may require some form of signal conditioning (such as amplification) before controlling an output or actuator (such as an audio speaker).

Most microcontrollers incorporate on-chip A/D converters, often with multiple input channels. The resolution and speed of these on-chip A/D converters are not usually at the upper end of the performance spectrum, but they are suitable for many applications, and the resulting single-chip solution is very convenient and economical. A small number of microcontrollers feature on-chip D/A converters, however, these are relatively uncommon. Digital-to-analog conversion is most often implemented with separate, single-purpose integrated circuits.

19.2 DIGITIZING CONTINUOUS SIGNALS

In addition to processing only digital data, a microcontroller executes commands in distinct intervals of time, as dictated by its instruction clock. These characteristics have several implications when converting between analog and digital signals. Figure 19.2 illustrates a representative analog voltage signal (or, more generally, the amplitude of a signal). A key characteristic of an analog signal is that it is **continuous** in both time and amplitude, no matter how closely it is examined. For the signal shown in Figure 19.2, we could theoretically zoom in to any magnification of the time scale (x-axis) or the voltage scale (y-axis), and never see anything other than a continuous signal.

An ideal A/D converter could perfectly measure and represent this signal, and conversely, a perfect D/A converter could produce this signal perfectly. However, because microcontrollers (and the A/D and D/A converters they control) operate only in discrete time intervals and express amplitude only in discrete increments, the results are only approximations of continuous analog signals.

Since microcontrollers execute instructions in discrete time intervals, they are only able to measure a signal periodically. Rather than monitoring a window of data (like that shown in Figure 19.2), microcontrollers are capable only of taking a series of snapshots of a

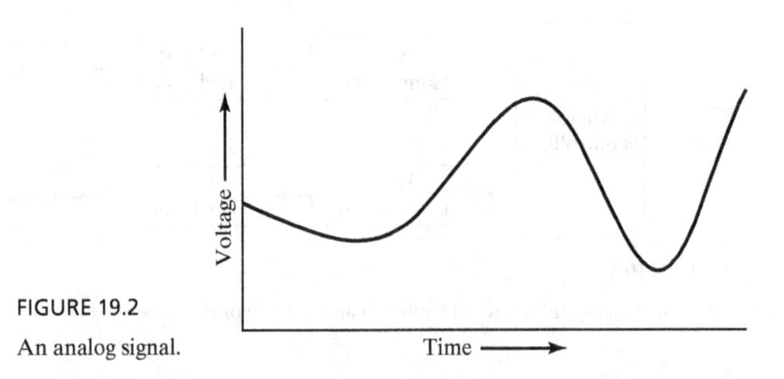

FIGURE 19.2
An analog signal.

signal at discrete points in time, as illustrated in Figure 19.3. This is referred to as **sampling**. With sampled data, the **sampling interval** (the amount of time between successive samples, labeled Δt) is a critical parameter, especially in cases where it is possible for important characteristics of the signal to occur quickly enough that they might be missed between samples or otherwise misinterpreted.

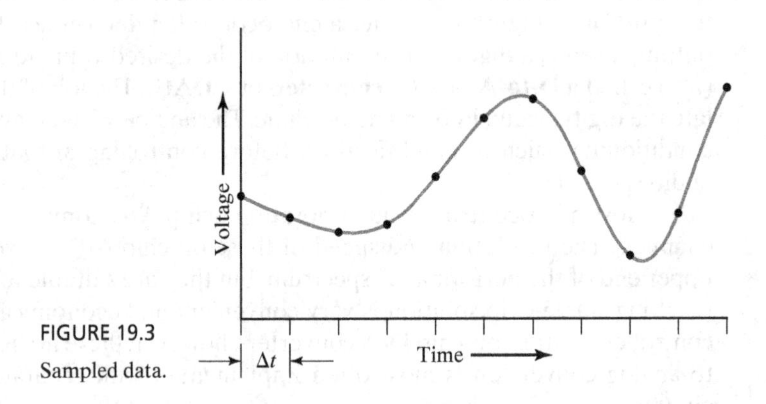

FIGURE 19.3
Sampled data.

Figure 19.3 represents a situation where samples are obtained at discrete intervals for a signal, and any voltage level may be represented. In this case, the sampled signal is continuous in amplitude, but discrete in time. We can flip this scenario around to consider a case in which the signal is continuous in time (it has a value at any and all times), but its amplitude must be one of a discrete number of steps. For example, the signal may have a value of 3 or 3.25 V, but it cannot be 3.14 V. This signal is continuous in time and discrete in amplitude, and is represented in Figure 19.4. The signal is said to have been **quantized**, and the difference

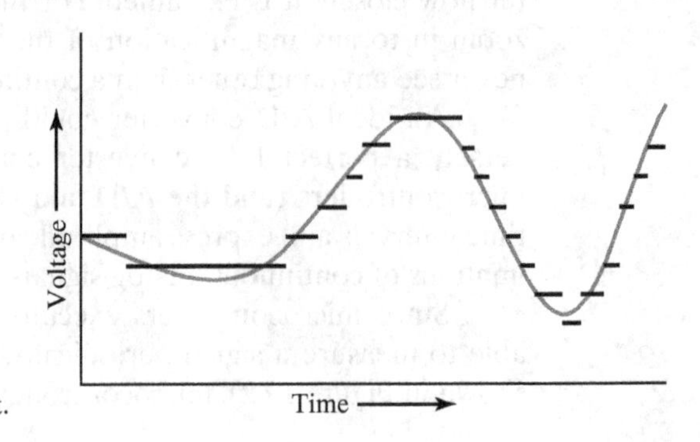

FIGURE 19.4
Quantized data: D/A converter output.

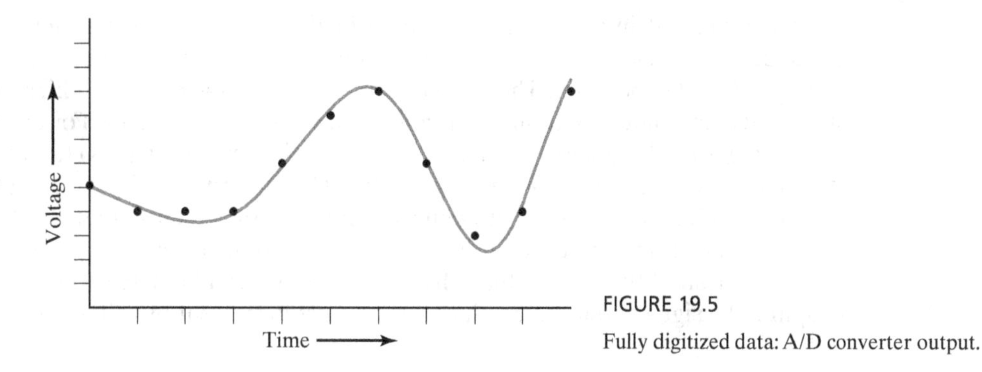

FIGURE 19.5
Fully digitized data: A/D converter output.

between the ideal value and the actual value is called **quantization error**. The **resolution** of the converter determines the size of the quanta. This is the type of output produced by real world D/A converters, and readers familiar with the field of digital control theory may recognize this as a **zero-order hold**, or **ZOH**.

Finally, we consider a case which is discrete with respect to both time and voltage: a fully **digitized signal** (Figure 19.5). This is how real world A/D converters represent analog signals: they sample a signal at discrete intervals, and quantize (or round) the amplitude to the nearest representable discrete value.

Figure 19.5 illustrates the limitations of representing an analog signal with a sequence of digital samples. As shown, the underlying analog signal is somewhat crudely approximated by the succession of digital values (the dots). The results can be improved dramatically by increasing the sampling rate (decreasing the sampling interval) and/or increasing the resolution (i.e., decreasing the minimum increment of amplitude values that can be represented).

The minimum sampling rate required for a given application is determined by the frequency content of the analog signal being converted. If the signal is not sampled at a high enough rate, important information will be missed and the frequency of the input signal can be misinterpreted. Consider the case illustrated in Figure 19.6, where an input signal with a comparatively high frequency is sampled at a lower frequency. The resulting series of samples could be interpreted as representing a sine wave at a much lower frequency than the actual input signal. This phenomenon is called **aliasing**.

There are two basic weapons in the designer's arsenal to prevent aliasing from occurring. The first tool is the **Nyquist Theorem**, named after Harry Nyquist (1889–1976), who showed that a signal must be sampled at a frequency that is at least twice that of the highest

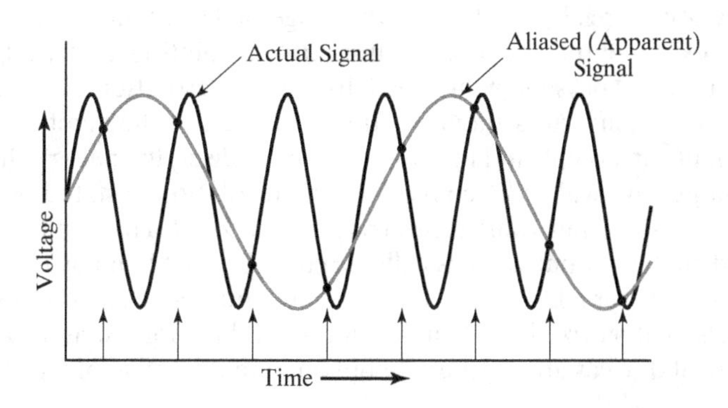

FIGURE 19.6
Aliasing.

frequency present in significant amplitude in the signal to prevent aliasing. This provides guidance for the selection of appropriate sample rates, which in turn drives the selection of an appropriate A/D converter. The second tool is the **antialiasing filter**, which removes undesirable frequency content in a sampled signal that could cause aliasing. For example, if a system is sampling a 100 Hz analog signal at 1 kHz, an antialiasing (low pass) filter would be used to attenuate any frequency content above 100 Hz. In order to be maximally effective, the antialiasing filter should attenuate signals at or above the **Nyquist limit** (the highest frequency signal we can accurately represent using a given sampling frequency, in this example, 500 Hz) so that their amplitude is less than the minimum that can be detected by the A/D converter sampling the signal—that is, less than the smallest increment of voltage that can be resolved.

19.3 A/D AND D/A CONVERTER PERFORMANCE

A full understanding of how well a given A/D or D/A converter performs first requires a solid grasp of the terms commonly used to describe their performance, and an understanding of how a theoretically ideal converter would perform, if one existed. In this section, we'll explore ideal converter behavior, and also the various sources of error—the ways in which real converter behavior deviates from the ideal. We begin by defining the most fundamental terms and concepts.

Resolution: For an A/D converter, the **resolution** is the number of discrete values that the digital output can take on. It is related to (but not the same as) the smallest change in the input voltage that can be distinguished. The range of allowable input voltages for an A/D converter is divided into discrete increments, also called **counts**. The number of counts returned by an A/D converter encodes the analog input value, so the output is also referred to as a **code**. When a converter has low resolution, the magnitude of each increment is large. Higher resolution converters have smaller increments. This situation is analogous to two rulers, one marked in centimeters and the other in millimeters. You can measure the length of an object with either ruler (provided the range of the rulers is sufficient), but the one marked in millimeters allows you to make measurements with 10 times the resolution as the one marked in centimeters. The resolution of an A/D converter is analogous to the number of marks on the ruler. For A/D converters, resolution is expressed in **bits**. Resolution, of course, is just a number, and could theoretically have any value and/or be expressed in any number base system, but for digital systems, bits are the natural choice. Converter resolution is such an important characteristic that it is the primary means of categorizing the devices, for example, a "16-bit A/D converter." Eq. (19.1) gives the expression for determining the number of counts for a given number of bits of resolution, N.

$$\text{Number of counts} = 2^N \tag{19.1}$$

Converters with any number of bits are theoretically possible, however, converters are currently commercially available in the range of 8 to 24 bits. An 8-bit converter, for example, has 2^8 possible output values, resulting in a resolution of 1 part in 256 (1:256). A 24-bit converter has 2^{24} possible values, or 1:16,777,216 resolution.

For D/A converters, the resolution is defined by the number of discrete values that the digital input can accept, and thus the number of discrete analog values that may be produced at the output. As with A/D converters, the resolution is stated in bits, and each successive increment in the input value is referred to as a count. The number of counts used to specify a desired analog output value is called a code. Eq. (19.1) holds for calculating the number of possible counts for a D/A converter, given the number of bits of resolution, N. D/A converters are also categorized by their resolution, in bits: for example, a "10-bit D/A converter." Commercial devices are most easily obtained in the range of 8 to 16 bits (1:256 to 1:65,536).

Lower resolutions (down to 4 bits) and higher resolutions (24 bits and more) are also available, however, these are much less common.

Least significant bit, also called an **LSB**: As we saw in Chapter 3, the LSB in a binary number is positionally the rightmost bit, and from a value standpoint the bit whose value is $2^0 = 1$. This is the bit which will change states when the total value of the binary number changes by 1. In the context of a given A/D converter, it is the smallest change in the input voltage that will change the digital output by 1. For an A/D converter, the magnitude in volts of each count is determined by the range of allowable input voltages for the converter (typically 0 V to $V_{\text{full scale}}$, but more generally the range between a higher reference voltage, V_{RH}, and a lower reference voltage, V_{RL}, which we will call V_{span}) and the number of bits of resolution (N). This quantity is referred to as a **least significant bit**, or **LSB**. To determine the magnitude (in volts) of an LSB, the full-scale input voltage of the converter is multiplied by the reciprocal of the number of counts [Eq. (19.2)].

$$\text{LSB} = \frac{V_{\text{span}}}{2^N} = \frac{V_{\text{RH}} - V_{\text{RL}}}{2^N} \tag{19.2}$$

For a given D/A converter, the definition of an LSB depends on the type of output the device has. Most converters have voltage output, however, current output is also common. Because of this, we must broaden our definition of the LSB slightly to allow for units other than volts: in a more general sense, an LSB is the magnitude of the change of a single converter count expressed *in the native units of the device under consideration.* In the simple case where a D/A converter output is a voltage, the definition is the same as for the A/D converter: an LSB is the magnitude (in volts) of each count, and it is determined by the range of allowable output voltages for the converter (V_{span}) and the number of bits of resolution, N [Eq. (19.2)]. For D/A converters whose native output is in the form of current, an LSB also has units of current. The equation for determining the magnitude (in amps) of an LSB for a device with current output is given by Eq. (19.3).

$$\text{LSB} = \frac{I_{\text{span}}}{2^N} \tag{19.3}$$

Example 19.1

What is the value (in volts) of 1 LSB on a 10-bit converter that is operated with reference voltages of 0 and 5 V?

$$\text{LSB} = \frac{V_{\text{RH}} - V_{\text{RL}}}{2^N} = \frac{5\text{ V} - 0\text{ V}}{2^{10}} = \frac{5\text{ V}}{1{,}024} = 4.88\text{ mV}$$

Example 19.2

If a D/A converter's output voltage should span the range from 0 to 5 V and exhibit a minimum voltage step of 100 μV, how many bits of resolution are required of the D/A converter?

$$\text{LSB} = \frac{V_{\text{RH}} - V_{\text{RL}}}{2^N} = \frac{5\text{ V}}{2^N} = 100\ \mu V$$

$$2^N = \frac{5\text{ V}}{100\ \mu V} = 50{,}000$$

$$N = \log_2(2^N) = \log_2(50{,}000) = 15.6$$

Therefore, we need a converter with at least **16 bits of resolution.**

Accuracy: For an A/D converter, **accuracy** describes how closely the actual input voltage matches the reported results of a conversion (i.e., the output code). Conversely, for a D/A converter, accuracy describes how closely the actual output voltage matches the commanded voltage (based on the input code). Accuracy is most often stated in terms of LSBs because this eliminates the need to consider how a particular converter is configured in a given circuit. For example, the 10-bit MAX149B A/D converter (Maxim Integrated Products) has a specified accuracy of ±1/2 LSB.

Monotonicity: This describes a characteristic wherein continually increasing values at the input result in continually increasing values at the output across the full range of operation, and likewise, continually decreasing values at the input result in continually decreasing values at the output. For this to be the case, the slope of the line defined by a device's transfer function does not change sign at any point. If the slope is zero anywhere, this opens up the possibility of missed codes (see below). Monotonicity is a highly desirable trait for converters to posses, since it prevents some potentially confusing scenarios, such as when an A/D input voltage increases but the output value decreases, or the output of a D/A decreases when in fact it was commanded to increase. A lack of monotonicity can be especially bedeviling in digital control systems, where systems can become unstable when the control effort is applied in the wrong direction. Monotonicity is a common characteristic in modern converters, and device data sheets usually loudly proclaim it.

Missed codes: For an A/D converter, this is a code (i.e., conversion value or specific number of counts) that is never returned at the output for any input voltage. This results when the transfer function has a local slope of zero in the vicinity of the missed code.

Maximum sampling rate: The maximum rate at which samples may be obtained by a given A/D converter, usually stated in units of Samples/second (S/s) or frequency (Hz). Commercially available A/D converters have maximum sampling rates ranging from a few hertz to hundreds of megahertz. More exotic examples can be obtained (at high cost) with sampling rates above 1 GS/s.

Settling time: The amount of time required for the output of a D/A converter to stabilize after commanding a change in the output. D/A converters are available with settling times as low as a few nanoseconds (10^{-9} s) or as high as hundreds of microseconds (10^{-4} s).

Temperature coefficient: The effect of changes in temperature on the results of a conversion (A/D or D/A). Ideally, the result of a conversion would be independent of temperature, however, in real converters, temperature affects performance in many ways that can affect the results. The variation is typically expressed as a maximum error in the output (in LSBs) over the range of operating temperatures. For example, a data sheet might state that the maximum temperature-induced error is 1/2 LSB over a temperature range of 0 to 70°C.

19.3.1 Ideal A/D Converter Performance

A/D converters sample an analog input voltage and effectively perform a rounding operation (they can only return a single numerical answer and aren't capable of interpolation), and the result returned (the "code") has a value nearest to the sampled input voltage. The code reported by an A/D converter is the numerator of a fraction of the range of input voltages for the converter. In general, the range of input voltages is set by the difference between the higher and lower voltage references, $V_{RH} - V_{RL}$, which may be externally supplied to the converter or, for some devices, generated internally. The lower reference V_{RL} often (but not always) has a value of 0 V. The denominator of the fraction is the maximum number of possible code values. When the resolution of the converter is stated in bits N,

$$V_{in} = \frac{\text{code}}{\text{max num codes}} \times V_{span} = \frac{\text{code}}{2^N} \times (V_{RH} - V_{RL}) \tag{19.4}$$

which can be rearranged to yield an expression for the output code as a function of the input voltage, Eq. (19.5):

$$\text{code} = \frac{V_{\text{in}}}{V_{\text{span}}} \times \text{max num codes} = \frac{V_{\text{in}}}{(V_{\text{RH}} - V_{\text{RL}})} \times 2^N \tag{19.5}$$

This relationship can be represented graphically as a transfer function, which is shown in Figure 19.7. Recall that a **transfer function** is the functional relationship between a device's input signal (in this case, the analog input voltage) and its output (the digital representation of the analog input).

Figure 19.7 shows the transfer function of an ideal A/D converter, with the analog input voltage on the x-axis and the digital output on the y-axis. In the representative ideal transfer function shown, the input voltage ranges from V_{RL} at 0 to V_{span} at 1, and the converter has a resolution of 3 bits (or 1 part in 8). In this simple example, a low resolution was chosen to improve the readability of the figure; in realistic applications, a 3-bit converter is unlikely to have adequate resolution to be very useful. Because the resolution is 3 bits, the result can take on one of 8 codes (2^N, where $N = 3$). In decimal, these are simply 0 through 7, which in binary is 000 through 111. These are the numbers along the y-axis. The task of the A/D converter is to identify the output code that most closely corresponds with the analog input voltage (the code that results in the smallest error), and return that value. If our example 3-bit converter has a V_{span} of 5 V, an LSB has a value of 5 V/(2^3) = 0.625 V. The output code 000 corresponds to input voltages between 0 and 0.3125 V—from the minimum acceptable voltage (0 V in this case) to 1/2 LSB. The binary code 001 corresponds to input voltages between 0.3125 and 0.9375 V (an interval of 1 LSB), and 010 (binary) corresponds to inputs between 0.9375 and 1.5625 V. It gets interesting at the top of the scale: it's typical in A/D converter design for the highest code to correspond to inputs within 1.5 LSBs of V_{span}. In our example, the binary code 111 will be returned for an input voltage between 4.0625 and 5 V (an interval of 1.5 LSBs). To prevent the larger error that would occur if the analog input were allowed to go all the way up to V_{span}, the maximum useable analog input voltage range is generally considered to be up to V_{span} – 1 LSB.[1] For an ideal A/D converter, the line drawn through the center of each LSB step is perfectly straight.

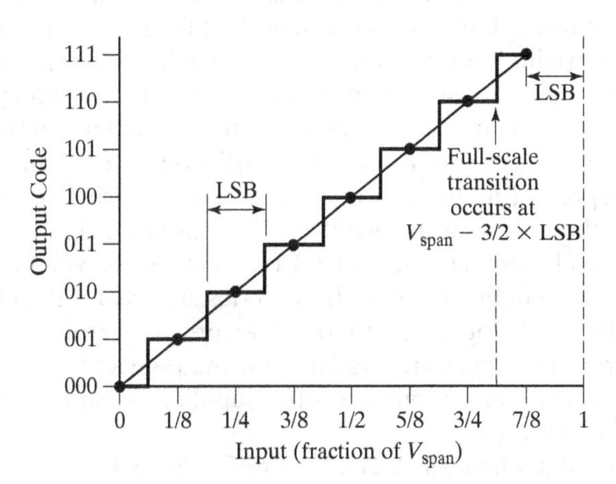

FIGURE 19.7

The ideal A/D converter transfer function.

[1] "Data Converter Codes—Can You Decode Them?," Kester, W., MT-009 Tutorial, Rev. A, Analog Devices, Inc., October, 2008.

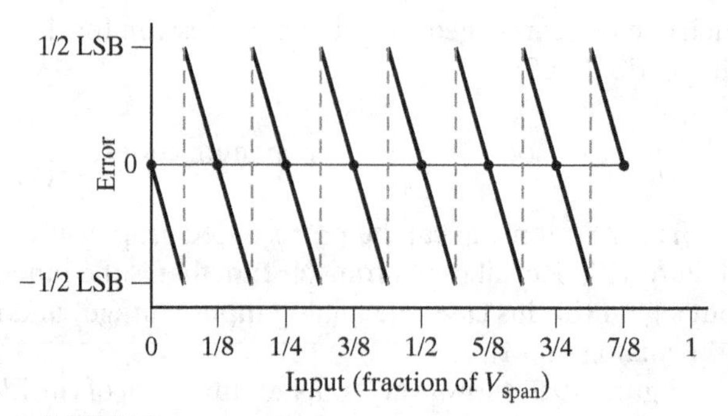

FIGURE 19.8

A/D converter quantization error.

19.3.2 Sources of Error for A/D Converters

For A/D converters, there are three major categories of error, and in this discussion, we will focus on two of the three categories. The first is the error that inevitably results from quantizing the data, called the **quantization error**. Recall that A/D converters round the results of conversions to within windows of 1 LSB, so the actual input voltage could be anywhere within ±1/2 LSB of a reported digital output code. This is shown graphically in Figure 19.8 for an 8-bit converter, and corresponds to the ideal transfer function above (Figure 19.7). Quantization error is inherent in the analog-to-digital conversion process and cannot be reduced or eliminated.

The second major category of error applies to conversions that occur at relatively low sample rates, and are governed by steady state, or DC, effects. There are several types of errors in this category, and we will explore how they cause deviations from the ideal A/D converter transfer function shown in Figure 19.7. The first two types of DC error are probably familiar to most readers who have experience taking measurements: these are offset error and gain error. The definitions of A/D converter gain and offset error are similar to those in other contexts, and may be more familiar as a result.

Figure 19.9a illustrates the effects of **offset error** for an A/D converter, which occurs when the line drawn through the midpoint of each of the steps intersects the x-axis at a location other than the origin. If a 0 V reference is available (or alternately, any nonzero reference voltage at the origin), software compensation of an offset voltage is straightforward to correct by adding or subtracting a constant value to the conversion result.

Figure 19.9b illustrates **gain error** for an A/D converter. Gain error occurs when the slope of the line representing the transfer function differs from the ideal slope, and results when the values of the codes (i.e., the width of the steps in the figure) are not equal to 1 LSB. In cases where the codes are less than 1 LSB, the actual slope will be greater (steeper) than the ideal slope, as shown in Figure 19.9b. If the codes are greater than 1 LSB, the actual slope will be less than the ideal slope. As with offset error, gain error is often straightforward to correct when reference voltages are available for the points at (or near) the origin and the full-scale voltage. These corrections are often applied in software to the digital values reported by the A/D converter.

Gain error results when all A/D codes differ from 1 LSB by a consistent amount. Assuming that all A/D codes are uniform is an idealization as well: for real devices, no single A/D code is exactly 1 LSB wide, and each has its own individual error, not necessarily the same as any other count. The result of these differences is a transfer function that is no longer truly linear, so this type of error is called **nonlinearity error**. There are two ways in which nonlinearity error is typically described.

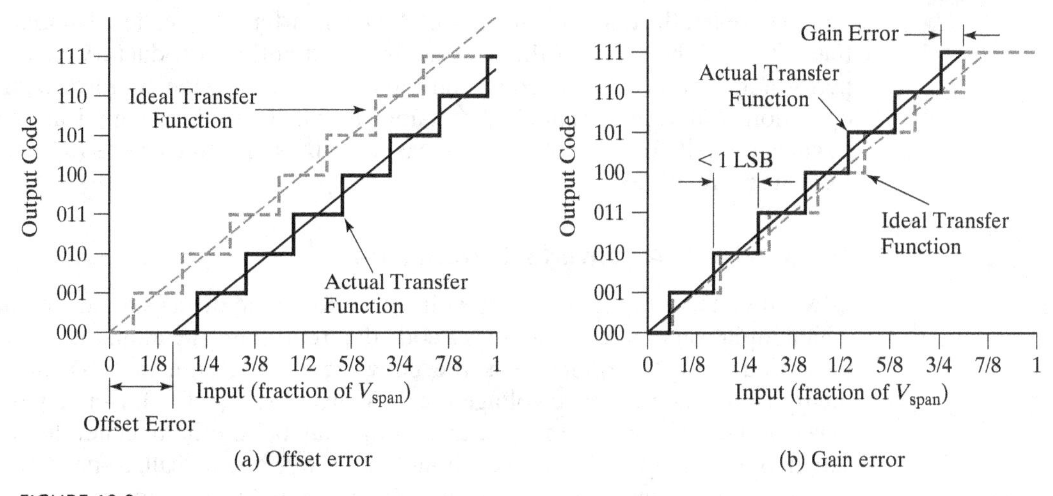

FIGURE 19.9

A/D converter offset error (a) and gain error (b).

The first is **differential nonlinearity error**, or **DNL error**, which is the difference between any single A/D code and the next. This is illustrated in Figure 19.10a. Note that in this admittedly exaggerated example, the line through the center of each code is far from straight. When differential nonlinearity error exceeds 1 LSB, it becomes possible to have a missed code—a code that won't ever be reported by the A/D. The second way of describing the results of nonuniform code widths is to consider the cumulative effects of all of the deviations in aggregate. This is called **integral nonlinearity error**, or **INL error**. Both types of error are typically specified in terms of LSBs.

In many cases, designers are interested in the combined results of all the various error sources, rather than the effect of any single contributor. Those with a "glass half full" outlook call this specification **absolute accuracy**, while those with "glass half empty" outlook call it **absolute error** or **total error**. These are equivalent terms. Whatever you decide to call it, it is the sum of the offset error, gain error, and integral nonlinearity error (differential nonlinearity error is already included in the integral nonlinearity error term).

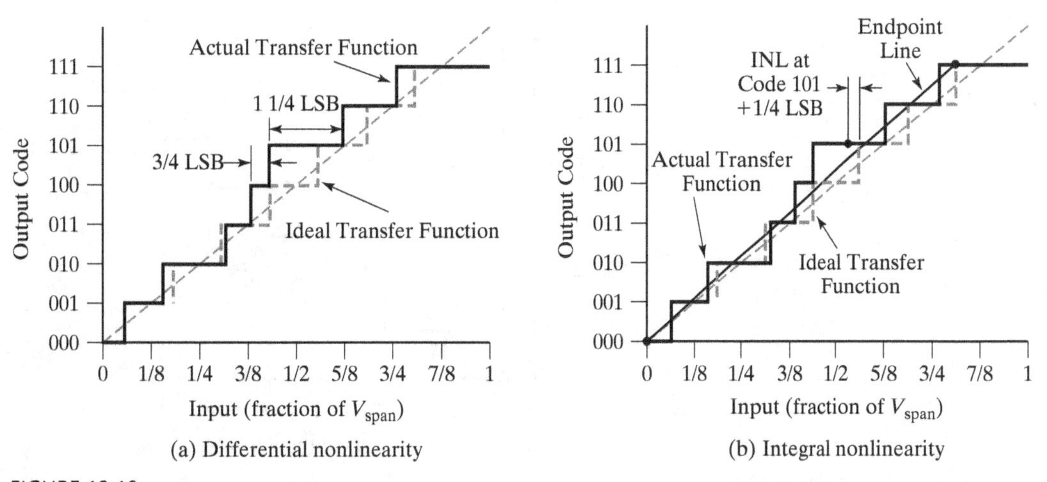

FIGURE 19.10

A/D converter differential nonlinearity (a) and integral nonlinearity (b).

There is a third major category of A/D converter error, called **dynamic error**, however, this is beyond the scope of the material included in this introductory chapter. The dynamic performance of an ADC becomes an important factor when characterizing or performing operations with rapidly changing (dynamic) signals. Consider an application where an audio signal is filtered, for example. In such cases, analysis of a converter's frequency domain performance is required.

19.3.3 Ideal D/A Converter Performance

D/A converters output an analog voltage, or for some devices a current, according to the digital input command. The input is a code that represents the numerator of a fraction whose denominator is 2^N (where N is the converter's resolution in bits) and this fraction is multiplied by the reference voltage for the converter, Eq. (19.6). Here again, the higher and lower bounds of the reference voltage (V_{ref}) may be internally generated by the device, or supplied externally to the device. Though the lower bound, called the lower reference voltage (V_{RL}), often has a value of ground (0 V), it may have a nonzero value, in which case it must be accounted for in the calculations. The upper bound is called the higher or upper reference voltage (V_{RH}). In general terms, $V_{ref} = V_{RH} - V_{RL}$.

$$V_{out} = V_{RL} + \frac{\text{code}}{2^N} \times (V_{RH} - V_{RL}) \tag{19.6}$$

Though V_{RL} is commonly 0 V, the possibility of using a nonzero value is available to increase the flexibility of the devices. When $V_{RL} = 0$ V, Eq. (19.6) reduces to Eq. (19.7), which applies in most applications.

$$V_{out} = \frac{\text{code}}{2^N} \times V_{ref} \tag{19.7}$$

D/A converters with current outputs are less common than those with voltage outputs. The main advantage of this type of device is that the designer has the ability to select an output stage op-amp that is best suited for a specific application. The equation governing the current at the output is usually the same as Eq. (19.6), with I_{out} replacing V_{out}. These devices

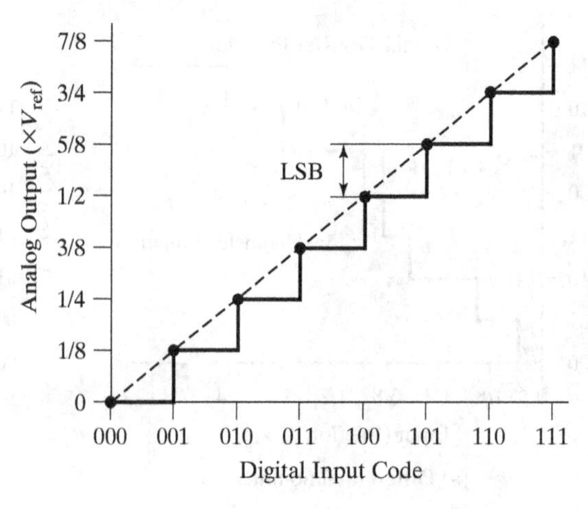

FIGURE 19.11

The ideal D/A converter transfer function.

usually provide an integral feedback resistor (which sets the gain for the output stage) that is matched in value and temperature stability with the other elements of the device.

Eq. (19.7) can be also represented graphically for the full range of inputs and outputs as a transfer function, which shows the relationship between the input code specified on the x-axis and the resulting analog output on the y-axis. A transfer function for an ideal 3-bit D/A converter with voltage output is shown in Figure 19.11.

19.3.4 Sources of Error for D/A Converters

Once you understand how real A/D converter behavior deviates from ideal, you are well positioned to understand the behavior of D/A converters, because the same types of DC errors that affect the performance of A/D converters also apply to D/A converters.

Figure 19.12a illustrates the effects of **offset error** for a D/A converter. This occurs when the transfer function intersects the y-axis at a location other than the origin. When a code of 0 is issued at the input of the converter, a nonzero output voltage (or current) results. Figure 19.12b illustrates **gain error** for a D/A converter. Gain error occurs when the slope of the line representing the transfer function differs from the ideal slope, and is a result of converter codes that do not have a value equal to 1 LSB. In cases where the counts are greater than 1 LSB, the actual slope will be greater (steeper) than the ideal slope, and when the counts are less than 1 LSB, the resulting slope will be less than the ideal slope (which is the case illustrated in Figure 19.12b). When a means of accurately measuring the output of the D/A converter is available, software compensation can sometimes be used to correct D/A offset and gain errors. However, in some cases, it isn't possible to force the transfer function to cross the y-axis at the origin. Consider, for example, the case illustrated in Figure 19.12a, where a code of 0 results in a voltage greater than 0. Here, there is no means of reducing the input code to a value below 0 that could bring the output down to a value of 0 V. To address issues like this, some form of hardware compensation would be necessary.

Pure gain error results when all of the D/A codes uniformly differ from 1 LSB by the same amount, but as we saw in the discussion above about A/D converters, this is an idealization as well. In general, each code has its own individual error which is not necessarily the

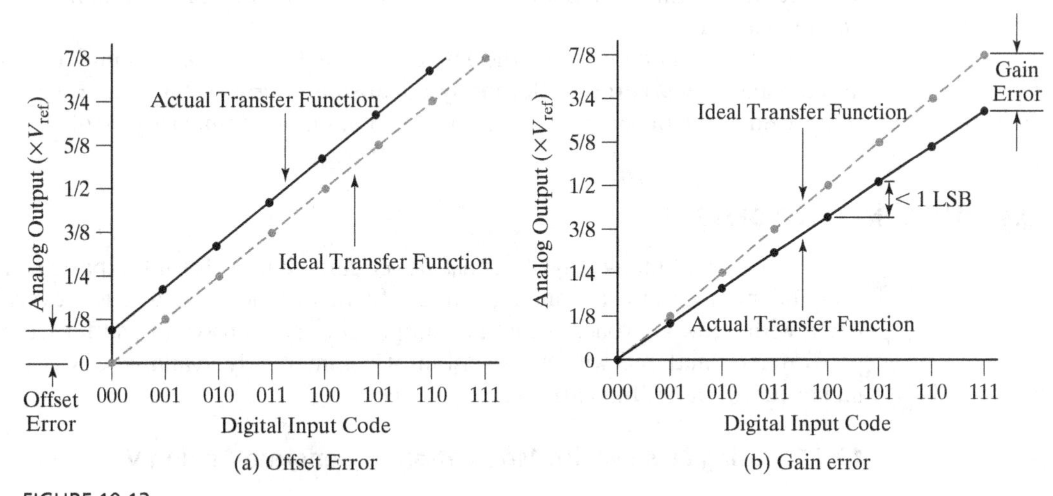

FIGURE 19.12

D/A converter offset error (a) and gain error (b).

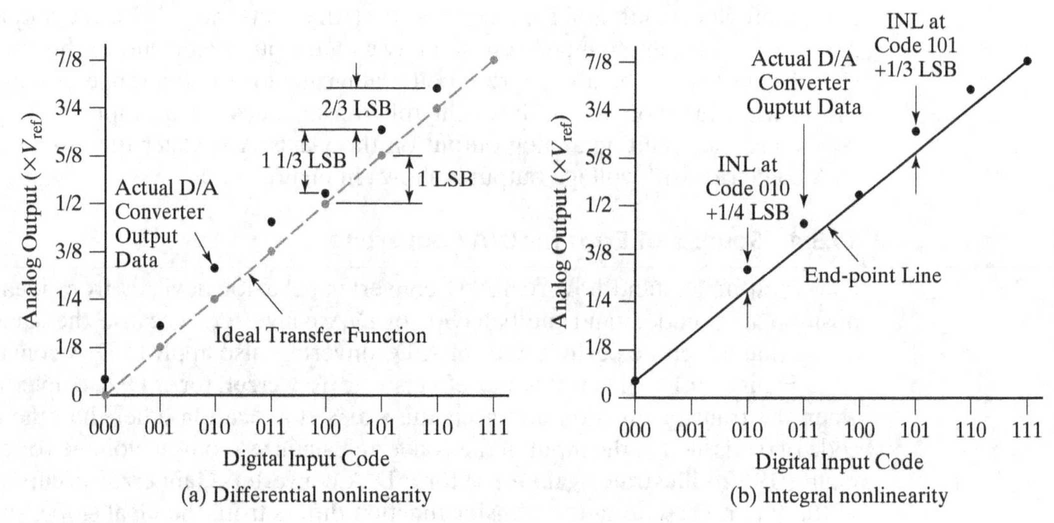

FIGURE 19.13

D/A converter differential nonlinearity (a) and integral nonlinearity (b).

same as the error for any other code. Because of this, the transfer function is not linear, and the resulting error is called **nonlinearity error.**

Differential nonlinearity error, or **DNL error**, is the difference between the value of any one code and the next code. This is illustrated in Figure 19.13a. When D/A converter differential nonlinearity error exceeds 1 LSB, the result is a nonmonotonic output. If this happens, an increase in the input code may result in a decrease in the output voltage, or conversely, a decrease in the input code may result in an increase in the output voltage. Monotonic behavior is highly desirable, but it is not required for some lower-criticality applications (such as toys and games). The **integral nonlinearity error**, or **INL error**, is the difference between the value of the output for any given input code and a straight line (Figure 19.13b). The line is defined either by connecting the two endpoints (in which case it is appropriately called the "endpoint line"), or by a best fit for all the data. Both differential and integral nonlinearity errors are expressed in terms of LSBs to eliminate any dependence on the value of V_{ref}.

As with A/D converters, the DC errors for D/A converters may be lumped together into the **absolute accuracy** (alternately, the **absolute error** or **total error**) specification. This is comprised of the offset error, gain error, and integral nonlinearity error.

19.4 D/A CONVERTER DESIGNS

There are a great many ways to construct circuits that perform the basic task of translating a digital number into an analog output. In this section, we discuss several of them. We begin with a few approaches that are simple and inexpensive but suffer from relatively low performance, and end with a discussion of commercially available devices that have much higher performance but carry a higher price tag.

19.4.1 Using Pulse Width Modulation to Generate Analog Voltages

Though there are relatively few microcontrollers that incorporate a true onboard D/A converter, nearly all microcontrollers incorporate timer subsystems. As we discussed in greater

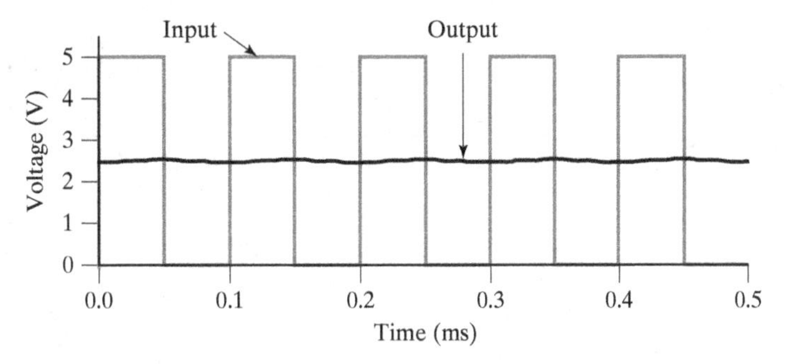

FIGURE 19.14

Using a 0-to-5 V, 10 kHz, 50% PWM signal to generate an analog voltage with an average of 2.5 V.

detail in Chapter 8, timer subsystems can be used to generate a digital pulse width modulation (PWM) signal, and many timer subsystems incorporate hardware capable of controlling a digital output so that it automatically generates a PWM signal. PWM is a technique where a signal is rapidly switched on and off and made to be active for a specified fraction of a fixed period and inactive for the remainder of that period. The fraction of the time that the signal is active is referred to as the duty cycle, which is expressed as a percentage of the total period. By varying the duty cycle, the time-averaged voltage can be controlled to values between logical 1 (100% duty cycle: when the signal is always active) and logical 0 (0% duty cycle: when the signal is never active). For example, with a 5 V logic circuit, an average voltage of 2.5 V may be created by establishing a PWM output with a 50% duty cycle—where the output is active and outputting 5 V 50% of the time, and inactive (at 0 V) the remaining 50% of the time. So far, this is nothing but a digital output rapidly switching on and off. By adding filtering to smooth the signal, a relatively stable analog voltage can be produced from a PWM signal. This is the very definition of a D/A converter, so we're definitely on the right track.

Figure 19.14 illustrates a case where a low-pass filter with a corner frequency of 100 Hz is used to smooth a 0-to-5 V, 10 kHz, 50% PWM signal in order to generate an analog voltage with an average value of 2.5 V. Because the corner frequency of the filter is a factor of 100 less than the PWM signal frequency, the 0-to-5 V swings are smoothed considerably, however, a substantial amount of ripple remains. The resulting ripple is 78 mV peak-to-peak, or about 1.6% of the amplitude of the original digital PWM signal. That may not sound too bad, but the performance of most true D/A converters is far better.

For any circuit, the amplitude of the ripple is at a maximum when the PWM duty cycle is 50%. The ripple can be reduced for all values of duty cycle by choosing a lower corner frequency for the low-pass filter. Further increasing the ratio between the PWM drive frequency and the corner frequency of the low-pass filter further reduces the ripple. This comes at a cost, however. The response time of the circuit, as indicated by the characteristic time constant τ, also increases as the corner frequency is decreased, so this method of generating an analog voltage will either inherently have large ripple or slow response. Neither of these are desirable characteristics for a D/A converter. The D/A converter designs we introduce in the next section perform considerably better in both respects but also add cost.

19.4.2 Summing Amplifier D/A Converters

A simple design based on a summing amplifier circuit is shown in Figure 19.15, and offers an excellent starting point for the discussion as we seek to achieve better D/A performance

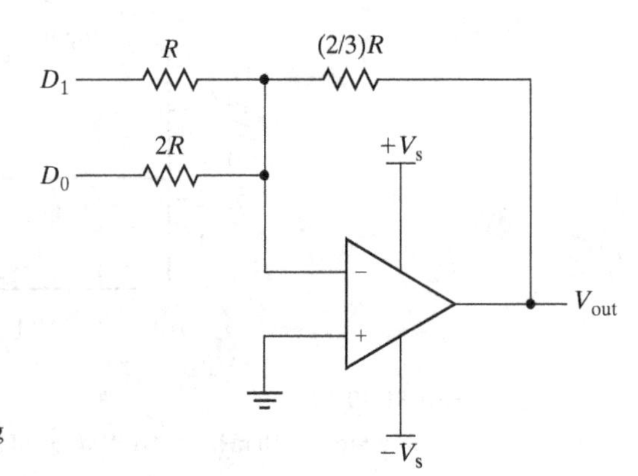

FIGURE 19.15

A simple 2-bit D/A converter (an inverting summing amplifier).

than can be obtained using PWM. This basic op-amp circuit was introduced in Chapter 11 on op-amps. It provides the basic function of a D/A converter, and delivers dynamic performance that is superior to what can be achieved with PWM. It also has some fundamental flaws that prevent its widespread use, and as a result, it is not used in commercially available devices. These flaws, however, are highly instructive, and will serve to motivate the discussion that follows this section about how commercially available D/A converters are configured.

Remember that the inputs of a D/A converter are digital (hence the name), and in Chapter 16, we described how the ideal 5 V logic signal represents a 0 (logical low) with 0 V, and a 1 (logical high) with 5 V. In Figure 19.15, the inputs are labeled D_0 (the LSB) and D_1 (the most significant bit, or MSB), and if we limit ourselves to introducing digital signals of either 0 or 5 V at either input, we can call these digital inputs. The resulting voltage at V_{out} ranges between 0 and −5 V based on the states at the inputs, and even though this is quite different from commercially available devices, it clearly meets the definition of a digital-to-analog converter.

To understand how it works and map out its transfer function, we begin by assuming that we're working with an ideal op-amp and that we can apply the Golden Rules of Op-Amps from Chapter 11. When the inputs are set to 00 (i.e., both inputs are connected to 0 V), the output will also be 0 V. We increment to the next code (01 in binary) by raising D_0 to a 1 (5 V), which causes current to flow through its input resistor (labeled $2R$). V_{out} responds by causing current to flow through the feedback resistor (labeled $2/3R$) so that the op-amp's inverting input is held at 0 V (since the noninverting input is connected to ground). The current through $2R$ must be the same as that through $2/3R$, so V_{out} has a value of −5/3 V, or −1.67 V. When we increment the input code again, to 10 in binary ($D_1 = 5$ V, $D_0 = 0$ V), the current through the resistor at the D_1 input, labeled R, must be the same as that through the feedback resistor $2/3R$, so V_{out} takes on a value of −10/3 V, or −3.33 V. When the input code is incremented to 11 ($D_1 = D_0 = 5$ V), we reach the maximum number we can represent with two input bits, and in this case $V_{out} = -5$ V.

This approach converts the ideal digital inputs to an analog output voltage, and is simple and effective. However, there are important differences between the way this design works and what is found in commercially available devices. Firstly, this circuit doesn't use a reference voltage, as described in Section 19.3.3. Instead, the output voltage is a function of the input voltages introduced at D_0 and D_1. Since we know that real digital signals are rarely, if ever, exactly 0 or 5 V, this approach isn't particularly practical. Secondly (and related to the lack of a true reference voltage), the output voltage for this circuit doesn't adhere to

Eq. (19.6) or Eq. (19.7) for D/A converter transfer functions. Instead, its output is described by the following equation:

$$V_{\text{out}} = -5\text{ V} \times \frac{\text{code}}{2^N - 1} \tag{19.8}$$

From this, we see that the transfer function has a negative slope, and the output voltage for this device is either zero or negative. The resulting plot of its transfer function doesn't look like the ideal case shown in Figure 19.11, and instead has points at 0 and −5 V, with evenly spaced intermediate points (Figure 19.16).

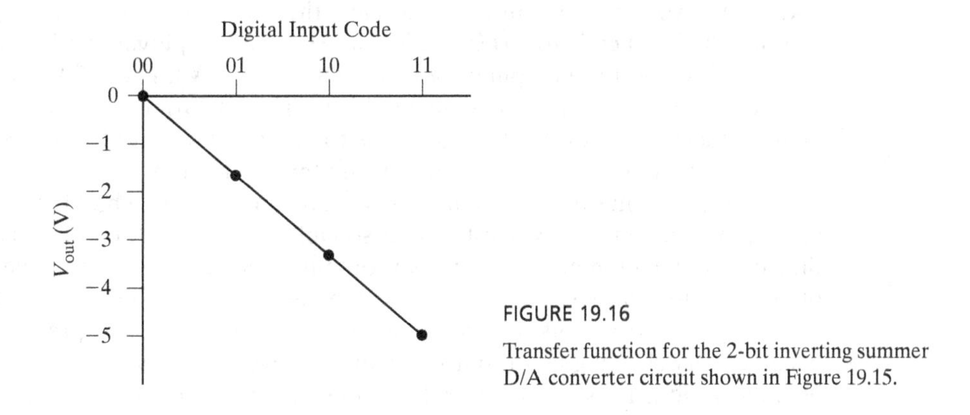

FIGURE 19.16

Transfer function for the 2-bit inverting summer D/A converter circuit shown in Figure 19.15.

There's nothing inherently wrong with this, however, these are *major* departures from the behavior of standard, commercially available D/A converters. This example serves to show how you might craft your own "home-made" circuit that provides the basic function of converting a digital input to an analog output.

If negative output voltages are undesirable but we're still willing to accept a nonstandard transfer function [i.e., one that doesn't adhere to Eq. (19.6) or Eq. (19.7)], we can modify the circuit in Figure 19.15 to that shown in Figure 19.17.

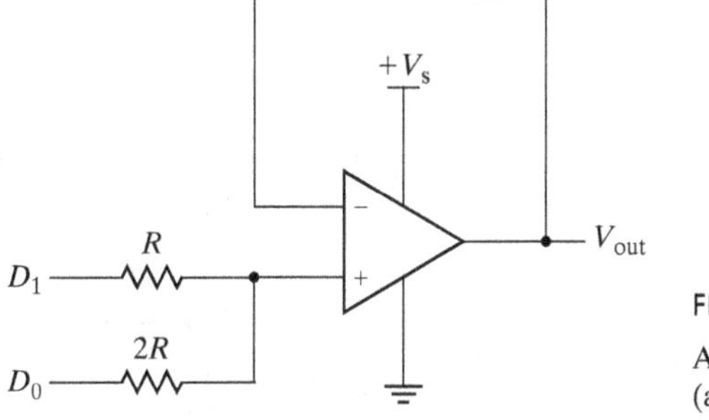

FIGURE 19.17

Another simple 2-bit D/A converter (a noninverting summing amplifier).

To analyze this circuit, we again assume ideal op-amp behavior and apply the Golden Rules. This is also a 2-bit D/A converter, and D_1 and D_0 are the digital inputs (D_1 is the MSB, and D_0 is the LSB). We will again introduce the digital codes at D_1 and D_0 to specify the desired values for V_{out} using 0 V to signify a logical 0 and 5 V for logical 1. Introducing an input code of 00 ($D_1 = D_0 = 0$ V) causes the voltage at the noninverting input (V_+) to be 0 V.

Since the output will do whatever is necessary to cause the voltage at the inverting input (V_-) to be the same as the noninverting input (0 V), V_{out} = 0 V as well. Incrementing the input code to 01 (D_1 = 0 V, D_0 = 5 V) creates a voltage divider circuit, with 5 V applied across the resistors $2R$ and R in series, and the op-amp's noninverting input connected to the point between the resistors. Recalling the equation for a voltage divider [Eq. (9.19)] and substituting in the appropriate names used in this example:

$$V+ = V_{D0}\frac{R}{2R + R} = \frac{5\text{ V}}{3} = 1.67\text{ V}$$

Since the output is connected in negative feedback (to the inverting input), V_{out} will take on the voltage that causes V− to have the same value as V+, so V_{out} = 1.67 V. Incrementing the input code to 10 (D_1 = 5 V, D_0 = 0 V) and employing similar reasoning results in V_{out} = 3.33 V, and for an input code of 11 ($D_1 = D_0$ = 5 V), V_{out} = 5 V. The resulting transfer function is −1 times Eq. (19.8)—again, this is different from the transfer function for commercially available devices. Our goal was to create a version of a summing amplifier D/A converter that doesn't invert the output, and this did the job.

The design has positive values for V_{out}, is simple, easy to build, inexpensive (requiring only an op-amp and a few resistors), reasonably fast (the settling time is determined by the time it takes the op-amp to achieve new output values), and is expandable to larger numbers of bits. Figure 19.18 shows an 8-bit implementation of a D/A converter based on this approach. It also reveals a major issue with this approach: the precision of the resistors becomes very challenging to control adequately over such a wide range of values. The circuit uses the various resistors to control the flow of current in an amount corresponding to the digital input codes, and its proper function depends on maintaining a precise factor-of-2 ratio between resistors at successive digit inputs. If each of these are 1% accurate (a commonly available tolerance), there is greater uncertainty in the amount of current flowing through the resistor (R) at the D_7 input (the MSB) than the current that flows through the resistor

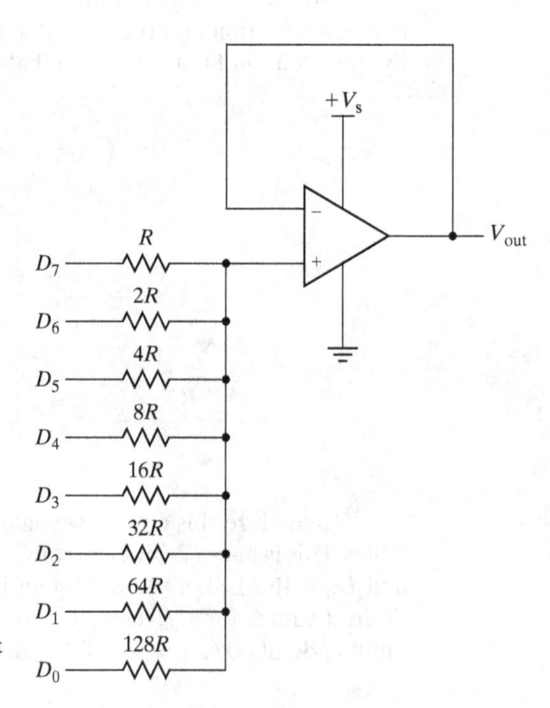

FIGURE 19.18

A noninverting summing amplifier circuit configured as an 8-bit D/A converter.

($128R$) at the D_0 input (the LSB) when it is connected to 5 V. The error from the uncertainty in the MSB is bigger than the difference between introducing a 1 or a 0 at the D_0 input. This could result in a nonmonotonic characteristic for the device, with very poor linearity—especially for the transition from 127 to 128 (decimal), which is 0111111 to 1000000 (binary).

19.4.3 String D/A Converters

Because of the challenge of tightly controlling resistor tolerances, commercially available D/A converters don't use the summing amplifier approach. Instead, they commonly come in one of two configurations: string D/A converters or *R*-2*R* ladder D/A converters. The simpler and less expensive configuration is the string D/A converter, which also has the lower performance of the two options. A **string D/A converter** uses a series of resistors, all of value *R*, connected in series (i.e., in a "string") between a reference voltage V_{ref} and ground. The heart of this architecture is essentially a voltage divider with many points at which the output voltage may be selected. The point at which an intermediate voltage is "tapped" off from this string is buffered and comprises the output of the D/A converter. A simplified string D/A converter design with 3-bit resolution is shown in Figure 19.19.

This design addresses both of the issues identified with the summing amplifier approaches we examined above. The first advantage of this approach is that it requires only a single resistor value, *R*. Its performance depends on matching the value of every resistor in the series from V_{ref} to ground as closely as possible. This is challenging, but not as difficult as creating a large number of precision resistors across a very wide range of values, as was

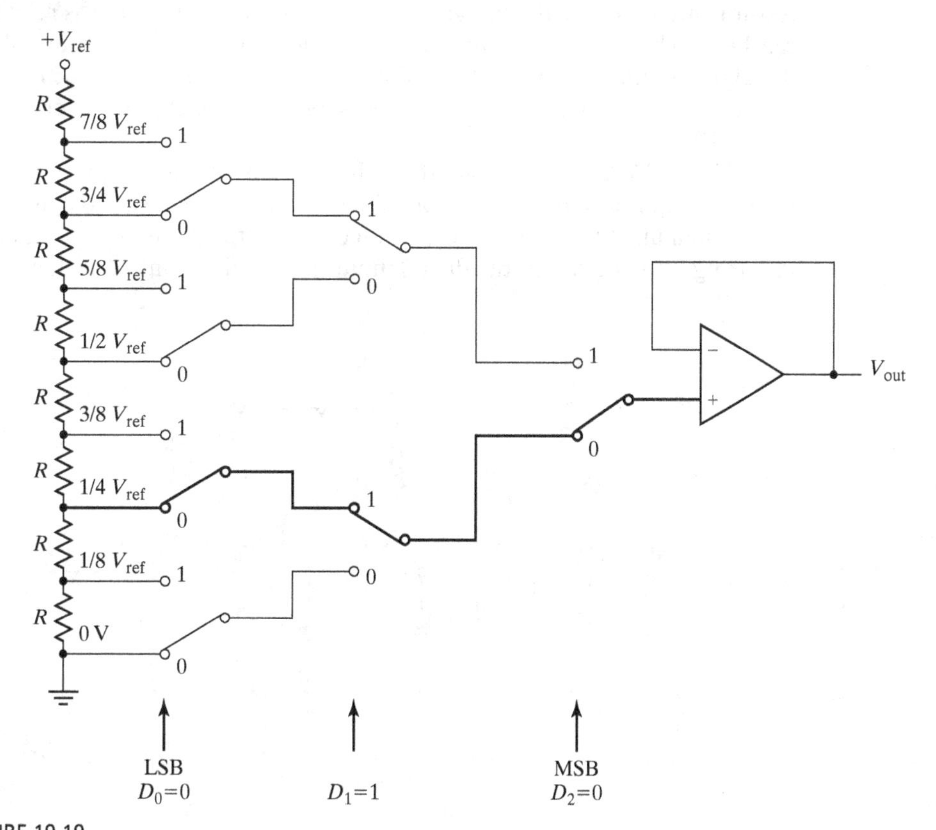

FIGURE 19.19

String D/A converter configuration with 3-bit resolution.

required for the summer-based approaches. Building D/A converters with higher resolutions involves only adding more resistors with the same value R and the switches needed to select the additional intermediate tap points.

The second important feature that makes the string D/A converter design a more viable option for real world use is that the architecture guarantees that the output is monotonic, since it is generated by picking off an intermediate point in a voltage divider comprised of a number of series resistors. Even if the value of each resistor varies considerably, moving up a step or down a step will always result in a respective increase or decrease in the output voltage.

Another important feature of this design that we haven't yet discussed is that the digital inputs (labeled $D_0 - D_2$ in Figure 19.19) control the state of switches, whose function is to select the point at which the output voltage is tapped. As long as the digital input voltages and currents meet the input requirements of the switches, the exact values won't affect the output voltage. The summer-based approaches scaled the digital input voltages themselves to create the output voltages, so proper function relied on introducing 0 V to represent logical 0 and 5 V to represent logical 1. This is not a realistic expectation for digital logic, as we discussed in Chapter 16.

19.4.4 *R*-2*R* Ladder D/A Converters

The string D/A configuration is conceptually straightforward, inexpensive to manufacture, and performs reasonably well, however, there are a few issues that limit it to applications that don't require high performance: it is difficult to achieve higher resolutions since 2^N resistors are required, the integral nonlinearity error that results is relatively large, and there can be glitches (noise) in the output voltage when a new output value is requested and the digital switches change states. For better performance, the ***R*-2*R* ladder D/A converter** architecture is preferred. A simplified 4-bit *R*-2*R* D/A converter circuit is illustrated in Figure 19.20.

The *R*-2*R* ladder configuration shares many of the good qualities of the string converter. It requires only two resistor values (R and $2R$—hence the name), which makes it possible to manufacture a higher resolution device with a monotonic output. As was the case for the string D/A converter, the digital inputs are used to control switches whose configuration

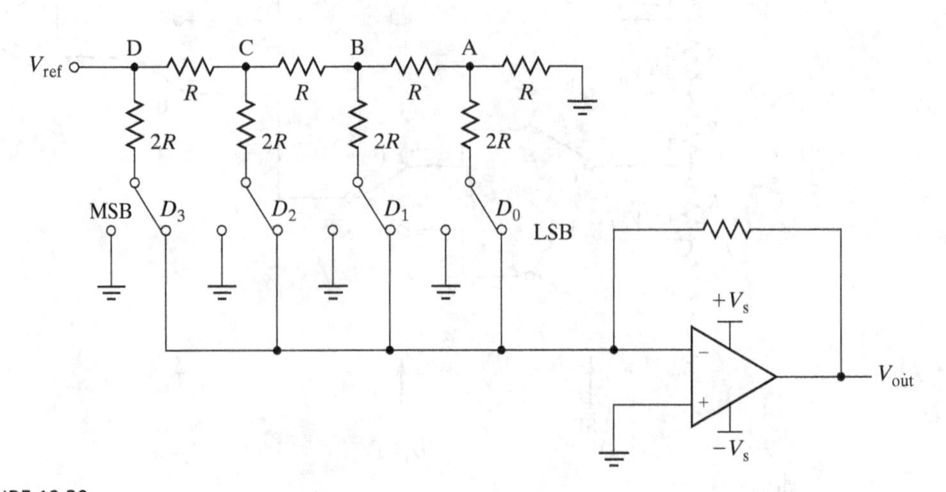

FIGURE 19.20

R-2*R* ladder D/A converter architecture (4 bits).

determines the device's output, so only the digital state (1 or 0) affects the analog output voltage and not the inputs' exact voltage and current values. An important improvement in the *R*-2*R* configuration is that each digital input bit controls a switch that connects a node in the circuit either to ground or to a point in the circuit held at 0 V by an op-amp. This point in the circuit is called a **virtual ground**, since it has the same potential as ground (0 V), but isn't physically connected to true ground in any way. The benefit of this feature is that the current flowing through all points of the circuit is constant, regardless of the state of any of the input switches. This minimizes glitches and noise.

To understand how the *R*-2*R* configuration works as a D/A converter, we first consider the voltages at the series of nodes (A–D) marked along the upper row of resistors in Figure 19.20. Starting at node A on the far right, we see that there are two resistors with value 2*R* connected in parallel to either ground or virtual ground. The resulting resistance between node A and 0 V is then *R*, and the position of the switch (D_0) doesn't affect this. Moving a step to the left and looking now at node B, we see 2*R* in parallel with *R*, which is in turn in series with the resistance from node A (recall that this was just *R*). Again, we have an equivalent resistance of *R* between node B and 0 V. Looking at node C, we see that the pattern repeats again: 2*R* in parallel with *R* in series with the resistance from node B to 0 V, which was *R*. Again, the resistance to ground is *R*. The same is true at the input node, D, and it would also be true if we added more bits of resolution.

With these results in mind, we are now prepared to consider the resulting flow of current in the circuit (Figure 19.21). The reference voltage at node D sees a combined resistance of *R* to 0 V (either actual or virtual ground), so the total current that flows is $I_D = V_{ref}/R$. This current is divided equally, with one half ($V_{ref}/2R$) flowing to the switch controlled by the MSB digital input (D_3) and the other half flowing to the rest of the network to the right. This half of the current flows first through node C, where it is divided in exactly the same way: half ($V_{ref}/4R$) flows down to the switch controlled by the digital input D_2, and the other half flows into the remaining portions of the network to the right. The process continues until there are no additional nodes. The resistor network thus produces a series of currents that

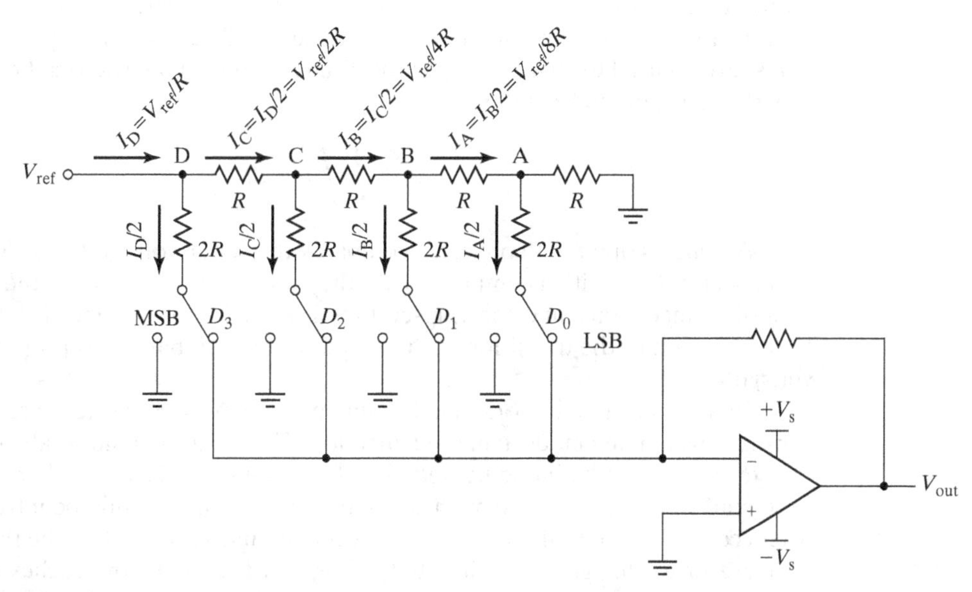

FIGURE 19.21

Current flow analysis for a 4-bit *R*-2*R* ladder D/A converter.

are divided evenly by factors of 2—exactly what is needed at the current summing node of the op-amp that serves as the output stage of the D/A converter.

Some *R*-2*R* ladder D/A converters integrate an op-amp at the output stage, in which case their outputs are in the form of an analog voltage. This is the case for the converter shown in Figure 19.21. While this approach is in many cases simpler to integrate and requires fewer components, it is less flexible than the alternative design, which omits the op-amp and gives its output in the form of a current. For the current output configuration, the designer is free to choose whichever output stage configuration best suits the requirements of a given circuit. There are many D/A converters to choose from with either current or voltage output—in fact, the biggest challenge is often selecting a single device from among the large number that would work perfectly well in a given circuit.

19.5 A/D CONVERTER DESIGNS

Just as we saw with D/A converters, there are many ways to construct A/D converters. Each approach brings its own particular strengths and weaknesses. In the following sections, we'll describe the layout of several architectures and their characteristics. The goal of this discussion isn't to enable readers to design their own A/D converters (though with additional study this would certainly be possible), but rather to help designers understand how each type of converter works and use this information to make informed choices when specifying A/D converters in their designs.

19.5.1 Single Slope and Dual Slope A/D Converters

One way to determine the value of an analog input voltage is to use the signal to influence some other physical quantity that is easy to measure. One such easily measured quantity is time: if an unknown voltage determines the time it takes for an event to occur, then measuring that time indicates the voltage at the input. This is the approach taken by single slope and dual slope A/D converters.

In a **single slope A/D converter**, a capacitor comprises the circuit element that performs the translation between voltage and time. The basic circuit layout of a single slope A/D converter is illustrated in schematic form in Figure 19.22. With this approach, a constant current source is used to charge a capacitor. The resulting voltage across the capacitor increases linearly, according to Eq. (19.9).

$$V = \frac{1}{C}\int I(t)\,dt \tag{19.9}$$

A single slope A/D converter compares the input voltage (V_{in}) to the voltage across the capacitor (V_C) with a comparator. At the moment when $V_{in} = V_C$, the output of the comparator changes state, and the elapsed time is stored. The amount of time from the start of the conversion to the transition at the output of the comparator is proportional to the input voltage.

Reasonably good "home-made" single slope A/D converters are straightforward to construct with many modern microcontrollers. The functional block labeled "Control Logic" in Figure 19.22 can be implemented in software on the microcontroller. The function of the FET could be implemented with an output line from the microcontroller. The software would control the output line to start conversions and keep track of the time required for the comparator to change state when the voltage on the capacitor reaches the value of V_{in}. A simple resistor can be used to charge the capacitor. Using a resistor instead of a true current

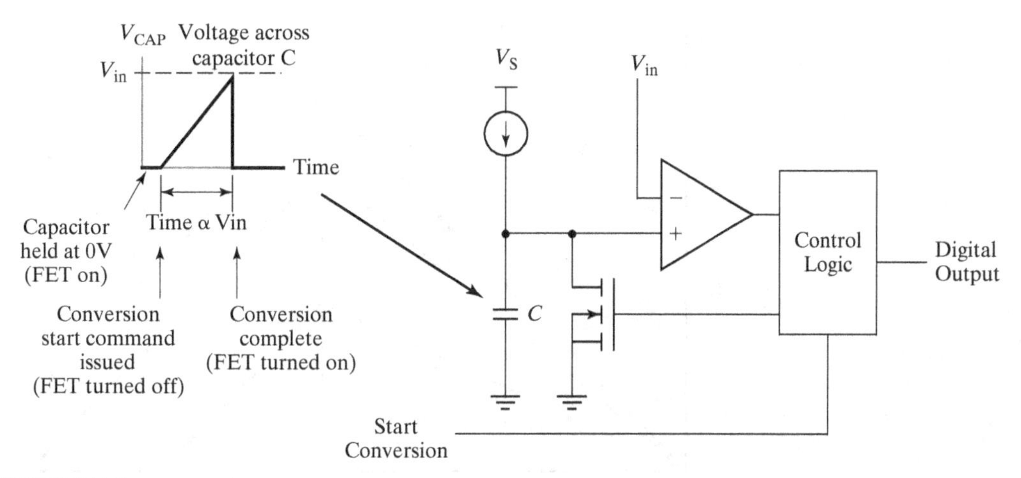

FIGURE 19.22

Single slope A/D converter block diagram.

source will produce a nonlinearity in the digitized value, but that effect can be linearized in the software on the microcontroller. In an effort to make it easy to incorporate this kind of functionality into designs, many manufacturers include an analog comparator and switchable pull-up resistors inside the microcontroller. With a microcontroller that has internal pull-ups and a comparator on-chip, the only external component necessary to make a functional A/D converter is a single external capacitor. This leads to a very low cost (though relatively low performance) way to provide for analog input to a microcontroller without the cost of adding a more complex A/D converter.

This approach works well; however, there are a few practical issues that prevent its widespread use. The biggest problem is that the conversion result depends directly on the value of the capacitor, *C*, which may not be known precisely, and may not be stable when the temperature varies. Another problem is that the input voltage may change during a conversion, which affects the result since it is compared with the voltage across the capacitor during the conversion.

The capacitance issue is addressed by using a **dual slope A/D converter**, which takes a similar approach with some key improvements. With a dual slope A/D converter, the capacitor is initially charged for a fixed amount of time T_0 with a constant current that is proportional to the input voltage, V_{in}, so the voltage achieved across the capacitor during this phase depends on V_{in}. After time T_0 has elapsed, the capacitor is discharged at a constant current over a period of time T_1 which is measured by the converter. The time required for the discharge process (T_1) depends only on the input voltage and not on the capacitance (Figure 19.23). Since the charge and discharge portions of the conversion are inversely proportional to *C* [Eq. (19.9)], *C* drops out of the expression.

Because dual slope A/D converters are relatively simple and don't require precision components (such as a precision capacitor with low temperature dependency), they are inexpensive to produce. Dual slope A/D converters are capable of achieving resolutions as high as 13 to 14 bits (1 part in 16,384), but several milliseconds are required to perform a conversion. This is considerably slower than most of the other architectures we discuss in the following paragraphs. Not all applications require high speed, however. For example, dual slope A/D converters are commonly used in digital multimeters, where their characteristics are a good match with the requirements. Above, we described a technique for constructing a single slope A/D converter using a microcontroller and an external capacitor. With slightly different

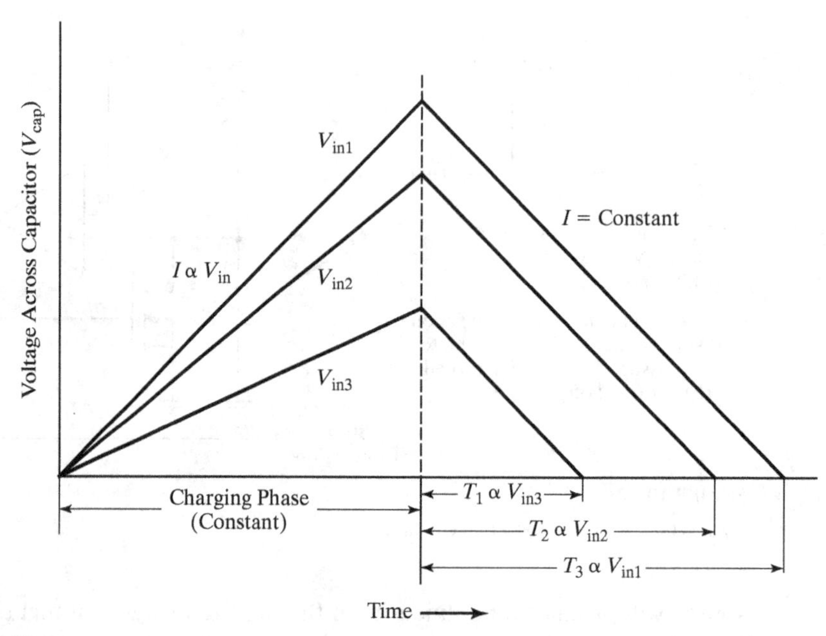

FIGURE 19.23
The voltage across the capacitor in a dual slope A/D converter.

software and hardware, a dual slope A/D converter can also be implemented using a microcontroller that switches a capacitor between a charging phase and a discharging phase. More and more small microcontrollers are incorporating internal pull-up resistors as well as analog comparators, which significantly reduces the part count for the resulting circuit.

19.5.2 Flash A/D Converters

The fundamental building block of flash A/D converters is the comparator. In Chapter 11, we introduced comparators, which are similar to op-amps but optimized to have an output that very rapidly takes on a value near 0 V (which can be used to represent a logical 0) when the voltage at the inverting input is greater than the noninverting input ($V_- > V_+$), and a value near its positive power supply voltage or some other voltage representing a logical 1 when $V_+ > V_-$. A comparator can be thought of as a 1-bit A/D converter: any input voltage

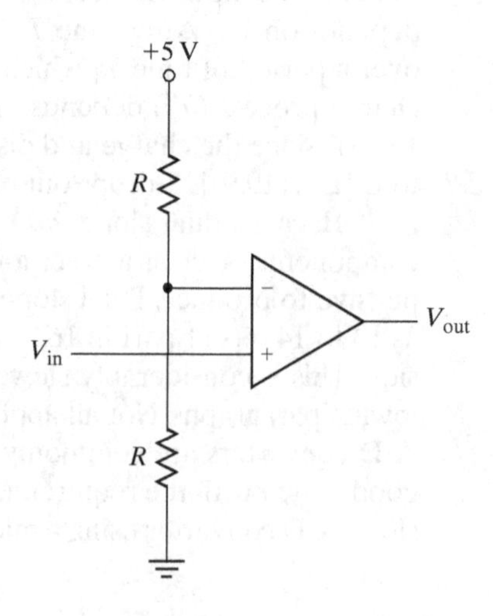

FIGURE 19.24
A 1-bit A/D converter (a comparator).

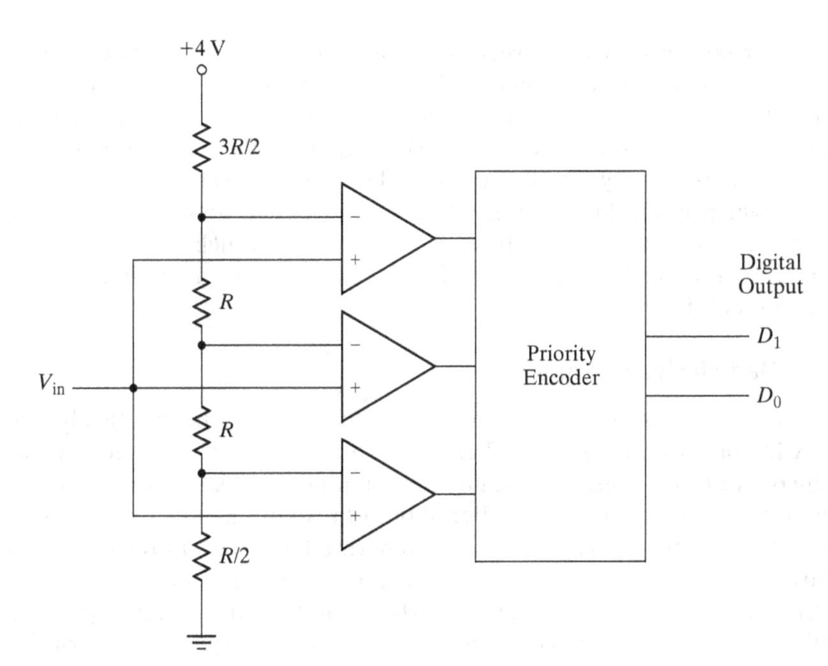

FIGURE 19.25

A 2-bit, comparator-based A/D converter.

above a given threshold results in an output of logical 1, and any input below the threshold results in an output of logical 0.

Figure 19.24 illustrates the case of a comparator serving as a 1-bit A/D converter with a threshold of 2.5 V. When $V_{in} > 2.5$ V, V_{out} will have a logical value of 1 ($V_{out} \approx 5$ V if the comparator output is open collector or open drain and pulled up to 5 V). Conversely, when $V_{in} < 2.5$ V, V_{out} will have a logical value of 0 ($V_{out} \approx 0$ V). In most applications where an A/D converter is required, resolutions of higher than 1 bit are usually desirable. To achieve higher resolution, several comparators can be used in conjunction, each with its own threshold voltage. A 2-bit, comparator-based A/D converter is illustrated in Figure 19.25.

In this configuration, a string of resistors (each with the same resistance, R, except for the two endpoints) creates the threshold voltages for each of the comparators by dividing up the full scale of the converter (in this example, between ground and $V_{ref} = 4$ V). The comparator on the bottom of the string in Figure 19.25 has a threshold voltage equal to

$$V_{ref} \times \frac{R/2}{4R} = 0.5 \text{ V}$$

Any input voltage below 0.5 V will cause a 0 to appear at this comparator's output, while input voltages above 0.5 V result in a 1. Similarly, the threshold voltage for the comparator in the center is 4 V × (3/8) = 1.5 V, and for the top comparator, the threshold is 4 V × (5/8) = 2.5 V. In general, for a given V_{in}, the comparators with threshold voltages less than V_{in} will have outputs with logical 1s, and those whose threshold voltages greater than V_{in} will have logical 0 outputs. The comparators' outputs are fed into a logic circuit called a priority encoder, which translates the results into a binary number representing the result.

This architecture is conceptually straightforward, and requires only resistors, comparators, and a bit of digital logic (the priority encoder). Most importantly, conversions are performed very quickly, and depend primarily on the time required for the comparators to reflect the difference between the input and threshold voltages, and the time required for the priority encoder to translate the output of the comparators into a binary number. Because of the high speed with which conversions can be performed, these are called **flash**

A/D converters. Flash A/D converters are indeed quick, capable of conversion rates above 1 gigasamples per second (GS/s). However, it is impractical to construct converters with high resolution using this approach because $2^N - 1$ comparators are required to implement a converter with a resolution of N bits. For higher resolutions, the number of comparators and resistors grows very rapidly. An 8-bit flash A/D converter requires $2^8 - 1 = 255$ comparators, which is the highest resolution that is easily obtainable in commercially available devices. Imagine a 16-bit flash A/D converter, which—if it were available—would require $2^{16} - 1 = 65{,}535$ comparators. This is beyond what can be implemented at a reasonable or competitive price.

19.5.3 Half-Flash A/D Converters

One architecture that achieves higher resolution than can be practically implemented in a flash A/D converter while only slightly sacrificing conversion speed is the **half-flash** (or **pipelined) A/D converter.** In these devices, separate flash A/D converters with relatively low resolution operate on half the number of bits of resolution of a full conversion at a time, and the result is combined back together to generate the final, full resolution output. A representative 8-bit half-flash A/D converter design is shown in Figure 19.26.

In this 8-bit half-flash A/D converter example, the analog input voltage is routed directly to a 4-bit flash A/D converter. The conversion is quickly performed (a hallmark of the flash converter architecture), and the result is used in two separate ways. First, it forms the 4 MSBs of the final output of the conversion. Second, the result is fed into a 4-bit D/A converter, whose output is subtracted from the original analog input, $V_{in} - V_{DAC}$. The result of this subtraction is the analog voltage that was rounded off when the initial 4-bit A/D conversion was performed. A second 4-bit flash A/D converter performs a quick conversion on this remainder to generate the lower 4 bits of resolution of the final conversion output. The 4-bit result from the first converter is combined with the 4-bit result from the second converter (the LSBs) to create the full 8-bit result. A full 8-bit flash converter requires $2^8 - 1 =$ 255 comparators and performs the conversion in one quick operation, while this approach requires only $2 \times (2^4 - 1) = 30$ comparators and completes the task in a little more than twice the time. The complexity of the device is dramatically reduced, and the time required to perform a conversion increases by a factor of approximately 2.

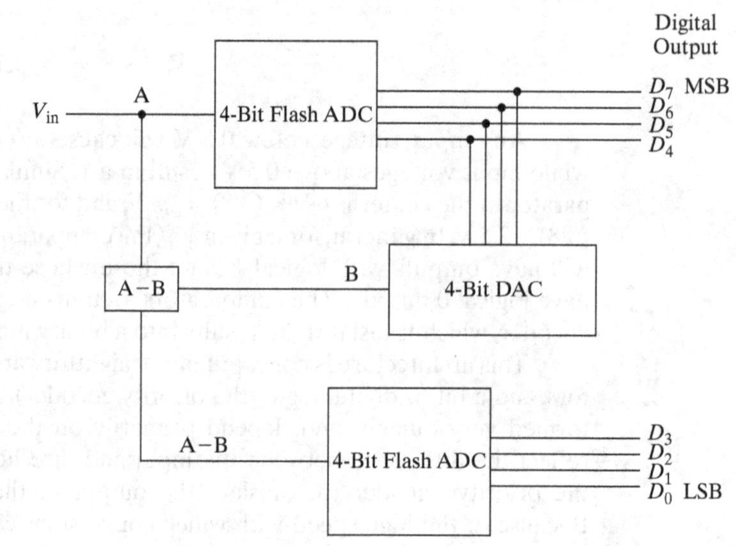

FIGURE 19.26

An 8-bit half-flash A/D converter.

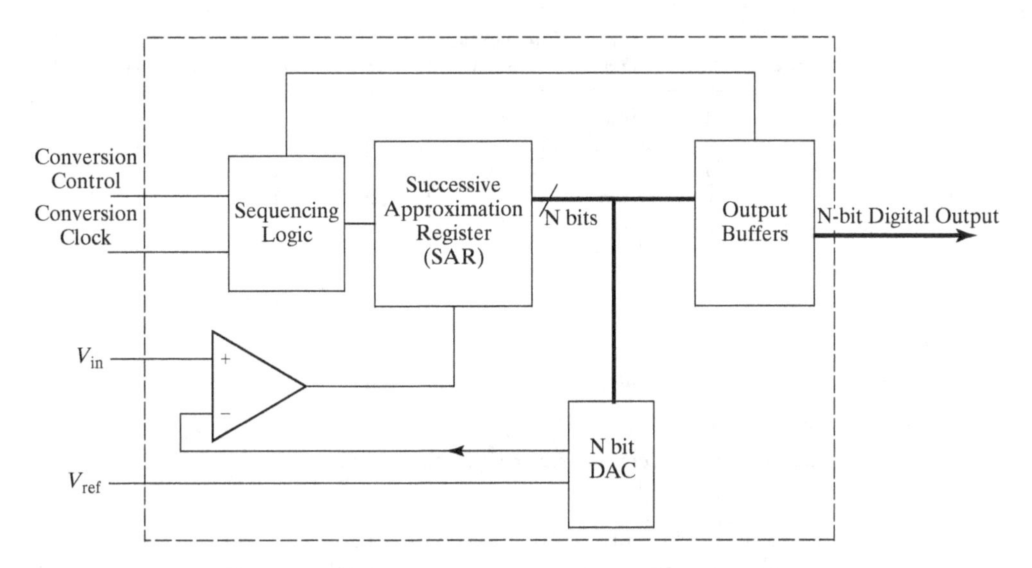

FIGURE 19.27

SAR A/D converter architecture.

19.5.4 Successive Approximation Register A/D Converters

Successive approximation register A/D converters, also called **SAR A/D converters**, employ a binary search algorithm to determine the value of an analog input and convert it to a digital representation at the output. SAR A/D converters start by guessing, or "approximating," what the input voltage is and then comparing this guess to the actual input voltage. Successive approximations are made based on the outcome of the previous comparisons, and the converter quickly narrows in on the correct answer. The guesses are selected based on a binary search algorithm: each guess is halfway between the current lower and upper bounds.

Figure 19.27 shows the basic architecture of a SAR A/D converter. The analog input voltage (V_{in}) enters the converter on the left side of the figure, where it is compared with a guess, which is the output of a D/A converter. The result of the comparison is fed to a block of logic called the successive approximation register, or SAR, which is responsible for performing the binary search and narrowing down the guesses. The result of the guesses and comparisons required for a conversion is the output of the converter, shown on the right side of the figure.

To get a feel for how the process works, we will step through the example illustrated in Figure 19.28. In this test case, the analog input voltage (V_{in}) is 1.37 V, and this will be converted to a digital result using an 8-bit SAR A/D converter with a 5 V reference. The process begins with the SAR guessing that the input voltage is 2.5 V, halfway between the minimum and maximum possible voltages (0 and 5 V). This guess is compared with the actual input voltage at the comparator, and the result is a logical 0, indicating that V_{in} is less than the guessed value. The comparator output is stored in the MSB of the result register. Next, the converter makes a second guess, which is again halfway between the minimum and maximum possible values based on what we know so far. The second guess is thus 1.25 V (the midpoint between 0 and 2.5 V). V_{in} (1.37 V) is greater than this second guess (1.25 V), so the output of the comparator is a logical 1, and this is stored in the next most significant bit of the result register (bit 6). The process repeats a third time, as the SAR guesses 1.875 V (halfway between 1.25 and 2.5 V). This time the result is greater than V_{in}, so the comparator's output is 0, which is stored in the result register at bit 5. The process repeats until all 8 bits of

Test	Clock	Bit Status	Comparator Output	MSB D7	D6	D5	D4	D3	D2	D1	LSB D0
1	1	Test	1	**1**							
	0	Reject	0	**0**	?	?	?	?	?	?	?
2	1	Test	1		**1**						
	0	Keep	1	0	**1**	?	?	?	?	?	?
3	1	Test	1			**1**					
	0	Reject	0	0	1	**0**	?	?	?	?	?
.		.	.	.	.	.	.	.	.	.	.
.		.	.	.	.	.	.	.	.	.	.
.		.	.	.	.	.	.	.	.	.	.
8	1	Test	1								**1**
	0	Reject	0	0	1	0	0	0	1	1	**0**

FIGURE 19.28

An 8-bit SAR A/D converter search algorithm example.

the result register are filled, and the final results are reported at the output of the device. For the sake of completeness, the final result is 01000110 in binary (70 in decimal).

For SAR A/D converters, the number of guesses required to perform a conversion is equal to the number of bits of resolution. The internal D/A converter also has the same resolution as the desired A/D output. All of the internal elements required to construct a SAR A/D converter are inexpensive to manufacture, and it is practical to make relatively high resolutions (up to about 18 bits) and relatively high conversion rates (up to about 10 MS/s) using this approach. However, since performing higher resolution conversions take more steps (one step for each bit of resolution), the higher the resolution, the slower the conversion, as a general rule. SAR converters currently occupy the broad middle ground for A/D converters with respect to both cost and performance, and as such are by far the most popular and common choice. Nearly all microcontrollers that include on-chip A/D converters use this architecture.

19.5.5 Sigma-Delta A/D Converters

Flash and half-flash A/D converters are capable of very fast conversions at low resolution, and SAR A/D converters are capable of moderate speed and resolution, but what if you need the utmost in resolution and accuracy? This is where **sigma-delta A/D converters** (alternately called **delta-sigma A/D converters**) excel. This type of converter is currently available in resolutions up to 24 bits (1 part in 16,777,216)—significantly better than any other architecture. Some examples of sigma-delta A/D converters are exceptionally slow (generally, the highest resolution parts), posting maximum conversion rates in the tens of samples per second. They are not all this slow, however: higher speed sigma-delta converters are capable of achieving conversion rates in the tens of megahertz for 12-bit resolution. A look at how these converters are constructed and how they perform conversions helps to understand how they achieve such high resolution.

Sigma-delta A/D converters make use of advanced digital signal processing techniques to achieve their high performance. They take a very large number of coarse samples very quickly (a technique known as **oversampling**) and average the resulting data. Figure 19.29 shows the fundamental building blocks of a basic sigma-delta A/D converter, which consists of a difference amplifier (the "delta" in the name), an integrator (the "sigma"), a comparator, and a 1-bit D/A converter (i.e., a switch that selects between either the high reference voltage, V_{RH}, or the low reference voltage, V_{RL}). As with the SAR A/D converter, stepping through an example of a conversion provides an excellent framework for learning how sigma-delta A/D converters work. Table 19.1 corresponds to the block diagram

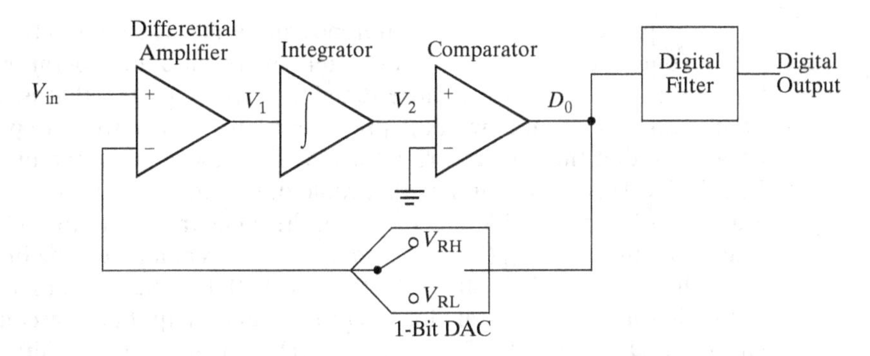

FIGURE 19.29

Sigma-delta A/D converter architecture.

in Figure 19.29. For this example, $V_{in} = 2.75$ V, $V_{RH} = 5$ V, and $V_{RL} = 0$ V. Before we begin the conversion, we assume that D_0 is a logical 0 and that this state corresponds to the output of the DAC being at 0 V.

TABLE 19.1 A sigma-delta A/D conversion example, with $V_{in} = 2.75$ V.

Clock	V_1	V_2	D_0	5 Clock Average of $D_0(V_{RH}-V_{RL})$	10 Clock Average of $D_0(V_{RH}-V_{RL})$	20 Clock Average of $D_0(V_{RH}-V_{RL})$
0	0	0	0			
1	2.75	2.75	1			
2	−2.25	0.5	1			
3	−2.25	−1.75	0			
4	2.75	1	1	3		
5	−2.25	−1.25	0	3		
6	2.75	1.5	1	3		
7	−2.25	−0.75	0	2		
8	2.75	2	1	3		
9	−2.25	−0.25	0	2	2.5	
10	2.75	2.5	1	3	3	
11	−2.25	0.25	1	3	3	
12	−2.25	−2	0	3	2.5	
13	2.75	0.75	1	3	3	
14	−2.25	−1.5	0	3	2.5	
15	2.75	1.25	1	3	3	
16	−2.25	−1	0	2	2.5	
17	2.75	1.75	1	3	3	
18	−2.25	−0.5	0	2	2.5	
19	2.75	2.25	1	3	3	2.75
20	−2.25	0	0	2	2.5	2.75
21	2.75	2.75	1	3	2.5	2.75
22	−2.25	0.5	1	3	3	2.75

The process begins at the difference amplifier, which takes the difference between V_{in} and the output of the 1-bit D/A converter. The result of the operation is $V_1 = 2.75$ V, which is then fed to the integrator (whose function is to integrate—that is, it sums the latest result from the difference amp with the previous results). Since we have just started a conversion, we assume that the previous result is 0 V. The new value at the output of the integrator is $V_2 = 2.75$ V. This is compared with a value of ground (0 V) at the comparator, whose output is a logical 1 (since 2.75 V > 0 V). When the comparator's output (D_0) is a 1, the 1-bit D/A converter output is V_{RH} (5 V, in this example). This completes the first cycle.

The second cycle is much like the first, with the exception that we've established some history in various functional blocks. The difference amplifier now compares V_{in} with the output of the D/A converter from the last cycle, which is 5 V. The difference is 2.75 V − 5 V = −2.25 V, and the integrator adds this to the previous value of 2.75 V (the result from the first cycle) to obtain a new value of 0.5 V. The comparator determines that this new result from the integrator (0.5 V) is greater than 0 V, so the comparator's output (D_0) is again a logical 1. For the sake of thoroughness, we'll step through the third cycle, however, at this point you should be able to predict the outcome. The difference amplifier output is again 2.75 V − 5 V = −2.25 V, and the new value in the integrator is −2.25 V + 0.5 V = −1.75 V. This time, the comparator's output is a refreshing change to logical 0, since −1.75 V < 0 V. The D/A converter output switches to 0 V (V_{RL}) in response, and this is in turn provided to the difference amplifier, where the cycle repeats many more times. The Digital Output (the result of the conversion) for this sigma-delta A/D converter is derived by digitally averaging the output of the comparator over a number of previous cycles. This averaging is performed by the Digital Filter block that takes the sequence of logical states (D_0) as its input. You can follow the successive steps in the rest of Table 19.1 to see how the digital output value converges on the correct answer. The table also shows how averaging over a larger window of previous results improves the answer (compare the results for the averages of 5 and 10 previous results with those for the average of 20 previous samples shown in the rightmost column of the table). In this example, the converter [using a 20 result moving average digital filter (Chapter 15)] converges on the correct result to within two decimal places in 20 cycles.

Though each sample and averaging cycle is performed very quickly, a very large number of cycles is required to achieve the higher resolutions possible with this architecture (e.g., 24 bits). It is common in commercial devices to oversample at a rate of 256 or 512 times the desired output rate so that 256 or 512 samples can be averaged to produce a single output value. The maximum output data rate is a classic performance trade-off between speed and resolution. Higher resolution requires more samples to be averaged, resulting in a slower output rate.

19.6 HOMEWORK PROBLEMS

19.1 You wish to sample a signal with frequency content between 0 and 1 kHz with an A/D converter. What is the minimum sampling rate required to ensure that aliasing will not inhibit your ability to accurately interpret the data?

19.2 You wish to use an 8-bit A/D converter with $V_{ref} = 5$ V to measure a signal (amplitude = 2 Vpp, frequency = 1 kHz) that includes substantial noise (amplitude ≤ 0.5 Vpp, frequency > 5 kHz). You will sample the signal at a rate of 10 kHz. Design an antialiasing filter that prevents the noise from affecting your readings, and alters the amplitude of the 1 kHz signal by 20% or less.

19.3 For an A/D converter with $V_{RL} = 0$ V and $V_{RH} = 5$ V, what is the value of an LSB if the resolution of the converter is 8 bits? 10 bits? 16 bits? 24 bits?

19.4 For an A/D converter with $V_{RL} = -5$ V and $V_{RH} = 5$ V, what is the value of an LSB if the resolution of the converter is 8 bits? 10 bits? 16 bits? 24 bits?

19.5 For a 12-bit A/D converter with $V_{ref} = 3.3$ V, what is the uncertainty contributed by quantization error?

19.6 For a 10-bit D/A converter with $V_{RL} = 0$ V and $V_{RH} = 5$ V, what is the output voltage if the input code is 721?

19.7 For a 16-bit D/A converter with $V_{RL} = -2.5$ V and $V_{RH} = 2.5$ V, what input code would be needed to produce an output of 0.701 V?

19.8 For an 8-bit A/D converter with $V_{ref} = 5$ V, what is the input voltage if the output code is 241 (decimal)?

19.9 For a 20-bit A/D converter with $V_{RL} = 0.25$ V and $V_{RH} = 3$ V, what is the input voltage if the output code is 725413 (decimal)?

19.10 Without adding any op-amps, how could you redesign the circuit shown in Figure 19.17 so that the output of the summing D/A converter ranges from 0 to 10 V?

19.11 Using the Internet, find examples of a flash A/D converter, a SAR A/D converter, and a sigma-delta A/D converter from the following manufacturers: Maxim Integrated Products and Texas Instruments. Create a table that compares their resolution, maximum sample rates, and price. *Hint*: It may simplify your search if you begin by locating the manufacturers' data converter selection guides.

19.12 Construct a table like the one shown in Figure 19.28 that illustrates each of the steps for an 8-bit SAR A/D converter if the input voltage is 1.095 V and $V_{ref} = 5$ V. What is the output code when the conversion is complete?

FURTHER READING

"Analog-to-Digital Converter Architectures and Choices for System Design," Black, B., *Analog Dialog* 1999; 33–38.

The Art of Electronics, Horowitz, P., and Hill, W., 2nd ed., Cambridge University Press, 1989.

"Comparing DAC Architectures," Baker, B., *EDN Magazine* January 18, 2007; 34.

"Data Converter Codes—Can You Decode Them?", Kester, W., MT-009 Tutorial, Rev. A, Analog Devices, Inc., October, 2008.

"Understanding Data Converters," Texas Instruments Application Report SLAA013, 1995.

"Understanding Flash ADCs," Application Note 810, Maxim Integrated Products, October 2, 2001.

CHAPTER

20 Voltage Regulators, Power Supplies, and Batteries

20.1 INTRODUCTION

In earlier chapters, we've discussed voltage and current primarily in the context of their capacity to relay information. Examples of this include the output signal from a sensor (e.g., a photo-transistor), the transfer function of an op-amp circuit (e.g., a noninverting amplifier), the response of a low- or high-pass filter, and the input/output characteristics of digital logic devices. In this chapter, we consider another important aspect of voltage and current: supplying power to circuits. These applications are fundamentally different than the material we've covered so far, in that they involve the delivery of a specific and stable voltage at higher levels of current—it is not unusual for circuits in mechatronic systems to require several amps or even tens of amps. As a result, the design of power circuits for mechatronic systems warrants its own discussion, and this is the subject of this chapter. After reading this chapter, you should be able to:

1. understand the performance requirements of power sources,
2. understand the operation and characteristics of voltage regulators,
3. understand how linear voltage regulators work and how to apply them,
4. understand how switching voltage regulators work and how to apply them,
5. understand the design, function and performance of linear and switching power supplies,
6. understand the fundamentals and characteristics of batteries,
7. design circuits that will meet the power supply requirements of a mechatronic system.

20.2 POWER REQUIREMENTS AND POWER SOURCES

Mechatronic systems are end users of electrical power, usually drawing power either from batteries or from an electrical outlet in a building connected to the electrical grid. Power from the electrical grid is generated mostly by harnessing an energy source to turn an electrical generator. In the United States, most of this energy is derived from combustion (e.g., coal- or natural gas-fired power plants), nuclear fission (nuclear power plants), or the flow of fluids (hydroelectric power plants). Batteries, on the other hand, derive electrical power from chemical reactions. Regardless of the origins of the power consumed, the great majority of mechatronic systems run on low voltage direct current (DC) at relatively low levels of current. Each application has its own specific set of requirements, but the most common voltages for mechatronic systems are in the range of 5 to 24 VDC, with more exotic applications going as low as 1.2 VDC and as high as 48 VDC. Current requirements commonly range from a few milliamps to a few amps, but in extraordinary cases, the limits can be much lower (a few microamps) or much higher (tens of amps).

For mechatronic systems to operate reliably and perform well, they require a power supply that provides a stable voltage and enough current to meet all component and system requirements under all possible operating conditions. Most common sources of electrical power, including electrical outlets and batteries, do not inherently satisfy these requirements, and additional circuitry is required. Power drawn from an electrical outlet is alternating current (AC), and at much higher voltages than most mechatronic systems require (nominally 120 VAC in the United States). Batteries, though their output is direct current (DC), have terminal voltages that vary depending on the type of battery and the state of charge: a fully charged battery will generally have a significantly higher output voltage than a nearly discharged battery. In the following sections, we discuss how to manage both approaches to supplying power for your mechatronic designs. In the interest of getting to the most useful material as quickly as possible, we begin the discussion with voltage regulation: how to create a stable voltage for your circuits from a less-than-ideal power source. With an understanding of how to achieve this, we then examine the most common means of supplying power: power supplies and batteries.

20.3 VOLTAGE REGULATORS

The function of a voltage regulator is analogous to the function of a pressure regulator in fluid flow. Where the pressure regulator controls the pressure of a fluid, a voltage regulator controls the voltage in a circuit. Recall from the electrical-fluid analogy developed in Chapter 9 that electrical voltage is analogous to fluid pressure, and the analogy holds when we consider the regulation of both quantities. Figure 20.1 illustrates how a fluid pressure regulator accomplishes its task. A fluid whose pressure may vary is introduced at the inlet (bottom left). The flow of fluid from the inlet to the outlet (center right) is controlled by a poppet, which is a valve whose position is determined by the pressure differential across a diaphragm. One side of the pressure differential is the outlet pressure acting on the area of the diaphragm, and the opposing side is a force selected by the user by means of a screw pushing a spring against the diaphragm. When the pressure at the outlet is too high (the force from the fluid pressure is greater than the spring force), the net force on the diaphragm pushes the diaphragm up. This closes the poppet and reduces or stops the flow of fluid from the inlet, and the pressure at the outlet drops. When the pressure at the outlet is too low (the force from the spring is greater than the force from the fluid pressure), the net force on the diaphragm pushes the diaphragm down. This opens the poppet and increases the flow of fluid from the inlet, and the pressure at the outlet increases. When the forces are balanced the pressure is exactly as desired. Under these conditions, the poppet is partially open and in the correct position to maintain the desired pressure at the outlet. Under steady-state conditions, the regulator doesn't have much work to do. However, when conditions are dynamic (inlet pressure or outlet flow rate variations), the regulator must react as quickly as possible to maintain the outlet pressure at the desired value.

A voltage regulator performs an analogous function in the electrical regime. The input voltage may vary, as well as the flow of current through the regulator, but the regulator does

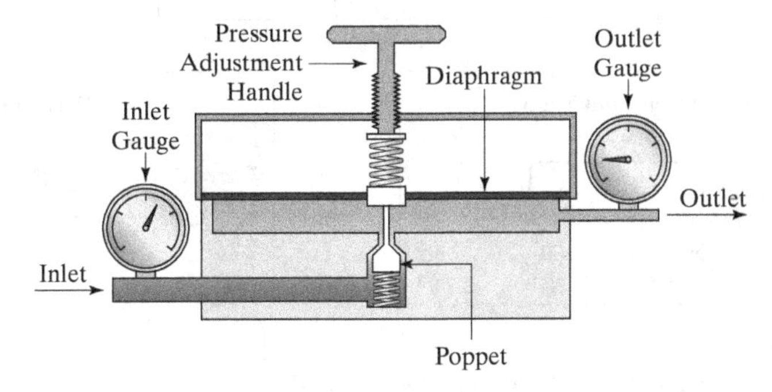

FIGURE 20.1
Fluid pressure regulator.

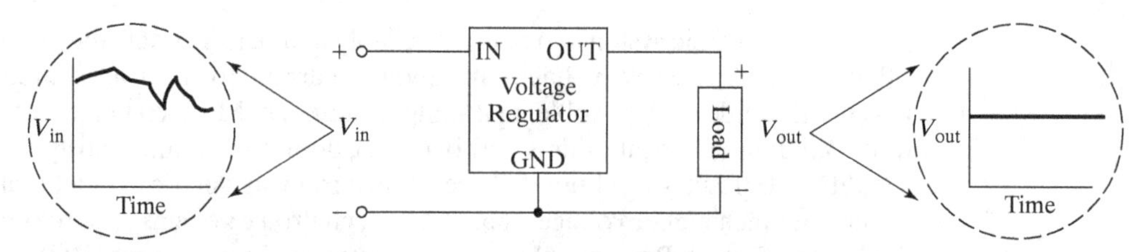

FIGURE 20.2
Ideal voltage regulator. (Courtesy of Texas Instruments Incorporated.)

everything possible to hold the output voltage constant. An ideal voltage regulator would be capable of maintaining a perfectly steady output voltage regardless of conditions at the input or changes in the amount of current flowing (Figure 20.2).

20.3.1 Voltage Regulator Terms and Definitions

As we've repeatedly seen throughout this text, no component is perfect—and voltage regulators are no exception. In this section, we introduce the many terms commonly used to quantify the ways in which voltage regulators fail to achieve perfection. As is the case for other components, these terms and definitions are used in regulators' data sheets so that their level of perfection (and imperfection) is quantified and can be used to compare performance between devices.

Line regulation: The ability of a voltage regulator to hold the output voltage constant when the input voltage changes. Figure 20.3 illustrates a typical test case, where the input voltage undergoes an abrupt change of approximately +1 V followed by another of −1 V. When this occurs, the output changes by a small amount. The specification for line regulation is given as a maximum variation observed in the output voltage when the input voltage is allowed to swing over some specified voltage range (usually the full range of permissible input voltages). Typical line voltage specifications range from a few millivolts to a few tens of millivolts.

Load regulation: The ability of a voltage regulator to hold the output voltage constant when the current flow rate through the load changes, which is also the current flow rate through the regulator. For example, the load might be a high-output LED that is turned on and off, drawing considerable current when on. Figure 20.4 illustrates this scenario. The current at the output (I_O) increases sharply from 0 to 150 mA, and then returns abruptly to 0 a short time later. The voltage at the output changes slightly during this transient. The specification for load

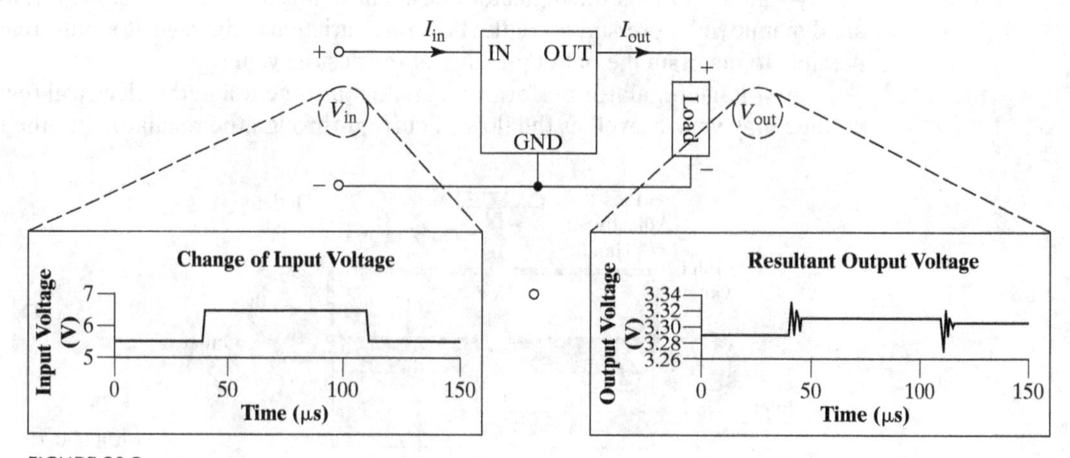

FIGURE 20.3
Line regulation. (Courtesy of Texas Instruments Incorporated.)

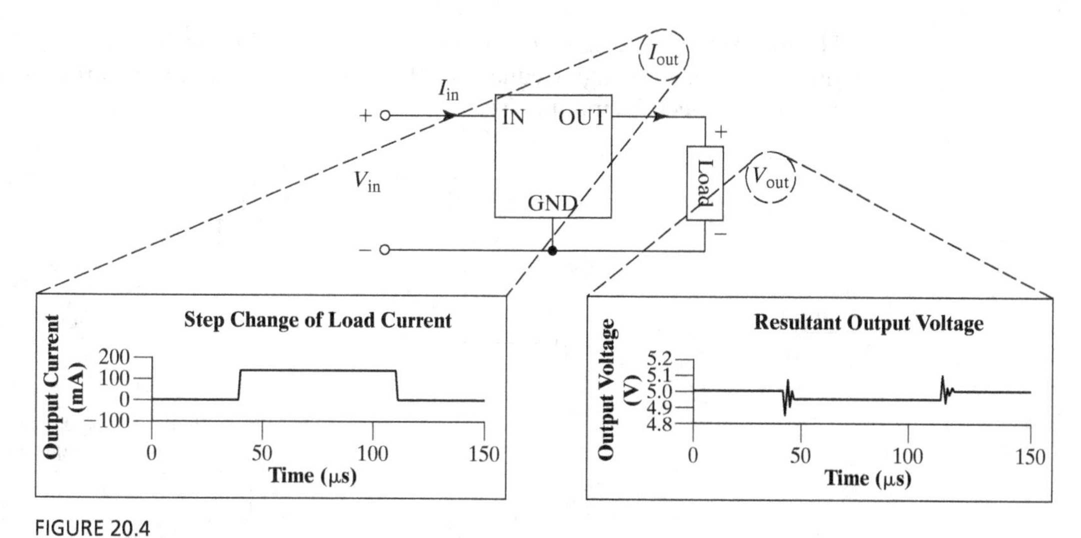

FIGURE 20.4

Load regulation. (Courtesy of Texas Instruments Incorporated.)

regulation is given as the maximum variation observed in the output voltage when the output load current swings over some specified current range. For load regulation specifications, a few millivolts to a few tens of millivolts are typical.

Output noise, also called **output ripple**: The amount that the output of a voltage regulator varies when the input voltage and the output current are both held constant (Figure 20.5). Even under these steady-state conditions, the output voltage of a regulator will vary slightly, and this amount is the output noise or output ripple. Data sheets may state noise in terms of peak-to-peak voltage or root-mean-square (rms) voltage. Typical values for output noise/ripple are in the range of tens to low hundreds of microvolts.

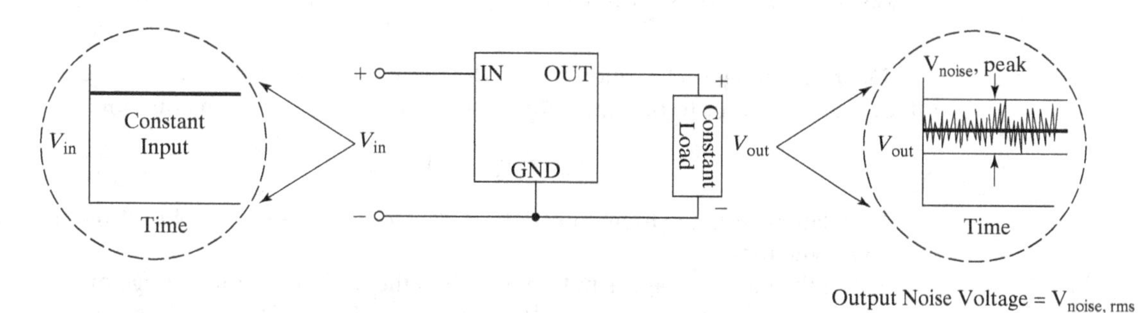

FIGURE 20.5

Output noise, also called output ripple. (Courtesy of Texas Instruments Incorporated.)

Power supply rejection ratio (or **PSRR**): The ability of a voltage regulator to attenuate changes in the input voltage across a range of frequencies. PSRR is related to line regulation (above), which describes a regulator's performance to step changes. PSRR describes how well a regulator is able to attenuate periodic disturbances over a range of frequencies, and is defined as:

$$PSRR = \frac{V_{out,\ ripple}}{V_{in,\ ripple}} \tag{20.1}$$

Figure 20.6 shows PSRR performance of a representative voltage regulator (the MAX1792 from Maxim Integrated Products, Inc.). PSRR is often expressed in units of decibels (dB), since the attenuation is typically very large, especially at low frequencies (below

1 kHz, for example). At higher frequencies, rejection of noise from the input power source diminishes, and at very high frequencies (hundreds of kilohertz), most regulators have a limited ability to reject the input noise.

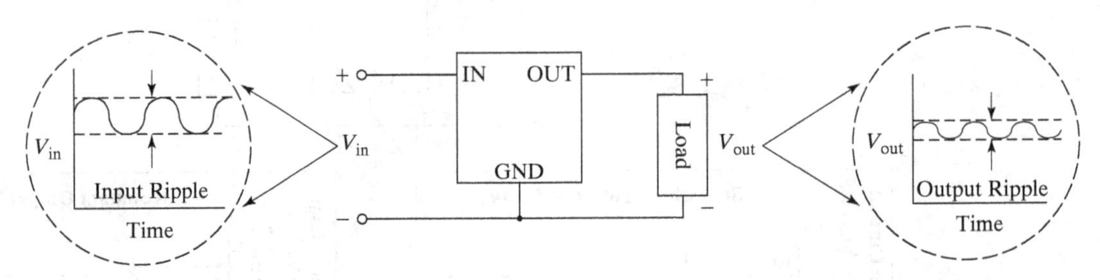

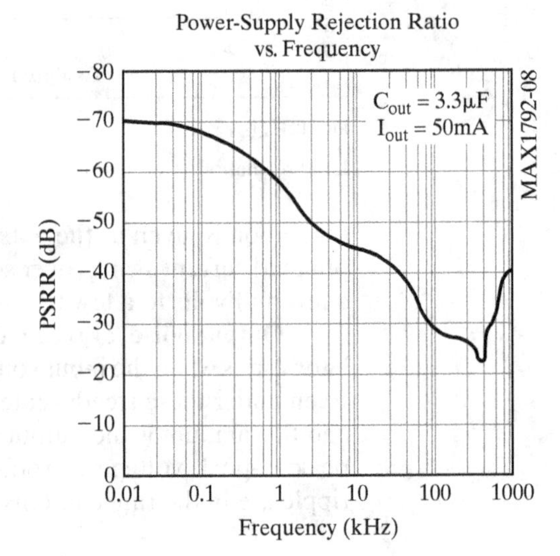

FIGURE 20.6

Top: Illustration of the power supply rejection ratio (PSRR). (Courtesy of Texas Instruments Incorporated.) Bottom: Characteristic PSRR for the MAX1792 voltage regulator. (Copyright Maxim Integrated Products. http://www.maxim-ic.com. Used with permission.)

Dropout voltage: The amount that the input voltage must exceed the desired output voltage for a regulator to operate within specifications. Stated as an equation:

$$V_{in} \geq V_{out} + V_{dropout} \tag{20.2}$$

Regulators designed to minimize the dropout voltage are called **low dropout regulators** (or **LDO** regulators).

When the input voltage is not greater than the desired output voltage plus the dropout voltage [i.e., Eq. (20.2) is not satisfied], the output voltage of the regulator will have a value of approximately $V_{out} = V_{in} - V_{dropout}$. Once Eq. (20.2) is satisfied, however, the output voltage (V_{out}) will be held at the desired regulated value. This is illustrated in Figure 20.7, which shows the operation of an LM7805 from National Semiconductor (a low-cost linear regulator with a 5 V nominal output) for input voltages between 0 and 15 V.

Efficiency: The ratio of the power output of a voltage regulator to its power input. Efficiency is expressed as:

$$\eta = \frac{P_{out}}{P_{in}} \tag{20.3}$$

No voltage regulator can achieve 100% efficiency—there's always a cost for including the benefits of a regulated voltage in your circuits. Some of the most efficient regulators achieve efficiencies of greater than 95% (see Section 20.3.3 on switching regulators), while some of the least efficient regulators (when used in the least efficient ways) may be 40% efficient or less.

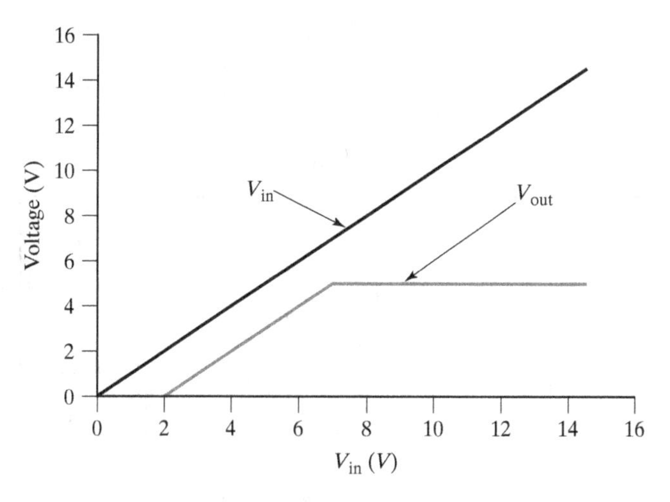

FIGURE 20.7

Input and output voltages for an LM7805 with a dropout voltage of 2 V.

20.3.2 Linear Voltage Regulators

The simplest type of voltage regulator to use is the **linear voltage regulator**. In their simplest form, they are three-terminal devices: one pin for ground, one pin for the input (supply) voltage, and one pin for the regulated output voltage. Figure 20.8 shows a few of the most common packaging options for linear voltage regulators. Larger packages (such as those shown in Figures 20.8a and b) with significant surface area are typically better able to dissipate heat and are rated for higher current flows. Smaller surface-mount packages (such as the SOT-89 in Figure 20.8c) are often in very good thermal contact with a printed circuit board and can also achieve very good heat dissipation.

Though linear voltage regulators can be obtained in a variety of shapes and sizes, connecting them in a circuit usually involves less variation. Figure 20.9 shows a typical linear voltage regulator circuit. In this case, a 7.2 V power source connected at V_{in} is regulated down to 5 V at V_{out}.

For specific regulators, capacitors may be required or recommended for stability at the input (C_{in}) and/or the output (C_{out}). For the most stable devices, C_{in} is only necessary if the regulator is physically located far from the input power source, and C_{out} is recommended (but not required) to improve transient response. In the most demanding applications (such as LDO regulators, which we discuss below), specific values and types of capacitors may be required.

(a) TO-220 package

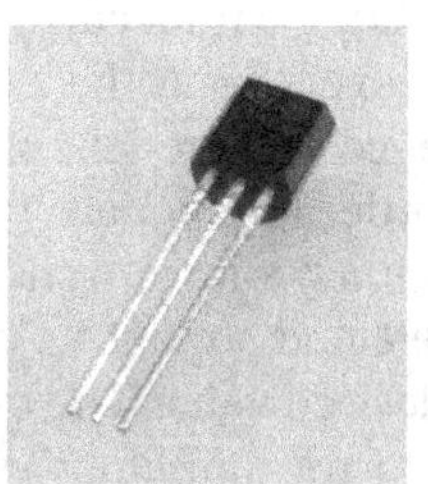

(b) TO-92 package

(c) SOT-89 package

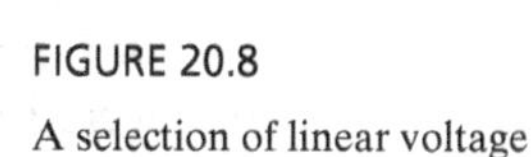

FIGURE 20.8

A selection of linear voltage regulator packaging options.

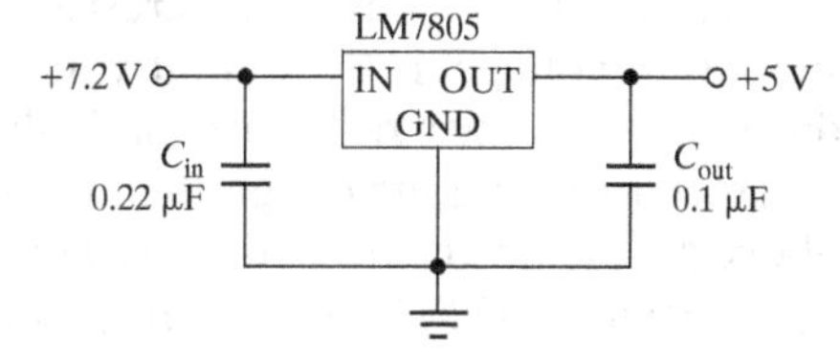

FIGURE 20.9

A simple linear voltage regulator circuit.

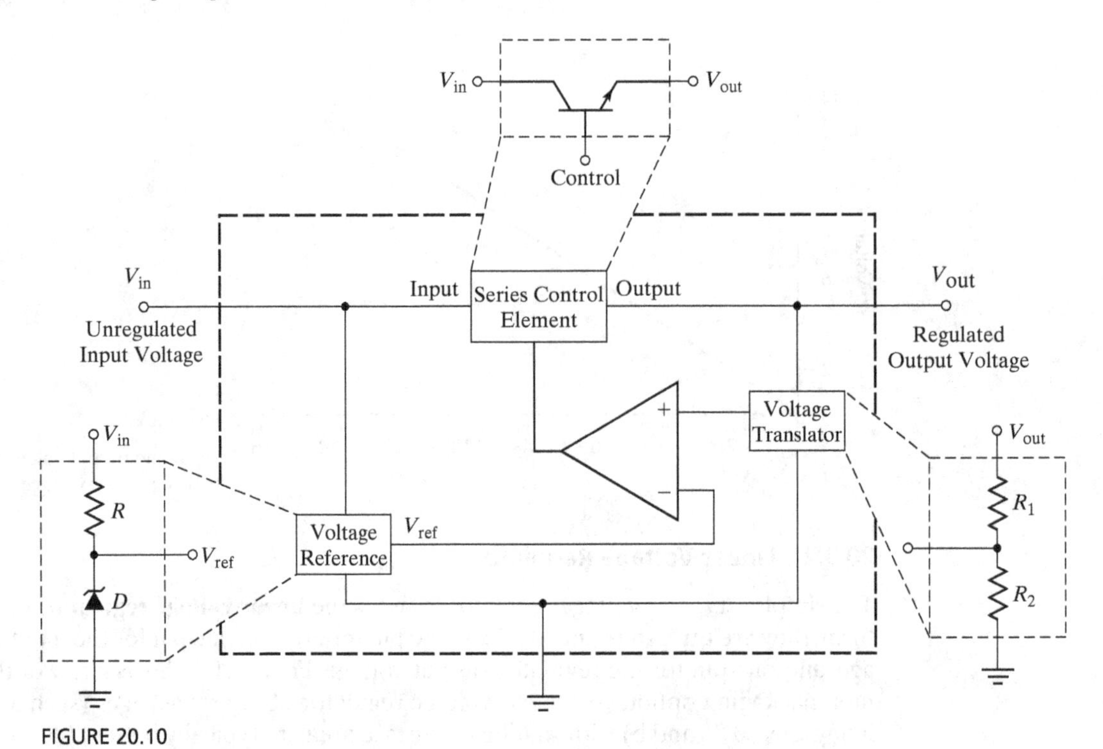

FIGURE 20.10

Linear voltage regulator block diagram.

Figure 20.10 illustrates the basic functional blocks of a linear voltage regulator. To create the steady, regulated output voltage, the regulator continuously compares the output voltage to an internal voltage reference, and modulates the voltage drop between the input and the output by adjusting a linear control element, here labeled the "**series control element**," also called the **pass element**. This is what distinguishes the regulator as linear: the series control element linearly and continuously adjusts the voltage at the output to match the desired value. The series control element is analogous to the poppet in the fluid pressure regulator (Figure 20.1); it may be thought of as modulating the flow of electrons by increasing or decreasing the effective resistance across its terminals. The internal voltage reference may be created using any of a number of methods, but one particularly straightforward means is to use a low voltage Zener diode with a current limiting resistor, as depicted. As long as V_{in} remains sufficiently high to reverse bias the Zener beyond its breakdown voltage, the circuit will provide a stable reference voltage (the Zener voltage). Typical reference voltages are low: 1.2 V, for example. In order to compare the output voltage to the reference voltage, we see in the block diagram that a "voltage translator" is used. This block consists of a voltage divider with the resistors selected so that its output is equal to the reference voltage when the output voltage has the desired value. The output of the reference and the divider are connected to the inputs of an "error amplifier," which determines the difference between the two values (the "error" of the output), and uses this value to control the series control element and adjust the output voltage. If the output voltage is determined to be too high, the error amplifier responds by increasing the effective series resistance across the control element, which then reduces the output voltage. If the output voltage is determined to be too low, the error amplifier reduces the effective resistance until the output voltage rises to the desired value. Using this approach, the output voltage continuously "hunts" for the desired value, with small excursions above and below the desired value. For many linear regulators, the bandwidth of the control loop is around 100 kHz. This corresponds closely to the frequency

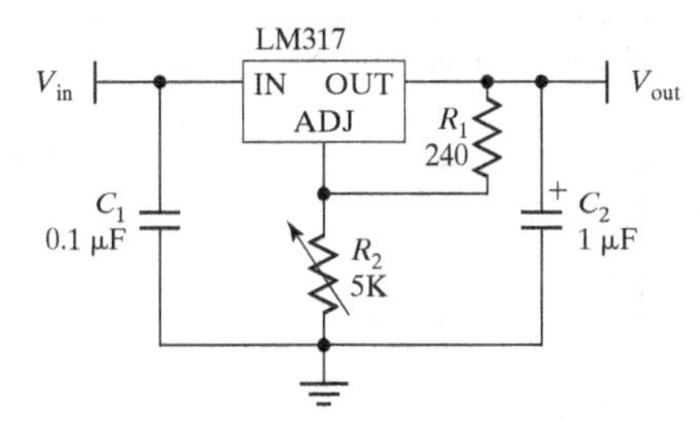

FIGURE 20.11

A circuit that uses an LM317 adjustable voltage regulator, with a user-configurable output voltage between 1.2 and 25 V.

content of the noise present in the output voltage and to the ability of the regulator to reject noise with frequency content approaching or exceeding its own control loop bandwidth—recall that the PSRR decreases as frequency increases.

Linear regulators are available with a fixed voltage output, in which case they are called **fixed regulators**. Examples of fixed regulators include the family of devices of which the LM7805 is a member. Other devices in the family are the LM7812 (with a 12 V output) and the LM7815 (with a 15 V output). Fixed output linear regulators are available with output voltages between 0.8 and 15 V. Linear regulators are also available with a user-configurable adjustable output, in which case they are called **adjustable regulators**. For example, the LM317 may be configured to deliver output voltages between 1.2 and 37 V. Adjustable regulators give the designer control over the "voltage translator" block shown in Figure 20.10. By selecting the resistors in the voltage divider, the designer determines the output voltage that the regulator will seek to maintain. A typical application circuit is shown in Figure 20.11.

20.3.2.1 Low Dropout Linear Regulators

The configuration of the series control element in Figure 20.10 determines the regulator's dropout voltage, that is, the minimum voltage drop that can be achieved between a regulator's input and output when the regulated output has the desired value (see the definitions in Section 20.3.1). The block diagram in Figure 20.10 shows an NPN BJT as the series control element, which has the disadvantage of a rather high dropout voltage (~2 V) but the advantage of being extremely stable. In this context, "stability" refers to the ability of the regulator's feedback control loop (comprised of the series control element, the voltage translator, and the error amplifier in Figure 20.10) to maintain a stable output voltage. To reduce the dropout voltage, alternate configurations of series control elements are used, and regulators that use this approach are called low dropout, or LDO, regulators.

The pass elements shown in Figures 20.12a and 20.12b are NPN BJT-based designs and are typical of standard voltage regulators (not low dropout). These designs have the advantage of being very stable, but have relatively high dropout voltages of approximately 2 V. The PNP-based design in Figure 20.12c has a much lower dropout voltage (around 0.6 V), and the FET-based designs (Figures 20.12d and 20.12e) are lower still, featuring dropout voltages in the range of 0.2 V.

The lower dropout voltage designs trade off stability to achieve the lower voltage drop and require designers to pay closer attention to the circuit design, in particular the capacitors at the input and output of the regulator (not just the capacitance, but the type of capacitor too). Real capacitors have an equivalent series resistance (ESR) associated with them that changes the behavior of the control loop that is at the heart of a voltage regulator. The total resistance between an LDO regulator's output capacitor and ground is called the **compensation series resistance** (**CSR**). In order for an LDO regulator to operate in a stable regime, the CSR must be within the range of values specified by the manufacturer. Values either above or below the specified range will result in unstable behavior, the result of which is an unstable output voltage—which defeats the purpose of

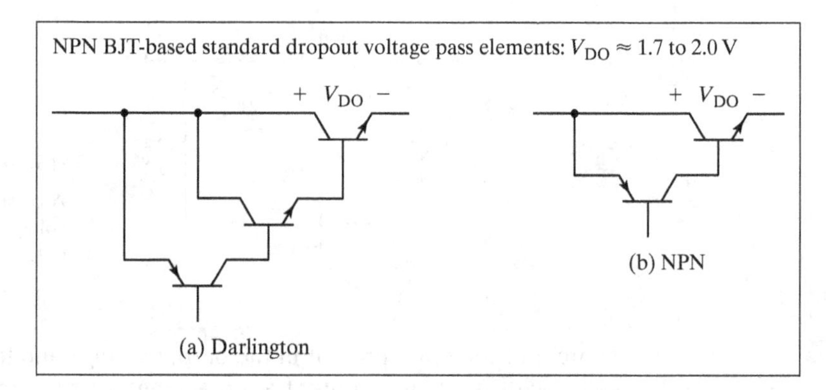

PNP BJT-based LDO voltage pass element: $V_{DO} \approx 0.6$ V

+ V_{DO} −

(c) PNP

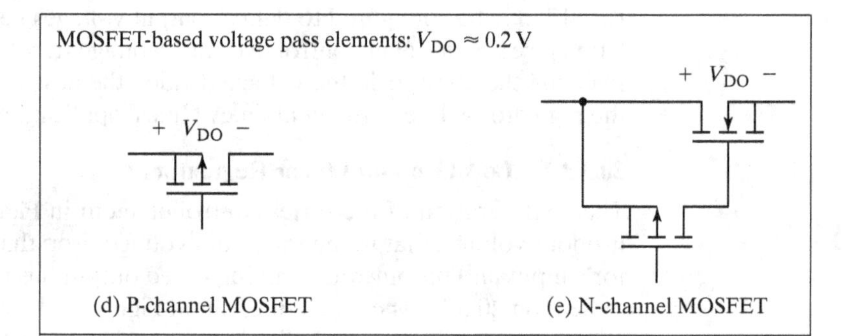

FIGURE 20.12

Linear regulator pass element configurations. (Courtesy of Texas Instruments Incorporated.)

adding a voltage regulator in the first place. The CSR is usually dominated by the ESR of the output capacitor (see Section 8.14.1). LDO regulator data sheets often specify the allowable range of CSR or ESR and provide a graph to describe operation over a range of output current values (Figure 20.13). Because of the graph's distinctive shape and the consequences of straying outside the specifications, the regions of stability and instability are sometimes described as the **tunnel of death**. The example shown is for the TPS7250, an LDO regulator from Texas Instruments with a fixed 5 V output. Graphs such as that shown in Figure 20.13 are not always provided by the manufacturer, but some guidance regarding the selection of capacitors to ensure a successful design will be given in some form.

20.3.2.2 Heat Dissipation

In Chapter 9 (Basic Circuit Elements), we showed that the power dissipated in a device (in the form of heat) is equal to the voltage drop across the device times the current flowing through the device: $P = VI$. This holds for linear voltage regulators and is often a primary design consideration since substantial power may be dissipated by these components due to the voltage drop from the input to the output. For linear regulators, we can further expand the power consumption expression to include a term for the device's internal circuitry:

$$P = (V_{in} - V_{out})I_{out} + V_{in}I_Q \tag{20.4}$$

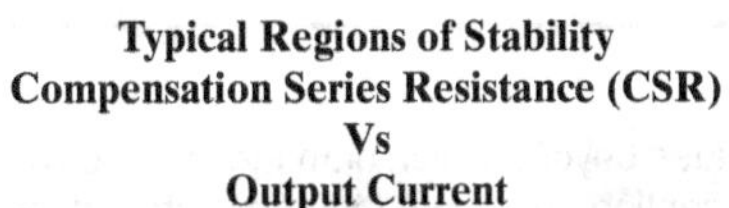

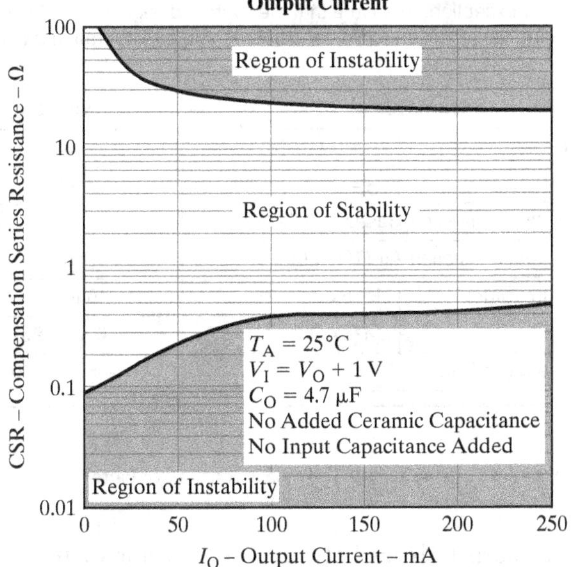

FIGURE 20.13

CSR values that result in stable and unstable operating regimes for the TPS7250 LDO regulator. (Courtesy of Texas Instruments Incorporated.)

The first term is the result of the current (I_{out}) flowing through the regulator to the load, with the voltage drop equal to the difference between the input (unregulated) voltage and the output (regulated) voltage. The second term is the power required by the regulator to function, and is equal to the input (unregulated) voltage times a quantity called the **quiescent current** (defined as the amount of current consumed by a device to function, which does not include driving any external loads). Given that the purpose of a regulator is to provide current to a load at a regulated voltage, the power dissipation can be large if the output current is substantial—generally more than a few hundred milliamps—and/or if the voltage drop between the input and the output is large. This is particularly an issue for linear voltage regulators, which simply (and somewhat inelegantly) burn off the difference between the input voltage and the desired output voltage (times the current) as heat. This functional approach results in low efficiency [defined as the ratio of the power out to the power in, Eq. (20.3)]. For linear regulators, efficiency is usually in the range of $\sim 40\% \leq \eta \leq \sim 80\%$.

The power converted into heat within a device is not an issue in and of itself—that's only part of a potential problem. Excessive *temperature*, rather than heat itself, can damage or destroy a device. At steady state, heat dissipation must match heat generation. Heat dissipation is driven by the temperature difference between the device and the world (ambient). Therefore, the greater the amount of heat to be dissipated, the greater the temperature difference between the device and the ambient. For a given ambient temperature, this means the hotter the device gets.

How much heat is too much? To determine the limits of operation, we need information about the device itself and the environment. Specifically, we need to know:

- the maximum allowable temperature at which the silicon die (the "**junction**") can function within specifications,
- the heat transfer characteristics between the junction and its exterior surface (the "**case**"),

Absolute Maximum Ratings

Absolute maximum ratings are those values beyond which damage to the device may occur. The datasheet specifications should be met, without exception, to ensure that the system design is reliable over its power supply, temperature, and output/input loading variables. Fairchild does not recommend operation outside datasheet specifications.

Symbol	Parameter		Value	Unit
V_I	Input Voltage	V_O = 5V to 18V	35	V
		V_O = 24V	40	V
$R_{\theta JC}$	Thermal Resistance Junction-Cases (TO-220)		5	°C/W
$R_{\theta JA}$	Thermal Resistance Junction-Air (TO-220)		65	°C/W
T_{OPR}	Operating Temperature Range	LM78xx	-40 to +125	°C
		LM78xxA	0 to +125	
T_{STG}	Storage Temperature Range		-65 to +150	°C

FIGURE 20.14

Temperature limits and heat transfer characteristics for the LM7805. (Courtesy of Fairchild Semiconductor.)

- the heat transfer characteristics between the exterior of the device (case) and the ambient environment (if no heat sink is used), or between the exterior of the device and the heat sink.
- (if a heat sink is used) the heat transfer characteristics between the heat sink and the environment.

This information is usually included in the Absolute Maximum Ratings section of a device's data sheet (Figure 20.14). In some data sheets, the thermal specifications may be scattered throughout multiple sections, and finding everything needed may be a bit of a challenge.

In electronic design, the most common approach for analyzing heat transfer is that of the **equivalent thermal circuit**, which is illustrated for a representative integrated circuit package in Figure 20.15. With this approach, a thermal system is modeled as an electrical circuit, with temperature analogous to voltage, heat flow analogous to current, and the heat transfer characteristics between two points in the system analogous to electrical resistance. The resistance to the flow of heat between two points is called the **thermal resistance** and is represented by the symbol Θ (sometimes R_Θ). Thermal resistance has units of °C/W—the change in temperature per unit power. During operation, heat is generated at the silicon die located

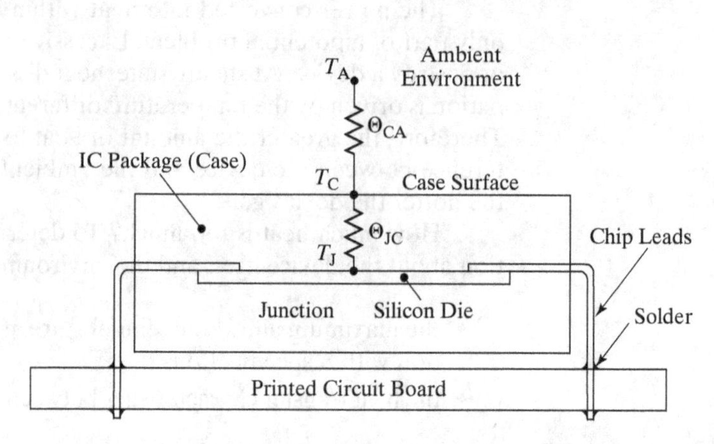

FIGURE 20.15

An equivalent thermal circuit analysis of the heat transfer from an integrated circuit's junction to the ambient environment.

at the center of the integrated circuit (the junction), and the temperature at this location is labeled T_J. The thermal resistance between the junction and the surface of the integrated circuit housing (the case) is given the symbol Θ_{JC} (the "thermal resistance between the junction and the case"). The temperature of the case is T_C, and the thermal resistance between the case and the ambient is Θ_{CA}. The ambient temperature is T_A (or sometimes T_∞).

The governing equation for heat transfer in a thermal circuit is analogous to Ohm's law ($V = IR$):

$$\Delta T = P\Theta \tag{20.5}$$

where ΔT is the difference in temperature between two points in the circuit, P is the power to be dissipated, and Θ is the thermal resistance between the points. For the thermal circuit illustrated in Figure 20.15, Eq. (20.5) may be rewritten as:

$$T_J - T_A = P(\Theta_{JC} + \Theta_{CA}) \tag{20.6}$$

This equation may be populated with known values and operating conditions and solved for unknown parameters. Eq. (20.6) is broadly applicable to practically all devices and scenarios, regardless of the type of device or whether a heat sink is used, as long as the appropriate values of temperature and thermal resistance are used. As an example, we can use the values for the Fairchild LM7805 described in Figure 20.14 to determine the maximum allowable power dissipation when the ambient temperature is 25°C and no heat sink is used. The data sheet specifies the thermal resistance from the junction to the ambient (given the symbol $R_{\Theta JA}$) as a single value (65°C/W), so we can write Eq. (20.6) in terms of P_{max} as:

$$P_{max} = (T_J - T_A)/(R_{\Theta JA}) = (125°\text{C} - 25°\text{C})/(65°\text{C/W}) = 1.54\text{ W}$$

The heat transfer from the case to the ambient (Θ_{CA}, that is) may be improved by adding a heat sink to the device. Heat sinks improve heat transfer (decrease the thermal resistance) between the case and the ambient by providing increased surface area and incorporating shapes that increase the heat transfer to the environment, such as fins. Two representative heat sinks are shown in Figures 20.16a and b. Some heat sinks even incorporate an integral fan, which can dramatically improve heat transfer by introducing forced convection. This approach is standard for heat sinks used to cool microprocessors in desktop computers, which in some cases must dissipate more than 100 W. A separate fan can be added to most circuits regardless of the heat sink designs used, which can dramatically decrease the thermal resistance through the use of forced convection. In order to ensure good thermal contact between a device and its heat sink, a thin layer of **thermal grease** (also called **heat sink compound**) is highly recommended (Figure 20.16c).

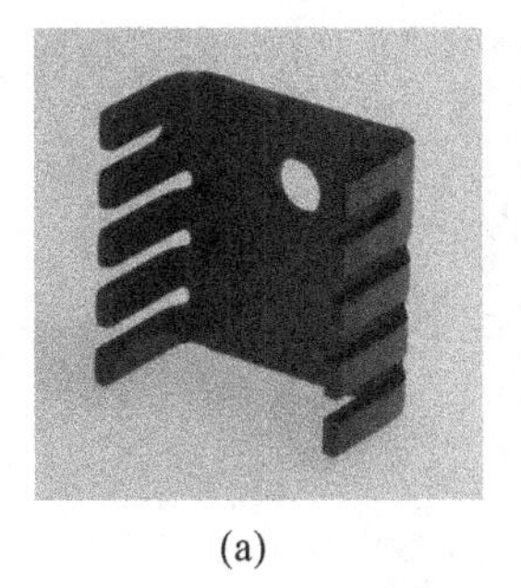

(a)

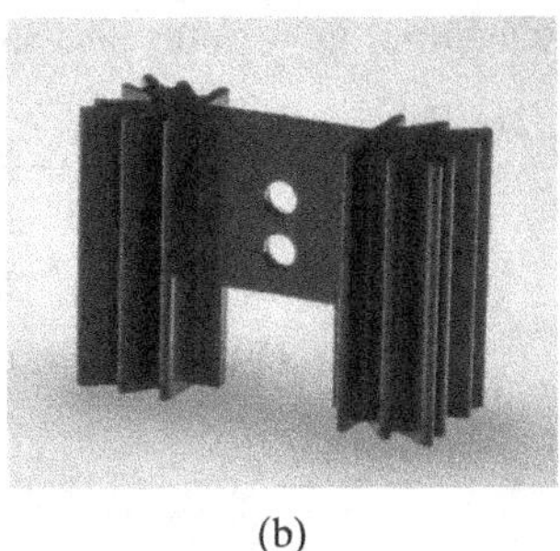

(b)

(c)

FIGURE 20.16

Representative heat sinks for TO-220 packages (a and b), and (c) heat sink compound (a.k.a. thermal grease).

Example 20.1

A 7.2 V nickel-cadmium battery pack is used to power a circuit that requires a regulated 5 V power supply. If an LM7805 linear voltage regulator with the specifications listed in Figure 20.14 provides the regulated 5 V, what is the maximum amount of current it can provide a) when no heat sink is used, b) when a heat sink with a thermal resistance of $\Theta_{CA} = 24.5°C/W$ is used (a typical value for a low-cost TO-220 heat sink), and c) when the battery pack voltage is doubled to 14.4 V?

a) For the case where no heat sink is used, the system is as shown in Figure 20.15, and we can solve Eq. (20.6) for the maximum power and then the maximum current:

$$\Delta T = P\Theta$$

$$P_{max} = (V_{in} - V_{out})I_{max} = (T_J - T_A)/(R_{\Theta JA}) = 1.54\ \text{W} \tag{20.7}$$

$$I_{max} = \frac{P}{\Delta V} = \frac{(T_J - T_A)/(R_{\Theta JA})}{(V_{in} - V_{out})} = \frac{(125°C - 25°C)/(65°C/W)}{(7.2\ V - 5\ V)} = 0.70\ A$$

b) When a heat sink with thermal resistance $\Theta_{CA} = 24.5°C/W$ is added to the LM7805, the equation representing the equivalent thermal model must now include the thermal resistance from the junction to the case and the heat sink to the ambient (labeled below as $R_{\Theta HA}$). We assume that the case and heat sink are in such intimate contact that the case to sink thermal resistance is negligible. We also assume that the thermal resistance between the heat sink and ambient is much lower than that of the case and ambient, so we will neglect the heat transfer contributions of the case in this analysis. This is a conservative assumption, since the case will actually contribute a small amount in a real system, and the overall junction temperature will be reduced slightly as a result. When these updates are made to Eq. (20.7), we get:

$$I_{max} = \frac{(T_J - T_A)/(R_{\Theta JC} + R_{\Theta HA})}{(V_{in} - V_{out})} =$$

$$\frac{(125°C - 25°C)/(5°C/W + 24.5°C/W)}{(7.2\ V - 5\ V)} = 1.54\ A$$

Elsewhere in the LM7805 data sheet, the maximum output current specification is given as 1 A, so if this heat sink is used, heat dissipation does not impose the upper limit on current.

c) When the battery pack voltage is doubled to 14.4 V, the value of V_{in} in the denominator of both expressions increases. The voltage drop across the regulator increases from 2.2 to 9.4 V, and this has a substantial impact on the heat dissipation. The resulting maximum current value with no heat sink is:

$$I_{max} = \frac{(T_J - T_A)/(R_{\Theta JA})}{(V_{in} - V_{out})} = \frac{(125°C - 25°C)/(65°C/W)}{(14.4\ V - 5\ V)} = 0.21\ A$$

And the maximum current value with the heat sink is:

$$I_{max} = \frac{(T_J - T_A)/(R_{\Theta JC} + R_{\Theta HA})}{(V_{in} - V_{out})} =$$

$$\frac{(125°C - 25°C)/(5°C/W + 24.5°C/W)}{(14.4\ V - 5\ V)} = 0.47\ A$$

20.3.3 Switch-Mode (Switching) Voltage Regulators

While linear voltage regulators use a linear control element to create a regulated output voltage, switched-mode (or "switching") voltage regulators employ a completely different approach. **Switching voltage regulators** make use of the energy storage characteristics of inductors and capacitors to translate an unregulated DC input voltage into a regulated DC output voltage. A digital controller is used to modulate (switch) the flow of current to an inductor or a capacitor to achieve the desired result—hence the name "switching" voltage regulator. Switching regulators are exceptionally flexible: depending on the configuration, the output voltage may be less than the input voltage, greater than the input voltage, or even have the opposite polarity as the input voltage. Switching regulators also typically achieve much higher efficiency (η) than linear regulators, and because they waste less power in the form of heat, they can more easily be configured to deliver high levels of current, such as 10 A or more. All of these beneficial attributes come at the cost of substantially higher output noise (ripple) and increased circuit complexity. An examination of the functional elements of switching voltage regulators helps understand why.

20.3.3.1 Charge Pumps

For low-current applications (up to about 100 mA), a common and economical approach is to use capacitors to temporarily store charge, then switch the connections to the capacitors so that the stored energy is delivered to boost or invert the output voltage. Devices that employ this approach are referred to as **charge pumps** or **flying capacitor voltage converters**. Figure 20.17 shows a basic configuration charge pump whose output voltage has the opposite polarity of the input voltage.

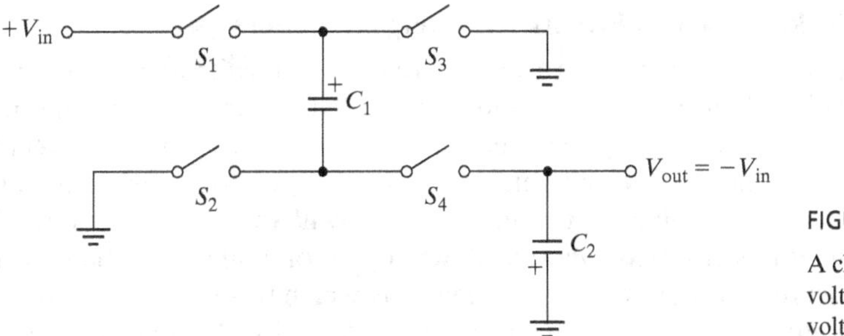

FIGURE 20.17

A charge pump (flying capacitor voltage converter) configured as a voltage inverter.

Inverting the voltage takes place in two steps. The first step involves closing the switches on the left (S_1 and S_2) while opening the switches on the right (S_3 and S_4), which charges the capacitor at the center of the diagram, C_1. Once C_1 is charged, the top (positive) terminal of C_1 has a potential of $+V_{in}$ relative to the bottom (negative) terminal. Said another way: the negative terminal of C_1 has a potential of $-V_{in}$ relative to the positive terminal—and this is the effect used to invert the input voltage at the output. To complete the voltage inversion, S_1 and S_2 are opened so that the input voltage is disconnected, and S_3 and S_4 are closed. This connects the positive terminal of C_1 to ground through S_3, and the negative terminal of C_1 to V_{out}. Since the negative terminal of C_1 is at a potential of $-V_{in}$ relative to the positive terminal, V_{out} has a value of $-V_{in}$ relative to ground: the voltage has been inverted. The capacitor at the output (C_2) smoothes the variations in V_{out} that would occur from repeatedly switching the capacitor terminals between the input and the output.

A rearrangement of the charge pump (or flying cap converter) shown in Figure 20.18 results in a device whose output voltage is twice the input voltage. The initial step of the process is similar to that of the voltage inverter: S_1 and S_4 close to charge the capacitor C_1. This places the positive terminal of C_1 at $+V_{in}$ relative to its negative terminal. For the second phase, however, S_2 connects the negative terminal of C_1 to $+V_{in}$ and S_3 now connects the positively charged terminal of C_1 to the output, V_{out}. Since there is a ΔV across C_1 equal to $+V_{in}$ and its negative terminal is now at $+V_{in}$, the result is $V_{out} = 2 \times V_{in}$. C_2 again serves to smooth the ripple in the output voltage that results when C_1 is switched between the input and the output. This general approach can also be used to generate higher multiples of the input voltages by sequencing additional stages of charge-storing capacitors and switches. This is not common because of the increased resistance from multiple switches in the output path.

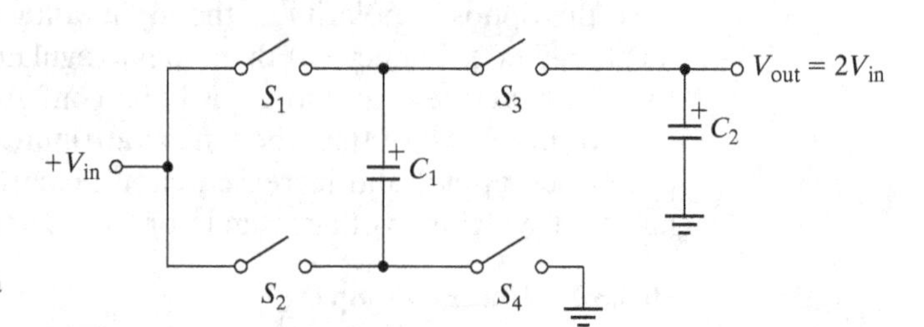

FIGURE 20.18

A charge pump (flying capacitor voltage converter) configured as a voltage doubler.

The output voltages of both the inverting and doubling configurations can also be regulated to values other than $-V_{in}$ or $2V_{in}$ by altering the timing with which the switches are opened and closed. For example, by allowing more or less charge to accumulate at the capacitor C_1 during the charging stage, the output of inverting charge pump can be regulated to values between ground and $-V_{in}$. For the doubling charge pump, the output can be regulated to values between V_{in} and $2V_{in}$.

20.3.3.2 Buck, Boost, and Flyback Switching Regulators

When currents higher than about 100 mA are required, switching voltage regulators that use inductors, rather than capacitors, as the storage element are more commonly used. The most common type of switching regulator used in these instances is the **buck regulator**, also called a **step-down regulator**. The input voltage of a buck regulator is higher than the desired regulated output, and the regulator is said to "step down" the voltage from the input to the output. The function performed by a step-down regulator is similar in many respects to that of a linear voltage regulator, however, the means by which this is achieved are completely different. Buck regulators generally have a circuit topography like that shown in Figure 20.19.

In this circuit, a controller determines the state of the switch, S_1, between V_{in} and the inductor, L. When the switch is closed (bottom left of figure), current through the inductor begins to flow as the magnetic field builds and creates a store of energy. When the switch is opened (bottom right of figure), the magnetic field of the inductor collapses, which (for a brief period of time) maintains the flow of current through to the inductor to V_{out}. The diode, D, holds the voltage on the left side of the inductor to one forward voltage drop below ground, while current continues to flow through the inductor from left to right (since current through an inductor cannot change instantaneously). Some of the current is supplied directly to the load at V_{out} and then flows through the diode D back to the inductor. The charge that makes up the rest of the current accumulates on the plates of the capacitor, C, raising its potential. Throughout the process, the capacitor C serves to filter and reduce variations in the output voltage, V_{out}. The controller for the regulator monitors V_{out} and modulates the state of the switch (usually through the use of pulse width modulation, PWM), so that V_{out} is held as close to the desired value as possible, even as V_{in} or the characteristics of the load powered by the regulator change.

The circuit topology may be rearranged to create a **boost regulator**, also called a **step-up regulator**. In this configuration, the input voltage is lower than the desired regulated

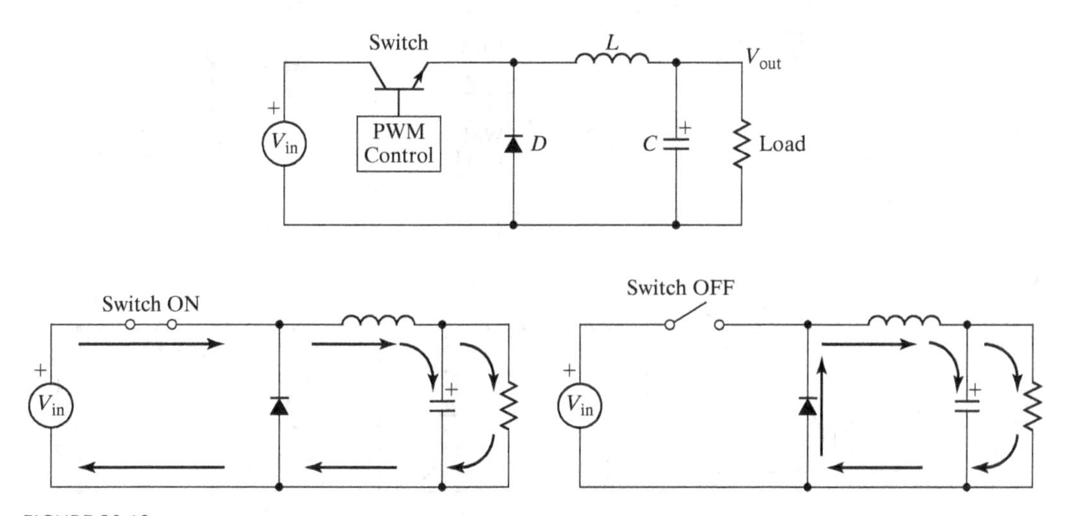

FIGURE 20.19

A switching voltage regulator with a buck (step-down) configuration. (Courtesy of National Semiconductor Corporation.)

output voltage, and the regulator is said to "step up" the voltage from the input to the output. A typical topology is shown in Figure 20.20.

For the boost regulator, when the switch (transistor) is closed (bottom left of figure), current begins to flow through the inductor and the magnetic field is established. When the switch is opened (bottom right of figure), the inductor's magnetic field collapses which causes the flow of current through the inductor to continue toward the diode and the voltage dramatically increases. The diode ensures that the current flows only from left to right in the circuit (which could happen during the part of the cycle when the switch is closed and current is increasing in the inductor), and the capacitor smoothes the voltage peaks and reduces the output voltage ripple. A digital controller monitors the output voltage (V_{out}) and modulates the state of the transistor switch to hold V_{out} at the desired value. The resulting output voltage established is greater than the input voltage: the voltage has been "boosted" or "stepped-up."

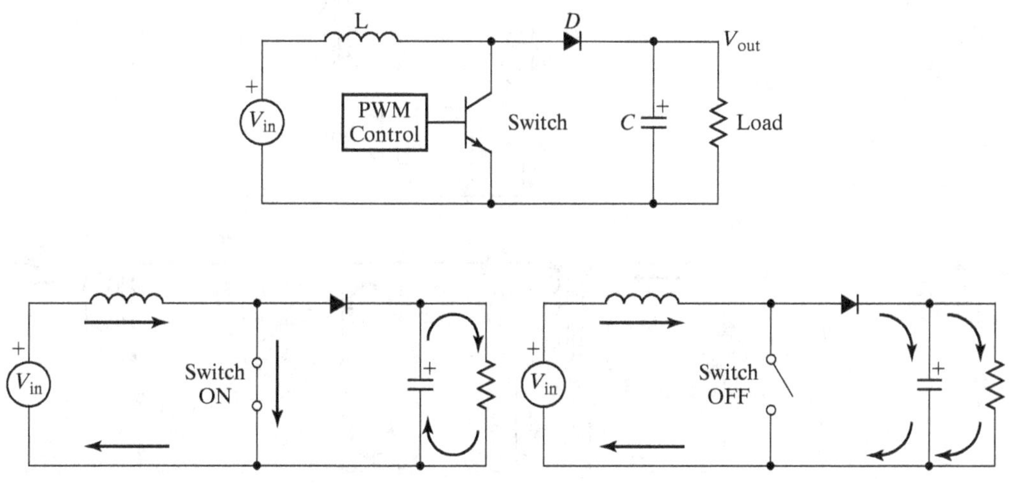

FIGURE 20.20

A switching voltage regulator with a boost (step-up) configuration. (Courtesy of National Semiconductor Corporation.)

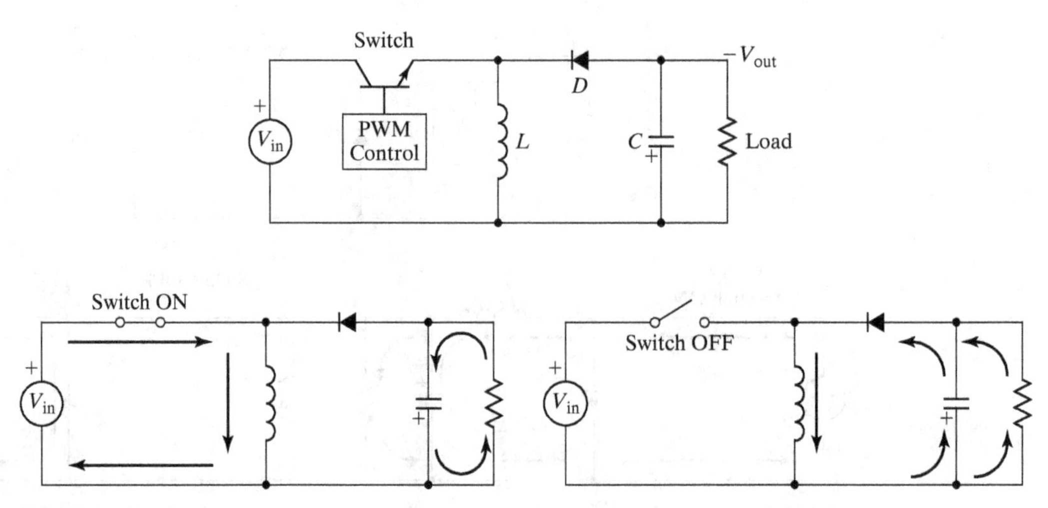

FIGURE 20.21

A switching voltage regulator with a configuration that inverts voltage polarity. (Courtesy of National Semiconductor Corporation.)

Another arrangement of the same components allows for a reversal of the polarity of the voltage, so that $V_{out} = -V_{in}$. A topology that performs this translation is shown in Figure 20.21.

When the switch is closed (bottom left of figure), current begins to flow through the inductor to ground and the magnetic field begins to build. When the switch is opened (bottom right of figure), the current flowing through the inductor to ground continues as before (driven by the inductor's collapsing magnetic field), in a loop from ground through the capacitor. As this occurs, the voltage at the top of the inductor decreases and the diode becomes forward-biased. The overall result is a drop in the voltage at the output to a negative voltage whose exact value ($-V_{out}$) is determined by the controller modulating the state of the switch.

A variation of the boost topology (and in some cases, the polarity reversing topology) is the **flyback regulator** (Figure 20.22). Instead of using an inductor as the energy storage

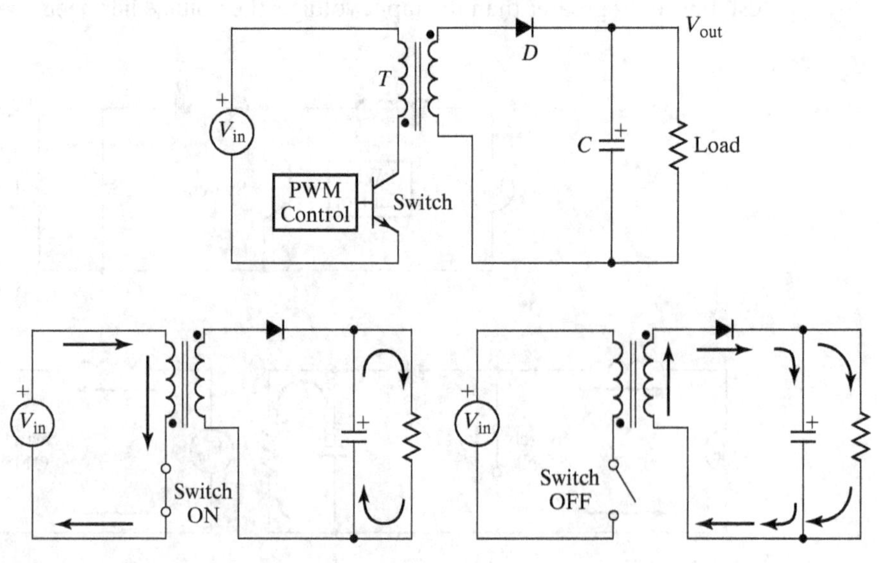

FIGURE 20.22

A switching voltage regulator with a flyback (transformer coupled) configuration. (Courtesy of National Semiconductor Corporation.)

element in the circuit, this configuration uses a transformer. Transformers are comprised of two inductors whose magnetic fields are shared in such a way that alternating current flowing in one (the **primary coil**) induces current to flow in the other (the **secondary coil**), with the voltage of the induced current in the secondary coil determined by the ratio of the number of turns in each inductor. Use of a transformer allows for a great deal of flexibility in the design, including an output voltage much greater than the input voltage [exploiting the transformer's ratio of turns between the secondary (output) coil and the primary (input) coil], and multiple outputs with different output voltages, some or all of which may have their polarity inverted.

When used to create an output voltage that is a multiple of the input voltage ($V_{out} = N \times V_{in}$), the transformer is arranged so that the polarity of the primary coil is the inverse of that of the secondary coil. This is indicated by the location of the dots near each coil in the transformer of Figure 20.22. When the switch is closed (lower left of figure), current and a magnetic field build in the primary coil of the transformer. When the switch is opened (lower right), the primary coil's collapsing magnetic field causes an increase in voltage at the bottom terminal, which induces N times the voltage increase in the secondary coil, where N is the ratio of the transformer's secondary turns to primary turns. The voltage induced in the secondary coil can be very large as a result. The rest of the circuit resembles the topography of the boost configuration above, where current flows in a loop between the secondary coil, through the diode, charges the capacitor and supplies current to the load connected to V_{out}.

Multiple outputs may be created by using a transformer that has multiple secondary coils coupled to a single primary coil, as shown in Figure 20.23. One issue with this approach is that the controller may close the loop based on only one of the output voltages, and the regulation of the other voltages suffers somewhat as a result. If the regulation of all of the output voltages is critical to the circuit, the outputs not monitored by the controller may be further regulated by some other means, such as a linear LDO regulator (Section 20.3.2.1). Output voltages with polarity opposite that of the input may be generated simply by using transformers whose primary and secondary polarities are not inverted relative to each other.

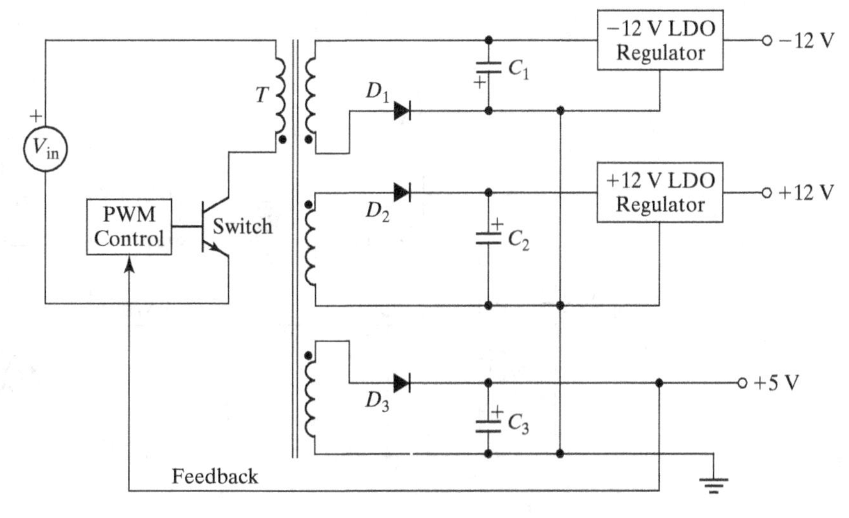

FIGURE 20.23

A switching voltage regulator circuit with multiple output voltages, including one with reversed polarity. (Courtesy of National Semiconductor Corporation.)

20.3.3.3 Practical Applications of Switching Voltage Regulators

In practice, designing with switching voltage regulators is more straightforward than it may seem after first exposure to their theory of operation. Because the performance of a switching regulator (regardless of the configuration used) depends on a great many factors, vendors of switching regulators work hard to simplify the design process. After all, the easier the devices are to use, the more devices they will sell. The easiest of the simplified approaches to use is that of a fully integrated solution, where the vendor has designed and assembled the entire circuit and provided it as a module to the user. These require only a connection to the input voltage and ground, and they provide a regulated output voltage. Texas Instruments provides a wide range of such modules, and a representative unit is shown in Figure 20.24. Such modules are relatively expensive, compared with designing and assembling a switching regulator from the elemental components, but the reduced engineering design time makes them attractive for lower volume products.

FIGURE 20.24

A fully integrated switching voltage regulator module from Texas Instruments.

Another highly successful approach was pioneered by National Semiconductor, which provides design and circuit modeling tools tied to their Simple Switchers product line of switching voltage regulators. The Web-based design software, called WebBench Tools®, requires straightforward high-level input from designers [e.g., input voltage range, output voltage(s), output current, ambient temperature, desired feature set], and then provides several candidate circuit designs, each optimized for cost, parts count, performance, efficiency, and so on. As a result, the designer's task is simplified to selecting the circuit that best fits the requirements of a given project. A typical step-down regulator circuit designed with the aid of WebBench Tools is shown in Figure 20.25.

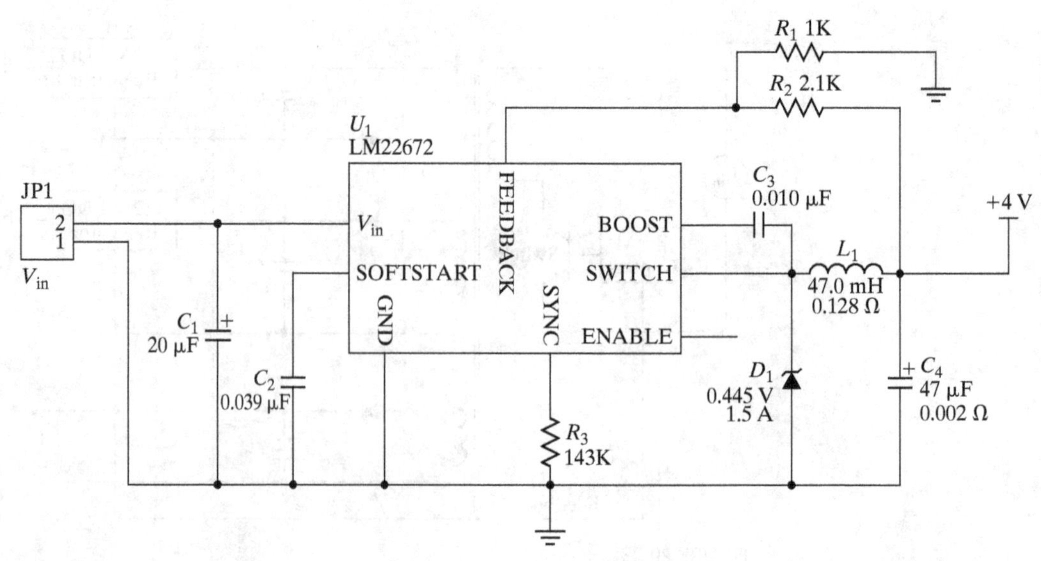

FIGURE 20.25

A step-down (buck) switching regulator circuit designed using WebBench Tools from National Semiconductor.

Because vendors have taken this approach, designing with switching voltage regulators is usually only slightly more complex than designing with linear voltage regulators. In a variety of applications, switching regulators provide performance that greatly exceeds that of linear regulators, such as creating an output voltage that is greater than the input voltage, creating multiple output voltages from a single input voltage in a single circuit, providing high levels of output current, and achieving high efficiencies. Typical switching voltage regulators achieve efficiencies in the 80–90% range, and the highest efficiency circuits can deliver better than 95% efficiency. The biggest drawbacks of using a switching voltage regulator are the increased parts count [the regulator, an inductor, and several capacitors and resistors (Figure 20.25)] and the higher voltage ripple or noise at the output. The output ripple makes switchers problematic in highly sensitive applications (e.g., high-resolution analog-to-digital converters, sensors, precision instrumentation). In general, however, switching voltage regulators have a great deal to offer in a wide variety of applications. A comparison of the performance of linear and switching voltage regulators is given in Table 20.1 to assist in choosing the most appropriate technology for a given design.

TABLE 20.1 A comparison of linear and switching voltage regulator performance. (Courtesy of National Semiconductor Corporation.)

Specification	Linear Voltage Regulators	Switching Voltage Regulators
Line Regulation	0.02–0.05%	0.05–1%
Load Regulation	0.02–0.1%	0.1–1%
Output Ripple	0.5–2 mV rms	18–71 mV rms
Efficiency	40–80%	> 80%
Power Density	0.5 W/in^3	2–5 W/in^3

20.4 POWER SUPPLIES

Linear and switching voltage regulators create the fixed, steady DC output voltage that most circuits require from an unregulated DC input voltage. This, of course, requires that you have access to a suitable DC input voltage, unregulated though it may be. The remainder of this chapter is devoted to a discussion of the two most common means of providing such DC input voltages: power supplies (this section) and batteries (Section 20.5).

The term **power supply** refers to devices that draw AC power from the electrical grid (120 VAC, 60 Hz in the United States) and convert it to a lower DC output voltage, suitable for supplying power (hence the name) to electrical circuits. Just as there were two major categories of voltage regulators (linear and switching), there are two major categories of power supplies: linear power supplies (which use linear voltage regulators at their outputs) and switching power supplies (which make use of switching voltage regulators). Since we've described how each type of voltage regulator works in the preceding sections of this chapter, the focus of this discussion is what happens between the AC power input and the voltage regulator—for example, how 120 AC line voltage can be translated to 5 VDC.

20.4.1 Linear Power Supplies

Linear power supplies create a stable, low-noise DC output voltage from an AC input voltage and incorporate a linear voltage regulator at their output stage. Linear power supplies tend to be more expensive, larger, and heavier than the alternatives (switching power supplies, Section 20.4.2) for a given quantity of output power. They are often used to power circuits where clean, low-noise power is a requirement: laboratory experiments, circuit prototypes, and measurement instrumentation, for example. Examples of commercial power supplies are shown in Figure 20.26.

(a) Bench-top, adjustable output supply for laboratory use

(b) Open-frame, fixed output

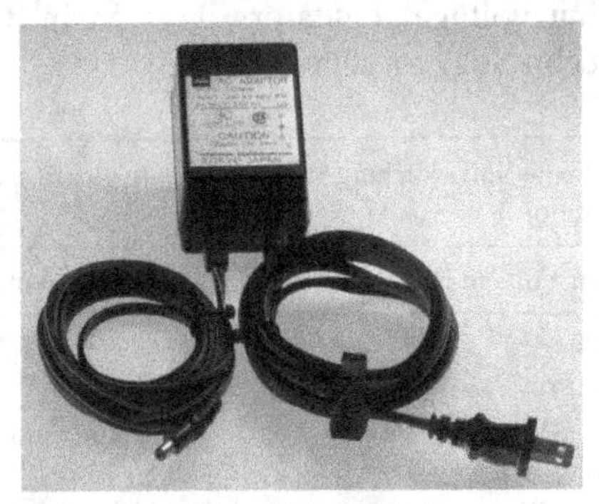

(c) Desk-mount

(d) Wall-mount, a.k.a. "wall wart"

FIGURE 20.26

A selection of linear power supplies.

Figure 20.27 is a block diagram that shows how linear power supplies work. The input to the power supply is AC power from an electrical outlet, called **line voltage**. In the United States, this is nominally 120 VAC, 60 Hz. Around the world, line voltage varies widely. In Japan, for example, it's 100 VAC at 50 Hz; in most of Europe, it's 230 VAC at 50 Hz. As a result, many electronic designs are specific to the country or countries where the devices will be used.

The first functional block of a linear power supply reduces the AC voltage from line voltage to a lower AC voltage, closer to the desired DC output voltage. A transformer is the standard means of reducing the voltage since it performs the task with minimal power losses. In Figure 20.27, the transformer has a turns ratio of 7:1 (secondary turns to primary turns), which reduces the voltage to 17 VAC. The next functional block is a rectifier, which converts the AC voltage to DC. Use of a full-wave rectifier is common, which allows the positive half of the AC sine-wave cycle to pass to the output unchanged but inverts the negative half of the cycle. The output of the rectifier resembles a succession of voltage humps. The next functional block uses a large capacitor to filter (or "smooth") the variations in the output from the rectifier, resulting in a quasi-steady—though still unregulated—DC voltage. The larger the value of this filter capacitance, the smoother the resulting voltage. The final stage is a linear voltage regulator that translates the unregulated DC voltage from the filter capacitor, ideally into a steady, low-noise regulated output voltage suitable for powering a circuit.

Figure 20.28 shows the schematic of a representative low-cost bench-top laboratory power supply, which closely parallels the block diagram of Figure 20.27. This supply accepts US line voltage (120 VAC, 60 Hz), and provides three regulated DC outputs: 5 V (fixed) at up to 3 A, 3 to 15 V (adjustable) at up to 1 A, and −3 to −15 V (adjustable) at up to 1 A. The

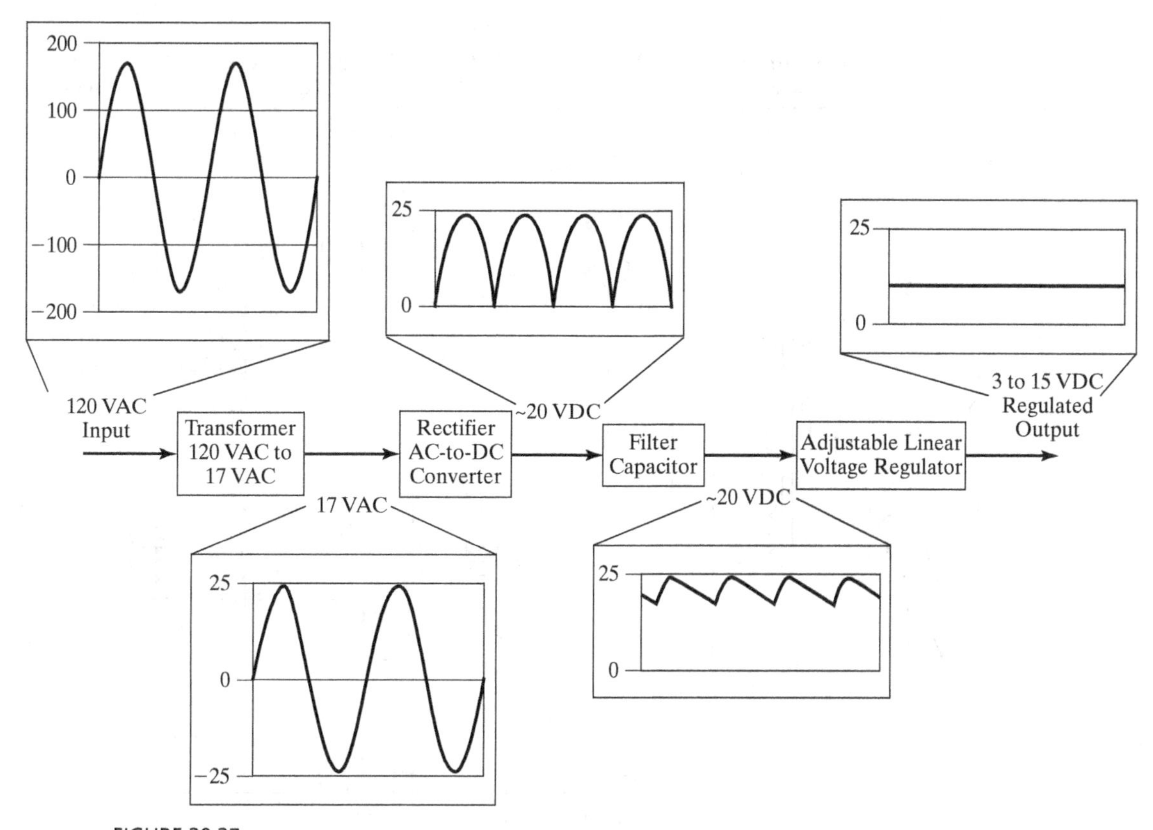

FIGURE 20.27

Linear power supply block diagram.

functional elements of the circuit are exactly as outlined in the block diagram (Figure 20.27), with a few specific elements added or rearranged for practical purposes. For example, AC power enters through an AC plug at the left of the circuit (P_1), and then flows through a fuse at F_1 (a wise precaution—a fuse will prevent a fire in the event of an anomaly that involves excessive current flow, such as a short circuit) and a switch (S_1). The next functional block and component is the transformer (T_1), which in this supply has a single primary and two secondary windings. The top secondary winding has a turns ratio of 7:1 and an output of 17 VAC, and the bottom secondary winding has a ratio of 13.3:1 and an output of 9 VAC. It's important to note that the two secondary windings of the transformer do not share a physical connection, which means that their respective grounds are not common (note the different ground symbols used for each section of the circuit). The voltages on the secondary side of the transformer have no specific relationship to each other until they are connected in some way by the user at the output. Many power supplies are intentionally configured this way (with floating outputs) to maximize their flexibility. Separate grounds are often connected together by the user, however, this is by no means required. Since the voltages are independent, they could potentially be "stacked" in a variety of arrangements. For example, the 5 V supply could be connected in series with the 15 V supply to achieve 20 V at up to 1 A.

The 9 VAC output of the transformer's bottom secondary is rectified with a full wave rectifier (D_5–D_8) and filtered by a 4,700 μF capacitor (C_7). Finally, this is regulated by an LM323 (U_3), a linear regulator with a fixed 5 V output that can provide up to 3 A. The 17 VAC output of the transformer's top secondary winding is separated into a positive voltage and a negative voltage through a clever arrangement of each half of the rectifier

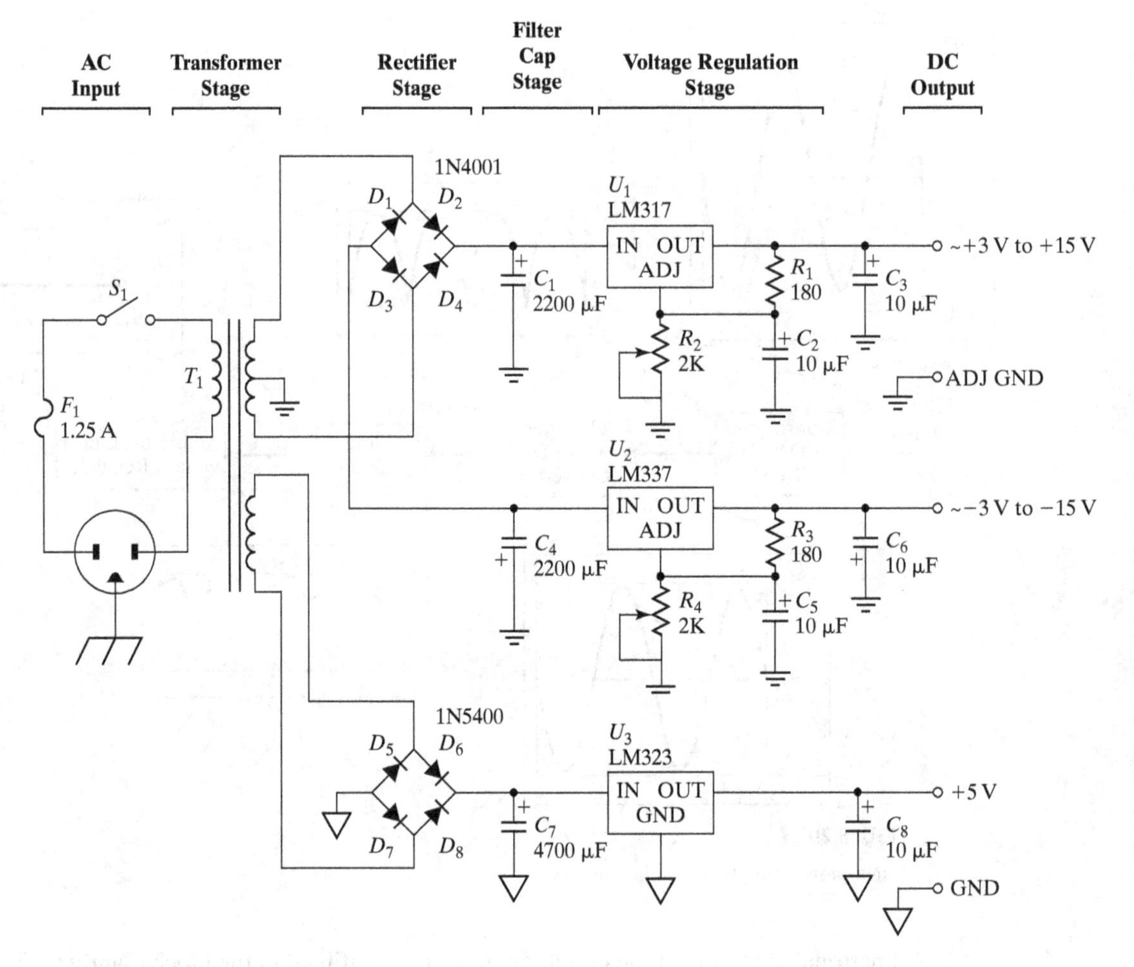

FIGURE 20.28

Schematic for a linear power supply with three outputs: 5 V at up to 3 A, variable +3 to +15 V at up to 1 A, and variable −3 to −15 V at up to 1 A.

bridge and the center tap of the transformer secondary. The positive and negative outputs of the rectifier are each filtered by a 2,200 μF capacitor (C_1 and C_4). The positive output is regulated by an LM317 (U_1), which is a 1 A adjustable linear regulator. A potentiometer (R_2) adjusts the output voltage between 3 and 15 V. The negative output is regulated by an LM337 (U_2), which is a 1 A adjustable linear regulator for negative voltages. Its output voltage is similarly adjusted via a separate potentiometer (R_4) to a value between −3 and −15 V.

The linear power supply shown in Figure 20.28 is quite basic, but incorporates elements common among linear power supplies. More costly and feature-rich linear supplies may include output voltages that are selectable via a serial interface or GPIB, very low noise, or very high stability. Top quality and feature-laden linear power supplies can be very expensive: many examples cost well over $1,000. Basic linear power supplies start in the $50–$100 range.

20.4.2 Switching Power Supplies

Linear power supplies are best for producing low-noise, stable output voltages. **Switching power supplies** (also called **switch-mode power supplies** or **SMPS**), on the other hand, have relatively noisy output voltages, but excel in areas where linear supplies don't: low cost, weight, size, and high power capacity are among their many strengths. For this reason, switching power supplies are the most common choice in applications where low-noise power is not a requirement. This category includes a great many devices, including personal computers, telephony equipment, and medical equipment, to name just a few. Figure 20.29 shows examples of commercial switching power supplies.

Because switching power supplies use switching voltage regulators at their output stage, the requirements for the circuits between the AC power input and the regulation stage are somewhat different from linear power supplies. Functionally, the differences are not great, however, the net result is a savings of cost, weight, and volume.

Switching voltage regulators operate at frequencies of at least 20 kHz (higher than audible frequencies for humans), and usually in the hundreds of kilohertz. The circuit's switching elements tend to introduce substantial **conducted emissions** (undesirable electrical noise fed back into the AC supply by a device), so the first functional block shown in Figure 20.30 is a filter to block this noise. A **line filter** prevents electrical noise from being fed back into the AC supply. If a line filter isn't used, the electrical noise may interfere with the function of other devices plugged into the same AC supply, such as a nearby computer, radio, or TV. The output of the line filter remains at 120 VAC and feeds into the next functional block of the power supply.

Recall that switching voltage regulators feature high efficiency—usually at least 80%, so unlike the linear power supply there is no inherent need to step down the input line voltage to a lower AC voltage prior to the regulation stage. This eliminates the need for a large, heavy, and costly input transformer, which comprised the first block of the linear power supply's block diagram (Figure 20.27). For a switch-mode power supply, the requirement for the voltage at the input of the regulation stage is simply that it be DC, so rectification and filtering stages are required. Since the line voltage (in the United States) is approximately 120 VAC rms, the output voltage of the rectification stage and the filter stage will be somewhere around 170 VDC (120 V/0.707). This can be fed directly into a switching voltage regulation

(a) Closed-frame

(b) Open-frame

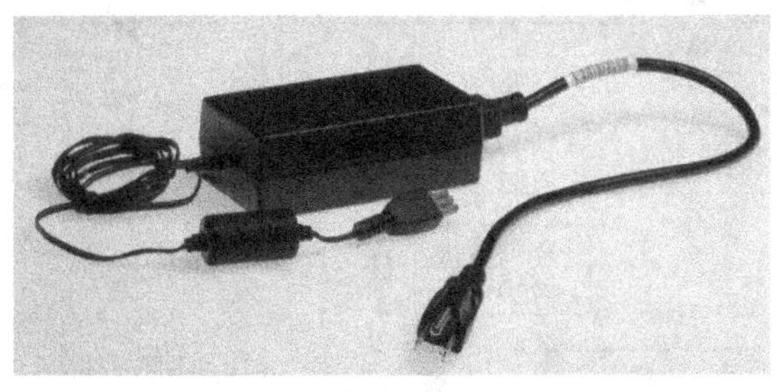

(c) Desk-mount

(d) Wall-mount ("wall wart")

FIGURE 20.29

A variety of switch-mode (switching) power supplies.

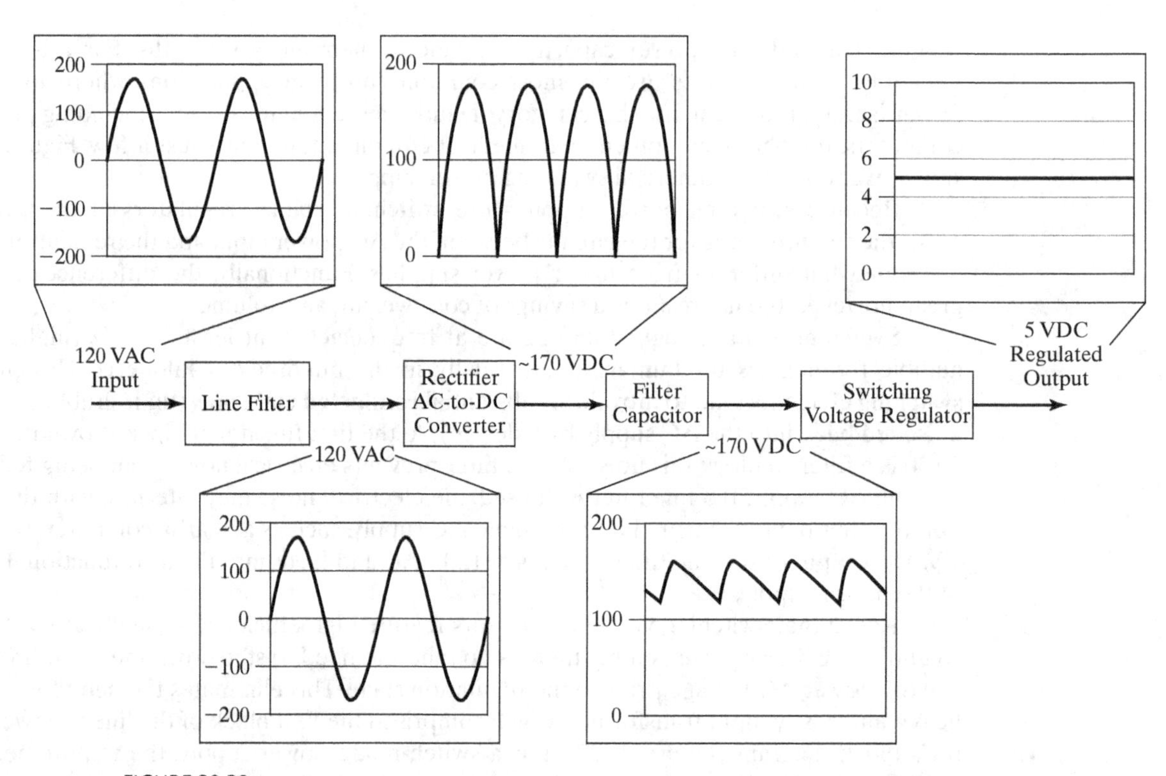

FIGURE 20.30

Switching power supply block diagram.

circuit, and a flyback configuration is especially well-suited for this application because of its use of a transformer to store and deliver energy to a load (Section 20.3.3.2), and the ability to efficiently create a regulated output voltage that is much less than the unregulated input voltage. Figure 20.31 is a simplified schematic that shows an implementation of the approach described in the block diagram.

In Figure 20.31, note that there is no electrical isolation between the input line voltage and the regulated output voltage. This is a potential safety hazard: a person completing a circuit between the output voltage and earth ground could potentially experience an electrical shock, and a short circuit between the output and earth ground could cause large amounts of

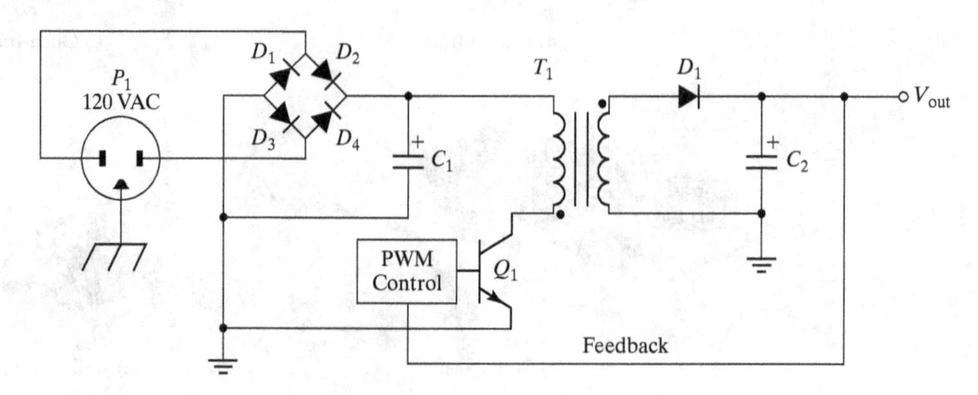

FIGURE 20.31

Simplified schematic for a line-powered switching power supply.

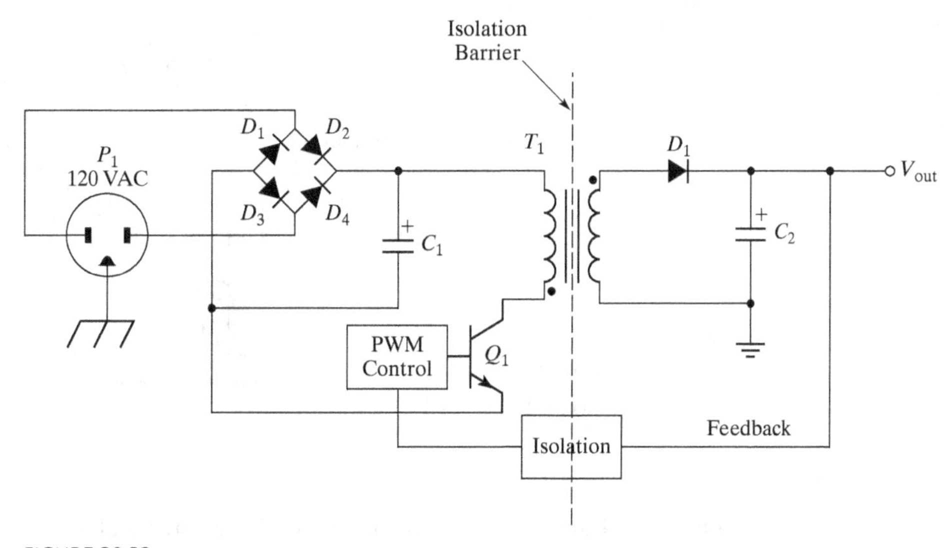

FIGURE 20.32

Simplified schematic for a line-powered switching power supply with electrical isolation.

current to flow. Because of this, electrical isolation is usually incorporated somewhere in the circuit. It could be achieved by using an input transformer like linear voltage regulators do, but it is less expensive and more elegant to isolate the feedback path from the output voltage to the control of the switching element at the primary side of the flyback transformer, as shown in Figure 20.32. This isolation can be achieved in a number of ways: optoisolation (the use of a paired LED and a photo-transistor, discussed in Chapter 21) is a common approach. Commercial switching power supplies are more complex than the circuit outlined in Figure 20.31, since they incorporate additional circuitry to optimize efficiency, short-circuit protection, and under- and over-voltage protection.

20.5 BATTERIES AND ELECTROCHEMICAL CELLS

A **cell** is the fundamental electrochemical element used to power portable devices that many of us use everyday: mobile phones, media players, portable computers, cars, toys—the list is long indeed. A **battery** is an assembly of cells, arranged either in series to provide greater voltage, in parallel for greater capacity, or a combination of series and parallel cells to achieve both greater voltage and capacity. In common usage, "battery" has become a general term that describes both cells and true batteries (combinations of cells), but in the strictest sense, this use of the word is imprecise. An example of a single cell and a battery is shown in Figure 20.33.

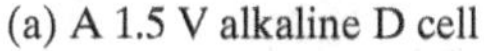
(a) A 1.5 V alkaline D cell

(b) A 9 V alkaline battery

FIGURE 20.33

A representative alkaline (a) cell and (b) 9 V battery (six alkaline cells in series).

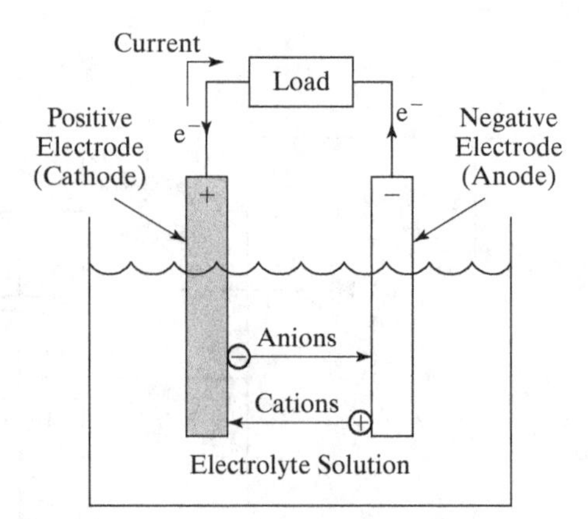

FIGURE 20.34
An electrochemical cell.

Figure 20.34 shows schematically how electrochemical cells are constructed. The essential components are a positive electrode (the cathode), a negative electrode (the anode), and an electrolyte solution that enables the positive and negative electrodes to chemically exchange cations (positively charged ions) and anions (negatively charged ions) during a chemical reaction. During use, the materials making up the anode and cathode react with the electrolyte and exchange electrons through a separate electrical circuit connecting the electrodes. The cell's stored chemical energy is exhausted when the chemical reaction nears completion. The materials selected for the electrodes and electrolyte determine the type and performance characteristics of the resulting cell, such as how much energy it can store, and whether it is rechargeable. In a typical alkaline cell, for example, the positive electrode is composed of manganese dioxide (MnO_2), the negative electrode is zinc (Zn), and the electrolyte is potassium hydroxide (KOH), often in gel form to reduce the likelihood of leaks.

To see how a cell works, we examine the chemical reactions occurring between each electrode and the electrolyte. Here again, the alkaline cell provides a great example. When current is drawn from a cell, the zinc anode combines with hydroxyl ions (OH^-) in the electrolyte to produce zinc oxide (ZnO), water, and (most importantly for our use of the device as a source of power) two electrons. A continuous reaction allows for the continuous production of electrons, and hence useful electrical current.

$$Zn + 2OH^- \rightarrow ZnO + H_2O + 2e \tag{20.8}$$

Meanwhile, at the cathode, the manganese dioxide cathode reacts with water in the electrolyte solution and uses two incoming electrons to produce manganite (MnOOH) and hydroxyl ions. Recall that the hydroxyl ions were needed by the zinc anode for its part of the reaction.

$$2MnO_2 + 2H_2O + 2e \rightarrow 2MnOOH + 2OH^- \tag{20.9}$$

The simplified reaction (i.e., just showing conditions before and after the reaction, and not showing the intermediate steps) is:

$$Zn + 2MnO_2 + H_2O \rightarrow ZnO + 2MnOOH \tag{20.10}$$

For an alkaline cell then, the zinc reacts and becomes zinc oxide, and the manganese dioxide becomes manganite, and two electrons are given off and reabsorbed. It's primarily the electrons that make the process interesting to us.

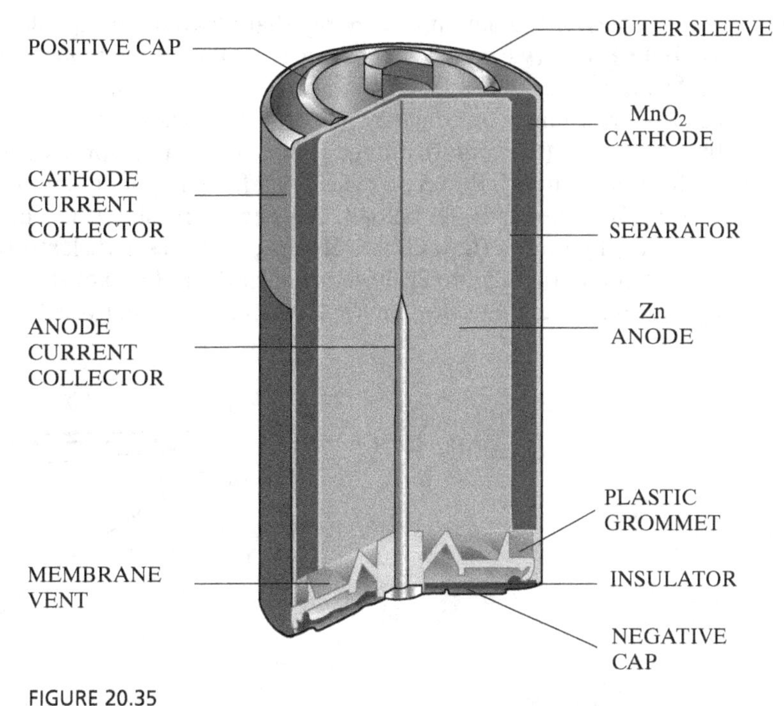

FIGURE 20.35

Anatomy of a Duracell alkaline cell. (Courtesy of Procter & Gamble Company.)

You aren't likely to run across any commercially available cells that resemble Figure 20.34—such a cell design would be bulky and prone to messy electrolyte spills. However, modern cells aren't all that different from the figure: they incorporate the same elements, but these are arranged in ways that make them highly reliable, robust, and unlikely to accidentally release the electrolyte solution. Figure 20.35 shows a cutaway view of a modern cell, in this case an alkaline cell made by Duracell.

The most notable differences are the compact concentric arrangement of the anode and cathode materials, and the use of a separator (a barrier which separates the anode and cathode physically and electrically, but is permeable to the electrolyte). A safety vent is also included at the base to allow for the controlled release of gases which can result from use of the cell beyond its specifications. Other than these clever design elements, however, the cell shown in Figure 20.34 is equivalent to the cell in Figure 20.35.

20.5.1 Battery Performance and Characteristics

If an ideal battery existed, it would store large amounts of energy, have a long service life, occupy little space, weigh little, cost little, and be unaffected by environmental factors. Your own experiences with battery-powered devices have probably convinced you that real batteries fall somewhat short of these ideals. When selecting a battery for a given application, designers must compare the strengths and weaknesses of the available battery technologies against the design goals, and select the best compromise. Would you prefer long battery life, or a very small device? Disposable or rechargeable batteries? In this discussion, we will define many of the terms used to describe battery performance which allow for meaningful comparisons. Then, in the next section, we look at a variety of batteries types, and how well they perform in each of these categories.

Capacity: The amount of energy that a battery is capable of storing, usually given the symbol **C**. Battery capacity is typically stated in units of amp hours (Ah) or milliamp hours (mAh). Capacity is measured by draining a fully-charged battery under constant current, constant resistance, or constant power conditions until the terminal voltage falls below a threshold level (the "**cutoff voltage**"—see below) that corresponds to the effective depletion of the stored energy. Battery capacity is affected by a number of factors, such as the rate of current flow (see C-Rate, below), cell chemistry, size (the quantity of reactants present), design configuration (how effectively the electrodes and electrolyte react), temperature, and storage conditions. Figure 20.36 shows the effect of two of these variables (different current flow rates and temperature) on a specific nickel-metal hydride battery.

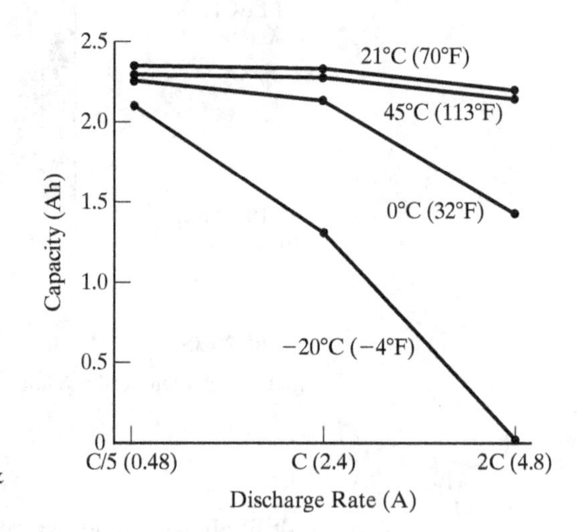

FIGURE 20.36

Influence of discharge rate and temperature on Duracell DR30 nickel-metal hydride (NiMH) batteries. (Courtesy of Procter & Gamble Company.)

C-rate: A measure of the rate at which current is drawn from a battery relative to its capacity. Stated as an equation:

$$\text{C-rate} = \frac{\text{Current}}{\text{Rated capacity}} \qquad (20.11)$$

A C-rate of 1 is a rate of current flow equal to the specified capacity of a battery and is written as "1 C." For example, for a 1,500 mAh battery, 1 C is 1,500 mA (or 1.5 A). A C-rate of C/10 for the same battery is equal to 1,500 mA/10 = 150 mA (a low level of current demand for this particular battery), and a C-rate of 2 C is 1,500 mA × 2 = 3,000 mA (a high level of current demand). In Figure 20.36, we see that draining the Duracell DR30 battery at 1 C at room temperature results in a capacity of about 2.3 Ah. This indicates that the battery can supply 2.3 A for 1 hour before the terminal voltage drops below the cutoff voltage.

Energy density: A measure of how much energy is stored in a cell or battery per unit mass, usually stated in watt hours per kilogram (Wh/kg).

Volumetric energy density: A measure of how much energy is stored in a cell or battery per unit volume, usually stated in watt hours per liter (Wh/L).

Discharge curve: A graph of a battery's voltage as a function of time under specific discharge conditions. A typical discharge curve has its maximum voltage at the initial value, and the voltage decreases as the battery is discharged. Often, but not always, there is a plateau region where the voltage is relatively constant, and finally a knee where the voltage drops off precipitously, signaling the depletion of the battery's stored energy. Most often, the discharge curve is shown for a given C-rate, although it may also be given for constant resistance load (in which case the current decreases over time since the cell voltage decreases as it discharges) or constant power. Like battery capacity, the characteristics of the discharge curve are strongly influenced by the temperature of the cell (lower temperatures slow down the chemical reactions in the cells, while higher temperatures speed things up, see Figure 20.37a) and by the amount of current drawn from the battery (higher current flow rates result in lower capacity, see Figure 20.37b).

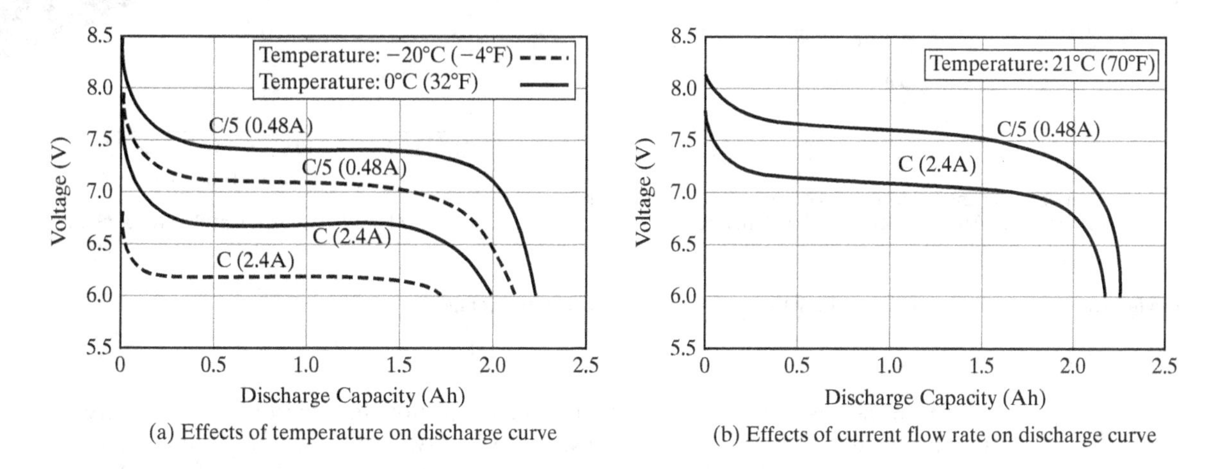

(a) Effects of temperature on discharge curve

(b) Effects of current flow rate on discharge curve

FIGURE 20.37

Discharge curves for the Duracell DR30 NiMH cell, showing the impact of (a) varying temperature and (b) current flow rate. (Courtesy of Procter & Gamble Company.)

Cutoff voltage, also called the **end voltage**: The voltage at which a cell is said to be effectively discharged. The cutoff voltage is a quantity that may be determined by cell chemistry (e.g., the nominal cutoff voltage of an alkaline cell is 0.8 V, corresponding to the point where 95% of its capacity has been consumed), or in a specific application (e.g., the cutoff voltage of a cell in a specific toy is 1.1 V, because the device will not function normally once the battery voltage drops below that threshold).

Open-circuit voltage: The voltage measured at the terminals of a cell or battery in the absence of any electrical load. This parameter is not particularly useful, because the cell or battery is not performing any useful work under these conditions, and the voltage will be lower when a load is connected.

Closed-circuit voltage: A cell's terminal voltage measured when subjected to a specified electrical load.

Midpoint voltage: A cell's terminal voltage measured when the cell is 50% discharged.

Self-discharge: The reduction in capacity that occurs when a cell is unused for an extended period. Self-discharge is a result of a low level of continuous chemical reaction occurring within a cell, even though no electrons are drawn from the cell terminals. The degree of self-discharge depends heavily on the cell chemistry. For example, alkaline cells typically lose less than 5% of their capacity per year when stored at room temperature, but NiMH cells will lose on the order of 20% *per month* under similar conditions. Storage temperature also has a large influence on the self-discharge rate, as shown in Figure 20.38. Here, the slower reaction speeds that result from lower temperatures have a favorable effect: less of the cell's energy is lost during storage at low temperatures.

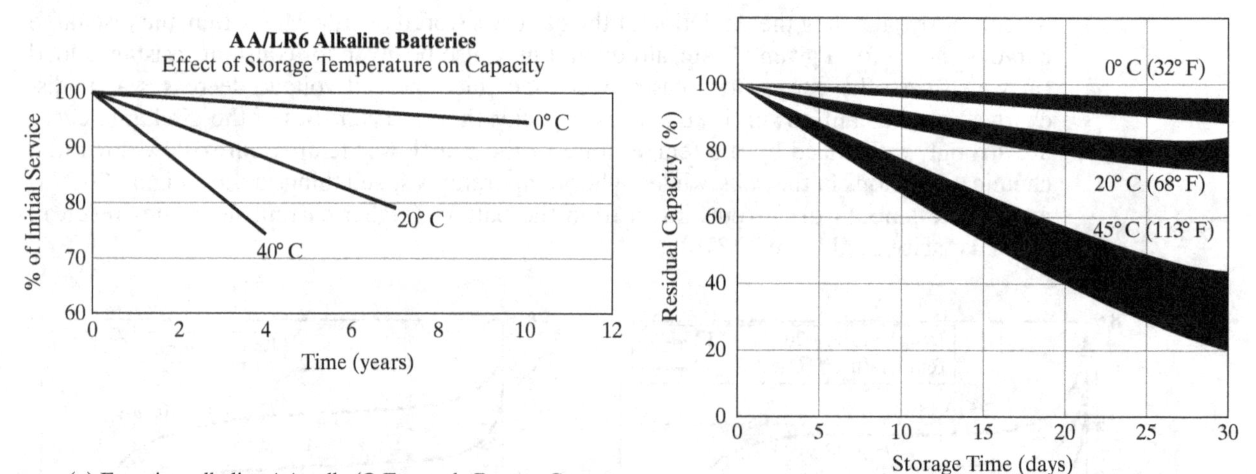

(a) Energizer alkaline AA cells. (© Eveready Battery Company, Inc., St. Louis, Missouri. Reprinted with permission.)

(b) Duracell NiMH cells. (Courtesy of Procter & Gamble Company.)

FIGURE 20.38
The influence of time and temperature on self-discharge rate during battery storage.

Internal resistance: The resistance to current flow in a battery-powered circuit due to the battery itself. A cell's internal resistance has two primary components: a smaller first component due to the ohmic characteristics of the electrical elements and connections within the battery (e.g., the terminals, internal connections, and wiring), and a larger second component called ionic resistance, which is a function of the chemistry and construction of the cell. To determine the overall internal resistance, a cell is first subjected to a low level of current flow (a few milliamps) and a terminal voltage measurement is taken. The cell is then subjected to a step increase in current flow to a much higher level of current flow (several hundred milliamps), and the terminal voltage is measured again after the terminal voltage has stabilized at a new value.

In Figure 20.39, the initial low level of current flow is 5 mA, and the higher level is 505 mA, a difference of 0.5 A. The terminal voltage has an initial value at 5 mA of 1.485 V and a final value of 1.378 V at 505 mA, a difference of 0.107 V. Using Ohm's law, the total internal resistance of the cell is $R = 0.107\text{ V}/0.5\text{ A} = 0.214\ \Omega$. Internal resistance is primarily affected by cell chemistry and cell design. Low internal resistance is a highly desirable quality in a cell, resulting in lower internal losses to heat, and higher maximum current flow.

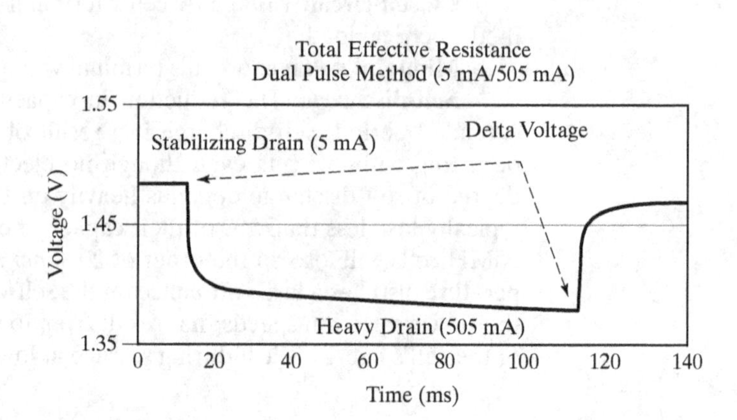

FIGURE 20.39
Measuring the internal resistance of a cell. (© Eveready Battery Company, Inc., St. Louis, Missouri. Reprinted with permission.)

20.5.2 Primary Batteries

Depending on the materials used to construct cells and batteries, the chemical reaction may be irreversible (in which case the cell is not rechargeable) or reversible (in which case the cell is rechargeable). Cells that are not designed to be recharged are called **primary cells**, and similarly, batteries constructed from primary cells are called **primary batteries**. Primary cells and batteries are disposable power sources: once the chemical reaction is exhausted, they must be replaced with new ones. Table 20.2 shows the chemistry and characteristics of the most commonly used types of primary cells.

The energy density characteristics of primary (nonrechargeable) batteries are typically much higher than those of rechargeable batteries (see Section 20.5.3). Though the use of primary batteries involves the inconvenience and cost of periodic replacement and disposal when their energy stores are exhausted, their low cost, wide availability, high energy density, and compactness make them attractive choices for many applications. Sometimes, they are the only practical choice, such as hearing aids.

Up until the 1940s, **carbon-zinc** batteries were the only commercially available primary batteries. Since the late 1960s, carbon-zinc has been largely replaced by **alkaline batteries**, which share similar chemistries but have higher energy storage capabilities—more than twice as much. This is achieved through the use of a different electrolyte solution and cell designs. Though alkaline batteries cost more than carbon-zinc batteries, they are a more cost effective choice because of their longer service life. As of mid-2009, alkaline batteries are the baseline primary battery choice for typical applications. Special design requirements, such as very long life or very compact power sources, may drive designers to select an alternative.

Other primary battery options include **silver oxide**, **zinc-air**, and a group called **lithium cells** (the anodes of which are comprised of lithium, while the cathodes may be any of a variety of different materials, such as manganese dioxide, sulfur dioxide, thionyl chloride, and others). Silver oxide and zinc-air cells feature much higher energy density than alkaline batteries and are common choices for compact devices. Zinc-air batteries deserve special mention because of their superior energy density and volumetric energy density characteristics. This type of

TABLE 20.2 Primary cell chemical composition and characteristics.

Chemistry	Anode	Cathode	Midpoint Voltage	Energy Density (Wh/kg)	Volumetric Energy Density (Wh/L)	Applications
Carbon-zinc	Zn	MnO_2	1.5	85	165	Toys, radios, flashlights
Alkaline	Zn	MnO_2	1.5	145	400	Consumer electronics, toys, radios, flashlights
Silver-zinc	Zn	Ag_2O	1.6	135	525	Watches, cameras
Zinc-air	Zn	Ambient air	1.5	370	1300	Hearing aids, pagers
Lithium-manganese dioxide	Li	MnO_2	3.0	230	535	Watches, cameras, thermometers
Lithium-sulfur dioxide	Li	SO_2	3.0	260	415	Flashlights, emergency equipment
Lithium-thionyl chloride	Li	$SOCl_2$	3.6	590	1100	Memory backup, security systems, measurement instruments

battery has a zinc anode, but carries no cathode material on board—it's essentially all "fuel" and no oxidant, and hence can achieve the highest energy density figures. It does, of course, have a cathode: the oxygen in the ambient air, which conveniently surrounds the cell in adequate concentration almost everywhere on earth. Small perforations in the housing of a zinc-air battery casing allow the free exchange of air with a porous carbon electrode to form the electrical contact. Zinc-air batteries are shipped with an adhesive cover over the air holes to keep the oxygen out and minimize self-discharge until the device is ready to use.

Primary cells with lithium anodes (Li/MnO_2, Li/SO_2, and $Li/SOCl_2$) also have high energy density characteristics, as well as a significantly higher cell voltage. Other cell chemistries shown have a midpoint voltage of around 1.5 V, whereas lithium cells are at least 3 V. In applications that require higher battery voltages, this reduces the number of cells needed. In particular, lithium-thionyl chloride ($Li/SOCl_2$) cells have high midpoint voltage (3.6 V) and energy density that competes well with the zinc-air cell. However, $SOCl_2$ is highly toxic and corrosive.

20.5.3 Secondary Batteries

Rechargeable cells are called **secondary cells**, and rechargeable batteries are **secondary batteries**. A secondary cell or battery is recharged when the chemical reaction is caused to proceed in reverse by introducing a flow of electrons (e.g., an electrical current) into the device. If the reaction were perfectly and completely reversible, rechargeable cells would have an infinite life. Sadly, this is not the case. However, many types of secondary cells/batteries can be recharged a great number of times (many types allow for several hundred discharge-charge cycles, and some can achieve over a thousand). Examples of secondary cells include nickel-cadmium (NiCd), nickel-metal hydride (NiMH), lead acid, lithium ion, and lithium polymer cells (Table 20.3).

TABLE 20.3 Secondary cell chemical composition and characteristics.

Chemistry	Anode	Cathode	Midpoint Cell Voltage	Energy Density (Wh/kg)	Energy Density (Wh/L)	Discharge-Charge Cycle Life	Applications
Lead acid, sealed lead acid (SLA)	Pb	PbO_2	2.0	35	70	200–300	Vehicles, tools, stationary equipment, medical instruments
Nickle-cadmium	Cd	Ni oxide	1.2	35	100	1000–1500	Toys, power tools, medical instruments
Nickle-metal hydride	MH	Ni oxide	1.2	75	240	500–1000	Toys, power tools, medical instruments, electric vehicles
Lithium ion	Li_xC_6	$Li_{(i-x)}CoO_2$	4.1	150	400	500–1000	Consumer electronics, portable computers, electric vehicles
Lithium polymer	Li	$LiCoO_2$	3.7	200	300	500–1000	Consumer electronics, portable computers, electric vehicles

FIGURE 20.40

An SLA battery (6 V, 1.3 Ah).

Whether you've had direct experience designing with the venerable **lead acid** battery (invented in the mid-1800s), it's likely that you use one each day. Lead acid batteries are standard in vehicle applications (cars, motorcycles, boats, buses, planes, and so on) to supply current to the electrical systems when the primary engine either isn't running or as a backup when the engine isn't rotating fast enough to generate adequate electrical power on its own. For example, the battery powers the starter motor and can power the radio and headlights when the car's engine is off. Lead acid batteries are very inexpensive and are capable of very high current flow rates, however, they have low energy densities, and both the lead and the sulfuric acid used in their construction are significant environmental hazards. Lead acid batteries designed for use in electronic devices are usually sealed to prevent the acid from leaking, and the acid used is in gel form, rather than liquid (in which case they are called **gel cells**). **Sealed lead acid** (or **SLA**) batteries aren't truly sealed—their internal pressure is valve regulated so that they may vent any gases produced during charging or discharging (Figure 20.40).

The performance characteristics of **nickel-cadmium** (NiCd) batteries aren't much better than SLA batteries in terms of energy density; however, they can be discharged and recharged many more times before their capacity decreases significantly. Typical SLA batteries can be recharged 200–300 times, compared with 1,000–1,500 times for NiCds. NiCd batteries also have very low internal resistance (on the order of 100 mΩ, depending on the design of the cell) and can tolerate very rapid discharge at very high currents. For these reasons, they are popular in the radio-controlled vehicle and cordless tool markets. A significant issue for NiCd batteries is the use of cadmium as the anode, since cadmium is a highly toxic heavy metal. Manufacture and disposal of NiCd batteries is problematic, as a result, and their popularity is fading.

Largely because of the nickel-cadmium battery's low energy density and toxicity issues, **nickel-metal hydride** batteries (NiMH) have become popular in applications where NiCds once dominated. Nickel-metal hydride batteries are superior to nickel-cadmium both in energy density (which is on the order of twice that of NiCd) and environmental concerns (the metal hydride anode is not hazardous). The number of charge cycles that a NiMH battery can provide, however, is at the low end of the range of NiCd. However, given that NiMH batteries store approximately twice the energy of NiCd, the balance remains tipped in their favor. Both NiCd and NiMH have very high self-discharge rates: approximately 20% per month at room temperatures (see Figure 20.38). Figure 20.41 shows a typical packaged three-cell NiMH battery pack.

Lithium ion (Li-ion) and **lithium polymer** (Li-poly) batteries also achieve high energy density values, in addition to having a higher cell voltage. These cells also lend themselves well to custom, noncylindrical shapes, and are often designed into devices where space is at a premium. As of mid-2009, Li-ion and Li-poly batteries are the preferred choice for mobile phones, media players, and portable computers. Designing with Li-ion and Li-poly batteries

FIGURE 20.41
A three-cell NiMH battery (3.6 V, 1150 mAh).

requires extra precautions and a great deal of care relative to SLA, NiCd, and NiMH batteries. Overcharging them can result in an explosion, as can short circuiting or physically damaging a battery. Also, deeply discharging a Li-ion or Li-poly battery can result in permanent damage. Because of the importance of proper technique when charging and discharging Li-ion and Li-poly batteries, protection and monitoring circuits are usually needed, and are often manufactured into the batteries themselves. This improves their safety profile and reliability, but also increases cost. Figure 20.42 shows a typical Li-ion cell phone battery.

20.5.3.1 Charging Secondary Batteries

Ideally, charging (or recharging) secondary batteries would be simple, quick, safe, and reliable. The degree to which these goals can be simultaneously achieved varies greatly by cell type, and in this section, we briefly describe the techniques used to restore the energy in a secondary cell.

The simplest approach is to charge the battery at a constant voltage with a low level of current. This is called **trickle charging**. With this technique, it takes a long time to fully recharge the battery (this depends on its capacity); however, minimal controls are required to monitor the cells and adjust the behavior of the charger. Trickle charging can be used with almost any type of cell or battery and is usually the most inexpensive (though time consuming) choice. To speed up the charging process, a more active approach is required. Usually, rapid charging requires a multistep approach. For example, a full charge cycle may involve a constant current stage, a constant voltage stage, a voltage limited stage, a temperature limited stage, and/or a trickle charge stage. There are other possibilities as well, such as monitoring the rate of change of cell voltage (dV/dt) and temperature (dT/dt).

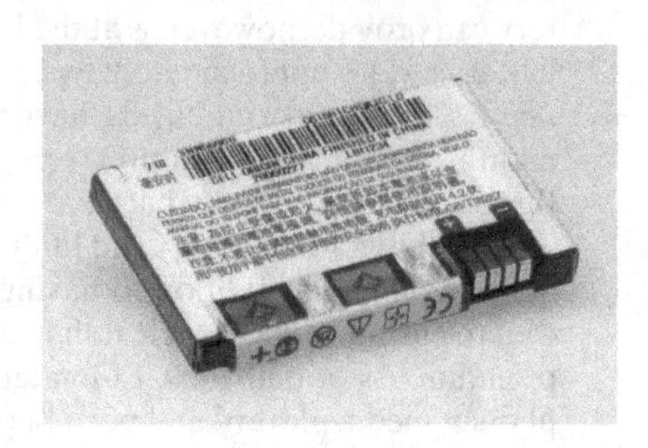

FIGURE 20.42
A Li-ion cell for a mobile phone.

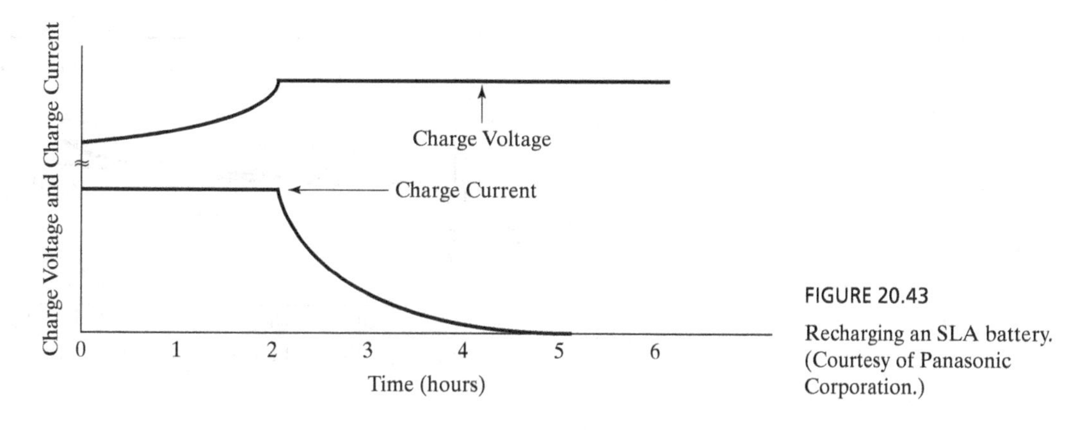

FIGURE 20.43

Recharging an SLA battery. (Courtesy of Panasonic Corporation.)

Recharging an SLA battery provides an excellent example, and a typical approach is shown in Figure 20.43. The first stage of the charge cycle is a **constant current** stage. In this example, the current is held at a relatively high current level for 2 hours. A typical choice for this stage is 0.4 C or higher for a given battery. During this time, the cell voltage rises steadily, but then begins to level off. When the cell voltage reaches a threshold (about 2.45 V/cell for SLA batteries), the charger switches to a **constant voltage** stage, where it maintains this cell voltage and modulates current flow. Once the current required to maintain the cell voltage decreases and approaches a low, nearly constant value, the battery is fully charged. At this point, charging may cease, or the charger may enter a third stage. The **float charge** (i.e., trickle charge) stage keeps the battery fully charged if significant time elapses between the end of the second stage and use of the battery.

Lead acid batteries are relatively forgiving of imprecise charging, and the terminal voltage is the only parameter needed to determine when to change stages of the charging cycle. Severely overcharging a lead acid battery at high C-rates can result in the production of O_2 and H_2 gas through electrolysis, and in high enough concentrations, the mixture is explosive. This is particularly undesirable in a battery filled with a sulfuric acid solution. So exercise caution when recharging lead acid batteries, and use the appropriate charging techniques.

Recharging nickel-cadmium or nickel-metal hydride batteries is similar in many respects. The major difference between recharging SLA batteries and NiCd or NiMH batteries comes at the end of the process: deciding when to terminate the charging cycle. To maximize the life of both NiCd and NiMH batteries, they should be fully recharged; performing an incomplete recharge cycle would be a conservative way to avoid damaging the devices, but this would come at the expense of a considerable fraction of the battery's usable life and value.

To rapidly charge NiCd and NiMH batteries, a constant current charge cycle is used. This can be done at high current rates (up to 1 C, depending on the battery). Overcharging either battery type with high currents can result in damage to the battery, so a means of determining when the batteries are fully charged is needed. Both NiCd and NiMH batteries exhibit distinctive voltage and temperature characteristics that can be used for this purpose. As shown in Figure 20.44, the cell voltage increases as the charge input nears 100% of capacity, and then begins to decrease just as it achieves 100%. For NiCd, the effect is more pronounced than for NiMH, but is observable in both cases. As the cell nears a full charge, the temperature also begins to increase rapidly, and if a means of temperature measurement is available, this can also be used to determine when to terminate the charge cycle. Some NiCd and NiMH battery packs have temperature sensors (often an NTC thermistor, see Chapter 13) manufactured into the packaging to ensure accurate measurements.

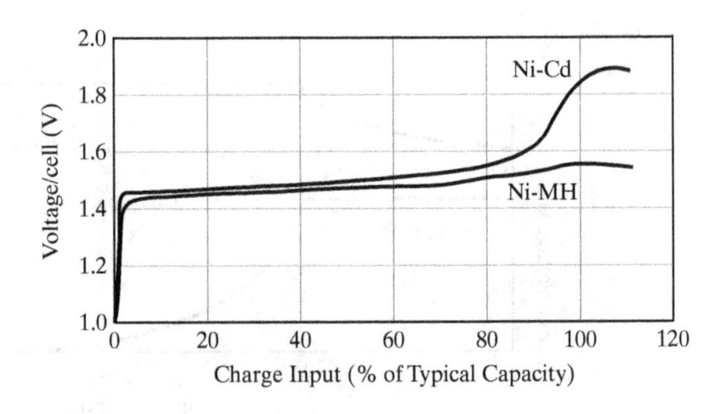

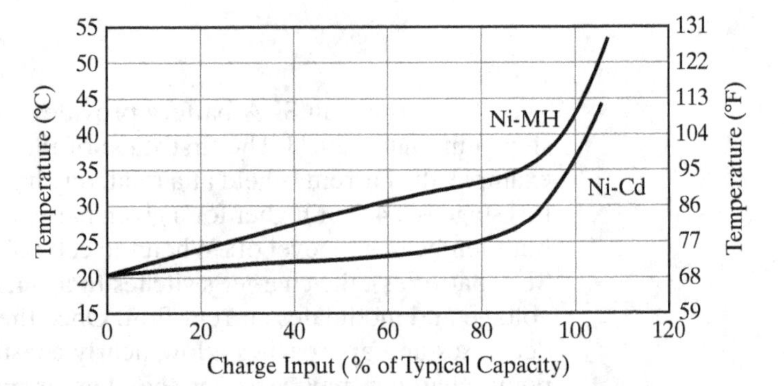

FIGURE 20.44

Voltage (top) and temperature (bottom) characteristics of NiCd and NiMH cells during charging. (Courtesy of Procter & Gamble Company.)

When the voltage peaks and then decreases by a threshold amount (labeled $-\Delta V$ in Figure 20.45), and when the cell temperature has either reached a threshold maximum value (labeled TCO) or the temperature change has reached a threshold rate (labeled dT/dt), the charger will terminate the charge cycle.

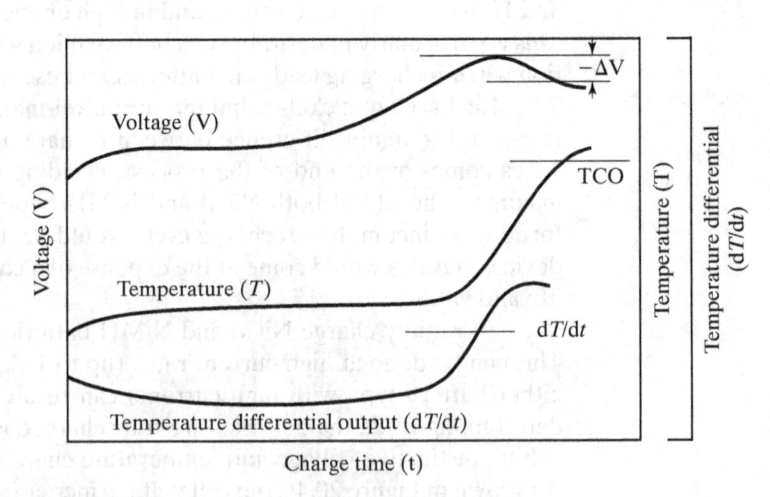

FIGURE 20.45

Determining when to terminate charging of a NiCd or NiMH battery using cell voltage and temperature. (Courtesy of Procter & Gamble Company.)

Both nickel-cadmium and nickel-metal hydride batteries exhibit a **memory effect**, though this is much more pronounced for nickel-cadmium. This occurs in batteries that have been repeatedly partially discharged. Each time the battery is partially discharged and then recharged, its discharge curve is altered such that the terminal voltage is reduced slightly. If terminal voltage is used as an indication of the state of charge of the battery (as it often is), this has the apparent effect of reducing the battery's capacity, as the lower voltage is reached sooner each time the battery is used. For this reason, the memory effect is also referred to as the **voltage depression** effect. The memory effect is not permanent; however, fully discharging and then recharging a battery a few times eliminates it.

Figure 20.46 illustrates the memory effect for a representative Duracell NiMH cell. For the discharge curve labeled cycle #1, the cell was discharged at 1 C to a terminal voltage of 1.0 V, and then recharged at 1 C until $-\Delta V = 12$ mV. For cycles #2 through 18, the cell was discharged at 1 C to a terminal voltage of 1.15 V (i.e., only partially discharged), and then recharged at 1 C until $-\Delta V = 12$ mV. During each of these discharge-charge cycles, note that the time required to reach the 1.15 V cutoff was progressively shorter. By cycle #18, the time had been reduced by about 0.16 hours (21%)—a very significant and noticeable amount. For cycles #19 through 21, the cell was fully discharged to 1.0 V each time and recharged as in the preceding cases. At the end of these three restorative cycles, the discharge curve for cycle #21 very closely resembles that of cycle #1: the memory effect has been almost entirely "erased."

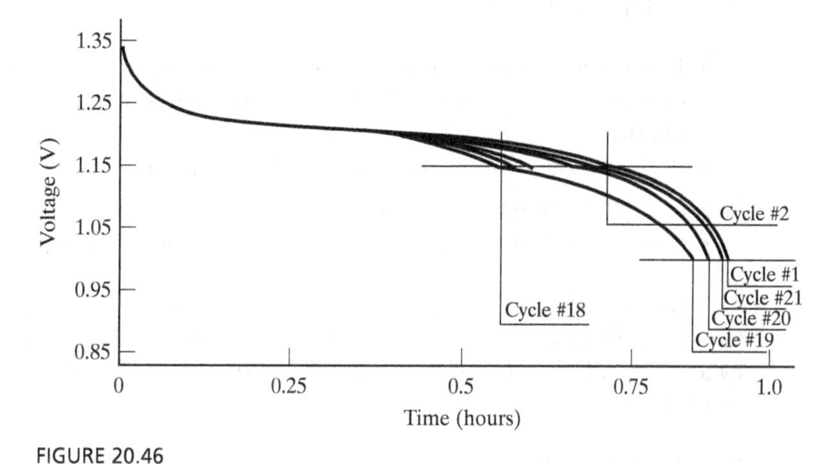

FIGURE 20.46

The battery memory effect, exhibited by NiCd and NiMH batteries. (Courtesy of Procter & Gamble Company.)

Charging and discharging Li-ion and Li-poly batteries are similar in many respects to the process for SLA batteries (see above). There are, however, a few important differences: the cells can be permanently damaged, or catch fire and burn vigorously, or even explode if used incorrectly. As a result, extra caution is highly recommended. Figure 20.47 illustrates a standard approach to recharging a Li-ion cell. The first two charging stages are the same for Li-ion and the SLA battery (above), but the third stage is different. The first is a constant

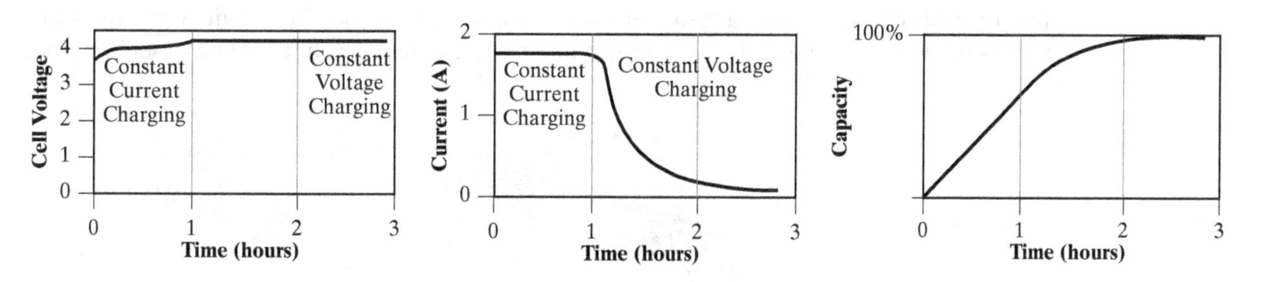

FIGURE 20.47

Charging characteristics of a SANYO UR18650F Li-ion cell. (Courtesy of SANYO Electric Co., Ltd.)

current stage, with a high level of current flow (about 1.75 A in the example). This stage continues until the cell voltage has reached a threshold of 4.2 V. Once this threshold is reached, a second constant voltage stage begins. The cell voltage is maintained at 4.2 V while controlling the flow of current. Once the current flow rate has decreased to a low threshold, the constant voltage stage ends and the cell is fully charged. The current threshold that triggers the end of the second stage can be battery specific, but 3% of rated current (0.03 C) is a typical choice. Once the second stage is complete, the cell is fully charged. For Li-ion and Li-poly cells, a continuous trickle charge can result in overcharging and permanently damaging the cells, so the topping charge may be omitted altogether, or may be supplied intermittently, such as 1 out of every 500 hours. This makes up for the self-discharge rate of Li-ion or Li-poly cells, and also prevents overcharging and damaging the battery.

Because there are several ways for the charging process to go awry with Li-ion and Li-poly cells, it is common to use a specialized integrated circuit, called a **fuel gauge chip**, to manage the process. Fuel gauge chips measure a cell's voltage, temperature, and current flow to determine the following:

- Remaining capacity (how much of its charge remains?)—they may even have models of how cells age over multiple discharge-charge cycles to improve accuracy over the life of a cell
- Cell under-voltage (is the cell over-discharged?)
- Cell over-voltage (is the cell over-charged?)
- Short-circuit conditions (is the current flow above a selectable limit?)

Many of these devices also manage the process of charging the cells, which is a natural function for them, given the information they track about a cell. Fuel gauge chips are offered by a variety of manufacturers, including Maxim Integrated Products, TI, Linear Technology, and Intersil.

20.5.4 Battery Safety and Environmental Issues

No discussion of batteries would be complete without a few words about the related safety and environmental issues. As stated above in the general definition, cells and batteries are devices that store chemical energy and convert it into electrical energy. Most batteries are capable of doing this very rapidly—much more rapidly than we intend, in some cases. Under anomalous conditions (like a short circuit), all of a battery's stored energy can be suddenly released, and this can result in a very rapid increase of battery temperature, a fire, or even an explosion. In addition to the obvious risk to the people nearby (imagine this happening to the mobile phone in your pocket), this also has an impact on the future performance of the battery and the device it powers. To mitigate the risk that these events pose, nearly all battery-powered devices incorporate a **fuse** in series with the battery. When current flow through a fuse exceeds its rated value, a fusing element heats up and breaks the circuit. Some fuses are single-use devices (Figure 20.48a), in which the fusing element gets hot enough to self-destruct

(a) Single-use fuse

(b) Resettable fuse

FIGURE 20.48

Two types of fuses.

and thereby break the circuit. Other fuses are multiple-use devices, such as the resettable fuse shown in Figure 20.48b. This is a positive temperature coefficient thermistor (a PolySwitch, made by Tyco Electronics) whose resistance increases sharply when its temperature rises above a threshold level (PTC thermistors are discussed in Chapter 13, Section 13.5.3.2).

It's good design practice to incorporate a fuse of some sort in all your designs. For many products, it is required by regulatory bodies. The primary function of a fuse is to prevent a fire or explosion in the event of unintentionally high current flow. Though your designs may never experience excessive current flow or anomalous conditions and the fuses might never be called into service to save the day, if and when they do, you (and your customers) will be glad you included them.

In addition to the risk of fires and explosions, some of the chemicals used to make cells and batteries are highly toxic and environmentally hazardous. Another good design practice is to avoid designing with the most toxic battery technologies, such as nickel-cadmium or lithium-thionyl chloride. Though battery recycling is available in most regions within the United States, it's not possible to ensure that the batteries you design into a device will be properly disposed of. And if you design a device that enters high volume manufacturing, it's quite realistic that you may literally create a mountain of toxic spent batteries when the devices are retired. A sure way of preventing this is to choose less toxic battery chemistries when possible. There are good alternatives to most of the more toxic battery chemistries (e.g., use NiMH instead of NiCd batteries), so in most cases it's not difficult to do the right thing.

20.6 HOMEWORK PROBLEMS

20.1 A 5 V LDO voltage regulator has a maximum dropout voltage specification of 0.21 V. If it is to provide a stable output of 5 V, what is the minimum voltage that must be supplied at the input of the regulator?

20.2 What is the efficiency of the LDO voltage regulator described in Problem 20.1 if the input voltage is the minimum voltage that results in a regulated 5 V output? What is the efficiency if the input voltage is increased to 9 V? 15 V?

20.3 An LM7805 linear voltage regulator is used to provide a stable output voltage of 5 V. What is the minimum voltage that must be supplied at the input of the regulator to ensure the output is at 5 V?

20.4 What is the efficiency of the LM7805 voltage regulator described in Problem 20.3 if the input voltage is the minimum voltage that results in a regulated 5 V output? What is the efficiency if the input voltage is increased to 9 V? 15 V?

20.5 For the circuit shown in Figure 20.49, the LED has a forward voltage drop $V_f = 2$ V. How much heat is dissipated in the regulator? In the resistor? In the LED? If the ambient temperature is 40°C, is the LM7805 operating within specifications (use the specifications in Figure 20.14)? If the resistor is rated at 1/4 W and the LED is rated for a maximum continuous current of 40 mA, are these components operating within specifications?

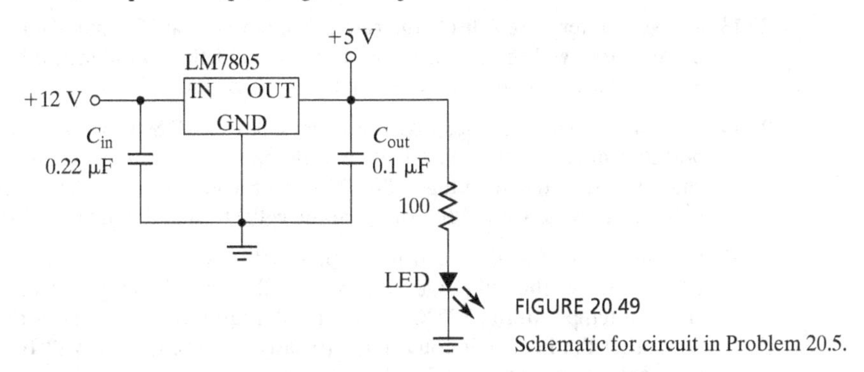

FIGURE 20.49
Schematic for circuit in Problem 20.5.

20.6 A switching voltage regulator circuit operates at 85% efficiency, and its input voltage is 24 V. If it provides a regulated output of 12 V at 2 A, what is the average current flowing into the input? How much power does the regulator circuit dissipate under these conditions?

20.7 Your bench-top laboratory power supply has three regulated outputs. The first is a fixed +5 V output, capable of supplying up to 3 A. The second is an adjustable 0 to +15 V output that can supply up to 1 A, and the third is an adjustable 0 to −15 V output, also capable of up to 1 A (Figure 20.50). The ground for the fixed output is independent of the ground for the adjustable outputs. To power a new prototype circuit, you need 18 V and a maximum of 0.8 A. Can you use this power supply to power the circuit? If so, show how it will be configured. If not, explain why not.

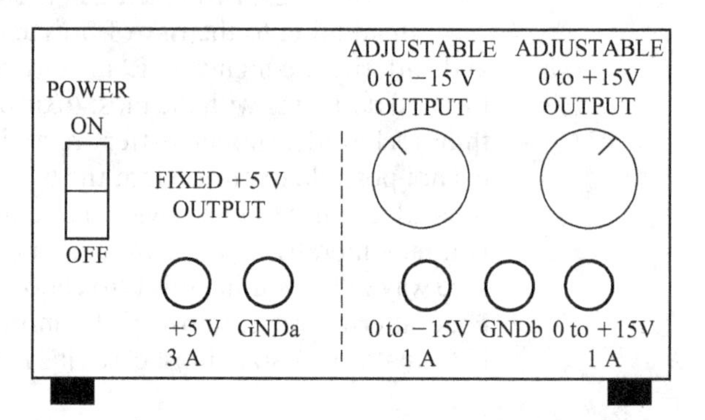

FIGURE 20.50

Power supply front panel for Problem 20.7.

20.8 You revise the circuit you created in Problem 20.7, and now it requires 28 V and as much as 0.5 A of current. Can you use the power supply described in Problem 20.7 for this new circuit? If so, show how you will configure it to meet the circuit's requirements. If not, explain why not.

20.9 Determine the minimum battery capacity (in mAh) required to power a portable device that consumes an average of 350 mA at 3 V for at least 8 hours. The battery voltage is 4.8 V and the device incorporates a 3 V regulator that operates at 85% efficiency.

20.10 Using Table 20.3, what is the minimum possible volume and mass of the NiMH battery pack needed to satisfy the requirements of Problem 20.9?

20.11 A particular car uses an internal combustion engine with 20% thermodynamic efficiency for locomotion, carries 12 gallons of gasoline (energy density = 44 MJ/kg) and gets 25 miles per gallon. If all other factors are assumed to be equal, what mass (in kg) and volume (in m^3) of lead acid batteries would be needed to provide an equivalent range for an electric version of the vehicle? What volume and mass are required if Li-ion batteries are used instead? Assume that the overall energy conversion efficiency for an electric vehicle is 85%.

20.12 What fraction of an alkaline cell's original charge remains if it is stored at room temperature for 5 years? What fraction of a NiMH cell's original charge remains if stored at room temperature for 5 months?

20.13 A given battery's self-discharge rate is 2% per year at 0°C and 5% per year at 25°C. How much more charge will the battery stored at 0°C hold than an identical battery stored at 25°C after 10 years? Is it a good idea to store batteries in a freezer?

20.14 Design a circuit that uses alkaline cells and an LM7805 linear voltage regulator to provide a portable device with 5 V. Alkaline cells have a terminal voltage of 1.5 V when new (fully charged), and an end voltage of 0.8 V. Design your circuit so that it provides a regulated output voltage of 5 V across the full range of the cells' terminal voltage (0.8 to 1.5 V).

20.15 For the circuit you designed in Problem 20.14, calculate the resulting heat dissipation in the LM7805 under the full range of possible cell terminal voltages when delivering 275 mA. For an ambient temperature of 25°C, is a heat sink required? If so, what is the maximum thermal resistance that the heat sink could have to satisfy the requirements? If not, what is the maximum allowable ambient temperature?

FURTHER READING

The Art of Electronics, Horowitz, P., and Hill, W., 2nd ed., Cambridge University Press, 1989.

Batteries for Portable Devices, Pistoia, G., Elsevier, Inc., 2005.

"Introduction to Power Supplies," Locher, R.E., Application Note 556, National Semiconductor, 2002.

OEM resources (www.duracell.com).

OEM resources (www.energizer.com).

"Switching Regulators," National Semiconductor Corp. (http://www.national.com/appinfo/power/files/f5.pdf).

"Technical Review of Low Dropout Voltage Regulator Operation and Performance," Lee, B., Texas Instruments Application Report LSVA072, 1999.

"Thermal Applications," Application Note TN-00-08, Micron Technology.

"Understanding Terms and Definitions of LDO Voltage Regulators," Lee, B., Texas Instruments Application Report SLVA079, 1999.

CHAPTER

21

Noise, Grounding, and Isolation

In many mechatronic systems, we can observe signals finding their way into unintended places. This is especially true when circuits driving actuators and circuits handling the signals from sensors are integrated into a single device. There, it is common to find that effects from the actuator drive signals are corrupting the signals in the sensor circuits. In the general case, we see the remnants of "large" signals appearing where they don't belong. Since this transfer is undesirable, we consider this signal crossover to be noise. In this chapter, we will deal with the mechanisms by which this transfer takes place and steps that we can take to minimize the noise that is received by a circuit. In addition, we will deal with the concept of electrical isolation and go into detail on one method for how to use optical isolation to keep two systems completely separate, yet still allow them to communicate. After reading this chapter, the student should be able to:

1. recognize the symptoms of conductively coupled noise, understand how it occurs, and how to reduce it,
2. recognize the symptoms of capacitively coupled noise, understand how it occurs, and how to reduce it,
3. recognize the symptoms of inductively coupled noise, understand how it occurs, and how to reduce it,
4. understand the purposes for and function of electrical isolation,
5. understand when and how to apply optical isolation.

21.1 NOISE COUPLING CHANNELS

When we talk about noise being transferred between systems, we speak of the **coupling** between those systems and the noise being coupled (and noise energy being transferred) from one system to another. Figure 21.1 shows the signal path in the general case of noise coupling.

To effectively deal with noise, you must be able to identify the manifestation of each of these blocks in your particular application. Once that is done, you can wage your war against noise on three fronts: 1) at the source, 2) by reducing the effectiveness of the coupling channel, and 3) by reducing the sensitivity of the receptor to the noise that makes it through the coupling channel.

Noise Source → Coupling Channel → Noise Receptor

FIGURE 21.1
The generic noise signal path.

The physical mechanism through which coupling takes place is referred to as the **coupling channel**. There are four principal coupling channels through which noise is transferred from one system to another, and each involves a transfer of energy through a different physical mechanism. In this chapter, we will deal in detail with the coupling channels that are most important in mechatronic systems. Briefly, the four channels are:

1. **Conductive coupling** is the only one of the coupling mechanisms that involves physical contact between the interacting systems. Conductive noise is the result of a difference in voltage potential (ΔV)—a mechanism that you are very familiar with at this point in the text.
2. **Capacitive coupling** does not involve physical contact, but instead relies on the action of changing electric fields to induce a current flow. The two principal driving forces are rate of change of voltage (dV/dt) at the noise source and the shared capacitance between two conductors. The latter is dominated by the shared conduction area between the two circuits and the distance between the conductors.
3. **Inductive coupling** also does not involve physical contact, depending instead on the action of a changing magnetic field to induce a current flow. The two principal driving forces here are rate of change of magnetic flux (dB/dt) at the noise source and the mutual inductance between two circuits. The mutual inductance is dominated by the shared loop area between the two circuits.
4. **Radiative coupling** is not only noncontact but, in fact, requires a significant separation between the two circuits. The phenomenon known as radiative coupling occurs when the distance between the source and the receptor is greater than the wavelength (λ) of the source signal ($\lambda = v/f$, where v is the speed of light and f is the frequency of the signal). This situation is referred to as "far field," where the effects of the electric fields and magnetic fields merge into an electromagnetic field. This is the mechanism through which radio signals are broadcast and received. For frequencies below 100 MHz, the wavelengths are greater than 3 m. Since we will generally be dealing with noise coupling *within* a device, where the distances are usually much less than the wavelength of the signals, radiative coupling will not play a role. However, when producing a mechatronic product, it will be necessary to obtain regulatory approval for the device. The regulations stipulate strict limits on radiated emissions from your device (how much of a "noise source" your device is to nearby devices). In these cases, understanding radiative coupling will be critically important.

Calculating the amount of energy that will be transferred between the noise source and the receptor through the noncontact coupling channels requires the solution of Maxwell's equations for the specific physical situation. Unfortunately, these solutions are complex for most real world situations and are strongly influenced by boundary conditions that are difficult to model. Rather than setting up and solving Maxwell's equations here, we will present a lumped component representation of the coupling channels. While this will not give us a tool that is numerically accurate at predicting the amount of noise, it *will* provide us with insight that will allow us to understand how the various aspects of the system contribute to the noise coupling process.

21.2 CONDUCTIVELY COUPLED NOISE

21.2.1 The Origins of the Conductive Coupling Channel

Conductively coupled noise is a result of Ohm's law: when current flows through resistances common to multiple subsystems, voltage drops attributable to each of the current flows result. When time-varying signals are involved, as they almost always are, resistance should

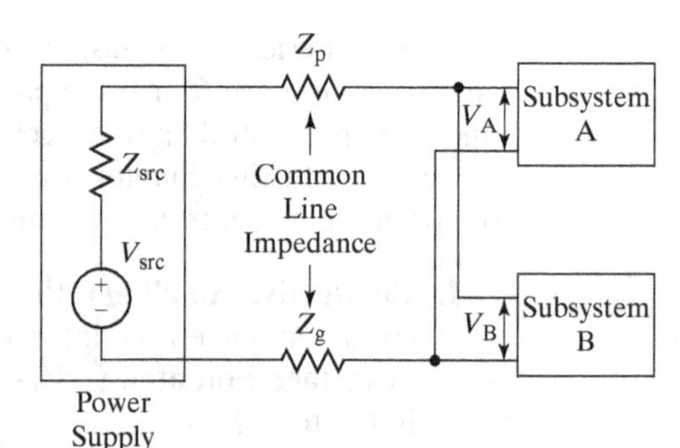

FIGURE 21.2

Noise coupling through shared impedances.

be replaced by impedance (which will include the effects of inductance and capacitance) to capture the effects of the changing signals. This situation is shown in block diagram form in Figure 21.2.

In this situation, the two subsystems, A and B, share common wiring back to a common power supply. Current flowing to Subsystem A affects Subsystem B through the voltage drops created across the common impedances (Z_p and Z_g) and the power supply source resistance, Z_{src}. While the DC resistance of the wiring and the power supply can be quite small, the magnitude of the current drawn by actuators can make the products ($I \times Z_p$ and $I \times Z_g$) significant. For example, if the current flow were 2 A (a relatively small current draw for a DC motor) and the common impedances were each 0.033 Ω (2 ft of 22 gauge wire), then Subsystem B would experience a 132 mV decrease in its power supply voltage (V_B) in response to the steady-state current drawn by Subsystem A. In addition to this DC effect, the rapid changes in current characteristic of digital logic transitions make the *impedance* (due to inductance) in the wires and power supply even more significant. It is very easy for the magnitude of these disturbances, seen at the other subsystem(s), to reach hundreds of millivolts. For example, if the inductance of the common conductors amounts to 1 μH and the current in Subsystem A changes by 10 mA in 20 ns (typical of high-speed digital logic circuits), then the voltage across the inductance would be $L(dI/dT) = 1\ \mu H \times (100\ mA/20\ ns) = 500\ mV$. While the noise margins of other digital logic may be able to tolerate this level of noise, analog circuits are much more critical. Variations in an analog voltage due to noise are indistinguishable from the underlying analog signal.

21.2.2 Reducing Conductively Coupled Noise

To begin the process of reducing the conductively coupled noise, let's fit the elements of Figure 21.2 into the format of Figure 21.1. This is shown in Figure 21.3.

Here, we see the source of the noise as Subsystem A—more specifically in this example, the changing currents drawn by the digital device(s) of Subsystem A. The noise is coupled through two conductive channels: the impedance of the power supply and the common impedance of the wiring from the power supply to the subsystems. To minimize the effects of the noise, we should look at how we can reduce the source of the noise, reduce the coupling, and reduce the sensitivity of the receptor to the noise.

21.2.3 Reducing Noise at the Source: Decoupling

The root cause of all conductively coupled noise is changing current being drawn through a common impedance. While chip designers seek to minimize the transition currents that are drawn by digital devices, the fact is that when outputs change state, charge must be moved around and that requires changing currents. This is not a phenomenon unique to digital devices. Analog circuits, too, generate changing currents as the input signal changes and can

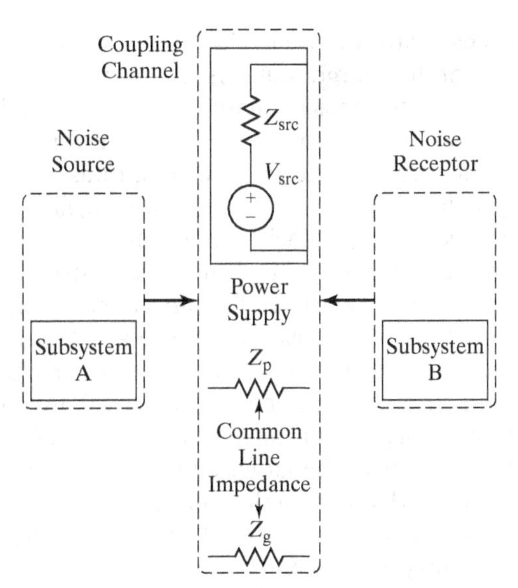

FIGURE 21.3

Identifying the noise path.

therefore be the sources of conductively coupled noise. While we cannot eliminate these transient currents, we can take steps to minimize how much of that current is drawn through the common impedances. We do this by providing a local store of charge in close physical proximity to the pins of the devices that will source and sink the transition currents. In this way, the rapid movement of charge necessary to make up the transition current is drawn from the local store rather than through the impedances shared with other parts of the device. The local store of charge can be added with the use of **decoupling capacitors** (sometime referred to as **bypass capacitors**).

To be most effective, decoupling capacitors must have specific characteristics and be positioned properly relative to the device demanding the current. The decoupling capacitor must be located as close as is physically possible to the power and ground pins of each and every IC in the circuit. To minimize the effects of the inductance of the leads of the capacitor, component leads should be as short as possible. The wires or **printed circuit board (PCB)** traces from the capacitor to the power and ground pins should be as direct and short as possible and, in the case of PCB traces, they should be wider than ordinary signal traces to reduce their impedance. The directness of the routing is important to minimize the area encompassed by the current loop formed between the decoupling capacitor and the IC. The reason to minimize this area is discussed in Section 21.4. When prototyping with a solderless breadboard, it is common to place the decoupling capacitor directly above each IC with the leads connected to the power and ground pins (Figure 21.4).

FIGURE 21.4

Decoupling capacitor placement on a solderless breadboard.

When choosing a capacitor for use in decoupling applications, it is important to keep in mind the characteristics of the charge transfers that need to take place. Transition events are generally short in duration (a few tens to hundreds of nanoseconds), with transitions that happen over very short periods of time (on the order of a few nanoseconds). The magnitude of the currents can be quite large, on the order of hundreds of milliamps, so the charge storage necessary to supply these currents can be substantial, and this will require quite a bit of capacitance. The ideal decoupling capacitor would have a large capacitance, maintain that capacitance to a high frequency, and be small and inexpensive. Unfortunately, no single capacitor with this set of characteristics exists. To best approximate the desired qualities, we decouple using several different types of capacitors placed at different locations in the circuit. For the capacitors that are closest to the ICs, monolithic ceramic capacitors are a good choice because they maintain their capacitance even at very high frequencies. This enables them to respond to the very sharp transitions in current that occur when a logic device switches state. Bulk capacitance, which will cover the larger amplitude but lower-frequency components of the transients, is best supplied with tantalum electrolytic capacitors, whose high capacitive density and moderate costs make them a good choice. Depending on the size of the circuit, these tantalum capacitors may be distributed around the board (for larger boards) or placed at the point where power is brought into the board (for smaller boards). The monolithic ceramic and tantalum capacitors are often supplemented with a single larger (100 μF or more) aluminum electrolytic capacitor where the power enters the board. For special cases (such as an H-bridge), several hundred μF are often recommended in the driver IC data sheets (typically 100 μF per amp of bridge current for H-Bridges). The recommended location for the capacitors is also directly across the power and ground pins of these high-current drivers.

There are basically two different ways to determine the amount of capacitance needed for decoupling: model the system or use "rules of thumb." Accurate system modeling can lead to a minimal number of capacitors, but this can be highly complex, time consuming, and in any case, is well beyond the scope of this book. On the other hand, using rule of thumb methods to choose the amount of decoupling capacitance for general purpose applications is very straightforward with only a few rules:

1. Include a 0.01–0.1 μF monolithic ceramic capacitor placed as close as possible to the power and ground pins of every IC. When wiring these capacitors to the power and ground pins, keep the connections as direct as possible.
2. Include a tantalum capacitor for every 5–10 monolithic ceramic capacitors used at the ICs. The size of the tantalum should be at least 5–10 times the *total* capacitance of the monolithic ceramic capacitors. Including capacitance in excess of this recommendation causes no harm, but produces little measureable improvement in performance.
3. Include a 100 μF aluminum electrolytic capacitor at the point where power enters the board or breadboard.

Despite the simplicity of the solution (adding a few capacitors across power and ground), the positive impact of decoupling capacitors is immense. You should never build a circuit, even for testing purposes, without including adequate decoupling capacitance.

21.2.4 Reducing the Coupling of Conductive Noise

In Figure 21.3, we see that the coupling channel for our prototypical conductive noise problem has two components: the output impedance of the power supply that serves both subsystems and the shared impedances of the power leads from the power supply to the subsystems, Z_p and Z_g. We could reduce the extent of the noise coupling by reducing either or both of these mechanisms. We will examine each of these approaches in turn.

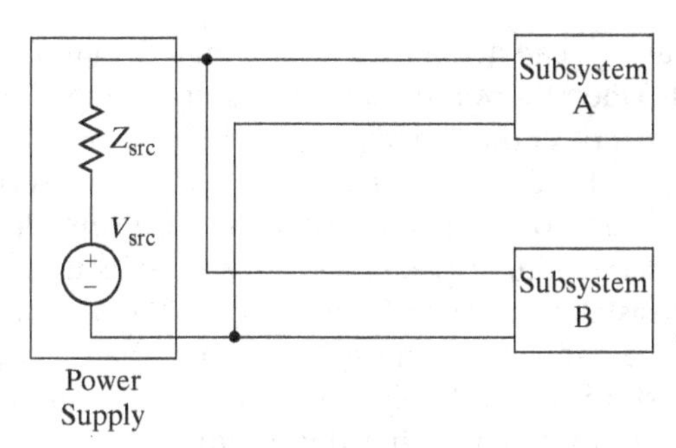

FIGURE 21.5

Eliminating the common impedance with separate wires.

First, we examine the wires connecting the subsystems to the power supply. Since the impedance of a wire is inversely related to the diameter of the wire, we could reduce the common impedance due to the wires by replacing the common wires with ones of a larger diameter. While this will work, it is not as cost effective as taking some of the extra copper that would have gone into the larger diameter wire and using it to run a separate set of wires from the power supply to the subsystems, Figure 21.5, which eliminates all but a tiny fraction of the shared impedance due to the common wire.

The output impedance of the power supply, Z_{src}, also represents a common impedance between the two subsystems. The output impedance of the power supply might be reduced by changing the design of the power supply, but if the power supply is based on IC voltage regulators this may prove challenging as the output impedances are typically already very low. If we have already implemented separate wiring to the two subsystems, then an alternative approach would be to provide separate voltage regulators for each of the subsystems, effectively eliminating the common source resistance.

A hybrid approach would be to use a common power conductor for both subsystems with separate ground return conductors and local voltage regulation at each of the subsystems. This approach reduces the number of conductors from four to three and uses the ability of the voltage regulator to reject noise on its input to reduce the noise seen by the subsystems.

21.2.5 Reducing Conductive Noise at the Receptor: Power Supply Filtering

When everything possible has been done to reduce the noise at the source and reduce the effectiveness of the coupling, the last line of defense is at the receiver. While the actions taken at this stage in the coupling process are not as targeted as reductions earlier, the amount of noise on the power conductor *can* be reduced by low-pass filtering the power conductor. Since introducing a series resistor to create a low-pass filter, as we did in Chapter 9, would create undesirable voltage drops at DC that vary with current flow, the preferred series element for a low-pass filter on the power line is an inductor. When combined with the required decoupling capacitors for the circuits being supplied with power, the result is a circuit like that of Figure 21.6.

One of the significant drawbacks to this approach is that the circuit of Figure 21.6 has a resonance at $f_r = 1/2\pi\sqrt{LC}$. When excited at the resonant frequency, the amplitude of the

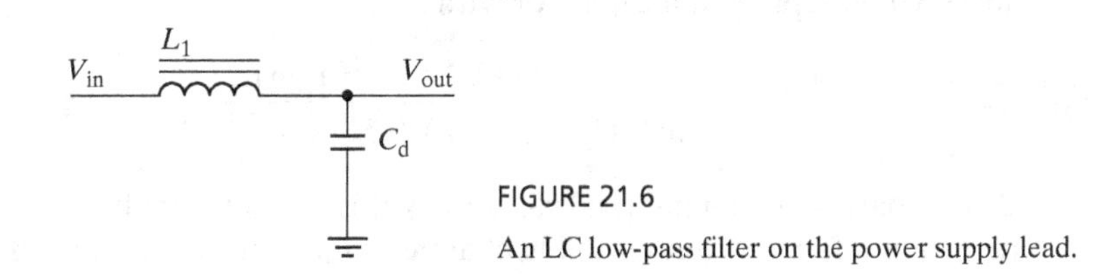

FIGURE 21.6

An LC low-pass filter on the power supply lead.

noise at f_r that appears at V_{out} can actually be larger than the amplitude present at the input! This situation will occur when there are sharp edges at the input (which contain a range of frequencies) and there is not enough damping in the system. It is therefore important to insure that the damping factor $\xi = R/2\sqrt{C/L}$ (where R is the series resistance of the real inductor) remains above 0.5 to insure that there is sufficient damping. This can require substantial amounts of decoupling capacitance. For example, if L_1 is a power inductor with $L = 1$ μH and a series resistance of 0.048 Ω (yielding a 48 mV drop when conducting 1 A at DC), then the decoupling capacitance must be greater than 434 μF in order to obtain a damping factor of at least 0.5. This is likely to be substantially more capacitance than would be indicated by the rules of thumb introduced in Section 21.2.3.

21.2.6 Best Practices for Reducing Conductive Noise

1. Include adequate bypass capacitors locally at every digital device and in bulk for the circuit as a whole.
2. Minimize the common impedances between subsystems by running separate power and ground wires for subsystems likely to either cause significant power supply fluctuations (high current loads or fast switching circuits) or be sensitive to fluctuations on the power supply (analog circuits).
3. When suggestions 1 and 2 are not sufficient, consider adding power supply filtering to the noise sensitive circuits.
4. When all else fails, consider isolation (Section 21.5).

21.3 CAPACITIVELY COUPLED NOISE

21.3.1 The Origins of the Capacitive Coupling Channel

Recall from Chapter 9 that a changing voltage on one plate of a capacitor can cause a displacement current to flow into or out of the other plate, even though there is no physical connection between the plates. Recall also that any two conductors separated by an insulator create a capacitor. This is true even for two wires. Therefore, a changing voltage in one wire can cause a current to flow in an adjacent wire. In fact, we can use Eq. (9.61) to develop an expression for the amount of capacitance C, in F/m, that will exist between two parallel circular conductors of diameter d spaced apart by a distance D (where $\varepsilon = 8.85 \times 10^{-12}$ F/m):

$$C = \frac{\pi\varepsilon}{\cosh^{-1}(D/d)} \tag{21.1}$$

For large spacing between the wires (i.e., $D/d > 3$), this reduces to

$$C = \frac{\pi\varepsilon}{\ln(2D/d)} \tag{21.2}$$

It is relatively easy for a pair of wires to develop small but significant amounts of capacitance between them. For example, a pair of 24 gauge (0.51 mm) wires, 10 cm long, spaced 1 cm apart would represent a capacitor with

$$C = \frac{\pi \times (8.85 \times 10^{-12}\text{ F/m})}{\ln[2(1 \times 10^{-2}\text{ m})/0.51 \times 10^{-3}\text{ m}]} \times 0.1\text{ m} = 0.76\text{ pF}$$

By comparison to commercial capacitors, this is a very small value, but, as we shall see, it can represent a fairly strong coupling channel under certain conditions.

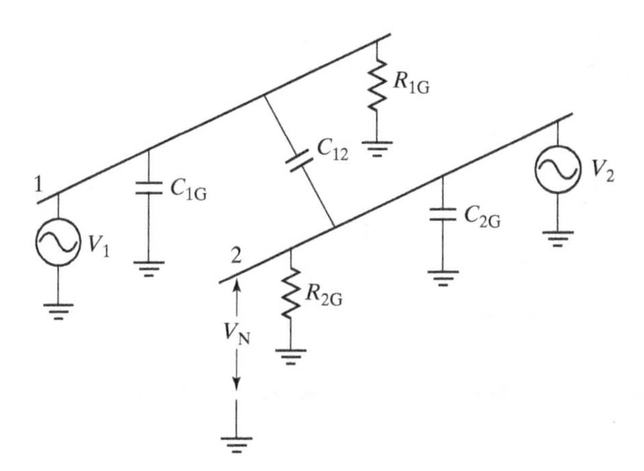

FIGURE 21.7

Physical schematic of two wires and the associated electrical elements.

When there is a capacitance between two wires, then a change in voltage on one wire will induce a current on the second wire according to Eq. (9.23), $I = C(dV/dt)$. The situation is shown in a physical schematic in Figure 21.7.

The conductor labeled 1 represents one of the two wires. It is driven at one end by a voltage source V_1. The far end of the wire has a resistance, R_{1G}, that represents the input impedance of the circuit being driven by V_1. There is a capacitance, C_{1G}, between wire 1 and ground, as well as a capacitance, C_{12}, between wire 1 and wire 2. The conductor labeled 2 represents the other wire in the pair. It too has a driving source (V_2), a capacitance with ground (C_{2G}), and a characteristic impedance (R_{2G}). The voltage that appears at the near end of wire 2 (V_N) will have components due to the driving source (V_2) and to the circuit response to the current injected through C_{12}. For example, if wire 1 was carrying a digital logic signal that swung from 0 to 5 V in 15 ns (typical for high-speed CMOS logic), it would produce a dV/dt of 5 V/15 ns = 333 V/μs. With a 0.76 pF capacitance between wires 1 and 2, this would induce a current of 0.25 mA in wire 2. If the voltage source V_2 were a relatively high output impedance sensor such that the characteristic impedance (R_{2G}) was 10 kΩ, then the logic transition would induce a 2.5 V transient in wire 2.

21.3.2 Reducing Capacitively Coupled Noise

We begin our exploration of how to reduce capacitively coupled noise by redrawing Figure 21.7 as an electrical schematic and labeling it using the nomenclature from Figure 21.1. In Figure 21.8, we have temporarily eliminated the source V_2 to focus on the response to the noise source.

Using Figure 21.8 as a guide, we can develop an expression for V_N which is only the response to the noise source, Eq. (21.3).

$$V_N = \frac{j\omega\left(\dfrac{C_{12}}{C_{12} + C_{2G}}\right)}{j\omega + \dfrac{1}{R_{2G}(C_{12} + C_{2G})}} V_1 \tag{21.3}$$

Notice that while R_{1G} and C_{1G} affect the currents that will flow in the noise source, they do not affect the voltage at the left-hand side of C_{12}, therefore, they do not enter into the expression for V_N.

Looking at Eq. (21.3), it is not immediately clear how the values of the components affect the noise voltage, V_N. We can improve the clarity of the relationship by considering

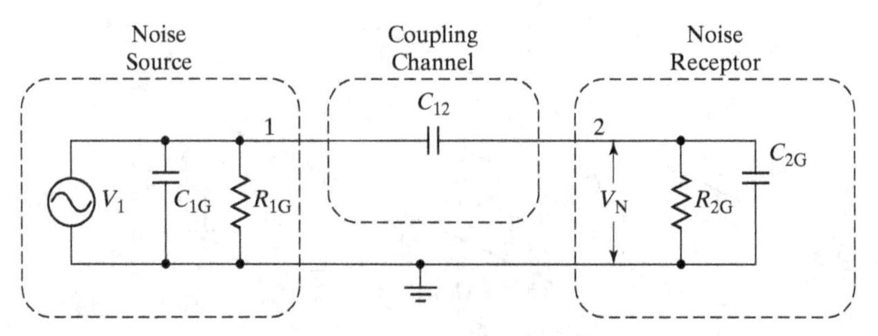

FIGURE 21.8

The elements of capacitive coupling in electrical schematic form.

simplifications for the conditions when the impedance R_{2G} is either much smaller or much larger than the impedance of the capacitances. When

$$R_{2G} \ll \frac{1}{j\omega(C_{12} + C_{2G})}$$

then Eq. (21.3) can be reduced to

$$V_N = j\omega C_{12} V_1 R_{2G} \tag{21.4}$$

The $j\omega C_{12}V_1$ term represents the amount of current injected into conductor 2, which flows across R_{2G} to produce V_N. Eq. (21.4) describes a noise voltage that rises linearly with increasing frequency.

On the other hand, when

$$R_{2G} \gg \frac{1}{j\omega(C_{12} + C_{2G})}$$

the response is dominated by the capacitive divider and Eq. (21.3) can be reduced to

$$V_N = \left(\frac{C_{12}}{C_{12} + C_{2G}}\right)V_1 \tag{21.5}$$

In this case, the noise voltage is independent of frequency, that is, a horizontal line in a plot of noise amplitude vs. frequency.

Eq. (21.4) represents the situation when ω is relatively low and therefore the impedance of the coupling capacitance is high. As the frequency of the noise source increases, Eq. (21.5) begins to dominate. We will use these relationships to help us decide how to reduce the effects of capacitively coupled noise. Looking at Eq. (21.4), we can see that the only capacitance that enters into the noise expression is C_{12} and that reducing any of ω, V_1, C_{12}, or R_{2G} will reduce V_N. Once the noise frequency rises to the point where Eq. (21.5) dominates, defined by

$$\omega = \frac{1}{R(C_{12} + C_{2G})} \tag{21.6}$$

our efforts to reduce the noise amplitude should be focused on decreasing V_1, where possible, decreasing C_{12}, and/or increasing C_{2G}.

21.3.3 Reducing Capacitively Coupled Noise at the Source

Examining Eqs. (21.4) and (21.5), we see that the only terms that originate with the noise source are the amplitude, V_1, and the frequency, ω; therefore, these are the only "knobs" we have to turn as we try to reduce the noise at the source. Since the magnitude of the signal on conductor 1 will be set by the requirements of that part of the circuit, it will be the rare instance where reducing the voltage amplitude will be an approach to reducing the noise that is available to us. The equations do show us, however, that reducing the source voltage V_1 (for example, by changing from 5 V logic to 3.3 V logic) will proportionately reduce the magnitude of the capacitively coupled noise generated by the logic signals.

The fundamental frequency of the noise source will largely be determined by the functional requirements for the circuit that generates the signal, and therefore also is not likely to be a parameter that can be altered to reduce the noise coupling. While reducing the fundamental frequency is usually not an option for reducing noise, it is worth noting that in the situations where it is a possibility (e.g., the frequency of a pulse width modulation drive signal), choosing a lower frequency will result in less noise. However, when looking at digital signals, the *fundamental frequency* is not the only one that should be considered. The very short rise and fall times associated with the digital outputs on high-speed CMOS microcontrollers means that these signals have significant high-frequency content even when the fundamental frequency is relatively low. The 15 ns rise and fall times typical of high-speed CMOS logic are unnecessarily fast in many situations. To illustrate this, consider that a 15 ns rise/fall time for 5 V logic corresponds to the same slew rates (dV/dt) as a 1 V sine wave at 53 MHz! These fast edges are often a significant source of capacitively coupled noise. The speed of these edges is not always required, and in these cases, it is possible and beneficial to slow the edges down and reduce the amount of noise generated. This is especially true in applications where digital signals are routed beyond the chassis of a product. In situations like this, it is common to place a ferrite bead (a bead of material with a high magnetic permeability) around the wire as it exits the housing. This creates a small series inductance with a large impedance to high-frequency signals while producing no DC voltage drop. The effect is to slow the rise and fall times without impacting the low-frequency behavior of the circuit, such as the fundamental frequency.

In situations where we have some latitude and can tolerate somewhat slower transitions (e.g., driving a power MOSFET to turn an external device on and off), it is to our advantage, from a noise standpoint, to purposely reduce the rise and fall times by introducing a low-pass filter in the drive signal path. This is particularly easy to do in the case of the gate drive on a MOSFET. Since the gate of the MOSFET inherently has significant capacitance, we can create a low-pass filter by simply adding a series resistor to the signal that drives the gate. When implementing this solution, the physical placement of the resistor is important. Consider the two situations shown in Figure 21.9.

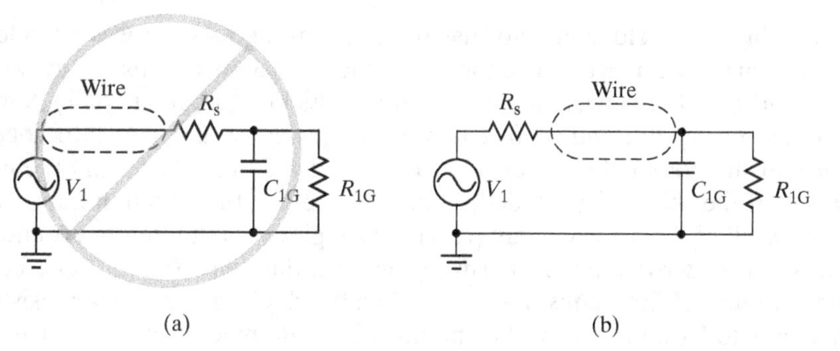

FIGURE 21.9

Position matters when placing a gate resistor.

In Figure 21.9a, the resistor is placed close to the MOSFET, and the wire between the driving voltage and the MOSFET carries a signal with the fast rise and fall times. In contrast, Figure 21.9b shows the resistor placed close to the driving source so that the line to the gate of the MOSFET carries the filtered drive signal. The wire carrying the slower transitions shown in Figure 21.9b has less high-frequency content, and as a result will be less of a source for capacitively coupled noise.

21.3.4 Reducing the Coupling of Capacitively Coupled Noise

The coupling channel consists entirely of the stray capacitance represented by C_{12} in Figures 21.7 and 21.8. The magnitude of that capacitance is determined by Eq. (21.7):

$$C = \frac{\varepsilon\varepsilon_0 A}{d} \tag{21.7}$$

where ε_0 is the permittivity of free space (8.85×10^{-12} F/m), ε is the dielectric constant (1 for air, which is commonly the bulk of the substance between conductors exchanging capacitively coupled noise), A is the shared area of the two conductors, and d is the spacing between them.

From Eq. (21.7) we can see that to reduce the coupling capacitance C_{12}, we should increase the distance between the conductors and orient them to minimize the shared area. We can minimize the shared area between the noise source conductor and the noise receptor conductor by **not** routing the conductors in parallel, and when the conductors must cross, making sure that the crossing happens at 90° angles.

21.3.5 Shielding

There is another approach to reducing the energy coupled into the receptor conductor that is not explicitly suggested by the material presented so far. This is to introduce a shield. A **shield** is a conductor that is typically positioned so that it surrounds the receptor conductor to the maximum possible extent. Electrically, this creates a situation like that shown in Figure 21.10.

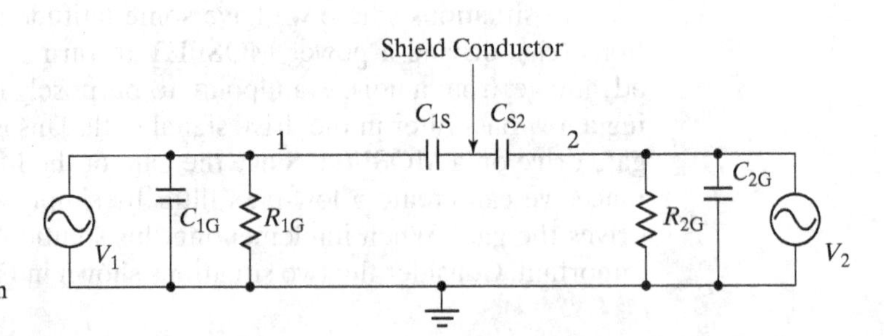

FIGURE 21.10

Electrical schematic of a system with floating shield (BAD!).

Adding the shield conductor has the effect of taking what was a single capacitor, C_{12}, and replacing it with two capacitors: C_{1S} (the capacitance between conductor 1 and the shield) and C_{S2} (the capacitance between the shield and conductor 2). Simply adding the shield has no benefit, and in fact may make the noise coupled into conductor 2 worse because of the potentially larger shared area between the shield and the noise conductor. However, if the shield is grounded, as in Figure 21.11, the situation improves dramatically.

Now, all of the noise current passing through C_{1S} is shunted to ground through a very low (essentially zero) impedance. This means that the shield side of C_{S2} is held at a constant ground potential. With a constant potential on the shield, any dV/dt that exists across C_{S2} will be due only to V_2, and there will be no noise current injected into conductor 2 through C_{S2}. If the shield were to completely enclose conductor 2, there would be no capacitively coupled

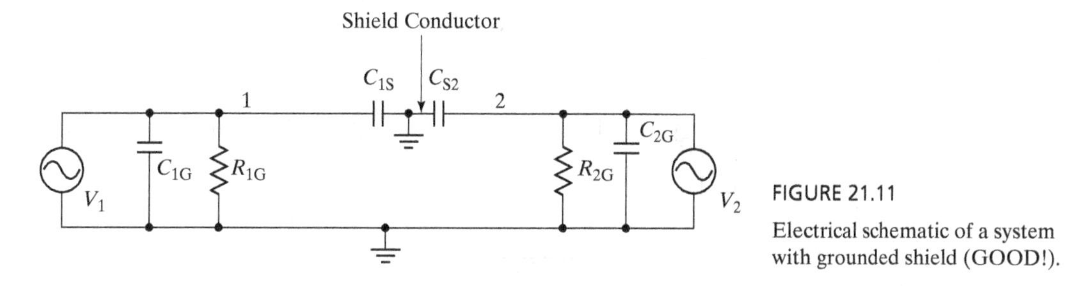

FIGURE 21.11

Electrical schematic of a system with grounded shield (GOOD!).

noise. In practice, shielding doesn't always completely surround the conductor (this depends on how the shield is made), and the conductor is typically exposed at the ends to make circuit connections, so the complete elimination of noise through shielding is not possible. Figure 21.12 shows what this might look like in practice, where the shield conductor would be made from a wire braid or metalized plastic film that would surround the signal conductor with an insulating layer in between.

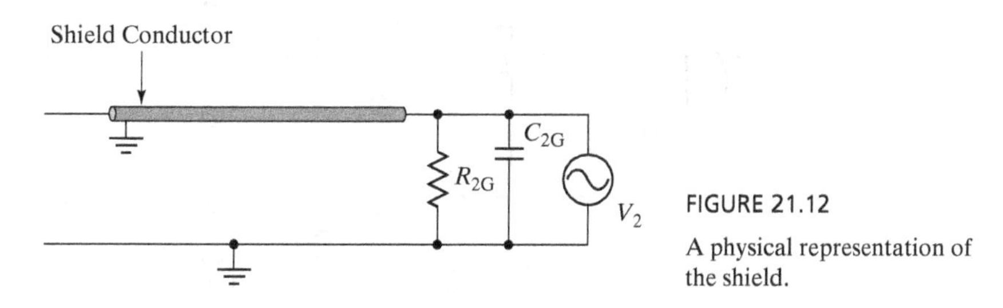

FIGURE 21.12

A physical representation of the shield.

The decision of how to connect the shield to ground has a significant impact on the shield's effectiveness. For maximum effectiveness, the shield should only be grounded at one end. If the shield were grounded at both ends (at physically separate locations), there is a high likelihood that an instantaneous voltage difference would exist between these two "grounds." If that were the case, then the shield would no longer be held at a constant potential, and the potentials on the shield would cause current to flow and comprise a new source of noise that will be injected into conductor 2. Since encasing conductor 2 in the shield has created a large shared area, the effectiveness of this new and unintentional noise coupling channel would be very high, resulting in large noise currents for relatively small amplitude potentials. Therefore, always make sure that a shield is grounded at only one end.

21.3.6 Reducing Capacitively Coupled Noise at the Receiver

In reviewing Eqs. (21.4) and (21.5) for opportunities to reduce the sensitivity of the receptor side of the coupling, we find that the only parameter associated with the receptor is the input impedance R_{2G} (using the nomenclature established in Figure 21.7 and those following). The larger the value of R_{2G} is, the greater the voltage that results from a given amount of injected noise current. Unfortunately, the size of R_{2G} is often governed by the size of the output impedance of the driving voltage. When the driving voltage originates from a sensor with a high output impedance, then there is no choice but to make the input impedance even higher in order to minimize voltage divider effects (i.e., loading). However, there are choices we can make for the location of signal conditioning circuits that can minimize the circuit's sensitivity to capacitively coupled noise. Consider the circuit of Figure 21.13a.

Here, we a have a photo-transistor mounted remotely from the trans-resistive stage. The coupled noise is modeled as a current source that injects current into the wire from the photo-transistor emitter to the trans-resistive stage input. At the op-amp, this injected current

BAD

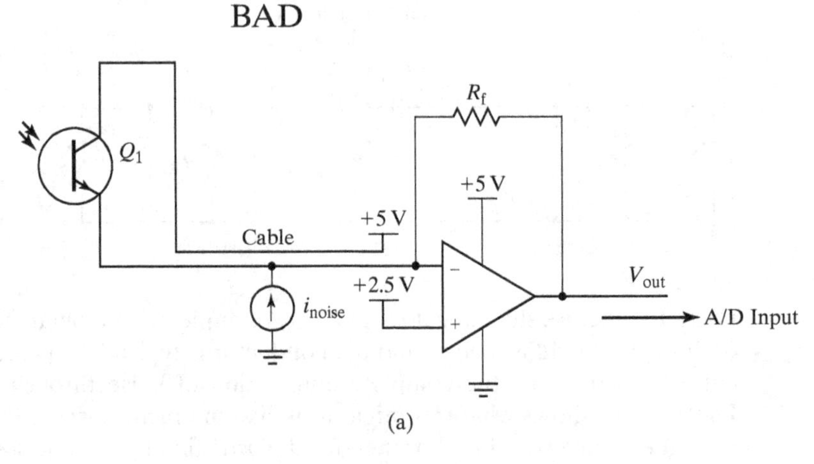

GOOD

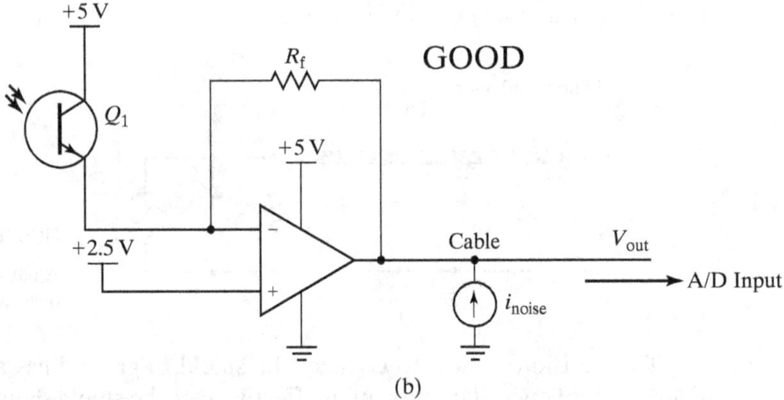

FIGURE 21.13

Photo-transistor sensor and possible trans-resistive stage locations.

is indistinguishable from the real photo-current and is amplified indiscriminately. The R_{2G} for this circuit will be the R_f of the trans-resistive stage.

Compare that with the situation shown in Figure 21.13b. Here, the trans-resistive stage has been located remotely with the sensor and the wire that is subject to noise coupling carries the output of the op-amp. In this case, the R_{2G} that the noise current flows across to create the noise voltage will be dominated by the output impedance of the op-amp, which will be quite small (a few ohms). Compared to the arrangement of Figure 21.13a, this output signal will show a substantially reduced sensitivity to the capacitively coupled noise because of the much smaller R_{2G}.

21.3.7 Best Practices for Reducing Capacitively Coupled Noise

1. Reduce the coupling capacitance by minimizing the shared conductor area between noise sources and sensitive circuits (don't make runs of parallel wires).
2. Where possible, reduce the rise and fall times of signals to minimize the generated dV/dt.
3. Minimize the impedance of circuits whose long runs of wire are vulnerable to noise coupling. This is most easily accomplished by providing local buffering of high impedance sensors so that the long wires are driven by the low output impedance of the buffer.
4. When suggestion 3 is not possible, consider shielding the sensitive circuits, taking care to ground the shield at only one point.

21.4 INDUCTIVELY COUPLED NOISE

21.4.1 The Origins of the Inductive Coupling Channel

Inductive coupling depends on changing magnetic fields to couple energy between subsystems. This is the same phenomenon that makes transformers work. For there to be inductive coupling, there must be a changing magnetic field and that field must intersect a conductive loop. Electrically, we can model the situation with a circuit like that of Figure 21.14.

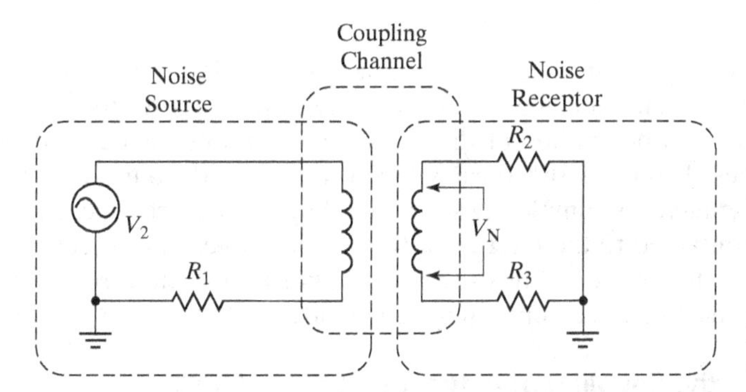

FIGURE 21.14

Electrical representation of the inductive coupling process.

For inductive coupling, we see a situation that is different than the two coupling situations discussed above (conductive and capacitive). Here, the noise source shares an element with the coupling channel: the loop area of the source circuit which is represented by the inductance on the noise source side. The noise receptor also shares the same type of element with the coupling channel in the form of the loop area of the receptor circuit. This is represented by the inductance on the receptor side. The coupling channel is the loop area that is shared between the circuit loops of the source and receptor.

21.4.2 Reducing Inductively Coupled Noise at the Source

The strength of the magnetic field generated by the noise source will depend on the current flowing in the noise source circuit. Presumably, this current is a requirement for the desired function of the source circuit (e.g., driving a motor), so, reducing this current is rarely a reasonable option. The other aspect of the noise source is the physical area of the loop formed by the conductors. This is something that is generally easily controlled. For example, the leads going to a motor should be kept as close together as possible. This will minimize the loop area created to the region between the two conductors. Better yet, if the leads are twisted together, there will be partial cancellation of the external magnetic field from the adjacent twists, reducing the external magnetic field (Figure 21.15b). Figure 21.15a shows a poorly executed

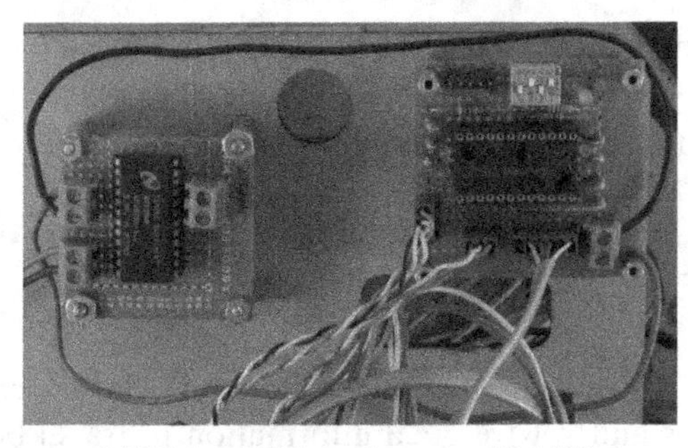

(a) Wiring with large loop area (BAD!)

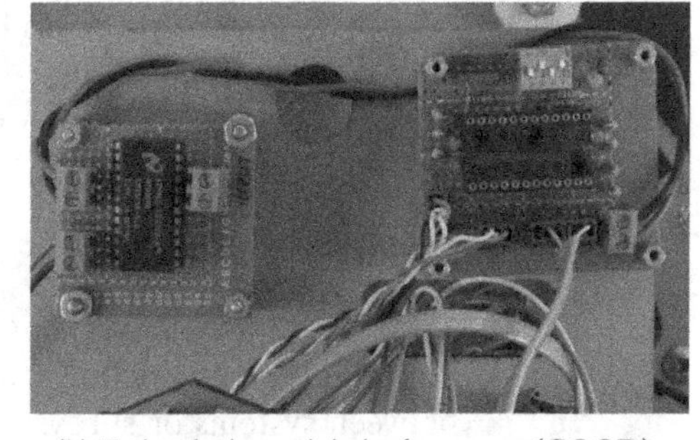

(b) Twisted wires minimize loop area (GOOD)

FIGURE 21.15

Wiring affects loop area.

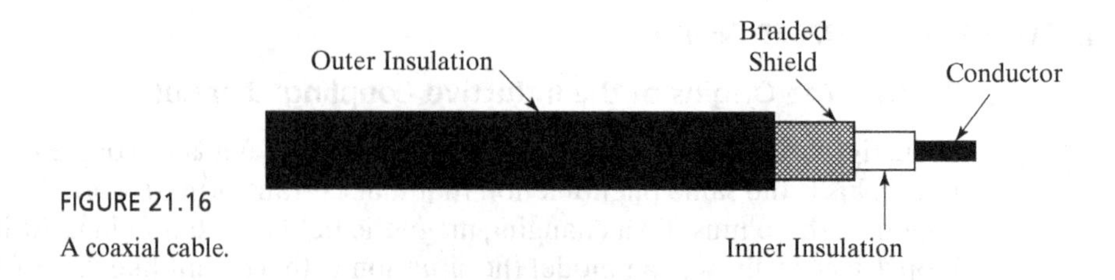

FIGURE 21.16
A coaxial cable.

circuit, where the wiring maximizes the loop area. Figure 21.15b shows a much better way of arranging the wiring and minimizing the loop area: twisting the wires together.

Shielding can also be applied at the source in situations such as driving a motor, by using coaxial cables (Figure 21.16) to carry the current to and from the motor. In a coaxial cable, the inner conductor is completely surrounded by an outer conductor. The loop area in this case will be contained to the interior of the coaxial cable, and the return current in the outside (shield) conductor will be flowing in the opposite direction from the current in the inner conductor canceling the external magnetic field.

21.4.3 Reducing the Coupling of Inductively Coupled Noise

Since the strength of a magnetic field falls off with the cube of distance, keeping the source and receiver as far apart as possible is a good strategy for reducing the coupling. This strategy and minimizing the shared loop area are the principal strategies for reducing inductive coupling between the noise source and the receptor. As was pointed out in Section 21.4.2, minimizing the loop area is best done by keeping the wires that make up the source as close to one another as possible, and doing the same for the wires that make up the receptor. This, in turn, is most easily accomplished by wiring both source and receptor circuits using twisted pairs of conductors.

21.4.4 Reducing Inductively Coupled Noise at the Receptor

Since the noise due to inductive coupling appears as a voltage across the shared loop, there is little that can be done electrically to reduce the coupling at the receptor. The best method is to keep the loop area of the receptor circuit as small as possible by wiring susceptible elements of the receptor circuit using twisted pairs of conductors.

Conquering inductively coupled noise reduces down to a single approach and technique: nip it in the bud by wiring all potential source and receptor circuits using twisted pairs of conductors.

21.4.5 Best Practices for Reducing Inductively Coupled Noise

1. Wire circuits prone to being inductive noise sources using twisted conductor pairs or coaxial cable.
2. Wire circuits sensitive to inductively coupled noise using twisted conductor pairs or coaxial cable.

21.5 ISOLATION

There are many situations in which it is imperative that there be no conductive contact between systems or subsystems and yet we need information to travel between those systems. The classic example of this situation arises in the design of line (wall outlets, or "mains") powered medical devices that must contact a patient directly. With a direct connection to a patient, even a very small undesired current can disturb the natural electrical signals in the body, such as heart rhythms. It is absolutely essential (as well as legally required) that under no condition does the device cause more than tiny amounts of current to flow in the patient. In the case of a line powered device, if there were a conductive connection, then a failure inside the device could expose the patient to line voltage (120 VAC in the US). This would be, quite literally, a deadly situation. As a result, in situations like this, we need a way to electrically isolate the patient from the line powered portion of the device yet pass signals from a patient to a device.

Isolation is achieved by creating devices that use a noncontact mechanism to transfer signals across an isolation barrier. The magnitude of the voltage that can be allowed to appear across the isolation boundary (**isolation voltage**) is one of the characteristics that distinguishes the different isolation devices. In examining noise coupling, we have seen two mechanisms, capacitive and inductive coupling, in which signals are transferred without conductive contact. Indeed, both of these phenomena have been harnessed to make chip scale (IC sized) devices that provide electrical isolation while allowing information to be passed across an isolation barrier. Inductive coupling has been used to transfer not only information but also power to the isolated portion of a circuit. While most isolation devices transfer digital signals across a boundary, devices are also available that transfer analog signals.

Medical devices are by no means the only places where electrical isolation is necessary or desired. It is also useful in situations where there may be a very large difference in voltage between the controlling signal and the circuit to be controlled, for example, controlling high voltage (> 100 V) circuits using the outputs from microcontrollers (which often have 0 to 5 V outputs). There will also be situations in which it is impossible to reduce the conductively coupled noise to an acceptable level. In such situations, it may be desirable to power sensitive circuits from a completely separate power supply and pass signals across an isolation boundary in order to prevent conductively coupled noise from appearing in the sensitive system.

21.5.1 Optical Isolation

While capacitive and inductive coupling are used to make devices to provide isolation, the most common coupling mechanism in use involves the transfer of signals via light: optical coupling. These devices, known as **opto-isolators**, have been available for many years and provide an inexpensive way to isolate circuit elements and subsystems. The essence of an opto-isolator is simply an LED and a photo-transistor packaged together and optically sealed to exclude all external light (Figure 21.17).

A circuit on one physical side of the device (which also corresponds to one side of the isolation barrier) controls the current through an LED. The light from the LED crosses a cavity within the package and falls onto a photo-transistor with connections on the other physical side of the package, on the other side of the isolation barrier. The schematic shows something new in photo-transistors: a base lead. The electrical base connection is provided to enable the connection of an external resistor between the base and the emitter to improve switching times. In the most common applications of the devices, it is left unconnected.

Designing with an opto-isolator is not much more difficult than designing an LED drive circuit and a photo-transistor sensor circuit, which we have already seen in Chapters 10 and 13. A series resistor is chosen (based on the drive voltage and the V_F of the diode) to

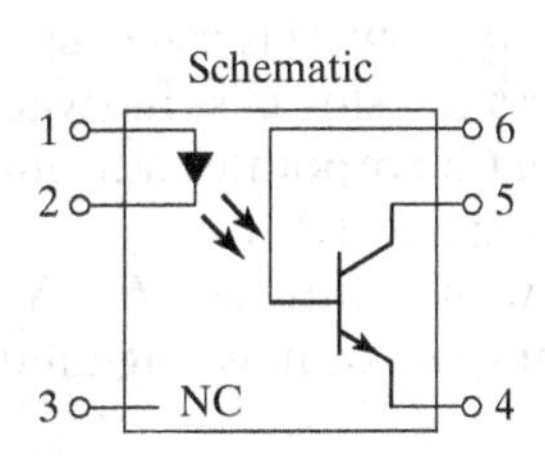

FIGURE 21.17
The opto-isolator. (Courtesy Fairchild Semiconductor.)

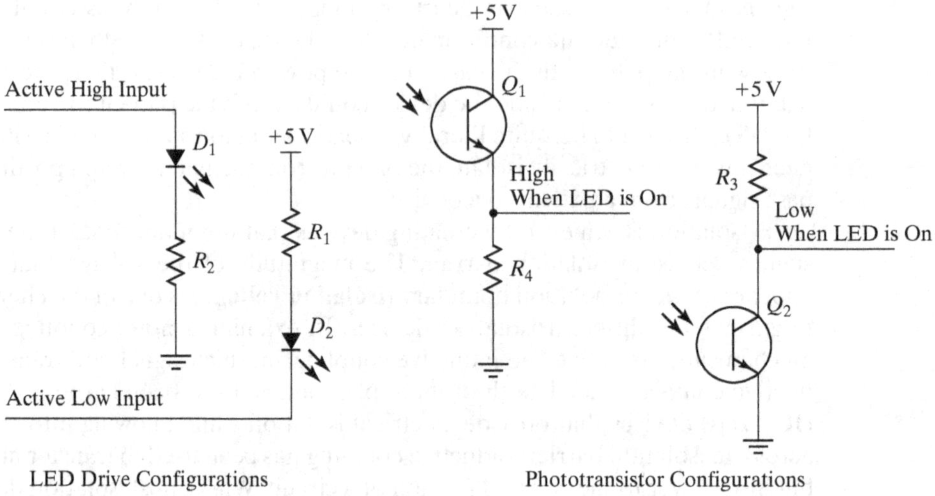

FIGURE 21.18

Opto-isolator drive and output configurations.

limit the current through the LED (for most opto-isolators, currents of less than 10 mA are used). On the output side, only a pull-up or pull-down resistor is necessary to convert the photo-current into a voltage (Figure 21.18).

The new thing that we will need to deal with involves specifying the relationship between LED drive current and the photo-current through the photo-transistor. The parameter of interest in a data sheet will be labeled as the **current transfer ratio** (**CTR**). It will be expressed as a percentage to describe the percent of the LED drive current that will be expected to flow in the photo-transistor. For example, if an opto-isolator is specified as having a 10% CTR and the LED side is being driven with a 10 mA current, then we would expect 10% of 10 mA, or 1 mA, of current to flow from collector to emitter in the photo-transistor. As with the transistor circuits that we discussed in Chapter 10, the photo-transistor is a valve, not a pump, so, the components connected externally to the photo-transistor (such as a pull-up or pull-down resistor) will determine the actual current that will flow, up to the amount determined by the CTR and LED drive levels.

By replacing the single photo-transistor of the opto-isolator with a photo-darlington transistor, it is possible to achieve CTRs in excess of 1,000%, at the expense of switching rate due to the slower switching times of the photo-darlington. Indeed, one of the shortcomings of opto-isolators in general is the speed of switching. The slow switching speeds have been a major driving force in the development of isolators based on other technologies that we mentioned above and we describe in more detail below (capacitive and inductive). Opto-isolators based on photo-transistors have turn-on (delay) and rise times on the order of a few microseconds. The turn-off delay and fall times are often a factor of 10 longer. When photo-darlington transistors are used in the output, these times increase by a factor of 10 or more. Compared to digital logic switching times, which are on the order of tens of nanoseconds, these switching times are very slow. This can cause problems when they are connected to digital device inputs. To address the slowness in switching, a family of opto-isolators was developed (the H11Lx series) that incorporates a Schmitt trigger input and open collector output within the opto-isolator (Figure 21.19).

These devices will operate with as little as 1.6 mA of current through the LED, a level easily achieved directly by the outputs of most digital devices. The output has rise and fall

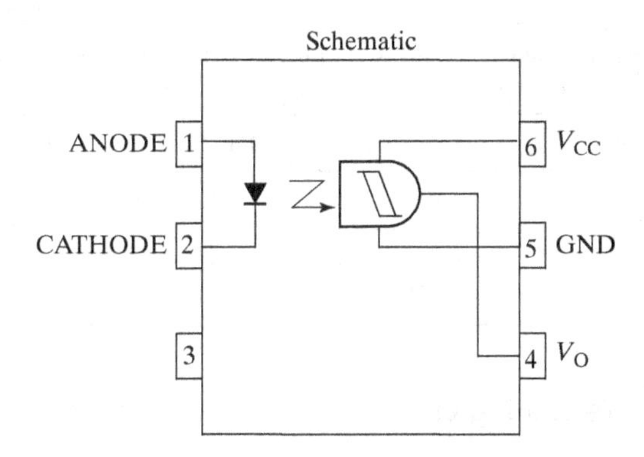

Truth Table

Input	Output
H	L
L	H

FIGURE 21.19
An opto-isolator with a logic output. (Courtesy Fairchild Semiconductor.)

times on the order of 100 ns. While this is not as fast as most digital devices, it is adequate to interface directly to digital logic with no problems due to slow rise and fall times. Devices such as these will allow data at rates up to about 1 Mbit/s to be transferred through the isolation boundary.

To achieve significantly higher speeds requires much greater complexity (and cost) within the device. By carefully controlling the drive current to the LED, substituting a photo-diode for the photo-transistor, and incorporating a trans-resistive amplifier and voltage comparator on the output, devices such as the HCPL-0721 from Avago Technologies can achieve speeds sufficient to pass 25 Mbit/s digital signals across the boundary. Because of the internal LED drive circuit and output comparator, there is no longer a direct relationship between the current input to control the LED and the amount of current that can be sourced or sunk at the output. As a result, the concept of CTR does not apply to devices such as these.

21.5.2 Capacitive Isolation

In the quest to build devices capable of ever-higher data throughput rates, several manufacturers have developed isolators that make use of capacitive coupling in order to transfer signals across the isolation boundary. As you may recall from Section 21.3.1, the displacement current induced by capacitive coupling is related to dV/dt. The implication of this is that we cannot use capacitive coupling to pass a DC signal directly across the isolation boundary. In order to pass DC signals, these capacitive isolators include internal circuitry to modulate the input signal (to guarantee the presence of dV/dt at the capacitor) and then more circuitry to demodulate it on the far side of the isolation boundary. In return for this increased complexity, isolators based on capacitive coupling have achieved data rates in excess of 100 Mbit/s.

21.5.3 Inductive Isolation

By taking advantage of advances in manufacturing technology, Analog Devices and Avago have each developed a line of digital isolators based on inductive coupling. These devices contain microscale transformers. Like capacitive isolators, these devices cannot pass DC and therefore must modulate low-frequency signals in order to pass them across the boundary. Unique to the inductive coupling approach is the ability to transfer significant amounts of energy across the isolation boundary. This ability makes it possible to power the isolated circuitry without a separate power source on the other side of the boundary. Capacitive coupling, on the other hand, requires a separate power source on each side of the boundary. From a data transfer speed perspective, these devices can reach speeds as high as 100 Mbit/s.

TABLE 21.1 Comparison of electrical isolation technologies. (Courtesy Avago Technologies Ltd., Texas Instruments Incorporated, Anolog Devices, Inc.)

Part Number	Coupling Technology	V_{cc}	Data Rate	Isolation Voltage
HCPL-0721	Optical	5 V	25 Mbit/s	3,750 Vrms
ISO721	Capacitive	3.3 V, 5 V	150 Mbit/s	4,000 $V_{(peak)}$
ADuM1100	Inductive	5 V	100 Mbit/s	2,500 Vrms
		3.3 V	50 Mbit/s	

21.5.4 Comparing Isolation Technologies

Table 21.1 summarizes the differences between the three isolation technologies we've covered in this chapter. As the table shows, they differ not only in the data rates achievable, but also in their ability to operate with lower power supply voltages and the magnitude of voltage difference that can be tolerated across the isolation boundary (per U.L. 1577).

21.6 HOMEWORK PROBLEMS

21.1 What are the dominant noise coupling channels within typical mechatronic systems?

21.2 True or False: When building a small circuit (1–2 ICs) on a breadboard, it is acceptable to omit the decoupling capacitors.

21.3 True or False: Analog circuits do not require decoupling capacitors.

21.4 True or False: When using shielded wire, the shield should be connected to ground at both ends so that it provides a common ground connection between the systems.

21.5 Figure 21.20 shows the block diagram for a mechatronic system with two power supplies, a microcontroller, an analog sensor block, and a motor driver block. The microcontroller and the analog sensor block need to be powered by 5 V, while the motor driver runs on 12 V. The motor driver takes its input (DI) from the microcontroller digital output (DO) and the microcontroller receives its analog sensor input (AI) from the sensor circuit output (AO). Draw a set of wires that will supply power and connect control and sensor signals between the blocks while minimizing the potential for conducted noise coupling between the subsystems.

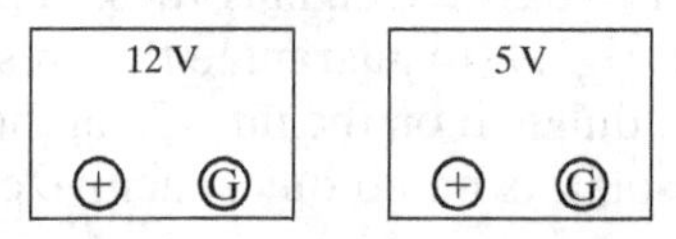

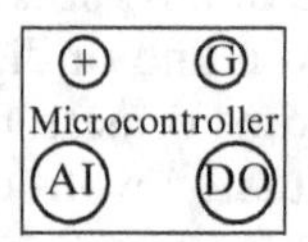

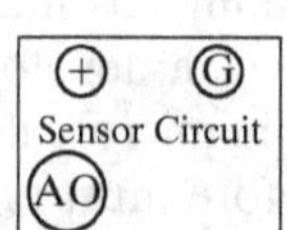

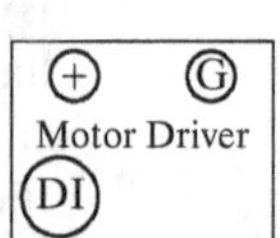

FIGURE 21.20
System block diagram for Problem 21.5.

21.6 True or False: Simply twisting wires together is an effective way to minimize inductively coupled noise.

21.7 Explain in your own words how optical isolation can reduce conductively coupled noise.

21.8 In the circuit of Figure 21.21, if the digital output is in the high state, would you expect the signal seen by the digital input to be in a high or low state? Explain why.

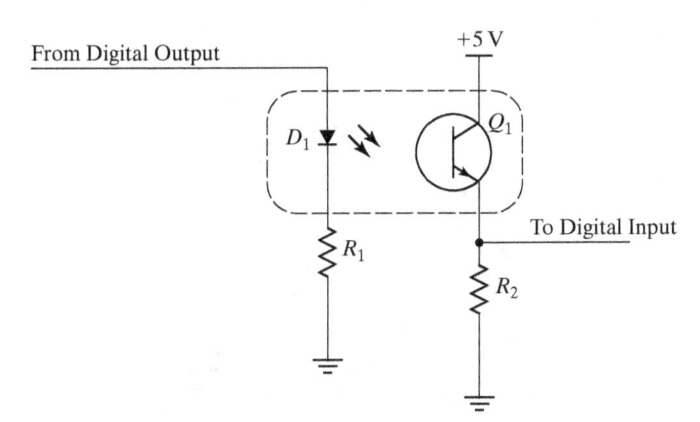

FIGURE 21.21

Circuit for Problems 21.8 and 21.9.

21.9 In the circuit of Figure 21.21, if the digital output swings between 0 and 5 V, D_1 has a V_f of 1.5 V, $R_1 = 3.3\ k\Omega$, $R_2 = 10\ k\Omega$, and I_{IH} on the digital input is 1 μA, what is the minimum CTR required of the opto-coupler in order to produce a high-state voltage of 3.5 V at the digital input?

21.10 In the circuit of Figure 21.22, if the digital output is in the high state, would you expect the signal seen by the digital input to be in a high or low state?

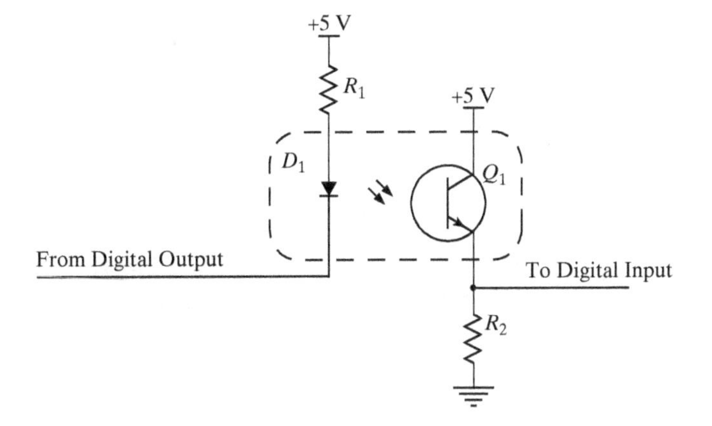

FIGURE 21.22

Circuit for Problems 21.10 and 21.11.

21.11 In the circuit of Figure 21.22, if the digital output connected to the cathode of the diode (D_1) swings between 0.4 V while sinking a maximum of 3.2 mA and 4.5 V while sourcing a maximum of 1 mA, D_1 has a V_f of 1.5 V, the CTR of the opto-coupler is 50%, and the digital input connected to the collector of the photo-transistor has specifications of $V_{IH} = 3.5$ V, $I_{IH} = 10$ μA, $V_{IL} = 1.5$ V, $I_{IL} = -10$ μA, choose values for R_1 and R_2 that will result in acceptable input voltages to the digital input.

FURTHER READING

Ciarcia's Circuit Cellar, Ciarcia, S., Vol. III, McGraw-Hill, 1982.

Noise Reduction Techniques in Electronic Systems, Ott, H. W., 2nd ed., John Wiley & Sons, Inc., 1998.

CHAPTER 22

Permanent Magnet Brushed DC Motor Characteristics

22.1 INTRODUCTION

Direct current (DC) motors comprise one of the most common types of actuator designed into electromechanical systems. They are a straightforward and inexpensive means of creating motion or forces. More often than not, you'll find yourself using motors to put the "*mech-*" into "*mechatronics.*"

Motors are actually complex assemblies that exploit the relationships between electrical current and magnetic fields in order to create useful torque and do work. And, like all real world components and complex assemblies, motors have several interesting characteristics, trade-offs, quirks, and (of course) pitfalls to avoid. Understanding the issues will enable designers to successfully select and use DC motors. In this chapter, we introduce the most common type of DC motors: brushed permanent magnet DC motors. We describe their basic steady-state characteristics, and how to select an appropriate brushed permanent magnet DC motor to meet the requirements of a design. Once you have mastered the concepts contained in this chapter, you should be able to:

1. identify the functional elements of brushed permanent magnet DC motors and describe their function,
2. describe how useful torque is generated from the interactions between the magnetic field established by the permanent magnet in the housing and the rotor's coils,
3. understand and use the characteristic constants for a DC motor, the torque constant K_T, and the speed constant K_e,
4. understand back-EMF,
5. be able to identify DC motor operating points for peak power and peak efficiency,
6. be able to specify an appropriate DC motor for a given application,
7. be able to specify and use a gearhead in combination with a DC motor for a given application.

22.2 SUBFRACTIONAL HORSEPOWER PERMANENT MAGNET BRUSHED DC MOTORS

The category of DC motor that is the least expensive, easiest to use, and thus the most popular, is the subfractional horsepower permanent magnet brushed DC motor. A representative example is shown in Figure 22.1. "**Subfractional horsepower**" refers to its limited power output, and distinguishes it from larger varieties of motors. "**Permanent magnet**" refers to the means used to establish one of the motor's two interacting magnetic fields. "**Brushed**" refers

FIGURE 22.1

A typical subfractional horsepower brushed permanent magnet DC motor.

to the method of commutation (the way in which coils are energized to establish the motor's other magnetic field). Finally, "**DC**" indicates that these motors operate on DC (direct current), rather than AC (alternating current).

Motors exploit the phenomenon described by Ampere's law (one of Maxwell's equations), which states that a flowing electrical current establishes a magnetic field. In the case of current flowing within a wire, a concentric magnetic field is produced that surrounds the wire (Figure 22.2a). By placing a current-carrying wire and its accompanying magnetic field within an external magnetic field, a force results from the interaction of these two magnetic fields (Figure 22.2b). We can determine the resulting force $F = I(\ell \times B_{ext})$ for a current I flowing in a length and orientation of wire ℓ immersed in an external magnetic field B_{ext}.

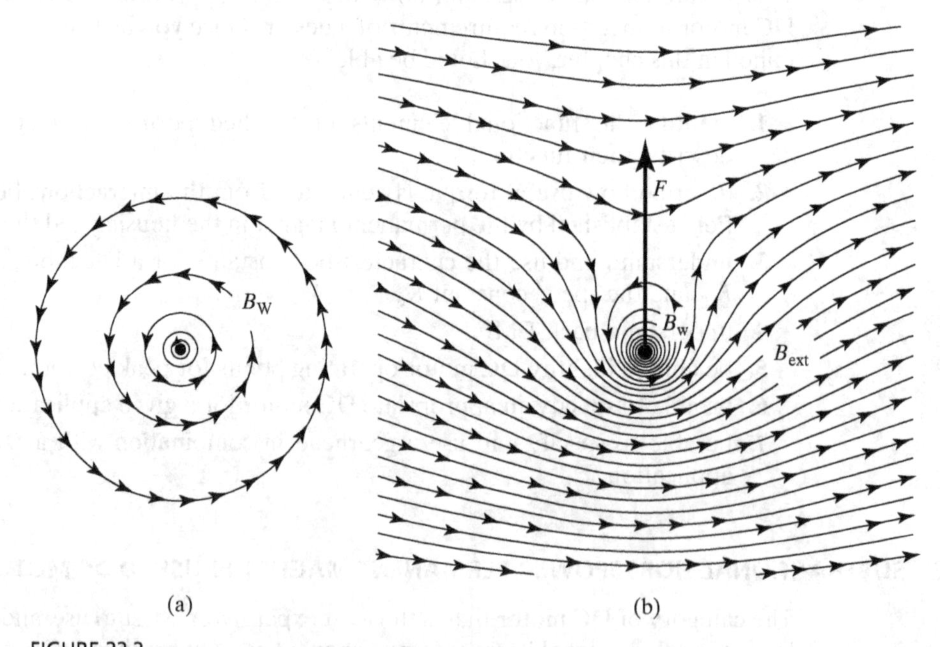

FIGURE 22.2

Magnetic field interactions and resulting forces within a permanent magnet DC motor.

The permanent magnet DC motor's design harnesses the force imparted to the wire to create torque on the motor's rotor, which is constrained by the motor's bearings so that the only motion permitted is rotation (see Figures 22.3 and 22.4).

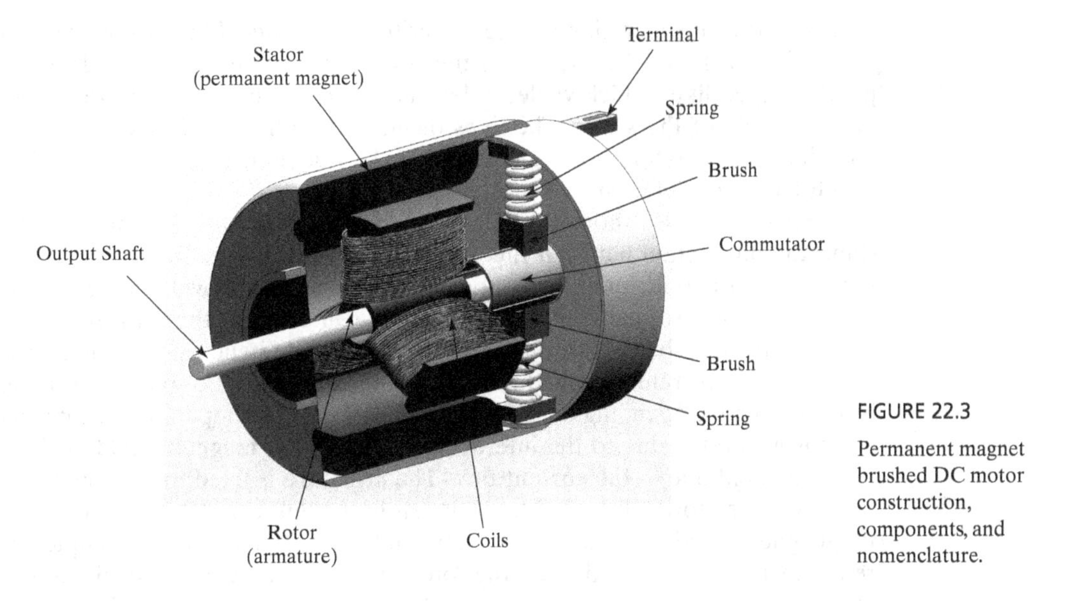

FIGURE 22.3

Permanent magnet brushed DC motor construction, components, and nomenclature.

Figure 22.3 shows a cutaway view of a typical permanent magnet brushed DC motor. The construction generally consists of a **stator** (the stationary element in an electromagnetic device), which is made up of powerful permanent magnets that generate a static, nonrotating magnetic field; a **rotor** which carries the **armature** (the moving element in an electromagnetic device, and in the case of a permanent magnet brushed DC motor, the element that incorporates the **windings** or **coils**) and the **commutator** (from Latin *commutare,* meaning "to change often"), and rotates in the **bearings** that support it; and a **housing** that holds the stator, rotor bearing supports, and brushes in a fixed relationship to one another.

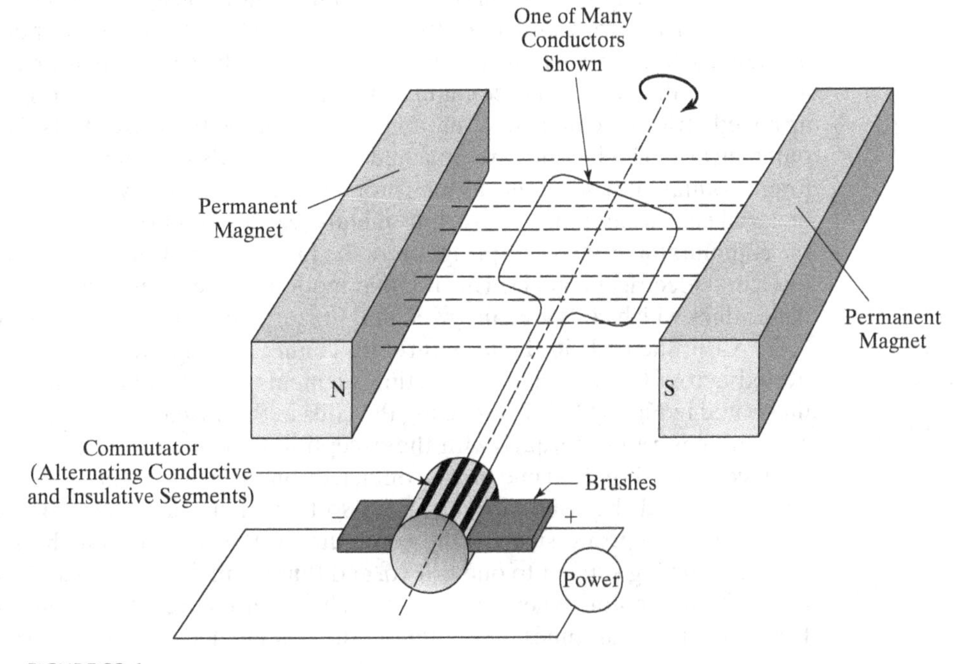

FIGURE 22.4

Permanent magnet DC motor stator and armature detail.

In terms of generating torque, the critical elements of the motor are the stator and the armature, which are the sources of the two interacting magnetic fields. The stator is commonly shaped like a thick-walled tube, and the rotor and armature fit in the hollow space in the middle of the stator. The lines of magnetic flux established by the stator run from one side of the stator to the other. Figure 22.4 demonstrates the magnetic flux lines in a simplified representation.

Figure 22.4 also shows a single winding of the armature, and makes it easier to understand the interaction between the armature and the stator. The armature contains a large number of wire loops, or coils, identical to the single one shown, arranged in a radial pattern around the rotor so that continuous torque is generated as the rotor rotates. Also, the additional loops contribute additional resulting forces, and hence more motor torque.

Causing current to flow through the looped wire in the coil causes a magnetic field to be established. Depending on the orientation of the coil loop in the stator's magnetic field, a force is generated due to the interaction between the magnetic field of the stator and the magnetic field due to the current flow. The armature is fixed to the rotor, and as a result, it reacts to these forces by rotating. If the entire system consisted of a single coil loop within the magnetic field of the stator as shown in Figure 22.4, inducing a simple, unchanging current flow in the coil would cause the rotor to turn until the magnetic fields are aligned, eliminating the forces that generate torque. This will happen when the direction of the force has no component perpendicular to the rotor's radius, and the rotor will come to rest at an equilibrium position. In order to spin continuously, permanent magnet DC motors are designed to switch the current flowing in the coils continuously, never coming to an equilibrium position. When the rotor approaches a point near equilibrium, the direction of the current flow in the coil is first stopped, and then reversed. This has the effect of moving the equilibrium point as the rotor spins. In much the same way that a carrot is dangled in front of a cart horse to encourage it to pull, the equilibrium point is always being moved out of reach of the rotor in order to keep it spinning continuously.

The commutator and **brushes** together perform the switching of the current in the coils of the armature so that the motor may rotate continuously. These are shown on the upper right side of Figure 22.3. The commutator-to-brush interface is the point where current is introduced into the coils. The commutator is part of the rotor/armature assembly, and rotates along with the rotor. The commutator is shaped like a smooth ring with longitudinally arranged strips of conductive material alternating with strips of insulating material on its outer surface. The brushes are attached to the motor's housing, and make contact with different commutator segments as the rotor spins. Brushes are typically made of a low-friction material such as graphite or precious metals (e.g., platinum), and are in sliding contact with the commutator as it spins. Springs press the brushes firmly against the commutator to maintain good electrical contact. This relative motion between the brushes and the commutator determines which coils are energized and the direction of the current flow.

All of the coils in the armature are connected together in series, with each coil connected across two adjacent conducting segments of the commutator. This arrangement is illustrated in Figure 22.5. Connecting the coils in this fashion allows all the coils to carry current and contribute torque (with the exception that one of the coils periodically has both leads connected to the same brush contact, during which time no current flows in that single coil). If instead the coils were arranged so that their leads were connected to conducting segments on opposite sides of the commutator, this would limit the number of coils that were conducting current to one or two, and thus significantly reduce the overall torque generated by the motor. When connected as shown in Figure 22.5, all but one of the coils contribute torque at all times. In the figure, the coil labeled "c" has no current and contributes no torque.

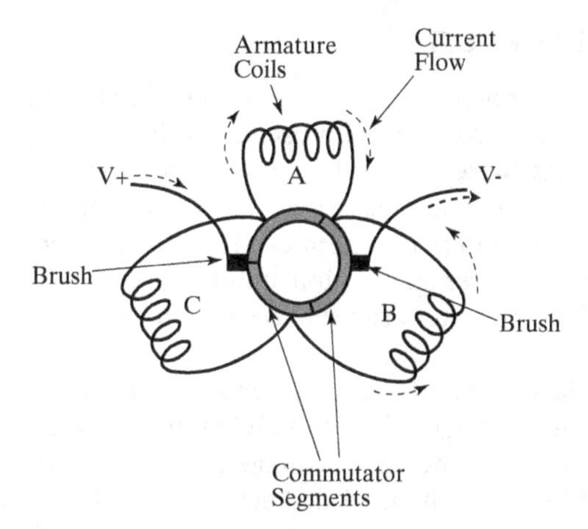

FIGURE 22.5

Electrical connections and layout of armature coils, commutator, and brushes.

Although the commutator and brushes are key elements that make a permanent magnet DC motor work, they are also the weakest link in the system. Brushes are sacrificial components that wear over time, and are the components of the motor that are most likely to fail. Also, because they are in sliding contact with the commutator, they cause "**brush drag**," or frictional forces, which are the unavoidable result of dragging one material over another. Overcoming these drag forces requires torque that is then not available to do work. These forces are therefore considered losses. In addition, since the brushes are pressed against the commutator with springs, this creates a dynamic system complete with resonances. As the rotor and commutator spin faster and faster, the brushes will eventually reach a condition where they are not able to follow the contours and stay in good electrical contact because of a phenomenon known as **brush bounce**. Ultimately, this limits the maximum speed of the motor. Increasing the spring force pressing the brushes against the commutator is one solution, but this increases frictional losses and accelerates wear on the brushes. Brushes are most commonly made out of graphite, which is a material that has a relatively high electrical resistance (1–10 Ω, typically) and is physically dirty. Graphite brushes cast off particles and dust as wear occurs. In addition, the brush–commutator interface is electrically noisy as the connections to the individual coils in the armature are continuously made and broken. These can be the source of significant electromagnetic interference (EMI) or noise in a system.

22.3 ELECTRICAL MODEL

Electrically, permanent magnet brushed DC motors can be modeled as a series of three basic electrical components: a resistor, an inductor, and a source of electromotive force (EMF), or voltage (Figure 22.6). This voltage source is commonly called the "**back-EMF**" or "**counter EMF.**" The origins of the resistive and inductive components are easy to see. The resistor in the model is a result of the finite resistance per unit length of the wire used to construct the coils in the armature. The inductor is a result of coils of wire that make up the armature windings. All coils of wire act as inductors. The back-EMF, on the other hand, takes a little more discussion to clarify.

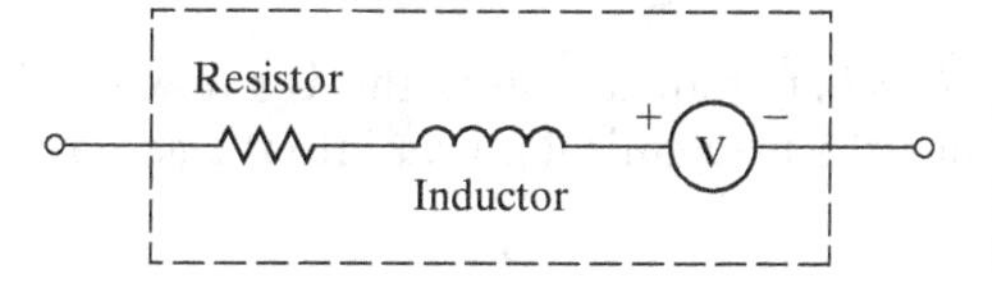

FIGURE 22.6

Electrical model of a permanent magnet brushed DC motor.

22.4 BACK-EMF AND THE GENERATOR EFFECT

Recall that the torque generated by a permanent magnet DC motor is the result of the current flowing in the armature coils in the presence of the stator's magnetic field. This effect is known as the **Lorentz force law** and is described by Ampere's law. **Faraday's law** is another of Maxwell's equations, and describes the result of moving a coil of wire through an external magnetic field: a voltage is generated. This is the principle used to generate electricity in a hydroelectric power generation plant, for example, where the potential and kinetic energy of the flowing water are used to spin a rotor and armature in the presence of a magnetic field.

Faraday's law is the effect that requires the inclusion of the back-EMF term in our electrical model of a permanent magnet brushed DC motor: the armature is spinning inside the field created by the stator. This induces a voltage (the back-EMF) across the coil as it spins. This voltage is opposed to the voltage placed across the coil that made the rotor spin in the first place. The somewhat surprising and nonintuitive result of this is that the motor acts as a generator *at the same time* that it acts as a motor. The fact that we are applying a voltage and causing the motor to turn doesn't eliminate the phenomenon described by Faraday's law. When a motor turns, it acts as a generator and the voltages add by superposition, though they have opposite signs. The effect of this voltage is to reduce the voltage drop across the motor's terminals and therefore the current flow in the motor's coils when the motor is rotating.

22.5 CHARACTERISTIC CONSTANTS FOR PERMANENT MAGNET BRUSHED DC MOTORS

As a motor turns faster, more back-EMF is generated since the coils in the armature are moving faster through the stator's magnetic field. The magnitude of the back-EMF is related to the rotational speed through a constant K_e, called the **speed constant** or **voltage constant:**

$$E = K_e \omega \tag{22.1}$$

where: E = back-EMF [V]

K_e = speed constant $\left[\frac{\text{V}}{\text{rad/s}}\right]$

ω = rotational speed [rad/s]

The value of K_e is determined by the construction, geometry, and materials properties of the motor. Quantities like the motor's physical dimensions, the number of turns in the coil windings, and the magnetic flux density of the stator all contribute to the value of K_e. We can continue to examine the generator effect to arrive at another useful relationship.

In this development, we will ignore, for the moment, the nonideal mechanical and electrical losses associated with the motor/generator operation. The largest of the effects we will assume are negligible in this discussion is the torque required to overcome friction in the motor. Because of friction, the torque generated by the motor may be treated as being made up of two terms: frictional torque and usable torque (or torque that is available at the motor's output shaft and may be used to drive a load):

$$T_M = T_L + T_f \quad \text{[Nm]} \tag{22.2}$$

For now, we assume $T_f \approx 0$. For most motors, this is a reasonable first approximation. If the losses are negligible, then the mechanical power into the generator, $T \times \omega$, will equal the electrical power out, $E \times I$.

$$P = EI = T\omega \quad \text{[W]} \tag{22.3}$$

We can combine Eq. (22.1) with Eq. (22.3) to yield:

$$K_e\omega\, I = T\omega \qquad [\mathrm{W}] \tag{22.4}$$

which can be simplified to:

$$T = K_e I \qquad [\mathrm{Nm}] \tag{22.5}$$

At this point, you may be wondering how we took a constant, K_e, with units of [volts/rotational speed] and tuned it into one with units of [torque/current]. The answer lies in the origin of the constant K_e. The constant originates in the relationship between current flow and magnetic fields described by Maxwell's equations. The constant can be expressed equivalently with units of [volts/(radians/second)] or [newton · meter/ampere]. To minimize confusion, we treat the constant in Eq. (22.5) as different and call it K_T, the **torque constant**. The value of K_T is determined by the same factors that determine the value of K_e: the construction, geometry, and materials properties of the motor. The motor's physical dimensions, the number of turns in the coil windings, and the magnetic flux density of the stator all contribute to the value of K_T, just as they did with K_e. With the substitution of K_T for K_e, Eq. (22.5) takes on its more common form:

$$T = K_T I \qquad [\mathrm{Nm}] \tag{22.6}$$

where: T = torque [Nm]
K_T = torque constant [Nm/A]
I = current [A]

This distinction between K_T and K_e is particularly useful in that numerically $K_T = K_e$ when compatible units are used (e.g., [volts/(radians/second)] and [newton · meter/ampere]). When the constants are expressed in inconsistent units (which is far more common in practice), a conversion factor must be applied to convert between them. Two of the most commonly needed are:

$$K_T\ [\mathrm{in\cdot oz/A}] = 1.3542 \times K_e\ [\mathrm{V/krpm}] \tag{22.7}$$

$$K_T\ [\mathrm{Nm/A}] = 9.5493 \times 10^{-3} \times K_e\ [\mathrm{V/krpm}] \tag{22.8}$$

While we have structured this discussion by examining the case of a generator, the direction of torque flow, either into or out of the motor shaft, has no effect on the fundamental interaction between the magnetic field and the electrons in the conductor. As a result, these equations [Eqs. (22.1) and (22.6)] are true when the motor acts as a generator as well as when it acts as a motor. One important result of this is that the maximum current flow, and therefore [by Eq. (20.6)] maximum torque, occurs at zero rotational velocity, when there is no back-EMF.

Example 22.1

A motor with $K_T = 5.89$ in · oz/A and a coil resistance of 1.76 Ω is driven with a supply voltage of 12 V. If the motor's friction torque is 1.2 in · oz, what is the maximum torque available for driving a load? How much current is flowing under these conditions?

By combining Eqs. (22.2) and (22.6), we can find the torque available at the output shaft:

$$T_M = T_L + T_f = K_T I$$

$$T_L = K_T I - T_f$$

Using Ohm's law to substitute for current:

$$\boxed{I = \frac{V}{R} = \frac{12\,\text{V}}{1.76\ \Omega} = 6.82\,\text{A}}$$

$$T_L = \frac{K_T V}{R} - T_f = \frac{(5.89\,\text{in}\cdot\text{oz/A})(12\,\text{V})}{1.76\ \Omega} - 1.2\,\text{in}\cdot\text{oz}$$

$$\boxed{T_L = 39.0\,\text{in}\cdot\text{oz}}$$

After the effects of friction are subtracted off the total torque produced by the motor under these conditions, 39.0 in · oz of torque is available at the motor shaft to drive a load. The current required to generate this torque is 6.82 A.

22.6 CHARACTERISTIC EQUATIONS FOR CONSTANT VOLTAGE

To more fully understand the torque and speed characteristics of a motor, we can start by examining what goes on when we place the motor into a circuit with a driving voltage (Figure 22.7).

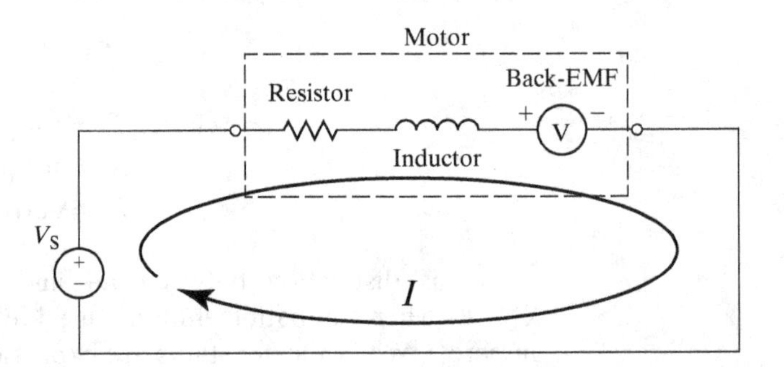

FIGURE 22.7
DC motor circuit with driving voltage.

We can use Kirchhoff's laws to write a loop equation to describe the steady-state current flow in this circuit:

$$V = IR + K_e\omega \qquad [\text{V}] \tag{22.9}$$

where: V = voltage [V]
I = current [A]
R = resistance of motor coils [Ω]
K_e = voltage constant $\left[\frac{\text{V}}{\text{rad/s}}\right]$
ω = rotational speed [rad/s]

The voltage drop across the motor's coils has an $I \times R$ term, as you would normally expect, *plus* the effects of the back-EMF generated by spinning the motor, expressed in the term $K_e\omega$.

Some implications of equations Eqs. (22.1), (22.6), and (22.9) are:

- The higher the rotational speed of the motor, the lower the current flow and therefore the lower the torque. This occurs because of the back-EMF.
- Maximum speed corresponds to zero current flow and therefore zero torque (we obviously can't achieve this with a real motor, which must, at the very least, overcome the force of friction).

- When $\omega = 0$ (a condition referred to as "stall"), $V = IR$ and current and torque will both be at a maximum.

By substituting Eq. (22.9) into Eq. (22.6), we can develop an expression relating torque to speed.

$$V = \frac{T}{K_T}R + K_e\omega$$

$$V - \frac{T}{K_T}R = K_e\omega$$

$$\omega = \frac{V}{K_e} - \frac{R}{K_T K_e}T \qquad \text{[rad/s]} \tag{22.10}$$

Eq. (22.10) shows that, for a given voltage V, torque and speed for a motor are linearly related. Often, this is graphically represented with a plot showing a family of lines relating T vs. ω for several constant values of voltage, V. Figure 22.8 shows a typical example.

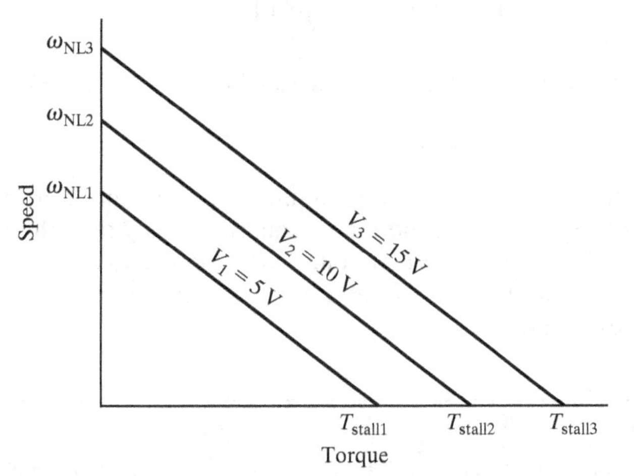

FIGURE 22.8

Typical torque vs. speed curves for a permanent magnet brushed DC motor.

There are a few aspects of Figure 22.8 that we should identify and label. The first is the y-intercept of each line. This is the maximum speed that the motor can achieve for a given voltage, which occurs for the idealized case where there is no torque generated. This is called the "**no-load speed,**" written as ω_{NL}, and is the V/K_e term in Eq. (22.10). Thus, for a permanent magnet brushed DC motor:

$$\omega_{NL} = \frac{V}{K_e} \qquad \text{[rad/s]} \tag{22.11}$$

The slope of the line given by Eq. (22.10) is the multiplier on T, which is $R/K_T K_e$. The slope term is also given its own symbol, R_M, and is called the "**speed regulation constant**":

$$R_M = \frac{R}{K_T K_e} \qquad \left[\frac{\text{rad/s}}{\text{Nm}}\right] \tag{22.12}$$

By substituting ω_{NL} and R_M into Eq. (22.10), we get an expression that is more easily identified as that of a straight line:

$$\omega = \omega_{NL} - R_M T \qquad \text{[rad/s]} \tag{22.13}$$

The x-intercept of the constant-voltage line represents the case where $\omega = 0$, which occurs when the motor is stalled. This is the point at which torque, and therefore current, is maximized. This is called the "**stall torque**," and given the symbol T_{stall} or T_S. The corresponding "**stall current**" is given the symbol I_{stall} or I_S). If we set $\omega = 0$ in Eq. (22.13), we can also express T_{stall} as:

$$0 = \omega_{NL} - R_M T_{stall}$$

$$T_{stall} = \frac{\omega_{NL}}{R_M}$$

Recalling from Eq. (22.11) that $\omega_{NL} = V/K_e$:

$$T_{stall} = \frac{V}{R_M K_e} \quad [\text{Nm}] \tag{22.14}$$

Then, recalling from Eq. (22.12) that $R_M = R/K_T K_e$ and then simplifying gives:

$$T_{stall} = \frac{K_T V}{R} \quad [\text{Nm}] \tag{22.15}$$

Simplifying further using Ohm's law, we once again obtain Eq. (22.6):

$$T = K_T I \quad [\text{Nm}]$$

Stall occurs whenever the motor is attempting to spin or move against a force that exceeds the amount of torque it can generate internally, *and* any time the motor is started from a resting position. This is an important point: *stall torque (and stall current) occurs any time the motor starts from a stop*. This is a critical point to consider when designing circuits to drive DC motors, as covered in Chapter 23.

Example 22.2

A permanent magnet brushed DC motor will be used to spin a cooling fan installed in a toy that would otherwise overheat and deform. The application requires a motor that can supply at least 225 mNm of torque at 2,000 rpm. Will a motor with a no-load speed of 9,550 rpm and a coil resistance of 2.32 Ω, powered by a 24 V battery pack, be adequate for the task?

Starting with Eq. (22.13) and rearranging it as an expression for T:

$$\omega = \omega_{NL} - R_M T$$

$$T = \frac{\omega_{NL} - \omega}{R_M} = \frac{K_T K_e(\omega_{NL} - \omega)}{R_{coil}}$$

We can determine K_e from Eq. (22.11):

$$K_e = \frac{V}{\omega_{NL}} = \frac{24\text{ V}}{9.55\text{ krpm}} = 2.51\text{ V/krpm}$$

and K_T from Eq. (22.8):

$$K_T = \left(9.5493 \times 10^{-3}\frac{\text{Nm/A}}{\text{V/krpm}}\right)(2.51\text{ V/krpm})$$

$$= 0.0240\text{ Nm/A} = 24.0\text{ mNm/A}$$

Substituting these values into our expression for torque:

$$T = \frac{K_T K_e(\omega_{NL} - \omega)}{R_{coil}}$$

$$= \frac{(24.0\,\text{mNm/A})(2.51\,\text{V/krpm})(9.55\,\text{krpm} - 2\,\text{krpm})}{2.32\,\Omega}$$

$$\boxed{T = 196\,\text{mNm}}$$

This motor will not provide adequate torque at 2,000 rpm for this application, so a different motor will need to be evaluated.

Given that a motor can operate anywhere between stall and no-load conditions, a couple of fundamental questions arise: are all operating points equally desirable and useful? *or* are some better than others? A close look at motor power and efficiency will help answer these questions.

22.7 POWER CHARACTERISTICS

For the purposes of this discussion, mechanical power is defined as $P = T\omega$. Recall from Eq. (22.2) that overall motor torque is made up of a friction torque term and a usable torque term, so the full expression for motor power output becomes:

$$P = T\omega = (T_f + T_L)\,\omega \qquad [\text{W}] \tag{22.16}$$

For this discussion, we will assume that the friction torque is relatively small and may be safely neglected. Again, for most motors, this is a reasonable first approximation. However, for any of these discussions, the effects of frictional torque may be explored by carrying the T_f term in the equations through and redeveloping the results.

As with the relationship between torque and speed, a motor's torque and power characteristics are usually presented graphically for lines of constant voltage, and drawn as a family of curves. Figure 22.9 shows a typical family of curves for a representative motor. The power output characteristic is parabolic in shape, having a maximum at $\frac{1}{2}T_{stall}$ for a given voltage.

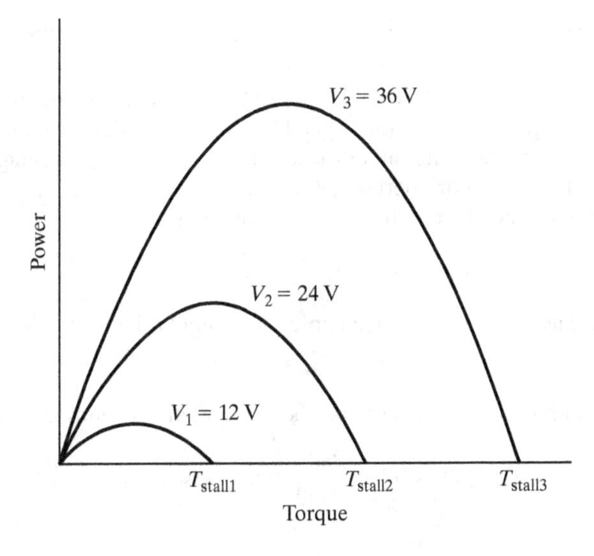

FIGURE 22.9

Typical torque vs. power output curves for a permanent magnet brushed DC motor.

To understand the shape of the curve and the position of the peak value, start from the statement that $P = T\omega$ [Eq. (22.16)]. By substituting Eq. (22.13) for ω, this can be rewritten as:

$$P = T(\omega_{NL} - R_M T) \quad [\text{W}] \tag{22.17}$$

Then, by combining terms and substituting $\omega_{NL} = V/K_e$ from Eq. (22.11), we arrive at an expression relating power to torque:

$$P = \frac{VT}{K_e} - R_M T^2 \quad [\text{W}] \tag{22.18}$$

Taking the derivative of Eq. (22.18) with respect to torque and setting the results equal to 0 yields the point of **maximum power.** The results of that exercise are that maximum power output for a permanent magnet brushed DC motor occurs when $T = ½T_{stall}$.

We can make use of this by starting with Eq. (22.18), and substituting $T = ½T_{stall}$ and $T_{stall} = V/R_M K_e$ from Eq. (22.14), to develop a relationship between P_{max} and applied voltage:

$$P_{max} = \frac{V^2}{2R_M K_e} - R_M\left(\frac{V}{2R_M K_e}\right)^2$$

Substituting $R_M = R/K_T K_e$ from Eq. (22.12) gives:

$$P_{max} = \frac{V}{K_e}\left(\frac{K_T V}{2R}\right) - R_M\left(\frac{K_T V}{2R}\right)^2$$

Finally, simplifying this gives the result:

$$P_{max} = \left(\frac{K_T}{4K_e R}\right)\cdot V^2 \quad [\text{W}] \tag{22.19}$$

These results show that P_{max} is proportional to V^2, since the term $(K_T/4K_e R)$ is a constant for a given motor. This is an important result: the mechanical power output of permanent magnet brushed DC motors changes as the square of the applied voltage. Changes in voltage have a substantial impact on a motor's power output.

Example 22.3

A motor with a terminal resistance of 0.316 Ω and $K_T = 30.2$ mNm/A is powered by a 12 V supply. Measurements show that the operating rotational speed is 3,616 rpm with a current flow of 1.79 A. How much power does the motor generate under these conditions? What percentage of the maximum possible power is this for the motor operating at 12 V?

The amount of power generated by the motor is given by Eq. (22.16):

$$P = T\omega = (T_f + T_L)\,\omega$$

In this example, we will assume that T_f is negligible, and simplify the expression:

$$P = T\omega$$

To maintain consistent units, we need rotational speed in radians/seconds instead of rpm:

$$3{,}616 \text{ rpm} \times 0.105\,\frac{\text{rad/s}}{\text{rpm}} = 380 \text{ rad/s}$$

Then, from Eq. (22.6), we can determine the torque produced by the motor, since we know both K_T and motor current, I:

$$T = K_T I$$

Substituting this back into the expression for power gives the solution:

$$P = K_T I \omega$$

$$\boxed{P = (0.0302\ \text{Nm/A})(1.79\ \text{A})(380\ \text{rad/s}) = 20.5\ \text{W}}$$

In order to determine the maximum power possible for this motor at 12 V, we will need to determine T_{stall}, from which we can determine ω at $\frac{1}{2}T_{stall}$ and calculate maximum power. From Eq. (22.15):

$$T_{stall} = \frac{K_T V}{R} = \frac{(0.0302\ \text{Nm/A})\ (12\ \text{V})}{0.316\ \Omega} = 1.15\ \text{Nm}$$

For this motor powered at 12 V, the rotational speed at $\frac{1}{2}T_{stall}$ is:

$$\omega = \omega_{NL} - R_M(\tfrac{1}{2}T_{stall}) = \frac{V}{K_e} - \frac{R}{K_T K_e}\frac{K_T V}{2R}$$

$$\omega = \frac{V}{K_e} - \frac{V}{2K_e} = \frac{V}{2K_e}$$

From K_T, we can determine K_e using Eq. (22.8):

$$K_e = \frac{0.0302\ \text{Nm/A}}{9.5493 \times 10^{-3}\ \frac{\text{Nm/A}}{\text{V/krpm}}} = 3.16\ \text{V/krpm}$$

Then, ω for $\frac{1}{2}T_{stall}$ is:

$$\omega = \frac{V}{2K_e} = \frac{12\ \text{V}}{2\,(3.16\ \text{V/krpm})} = 1{,}897\ \text{rpm}$$

Finally, maximum power is:

$$P_{max} = \tfrac{1}{2}T_{stall}\omega = (1/2)(1.15\ \text{Nm})(199\ \text{rad/s})$$

$$P_{max} = 114\ \text{W}$$

When the motor is generating 20.5 W, the percentage of maximum power is:

$$\boxed{\frac{P}{P_{max}} = \frac{20.5\ \text{W}}{114\ \text{W}} \times 100 = 18.0\%}$$

22.8 DC MOTOR EFFICIENCY

An additional quantity of great interest is that of **motor efficiency**, η. In this analysis, efficiency is defined as the ratio of mechanical power produced by the motor to electrical power consumed by the motor:

$$\eta = \frac{P_{out}}{P_{in}} = \frac{T_L \omega}{VI} = \frac{(T_M - T_f)\omega}{VI} \tag{22.20}$$

Efficiency is maximized at the point in the operating curve when a balance is struck between generating the most useful work while dissipating a minimum of power as I^2R losses (heat) and friction losses. Figure 22.10 shows the typical efficiency characteristics of a permanent magnet brushed DC motor, and its relationship to power, torque, speed, and current.

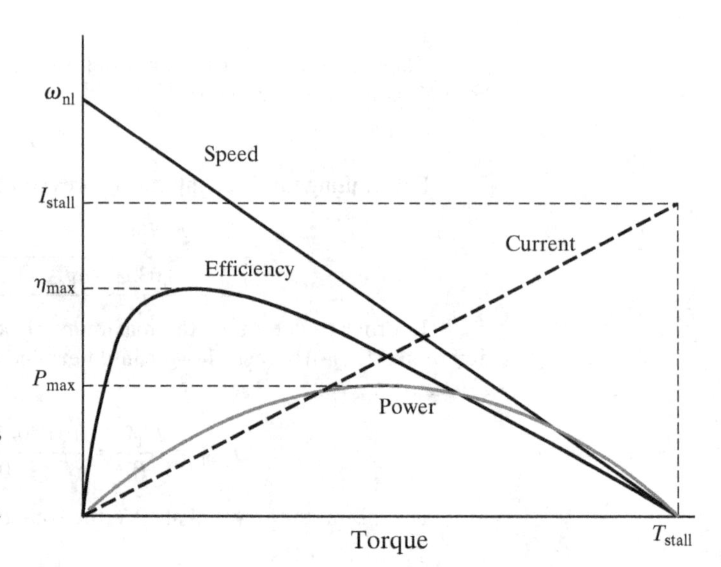

FIGURE 22.10

Composite DC motor characteristics, showing the interrelationship between efficiency η, torque T, current I, speed ω, and power P for a given voltage.

At high torques, with correspondingly high currents, I^2R losses (dissipated through heating of the coils) are high and therefore efficiency is low. At very low torques, in spite of the high rotational speed ω, little useful mechanical power is produced and most or all of the power is consumed overcoming friction, which also leads to low efficiency. Peak efficiency must then occur somewhere in between. In general, maximum efficiency operation occurs at relatively high ω, and low values of torque and current leading to minimal I^2R losses. The point of maximum efficiency operation varies from motor to motor.

In most applications, it will be desirable to operate DC motors in the region between the point of maximum efficiency and the point of maximum power. Note that the slope of the efficiency curve falls away from the maximum much more gradually in this region, rather than dropping rapidly to zero for higher speeds. Also, most motors are not capable of running continuously with high current levels, such as those that occur at torque levels above the point corresponding to maximum power because they will overheat. The specifications of any given motor should always be consulted for this type of information.

For operating conditions that result in the **maximum motor efficiency**, a motor will generate the most mechanical output power for a given power input. We will solve for the motor current, I, that results in the efficiency being maximized [1]. For this analysis, frictional effects definitely may *not* be safely neglected, since they dominate the efficiency of the motor for conditions involving low torque/high speed operation.

It will be useful to restate efficiency in order to solve for the current, I, that maximizes η. The general equation for motor efficiency is stated above in Eq. (22.20). We can restate V using Ohm's law as follows:

$$V = I_S R \quad \text{[V]} \tag{22.21}$$

since V, I_S, and R are all constants. Substituting this back into the P_{in} term of Eq. (22.20) gives:

$$\eta = \frac{(T_M - T_f)\omega}{I_S I R}$$

Substituting the expression for ω from Eq. (22.10) gives:

$$\eta = \frac{(T_M - T_f)\left(\dfrac{V}{K_e} - \dfrac{RT_M}{K_e K_T}\right)}{I_S I R}$$

and replacing total motor torque produced $T_M = K_T I$ [Eq. (22.6)] and $V = I_S R$ [Eq. (22.21)] gives:

$$\eta = \frac{(K_T I - T_f)\left(\frac{I_S R}{K_e} - \frac{K_T I R}{K_T K_e}\right)}{I_S I R}$$

Simplifying this gives:

$$\eta = \frac{(K_T I - T_f)\left(\frac{I_S R - I R}{K_e}\right)}{I_S I R}$$

The frictional torque term, T_f, may also be expressed as:

$$T_f = K_T I_{NL} \qquad [\text{N} \cdot \text{m}] \tag{22.22}$$

since the no-load current, I_{NL}, is the amount of current required to overcome the force of friction only, without generating any additional useful torque—the definition of the no-load condition. Substituting this into our expression for efficiency results in the following:

$$\eta = \frac{(K_T I - K_T I_{NL})\left(\frac{I_S R - I R}{K_e}\right)}{I_S I R}$$

Finally, since $K_T = K_e$ in consistent units, efficiency can be expressed as:

$$\eta = \frac{(I - I_{NL})(I_S R - I R)}{I_S I R}$$

$$\eta = \frac{(I - I_{NL})(I_S - I)}{I_S I}$$

$$\eta = 1 - \frac{I}{I_S} - \frac{I_{NL}}{I} + \frac{I_{NL}}{I_S} \tag{22.23}$$

$$\eta = \left(1 - \frac{I_{NL}}{I}\right)\left(1 - \frac{I}{I_S}\right) \tag{22.24}$$

With Eqs. (22.23) and (22.24), we have expressions for efficiency only as functions of motor current (I), no-load current (I_{NL}), and stall current (I_S). In order to find the current that results in the operating point of maximum efficiency, take the derivative of Eq. (22.23) with respect to current (I), set the results equal to 0, and solve for I.

$$\frac{\partial \eta}{\partial I} = \frac{\partial}{\partial I}\left(1 - \frac{I}{I_S} - \frac{I_{NL}}{I} + \frac{I_{NL}}{I_S}\right) = 0$$

$$\frac{I_{NL}}{I^2} - \frac{1}{I_S} = 0$$

$$I = \sqrt{I_{NL} I_S} \tag{22.25}$$

Substituting the result in Eq. (22.25) back into Eq. (22.24) gives us the expression for maximum efficiency we were after:

$$\eta_{max} = \left(1 - \frac{I_{NL}}{\sqrt{I_{NL} I_S}}\right)\left(1 - \frac{\sqrt{I_{NL} I_S}}{I_S}\right)$$

Simplifying this gives the more compact result:

$$\eta_{\max} = \left(1 - \sqrt{\frac{I_{\mathrm{NL}}}{I_{\mathrm{S}}}}\right)^2 \tag{22.26}$$

Recalling from Eq. (22.22) that $T_{\mathrm{f}} = K_{\mathrm{T}} I_{\mathrm{NL}}$ and $V = I_{\mathrm{S}} R$ from Eq. (22.21), we can rewrite Eq. (22.24) in terms that will allow us to draw a few additional conclusions:

$$\eta_{\max} = \left(1 - \sqrt{\frac{T_{\mathrm{f}} R}{K_{\mathrm{T}} V}}\right)^2 \tag{22.27}$$

This expression for maximum efficiency shows that increases in friction decrease efficiency, as do increases in resistance.

After stepping through this development, the trade-offs between speed and torque that affect efficiency should be clear; both from the final results in Eq. (22.27) and the initial expression for efficiency given in Eq. (22.20):

- Very high speed/low torque operation (e.g., at or near the no-load condition) is dominated by friction and hence not efficient. Eq. (22.20) illustrates this point, as T_{M} goes to 0 at the no-load condition.
- Very low speed/high current operation (e.g., at or near the stall condition) is dominated by heating of the coils through I^2R losses and hence inefficient. Eq. (22.20) also illustrates this point, as ω is 0 at stall conditions.
- The point of maximum efficiency operation lies between these two endpoints, and I^2R losses typically exert a more dominant influence. This results in maximum efficiency for relatively high speed/low torque operation (e.g., somewhere near the no-load condition). Eq. (22.27) illustrates the balance between efficiency at stall and at no-load conditions.

Example 22.4

If the no-load current for the motor in Example 22.3 is 137 mA, what is the efficiency of the motor running under those conditions? What rotational speed results in maximum efficiency operation? What is the maximum efficiency?

In addition to the no-load current, we need to know the stall current, I_{S}. From Eq. (22.5):

$$I_{\mathrm{S}} = \frac{T_{\mathrm{S}}}{K_{\mathrm{T}}} = \frac{1.15\,\mathrm{Nm}}{0.0302\,\mathrm{Nm/A}} = 38.0\,\mathrm{A}$$

Using Eq. (22.24) to find the efficiency of the motor for these conditions:

$$\eta = \left(1 - \frac{I_{\mathrm{NL}}}{I}\right)\left(1 - \frac{I}{I_{\mathrm{S}}}\right) = \left(1 - \frac{0.137\,\mathrm{A}}{1.79\,\mathrm{A}}\right)\left(1 - \frac{1.79\,\mathrm{A}}{38\,\mathrm{A}}\right)$$

$$\boxed{\eta = 88.0\%}$$

The current that results in maximum efficiency is given by Eq. (22.25):

$$I = \sqrt{I_{\mathrm{NL}} I_{\mathrm{S}}} = \sqrt{(0.137\,\mathrm{A})(38\,\mathrm{A})} = 2.28\,\mathrm{A}$$

This corresponds to a rotational speed at 12 V as given by Eq. (22.13):

$$\omega = \omega_{\mathrm{NL}} - R_{\mathrm{M}} T$$

Substituting T from Eq. (22.6), ω_{NL} from Eq. (22.11), and R_M from Eq. (22.12) into Eq. (22.13), and recalling that $K_e = 3.16$ V/krpm from Example 22.3, gives:

$$\omega = \frac{V}{K_e} - \frac{R(K_T I)}{K_T K_e} = \frac{V - RI}{K_e} = \frac{12\,\text{V} - (0.316\,\Omega)(2.28\,\text{A})}{3.16\,\text{V/krpm}}$$

$$\omega = 3{,}567 \text{ rpm}$$

Finally, Eq. (22.26) gives the maximum efficiency for this motor as:

$$\eta_{max} = \left(1 - \sqrt{\frac{I_{NL}}{I_S}}\right)^2 = \left(1 - \sqrt{\frac{0.137\,\text{A}}{38\,\text{A}}}\right)^2$$

$$\eta_{max} = 88.3\%$$

It turns out that the motor in Example 22.3 was running very close to maximum efficiency. For most practical purposes, a gain of 0.3% is negligible.

22.9 GEARHEADS

As shown in Section 22.8, the most efficient operating currents (and therefore torques) for permanent magnet brushed DC motors lie above no-load current, and below the condition for maximum power, $\frac{1}{2}I_{stall}$ or $\frac{1}{2}T_{stall}$. This region is especially desirable if a motor is to be in continuous or high duty cycle use, as most motors are not rated for long periods of high torque output. The coils of the armature will eventually reach high temperatures and fail.

Given that most motors have output shaft speeds of several thousand rpm and little torque near the point of maximum efficiency, and given that many applications for permanent magnet brushed DC motors require substantially lower speeds and higher torques, **gearheads** (also called **gearboxes**) may be required. Gearheads are an inexpensive and compact means of decreasing motor output shaft speeds and increasing available torque. In theory, they could also be constructed to increase the output shaft speed and decrease the available torque, however, this is seldom—if ever—useful. In practice, gearheads are used to reduce the speed of the output shaft to more useful ranges.

The relationship between the input shaft speed for a gearhead and its output shaft speed is called the gear ratio, and is given by:

$$\text{Gear ratio} = N = \frac{\omega_{in}}{\omega_{out}} \tag{22.28}$$

The number of teeth in the gears used to construct a gearhead determines the **gear ratio**. Gear ratios are often expressed as N:1, for example, "10:1" or "250:1." The range of possible gear ratios is very wide, from near 1:1 with spur gears (such as Figure 22.11a) to several-thousand-to-one with planetary gears (Figure 22.11b).

Ideally, all the power produced by the motor and introduced to the gearhead at the input shaft would be available at the gearhead output shaft. This would be the case if the gearhead were frictionless and 100% efficient. Unfortunately, this is never the case for real gearheads. Taking losses into account, the expression for **gearhead efficiency** is given by:

$$\eta = \frac{P_{out}}{P_{in}} = \frac{T_{out}\omega_{out}}{T_{in}\omega_{in}} \tag{22.29}$$

In commercially available gearheads, it is not unusual to see efficiency ratings below 50%, especially for inexpensive gearheads and those with very high gear ratios. For smaller

(a) (b)

FIGURE 22.11

Typical gearheads used with brushed permanent magnet DC motors: (a) spur gearhead and (b) planetary gearhead.

gear reductions (4:1, say), efficiency may be in the 90% range. In any application, gearhead efficiency is a major consideration and can greatly affect system performance.

As speed is decreased across a gearbox, torque is increased. Combining Eqs. (22.28) and (22.29) gives an expression for the torque available at the output shaft of a gearbox:

$$T_{out} = \eta T_{in} N \tag{22.30}$$

Example 22.5

A mobile robot requires a drive motor that can supply at least 75 in · oz torque at 50 rpm. The gearhead that has been chosen for the application has a ratio $N = 24$ and efficiency $\eta = 63\%$. Since the next step will be to select the motor that will drive the gearhead, what speed and torque operating points will be required of a motor under these conditions?

Since the ratio and the required output shaft speed of the gearhead are specified, we can determine the required rotational speed of the motor from Eq. (22.28):

$$N = \frac{\omega_{in}}{\omega_{out}}$$

$$\boxed{\omega_{in} = N \cdot \omega_{out} = (24)(50 \text{ rpm}) = 1{,}200 \text{ rpm}}$$

Then, since the efficiency and the required output torque are specified, we can determine the torque required from the DC motor from Eq. (22.30):

$$T_{out} = \eta T_{in} N$$

$$\boxed{T_{in} = \frac{T_{out}}{\eta \cdot N} = \frac{75 \text{ in} \cdot \text{oz}}{(0.63)(24)} = 4.96 \text{ in} \cdot \text{oz}}$$

The motor will be required to produce at least 4.96 in · oz torque at 1,200 rpm. The final step in specifying a drive system (a motor mated to the gearhead) for this application requires verifying that the motor is capable of supplying 4.96 in · oz at 1,200 rpm when driven by the available power supply with or without the use of pulse width modulation. Consideration should also be given to whether this is within the motor's continuous operation regime (in which case it may be operated under these conditions indefinitely), or the motor's intermittent operation regime (in which case the length of time that the motor delivers the required torque and speed before overheating becomes an important factor).

22.10 HOMEWORK PROBLEMS

22.1 The mechatronic system shown in Figure 22.12 is designed to periodically hoist a 10 oz. mass above a platform where it is normally resting. The spool has radius 3/8 in., and is directly connected to the output shaft of the motor. If the motor has a stall torque of 29.5 in · oz at 15 V, what is the minimum voltage required to hoist the mass?

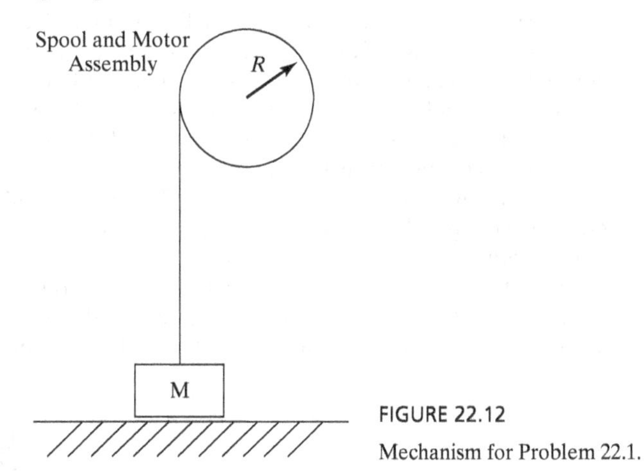

FIGURE 22.12

Mechanism for Problem 22.1.

22.2 As you stroll the isles of the local flea market, you come across a booth stocked with surplus permanent magnet brushed DC motors. Your eyes widen with excitement as you notice a particularly shiny gear motor priced at just \$1.50. Whipping your trusty multimeter out of its belt holster, you measure the winding resistance to be 18.9 Ω. Next, you pull a small torque wrench out of your fanny pack and measure the stall torque, which is 2.8 Nm when powered by the 12 V battery you keep handy for just such occasions. The gearhead is marked "100:1." The application you have in mind for this motor requires that the gearhead output shaft deliver 400 mNm at 35 rpm when driven at 15 V. You assume that the frictional losses in the motor and gearbox are negligible. Determine the appropriateness of the motor by answering the following:

(a) Will the motor and gearhead meet the requirements for torque and speed at 15 V? If not, what drive voltage would enable you to meet the design requirements?

(b) What is the current required to operate at the design point?

22.3 A motor with $K_T = 105$ mNm/A, $R_{coil} = 10\ \Omega$, and ω_{NL} (at 48 V) = 4,320 rpm will be operated with a 48 V supply. If this motor is connected to a mechanism that has frictional torque losses of $T_f = 55$ mNm, what will its output shaft rotational speed be?

22.4 For the system shown in Figure 22.13, electrical contact A is attached to a fixed reference point via spring A with spring constant K_A. The spring is spooled onto the shaft of a DC motor in order to move the electrical contact from its initial location, until it makes contact with electrical

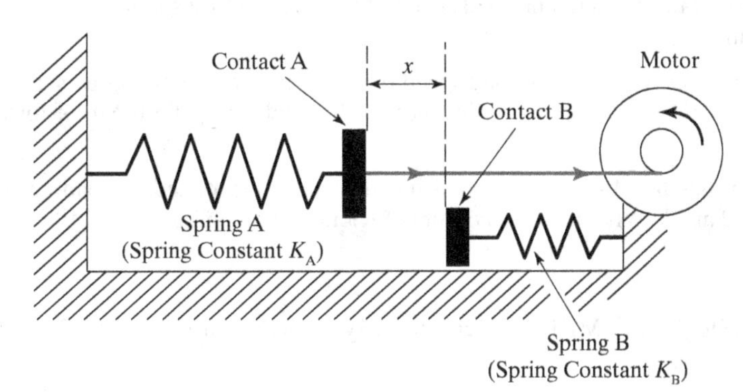

FIGURE 22.13

Mechanism for Problem 22.4.

contact B that is also attached to a fixed reference point via spring B that has a spring constant K_B. The first spring constant is $K_A = 100$ N/m, the second spring constant is $K_B = 15$ N/m, and the motor's shaft diameter is 0.5 in. The motor is powered at 15 V, has a no-load speed $\omega_{NL} =$ 4,080 rpm (at 15 V), and coil resistance $R = 9.73\ \Omega$. What is the value of the initial distance, x, between the contacts to ensure that contact B is displaced 1 mm when the motor is switched on and pulls contact A into contact B and compresses spring B?

22.5 You wish to design a new ultrahigh quality, portable, battery-powered coffee burr grinder for backpacking espresso fanatics. The grinding elements you have selected (Figure 22.14) are adjustable so that a coarse grind results when the burr cones are moved far apart from each other, a fine grind results when the cones are moved close together, and any intermediate point may be selected by the user. The manufacturer of the burr cones claims that the torque required to grind coffee ranges from 0.1 Nm (coarse grind) to 0.5 Nm (fine grind), and that the burr cones only function when rotating between 6 and 10 rpm. You will use a 12 V battery and a motor with $\omega_{NL} = 13{,}900$ rpm, $T_{stall} = 28.8$ mNm, $I_{stall} = 3.55$ A, coil resistance $R = 3.38\ \Omega$, maximum continuous current $I = 0.614$ A, $K_T = 8.11$ mNm/A, and $K_e = 0.847$ V/krpm. There are three gearheads available to you for this design: the first has an 850:1 ratio with 65% efficiency, the second has a ratio of 1,621:1 with 59% efficiency, and the third has a ratio of 3,027:1 with 59% efficiency. Which gearhead satisfies all the constraints (including continuous operation of the motor)?

FIGURE 22.14

Burr coffee grinder elements for Problem 22.5.

22.6 A motor with $K_T = 16.1$ mNm/A, $R_{coil} = 1.33\ \Omega$, and ω_{NL} (at 18 V) = 10,300 rpm will be operated with an 18 V supply. The maximum permissible continuous torque specification is 24.2 mNm. What rotational speed does this correspond to?

22.7 Starting with the expression for motor power output given in Eq. (22.18), show that maximum power is developed at $\frac{1}{2}T_{stall}$.

22.8 What are the roles of the commutator and brushes in a brushed DC motor?

22.9 A motor has a measured $R = 14.5\ \Omega$, and a measured stall torque of 4.47 mNm when operated on 9 V. What is the expected no-load speed of this motor? You should ignore motor friction for this problem.

22.10 A motor has a measured no-load speed of 11,500 rpm, and a measured stall torque of 4.47 mNm. What is the expected speed of this motor when delivering 1.5 mNm of torque into an external load?

22.11 A motor has $R = 14.5\ \Omega$, a no-load speed of 11,500 rpm, and no-load current of 12 mA when operated at 9 V. What is the maximum efficiency when operated on 9 V?

REFERENCE

[1] Lecture: *DC Motors,* ME112—Mechanical Systems Design, Gerdes, C., Stanford University, 2003.

FURTHER READING

DC Motors Speed Controls Servo Systems: An Engineering Handbook, 5th ed., Electro-Craft Corporation, 1980.

Design with Microprocessors for Mechanical Engineers, Stiffler, A.K., McGraw-Hill, 1992.

Maxon Precision Motors, Inc., USA, *www.maxonmotorusa.com.*

Mobile Robots, Inspiration to Implementation, Jones, J., Flynn, A., and Seiger, B., 2nd ed., A K Peters, Ltd., 1999.

CHAPTER 23

Permanent Magnet Brushed DC Motor Applications

23.1 INTRODUCTION

In Chapter 22, our discussion was focused on motor construction and on characterizing motor behavior and performance. These are essential topics for determining what specifications are required for a motor in a given application, and for selecting the right motor for each occasion, but they neither describe how to integrate a motor into a system once it is selected nor how to control it once it is integrated. This chapter focuses entirely on the practical issues related to designing circuits with permanent magnet brushed DC motors and successfully applying them. The practical information in the following discussion is just as critical as the more analytical discussion in Chapter 22, and will make the difference between a successful system design and a well-specified but unsuccessful one.

In this chapter, you will learn:

1. non-steady-state electrical characteristics of motors: how they respond when they are switched on and off,
2. what inductive kickback is and how to manage it,
3. how to implement unidirectional and bidirectional control of brushed DC motors,
4. what components are useful for designing motor control circuitry and how to select them,
5. how to implement speed control of motors using pulse width modulation (PWM).

23.2 INDUCTIVE KICKBACK

A very simple implementation of a permanent magnet brushed DC motor that can be switched on and off is one in which one lead is connected to a power source and the other lead is connected to a switch to ground, as in Figure 23.1a. This has the disadvantage of requiring physical interaction with the switch to power the motor on and off. In order to switch the motor electronically, you might be tempted to replace the switch with a transistor (one of many possible approaches), as shown in Figure 23.1b.

When the switch is closed or the transistor is on, current flows from the power source, through the motor, through the switch or transistor, and finally to ground. When the switch is open or the transistor is off, no current may flow. Unfortunately, the approach shown in Figure 23.1b has a terrible fatal flaw, and *after a few cycles of switching the motor on and off, the transistor will most likely be destroyed* by the effects of inductive kickback—the inevitable effect of trying to abruptly turn off the current flowing in an inductor, such as those made up by the motor's coil windings.

One of the primary practical issues that must be addressed when designing drive circuitry for permanent magnet brushed DC motors (and, in fact, any load with significant

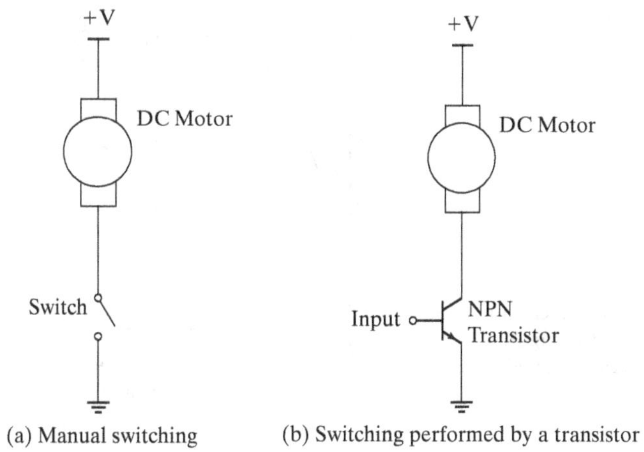

(a) Manual switching

(b) Switching performed by a transistor

FIGURE 23.1

Ill-advised ways of controlling current through a motor (no protection against inductive kickback).

inductance) is a phenomenon known as **inductive kickback**. Inductive kickback is a large voltage spike that occurs whenever the current flowing through an inductor is abruptly switched off, and can result in the destruction of drive circuit components if not handled correctly. Fortunately, inductive kickback can be managed effectively with the addition of a few inexpensive components.

Intuitively, what is happening is that we have built a magnetic field around the coil established by the current flow while it was energized. When we shut down that current flow, the magnetic field collapses. The collapsing magnetic field moves through the coils of the inductor producing a generator effect that creates the voltage we see when we turn off the current flow. The resulting voltage generated can be quite large. Recall that the voltage potential across an inductor is given by:

$$V = L\frac{\partial I}{\partial t} \quad [\mathrm{V}] \tag{23.1}$$

The current flowing through an inductor cannot be instantaneously changed. Depending on the magnitude of the inductance L, an abrupt increase in the voltage applied across the leads of an inductor causes the current I to ramp up in an exponential fashion, as shown in the time period between 0 and about 0.4 ms in Figure 23.2. In this portion of the figure, the applied voltage is initially 0 V, until time $t_0 = 0$ ms, when the voltage is abruptly switched to 12 V. The resulting current doesn't switch on immediately in response to the change in voltage, rather it rises exponentially to its final value which is determined by Ohm's law ($V = I \times R_{\text{inductor}}$). The time constant characterizing the current rise time in an inductor is:

$$\tau = \frac{L}{R} \quad [\mathrm{s}] \tag{23.2}$$

During the relatively lengthy time (on an electrical timescale, that is) that the current is rising, the inductor is establishing a magnetic field. This accompanying magnetic field is what affords the inductor its characteristic behavior: changes in current are resisted by the magnetic field as it adjusts to the changes. Energy is traded back and forth between the current flow and the magnetic field while changes occur. If the current flow increases, the energy in the magnetic field must also increase, and energy will be drawn from the current flow to increase the magnetic field energy. If the current flow decreases, energy from the magnetic

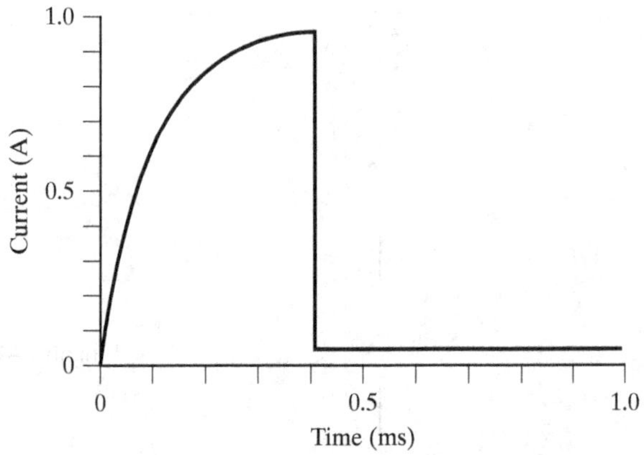

FIGURE 23.2

Current response in an inductor for a step increase in applied voltage at t = 0 ms and a step decrease in applied voltage at t = 0.4 ms.

field must also decrease, and energy will be contributed from the field to the current flow. In this way, *inductors act in such a way that they resist changes in current flow, whether that change is an increase or a decrease.*

In the case where the applied voltage across an inductor is abruptly switched off (which happens every time a motor is turned off), the magnetic field established by an inductor (e.g., the motor's coil windings) will contribute its stored energy in an effort to force current to continue to flow. The energy stored in the inductor's magnetic field is given by the expression:

$$E = \frac{1}{2}I^2L \qquad [\text{J}] \tag{23.3}$$

For a brief time (a few microseconds, depending on the coil's inductance and resistance), the current flowing in the inductor will continue, first at a level equal to the current that was flowing immediately prior to the voltage being switched off, and then decaying rapidly as the energy from the magnetic field is exhausted. The rapid decrease in current shown in Figure 23.2 at t = 0.4 ms illustrates the current decay characteristics that result when the applied voltage is switched off. The time required for the magnetic field to collapse is short relative to the time required to establish the magnetic field (i.e., the rise time of the current), but it is significant.

In order to maintain this current flow in the absence of any external applied voltage, the inductor establishes a very large voltage differential across its terminals, using the energy from the magnetic field. For a brief time (equal to the amount of time it takes for the current to cease flowing, typically a few microseconds), the voltage across the inductor will spike to very high levels, often reaching well above 1 kV. Figure 23.3 shows the resulting voltage response. This effect is known as inductive kickback, with the developed voltage referred to as either the **kickback voltage** or **flyback voltage**. The energy stored in the inductor's magnetic field must go somewhere, either by being bled off through some leakage mechanism in one of the system components or through arcing. In all cases, the energy won't go anywhere until there is current flow. As a result, the maximum voltage achieved by the inductive kickback spike is only limited by the first opportunity for some current to flow.

Inductive kickback, which results in voltage spikes that may be over 1 kV, can exceed the specifications of many of the electronic components used to control motors. Even in a very simple configuration, such as the one shown in Figure 23.1a, the inductive kickback can cause problems, since most manual switches are not rated for more than a few hundred volts across their terminals. Most switches will function well for several cycles, but under these extreme circumstances, arcing will occur which will eventually damage the switch. If, however,

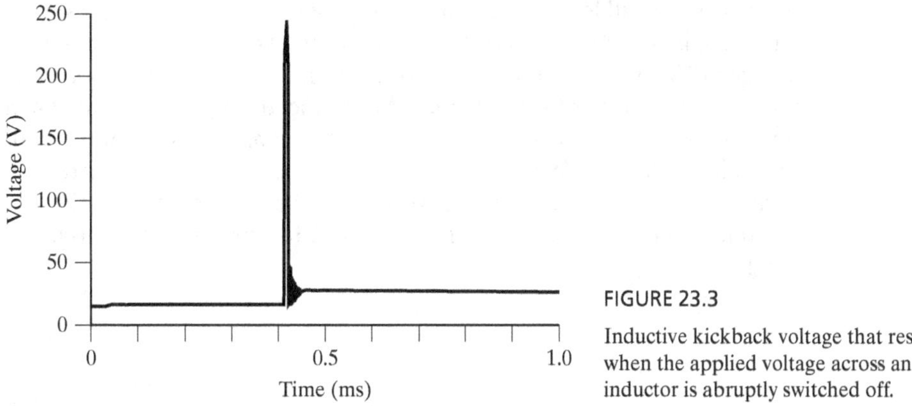

FIGURE 23.3

Inductive kickback voltage that results when the applied voltage across an inductor is abruptly switched off.

a typical field-effect transistor (FET) or a bipolar junction transistor (BJT) is used to control a motor, such as the circuit shown in Figure 23.1b, the transistor will soon be destroyed by exposure to the inductive kickback, since most silicon-based components aren't rated to withstand more than 30–150 V across their terminals.

Fortunately, dealing with inductive kickback is straightforward. In general, the strategy is to add components to the control circuit that provide a low-voltage current path, thereby reducing, or **snubbing**, the magnitude of the kickback voltage to safe levels.

The most straightforward means of reducing inductive kickback to safe levels is to place a diode across the terminals of the motor in a reverse-biased orientation. This limits the voltage spike caused by inductive kickback to the forward voltage drop of the diode (about 1 V under these high current conditions) plus the positive voltage supply (V_s). A diode used in this configuration is called a **kickback diode** or **flyback diode** or **diode snubber**. Figure 23.4 shows a kickback diode in place across the terminals of a DC motor.

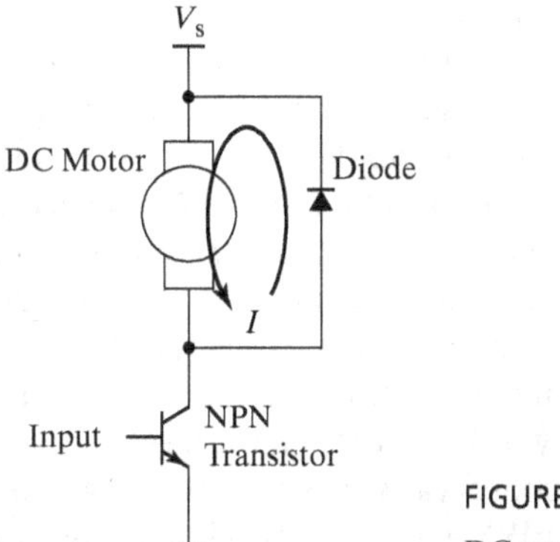

FIGURE 23.4

DC motor circuit with kickback diode.

For normal operation of the motor, the diode is reverse-biased and no current will flow through the diode. The diode only conducts when the collector voltage of the transistor exceeds the diode's forward voltage drop of ~1 V, which occurs only during inductive kickback.

With the kickback diode in place, the voltage spike is effectively limited to the positive voltage supply level plus the forward voltage drop of the diode, $V_C = V_s + V_f$. Once the voltage reaches this value, the energy stored in the magnetic field of the solenoid will be dissipated as a current that circulates in the direction of the arrow in Figure 23.4. With the addition of the diode snubber, the voltage measured at the lower motor terminal (the collector of the transistor) will look like Figure 23.5, which shows an electrical simulation of the circuit in Figure 23.4. During the time that the collector voltage is elevated above the power supply voltage (12 V in the simulation), current circulates through the loop created by the coil of the motor and the forward-biased diode. Mechanically, this means that when we shut off the current flow to the motor by switching off the transistor, the motor will not stop producing torque immediately. There will be a delay as the recirculating current prolongs the decay of the magnetic field. The diode has protected the transistor from the large voltage spike but introduced a delay in turning off the motor. In our simulated circuit, this delay is approximately 180 μs.

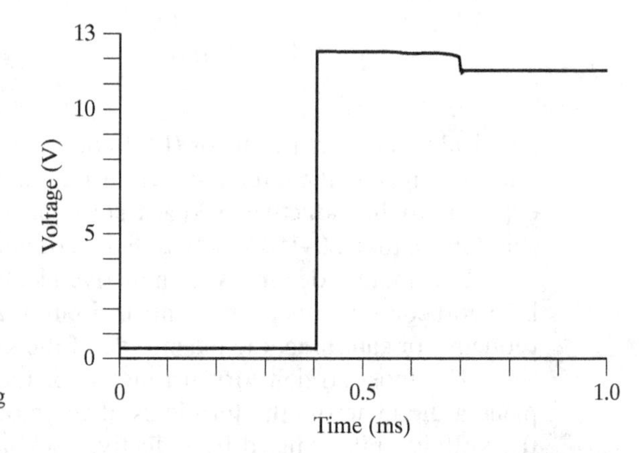

FIGURE 23.5
Kickback diode snubbing inductive kickback.

When selecting a kickback diode, two characteristics should be considered. First, the diode should be able to switch very rapidly from the forward-biased condition, when it is conducting current (as when inductive kickback is occurring and a voltage spike is developing across the terminals of the motor), to the case where it is reverse-biased and blocking the flow of current (as occurs when the motor is powered on and operating normally). This specification is called the **reverse recovery time** and is usually given the symbol t_{rr}. Reverse recovery time specifications vary widely for different types of diodes. Diodes that are appropriate for this application will often be labeled **fast recovery** or even **ultrafast recovery**. The faster, the better—diodes with reverse recovery times of a few hundred nanoseconds or less are good candidates. The second important diode specification to consider is the maximum allowable current flow. The maximum current that the diode will experience under normal conditions is the maximum current through the motor: the stall current. A very conservative choice would be to specify a diode that can tolerate the stall current continuously. In practice, this is unnecessarily conservative: even if the motor is switched on and off very rapidly (which happens any time PWM speed control is used, as discussed in Section 23.4), the diode will only experience these peak current levels a small fraction of the time. A diode that has a peak intermittent current specification that exceeds a motor's stall current will usually suffice.

There are several alternative methods of snubbing the inductive kickback. A single kickback diode is the simplest solution, but higher overall system performance can be achieved using slightly more complex arrangements that shorten the time required to dissipate the energy stored in the motor coils' magnetic fields.

In the model using the simple kickback diode illustrated in Figures 23.4 and 23.5, the duration that the kickback diode is forward-biased and conducting is about 180 μs. During this time, the energy from the collapsing magnetic field causes current to flow in a loop that includes the motor and the kickback diode. This energy is dissipated both in the motor, since torque is generated by the current flow and $I^2 \times R$ losses occur in the coil windings, and as $V \times I$ losses across the diode. This approach is straightforward and gets the job done, but it results in maximum voltage levels and power dissipation levels that are well below the limits for most components. An expression for the dissipation of this energy would be:

$$E_{\text{stored}} = \int \text{Power}(t)\, dt \qquad [\text{J}] \tag{23.4}$$

Restating the expression with the assumption that the power dissipation has an average value:

$$E_{\text{stored}} = \text{Average power} \times \Delta T \qquad [\text{J}] \tag{23.5}$$

Minimizing ΔT, the time required to dissipate the energy, while ensuring that the kickback voltage never rises above the specified maximum levels for any of the system components, and that the average power dissipation limits of all components is never exceeded, will result in improved system response. By limiting the kickback voltage and power dissipation to the highest possible *safe* values, the duration required to dissipate the energy will be minimized, and this aspect of system performance will be optimal.

One means of implementing this strategy is to add a resistor to the diode in the kickback loop, as shown in Figure 23.6a. The additional resistance increases the energy dissipation rate by increasing the resistance R in the $I^2 \times R$ loss term. The value of the resistor should be selected so that the voltage at the transistor's collector terminal can't exceed the device's maximum collector-to-emitter voltage specifications [labeled $V_{\text{CEO(SUS)}}$ in most BJT data sheets]:

$$I_{\text{peak}}\, R + V_{\text{f}} + V_{\text{s}} \leq V_{\text{CEO(SUS)}} \tag{23.6}$$

Figure 23.6b shows the results of a simulation of this approach and predicts that the kickback spike will be dissipated in about 100 μs—almost a factor of 2 faster than a diode alone.

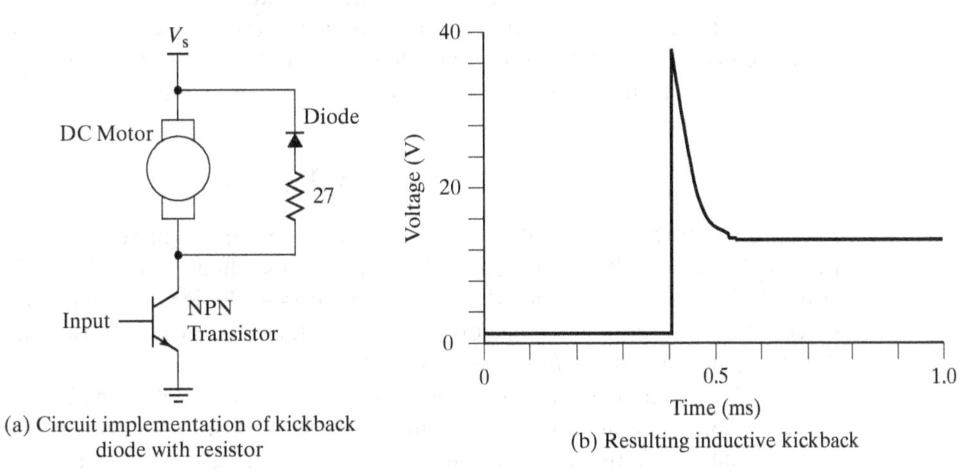

(a) Circuit implementation of kickback diode with resistor

(b) Resulting inductive kickback

FIGURE 23.6

Kickback suppression with a diode in combination with a resistor.

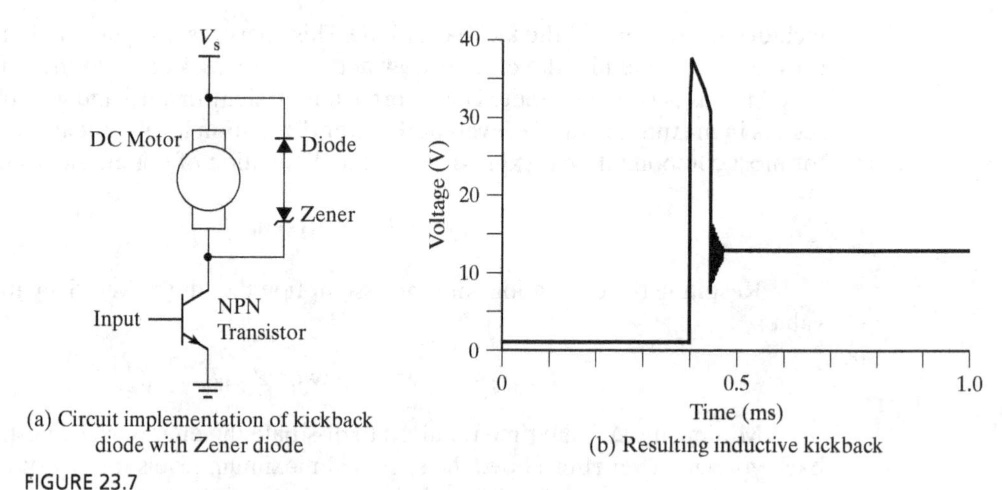

(a) Circuit implementation of kickback diode with Zener diode

(b) Resulting inductive kickback

FIGURE 23.7

Kickback suppression with a diode in combination with a Zener diode.

Since the maximum voltage observed using this approach is higher than that seen with a simple kickback diode, care must be taken when selecting the value of the resistor in the kickback loop so that the maximum voltage specifications of the other components in the system are not exceeded. Also, the power rating of the resistor must be adequate for the task.

Further performance gains can be realized by replacing the resistor illustrated in Figure 23.6 with a Zener diode arranged as shown in Figure 23.7a. Note that it is forward-biased with respect to the overall power supply, but that since the standard diode is reverse-biased, no current will flow in the kickback loop under normal operating conditions. However, when the kickback voltage exceeds the Zener's reverse-breakdown voltage plus the forward drop of the standard diode, both diodes will begin to conduct. The Zener diode will hold the voltage drop across its anode and cathode to the reverse-breakdown voltage (12 V for the Zener diode used in the simulation shown in Figure 23.7b), and the standard diode in forward conduction will have a voltage drop of approximately 1 V. One benefit of this approach is that the diodes allow the designer to very accurately and confidently set the maximum voltage that the kickback surge will attain. Similar to the selection process for the diode-resistor snubber circuit described above, the Zener diode should be selected so that the voltage at transistor's collector terminal won't exceed the device's maximum collector-to-emitter voltage [$V_{CEO(SUS)}$] specifications:

$$V_Z + V_f + V_s \leq V_{CEO(SUS)} \tag{23.7}$$

Zener diodes are available with a very large variety of reverse-breakdown voltages, ranging from 2 to 200 V, so there are many options when setting the limit for the kickback voltage. Standard and Zener diodes are not ohmic devices—that is, they have no "resistance," only an inherent voltage drop when conducting. Because of this, the energy stored in the inductor's magnetic field is dissipated as $I^2 \times R$ losses across the motor windings, as $V_{diode} \times I$ losses across the diode, and as $V_{Zener} \times I$ losses across the Zener diode. As shown in Figure 23.7b, this approach results in the dissipation of the energy in approximately 50 μs, or more than three times faster than the case for the snubber diode alone.

The selection criteria for the standard diode and the Zener diode in this application are essentially the same as the criteria for the configuration with the standard diode alone. For the standard diode in this case, select a fast recovery diode (one with a short reverse recovery time specification) and ensure that it can safely handle the motor stall current. The typical choice is to select a diode with a peak intermittent forward current specification equal to or greater than the motor's stall current. For the Zener diode, ensure that the power dissipated when the diode is conducting in the reverse-breakdown mode is less than the diode's rated power dissipation. This process is a little more involved for Zener diodes than for standard diodes, as Zener diodes that have peak intermittent current ratings in excess of typical motor stall currents are not as easy to find or as inexpensive. For Zener diodes, a more detailed analysis of average power dissipation is required to ensure that its power dissipation specifications are not exceeded.

Recall from Eq. (23.3) that the energy stored by an inductor is $E = ½I^2L$. From this, we can develop an expression for the average power dissipation resulting from kickback that takes into account how frequently a motor is switched on and off. In practice, motors may be switched on and off very frequently, as when PWM drive is used. PWM is discussed in Section 23.4.

$$P = \left(\frac{1}{2}I_{\text{stall}}^2 L\right) \times \text{Frequency} \qquad [\text{W}] \tag{23.8}$$

Note that this is the average power that will be dissipated during kickback, and that the dissipation will be distributed among the motor, the standard diode, and the Zener diode. A Zener diode that is rated for average power dissipation in excess of the levels calculated with Eq. (23.8) in a specific application will be a good choice, with a comfortable margin of safety.

A Zener diode can also be used to directly limit the voltage across the transistor. The example circuit used throughout this discussion has used an NPN BJT to control a DC motor. There is a limit to the voltage that can be applied across the BJT's emitter (the terminal connected to ground in Figure 23.8a) and collector (the terminal connected to the DC motor in Figure 23.8a) without damaging the component. Most NPN BJTs can tolerate a maximum **collector-emitter voltage** of 30 to 60 V. When the motor is on and running steady state, the collector-emitter voltage will be very small (about 0.2 V, depending on the specifications of the BJT). When the motor is off and in steady state, the collector-emitter voltage will be the power supply voltage, V_s. When the motor is switched off and inductive kickback is occurring, much higher collector-emitter voltages will be observed, and the BJT is likely to be destroyed if

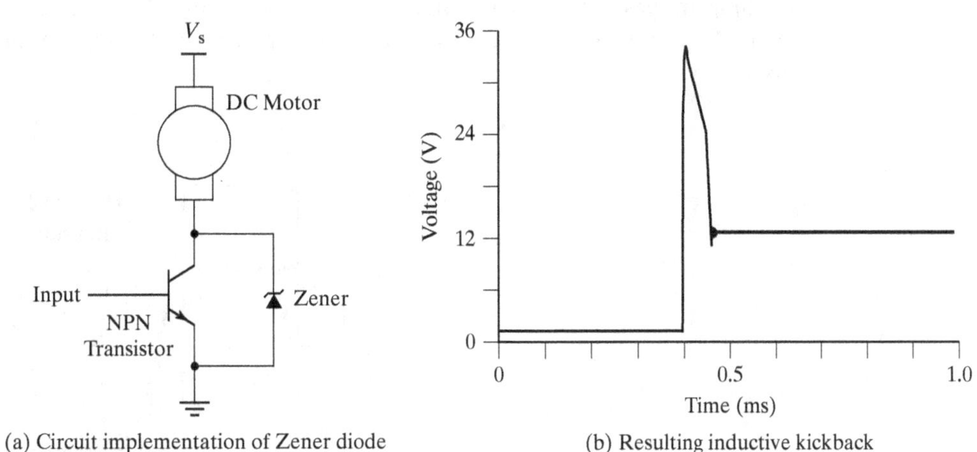

(a) Circuit implementation of Zener diode

(b) Resulting inductive kickback

FIGURE 23.8

Kickback suppression with a Zener diode across the switching element.

no snubbing is implemented. Placing a Zener diode across the collector and emitter of the transistor in a reverse-biased orientation will limit the voltage across the transistor to the Zener's reverse-breakdown voltage. In Figure 23.8a, a Zener diode with a reverse-breakdown voltage of 34 V has been selected. As before, care must be taken when selecting the Zener diode to ensure that it is capable of dissipating the heat that will be generated when it is conducting during kickback. To ensure that a particular Zener diode is suitable, Eq. (23.8) may be applied to calculate the average power dissipation requirements. Select Zener diodes with power rating specifications above the calculated average power dissipation.

A drawback to this approach, despite its simplicity, is that whenever inductive kickback occurs, the power supply ground will be required to absorb a short but significant current spike. Depending on the quality of the power supply and the wiring used to connect the motor and Zener, this could cause substantial noise in the power supply's ground. While it's not likely to cause a problem for the motor or the transistor, other components connected to the power supply, such as analog circuitry or microcontrollers, may not tolerate the results well. (See Chapter 21: Noise, Grounding, and Isolation for a discussion of these issues and remedies.)

23.2.1 Inductive Kickback Summary

Turning off the current flow through a motor as rapidly as possible will result in improved performance; however, the complexity and cost of the design must also be considered when deciding how to deal with inductive kickback. Figure 23.9 compares the time required to stop the current flow using each of the techniques discussed above. The goal is to get as close as possible to the case where no kickback snubbing is performed, which is the fastest, without allowing the kickback to destroy any circuit components. Each of the kickback snubbing techniques presented will protect the circuit components, but each technique has a different effect on the turn-off time of the kickback current and differs in cost and complexity. The diode-only approach has the benefit of simplicity and low cost, but results in the slowest turn-off time, and therefore the least optimal motor performance. This approach is perfectly adequate for a wide variety of applications, however, improvements are possible. Adding a Zener diode to the standard diode across the motor leads, for example, results in the fastest current turn-off time, and therefore the most desirable motor performance. The other configurations discussed (the reverse-biased diode plus a resistor in a loop around the motor leads, and the Zener diode across the collector and emitter of the switching transistor) result in current turn-off times that lie between these extremes. The best method of snubbing inductive kickback depends on the application, but *some* method must *always* be used.

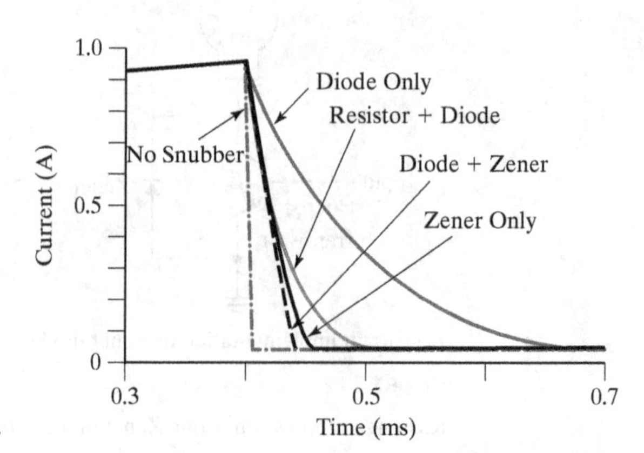

FIGURE 23.9
Inductor current decay comparison.

23.3 BIDIRECTIONAL CONTROL OF MOTORS

Any of the circuits discussed above that include kickback suppression are good choices for unidirectional control of DC motors, but what about applications that will require a motor to spin in both directions (clockwise—CW, and counterclockwise—CCW)?

For the circuits shown in Figures 23.4, 23.6, 23.7, and 23.8, current can only flow from the positive terminal of the power source to ground, so the motor can only turn in one direction. Enabling a motor to turn either CW or CCW requires the capability to cause current to flow through the motor in either direction. A different, more complex drive circuit will be required to accomplish this. One possibility is to make use of a "dual" power supply, that is, one with both a positive voltage supply and a negative voltage supply. By switching one on or the other, current can flow through the motor in either direction. Figure 23.10 illustrates this configuration. Note that kickback diode protection must be included, preventing inductive kickback from exceeding the maximum collector-emitter voltage for each of the transistors.

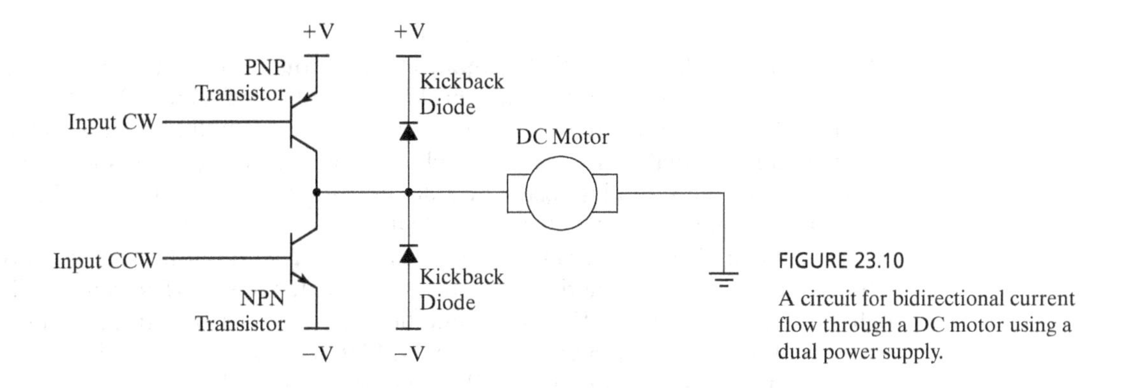

FIGURE 23.10

A circuit for bidirectional current flow through a DC motor using a dual power supply.

For this circuit, if the voltage at the Input CW node is at least 0.6 V below the positive voltage supply (+V) and the Input CCW node is held at the negative supply voltage (−V), the PNP transistor will be on and conducting, and the NPN transistor will be off, resulting in current flowing from the +V terminal through the motor to ground. This will cause the motor to spin in one direction (our labeling convention indicates that it will spin CW).

We can reverse the direction of current flow through the motor by changing which transistor is on. If Input CW is held at +V, and Input CCW has a value of at least 0.6 V above −V, the PNP transistor will be off and the NPN transistor will be on and conducting. Under these conditions, current will flow from ground (0 V) to the negative supply terminal (−V), which has a voltage potential lower than that of ground, and the motor will spin in the CCW direction.

Turning on both transistors at the same time must be avoided in this circuit, as it would result in a short circuit from +V to −V through the transistors. This would likely destroy the transistors and/or the power supply.

For this example, current flows from left to right across the motor if the upper transistor is on, and from right to left across the motor if the lower transistor is on. This is one way to accomplish bidirectional current flow, but in practice, it is not common to have a high current negative voltage supply available. It would be more practical to design a circuit capable of causing current to flow in either direction through a motor using only a single positive power supply. The most common way of accomplishing this is through the use of a network of transistors, arranged in what is known as an **H-bridge** configuration. The shape of the circuit resembles an "H," which gives rise to the distinctive name. Figure 23.11 illustrates an H-bridge.

Recall that the objective of this circuit is to enable current to flow through the motor in either direction (left-to-right or right-to-left) using only a single positive power supply

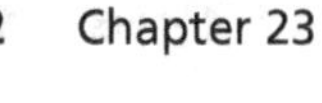

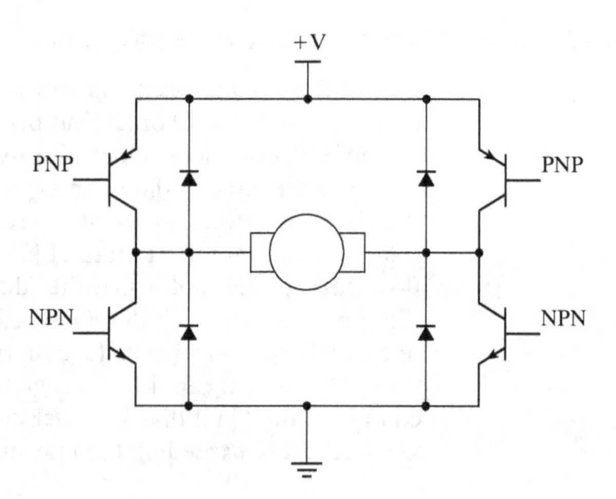

FIGURE 23.11
A typical H-bridge circuit, including kickback diodes to protect against inductive kickback.

(shown in Figure 23.11 as +V). The diodes that surround the motor are in place to protect the transistors from the inductive kickback, as described in Section 23.2. Since current may flow in either direction across the motor, the high voltage inductive kickback can occur on either motor terminal and must be snubbed to prevent the destruction of the transistors. Though any of the snubbing techniques discussed above are applicable with H-bridges, by far the most common practice for brushed DC motors is the use of standard diodes.

Examining Figure 23.12a first, we see that we have established the situation where the left PNP transistor is on, the right PNP transistor is off, the left NPN transistor is off, and the right NPN transistor is on. For these conditions, current can flow from the positive voltage terminal of the power supply, through the left PNP transistor, through the motor from left to right, and finally through the right NPN transistor to ground.

In Figure 23.12b, the opposite conditions have been established: by reversing the state of each of the transistors, current can now flow from the positive terminal of the power supply, through the PNP transistor on the right side, through the motor (this time from right to left), and finally through the left NPN transistor to ground.

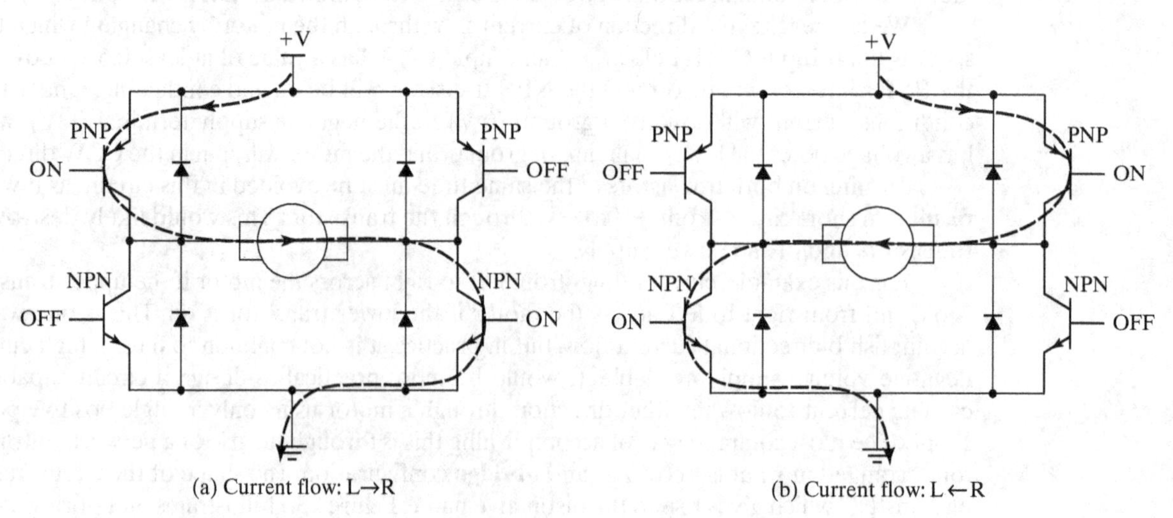

(a) Current flow: L→R

(b) Current flow: L←R

FIGURE 23.12
Using an H-bridge to control the direction of current flow in a motor.

Other possibilities exist for combinations of transistors states, of course. For instance, if both transistors on either the left side or the right side of the "H" are switched on, current will flow without any resistance, directly from power to ground, Figure 23.13. In this case, a short circuit is effectively established between power and ground, since there is no load or other component in the path to resist the flow of current. This results in very high levels of current flow. If it is sustained for any significant length of time, one of the components involved is very likely to be destroyed by the excessive current levels. This unrestrained current is referred to as **shoot-through current**. Shoot-through current can occur under normal circumstances, such as while an H-bridge is in the process of switching from one desirable configuration to another (for instance, when reversing the direction of the current flow through a motor). This occurs because the transistors don't have symmetric turn-on and turn-off times (turn-off time usually being longer than turn-on time). Under these conditions, brief bursts of shoot-through current may be experienced. Shoot-through current happens, but it is NOT desirable. The designers of H-bridges take great care to ensure that shoot-through currents are minimized, often including intentional delays, called **dead time**, into the switching of the transistors to minimize the overlap time, when the transistors on one side of the H-bridge are both conducting, that results in shoot-through. The need to control shoot-through currents is one of the significant issues that makes the design of successful H-bridges a nontrivial task.

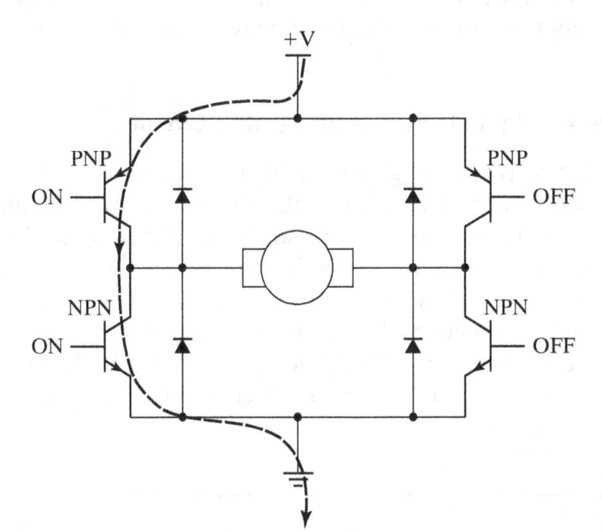

FIGURE 23.13

Shoot-through current in an H-bridge, where current flows unimpeded from power to ground, usually with disastrous effects.

Another, more useful, combination of transistors to turn on configures an H-bridge to dynamically brake, or resist the rotation of, a motor. When the leads of a spinning motor are shorted together, the current that is generated by the back-EMF (recall that the spinning motor acts as a generator) causes current to flow in a loop, as shown in Figure 23.14. The direction of current flow during **dynamic braking** depends on the direction of the motor's rotation. Current will flow from one of the motor's terminals, through one of the kickback diodes, around to one of the transistors that are in the "on" state, and finally around to the motor's other terminal. The internal resistance of the motor coils, the forward voltage drop across the diode, and the saturation voltage of the transistors serve as the load for this current, and the electrical power being generated by the rotating motor will be dissipated as heat. In the absence of a load caused by dynamic braking, the only forces acting to slow the rotation of a motor are the frictional resistance of the motor bearings and other mechanical systems coupled to the motor. Dynamic braking can be performed with an H-bridge

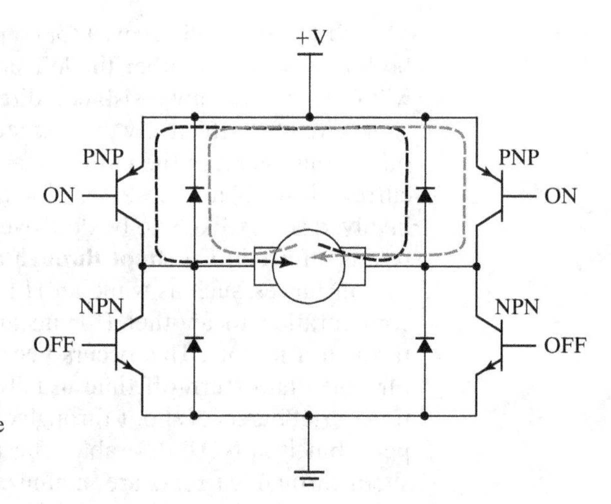

FIGURE 23.14
An H-bridge configured to short both leads of motor together, creating dynamic braking. Depending on the direction of rotation of the motor, the current flow will be either CW or CCW, as shown.

by turning on both the upper transistors while turning off both the lower transistors, or by turning off both the upper transistors while turning on both the lower transistors. This effectively shorts both leads of the motor together and results in dynamic braking. This is useful when switching a motor off in circumstances where it is desirable for the motor to stop rotating more quickly than would otherwise occur if it were simply allowed to slowly decelerate and coast to a stop.

23.3.1 Commercially Available H-Bridge Integrated Circuits

Bidirectional switching of current through a load, such as a motor, is such a common need in electronic design that there are numerous integrated circuits (ICs) in the market that implement H-bridges in a single convenient package, so that designers don't have to create one from discrete components. H-bridges are available in a wide variety of current drive capabilities, package sizes, and features. They all perform the same basic task, however: switching current through a load in two directions. Some integrated solutions you may wish to investigate are the Texas Instruments SN754410NE, the ST L293, the ST L298, and the National Semiconductor LMD18200, several of which are discussed in detail in Chapter 17: Digital Outputs and Power Drivers.

Example 23.1

Design drive circuitry based on an L293B H-bridge IC (manufactured by STMicroelectronics) that is capable of controlling a Maxon RE-13 (Part #: 118638) motor using a 12 V power supply. The motor has terminal resistance $R = 14.1\ \Omega$ and coil inductance $L = 0.48$ mH. It will be switched on and off at a maximum rate of 1 kHz. To protect the circuit from excessive inductive kickback voltages, use standard diodes. Ensure that the specifications for all components are met by the design.

In order to satisfy the constraints, the L293B must be capable of switching the maximum current possible for this motor powered at 12 V. Since the terminal resistance is given, we can solve for the stall current ($\omega = 0$):

$$I_{\text{stall}} = \frac{V}{R} = \frac{12\ \text{V}}{14.1\ \Omega} = 0.851\ \text{A}$$

The specifications for the L293B show that it is capable of handling up to 1 A per channel (continuous). Since the stall current is the maximum current the system will experience, the L293B will be able to safely handle controlling this motor.

The next issue to address is the inductive kickback. The problem specifies the use of standard diodes, so we must identify a candidate that has a suitable reverse recovery time specification and appropriate current capabilities. A diode that is very commonly employed as a kickback diode is the 1N4935, which is identified as a "fast recovery diode" in the list of features found in its data sheets. Recall that kickback diodes should have a reverse recovery time specification of a few hundred nanoseconds or less, and the 1N4935 has $t_{rr} < 200$ ns. The requirement for low reverse recovery time has been met. Recall also that good design practice dictates that the diode's specification for peak intermittent current should be above the maximum current that the circuit will encounter, which is the stall current calculated above (0.851 A). The peak intermittent current specification for the 1N4935 is 10 A, which is more than 10 times the stall current expected for this circuit. The 1N4935 is therefore a good choice in this respect. The issue at the heart of the specifications for peak (and continuous) current is power dissipation: the energy stored in the motor's coil is dissipated during kickback by the standard diode and the motor's own coil resistance. We can analyze the power dissipated by the snubbing circuit from Eq. (23.8), where we find:

$$P = \left(\frac{1}{2}I_{\text{stall}}^2 L\right) \times \text{Frequency} = \frac{1}{2}(0.851\text{ A})^2(0.48\text{ mH})(1\text{ kHz}) = 0.174\text{ W}$$

The total power dissipated during kickback is thus 0.174 W, which will be distributed between the standard diode and the resistance of the motor's coils. A conservative approach is to assume that the power dissipated in the diodes dominate the power dissipated by the motor's coils. The most conservative assumption is that *all* the power is dissipated in the diode; using this reasoning, any standard diode capable of dissipating at least $P_D = 0.174$ W will suffice. The data sheet for the 1N4935 doesn't directly state a maximum power dissipation specification; however, we know that its maximum continuous forward current is 1 A, and the maximum forward voltage drop at 1 A is 1.2 V. From this, we can infer that the maximum continuous power dissipation capability of the 1N4935 is $P_{D\,max} = VI = 1.2\text{ V} \times 1\text{ A} = 1.2$ W. This is much greater than the 0.174 W it will dissipate in this circuit, so we've verified that the 1N4935 is an acceptable choice. Thermal analysis of the 1N4935 under these conditions also shows that the junction temperature remains well below the maximum allowable levels, further supporting the choice of the 1N4935.

At this point, all of the components have been selected and the schematic may be drawn (Figure 23.15).

The inputs "Half1"and "Half2" each control one side, or leg, of the H-bridge. Each side of the H-bridge is called a **half-bridge**. A logic-level high to an input (such as the one labeled "IN1") connects the corresponding output (OUT1) to the motor supply voltage, minus a voltage drop due to

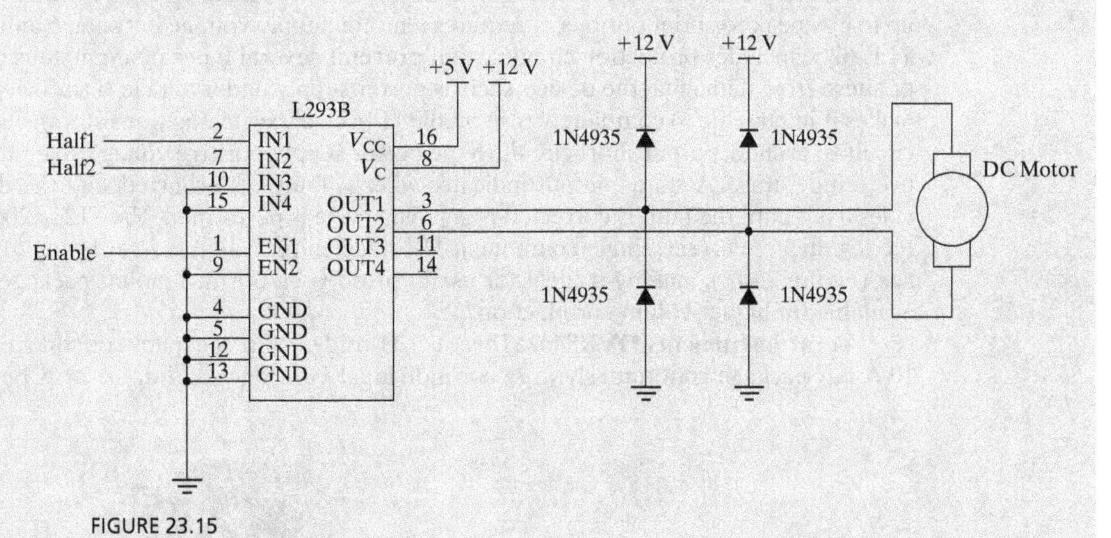

FIGURE 23.15

H-bridge circuit for Example 23.1.

losses (the saturation voltage). This drop is V_{CEsatH} (collector-emitter saturation voltage, high) and is specified to have a typical value of 1.4 V in the L293B data sheet. A logic-level low to an input connects the corresponding output to ground, again with losses due to the saturation voltage across the active switch. This is the low side saturation voltage (V_{CEsatL}) and is specified to be typically 1.2 V in the data sheet. Setting Half1 to a logic-level high and Half2 to a logic-level low will drive current through the motor in one direction and cause the motor to turn. Reversing the inputs drives the current in the opposite direction, so the motor also turns the other direction. Setting Half1 and Half2 both to logic-level low at the same time (or both to logic-level high) connects the terminals of the motor together and causes dynamic braking. The input called "Enable" controls current flow through the half-bridges. Setting the Enable pin low inhibits current flow and turns the motor off, allowing it to coast to a stop.

23.3.2 H-Bridges for Higher Current Applications

Many mechatronic devices are well-served by small, low-torque motors, and these often require only small amounts of current (e.g., 1 A or less). However, this doesn't apply in all situations: some applications call for substantial amounts of current to flow bidirectionally through a DC motor (or, more generally, any type of load). Circuits capable of managing such current levels are often little different than those designed for lower currents, but the components selected must be capable of managing the higher currents. There are a great many H-bridges available that are capable of handling between 1 and 2 A (some of which were enumerated in Section 23.3.1), and similarly there are many capable of managing 2 A up to about 10 A, as well as a category of H-bridge driver chips that allow for still greater current levels by moving the switching transistors off-board.

Once a designer determines how much current must be supplied to a motor (or other load) for a given application, the drive components for the motor must be capable of managing the required current levels. Identifying candidate components involves searching the product offerings of electronic component distributors and/or manufacturers, and then following their guidelines for the proper use of the devices. In the remainder of this section, we present a few representative devices.

Infineon TLE-5206: This device incorporates a single H-bridge driver and is thus able to drive a single motor bidirectionally. It is capable of delivering up to 5 A continuously, and up to 6 A peak for brief periods. It requires a motor supply voltage between 6 and 40 V. The TLE-5206 includes protection circuitry that prevents several types of anomalous conditions or abuse from damaging the device, such as power supply undervoltage (insufficient voltage makes it impossible to completely turn on the H-bridge's switching transistors), output short circuit to ground, output short circuit to the power supply positive voltage, overcurrent, and overtemperature. A status output indicates when a fault has occurred, and the device disables itself until the fault is corrected or a power cycle is performed. The TLE-5206 is one of the few higher-current, single component H-bridge chips available in a through-hole package (Figure 23.16), making it ideal for use in prototypes. Surface-mount packages are also available for higher-volume production.

Texas Instruments DVR8402: This dual H-bridge chip is capable of controlling up to 10 A per package continuously (5 A per individual H-bridge) and up to 24 A per package

FIGURE 23.16

The Infineon TLE-5206, a 5 A H-bridge.

for brief intervals. The two H-bridges can also be wired together in parallel to provide a total of 10 A (continuous) to a single motor, or more generally, another type of load. Two separate power supplies are required by the DVR8402: a 0–50 V supply for the motor, and an 11.4–13.2 V supply for the internal logic circuits. Independent control of each of the four half-bridges on the chip is provided, enabling the use of dynamic braking. The internal MOSFETs used in the H-bridge circuits have a low $R_{DS(ON)}$ specification of 90 mΩ (recall that we discussed MOSFETs and the $R_{DS(ON)}$ specification in Chapter 10). This results in an impressively low 0.45 V drop across each transistor when the maximum 5 A is flowing. The DVR8402 provides protection against overtemperature and overcurrent conditions, making it more difficult (but certainly not impossible) to damage. This device is available only in a 36-pin surface-mount package, which can make it more challenging to use in prototype circuits.

Allegro A3985: This device isn't actually an H-bridge (a fully integrated device to which a motor or other load may be directly connected). Rather, this is an **H-bridge driver**: a type of device that combines control logic and gate driver outputs designed to be connected to the gates of external N-channel MOSFETs in both the high-side and low-side positions to create a full H-bridge. The designer specifies and connects the external MOSFETs separately. Moving the switching elements off-chip removes the most significant sources of heat within the device, and provides the designer with much greater flexibility and control over cost and performance of the motor-driving circuit. H-bridge driver chips allow for the control of much higher currents than single-chip solutions, provided the MOSFETs mated to the driver are capable of controlling the required currents. The A3985 incorporates two separate H-bridge driver modules, so each chip can control two motors. It requires a motor supply voltage between 12 and 50 V, is controlled via an SPI serial communications link (which we described in Chapter 7), and is available only in a surface-mount package. The A3985 is highly configurable, which allows designers to set it up to match the requirements of a wide variety of applications, but this often results in additional (sometimes extreme) effort to bring the circuit up and get everything working properly. However, if a circuit you're designing requires a 20 A motor, H-bridge driver chips with external MOSFETs may be the best choice.

23.4 SPEED CONTROL WITH PULSE WIDTH MODULATION

Another essential aspect of controlling motors is the ability to change the speed of rotation and the amount of torque produced. One simple but usually impractical means of achieving this is to adjust the supply voltage up and down as required. However, it is more effective and common to use the concept of PWM. This is a control technique where power to the motor is switched on and off rapidly, at rates high enough that the effects of the switching are negligible. The resulting effective voltage is then the average fraction of the time the power is on. This technique is also used in many other applications and is discussed elsewhere in the text.

Figure 23.17 illustrates the concept. The drive signal is switched on and off with a given period and is in the "on" state at voltage V_{ON} for a fixed fraction of the period. This "on" time is referred to as the **duty cycle** and is stated as a percentage, calculated as:

$$\text{Duty cycle (\%)} = \frac{\text{“On” time}}{\text{Period}} \times 100 \tag{23.9}$$

Figure 23.17 shows a duty cycle of 50%, with the resulting average *perceived* voltage of this PWM signal equal to 50% of the maximum voltage. The frequency of the PWM drive signal is calculated by taking the reciprocal of the period:

$$\text{PWM frequency} = \frac{1}{\text{Period}} \tag{23.10}$$

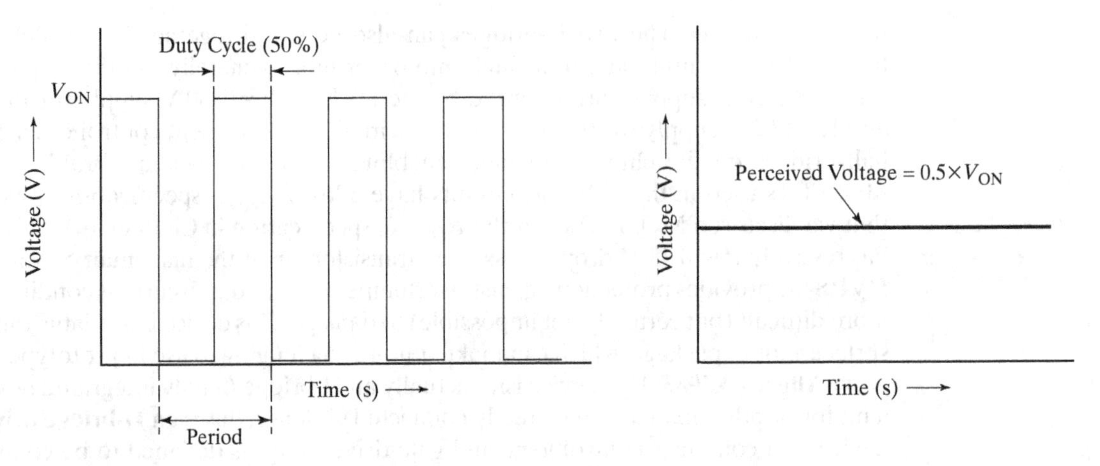

FIGURE 23.17
PWM with 50% duty cycle.

Changing the duty cycle of a PWM signal changes the average, or perceived, voltage level. For instance, adjusting the duty cycle shown in Figure 23.17 so that it is in the "on" state for 80% of the period will change the resulting perceived voltage as shown in Figure 23.18.

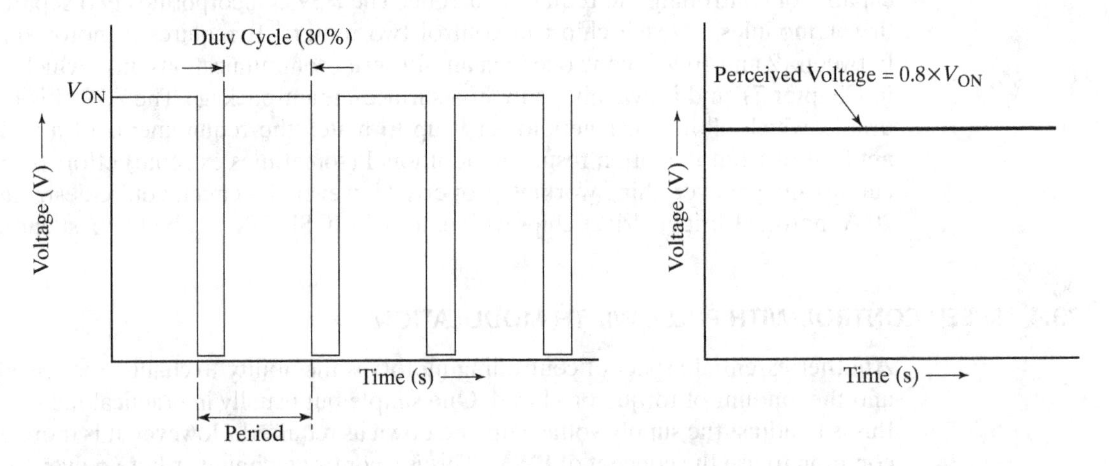

FIGURE 23.18
PWM with 80% duty cycle.

By increasing the duty cycle to 80%, the perceived voltage increases to 80% of V_{ON}. If PWM is implemented at a frequency that is too low, the result will be a jerky, stop-start response. Instead, the desired result is to approximate the "perceived voltage" as closely as possible with a minimum of perceptible ripple. When driving a permanent magnet brushed DC motor with PWM, the smoothing or filtering is very effectively performed by the physical inertia of the mechanical system and the electrical inductance of the motor windings. When done properly, the switching occurs too rapidly for the mechanical system to follow. For applications other than controlling the speed of a DC motor (e.g., approximating an analog output voltage with a microcontroller digital output pin), filtering circuits may be used to perform the required smoothing of the PWM output.

A typical mechanical time constant for a permanent magnet brushed DC motor is a few milliseconds, without a gearhead or other components attached. When selecting the frequency of the PWM drive signal, the designer must take into consideration the physical characteristics of the mechanical drive system, as well as the electrical characteristics of the motor. Usually, a fairly wide range of PWM frequencies will give reasonable results.

At the lower limit of acceptable PWM frequencies, the mechanical time constant of the physical system (i.e., the amount of time the system requires to respond to a change in an input) will dominate. In the extreme case where a very long period is chosen—much longer than the mechanical time constant—the motor will very obviously start and stop. Consider the case of PWM period of 2 seconds (which is a 0.5 Hz PWM frequency): the motor will almost certainly accelerate and decelerate noticeably during each period. This is the undesirable **torque ripple** that will become much less noticeable if higher PWM frequencies are used.

The same effect that leads to torque ripple also affects the linearity of the response of the motor to changes in duty cycle. As an example, Figure 23.19 shows the speed of a geared DC motor vs. drive duty cycle at several PWM frequencies.

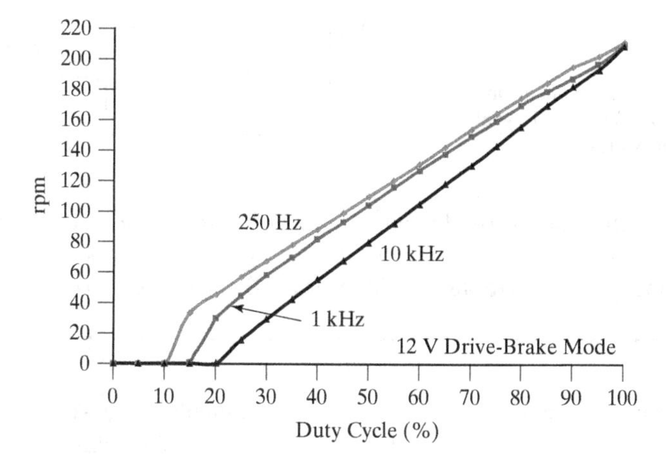

FIGURE 23.19

Effect of PWM frequency on the linearity of the duty cycle–speed relationship.

Increasing the PWM frequency results in better linearity of the relationship between duty cycle and motor speed. Usually, PWM frequencies in the range of 100 to 20,000 Hz will give workable results.

Regardless of the PWM frequency, torque ripple (and, by extension, **current ripple**) will occur to some degree. Higher PWM frequencies result in less ripple, but since current flow in the inductive motor coils is never able to follow the crisp edges of the PWM drive signal, ripple is unavoidable. Figure 23.20 illustrates how a steady state, low duty cycle PWM signal switching between a 0 V and V_{dc} with an average value of E induces a current flow in the motor's windings that has a minimum value of I_m and a maximum value of I_M. Shorter periods, T, will reduce the magnitude of the ripple, and hence the difference between I_m and I_M. Again, the rise and fall times of the current levels are governed by the motor winding's

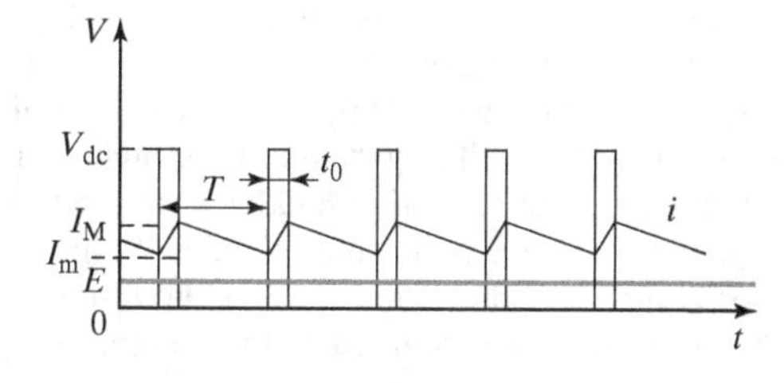

FIGURE 23.20

Current ripple characteristics for a steady state, low duty cycle PWM drive signal. (Courtesy of Maxon Motor AG.)

inductive time constant, $\tau = L/R$, Eq. (23.2). The amount of ripple current that can be tolerated will differ depending on the requirements of each application. One manufacturer, Maxon Precision Motors, recommends that ripple current be limited to 10% of their motors' maximum continuous current specifications [1].

When the PWM duty cycle is in the process of changing and isn't in steady state, the current flowing in the motor's coils will rise or fall in response to the changing average perceived voltage. The current can't immediately reflect instantaneous change in the duty cycle, just as it can't perfectly follow the crisp edges of the PWM signal as it switches on and off. Figure 23.21 shows how the motor current responds to changes in PWM duty cycle, again dictated by the time constant of the motor's coil windings.

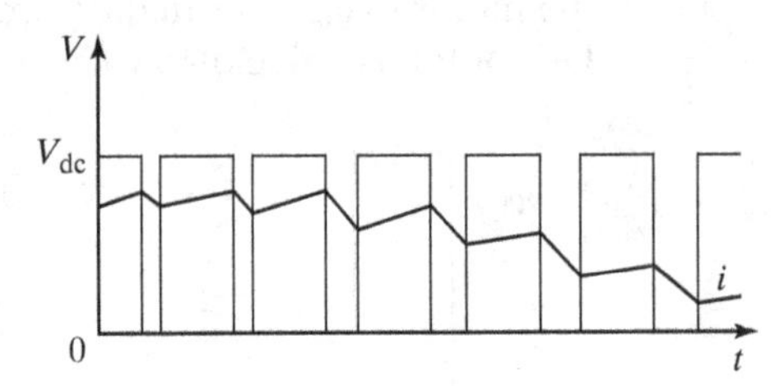

FIGURE 23.21

Motor current levels changing in response to a non-steady-state PWM signal. (Courtesy of Maxon Motor AG.)

PWM speed control can be readily combined with position or velocity feedback to create a servomotor controller (also somewhat inaccurately called a servomotor amplifier). We'll cover this concept in more detail in Chapter 28: Basic Closed-Loop Control.

Example 23.2

A motor will be required to spin continuously CW, and must supply 0.25 in · oz torque at 5,000 rpm when powered by 12 V. The motor has $K_T = 1.657$ in · oz/A, $K_e = 1.230$ V/krpm, and terminal resistance $R = 21.3\ \Omega$. Given that bidirectional control is not required, use a DS3658 IC, manufactured by National Semiconductor, and described as a "high-current peripheral driver" in the data sheet, to drive the motor.

1. Design a circuit to drive the motor, and show that none of the component specifications will be violated.
2. Specify the PWM duty cycle that will be required to obtain the specified torque and speed. Is there any uncertainty about what the resulting speed and torque will be? If so, where is this uncertainty introduced?

To answer the first question, we need to understand the function of the DS3658. It incorporates four independent switches, each of which can connect a load to ground or disconnect it from ground (low-side drive, Chapter 17), and each of which is controlled by a corresponding logic-level input. In the case of a motor or similar load, when one terminal of the load is connected to a power supply and the other terminal is connected to one of the DS3658's output pins, the motor will be turned on when the switch connects the output to ground and turned off when it is disconnected. One detail to look out for, though: the output pins are not perfect and can't connect a load to ground without some small losses. These losses manifest themselves as a slight rise in voltage from ground to the level at the output pin when it is on. This voltage rise is called the "saturation voltage," and is labeled on the data sheet as "Output Low Voltage."

The circuit we need to design must take into account that we will be using PWM drive to control the speed and torque of the motor (so this will require that we switch the motor on and off rapidly), we must protect the components from inductive kickback, and we must ensure that the components are capable of switching the current flowing through the motor. The circuit below, Figure 23.22, will be effective for controlling the switching, but the remaining specifications will need to be checked to ensure that none are being violated by the design.

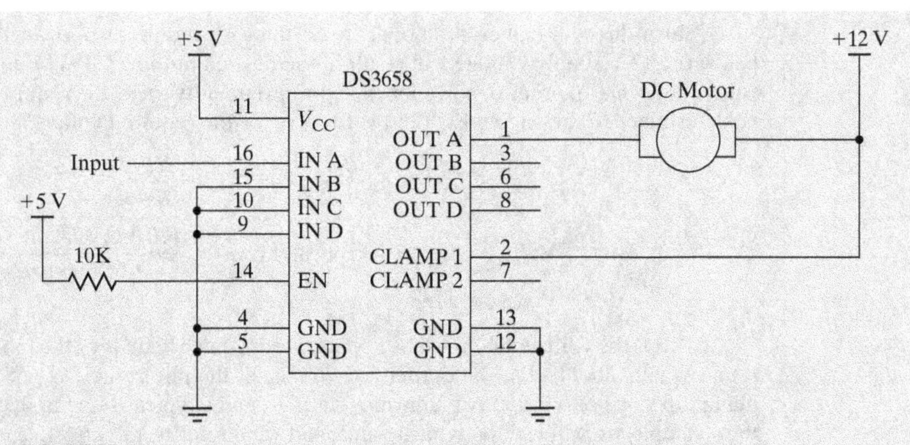

FIGURE 23.22

Motor driver circuit for Example 23.2.

Examining the data sheet for the DS3658 shows that each of the chip's four channels can safely switch 600 mA. Conveniently, the chip has built-in kickback diodes (CLAMP 1 and CLAMP 2), each capable of handling up to 800 mA of peak current. When making use of the incorporated kickback diodes, they should be clamped to supply voltages less than or equal to 35 V. The "output low voltage" specification is 0.35 V typical, 0.70 V maximum, and no minimum is given. Since we can't know in advance what the actual output low voltage will be for our system, we will assume it is the "typical" value, and will take into consideration the maximum and minimum values if they are going to have an effect. In order to determine whether the design meets the specifications or not, the second part of the problem will need to be solved.

First, it will be useful to understand if the torque and speed requirements can be met without using PWM. This is, effectively, the case where PWM with 100% duty cycle is used. Once we have established that the system requirements call for a duty cycle less than 100%, we know we can create a successful design. The maximum current the system will need to safely manage is the motor's stall current, which will occur any time the motor is switched on, so we will need to ensure that the system can do this safely. With a 12 V power supply and including the effects of the typical losses across the DS3658, the stall current is easy to determine using Ohm's law ($V = I_{stall} \times R$). Here, the voltage across the motor is the power supply voltage (12 V) minus the output low voltage (0.35 V):

$$V_{motor} = V_{supply} - V_{sat}$$

$$I_{stall} = \frac{V_{motor}}{R} = \frac{12\text{ V} - 0.35\text{ V}}{20.3\ \Omega} = 0.574\text{ A}$$

This is close to the specified maximum allowable current flow for a DS3658 channel, 0.6 A, but is within the specifications. Also, the clamp diodes can safely handle up to 0.8 A, so this is within their capabilities as well. However, this is not the worst case. The worst case will be when the output low voltage has a minimum value (resulting in the maximum current flow). Since no minimum value is stated in the data sheet, the safest assumption is that it can have a minimum value of 0 V. The results of that calculation are that $I_{stall,\,max}$ is 0.591 A, which is still below the chip's maximum current capabilities. The DS3658 can safely control this motor under these conditions.

For 100% duty cycle, the speed and torque produced can be found using Eq. (22.10) and entering the conditions that are given to solve for the rotational speed that will result:

$$\omega = \frac{V_{motor}}{K_e} - \frac{R}{K_T K_e} T$$

$$\omega = \frac{(12\text{ V} - 0.35\text{ V})}{1.23\text{ V/krpm}} - \frac{(20.3\ \Omega)(0.25\text{ in}\cdot\text{oz})}{(1.657\text{ in}\cdot\text{oz/A})(1.23\text{ V/krpm})}$$

$$\omega = 6{,}981\text{ rpm}$$

Since this is a higher rotational speed than we require, we know that an average voltage below the 12 V supply voltage will result in the desired output. A PWM duty cycle less than 100% will achieve this. In order to solve for the appropriate duty cycle, the required average voltage must be determined. Rearranging Eq. (22.10) to solve for the required voltage gives:

$$V_{motor} = K_e\left(\omega + \frac{RT}{K_e K_T}\right) = K_e\omega + \frac{RT}{K_T}$$

$$V_{motor} = (1.230\ \text{V/krpm})(5\ \text{krpm}) + \frac{(20.3\ \Omega)(0.25\ \text{in}\cdot\text{oz})}{1.657\ \text{in}\cdot\text{oz/A}}$$

$$V_{motor} = 9.21\ \text{V}$$

This is the voltage that must be applied across the terminals of the motor, but recall that there is uncertainty about what the output low voltage at the pin of the DS3658 will be. It could have a minimum value of 0 V, a maximum value of 0.7 V, and a typical value of 0.35 V. For the purposes of this example, we will use the typical value, and we'll refer to the effective, average voltage that we will achieve by applying PWM to the supply voltage as "V_{ave}."

$$V_{motor} = V_{ave} - V_{sat}$$

$$V_{ave} = V_{motor} + V_{sat} = 9.21\ \text{V} + 0.35\ \text{V} = 9.56\ \text{V}$$

In order to achieve an average voltage of 9.56 V, the 12 V power source should be switched on with the duty cycle determined as follows:

$$\text{Duty cycle} = \frac{V_{ave}}{V_{supply}} = \frac{9.56\ \text{V}}{12\ \text{V}} \times 100\% = 79.7\%$$

To sum up: we've insured that all the specifications are satisfied, and we will use a PWM duty cycle of 79.7% to achieve the desired operating point. There is uncertainty associated with the DS3658's output low voltage, and this will affect the speed and torque generated. Great care should always be taken to choose the output low voltage specification that results in the most conservative design. Practically always, this is one of the extremes, the maximum or minimum, and not the typical value.

23.5 HOMEWORK PROBLEMS

23.1 Why is it necessary to include a snubber circuit when switching inductive circuit elements (e.g., brushed DC motors)?

23.2 Name at least two benefits of using higher PWM frequencies.

23.3 For each of the snubbing methods described in this chapter,

(a) rank them in order of slowest current decay time to fastest.

(b) draw a schematic showing how each is connected, label the peak voltage drops across each of the components that comprise the snubber circuit, and write the expression for the peak voltage at the connection to the switching element.

23.4 A sensor will be mounted on a small platform that will be required to rotate continuously at a slow, steady rate. A motor with the following specifications is proposed for turning the platform: $R = 102\ \Omega$, $\omega_{NL} = 427.3$ rad/s at 36 V, $T_{stall} = 29$ mNm at 36 V. The application calls for the motor to deliver 5 in · oz of torque at 300 rpm. The system is to be powered by a 12 V supply and switched by a BJT rated to handle a maximum of 500 mA continuously with a collector-emitter voltage drop of $V_{CE} = 0.2$ V.

(a) Can the BJT safely switch the required current?

(b) Is it possible to meet the design requirements for torque and speed with the given motor and power supply? Justify your answer and show all calculations performed to reach your conclusion.

(c) What is the average current required when running at the design point?

(d) What applied voltage is required when running at the design point?

23.5 A Pittman™ 14203-48 DC motor is to be used in an application where it is required to produce torque of at least 10 in · oz when stalled. The motor has a no-load speed of $\omega_{NL} = 3{,}330$ rpm (at 40 V), stall torque $T_S = 161$ in · oz (also at 40 V), torque constant $K_T = 18.8$ in · oz/A, speed constant $K_e = 13.9$ V/krpm, and resistance $R = 5.53\ \Omega$. Design a circuit to drive the motor using an ST L298N H-bridge with two separate power supplies: +9 V for driving the motor and +5 V for powering the logic. Incorporate kickback diode protection. The design should take in logic-level signals for enabling the motor (turning it on and off) and changing the direction of the motor's rotation. Show how all specifications of the motor, the kickback diodes, and the L298N are satisfied.

23.6 A motor with a stall current of 0.9 A and coil inductance of 12 mH is driven with PWM frequency of 500 Hz. The kickback protection for the drive circuit consists of a fast recovery diode (a 1N4935) in combination with a 24 V, 5 W Zener diode (a 1N5359B). What is the peak power dissipated by the standard diode? By the Zener? Does this circuit operate within the specifications of these diodes?

23.7 Design a circuit that performs bidirectional control of a DC motor that uses only the following components: a two-position, ON-ON type, double-pole, double-through (DPDT) switch; a DC motor; and standard diode(s). Show a wiring diagram of your circuit and be sure to include kickback protection with the standard diodes.

23.8 When implementing PWM in a circuit using a 12 V power supply, what is the average output voltage when the duty cycle is 42%? If the output voltage is to be increased so that its average value is 8.79 V, what duty cycle is required?

23.9 In a general and qualitative way, sketch the current response vs. time for a motor with $L = 15$ mH and $R = 5\ \Omega$ ($\tau = L/R = 3$ ms) under PWM control for the following scenarios. Include a sketch of the PWM drive signal on a separate axis for reference. Do not perform any calculations or analysis to answer these questions. Assume that the motor is stalled, and a snubbing technique is used that results in a current fall time that is roughly equivalent to the rise time.

(a) PWM frequency = 25 Hz, duty cycle = 90% (period = 40 ms, on time = 36 ms, off time = 4 ms)

(b) PWM frequency = 500 Hz, duty cycle = 40% (period = 2 ms, on time = 0.8 ms, off time = 1.2 ms)

(c) The motor is initially off. At time t_0, a 20 kHz drive signal is enabled at 60% duty cycle (period = 50 μs, on time = 30 μs, off time = 20 μs). Illustrate the current transient response.

(d) The motor is initially running at 100% duty cycle. At time t_0, the PWM duty cycle is changed to 25% at 20 kHz (period = 50 μs, on time =12.5 μs, off time = 37.5 μs). Illustrate the current transient response.

(e) The frequency of the PWM drive signal is increased steadily from 25 Hz to 1 kHz, while the duty cycle is held constant at 20%. (The initial period = 40 ms, initial on time = 8 ms, and initial off time = 32 ms. The final period = 1 ms, final on time = 0.2 ms, and final off time = 0.8 ms.)

23.10 A motor with $K_T = 1.657$ in · oz/A, $K_e = 1.230$ V/krpm, and terminal resistance $R = 20.3\ \Omega$ is driven at 14 V. How fast will the motor spin under a constant 0.15 in · oz load if it is driven by PWM with a 25% duty cycle and a frequency well above the motor's electrical time constant? What is the speed if the duty cycle is increased to 85%?

23.11 Design kickback protection circuitry that incorporates a standard diode and a resistor for a motor that has coil inductance of 50 mH and a stall current of 900 mA when driven by a 12 V power supply. The motor is controlled by an NPN BJT. Limit the inductive kickback voltage spike to 36 V at the collector of the BJT. Draw a schematic diagram of the complete circuit. Under what conditions can the diode and the resistor safely manage the current and resulting power dissipation? Is the maximum PWM frequency a reasonable choice?

REFERENCE

[1] "PWM-Scheme and Current Ripple of Switching Power Amplifiers," White paper publication, Maxon Precision Motors, Inc., August 29, 2000.

FURTHER READING

"Driving DC Motors," Maiocchi, G., AN281 STMicroelectronics, 2003.

"How to Drive DC Motors with Smart Power ICs," Sax, H., AN380 STMicroelectronics, 2003.

Mobile Robots, Inspiration to Implementation, Jones, J.L., et al., 2nd ed., AK Peters, Ltd., 1999.

CHAPTER

24

Solenoids

24.1 INTRODUCTION

Solenoids, like the motors that we cover in Chapters 22, 23, 25, and 26 are based on electromagnetic force. In this chapter, we will examine the construction of DC solenoids, their performance characteristics, and the kinds of electronic circuits necessary to drive them. When you have completed this chapter, you will know about:

1. the range of forces and strokes achievable with solenoids,
2. the limitations on the length of time solenoids are energized (their "on time"),
3. how to maximize solenoid performance through the design of the electronic drive circuits.

24.2 SOLENOID CONSTRUCTION

Solenoids and DC motors are essentially electromagnets. Electromagnets make use of the fact that a current flowing in a wire produces a magnetic field surrounding that wire. The strength of the field surrounding a lone, individual wire isn't terribly impressive, but by arranging a length of wire into a coil, the magnetic field can be concentrated to the point that you can do useful work with it. The simplest application of this phenomenon in an actuator is what is called a solenoid. The solenoid is nothing more than an electromagnet and a movable iron or steel core that can respond to the magnetic field created by the electromagnet. Typical **tubular solenoids**, such as the one shown in Figure 24.1, have an internal construction something like the cross-sectional drawing shown in Figure 24.2. The **C- and D-frame solenoids** shown in Figure 24.3 make use of a simpler construction technique that is less expensive while still delivering good functionality.

In all of these solenoids, the plunger is the core and is also referred to as the **armature**, the moving part of an electromagnetic device. In a typical **pull-type solenoid** application, the plunger is pushed or pulled out of the housing of the solenoid by a spring when the coil of the solenoid is not energized. When the solenoid is turned on and the coil is energized, the plunger is attracted to the center of the electromagnet coils and drawn into the housing of

FIGURE 24.1

An example of a tubular solenoid.

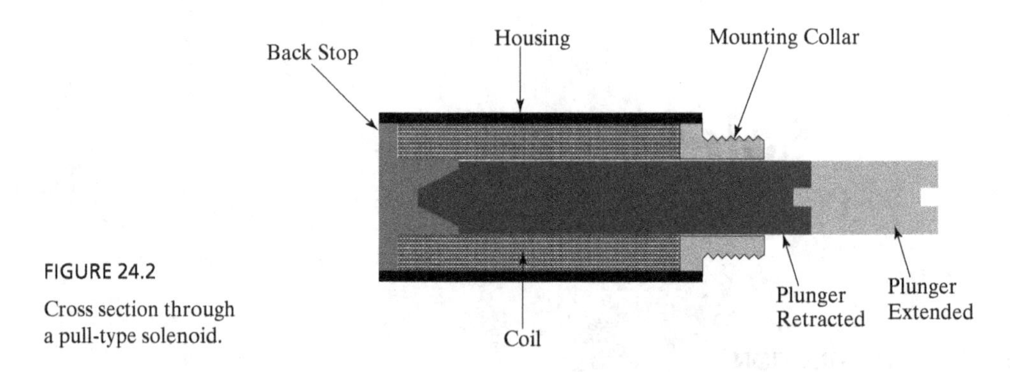

FIGURE 24.2

Cross section through a pull-type solenoid.

FIGURE 24.3

C- and D-frame solenoids.

the solenoid. To construct a **push-type solenoid**, a solenoid that creates a push force when the coil is energized, a pin is added to the plunger so that it will move out of the housing of the solenoid as the plunger is drawn toward the center. This type of construction is illustrated in Figure 24.4.

By combining an inclined plane to create screw threads with the linear motion of the plunger, it is possible to create a **rotary solenoid** that produces rotation over a limited angular displacement. The function of these devices is similar to the way that back-drivable lead screws work. In this style of rotary solenoid, shown in Figure 24.5, there are typically three spiral grooves, called "ball races" that are cut or stamped into both the housing and the plunger. The depth of the races increases along the arc of the spiral. These ball races carry ball bearings that allow the housing and the plunger to rotate relative to one another while supporting axial loads. As the plunger is pulled into the housing, the ball races guide the plunger through an angular, as well as axial, displacement. The screw thread and ball races provide both a means of converting linear motion to rotary motion and a bearing system to support loads during the rotary motion.

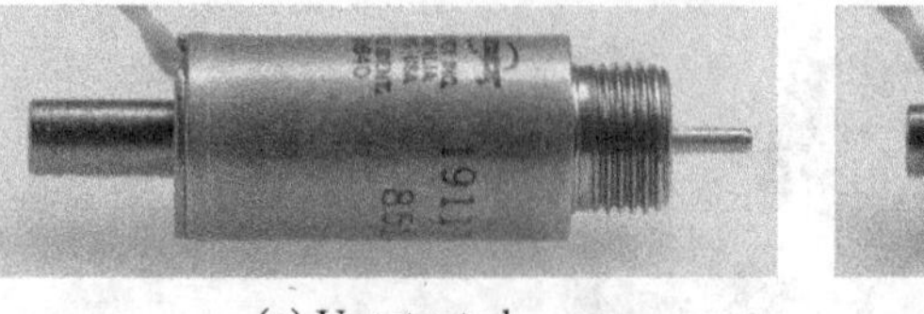

(a) Unactuated

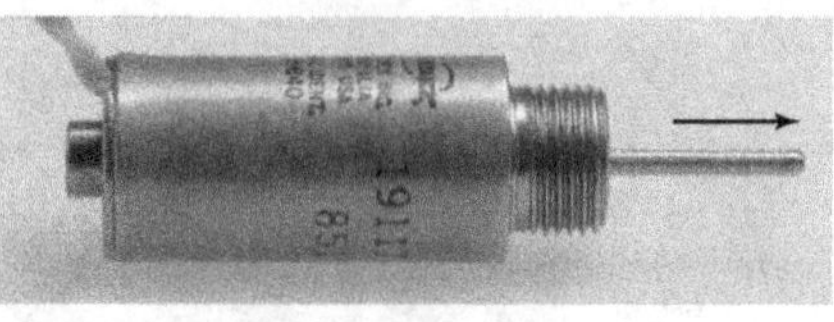

(b) Actuated

FIGURE 24.4

A push-type solenoid.

FIGURE 24.5

A rotary solenoid based on inclined ramp.

"True" rotary solenoids are also available, though more expensive. In these devices, the coil is arranged in a segment of solenoid body and a segment shaped plunger is drawn into the coil to create the rotary motion, as shown in Figure 24.6. For these devices, no axial motion accompanies the radial motion.

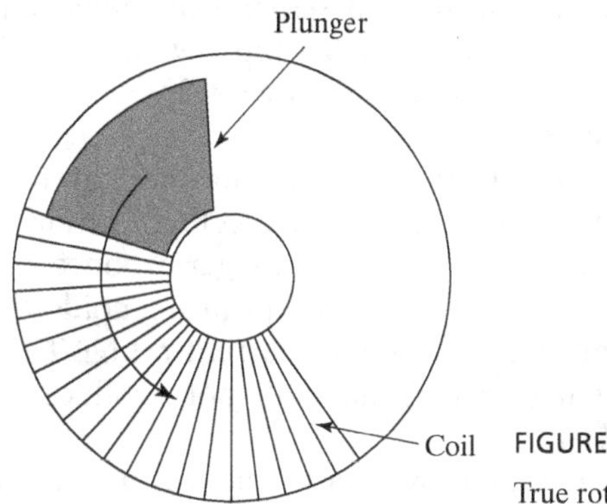

FIGURE 24.6

True rotary solenoid.

24.3 SOLENOID PERFORMANCE

The strength of the magnetic field produced by the coil determines the force produced by the solenoid. The strength of the magnetic field, B, in a simple electromagnet is a function of the number of turns in the coil, N, the length of the coil, L, the amount of current flowing through the coil, I, and the magnetic permeability, μ, of the core material.

$$B = \mu \frac{N}{L} I \tag{24.1}$$

For a given solenoid housing, there may be a choice of several different coil winding configurations. These windings differ in the numbers of turns and wire diameters and therefore the wire's resistance. These differences allow the manufacturer to provide solenoids suitable for a range of operating conditions.

Typical useful strokes range up to 15 mm for small (~30 cc, or cubic centimeters) solenoids while delivering forces on the order of 5 N (1.12 lbf). These same small solenoids are capable of producing 25–35 N (5.6–7.9 lbf) holding forces in their fully retracted position.

One of the key limitations in solenoids is the heat generated in the coil while the solenoid is actuated. The force generated by the solenoid is proportional to I, the current through the solenoid. Meanwhile, the heat generated in the resistance of the coil is I^2R. As a result of the heat input, most solenoids are rated for a maximum continuous power dissipation. At current levels through the coil that result in less than or equal to the specified maximum power dissipation rating, the solenoid may be energized continuously. If the application requires a higher force and a correspondingly higher current, the designer must limit the time that the solenoid is energized in order to allow it to cool. This limitation is normally expressed in two terms, a maximum duty cycle and a maximum on time. The **maximum duty cycle** describes the fraction of the time (in percent) that a solenoid can be energized:

$$\text{DC}(\%) = \frac{\text{On time}}{\text{Total time}} \times 100 \tag{24.2}$$

The **maximum on time** specifies the longest continuous interval that the solenoid can be energized without exceeding its maximum temperature. For example, a solenoid rated at 7 W continuous could be operated at 70 W with a 10% duty cycle (on for 1/10th of the total time). That same solenoid might have a maximum on-time limitation of 7 seconds at 70 W implying that the solenoid could be energized at 70 W for 7 seconds but would then need to be off for at least 63 seconds to remain below the 10% duty cycle. Note that achieving these power levels requires substantial amounts of voltage and/or current. A solenoid operated at a power level of 70 W from a 12 V supply would draw 70 W/12 V = 5.8 A. A solenoid rated for a continuous 7 W power dissipation at 12 V would need to be energized by 120 V in order to achieve a 70 W power dissipation.

There are limits to the improvement in peak force that can be generated by driving higher currents for shorter duty cycles. At high currents, the materials in the solenoid reach a point of magnetic saturation and no longer produce further increases in force proportional to further increases in current. For typical solenoids, this point is reached at approximately 10 times the maximum continuous current rating. Therefore, duty cycles below 10%, with correspondingly higher currents, will not produce force increases proportional to the increases in current.

The speed of actuation of a solenoid is increased by increasing the drive voltage. In contrast to the current, the only inherent limitation on the voltage at which a solenoid may be operated is determined by insulation characteristics of the wire used to create the coil windings. As a result, a given solenoid may be operated at a wide range of voltages, as long as the power dissipation limits are observed.

The way that these solenoid operating conditions are compared is in terms of a unit referred to as the **ampere-turns** of the solenoid. This is simply the product of the number of turns in the coil and the amount of current driven through the coil at a particular actuation voltage. As Eq. (24.1) shows, if the size of the coil and material of the plunger remain the same, the product of current and turns will indicate the relative magnetic field strength, and therefore the force generated by the solenoid. This relationship allows a designer to predict the force from a solenoid energized at something other than its continuous power dissipation rating.

There is a highly nonlinear relationship between the force provided by the solenoid and the extension, or **stroke**, of the plunger. A typical force vs. stroke plot is shown in Figure 24.7. While the force can be quite high for very short strokes near the fully retracted position, the force available falls off very rapidly as the plunger is moved further out of the housing.

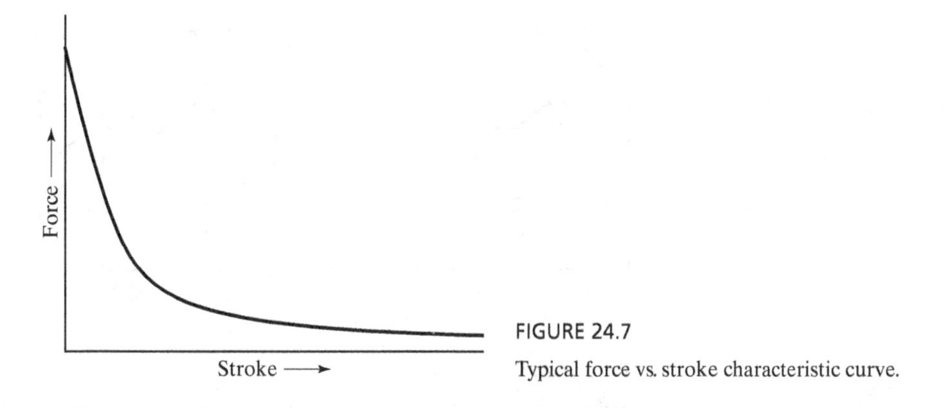

FIGURE 24.7

Typical force vs. stroke characteristic curve.

The shape of the force vs. stroke characteristic can be altered by varying the geometry of the plunger and the back stop within the housing. By tapering the end of the plunger, it is possible to trade peak force near the stop for a flatter force vs. stroke curve, as seen in Figure 24.8.

One way that manufacturers have come up with to address the need for high forces with long strokes is to provide a solenoid with two windings. One of the windings has a relatively lower resistance, resulting in higher current flows and correspondingly higher forces. The second winding has a higher resistance with lower currents. To retract the solenoid, both coils are initially energized, resulting in high retraction forces, even at a relatively large stroke position. After the solenoid has retracted, the current through the low-resistance winding is stopped, leaving only the high resistance, lower current winding energized. If the plunger has retracted to the home position, the force provided by the high-resistance winding alone will still be quite high and, depending on the application, sufficient to hold the solenoid in the retracted position. Because the low-resistance coil has been de-energized, the heat generated is greatly reduced.

In situations where the size of the solenoid is critical and it is impractical to provide two windings, a similar effect can be achieved through an electronic drive technique known as **peak and hold**. In this approach, there is only a single, low-resistance coil in the solenoid. A power supply will be used that generates a drive voltage that, if simply turned on, would generate a current level well above the maximum continuous current specification for the solenoid. The drive electronics are designed such that when the solenoid is commanded to actuate, the full drive voltage is applied to the coil to generate high forces and a quick

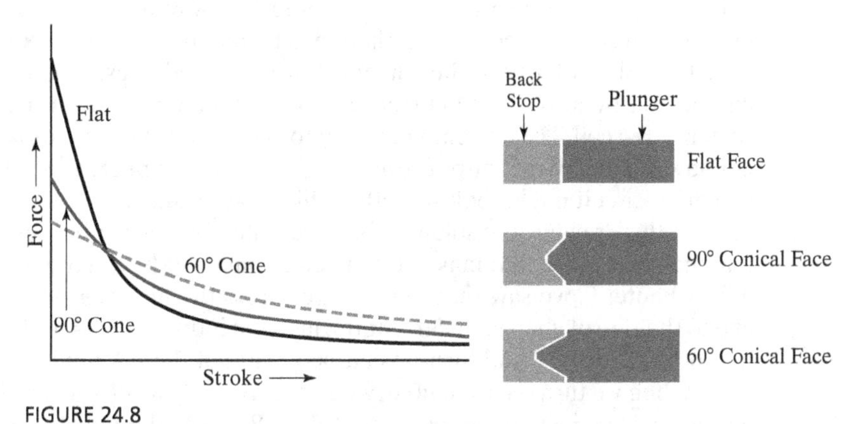

FIGURE 24.8

Changes in the force vs. stroke characteristic achieved by altering the plunger geometry.

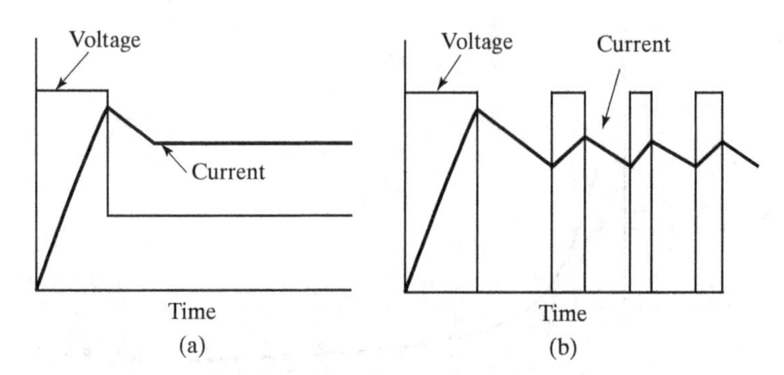

FIGURE 24.9

Voltage and current wave forms during peak and hold operation using (a) linear drive approach and (b) chopper drive approach.

retraction of the solenoid. After the solenoid has retracted, the drive electronics reduce the voltage to the coil by either increasing the voltage drop across the switching element (called **linear drive**) or by pulsing the drive voltage to the coil to produce a lower average drive voltage (called **chopper drive**). These techniques are illustrated in Figure 24.9.

With either linear drive or chopper drive, the actuation sequence begins with the full power supply voltage continuously applied to the solenoid. This is referred to as the "peak" phase. The end of the peak phase can be determined simply by a fixed time delay or, for more critical applications, by monitoring the actual current through the coil and reducing the effective drive voltage after the coil current has risen to the desired peak value. After the peak phase is complete, the system enters into the "hold" phase. Here the linear drive system differs from the chopper drive. In the linear drive system, a linear control element (typically a transistor operated in the linear region) is used to reduce the voltage applied to the solenoid. In a chopper drive system, the full power supply voltage is applied to the solenoid but not continuously. Instead, it is pulsed on and off at a high frequency and low duty cycle to produce the effect of a lower drive voltage.

24.4 DRIVING A SOLENOID

All that is necessary to actuate a solenoid is a way to start and stop the flow of current through the solenoid coil. Since there are no components incorporating a permanent magnetic field, the polarity of the current flow in the coil doesn't matter—the result of energizing the coil by introducing current in either direction is to retract the plunger toward the center of the coil. The current necessary to actuate most solenoids is large enough that it cannot be supplied by the output from a logic device or a microcontroller. The simplest version of a drive circuit might look something like Figure 24.10.

In this drawing, the solenoid is represented by the symbol for an inductor to remind us that the coil of wire that makes up the electromagnet has a large inductance associated with it. In Chapter 9, we saw that the voltage across the inductor was related to the inductance and to the rate of change of current through the inductor: $V = L\, dI/dt$. This has implications for both the turn-off and turn-on characteristics of the solenoid.

When we turn on a solenoid, we suddenly apply a voltage across its terminals. This does not produce an instantaneous current flow. Because the inductor resists changes in current flow through it, it takes some time for the current in the coil to build up. How long? As we

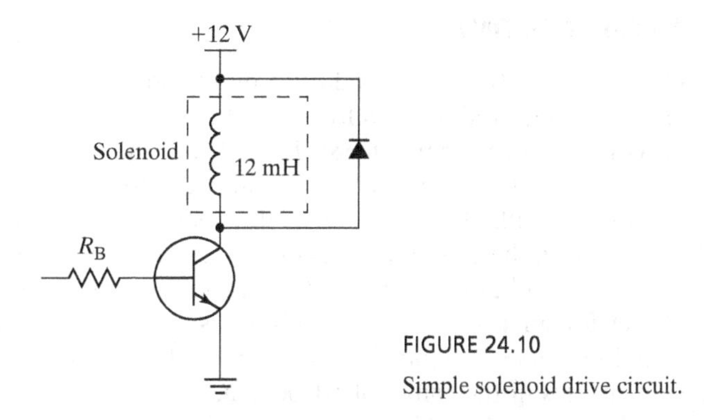

FIGURE 24.10
Simple solenoid drive circuit.

saw in Chapter 9, the time constant, τ, for a circuit like this is L/R. The L term is the inductance of the coil. The R term is the resistance of the wire that makes up the solenoid's coil. For a solenoid that has an inductance of 12 mH and a resistance of 11.4 Ω, the turn-on transient for the current through the inductor would look something like that shown in Figure 24.11, with the current taking approximately 400 μs to reach the peak value.

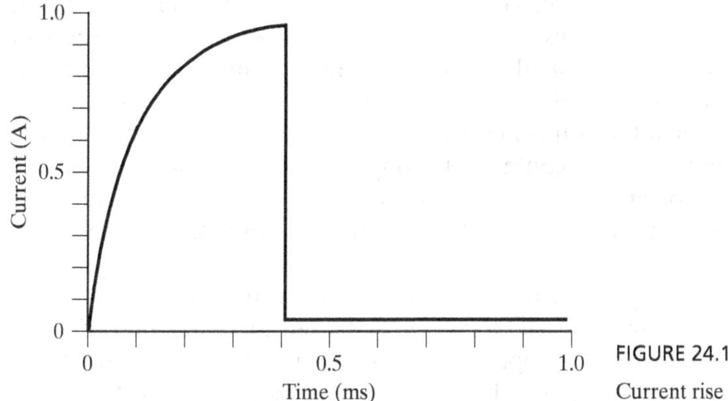

FIGURE 24.11
Current rise in a solenoid as it is turned on.

For many applications, this time that the current takes to rise is so short, relative to the total time the coil is energized, that it creates no problems. In some applications though, the solenoid must be activated very quickly and this rise time is an issue. In these cases, the peak and hold drive technique discussed in Section 24.3 may be of use. By using a peak and hold driver with a much larger voltage than could be applied to the solenoid continuously, it is possible to dramatically shorten the rise time. For example, moving from a 12 V drive voltage to a 42 V drive voltage would shorten the time to reach 1 A of current from 400 μs to about 114 μs for the 12 mH, 11.4 Ω inductor in the example above.

The inductive nature of a solenoid coil also comes into play when we turn the solenoid off, and here, it is cause for concern. When we turn off the current flow through an inductor, the inductor responds by creating a large and potentially destructive voltage across the terminals of the inductor, called the **kickback voltage**. This is exactly the same effect that we saw and covered for the DC motor in Section 23.2. As with the DC motor, it is very important to always include a snubber element in any solenoid drive circuit. Because of the large transient voltages created when the coil is switched off, it is common in industrial automation applications to use an opto-isolated driver (often referred to as a solid-state relay) to switch solenoids. When solenoids are used inside products, the drive circuits are typically implemented with discrete components rather than the larger, more expensive prepackaged driver blocks used in industrial automation.

24.5 MECHANICAL RESPONSE TIME

In addition to the time that it takes to build and eliminate the current flow in the coil, additional time is required to actually move the plunger. The plunger and whatever physical system is connected to it have mass, after all, and therefore cannot move instantaneously. The force created by the electromagnet, minus the force used in overcoming the return spring and friction is applied to the mass of the system to accelerate it into motion. Unfortunately, when specifying the characteristics of the return spring, the system designer will be forced to make a trade-off between the speed of actuation and the speed of return. In general, actuation times for small solenoids with plungers that have a mass of a few grams and a stroke of a few millimeters are typically in the range of a few milliseconds. For larger solenoids with plunger masses in the range of 50–60 g, actuation times are in the low tens of milliseconds. Solenoids that produce higher forces are, in general, larger and slower than solenoids producing lower forces.

24.6 APPLICATIONS

The best applications for solenoids are those that require only a small amount of linear travel. A common application is one in which the solenoid inserts or retracts a locking pin. The large forces in the system are carried by the locking mechanism and the solenoid need only move a relatively small part to effect the locking or unlocking action. Solenoids also appear as the actuating element in an electrical relay (Figure 24.12).

Here, when the control current flows through the coil, the electromagnet attracts the armature and closes the contacts allowing current to flow in the controlled circuit. The relay allows a control current signal to control a separate controlled circuit. One very convenient additional benefit of this approach is that the current paths, control and controlled, are electrically isolated from one another.

Another very common application across many industries is that of the solenoid valve. In these devices, the solenoid plunger is attached to a pneumatic or hydraulic valve to provide the actuating force to open or close the valve. To maximize the force available for actuation, it is not uncommon to have two separate solenoids in the valve instead of a return spring. One solenoid is used to open the valve and the other is used to close the valve. We cover solenoid valves in greater detail in Chapter 27.

Probably the type of solenoid valve produced in the highest volume is the automotive fuel injector. In this device, a solenoid operates a spring loaded needle valve to deliver a small quantity of fuel to each cylinder in an engine. This application is one instance in which the mechanical performance of the valve is critical. As a result, the drive circuits for fuel injectors often apply the peak and hold solenoid drive technique so that they can achieve the shortest possible opening times for the fuel injectors.

One of the most entertaining applications of solenoids is in arcade pinball machines. In these devices, solenoids are employed to perform a great number of tasks, mostly relating to moving the ball around the playing area: they actuate flippers and bumpers, eject balls, ring bells, and register new high scores with a bone-jarring series of knocking noises.

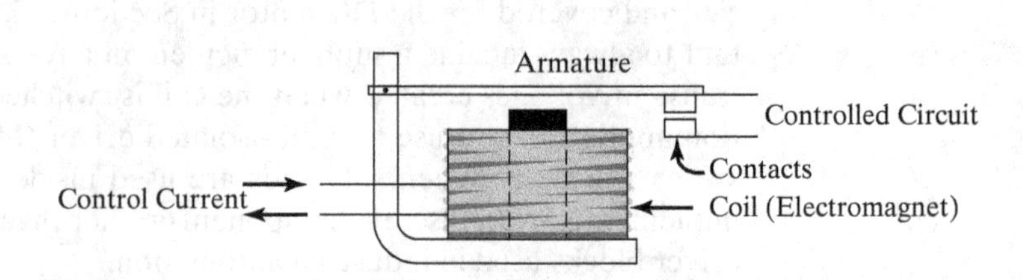

FIGURE 24.12

A solenoid as the actuating element of a relay.

24.7 HOMEWORK PROBLEMS

24.1 Describe the general characteristics of applications in which solenoids may be a good choice.

24.2 If solenoids prove to be a poor choice for meeting the requirements of a given application, name at least three alternative means of creating linear motion.

24.3 Given the performance characteristics for four solenoids shown in Figure 24.13, answer the following questions:

(a) Which solenoid(s) would be suitable for a continuous duty application that needed to provide more than 2.5 N of force at a displacement of 5 mm?

(b) Which solenoid(s) would be suitable for the same application if a 4:1 peak and hold driver (i.e., the initial phase of actuating the solenoid is driven with four times the voltage corresponding to a continuous duty cycle) were used to drive the solenoid?

(c) The construction of solenoid #4 differs from that of the others shown. How is it different? What impact does the difference have on its force vs. stroke characteristics?

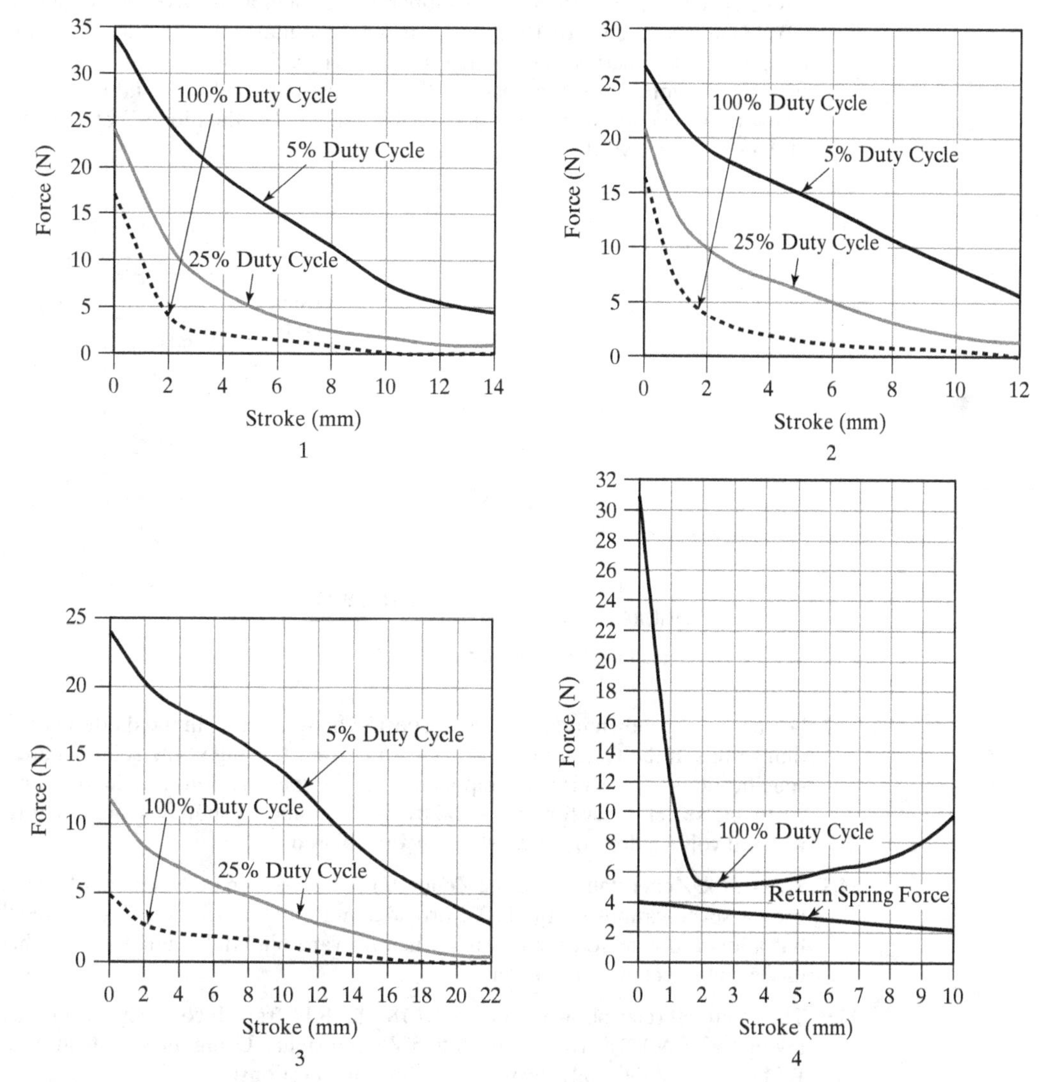

FIGURE 24.13

Data for Problem 24.3.

24.4 The performance requirements for an automotive fuel injector are stringent. For example, calculate the total time available for a fuel injector to introduce fuel into an automotive cylinder when a four-stroke engine is turning at 6,000 rpm. How does this compare to the mechanical actuation times of typical solenoids?

24.5 Design a peak and hold driver that will take two separate 0–5 V digital inputs and control the current through a solenoid to a 4:1 peak:hold ratio. The two inputs are called Peak/$\overline{\text{Hold}}$ and Hold/$\overline{\text{Off}}$. When the Peak/$\overline{\text{Hold}}$ input is in the logical high state, the Hold/$\overline{\text{Off}}$ input will be in a logical low state, and the current through the solenoid should be at the peak value. When the Hold/$\overline{\text{Off}}$ input is in the logical high state, the Peak/$\overline{\text{Hold}}$ input will be in the logical low state and the current should be at the hold value. When both inputs are in the logical low state, the solenoid should be off (no current flowing). The two inputs will never be in the logical high state at the same time. The solenoid has a resistance of 12 Ω and will be operated from a 12 V power supply. Choose standard 5% resistor values, but do not account for the 5% variation in your answer when assessing the current ratios.

24.6 The circuit shown in Figure 24.14 uses a standard diode (whose maximum average power dissipation specification is 0.3 W) as a snubber. If the solenoid is driven with pulse width modulation (PWM) at a frequency of 10 Hz, what is the minimum coil resistance R_c that the solenoid may have without violating the diode's power dissipation specification? You may assume that $V_{CE(sat)}$, the voltage drop across the TIP31 transistor's collector and emitter, is 0.5 V when the transistor is on and saturated. You may also assume that the diode D has a forward voltage drop of 0.6 V when forward-biased.

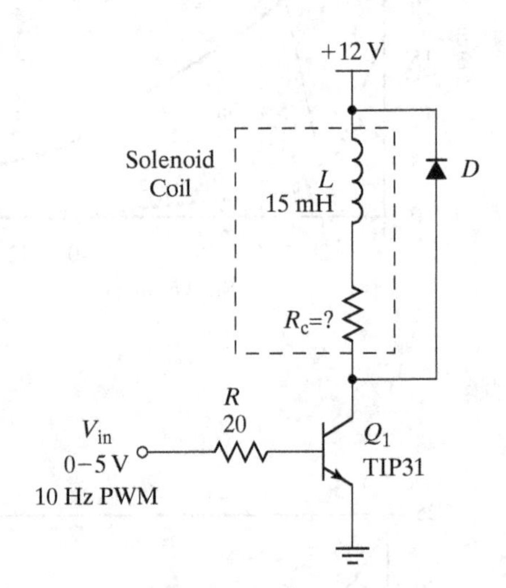

FIGURE 24.14
Circuit for Problem 24.6.

24.7 For the circuit shown in Figure 24.15, specify the resistor R_s in the diode-resistor snubber circuit that results in the fastest possible decay of current flowing in the solenoid and does not violate specifications for any circuit component. Use the specifications provided for the TIP31 in Figure 24.16 and select a resistor with standard 5% tolerance. You may assume that the diode D has a forward voltage drop of 0.6 V when forward-biased.

24.8 Specify the Zener voltage for the Zener diode (D_2) shown in the circuit of Figure 24.17. You may use the specifications for the TIP31 provided in Problem 24.7. You may assume that the diode D_1 has a forward voltage drop of 0.6 V when forward-biased. Include the data sheet for the Zener diode you select in your answer.

24.9 The electrical relay shown in Figure 24.18 (the RT1 from Tyco Electronics) is capable of switching up to 250 VAC at 16 A with a 0 to 5 V input signal. Using the specifications provided for the RT114005 (5 V DC coil), answer the following questions:

(a) When controlling the coil with a 5 V drive signal, how much current will flow? What is the threshold voltage at which the contacts will close? What is the threshold voltage at which the contacts will open? (continued on page 587)

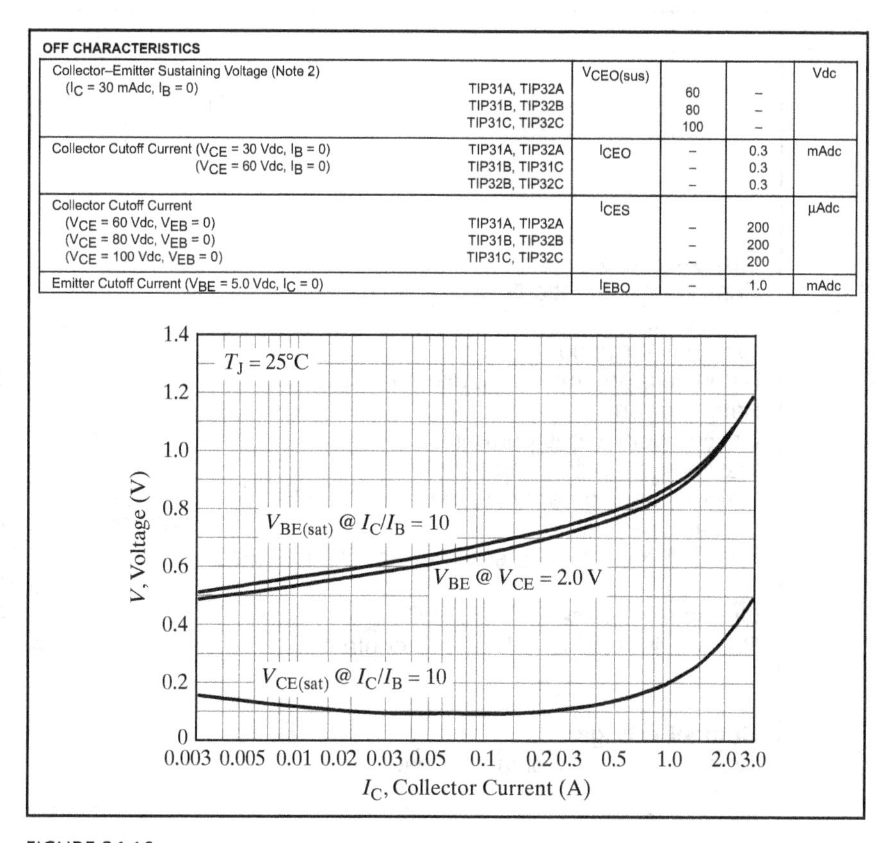

OFF CHARACTERISTICS

Characteristic	Device	Symbol	Min	Max	Unit
Collector–Emitter Sustaining Voltage (Note 2) (I_C = 30 mAdc, I_B = 0)	TIP31A, TIP32A TIP31B, TIP32B TIP31C, TIP32C	$V_{CEO(sus)}$	60 80 100	– – –	Vdc
Collector Cutoff Current (V_{CE} = 30 Vdc, I_B = 0) (V_{CE} = 60 Vdc, I_B = 0)	TIP31A, TIP32A TIP31B, TIP31C TIP32B, TIP32C	I_{CEO}	– – –	0.3 0.3 0.3	mAdc
Collector Cutoff Current (V_{CE} = 60 Vdc, V_{EB} = 0) (V_{CE} = 80 Vdc, V_{EB} = 0) (V_{CE} = 100 Vdc, V_{EB} = 0)	TIP31A, TIP32A TIP31B, TIP32B TIP31C, TIP32C	I_{CES}	– – –	200 200 200	μAdc
Emitter Cutoff Current (V_{BE} = 5.0 Vdc, I_C = 0)		I_{EBO}	–	1.0	mAdc

FIGURE 24.16

Data for Problem 24.7. (Used with permission from SCI LLC, DBA ON Semiconductor.)

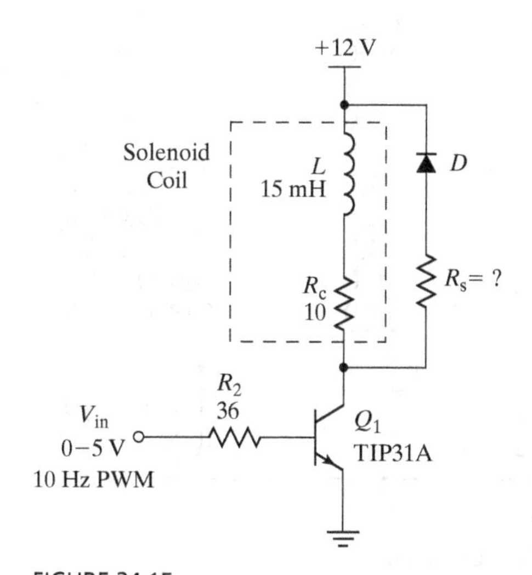

FIGURE 24.15

Circuit for Problem 24.7.

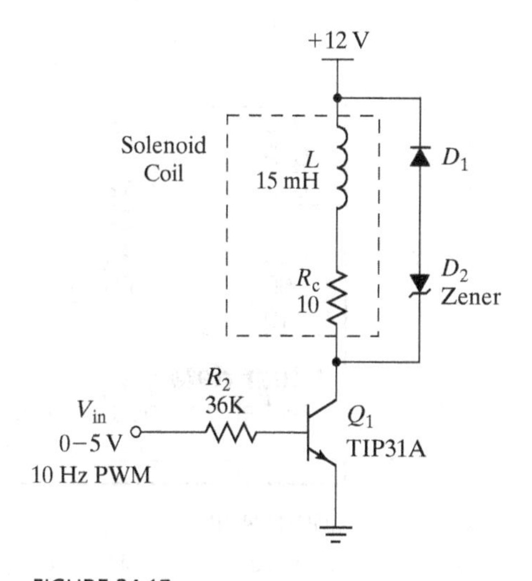

FIGURE 24.17

Circuit for Problem 24.8.

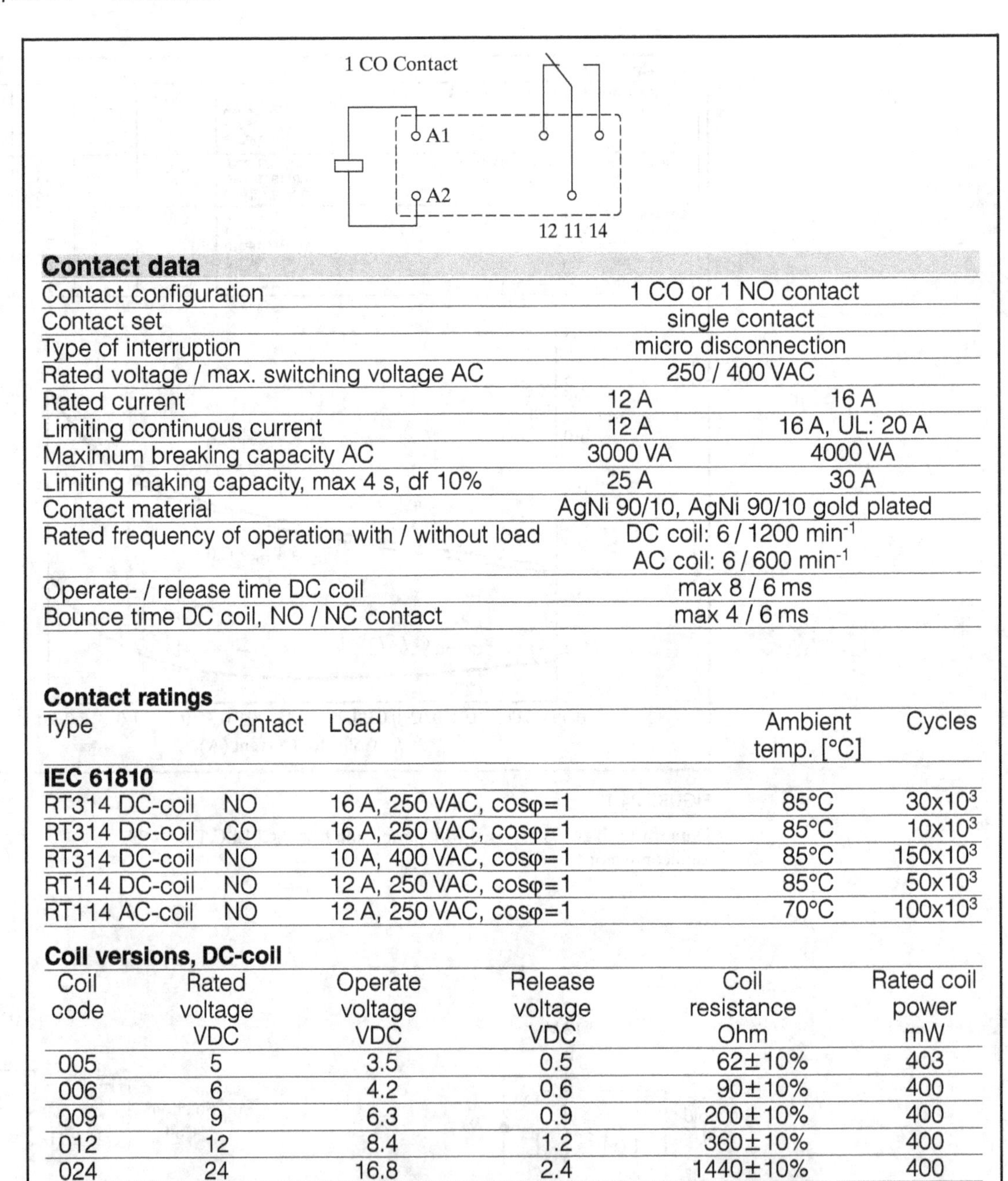

Contact data

Contact configuration	1 CO or 1 NO contact	
Contact set	single contact	
Type of interruption	micro disconnection	
Rated voltage / max. switching voltage AC	250 / 400 VAC	
Rated current	12 A	16 A
Limiting continuous current	12 A	16 A, UL: 20 A
Maximum breaking capacity AC	3000 VA	4000 VA
Limiting making capacity, max 4 s, df 10%	25 A	30 A
Contact material	AgNi 90/10, AgNi 90/10 gold plated	
Rated frequency of operation with / without load	DC coil: 6 / 1200 min^{-1} AC coil: 6 / 600 min^{-1}	
Operate- / release time DC coil	max 8 / 6 ms	
Bounce time DC coil, NO / NC contact	max 4 / 6 ms	

Contact ratings

Type	Contact	Load	Ambient temp. [°C]	Cycles
IEC 61810				
RT314 DC-coil	NO	16 A, 250 VAC, cosφ=1	85°C	$30x10^3$
RT314 DC-coil	CO	16 A, 250 VAC, cosφ=1	85°C	$10x10^3$
RT314 DC-coil	NO	10 A, 400 VAC, cosφ=1	85°C	$150x10^3$
RT114 DC-coil	NO	12 A, 250 VAC, cosφ=1	85°C	$50x10^3$
RT114 AC-coil	NO	12 A, 250 VAC, cosφ=1	70°C	$100x10^3$

Coil versions, DC-coil

Coil code	Rated voltage VDC	Operate voltage VDC	Release voltage VDC	Coil resistance Ohm	Rated coil power mW
005	5	3.5	0.5	62±10%	403
006	6	4.2	0.6	90±10%	400
009	9	6.3	0.9	200±10%	400
012	12	8.4	1.2	360±10%	400
024	24	16.8	2.4	1440±10%	400
048	48	33.6	4.8	5520±10%	417
060	60	42.0	6.0	8570±12%	420
110	110	77.0	11.0	28800±12%	420

Other data

Mechanical endurance DC coil	> 30 x 10^6 cycles
AC coil	> 10 x 10^6 cycles

FIGURE 24.18

Relay and data for Problem 24.9. (Courtesy of Tyco Electronics Corp.)

(b) How many switching cycles are the electrical contacts rated for? How many actuation cycles are the mechanical elements rated for? Which is most likely to fail first?

(c) What is the highest voltage and current you can control with this relay?

(d) How quickly can the contacts switch on or off? Is it reasonable to use this relay to implement PWM control? Why or why not?

(e) You are to use this relay to build a motion controlled porch light. Design a circuit which controls power to a standard 120 VAC/60 W light bulb with an input signal that is 0 V when the light bulb is off and 5 V at up to 100 mA when the light bulb is to be on.

24.10 You are building a test fixture for a piece of electronic equipment, and you need to characterize the life of a momentary switch located on the device's front panel. You decide to use the solenoid with the characteristics shown in Figure 24.19 to depress the switch repeatedly and count the number of cycles until failure. If you position the solenoid plunger 0.220 in. away from the switch and the switch deflects 0.020 in. when pressed, what is the maximum force you can generate with the solenoid in the fixture? If the switch requires 5 oz., what is the corresponding maximum duty cycle you can use with this solenoid?

24.11 If the solenoid from Problem 24.10 has 965 turn windings and a DC resistance of 20 Ω, design a drive circuit to control the solenoid at the maximum permissible current for 10% duty cycle. The control signal will be a 0–5 V output from a microcontroller ($I_{oh} = -100\ \mu A$). Specify the components and necessary power supply voltage chosen from the set of 5, 12, 15, and 24 V.

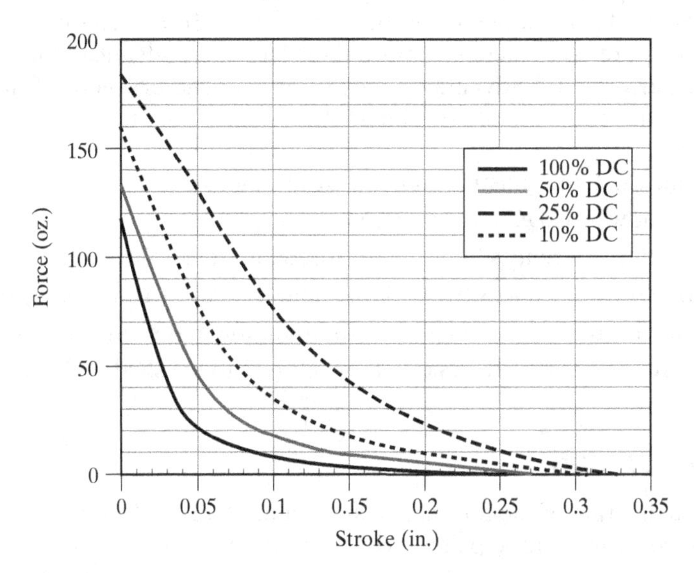

FIGURE 24.19
Data for Problem 24.10.

FURTHER READING

"Introduction to Solenoids," http://w2s.ledex.com/ledx/ds/lx000/lx0002e.lasso?pcode=L451 on 3/30/10.

"Solenoid Design and Operation," http://www.bcrn.com/bicronusa/images/solpdf/soldesop.pdf on 3/30/10.

"Technical Data," www.ledex.com/ltr2/access.php?file=pdf/2801.pdf on 3/30/10

CHAPTER

25

Brushless DC Motors

25.1 INTRODUCTION

In the brushed permanent magnet DC motor, the brushes and commutator work together to keep the rotor turning when a voltage is applied. This yields a motor that is very easy to drive with a convenient linear relationship between available torque and motor speed. However, the interface between the brushes and the commutator is a source of problems as well. The mechanical drag that results from the physical contact between these components reduces the available torque, creates acoustic and electrical noise, creates debris, and limits the maximum speed of the motor. The **brushless DC (BLDC)** motor addresses each of these issues and brings some added benefits as well. In this chapter, we will examine how BLDC motors are constructed, explore their basic and advanced modes of operation, and discuss the performance advantages they offer over the brushed DC motor. After reading this chapter, the reader should be able to:

1. describe the basic construction of a BLDC motor,
2. explain how a BLDC motor operates,
3. explain the differences between sensored and sensor-less commutation,
4. describe the differences between the delta and star (Y) winding configurations,
5. explain difference between block commutation and sinusoidal commutation,
6. compare the characteristics of BLDC motors to brushed DC motors.

25.2 BLDC MOTOR CONSTRUCTION

A simplified diagram of the internal arrangement of a BLDC motor is shown in Figure 25.1. This motor consists of three pairs of coils wound on the pole pieces of the fixed housing of the motor (the **stator**), and a **rotor** that incorporates a permanent magnet. This diagram shows a rotor with two magnetic poles, though rotors with four poles are also common. We saw these same basic elements (a magnet and an electromagnet) in the brushed DC motor in Chapter 22. Indeed, the interaction between the permanent magnet and the electromagnets formed by the coils is the source of torque in both types of motors, and the torque–speed relationships are similar as well. The difference between the motors is that the BLDC motor is turned "inside-out" relative to the brushed DC motor. Placing the coils onto the stator means that it is no longer necessary to supply current across a moving boundary, which was the function of the brushes and commutator. The commutator of the brushed DC motor also served the purpose of energizing the electromagnet coils in the correct sequence to provide a torque that resulted in continuous rotation of the rotor. For the BLDC motor, which has no

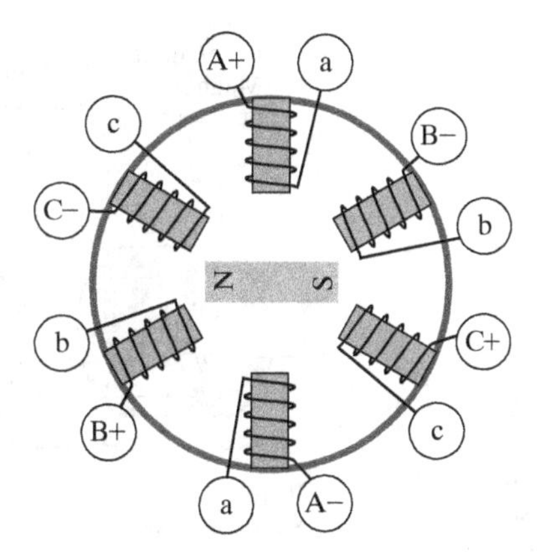

FIGURE 25.1

A simplified representation of the core of a BLDC motor.

commutator, the commutation process (sequencing of drive to the coils) must be performed by separate drive electronics.

The motor shown in Figure 25.1 is a three-phase motor with three pairs of electromagnet coils. The points labeled with matching lower case letters are connected together to join the diametrically opposite pairs to produce three distinct coils (one for each phase). Physically, they are arranged so that they are offset from one another by 60 degrees. Electrically, these coils are arranged in one of the two possible arrangements shown in Figure 25.2. The number of **phases** in a BLDC motor is the number of individually controllable coils. Even though there are six separate coils shown in Figure 25.1, they are wired and controlled in pairs, so this is actually a three-phase motor.

Windings configured as shown in Figure 25.2a are considered to be **Wye connected** (or simply **Y**). This arrangement is sometimes also referred to as **star connected.** Figure 25.2b shows an example of a **delta connection.** We will describe the impact of the wiring arrangements in more detail in Section 25.4, when we talk about the operation of the BLDC motor.

The last component of the construction of a BLDC motor is one that is commonly used, but optional. In order to perform the switching of the coil currents at the correct time to produce smooth rotation, we need to know something about the position of the rotor

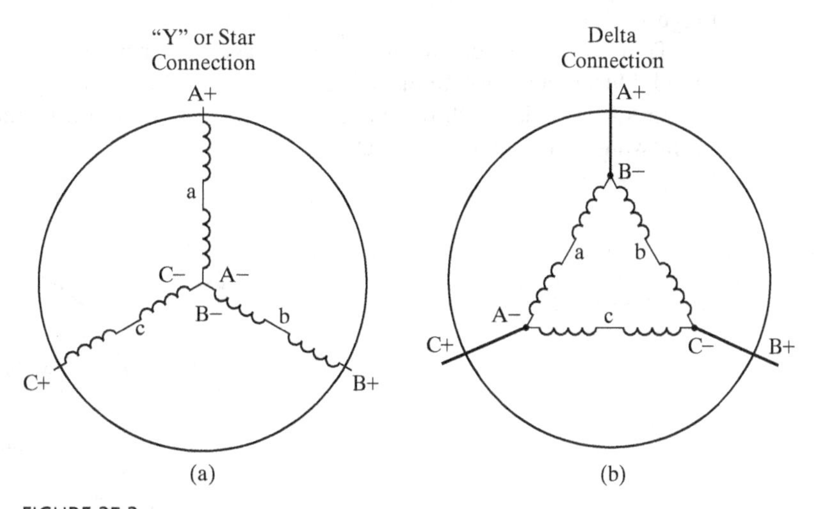

FIGURE 25.2

Brushless wiring configurations.

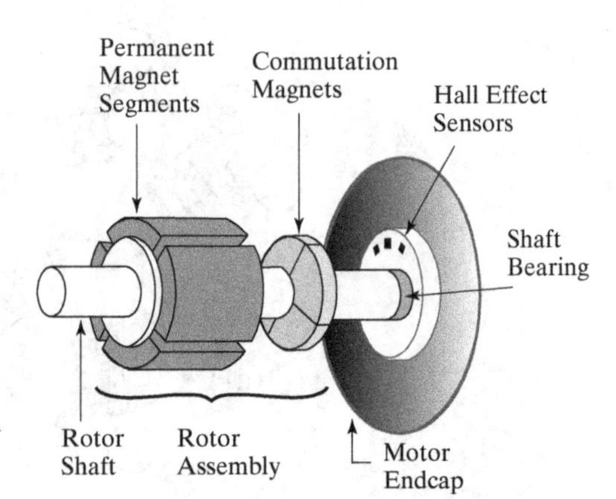

FIGURE 25.3

Hall effect sensors and separate commutation magnet on rotor for position sensing. (Courtesy of Bodine Electric Co.)

relative to the coils in the stator. The most common way of doing this is to incorporate a coarse position sensor into the motor, as shown in Figure 25.3.

In Figure 25.3, the motor's position sensor is formed by the commutation magnets and the Hall effect sensors mounted on one end of the motor. In this construction, the commutation magnets are associated with the rotor position sensing, and don't contribute to the motor's torque. The position of the commutation magnets relative to the rotor's permanent magnets is chosen such that whenever one of the Hall effect sensors is triggered by the commutation magnets, the corresponding stator coil will be activated.

Recall that sensors for determining the rotor's angular position were "optional," though commonplace. Some BLDC motors have no integral position sensors. Operating sensor-less motors requires techniques different from those used to drive motors with position sensors. Both approaches are described in the following sections.

25.3 BLDC MOTOR OPERATION

Before we get into the details of the operation of the motor, we need to distinguish between electrical cycles and mechanical cycles. For a three-phase motor, the electrical cycle is shown in Figure 25.4.

There are six stages to the electrical cycle. During each stage, one pair of the leads is excited. During a complete electrical cycle, each of the three phase windings is excited during four stages: twice with the current flowing in one direction and twice more with the current flowing in the opposite direction.

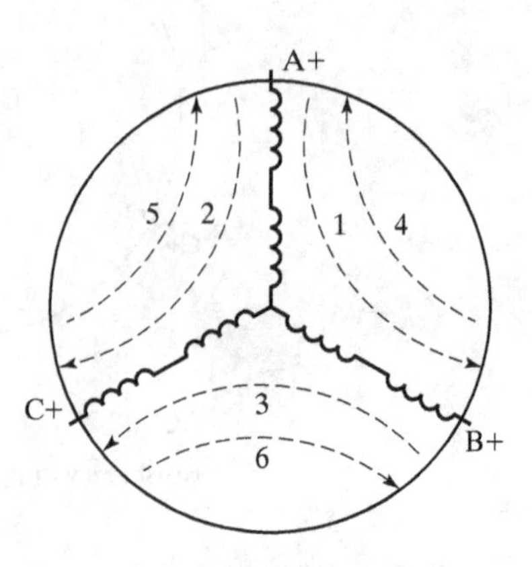

FIGURE 25.4

Current flows during the six stages of the electrical cycle for a three-phase motor.

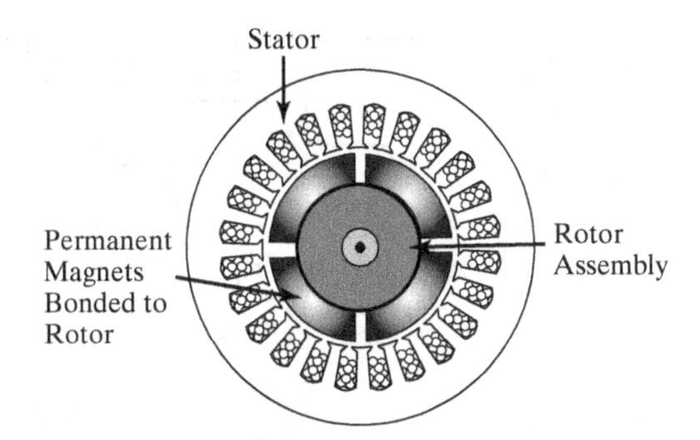

FIGURE 25.5

The core of a BLDC motor with 24 electromagnetic poles. (Courtesy of Bodine Electric Co.)

In the simplified motor of Figure 25.1, one electrical cycle (the sequence of all six stages shown in Figure 25.4) corresponds to one mechanical cycle—that is, one complete revolution of the rotor. In most commercially available motors, there are more than six internal poles, and the windings for the three phases are distributed around the periphery of the stator. This is why we have described Figure 25.1 as depicting a "simplified motor." Figure 25.5 shows the interior of a more representative BLDC motor, with 24 pole pieces on the stator and four magnetic poles on the rotor.

Using more electromagnetic poles results in smoother rotation with less torque ripple, and requires that the drive electronics cycle through the six stages of the electrical cycle multiple times (four times, in the case of the motor shown in Figure 25.5) to complete a revolution of the motor shaft.

Much as a sine wave goes through a complete cycle in 360 degrees, we speak of the electrical cycle as consisting of 360 degrees. Using this terminology, the interval between changes in drive to the coils is 60 electrical degrees. The simplified motor of Figure 25.1 would move through 60 mechanical degrees as the drive moved through 60 electrical degrees. However, if there were 24 rather than 6 poles to the stator (as in the motor of Figure 25.5), the motor would move through only 15 mechanical degrees for the same 60 electrical degrees.

Basic operation of the motor requires that we sequence through energizing the coils (the electrical cycle) in order to pull the rotor along and sustain the rotation. To change the direction of rotation, we reverse the *sequence* in which we energize the coils. Once you have decided which direction you want the motor to rotate, the key question becomes: "Which coil should I be energizing right now?" To answer that question, you need to know which of the two basic modes of operation for a BLDC motor you are going to use. The mode you choose is determined by whether or not your motor includes position sensing.

25.3.1 Sensored Commutation

When using sensored commutation (i.e., commutating a motor with integrated position sensors), the outputs of the position sensor are used to control the drive currents to the three coils. The position sensors are most commonly Hall effect sensors, though sinusoidal commutation (see Section 25.4) requires the use of an encoder with higher resolution than simple Hall effect sensors.

The number of poles on the commutation magnets and the mounting locations of the Hall effect sensors are such that each of the three sensors is active for 180 electrical degrees and the three sensors are offset from one another by 60 electrical degrees. This results in an output pattern like that shown in Figure 25.6.

We see that six unique combinations of the outputs from the three sensors are present, with a change in the pattern every 60 electrical degrees. These three sensor outputs can be

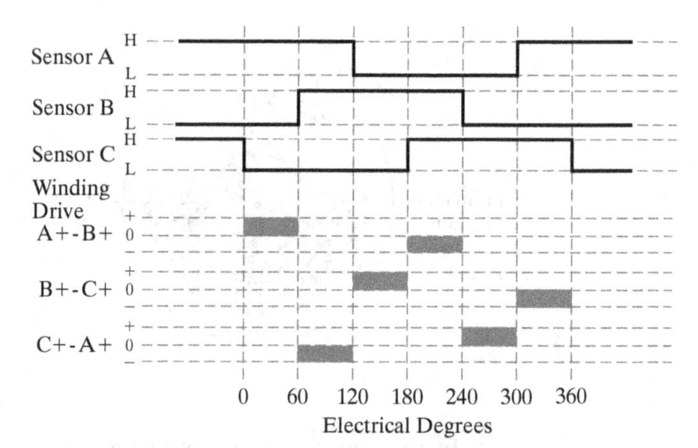

FIGURE 25.6

Hall effect sensor and coil drive waveforms.

logically combined to determine, at any given point in time, which of the three winding pairs should be energized and in which polarity.

25.3.2 Sensor-less Commutation

As we shall see, the term "sensor-less commutation" is a bit of a misnomer. While there are no Hall effect sensors—in fact no added sensors at all—**sensor-less commutation** makes use of elements within the motor itself as sensors. Recall that Faraday's law states that a moving magnetic field intersecting with a conductor induces a voltage. This is the same as the "back-EMF" that we discussed for the brushed DC motor. Recall also that back-EMF is created as the rotor of the motor rotates, inducing a voltage across the coils. In the case of the BLDC motor, the coils are located on the stator. If we sense this voltage within the coils, we are able to deduce when a pole of the rotor's permanent magnets passes by a coil of the stator. This is the essence of "sensor-less" commutation. In effect, the coils are used to sense the rotor's magnetic poles.

To work at all, sensor-less commutation requires that the magnet of the rotor be moving fast enough to induce a detectable voltage in an unenergized coil of the stator. If this is the case, how do you get the motor started? Under starting conditions, the speed of the rotor is zero, after all, so there is no back-EMF. There are a couple of common approaches that have been taken to getting the motor moving fast enough so that the back-EMF can be detected and used for commutation.

The simplest way is to start the motor slowly. If commutation begins slowly enough that you are sure that the rotor will have time to align itself with the moving magnetic field, you can then accelerate the commutation rate and "pull" the rotor up to a speed that will produce enough back-EMF. At that point, you can transition to commutation based on the signals derived from the back-EMF. This approach works well, but has a potential anomaly at startup. If you don't know anything about the position of the rotor at startup, then it is possible that the first coil that is energized in order to drag the rotor up to speed will actually cause the rotor to move backwards slightly. For many applications, this slight hiccup at startup is not significant, and this approach can be an appropriate choice.

For applications that absolutely cannot tolerate rotation in the wrong direction at any time, there is another more complex approach. This approach makes use of the fact that the presence of the rotor in proximity to a coil will cause the inductance of that coil to change. In operation, this technique begins by injecting a small current into each of the coils. The injected current is used to measure the inductance of the coils. From this measurement, it is possible to identify the first coil to energize so that the rotation can be guaranteed to be in the desired direction. This approach is very specific to the characteristics of a given motor. While the approach is generally applicable, the specific values of inductance and their relationship to the position of the rotor will vary between motor designs.

25.4 DRIVING A BLDC MOTOR

The kind of circuit that you need to drive a BLDC motor depends on the wiring type for the particular motor. Figure 25.7 shows the wiring and typical drive for a Wye motor.

This approach connects a half-bridge (see Section 17.6) to each of the three motor windings. The motor is driven by energizing two half-bridges at a time in order to drive current through a pair of windings. By changing which pair of half-bridges is active, as well as which is sourcing and which is sinking, it is possible to drive current in both directions through each of the winding phases. With this approach, two winding phases are always active with opposite magnetic polarities. This will produce the maximum torque output from the motor at the expense of a high voltage required to drive the two phases in series and a higher inductance of the series coils.

For applications that require less torque and are particularly sensitive to cost, it is possible to simplify the drive electronics by using a Wye motor with a **center tap**, as shown in Figure 25.8.

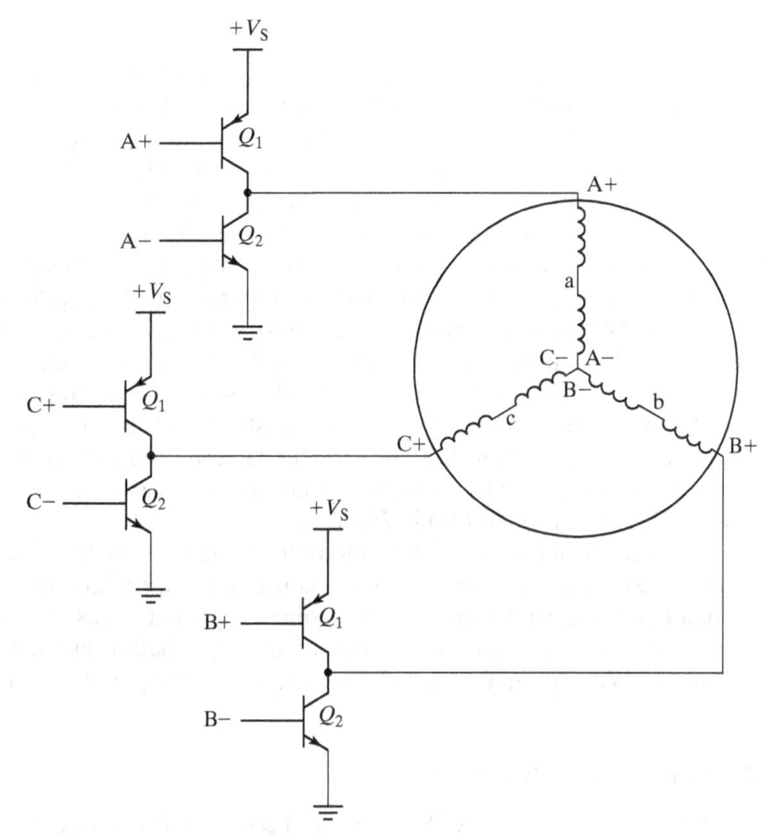

FIGURE 25.7

Drive circuit for a Wye motor.

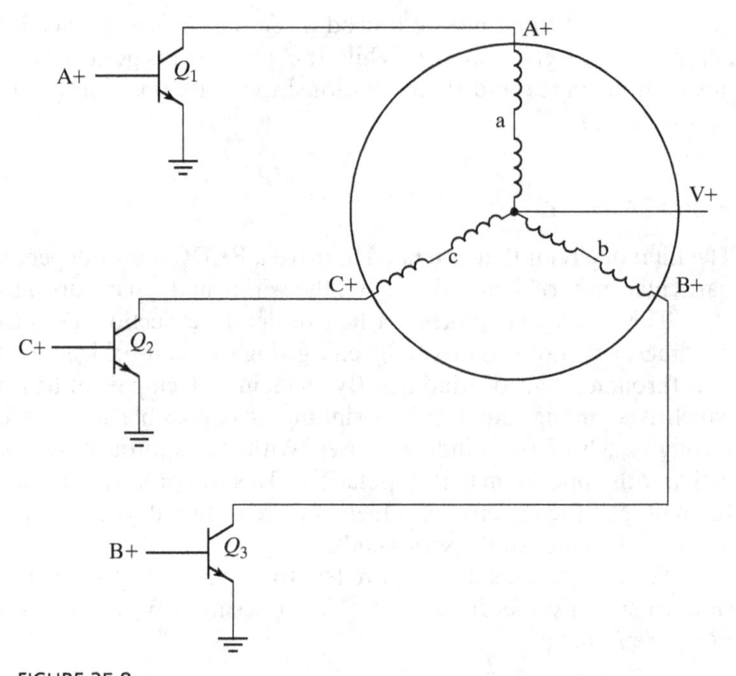

FIGURE 25.8

Drive circuit for a center-tapped Wye motor.

This drive approach trades away torque for simplicity of drive. With this approach, a single transistor (rather than a half-bridge) is used to drive each phase, and only one of the three phases is active at any instant in time, and the windings only produce a single polarity of magnetic field. As a result, motors using this winding and drive approach produce lower torque than that achievable using half-bridge drive electronics with the Wye motor.

The delta wiring shown in Figure 25.9 uses the same kind of half-bridge drive electronics used to achieve maximum torque with the Wye motor. This wiring configuration makes a different set of trade-offs between drive requirements and performance.

The Wye motor drives a pair of phase windings in series using two active half-bridges at a time. The delta drive uses a pair of half-bridges to drive one coil at a time (mostly). If the half-bridge at A is sourcing and the half-bridge at B is sinking, there are two current paths: A-B and A-C-B. Because of the higher resistance of the A-C-B path, most of the current will flow A-B, but some will flow A-C-B. The performance impact of the difference in wiring is most easily seen by comparing the specifications for two motors that differ only in how the windings are connected (Table 25.1).

At the same drive voltage, the delta wound motor has a higher no-load speed and a higher stall torque. Because its torque constant is approximately $1/\sqrt{3}$ that of the Wye (star) wound motor, it will require correspondingly higher currents to produce the same torque output. The motor configured with delta winding is better suited to high-speed applications, while the Wye wound motor will generate higher torque for a given amount of current.

25.5 COMMUTATING A BLDC MOTOR

In Section 25.3.1, we described sensored commutation based on the use of Hall effect sensors. In the approach described there, the coils of the motor were switched on and off as the outputs of the Hall sensors changed. This approach is referred to as **block commutation.**

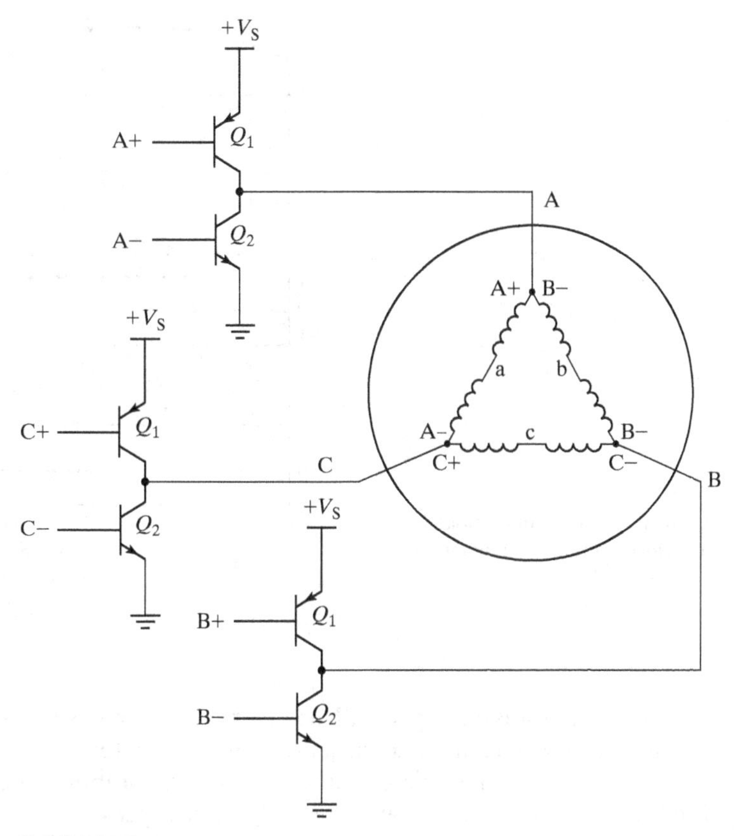

FIGURE 25.9

Drive configuration for a delta wound motor.

TABLE 25.1 Comparing specifications for Wye and delta configurations. (Courtesy of MicroMo Electronics Inc.)

Coil Connection	Δ Delta	Y Star	Units
Nominal Voltage	24	24	Volts
Terminal Resistance, phase-phase	0.24	0.69	Ω
Efficiency	86	85	%
No-Load Speed	9550	5450	rpm
No-Load Current	0.554	0.217	A
Stall Torque	2406	1455	mNm
Speed Constant	401	228	rpm/Volt
Back-EMF Constant	2495	4384	mV/rpm
Torque Constant	23.83	41.86	mNm/A
Terminal Inductance, phase-phase	76	220	μH
Mechanical time constant	5	5	ms
Rotor Inertia	130	130	gcm^2

While it is relatively simple and effective, it does have its drawbacks. The biggest drawback is that the abrupt nature of the changes in drive and the relatively coarse position information available result in variations in available torque as the motor moves through a single rotation. This variation is referred to as **torque ripple** and is undesirable.

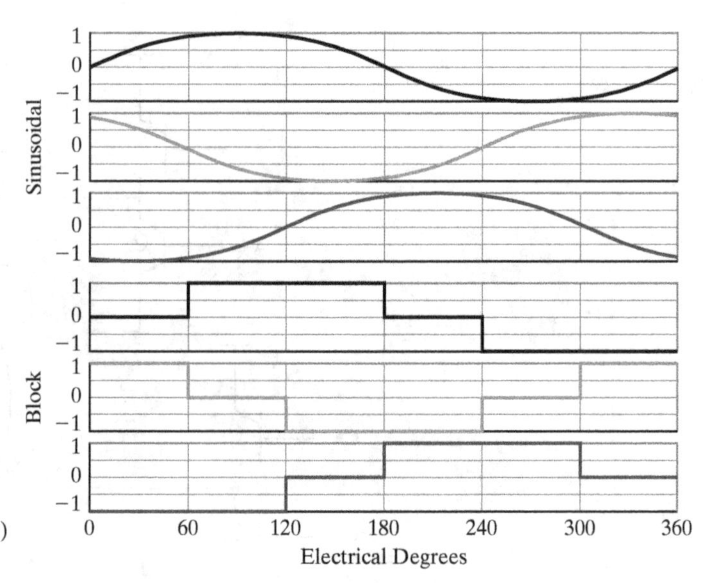

FIGURE 25.10

Current waveforms in sinusoidal (top) and block (bottom) commutation.

If the motor is equipped with a higher resolution position sensor, it is possible to use an alternative drive technique to virtually eliminate the torque ripple. To do this, the drive voltages to the coils are modulated rather than being abruptly switched on and off. As the rotor rotates, the drive voltages are smoothly varied between the coils, producing a constant torque output independent of position. Because the required variation in drive current is sinusoidal, this technique is referred to as **sinusoidal commutation** and results in current waveforms that look like those shown in Figure 25.10.

Because the current in each coil is a function of the combination of the applied voltage and the back-EMF, achieving good sinusoidal commutation is a nontrivial task. The current state of the art employs current feedback from the coils and sophisticated digital signal processing (DSP) chips to perform the calculations and generate the required waveforms. The result of all of this extra effort, however, is excellent torque control all the way down to zero speed. Despite the complexity of implementing sinusoidal commutation, the cost of the required electronics continues to fall, and as a result, it is seeing wider application.

25.6 BLDC MOTOR DRIVER INTEGRATED CIRCUITS

Because of the increased drive complexity of BLDC motors, most applications using small BLDC motors employ some sort of dedicated driver IC. These integrated solutions generally include an interface to Hall effect position sensors as well as the drive sequencing circuits, and require only enable and direction (and sometimes brake) inputs from the microcontroller. A block diagram for a representative device, the Allegro Microsystems® A4931, appears in Figure 25.11. Devices such as the ST Microelectronics® L6235 include all of these functions plus the actual output drive transistors, while other solutions, such as the Allegro Microsystems A4931, include high-voltage drivers but assume that the power transistors will be provided external to the controller. With the driver transistors included in the package, the coil drive currents are usually limited to the 2–3 A/coil range by the power dissipation capabilities of the IC package. When the output transistors are moved external to the driver,

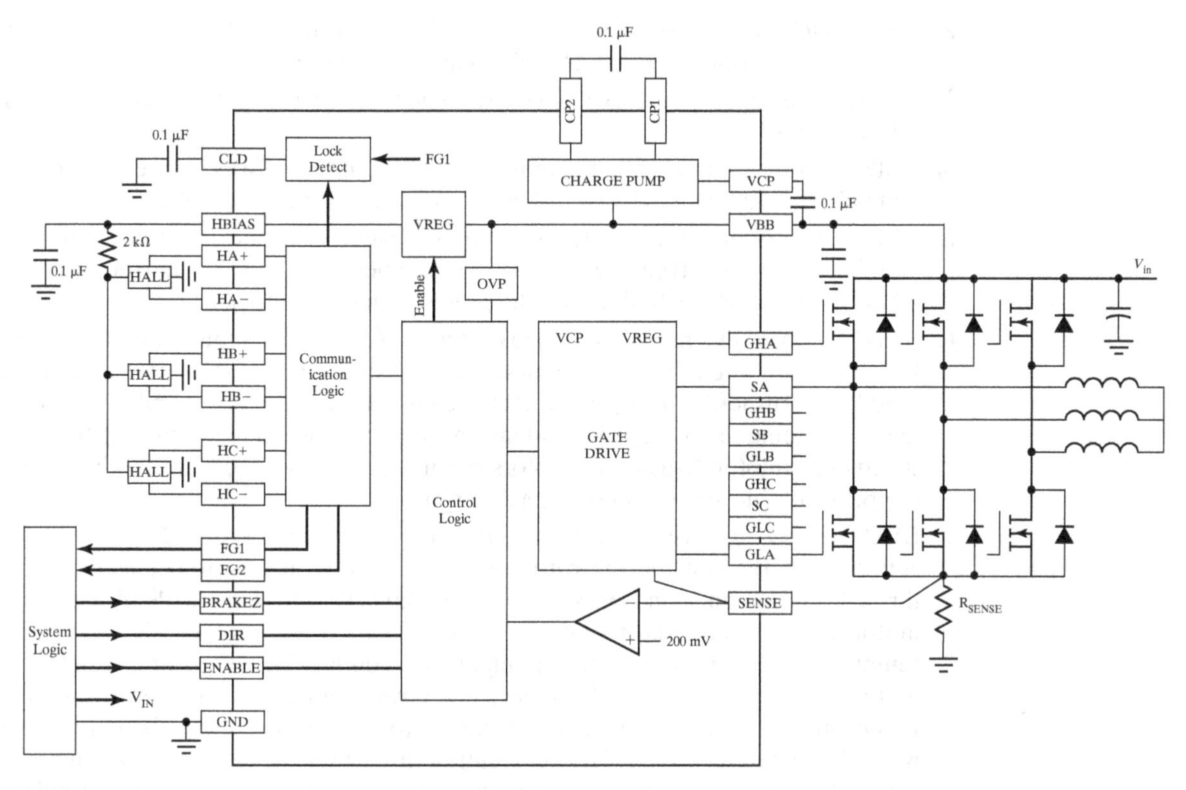

FIGURE 25.11

Block diagram of a BLDC motor controller (Allegro Microsystems A4931). (Courtesy of Allegro Microsystems Inc.)

the choice of transistor and mounting and cooling methods can allow for much larger drive currents.

Most of the available BLDC motor driver chips employ simple block commutation and Hall sensor feedback, though a few, such as the Toshiba® TB6551, provide for sinusoidal commutation. The highest performance BLDC motor controllers use an encoder to provide for fine position feedback. These high-performance controllers are not single integrated circuits, but employ a DSP to perform the relatively complex calculations necessary to determine and control the required phase currents and timing that result in smooth torque output over a wide range of speeds. The complexity of the interaction between the mechanical and electrical dynamics of the motor and the controller means that these types of controllers are not general purpose controllers, but are instead matched to a particular motor design.

25.7 COMPARING BRUSHED AND BRUSHLESS DC MOTORS

From a functional standpoint, brushed and brushless DC motors are very similar: they both exhibit a linear relationship between torque and speed. However, that is about the only common characteristic they share. The brushless motor requires more electronics to drive it than does the brushed motor, which increases the overall system complexity and cost. In return for this increased complexity and cost, the BLDC motor offers many advantages:

1. The BLDC motor is quieter acoustically. Without brushes, the bearings are the only moving surfaces in contact and they produce much less noise than the brushes.

2. The BLDC motor is quieter electrically. Without brushes, there is no arcing at the brush/commutation interface. This eliminates the attendant electrical noise.
3. BLDC motors are safer in explosive environments because without brushes there is no source of sparks.
4. BLDC motors are more reliable than brushed DC motors. Brushes are a wear element that will eventually fail. There is, of course, no such part in a brushless motor.
5. BLDC motors are more efficient than brush-type motors. The brush/commutator interface adds a relatively high electrical resistance element in the current path as well as mechanical friction, both of which reduce the motor's efficiency.
6. BLDC motors have much higher maximum speeds. The maximum speed of a brushed DC motor is limited by the dynamics at the brush/commutator interface. At very high speed, the brushes tend to bounce off the commutator segments (similar in many ways to the phenomenon of valve float in internal combustion engines), limiting the ultimate speed of the motor. Brush-type motors are limited to about 10,000 rpm, while brushless motors can achieve speeds up to 100,000 rpm.
7. BLDC motors are easier to cool than brushed motors. In a brushed motor, the heat is generated in the coils on the rotor, with very limited means of exhausting that heat through the bearings and brushes or motor housing. Since the coils of the brushless motor are stationary and mounted on the housing of the motor, it is much easier to remove the heat. This leads to a number of secondary benefits. For example, BLDC motors can be more compact for the same power output, and it is also possible for a BLDC motor to produce higher continuous torques at lower speeds than is possible with a brushed-type motor. The net result of this is that BLDC motors, when equipped with a high-resolution encoder, can sometimes be used in direct drive applications because their high torque/low speed capabilities eliminate the need for a gearbox.

Despite all of the positives, the added cost and complexity of the drive electronics required to operate BLDC motors will insure that there is a continued market for both brushed and BLDC motors for the foreseeable future.

25.8 HOMEWORK PROBLEMS

25.1 How do the terminal-to-terminal inductances compare between a Wye-connected and a delta-connected motor?

25.2 How would you expect the difference in inductance between a Wye-connected and a delta-connected motor to be manifested in the performance of the motor? Describe the differences in expected performance.

25.3 What is the relationship between mechanical and electrical degrees for a BLDC motor with 48 internal poles?

25.4 If the coils in Figure 25.1 were not wired in pairs, how many phases would this motor have?

25.5 If the coils in Figure 25.1 were not wired in pairs, how many stages would the electrical cycle contain?

25.6 Explain why a high-resolution encoder is necessary to achieve the smoothest torque output when using sinusoidal drive.

25.7 Using the types of subsystems commonly available on a microcontroller, how would you generate (in general terms) the varying drive currents necessary to implement sinusoidal drive?

25.8 If a microcontroller with circuitry to measure coil currents were used to implement sensor-less commutation, explain one way that you could use the microcontroller to measure the inductance of the coils.

25.9 Cite three reasons for why a BLDC motor would be a superior choice over a brushed motor in a laptop computer cooling application.

25.10 What is the alternative to block commutation for a BLDC motor? Why would a designer choose block commutation?

25.11 How many output transistors are necessary in the drive circuit for delta wound motor? Is there an alternative winding and/or wiring layout that requires fewer transistors?

FURTHER READING

"AN1130 An Introduction to Sensorless Brushless DC Motor Drive Applications with the ST72141," STMicroelectronics.

"AN885 Brushless DC (BLDC) Motor Fundamentals," Microchip Technology, Inc., 2003.

"AN970 Using the PIC18F2431 for Sensorless BLDC Motor Control," Microchip Technology, Inc., 2005.

"Catalog S-14," Bodine Electric Company.

"Technology Short and to the Point," Maxon Motors, Inc., USA.

CHAPTER 26

Stepper Motors

26.1 INTRODUCTION

The construction of a brushed permanent magnet DC motor is optimized for continuous rotation. The brushes and commutator work together to keep the rotor turning with a constant voltage applied. This is nice if what we want is a rotor that turns more or less continuously at some controllable speed. What about if we want to use a motor to position something, rather than have it move continuously? You can do that with a brushed or brushless DC motor, but you will need to add other components to the basic motor in order to perform position control, such as some form of position sensor to detect where the motor is relative to where you want it to be and a controller to read the position from the sensor and adjust the drive to the motor to get it to hold that position. This results in what is called closed-loop control (Chapter 28). If our goal is position control with less hardware, then there is another kind of motor that we can look to: the stepper motor. The stepper motor, as its name implies, works by taking discrete steps rather than rotating continuously as ordinary DC motors do. If you design within the specifications and constraints for a stepper motor, no additional position feedback is usually necessary, meaning you can use them to implement **open-loop** (i.e., without feedback) position control. This makes stepper motors ideal for a wide range of positioning applications, especially those with limited demands for fast response and extreme accuracy.

Once you have mastered the concepts contained in this chapter, you should be able to:

1. describe the different types of stepper motor constructions,
2. describe the different types of wiring used in stepper motors,
3. understand the different types of drive sequences used with stepper motors and the advantages and disadvantages of each,
4. describe the various regions of the torque–speed relationship for a stepper motor,
5. describe the dynamic behavior of the stepper motor when taking a step,
6. understand the differing electronic approaches to driving stepper motors and their impact on motor performance.

26.2 STEPPER MOTOR CONSTRUCTION

The basic **permanent magnet (PM) stepper motor** is depicted in Figure 26.1. The motor consists of a stator, with two sets of coils wound on the pole pieces of the fixed housing, and a rotor that incorporates a permanent magnet. This should sound pretty familiar: it is the same basic structure as the permanent magnet brushless DC motor that we covered in the last chapter. The difference here is that we want the stepper motor's rotor to rotate to a given position, stop, and hold that position.

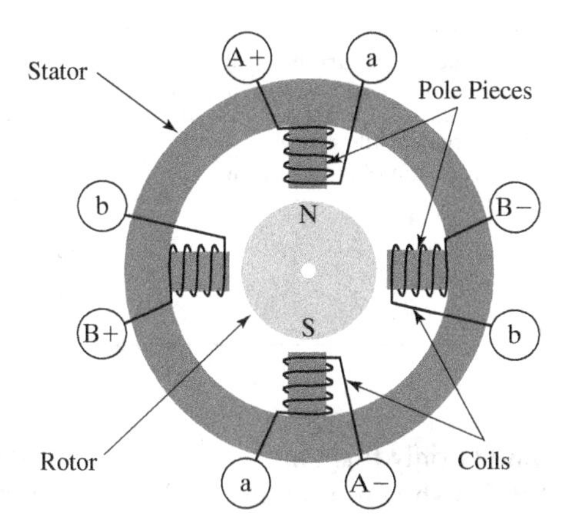

FIGURE 26.1

Basic layout of a PM stepper motor.

To show how the motor works, we'll go through the actions required to cause it to make a series of steps. The sequence shown in Figure 26.2 corresponds to the discussion.

We start by running current through coils A and B (note that they're connected, so our only options are to energize them both or de-energize them both), such that the inner pole of the upper coil (A) becomes a south magnetic field and the inner pole of the lower coil (B) becomes a north magnetic field (Figure 26.2i). The permanent magnet rotor will rotate to align its south and north fields to the respective north and south fields of the electromagnet. If we continue to allow the current to flow through the coils A and B, then the rotor will move into alignment and stop, holding that position. If we now remove the current from coils A and B and energize coils C and D (again, these are connected, so we can't switch them on individually), the rotor will rotate to realign itself with the fields from these electromagnets and move to the position shown in Figure 26.2ii. If we then remove the current from coils C and D and

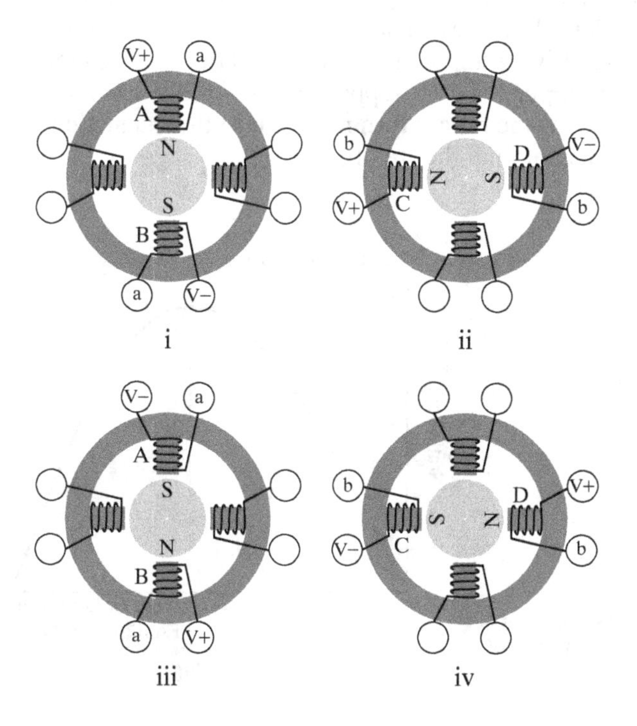

FIGURE 26.2

A PM stepper motor rotating through a sequence of steps.

TABLE 26.1 Sequence of current through coils resulting in rotation of PM stepper motor rotor (corresponds to Figure 26.2).

Coils Energized (the arrow indicates the direction of current flow)	Position
A → B	Figure 26.2i
C → D	Figure 26.2ii
B → A	Figure 26.2iii
D → C	Figure 26.2iv

reapply current to coils A and B, only this time with the current flowing in the opposite direction, we will create the situation shown in Figure 26.2iii. We can repeat this pattern and the rotor will move in a full circle as the coils are energized in the sequence shown in Table 26.1.

The stepper motor in the example has a four-step cycle and completes one revolution every four steps, so each step corresponds to 90° of rotation.

With current applied to a set of coils, the stepper motor will move to an equilibrium position and stop. Once the rotor has come to rest, it must no longer be producing any torque. If it were producing torque, it would move further, after all. If an external torque displaces the rotor from the rest position, the torque necessary to cause a displacement increases as the angle grows. That is, moving the first 0.1° of displacement requires less external torque than moving the next 0.1°. In this way, the magnetic force that produces the holding torque is somewhat like a spring. However, unlike a spring, the increase in the required torque follows an approximately sinusoidal form up to a peak and then back down to zero as you move further from the resting position of the rotor. This is shown in Figure 26.3.

The peak of the curve in Figure 26.3 is the maximum torque that you can apply without causing the rotor to break free of its position and rotate. This peak value is referred to as the **holding torque** for the stepper motor. Much like a DC motor produces the maximum torque when the rotor is stalled in one position, the holding torque is the maximum torque that the stepper motor can resist. The useful torque that the motor can produce in motion, for example, to move a load, is always less than the holding torque. We'll look at this in more detail when we talk about driving stepper motors a little later in the chapter.

Without any power applied, a PM stepper motor still has some holding torque. This holding torque, referred to as the **detent torque**, is due to the magnetic attraction between the rotor (a permanent magnet) and the stator poles.

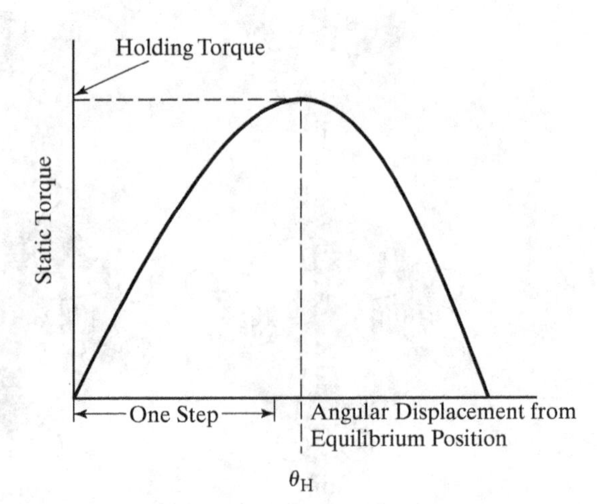

FIGURE 26.3

Torque vs. angular displacement in a PM stepper motor. (Courtesy of Oriental Motor Co., Ltd.)

Most PM stepper motors have many more than four steps per revolution (90°/step). However, because of the difficulty in making multipole magnets for the rotor, PM stepper motors have relatively large step angles. The number of steps per revolution is a function of the number of magnetic poles on the rotor and the number of independently controllable coils, referred to as **phases**, on the stator. For a PM stepper motor, the number of steps per revolution, S, is

$$S = 2pN_{\text{poles}} \tag{26.1}$$

where p is the number of phases. We can increase the number of steps per revolution by increasing either the number of magnetic poles on the rotor or the number of phases in the stator, or both. In our simplified example, $p = 2$ and $N_{\text{poles}} = 1$, so the number of steps per revolution is 4. The most common step size for PM stepper motors is 15° per step or 24 steps per revolution. A two-phase stepper motor with 15° steps would need six permanent magnet poles to its rotor.

26.3 THE VARIABLE RELUCTANCE STEPPER MOTOR

The PM stepper is only one of several ways to create a stepper motor. Figure 26.4 shows an example of what is called a **variable reluctance (VR) stepper motor**. This is also sometimes referred to as a **switched reluctance motor**. **Reluctance** is the opposition that a material offers to the magnetic lines of force, somewhat analogous to magnetic resistance. The VR motor makes use of the fact that a piece of iron placed in a magnetic field will always align in the minimum reluctance position—thus presenting the minimum resistance to the magnetic field.

The rotor of this motor is not a magnet, but is instead made of some magnetically permeable material, typically soft iron, laminated steel, or solid steel. The key to this motor's function is simple magnetic attraction and the relationship between inner and outer teeth on the rotor and stator. Notice how, in the position shown, four of the inner teeth exactly align with teeth and coils on the stator. This is called the minimum reluctance position and it is the

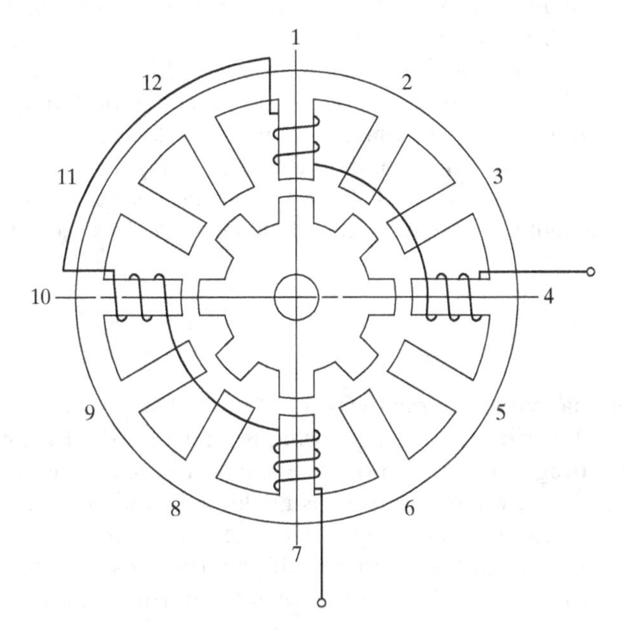

FIGURE 26.4

Construction of a VR stepper motor. (Courtesy of Oriental Motor Co., Ltd.)

position that the rotor will take if the coil wound on poles 1, 4, 7, and 10 is energized. Intuitively, what is happening is that the magnetically permeable material of the rotor is attracted to the electromagnets on the stator. The minimum reluctance point is the closest that it can get in aligning to the magnetic flux field lines.

If we energize the four outer teeth adjacent and to the left of the coils shown, shifting one position counterclockwise (CCW) (i.e., poles 3, 6, 9, and 12), the rotor will rotate *clockwise* (CW) as the rotor teeth align with the closest energized stator teeth. Moving the current to the next set of poles around the stator causes the next step. For the variable reluctance type motor, three phases are the minimum necessary to construct a motor whose direction of rotation can be controlled. While three-phase VR stepper motors are the most common, motors with four and five phases are also available and provide increased numbers of steps per revolution.

Unlike the PM stepper motor, when no power is applied, there is no detent torque in the variable reluctance motor. Without power, there is no electromagnetic field in the stator and with only a permeable metal rotor, there is no magnetic field from a permanent magnet, hence no detent torque.

For VR stepper motors, the number of steps per revolution is a function of the number of teeth on the rotor and the number of phases,

$$S = pN_{\text{teeth}} \tag{26.2}$$

In this example shown in Figure 26.4, there are eight teeth on the rotor and three phases, yielding $3 \times 8 = 24$ steps per revolution (15°/step). As with the PM stepper motor, we can increase the number of steps per revolution by adding teeth to the rotor or phases to the stator, or both.

The variable reluctance motor is finding increased application in situations where position control is not the goal. They are beginning to find favor as an AC motor replacement in applications where efficiency is of great interest. In these applications, the motor is sometimes called a switched reluctance motor and is operated much like a sensorless, brushless DC motor (Chapter 25). This development is fueled by the availability of inexpensive microcontrollers with considerable computing horsepower, and inexpensive MOSFETs. The simplicity of the VR stepper motor is such that manufacturers are very nearly to the point that the cost savings in the manufacturing of the VR stepper motor makes up for the cost of the electronics necessary to drive the motor. The key manufacturing advantage of the VR stepper motor is that its rotor does not require coil windings, unlike a conventional AC motor. It does require some sophisticated control in order to produce smooth output torque, but the knowledge of how to implement that control is now available and the cost of the computer necessary to implement the control is continuously decreasing. The efficiency of the resulting motor can be as high as 90%, a 25–30% energy savings over the typical AC motor. The first applications are in high use, high reliability cycled load devices such as refrigerators and A/C compressors.

26.4 THE HYBRID STEPPING MOTOR

If we combine the permanent magnet in the rotor of the PM stepper motor with the toothed rotor configuration of the VR motor, we get what is known as the **hybrid stepping motor**. Figure 26.5 shows a photograph of the rotor from a hybrid stepper motor.

Unlike the PM design, where the rotor is made up of radially oriented magnets, the hybrid stepper motor's rotor is axially magnetized. There are two sets of teeth on this rotor, one at the north pole and one at the south pole. If you look closely at the teeth on the two ends of the rotor, you will see that the set of teeth at the north pole is offset from the set of

FIGURE 26.5

The rotor of a hybrid stepper motor.

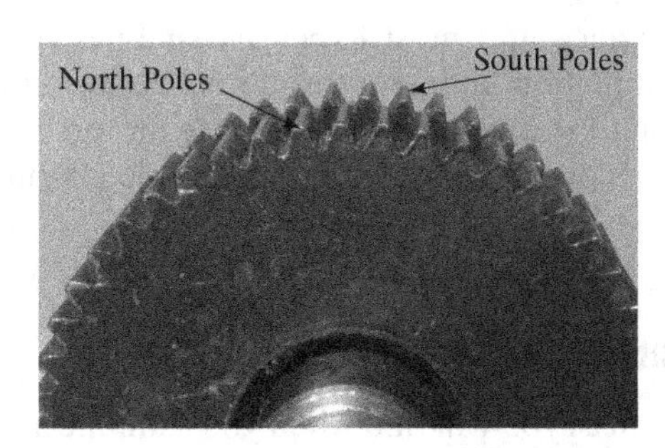

FIGURE 26.6

An end-view of a hybrid stepper motor rotor, showing the offset between the teeth at the north and south poles of the rotor.

teeth at the south pole by half the tooth pitch. This is clearly shown in the end-view shown in Figure 26.6.

Offsetting the teeth in this way improves (decreases) the step size by providing new equilibrium points at twice the raw tooth pitch.

Because the rotor is magnetized, this motor, like the PM stepper motor, has detent torque. You can easily feel this by turning the shaft of such a motor with your hand and noting how the shaft jumps from detent to detent, giving a feeling of roughness to the rotation.

The hybrid stepper motor can be manufactured with much smaller step sizes than a PM motor because it depends on the rotor tooth arrangement rather than the number of magnetic poles for setting the step size. Typical hybrid stepper motors have 200 or 400 steps per revolution (1.8° or 0.9° step sizes).

26.5 COMPARING STEPPING MOTOR TYPES

The least expensive of the stepper motor types is the VR stepper motor. The low cost is due to the fact that the rotor does not contain a magnet. The lack of a magnet has another advantage as well as a significant drawback. Since the rotor does not contain a magnet, it can be made not only less expensively but also lighter (lower mass) than a motor with a permanent magnet rotor. The lower mass allows the rotor to accelerate more quickly for a given torque produced by the motor. The drawback to a variable reluctance motor in a positioning application is that, without a magnet in the rotor, there is no detent torque. This means that when it is not powered, the motor is completely free to rotate due to external forces. This is a distinct liability in a positioning application where removing the power would leave the load completely free to move.

More expensive than the VR stepper motor is the PM stepper motor. These motors trade simplicity and low cost for a relatively large step angle. The most common step angles for PM stepper motors are 15° (24 steps/rev) and 7.5° (48 steps/rev). These motors are most commonly used in situations where there will be a gear stage between the motor and the element driven. The gearing will be such that there will be several revolutions of the motor for each revolution of the output, effectively increasing both the torque and the step resolution as seen at the output. The most common applications in which you are likely to encounter these motors are in optical scanners, where they are used to move the scanning assembly, and ink jet printers, where they move the printhead.

The descriptions of the PM and VR stepper motors that we have presented here are what are referred to as **single-stack motors.** It is possible to gang two or more of these motors onto a single shaft and treat it as a single motor. This would be referred to as a **two-stack motor**. More than two motors may certainly be ganged, as well. Ganging two or more motors onto a single shaft has the advantage of producing more torque at the shaft. Perhaps not so obvious is the fact that if you offset the stacks, it is possible to increase the number of steps per revolution as well.

The highest performance and most expensive stepper motor is the hybrid stepper motor. Because of its combination of detent torque, efficiency, and small step angles, it is the most commonly used form of the stepper motor. These motors find application in fine positioning applications such as precision scanning, diagnostic instrumentation, and laboratory automation systems.

26.6 STEPPER MOTOR INTERNAL WIRING

One of the obvious differences you will find when you examine a selection of stepper motors is that they have differing numbers of wires coming out of them. This reflects the differing wiring and phase (coil) arrangements inside the motors. In the examples that we have been dealing with so far, there have been either two or three sets of coils. In the terminology of stepper motors, these are two-phase and three-phase motors. Each independently controllable coil, even if it is wound on multiple pole pieces, represents a single phase. While two-phase motors are the simplest and most common, it is possible to purchase stepper motors with three, four, five, or more phases. These motors trade increased complexity in the drive electronics for such characteristics as decreased step size and increased torque.

You can tell much about a stepper motor from looking at the wires coming from it. Figure 26.7 shows schematics of some possible wiring configurations that you are likely to encounter.

Figure 26.7a is the kind of two-phase motor that we have discussed in the PM stepper motor example earlier in this chapter (Figures 26.1 and 26.2). It has four wires that connect to two independent coils; hence it is a two-phase motor. Figure 26.7b is a three-phase motor like the VR example (Figure 26.4). In both of these cases, there is a single coil on each pole piece and the winding style is referred to as **monofilar** wound.

The motors in Figures 26.7c and 26.7d are wound (and drawn) very differently from the other motors. To understand the significance of this style of winding, first compare Figures 26.8a and 26.8b.

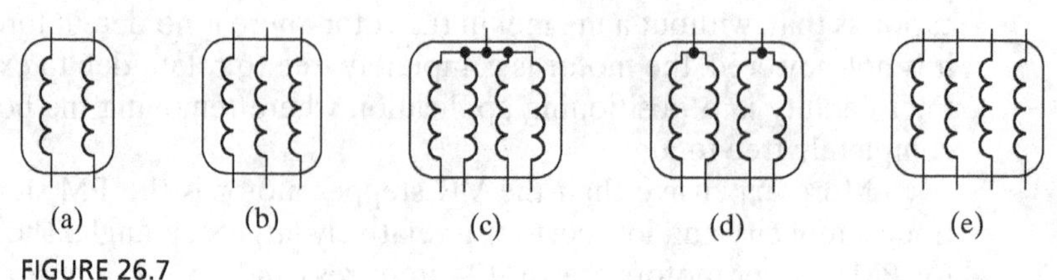

FIGURE 26.7

Wiring schematics for a selection of stepper motors.

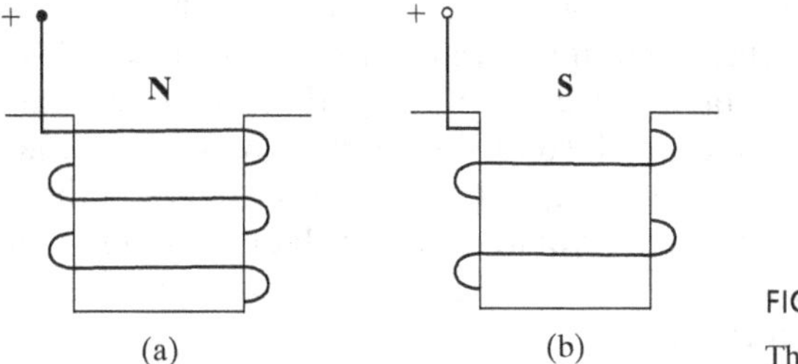

FIGURE 26.8
The two possible winding directions.

This figure shows the two possible ways that you might wind the coil onto a pole piece of a motor. If current were passed through the coil on the left (Figure 26.8a) from top to bottom, the top of the coil would become the north pole. If we passed current from the top to the bottom of the coil on the right (Figure 26.8b), the top of the coil would become the south pole. The direction of the coil winding has reversed the magnetic polarity, even though the direction of current flow was the same.

The motors shown in Figures 26.7c and 26.7d are wound so that windings like those shown in both Figures 26.8a and 26.8b appear on each pole piece. This winding style is referred to as **bifilar** and allows the magnetic polarity of a pole piece to be changed by choosing which of the two coils is energized. To simplify wiring to this style of motor, the input terminals for the whole motor (Figure 26.7c) or for each coil (Figure 26.7d) are brought out on a single wire.

The diagram of the stepper motor depicted in Figure 26.7e is actually ambiguous as to the number of phases. From an electrical standpoint, it clearly has four coils, qualifying as a four-phase motor, but there are two very different internal constructions that might result in this same schematic.

Figure 26.9a shows the internal construction for what is referred to as a **universal wound two-phase stepper motor**. The construction is similar to the bifilar wound motors in that there are two windings on each pole piece. In this case, each pair of windings is wound on four pole pieces. Bringing out both wires from the two windings of each phase (for a total of eight leads) allows a great deal of flexibility in how the motor is driven. A motor like this can be wired to act like a bifilar wound motor or a monofilar wound motor with three different coil resistances (R using a single coil, $2R$ using the two coils in series, and $1/2R$ using the two coils in parallel). There are, however, only two distinct phases, A and B, inside the motor.

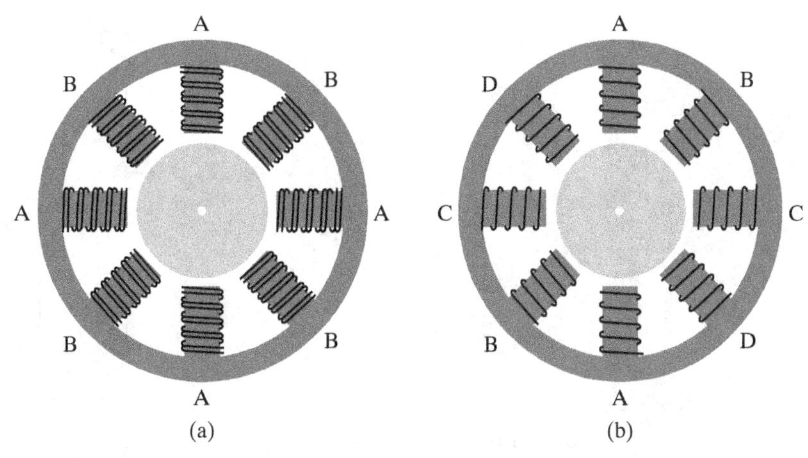

FIGURE 26.9
Comparison of a universal wound two-phase motor (a) and a four-phase motor (b).

Figure 26.9b shows a motor that is electrically equivalent to that in Figure 26.9a but the four windings are wound on four separate pairs of pole pieces. This is a true four-phase motor.

You could develop a procedure for experimentally determining if the motor that you had was a two-phase universal wound motor or a four-phase motor by noting that for a universal wound motor, if you energize the two coils wound on the same pole pieces with current of opposite polarity, they would cancel each other and not create any torque. Since the coils on a four-phase motor all lie on different pole pieces, energizing two coils will always produce a holding torque.

26.7 DRIVING STEPPER MOTORS

Driving stepper motors and getting them to actually take steps involves two distinct parts: the drive electronics that produce the current flows in the coils and the sequencing electronics that determine the order in which the coils will be energized. Let's start by taking a look at the electronics required to produce the current flows.

Figure 26.10 shows the drive circuit required for a two-phase, monofilar wound stepper motor.

In order to produce both magnetic field orientations in each coil, we need to be able to reverse the direction of current flow in each coil. This is similar to the need to reverse polarity to control the direction of a brushed DC motor (Chapter 23). The need to produce a drive of two polarities gives rise to another term used to describe this type of motor: a **bipolar stepper motor**. It takes two full H-bridges to drive the two phases of this motor. In general, bipolar stepper motors require a full H-bridge for each phase.

Figure 26.11 shows the drive circuit required for a two-phase, bifilar wound stepper motor. Either Q_a or Q_b will be turned on to energize phase A. Since the two windings of phase A are wound in opposite directions on the same pole pieces, turning on both Q_a and Q_b at the same time would cause the resulting magnetic fields to cancel, and thus isn't something you'd normally want to do. Since the bifilar windings allow us to change the orientation of the magnetic field without changing the direction of current flow, these motors are often referred to as **unipolar stepper motors**. The bifilar windings reduce the complexity of the drive circuit to four transistors.

Since the drive circuit for the unipolar stepper motor requires half the transistors of bipolar drive, you might ask "Why not simply make all stepper motors as unipolar drive?" To understand the answer to that question, you need to think about how efficiently the windings

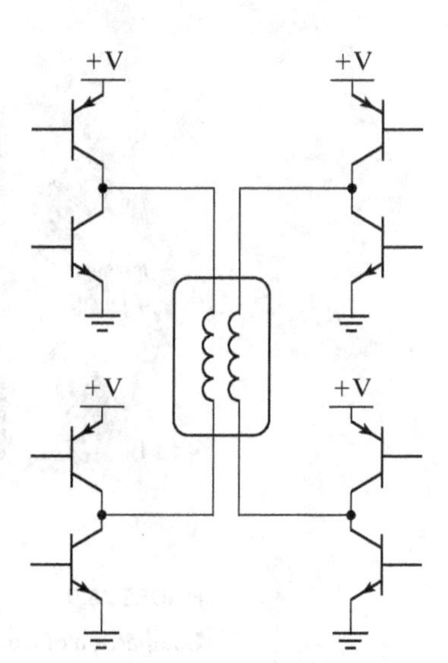

FIGURE 26.10

Drive circuit for a monofilar wound, two-phase stepper motor.

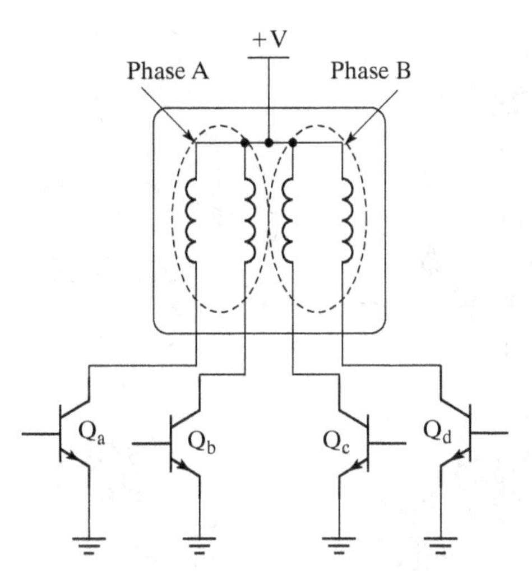

FIGURE 26.11

Drive circuit for a bifilar wound stepper motor.

in these motors are being used. Recall from Chapter 24 that, for a given current, the strength of the magnetic field and therefore the force or torque created is a function of the number of turns in the coil. This would lead us to put as many turns as possible into the available volume within the motor in order to maximize the torque it can generate. In the monofilar wound, bipolar driven stepper motor, all the turns for a given phase are always active. By contrast, in a bifilar wound, unipolar driven stepper motor, there are never more than half the turns active at any one time. As a result, for a given motor design, the torque density (torque produced for a given physical size) will always be better with a bipolar driven stepper motor than for a unipolar driven stepper motor. The bifilar wound stepper motor finds wide use in applications where the ultimate in volumetric efficiency is not required and the lower cost of the simplified drive circuit is important. It is also popular as a standard motor configuration because of the drive flexibility that it provides. In that respect, the six-wire configuration of Figure 26.7d offers an advantage over the similar configuration of Figure 26.7c. Because the central lead of the two coils has been brought out separately, the six-wire stepper motor can be treated as a bipolar motor with two independent coils by ignoring the two center taps. Motors like this can be driven either in a unipolar mode (as shown in Figure 26.11) or in a bipolar configuration (like that shown in Figure 26.10). The same stepper motor driven in a bipolar drive circuit with the same input electrical power will typically produce 30–40% more torque than that motor driven with the same input power in a unipolar drive circuit. This is a direct result of engaging more of the coils in the production of the magnetic field. The drawback to using the six-wire motor in a bipolar drive circuit is the significant (4x) increase in coil inductance that that circuit produces. As you will see in Section 26.12, this has a detrimental effect on the maximum stepping speed.

26.8 STEPPING SEQUENCES FOR STEPPER MOTORS

While the drive circuits will provide current flow through the coils of the stepper motor, it is the sequence in which those coils are energized that determines whether the motor rotates and moves to a new position or simply oscillates back and forth between two adjacent positions. By choosing the sequencing of the coils, we can control both the size of the step that will occur and the resting position in the new step location.

The simplest drive sequencing follows the example that we used to introduce the PM stepper motor (Figure 26.12).

This drive technique, with only one phase energized at a time, is known as **wave drive**. Figure 26.13 shows how the transistors are energized to implement wave drive for both bipolar and unipolar stepper motors.

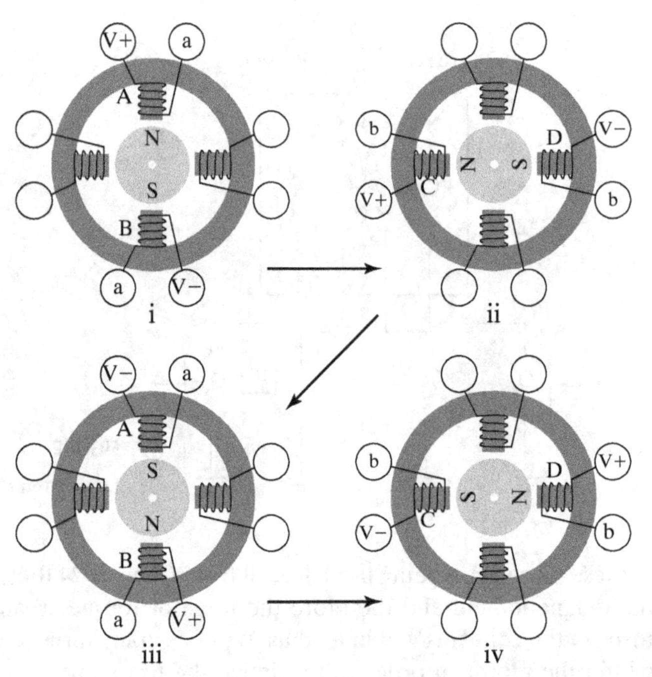

FIGURE 26.12

Stepping sequence for a PM stepper motor with wave drive (one phase on at a time).

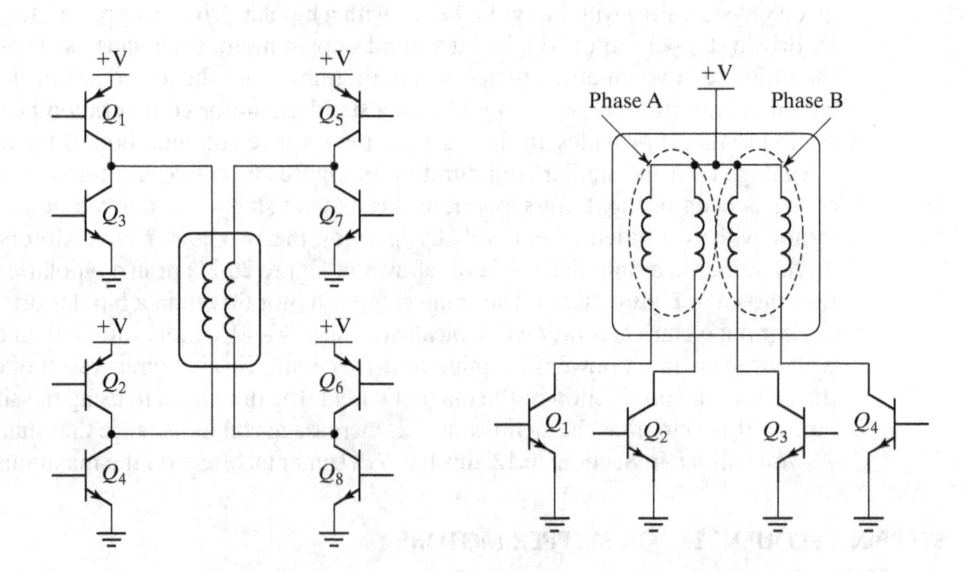

Bipolar Drive (CW ↓)

Step	Q_1+Q_4	Q_2+Q_3	Q_5+Q_8	Q_6+Q_7
1	On	Off	Off	Off
2	Off	Off	Off	On
3	Off	On	Off	Off
4	Off	Off	On	Off
1	On	Off	Off	Off

Unipolar Drive (CCW ↑)

Step	Q_1	Q_2	Q_3	Q_4
1	On	Off	Off	Off
2	Off	Off	Off	On
3	Off	On	Off	Off
4	Off	Off	On	Off
1	On	Off	Off	Off

FIGURE 26.13

Energizing sequences for wave drive.

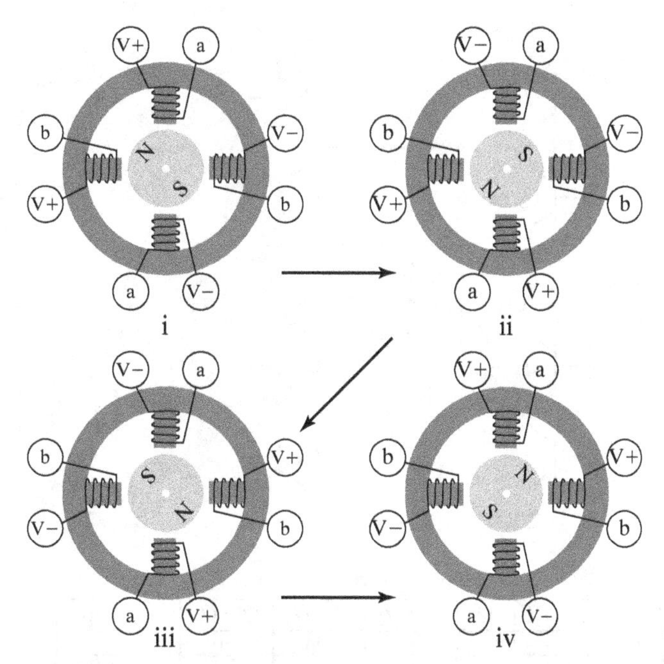

FIGURE 26.14

Stepping sequence for a PM stepper motor with full-step drive (two phases on at a time).

If the transistors are energized in the sequence reading down the table from top to bottom, the motor will rotate in a clockwise direction. If they are energized in the sequence reading up the table from bottom to top, rotation will be in the counterclockwise direction.

The most common drive mode for stepper motors is referred to as **full-step** drive. This drive mode involves energizing both of the phases at the same time. Figure 26.14 shows the rotor behavior when using full-step drive.

The figure shows that full-step drive moves through the same step angle (in this case, 90°) as the wave drive configuration. What is different is the equilibrium position at each step. The equilibrium position for full-step drive is shifted by half the step angle relative to the equilibrium position for wave drive.

What you can't see from the diagrams is that there is also a difference in the torque produced by two-phase-on drive. With two phases energized at the same time, the current drawn is twice that of one-phase-on and that might lead you to conclude that you would get twice as much torque. However, because the equilibrium position is not as closely aligned with either pole piece, you do not get the full benefit of twice the current. The actual torque produced by two-phase-on drive is greater by a factor of 1.4 as compared to the one-phase-on drive. It is worth noting here that one of the benefits of going with a larger number of phases is that you can energize more phases at the same time. For example, with a five-phase stepper motor, you can operate with three phases on and obtain 1.6 times as much torque as with a single phase on.

Figure 26.15 shows how the transistors are energized to implement full-step drive for both bipolar and unipolar stepper motors.

It is possible to combine both wave drive and full-step drive into a single sequence. If we do this, we find that the rotor position alternates between the equilibrium positions for full-step drive and the equilibrium positions for wave drive. For this drive technique, when the motor has taken two steps, it has only traversed the angular displacement of a single full-step.

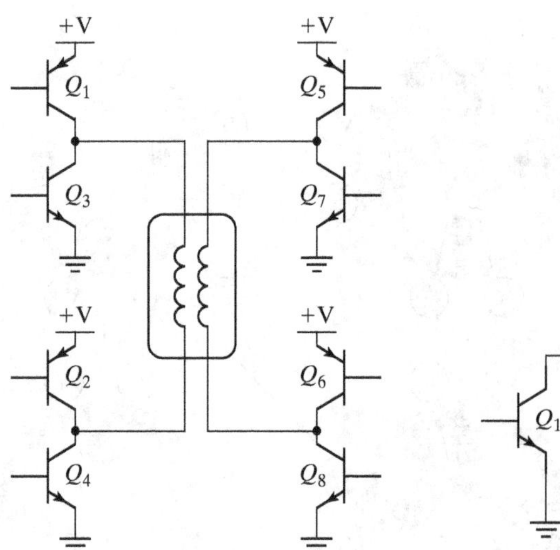

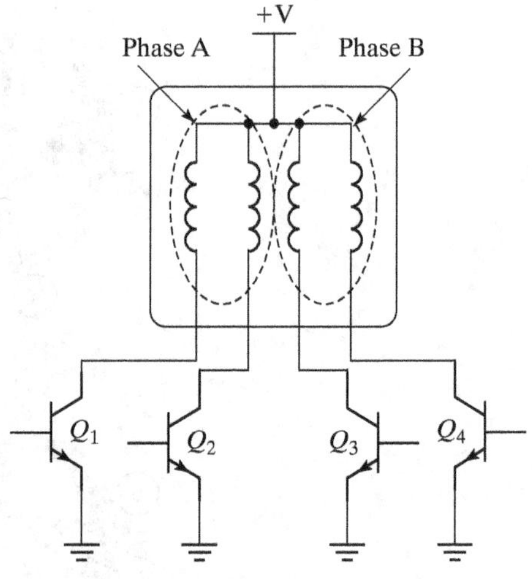

Bipolar Drive (CW ↓)

Step	Q_1+Q_4	Q_2+Q_3	Q_5+Q_8	Q_6+Q_7
1	On	Off	On	Off
2	On	Off	Off	On
3	Off	On	Off	On
4	Off	On	On	Off
1	On	Off	On	Off

Unipolar Drive (CCW ↑)

Step	Q_1	Q_2	Q_3	Q_4
1	On	Off	On	Off
2	On	Off	Off	On
3	Off	On	Off	On
4	Off	On	On	Off
1	On	Off	On	Off

FIGURE 26.15

Energizing sequences for full-step drive.

As a result, this drive mode is known as **half-step** drive and, for a given motor, produces twice as many steps per revolution. Figure 26.16 shows the energizing sequence for half-step drive.

Half-step drive inherits characteristics from both full-step and wave drive. In particular, it exhibits the torque characteristics of both modes. When a single phase is energized and the rotor is at a wave drive equilibrium position, it exhibits the torque characteristics of wave drive. When two phases are energized and the rotor is at a full-step position, it exhibits the larger torque of the full-step mode. Because the torque available oscillates between these two values, designers must content themselves with designing to the lower torque level of the wave drive mode when using the half-step drive mode. For a given torque requirement, it will be necessary to use a larger motor if it will be operated in half-step mode.

We can combine the half-step drive mode with pulse width modulation (PWM) to produce the last drive mode that we will examine. To understand this mode, take a look at Figure 26.17a.

This is the normal rotor position for full-step or two-phase-on drive. Figure 26.17b shows what happens to the rotor position if we change the drive to the phase wound on pole pieces A and B so that it is driven with PWM at a 50% duty cycle. The magnetic field associated with that phase has been weakened due to the decreased current flowing in the coils, causing the rotor to rotate to a new equilibrium closer to the direction of the C-D pole pieces. We have created a new equilibrium position between the wave drive position and the full-step position. By sequencing the drive to a phase through steps of OFF → 50% → 100% → 50% → OFF, rather than simply ON → OFF, we can double the number of steps again, creating, in effect, a quarter-step drive mode. This process can be extended to arbitrarily small changes in duty cycle, creating what is known as **microstepping**. This approach can produce thousands of "microsteps" per revolution with a motor having only a few hundred

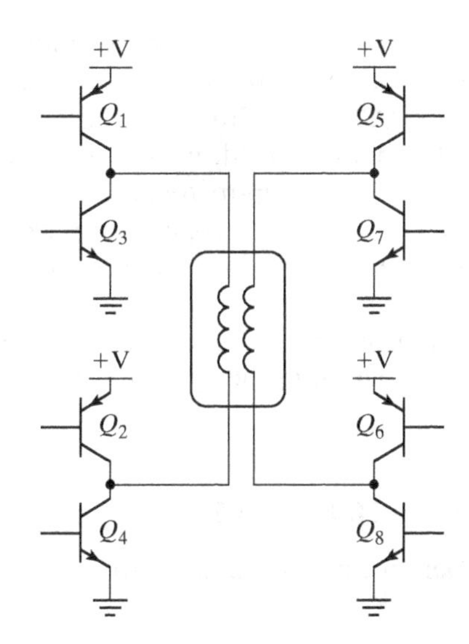

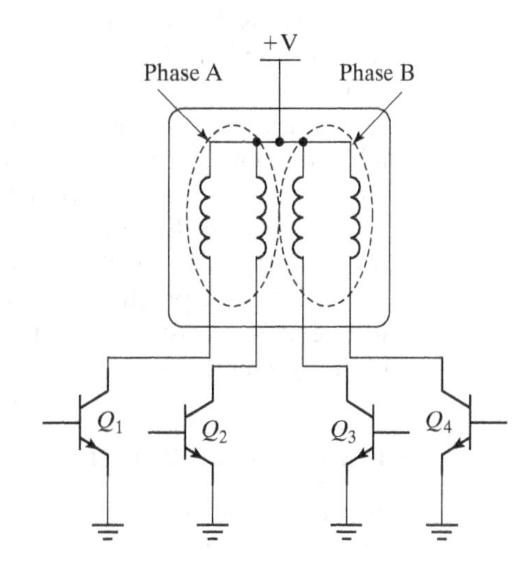

Bipolar Drive (CW ↓)

Step	Q_1+Q_4	Q_2+Q_3	Q_5+Q_8	Q_6+Q_7
1	On	Off	On	Off
2	On	Off	Off	Off
3	On	Off	Off	On
4	Off	Off	Off	On
5	Off	On	Off	On
6	Off	On	Off	Off
7	Off	On	On	Off
8	Off	Off	On	Off
1	On	Off	On	Off

Unipolar Drive (CCW ↑)

Step	Q_1	Q_2	Q_3	Q_4
1	On	Off	On	Off
2	On	Off	Off	Off
3	On	Off	Off	On
4	Off	Off	Off	On
5	Off	On	Off	On
6	Off	On	Off	Off
7	Off	On	On	Off
8	Off	Off	On	Off
1	On	Off	On	Off

FIGURE 26.16

Energizing sequences for half-step drive.

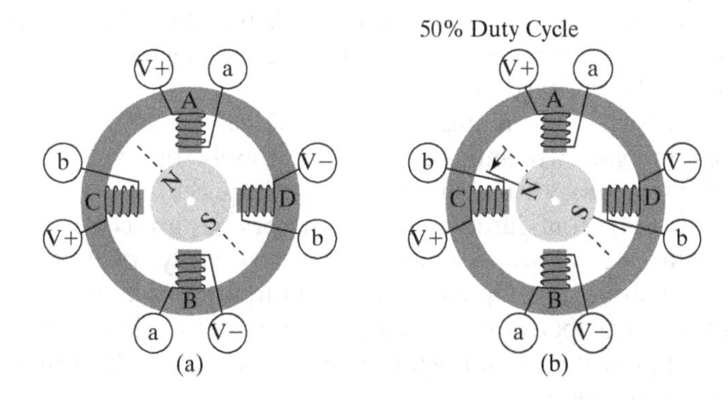

FIGURE 26.17

The effect of PWM on the position of the rotor.

natural steps per revolution. The main drawback to this approach is that as the steps are made smaller, the torque available to move from one step to the next falls. The incremental torque available to move from one microstep to the next is

$$T_{\text{inc}} = T_{\text{hold}} \sin\left(\frac{90}{N}\right) \tag{26.3}$$

where T_{inc} is the incremental torque available, T_{hold} is the stepper motor holding torque, and N is the number of microsteps per physical step. If you created a 16 microsteps per step system, the available torque would be only 9.8% of the holding torque of the motor.

While any stepper motor can be microstepped, in motors specifically designed for microstepping, particular attention is paid to the step-to-step uniformity of the torque vs. displacement characteristic. To achieve equal angular steps, it is necessary to change the drive duty cycle according to a sine function. Any deviations from a sine function in the motor response (and there will always be at least some) will reduce the accuracy of the microsteps. The number of practical microsteps that a system can resolve will depend on the native torque characteristics of the motor, the friction in the system, and the torque requirements of the system the motor is driving.

26.9 GENERATING THE DRIVE SEQUENCES FOR STEPPER MOTORS

At a conceptual level, driving a stepper motor requires a controller with two basic elements, shown in Figure 26.18.

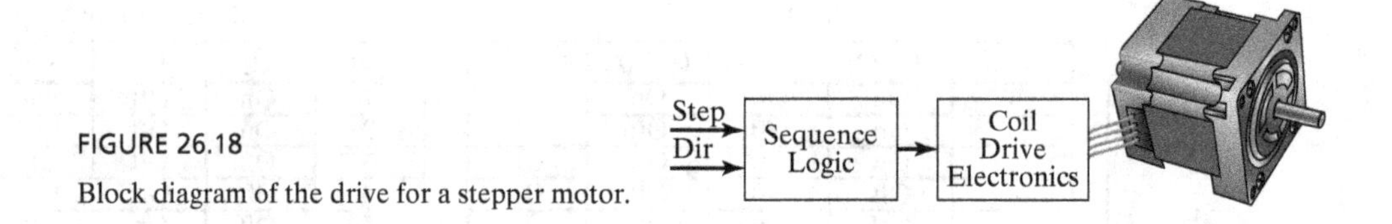

FIGURE 26.18

Block diagram of the drive for a stepper motor.

The sequencing logic determines and outputs the logic-level drive commands that are required to turn on and off the various phases of the stepper motor to obtain the desired rotation of the rotor. The logic output will depend on the preferred drive mode (full-step, wave drive, and so on) and other inputs that indicate when it is time to cause the motor to take a step and the desired direction of rotation. The Coil Drive Electronics block contains the power electronics needed to switch the relatively high levels of current required by the motor, typically ranging from a few hundred milliamps to a few amps. This block is responsible for generating the current flows through the coils based on the logic commands from the sequencing logic. In the last section, we covered the sequencing logic, now we will move on to generating the coil drive currents.

As we saw above, when examining the wiring of stepper motors, there are two basic drive configurations: bipolar drive and unipolar drive. Bipolar drive requires a full H-bridge for each phase of the motor, while unipolar drive requires two transistors per phase. Integrated circuits that are configured as two full H-bridges are commonly available, and appropriate for driving bipolar stepper motors. Similarly, arrays of transistors suitable for providing the current drive to unipolar stepper motors are widely available. It is also possible to purchase ICs to perform the sequencing logic function, either as stand-alone components or as complete component with the integrated drive electronics to provide the high current outputs to the motor.

Figure 26.19 shows an example of a sequencing chip, the STMicroelectronics L297, connected to a dual H-bridge chip, the L298, to drive a two-phase bipolar stepper motor.

By contrast, both the logic and current drive functions are integrated in the ON Semiconductor MC3479, a driver for bipolar stepper motors requiring up to 350 mA/phase. For providing the current drive to a unipolar stepper motor, there are a range of IC transistor arrays and devices, known as peripheral drivers, available. Examples include the ULN2003 (Chapter 17) suitable for currents up to 350 mA/phase and the SLA7032 suitable for up to 1,500 mA/phase.

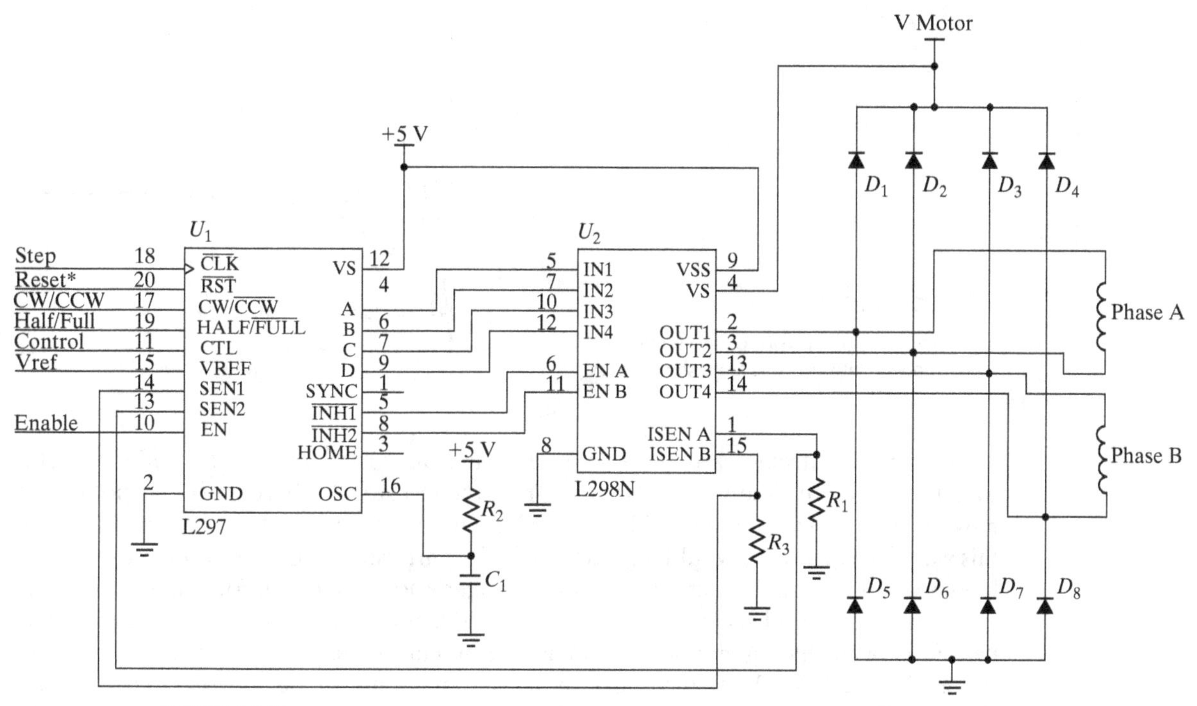

FIGURE 26.19

The L297/L298 combination used to drive a stepper motor.

In many applications that include a microcontroller, there are sufficient resources, in terms of I/O lines and code space, to generate the sequencing logic in software. In these cases, all that is necessary to drive the stepper motor will be the coil drive electronics. A few special purpose microcontrollers exist, such as the Freescale MM908E625, that integrate a microcontroller and power electronics into a single package so that the microcontroller can directly drive the motor.

26.10 STEPPER MOTOR DYNAMICS

An ideal stepper motor would move instantaneously from one step to the next and stop at the new position with no overshoot and no settling time. Real stepper motors are not so perfect. Because of the inertia of the rotor and the fact that the forces are applied through magnetic fields that have compliance, the movement from position to position does not achieve this ideal. When a real stepper motor is commanded to move from one position to the next, the rotor is first accelerated toward the new position, then decelerated as it moves past the intended position, and then accelerates back toward the intended position. This process continues with smaller and smaller excursions from the intended position until it finally settles at the new position. If we plotted the angular position vs. time for a single step, it looks like Figure 26.20. Though no pun is intended here, this behavior is typically referred to as the **step response** of a system. The system exhibits a rise time, overshoot, a damped response, a decaying oscillation, and a settling time. These are general industry terms used to describe the response of a wide variety of dynamic systems. In this application, it is literally the *step* response for a stepper motor.

As long as we are only attempting to step very slowly, these dynamics will not have a noticeable detrimental effect on the operation. As we seek to step the motor faster and faster, we will begin to take the next step before the motion of the last step has completely

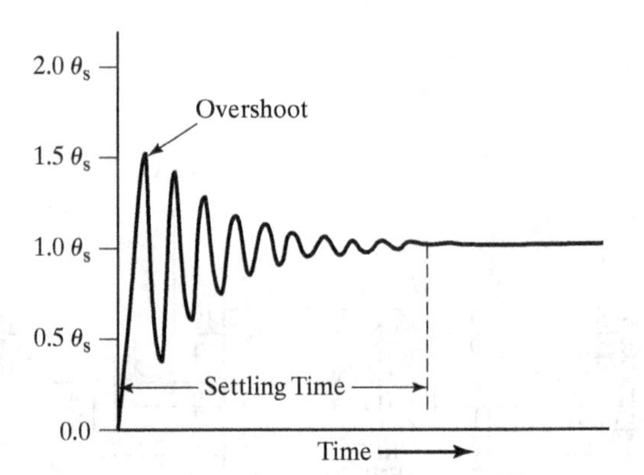

FIGURE 26.20
The dynamics involved in a stepper motor's rotor moving from step to step.

settled. In the extreme case, we would be asking the rotor to take a second step before it even had a chance to complete the move from the last step. In this case, it may not be able to move to the second step and will end up skipping one or more of the commanded steps. If this were to happen, we would have defeated the purpose of using a stepper motor for open-loop position control because we would no longer know where the rotor was positioned.

The details, such as frequency and amplitude, of the oscillations will depend on a number of factors including motor and load inertia, and mechanical and electrical damping in the system. Figure 26.21 shows the effect on the step of adding an inertial load to the motor. The primary effect that we see is the reduction in the natural frequency of the oscillation due to the added inertia.

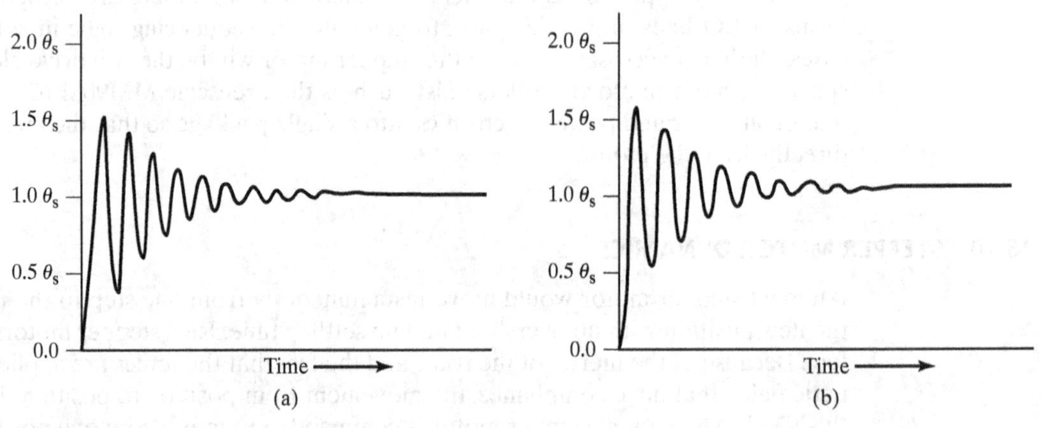

FIGURE 26.21
Change in step response due to added inertia: (a) with no added inertia and (b) with 150 g-cm^2 added inertia. (Courtesy of Oriental Motor Co., Ltd.)

Figure 26.22 shows the effect of adding a friction load to the motor with inertial load combination from Figure 26.21b. The friction load increases the damping in the system, resulting in much less overshoot and quicker settling.

Damping can also be added by changing the way that the motor is driven electrically. Figure 26.23 shows the difference in step response between one-phase-on (wave drive) and two-phase-on (full-step) drive modes.

The one-phase-on step response shows higher amplitude and longer ring down of the oscillations, indicating less damping than two-phase-on drive. The reason for the increased

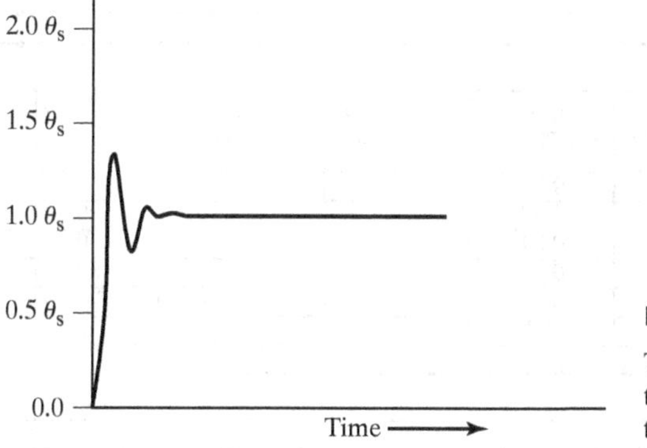

FIGURE 26.22

The effect of adding friction damping to the system of Figure 26.21b. (Courtesy of Oriental Motor Co., Ltd.)

damping in two-phase-on drive can be understood by recalling the back-EMF effect described in Chapter 22 (brushed DC motors). As the magnetic fields associated with the permanent magnet in the rotor move past the coils of the stator, they induce voltages in those coils. Those induced voltages are in opposition to the external drive voltages that might be applied. The applied voltages are resisting the velocity-induced effects of back-EMF. Because the amplitude of the back-EMF reaction is related to the velocity of the moving magnetic field, the effect is to increase the damping in the system.

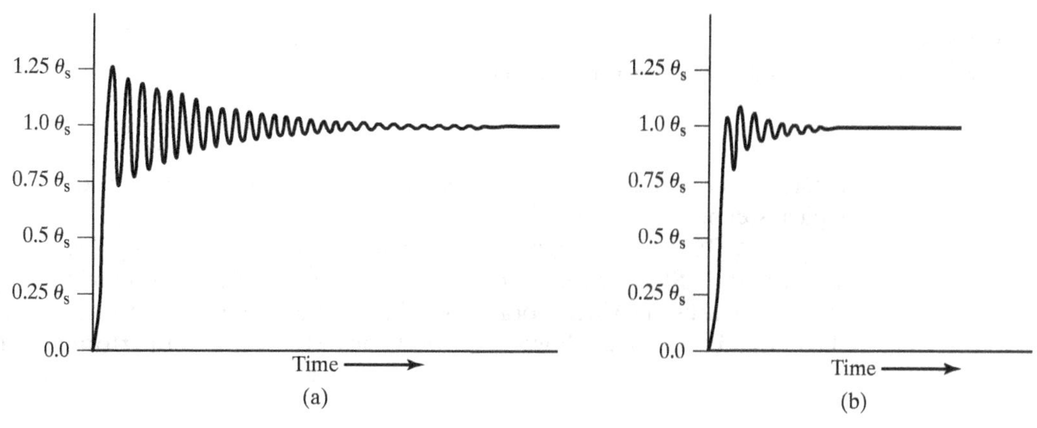

FIGURE 26.23

Effects of one-phase-on and two-phase-on drive on the dynamic step response. (Courtesy of Oriental Motor Co., Ltd.)

26.11 STEPPER MOTOR PERFORMANCE SPECIFICATIONS

The most fundamental specification for a stepper motor is its step size. In addition to the step size, the manufacturer will quote step error and positional accuracy. These are often associated with an entire family of motors, since they result from the motor design and manufacturing capabilities.

The step error is the maximum angular deviation between an actual step angle and the theoretical, ideal step angle for that motor. It is important to understand that this specification is for a single step; therefore, it is not additive across multiple steps.

The positional accuracy is a measure of the maximum possible angular position error between any two step positions. The positional accuracy is defined as the difference between the largest positive step error and the largest negative step error. For example, if you were at a step that exhibited a +0.04° step error and moved to a position that exhibited a –0.02° step

Motor data											
Motor order number	Type	Nominal voltage (V)	Current (A/ Phase)	Resistance (O/ Phase)	Inductance (mH/ Phase)	Holding torque (mNm)	Detent torque (mNm)	Rotor inertia (g-cm²)	Overall length (L=mm)	Weight (g)
1423-022-40343	2SQ-022BA34	2.2	0.55	4	5.5	95	4	19	34	195
1423-006-40343	2SQ-060BA34	6	0.67	9	10	120	4	19	34	195
1423-091-40343	2SQ-091BA34	9.1	0.24	38	74	140	4	19	34	195
1423-012-40343	2SQ-120BA34	12	0.48	25	27	150	4	19	34	195
1425-063-40343	2SQB-063BA34	6.25	0.25	25	35	71	4	18	34	195
1425-032-40343	2SQB-032BA34	3.2	0.7	4.5	5.3	80	4	18	34	195
1425-055-40343	2SQB-055BA34	5.5	0.36	15	18	70	4	18	34	195

General data	
Step angle	1.8° (2SQ), 3.6° (2SQB)
Positional accuracy	± 5% max. noncumulative
Number of phases	2
Temperature rise	70 deg. max.
Insulation resistance	50 MO min. at 500VDC
Dielectric strength	AC500V for one minute
Insulation class	B
Lead wire size	AWG #22 L= 150m/m
Number of leads	4
Operating ambient temperature	-20°C to +50°C
Material of brecket	Aluminium

FIGURE 26.24

Basic specifications for a typical hybrid stepper motor. (Courtesy of Portescap.)

error, the positional accuracy between those two steps would be 0.06°. Figure 26.24 shows a typical stepper motor data sheet.

Like DC motors, stepper motor characteristics are typically described with torque vs. speed curves. Stepper motor curves are not as simple as permanent magnet DC motor curves, and must also incorporate more information to fully characterize the performance of the motor. Figure 26.25 shows the prototypical stepper motor performance graph.

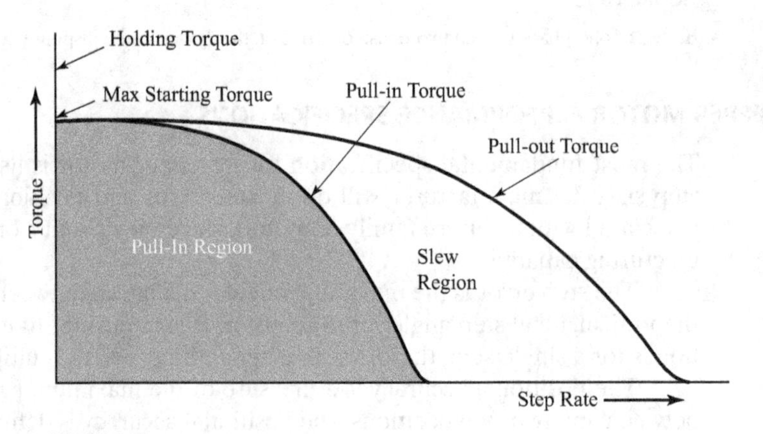

FIGURE 26.25

Speed vs. torque for a stepper motor.

The first thing that stands out is that the graph shows two lines for a single motor. The lower curve, which bounds the **pull-in region**, sometimes labeled as the **start-stop region**, indicates the range of torque-speed points where the motor can be started and stopped at will without missing steps. For torque-speed combinations in this region, the motor can be instantaneously taken from stopped to stepping at a given rate.

The upper curve defines the **pull-out torque** line, which shows the upper limit of torque-speed points that the motor can achieve. Operation to the right or above this line is not possible without missing steps. The region between the pull-in region and the pull-out torque line is referred to as the **slew region**. The motor can be operated in this region, but it may not be stopped or started in this region without losing steps. To operate in the slew region, the motor must be started in the pull-in region and then accelerated into the slew region. To stop a motor that is operating in the slew region, the motor must first decelerate back into the pull-in region.

Notice that the holding torque is somewhat higher than the maximum starting torque. This difference results from the fact that some torque is necessary to accelerate the motor to the new position. The motor can hold a given position against an opposing torque, but not have enough excess torque to move to a new position against that same torque.

Another prominent feature of the curves in Figure 26.25 is the steep fall-off in torque at the higher step rates for both the pull-in region and pull-out torque. To understand what causes that effect, you need to remember that the coils of the motor have significant inductance and that you cannot instantaneously start and stop current flow through an inductor. Figure 26.26 shows the current rise in the coil of the motor, which correlates with the production of torque.

At the slower stepping rates, the current has ample time to rise to the maximum value where the motor produces the best torque. As the step rate is increased, the rise time of the current in the inductance of the coil limits the peak value that the current attains before a new step is requested, and therefore the torque is reduced. The faster the step rate, the shorter the time available for the current to rise and the lower the torque produced.

$$\tau = \frac{L}{R} \tag{26.4}$$

As we saw in Chapter 9, for an inductor, the rise time of the current is characterized by a time constant given by L/R, where L is the inductance of the coil and R is the resistance of the winding.

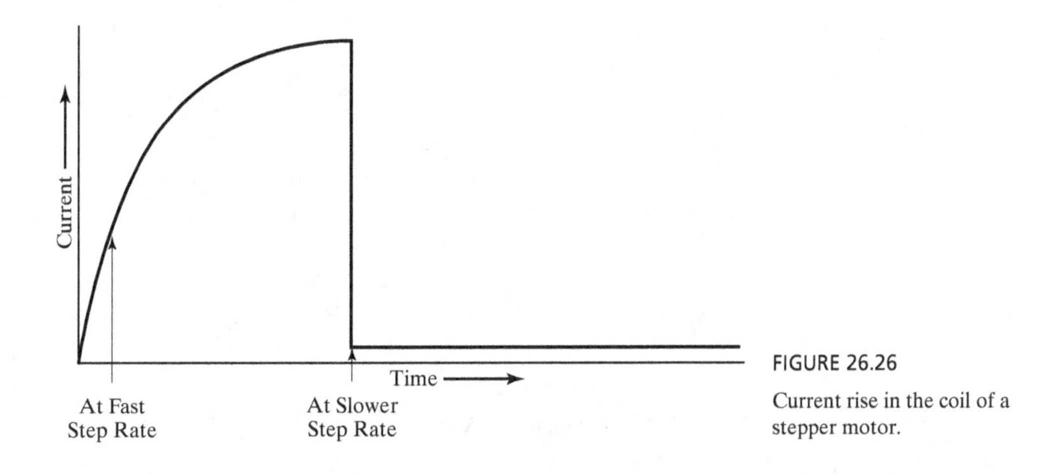

FIGURE 26.26

Current rise in the coil of a stepper motor.

26.12 OPTIMIZING STEPPER MOTOR PERFORMANCE WITH DRIVE ELECTRONICS

If we could find a way to decrease the rise time of the current in the motor coils, we could allow the motor to produce increased torque at higher step rates. One additional benefit of this improved rise time is that the maximum stepping rate will be increased—often dramatically. If we look at the expression for the time constant of the current rise, we see that we could reduce the time constant by reducing the inductance of the motor or increasing the resistance of the coils. Neither of these is possible for a motor that we already have in hand. What we could do, though, is to increase the resistance in the coil *circuit* by adding a series resistor external to the motor, as in Figure 26.27.

This configuration is known generically as ***L/nR*** **drive**, often shortened to just ***L/R*** **drive**. The most common values for the external resistance are one and three times the resistance of the coil, yielding a factor of 2 or 4 reduction in the time constant. In order for this approach to work, the excitation voltage to the motor, shown as +V in Figure 26.27, must be raised so than at steady state, the same current flows through the winding of the motor. If this were not done, then the speed and holding torque would be reduced, resulting in worse performance than we started out with. In the case of *L/2R* drive, twice the original voltage is required to maintain the original current. For *L/4R* drive, this would mean raising the voltage to four times the original voltage.

To understand why adding this external resistance results in a faster rise time, consider what happens on a very short time-scale after the voltage is applied (Figure 26.28).

At the instant the voltage is applied (Figure 26.28a), there is no current through the external resistor or the resistance of the coil, and the full voltage appears across the inductor, driving the rise in current. Since with *L/2R* or *L/4R* drive we are applying a larger voltage, it stands to reason that the rate of current rise would be higher than it would be at a lower voltage without the external resistance. As the current starts to rise (Figure 26.28b), a voltage drop will develop across the external and internal resistance, reducing the voltage that appears across the inductance and reducing the rate of current rise. At steady state (Figure 26.28c), the added resistance effectively cancels the increased drive voltage and the same current flows through the motor with a simple drive or *L/2R* or *L/4R*.

The drawback to this approach—and it is a significant drawback—is higher power consumption. If the same current, *I*, flows at steady state in all cases, then the power dissipated in

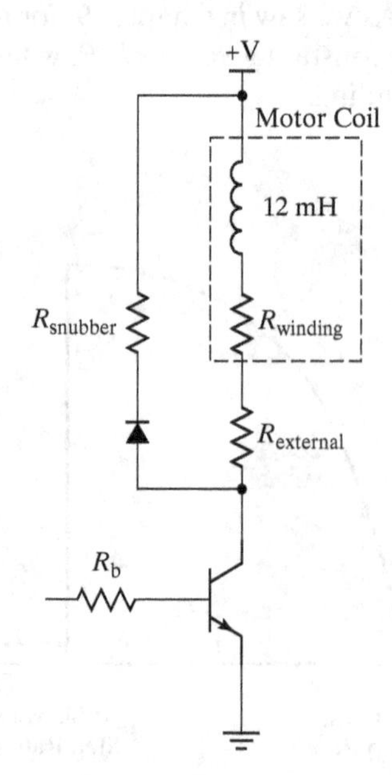

FIGURE 26.27
Circuit for *L/nR* drive.

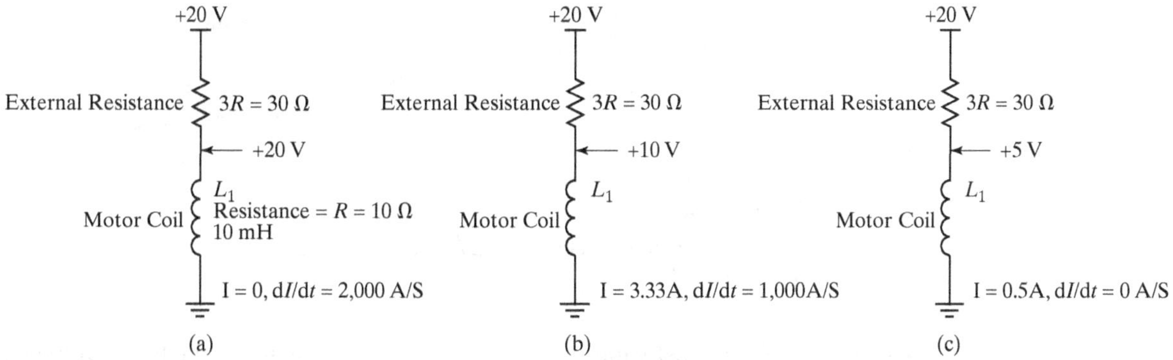

FIGURE 26.28

Voltages as the current builds.

the simple case would be VI. For $L/2R$ drive, it would be $2VI$ and for $L/4R$ drive, it would be $4VI$. At steady state, all of the extra power dissipation would be wasted as heat in the external resistors. But despite the power dissipation issues, this is a popular solution for improving stepper motor performance because it is so simple to implement. Figure 26.29 shows the kind of improvements that are possible by going to $L/2R$ or $L/4R$ drive, as compared to drive with no external resistance. Notice that there is no change in holding torque. At steady state, the same current flows in all three cases, therefore, the steady state holding torque will be the same.

If the increased power dissipation for $L/2R$ or $L/4R$ drive presents a problem in a particular application, but you would still like to improve the high speed performance of the stepper motor, there are other techniques available that trade increased complexity in the electronics for improved efficiency.

We could achieve a similar improvement in high-speed torque with less power consumption, if we could change the drive voltage to the motor coils. We could use a high drive voltage to get the current to rise to the desired level quickly, then drop the drive voltage back to a lower level to maintain that current. This approach is known as **dual voltage drive**. It is a viable approach when the drive voltages necessary are already available in the system. However, if the lower drive voltage needs to be created from the higher voltage via a linear regulator, all that we will have achieved is having moved the power dissipation from the resistors in $L/2R$ drive to the linear regulator of the lower voltage power supply.

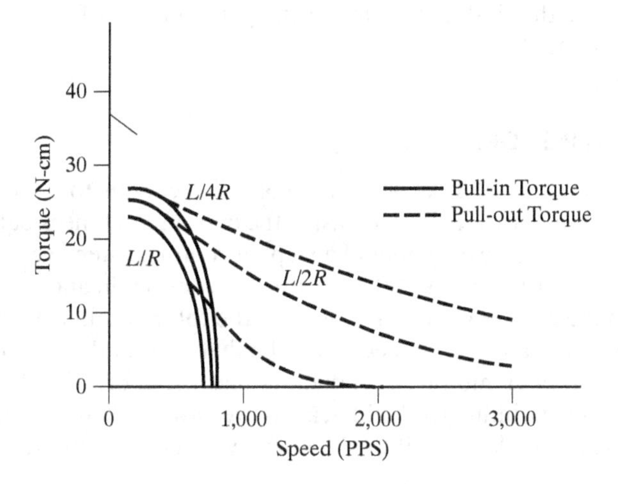

FIGURE 26.29

Comparing simple (L/R) drive with $L/2R$ and $L/4R$ drive. (Courtesy of Oriental Motor Co., Ltd.)

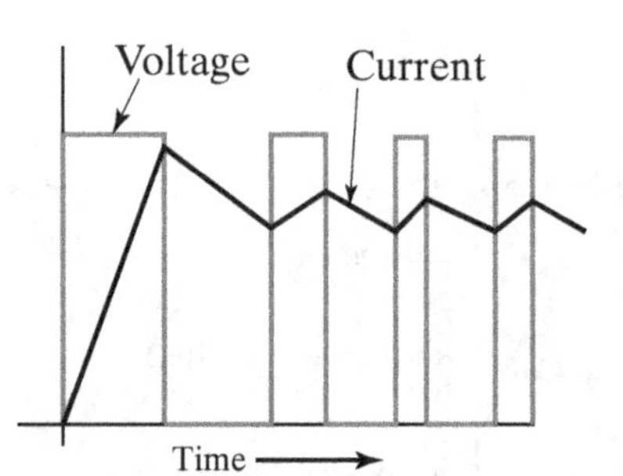

FIGURE 26.30

Voltage and current waveforms with chopper drive.

An alternative, and vastly more power efficient, way to achieve the lower voltage is by modulating the duty cycle of the higher drive voltage. In this case, the lower voltage is actually zero—the higher drive voltage is switched between on and off. This is known as **chopper drive** and is essentially identical to the chopper drive peak and hold solenoid drive technique discussed in Section 24.3. The voltage and current waveforms are shown in Figure 26.30.

When first turned on, the full voltage is applied to the coil, resulting in a rapid current rise. Because there is no external resistor in the circuit, the rapid rate of current rise continues until the desired coil current is reached. When the current reaches the desired level, the full-drive voltage is pulsed on and off to create the desired average current level in the coil. Because we are not using a resistor to dissipate the excess power but instead leveraging the time averaging ability of the inductor, the efficiency of this approach is substantially better than *L/nR* drive. The absence of the resistor also results in a more rapid rise in the coil current than is possible with *L/nR* drive. Thus, replacing a simple drive with a chopper drive can yield factors of 3–10 improvement in the maximum stepping rate possible with a motor.

There are two approaches that are commonly used to control the modulation in chopper drive systems. The simplest is to rely upon an open-loop approach, and to calibrate the initial on time and sustaining duty cycle. If the specific motor that will be driven and the drive voltage are known in advance, then the rise time of the current and the necessary on-off duty cycle can be determined a priori and built into the control electronics. If the specific motor and drive voltage are not known, then a closed-loop approach is required, and the current through the motor must be sensed. The current measurement would be used to determine when to transition from continuous drive to the duty cycle drive and also to dynamically adjust the duty cycle necessary to achieve the desired current limit (see Figure 26.19).

While chopper drive is much more power efficient than *L/nR* drive, the high frequency switching of the drive voltage necessary to modulate the current causes this approach to generate more electrical noise. This noise is an issue that needs to be dealt with at the system level in order to ensure that it does not impair the performance of other parts of the system or other nearby equipment.

26.13 THE ROLE OF SNUBBING IN PERFORMANCE

While the content of Section 26.12 concentrated on drive circuits to improve the rise time of the current in the stepper motor coils, decreasing the current fall time is equally important. As we saw in Section 26.8, when we command a stepper motor to step, we must energize a new coil as well as cause an existing current flow to stop (or reverse) in another coil. While current continues to flow in the coil we wish to de-energize, the rotor is held in its current position. To achieve the best stepping rate, we will need to make that current decay as quickly as possible.

The coils of a stepper motor are inductors and therefore the drive electronics will require protection against inductive kickback. As we saw in Chapter 23, there are a range of choices of circuits available to limit the kickback voltage. The simplest of those circuits is

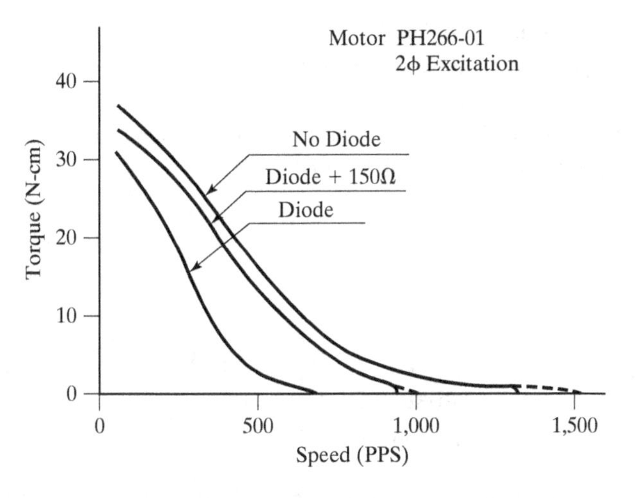

FIGURE 26.31

Stepper motor performance with different snubber circuits. (Courtesy of Oriental Motor Co., Ltd.)

a reverse-biased diode placed across each coil. In the case of the stepper motor, this solution has the undesirable characteristic of the longest decay time for the current. During this decay time, the motor will be held in its current position by the continued current flow through the coil. To shorten the decay time and achieve the maximum stepping rate, we want to use a snubbing circuit that minimizes the decay time while still protecting the drive electronics. In Chapter 23, we introduced a number of methods for achieving this. Figure 26.31 shows the improvement in the performance of a stepper motor possible by using a snubber circuit with a faster decay time.

26.14 HOMEWORK PROBLEMS

26.1 Show the three different wiring configurations with three different coil resistances for a universal wound motor driven in a bipolar drive circuit.

26.2 Show how you would wire a universal wound motor to use it in a unipolar drive circuit.

26.3 If a stepper motor were specified as having a maximum step error of ±1°, what is the worst case positional accuracy for that motor?

26.4 Name the three types of stepper motor construction described in this chapter.

26.5 For the motor shown in Figure 26.32,

(a) how many phases does the motor shown below have?

(b) is it bipolar or unipolar?

(c) draw a basic schematic that shows how to connect the motor. For your answer, use a representative power supply and generic transistors.

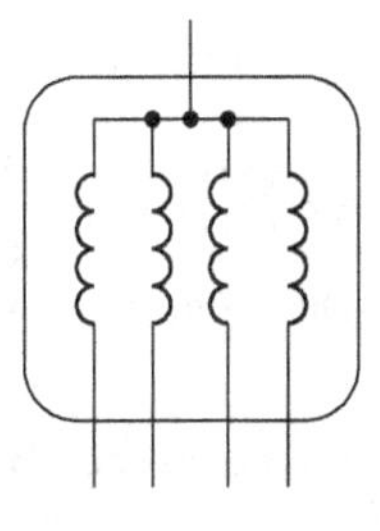

FIGURE 26.32

Motor for Problem 26.5.

26.6 For the motor shown in Figure 26.33,

(a) how many phases does the motor shown below have?

(b) is it bipolar or unipolar?

(c) draw a basic schematic that shows how to connect the motor. For your answer, use a representative power supply and generic transistors.

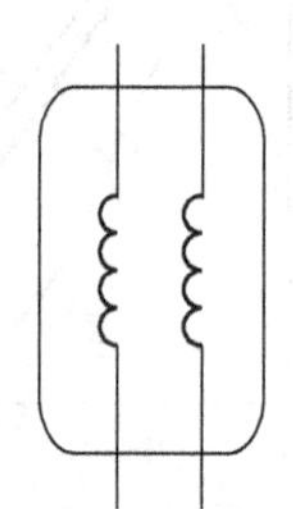

FIGURE 26.33

Motor for Problem 26.6.

26.7 The motor shown in Figure 26.34 can be wired and used in two different configurations.

(a) Describe the two different configurations (how many poles, bi- or unipolar, bifilar or not bifilar)?

(b) Draw a basic schematic showing the two ways the motor can be wired. For your answer, use a representative power supply and generic transistors.

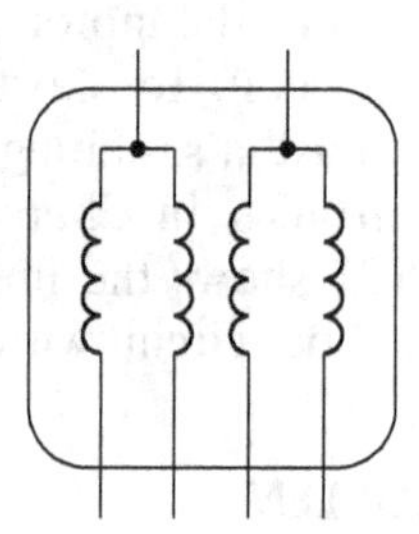

FIGURE 26.34

Motor for Problem 26.7.

26.8 For the motor whose torque-speed characteristics are shown in Figure 26.35, answer the following questions:

(a) What is the maximum step rate?

(b) If the motor is initially not rotating, will it be possible to immediately begin stepping the motor at a rate of 200 pps (pulses per second) while generating 20 mNm of torque?

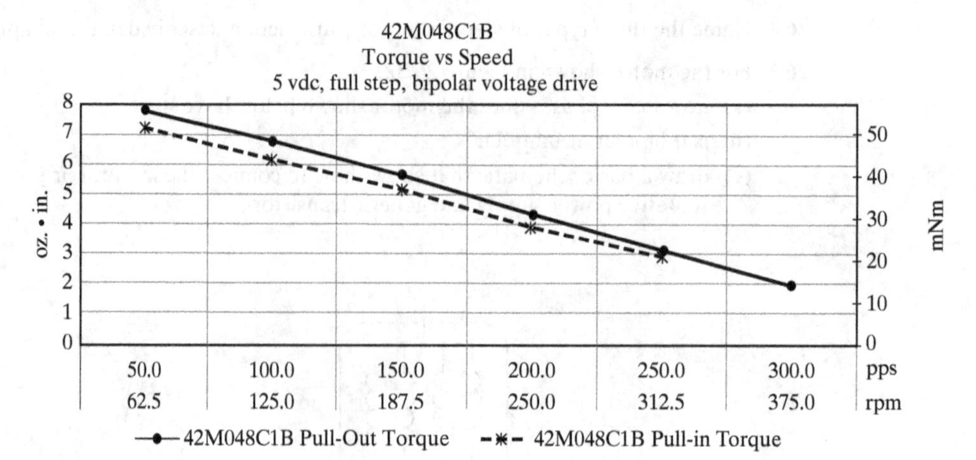

FIGURE 26.35

Motor specifications for Problem 26.8. (Courtesy of Portescap.)

(c) What is the maximum starting torque?

(d) What is the fastest the motor can be stepped while generating 40 mNm of torque?

(e) How much torque can the motor generate if it immediately begins stepping at 250 pps?

(f) What is the maximum starting step rate?

(g) A motor is required to generate 20 mNm of torque and step at a rate of 250 pps. Describe a control scheme that allows you to achieve the desired performance.

26.9 Name the three types of drive modes for stepper motors. For each type of drive mode, state whether the torque produced is consistent from step to step, or if it varies. If it varies, state why it varies.

26.10 Which motor characteristic(s) limit its maximum step rate?

26.11 The maximum step rate for a motor is related to the rate at which current begins to flow in the motor's coils and the rate at which it stops flowing. For the 1423-012-40343 motor whose specifications are given in Figure 26.24,

(a) calculate the characteristic rise time, τ, of the current in the motor's coils.

(b) calculate the new value of τ, when you implement L/R drive by adding an external resistance of three times the motor's internal coil resistance. What drive voltage is needed to achieve the same steady-state current that would result without L/R drive when the motor is driven with its rated voltage? How much power is dissipated (i.e., wasted) in the external resistors?

(c) if you implement chopper drive with a drive voltage that is three times the rated voltage for the motor, how much time is required for the current to reach the level achieved in the characteristic rise time calculated for part(a)?

26.12 Show how you would wire the stepper motor from Figure 26.7d to be driven using bipolar drive.

FURTHER READING

"Stepper Motor Handbook," Airpax Corporation, 1982.

Stepping Motors and their Microprocessor Controls, Kenjo, T., 2nd ed., Oxford University Press, 1994.

"Technical Information on Stepping Motors," Oriental Motors U.S.A., Corp.

CHAPTER 27

Other Actuator Technologies

27.1 INTRODUCTION

In the preceding chapters, we have focused exclusively on actuators that generate force or motion by harnessing interactions between magnetic fields and electric currents. Permanent magnet brushed DC motors (Chapters 22 and 23), solenoids (Chapter 24), permanent magnet brushless DC motors (Chapter 25), and stepper motors (Chapter 26) are all variations on this theme. However, it is not always the case that an actuator from this list will best meet a design challenge, and an alternate means of actuation must be sought. For instance, a solenoid might not be the best choice for an application where linear motion with a 0.0001 in. stroke is required. And both brushed and brushless DC motors are heavy because of the large coils and permanent magnets they incorporate: what if you needed actuators that are much less massive? There are many other physical phenomena that have been put to use in actuators to accomplish useful work, and in this chapter, we survey several of them.

In this chapter, you will learn about:

1. pneumatic and hydraulic systems,
2. piezo actuators,
3. shape memory alloy (SMA) actuators.

27.2 PNEUMATIC AND HYDRAULIC SYSTEMS

A means of actuation fundamentally different than any we've discussed up to this point in the text is derived from pressurized fluids. The working fluids may be either a gas, such as air in pneumatic systems, or liquids, such as hydraulic oil in hydraulic systems. The theory of operation of fluid drive systems is the same, though, regardless of the phase of the working fluid: pressure differentials are established to create flow, and create useful force and motion in the process.

A schematic diagram of a typical fluid actuation system is depicted in Figure 27.1. The system is comprised of a compressor, reservoir or tank, and a pressure regulator feeding working fluid at a controlled pressure to a valve and an actuator. The compressor, reservoir, and regulator serve to ensure that the pressure and flow characteristics of the working fluid are reasonably well-known and constant, so that actuating the solenoid valve gives the desired result at the actuator. The actuator shown in Figure 27.1 is a piston that provides linear motion. The force available from the actuator is the product of the working fluid pressure and the piston's surface area, minus losses from friction: $F = P \times A - F_f$.

For a typical pneumatic piston actuator, the working pressure might be 20 psi and the piston diameter might be 1 in., resulting in a force of 15.7 lb. Piston actuators can be easily obtained with a **stroke** (linear range of motion) ranging from a fraction of an inch up to

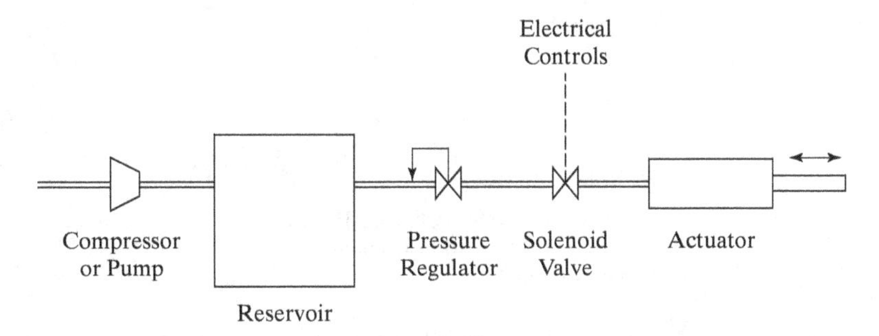

FIGURE 27.1

Schematic representation of a fluid drive system that incorporates a solenoid valve to control the working fluid.

about 10 in. Also, because of the limited number and simplicity of the moving parts, pneumatic systems can survive very large numbers of cycles before wearing out. Solenoid valves are often rated for tens to hundreds of millions of actuation cycles, and as long as everything is kept clean, the seals on actuators can provide very long service life as well.

In contrast to pneumatic systems, where pressures don't normally exceed 100 psi, the working pressures of hydraulic systems can be much higher: 3,000 psi is typical. This can result in far greater actuation forces, and hydraulic actuators are capable of creating thousands of pounds of force over very long strokes. Along with these much higher working pressures, the components must be capable of withstanding high pressures and loads, and it becomes critical to ensure that the working fluids don't leak. Whereas air leaking from a pneumatic system isn't usually too dangerous, 3,000 psi fluid shooting out of a leaking hydraulic component can be a memorable (and hazardous) event!

27.2.1 Solenoid Valves

Admittedly, this category of actuator still makes use of electromagnetics to create motion: a standard solenoid is used to actuate a **solenoid valve**. From an electrical standpoint, these are much the same as actuators we have covered in the preceding chapters. The difference here is that the solenoid valve is a component within a larger system comprising an actuator, and within these systems, solenoid valves are used to modulate the pressure and flow of the fluid that does the real work. The effort required to control the solenoid valve is small compared to the forces and motion that can be generated by the working fluid and its associated actuators.

The theory of operation of solenoid valves is refreshingly straightforward: a linear solenoid moves a plunger or spool relative to one or more orifices in order to control the flow of fluid through a valve body. The routing of the fluid and the resulting actions can get complex, but at the most basic level, the actuation of a solenoid valve doesn't get significantly more complicated.

27.2.1.1 Direct Acting Solenoid Valves

The simplest example of a solenoid valve is a direct acting 2-way valve. For the **direct acting** category of solenoid valve, the solenoid's plunger acts directly as the valve's metering component, and the force to actuate it must overcome the working pressure in the system to either open or close (depending on whether the valve is normally open or normally closed—both are readily available and commonplace). An example of a 2-way direct acting solenoid valve is shown in Figure 27.2.

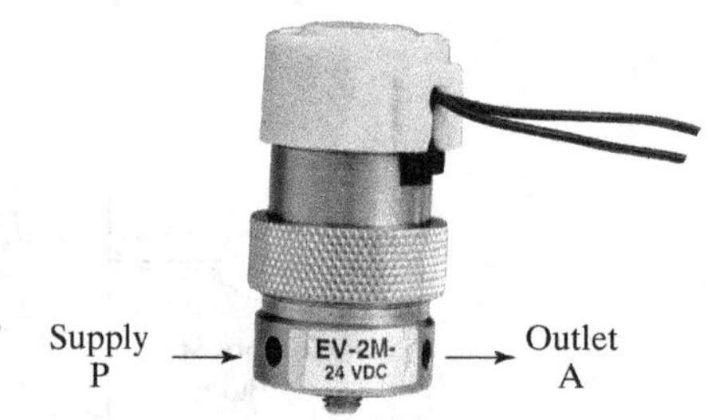

FIGURE 27.2

2-way solenoid valve (Clippard EV-2). (Courtesy of Clippard Instrument Laboratory Inc.)

A **2-way** valve simply allows or stops the flow of the working fluid from the pressure port (labeled "P") to the outlet, or actuator, port (labeled "A"). A useful electrical analogy is that of a momentary single pole, single throw (SPST), ON-OFF electrical switch, which either makes or breaks a single circuit.

Figure 27.3 shows a cutaway view of a normally closed, 2-way direct acting solenoid valve that illustrates how it works. On the left, the valve is shown in the closed configuration. In this state, no current flows through the solenoid coil, and plunger is held tight against the valve seat by the fluid pressure from port P and the core spring. When the solenoid is energized, the plunger is pulled up toward the center of the solenoid coil, allowing flow from the pressure port (P) to the actuator port (A). To do this, the solenoid must overcome the force from the core spring and the fluid pressure acting on the plunger.

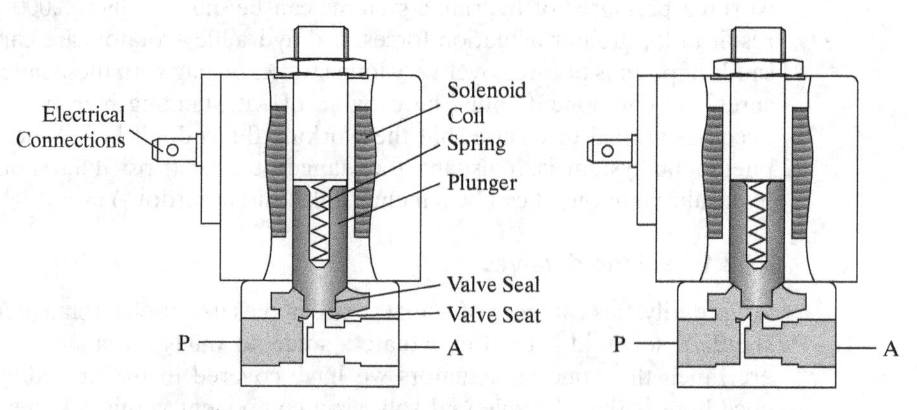

FIGURE 27.3

Cutaway view of a normally closed, direct acting 2-way solenoid valve.

Figure 27.4 represents the states of the 2-way solenoid valve in a simpler schematic format. On the left side, the valve body blocks the path between the inlet (P) and outlet (A) ports, and on the right, the valve body is in a position that allows flow.

Carrying the analogy to electrical switches further is helpful: just as switches are available with many combinations of poles and throws (e.g., single pole/double throw, double pole/double throw, and so on), equivalent solenoid valves are often available. For example, a **3-way solenoid valve** is analogous to a single pole, double throw (SPDT) ON-ON electrical switch, and controls the flow so that the valve outlet port connected to the actuator is

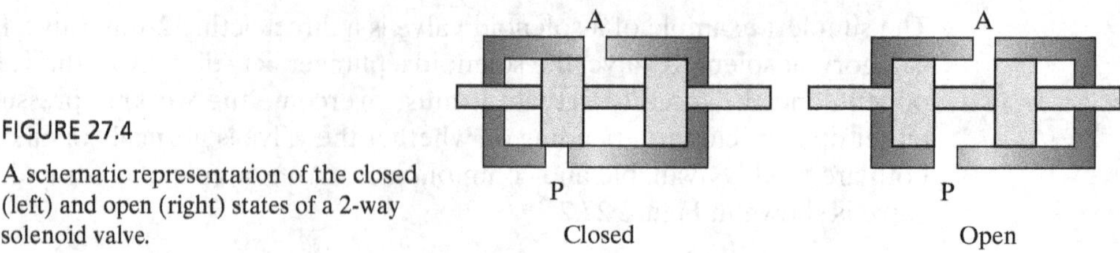

FIGURE 27.4

A schematic representation of the closed (left) and open (right) states of a 2-way solenoid valve.

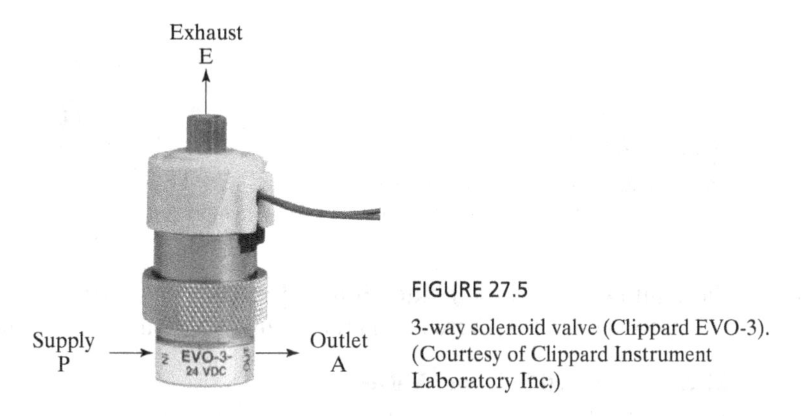

FIGURE 27.5

3-way solenoid valve (Clippard EVO-3). (Courtesy of Clippard Instrument Laboratory Inc.)

switched between the supply inlet port and an exhaust port. An example of a 3-way direct acting solenoid valve is shown in Figure 27.5.

This type of valve is used to pressurize the actuator when the solenoid valve is energized, and to release the pressure and relax or retract the actuator when the valve is de-energized. A cutaway view of a normally closed, direct acting 3-way solenoid valve is shown in Figure 27.6, where the inlet is labeled port P, the outlet is port A, and the exhaust is port E. When the solenoid is off, the plunger is pressed against valve seat 1 and pulled away from valve seat 2 by a spring at the plunger's center. Thus, the valve between the pressure port (P) and the actuator port (A) is closed, while the valve between the actuator port and the exhaust port (E) is open. The exhaust port is typically open to atmospheric pressure, so this has the effect of venting the actuator. When the solenoid is enabled, the plunger is pulled toward the middle of the solenoid coil, which reverses the states of valves 1 and 2: the pressure port is connected to the actuator port, and the connection between the actuator port and the exhaust port is closed.

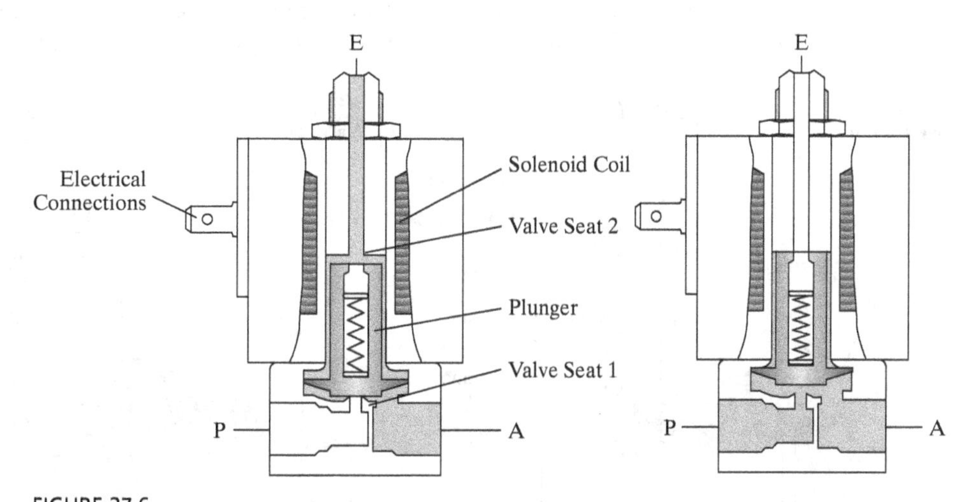

FIGURE 27.6

Cutaway view of a normally closed, direct acting 3-way solenoid valve.

Figure 27.7 is a schematic representation of this type of valve, and shows how the outlet/actuator port (A) is connected to either the pressure inlet port (P) or to the exhaust port (E), depending on the position of the plunger.

There are a great variety of solenoid valve constructions, with 2-way and 3-way valves representing the simplest examples. Solenoid valves with 4-way valves are also common, and

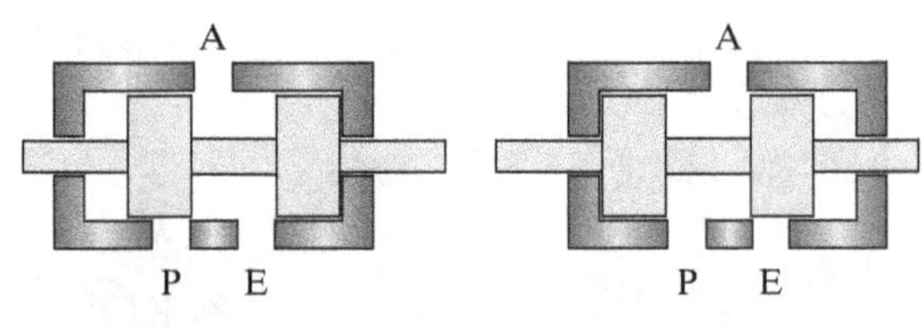

FIGURE 27.7

A schematic representation of the exhaust (left) and pressurizing (right) states of a 3-way solenoid valve.

these allow for powered extension and retraction of actuators. While these are the most common valve types, there are a great many other variations available.

27.2.1.2 Piloted Solenoid Valves

For applications requiring high pressures and/or flow rates, the resulting forces on a direct acting valve solenoid's plunger become very large. Increasing pressure or surface area requires ever-larger solenoids, which eventually reaches a practical limit with either size or the amount of current required to generate the necessary forces. Instead of the direct approach, **piloted solenoid valves** cleverly harness the pressure of the working fluid itself to create the forces needed to open and close. Piloted valves employ a small solenoid valve, called the **pilot valve**, to control the application of the working fluid's pressure across a diaphragm at the **main valve** orifice. The diaphragm incorporates perforations that allow a small amount of fluid to flow to the pilot valve. Consider the cutaway view of a piloted 2-way solenoid valve shown in Figure 27.8.

For the valve shown, when the pilot valve's solenoid coil (in the upper right part of the diagram) is not energized, the plunger is pressed against the valve seat by the plunger's spring and the pressure of the working fluid. In this case, the pilot valve is closed, and the pressure from the working fluid introduced at the high-pressure port (P) acts on the top of the diaphragm at the main valve orifice, holding it closed. A compression spring also applies downward force at the top of the diaphragm to ensure that the valve stays closed when the pilot valve's solenoid isn't energized. In this case, no flow is allowed to the outlet port A.

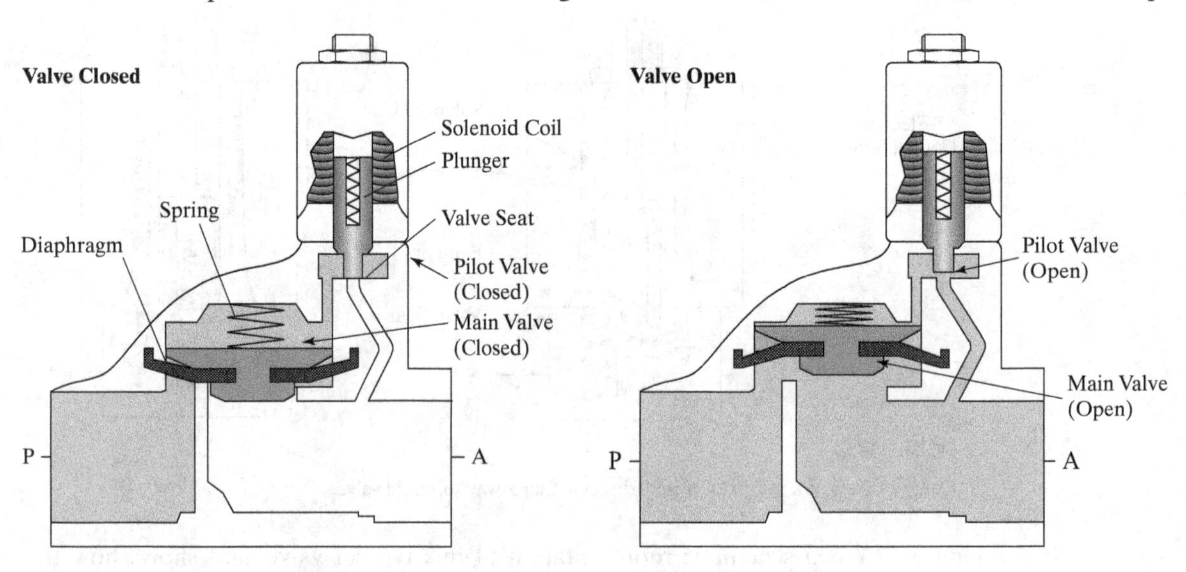

FIGURE 27.8

Cutaway view of a normally closed, piloted 2-way solenoid valve.
Left: Solenoid off, valve closed (no flow). Valve plunger is down, closed tight against valve seat.
Right: Solenoid on, valve open (flow). Plunger is up, pulled away from valve seat by solenoid.

When the pilot valve's solenoid is energized and opened, a small amount of fluid is allowed to flow, which is sufficient to pressurize the underside of the main valve's diaphragm. A pressure drop occurs across the orifices in the diaphragm and the pilot valve as fluid flows through the pilot valve. The resulting pressure acting on the top side of the diaphragm is lower than the pressure acting on the bottom side from the high-pressure port, and this causes the valve to open. In this way, a small solenoid can be used to control a large pressure and flow. Keep in mind that piloted valves have a high-pressure port and a low-pressure port, and that these must be hooked up correctly in order for the valve to function properly. They only adhere to the SPST switch analogy when the ports are properly connected. This is not always the case with direct acting solenoid valves, which can often be hooked up in either orientation. The schematic representation of the flow control across this piloted 2-way valve is the same as that shown for the direct acting 2-way valve in Figure 27.4.

27.2.2 Servo Valves

Solenoid valves can be switched between off and on, but are not intended to be operated at intermediate positions. They're great for such on/off applications, but there are many applications that require flow of a working fluid to be metered and adjusted somewhere between fully on and fully off. This is the job of the **servo valve**, also called an **actuator valve** in the industry. This type of valve is most commonly used in process control industries, and is not frequently used to control the flow of working fluids to standard pneumatic or hydraulic actuators.

The name "servo valve" indicates closing a feedback loop around a parameter of interest, usually the position of the valve element that meters flow, and implementing closed-loop feedback control to achieve fast and accurate response of the valve. Figure 27.9 shows a classical feedback control block diagram representation of a servo valve. Other parameters that can be fed back and controlled include pressure and flow rate, as well as the velocity or acceleration of the valve's metering components.

Such valves do not necessarily need to function under automatic feedback control: they may also be run open loop (without an automated controller). In this case, the "actuator valve" term is more apt, as the result is simply a metering valve whose metering element is controlled by an actuator, rather than manually.

Servo valves are supplied by a number of manufacturers, and a variety of constructions are available. One popular design is essentially the same as a solenoid valve, with a solenoid's plunger coupled to the valve's metering component. In this type of application, the solenoid is arranged to act against a relatively stiff spring which provides the return force needed to return the valve to its de-energized state. Activating the solenoid with 100% of rated current will fully actuate the valve (either fully open or fully closed, depending on the valve type), while supplying it with values between 0% and 100% will cause it to move to a corresponding intermediate position. **Pulse width modulation** or **PWM** (see Chapter 23), is typically used to achieve intermediate levels of current, and is created by rapidly pulsing the supply voltage on and off while varying the percentage of the time the voltage is on. This is usually more efficient and convenient than creating a continuously variable voltage supply for the solenoid. For example, a particular servo valve may be powered by a fixed 24 V

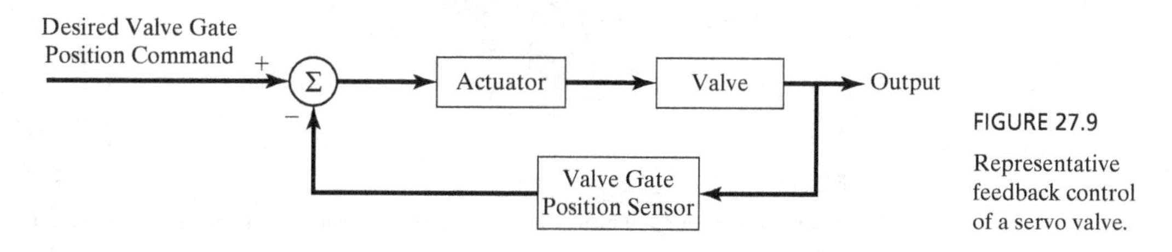

FIGURE 27.9

Representative feedback control of a servo valve.

power supply, and in order to get the valve to open halfway, the supply is turned on and off at a **PWM frequency** of 10 kHz, and during each cycle, it will be on 50% of the time and off the remaining 50% of the time. The percentage of time that the signal is "active" (in this case, current flowing through the solenoid) is the **duty cycle**.

Other common designs of servo valves include:

- Dual-solenoid valves, in which one solenoid actuates the valve's metering components in one direction, and the other pushes it back in the opposing direction.
- Gear motor servo valves (Figure 27.10), in which a DC or AC motor's output shaft is geared down to increase the torque and reduce the speed to a range appropriate for the application. The position of the valve's metering component is measured directly with a potentiometer or LVDT (linear variable differential transformer).
- Torque motor servo valves, in which an arrangement of opposing rotary solenoids is used to rock a lever arm and actuate a valve's metering components.

The best design for a given application depends mostly on the flow rates and speed of response required. The control electronics that are coupled to the valve's actuators are also a major concern for the designer, and interfacing these components to the rest of a system can be one of the most challenging parts of incorporating a servo valve into your system. Since servo valves are most common in process control industries, the standards that have evolved for integrating the controllers are best suited to these applications.

27.2.3 Pneumatic and Hydraulic Actuators

Controlling the flow in pneumatic or hydraulic systems with valves allows you to contemplate doing useful work, but it's only half the battle: to finish the job, you also need actuators. In this section, we present a few of the most common options.

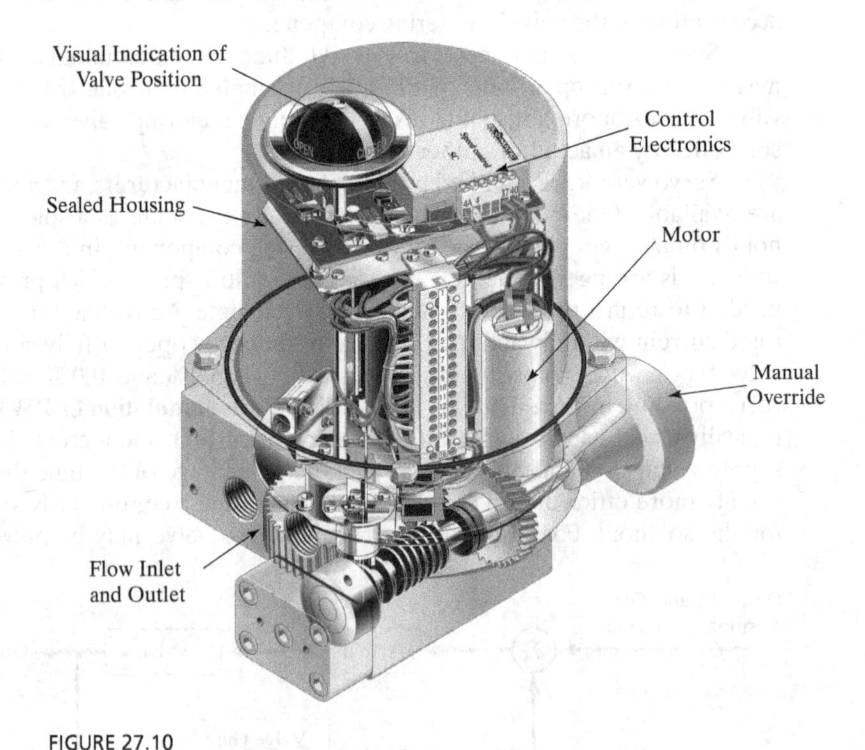

FIGURE 27.10

A servo valve actuated by a geared motor. (Courtesy of Emerson Valve Automation.)

Linear actuators: The most common pneumatic and hydraulic linear actuators are based on the principle of pressure acting upon the face of a piston within a cylinder, resulting in the generation of force and linear motion.

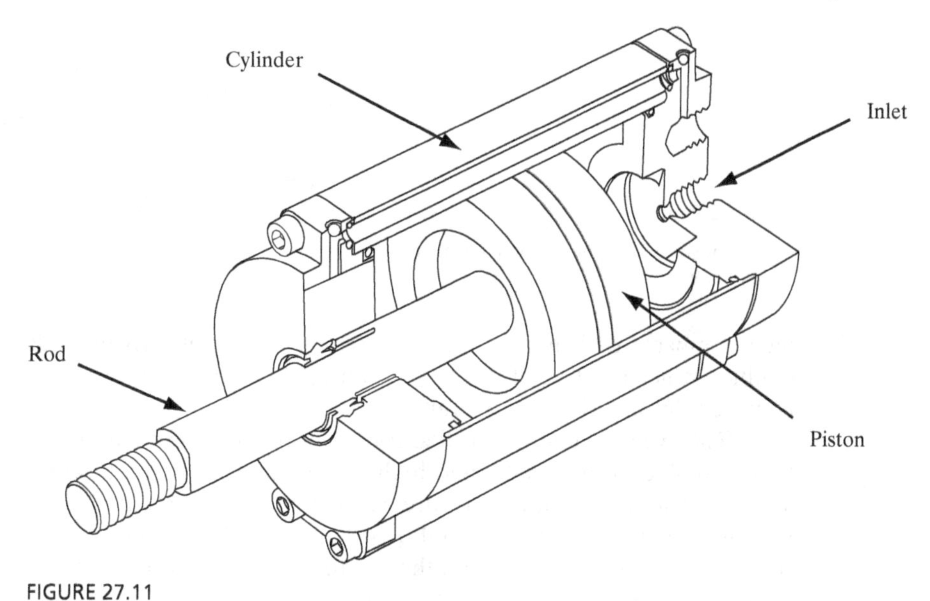

FIGURE 27.11

Cutaway view of a typical piston-based linear actuator.

The motion of the piston within the cylinder can be transferred to external components by means of a rod, as depicted in Figures 27.11 and 27.12a, or a **slide** (also called a **rodless cylinder**), shown in Figures 27.12b and 27.12c. For cylinders that include rods, they may be unconstrained so that they are able to rotate relative to the cylinder, or they can be constrained so that rotation is prevented. Damping may be incorporated into the cylinder design to reduce shock during accelerations and decelerations, or the cylinder may be undamped. In a rodless cylinder, the piston is located inside a cylinder with pressure acting on both sides, so that it can slide in either direction in response to changes in the differential pressure. The piston is coupled to a mounting block either through a slot (Figure 27.12b) or using strong magnets (Figure 27.12c). The major advantage of slides or rodless cylinders is

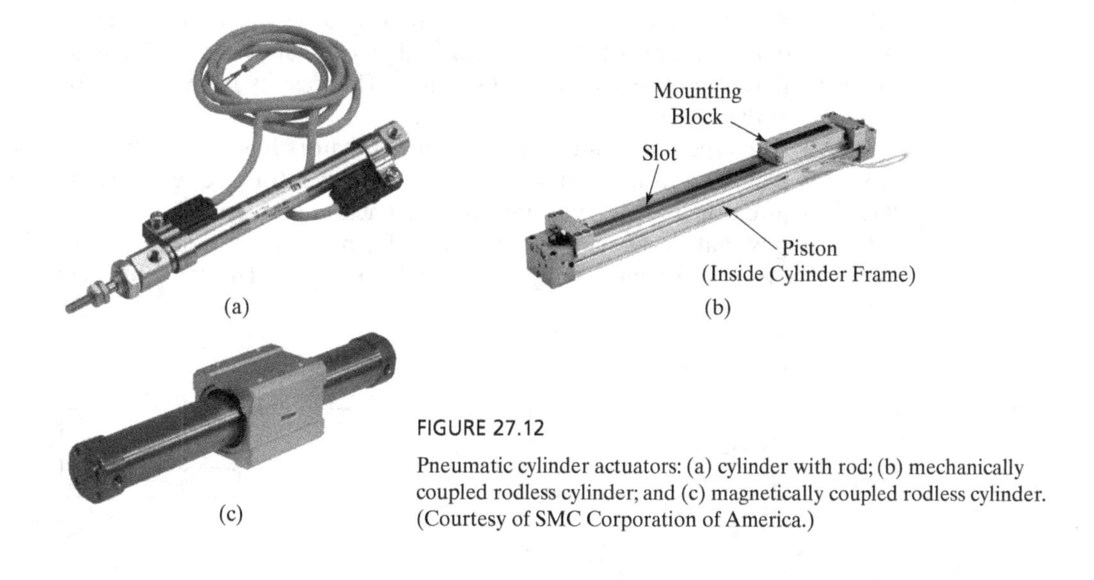

FIGURE 27.12

Pneumatic cylinder actuators: (a) cylinder with rod; (b) mechanically coupled rodless cylinder; and (c) magnetically coupled rodless cylinder. (Courtesy of SMC Corporation of America.)

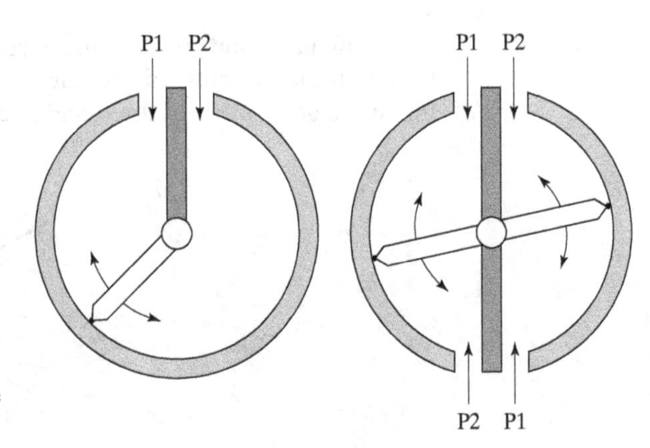

FIGURE 27.13

Layout of single (left) and double (right) vane rotary actuators.

their compactness. When a traditional piston and rod is fully extended, its length effectively doubles, which can be difficult to accommodate in many applications. The position of a slide, however, doesn't affect its overall size.

Rotary actuators: Controlled rotary motion can be created from the pressure differentials available in pneumatic and hydraulic systems in a great variety of ways. The simplest and most widely available types are called rotary vane actuators, and are based on a rotating version of a piston. Figure 27.13 shows the layout of such actuators: the figure on the left is a single vane rotary actuator, and the figure on the right is double vane. Figure 27.14 shows a physical example of a rotary vane actuator.

FIGURE 27.14

A single vane rotary actuator. (Courtesy of SMC Corporation of America.)

In the simplest case of the single vane actuator, the output shaft is located at the center of the device, and a vane that serves as the piston rotates through the volume within the cylinder when a pressure differential is established between the two chambers defined by the vane. In the case of the double vane (right), two vanes are used and hence the pressure differential acts upon twice the surface area. This results in increased force and speed of actuation at the expense of range of motion.

All vane type pneumatic and hydraulic actuators have a specific and limited range of motion, with angles from about 45° to 280°. High levels of torque are available (up to about 80 kNm) at reasonably high mechanical efficiencies (80%–95%).

Designs that derive rotational motion from linear motion are also widely available, such as the rack-and-pinion design shown in Figure 27.15. This type of actuator translates the

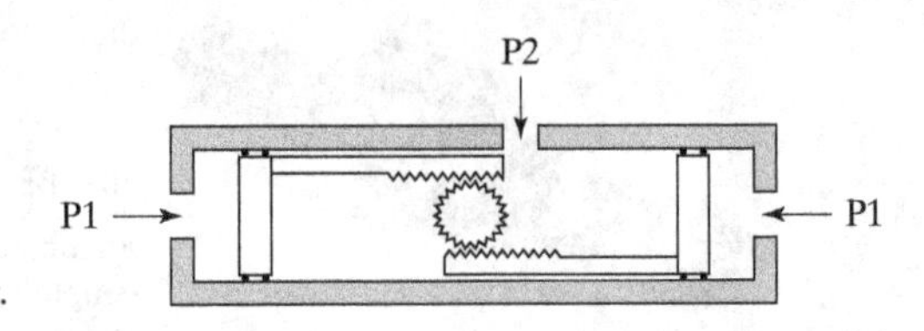

FIGURE 27.15

Rack-and-pinion rotary actuator.

linear motion from one or two piston/cylinder pairs (shown at either end of the figure) into rotary motion of the output shaft at the center. These types of rotary actuators allow for more than a single rotation of the output shaft, are capable of providing very high output torque (as much as 5 MN·m!), and have high mechanical efficiencies (~85%–90% for single rack designs and ~90%–95% for double rack).

Other types of rotary actuators worth considering include helical spline, enclosed piston crank, bladder, helix, scotch yoke, and piston chain designs (Figure 27.16). Manufacturers usually supply a variety of different types of actuators, so investigate the specifications and prices to determine which one will best meet your requirements.

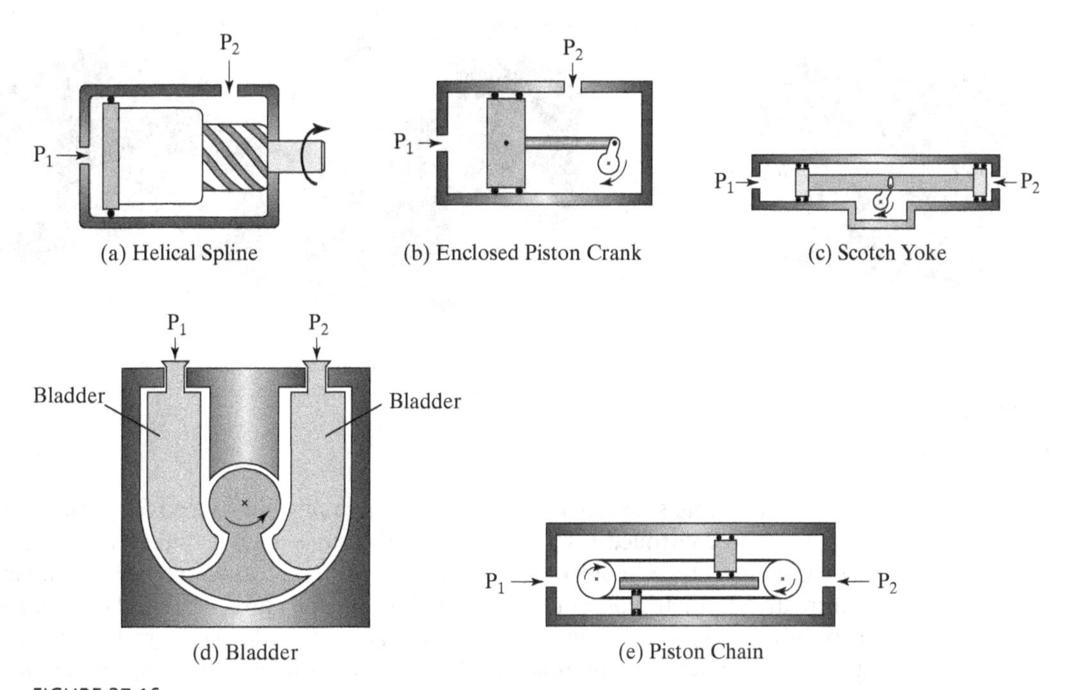

FIGURE 27.16

A variety of rotary actuator types.

Grippers: Grippers are special purpose assemblies of actuators and linkages that are used to grasp and manipulate workpieces. In a limited way, they mimic the human hand's ability to pick up and hold objects. They are available in 2-finger configurations where the finger motion is parallel, angular, angular with a wide range of motion (e.g., 180°); 3- and 4-finger configurations for grasping and centering items, in-and-out motion (called escapement actuators); and toggle grippers that maintain their gripping force after the drive pressure is released. Pictures of many of these options are shown below in Figure 27.17.

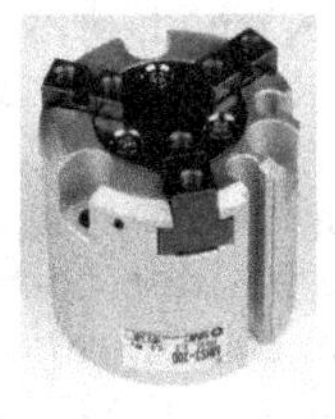

FIGURE 27.17

A variety of pneumatic grippers (from left): angular, parallel 2-finger, rotary 3-finger, parallel 3-finger, escapement, toggle. (Courtesy of SMC Corporation of America.)

27.3 RC SERVOS

Solenoid valves and servo valves aren't the only types of "other" actuators that use electromagnetic actuators (solenoids and motors) as a means of actuation. In this section, we'll transgress even further to bring RC servos to your attention. These "other" actuators are really nothing more than a standard permanent magnet brushed DC motor, coupled to a low-precision position feedback system. They are easy to use, simple to interface with, readily available, and inexpensive. As a result, they are very commonly used in mechatronic applications. Figure 27.18 shows exterior and interior views of a typical RC servo.

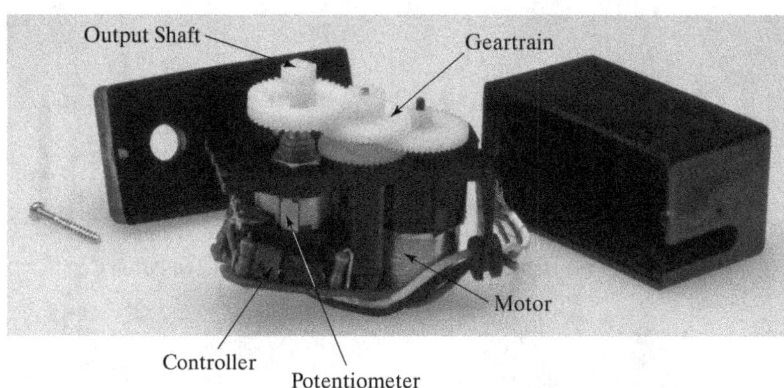

FIGURE 27.18
RC servo motors.

As the name suggests, **RC servos** were originally introduced for position control applications in radio controlled ("RC") device applications, such as model cars and airplanes. They are commonly used to control steering angle in cars, and elevator, rudder, and aileron position in airplanes. In their most common embodiments, they are comprised of a plastic housing that incorporates a motor coupled to a high ratio gear reduction stage, resulting in low output shaft speed (i.e., tens of rpm) and relatively high torque (usually in the range of about 10–200 oz. • in./0.071–1.4 Nm). In many cases, RC servos are considered as an alternative to stepper motors as a means of implementing position control, often because of the relatively low torque capabilities of most stepper motors. However, the output torques of typical stepper motors and RC servos are generally in the same range, so good practice dictates that the specifications for each type of motor be considered before selecting one approach or the other. Inexpensive RC servos, in particular, often have output torque at the low end of the typical range (~10 oz • in.).

The position of the servo's output shaft is measured by a potentiometer and used for closed-loop position control. A convenient mounting flange is typically attached to the output shaft. The majority of RC servos are not capable of continuous rotation, instead having a range of motion less than 360° with physical stops at their limits of travel. Some, however, do allow for continuous rotation or can be user-modified to achieve it. Lower cost RC servos incorporate inexpensive, small, brushed permanent magnet DC motors and plastic gears, while higher cost options use brushless DC motors, and some are available with metal gears.

RC servos require power connections for the motor and control electronics, as well as a position command input. The specification for the position command input is common among RC servos. The position command is a 0–5 V digital pulse whose duration corresponds to the desired angular position. The duration of the pulse varies between 0.5–3 ms, with 1.5 ms corresponding to the approximate halfway position. The time between pulses is typically between 20 and 30 ms, but wide variation is usually allowable (Figure 27.19).

As a result of their convenient packaging, low cost, and high availability, RC servos are used in a very wide variety of applications, well beyond the radio controlled device realm.

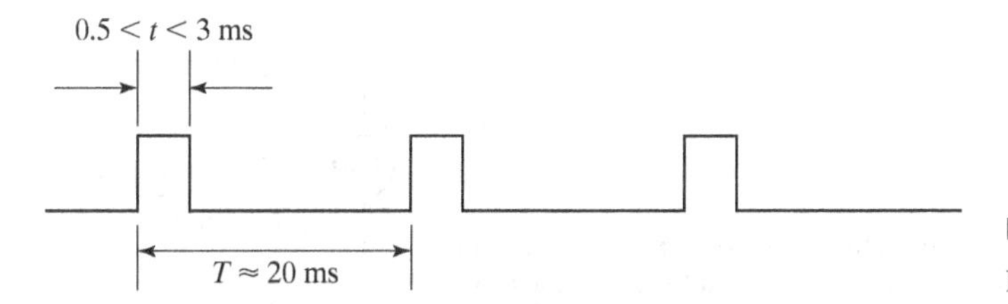

FIGURE 27.19

RC servo position control pulse train.

Stickybot: a wall-crawling robot. (Courtesy of Stanford University.)

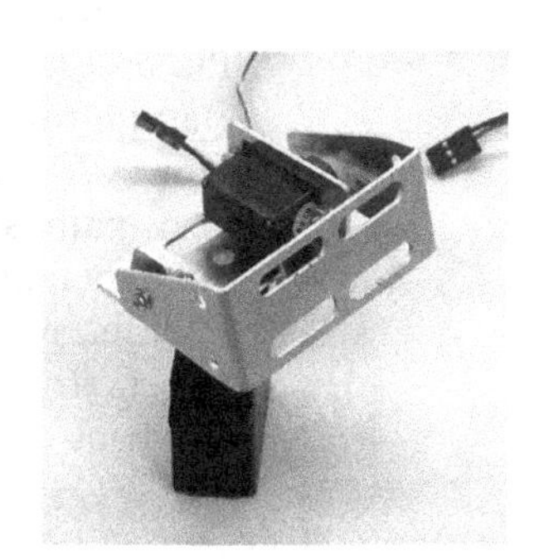

Miniature pan-and-tilt camera mount. (Courtesy of SuperDriod Robots, Inc.)

Snake robot. (Courtesy of Dr. Gavin S. P. Miller, www.snakerobots.com.)

FIGURE 27.20

A selection of novel RC servo applications.

For example, they have been incorporated into wall-crawling robots, such as Stanford's Stickybot, snake robots, and pan-and-tilt camera mounts (Figure 27.20).

27.4 PIEZO ACTUATORS

Piezo actuators derive their function from a physical phenomenon called the **piezoelectric effect**, discovered by Pierre Curie (husband of Marie Curie) and his brother Jacques in the 1880s. They determined that certain naturally occurring crystals (e.g., quartz, tourmaline, topaz, cane sugar, and Rochelle salt) reacted to the introduction of an electric potential by changing shape, and conversely when subjected to mechanical strains, the materials created an electric potential. It is the former characteristic that is exploited to create actuators from

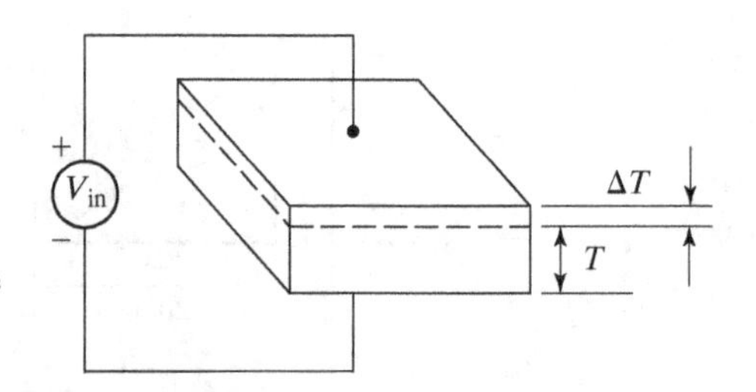

FIGURE 27.21

Diagram of a piezoactive material deforming (ΔT) under the application of electric potential ($V_{in}+$ and $V_{in}-$) to the top and bottom surfaces.

piezoelectrically active materials: the application of electric potential across the material causes it to deform, and the resulting motion and forces are used to create an actuator (see Figure 27.21).

Piezoelectric ceramics, developed during WWII, exhibit much higher piezoactivity: on the order of 100 times greater strain per volt than previously known crystalline materials. Barium titanate and lead zirconate titanate (**PZT**) are examples of piezoelectric ceramics. However, even these materials don't deform much (less than 0.2%, typically). Their response also exhibits hysteresis—that is, the dimensions of the material at a given voltage are different depending on the voltage history. The displacement resulting from an increasing voltage is different from the displacement from a decreasing voltage (Figure 27.22). Correction factors (open-loop compensation) or independent position measurements (closed-loop feedback control) can be used to compensate for this behavior.

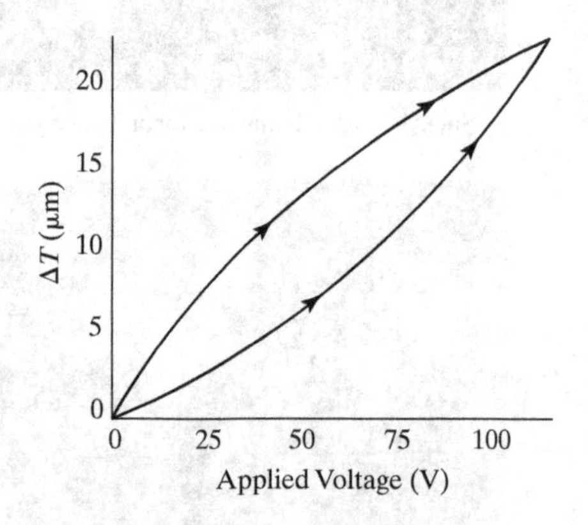

FIGURE 27.22

Typical response of a piezo element to an applied voltage: small displacements (y-axis) result from high applied voltages (x-axis). Note that the response exhibits hysteresis.

Piezo actuators are used mostly to create motion on a very small scale, though it can be very accurate and repeatable and achieve very impressive resolution. Ranges of motion from nanometers to hundreds of microns are achievable, with subnanometer resolution. Precise control of a piezoactive element is limited primarily by how well you can control the applied voltage. In cases where the ultimate in precision and resolution are required, electrical noise, component drift, temperature effects, and so on are responsible for limiting system performance.

Piezo actuators have several very impressive characteristics, including:

- they can generate very high forces (thousands of pounds),
- they have very high positional resolution (subnanometer, in the absence of frictional effects),
- they have very fast response (bandwidths in the tens of kilohertz),
- they don't generate, and aren't effected by, magnetic fields,
- they consume very little power, especially when holding a position,
- they have very long service lives (several billion cycles are typical),
- they can be used in severe environments (cryogenics, clean rooms, explosive atmospheres, and so on).

However, there are a few major limitations that prevent piezo actuators from enjoying widespread use:

- piezo actuators are relatively expensive (~$100 to several thousand dollars),
- the maximum range of motion of stacked (or series) piezo actuators is very small (in the hundreds of micrometers),
- the maximum range of motion of bimorph (or bender) piezo actuators is less limited but still small (up to a few millimeters), but these generate lower forces,
- piezo actuation elements are very brittle, and do not tolerate tensile or shear loading well,
- piezoactive materials exhibit substantial hysteresis, which can be mitigated with open-loop compensating techniques or closed-loop control,
- relatively high voltage is required to achieve the full range of motion from piezo actuators (hundreds to thousands of volts),
- the precision, stability, and noise characteristics of the electronics used to apply a voltage across a piezo element ultimately determine how accurately it can be controlled.

Despite the impressive capabilities and simplicity of piezo actuators, they are reserved for fairly exotic applications. For example, they are commonly used in optics, MEMS, laboratory environments, and semiconductor manufacturing.

27.4.1 Types of Piezo Actuators

Several configurations of piezo actuators have been devised to create distinct classes of actuation that are useful in a variety of applications. In this section, we describe a number of the most common piezo actuators.

27.4.1.1 Stacked Piezo Actuators

The most straightforward and common configuration is that of the **stacked**, or **serial**, **piezo actuator**. This type of actuator adds the combined deformations of many piezo elements together to result in an increased range of motion. For example, if a single piezo element is capable of deforming as much as 1 μm, then stacking five of them together (as shown in Figure 27.23) will result in a range of motion of 5 μm. These are available as simple

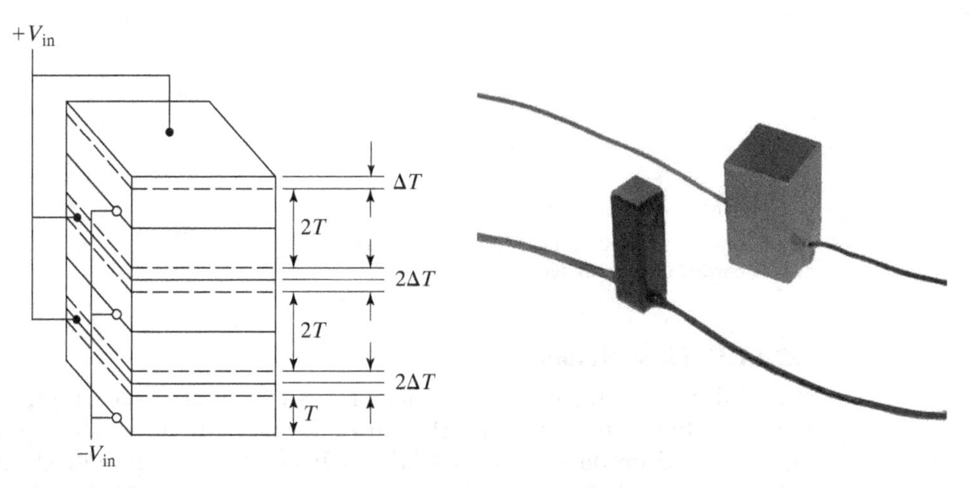

FIGURE 27.23

Stacked, or serial, piezo actuators. Left: diagram showing construction and resulting actuation. Right: typical piezo stack actuators. (Courtesy of Piezo Systems, Inc.)

laminated assemblies (as shown) and also in protective housings. Housings may also incorporate elements that prevent improper loading (tensile and shear) of the piezo elements to prevent damage.

27.4.1.2 Bimorph Piezo Actuators

The second major type of actuator construction is the **bimorph**, or **bender**, **piezo actuator**. This construction stacks two piezo elements side by side, and applies the voltage differentially across them (one element gets bigger while the other gets smaller) to cause it to bend. This is illustrated in Figure 27.24.

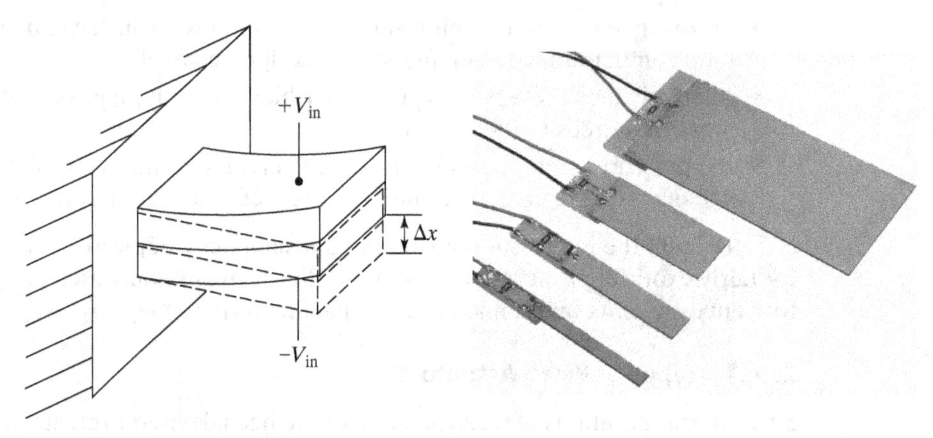

FIGURE 27.24

Bimorph, or bender, piezo actuators. Left: diagram showing typical construction, application of voltage, and resulting motion of bimorph actuator. Right: examples of bimorph actuators. (Courtesy of Piezo Systems, Inc.)

In addition to general purpose actuators, bimorph actuators have been used to create small fan elements. Though expensive relative to conventional fans, they consume little power, deliver high flow velocities and volume flow rates, have long service lives, and allow designers to target "hot spots" with a very precisely aimed air stream (Figure 27.25).

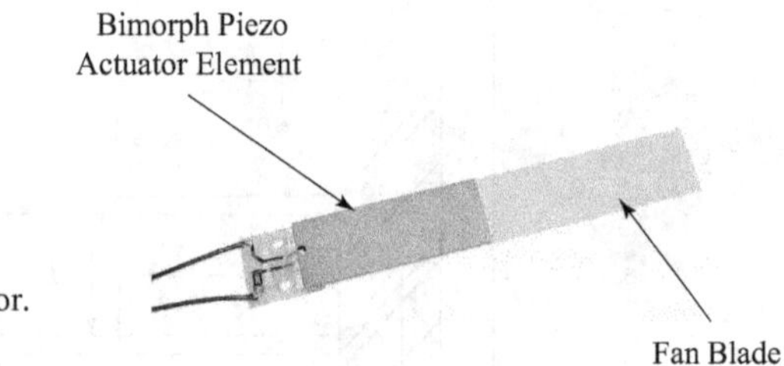

FIGURE 27.25

A simple fan based on a bimorph piezo actuator. (Courtesy of Piezo Systems, Inc.)

27.4.1.3 Piezo Motors

A third major category of piezo actuator is commonly called a **piezo motor**. Within this category, there are two approaches to achieving motion: so-called "inchworm®" drives (a trademarked product name of Burleigh Instruments®, who introduced the devices) and ultrasonic motors. **Inchworm drive** actuators use a finger or multiple fingers to push and pull either a slide element (for linear motion) or a rotor (for rotational motion). Essentially, the fingers "walk" the plunger or rotor in successive grasp-push-release actions to create

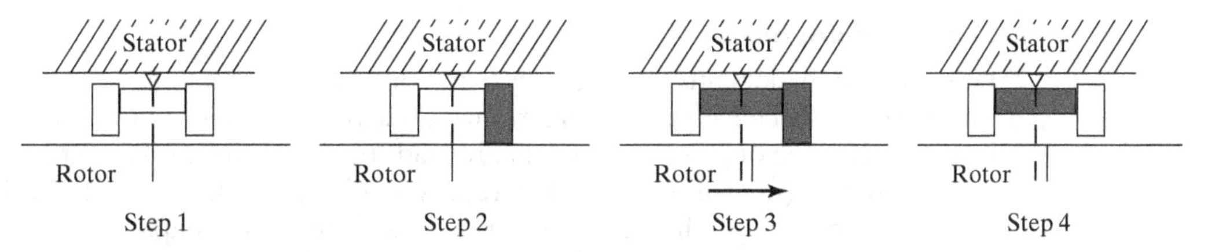

FIGURE 27.26

How a piezo motor creates motion. 1) Unpowered—the rotor of the piezo motor may rotate freely. 2) Grip—a piezo element on the right side pushes securely against the rotor. 3) Pushed—the central element of the actuator expands to rotate the rotor clockwise. 4) Release and return—the right-hand piezo element relaxes to release the rotor, and the central piezo element relaxes to prepare for the next sequence.

pseudo-continuous motion. This sequence is shown in Figure 27.26. The motion resulting from the inchworm approach has a discontinuous characteristic, similar to the operation of a stepper motor (Chapter 26), but with much finer resolution.

The inchworm piezo motor approach allows for large ranges of motion (not fundamentally limited, but often in the tens of centimeters), and each step can be controlled precisely via analog electronics to provide very high resolution positioning (subnanometer). Rotational actuators are capable of rotating continuously. These piezo motors are capable of jogging at high feed rates, such as 100 mm/s, and can provide impressively high actuation forces—as high as hundreds of newtons.

Ultrasonic motors employ an entirely different approach to creating relative motion between a stator (the stationary housing of the actuator) and the moving rotor or slide. In these devices, a wave is induced in the piezo actuation elements. One side of the piezo elements is attached to the stator, and the other side is pressed up against the rotor or slide. By controlling the characteristics of the waves induced in the piezo component, relative motion between the stator and the rotor or slide is created. Actuation forces are transmitted from the piezo element to the rotor or slide through friction. The concept is illustrated in Figure 27.27.

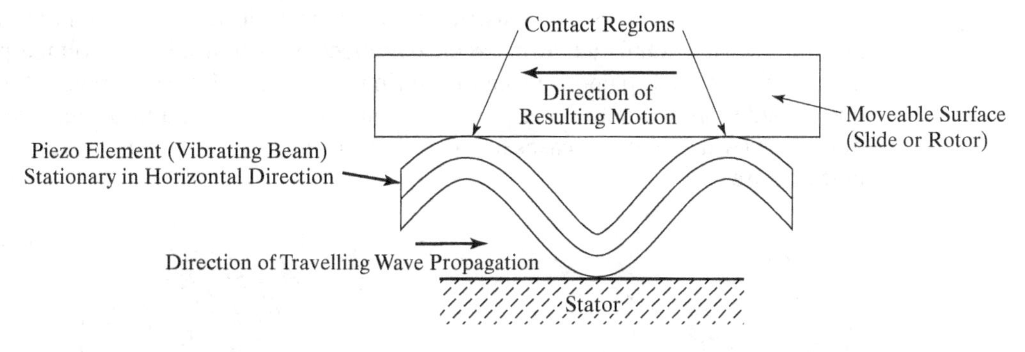

FIGURE 27.27

A traveling wave ultrasonic motor.

Traveling wave ultrasonic motors have been commercially available since the late 1980s, and were first introduced by Canon®, Inc. to automatically focus camera lenses. Since then, these actuators have been put to many other general purpose uses. When a traveling wave is induced in a piezo beam element, its points of contact with the movable surface (rotor or slide) have elliptical motion, which results in relative motion between the movable surface and the piezo. The process occurs very quickly, as the drive frequency is usually in the tens of kilohertz (well above the audible range for humans). Unlike the motion resulting from the inchworm drive approach, ultrasonic motors have smooth continuous motion.

27.4.1.4 Piezo Buzzers

Piezo actuation elements are also very commonly used to create devices that produce sound waves. These are called **piezo buzzers.** Buzzers actuated by piezo elements are compact, and many are capable of producing piercingly loud sounds (90 dB or more) while requiring only minimal power (tens of milliwatts). A representative piezo buzzer is shown in Figure 27.28. Piezo buzzers are used in most electronic devices that emit simple sounds as part of a user interface. For example, they are often incorporated in computers, laboratory instrumentation, toys, and home appliances to produce user feedback in the form of simple beeps and buzzes. The attributes of piezo actuators are a very good match for creating sound waves: the materials' fast response times allow for the production of sound frequencies from below 100 Hz to well above 10 kHz, and piezo materials' small deformations—a big limitation in many actuation applications—are well-suited for the task of vibrating a diaphragm to produce sound waves.

FIGURE 27.28
Typical piezo buzzer with plastic housing and wire leads.

Figure 27.29 shows the construction of typical piezo buzzer elements. A thin ring of piezo material is sandwiched between two electrodes that allow a voltage potential to be applied across the top and bottom of the ring. The piezo element is bonded to a metal plate on one side (shown below the piezo element in the figure), and a thin diaphragm on the other side (which in many cases forms one of the electrodes, shown above the piezo element in the figure).

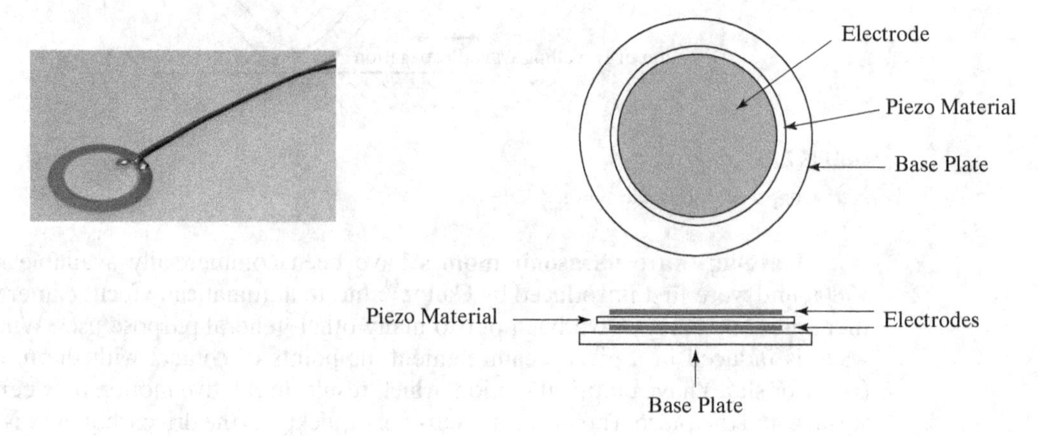

FIGURE 27.29
Piezo buzzer element construction.

As with all the piezo actuator technologies described above, applying a driving voltage across the piezo material causes it to deform. Because of the way the material is constrained by the metal plate and the diaphragm, the assembly assumes concave and convex shapes depending on the polarity of the applied voltage. This is illustrated in Figure 27.30. Alternating the polarity of the applied voltage causes the buzzer element to vibrate back and forth as it alternates between a concave and a convex shape, and this vibration produces pressure variations in the surrounding air—that is, sound waves. As long as the vibrations are within the frequency rage of human hearing (approximately 20 Hz to 20 kHz) and of sufficient intensity (the piezo actuator and diaphragm vibrate with sufficient amplitude), the sound waves produced will be audible.

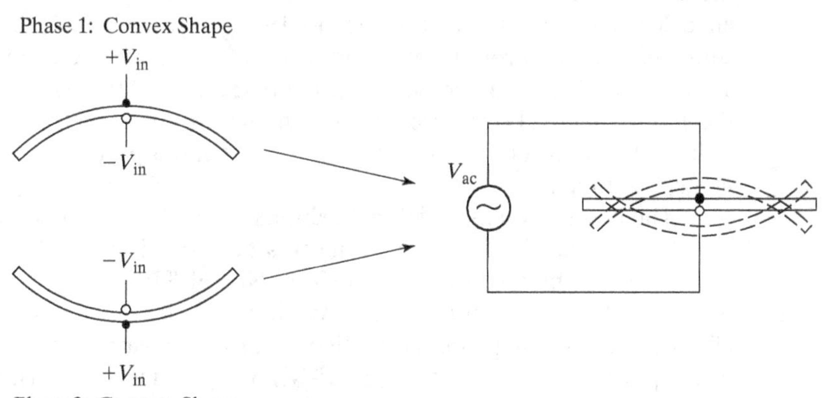

FIGURE 27.30

A piezo buzzer element producing sound waves when alternately driven into a convex shape (Phase 1) and a concave shape (Phase 2) by changing the polarity of the driving voltage.

The drive voltages required to produce audible tones are typically in the range of a few volts peak-to-peak. Higher drive voltages cause greater deformations of the piezo actuator and thus louder sounds, but many buzzers are readily available for which 5 V peak-to-peak (conveniently corresponding to standard logic-level voltages and power supplies) produces a very useful tone. Only a small amount of current is required by most piezo buzzers, in the range of a few milliamps. These minimal constraints allow for relatively simple drive circuits that are required to produce 1) an oscillating signal (a simple square wave works well, as does a sine or sawtooth wave), 2) an amplitude sufficient to create adequate deformation of the piezo element (a few volts peak-to-peak), 3) a few milliamps of current, and 4) frequencies somewhere within the audible range for humans. Such signals may be produced directly from the output pin of a microcontroller (such as the PWM output of a PIC12F609, described in Chapter 2), a 555 timer chip (Chapter 18), or one of the many driver circuits recommended by the manufacturers of piezo buzzers. A category of buzzer is also available that incorporates the drive electronics and the piezo buzzer element in a common housing, and requires only a connection to power to emit sound. The notes created by such buzzers are limited to whatever the internal drive circuit produces, but incorporating these devices into a circuit is exceptionally straightforward: simply connect power and ground, and they produce a tone.

In addition to the piezo-based actuators described above, there are many other available configurations, as well as kits that enable would-be piezo actuator designers to experiment with constructing shapes and configurations for custom applications. When high forces, very fine resolution, and small displacements are required—and high voltages aren't prohibitive—piezo actuators are an excellent option.

27.5 SHAPE MEMORY ALLOY ACTUATORS

Shape memory alloys, commonly called **SMAs**, exhibit remarkable mechanical characteristics that result from changes in the material's crystalline structure caused by temperature changes. With careful processing, these materials can be configured to return to a predetermined shape at a specific temperature, often with substantial force. The desired default shape of a piece of SMA is "taught" to the alloy by constraining it in that shape while the temperature is raised to the point where annealing occurs. After the material has been annealed and cooled below a threshold temperature (which depends on the SMA's constituents and their proportions), it may be deformed into new shapes, and when the temperature is raised to a higher threshold temperature (again, depending on the composition of the material), it will very forcefully reconfigure itself and return to its default shape. Subject to limitations on temperature and strain levels, the process may be repeated a very large number of times.

The best known material that exhibits this behavior is a mixture of nickel and titanium, sometimes called **NiTi**. This material was developed by the Naval Ordinance Laboratory, which dubbed the material **Nitinol** (from **NI**ckel, **TI**tanium, **N**aval **O**rdinance **L**ab). Nitinol is one of the best performing SMAs, though it is also among the costliest. Many other alloys exhibit the shape memory effect, with widely varying characteristics and cost, including: copper-aluminum-nickel (Cu-Al-Ni), copper-zinc (Cu-Zn), copper-tin (Cu-Sn), gold-cadmium (Ag-Cd), cobalt-nickel-aluminum (Co-Ni-Al), cobalt-nickel-gallium (Co-Ni-Ga), and nickel-iron-gallium (Ni-Fe-Ga). A wide variety of SMAs are commercially available and easy to purchase, however, Nitinol is by far the most commonly used and readily available.

The mechanism of action of SMAs is a phase change of the molecular structure of the material. This isn't the typical solid-to-liquid phase change that springs immediately to mind, however—the material remains very much a solid throughout the process. Rather, the phase change occurs within the crystalline structure of the alloy (Figure 27.31), and the different phases (**martensite** vs. **austensite**) have very different properties.

At low temperatures, SMA materials' crystal structure is entirely martensitic, resembling a series of connected parallelograms. Above a **transition temperature**, the crystal structure changes to austensite, which is cubic in shape. The transition temperature is a material property that is determined by the ratio of the constituent elements of the alloy. For Nitinol, the ratio of nickel/titanium is very close to 50%/50%. By adjusting the mixture within a 1%

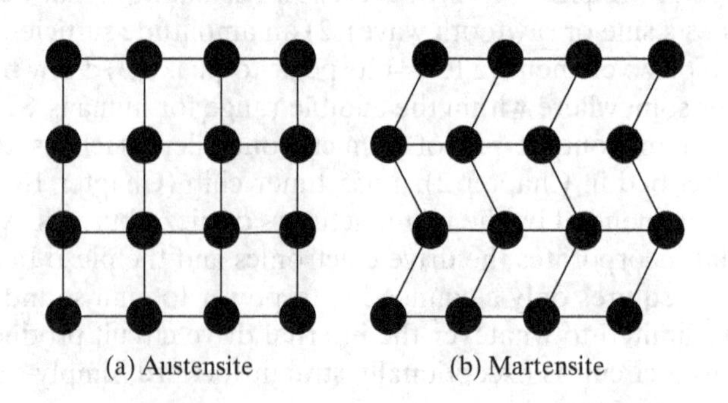

FIGURE 27.31

Differences in crystalline structure for SMA materials: (a) austensite above the transition temperature and (b) martensite below the transition temperature.

window, the transition temperature can be selected to be anywhere between −100°C and 200°C. Getting the right transition temperature requires that the composition of the alloy be very tightly controlled.

The transition from martensite to austensite or austensite to martensite is highly hysteretic, with the phase transitions occurring at different temperatures depending on whether the temperature is increasing or decreasing.

The default shape of the material—the shape that the SMA wants to return to as it transitions to austensite—is taught to the alloy (also called "training" or "programming") by constraining the material in the desired shape during an annealing process. For Nitinol, the annealing temperature is about 540°C, and the process is complete within about 5 minutes. Once the material is trained, the behavior across the full range of temperatures is fully determined:

1. Below the transition temperature, the structure is entirely martensite and the material may be deformed as desired (up to a limit: about 8% strain for some alloys).
2. As the temperature of the material is increased, the structure of the material begins to transition to austensite at the point called the **austensite start temperature** (A_S). This is shown in Figure 27.32 at the start of the transition labeled "Heating Phase." As the material transitions to austensite, it returns to the shape it held when it was annealed. Large forces can be generated as the crystalline structure changes.
3. Further increasing the temperature increases the ratio of austensite, until the entire structure is austensite. The temperature at which the transition is complete is called the **austensite finish temperature** (A_F), as shown at the top right in Figure 27.32.
4. Increasing the temperature above A_F has no effect on the material's structure, until high temperatures near the annealing temperature are reached. Returning the material to the annealing temperature will "retrain" the material, so it's something to avoid. (As an aside, between A_F and the annealing temperature, the material is said to be "superelastic," where very high strains—up to 10% in some cases—can be imposed without causing plastic deformation. It's not an actuator, but it's something very interesting and worth mentioning here.)
5. If the A_F temperature has been reached and the temperature is then decreased, the structure will begin transforming back into martensite at the **martensite start temperature** (M_S), shown at the top center of the figure. Note that this is a lower temperature than the austensite finish temperature, due to the hysteresis.
6. As the temperature is further decreased in the "Cooling Phase," the fraction of martensite continues to increase until the material is comprised entirely of martensite. As cooling occurs, the SMA maintains its trained shape, though the crystalline structure is changing. The temperature at which this transformation is complete is called the **martensite finish temperature** (M_F).

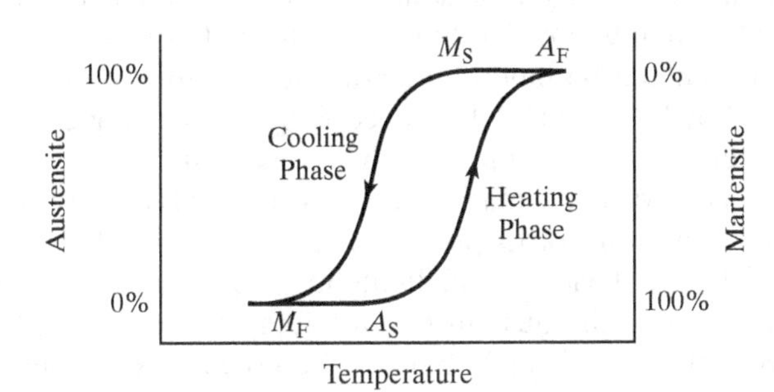

FIGURE 27.32

Transition temperatures and hysteretic phase change behavior of SMAs.

SMA actuators exploit the forces generated as the fraction of austensite increases, and the material does everything it can to return to its annealed shape. The SMA can be conveniently heated by causing electric current to flow through it, though any means of raising the temperature will have the desired effect. One common approach to building SMA actuators relies on small diameter wires that are trained with a default length. After they have been stretched so that they are longer than their default length, they are heated by flowing an electric current through them. Turning the current on heats the SMA and causes it to contract forcefully, and turning off the current allows it to cool and relax. An external force is required to stretch the material as it cools, otherwise the material will remain essentially in its trained shape after it returns to martensite, making it a pretty boring actuator after that point (turning on the current will simply heat the device without resulting in any further motion).

We mentioned above that the strain level that can be imposed when an SMA is below the transition temperature has a limit. Above this limit, permanent deformation of the material will result. As an example, for Nitinol, this maximum strain is about 8%. Up to this degree of strain, NiTi will completely return to its trained shape when heated above A_F. If more than 8% strain is applied, the material won't be able to fully return to its annealed shape. As you might guess, the material has a very short lifetime when operated at or near the maximum allowable strain: it begins to change properties and fatigue after only a few tens of cycles. To achieve longer lifetimes, such as millions of cycles, strain should be limited to 2%–5% (as with most materials, the lower, the better).

The speed of response of all SMAs is dominated by the time it takes to heat and cool the material. In general, this is quite slow compared with many other actuation technologies. Whereas the contraction of an SMA actuator can usually be accomplished quite quickly by flowing electrical current through it to generate heat internally according to $P = I^2R$, cooling the material can take much more time. How much time depends on the shape, mass, and specific heat of the SMA, and the heat transfer rates from conduction and convection (plus perhaps a small contribution from radiation). Optimizing all the variables still results in comparatively slow response times: a 1 Hz cycle time (a "cycle" is a contraction and subsequent relaxation of the SMA) is about the best you can expect.

In contrast to the slow response times, the forces that SMAs are capable of generating are very large. The power of the molecular structure's phase change is harnessed to create the motion, and this is large relative to the size of the actuating element: for Nitinol, the maximum output is about 600 MPa (600 MN/m^2). That is, a piece of Nitinol with cross-sectional area of 1 m^2 (10.76 ft^2) is capable of generating a force of 600 MN (224,809 lb$_f$)! Looking at this another way, to generate a force of 10 lb$_f$ requires a Nitinol wire with a cross-sectional area of only 1.15×10^{-4} in.^{2}, and a diameter of 0.012 in.

The simplest SMA actuator elements are available off-the-shelf in the form of wires. Different alloys and sizes are available, allowing you to choose from a range of actuation force, transition temperature, speed of response, and longevity. These off-the-shelf wires have the advantage of having been processed and "trained" by the manufacturer, so they're ready to use. SMA materials are also available in numerous forms of raw stock: sheet, bar, rod, and tube stock are all common and reasonably easy to find. The transition temperature is something you specify when you order, but the material hasn't been trained (annealed) yet. The raw SMA material can be modified to fit your requirements, and then annealed. Figure 27.33 shows a hobby kit called "Solar Space Wings." The wing elements are actuated by SMA wires, and power for the kit is provided by a solar cell.

Numerous products have been manufactured using SMA as an actuating element, but you may not have been aware of its presence. A classic example is an automatic, non-powered opening and closing device for skylights in greenhouses. When the temperature inside the greenhouse exceeds the austensite start temperature (A_S), the SMA actuator opens a skylight to vent the hot air and cool off the greenhouse. When the temperature

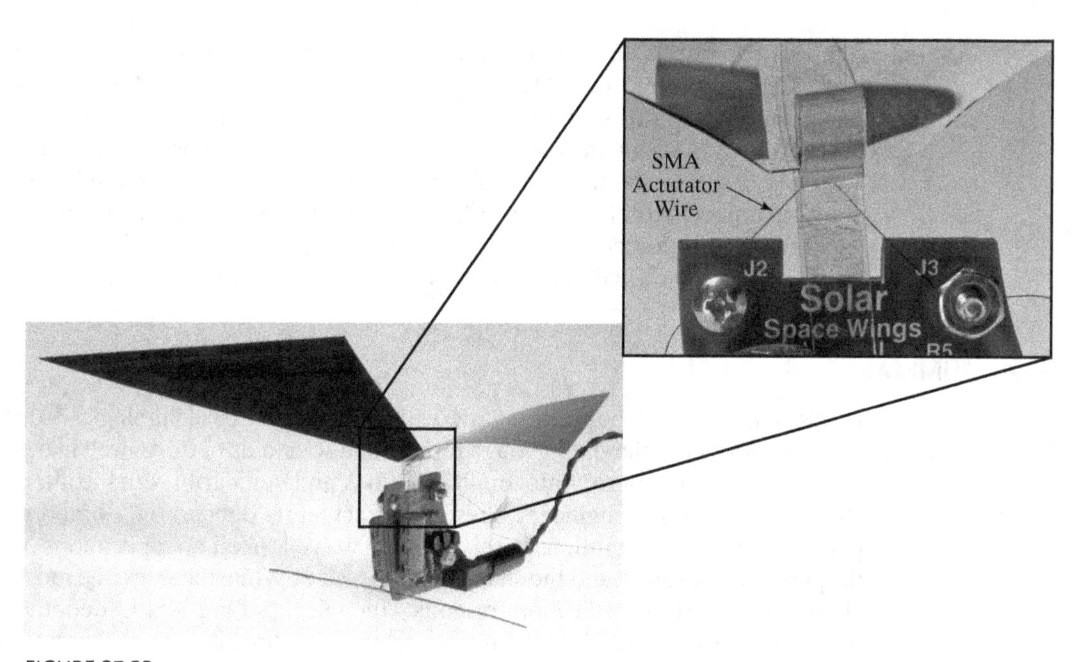

FIGURE 27.33

"Solar Space Wings," a solar-powered hobby kit actuated by SMA wires.

decreases below the martensite start temperature (M_S), the actuator closes the vent. It's an elegant way to sense temperature and actuate a controller (the window) without requiring external power or software.

A similar approach has been used to create devices that protect users from exposure to hazardous temperatures. For example, SMA actuators have been used to create an automatic locking mechanism on a self-cleaning oven door. When the self-cleaning cycle is initiated and the oven heats up beyond the latch's SMA transition temperature, the latch engages and prevents anyone from opening the oven (and potentially burning him- or herself) until the cycle ends and the oven cools to a safe level. Along similar lines, SMA actuators have been used to create antiscald valves that protect people from accidentally turning on a shower or sink with the water temperature set to unsafe temperatures. When water above an SMA element's transition temperature starts to flow through the valve, the SMA closes the valve and prevents injury. SMA has also been used in resettable fire sprinkler valves, and in circuit breaker elements in lithium ion batteries, which have a nasty reputation for catching on fire when they overheat.

One place that it's especially inconvenient to wire up an external power supply is inside the human body. Fortuitously, Nitinol is highly biocompatible, and suitable for use in devices that are permanently implanted in the body. It is also convenient that the internal temperature of the body is predictable (98.6°F, 37°C), so the transition temperature can easily be selected to be well below body temperature, ensuring that an SMA device will exert a strong force to achieve its trained shape when it's inside the body. Because of these characteristics, Nitinol has been used in numerous medical devices, including stents (small tubes used to prop open clogged arteries, veins, intestines, and other body lumens), tensioning wire for orthodontic braces, clips, and surgical retractors, to name a few.

Of interest to mechatronic engineers, integrated off-the-shelf SMA-based actuators designed to replace DC motors and solenoids are emerging. They offer high energy density (usable work output per unit mass), small package size, high actuation forces, and long life (depending on how you use them, of course). The response time, however, is relatively slow compared to motors and solenoids, the actuators they're intended to displace in the marketplace. One example is the DM01 linear actuator from Miga Motors®. It's available in versions that can generate either push or pull forces of up to 4.5 lb, has a stroke of 0.5 in., weighs about 25 g, and has dimensions of 3.125 × 0.864 × 0.286 inches (L × W × D). It is capable of actuating to a new commanded position within about 25 ms, and will relax to its passive state if unconstrained within 3–15 s in still air. According to the manufacturer, improved cooling, for example, with forced convection, can improve this by a factor of up to 100. These characteristics would be difficult or impossible to match with traditional motors or solenoids.

27.6 SUMMARY

In this chapter, we've looked at a variety of "other" actuator technologies—that is, alternatives to DC motors and solenoids. Though motors and solenoids perform well in a great many applications, they're not appropriate for all situations, and familiarity with the alternatives is invaluable for mechatronic engineers. These other actuators depend on a broad range of physical phenomena to create motion. In this chapter, we've looked at applications of fluid dynamics, the piezoelectric effect, and the shape memory effect. While these are the most commonly used alternatives, there are still more options—though these are less frequently used outside of research environments. We chose not to include actuators like magnetorestrictive, ferrofluidic, and electroactive polymer actuators in this chapter, but they are also worthy of independent investigation and might provide the perfect means of actuation in a given application.

Table 27.1 enumerates the technologies we've discussed in the sections above, and compares their characteristics. It is intended to help you quickly assess whether one or more of these technologies would be a good fit for an actuation challenge you are facing.

27.7 HOMEWORK PROBLEMS

27.1 Describe the basic difference between a solenoid, as introduced in Chapter 24, and the solenoid valve introduced in this chapter.

27.2 How does a direct acting solenoid valve differ from a piloted solenoid valve?

27.3 How does a solenoid valve differ from a servo valve?

27.4 Would you expect a single vane or double vane rotary actuator to be smaller for a given amount of torque produced? Explain why.

27.5 A servo motor capable of a 270° rotation responds to a range of pulses between 0.5 and 2.5 ms to move between the two extremes of its rotation. What timing resolution is necessary in the pulse generation in order to provide position control in 1° increments?

27.6 An engineer wants to use a microcontroller's 8-bit hardware PWM subsystem to generate the pulses to position the motor from Problem 27.5. If the clock for the PWM subsystem is chosen so that the maximum PWM period is 20 ms, what duty cycles (0–255 for the full 8-bit PWM) should be requested to place the servo at its minimum and maximum positions?

27.7 For the combination of motor and PWM drive system from Problem 27.6, what is the best expected angular resolution for the system (i.e., how much will the motor move when the PWM count changes by 1)?

27.8 A particular positioning application requires motion on the order of 1 cm to be produced with μm resolution while producing 10 N of force. Suggest an appropriate actuator technology to fit this application.

27.9 Explain the process that you would go through to produce an actuator that produced a pull-force using a length of Nitinol wire. Start with the raw, untrained wire and describe the processes through to actuation.

TABLE 27.1 A comparison of actuator technologies discussed in this chapter.

Actuator Type	Physical Basis	Control Mechanism	Popularity	Maturity	Ease of Use	Ease of Procurement	Range of Motion	Maximum Bandwidth	Pros	Cons
Solenoid valves	Electromagnetics	Electrical current, on/off	+++	+++	+++	+++		1 kHz	Fast response Long life	
Servo valves	Electromagnetics	Electrical current, on/off, PWM, or analog	+	+++	+++	+		1 kHz	Good for process control applications Long life	Few suppliers
Pneumatic actuators	Fluid dynamics ($F = P \times A$)	Valves (see above)	+++	+++	+++	+++	0.1–10s of inches	1–10 Hz	Large forces Large range of motion Long life	System complexity
Hydraulic actuators	Fluid dynamics ($F = P \times A$)	Valves (see above)	+	+++	++	+++	0.1–10s of inches	1–10 Hz	Very large forces Large range of motion Long life	High working pressures (~3,000 psi) System complexity
Piezo actuators	Piezoelectric effect	Voltage	–	++	–	+	1 nm–10s of μm	10 kHz	High positional resolution and accuracy Low current Low power Holding position requires no power Unaffected by magnetic fields	Small range of motion (μm) High voltage (~1 kV) Expensive
SMA actuators	Molecular crystalline structural changes	Temperature (heat transfer)	–	++	++	++	Depends on configuration	1 Hz	Large forces Long life High energy density Does not (necessarily) require electrical power Simplicity of design Biocompatible (NiTi)	Slow response (~1 Hz) If electrically heated: high current/high power Expensive Complex manufacturing process

27.10 From the list of actuator technologies below (none of which were discussed in this chapter), select one and 1) briefly describe its theory of operation, 2) cite at least one application, and 3) complete a new row for Table 27.1 corresponding to the actuator technology you selected (include justifications for your answer to part 3).

(a) Magnetorestrictive actuators

(b) Magnetorheological actuators

(c) Dielectric electroactive polymer actuators

FURTHER READING

Actuator Design Using Shape Memory Alloys, Waram, T., 2nd ed., 1993.

"Introduction to Piezo Transducers," Catalog #7C, 20–23, Piezo Systems, Inc., 2008.

"Introduction to Piezoelectricity," Catalog #7C, 57–59, Piezo Systems, Inc., 2008.

Introduction to Pneumatics and Pneumatic Circuit Problems for FPEF Trainer, Groot, J.R., et al., Fluid Power Education Foundation.

Introduction to Ultrasonic Motors, Toshiiku, S., et al., Oxford University Press, 1993.

Muscle Wires Project Book, Gilbertson, R., 3rd ed., Mondo-tronics, Inc., 2000.

CHAPTER 28

Basic Closed-Loop Control

28.1 INTRODUCTION

In this textbook on mechatronics, we devote only a single chapter to basic closed-loop control. You might very reasonably ask "Isn't that what mechatronics is all about?" To some degree, the goal of basic closed-loop control isn't much different from the overall goal of a mechatronic system. The distinctions are in the details. This chapter is about the details of strategies and algorithms that you can employ in order to control a mechatronic system.

The majority of this chapter will deal with techniques for implementing what is called "closed-loop control." Closed-loop control is also often referred to as "feedback control." This is because the essence of closed-loop control is the measurement of the output and the "feeding back" of that information to the controller in order to use the information to better achieve the desired output. We discuss the implementation of closed-loop control using on-off ("bang-bang") control as well as linear control based on proportional, integral, and derivative terms.

We also introduce an approach called "open-loop control." In contrast to closed-loop control, open-loop control techniques do not make use of information about the state of the system output to influence the control signal; rather, they rely upon assumptions about the response of the system to the control signal. We will discuss the pros and cons of open-loop and closed-loop control and why you might prefer one over the other in a given application. We will also introduce some ad hoc methods for control of systems that are not well served by traditional control methods.

28.2 TERMINOLOGY

Control systems are often described in terms of block diagrams that capture the essence of the various parts of the system. Figure 28.1 shows the simplest of these kinds of block diagrams, a system with one input and one output.

The system in the diagram is labeled the "**Plant**," a nod to the history of closed-loop control in industrial chemical processes. The **control input** is the parameter that we change to affect the plant. In the case of the DC motor control that we will be using as an extended example throughout this chapter, this would be the duty cycle drive of the voltage to the motor. We will often talk about the **control effort**, which is just another way of saying the magnitude of the control input. The **output response** is the parameter that we are interested in monitoring and controlling. For our DC motor example, this will be the speed of the motor.

In discussing control, we will be focusing heavily on how the system responds to the control input as well as to disturbances from outside the system itself. One of the most common ways of presenting and talking about these responses is to look at the **step response** of a

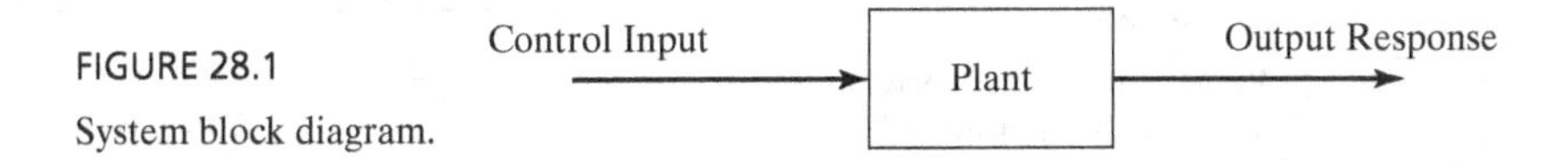

FIGURE 28.1
System block diagram.

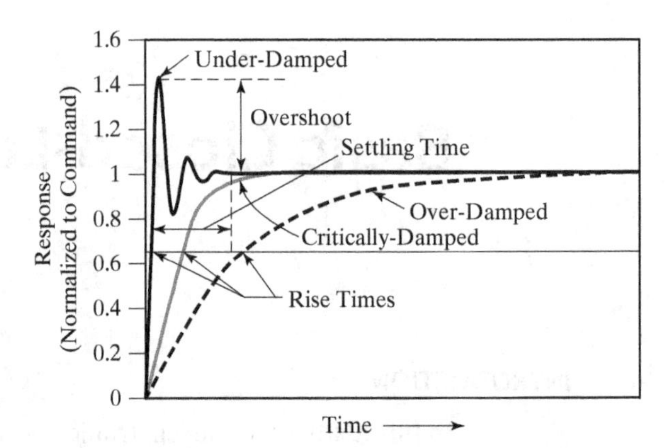

FIGURE 28.2

Step responses of overdamped, underdamped, and critically damped systems, with terminology labeled.

system. The step response is simply the response of the system to an instantaneous change in the command, or control input. Figure 28.2 shows step response plots for three types of system response: underdamped, overdamped, and critically damped. The step response features that we will use to characterize the different kinds of system responses are marked.

In this figure, the control input was changed from 0 to 1 at time $t = 0$ for three systems whose dynamic responses are characteristic of overdamped, underdamped, and critically damped systems. The **overdamped** response rises somewhat slowly to the new value, and always approaches from the direction of the value of the system output prior to time $t = 0$. The **underdamped** response exhibits a much faster **rise time** but **overshoots** the target value and exhibits a decaying oscillation in the response before it settles to a stable value (the **settling time**). The **critically damped** response walks the fine line between over- and underdamping. It exhibits a quick rise time with no overshoot of the final stable value. In practice, it is difficult and often unnecessary to achieve true critical damping.

28.3 OPEN-LOOP CONTROL

In most contexts, the mention of motor control is assumed to mean closed-loop control, but this is not necessarily the case. Before we get into the more traditional topic of closed-loop control, let's take a look at the simpler alternative: **open-loop control**. Using open-loop control, you map the response of a system to the control variable and then use that mapping to produce the desired outputs during operation. As an example, consider the control of a brushed permanent magnet DC motor that you wish to operate at constant speed. As we saw in Chapter 23, if we use drive-brake mode and a high enough pulse width modulation (PWM) frequency, we can obtain a well-behaved linear relationship (albeit with an offset) between PWM duty cycle and the motor's rotational speed (Figure 28.3).

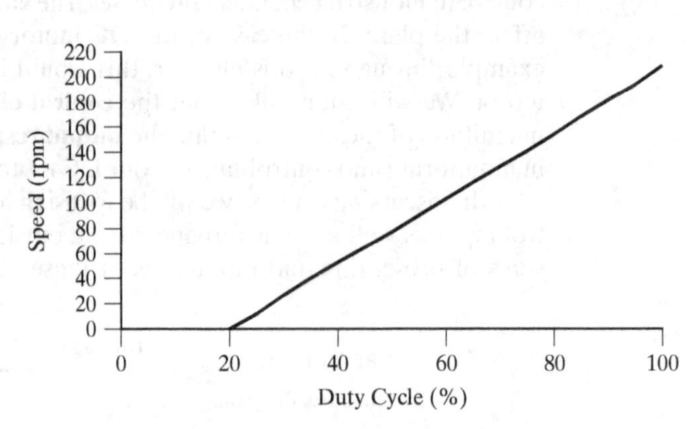

FIGURE 28.3

PWM duty cycle vs. speed for a sample DC motor.

This relationship between duty cycle and motor speed holds so long as the load on the motor remains constant. This may be a reasonably valid assumption in some instances. To set the speed of the motor when we know the relationship between duty cycle and speed, we simply calculate what the required duty cycle should be based on the desired speed and then output that duty cycle to the electronics driving the motor. If the relationship between duty cycle and motor speed is more complex than the simple linear relationship, we can use piecewise linear approximations to look up, or a polynomial curve fit to calculate, the correct duty cycle based on the desired speed.

While this will work to some degree, it is easy to see how it might not result in the desired speed. If the characteristics of the actual motor differ in any way from the characteristics of the motor used to create the map, then the resulting speeds will differ as well. If the loading conditions on the motor differ from the mapping conditions or if the load changes over the life of the device, then the mapping will no longer be valid or produce the desired motor speed.

Despite the shortcomings of open-loop control, the resulting system response may actually be "good enough," depending on the system requirements. As an example of this, consider the situation where the motor drives a fan that is being used to cool a heat sink on a power transistor. Based on the load being driven by the transistor (which is under the control of the microcontroller), we want to increase or decrease the speed of the cooling fan motor. If we know that we are increasing the transistor load, it is easy to increase the speed of the cooling motor in proportion to the increase in load. For a price-sensitive product, it may be prohibitively expensive to measure the speed of the motor (or better yet, the volume of the resulting airflow), and feed that measurement back to control the motor. In addition, it is not critical that the motor speed matches the commanded speed exactly. If it is close ($\pm 10\%$ might do it), this would be "good enough." Since the load that the fan presents to the motor will not change with time and the application does not require high accuracy, the deviations in speed that we might get with open-loop control may well be small enough to trade-off in favor of keeping the price of the product down.

28.4 ON-OFF CLOSED-LOOP CONTROL

When open-loop control is just not good enough, we need to do something more elaborate in order to meet the performance goals. The most common solution is to turn to some form of **feedback** and **closed-loop control** to improve the performance. The simplest form of closed-loop control is based on what is called "**on-off**" or "**bang-bang**" control. The name comes from the fact that with this approach, the control effort is either fully on or fully off, and there is no attempt made to modulate the control effort to any intermediate value. This is not to say that the desired control point cannot be set to any value that we would like—on the contrary, we can usually pick any value we like. What is important is that the response of the system to the control effort must be relatively slow compared to the rate at which the control effort can be turned on and off.

This type of control is very common in thermal control systems such as the heating/cooling systems in buildings, refrigerators, and baking ovens. In the case of heating and cooling systems and refrigerators, the "actuators" (the furnace or compressor on the refrigerator or A/C system) are such that they cannot be modulated. They are either on or off. In the case of ovens used for baking, the costs associated with implementing modulated control are not justified by the application.

In theory, on-off control systems can use a single set-point to control the actuator. If the measured parameter is below the set-point, the actuator is activated and if the parameter is above the set-point, the actuator is deactivated. Unfortunately, this can lead to chatter (rapidly turning the actuator on and then off) when the measured parameter is very close to the set-point and moves slightly above and below the set-point. In practice, on-off control systems work best by setting two thresholds: one above and one below the desired set-point. For a heater, if the ambient temperature is below the upper threshold, the actuator (furnace) will be turned on. It will remain on until the ambient temperature crosses the upper threshold, at which point it will be turned off. The furnace will remain off until the temperature again falls below the lower threshold, at which point the cycle will begin again. You might recognize this behavior as similar to that of the comparator with hysteresis that you encountered in Chapter 11. The two thresholds result in a path-dependent behavior of the furnace and therefore hysteresis in its response to temperature changes. In the terminology of controls engineers, the gap between the two set-points is referred to as the **dead-band**. The state of the furnace will only change if the temperature strays outside the dead-band.

On-off control systems are viable solutions when a number of conditions are met. First and foremost is that the system requirements allow for the variations in the output response that are inherent in on-off control. In the case of a home heating system, we humans are not so sensitive to temperature that slow swings of a few degrees are troublesome. Baking in an oven is also generally tolerant of small swings in temperature. You could contrast these situations to the temperature control necessary for a chemical reaction chamber, where small changes in temperature can produce large changes in the products of the reactions. In such a situation, on-off control would probably not be acceptable.

The second condition that must be met in order to consider using on-off control is that the combination of the dynamics of the system and the "strength" of the actuator be such that the peak-to-peak variation in the controlled parameter can be limited enough to meet the requirements of the system. To understand this condition, consider the case of an oven with a very large heating element. Because of the thermal capacity of the heating element and the fact that heat does not travel instantaneously, it will take some time after the heater is turned on before the temperature in the oven stops falling. During this time, the power from the heater is going into raising the temperature of the heating element well above the temperature of the oven so that heat can be transferred into the air in the oven. The temperature in the oven will initially continue to fall, then change to rising, as more and more heat is transferred into the air. When the temperature at the sensor reaches the upper threshold, it will turn off the power to the heating element. However, the temperature in the oven will continue to rise because the heat capacity of the heating element (which is at a temperature higher than the oven temperature) will continue to add heat to the oven air for some time after the power is turned off. This will result in the actual temperature overshooting the upper threshold during heating in much the same way that it undershot the lower threshold during cooling. How much it overshoots and undershoots will be a function of the thermal dynamics of the system, governed largely by the heat capacity of the heating element and the rate of heat transfer to the air in the oven. For an on-off control system to be viable, it must be possible to limit the excursions to a small enough amplitude to meet the needs of the application. In the case of our oven, for example, we wouldn't want to oscillate between 300° and 400° to achieve an average 350° temperature.

28.5 LINEAR CLOSED-LOOP CONTROL

If the system requirements cannot tolerate the excursions that are inevitable with on-off control, we turn to some form of feedback and closed-loop control in which we modulate the output to intermediate values in order to achieve finer control. This is the domain of **linear closed-loop control**. We'd like to note that entire classes—indeed entire series of classes—are devoted to this subject. Our treatment of it will cover a few pages, with the disclaimer

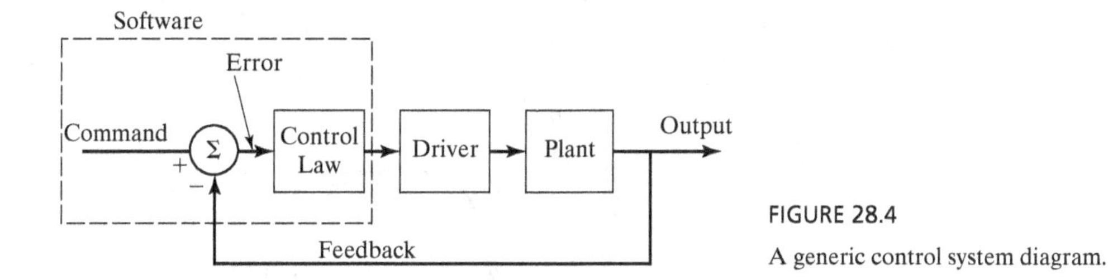

FIGURE 28.4
A generic control system diagram.

that we intend this to serve as an appetizer for a vast feast of a topic. Numerous excellent texts [1–3] are available for students who wish to learn more, but this is outside the scope of this text.

In this section, we are going to show you how you might implement linear closed-loop control of a system by incrementally developing an example of controlling the speed of a DC motor. There is nothing unique to motor control in what we will develop. You might perform a similar process for any system that you would like to control and get similar results. This system (the DC motor) simply serves as a concrete and intuitive example and one that can provide real data which we can use to evaluate the algorithms that we develop.

Closed-loop control systems are often represented schematically with diagrams like Figure 28.4.

In the example we will develop, the elements inside the dashed box are implemented in software inside the microcontroller, but this is not necessarily true of all control systems. This could also be implemented in hardware. The Driver box contains the power electronics necessary to drive the Plant, which is the thing that we are trying to control. In this case, the Plant is the DC motor. The output of the Plant, the parameter that we are seeking to control, is the speed of the DC motor.

28.5.1 Getting Started

If you needed to develop an algorithm for controlling something, you might logically consider starting with a thoughtful analysis of how you would personally go about manually controlling the device. That provides an excellent intuitive foundation, so let's do that here. This is the scenario: you are manually controlling the speed of a DC motor with gearbox. You have control of a knob that affects the voltage applied to the motor. The feedback that you have is an indication of how fast the motor is currently turning. Take a moment and think through the set of rules, your **control law**, that you might come up with to describe how to vary the knob to get the motor to run at the desired speed.

One of the things that you may have come up with is to look at what the motor shaft speed is relative to where you want it to be and use the difference to decide what the knob setting ought to be. The further away from the desired speed the motor is, the more control effort you should exert to get the actual motor speed to match the desired speed. If the motor is spinning too slowly, you should increase the effort (increase the voltage) and if it's too fast, you should reduce the effort (decrease the voltage). Mathematically, you might express that algorithm as:

$$\text{Control Effort} = (\text{Where We Want to Be} - \text{Where We Are}) \times \text{Scaling Factor} \quad (28.1)$$

The difference is arranged to give a positive result when we need to increase the speed of the motor. The difference between *WhereWeWantToBe* and *WhereWeAre* could be thought of as the **error** in our speed. Indeed, this turns out to be a central quantity in closed-loop control, so let's give it the name *Error*. In the parlance of controls, the scaling factor is

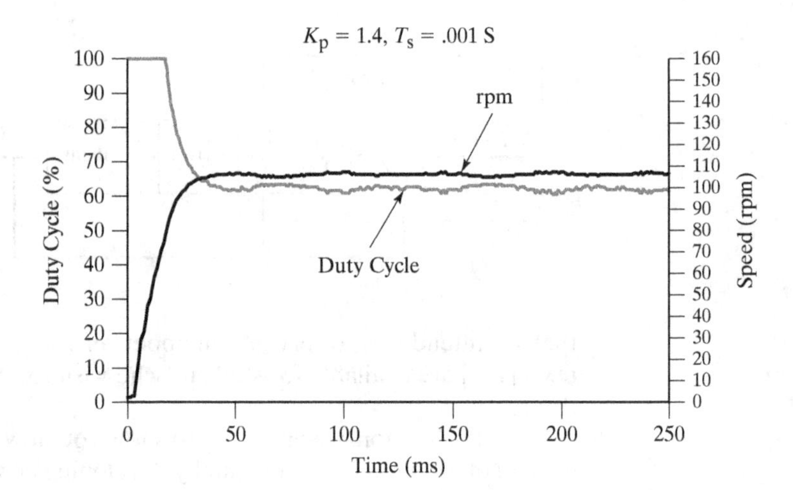

FIGURE 28.5

DC motor step response using only proportional gain.

referred to as the **gain**. In this case, the gain is applied directly to the error yielding an effort that is proportional to the error, so we call this the **proportional gain**. It is usually given the mathematical symbol $\boldsymbol{K_p}$.

We could implement this control law with a snippet of C code like this:

```
RPMError = TargetRPM - CurrentRpm;
RequestedDuty = (RPMError * ProportionalGain);
SetDuty(RequestedDuty);
```

We ran this code on our DC motor-gearbox test setup with a desired output shaft speed (TargetRPM) of 150 rpm. The results are shown in Figure 28.5.

This is a plot of motor speed and duty cycle (our control effort) as a function of time as the motor is started. This is referred to as the system's **step response** because at time = 0, the commanded speed stepped instantaneously from 0 to 150 rpm.

By examining this plot, we can see a number of things about the performance yielded by this control law. Notice that initially, the duty cycle is at 100%. Here, the error multiplied by the proportional gain was enough to force the control effort to its maximum value. As the motor accelerated toward the target rpm, the error dropped and at a time of about 20 ms, the (RPMError × ProportionalGain) term no longer resulted in 100% duty cycle, and the commanded duty cycle started to drop. Shortly after that, we reach a stable equilibrium point where the commanded duty cycle achieves a (relatively) constant speed. Note that there is a little bit of jitter (±1.1 rpm) in the motor's speed. While it may be almost constant, it is, unfortunately, not the speed we were looking for. We commanded 150 rpm, but only got about 106 rpm. You might suggest increasing the gain in order to reduce the rpm error. That sounds like a good idea, but when we tried it, we didn't get the desired results. Increasing the proportional gain too much produces an unstable system with a step response like that in Figure 28.6.

Here, we increased the proportional gain in an attempt to reduce the steady-state rpm error. While the average speed of the motor has increased to about 145 rpm (still less than the commanded 150 rpm), the fluctuations in the speed have also increased dramatically. By looking at the duty cycle trace, we can see that those variations in speed are the direct result of the large swings in the commanded duty cycle. This system is unstable, with the command exhibiting large-scale oscillations.

A quick look at our control law can explain why this approach will never produce a system that achieves the desired rpm. If the error goes to 0 (meaning that we have achieved the target rpm), the control effort goes to zero as well. Without any control effort,

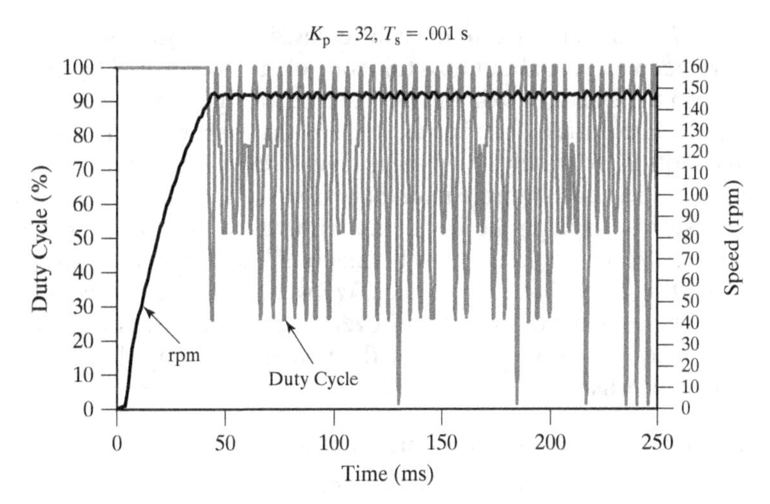

FIGURE 28.6

DC motor step response using only proportional gain, with the proportional gain set too high.

the controller will not be able to maintain the commanded motor speed and therefore a condition of zero error. In this system, a closed-loop control law based only on proportional gain will never achieve the desired rpm.

28.5.2 Getting Smarter

With the benefit of the experience we gained in the section above using only proportional gain, let's return to the mental exercise of approaching the development of the control system in ways that mimic how we might personally control the motor manually. We might reasonably say: "When I saw that I was not getting to the desired rpm, I would start ramping the voltage up so that I would eventually get there." It turns out that this is a very good next step in improving our control. What we need is a way to look back in time and notice that we are constantly failing to meet the desired rpm (a persistent error) and make an adjustment to the commanded duty cycle to drive the error to zero. One way that we might do that would be to integrate all the past errors and use that integrated error to add to our control effort. While there are several reasonable approaches to implementing a discrete integrator, one of the simplest, and most intuitive, is known as the trapezoidal rule and is shown in Eq. (28.2).

$$\text{Current Integral Error} = \Delta t \times [(\text{Error at Last Step} + \text{Current Error})/2] \quad (28.2)$$

If we are sampling at a constant rate (a requirement for this approach to control), then the Δt term will be a constant and we can pull both the Δt and the division by 2 out of the calculation (actually, move them to the gain term), leaving the integral as simply the summation of the errors. This is a way to condense a history of error into a single number that we can use for control. We might express this new approach mathematically as

$$\text{Control Effort} = (\text{Error} \times \text{Proportional Gain}) + (\text{Integral Gain} \times \Sigma\text{Error}) \quad (28.3)$$

The summation is the discrete equivalent of integration, so we refer to the scaling factor that we use with the summation as the **integral gain**. This type of control, with a proportional term and an integral term, is commonly referred to as **PI control** or **proportional integral control**. The gain on the summation of the error is usually given the symbol $\boldsymbol{K_i}$.

With this addition to the control law, it is time to look at an implicit assumption that we have been making about how this control law is executed by the microcontroller. For a summation to be an approximation of an integral, there must be some other variable over which we are integrating. In this case, time is that other variable. What has been implicit in the assumptions represented by these control laws is that they are executing on a very regular basis with a constant time interval between updates. This is a very important point. Without precisely controlling the frequency with which the control law is executed and the output updated, time would need to be made an explicit part of the control law and that would make things much more complex. We want to avoid that complexity, so we will arrange to have the control law execute on a very regular basis (in this case, every 1 ms).

The C code to implement the combined proportional plus integral (PI) control law would look like:

```
RPMError = TargetRPM - CurrentRpm;
SumError = SumError + RPMError;
RequestedDuty = ProportionalGain*(RPMError + (IntegralGain * SumError));
SetDuty(RequestedDuty);
```

Note that the RequestedDuty expression used in the code snippet is a rearrangement of Eq. (28.2). It makes the proportional gain act not only on the error term but also on the integral term. This is one of several common expressions used to implement closed-loop control, and it will have advantages later (in Section 28.5.5) when the time comes to determine the necessary gains for a system.

When we tried this program on our experimental system, we obtained the response shown in Figure 28.7.

The initial response of the system is very similar to that obtained with proportional-only control, since in this region, the proportional-only control generated enough control effort to force the duty cycle to its maximum (often referred to as **saturation** of the actuator). By the time the proportional control effort starts to drop off (after about 20 ms), the integral term has had enough time to accumulate an integrated error and it starts to contribute significantly. The result of the integral term is the slow rise of the motor speed to the desired set-point. Our new PI control is working very well on the test system under this set of conditions. However, one of the drawbacks to integral control is that it tends to destabilize a system. If we add integral control to a system that was only marginally stable with proportional-only control, we

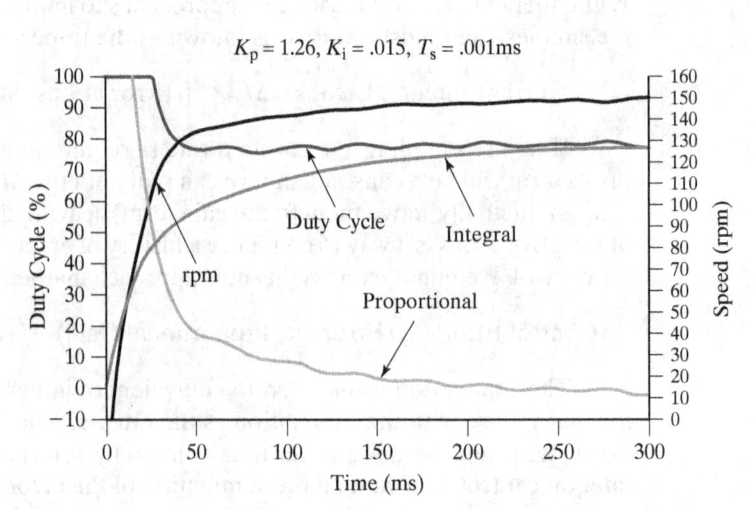

FIGURE 28.7 Step response with PI control.

would have to reduce the proportional gain in order to keep the system stable. We'll probe this issue in the next section when we consider disturbance rejection, and then we'll revisit the effects of adding integral control a little later when we talk about choosing gains.

28.5.3 Disturbance Rejection

Now that we have a system that can reach a stable commanded speed, we can start to look at what happens to the system if we introduce a disturbance. This is really the best reason for implementing closed-loop control in any system: it allows the system to detect and adjust for unpredictable changes in the operating environment in order to maintain the commanded output. The ability to overcome these external disturbances is known as **disturbance rejection**. Figure 28.8 shows the response of our DC motor example system to an external drag force being repeatedly applied and removed from the system. In this case, we gently pinched the wheel attached to the motor to create the disturbances.

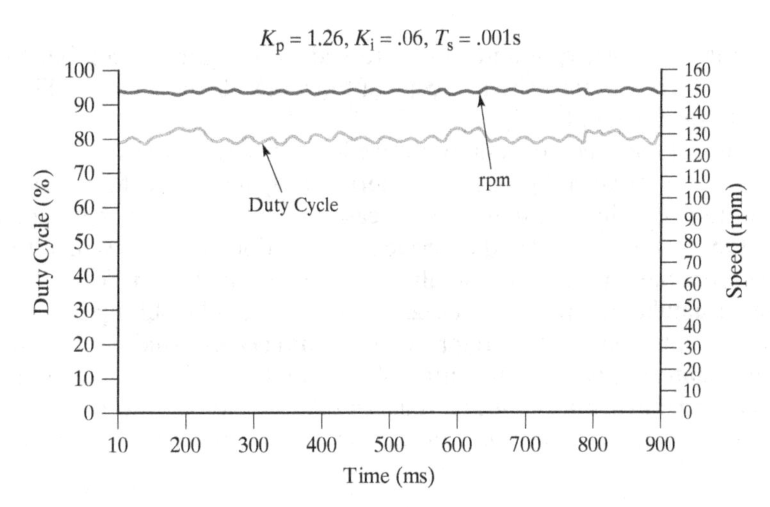

FIGURE 28.8
Disturbance rejection with PI control.

If we just looked at the rpm trace, it would be hard to tell that there had even been any disturbance. The system is doing a pretty good job maintaining the set-point! Looking at the duty cycle (the control effort), we can see when the external load was applied (150–250 ms and 575–650 ms), since it resulted in an increase in duty cycle in order to maintain the commanded rpm.

The system seems to be working pretty well under the conditions imposed so far, but to insure stability, we should also look at what happens when we push it beyond its capabilities. In terms of disturbance rejection, this means applying an external load that is larger than the controller can compensate for, even by saturating the actuator and going to 100% duty cycle. When we subjected our test motor setup to such large disturbances with the same PI control system described above, we found that the system responded as shown in Figure 28.9.

Just past the 200 ms point in the test, we applied the large external load and observed that the control effort quickly saturated at 100% duty cycle while the speed dropped below the target, indicating that we exceeded the ability of the control system to respond to the disturbance. This is exactly as we expected, since we intentionally introduced a large disturbance to observe the response of the system under these conditions.

The problem with our simple PI control approach was revealed when we suddenly removed the load (approximately, 375 ms into the test). At that point, the motor speed

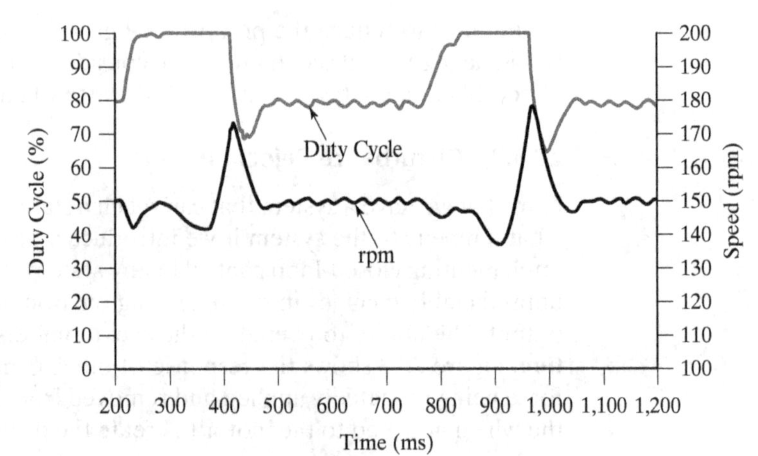

FIGURE 28.9

Step response of a PI control system to overload from a disturbance.

overshoots the target rpm and we are forced to wait for the combination of integral and proportional control [mostly the decay of the integral control term (Figure 28.10)] to bring us back to the set-point.

Why did the speed overshoot? If we look closely at how the algorithm works and the values of the integral and proportional terms, we can see. For the integral term, we are summing up the error in an attempt to increase the control effort that is being exerted on the motor. Once we reach 100% duty cycle, we have done all that we can to overcome the disturbance, but the error kept accumulating. When we finally removed the disturbance, we had a large accumulated error that caused a large command resulting in overshoot of the target speed. This phenomenon is referred to as **integrator windup**. The solution is to build **integrator antiwindup** into our control algorithm. The easiest way to do this is to monitor the commanded control effort. Once it reaches 100%, we should stop integrating the error, since we are already doing everything that we can to reach the target. We should also do the same

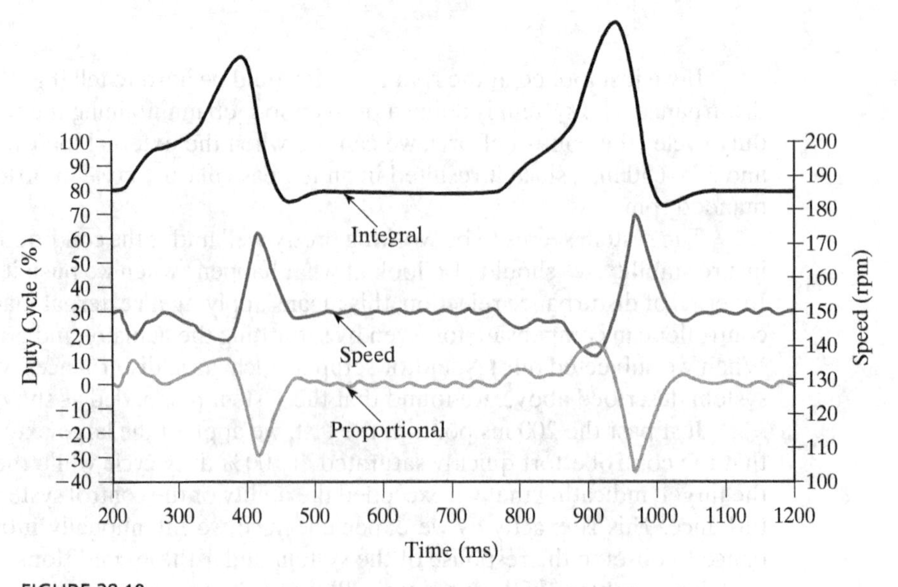

FIGURE 28.10

Integrator windup showing proportional and integral contributions.

thing in cases where we have commanded 0% duty cycle, since, at that point, we have done all that we can to slow the motor down. With that modification, the algorithm looks like:

```
RPMError = TargetRPM - CurrentRpm;
SumError += RPMError;
RequestedDuty = ProportionalGain*(RPMError + (IntegralGain * SumError));
// add Anti-Windup for the integrator
    if (RequestedDuty > 100) {
        RequestedDuty = 100;
        SumError-=RPMError;     /* anti-windup */
    }else if (RequestedDuty < 0){
        RequestedDuty = 0;
        SumError-=RPMError;     /* anti-windup */
    }
SetDuty(RequestedDuty);
```

The disturbance rejection performance of this algorithm looks like that shown in Figure 28.11 where we no longer see the large speed overshoot when the load is removed.

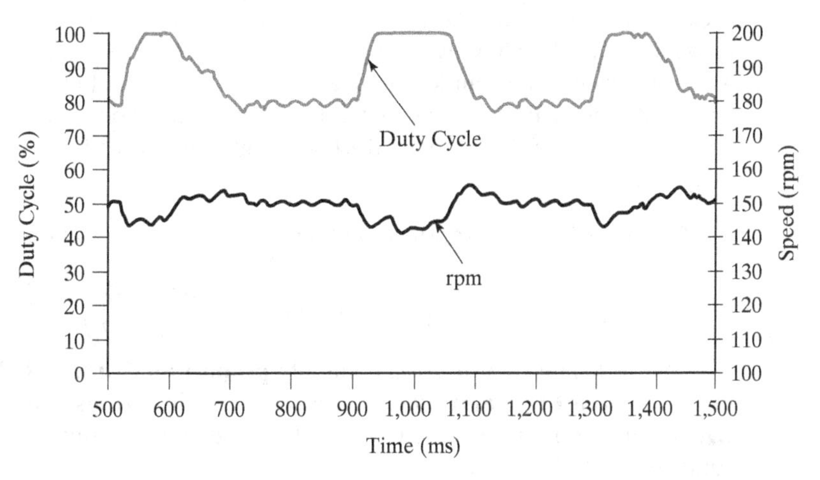

FIGURE 28.11

PI control performance with integrator antiwindup implemented.

28.5.4 Improving Performance Further by Adding Derivative Control

The system response achieved so far, as shown in Figures 28.7 and 28.11, is reasonably good—certainly better than we were able to achieve with open-loop control or proportional gain alone—but what if we'd like to improve the step response of the system further by decreasing the amount of time it takes the motor to reach the target speed when given the initial step input change? We have two gains that we can adjust, the proportional gain and integral gain. What happens if we increase the integral gain, trying to get the system to approach the set-point more quickly? If we increase the integral gain, we would get a family of response curves like that shown in Figure 28.12.

Initially, adding integral gain improves the rise time. However, we quickly reach a point where adding more integral gain begins to result in overshooting the target. If we continue

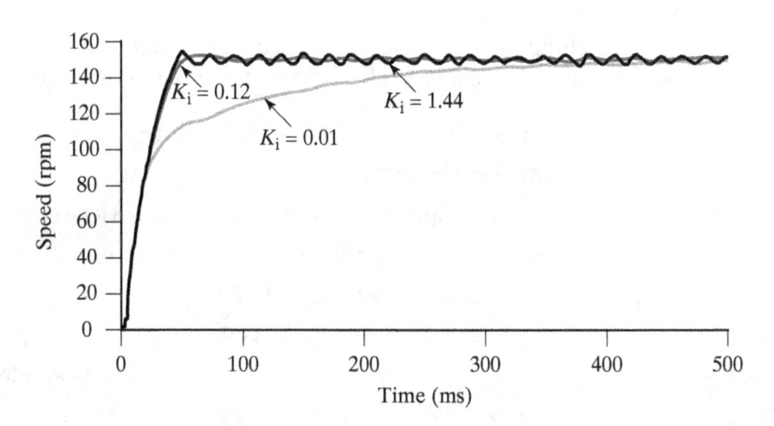

FIGURE 28.12
The effects of increasing integral gain for $K_p = 1.26$.

to increase the integral gain, we don't improve the rise time, but we drive the system into oscillation.

Adding more integral gain looks like it has the potential to improve the response, if we can just get rid of that pesky oscillation. We need a different approach, so let's go back to pretending to be the controller ourselves and see if we can come up with a way to deal with the overshoot that we get from adding more integral gain.

What if we monitor when we are approaching the set-point very quickly and start backing off on the control effort to keep from overshooting? In this way, we would be looking forward in time to where we think we are going and make adjustments now to better put us where we want to be. Our key indicator here is the rate of change of the rpm. If we added a term that looked at rate of change of the error (i.e., the derivative of the error) and used it to reduce or increase the control effort, we might end up with a control law that looked something like:

$$\text{Control Effort} = \text{Proportional Gain}\left(\text{Error} + (\text{Integral Gain} \times \Sigma\text{Error}) + \left(\textit{Derivative Gain} \times \frac{\mathrm{d}\textit{Error}}{\mathrm{d}T}\right)\right) \tag{28.4}$$

With this term, we have a complete **proportional, integral, derivative controller**, commonly called a **PID controller**. The derivative gain term is commonly assigned the symbol $\boldsymbol{K_d}$. For our example system, this results in the step response shown in Figure 28.13 which compares PI to PID control.

In this case, the improvement is not dramatic. If the system had less inertia, the effects of the derivative control would have been more pronounced. Adding the derivative term has

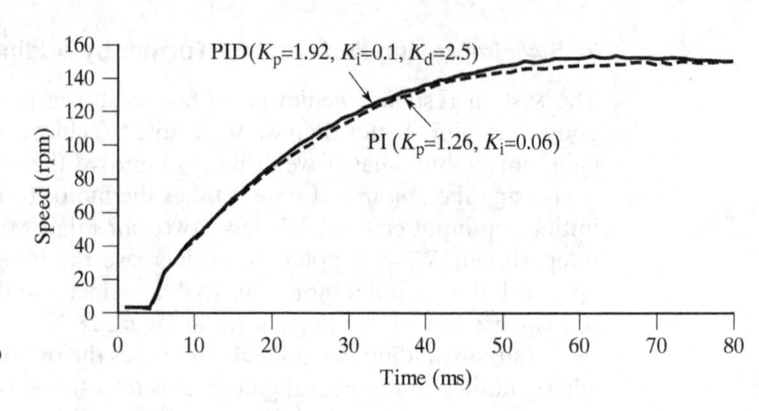

FIGURE 28.13
Step response with PID and PI controllers.

allowed us to increase the proportional and integral gains without driving the system unstable and resulted in a system that responds more quickly (e.g., it has a faster rise time) with only minor overshoot. In general, adding derivative control adds stability to a system. In this case, the added stability compensates for the destabilizing effect of integral control. The potential drawback to derivative control is that it tends to accentuate any noise present in the feedback signal. Small but rapid changes in the signal will be magnified by the derivative gain. To compensate for this tendency, it is common to implement a simple low-pass digital filter (Chapter 15) on the derivative term before using it in the control law.

To implement derivative control, we need to keep track of what the error was the last time we updated the control effort so that we can develop a term that is the rate of change of the error. Once again, implicit in this formulation is that the updates are happening at a very regular time interval so that we can treat the simple difference in the error as a rate of change (derivative). This parallels the development of the integral as a simple summation of the error in Section 28.5.2. In C code, this would look like:

```
RPMError = TargetRPM - CurrentRpm;
SumError += RPMError;
RequestedDuty = ProportionalGain*(RPMError + (IntegralGain * SumError) +
  DerivativeGain * (RPMError-LastError)));
if (RequestedDuty > 100) {
     RequestedDuty = 100;
     SumError-=RPMError;    /* anti-windup */
}else if (RequestedDuty < 0){
     RequestedDuty = 0;
     SumError-=RPMError;    /* anti-windup */
}
LastError = RPMError; /* update memory for derivative term*/
SetDuty(RequestedDuty);
```

28.5.5 Choosing the Right Gains

At this point, it should be getting pretty clear that selecting the proper values for the various gain terms is a key issue in creating a successful control system. This activity is often referred to as **tuning**. As we have shown in our running example with the DC motor, too much gain can drive the system into oscillation. Too little gain, and the performance is not very impressive, taking a long time to stabilize at the desired set-point. So, just how do you go about choosing the optimum gains? It turns out that this is not an easy question to answer.

There are three very broad approaches to finding the proper gains to use for a given system. Depending on the complexity of the system and your ability to carry out experiments on it, one of the three (or more likely a combination) will be the best choice. It will depend on your understanding of the dynamics of the system, your ability to carry out the necessary analysis, and your ability to experiment with the system.

28.5.5.1 Iterative Trial and Error

After the advent of feedback control but before we understood how to analyze the interactions between the system and the controller, trial and error was the only means available to develop a set of gains. As you might expect, it is the least reliable and predictable of the approaches in terms of its ability to produce an "optimum" set of gain values. To have a good chance of success requires experience on the part of the tuner and a set of requirements that

TABLE 28.1 The effects of adding the P, I, and D terms to a controller.

Control Term	Rise Time	Overshoot	Settling Time	Steady-State Error
K_p	Decreases	Increases	No Change	Decreases
K_i	Decreases	Increases	Increases	Eliminates
K_d	No Change	Decreases	Decreases	No Change

allow for less than ideal performance from the controller. The trial and error approach relies on understanding the kinds of effects that the different gains have on the various aspects of the system behavior. While this approach is not rigorous, it can yield acceptable results. And, perhaps more importantly, an understanding of the interactions of the gains is a useful bit of knowledge to have in your tool kit.

Table 28.1 shows, in broad terms, the effects of adding each of the control terms to the control law on various aspects of the closed-loop response. It can be used as a guide to which terms might be necessary or useful in a controller. Since the individual terms interact so strongly, it is only a general guide.

The way to read Table 28.1 is to say, for example, that adding derivative control to a P or PI controller would have no effect on the rise time or steady-state error of the controller while decreasing the overshoot and settling time.

On the other hand, if you are already working with a PID controller, Table 28.2 provides a guide to the effects to expect from increasing each of the gains in the control law. Again, because of the strong interrelationships between the gains, this is only a general guide.

The only unusual aspect of this table is the effect of K_i on the settling time. As long as the response remains overdamped (does not overshoot the target), increasing K_i will decrease the settling time at the desired value. Once the response transitions to an underdamped response (initially overshooting the target), increasing K_i will increase the settling time.

28.5.5.2 Ziegler-Nichols Tuning

In 1942, J. G. Ziegler and N. B. Nichols studied the problem of finding an appropriate set of gains for a PID controller. They developed a set of guidelines [4] for making a "best first guess" at what the values ought to be for each of the gains.

The plant is operated in open-loop control during tuning. The plant is subjected to a step function change in control effort, and the response of the system is measured. For many plants, this will result in a step response graph like that shown in Figure 28.14.

Here, the time d is a "pseudo-delay" from the time the command changed. T represents the process time constant and is arrived at by drawing a line tangent to the steepest part of the transition curve and extending that line upwards and downwards. T is the time difference between where the tangent intersects the original steady-state plant output value and where the tangent intersects the final plant output value. Both d and T should be expressed in terms of the number of control loop execution times in order to normalize out the effects of

TABLE 28.2 The effects of increasing the P, I, and D gains in a PID controller.

Control Term	Rise Time	Overshoot	Settling Time	Steady-State Error
K_p	Decreases	Increases	Increases	Decreases
K_i	Decreases	Increases	Decreases, then Increases	Eliminates
K_d	Increases	Decreases	Decreases	No Change

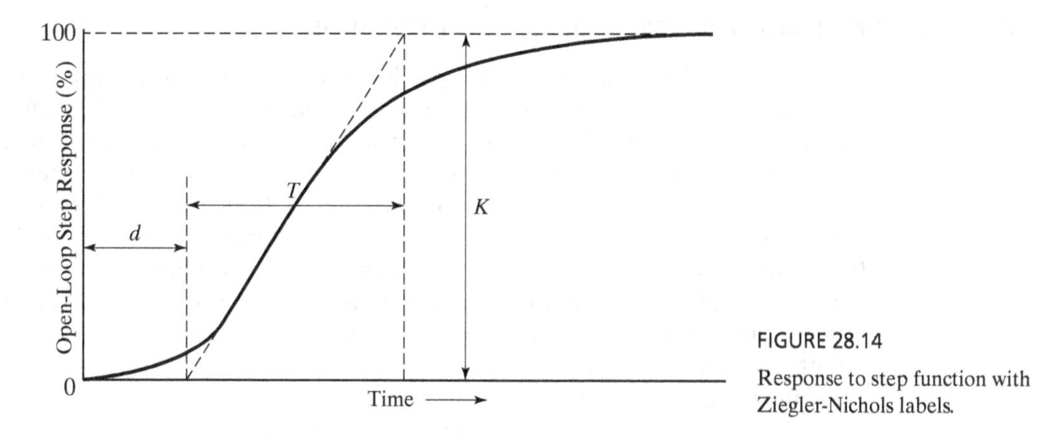

FIGURE 28.14

Response to step function with Ziegler-Nichols labels.

changing how fast the control loop executes. The value K is a measure of how much the plant output changed for a given command change. The ratio

$$G_p = \frac{K}{\text{Change in Control Effort}} \tag{28.5}$$

is referred to as the process gain (G_p).

In the Ziegler-Nichols open-loop tuning method, once you have measured these parameters, you can use them to calculate the PID gains using the formulas below:

$$K_p = 1.2\frac{T}{dG_p} \tag{28.6}$$

$$K_i = \frac{0.5}{d} \tag{28.7}$$

$$K_d = 0.5d \tag{28.8}$$

For our DC motor system example, we found the process gain (G_p) to be 2, the pseudo-delay (d) to be 5 ms, and the process time constant (T) to be 16 ms. The values of d and T were normalized to the control loop rate (1 ms) and inserted into Eqs. (28.6)–(28.8) to yield the Ziegler-Nichols gain values of $K_p = 1.92$, $K_i = 0.1$, and $K_d = 2.5$ for a system with a 1 ms control loop rate. These constants yielded a closed-loop system with the performance shown in Figure 28.13.

28.5.5.3 Tuning Based on Physical Models

The Ziegler-Nichols approach to tuning the PID gains is based on the assumption that the physical response of the plant can be modeled as a first order system (see Ref. [1] for details on the order of a system) plus a delay, or dead time. This may or may not be a good assumption for a given plant. When the plant dynamics are complex or unknown, this may be a reasonable assumption to make, or at least a good place to start. However, in many cases, it is possible to develop an analytical model of the plant based on the known dynamics of the system. Doing this requires some mathematics (the Laplace transform) that are beyond the scope of this book, but the material is common at some point in most engineering curricula. If you haven't encountered this yet, take heart: once you learn it and a bit of control theory, you will be able to skip most of the empirical tuning in favor of an analysis that will lead to calculated values for the gains. To learn more about how you can apply these analytical techniques to the calculation of the gains see References [1–3].

28.6 SYSTEM TYPE AND THE NEED FOR INTEGRAL CONTROL

In Section 28.5.1, we showed that proportional control alone would not work to control the speed of a DC motor. While proportional control alone does not work for that system and, in fact, a whole class of similar systems, that is not the case for all systems. There is another class of systems in which proportional control alone can work. Controls engineers would speak of the speed control of the DC motor as representing a Type-0 system. This is a reference to an exponent that appears in a mathematical representation of the behavior of the system. Type-0 systems are characterized by having a constant nonzero error when using proportional control alone. Using a more intuitive approach, we can think of Type-0 systems as those systems where a constant command input results in an output that stabilizes at some value. Another example of such a system would be an oven. For a given amount of heat input to the oven, the temperature in the oven will stabilize at some temperature.

The behavior of the speed of a DC motor and the temperature of an oven is different than, for example, a DC motor where the output of interest is *position* rather than velocity. A constant command input to a DC motor produces an output *position* that is constantly increasing. The position of a DC motor represents not a Type-0 system, but a Type-1 system, and this class of systems *can* be controlled with only proportional control. In fact, many positioning tasks can be controlled using P, or for higher performance, PD controllers and still result in systems with zero steady-state error without an integral control term in the controller.

28.7 SELECTING THE CONTROL LOOP RATE

So far, we have danced around the issue of the timing of the control loop by noting that it must be constant and that the time between executions of the control loop will affect the values that we use for the gains, but we haven't actually told you how to choose a rate. We'll now tackle that question, which has no easy answer.

You might reasonably ask: "What is the minimum rate at which I must run the control loop?" The answer is "That depends on your performance requirements." In an absolute sense, there is no lower limit on the rate at which we run the control loop. We can run it very slowly, at the expense of only being able to respond very slowly to changes in the system or its environment. If we wish to have a system whose response time approaches the maximum achievable, then we will need to operate our control loop at a rate comparable to or faster than the time constant of the system. In the case of our DC motor example, the mechanical time constant of the system (time to change speed to 63% of its final value) was found to be 36 ms. All of the closed-loop data that we have presented so far in this chapter was collected with a control loop rate of 1 ms. At this loop rate, the system is fairly resilient to changes in the values of the gains. If we slow the loop rate down, it becomes much more sensitive to the gains. Figure 28.15 shows the effects of changing the loop times on a PID controller for our example DC motor, each using gains calculated based on the Ziegler-Nichols tuning method.

At 1 and 5 ms loop rates, the systems are stable with reasonable step responses. At 10 and 20 ms loop rates, the systems are marginally stable and unstable, respectively. This is not to say that it is not possible to stabilize the system at slower loop rates. Figure 28.16 compares the Ziegler-Nichols gains at 1 and 5 ms with a set of hand-tuned gains at 10 ms.

From this, we can see that it is possible to produce very similar results at lower loop rates, at the expense of the time and effort of hand tuning the gains, since the Ziegler-Nichols based gains turned out to be marginally stable at this loop rate.

From these results, we can develop some rules of thumb for choosing the lower bound of the loop rate. At loop rates that are 1/7–1/10 the mechanical time constant of the system,

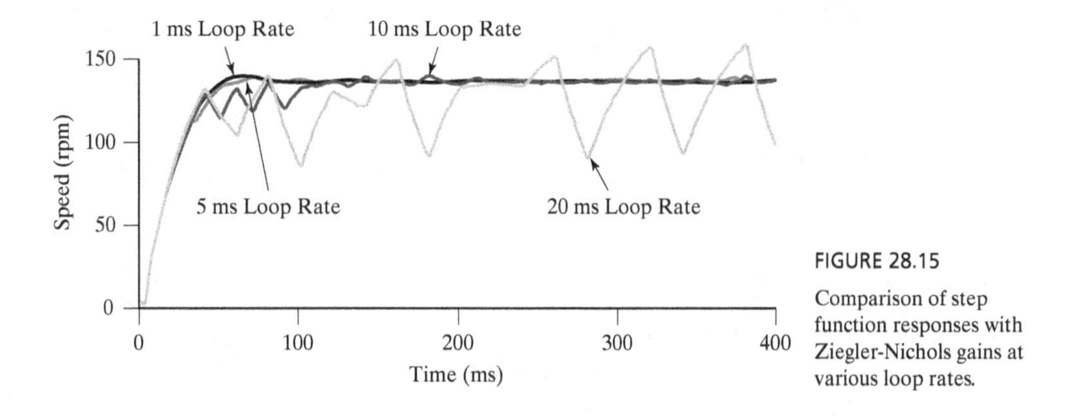

FIGURE 28.15

Comparison of step function responses with Ziegler-Nichols gains at various loop rates.

it is possible to use the Ziegler-Nichols method for choosing a set of gains that will produce a workable system. At loop rates 1/3–1/4 the mechanical time constant, it is still possible to produce performance similar to the higher loop rates, but at the cost of hand tuning the gains.

What about an upper limit on the loop rate? Is there one? In general, the faster the loop rate, the better the output control, as long as the system has a chance to respond between changes to the control input. In our system, we are using a 10 kHz frequency on the PWM drive for the motor. It would not make sense (and be counterproductive) to try and change the duty cycle before at least one full cycle of the PWM waveform had been output. Another situation in which there would be an upper limit would be that in which the system that we are controlling had a true time delay (as opposed to the pseudo-delay of the Ziegler-Nichols method). In the case of a system with a time delay, we wouldn't want to make a new change to the control effort until the last change that we made had a chance to propagate through the delay of the system and be reflected in the output that we are monitoring for our feedback.

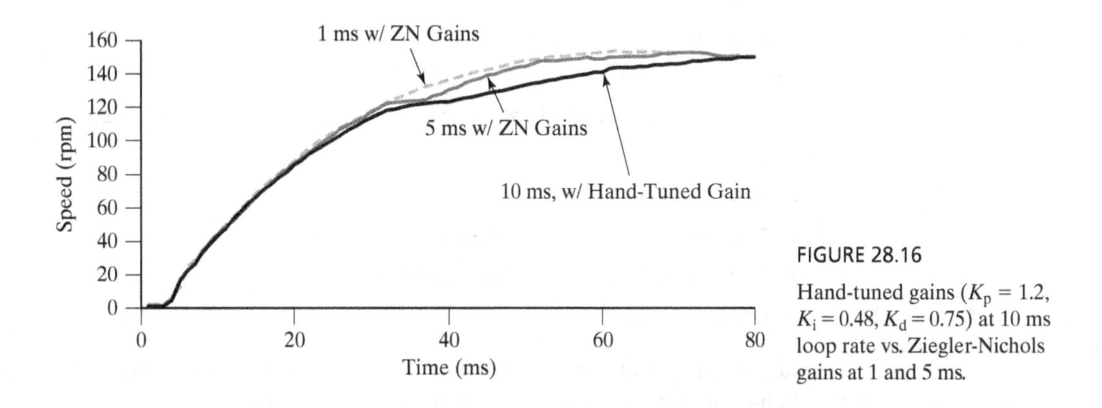

FIGURE 28.16

Hand-tuned gains ($K_p = 1.2$, $K_i = 0.48$, $K_d = 0.75$) at 10 ms loop rate vs. Ziegler-Nichols gains at 1 and 5 ms.

28.8 AD HOC METHODS

The linear closed-loop methods that we have discussed so far (P, PI, and PID) all assume at least an approximately linear relationship between the control effort and the response of the plant. There are more elaborate control theories for controlling systems in which there is a continuous but nonlinear relationship. But what about those systems where the control

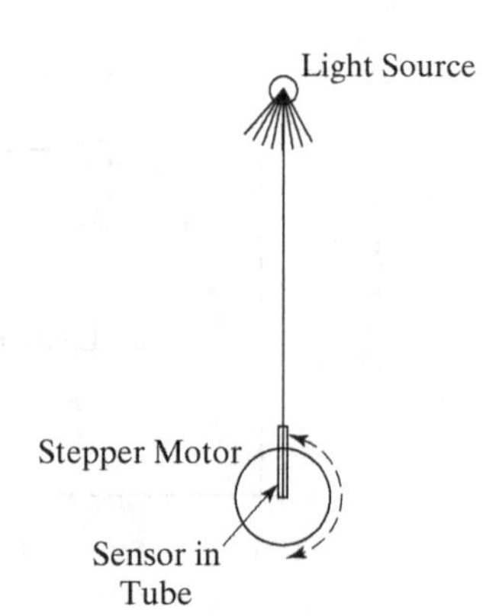

FIGURE 28.17

A stepper motor-based aiming/tracking system.

effort is discrete? An example of such a system might be an aiming system based on a stepper motor for positioning. The feedback signal would be the amplitude of the signal from a photo-sensor. Such a system might look schematically like that in Figure 28.17.

Here, we need an algorithm that accounts for the discrete nature of the steps from the stepper motor and for the time delay between when a step command is issued and when the system stabilizes at the new position. This time delay will become the maximum loop rate at which we run the control algorithm. We cannot reasonably make new changes until the effects of the last change can be measured. The algorithm itself should make use of the fact that the output from the sensor will be at a maximum when the sensor is directly aligned with the light source and the amplitude will fall off when pointed even a little bit on either side of the correct alignment. If we assume that the control algorithm will be called at an appropriate rate, we might express such an algorithm in pseudo-code as:

```
Persistent local variables: LastLightLevel, LastDirection
AimingAlgorithm:
    Read light sensor amplitude as CurrentLightLevel
    If CurrentLightLevel is less than LastLightLevel
          Change CurrentDirection to opposite LastDirection
    Else
          Leave CurrentDirection the same as LastDirection
    Endif
    Command 1 step in CurrentDirection
    Set LastDirection to CurrentDirection
    Set LastLightLevel to CurrentLightLevel
End AimingAlgorithm
```

Once aligned, this algorithm will result in an oscillation among three positions: perfectly aligned and one step either side of the correct alignment.

An analogous situation can also exist in systems when the control effort can be continuous (or approximately continuous) but the feedback is discrete. An example of such a system might be a mobile platform that is attempting to follow a strip of black tape on a light colored floor using reflective optical sensors. The response of the sensors is such that they are effectively digital in nature, indicating only the presence or absence of the tape. One possible arrangement of sensors for such a system is shown in Figure 28.18.

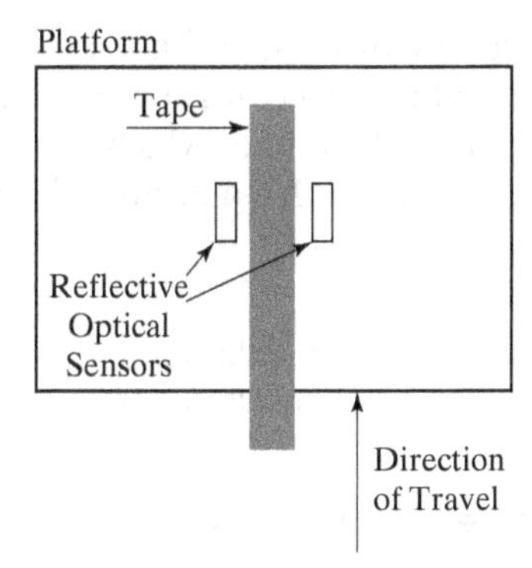

FIGURE 28.18

Possible sensor arrangement for tape following platform.

Here, the goal is to keep the left and right sensors off the tape. The platform begins with the sensors on either side of the line. If the platform moves straight and the line is straight, this situation will remain. If either the platform drifts slightly in one direction or the other or the tape line turns, then one of the sensors will move over the tape, indicating the need to turn the platform. The rate at which we turn the platform will determine how sharp a radius in the tape line that the platform will be able to follow.

The simplest approach to this problem is a solution that mimics the on-off control that we presented in Section 28.4. Here, there would be three possible steering states: Straight, TurningLeft, and TurningRight. The turning rates for TurningLeft and TurningRight would be chosen in order to navigate the tightest turns in the tape track. The algorithm to implement this approach would be fairly simple:

```
SimpleTapeTracking:
    If LeftSensor OffTape and RightSensor OffTape
          DriveStraight
    Else
          If LeftSensor OnTape
                TurnLeft
          Else RightSensor must be on tape
                TurnRight
          Endif
    Endif
End SimpleTapeTracking
```

When implemented, what we would find is that the platform would waddle along the line constantly bouncing off of the tape on alternating sides. The frequency of the waddling would be related to the tightest turn radius: the tighter the turn radius, the higher the frequency of waddling when following a straight line.

An algorithm like this, while workable, wastes a lot of time turning when the line is straight. We could improve on this if we had some way of inferring *how much* we needed to turn rather than always turning at a rate appropriate for the tightest turn. Since our sensors are digital, we need to look to another input. In this case, we can use the time since the last correction as an indicator of how much we should turn. If it has been a long time since the last correction, then we are probably drifting and need only make a shallow turn. On the other hand, if we were turning left at the last control input and we still need to turn left, then we should make a much tighter turn. The rate at which we wish to turn is inversely proportional to the time since the last correction. An algorithm in pseudo-code to implement this approach might look like:

```
Persistent local variables TimeOfLastCorrection,
DirectionOfLastCorrection, RateOfLastCorrection
BetterTapeTracking:
    If LeftSensor is OffTape and RightSensor is OffTape
         Continue in last direction (no new correction)
    Else we need to make a correction
         DeltaTime = CurrentTime - TimeOfLastCorrection
         If LeftSensor is OnTape
            DirectionOfCorrection is Left
            If DirectionOfLastCorrection is Left
                  RateOfTurn is RateOfLastCorrection +
                                 ScaledInverseOf(DeltaTime)
            Else
                  RateOfTurn is ScaledInverseOf(DeltaTime)
            Endif
         Else RightSensor is OnTape
            DirectionOfCorrection is Right
            If DirectionOfLastCorrection is Right
                  RateOfTurn is RateOfLastCorrection +
                                 ScaledInverseOf(DeltaTime)
            Else
                  RateOfTurn is ScaledInverseOf(DeltaTime)
            Endif
         Endif
         Issue SteerVehicle command based on DirectionOfCorrection
                                               And RateOfTurn
         RateOfLastCorrection = RateOfTurn
         DirectionOfLastCorrection = DirectionOfCorrection
         TimeOfLastCorrection = CurrentTime
    Endif
End BetterTapeTracking
```

The nature of the SteerVehicle command will depend on the design of the drive and steering system for the mobile platform, and the scaling of the inverse function to create the steering will depend on the nature of the steering/speed control mechanism on the platform. Appropriate scaling will need to be determined experimentally.

28.9 HOMEWORK PROBLEMS

28.1 Which PID term(s) contribute(s) instability to the system when the gain is too large? Which one(s) contribute(s) stability?

28.2 Write pseudo-code for a function to implement a bang-bang control law as described in Section 28.4.

28.3 Explain, in your own words, why proportional-only control can never reach the set-point.

28.4 A system that implements PI control exhibits a step response like that shown in Figure 28.19

(a) How would you modify the gains to eliminate the oscillatory behavior while preserving as much of the rise time as possible?

(b) If you were simply to add a derivative term (with some nonzero gain) to this system (keeping the other gains the same), how would you expect the response to change?

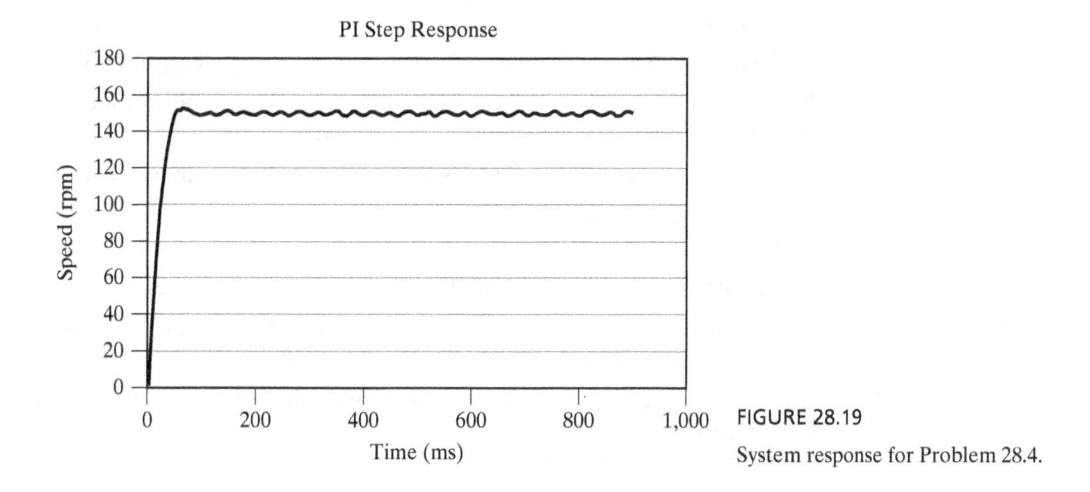

FIGURE 28.19

System response for Problem 28.4.

28.5 Lab-mates have collected some step response data from a system that they are trying to control using a PI controller. The plots of rpm and duty cycle are shown in Figure 28.20. They have asked you whether or not you think that they can increase the gains to improve the response of this system, and if so, which gain should they try increasing first?

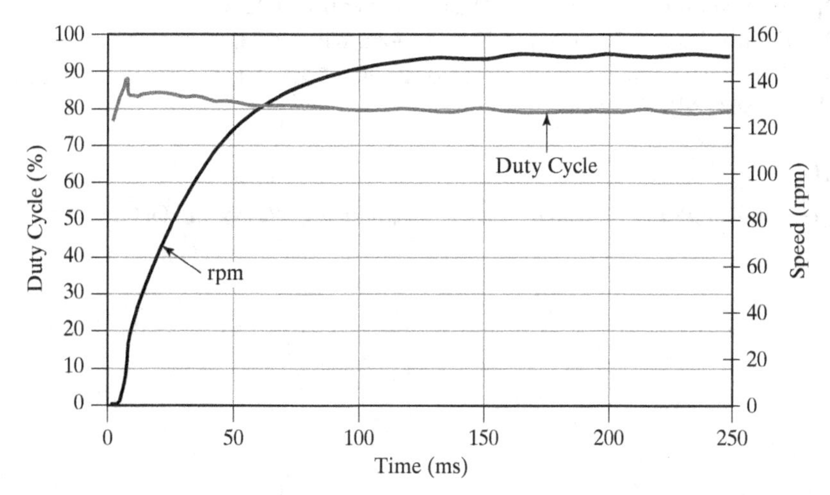

FIGURE 28.20

System response for Problem 28.5.

28.6 A particular mechanical system with a time constant of 100 ms is to be controlled using a PID controller running at a 25 ms loop rate. Would you expect to be able to use the Ziegler-Nichols formulas to calculate stable gains for this system? Explain your answer based on the guidelines from this chapter.

28.7 The results of an open-loop step response for a particular system yield the following data:

Pseudo-delay = 8 ms

$T = 25$ ms

Change in output = 20 rpm for a 10% change in commanded duty cycle.

If the system is to be controlled using a PID controller running at a loop rate of 1 ms, what should the Ziegler-Nichols gains be for K_p, K_i, and K_d?

28.8 The results of an open-loop step response for a particular system yield the following data:

Pseudo-delay = 2 ms

$T = 12$ ms

Change in output = 30 rpm for a 10% change in commanded duty cycle.

If the system is to be controlled using a PID controller running at a loop rate of 1 ms, what should the Ziegler-Nichols gains be for K_p, K_i, and K_d?

28.9 The results of an open-loop step response for a particular system yield the following data:

Pseudo-delay = 30 ms

$T = 120$ ms

Change in output = 30 rpm for a 15% change in commanded duty cycle.

If the system is to be controlled using a PID controller running at a loop rate of 5 ms, what should the Ziegler-Nichols gains be for K_p, K_i, and K_d ?

28.10 A particular actuator-driven mechanical system is found to have a time constant of 30 ms from actuation to stable response. Suggest a reasonable control loop rate to use in controlling this system. Explain your reasoning.

28.11 Only certain systems are good candidates for bang-bang control. What is the prime characteristic to consider when evaluating whether or not a system is suitable for use with bang-bang control?

REFERENCES

[1] *Feedback Control of Dynamic Systems,* Franklin, G.F., Powell, J.D., and Emami-Naeini, A., Prentice Hall, 2005.

[2] *Modern Control Engineering,* Ogata, K., Prentice Hall, 2001.

[3] *Control Systems Engineering,* Nise, N., John Wiley, 2003.

[4] "Optimum settings for automatic controllers," Ziegler, J.G., and Nichols, N.B., *Transactions of ASME,* 1942; (64)759–768.

FURTHER READING

PID Without a PhD, Embedded Systems Programming, Wescott, T., October 2000.

CHAPTER

Rapid Prototyping

29

29.1 INTRODUCTION

Mechatronic projects are often complex, far-reaching in scope, ambitious, and time constrained. They can also be a tremendous amount of fun. Because of this, mechatronic engineers are known for the zeal with which they address design challenges. Many have found that a key element of a successful project is the ability to proceed quickly from a concept to a functional, finished system. This process incorporates significant risk—it may indeed be *the* riskiest portion of a project, as the outcome of this stage is the definition of exactly *how* the requirements of a project will be tackled. In the unfortunate event that a poor path is selected (and this is not at all uncommon), the remainder of the project can be significantly more difficult and less pleasant, and the fate of the project itself can fall into question. Since it's so common to find yourself marching down a path that isn't working out, the best approach is to get to that realization as quickly as you possibly can, so that you have time to backtrack, change directions, and try something new. Building your prototypes *rapidly* gives you the best chance of success in this regard. Hence, the title of this chapter: Rapid Prototyping.

The goal of this chapter is to introduce the reader to (or, in some cases, remind the reader of) several prototyping techniques that produce results from which meaningful conclusions can be drawn regarding the promise of a given design approach. These techniques share the common trait that they produce results and prototypes quickly and relatively inexpensively.

In this chapter, you will learn:

1. prototyping philosophies: build in hardware vs. model in software,
2. for mechanical systems:
   a. what solid modeling is, and what its benefits are,
   b. prototyping with foamcore and other art store supplies,
   c. 2½-dimensional prototyping: laser cutters and less expensive laser-cutter-inspired techniques for creating complex and robust assemblies,
   d. tab and slot construction,
   e. leveraging the toy world: LEGO, Erector Sets, and K'NEX,
3. for electrical systems:
   a. components and materials,
   b. prototyping techniques: breadboard, wire wrap, perf board, and printed circuit boards (PCBs),
4. the necessities and benefits of testing and iterating,
5. suppliers and sources.

29.2 WHY MAKE PROTOTYPES?

When presented with a design challenge, a common impulse is to conceive a solution and then immediately set about building that solution in its final form. After all, if you get it right the first time, you'll quickly be finished! Right? Wrong! As an example, consider a race car chassis designer, who thinks of a new design for a tubular welded frame, and then rushes to the shop to begin cutting and notching tubing, and welding it together. On the third day of the fevered construction of this hypothetical chassis, the designer finds that the chassis doesn't have a mounting point for the transmission, and that the front wheels won't fit. The existing frame is useless, and has to be hauled away to the scrap yard. Three days of intense work in the shop have been wasted, and worst of all, they'll have to be repeated when a new design is started from scratch. This could have been avoided by spending a few hours or days creating a simple mockup using dowels and glue, or investing the time to create a 3-D CAD solid model of the design.

This impulse to rush forward and build exists in practically all disciplines: software development, circuit design, machine design, creative writing, technical writing, and so on. While this approach can *sometimes* produce good results, a more common result is that a lot of time is spent on an unusable first iteration of the project. This phenomenon is so widespread that Frederick Brooks, in his well-known book "The Mythical Man-Month," writes "plan to throw one away; you will, anyhow."

The key idea is that the first iteration, the first prototype, usually has a significant number of serious flaws and oversights, the result of things the designer had not considered. These almost inevitable shortcomings imply that the first prototype should be completed as quickly as possible, incorporating only the bare minimum of physical and esthetic integrity necessary to reveal these flaws and oversights. This will position the designer to begin the next prototype, the next iteration, informed by the lessons learned from the first prototype. The number of iterations is likely to be greater than 1—sometimes much greater. As the iterations continue, changes between iterations usually become less substantial, and the design stabilizes and eventually meets the full set of requirements. Sometimes, however, the prototypes indicate that a particular approach has no promise at all—there are just too many serious flaws to overcome. In both cases, the prototypes are key to the success of the designer and the project.

The ultimate question is: How do you create a prototype as quickly as possible that doesn't reveal more about the weaknesses of the prototyping techniques or simulation tools than the design under evaluation? This is no small feat, and finding the perfect balance is a lifelong pursuit. In this chapter, we share information about a few approaches and tools that we've found to be effective, and hope that they will also speed you towards working systems in your design work. The goal of this discussion is primarily to expose you to practical ideas and techniques, rather than teach you much about how to employ them. In this way, you will have seen them and placed them in your mental inventory of things to consider. Once you determine that one or more could be useful, and wish to learn more, a text or other resource specific to each technique would be your next step.

29.3 PROTOTYPING PHILOSOPHIES: BUILD OR SIMULATE?

For both electrical and mechanical systems, there are a wide variety of options for simulating designs and also for rapidly prototyping designs. The reasons to choose one approach over another depend on the circumstance of the project, and there isn't a consistent choice that can be applied in all cases. For instance, it may be the best choice in one case to quickly build a model of a concept to see if it holds promise and see if anything basic has been overlooked. In another case, it may be best to construct a 3-D CAD solid model of an assembly and constrain it so that it incorporates the envisioned geometries or ranges of motion. With either approach, you will learn more about the potential performance of a final design: it may work flawlessly as originally conceived (a rare but delightful event), or it may reveal the flaws in your logic. Parts may not work unless they are somehow able to violate the laws of physics, or perhaps materials properties are exceeded. Also, it may be too difficult to assemble or manufacture. Building hardware and software models can reveal all these things and more.

Given that there is substantial room for error with either approach, the best answer is to use both approaches to the maximum possible extent. The extent to which this is possible will be governed by the resources you have available to you during the project: How much time do you have? How many people are on your team and what are their skill sets? How familiar are you or your teammates with the simulation and modeling tools at the beginning of the project? How much experience do you have building models and using shop tools? Will you have to learn how to use the tools (simulation or shop) during the project or have you already mastered them? All of these factors will determine which approach or approaches are most appropriate for your project.

One substantial benefit of simulation and modeling is that iterations of simulation prototypes are inexpensive—both with respect to time and material costs. The up-front investments for simulation may be high, however, and include learning how to use the simulation tool and the time required to input the relevant characteristics of your design into the model. But this initial period of modeling and simulation is indeed an *investment* in many senses of the world: it has the potential to pay off richly. The time required to build the model may be substantial, but the information gained during that process can be highly valuable.

For instance, during the construction of a solid model of a four-bar linkage, you may discover that the method you were planning to use to create the hinge joints will not permit a particular member to rotate 360°, as the design requires. Changing the design of the joint in the solid model can be done very quickly and easily compared with a physical model. In the case of circuit design, a simulation of a digital logic circuit may be able to tell you whether a scheme you have created for managing the inputs to a motor drive circuit incorporates all the correct logical inversions and truth tables. Simulating your circuit with **SPICE** (**Simulation Program with Integrated Circuit Emphasis**), or any of several other readily available circuit simulation applications (which we introduced in Chapter 9, Section 9.10), might take a few hours to input and simulate. Compare both of these examples with a "build-it-first" approach, which would have required you to obtain components and supplies, either through the mail or by making trips to local vendors, and then building and testing the designs—possibly to find that you need additional components. The entire cycle could take days, and is highly likely to be repeated several times!

Unless the assignment is purely a paper exercise or simulation, you will ultimately have to construct your design and demonstrate that it works. The phrase "the proof is the pudding" is particularly apt with prototyping—not that there's much in common between mechatronics systems and pudding, but only when a system has been realized and demonstrated do you *know* it will actually work. The goal should be to prototype *and* simulate your designs as quickly as possible, allowing for as many iterations as possible, given the time and resources available.

29.4 RAPID PROTOTYPING OF MECHANICAL SYSTEMS

29.4.1 Solid Modeling Tools

For physical/mechanical systems, there really is no substitute for a scale drawing or a solid model. There are very few systems simple enough that you can get away with a hand sketch, or with no sketch at all, without finding a large number of surprises and show-stoppers.

Correcting surprises in real-time, as you build, will tax your creativity heavily, and doesn't often lead to a finished design that works well. It becomes too difficult psychologically to undo any of the work completed up to the point where problems were discovered, and the resulting design is very reactive. The story of the twists and turns taken during the construction-design process is sometimes apparent to even the casual observer. Instead, invest the time to model the system using at least a 2-D, but preferably a 3-D or solid modeling CAD program. Popular examples of these kinds of programs are Autodesk Inventor (from AutoDesk Inc.), SolidWorks (from SolidWorks Corp.), SolidEdge (from UGS), and SketchUp (*free* from Google), all of which are powerful solid modeling programs and have large numbers of users. They are intuitive and relatively easy to learn, and are available free or at low cost to students at many schools. Other high-end, professional-grade options include Unigraphics and Pro-Engineer. Lower cost, lower capability options are also available, such as QuickCAD (from Autodesk), Design Xpress (from Alibre), and TurboCAD (from IMSI), to name but a few. An Internet search will turn up these and many more—the authors are not recommending any particular package, but we are recommending the use of this *type* of tool.

In addition to enabling designers to explore how their designs will fit together and look with a CAD model, assemblies can be defined to allow specific parts to move relative to other components in the design, and this motion can be constrained. Animations of the device in motion or in use can be created with relative ease, so you can directly visualize how your design will look when it's in use.

Many solid modeling applications incorporate **finite element analysis** (**FEA**) tools, enabling designers to better explore quantities such as stress and strain, heat transfer, fluid flow, and electromagnetic fields for components or systems with complex geometries, where analytical approaches might be overly complex or altogether impossible. SolidWorks, for example, accommodates a wide variety of "plug-ins"—optional software modules that you can add to the application to analyze the physical behavior and response of the models you've built. These add cost (sometimes more than the cost of the solid modeling program itself), but the insights they can contribute are sometimes essential.

Solid modeling software is especially powerful when used in combination with some of the rapid prototyping techniques described below, such as **LaserCAMM** and **FDM** (**fused deposition modeling**). These techniques work directly with file types that can be created with solid modeling software, and can create 2-D and 3-D parts very quickly that match the modeled parts within a few thousands of an inch. Solid models also work very well in conjunction with the cheapest and easiest of physical modeling techniques, such as foamcore and hot glue, discussed in Section 29.4.3.

29.4.2 Modeling System Dynamics

Solid modeling programs can help you explore the shape of components and the interactions of multiple components in assemblies. Another category of software enables you to model a system's dynamics, including applications such as MATLAB and its companion system-modeling package Simulink (The MathWorks, Inc.) and Working Model (Design Simulation Technologies, Inc.). If dynamics will play a major role in the success of a system you are designing, then you will likely find tools like these to be indispensible. When model-based, closed-loop feedback control will be implemented, modeling software is almost always used.

29.4.3 Foamcore, X-ACTO Knives, and Hot Glue

When attempting to build a prototype and simultaneously optimize the amount of time spent, the number of new processes that must be learned, the cost of the prototype, and the learning that can be gleaned about the design rather than the weaknesses of the prototyping technique, it's hard to beat foamcore. **Foamcore** is a low-tech, widely-available composite

FIGURE 29.1

Typical foamcore sheet and detail of construction.

sandwich of paper (the "bread") and foam (the "meat"). A typical foamcore sheet is shown in Figure 29.1. It's stocked by most hobby shops, framing stores, and art supply stores. It is easy to work with: it can be cut with a hobby knife or other sharp blade, bonds easily with a variety of glues, and can be marked with paints and inks. It is inexpensive: somewhere between $5 and $10 per 24 × 32 in. sheet with 3/16 in. thickness (a wide variety of sizes and thicknesses are readily available).

It's essential that when you cut foamcore, you use a very sharp blade. This helps ensure that the resulting foamcore assemblies and glue joints are clean and strong. A high-quality and reasonably priced choice is a hobby knife, such as an X-ACTO knife. A very simple hobby knife, such as the one shown in Figure 29.2, will do the job in most cases. More extensive hobby knife kits, comprised of several handles, several different types of blades, and a

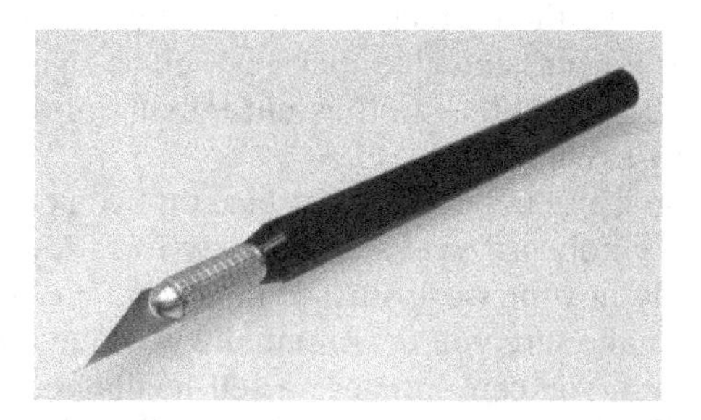

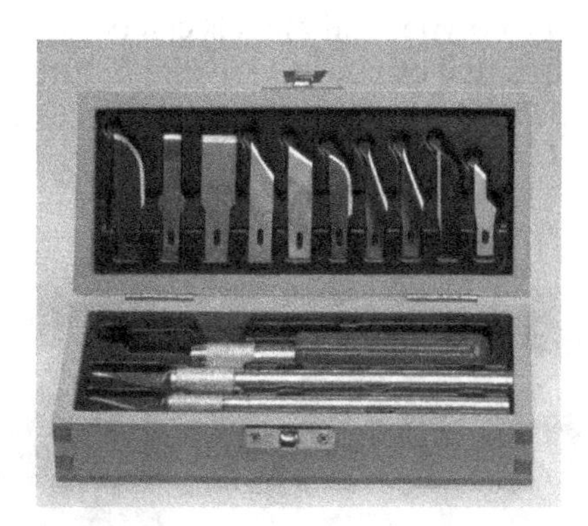

FIGURE 29.2

Hobby knives and blades. Top: inexpensive and simple handle. Bottom left: boxed set with handle and blade selection. Bottom right: extra blades in a plastic dispenser that incorporates a safe place to dispose of old, dull blades.

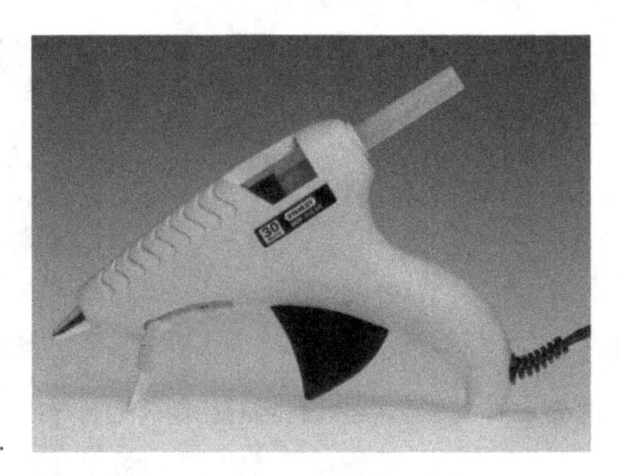

FIGURE 29.3
Hot-melt glue gun.

sturdy box, are also available for around $20. There are many other brands of hobby knives that work very well in this application. Regardless of which knife you choose, you will need to get a large supply of new blades—it doesn't take long before a blade becomes too dull to make clean cuts. Install a new blade at the first sign of dullness. Dispose of dull blades properly—they are still very sharp and can cause serious injuries if they're discarded carelessly.

Although a wide variety of adhesives bond well to foamcore, the most convenient choice is hot-melt glue. Hot-melt glue is quick, bonds in seconds, is reasonably removable when modifications are needed, quick to learn, and inexpensive ($10 is adequate to get a reasonably high-quality glue gun and several glue sticks). Figure 29.3 shows a typical hot-melt glue gun.

There are a few different types of glue guns and glue sticks, so care should be taken that you're buying the correct type for hobbyist or model-making applications. There are high-temperature guns and glue sticks, as well as high-strength glues, that may work perfectly well but will be overkill, as well as unforgiving of any errors. Stick with the quick-cooling low-temp, low-strength options.

You'll also need a work surface that you can gouge repeatedly with an X-ACTO blade. Leftover cardboard works adequately for a few cuts, but after that the slices you've made previously in your work surface begin to take the blade places you don't intend, so you'll need to make sure you're continually working on a fresh portion of cardboard cutting mat. Alternately, you can purchase a self-healing cutting mat (Figure 29.4), made precisely for this application. Depending on the size, brand, and quality, they cost between $5 and $50.

Construction of foamcore prototypes and models can be accomplished in a great variety of ways. Virtually any technique that can be used to join or form sheet metal or wood has an analog with foamcore, Figure 29.5 shows several possibilities. Comparisons to welding are

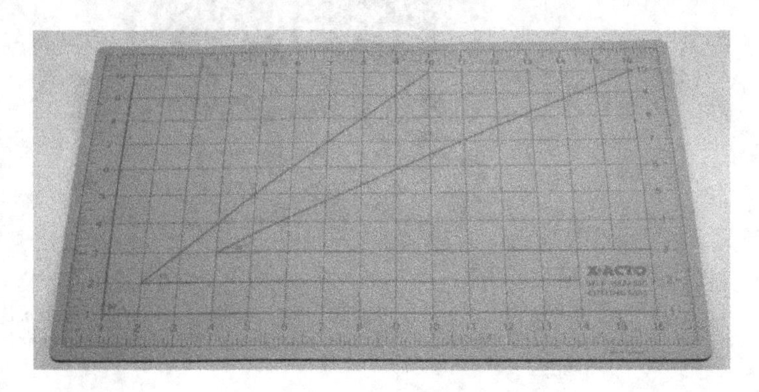

FIGURE 29.4
Self-healing cutting mat.

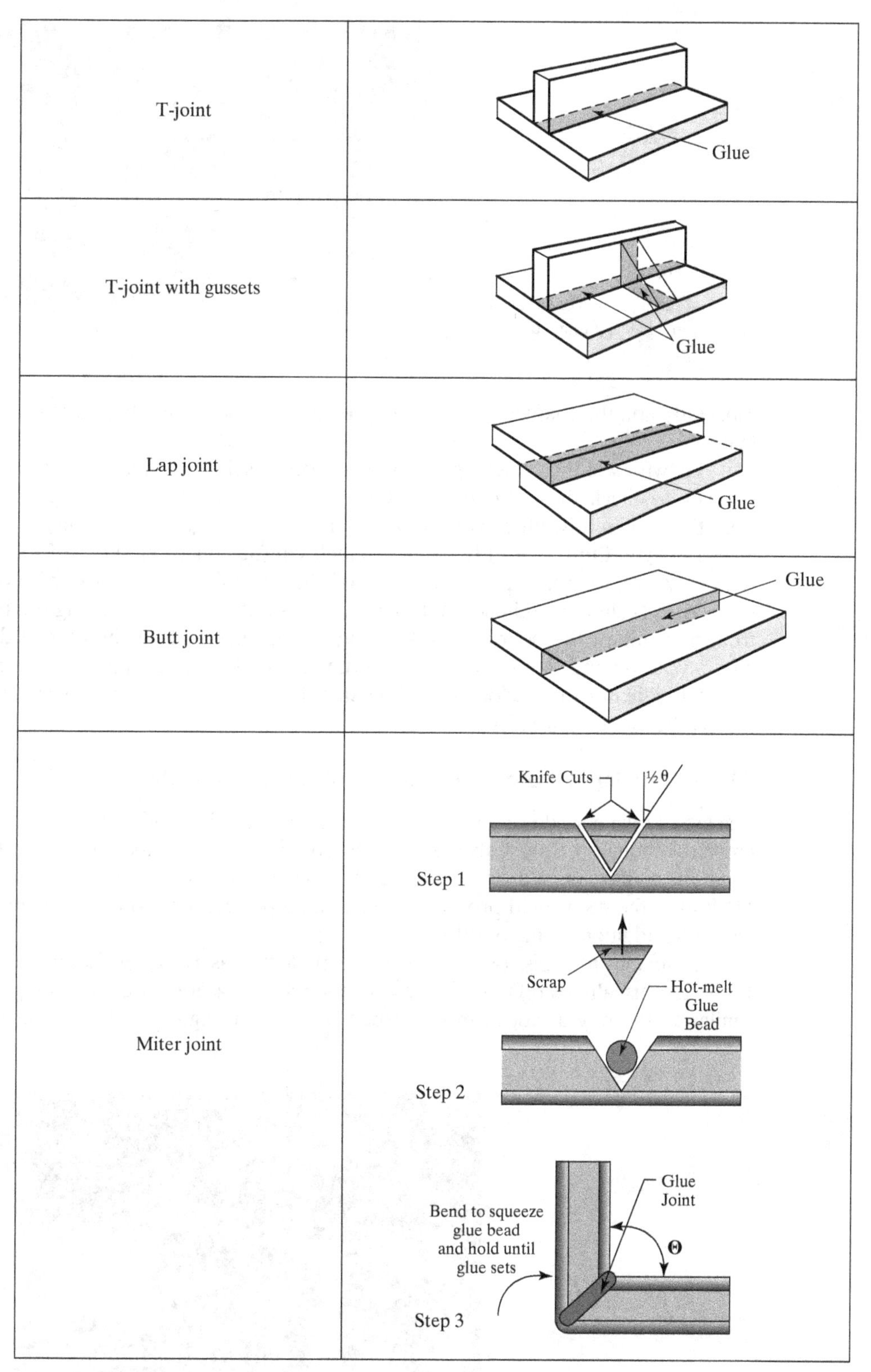

FIGURE 29.5
Examples of several effective foamcore jointing techniques.

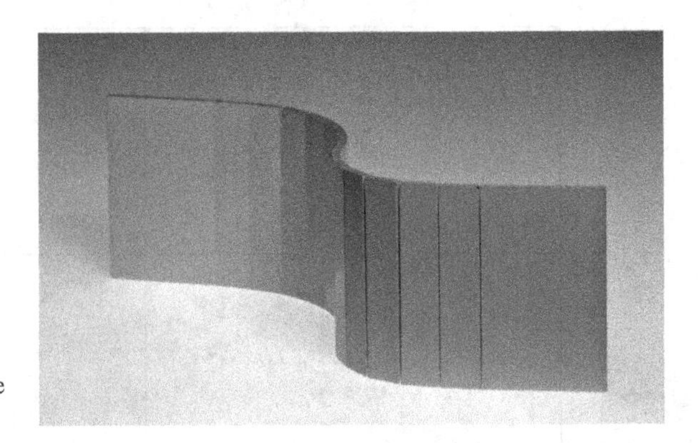

FIGURE 29.6
Example of a complex curve made using foamcore.

especially apt: the foamcore is the sheet metal and the hot-melt glue is the welding rod/fill material.

As with welding, "tack" glue spots are very useful for holding pieces in position temporarily to check fit and alignment. Once you're sure everything is positioned where you want them, then fully glue everything together with a larger shot of glue.

Lest you think that only straight panels can be constructed out of foamcore, it is also relatively easy to create complex curves of almost arbitrary complexity. To do this, create a series of cuts similar to those used to construct a miter joint. Removing the paper and foam from one side of the board while keeping the paper intact on the other side of the board allows you to bend the work piece into the desired shape and then lock it in position with hot-melt glue in strategic locations. An example of a complex shape constructed from foamcore is shown in Figure 29.6.

29.4.4 2-D Rapid Prototyping: Laser Cutting/LaserCAMM

The use of foamcore doesn't necessarily require that a solid model of the concept under test be created—though it almost always benefits from it. If a solid model has been built, or even if components have been drawn using a 2-D CAD program with specific file formats (typically a DXF file), there's a rapid prototyping technique perfect for turning diagrams into 2-D parts with little additional time or effort: laser cutting.

Laser cutting tools, for example, those made by LaserCAMM, LaserCut, Inc., and Epilog Laser, Inc., usually read DXF file formats, and move a laser over a flat work piece in a manner similar to the way a plotter moves a pen over paper. Figure 29.7 shows a photograph of a

FIGURE 29.7
LaserCAMM laser cutter.

FIGURE 29.8

A laser cutter in the process of making parts. Note the bright spot on the work piece, indicating where the cut is being made.

typical laser cutter. Most laser cutters can be used with foamcore, acrylic, fiberboard, ABS, Delrin, nylon, Masonite, and many other materials. Laser cutters cut by burning through the material with a powerful laser directed to the cutting point via moveable mirrors (Figure 29.8). Some cutters can even handle steel, aluminum, titanium, and other metals. Care should always be taken, however, as there are some materials that are definitely not OK to laser cut, and you should check with the manufacturer or vendor before trying to cut anything. For instance, it is not usually possible to laser cut polycarbonate—it can burst into flames.

Complex assemblies made from components that are 2½-D (2-D plus constant thickness) can fulfill a great many requirements. More complex shapes can be created by taking successive cross sections: by cutting and stacking a succession of slices, arbitrary 3-D shapes can be assembled, though there will be "steps" between the layers. Laser cutting is so easy, fast, and accurate; it has become one of the most commonly used techniques in our labs, and each year, most of the robots and other devices built by students incorporate several laser cut parts. The robot shown in Figure 29.9 is an examplary student project made with laser cut acrylic.

If you don't have access to a laser cutter, don't despair: there are a great number of vendors that perform laser cutting services and supply materials. Usually, you can upload part files at the vendor's website in DXF format, after which they will source the material you specify, cut the part, and ship it to you.

FIGURE 29.9

A robot built primarily with laser cut acrylic sheets. (Courtesy of Tim Wong.)

29.4.5 2-D Rapid Prototyping, Cheap!

If you don't have access to a laser cutter, and you don't have the $20,000 or more that it takes to procure one, and you don't have the resources or time to have a vendor laser cut parts for you, there is another method for creating 2-D parts that can quickly yield reasonable results. It requires considerably more effort, however.

Once you have created 2-D part drawings of components you would like to fabricate (preferably these have been derived from a 3-D solid model of your assembled device), print them on regular paper at 1:1 scale using a printer with reasonable print resolution. Larger parts may require a panelized printout spread over several pages or larger-format paper. With a pair of scissors or a hobby knife, roughly cut the excess paper away from the border of your part, and stick the remaining part drawing onto the flat material of your choice (acrylic, foam-core, Masonite, and so on). This can be done with a large variety of adhesives, like clear tape, glue stick, or spray-mount. The objective is to make sure that the portions of the paper that have the part outline remain firmly affixed to the material underneath, even as you cut, scrape, saw, file, and sand it. Now that you have your material with a high-fidelity part outline bonded to it, roughly cut the part out using the tool you feel is most appropriate for the scale and complexity of the part. A band saw, scroll saw, or jigsaw might work well, or a Dremel rotary tool. Perhaps the part can be rough-cut manually, with a coping saw, or a hobby knife with a saw blade installed. In this step, you are attempting to cut close to the final part outline, but not quite on the line. The next steps are to do the finishing cut and smooth the edge of the material. If you need to follow the part outline precisely, you can use progressively finer files to get closer to, and then eventually reach, the part outline. Finish with sandpaper if needed.

This process takes longer than laser cutting, to be sure, and it is less accurate. However, if you are resource and/or budget constrained, this is a reasonably quick and effective way to create parts that get the job done.

29.4.6 Tab and Slot Construction

A simple means of greatly increasing the strength of many jointing techniques is to use **tab and slot construction**. Rather than relying solely on adhesive bonds between the surfaces of two adjacent pieces in an assembly, using tabs and slots interlocks the pieces. The result is an assembly that does a vastly better job of distributing loading to its component pieces and, when well executed, reduces the loading on adhesives and fasteners. To implement tab and slot construction, simply look for opportunities to create interlocking features in the points, lines, and surfaces where parts meet. You can use tab and slot construction with practically any material and fabrication technique. It works especially well with foamcore, laser cut sheets (acrylic, MDF, plywood, and so on), and the cheap laser-cutter-inspired 2-D techniques described in the preceding section, and it can also be helpful when incorporated into 3-D rapid prototyping techniques described below.

Figure 29.10 shows an example of a design that harnesses the strength of tab and slot construction. In the assembly shown, foam balls are gravity fed down a length of plastic pipe until they contact a wheel attached to a motor that spins the wheel at several thousand rpm. When the ball makes contact with the wheel, the ball shoots out of the tube at high velocity toward a target. The laser-cut rectangular structure in the middle supports the pipes and the motor, and bears significant loads when the robot moves and during ball delivery. The design survived the rigors of integration, testing, debugging, and competition without failure, and this can be attributed at least in part to the use of tab and slot construction.

29.4.7 Toys

Sometimes, creating a great prototype doesn't require you to fabricate a single thing. There's a large category of building materials that are readily available at reasonable cost in almost every neighborhood: **toys**. Though it might not be immediately apparent, toys—especially

FIGURE 29.10

An example of an assembly that incorporates tab and slot construction.

construction toys such as LEGO, Erector Sets, and K'NEX—are a powerful and fun way to explore design concepts very quickly. The basic building blocks for these types of toys are designed to be rapidly assembled, flexible, and reconfigurable. And, best of all, they require absolutely no fabrication!

Toys may not be the best approach for exploring the form or integration of an assembly, but because they often include basic mechanical components such as wheels, shafts, and gears, you can quickly test ideas related to machine design and draw high-quality conclusions about how a final design concept will ultimately work. One especially compelling example from a previous class project involves a robot designed by a student team that decided to use a catapult to fling Nerf Ballistic foam balls. The project assignment required delivering such a ball into a target basket from a distance 3 to 6 ft away. The team quickly created a prototype of their idea, using a brushed DC motor, wire ties, a few LEGO components, and a plastic spoon for holding the ball. This first prototype (Figure 29.11) worked so well and proved so robust that the assembly was simply incorporated into their final design. Early prototypes don't usually work out this well, but it's a wonderful thing when they do. We document another example of this in Chapter 32, in the discussion about the development of Team InTheRuff's foam ball dispenser.

FIGURE 29.11

Nerf Ballistic ball catapult constructed from LEGO and a plastic spoon.

FIGURE 29.12
Prototypes made with K'NEX components (left) and Erector Set components (right).

Though LEGO have seen the most extensive use in our labs, K'NEX, Erector Sets, and many other options are readily available and very flexible alternatives. Each type of toy has its own approach for creating structures and assemblies (Figure 29.12), and one may be better suited for the purpose of your prototype than others. It's up to you to determine which toy and technique will work best and show you the most.

29.4.8 3-D Rapid Prototyping: SLA, SLS, FDM, and Soft-Mold Castings

When 2-D parts aren't adequate and a 3-D prototype is needed, there are a number of ways of creating high-fidelity rapid prototypes. As you might expect, it's more expensive and takes longer to create a 3-D prototype, but this approach can potentially result in an early prototype that is very close in look, feel, and function of a final design that might be made from parts that will take much longer to make. All of the processes described below require that a solid model of the prototype be created first and saved in a specific file format (typically an STL file), and then uploaded into the machine that will make the part. Options for rapid 3-D prototyping techniques include:

- *Stereolithography (SLA)*: With this technique, an ultraviolet laser is directed into a vat of photo-curable epoxy resin. At the beginning of this process, a platform that can be raised and lowered within the vat of resin is positioned near the top surface, and the laser traces out the first "slice" of the 3-D part to be made. The thickness of each slice is configurable, with a typical value around 0.005 in. and a minimum of about 0.002 in. Once the first slice, or layer, of the model is complete, the platform is lowered one slice-depth into the vat, and the next section of the prototype is traced by the laser. This layer is "grown" on top of the previous layer, and in this way, nearly arbitrary 3-D shapes may be built (with limitations on feature size and support). You can choose from "standard" material and finish (a milky, translucent material, as shown in Figure 29.13), or "water clear." Parts made using SLA do not have tremendous structural integrity; they are generally delicate and brittle. Delivery times for SLA parts can be as fast as one day, but this greatly affects the cost. If you are able to wait a few days for your parts, the cost drops and can become quite reasonable. Depending on the amount of material required, and the time required by the SLA machine to make it, parts can cost as little as $50. The maximum cost is pretty much limitless if you need a large, high-resolution part, one-day turnaround, and overnight delivery! In general, the SLA process provides the lowest cost and fastest turnaround time, but the parts tend to be delicate and cannot withstand elevated temperatures.

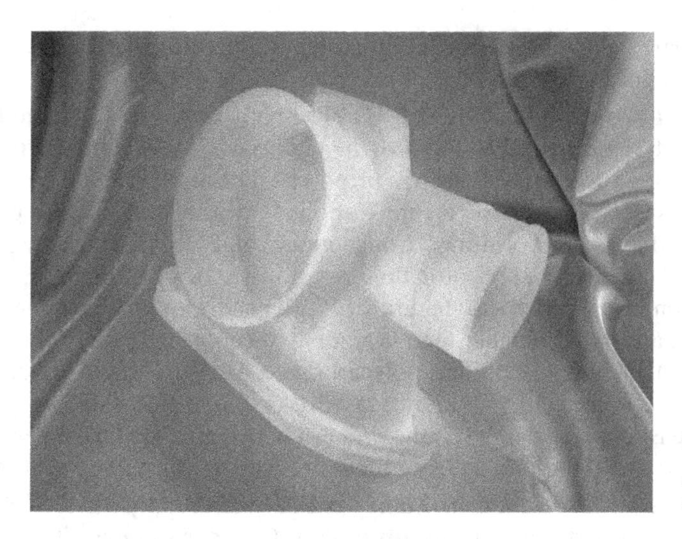

FIGURE 29.13

An SLA part. (Courtesy of Quickparts.com, Inc.)

- *Selective laser sintering (SLS)*: This process is similar to the SLA process described above, however, instead of using a UV laser to cure layers of resin to grow a model, SLS uses a laser to sinter (i.e., bond together by heating without melting), a powdered plastic material that has been spread in a thin, uniform layer on a platform. The process begins when the SLS machine spreads out the thin layer of material, and then the laser traces out the first layer of the part to be grown. When that is complete, the platform is lowered the depth of one layer, and a new layer of powder material is distributed on top of the part being grown. This layer is supported elsewhere by the leftover, unsintered powder material surrounding the growing part. The laser then traces out the next layer to define its shape and bond it to the previous layer. The process repeats until the part is complete (Figure 29.14). The primary advantage of SLS over SLA is that the parts are considerably stronger, and a variety of materials can be used that approximate the strength of common materials, such as polycarbonate and nylon. In addition, materials that are chemically resistant are available, as are flexible materials suitable for creating living hinges and snap features. SLS parts can be put under considerable loads and handled with lower risk than SLA parts. The resolution achievable with SLS parts is about the same as SLA parts, however, the surface finish is usually inferior.

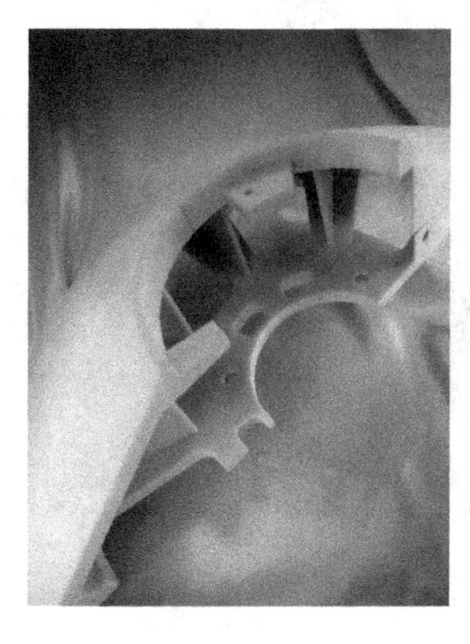

FIGURE 29.14

An SLS part. (Courtesy of Quickparts.com, Inc.)

SLS part delivery times are a little longer (three to five days for some vendors), and cost is a little higher than SLA parts.

- *Fused deposition modeling (FDM)*: The approach used to create FDM parts is unlike those of SLA or SLS. With FDM, the material is heated and extruded through a nozzle to develop successive layers of a prototype. The model is built on a platform that is lowered as each successive layer is applied by the nozzle. This process is sometimes referred to as "3-D printing." Conceptually, an FDM machine (Figure 29.15, left) is like a CNC cake icing applicator that lays down a bead of melted plastic. Similar to SLS components, FDM parts (Figure 29.15, right) are tough, have high structural integrity, and can be used in higher temperature applications. A variety of common materials can be used, including ABS and polycarbonate. The resolution and surface finish of FDM parts are inferior to SLS and SLA parts, as the bead of melted plastic created by the nozzle is relatively large (on the order of 0.010 in.). FDM parts generally cost more than SLA or SLS parts, and delivery times are on par with SLS (e.g., three days or more).
- *Castings from soft tooling*: Many vendors who supply SLA, SLS, or FDM parts also supply cast parts made from **soft tooling**. Cast parts made with soft tooling are a natural fit for vendors of SLA, SLS, or FDM parts because the process of creating the **castings** begins by making a "positive" of the prototype (i.e., something with the desired final shape), often using one of the rapid prototyping methods described above. Once the positive is in hand, a mold with a "negative" (a cavity that conforms to the part's outer contours) is created by placing the part within a container and filling the space around the part with a liquid that hardens in a short time (a few hours to a day). Once the mold has fully cured, the part model is removed, leaving a cavity. This cavity can then be filled with a suitable casting material, such as a self-curing urethane. Only a few materials can be used with this process (nylon and urethanes are the standards), but you can choose from a wide range of material durometers. Values between 25 A (very soft) and 80 A (very hard) are available, which gives designers considerable flexibility. Before a mold is made, of course, some thought must be given to how the part will be extracted from the mold, such as where parting lines will be, and how complex the geometry can

FIGURE 29.15

An FDM machine and example FDM parts.

be and still have the mold release the part at the end of the process. Not all shapes are moldable, but much is possible within the limitations of the process. (For a good example and many pictures of the process, see [1] in References.)

Lead time for cast parts is in the one- to two-week range, largely due to the need to first create the part model, and then the multiple curing times required for the mold and castings. For each mold made, one to two parts a day can be cast, since each one must cure in the mold. This process is slow enough that it's arguable whether it truly qualifies as a "rapid prototyping" technique (especially considering the very short timelines of most student projects), but it is a reasonably quick and economical way to produce several copies of the same part.

29.5 RAPID PROTOTYPING OF ELECTRICAL SYSTEMS

Conveniently, the basic principles of prototyping mechanical systems apply equally well to electrical prototypes. The goal is to quickly and inexpensively determine if a candidate design you are considering holds promise (in this case, an electrical system), and is deserving of the time and resources required to take it to a finished project.

For an electrical system, we recommend the following four steps:

1. Create a schematic of the circuit you plan to build.
2. Decide if you can effectively and efficiently model your system/circuit, or at least parts of it, using software simulations.
3. Build a prototype of your circuit and test it thoroughly.
4. Iterate—revise your schematic, rerun simulations, and update your circuit prototype.

Compare this list to the steps we recommend for creating mechanical prototypes (summarized from the discussion above):

1. Create a solid model representation of the part or assembly you plan to build.
2. Decide if there are aspects of the device's physical characteristics you can effectively and efficiently model, such as stress-strain, heat transfer, fluid flow, and so on.
3. Build a prototype of your component or assembly and test it thoroughly.
4. Iterate—revise your solid model, rerun simulations, and update your mechanical prototype.

The process is identical, but the specific steps, tools, and techniques are unique to each discipline. In this section, we'll discuss several techniques that enable you to quickly and inexpensively follow the guidelines with respect to electrical systems.

29.5.1 Schematic Capture and Circuit Simulation Tools

The first step in any design, especially electronic, is to document it with an accurate representation showing the relevant details. For mechanical systems, this is a solid model. For electrical systems, this is a schematic. Just as it can be tempting (and ill advised) to skip this step for a mechanical design, it can be tempting to skip this step in electrical design as well. However, this step is essential! If the circuit is anything more than trivially complex, you will need to refer to the schematic to build and troubleshoot it. If you would like to show the circuit to anyone (such as a team member or a supervisor), you will need a medium with which to communicate the design—one that is superior to verbal descriptions and hand-waiving. If you take a break while building the circuit, or if you build it and then leave it on the

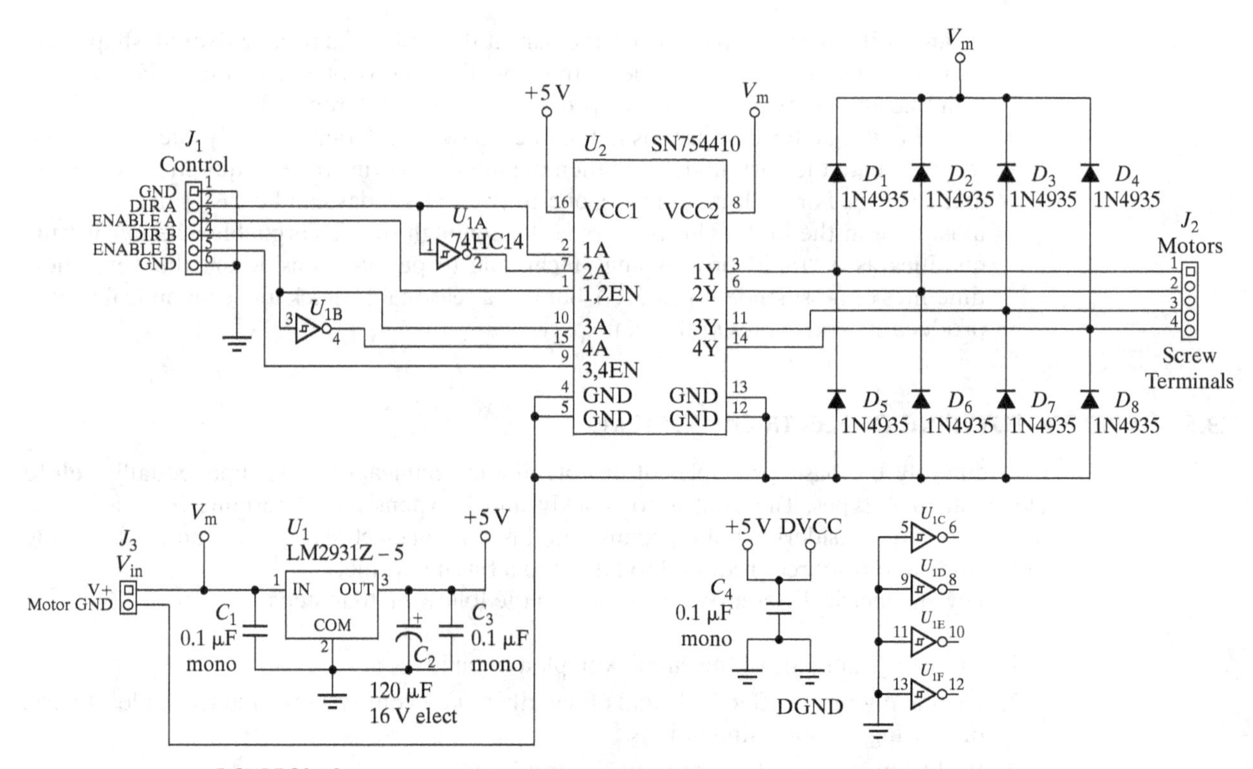

FIGURE 29.16

Circuit schematic.

workbench for any length of time and then return to the project, you will need to refamiliarize yourself with the circuit. The best way to accomplish all of these is with a complete and accurate schematic (Figure 29.16).

Schematics can be captured perfectly well with the use of pencil and paper. There are also a large number of excellent schematic capture computer programs available that improve readability and make the design easy to edit. Prices range from many thousands of dollars for professional tools (e.g., Altium Designer, OrCAD, Allegro, Multisim), to a few hundred dollars for lower-end pro tools (e.g., AutoTRAX, Electronics Workbench, Eagle), to free and open-source tools (e.g., ExpressPCB, FreePCB, gEDA, KiCAD). Despite being free, several of these software tools have surprisingly extensive capabilities and provide viable alternatives to the expensive professional tools. Upgrades and improvements are released frequently. Many of the commercial schematic capture applications are available in free versions that are either time-, capability-, or design size-limited. You should use whichever tool is readily available to you, or the one that best fits your budget.

Since schematic capture is the first step in the design process, you should select a capture program with later steps in mind. If you will need to run simulations of your circuit (see also Chapter 9, Section 9.10), be sure that your schematic capture program accommodates that. Most schematic capture tools incorporate modeling capabilities directly, and are straightforward to use. If you're going to create a **printed circuit board (PCB)** of your design, it would be a good idea to choose a schematic capture program that integrates easily with the board layout program you intend to use. Often, they are bundled and work together seamlessly.

Before you build the circuit you've defined in a schematic, you may wish to explore its behavior and performance using a circuit simulation program. Why not confirm that a circuit functions as intended while you're working on the schematic, rather than when you slowly and painstakingly build it? Most good commercial schematic capture programs, and some that are available for free (e.g., LTspice from Linear Technology), incorporate some variant of SPICE circuit simulation, so simulating a circuit for which you've just finished a schematic often involves only a few more steps. Testing new ideas and iterating in simulation are much faster than building systems on a workbench and testing with voltmeters and oscilloscopes. To be sure, it takes time to learn how to use a new simulation tool, and there is overhead required to set up and execute a simulation, but the benefits and time savings can be substantial. Though you can in principle simulate any circuit, there are types of circuits that are particularly rewarding to simulate [e.g., circuits that include moderately or highly complex logic (sequential and combinatorial) and analog filters (passive and active)]. In addition, you can explore issues like part tolerances and temperature effects on overall system function.

29.5.2 Circuit Prototyping: Breadboards, Wire Wrap, and Perf Boards

Once you're ready to build the first prototype of a circuit, there are a few choices. The major difference between the options is the time required to assemble the circuit, which corresponds directly with how robust the prototype is.

The quickest and easiest way to assemble a prototype of a circuit is with the use of a **solderless breadboard**, which allows you to connect leaded components (wires, resistors, capacitors, ICs, and so on) without the need for solder or tools. Breadboards come in many sizes and configurations, but share common characteristics: they are comprised of grids of holes, into which the leads of electronic components may be inserted (Figure 29.17).

The holes have hidden spring features that ensure good electrical contact and mechanically retain components. Horizontal rows of holes are electrically connected to facilitate connecting several items together, and vertical columns of holes along the outer boundary of the horizontal rows are electrically connected together to enable their use as power rails, since it is common to distribute power and ground throughout a circuit (Figure 29.18).

Breadboards are best for making quick prototypes, since a circuit can be quickly assembled, tested, and reconfigured. The flipside of this is that circuits are very easy to disassemble too, which can sometimes happen by accident. Parts can be dislodged and disconnected—sometimes without any visible indication that something is wrong. Breadboards are quick and easy, but they're not particularly robust. When assembling circuits using a breadboard, it's

FIGURE 29.17

A breadboard with a prototype circuit. (Courtesy of Tim Wong.)

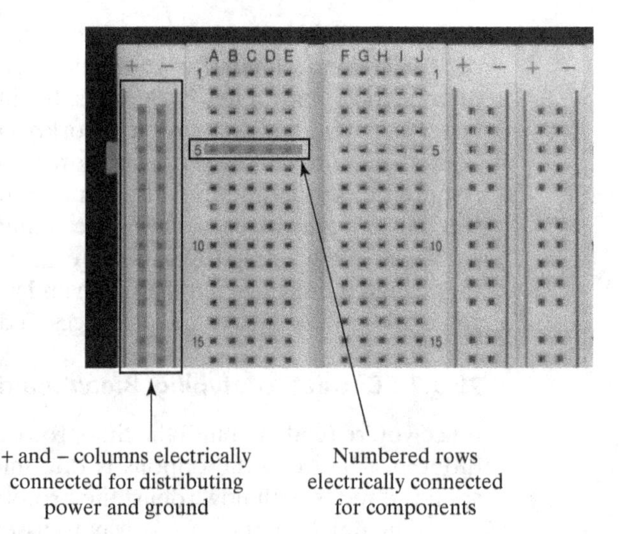

FIGURE 29.18
Breadboard hole connectivity.

generally good practice to avoid stacking wires on top of components, and allowing wires to become intermixed, as this can result in a circuit that is very difficult to debug or modify. Keep in mind that when working with prototypes at the earliest stages, bugs and revisions are almost certain to occur, so managing the design so that it's well organized and clean will almost certainly result in quicker overall success. The left side of Figure 29.19 shows a good example of a well-made breadboard circuit. Note the easy access to all the pins of each IC, and the way the wires run close to the breadboard surface instead of lofting up over the ICs and other components. The right side of Figure 29.19 shows an example of an implementation ready for disaster. The long wires and component leads will be knocked out of the holes very easily, and serve as excellent (i.e., insidious) antennae for any nearby sources of electrical noise.

Wire wrap is another circuit prototyping method, and it results in more robust prototypes than those possible with breadboards. So robust, in fact, that the guidance computers used in the Apollo space missions were constructed using this technique [2]. The robustness of wire wrapped prototypes comes at a cost, however: it takes considerably longer to create a wire wrapped prototype, and it is harder to debug and reconfigure than prototypes made with a breadboard.

FIGURE 29.19
Examples of breadboard construction technique. Left: good technique. Right: poor technique. (Courtesy of Salomon Trujillo.)

Wire wrapped circuits make use of component and IC sockets with lengthy square leads that extend through an insulated perforated board. The socket leads are long enough to wrap a short (~1 in.) length of 30 AWG wire a number of times (5–7 wraps of bare wire with the insulation removed, followed by 1–2 additional wraps that still have the insulation intact). The component socket leads are usually long enough for three such wire connections to be made, though they are available in differing lengths (1-, 2-, and 3-wire options are common). Special tools are required to properly strip the thin insulation off the wire-wrap wire and to correctly twist wire with the correct tension around the square socket leads. Figure 29.20 shows a manual wire wrapping tool which incorporates a wire stripper in the handle. Below the tool, a length of 30 AWG wire-wrap wire is shown with the insulation stripped off one end.

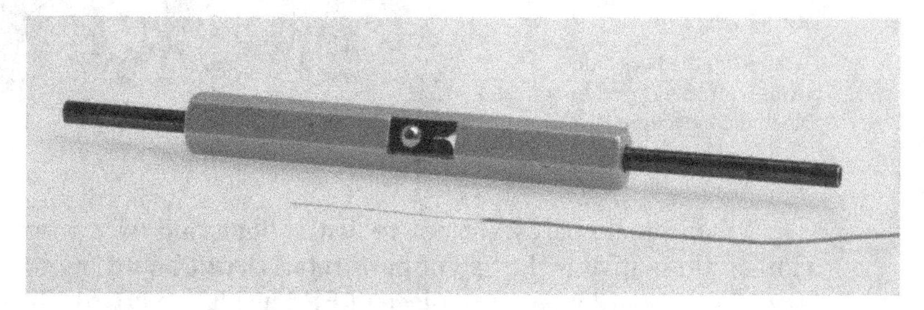

FIGURE 29.20

Wire wrap tool and a length of the 30 AWG solid-core wire used with wire wrap method.

Since the wire is wrapped on the underside of the components, it's essential to keep in mind that all the part pin-outs are flipped. Since this very consistently causes mistakes when adding the wiring, plastic templates are available to help keep track of the correct pin numbers for common socket sizes. Figure 29.21 shows a good example of a circuit built using wire wrap, and includes several of these numbering templates.

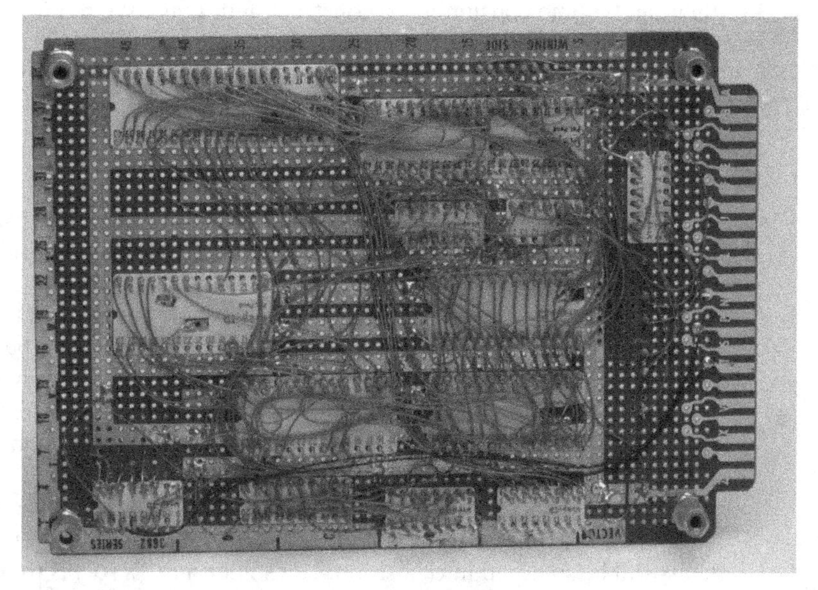

FIGURE 29.21

The bottom side of a wire wrapped prototype circuit.

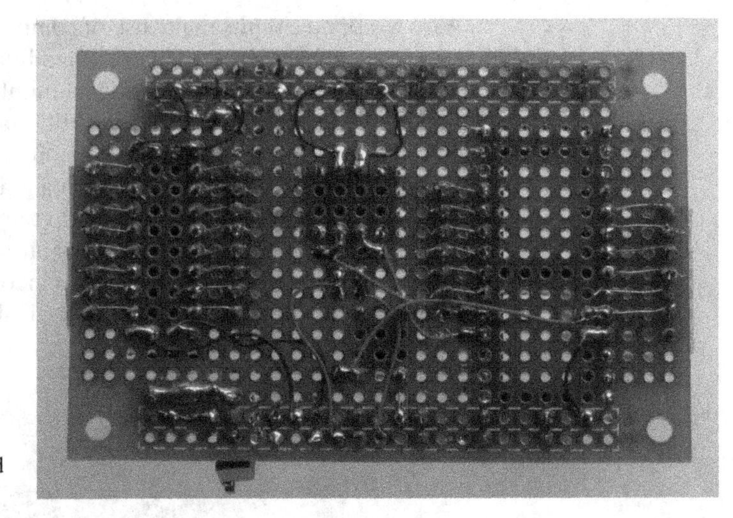

FIGURE 29.22

Detail from the back side of a circuit built using "perf board" and soldered connections.

When breadboards are not robust enough and wire wrapped circuits are too bulky, another alternative is the use of **perforated circuit board**, more commonly known as "**perf board**." Perf board is simply a piece of FR4 insulating board (the fiberglass material used to manufacture most PCBs), with a grid of holes drilled on 0.1 in. centers. The holes are plated with copper and tin to facilitate solder connections. Component leads can be inserted into the holes, and wires can be soldered to the back side of the board to establish the desired interconnections. Perf boards are available at most electronic supply stores. Building perf board prototypes requires a soldering iron, wire cutters, and needle-nose pliers, at a minimum. Other tools may also be required, depending on the nature of the prototype. Note that perf board circuits also require the "flip" of the board that causes so much confusion about the location of the pins, so care must again be taken during their construction. Figure 29.22 shows detail from a circuit built using perf board.

Circuits built using perf board and soldered connections can be very robust, however, they can be time consuming to construct and revise. We don't recommend this approach for initial prototypes, when numerous revisions are anticipated. Perf board is better suited in later stages of a project, when physical robustness is necessary and the circuit is well understood and unlikely to change significantly going forward.

29.5.3 Prototype PCBs

A PCB (Figure 29.23) is comprised of an insulating substrate, usually a fiberglass composite called FR4, with copper laminated to its surfaces. Additional insulating layers and copper layers can be added, up to a limit of about 20 copper layers. Holes can be drilled through the board to allow for insertion of leaded, **through-hole components**, and to facilitate connections between conductive layers of the board, and these holes can be plated with copper and tin during the board manufacturing process. **Surface-mount components**, by contrast, are intended to be soldered to the surface of the PCB, and do not require holes to accommodate their leads. To create the desired interconnections between components, the copper is selectively removed from the board through an etching process to create networks of point-to-point connections. A **solder mask** (a translucent insulating covering, often green in color) can be applied over the board and copper to cover and protect everything except where components will be soldered into position. The solder mask's role is to protect against unintentional connections (i.e., short circuits) between traces on the PCB, and to facilitate mass production. Labels and markings can be added on top of the solder mask using a silk screening process, which helps communicate proper part placement and orientation, as well as indicating board

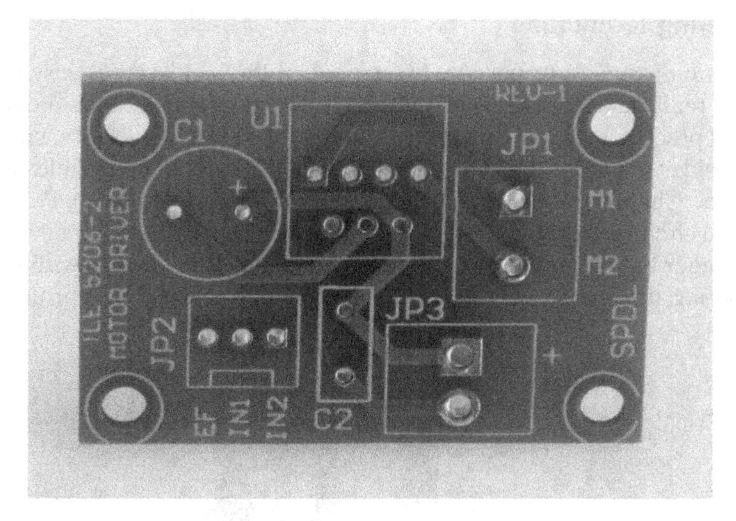

FIGURE 29.23

A PCB. Note the plated through-holes, the connecting copper traces under the solder mask, and the silk screen markings indicating part references and orientations.

function, revision level, date, copyrights, and so on. The beneficial characteristics of PCBs are manifold: the circuits are very robust once all the components are soldered in place; the process is repeatable, both in the manufacturing of the PCB and the assembly of the components; and circuit performance is highly consistent unit-to-unit. On the negative side, it can be difficult to modify a PCB if changes are required, and cost and design/delivery times may exceed the limitations of your project.

Until recently, PCBs could not have been considered an option for "rapid prototyping." The effort to design them was significant, the cost was prohibitively high, and the delivery times were lengthy. Because of this, PCBs were better suited for the later stages of a project, when they are absolutely required. By the end of the 1990s, however, several PCB manufacturers began offering low-cost, quick-turn options that made PCBs viable competition to wire wrap and perf board prototypes. Since that point, the time required to complete a prototype PCB design may be a very good investment, especially if a higher-production version of the same circuit is likely at some future point. As an example, a PCB layout for a simple circuit (say, less than 35 components) can be completed in one to two days, and the board can be ordered, manufactured, shipped, and received within three additional days, at a cost of $100 or less. This may not suit all prototyping needs, such as student assignments, but it makes tremendous sense for many projects, where the delay of three to four days will result in far better performance and reliability of the electronic subsystems. Figure 29.24 shows an example of an assembled PCB.

FIGURE 29.24

A PCB with components installed, often referred to as a printed circuit assembly, or PCA.

29.5.4 Soldering Technique

Though it is often assumed (especially by engineers) that soldering is easy and requires no training, this is really not the case. Making high-quality and reliable solder bonds is a skill that takes time and practice to acquire. Bad soldering is all too easy to achieve, unfortunately, and poor quality solder joints can cause you to spend considerable time troubleshooting a prototype circuit (see Chapter 31 on troubleshooting if this happens to you). Poor solder quality can cause open circuits, short circuits, intermittent connections, high-impedance connections, and a host of other baffling problems. Rather than wasting time dealing with the aftermath of bungled soldering, it behooves you to learn the basics before building a prototype.

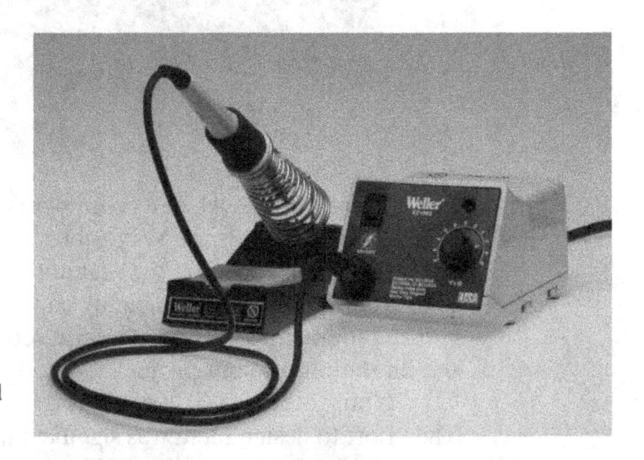

FIGURE 29.25

A high-quality, temperature-controlled soldering station.

The first critical element of success is to use good tools and materials. To get the best possible solder joints, it is very helpful (but not absolutely required) to have a temperature-controlled soldering iron (such as the one shown in Figure 29.25) with a tip that is in good condition. A new tip is clean and shiny (Figure 29.26a), while a tip that should be discarded is dull with a rough or even pitted surface (Figure 29.26b). When in doubt, replace the tip—they typically cost only a few dollars, and this has a big positive impact on solder joint quality.

Even a new tip can be used to make poor solder joints when it's dirty. Clean the tip frequently when the iron is hot. The best and simplest way to do this is with a wet sponge. Most high-quality soldering iron stands incorporate a space to hold a sponge (such as the one shown in Figure 29.25). The best results are obtained when all sides of the tip are wiped clean

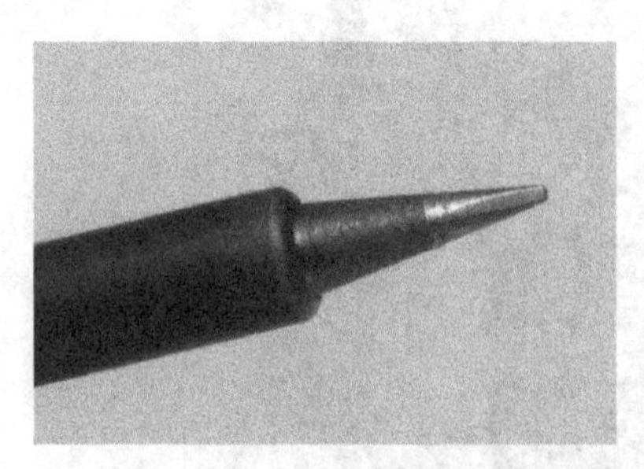

(a) A bright, shiny new soldering iron tip

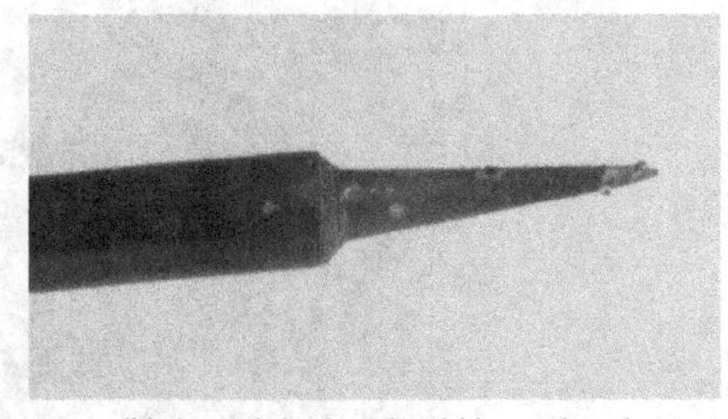

(b) A used tip that should be replaced

FIGURE 29.26

Soldering iron tip condition is critical to good solder quality.

on the sponge before *and* after each solder bond is made. With practice, this becomes a reflex that you'll perform without having to think about it consciously.

The choice of solder also has a major impact on the quality of the solder bond. Older solder is a mixture of tin and lead (a toxic metal), while newer solders are lead-free. These newer formulations work well in most applications, and obviate concerns about dealing with toxic materials in your workplace, on your hands and clothes, and so on. We highly recommend embracing these newer materials. For most electronic prototyping, solder with an outer diameter of 1/16 in. or 1/32 in. works best.

With the right tools and materials, you're well-positioned to create excellent solder joints. Technique is all that stands between you and success. The technique that allows you to achieve good quality solder bonds is to melt a small dab of solder onto the tip of the iron, and then to touch that melted spot of solder onto the component. With this small bridge of solder acting as a conduit for the iron's heat, the component lead will rapidly heat up to the point where larger quantities of solder will melt directly onto the component and the object it will be soldered to—a PCB, wire, or other component. In an ideal solder bond, solder will flow freely onto the surfaces you wish to connect together. The solder wets the surfaces and creates a curved meniscus, as shown in the left side of Figure 29.27. The resulting surface finish is bright and shiny, with no pits, gaps, discolorations, inclusions, or other visible defects. The right side of Figure 29.27 shows a cross-sectional view of an ideal solder joint, where the smooth meniscus is a result of good surface wetting, the proper quantity of solder, and good flow. If too much solder is applied, a blob results instead of a meniscus, so carefully controlling the amount of solder is also important.

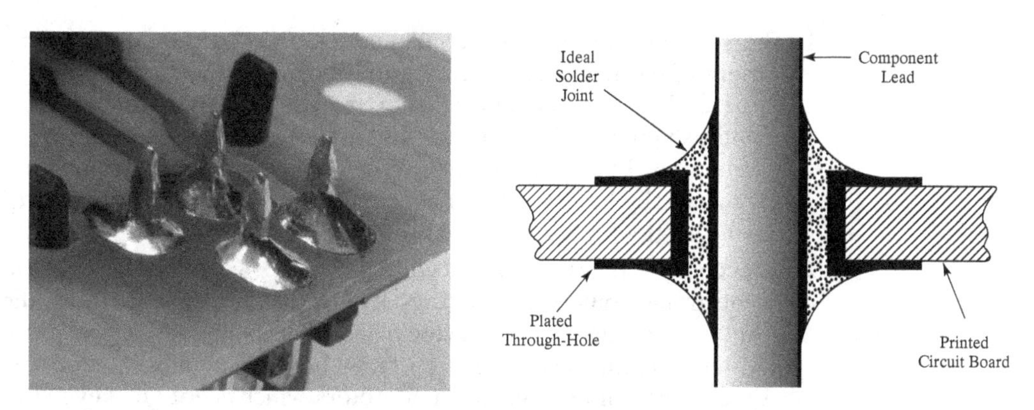

FIGURE 29.27

Characteristic appearance (left) and cross-sectional view (right) of good solder joints.

Poor solder joints look distinctly different from those in Figure 29.27. Figure 29.28 shows two of the most commonly occurring types of bad solder joints. Figure 29.28a shows a cold solder joint, which can result from an iron whose temperature is too low or solder that has cooled and then been reheated in place. Figure 29.28b shows a solder joint made with an iron whose temperature was too high. Both are easy to identify by their dull, rough, pitted, and sometimes even cracked appearance. Poor solder joints should be completely removed (either with a solder-sucker tool or solder wick) and replaced with a new solder bond. Bad solder isn't salvageable—it must be removed and replaced.

There are many excellent instructional websites and videos available on the Internet. Watching a skilled solderer demonstrate how to make good solder joints and describe the key steps as the process is performed is often very helpful and highly recommended.

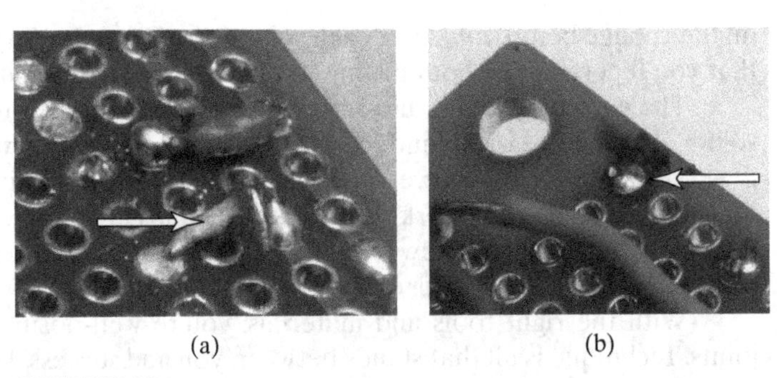

FIGURE 29.28

Examples of poor solder quality: (a) a cold solder joint and (b) a solder bond made with an iron at excessive temperature.

29.6 SUPPLIERS AND RESOURCES

There are a great many vendors who supply the components, materials, and services discussed in the chapter above. Much of what you need may be available locally, within a few blocks of your home, school, or workspace, though some items you'll need to order and have shipped to you. A Web search will result in a long but unscreened list of suppliers, so here we offer our top recommendations in each category.

For electronic components, tools, services, and supplies:

1. Digikey Corporation (www.digikey.com, 800-344-4539)—Digikey carries a very wide variety of electronic components and tools, and the vast majority of the items they carry are kept in stock and can ship immediately. They have an exceptional record of reliably shipping orders on time, and are especially good when you need components the next day. Their pricing is not particularly low, but they focus more on convenience: they generally have what you need and can get it into your hands quickly—essential characteristics for a supplier of prototype components.
2. Jameco Electronics (www.jameco.com, 800-831-4242)—Jameco also has a large variety of electronic components and tools, though not as extensive as Digikey's selection. Almost everything is kept in stock, and Jameco also reliably ships orders out quickly, and their pricing is very good on many items. They have a good selection of motors, especially DC motors and stepper motors, which is not Digikey's strong point. Jameco is a great place to order parts for stocking parts bins. For example, they provide kits with selections of resistor values, capacitors, hardware, and so on.
3. Newark (www.newark.com, 800-463-9275)—Newark has a huge selection of components and tools available, significantly larger than Digikey, even. You may find that you prefer Newark as a primary vendor—it will depend on the types of components you tend to buy, and whether Digikey or Newark does the best job stocking and shipping them to you.
4. Electronic Goldmine (www.goldmine-elec.com, 800-445-0697)—Electronic Goldmine is one of the few remaining sources of surplus electronic equipment. They carry many standard components (resistors, capacitors, logic chips, wire, motors), as well as a seemingly random yet delightful array of opportunistic finds (missile gyros, Geiger counters, solar cells). As with most surplus sources, prices are generally excellent.
5. All Electronics (www.allelectronics.com, 888-826-5432)—All Electronics is another excellent source for surplus electronic components and test equipment at low prices. They are similar in many ways to Electronic Goldmine, however, it is often worth

checking both vendors for specific items (and other surplus sources as well), since their stock depends on what's available to them at any given time.

6. Sparkfun (www.sparkfun.com, 303-284-0979)—Though Sparkfun sells some surplus electronic components, their primary focus is their large selection of evaluation boards for microcontrollers, GPS modules, Bluetooth, ZigBee, and others. In addition, they've created a spin-off PCB production company (www.batchpcb.com) that consolidates orders from many customers into larger, panelized PCB orders. By having these produced by subcontracted board manufacturing vendors, the resulting price is much lower for low quantity, small prototype boards.
7. RobotShop (www.robotshop.com, 866-627-3178)—A comprehensive source for fully assembled robots, as well as all the components necessary to design and construct your own robot. RobotShop sells robots for cleaning your house, feeding your dog, and cleaning your pool. It also supplies motors, servos, shaft couplers, wheels, casters, sensors, microcontrollers, motor driver boards, and a great many other components. Not just a great source of parts; it's a great place to look for ideas and inspiration too.
8. Alberta Printed Circuits (www.apcircuits.com, 403-250-3406)—AP Circuits was one of the first PCB board houses to provide ultralow cost, quick turnaround PCBs. The cheapest, fastest option available from AP Circuits is called "Basic Prototype," which does not include solder mask or silk screen, but for quick cheap prototypes, you can generally live without them as long as you include some labels in copper on the component side of the board. All orders for this prototype service are one-day turn, meaning they will fabricate your board in one day, and ship it to you the next day via overnight delivery. So the total time required to receive an order is something less than three business days. Prototype costs depend on the size of the boards, the number of holes, and the number of boards you're ordering, but most orders end up costing less than $150, including shipping and tax.

For mechanical materials, components, tools, services, and supplies:

1. McMaster-Carr (www.mcmaster.com, 562-692-5911)—McMaster-Carr has a huge catalog and offers an incredible variety of parts, including hardware, plumbing, tools, and materials. An exhaustive list would require several pages. In addition to being a one-stop shop, McMaster-Carr focuses on quick delivery, accuracy, and reliability, and have become the market leader for mail order materials and shop supplies. The description above for Digikey is equally apt for McMaster-Carr: they generally have what you need and can get it into your hands quickly.
2. Tower Hobbies (www.towerhobbies.com, 800-637-6050)—Hobby shops are an excellent source of ideas for how to build things, as well as mechanical components like motors, gears, servos, controllers, wheels, shafts, glues, rechargeable batteries, battery chargers, and so on. Tower Hobbies is one of the largest hobby shops with a good online presence. They also have a line of "Tower Hobbies" branded products, which tend to be less expensive and high quality.
3. The Robot Store (www.robotstore.com, 800-374-5764)—This company focuses solely on supplying components appropriate for the design and construction of small robots, so it's a natural fit for a discussion about rapid prototypes of mechatronic systems. They were acquired by Jameco Electronics in 2005, though their focus is the same as it was prior to the acquisition. They carry robot kits, assembled robots, and components: motors, shape memory alloy actuators, gears, shafts, couplers, wheels, and so on. The prices are somewhat high, but they have a very good selection of appropriate parts.

4. Amazon.com (www.amazon.com)—Though Amazon.com supplies many other things, the primary reason for including them on this list is that they are a great source for toys. Of course, toys are one of the things you'll probably be able to find close to your home or workplace, so you may not need to order them and pay for shipping. Prices are generally excellent, and the variety of toys that they stock is quite large. Even if you don't purchase from Amazon.com, they are a great place to go to look for ideas and compare prices.
5. Quickparts.com (www.quickparts.com, 877-521-8683)—Quickparts is an excellent vendor for the 3-D rapid prototype services described in this chapter. They provide SLA, SLS and FDM parts, and cast urethane parts, as well as prototypes made with a few other techniques not described here. They can get your parts made quickly at a reasonable cost, and are very helpful if you have questions or need to interact with them to make sure your order will come out the way you intend.

29.7 HOMEWORK PROBLEMS

29.1 Describe a situation in which you might want to build a foamcore prototype rather than build a solid model to test out a concept.

29.2 Describe a situation in which it might make more sense to make the first prototype as a solid model rather than a physical model.

29.3 True or False: Because of its sheet nature, foamcore is only suitable for creating planar shapes.

29.4 What is the purpose of using "tab and slot" construction?

29.5 What are the advantages of SLS over SLA prototyping? What are the drawbacks?

29.6 Contrast hard tooling with soft tooling. List advantages and disadvantages for each.

29.7 True or False: Cast parts made from soft tooling are limited to low-durometer (soft) materials.

29.8 List at least three distinct functions that a schematic diagram serves in the design/build process.

29.9 In the context of electronic design, what is SPICE?

29.10 What are some drawbacks to circuit prototypes made using solderless breadboards?

29.11 What is the major advantage of a wire wrapped prototype over a solderless breadboard prototype?

REFERENCES

[1] "Castings made with soft tooling," www.omnica.com.

[2] Entry for "*Apollo Guidance Computer* (AGC)," http://www.wikipedia.org, 5/7/07.

CHAPTER 30

Project Planning and Management[1]

30.1 INTRODUCTION

Mechatronics is an inherently multidisciplinary field, combining complex mechanical and electronic systems, often under software control. This is what makes mechatronics simultaneously so rewarding and so challenging. Mechatronic systems typically require a high degree of integration and testing—so much so that without a considerable amount of thought given to organizing your efforts, rapid progress, and even eventual success can prove elusive. To complicate things further, most worthwhile projects involve a variety of constraints: limits on time, resources, finances, manpower, technology, the market, and so on. Fortunately, there are a number of techniques and methodologies that address these issues, and even a brief discussion of them will benefit designers and managers in search of some control over their activities.

This chapter introduces several strategies for getting organized and mapping out the execution of complex projects, not just those involving mechatronics. A variety of methodologies will be covered, since no single approach is perfect for any real project. The approaches introduced here will give readers tools to analyze projects in different ways and view them from different angles, each revealing new and potentially useful insights about how they might be brought to completion. These tools apply to any activity: mechatronic or otherwise, simple or complex, short or long. Regardless, project planning and management skills can be invaluable. Many will sound like simple common sense, and indeed some are. The structure they lend to the process and the emphasis on advance planning are what make them such valuable and powerful tools.

In this chapter, you will learn:

1. how increasingly complex systems require active management,
2. the importance of focusing on design requirements and specifications,
3. techniques for quickly generating design alternatives and analyzing design trade-offs,
4. how to implement several project management tools, such as Gantt charts and network diagrams,
5. what concurrent design is and how to use it to structure design activities,
6. what systems engineering is and how to use it to structure design activities,
7. tools that help facilitate efficient teamwork.

[1]The material for this chapter is drawn from lectures given at Stanford University 1997–2003 by Dr. Christopher Kitts, currently Director of the Robotic Systems Laboratory and Associate Professor of Mechanical Engineering at Santa Clara University.

30.2 INCREASING PROJECT COMPLEXITY AND THE NEED FOR MANAGING THE PROCESS

Life wasn't always so complicated. Before the industrial age, for example, it was possible to work in a discipline that a single individual could be reasonably expected to master. Carpenters, blacksmiths, stonemasons, and other craftsmen defined the leading edge of their day's technology. Many of these careers continue to exist today, but they're no longer what you would call high technology. A potter can master every element of his or her craft, from identifying and preparing clay, molding it into a useful or decorative shape, applying glazes, preparing a kiln for firing the results, and finally producing finished ceramic items. Often, potters will even market their own goods directly to end users. The same can be said of farriers (horseshoers), who often make their own horseshoes in their own forge, custom-shaping them to suit the needs of individual horses. They also sell directly at the locations of their customers, going barn to barn to assess a horse's needs and apply new shoes. These traditional trades, however, developed centuries ago and haven't changed much over the years.

Contrast "craft" production, such as pottery and horseshoeing, with the design and manufacture of modern high technology systems. It would take an exceptionally well-trained and gifted individual a long time to design and construct, for example, all the components that comprise a modern drip coffee maker: an assembly of several molded plastic parts, plumbing, a heating element and control circuitry, and a glass or metal canister for holding the coffee. Or a microcontroller, which requires not only the design of the features to be etched or deposited onto a silicon die that gives it its function, but also the design and construction of equipment capable of creating the submicron scale features needed. A modern automobile requires expertise in engine design, electronic engine control, aerodynamics, and suspension dynamics, among many others. A Boeing 777 entails mind-boggling complexity, and the space shuttle more still. Teams of people, often groups of teams, and substantial time are needed to design and manufacture systems this complex. These people need management and guidance in order to ensure that the results of their efforts are successful. This is the case with mechatronic systems as well.

30.3 PLANNING AND EXECUTING A PROJECT

There are many industries which have recognized the need to focus attention on project management. In some cases, the goal is to simply implement best practices and ensure discipline. In other cases, it's not only a good idea, it's required by federal law (for example, medical device developers are required to implement "design control," a formalized project management process that is enforced by the FDA). Students who have spent time working within large organizations, FDA-regulated companies, or the aerospace industry will likely find the process outlined in Figure 30.1 familiar. It shows each of the phases of a typical project. This is a field for which there is no single right answer, of course. The process outlined and discussed here isn't guaranteed to be optimal for all projects and situations. However, many smart and experienced people have spent considerable time pondering the question of how projects should be managed, and many concluded that something like the process shown in Figure 30.1 works reasonably well in most cases.

An interesting aspect of Figure 30.1 is that there are several steps at the beginning and end of the process that don't typically jump to designers' minds when asked to create a system or device. The initial steps involve careful planning. The final steps explicitly show that *something* happens to systems and devices after a design is complete, and it's worth considering. The step that most designers reflexively (and sometimes exclusively) think about when a project begins is the one right in the middle of the figure, labeled "Design." However, if the other steps are ignored or given short shrift, the entire project is jeopardized. One of

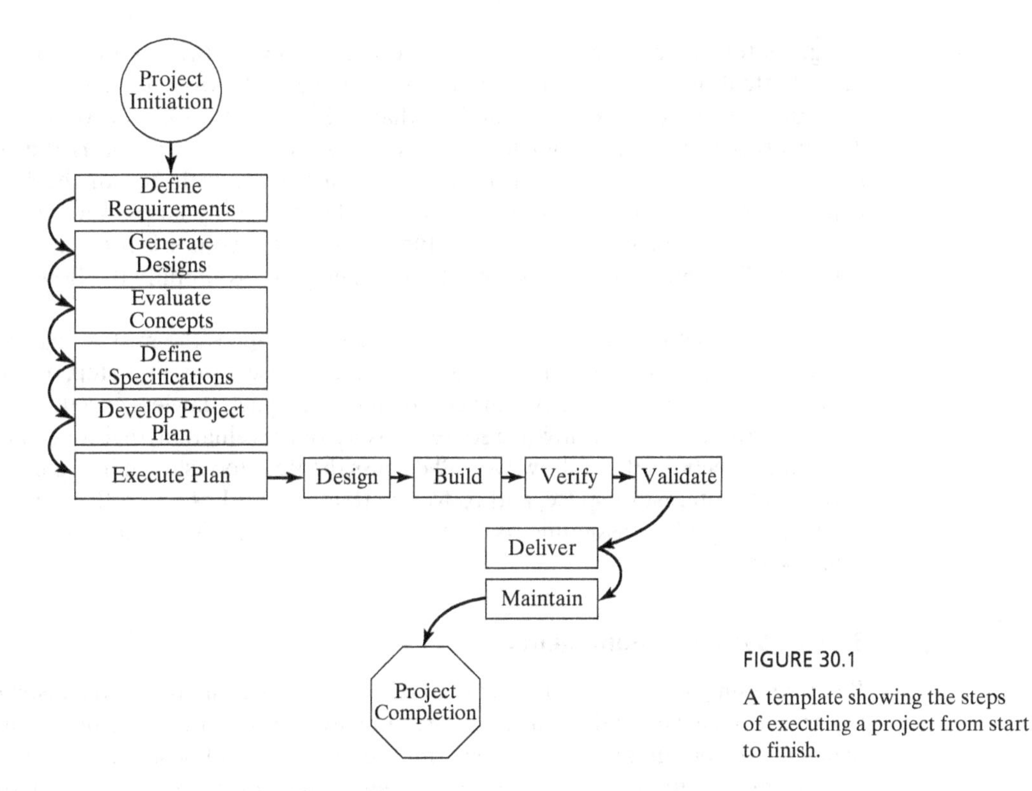

FIGURE 30.1

A template showing the steps of executing a project from start to finish.

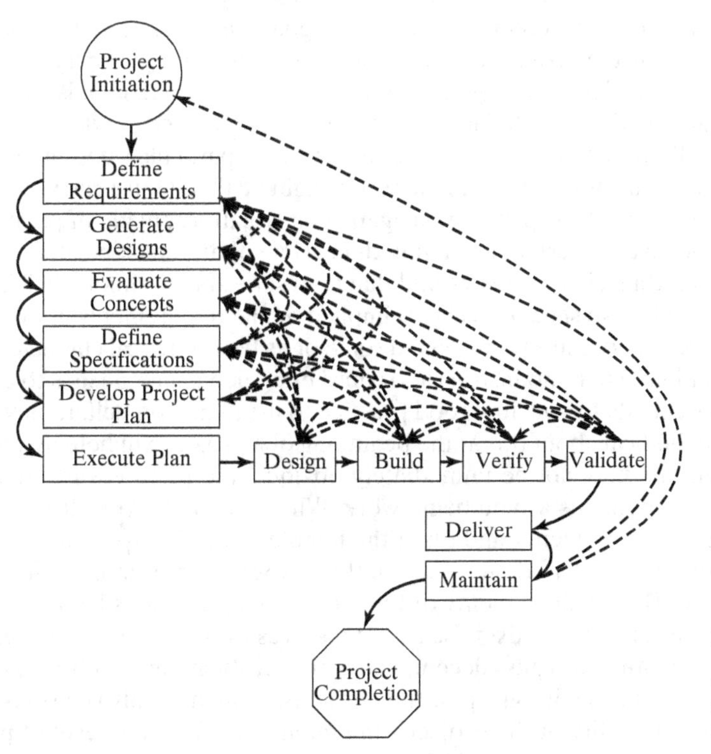

FIGURE 30.2

A more realistic process diagram, showing the true, iterative nature of real project execution.

our goals for this chapter is to discuss the importance of the "other" steps, and show how careful attention to them benefits both the design and the overall results.

In showing a process graphic like that in Figure 30.1, we feel we must admit that no design process ever turns out to be as step-wise, linear, and clean as the figure suggests. Design is a messy and iterative process, and in reality, each box in the flowchart has an implied arrow looping backward to each of the preceding steps. Showing all of the possible loops and return paths in a readable figure isn't easy, nor is it particularly helpful (see Figure 30.2), so most discussions about the design process simplify the flowchart in the interest of readability.

Leaving out the dashed, backward-pointing lines gives the misleading impression that a well-executed project won't involve going back to revise a previous step, or redefine requirements, or throw out a design, or start completely over. Such flawless execution is rarely, if ever, achieved! All real projects involve some measure of reevaluation, revision, and repetition.

In the sections that follow, we will discuss the key steps shown in Figure 30.1. We'll omit a few in the interest of space, and to focus attention on those with the highest relevance to mechatronic project assignments likely to be encountered in undergraduate or graduate coursework.

30.3.1 System Requirements

If you're going to have any chance at all of successfully designing and constructing a mechatronic system, it's essential that you start out with a firm understanding of what the requirements for the design are. You can only hit the target if you know what the target is!

In many situations, other groups or individuals are responsible for defining what it is you are to design and build. Your mechatronics instructor may assign a team project. The marketing department within your company may define a new product for you or your team to create. In these scenarios, they will most likely begin by drawing up a set of **system requirements.** In this context, **requirements** are the key goals and necessary functions of a system; the characteristics that the system must exhibit in order to fill a particular need. Requirements should not to be confused with **specifications**, which deal with demands placed on the particular designs and solutions created to meet the requirements (and are detailed further in the section below). Requirements should not assume any particular design or implementation. Typical requirements for a given mobile robot might be that it must carry its own power source on board, that it must be capable of navigating on a platform made of painted particle board, that it must not exceed certain maximum dimensions, that it must be able to detect a blinking red light with certain characteristics, and that it must be able to shoot a ball into a goal according to a set of rules. No specific design is implied by the requirements in this list, just the basic tasks the mobile robot must perform and some of its basic physical characteristics.

In situations where no instructor or marketing department exists to define and deliver system requirements to you, it is good practice to list them yourself. It may sound simplistic, but being methodical about this at the beginning of a project will help in many ways: it keeps options open that may not be immediately obvious, and helps ensure that you don't overlook major requirements as you begin work. When you make your list of requirements, list *only* the actual requirements, and resist the temptation to interpret, interpolate, or extrapolate new requirements based on what you think would be better, more interesting, or somehow improved. There will be plenty of time for these innovations later in the process.

The main value of this exercise is that it serves to structure your thoughts as you begin to work on creating designs, decomposing the problem into subsystems, and assigning resources. It facilitates your grasp of the entire assignment. It also clarifies the base level of acceptable functionality of the project—the minimum allowable level of performance. This is something you may find yourself thinking about quite often, especially as deadlines loom.

In fact, the term "**base level of functionality**" is such a key concept that it's become somewhat idiomatic around the authors' labs and classrooms. You are likely to find yourself coming back to this phrase repeatedly.

This exercise also serves to expose areas where the requirements are incomplete, ambiguous, or incorrect. Figuring this out as early as possible in the process has tremendous benefits. It gives you the ability to address the questions that arise. Perhaps someone with the authority to define the project further will need to clarify an issue. Perhaps you will need to make note of an issue and decide how to address it yourself. In any case, it takes time to find the best solution, and you will benefit from starting the process as early as possible.

Another benefit of listing the requirements of a project is that the exercise may reveal unobvious areas of freedom, or **loopholes**, that enable you to design a solution that fulfills the requirements perfectly in an especially clever or efficient way. Look for loopholes! Loopholes are legitimate and good. They encourage creativity and often make a project more fun. After you've listed the requirements, create a list of any loopholes you've been able to identify. Update your loophole list when new ones occur to you.

Cost limits or constraints are often among the most important requirements in the design of a real product, and are also usually a major factor in class assignments and projects. Changes in other requirements, discovery of loopholes, and changes in the overall market can all create opportunities for lower cost or impose requirements for lower cost. An in-class project will usually have an explicit cost limit, and this can eliminate some options in a design. When examining possible solutions to a requirement, cost estimates should be made early and continually revised throughout the process of design and development.

In real-world settings, it is not unusual for requirements to be revised frequently over long periods of time, as user needs and design constraints are better understood and loopholes are exploited. An iterative cycle is often established, which involves designers probing the limits of the requirements while they work to address them, as those responsible for generating the requirements perform more research or reinterpret existing data to address new questions. In academic settings, system requirements (e.g., assignments) that are complete, unambiguous, and consistent are much more common—in part because they are often written with a specific result in mind.

30.3.2 Generating Design Candidates and Alternatives

Once you have a firm grasp of the system requirements, it's time to consider exactly how you will go about meeting them. At this stage in the process, the challenge is to select viable approaches out of a great many candidate solutions. Depending on the nature of the design challenge, there may be no readily identifiable solutions, and some basic research must be performed or experts consulted in order to generate potential solutions.

Several tools and techniques have been developed to assist in these activities. Each is based on a particular means of focusing attention on at least the following, and sometimes considerably more:

- decomposing the challenge or problem so that the overall solution consists of solving several smaller (and, the theory goes, more manageable) challenges or problems,
- creativity, and the great benefits of generating as many solutions to a particular need as possible,
- documenting the process and the ideas, so that ideas that were not initially given much consideration can be easily recalled and reconsidered.

In this section, we'll discuss two tools that are particularly useful for mechatronic systems: morphological charts and brainstorming. As mentioned above, there are a great many more techniques that are worth considering that we will not cover here.

30.3.2.1 Brainstorming

Brainstorming is the activity of coming up with a great many different ways of accomplishing a set of requirements or goals. The technique is very popular, and perhaps already familiar to many readers. It shares many goals and characteristics with morphological charts (which we discuss in the next section), and the two are a natural fit—the ideas produced in a brainstorming session can go directly into a morphological chart, and a morphological chart can be used as the organizational basis of a brainstorming session. There are many formalized ways of conducting brainstorming sessions, however, most focus on the primary objective of generating copious numbers of ideas. At its core, brainstorming involves adhering to a few simple rules that seek to optimize idea "yield." The ability to keep it simple and dive right in without extensive training is one of the key reasons for brainstorming's popularity.

The simple rules of brainstorming are as follows:

1. **Generate *lots* of ideas, *fast*.** The primary goal of brainstorming is *quantity,* not quality. Sorting through the ideas generated is an exercise that should be performed sometime after the formal conclusion of a brainstorming session.
2. **Reserve judgment**. No matter how ridiculous an idea seems, go with it. Record it and move to the next ideas. Wacky ideas may prove to stimulate trains of thought that prove to be more reasonable. Under no circumstances should any participant rebut an idea or talk about why an idea won't work. Most ideas that are tossed out in brainstorming sessions are half-baked, and if the group gets bogged down dissecting them and proving that they won't work, additional ideas can't follow and the unfortunate person that contributed the idea feels persecuted.
3. **Appoint a moderator to direct the group's activities**. This participant bears great responsibility: he or she must keep the discussion moving in productive directions and enforce the rules of brainstorming while encouraging everyone to toss in as many ideas as possible. A well-managed brainstorming session is a fun and exciting activity. A poorly managed session can be a terrible experience. The performance of the moderator is usually the deciding factor, and it takes practice to get good at it.
4. **Record *all* ideas proposed in the session**. Capture the discussion and the sketches to the extent possible without slowing down the action. One of the participants other than the moderator should be responsible for capturing the ideas and results of the session. Often, it is useful to record who contributed each idea, especially in situations where patents or other intellectual property are likely to be created during the session.

While brainstorming can be done alone, the synergies that happen in group settings won't occur and this will make it less effective. Small groups work best—when too many people are involved, it's easy for chaos to ensue, or for less enthusiastic attendees to sit back and contribute little or nothing.

The overall goal of the exercise is to generate as many ideas as possible, without worrying about whether any of them are good or not. It's very likely that most of them aren't feasible, aren't realistic, are too expensive to realize, or have some fatal flaw that will quickly eliminate them at the next level of evaluation. However, one or more ideas may be highly creative, novel, and accomplishable. At the very least, your group should be able to generate a long list of candidate solutions. In the best-case scenario, you will come up with some great ideas that address the requirements with ingenuity and elegance.

30.3.2.2 Morphological Charts

A **morphological chart** is simply a list, usually in graphical form, of candidate solutions to the required system functions. The term "morphological" refers to the study of form, shape, or structure. The creation of a morphological chart involves creating a list of each of the functions that the system must perform or features it must possess, and then enumerating as many ways as possible of performing the required functions or creating the necessary features. Structurally, the morphological chart should be arranged so that each function or feature is at the beginning of a row, and quick sketches (or, if necessary, text descriptions) of the candidate solutions fills in the row, working left-to-right, as shown in Figure 30.3. Though it is simplistic, the exercise of creating a morphological chart achieves all three of the goals set out above, namely the decomposition of the problem, the creation of a great number of design solutions, and the documentation of these ideas.

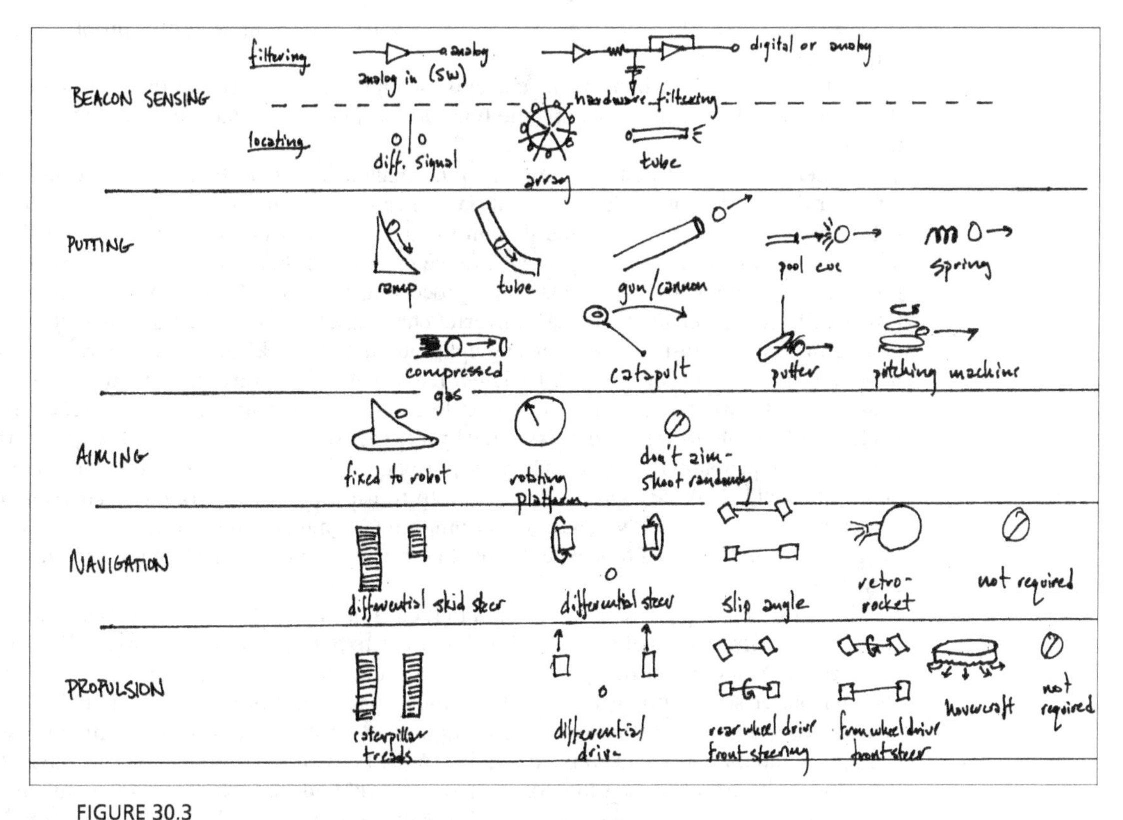

FIGURE 30.3

A morphological chart showing the required functions and features of the robot assigned in Example 30.1, and potential solutions.

Example 30.1

Your class has been assigned the following project:

Design a self-contained robot that is capable of putting a golf ball into one of four holes that are each 6 in. in diameter. The object of the game is to sink one golf ball in each of the four target holes as quickly as possible. The playing field is 10 ft $\times$ 10 ft, and is constructed of perfectly level particle board. The four target holes are randomly located around the playing field, and you will have no advance knowledge of their placement. Above each target is a beacon with an infrared LED that is modulated at 1 kHz with a duty cycle of 25%. When a ball is successfully sunk in a target, the modulated beacon above that target will be turned off. Your robot will be randomly placed on the playing field at the beginning of the game, but it will not be placed in such a way that two or more targets line up (four distinct targets will always be visible at the beginning of a game).

For this example, create a morphological chart that explores what is required of your robot and all the ways you can think of to meet the requirements.

The creation of a morphological chart begins with the enumeration of the functions and features that the design must fulfill. For this example, we will decompose the robot into subsystems that each perform one of the required tasks, and structure the list as follows:

1. beacon sensing
2. putting a golf ball
3. aiming a shot
4. robot navigation
5. propulsion mechanism

Keep in mind that there are a great many ways of decomposing this problem—there is no single "right" answer.

Next, we will draw in as many concepts and approaches as we can think of that can perform these functions. Figure 30.3 contains the finished morphological chart, followed by a discussion of the entries.

There are many potential solutions to the challenge of identifying a target, each relying on a different design. We found that the solutions we came up with for the category "beacon sensing" indicated that we needed to divide this into two functional categories: signal processing/filtering and directional beacon locating. Already, we've discovered that our first pass at decomposing the problem wasn't quite adequate. For signal processing, we might feed the sensor's output more or less directly into an analog-to-digital converter channel, and process the signal in software to determine if the photo-transistor were sensing light modulated at 1 kHz, 25% duty cycle. Alternatively, we could filter, rectify, and amplify to detect the signal in hardware, eliminating the need to write additional software. For directional beacon locating, we came up with several schemes: the first made use of two photo-transistors separated by a barrier so that the signal from the photo-transistors would be of equal amplitude only when they are pointed directly at a beacon. A more brute-force approach would be to use an array of eight photo-transistors, each separated by a barrier, so that the detectors aimed at beacons will have larger amplitude signals. In addition, a single sensor located at the end of a relatively long tube would probably receive light only when the tube is oriented directly in line with a beacon.

For the "putting" function, we found that each of the candidate designs centered around the storage and delivery of energy to a golf ball in order to putt it towards the target. We came up with schemes that made use of potential energy (ramps and tubes made into ramps), momentum (pool cue and putter shot), spring loaded release mechanisms, and pressure differential/gas expansion approaches (gunshots and cannons). As for aiming these shots, we thought of three possible ways: having the shooting mechanism fixed relative to the body of the robot so that the navigation function also took care of aiming the golf ball, or we could add a separate component (and hence a degree of freedom) upon which we would mount the putting mechanism. Or, we could dispense with aiming altogether, and randomly release balls in the hopes that one of them might eventually fall into a target.

Navigation and propulsion are closely linked. We could choose tank-like caterpillar treads to propel the robot, and differential skid steer to navigate. We might prefer to use two wheels and a caster in a similar approach that doesn't require the purchase or construction of appropriate tank treads. In this case, we might steer by differentially driving the wheels. We could choose a propulsion mode traditionally used with cars, such as rear-wheel-drive, front-wheel-steering, or vice versa. Thinking beyond the traditional, we could create a hovercraft and use air directed out the sides with fans or jets to propel and navigate the craft. Finally, we realized that there is no requirement for our robot to be mobile at all. We could create a robot that sat wherever it was placed at the beginning of a game and successfully fulfilled all the requirements without taking on the challenge of propulsion and navigation. Though it might not be the most exciting robot to watch, it is likely to have an excellent chance of successfully putting the golf balls into the targets in a short time, and could ultimately be the best performer. Also, by choosing this approach, we would be free to focus considerably more time and effort on the sensing, aiming, and putting functions. As a result, these subsystems are more likely to perform well and will probably be reliable much earlier.

30.3.3 Design Concept Evaluation: Prototypes and Iteration

In the best cases, a number of solutions have been conceived that look likely to meet project requirements and specifications. Choosing which candidate designs to move forward with becomes the next challenge. One of the best ways of objectively evaluating design ideas is to quickly construct prototypes of them. If you choose to prototype and compare a number of different ideas, you may find that one or more design candidates exhibit much greater potential (or deal-breaking shortcomings) than others. Often, this is not obvious when comparing the concepts on paper or in your mind's eye. We considered the subject of prototyping in much greater depth in Chapter 29, but the power of prototyping and iteration deserves brief mention here as a means of reducing uncertainty in the design and project management process.

Prototypes can show you where you have overlooked problems, or neglected to consider some of the more inconvenient laws of physics. On the positive side, prototypes can serve to validate your initial ideas and demonstrate the potential of a given approach. In either case, the goal for the earliest stages should be to build the first prototypes *quickly,* so that the first batch of problems and oversights can be identified as early as possible. If the first prototype versions reveal insurmountable problems, then shift effort to other concepts. If you identify issues that you believe are solvable, quickly iterate on the prototype and incorporate new solutions to the problems you found. You are likely to find a new batch of problems with this new prototype, even if you completely solved the initial problems. Iterating again (and again, and so on) will ideally bring you all the way to a working prototype that will serve as the basis of your final design. The experience gained during the first few iterations of a device prototype will serve to inform your estimates of how difficult the project is likely to be, what resources are needed, and how much time it is likely to take to complete. These topics all relate to project management, and we touch on these in later sections of this chapter.

30.3.4 Specifications

Once you have a firm grasp of the requirements and have generated possible solutions that address them, you are ready to generate the **specifications** for your system. It's common for confusion to exist over the difference between requirements and specifications. Requirements, as described above, are the key goals and functions of the systems: the characteristics or uses that the system must exhibit in order to fill a particular need. Specifications, on the other hand, are the particular and quantifiable metrics that enable a system to meet the overall requirements. For example, a particular mobile robot may be required to navigate a playing field constructed of level particle board. Specifications for the propulsion system of your design may be that it must be capable of moving the robot in a straight line with a maximum velocity of at least 25 cm/s, and that it must be capable of turning corners with a minimum radius of 5 cm. If that mobile robot must putt golf balls into 6 in. diameter holes in the playing field, the aiming system must have specifications governing its accuracy and resolution. For instance, it may need to be accurate to within ±2.5°, and be able to adjust the aim in 0.5° or finer increments. The list of requirements informs you of what the overall goals are. The specifications inform you of exactly how your product must perform to achieve the goals. Both are critical.

Defining the system specifications is especially critical when the system is broken down into subsystems and different individuals are responsible for designing and building the subsystems. In this case, the specifications spell out not only what each subsystem must do and how it must be done, but also how it fits together with all the other subsystems with which it must interact. These points of interaction, the **interfaces**, are highly important. If the interfaces are not well defined and well managed, the experience of bringing the subsystems together for the **integration** of the full system will be tedious and unpleasant. Problems as simple as mismatched bolt hole patterns or as complex as broken communications interfaces and protocols all have the same result: going back and reworking one system to fit with another. Careful planning and the generation of a good set of specifications will help prevent this type of nasty surprise.

As with the list of requirements, a complete list of your systems' specifications helps identify ill-defined or contradictory areas, and enables you to address them as early as possible. Again, it's always helpful to identify and address the problems at the planning stage. This may prevent wasting time and effort on designs that get scrapped when specifications are either better understood or clarified.

30.4 MANAGEMENT TOOLS

In addition to defining system requirements and specifications, and defining viable design candidates and alternatives to meet them—that is, defining what exactly you are going to be working on—the challenging task of managing the project from start to finish must also receive considerable attention. Complex projects require the coordination of resources (people, time, money, facilities, and so on) in order to be successful. Granted, they can eventually be accomplished in the absence of any active guidance or management, but the process is inefficient and the outcome is suboptimal. Management tools will make the difference.

The remainder of this chapter deals with tools that will assist with the organization of your thoughts and your projects. Several approaches are briefly presented, each of which is the subject of extensive thought and study elsewhere. Indeed, entire courses and even graduate degrees are devoted to the study of project management, concurrent design, and systems integration—we won't attempt anything more here than to discuss the most basic foundations for each.

30.4.1 Project Management

Project management is a term you've probably heard before, but in the context of this chapter, it has a specific meaning. **Project management** is the activity of organizing available resources to accomplish the tasks required to meet an objective [1]. It is the responsibility of the project manager to examine the starting point and the desired endpoint and determine how to navigate between the two points with the resources available (or, if the resources are inadequate, to determine what additional resources will be required).

The most important tasks relating to project management are:

- Define and relate tasks and subtasks
- Budget resources
- Create and maintain a schedule
- Control the processes

Once the project manager (if one has been appointed—it may also fall to the group as a whole to serve this purpose) has a firm grasp of these issues, there are several organizational tools that are well established that represent the flow of time, materials, effort, and other resources that go into the execution of the project, and help the manager to gauge how well the project execution matches the plan.

Probably the most popular project management tool that incorporates scheduling and resource allocation information is the **Gantt chart**. A Gantt chart consists of a number of rows, each representing a task or subtask. The leftmost entry in each row contains the task's name, and to the right of this is a timeline that indicates when each task will start and how long it will take to complete. The starting point and ending point of each task line are typically indicated with triangles that point downward. Milestones, such as the completion of a task or the generation and delivery of a report, should be included, and are usually indicated with a diamond symbol. Deadlines should be included as well—as they often drive the overall time available to complete the project. Deadlines are also usually indicated with a diamond symbol. Tasks that are interrelated in some way are connected with vertical lines. These may be used to indicate when one task can only begin, once another task has been completed, for example. Or, they may be used to indicate when several tasks will begin simultaneously. Vertical lines are used to show the interdependencies and timing that relate the various tasks. In short, any time-based information that is useful in understanding the schedule's constraints and requirements should be incorporated into the Gantt chart. An example Gantt chart is shown in Figure 30.4. This example isn't terribly long or complex—often real-world Gantt charts for complex projects can span many pages (both for the number of tasks enumerated and the time required for the completion of the project). Gantt charts are scalable to fit the needs of the specific project.

In addition to the tasks and timelines, resources may also be noted and incorporated into the schedule. The person or group assigned to a task can be noted in a separate column or next to the timeline, for example. If a facility or instrument will need to be managed (such as a machine shop or a soldering iron), that may also be incorporated into the chart, or a separate chart may be constructed for that purpose. The power of the Gantt chart is that it clearly communicates what tasks are required, and how much time is required to complete each task and the entire project. The Gantt chart makes it immediately clear which tasks are likely to require the most time, and encourage those managing the process to consider which, if any, tasks may be performed in parallel. Often, some tasks cannot begin before certain other tasks are complete, but when opportunities exist to complete a task that doesn't have these constraints, they may be scheduled for early completion if the necessary resources exist.

Though Gantt charts are powerful tools, they are not perfect. First, they presume that the manager knows—before the project really starts—what the required tasks are, how long they will take, and how they interrelate. Sadly, this is *never* the case. The experience of the manager figures heavily in the accuracy of each of these issues, but ultimately the manager uses his or her experience and judgment to simply estimate (i.e., guess) what the tasks are and how long they will take. This doesn't mean that Gantt charts are useless—far from it! This indicates that the project manager and the team should expect to continuously revise the schedule as new information and certainty develops while the project progresses. The manager and the team should be continuously monitoring the progress of the project and comparing it to the schedule to determine where there are signs of trouble or reassurance. If changes need to be made (and they usually do), the manager must be ready to recognize the need and act on it. This is one of the primary tools for the project manager to monitor and control the process. This comparison between the plan and the unfolding reality is the feedback that allows the management to exert control over the process. When everything works well, this means that the project will be brought to completion, on budget and on time.

Several popular software products exist for creating and maintaining schedules, such as Microsoft Project and FastTrack Schedule, as well as a number of shareware and freeware programs. However, aside from automating the generation of timelines, linking tasks, and the ease with which they can be updated and edited, these programs offer little that can't be done easily on paper or with a spreadsheet.

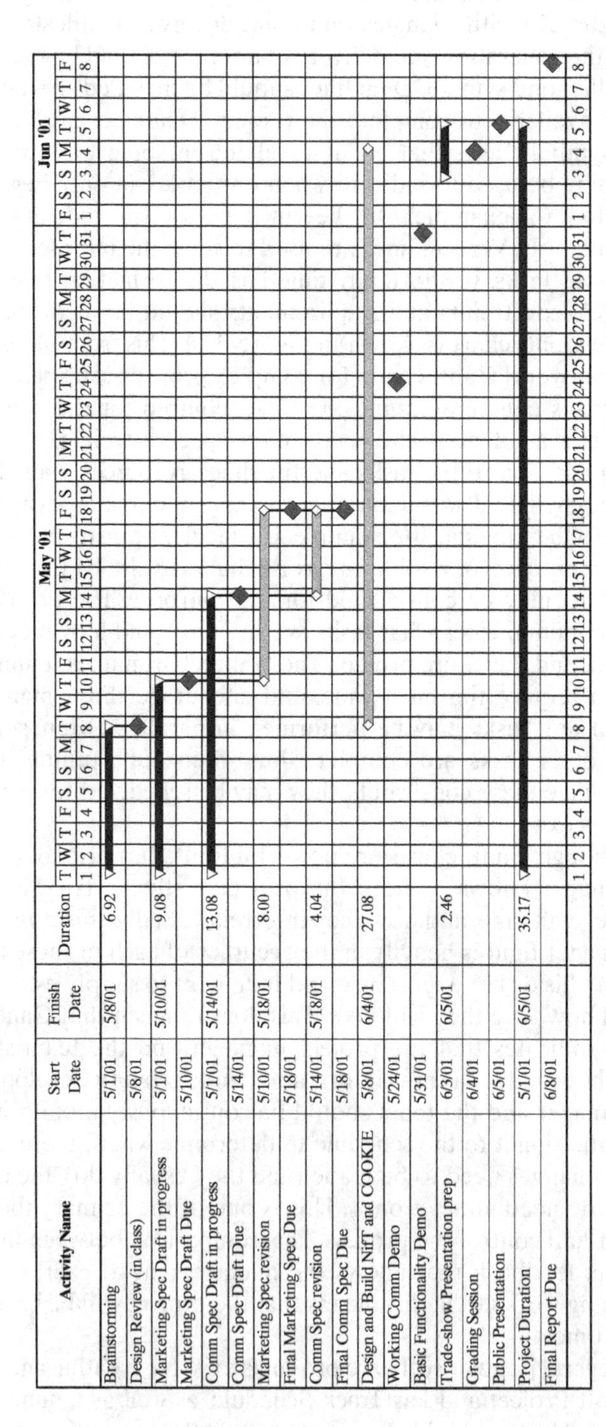

FIGURE 30.4
A Gantt chart.

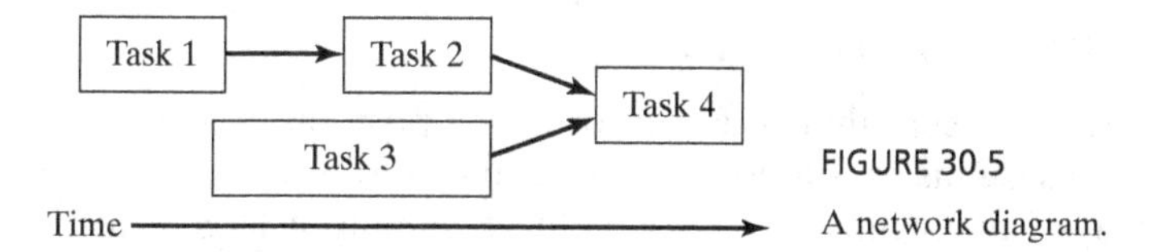

FIGURE 30.5
A network diagram.

Another graphical representation of tasks and their interrelation is the **network diagram**. Network diagrams show each of the required tasks as separate boxes and employ arrows that indicate predecessor tasks and successor tasks. The box representing each task may also be drawn in such a way that it communicates the time required for completion, but this is not always done. Figure 30.5 shows a representative network diagram.

Network diagrams very clearly show the dependencies between tasks, and the time required to complete each. One especially useful concept that they inherently highlight is the project's **critical path**, the task or chain of tasks that will take the most time to complete, or upon which many successor tasks depend. Gantt charts, while useful, do not always show the critical path as clearly.

Additional project management tools have been developed and are in widespread use, but Gantt charts and network diagrams should provide an excellent foundation from which to work. Also, there are a great number of references available for additional information and depth in project management.

Aside from the ability to plan tasks and timelines in detail, project management tools serve to allow managers to compare a plan to the reality of how a project is unfolding. Good managers continually work to gauge whether their team will be able to complete a project on time and within budget. Since it is exceptionally rare for a plan to be accurate and complete the first time (or even after multiple revisions), this comparison of projection vs. reality is a key tool for keeping track of where project risks are, and when to start exerting management control to rectify issues with a task that isn't quite working out as envisioned. Controlling the process, even as it continually changes and defies your plans, is a continuous process of its own. Each of the inputs is available for adjustment. Additional resources may be brought to bear in the form of additional personnel, expanded budgets or new facilities, and tools. Depending on the constraints, any or all of these may be available. The schedule can be rearranged to accommodate the new developments and realities. And the goals of any of the tasks or the entire project may be changed to make them easier to achieve: either by **de-scoping** (i.e., simplifying—or seeking something closer to the "base level of functionality" to which we refer above) or by switching to a design alternative that was not originally chosen. Often a design team will choose to move forward with a design that exceeds requirements in ways that may have appeal. This is a natural tendency, and especially hard to resist at the beginning of a project, when time and resources seem plentiful. The project manager must keep track of these "features" and be prepared to eliminate them as the project moves forward. De-scoping is common and often necessary, and these "features" should be high on the list of items that can be revised without jeopardizing the team's chances of meeting the overall requirements. The timing of de-scoping is also critical. If done too late, it may require more time and effort than it saves. De-scoping at the right time is often the key to delivering a product that meets the baseline requirements on time and within budget. Project managers need to constantly track the project's progress and resources, and watch for moments when de-scoping will be helpful.

Flexibility and freedom with the resources, tasks, and final configuration of the project should be continually evaluated and optimized. In this way, the project manager can use the available feedback and exert management controls to ensure the best outcome. This is the essence of project management.

30.4.2 Systems Engineering

One concept that is useful for decomposition of a complex system or project into simpler subsystems or components that can be more easily addressed is that of **systems engineering**. Systems engineering deals with examining a large scale or multidisciplinary project and dividing it according to functional subsystems so that the appropriate personnel and resources may be applied [1]. Systems engineering affords project planners a useful organizing principle to apply when struggling to generate tasks and schedules. Complex systems that incorporate subsystems that each require the efforts of highly specialized people are especially suited to this approach. Mechatronic systems are an inherently good fit with systems engineering, since they incorporate mechanical, electrical, and software components and can reach high levels of complexity.

Example 30.2

For the robot project assigned in Example 30.1, decompose the specified robot into functional subsystems that can each be designed and built by independent teams, brought together, and ultimately integrated into the full system.

One possible decomposition is proposed in Figure 30.6.

The system as shown breaks down very much along the same lines as those used in Example 30.1 for the creation of the morphological chart, with a few additions. (Indeed, a systems engineering approach is an excellent way to begin thinking about the functions required for the morphological chart.) In addition to Navigation, Propulsion, Aiming, Putting, and Sensing, we will also require a Structural subsystem (the physical design and configuration of the entire robot) and a Strategy subsystem (which might also be called "Programming," as the software subsystem will incorporate strategy as well as monitoring and controlling several robot functions). This is only one of many possible decompositions of the system. Again, there is no "right" answer here, and alternate solutions may be just as good as, or better than, this one. Undoubtedly, we would learn more as we begin work on the project.

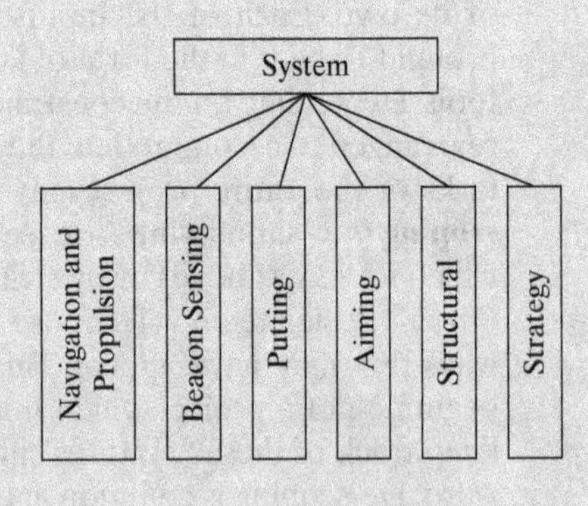

FIGURE 30.6

Decomposing a complex design using the principles of systems engineering.

Much of the risk involved in taking a systems engineering approach comes near the end of the project, during the integration of the completed subsystems. Because different people or groups may separately develop the components that make up the whole system, problems often arise when the subsystems are combined and they don't interact as intended. The time to address this issue is actually at the *beginning* of the project, not the end: great care must be taken at the beginning with the specifications. Special attention must be paid to the interfaces between each of the subsystems. Since the subsystems interact and interconnect either physically, electrically, or both, planning and communication are required to ensure that everything comes together and functions during system integration. During the execution of the project and the subsystems, if changes are made midstream and interfaces are affected, these changes must be communicated and the relevant interfaces updated. With a systems engineering approach, you will need to devote significant energy to specifying the subsystems and their interfaces, and verifying the function of each subsystem before it is integrated.

30.4.3 Concurrent Design

Another organizing principle useful for structuring project management is to consider all the life cycle phases that your project or product will go through, and this is the focus of **concurrent design**. Rather than simply insuring that it meets requirements and satisfies specifications, a successful project performs well in all phases of its life cycle, from the very beginning (we call it "analysis" here, though there are a number of ways to break this down), to the middle phases such as development (turning the concept into the finished design or product) and production [making a number of copies for distribution or sale to end users, from 1 (e.g., a satellite) to 1,000,000,000 (such as a microcontroller)], and finally into operations (which may include several elements such as repair and maintenance, upgrades or updates, customer satisfaction, and so on). During each of these phases, your product may need to be assembled, disassembled, tested, repaired, shipped—and a great number of other activities. Thinking carefully about each of these activities during the analysis and design phase of your project will make it much more successful over the entire course of its existence. This is ultimately how it will be judged, so paying attention to the full cycle will pay off handsomely. The application of the principles of concurrent design has popularized such phrases as **design for manufacturability (DFM)**, **design for assembly**, **design for test**, **design for serviceability**, and others.

An additional excellent reason for incorporating life cycle considerations into your design work is provided below in Figure 30.7, which demonstrates that the total cost involved in an average project or product is determined at the beginning, during analysis and development, though the payout of that cost occurs much later [2]. This indicates that the ultimate total cost of a project is largely in the hands of the designers and managers involved in analysis and development. If they are able to make the product easy and inexpensive to produce and operate, for example, the overall cost of the product will be substantially reduced.

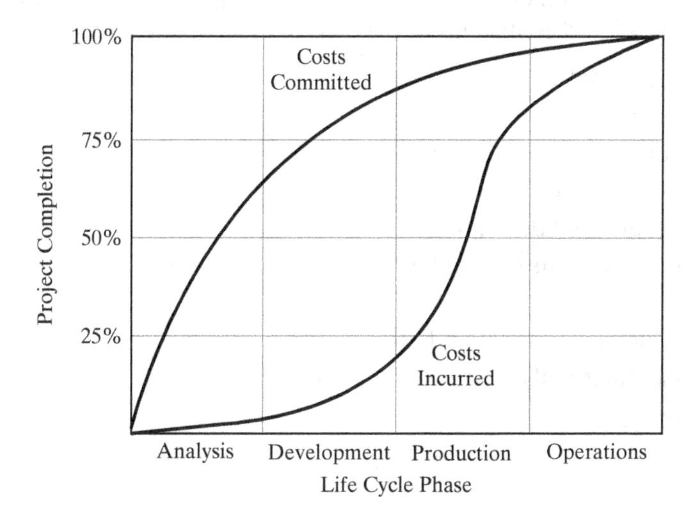

FIGURE 30.7

A good reason to consider all phases of a product's life: early design decisions have a dominant impact on total project cost.

In addition to designers simply thinking about ways of incorporating these issues in the design process (which has substantial value, of course), individuals or groups that are more expert in the other phases of a life cycle should also be included in the design process. For example, rather than conjecturing about how a design change might make a device easier to die cast, bring in an expert who has made a career of that particular field, and find out what he or she thinks is the optimal design. Manufacturing engineers, assembly technicians, repair technicians, sales people, and anyone else with a stake somewhere in your project's life cycle should all be involved in the design process, and, where possible, their input should be incorporated.

Whatever you are designing, it is good practice to consider which life cycle phases your product will experience, and which of these will optimize its success. In the case of the robot project described in Example 30.1, the success of the project is likely to depend on a design that accommodates extensive testing and modifications. It may only exist in the "operation" phase of life for a few brief minutes during a presentation or competition, but it may spend weeks bumping around a table or lab while its program, sensor circuits, propulsion system, navigation, putting system, and strategy are developed. Designing the robot to make the testing phase inexpensive (which is to say, easy, since a student team's most precious resource is time) will carry great benefits.

30.5 COMMUNICATION AND DOCUMENTATION

Documenting and communicating the state of your project and its subsystems to the others working toward its completion is also highly important. Under intense time pressure (such as a class project assignment), this is another area that is all too often neglected. Unfortunately, neglecting communication and documentation will usually end up costing you and your team far more time when something unexpected occurs. Rather than thinking that you don't have time to take care of it, consider how much time it will take to backtrack, generate the necessary documentation retrospectively, and then solve whatever hiccup is occurring. You can't afford *not* to document and communicate the state of your project!

At a minimum, it is good practice to keep a logbook that contains the evolution and current state of your design. Any member of your team should be able to turn to the logbook at any time to determine the status of the project and for specifics about the system's various subsystems and interfaces. In addition to traditional paper logbooks, there are a large and growing number of online tools (e.g., shared documents, wikis) that serve this purpose very well. These can substantially streamline the process of keeping the record up-to-date. As long as the information is easy to update and is accessible to you and your team whenever and wherever it's needed, it doesn't matter whether it's in hard copy or soft copy form. Regardless of the medium selected, a good logbook will contain up-to-date information about at least the following:

- an up-to-date schedule
- system requirements
- system and subsystem specifications
- concepts generated and selected
- sketches, diagrams, and descriptions
- schematics for all circuits (current *and* previous versions)
- code listings
- guides to the locations of files and programs
- notes from meetings and decisions that are made
- current status

Regular meetings are essential as well, though there's no substitute for having a written record of your project's progress that all can refer to at any time.

30.6 PITFALLS AND SUGGESTIONS

There are an infinite number of ways to run into difficulty with the design and construction of mechatronic systems, and indeed with projects in most other disciplines. In this chapter, emphasis has been placed on organizing principles, tools, and methodologies that help to get organized and improve chances for a successful outcome.

In an effort to show that the techniques described above are relevant to you and have real value, we encourage you to consider projects you've completed in the past. The following list is based on often-heard statements about what has gone awry with projects, all of which relate to the concepts and project management tools discussed in this chapter [1]. These are not hypothetical—we've heard these repeated numerous times by teams in our labs. The goal here is not to dredge up bad memories, but rather to encourage you to think about how to alter the course of your next project. Use the tools presented in this chapter, and you should find yourself saying or thinking these phrases far less often.

If you find yourself saying:	**You probably should have focused more on:**
"We ran out of time."	Managing critical path Task awareness Feedback/management controls De-scoping
"I need the computer too."	Resource scheduling Schedule coordination
"It had to do that too?"	Generating requirements . . .
"We tried for this elaborate and very cool scheme, but we ran out of time so we had to give it up, and in the end, we weren't able to get some of the required functions to work."	. . . and sticking to the requirements De-scoping
"There must be a better way."	Generating alternatives
"But how do I make that?"	Design for fabrication
"I'm not sure what's wrong."	Design for testing Document the design Subsystem verification Incremental integration
"It worked once . . ."	Testing and integrating early

30.7 HOMEWORK PROBLEMS

30.1 What is the focus of the systems engineering approach to structuring a project? What kinds of people are essential when using this methodology?

30.2 What is the focus of the concurrent design approach to structuring a project? What kinds of people are essential when using this methodology?

30.3 Are the project management, systems engineering, and concurrent design approaches mutually exclusive (i.e., if you use one of these approaches, are you precluded from using others)? Explain.

30.4 Using a systems engineering approach, enumerate at least 20 subsystems for the Ferrari 599 shown in Figure 30.8.

FIGURE 30.8

A Ferrari 599.

30.5 You are tasked with designing the successor to the Ferrari 599 (Figure 30.8), and wish to incorporate as much input from different domain experts as possible. Name a minimum of 10 types of experts you will include on your team.

30.6 Create a Gantt chart that reflects the syllabus for a course you are currently taking. Include interdependencies, start and end dates for all assignments, due dates, and milestones.

30.7 Create draft requirements for a laptop computer targeting the market segment of college students enrolled in an engineering program.

30.8 Which phase of the following products' life cycles would you anticipate would cost the most for the producer? Support each answer with a brief explanation.

(a) Disposable surgical tool

(b) Automobile

(c) Mars rover

(d) Communications satellite

(e) Ream of paper

(f) Desktop personal computer

30.9 You are tasked with leading the team that will design the next version of the Apple® iPod®.

(a) Create draft requirements for the device.

(b) Using a systems engineering approach, decompose the iPod into subsystems.

(c) Identify as many of the product's life cycle stages as you can think of.

(d) Identify the types of experts whose input will help minimize the expense of each of the iPod's life cycle stages.

30.10 You are assigned a class project in which you are to design a robot that putts 10 golf balls, one at a time, into one of three targets on a 4×8 ft. playing field. There is a 2 in. drop-off at the boundaries of the playing field, and your robot must stay on the playing field for the duration of the game. Each target is active for 10 seconds, after which time a new, randomly selected target becomes active. The active target is identified by a beacon that modules an infrared (IR) light at a frequency of 2 kHz with a 50% duty cycle. The wavelength of light emitted is easily sensed with an IR photo-transistor. Each round of the game lasts for 2 minutes. The robot that successfully putts the most golf balls into active targets wins. Each design team will be comprised of four students, and you will have two weeks to complete your design.

(a) Create a list of the requirements for a robot that can successfully compete in this putting game.

(b) Identify areas where the project specifications are vague.

(c) Identify loopholes in the project specifications.

(d) Create a Gantt chart describing how your team will go about completing the project.

(e) Identify the subsystems required by your robot.

(f) Create a morphological chart with as many candidate designs as possible for each of the subsystems identified in (e) above.

30.11 For your mobile phone, answer the questions below. If you do not own a cell phone, borrow one and explore its behavior to answer this question.

(a) List the functions that your phone is able to perform.

(b) From the list you generated to answer part (a), identify the set of functions that you consider to comprise the base level of functionality. What fraction of the phone's total function is essential to you?

30.12 Create a Gantt chart that describes your career goals for the next 15 years. Be sure to include important milestones.

REFERENCES

[1] Lecture: *Project Management,* ME118/318 and ME210, Kitts, C., Stanford University, 1998–2003.

[2] *Concurrent Engineering, What's Working Where,* Backhouse, C. J., Brookes, N. J., Gower Publishing Limited, 1996.

FURTHER READING

Conceptual Blockbusting: A Guide to Better Ideas, Adams, J. L., 4th ed., Basic Books, 2001.

Mechanical Design Process, Ullman, D.G., 3rd ed., McGraw-Hill, 2003.

CHAPTER 31

Troubleshooting

31.1 INTRODUCTION

All too often, newly minted mechatronic designers experience the following scenario: after completing the hard work of designing and assembling a mechatronic subsystem or even integrating an entire system, testing commences and surprises emerge. In this context, the term "surprise" is a euphemism for "problems," "bugs," and "mistakes"—issues the designers didn't foresee that cause them to scratch their heads and wonder what's actually happening, and what's causing the observed behavior. Although it's theoretically possible to build a system that works exactly as desired the first time power is enabled, it's exceedingly unlikely—it's a near certainty that the actual function of the system will deviate from the plan in a number of ways, even when great care was exercised throughout the previous stages of the project. The process of identifying and solving these problems is called **troubleshooting**. Troubleshooting mechatronic systems is especially challenging, because the systems tend to be complex, and the cause of undesirable system behavior could be in any of several possible places. It could be a design flaw somewhere within the electrical hardware, mechanical hardware, or software. It might be an implementation flaw: is the wire connecting the sensor to the op-amp sensitive to noise from a nearby solenoid? Is it even plugged in? This chapter will provide some tips and advice to help get through this process successfully.

Much of the advice presented here may seem obvious, and even a little preachy. The truth is that great troubleshooting technique really has to develop through first-hand experience in the lab. However, there are some things that can be built into the design of mechatronic systems that make troubleshooting easier, and there are rational approaches that can help in the identification and solution of problems. In this chapter, we'll describe some practices that can help get you through the troubleshooting process. To help drive the points home and lighten the mood, we've chosen to present this material in a David Letterman-esque "Top Ten" format, though we start with the most important point first.

Like it or not, it is simply one of the harsher realities of life that an integrated subsystem or completed system is unlikely to work perfectly the first time it is tested. While there are measures that can be taken to improve the odds of something working the first time—indeed, the old truism "an ounce of prevention is worth a pound of cure" is especially applicable to troubleshooting—the emphasis in this chapter is on getting through the troubleshooting process quickly and efficiently.

One of the most important points about troubleshooting that mechatronic designers need to accept that it's not a failure on your part when something you've created doesn't come together perfectly the first time. Most mechatronic systems are too complex for that to be possible. Rather, troubleshooting should be considered a normal stage in any project. You should design and implement your systems with the goal of working the first time, but allow for the likelihood that there will be trouble, and that you'll have to spend a significant amount of time tracking it down and correcting it. Plan for it; incorporate it into your schedule. With the right mindset, troubleshooting can actually be one of the most engaging and rewarding elements of a project—for this is where some of the greatest challenges lie (solving the mystery of *why* it is not working properly) and (when you're successful) some of the greatest rewards.

31.2 STARING INTO THE VOID

So you've completed the initial system/subsystem assembly, and your device just isn't working as expected. In order to fix the problems, you must first identify the problems. Being good at troubleshooting is actually mostly about being good at finding problems. Once a problem has been identified, it is often, though not always, straightforward to fix. The identification process is where a great deal of time and effort can be spent, and also a process in which there's substantial risk of damaging your system—before it's even working properly.

In many mechatronic systems, the entire system as a whole is very complex and includes a great many mechanical, electrical, and software elements. The total number of electrical circuits and components, the lines of code, and the mechanical parts can be very large indeed—in the thousands, or even millions—and thus so can the number of possible sources of problems. When something isn't working and it isn't obvious what the problem is, confronting this enormous list of possible causes can be daunting.

Ultimately, your goal is to identify the source of the problem. If you are looking at many thousands of details, spotting what's wrong is like looking for a needle in a haystack. You can stare at it all day and never find the needle. You'll have an easier time finding the problem if you isolate variables, and then perform simple tests to determine if you're on the right path.

Without a sound methodology to rely on, it may be tempting to reach into the system (hardware and/or software) and make a change based on a guess to see if it helps or hurts. The biggest problem with this approach is that it is very unlikely that you'll accidentally pick the one detail out of thousands that is wrong. It is much more likely that your first attempt at this will end up changing some detail that was correct into yet another incorrect detail. You may very well have created a new problem. There may be a discernable change in the system performance, but this is not progress: you haven't eliminated any bugs, you're just altering the system's behavior. Instead of this "lottery ticket" approach, we recommend something far more likely to converge on a solution, the *Number One Mechatronic Troubleshooting Tip*.

#1 Treat each test as an experiment: start with a hypothesis, and look for confirmation of the hypothesis in the result of the experiment.

The goal of the best scientific experiments is to isolate a single variable at a time and alter it in order to observe its effects on the overall system. The same is true with troubleshooting: each test should be as specific as possible, and you should begin each test with an expectation of the results of altering an element of your system. The "elements," or variables, that you vary may be electrical, software, or mechanical—whatever you believe might explain the unexpected behavior. When you successfully identify the element that is the source of the problem you're working to eliminate, fixing it is often straightforward. That's not always the case, but at least you have narrowed down the problem considerably, and have increased your understanding of it many fold.

When properly applied, the #1 Troubleshooting Tip prevents you from engaging in a great deal of unproductive testing. Take, for example, the following scenario. Suppose you have a system that is required to shoot a foam ball at a target when it detects an infrared light beacon, and suppose that the system inexplicably fails to shoot any balls. It can be tempting to make small changes to some part of the system (e.g., "maybe that op-amp isn't working"), and then to put it all back together again to see if it will suddenly begin to shoot the ball at the right moment. There are many problems with this approach, including the lack of specificity of the variable you changed and the lack of a specific expected result. There could be several things wrong with the system, and no single change will likely cause the desired total behavior.

Instead, our recommendation is to construct tests that look for changes in the system behavior that are immediately linked to the variable you change. So, instead of looking for the whole system to launch the ball after making, say, a change in the gain of the signal amplification circuit, we recommend that you look at the signals in the circuit just after the point where the change was made. Examine the signals before and after making the change. If you understand the circuit, you should be able to predict the result of the change. If things do not behave as expected, this means that either you don't understand this part of the circuit or that it is not behaving as it should. In either case, this is a clue that you need to dig in deeper at this location.

While you are poking, prodding, and testing your system, it is extremely important to maintain a thorough log of each test, including exactly what you did and what you learned from it (if anything). It's also crucial to restore any changes before making any more changes—hence enabling you to observe the effects of changing one variable at a time. In the case of software, this can be facilitated by frequently updating version numbers and archiving previous code versions so that it's possible to retrace your steps in the event that things suddenly become a lot worse. In the case of electrical hardware, it is essential to start with a complete and accurate schematic of the original circuits, so that you can make notes and revisions as you make changes. The overall concept is to leave a trail as you proceed. You may discover that you've gone down a blind alley and need to get back to a previous circumstance, and your notes will be the only thing marking your path. If you find yourself starting your 210th test, your memory will not serve you as well as you might like: you're bound to lose track of whether you've already tried a particular test before or what the results were. For these reasons, we offer the *Number Two Mechatronics Troubleshooting Tip*.

#2 Record all troubleshooting tests in a log book, and dutifully document any changes that you make to the system and the results that you obtain.

A word of caution: reassembly and restoration of a system after a test is a time fraught with risk. This is a very common time to accidentally plug wires in backwards, reverse power and ground connections, attach a connector incompletely so that the connections are intermittent, break screws, bend or decapitate electrical components, and so on. These mistakes often result in spectacular and catastrophic failures when you turn the system on again after reassembly—just when you're least emotionally equipped to handle it. You should take great care every time you reassemble your system. It also helps to design your system so that it's impossible, or at least very difficult, to assemble it incorrectly. For example, lockout features that ensure a part can only be installed in the correct orientation, electrical connectors that are keyed and shrouded so they prevent incorrect or incomplete hookup, and extensive error checking in your software (e.g., Is the sensor connected when the software boots?) all help you restore your system without inadvertently destroying anything.

Log books are also the appropriate place to record troubleshooting procedures that turn out to be generally useful. When the #1 Troubleshooting Tip has been applied well and a particular test or set of tests is likely to be used repeatedly, record the steps of the test procedure in your log book. After all, scientific experiments should be repeatable, and the results should also be reproducible by others. Recording the steps required to diagnose a problem or verify proper function of a subsystem allows members of a team other than the designer to perform the test. The designer doesn't have to be physically present in order for progress to occur. The practice of writing down troubleshooting procedures is used extensively in industry, where someone other than the designer of a system is likely to be the one who builds it and brings it to life.

31.3 AN OUNCE OF PREVENTION IS WORTH AN ALL-NIGHTER OF TROUBLESHOOTING

In this section, we stress the benefits of planning for troubleshooting when engaged in earlier phases of a project, such as design and assembly. We know from our own experiences that the design, assembly, and integration of complex mechatronic systems will not be perfect, so why not make the inevitable troubleshooting easier by planning for it when you're designing circuits, software, and mechanical components? **Insanity** was described by Albert Einstein as the repetition of the same activity over and over while expecting a different result; and this is certainly applicable in troubleshooting. Since we know we're going to be dealing with some troubleshooting, the *Number Three Mechatronics Troubleshooting Tip* is:

#3 When integrating a mechatronic system, add each subsystem incrementally, and then immediately test and verify the function of each new subsystem and the overall system.

One very common—and very bad—assumption is that it's reasonable to assemble everything all at once, turn in on, cross your fingers and say your favorite incantation, and hope that it all just works. This is the mechatronic equivalent of the "Hail Mary" pass in football. You can actually be pretty sure it won't work out. Because of this, it's always a wiser plan to start small and simple and build incrementally from humble—but wonderfully functional—origins.

As an example, consider a mechatronic system that incorporates a microcontroller running control software, a battery and a voltage regulator, a sensor circuit that reads and reports the level of a particular wavelength of infrared light, and a mobile platform that has limit switches in bumpers that trigger when it runs into objects and several motors with wheels. Rather than putting this all together and then switching the power on to find out what will happen, it's a much wiser plan to build the system slowly and incrementally, and verify everything after each step. Start with a subsystem that you know works the way you want it to (based on extensive and definitive testing), and then add one more subsystem. For our example system, you might start by verifying that the microcontroller executes a very simple program first, when it is powered by a bench-top power supply. Then you could switch to regulated battery power and (after verifying that the expected voltage is present on the correct pins in the microcontroller socket) verify that the microcontroller subsystem continues to work. Next, you might hook up the drive circuits for the motors, and add the software that will exercise the motors. Test and verify that you are able to control the motors correctly. You can repeat similar steps with the light sensor and the bumper limit switches. At the end of the process, the result should be a fully functional system. You will probably find bugs in every subsystem, and at the interface between each of the subsystems. But, because you are adding them one at a time, and because you're confident that everything else worked before you added each subsystem, you have greatly reduced the number of places you need to consider as you look for the source of undesirable behavior. It's a much smaller haystack!

While you're in the process of incrementally adding and testing subsystems, you can successfully limit the number of potential sources of problems, but devising tests to determine how well each subsystem is working can be challenging and time consuming. The *Number Four Mechatronic Troubleshooting Tip* can considerably streamline the process:

#4 Include elements that provide observability of the internals of your system.

When you power up a subsystem or an integrated system and the behavior doesn't match your expectations, where should you look first for problems? If you've incorporated elements that make it convenient to observe key aspects of your system's function and performance, you should be able to quickly gather a great deal of information. For example, statements in your software that indicate status can provide windows that allow you to observe the inner workings of your system; like sensor readings, what state your finite state machine is currently in, which events are detected, or other compact pieces of data. The information can be used to affect an output pin, displayed on a computer monitor or other display, or logged to a file. In the C programming language, the use of the `#ifdef` and `#endif` keywords facilitate quickly and conveniently enabling or disabling these types of debugging features. For example:

```
#define DEBUG
unsigned int ReadLightSensor(void) {
    unsigned int LightLevel;
    #ifdef DEBUG
    PTT |= BIT0HI;
    #endif

    LightLevel = ReadADC(LIGHT_SENSOR_CHANNEL);

    #ifdef DEBUG
    PTT &= BIT0LO;
    #endif

    return(LightLevel);
}
```

For this example code, if DEBUG is `#defined` (as is the case in the code shown), the commands that appear between the `#ifdef` and `#endif` statements will be compiled into the code and will execute at run time. If DEBUG has not been `#defined`, the commands are ignored when the code is compiled. With simple software "switches" such as this one, the code can easily be toggled between running in a debugging mode or a more streamlined, quieter mode. In this simple example, when DEBUG is defined, an output pin (pin 0 on Port T) is changed to a 1 when the function ReadLightSensor() begins executing, and changed to a 0 when the function ends. By connecting an oscilloscope probe to Port T pin 0, this enables you to monitor when this function executes, how long it takes to execute, and how the timing of its execution relates to other aspects of the system. Recompiling the code with the `#define DEBUG` statement deleted or commented out eliminates the toggling of the pin. Though this example is simple, techniques like this allow for much greater observability into what is happening as the code executes and responds to stimulus.

Another means of further improving observability of the software is to add print statements that are displayed in a terminal window on a host computer's monitor. In the C programming language, one way of doing this is through the use of `printf` statements. As above, this can be combined with the `#ifdef` approach to relay much greater amounts of information.

```
#ifdef DEBUG
printf("/rEntering ReadLightSensor()");
#endif
LightLevel = ReadADC(LIGHT_SENSOR_CHANNEL);
#ifdef DEBUG
printf("   LightLevel=%d   Exiting ReadLightSensor()", LightLevel);
#endif
```

In this revised example, every time the ReadLightSensor() function executes, a line of text prints in the terminal window that reads "Entering ReadLightSensor() LightLevel=854 Exiting ReadLightSensor()" (for the sake of this discussion, we arbitrarily assume the light sensor reading was 854). We know what code is executing, and we know what light level was read from the light sensor. This is far more information than we obtained from the oscilloscope probe connected to the toggling output pin, and with this additional observability, we are likely to be quicker and more effective troubleshooters. Whenever using print statements for debugging, you should be mindful of the fact that these characters are being printed through a serial port (Chapter 7) and that the time required to transmit each character is potentially significant to the function of your program. At 9,600 baud (a very common serial terminal rate), each character takes 1.04 ms to transmit to the terminal. At the maximum baud rate for a PC (115,200 baud), each character still takes 87 μs. In the example above, if the serial port is operating at 9,600 baud, the debugging messages shown would take 75 ms to print. On the surface, 75 ms may not sound like much time, but in the context of the execution time of your program, it is indeed very long, and can impact the behavior of the system.

It isn't necessary to rely solely on information printed on a terminal window while the code executes. Sometimes this is too much information, or the process of serially sending the data to the host computer consumes excessive processor time. One alternative is to fill a circular data buffer with debugging information as the code executes. The recorded data may then be sent serially en masse to the host computer upon request (such as when an intermittent glitch is observed) to see what was happening recently in the software during a period of interest. Most terminal programs have the ability to log (save) the information that they read in and display, so the data may optionally be stored and pondered further or otherwise postprocessed.

Rather than (or in addition to) software indications of system status and events, consider adding features in the hardware that can serve a similar purpose. For example, if you know (or suspect) that you will frequently wonder about the voltage or logical level at a specific point in your circuit, add a test point there (a hook or pin where you can conveniently and securely attach a scope probe or DMM lead). LEDs could be included that indicate logic high or low voltages or sensor readings above or below a given threshold. The main value of features such as these is that they allow you to understand what is happening at various locations in the code and the hardware, even if the majority of the overall system is not functioning.

As an example, consider a subsystem that is designed to detect infrared light signals from a beacon and use these signals to guide a vehicle toward the beacon. If the first test shows that the subsystem is unresponsive to the beacon, the problem could be in the software, or it could be in the sensor hardware, or it could be something more fundamental, such as a dead battery. Troubleshooting such a problem can be complicated, because it's difficult to know where to begin. However, a designer thinking about the Troubleshooting Tip #4 might have included a test point at the output of the sensor amplifier stage, as well as an LED indicator at the output of the photo-sensor signal conditioning system that would light up when the sensor detects the beacon. A "fuel gage" array of LEDs might be added to indicate the status of the system batteries. Also, the code might include some print statements that indicate when the software detects that the signal of interest has been detected. In this case, by watching for the hardware and software indicators, it is possible to quickly decide if the problem is in the sensor hardware or in the operation software, or perhaps elsewhere.

Figure 31.1 shows a sensor circuit with a test point (TP1) and a debugging LED indicator built-in. These circuit elements can be ignored during normal operation, but can be very useful in the early testing of the system. They are often useful later too, when something suddenly stops working and you're not sure what the source of the problem is. In order to enable this opportunity, the designer must plan for the inclusion of these components and any necessary drive circuitry, and must also position them where they can be easily accessed or viewed during testing and operation of the subsystems and the completed system.

A very effective methodology for troubleshooting a circuit such as the one shown in Figure 31.1 is to begin by) verifying that power and ground are properly connected everywhere in the circuit where they are needed, and then tracing signals from their origin to their final destination, and verifying that they have the proper characteristics at each stage. For the example beacon detection circuit, use an oscilloscope or DVM to verify that +5 V is connected to the collector of the photo-transistor (Q_1), the power pin of the LM6144 (U_1), the

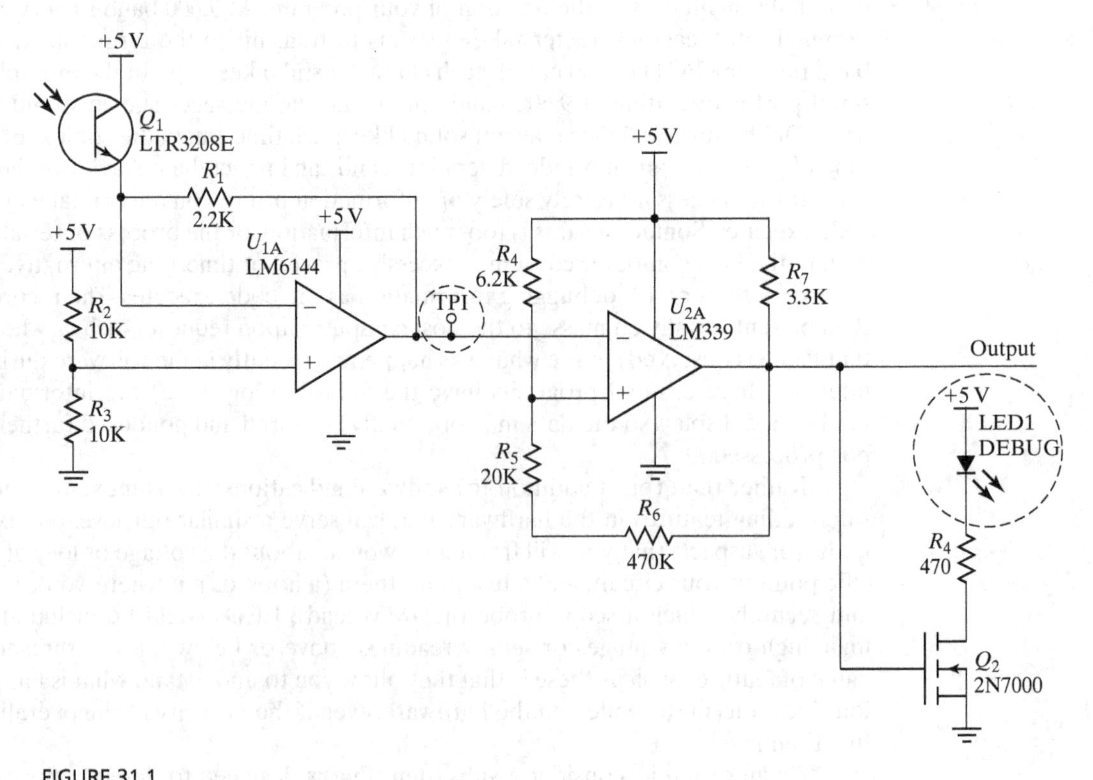

FIGURE 31.1
An infrared light beacon detection circuit with a test point (TP1) and a built-in debugging LED.

voltage divider at the noninverting input of the LM6144 (R_2), the power pin of the LM339 (U_2), and the anode of the debugging LED (LED1). Repeat the process for ground, which should be connected to the negative supply pin of the LM6144, R_3 of the voltage divider, the ground pin of the LM339, and the source of the 2N7000 (Q_2). To make sure you've verified all of the power and ground connections, it's helpful to highlight them on a printout of the schematic as you test. Most schematics are significantly more complex than the simple example shown, and it's easy to lose track of what you've tested and what you haven't. Once you're convinced that power and ground are connected correctly, trace the signals from their origins to their destinations. Check the output voltage of the R_2-R_3 voltage divider connected to the noninverting input of U_{1A}. Is it 2.5 V, as expected? Check the voltage at the test point TP1. Does it vary with the level of infrared light detected as you intended? If not, the problem lies somewhere in the sensor stage. If it does, progress downstream to the next stage: the comparator (U_{2A}). Does the output of the comparator change states when the input voltage reaches the correct trigger levels? Does the output achieve the correct voltage when in the high and low states? This is the kind of relentless troubleshooting from which trouble has no escape! The observability built into a system greatly facilitates this approach, and speeds you on your way to identifying issues.

It may seem excessive to build in indicators of this sort for all systems. In most cases, this is true. However there are frequently subsystems that prove to be more sensitive during early explorations and prototyping. The experience you have with such subsystems during the early stages of a project is usually a very good predictor of the problems you'll have throughout the life of the device. It is an excellent idea to pay attention to these more sensitive functions and to add elements during the design phase that allow you to observe their inner workings.

In many designs, there is also a need for calibration of components or operations after assembly is complete, perhaps because of uncontrolled environmental factors (e.g., ambient light, temperature, humidity, pressure), or because of system adjustments needed after rough handling. Since you are already aware that calibration will be necessary, we offer the *Number Five Mechatronics Troubleshooting Tip*:

#5 Design your systems so that all elements which may require adjustment are accessible during operation.

Consider a sensor system that you know will require frequent adjustment, such as the gain of the beacon sensor circuit example above. Rather than designing the system with a fixed gain resistor value that will frequently fail to provide the desired performance, replace it with a potentiometer as shown in Figure 31.2, or a DIP switch that allows you to select from a few alternative fixed resistances. Moreover, the trim pot or switch needed to adjust that parameter should be located in a place where it can be conveniently and accurately adjusted during operation. In cases where a parameter value will need adjustment in software, its value should be established in a single, obvious location in the code, such as a `#defined` constant at the beginning of a software module, and it should be well named so that it is clearly identifiable and easy to distinguish from other parameters. If appropriate, allowing for adjustability of the parameter at run time, rather than at compile time, may save considerable time, especially if you have a means of retaining the most recently-established value (e.g., in non-volatile memory).

The #5 Troubleshooting Tip applies equally to individual subsystems and the overall system design. It is often the case that obvious misbehavior only appears when the system is completely assembled. When this happens (and it will!), a typical first step is to back up a

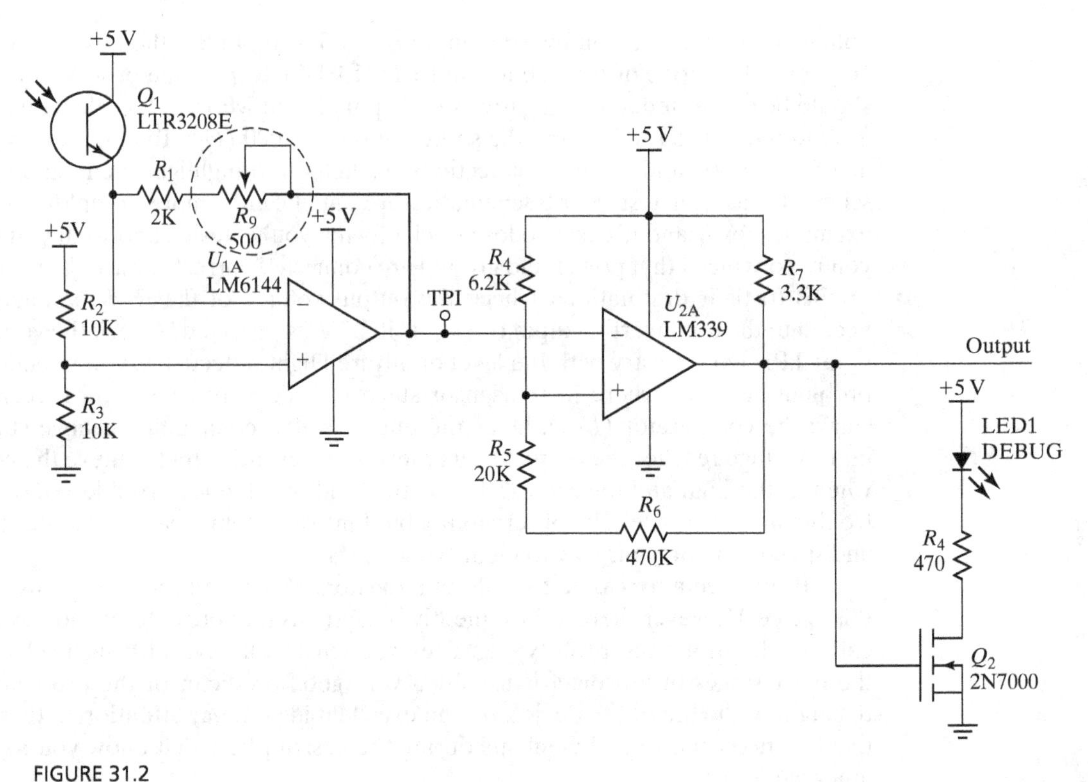

FIGURE 31.2

Incorporating a potentiometer to the gain resistor of the photo-transistor makes frequent calibration quick and convenient.

step or two and partly disassemble the system to allow access to individual hardware subsystems, or to partly disassemble the software to allow individual testing of modules. The *Number Six Mechatronics Troubleshooting Tip* relates to this scenario:

#6 Design the system so that it is easy to disassemble, and so that it may be operated in various states of disassembly.

The final structural design and the final software design may be optimized for many characteristics or features. These features are selected so that they best meet system requirements for full operation. However, before full operation is possible, it is necessary for each of the subsystems to survive the troubleshooting process. Repeated assembly and disassembly of devices will take its toll. Our strong advice in this regard is to place a high priority on planning for many assembly-disassembly cycles taking place. Design the system so that it's easy to disassemble and reassemble. Select components like fasteners, joints, and connectors that won't fatigue, collapse, liquefy or otherwise fail if many disassembly-assembly cycles are required to complete the debugging process.

In the hardware domain, this can mean providing wiring harnesses or mechanical structures that can allow the components to be disassembled and spread over the bench and still be turned on and operated in this condition. In the design of the system, this means making sure that the electrical wiring harnesses are long enough and flexible enough to allow them to be fully connected even when the system is partly disassembled. Such an extra length of wire in a harness is referred to as a **service loop**. It might mean providing mechanical support structures that can be moved laterally or vertically some fixed distances and then anchored to allow testing. It might mean providing detachable couplings for mechanical actuators so that the motors can be operated without spinning the wheels. Each design is different, so consider which opportunities are likely to have the biggest payoff as you design and build.

While it's probably easier to think about how the #6 Troubleshooting Tip can be applied in the hardware domain, it is equally important to apply it in the software domain. In software, we recommend designing modules so that they may be easily operated stand-alone (using a test harness as described in Chapter 6), or in combination with other modules as needed, prior to building the full hardware or software system. For example, it may become useful to test only a subset of the software with a specific hardware subsystem. This is much easier if you've created a software module specific to the hardware module that allows you to perform stand-alone testing and functional verification. Indeed, this allows you to thoroughly test both the hardware and software related to this subsystem prior to adding it to the milieu (strictly adhering to the #3 Troubleshooting Tip of incremental integration, of course) and consequently ratcheting up the complexity.

One of the most frequent and bedeviling questions that arises during the course of most projects is whether some new and baffling behavior is the result of software bugs or electrical hardware problems. Being able to rule out one domain or the other as the root of the problem has tremendous value, and you can simplify this process considerably by anticipating and planning for it well in advance. If you followed the guidance from the #3 Troubleshooting Tip, you'll build your system software and hardware incrementally, and test everything thoroughly at each step of the integration. This leaves you with a suite of simple software routines (test harnesses) that do little else than verify subsystem functionality and interconnections by turning actuators on and off, reading sensors and reporting the values, and so on. When a fully integrated system begins to misbehave and the source of the error isn't understood, reverting to these simple test harness routines can allow you to quickly determine if the problem is in software or hardware. If the test harness is still able to, for example, make a motor spin clockwise and counterclockwise, this allows you to quickly verify that the microcontroller outputs and inputs are working properly, the wiring is correct, and the electrical components are still functioning. In this case, the source of the problem is likely in your software. However, if the simple test routine is no longer able to turn the motor, then the problem is likely in the electrical hardware. You may have forgotten to plug a connector in, or maybe a connector has shaken loose because of vibration. A fuse might have blown. The setscrew on a shaft coupler on your motor may be loose, or the coupler may have jammed against the edge of a bearing housing. Whatever the problem is, using the test harness routines allowed you to eliminate the software as the probable cause and focus your attention on the electrical and mechanical elements.

A reviewer of a draft of this book contributed another excellent suggestion: early in the process of bringing a new mechatronic system to life, take the time to generate a simple program whose sole purpose is to read and report the value of inputs and outputs, and turn actuators on and off. The reviewer calls this utility the "sanity check," and it can be used any time there's a question about the hardware, the software—or your sanity. Over the course of a project, many sanity checks are sure to be helpful, and the investment of time required to create it will pay handsome returns.

The main message in this section is to remember that some subsystem testing will be necessary after full integration. Good system designs are those which facilitate observation and diagnosis of subsystem functionality while the system is fully integrated, provide access to and adjustment of parameters expected to require adjustment, allow for disassembly and individual testing of subsystems, and incorporate a simple reassembly process. Troubleshooting Tips #4, #5, and #6 are all about enabling these activities to take place easily and efficiently.

31.4 TROUBLESHOOTING ATTITUDE

In addition to the above recommendations on preparations and procedures, we want to close with some advice on the attitude of the troubleshooter:

#7 Know your limitations.

In the movie "Magnum Force," Clint Eastwood, reprising his role as Dirty Harry, has just killed off the last of a band of renegade vigilante policemen and then discovered that his chief (played by Hal Holbrook) was the leader of the renegades. After gratuitous fighting in which the chief gets the upper hand, he drives away in a car in which Harry has planted a bomb. After the bomb explodes, Harry says, "A man's got to know his limitations." This is every bit as true in troubleshooting as it is in action movies. When troubleshooting, this advice can be extremely useful, and it can even save lives (consider the consequences of inadequate troubleshooting in safety-critical applications like medical devices or the space program).

The thing to keep in mind is that the most challenging aspects of troubleshooting seem always to arise when troubleshooters are least capable of rising to the challenge. Frequently, troubleshooting occurs late at night, close to a deadline, after too many days of junk food, too many caffeinated beverages, and moments of extreme stress among a team. The last three of our Mechatronics Troubleshooting Tips deal with these circumstances.

#8 Avoid troubleshooting alone.

Don't try to be a hero—this just adds risk. Because of the stress often present at the troubleshooting stage of a project, it is especially helpful not to have individuals working alone at this time. When individuals work alone, the previous tips are often neglected, and a very complicated series of mistakes can happen. It is understandably tempting for individuals to try to tackle and solve the big problems in a system. However, in keeping with the #8 Troubleshooting Tip, it is very important to recognize that having some help can lead to faster recognition of errors and quicker, higher-quality fixes. In addition, the participation of additional team members can lead to very useful checks and balances against certain risky experiments or ill-conceived and dramatic design changes. Perhaps most importantly, the best kind of troubleshooting should resemble the scientific method: it should involve a hypothesis, an experiment and an expected result (see the #1 Troubleshooting Tip). The thinking and reasoning that should happen during troubleshooting are easier and more productive if they're discussed among reasonable and informed people while troubleshooting is in process.

There have been instances where a lone troubleshooter, working late at night, went a little crazy and caused additional problems in other parts of the system that proved to be much harder to diagnose, and ultimately caused his team's project to fail to meet specifications within the deadline. In combination with the #8 Troubleshooting Tip, we present the *Number Nine Mechatronics Troubleshooting Tip,* which can help ensure that this doesn't happen to you:

#9 Avoid doing more harm than good while troubleshooting.

While engrossed in troubleshooting, it is easy to lose sight of the impact of experiments on the entire system. For example, a thousand experiments to improve the accuracy of a ball launching mechanism may lead to the fatigue failure of key elements of the mechanism. Or, efforts to locate a wiring problem might accidentally lead to power connections being reversed, causing the destruction of some or all of the electronics. Or, adjustments in the way variables are passed between a pair of routines might cause widespread and perplexing problems with other functions throughout the entire code.

In all cases, the troubleshooting team must be thinking about worst-case scenarios during the troubleshooting process, and try very hard to avoid going too far backwards too quickly. Because of the sleep deprivation and long hours that often accompany the troubleshooting process, awareness of this is especially important. Even if things have been going slowly for a while, there is always the potential for a swift and brutal setback. These setbacks seem to happen the most frequently at the deepest, darkest hours, close to the deadline, and when personnel and other resources are stretched to the extreme. With this in mind, we offer one final Tip:

#10 Never underestimate the value of taking a break.

A good night's sleep, a decent meal, a shower, and a nice walk outdoors may seem like a real luxury when engrossed in troubleshooting. We all want to stick to the problem until it is solved. However, it is important to keep in mind that troubleshooting can be like solving a complex puzzle, where you need your wits about you, and where the key moment will happen quickly but only if you're properly receptive to the insight or observation. By taking a little time to detach from the details and tend to some other needs, you'll often find yourself in a much better state of mind for puzzle solving, and—somewhat paradoxically—you'll often get to the desired result more quickly. In addition, by taking a break, you can certainly avoid the mistakes that so often lead to setbacks that we mentioned in many of the above tips. Generally speaking, when nothing is happening, nothing bad is happening, and that is often much better than what *could* be happening if something *were* happening.

31.5 FINAL THOUGHTS

Troubleshooting is inevitable. Of all the stages of a project, it can be the one stage that seems most difficult to plan and organize, but the wise and experienced mechatronic engineer considers troubleshooting a natural part of every project and incorporates it into the overall schedule. Troubleshooting can be the worst part of a mechatronic project process and can be the part of the process where disaster rises up to doom a project.

However, troubleshooting is to be expected, and should ideally be approached as an organized and logical process. The content in a mechatronics course is largely in the form of simple concepts, linear equations, and limited challenges. One excellent description of the field of mechatronics is: "the art of getting 1,000 simple details 100% correct." In addition to the understanding of various equations and data sheets, one of the most valuable skills that students take away from a mechatronics class is how to troubleshoot.

So, when you begin testing your new system or subsystem and you discover that it's misbehaving, smile knowingly, take a deep breath, and review the Top Ten Mechatronics Troubleshooting Tips. What comes next may well be the defining moment of your project experience. It is also one of the most important learning opportunities in this area of study. Make the most of it!

31.6 HOMEWORK PROBLEMS

31.1 What is the first step in the troubleshooting process?

31.2 How is good troubleshooting like the scientific method?

31.3 While troubleshooting, it is important to maintain a ____ ____. (two words)

31.4 When assembling a mechatronic system for the first time, the methodology should be I______ I______. (two words each beginning with the letter I)

31.5 Name at least 3 steps that you can take during the design/build phase that will make troubleshooting easier.

31.6 What is the key aspect of the physical location of an adjustment potentiometer?

31.7 What is a "service loop" and what is its role in troubleshooting?

31.8 Describe some of the advantages of not troubleshooting alone.

31.9 What kinds of things are more likely to happen while troubleshooting when you are extremely tired?

31.10 True or False: It is possible to build a system that functions 100% correctly the first time that it is assembled.

CHAPTER 32

Mechatronic Synthesis

32.1 INTRODUCTION

Case studies are the traditional medium for demonstrating how the kinds of concepts that we have covered in this book get applied to the development process. The major drawback to case studies is that they typically focus on the product of the process and so miss the messy details and missteps that went into the product's development. The result is that the product seems to spring forth fully formed with no real sense of how it evolved. This can be very confusing to students, who get the impression that projects progress cleanly from concept to implementation. This is almost never the case and a myth that we hope to dispel in this chapter.

Perhaps the best way to demonstrate the mechatronic synthesis process would be to follow a real commercial project from the initial concept stage through to production. Unfortunately, that is not a realistic possibility. Companies guard their design and development process almost as closely as they do more traditional intellectual property, such as trade secrets and patents. The alternative that circumvents both of these issues is to follow the design, development, and implementation of two representative student projects that were the result of two teams' work in response to a single project assignment in an introductory mechatronics course for undergraduate seniors taught at Stanford University. Here, we have full access to the projects at every stage of development, so we can offer not only a glimpse of the development process but also provide commentary on the process.

An academic project like this is similar to what would be considered a "proof-of-concept" prototype in industry. The purpose of building this level of prototype is to prove out ideas and discover where omissions, oversights, and/or incorrect assumptions were made in the initial concept. In the interests of brevity, we have selected only the most important elements of the material prepared and submitted by the teams during the project. The subset that we present here was chosen because it illustrates the kinds of documents that were created and gives a reasonably full—and accurate—picture of the development process without getting bogged down in excessive minutiae. The parallels between some of the sections in this chapter and Chapter 30 (Project Planning) are not accidental. A major reason for assigning the project is to give the students an opportunity to practice applying those principles in their own work.

The goals of this chapter are to:

1. give readers the opportunity to see how a typical student mechatronic project grows and develops,
2. see how initial concepts change based on the realities of the time, resources and materials available,
3. look closely at both the early sets of concepts and the finished products to see how the project evolved,
4. show how two design teams applied the material covered in this book to build a mechatronic system that meets a set of predefined goals.

32.2 THE PROJECT DESCRIPTION

The project assigned to the design teams was titled "The Autonomous Golf Putter." The project description summarizes the goal as follows: "The basic purpose of the machine is to 'putt' golf balls into a 'hole' similar to that found on a golf green." The description goes on to elaborate on many aspects of the project. The complete project description appears as Appendix C, and you should take the time to read it through before continuing with this chapter.

The project description specifies the playing field, the rules of the game, and the constraints placed on the machines that will be built. As such, it is representative of the kind of document you might receive from a client or a marketing group that lays out what they would like the finished product to be able to do. As a class project description, it is a bit more detailed than you might find in an industrial setting and addresses issues (such as grading) that are not likely to be found in other circumstances. What it shares in common with an industrial project description is a focus on outcomes and constraints without any presumption of *how* the problem will be solved. When preparing this kind of document it is important that the writer be diligent not to allow a presumed solution path to color the description.

32.3 SYSTEM REQUIREMENTS

After reading through the project description and consulting with the client/marketing/professor to clarify points, one of the early tasks for the design team is to prepare a set of **system requirements**. Ideally, these are a succinct condensation of the relevant portions of the project description, along with requirements that are implied in the description but not stated explicitly. This is an exercise that requires great care in its execution. Just as the original description should not presume a solution, neither should its distillation into a set of requirements. In addition, this is one of the points in the process where it is possible for elements to sneak onto the list that are not actual requirements, but instead are ideas that the design team feels would be interesting or cool to include. This is such a common phenomenon that it has several names, such as **feature creep** and **feature-itis**. This is something to be avoided, both here and again when the requirements and the concepts for addressing the requirements are translated into a set of specifications.

For this project, one of the student teams (Team Zero) formulated the set of requirements shown in Figure 32.1 based on their reading of the project description.

Exercise:

Take a few minutes to compare Figure 32.1 against the project description in Appendix C. Do you see any requirements in the project description that did not make it into Team Zero's requirements list? Are there any requirements listed that don't have a root in the project description?

As prepared, Team Zero's proposed requirements list intermingles a number of different types of information. Some of the items belong in the requirements list and some do not. The section labeled "Absolute requirements" lists requirements that can all be traced to items in the project description. The section labeled "Environment and game play" mixes descriptive information about the field and game play with some actual requirements ("The putting of all balls must be completed within 2 minutes from start of round"). The section labeled "Warning" includes a true requirement ("The ball must not bounce outside the bin"), appropriately derived from a rule listed in the project description, which states "a ball must remain in the hole to count." The next line in this section, however ("The circuit must

Requirements

Environment and game play

Putting Green 4' × 8'. The sides not containing bins are bounded by a 3" wall
At one end of putting green is a series of bins, with edges below the surface
The goal is to putt into the bin labeled "a" (the hole)
The side-to-side center of the bin labeled "a" will be marked by a pole carrying a beacon emitting modulated IR light at a height of 12" above the putting green
Edge of the putting green adjacent to the bins marked by a strip of 1" black tape
Putter will be placed in a random orientation between 3' and 8' from the beacon pole
Putter loaded with golf balls at the beginning of each round
The putting of all balls must be completed within 2 minutes from start of round
When a golf ball enters the bin labeled a, the beacon extinguished for ~1/2 seconds
Once positioning/putting has started, may not touch the putter
The operator is allowed to input a single correction entry after every putt. The input may only take the form of the bin number of the hole into which the last putt fell. The entry is optional.
Scoring: a = 1, b = 2, c = 3, d = 4, not falling into bin = 4
Putter may not touch the pole that carries the beacon
A ball must remain in the hole to count (must not bounce)
The operator may not move the putter once it has been placed by the judges
The putter may not alter the playing field in any way

Absolute requirements

Must be a stand-alone entity
Connected only to power and ground
Occupy a volume not to exceed 13" × 13" (l × w) and 12" (h) when initiated
The machine contains the complete supply of golf balls (at least 5)
Must not be based on a commercial or pre-existing platform
Have an easily accessible power switch
Any exterior corners must have a radius of at least 1/4"
The putter should be safe both to users and spectators: no ballistic objects
All liquids, gels, and aerosols must be in three-ounce or smaller containers. All liquids, gels, and aerosols must be placed in a single, quart-size, clear plastic bag. Each putter can use only one, quart-size, zip-top, clear plastic bag
Putters may alter the Space-Time continuum only during the public presentations

Warning

The ball must not bounce outside the bin
The circuit must not be put into saturation if the beacon signal is too strong

FIGURE 32.1

Team Zero's proposed list of requirements.

not be put into saturation ...") is a requirement that has no root in the description and, in fact, is not a true requirement in the sense that it is possible to imagine a solution that does allow the sensor input to saturate at some point. Here, the team has assumed a solution path and created a requirement based on that path. This is not a good idea because it may eliminate viable (and potentially superior) solutions and strategies that could fulfill the actual requirements but would not satisfy this invented requirement.

32.4 DESIGN CANDIDATES AND ALTERNATIVES

Once the design team has a grasp of the project requirements, the next step is to generate possible solution candidates. As pointed out in Chapter 30, brainstorming can be an effective process for achieving this. The brainstorming list is typically winnowed down to a set that looks like they all have potential. The proposals that we show here are the concepts that made it past this "first cut" winnowing.

For the projects that we are considering here, Team Zero and the second design team (Team InTheRuff) generated a set of initial concepts that they brought before the rest of the class and teaching staff for review and comment.

32.4.1 Team Zero's Concepts

Team Zero took an approach where each team member prepared his or her favorite concept sketches which were then used as input into an integrated concept.

The proposal in Figure 32.2 imagines a mobile platform equipped with a scanning sensor to locate the beacon that marks the hole. The platform would identify the hole via the beacon, drive to some specific range from the beacon, and then roll the balls into the hole.

This second concept also incorporates a mobile platform. This drawing provides more detail on a number of the subsystems. In the upper-left portion of Figure 32.3, we can see that the shooting system now uses a gravity-driven approach, with the ball falling a short distance to impart kinetic energy and then rolling down a ramp which controls its direction, as opposed to the impulse system that seems proposed in Figure 32.2. Figure 32.3 also includes more detail on the ball storage subsystem. This proposal, like that of Figure 32.2, imagines that the platform will be equipped with bumper sensors to detect collisions with the perimeter wall that surrounds the playing area. The bottom portion of the figure includes details on a debugging board and how the aiming system might be configured to allow the beacon sensor to be aligned with the ball delivery subsystem.

The third concept, Figure 32.4, again shows a mobile platform, but adds a number of unique components to the basic elements that it shares with the other concepts. This concept imagines the use of omni-wheels, special wheels that incorporate a series of smaller orthogonal rollers along their perimeter that allow the wheels to roll both along and perpendicular to the major axis of the wheel (i.e., the axis of an axle that the wheel would normally rotate about). A representative omni-wheel is depicted in Figure 32.5. The design in Figure 32.4 also adds a compass sensor to be used in aligning the motion to the field and an ultrasonic distance sensor to detect the edge of the playing field and the point at which the balls will be dropped. This proposed design drives all the way up to the hole rather than to some predetermined distance from which the balls would be rolled or bounced into the hole.

Team Zero selected elements from these individual concepts and integrated them into a team design, which is shown in Figure 32.6. The ball storage and release system has taken on the name "pinwheel" but it remains a circular disk element that holds the balls and rotates to allow one ball at a time to fall through a hole and then be released. It is also worth noting that the team was beginning to think about other issues such as the project budget and a state machine to implement the software for the putter.

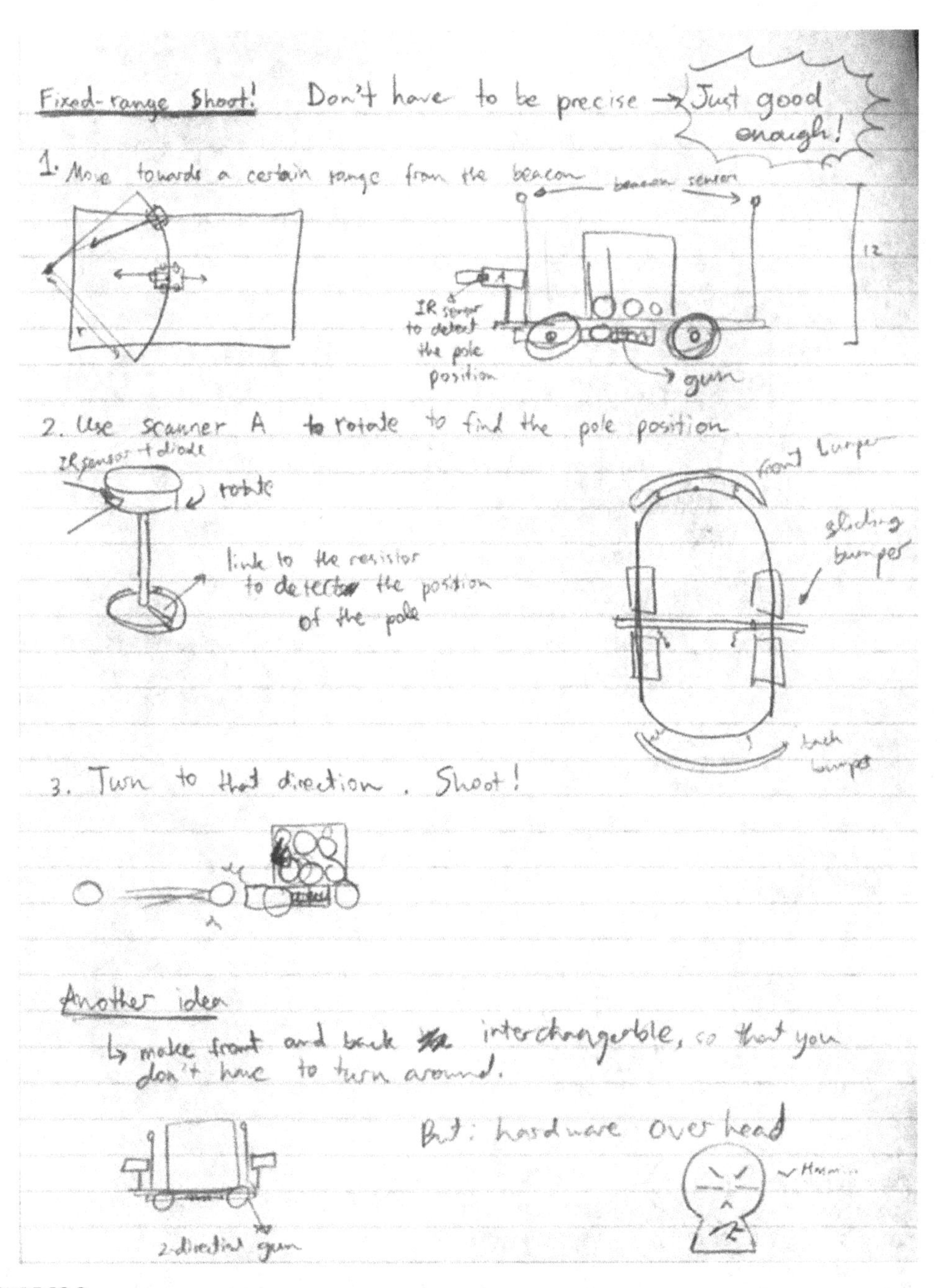

FIGURE 32.2
Design concept #1 from Team Zero.

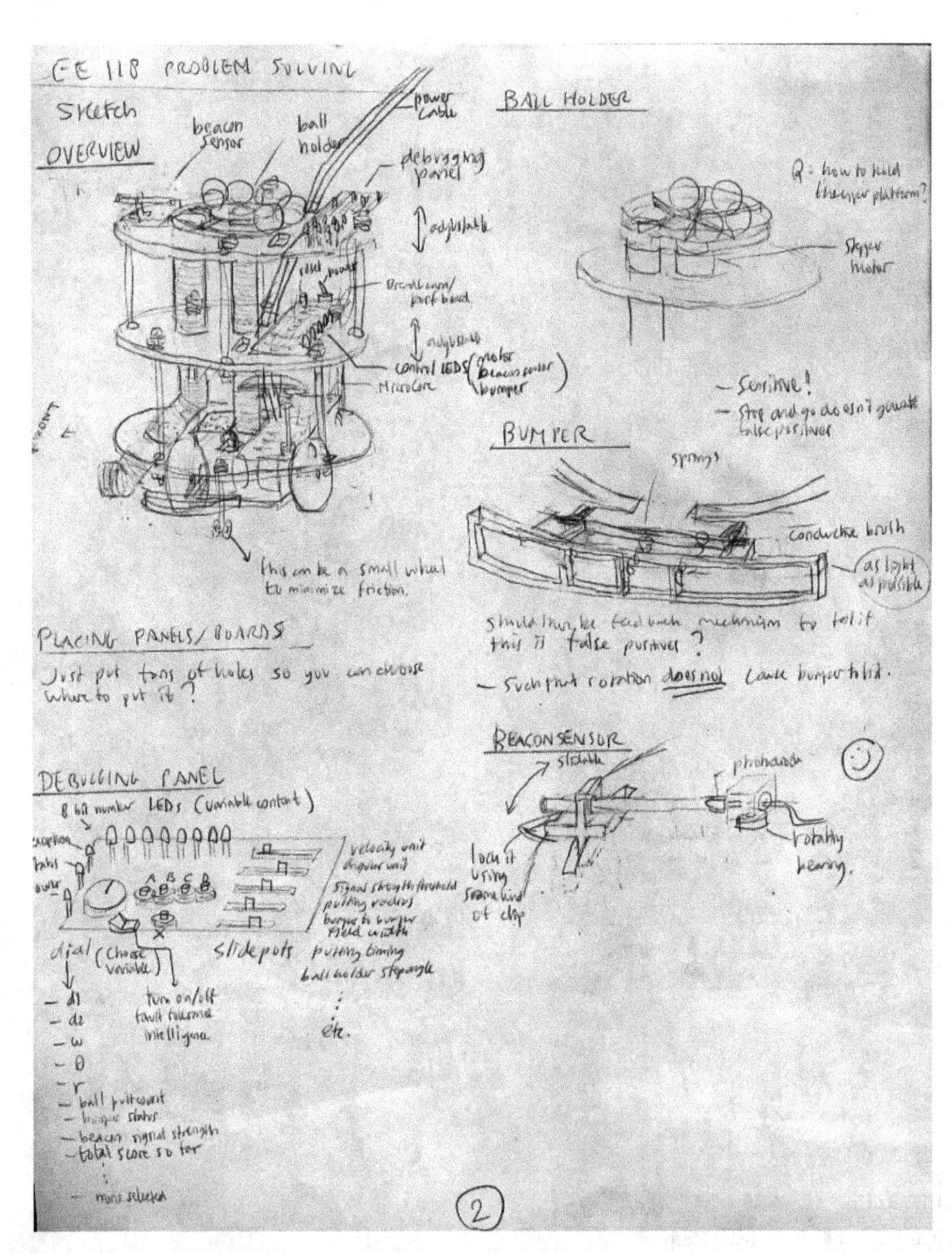

FIGURE 32.3

Design concept #2 from Team Zero.

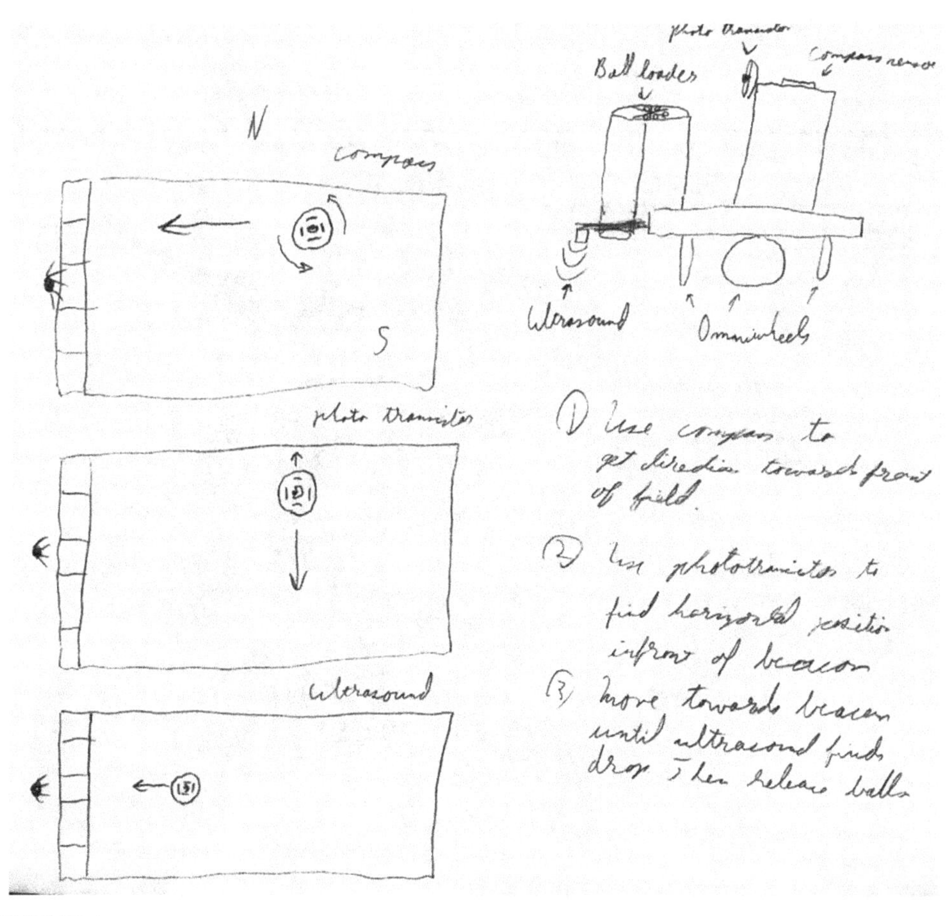

FIGURE 32.4

Design concept #3 from Team Zero.

FIGURE 32.5

An omni-wheel.

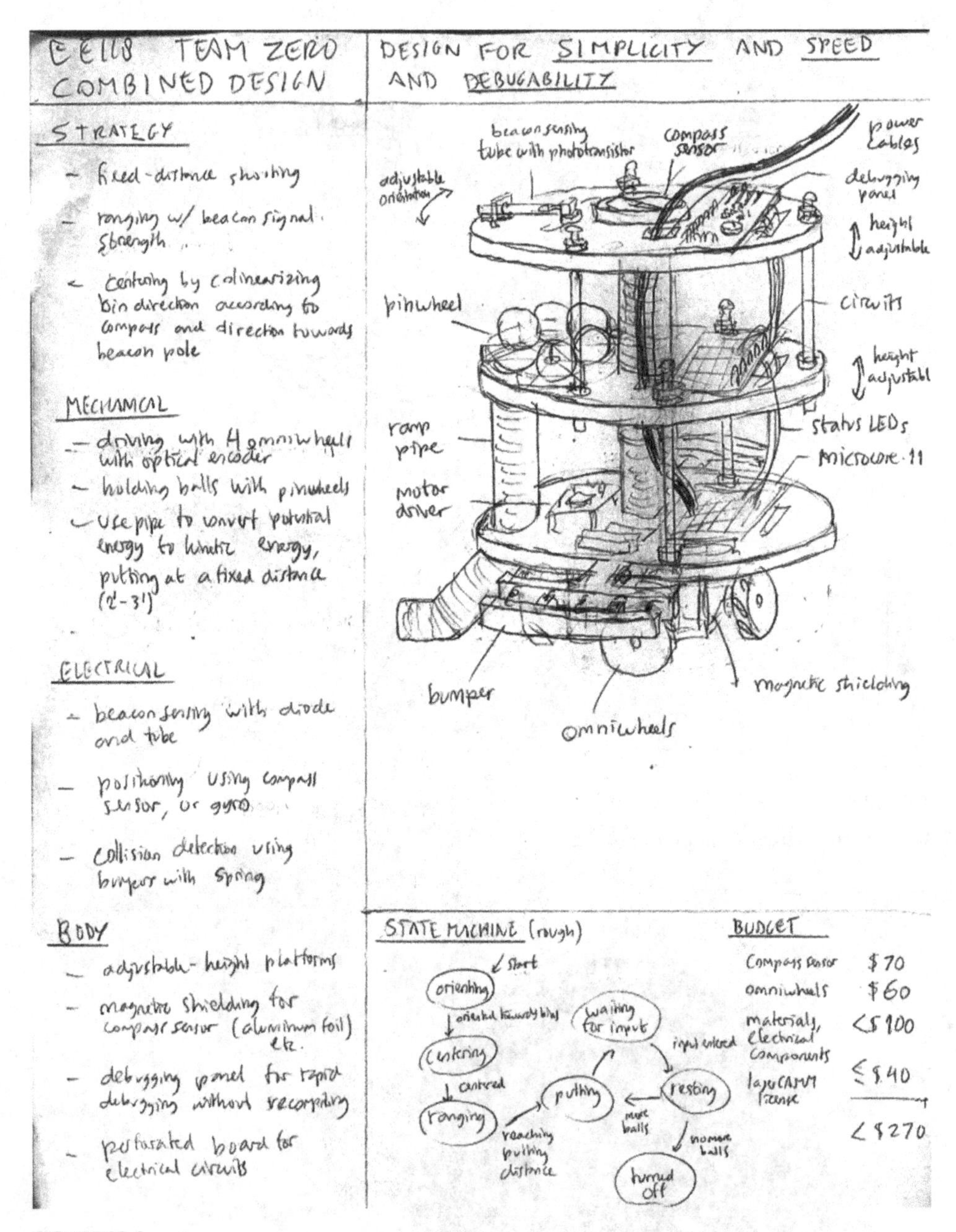

FIGURE 32.6

Team Zero's integrated concept, #4.

32.4.2 Team InTheRuff's Concepts

In contrast, Team InTheRuff condensed the results of their brainstorming ideas into a single concept that they presented at the design review. The concept shown was that of a mobile platform that would drive up to the black tape line in front of the hole before rolling the balls into the hole. The basic idea is depicted in Figure 32.7. Their mobile platform concept uses two drive motors for locomotion. The golf balls are carried in a cylindrical tube on which is mounted the photo-sensor that will be used for navigation.

This team paid particular attention to the details of how the balls might be released in a controlled fashion, so that (if desired) one ball at a time could be released. To affect that function, they came up with two different mechanisms, the first of which is shown in Figure 32.8.

A stepper- or servomotor-driven pawl pivots back and forth in such a way that the release of one ball blocks the progress of the ball immediately behind it. The central portion of the sketch gives a sense of scale and shows that they imagine a long ramp to hold many golf balls that could be fed into the ball release mechanism.

The team labeled the second ball release mechanism the "Gatling gun," shown in Figure 32.9.

In this concept, the four short sections of tube, each sized to hold exactly one ball, are attached to a central rotating plate. As the plate turns, the entry end of one tube (on the right) is exposed to the supply of balls, and a ball will drop into the tube if it is empty. When the plate rotates further, it moves through a position where both ends of the tube are blocked by elements that aren't shown in the diagram. Finally, when it reaches the position on the bottom, the exit (left) side of the tube is exposed and the ball is allowed to roll out.

The overall operation of the InTheRuff team's solution is captured in storyboard form in Figure 32.10. We can see that they plan to use a photo-sensor to locate the goal, drive to edge of the hole, then use the black tape at the edge of the hole for final alignment before releasing their golf balls into the hole.

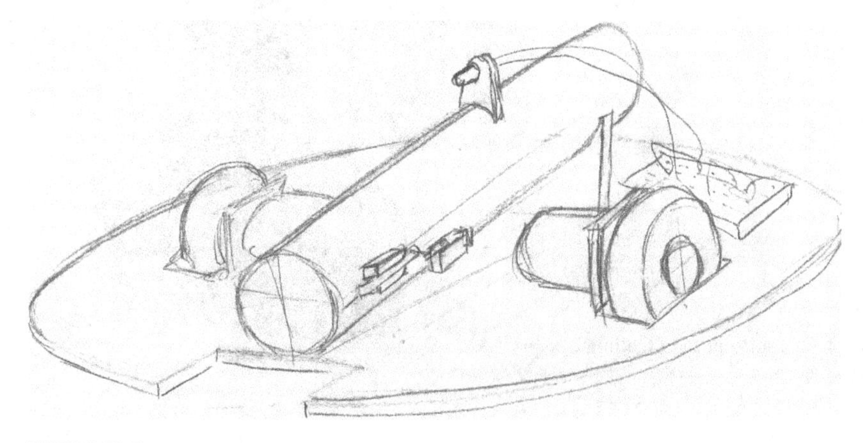

FIGURE 32.7

Team InTheRuff's platform concept.

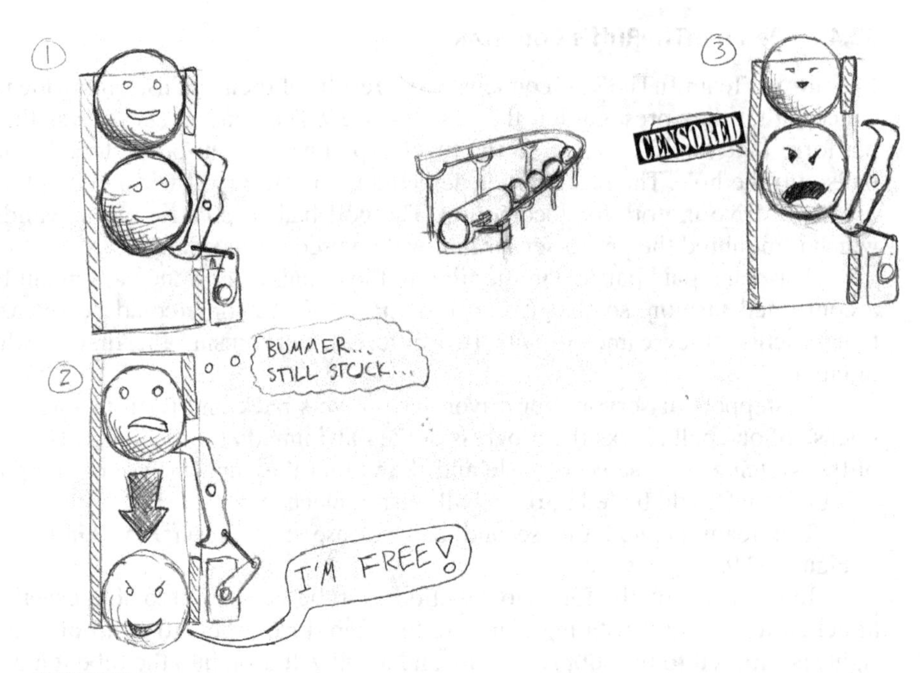

FIGURE 32.8

Team InTheRuff's proposed ball delivery mechanism.

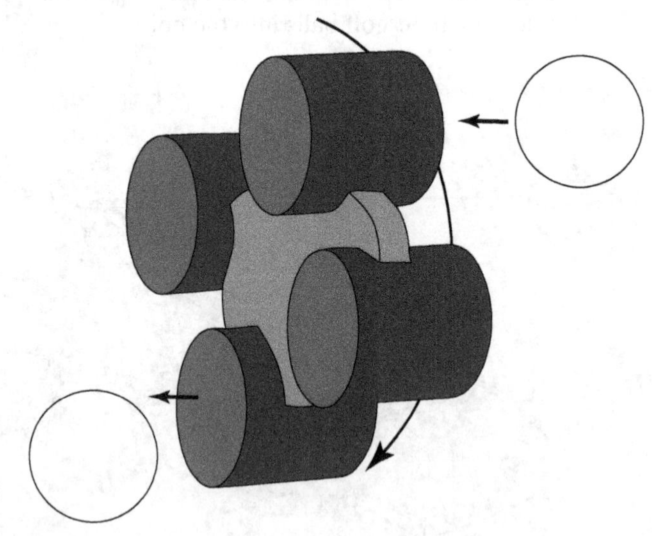

FIGURE 32.9

Team InTheRuff's concept for a Gatling gun ball release system.

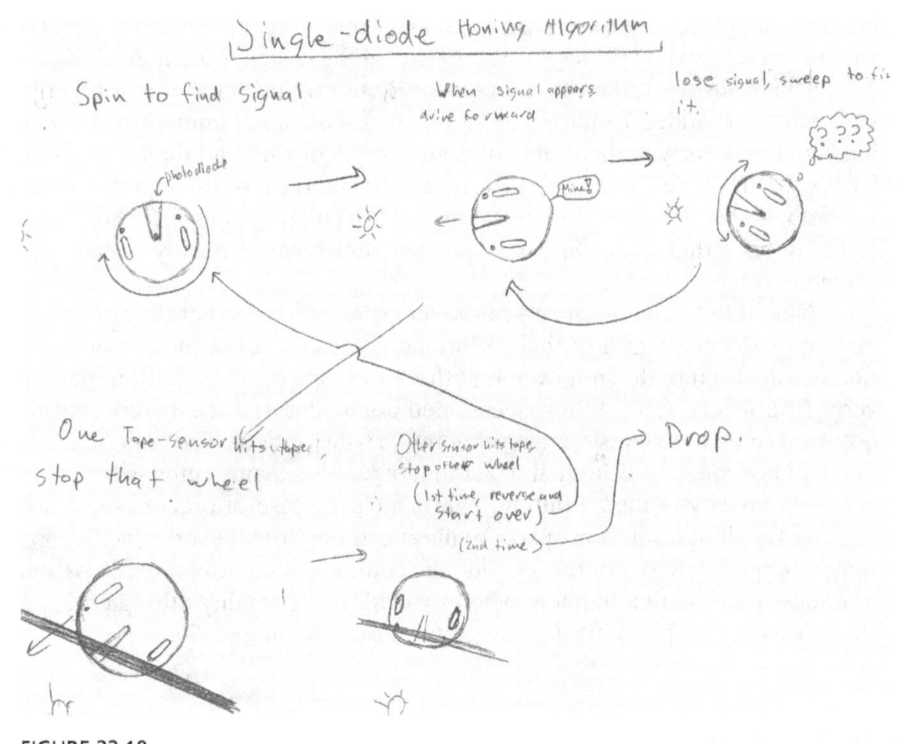

FIGURE 32.10

Team InTheRuff's proposed operation sequence diagram.

32.4.3 Review of the Design Candidates

During the design review, the teams were challenged to justify the features of the solutions that they had proposed. For example, Team Zero's integrated concept included omni-wheels. They were asked by the teaching staff to identify which requirement indicated that the omni-wheels were necessary (there aren't any). During the ensuing discussion, it became clear that the team had not realized the complexity of the drive software necessary to allow the platform to move holonomically (i.e., to move in any direction without first rotating), which is the advantage that omni-wheels provide over conventional wheels. Really, they had become enamored of the idea that their robot putter would move just like the "real" robotic platforms that they had read about, such as the one shown in Figure 32.11. By the end of the discussion, Team Zero admitted

FIGURE 32.11

Commercial robotic platform employing omni-wheels. (Courtesy of Wow Wee Ltd.)

that the omni-wheels were not a requirement and were, in fact, feature creep and in conflict with their own guiding principle of simplicity in the design (see Figure 32.6).

The inclusion of the compass sensor is another instance of feature creep. This particular sensor was added to allow the putter to follow an alignment algorithm that had been imagined very early in the brainstorming. This algorithm had their robot's platform aligning itself with the "hole" by driving in a straight line parallel to the row of bins at the goal end of the field. It was this algorithm that required the compass sensor (or some other way to align to the field) rather than the need for alignment being driven by some requirement of the problem.

One of the striking aspects of the teams' solutions is that they both decided to make a mobile platform integral to their solutions. A close reading shows that the project description doesn't require this, nor even hint that this might be a good solution, yet both teams converged on this concept. When questioned about this, the teams reported that it had never occurred to them to consider anything but driving options. The idea of a "sit and spin" solution had not come up, despite the fact that one of the teams emphasized that they were striving for simplicity in their solution. This is a classic case of preconceived notions artificially limiting the design solution space. While it can be difficult, some of the most creative solutions emerge when we operate outside the confines of the way things have been done before. The self-censoring that happens when we evaluate feasibility during brainstorming is one of the processes that prevents us from getting "outside the box."

32.5 MORPHOLOGY CHARTS

To explore the ways in which their concepts might be combined, Team Zero prepared and presented the morphology chart shown in Figure 32.12 in the design review.

In reviewing the chart, we can see that the team came up with a wider range of possible solution components than were shown in their individual concepts. In the shooting strategy, they also considered a "drive up and drop" approach using the walls and black tape as guides. Two of the ranging strategies (measure RMS amplitude and detect tape) are straightforward approaches, while the more elaborate "trigonometry" idea involves measuring the angle to the hole beacon from two places on the field. For this to work, the robot would need to know how far it moved between the two angular measurements in order to know the baseline for the triangulation. This ranging approach requires the existence of another sensor (a "distance traveled" sensor). A sensor that could make that measurement is shown in the positioning portion of the chart, but what is not captured is that the choice of the triangulation ranging approach requires the inclusion of something like the optical encoder to measure distance traveled.

The wheels section of the chart captures the two wheeled and omni-wheels approaches from the concept sketches and also includes a tank-tread approach, as well as indicating that the placement of the wheels relative to the "front" (the direction of putting) represents yet another design choice. The sketch for the omni-wheel solution shows four omni-wheels, despite the fact these are most effectively used in a configuration with three wheels (Figure 32.11).

For holding the balls, the solutions proposed are limited to carrying the required minimum number of balls. The team has apparently decided to ignore the competition round where the goal is to putt as many balls as possible in the allotted time. There was no explicit requirement that the putters hold more than five balls, so this is a perfectly legitimate decision.

The collision sensor section seems to be another instance of a "requirement" that evolved from other design choices, rather than one that is given in the project description. As it turns out, only one of the shooting strategies proposed would make use of driving along

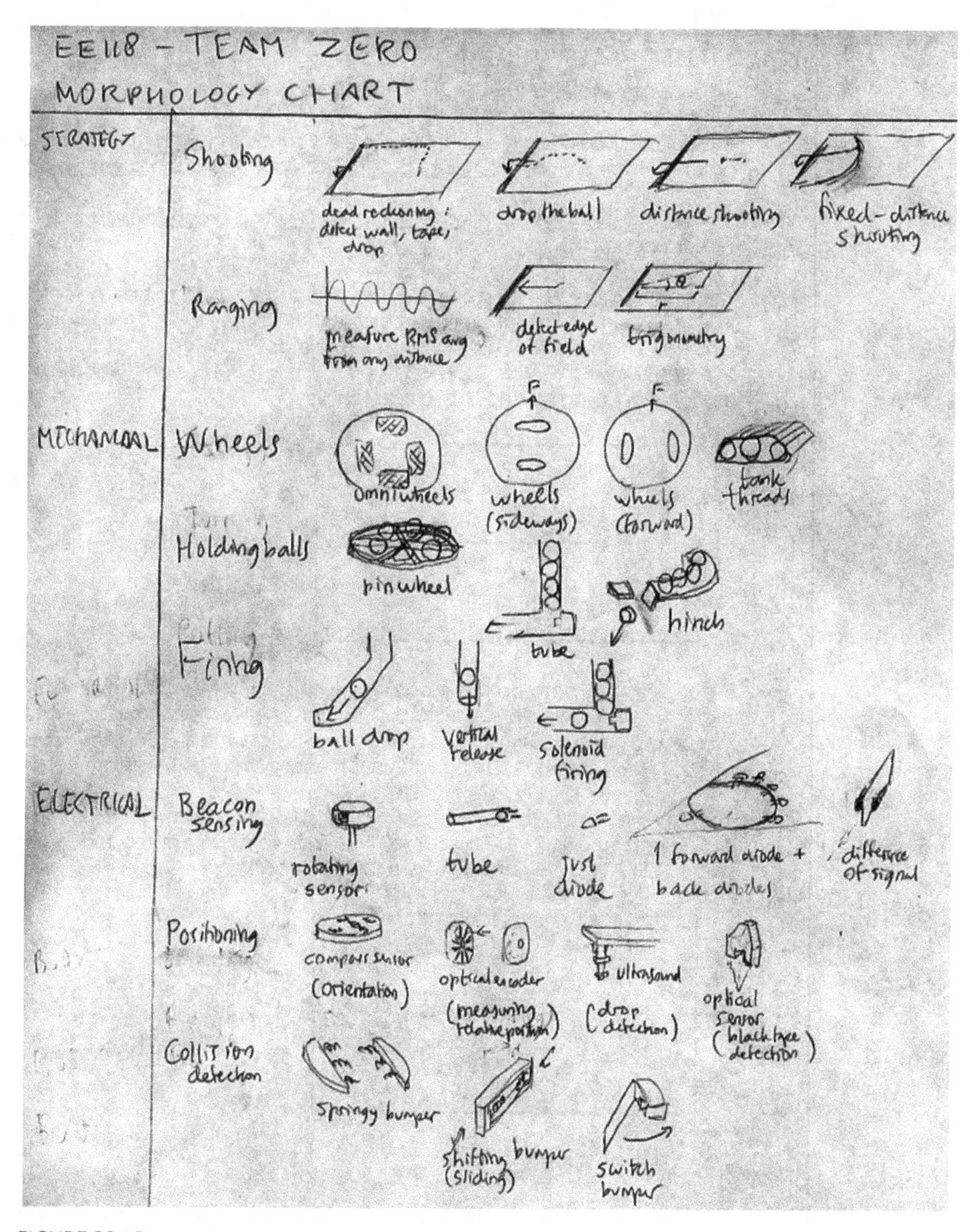

FIGURE 32.12

Team Zero's morphology chart.

the wall and therefore possibly need a collision/contact sensor. Despite this, two of the initial concepts and the final combined concept all include a collision sensor.

32.6 DESIGN CONCEPT EVALUATION: PROTOTYPES AND ITERATION

After the concept phase, the teams built quick prototypes and ran experiments to test and validate portions of their designs. One portion of those results is presented in Figure 32.13.

This figure presents the results from a set of well thought out and relevant tests and analysis. However, the "Lessons learned" section at the bottom of Figure 32.13 seems to be

PLAYING FIELD AND PUTTING

Based on our experiments on the playing field, we found that:

- The golf ball has a really small rolling friction coefficient with the putting green, so that a potential energy from the height of 3.25" can putt the ball more than 8' away.
- The beacon sensor is exactly 11" above the putting green, so we have to put our beacon sensor accordingly.
- The putting green tilts by a tiny amount, causing the ball trajectory to not follow a straight line. We observe a consistent deviation about 6"-8" to the right of the expected destination point when putting from 4'-8' away. However, the deviation from its track happens only when the ball starts slowing down. So, we can putt with enough velocity such that the effect of the deviation is negligible (no more than 1").

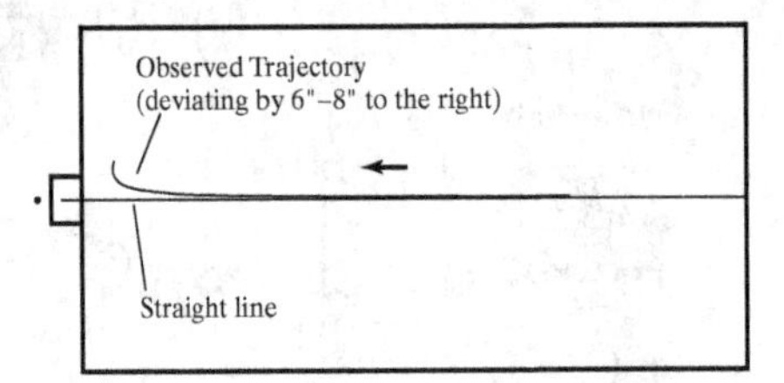

- When gliding exactly on the separating wall of 2 bins, the ball may easily fall outside the playing field. However, this case should be rare given accurate aiming. When the ball is aimed toward a bin, it takes a lot of energy before making the ball jump over the bin, so we are not worried about putting with much energy from short distance.
- There are regions in the playing field in which putting from those regions will not guarantee the ball to enter bin A. In the figure below, the regions are the shaded regions. In order to reliably putt to bin A, we will need to reposition our robot.

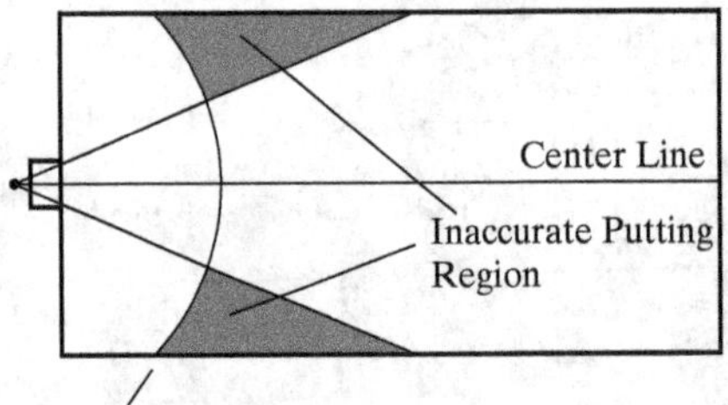

- The table in which the putting green is mounted is movable. This may affect the accuracy of our compass sensor calibration negligibly, as long as the putting green is not intentionally rotated.

Lessons learned:

- Our robot must have a centering mechanism, without which it cannot guarantee to putt into bin A every time. However, it is not important that we get to the center line exactly, because putting at an angle can still be very accurate.
- Our robot must putt from a reasonably close distance (i.e. less than 4' away) to minimize the effect of aiming inaccuracy due to tilted playing field. Thus, our robot must be able to move if it is initially set too far to the beacon.

FIGURE 32.13

Testing results from Team Zero.

colored by the initial design concepts and some inaccuracies in the analysis of the field. The second lesson seems at odds with the results reported in points 1, 2, and 3. Point 1 indicates that it is relatively easy to impart enough kinetic energy into the ball to allow it to putt from more than 8 ft away. The results reported in point 2 indicate that with sufficient velocity, the aiming inaccuracies are minimal (it also ignores that obvious solution of leveling the field). Finally, point 3 reports that it takes quite a lot of speed to get the ball to jump out of the bin at the back and that the team is not worried about putting with too much energy. Taken together, these points would seem to argue that putting from anywhere within the initial placement area should work.

Point 4 reports on the results of a geometric analysis of the problem. While the use of analysis to determine which approach would be best is admirable, unfortunately shortcomings in the way they set up the analysis led them to an incorrect conclusion. There will indeed be a region of the field from which a ray drawn directly to the beacon will not enter the required bin at the edge of the field. However, that region is much smaller than their diagram indicates and mostly includes portions of the field that are not part of the initial placement area. The actual position of the beacon emitters is not set back behind the bins, as shown in Figure 32.13, but immediately above the back edge of the bin, as shown in Figure 32.14. The effect of getting this important distinction right is to widen the acceptable shooting range and reduce the size of the impossible regions. In addition, the footprint of the putter must be accounted for in determining where the putter can be placed according to the rules. The combination of these two factors yields a more accurate geometric analysis shown in Figure 32.14.

Team InTheRuff performed a similar set of tests and came to the conclusion that it *was* possible to build a "sit and spin" solution for the project. Adhering to their ideal of simplicity, they abandoned the mobile platform that they had initially proposed and converted their design to putt from wherever it was initially placed.

During this phase of the project, Team InTheRuff built two quick prototypes to evaluate how accurately they could determine the location of the beacon, and to test the pawl-based ball release mechanism.

On the left in Figure 32.15, we can see the beacon locating prototype. It consists of an aluminum tube to narrow the field of view of a photo-transistor. It was mounted to a rotating platform built of LEGO blocks and driven by a LEGO drive motor and worm gear. The team built a simple software test harness to drive the motor back and forth to locate the peak amplitude from the photo-sensor.

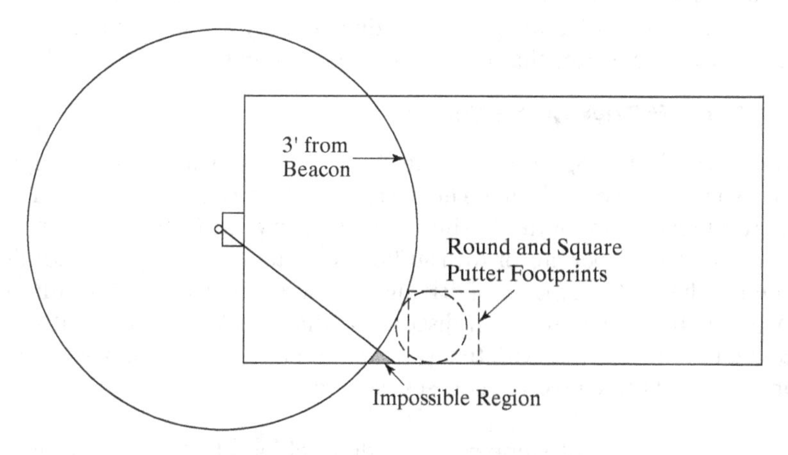

FIGURE 32.14

More accurate geometric analysis of possible putter placement.

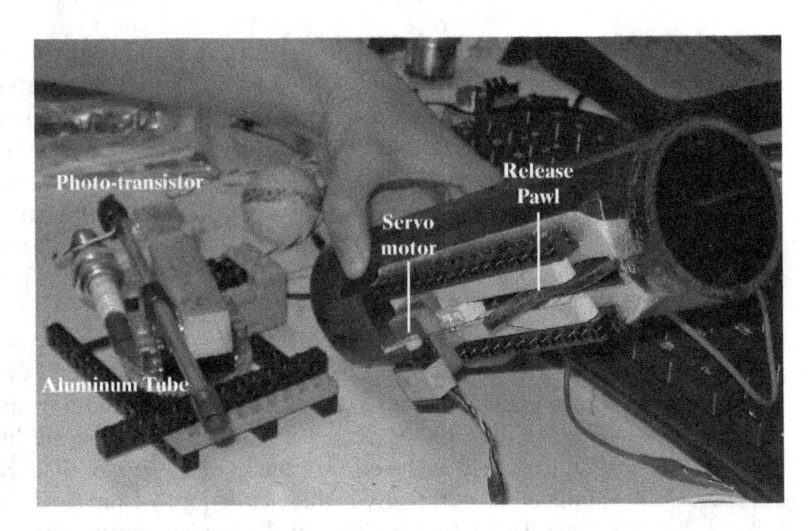

FIGURE 32.15
Team InTheRuff concept evaluation prototypes.

On the right in Figure 32.15 is the first prototype of the ball release mechanism sketched in Figure 32.8. The tube is a piece of ABS plumbing piping. The pawl was cut from a piece of thin plywood and mounted using several LEGO blocks and an axle. The pawl is driven back and forth between the two extremes of travel by a small RC servomotor. With a simple piece of test code to drive the servomotor to the two positions, the team was able to verify and tune the functionality of the ball release mechanism and then use that mechanism in further testing to determine the required tilt necessary to impart sufficient energy to the ball. The ball release system prototype provided a very repeatable platform on which to run subsequent range-testing experiments.

32.7 IMPLEMENTATION PHASE

The implementation phase in most projects is characterized by a range of different experiences: a lot of hard work, discoveries that things don't work the way that you had expected, and issues in integrating subsystems. The longer the time available for the project, the more attention that can be given to trying to anticipate and resolve all of the possible issues ahead of time. The reality for the very short, three week time span of the project described here is that there are many lessons learned on the fly as things are built and tested. The reason that we build a prototype is to discover where omissions, oversights, and/or incorrect assumptions were made in the initial concept. As such, the implementation phase is full of discoveries about things that were not anticipated or didn't actually work. In this section, we take a look at a few of these discoveries that were made by the teams that we are following.

32.7.1 InTheRuff Drive Motor Choice

Shortly after the first design review, Team InTheRuff converted their design from a mobile platform to a "sit and spin" design. The design that they came up with made use of a stepper motor to rotate the entire putter to simultaneously find the beacon and aim the putter. To do this, they needed a stepper motor with sufficient torque. The stepper motor driver board that is provided to the class is capable of driving bipolar, two phase motors with a current of up to 350 mA/phase. In testing, the team discovered that this level of current was insufficient to produce the required torque with the available stepper motors. They were faced with a decision and identified three possible courses of action:

1. **1.** Use a higher current stepper motor drive chip with the existing motor. This would require them to build the driver circuit from scratch.

2. Switch to a geared DC motor that would provide sufficient torque, but doesn't incorporate a means of controlling rotational position.
3. Use a higher current dual H-bridge board that was available to drive the existing motor at a higher current level and manually generate the phase drive signals in software.

Option 1 was identified as requiring the most physical work (the acquiring of parts and building of the driver circuit) but the least software, since they could use software pulse libraries provided to the class, just as they had initially intended. Option 2 would not require any new electronic hardware, since there were larger H-bridge modules available, nor new software since there were PWM libraries available. However, this solution was identified as somewhat risky because it was unknown how accurately and repeatably they would be able to aim the putter in the absence of any position control. In studying the third option, the team realized that while it would require adding software, they could isolate the changes to the software to a single function that would step a pair of output lines through the sequences needed to step the motor. In the end, the team decided that the lowest risk path was option 3.

32.7.2 Team Zero Ball Release Motor Choice

Team Zero also encountered a situation in which the stepper motor that they had initially planned to use had insufficient torque. In this case, though, the motor was to be used to turn a "pinwheel" on the ball release mechanism. As part of exploring possible solutions to the stepper motor's inadequate torque, the team built up a prototype of the ball release mechanism using a geared DC motor to drive the release wheel. This proved to have sufficient torque and enough damping and repeatability to be used by simply timing how long the motor was driven. As long as the wheel was placed in the correct initial position, no feedback was required to be able to release a full complement of balls one by one. Since this provided the simplest possible solution, the team chose to go with the geared DC motor for the release drive mechanism.

32.7.3 Team Zero Beacon Sensor Circuit Evolution

In planning for the beacon sensor circuit, Team Zero wanted to be able to determine the amplitude of the sensed signal so that they could use the information to infer their distance from the goal. To achieve this, they designed a circuit that used a photo-transistor amplified by a trans-resistive stage, which was then followed by a simple peak detector circuit (Chapter 14) to produce an analog output related to the amount of (IR) infrared light detected by the sensor. The initial design is shown in Figure 32.16.

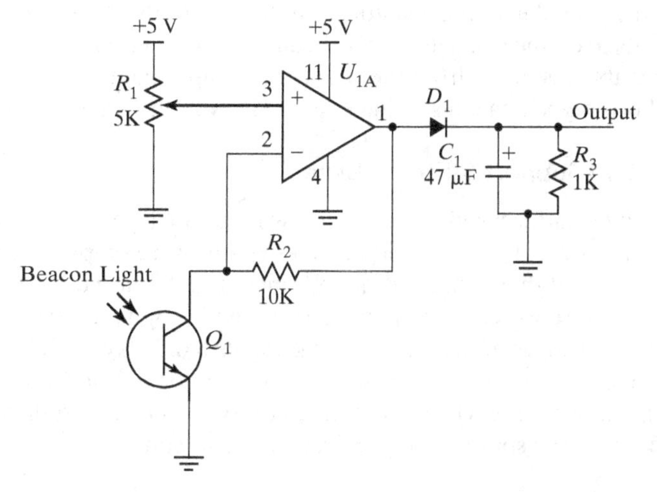

FIGURE 32.16

Initial beacon sensor design.

The output from the trans-resistive stage feeds into the peak detector made up of D_1, C_1, and R_3. The time constant of the RC is set at 47 ms. This is long compared to the approximately 1 ms period of the modulated beacon signal, but short in comparison to the speed at which the platform is expected to turn. This produces an analog output voltage with a relatively small amount of ripple due to the beacon modulation and an adequate fall off time as the sensor is turned away from the beacon. While this circuit worked under some conditions, it did not produce the range of output values that had been hoped for and it proved impossible to find a single value for the trans-resistive gain (R_2) that would produce a usable signal at the maximum range and not saturate at close range. The team eventually realized that the diode in the peak detector circuit was limiting the circuit's ability to follow low amplitude signals. If the output from U_1 was less than the forward voltage drop of the diode (V_f), then no signal would reach the output and no voltage would be applied to the RC. To address this issue, the team revised the design as shown in Figure 32.17.

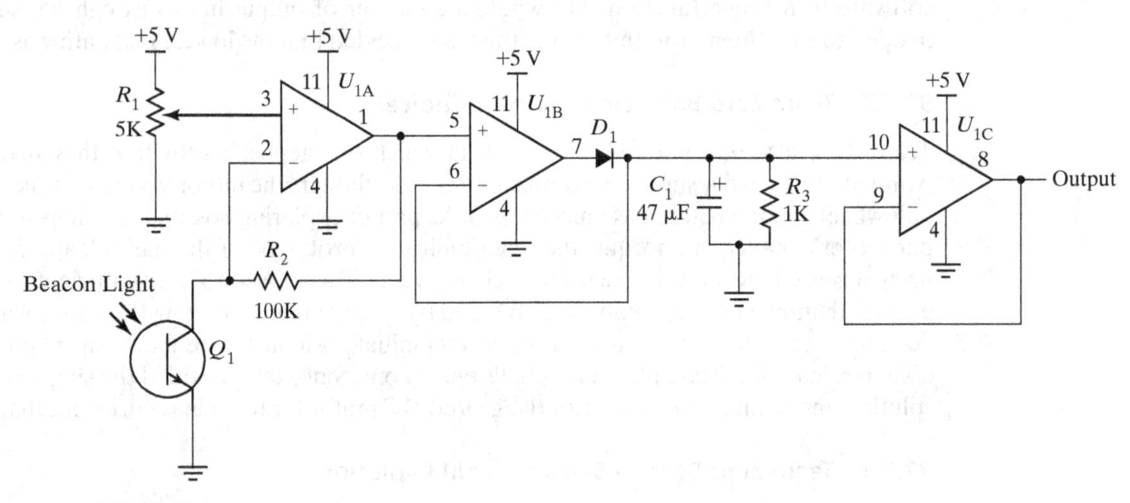

FIGURE 32.17

Revised signal conditioning circuit.

This design is composed of three distinct portions. The first is the trans-resistive stage, followed by an active peak detector in which the diode is placed within the feedback loop of an op-amp. This produces an apparent "ideal diode" with effectively no forward voltage drop, since the op-amp's output compensates for the real diode's drop. The final buffer stage isolates the voltage on the peak detector capacitor from the load of the A/D converter, minimizing the droop of the voltage on the capacitor. This circuit produced a usable output voltage across the full range of possible distances from the beacon, with approximately 1 V greater output span than the initial prototype beacon sensing circuit. This was the final circuit used in the project.

32.7.4 Team Zero Support Stiffness Issue

Team Zero's mobile platform rode on two drive wheels with a pair of balance struts at the front and a single balance strut at the rear. Initially, these struts were simply long screws that were fitted with feet at the end that contacted the playing field. When the platform was fully loaded with balls, the team discovered that the drag with the playing field caused the struts to deflect to the extent that the edges of the feet were digging into the playing field (Figure 32.18a). This caused uneven motion and in some cases the drive motor did not have sufficient torque to overcome the additional drag. The team decided to try solving the problem by adding bending stiffness to the struts. The solution they arrived at was to add sections of PVC pipe concentric

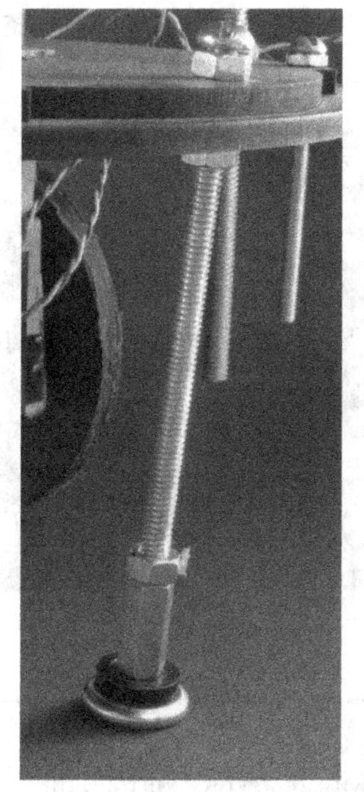

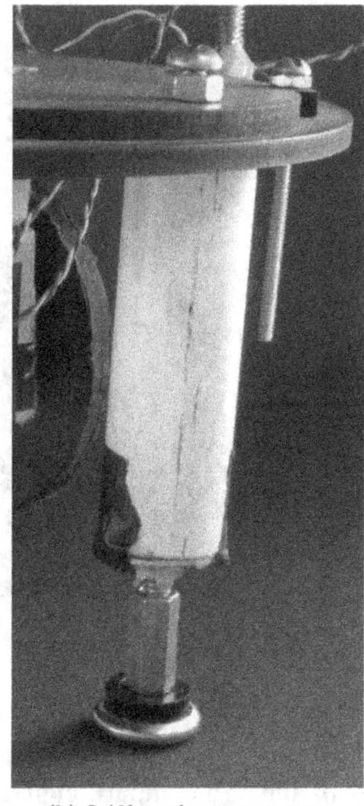

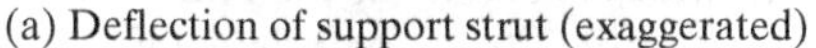

(a) Deflection of support strut (exaggerated)

(b) Stiffened support strut

FIGURE 32.18

Solving an unforeseen problem discovered during implementation: bending moments applied to support struts by friction.

with the threaded rod forming the strut. This increased the area moment of inertia, which in turn increased the bending stiffness, without adding much mass (Figure 32.18b).

32.7.5 Team Zero Compass Sensor Failure

When under time pressure, shortcuts are an inevitable temptation. However, it is often the case that those shortcuts result in situations that invite failure. Such was the case for Team Zero late at night two days before the project deadline. A power supply connection that was made quickly (using alligator clips) rather than robustly (using insulated and polarized connectors) allowed the 12 V power lead to contact the 5 V power lead. The power supply had protection circuits that kept it from being damaged. The microcontroller board also had protection devices that prevented any damage. Unfortunately, none of the analog circuits that the team had built had any protection. The shorting of the power supply leads damaged all of the analog signal conditioning circuits and destroyed the compass sensor. While replacement op-amps were available, a replacement compass sensor was not. In the final two days of the project, the team was forced to rewrite the code for aiming so that the project could work without the compass sensor. They did this by implementing an algorithm that repeatedly aimed based on finding the direction to the brightest beacon signal, then drove forward a bit, then aimed again then drove forward again, and so on—until the absolute value of the peak beacon signal was such that the putter was deemed to indicate that the device was at the right distance to putt (an empirically determined value).

32.8 THE COMPLETED DESIGNS

32.8.1 Team InTheRuff

The final version of Team InTheRuff's putter is shown in two views in Figure 32.19.

FIGURE 32.19

Team InTheRuff final putter.

Exercise:

Take a few minutes to compare Figure 32.19 with the sketch of the initial concept in Figure 32.7. How many characteristics can you identify that made it from the original mobile platform concept into the "sit and spin" solution of the final design?

In the final design, the platform sits atop a lazy Susan bearing, with three wheels mounted near the outer edge that enable the drive system to rotate the platform and prevent tipping.

Two of the wheels are not driven and just provide tipping support. The third wheel (Figure 32.20) is driven by a stepper motor to rotate the platform. The stepper motor produced 200 steps per revolution. With the chosen wheel size and placement, this combination provided an angular positioning resolution of approximately 0.5° per step. The team had calculated earlier that the angular width of the target bin was approximately 2° at the maximum distance, so this angular resolution allowed for four steps across the width of the bin.

FIGURE 32.20

Team InTheRuff stepper motor drive detail.

FIGURE 32.21
Team InTheRuff ball release mechanism detail.

The final ball release mechanism that InTheRuff used should look familiar (Figure 32.21). It is the original prototype that was built early in the process. It proved so successful that it was transplanted, with only a few modifications, into the final putter.

The aluminum tube, sensor, and decorative spark plug also made it from the early prototype (Figure 32.15) into the final putter.

The use of a stepper motor to aim the platform allowed for a very different approach to both signal conditioning and algorithm than that chosen by Team Zero. The signal conditioning circuitry used by Team InTheRuff is shown in Figure 32.22.

Careful study of the circuit reveals that the output of the circuit is a digital signal indicating whether or not IR light of sufficient intensity is detected by the sensor. The software algorithm uses this input by rotating the platform first in one direction until the edge of the beacon field of view is found. It then sweeps back to find the other edge of the field of view while counting the number of steps required to get from one edge to the other. The beacon is assumed to lie at the center of the field, so the final stage of aiming is to rotate the platform backward half the number of steps in the beacon field of view. This simple approach uses a

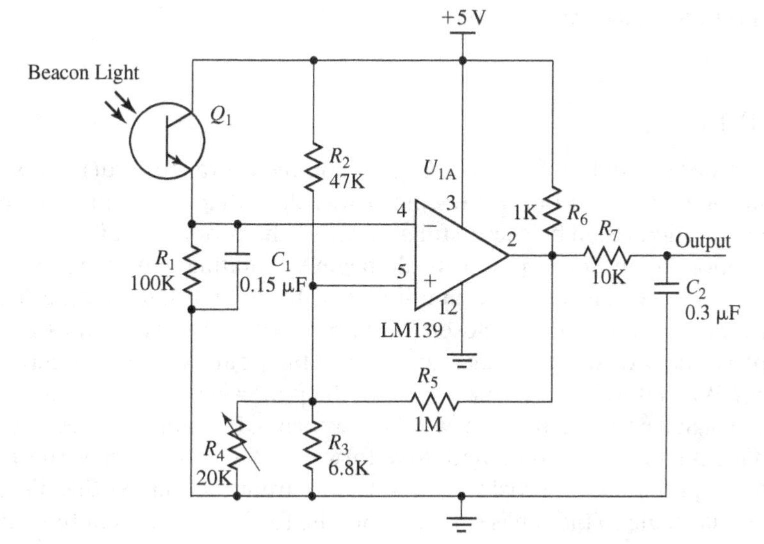

FIGURE 32.22
Team InTheRuff beacon signal conditioning circuit.

sensor circuit that is either off or saturated. Though Team InTheRuff did not originally state that they were striving for simplicity in their design, they ultimately achieved it with several of their subsystems: for example, their "sit and spin" platform and a digital beacon sensor.

32.8.2 Team Zero

The final version of the putter constructed by Team Zero is shown in Figure 32.23.

FIGURE 32.23
Team Zero final putter.

Exercise:

Take a few minutes to compare the photographs in Figure 32.23 with the sketch of the initial concept in Figure 32.6. Identify those aspects of the design that are visibly the same and different between the concept sketch and final implementation.

Team Zero did a remarkable job of adhering to their original vision, despite the many twists and turns they encountered along the way.

32.9 PERFORMANCE RESULTS

The machines were tested during a grading session over a range of distances from the bin and positions on the field. From this testing, it was clear that either a mobile platform or a "sit and spin" solution could be successful at meeting the project requirements.

In operation, Team Zero's putter would begin by rotating in place to locate the beacon signal. The software would read the amplitude of the signal and then decide whether to shoot from that distance or move closer to the bins. When the putter was placed near the beacon, it proceeded to rotate in place to aim and then putt, which is effectively a "sit and spin" strategy. When it was placed further away, the putter would first locate the peak of the beacon sensor signal and then drive forward a few inches. It would then recenter itself on the beacon signal, assess the amplitude, and drive forward a few more inches. This process would repeat until the putter was approximately 3 ft away from the bin. At this distance it would putt. Because the angle to the bin is smaller from the furthest reaches of the field, this putter proved more accurate than Team InTheRuff's when initially placed far away. Even when placed at the furthest distances, the putter made all five putts. The process of moving toward the beacon consistently moved the putter toward the center of the field and better aligned with the bin. When initially placed close to the bin at the edge of the field, the angle with the bin was not as favorable and the variability in the ball launch resulted in three balls putted into the correct bin and one in each of the bins on either side. While the aiming process was somewhat slow when starting from long distances, the putter always completed the five putts in much less than the allotted 2 minutes.

The putter built by Team InTheRuff proved to be extremely accurate and reliable. The putter made every putt attempted from all of the tested positions. The team had taken advantage of the rules and included the ability to input a correction in case of an inaccurate putt, but their device was so accurate that they never needed to make use of the feature.

32.10 GEMS OF WISDOM FROM THE STUDENTS

The teams were asked to include a "Gems of Wisdom" section in their final reports. The purpose of this section is to have the students be reflective and capture the big-picture lessons learned. It also gives the students an opportunity to pass on advice to future generations of students taking the class, in the hopes that they might help make their lives easier and their project results better. In the interest of full disclosure, we admit that we have deviated a bit from the focus on these two design teams and now offer a range of advice culled from several years' project reports.

1. A good way to begin your design plans is to focus on making your core functionality definitely achievable with the time and schedules that your team has. Add maybe one or two stretch goals that you can work on if you have extra time.
2. Strive for simplicity. Our project used two actuators and one sensor, each with a well-defined interface. This allowed us to test each component in isolation, yet be confident that they would work with the final integration.
3. Plan to make a project that JUST MEETS REQUIREMENTS. Don't go haywire! Reality doesn't hit you in the planning stage. Try to achieve a base set of goals that meet all specs, and WORK HARD TO GET THESE DONE EARLY. Once you get these done, then start pushing the envelope. We finished our base project (and if we had to, could have turned it in) early, and then added in the last week.
4. Design first, before you actually start implementing it.
5. Divide the labor. Two people working on one task is often slower than one person working on one task, which is slower than two people each working on half of a task. Instead of insisting that the whole group be present every time we wanted to work, trying to match differing schedules and work styles, we broke the project up into specific tasks with well defined interfaces. This allowed us each to work most efficiently, meaning alone, then incorporate our sections successfully.
6. Fail quickly. The quicker we fail, the more lessons we can get early in the process, and the better the final design will be. We learned that it is very hard to get everything right (or even 50% of everything right) in the first prototype. If we had a working prototype before constructing the final robot, we would have identified our main problems earlier and eventually achieved a better performance.
7. Validate assumptions. One of the main lessons we learned was that in the design phase, we should identify the assumptions we are making about our design and validate them as early as possible. Our initial design assumed that the putting mechanism would be

accurate, and that proper aligning/positioning with the beacon would best improve performance. However, our prediction was the opposite of what we discovered; the robot could align reasonably well with the beacon, but the putter was inaccurate to a fair degree. Even with only one beacon sensor and our simple positioning algorithm, our robot was able to make most balls into the correct bin. The misses were primarily attributable to variability in the putting mechanism. Thus, the addition of the compass sensor would have been unlikely to help at all. Had we focused on putter accuracy over positioning algorithms, we would have likely performed better. In the future, it would be advisable to keep an eye out for such bottlenecks in performance. However, because our robot was constructed later than anticipated, there was little that could have been changed.

8. Focus on repeatability. That is what robots do best, so by removing anything that was not deterministic, like movement of the robot, putting accurately became a simple matter of calibration.
9. Make the design adjustable. Consider making holes slots. Consider the use of pots.
10. Make the design maintainable. Secure components/cables in a manner that they can be replaced/rerouted. Make judicious use of cables, connectors, switches. Inclusion of status indicators. Have spare parts on hand.
11. Make sure the robot is easy to disassemble.
12. Make your circuits easily accessible when the robot is fully assembled.
13. Make the checkpoints—it will make your life incredibly easier to know that your core functionality is taken care of, especially if your team will be gone for Thanksgiving Break. Do not forget to factor in Thanksgiving Break—plan ahead of time depending on your team's schedules.
14. Do not underestimate the time it takes to integrate. This includes final wiring, building, decorating, and debugging, so it is very very significant.
15. This is a time-limited engineering project; don't over-design and under-deliver. A good enough circuit or algorithm is good enough. You don't have time to make it perfect, and it's usually not worth it to make it perfect.

ACKNOWLEDGMENTS

This chapter would not have been possible without the permission granted by the members of Team Zero (Computer Scientists count from Zero) and Team InTheRuff: Adam Bernstein, Ho Lum Cheung, Nancy Dougherty, Derianto Kusuma, Jordan LeNoach, Matthew Norcia, Kanya Siangliulue, and Wesley Zuber.

Resistor Color Code and Standard Values

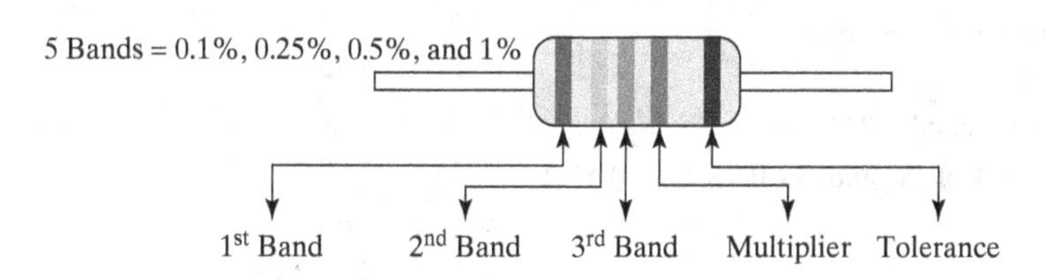

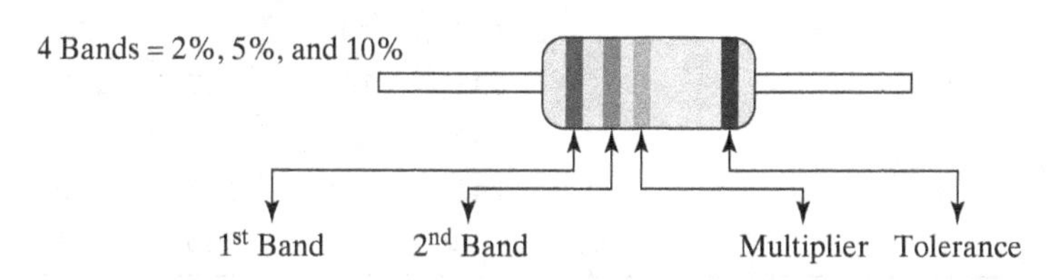

Color	1st Band	2nd Band	3rd Band	Multiplier	Tolerance
Black	0	0	0	1 Ω	Color
Brown	1	1	1	10 Ω	±1%
Red	2	2	2	100 Ω	±2%
Orange	3	3	3	1 kΩ	
Yellow	4	4	4	10 kΩ	
Green	5	5	5	100 kΩ	±0.5%
Blue	6	6	6	1 MΩ	±0.25%
Violet	7	7	7	10 MΩ	±0.1%
Grey	8	8	8		±0.05%
White	9	9	9		
Gold				0.1 Ω	±5%
Silver				0.01 Ω	±10%

5% RESISTORS

Standard Values

Standard 5% resistor values are found by multiplying one of the values in the list to the right by a power of 10 between −2 and 7.

The mantissa of the resistance value is marked by the first two color bands on the body of the resistor with the exponent being indicated by the third color band.

Examples:

red,red,red $= 22 \times 10^2 = 2.2$ kΩ

brown,black,yellow $= 10 \times 10^4 = 100$ kΩ

10	22	47
11	24	51
12	27	56
13	30	62
15	33	68
16	36	75
18	39	82
20	43	91

1% RESISTORS

Standard Values

Standard 1% resistor values are found by multiplying one of the values in the list to the right by a power of 10 between −2 and 7.

The mantissa of the resistance value is marked by the first three color bands on the body of the resistor with the exponent being indicated by the fourth color band.

Examples:

red,red,black,red = $220 \times 10^2 = 22\ k\Omega$

brown,black,black,yellow = $100 \times 10^4 = 1\ M\ \Omega$

100	147	215	316	464	681
102	150	221	324	475	698
105	154	226	332	487	715
107	158	232	340	499	732
110	162	237	348	511	750
113	165	243	357	523	768
115	169	249	365	536	787
118	174	255	374	549	806
121	178	261	383	562	825
124	182	267	392	576	845
127	187	274	402	590	866
130	191	280	412	604	887
133	196	287	422	619	909
137	200	294	432	634	931
140	205	301	442	649	953
143	210	309	453	665	976

Sample C Code

APPENDIX

B.1 MOVING AVERAGE

```
/*****************************************************************************
Function
   MoveAvg8

Parameters
   NewValue   int, the new value to enter into the moving average

Returns
   int, the value of the moving average after entering the NewValue

Description
   Implements an 8-point moving average using an 8-entry buffer and
   an algorithm that keeps a sum and subtracts the oldest value from
   the sum, followed by adding in the new value before dividing by 8.
   A very literal implementation.

Notes
   while the buffer is initially filling, it is hard to really call the
   the average accurate, since we force the initial values to 0.

Author
   J. Edward Carryer, 12/06/09 15:09
*****************************************************************************/

int MoveAvg8( int NewValue)
{
  static int RunSum = 0;
  static int Buffer[8] = {0,0,0,0,0,0,0,0};
  static unsigned char Newest = 0;
  static unsigned char Oldest = 1;

  // update the running sum: remove oldest value and add in the newest
  RunSum = RunSum - Buffer[Oldest] + NewValue;

  // put the new value into the buffer
  Buffer[Newest] = NewValue;

  // update the indices by incrementing modulo 8
  // for a binary modulo, the AND operation is faster than using %
  Newest = (Newest + 1) & 0x07;
  Oldest = (Oldest + 1) & 0x07;

  // now return the result by dividing by 8 (shortcut: shift right by 3)
  return(RunSum >> 3);

}
```

B.2 A SAMPLE PROGRAM BASED ON AN EVENTS AND SERVICES FRAMEWORK

```
/*************************************************************************
Module
   Rchdemo2.c

 Description
   Test/Demonstration module for implementing Cockroach functionality
   using Software Events and Services Framework.

Notes
   This is intended to replace old "sestest.c" and "rchdemo.c" demo
   programs.

History
When           Who What/Why
-------------- --- ---
01/10/01       jec  modified light on & off event to only read the sensor
                    once
12/18/00       te  initial code
*************************************************************************/

#include <stdio.h>
#include "me118.h"
#include "roachlib.h"
#include "timer.h"
#include "ses.h"

/*-------- Module Defines ----------------*/
#define HI_LIGHT_THRESHOLD 80
#define LO_LIGHT_THRESHOLD 60
#define TEN_SEC 2440
#define TIME_INTERVAL TEN_SEC

/*-------- Global Variables ----------------*/

/*-------- Function Prototypes ----------------*/
uchar TestForKey(EVENT_PARAM);
void RespToKey(SERVICE_PARAM);
uchar TestForLightOn(EVENT_PARAM);
void RespToLightOn(SERVICE_PARAM);
uchar TestForLightOff(EVENT_PARAM);
void RespToLightOff(SERVICE_PARAM);
uchar TestForBump(EVENT_PARAM);
void RespToBump(SERVICE_PARAM);
uchar TestTimerExpired(EVENT_PARAM);
void RespTimerExpired(SERVICE_PARAM);

/*-------- Module Code ----------------*/

void main (void)
{
  /* initialize the SES module to operate in round robin mode, with no
  timer delay */
  SES_Init(SES_ROUND_ROBIN, SES_NO_UPDATE);
```

```
  TMR_Init(TMR_RATE_4MS); //initialize timer module
  RoachInit();            //initialize roach library

  /* register each pair of Events and Services routines*/
  SES_Register(TestForKey, RespToKey);
  SES_Register(TestForLightOn, RespToLightOn);
  SES_Register(TestForLightOff, RespToLightOff);
  SES_Register(TestForBump, RespToBump);
  SES_Register(TestTimerExpired, RespTimerExpired);

  TMR_InitTimer(0, TIME_INTERVAL); //initialize timer 0 for use
  /* enter an infinite loop and handle the events */
  while (1)
    {
      SES_HandleEvents();
    }
}

/*************************************************
 * Function: TestForKey
 * Creator: KW                  Date: 1/06/00
 * Rev: TE                      Date: 12/18/00
 *
 * This is an event checking routine which returns true if a key
 * has been pressed. The pressed key is stored as the shared variable
 * for use by the service routine.
 *
 ****************************************************************/
uchar TestForKey(EVENT_PARAM)
{
  unsigned char theKey;

  char EventOccured = kbhit();
  if (EventOccured)
    {
      theKey = getchar();
      SET_SHARED_BYTE_TO(theKey);
    }
  return (EventOccured);
}

/***************************************************************
 * Function: RespToKey
 * Creator: KW                  Date: 01/06/00
 * Rev: TE                      Date: 12/18/00
 *
 * This is the service routine for a keypress.  It simply echoes the
 * character pressed to the screen.
 *
 ****************************************************************/
void RespToKey(SERVICE_PARAM)
{
  putchar (GET_SHARED_BYTE());
  putchar ('.');
}
```

```
/*****************************************************************
 * Function: TestForLightOn
 * Creator: J. Edward Carryer   Date: 04/11/94
 * Rev: TE                      Date: 12/18/00
 *
 * This is an event checking routine which returns true if the current
 * reading from the light sensors has changed from the last reading and it
 * exceeds a minimum "brightness threshold," implementing a hysteresis band
 * to prevent bouncing triggers.
 *
 *****************************************************************/
unsigned char TestForLightOn(EVENT_PARAM)
{
  static uchar Threshold = HI_LIGHT_THRESHOLD;
  static uchar LastLight = 0;
  uchar ThisLight = LightLevel();

  char EventOccured = ((ThisLight > Threshold) &&
                       (LastLight <= Threshold));
  if (EventOccured)
    Threshold = LO_LIGHT_THRESHOLD;
  /*      provide for hysteresis around the switching point to eliminate
          false triggers by DECREASING threshold level.  */

  else if (ThisLight <= Threshold)
    Threshold = HI_LIGHT_THRESHOLD;

  LastLight = ThisLight;
  return (EventOccured);
}

/*****************************************************************
 * Function: RespToLightOn
 * Creator: J. Edward Carryer   Date: 04/11/94
 * Rev: TE                      Date: 12/18/00
 *
 * This is the service routine for a "light on" event. If the light
 * has turned on, it simply prints "ON" to the screen.
 *
 *****************************************************************/
void RespToLightOn(SERVICE_PARAM)
{
  printf("\n ON");
}

/*****************************************************************
 * Function: TestForLightOff
 * Creator: J. Edward Carryer   Date: 04/11/94
 * Rev: TE                      Date: 12/18/00
 *
 * This is an event checking routine which returns true if the current
 * reading from the light sensors has changed from the last reading and
 * it is below a maximum "darkness threshold," implementing a hysteresis
 * band to prevent bouncing triggers.
 *
 *****************************************************************/
```

```
uchar TestForLightOff(EVENT_PARAM)
{
  static uchar Threshold = LO_LIGHT_THRESHOLD;
  static uchar LastLight = 0;
  uchar ThisLight = LightLevel();

  char EventOccured = ((ThisLight < Threshold) &&
                       (LastLight >= Threshold));

  if (EventOccured)
    Threshold = HI_LIGHT_THRESHOLD;
  /*      provide for hysteresis around the switching point to eliminate
          false triggers by INCREASING threshold level.  */

  else if (ThisLight >= Threshold)
    Threshold = LO_LIGHT_THRESHOLD;
  LastLight = ThisLight;
  return (EventOccured);
}

/************************************************************
 * Function: RespToLightOff
 * Creator: J. Edward Carryer    Date: 04/11/94
 * Rev: TE                       Date: 12/18/00
 *
 * This is the service routine for a "light off" event. If the light
 * has turned off, it simply prints "OFF" to the screen.
 *
 ************************************************************/
void RespToLightOff(SERVICE_PARAM)
{
  printf("\n OFF");
}

/************************************************************
 * Function: TestForBump
 * Creator:  TE                  Date: 12/18/00
 *
 * This an event checking routine. It returns true if one of the bumper
 * bits is low and the last check of the bumpers was not the same. This
 * prevents a single bump from registering multiple times. The bumper
 * reading is shared for use in the service routine.
 *
 ************************************************************/
uchar TestForBump(EVENT_PARAM)
{
  static uchar lastBump = 0x0F;
  uchar bumper;
  char EventOccured =  (((bumper=ReadBumpers()) != 0x0F) &&
                         (bumper != lastBump));

  if (EventOccured)
    {
      SET_SHARED_BYTE_TO(bumper);
      lastBump = bumper;
    }
```

```
  return (EventOccured);
}

/*****************************************************************
 * Function: RespToBump
 * Creator: TE                    Date: 12/18/00
 *
 * This is the service routine that is called when a bump is detected.
 * When a bump occurs, the actual bumper that is hit is printed to the
 * screen.
 *
 ****************************************************************/
void RespToBump(SERVICE_PARAM)
{
  unsigned char bumper;
  bumper = GET_SHARED_BYTE();

  // display which bumper(s) were hit
  switch (bumper)
    {
    case (0x0e):
      printf("Front Right...\n");
      break;
    case (0x0d):
      printf("Front Left...\n");
      break;
    case (0x0b):
      printf("Back Left...\n");
      break;
    case (0x07):
      printf("Back Right...\n");
      break;

    case (0x0c):
      printf("Both Front ...\n");
      break;
    case (0x03):
      printf("Both Back...\n");
      break;
    case (0x06):
      printf("Both Right...\n");
      break;
    case (0x09):
      printf("Both Left...\n");
      break;

    default:
      printf("What's this-> %x ?\n", bumper);
      return;
    }
}

/*****************************************************************
 * Function: TestTimerExpired
 * Creator: TE                    Date: 12/18/00
 *
```

```
 * This is an event checking routine which returns true if timer 0
 * has expired.
 *
 *
 ***************************************************************/
unsigned char TestTimerExpired(EVENT_PARAM)
{
  return(TMR_IsTimerExpired(0));
}

/****************************************************************
 * Function: RespTimerExpired
 * Creator: TE                      Date: 12/18/00
 *
 * This is the service routine that is called when timer 0 expires.
 * It displays the number of times that the timer has expired and
 * restarts timer 0. It also displays the current light level.
 *
 ***************************************************************/
void RespTimerExpired(SERVICE_PARAM)
{
  static Time =0;

  printf("\n %d",++Time);
  printf("Light level: %d\n", LightLevel());

  TMR_InitTimer(0, TIME_INTERVAL);
}
```

B.3 A SAMPLE TEMPLATE USING A STATE MACHINE WITH EVENTS AND SERVICES

```
/************************************************************************
 Module
   sesstate.c

 Revision
   1.0.0

 Description
   Test/Demonstration module for the Software Events and Services
   Framework driving a state machine. This module demonstrates how to
   combine the Events and Services framework with a state machine
   based design. This example is intended only to provide you with
   the general outline of what a solution should look like. It does
   nothing functional.

 Notes

 History
 When           Who  What/Why
 -------------- ---  -----------
 01/09/02 13:45 jec  Created from sestest.c.
***********************************************************************/
```

```
/*--------------------- Include Files ---------------------*/
#include <stdio.h>
#include <me118.h>
#include <ses.h>

/*--------------------- Module Defines ---------------------*/
#define EVENT1 1
#define EVENT2 2
#define EVENT3 3
#define STATE0 0
#define STATE1 1
#define STATE2 2

/*--------------------- Module Types -------------------------*/

/*--------------------- Global Functions ---------------------*/

/*--------------------- Module Functions ---------------------*/
uchar TestEvent1(EVENT_PARAM);
uchar TestEvent2(EVENT_PARAM);
uchar TestEvent3(EVENT_PARAM);
void DemoStateMachine(SERVICE_PARAM);/*this will be the one service routine*/

/*--------------------- Module Variables ----------------*/

/*--------------------- Module Code ---------------------*/
void main(void)
{

  puts("Starting...\n");

  SES_Init(SES_ROUND_ROBIN, SES_NO_UPDATE);

  SES_Register(TestEvent1,DemoStateMachine);
  SES_Register(TestEvent2,DemoStateMachine);
  SES_Register(TestEvent3,DemoStateMachine);

  while (1)
    SES_HandleEvents();

}

/****************************************************************************
 Function
   TestEvent1

 Parameters
   EVENT_PARAM    standard parameter for Event checker, will be
                  used to pass the event code to the Service routine.

 Returns
   unsigned char  non-zero if the event was detected.

 Description
   Dummy event checker. Event Code for this event is placed in the
   shared variable and TRUE (1) is returned to announce that the event
   has occurred. This will cause SES to call the DemoStateMachine()
   function.
```

```
 Notes
   None.

 Author
   J. Edward Carryer, 01/09/02 13:50
***************************************************************************/
uchar TestEvent1(EVENT_PARAM)
{

  static unsigned char SharedEvent; /*used to pass the event code */
  /* between the event checker and the service routine,    */
  /* which is the DemoStateMachine. It *must* be  */
  /* static to preserve its value after the       */
  /* function terminates.                         */

  SharedEvent = EVENT1;
  SET_SHARED_BYTE_TO(SharedEvent);/*leave value where the service routine */
  return 1;                        /* can find it */
}

uchar TestEvent2(EVENT_PARAM)
{
  static unsigned char SharedEvent;

  SharedEvent = EVENT2;
  SET_SHARED_BYTE_TO(SharedEvent);/*leave value where the service routine */
  return 1;                        /* can find it */
}

uchar TestEvent3(EVENT_PARAM)
{
  static unsigned char SharedEvent;
  SharedEvent = EVENT3;
  SET_SHARED_BYTE_TO(SharedEvent);/*leave value where the service routine */
  return 1;                        /* can find it */
}

/***************************************************************************
 Function
   DemoStateMachine

 Parameters
   SERVICE_PARAM  standard parameter for Service routine, will be
                  used to get the event code from the event checking
                  routines.

 Returns
   None.

 Description
   Dummy state machine. Set up as the service routine for all the
   relevant events, it looks at the Event code passed to the service
   routine to identify the current event and uses that along with the
   current state to implement a dummy state machine.

 Notes
   None.
```

```
 Author
   J. Edward Carryer, 01/09/02 13:59
*********************************************************************/
void DemoStateMachine(SERVICE_PARAM)
{
  static unsigned char CurrentState = 0;
  unsigned char CurrentEvent;

  CurrentEvent = GET_SHARED_BYTE(); /* get the event code passed from  */
                                    /* the event checking routines     */
  switch (CurrentState) /*implement state machine with nested switch */
    {
    case STATE0 :
    {
      switch (CurrentEvent)
        {
        case EVENT1 :
          CurrentState = STATE1; /* change states */
          break;
        case EVENT2 :
          CurrentState = STATE2; /* change states */
          break;
        case EVENT3 :
          CurrentState = STATE0; /* change states */
          break;
        }
    }
    break;
    case STATE1 :
    {
      switch (CurrentEvent)
        {
        case EVENT1 :
          CurrentState = STATE2; /* change states */
          break;
        case EVENT2 :
          CurrentState = STATE0; /* change states */
          break;
        case EVENT3 :
          CurrentState = STATE1; /* change states */
          break;
        }
    }
    break;
    case STATE2 :
    {
      switch (CurrentEvent)
        {
        case EVENT1 :
          CurrentState = STATE0; /* change states */
          break;
        case EVENT2 :
          CurrentState = STATE1; /* change states */
          break;
```

```
        case EVENT3 :
          CurrentState = STATE2; /* change states */
          break;
        }
    }
    break;
    }

}

/*-------------------- Footnotes ---------------------*/

/*-------------------- End of file -------------------*/
```

Project Description for the Chapter 32 Case Study

EE118 Introduction to Mechatronics

EE 118 Winter 2009 Project: The Autonomous Golf Putter

Project Presentations on March 9 2009 starting at 7:00 PM

Goals:

The goal of this project is to provide you with an opportunity to apply your knowledge to solve an open-ended problem. The basic purpose of the machine is to 'putt' golf balls into a 'hole' similar to that found on a golf green. For the purposes of this project, 'putt' means to get the golf balls into the hole.

Purpose:

The underlying purpose of this project is to give you some experience in integrating all that you have learned. The avenue through which you will gain this experience is the design and implementation of an autonomous golf putter.

Specifications

The Field:

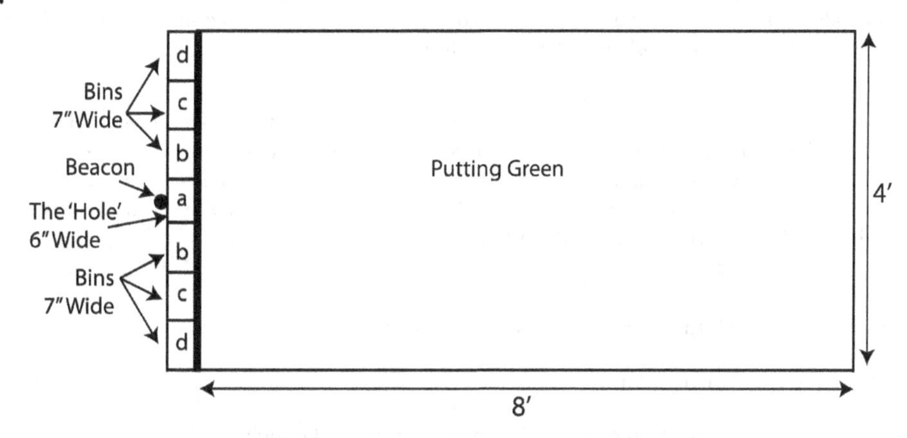

Fig. 1 Top View of Playing Field

- ☐ The Putting Green, as shown in Fig. 1, measures 4'x8'. On the three sides not containing bins, the Putting Green is bounded by a 3" high wall.
- ☐ At one end of the Putting Green will be a series of bins. The goal is to 'putt' your golf balls into the bin labeled **a** (the 'hole'). The edges of the bins will be below the surface of the putting green.
- ☐ The side-to-side center of the bin labeled **a** will be marked by a 'pole' carrying a beacon emitting modulated IR light at a height of 12" above the Putting Green.
- ☐ The edge of the Putting Green adjacent to the bins will be marked by a strip of 1" black tape.

The Putters:

- ☐ Your Putter must be a stand-alone entity, capable of meeting all specifications while connected only to power and ground.
- ☐ Your Putter is required to occupy a volume not to exceed 13"x13" in horizontal dimensions and 12" in height when initiated. Your machine must contain the complete supply of golf balls to be used during the event. At least five are required, but there is no upper limit, except that imposed by the total volume of the machine.
- ☐ Each Putter must be constructed as part of EE118. It may not be based on a commercial or otherwise pre-existing platform..
- ☐ Each Putter will have an easily accessible power switch.
- ☐ Any exterior corners on the Putter must have a radius of at least 1/4".

Game Play:

- ☐ The Putter will be placed on the Putting Green in a random orientation determined by the judges at a distance, also determined by the judges, between 3' and 8' from the beacon pole..
- ☐ The Putter will be loaded with golf balls at the beginning of each round.
- ☐ The putting of all balls must be completed within two minutes from the start of the round.
- ☐ When a golf ball enters the bin labeled **a**, the beacon will be extinguished for approximately $\frac{1}{2}$ sec.
- ☐ Once the positioning/putting process has been started, the operator may not touch the machine. The operator will be allowed to input a single correction entry after every putt. The input may ONLY take the form of the bin number of the hole into which the last putt fell. The entry of any correction is optional.
- ☐ Scoring will be based on the number of 'strokes' assessed for each ball putted. The bins on either side of the bin labeled **a** (the 'hole') will result in more strokes than the 'hole'. The further a bin is from the hole, the greater the strokes assessed. The bins labeled **b** count as 2 strokes. The bins labeled **c** count as 3 strokes. The bins labeled **d** count as 4 strokes. Any ball not falling into any of the bins will be considered to have fallen into the highest valued bin.

Rules:

- ☐ Your machine may not touch the pole that carries the beacon.
- ☐ A ball must remain in the hole to count.
- ☐ The operator may not move the Putter once it has been placed by the judges.
- ☐ Your Putter may not alter the playing field **IN ANY WAY**.

Safety:

- ☐ The machines should be safe, both to the user and the spectators. For this machine, that requires that no object associated with the machine become ballistic. This specifically includes the golf balls to be putted.
- ☐ All liquids, gels and aerosols must be in three-ounce or smaller containers. All liquids, gels and aerosols must be placed in a single, quart-size, zip-top, clear plastic bag. Each Putter can use only one, quart-size, zip-top, clear plastic bag.
- ☐ Putters may alter the Space-Time continuum only during the public presentations.

Check-Points

Project Milestones

Deliverables

First Review
2/17/09

During class-time on 02/17/09 we will conduct a design review. Each group should prepare a few sheets of paper showing your ideas. At least 5 concepts, with sketches, morphology chart, time schedules, and personnel assignments should be presented. These should be scanned into a no-frills Powerpoint file (landscape, 4:3 format, .ppt, not .pptx) for projection in class. You will walk us through your ideas. The other members of the class and the teaching staff will be on hand to hear about your ideas and provide feedback and advice.

Second Review
2/24/09

Calculations, system block diagram, preliminary test results.
Presented to teaching staff.

Third Review
3/5/09

Working versions of all systems, working software to test all systems, initial integration of systems.
Presented to teaching staff.

Final Presentations
3/9/09

Finished, operational machines

Report

A report describing the technical details of the machine will be required. The report should be of sufficient detail that a person skilled at the level of EE118 could understand, reproduce and modify the design. These reports will become entries into a design database that will be made available to future classes of EE118. The report will be delivered in two parts:

☐ On February 24, 2009, you will turn in a set of schematics, textual descriptions and software design documentation that describes the state of the design at that point in time. It need not be tested, nor even correct. It must be turned in as soft copy.

☐ On March 13, 2009 the final report is due. This should be a document whose structure is complementary to what was turned in on February 24. This document should describe the final mechanical design, circuits and software implemented. It should specifically address how these final designs differ from the original design. The intent is to capture some of the lessons learned in moving from the original design to the final design. This report must include a soft copy of the entire report in HTML.

Evaluation

Performance Testing Procedures:

All machines will be operated by a member of the team. Five attempts (putts) will be allowed at each level except Level 3. Putts not falling into the 'hole' will be scored according to the number of the bin into which they fall.

Level 1. This level will consist of two (2) rounds of five (5) putts each. Scoring for grading purposes will occur at this level.

Level 2. In this round the first shot is worth half as much as the succeeding shots.

Level 3. At this level, scoring is the reverse of normal golf: the higher the score the better. At this level, only balls entering the 'hole' count towards the score.

Level 4, This level is a Sudden Death Play-off.

Grading Criteria:

Each group will receive 5 grades for this project.

☐ 1) Concept (20%) This will be based on the technical merit of the design and coding for the machine. Included in this grade will be evaluation of the appropriateness of the solution, as well as innovative hardware, software and use of physical principles in the solution.

☐ 2) Implementation (20%) This will be based on the prototype displayed at the evaluation session. Included in this grade will be evaluation of the physical appearance of the prototype and quality of construction. We will not presume to judge true aesthetics, but will concentrate on craftsmanship and finished appearance.

☐ 3) Report (10%) This will be based on an evaluation of the written report. It will be judged on clarity of explanations, completeness and appropriateness of the documentation.

☐ 4) Performance (20%) Based on the results of the performance testing during the Final Presentation.

☐ 5) Interim Reviews (30%) Based on the three project milestone reviews.

☐ **Note** : This is a Mechatronics Project Design Activity. Grading in this class is based on complete system design and function. Therefore, a "beautiful" mechanical or electronic sub-system or elegant code is not a successful project if any of the other parts of the machine fail. Be sure to allocate resources (time and people) to all aspects of this project.

Index

0104, TXB0104, 391–394
1's complement, 33
1250, MMA1250, 277–278, 280
12F609, PIC12F609 microcontroller, 24–26
14490, ON14490, 284
16-bit microcontrollers, 13
16C84, PIC16C84 microcontroller, 117
16F690, PIC16F690, specifications, 406
2's complement, 33
2003, ULN2003 specifications, 407, 408
232, EIA-232 standard, 119
232, RS-232 standard, 118–119
2380, FDMS2380, 407
293B, L293B, 411–413
297, L297, 614–615
2-D part drawings of components, 682
2N7000 data sheet, 222
−3 dB point, 171, 339
3000, MAX3000E, 391
31, TIP31/32 data sheet, 220
32, TIP31/32 data sheet, 220
324, LM324, 260–263, 266
32-bit microcontrollers, 13
339, LM339, 402, 404
3479, MC3479 stepper motor driver, 379, 381, 383, 387
3985, Allegro A3985, 567
485, RS-485 standard, 119–120
4-bit microcontrollers, 12
5206, TLE-5206, 566
5305, IRF5305 data sheet, 228
555 timer
  astable configuration, 432–434
  inside of a, 432
  monostable configuration, 434–435
  uses, 435
6144, LM6144, 266–267
6484, LMC6484, 266–267
6816, MAX6816, 284
7400, SN7400, 399
74HC04, 400
802.11, WiFi (IEEE 802.11a/b/g/n), 127
802.15.4, Zigbee (IEEE 802.15.4), 127, 389–390
8402, DVR8402, 566–567
8-bit microcontrollers, 12–13
808, VN808CM, 408–409
9S12C32, Freescale MC9S12C32 16-bit microcontroller, 21–24, 130, 371–374, 385

A

Absolute accuracy
  for A/D converter, 447
  for D/A converter, 450
Absolute error, 447, 450
Abstraction, 76–77
Acceleration, measurements of, 272–273
Accelerometers, 272–273
AC-coupled square wave signal, 349
AC coupling, 194
AC coupling capacitor, 336
Accuracy
  for MMA1250 accelerometer, 278
  of sensor, 274
Active filters
  phase delay, 352–353
  response characteristics, 353–354
  shape of a signal waveform, 353
  simplified Sallen-Key design procedure, 356–360
  topologies, 354–360
  transient response, 354
Ada language, 48, 51
A/D converter
  clock, 142
  flash, 460–462
  sigma-delta, 464–466
  single slope and dual slope, 458–460
  successive approximation register, 463–464
Address bus, 14, 17
A/D or D/A converter performance
  accuracy, 444
  ideal, 444–445
  least significance bit, 443
  maximum sampling rate, 444
  missed code, 444
  monotonicity, 444
  resolution, 442–443
  settling time, 444
  sources of error, 446–450
  temperature coefficient, 444
Alberta Printed Circuits, 697
Aliasing, 441
Allegro A3985, 567
All Electronics, 696–697
Alternating current (AC), 147
Amazon.com
Amperes, 146
Ampere's law, 532
Amplification
  multistage, with AC coupling, 339
  multistage, with DC coupling, 338
Amplifier
  difference, 240–241
  instrumentation, 241
  inverting, 235–236
  noninverting, 233–235
  summing, inverting, 241–242, 451–453
  summing, noninverting, 453–455
  trans-resistive, 242–243
  unity gain buffer, 238–239
Amplitude modulation (AM), 121–122
Amplitude shift keying (ASK), 121
Analog comparator subsystem, 26
Analog-to-digital (A/D) converter subsystem, 141–143
Analog-to-digital converter (A/D converter), 9, 17, 23, 271, 439
  Freescale MC9S12C32 16-bit microcontroller, 24
AND operator, 40–41
ANSEL(H) register, 132
Antialiasing filter, 442
Application specific integrated circuits (ASICs), 8–9
Arabic numerals, 32
Arithmetic logic unit (ALU), 12–13, 34
ARM architecture, 13
Armature, 533
ASCII character code, 10
Assembler, 48
Assembly language, 47–48
Assembly time, 46
Associative law, 40
Asynchronous communication, 16, 23
Asynchronous serial communication, 108, 111–117
Attenuation, 170
Automobiles, 1
Autonomous Golf Putter project
  candidates and alternatives, 734–742
  completed design, 749–752
  concept evaluation, 743–746
  description of, 732
  game play, 770
  implementation phase, 746–749
  morphology charts, 742–743
  performance results, 752–753
  purpose, 769–770
  putters, 770
  report, 771–772
  rules, 771
  safety, 771
  student experiences, 753–754
  system requirements, 732–733

B

Back-EMF, 536
Balanced differential signaling standard, 119
Band-pass filter, 340
Band-reject filter, 341
Bandwidth, 281
  of a sensor, 276–277
Base-2, 11
Base of BJT, 206
BASIC programming language, 33, 37, 40, 48–50
BASIC Stamp language, 51
Battery, 493
  alkaline, 499
  capacity, 496
  charging of, 502–506
  closed-circuit voltage, 497
  constant current stage, 503
  constant voltage stage, 503
  C-rate, 496
  cutoff voltage, 497
  discharge curve, 496–497
  energy density, 496
  float charge, 503

Battery, *(continued)*
internal resistance, 498
lead acid, 503
Li-ion and Li-poly, 501–502
memory effect, 504–505
midpoint voltage, 497
NiCd and NiMH, 501, 503
open-circuit voltage, 497
performance and characteristics, 495–498
primary, 499–500
safety and environmental issues, 506–507
secondary, 500–502
self discharge rate, 497
silver oxide, 499
SLA, 501, 505
trickle charging, 502
voltage depression effect, 505
volumetric energy density, 496
zinc-air, 499
Beginners All-purpose Symbolic Instruction Code. *See* BASIC programming language
Bessel filters, 354
Binary digit (bit), 11
Bipolar junction transistors (BJT), 205–212, 555
NPN, 559
reading data sheet, 219–221
Bit 1, 11
4-bit binary number, 34
Bit-parallel communications, 107, 117–118
Bit rate, 107
Bit-serial communications, 107–108
Bits per second (bps), 111
Bit time, 111
Bit-wise operators, 40–41
Blocking code, 58
Bluetooth (IEEE 802.15.1), 127
Bode magnitude plot, 162
Bode phase plot, 352
Boolean algebra, 29, 40
Boost regulator, 482–483
Boot-loader program, 18
Boron, 199
Bounce time, 284
Brainstorming, 704
Brush bounce, 535
Brush drag, 535
Brushed permanent magnet DC motor subfractional horsepower, 531–535
Brushes, 534
Brushless DC (BLDC) motors
brushed DC motor vs., 597–598
commutation of, 594–596
construction, 588–590
delta connection, 589
driving, 593–594
Hall effect sensors, 590
motor driver ICs, 596–597
operation, 590–593
phases in, 589
star connection, 589
wye connection, 589
Buck regulator, 482
Buffering, 239
Bus, 13, 109
Butterworth filters, 354
Bypass capacitors, 513
Bypassing, 194
Byte, 36
Byte codes, 52

C

CAN (Controller Area Network), 16
Canister load cells, 304
Cantilever load cells, 304
Capacitively coupled noise
origin, 516–517
reduction, best practices, 522
reduction at receiver, 521–522
reduction at source, 519–520
reduction of, 517–518
reduction of coupling mechanism, 520
shielding for, 520–521
Capacitive coupling, 511
Capacitive isolation, 527
Capacitive proximity sensors, 314
Capacitive sensors
oscillator circuit, 291–292
step input, 290–291
Wheatstone bridge circuits, 292
Capacitors, 155–159
aluminum electrolytic, 188–189, 193–194
ceramic, 193
ceramic disk, 187
choosing of, 194–195
construction basics, 185–186
dielectric absorption (DA), 185
dynamic behavior of, 156
EIA tolerance codes, 191
electric double layer, 190–191
equivalent resistance for, 158
equivalent series inductance (ESL), 185
equivalent series resistance (ESR), 185
film, 190, 194
and filtering, 194
in hydraulic models, 156
labeling of, 191–193
metalized film, 190
monolithic ceramic, 188
multilayer ceramic (MLC), 188
polar and nonpolar, 186–187
principle of, 157
real, 185–186
in parallel, 157
in series, 157
simple parallel plate, 156
super, 190–191
tantalum, 189–190, 193–194
and time-varying signals, 158–159
water analogy for, 155
Cast operation, 37
C code, 757
CdS photo-cells, 300–301
Central processing unit (CPU), 8, 46–47
arithmetic logic unit (ALU), 12–13
number representation in, 11–12
Char, 36
Charge pumps, 481–482
Charge (Q), 146
Chebyshev filters, 354
Chopper drive
of solenoids, 580
of stepper motors, 622
Circuits
basic RC circuit configurations, 162–163
capacitors, 155–159
comparator, 246–248
complete, 147–148, 148
high-pass RC filter behavior in frequency domain, 172
high-pass RC filter behavior in time domain, 166–167
high-pass RC filter with DC bias, 173–174
inductors, 159–161
Kirchhoff's Current Law (KCL), 149–150
Kirchhoff's Voltage Law (KVL), 149–150
low-pass RC filter behavior in frequency domain, 169–171
low-pass RC filter behavior in time domain, 163–165
measuring current, 177–178
measuring voltage, 176–177
multitransistor, 217–218
open, 148
operational amplifiers (op-amps) 233–244
power supply outputs, 149
radio frequency (RF) and switching power supply, 161
real capacitors, 184–195
real resistors, 178–184
real time measurements, 176–178
resistance, 150–154
RL, behavior in time domain, 167–169
short, 148
simulation tools, 174–175
symbols, 148
Thevenin equivalent, 154–155
time and frequency domains, 161–162
transistor, 217–218, 224–225
voltage sources, 175–176
Clear a bit in a variable, 41
Clock frequency (f_{CK}) specification, 380
Closed-loop control
ad hoc methods, 667–670
bang-bang control, 653
choosing of gains, 663–665
control law, 655–658
derivative controls, 661–663
disturbance rejection, 659–661
implementing a discrete integrator, 657
implement the combined proportional plus integral (PI) control law, 658
linear closed-loop, 654–665
loop rate selection, 666–667
on-off closed-loop, 653–654
open-loop, 652–653
system type and need for integral control, 666
terminology, 651–652
Ziegler-Nichols approach to tuning the PID gains, 664–665
Closed-loop negative feedback, 232
CMOS logic families, 426–427
CMOS technology, 399
Cohesiveness, 80
Cold reference compensation, 307

Collector, 206
Combinatorial logic, 416–417
AND and OR gates, 417
decoder, 420–421
describing microcontroller subsystems using, 418–419
digital comparator, 419
digital multiplexer, 419–420
NOT operation, 417
truth table, 417–418
Common mode gain, 345
Commutative law, 40
Commutator, 533
Comparator, 245–248
packaging, 265
reading data sheet of, 264–265
with hysteresis, 247
Compiled language, 50
Compilers, 50–51
Compile time, 46
Compiling process, 50
Complement notation, 33
1's complement, 33
2's complement, 33
Complete circuit, 147
COM ports, 118
Concurrent design, 713–714
Conductive coupling, 511
Conductively coupled noise, 511
decoupling, 512–514
origins of, 511–512
reducing, best practices for, 516
reducing at receptor, 515–516
reducing coupling of, 514–515
reduction of, 512
Conductors, 198
Constant current circuit, 287
Continuous conversion mode, for A/D converters, 143
Control registers, 16, 129–130
Converters
A/D or D/A performances, 442–450
digitizing continuous signals, 439–442
interfacing between digital and analog domains, 438–439
resolution of, 441
Convolution, 363
Core, 8
Corner frequency, 171
Corner frequency (F_C) of filter, 171, 339
Coulomb (C), 146
Counters, 16, 133
Counter/timer subsystem, 133
Counting and representing numbers in different bases, 29–33
Coupling, in software, 80
Coupling channels, 511
C programming language, 39–40, 51
C++ programming language, 51
Critical path, 711
Crystalline silicon, 199
Current, 145–147
Current gain of a transistor, 208
Current ripple, 569
Current source, 147, 287
Current transfer ratio (CTR), 526
Cyclic redundancy checks (CRCs), 112

D

D/A converter
generation of analog voltages, 450–451
R-2R ladder, 456–458
string, 455–456
summing amplifier circuit as, 451–455
Dark current, 295
Darlington pair, 210–211
Data, 10
Data bus, 13
Data direction register (DDR), 130
Data transfer rate, 111
DC coupling capacitor, 336
Dead time, 411
Debugging, 63, 67, 722–726
Decimal system, 29
binary representation of, 39
Decomposition, 75–76
Decoupling, 194
Decoupling capacitors, 513
De-scoping, 711
Design for assembly, 713
Design for manufacturability (DFM), 713
Design for serviceability, 713
Design for test, 713
Detailed design, of software, 81
Detent torque, 602
Dibit, 122
Difference amplifier, 240–241
Differential nonlinearity error (DNL)
for an A/D converter, 447
for a D/A converter, 450
Differential signals, 344
Digikey Corporation, 696
Digital filters
moving average, 361–362
signal processing, 362–364
synchronous sampling, 364–365
Digital input/output (I/O) pins, 8, 16, 130
Digital inputs
current requirements, 373–374
pull-ups and pull-downs, 374–378
timing requirements, 378–381
voltage requirements, 370–373
Digital I/O system, 8–9. *See also* Digital inputs; Digital outputs
absolute maximum ratings, 370
clock input waveforms, 380
evaluation of compatibility, 384–386
false state, 367
hold time, 369
ideal behavior for digital devices, 368
interconnecting incompatible devices, 389–394
logical states, 367
propagation delay, 369
pull-ups and pull-downs for disconnectable and indeterminate inputs, 386–389
reading of data sheet, 369–370
real behavior for digital devices, 368–369
recommended operating conditions, 370
rise times and fall times, 369
setup time, 369
true state, 367
worst-case design principles, 370
Digital outputs
half-bridges and full-bridges, 409–411
high-side drivers, 408–409
low-side drivers, 405–408
maximum thermal resistance of a heat sink, 414
open-collector/open-drain, 401–403
rise over ambient (ROA) temperature, 413
thermal design issues, 411–414
three-state, 403–405
timing specifications, 383–384
totem-pole style, 398–401
voltage and current specifications, 381–383
Digital signal processors (DSPs), 7–9, 19–20, 26–27, 352, 360–365
Digital-to-analog converter (D/A converter), 17, 439
Digital voltmeter (DVM), 176
Digitized signal, 441
Diodes
hydraulic analogy for, 199, 200
forward bias, 200
forward voltage (V_f), 200
magnitude of V_f, 201
leakage current, 200
light emitting (LED), 204
photo, 205, 293–295
purpose of, 199
reverse-breakdown voltage (Vr), 200
reverse recovery period, 201
Schottky, 201–202
voltage vs. current (or *V-I*) characteristic for, 200
Zener, 202–203, 474, 559–560
Diode snubber, 555
Diode thermometer, 310
Direct current (DC), 147
Distributive law, 40
Divide-by ratio, 133
Dopants, 199
Doping, 199
Drift, 258
Driptech irrigation tubing manufacturing system, 1–2
Duty cycle, 567
DVD player, 57–58, 123
Dynamic braking, 563
Dynamic error, 448
Dynamic performance of an ADC, 448

E

EC++ programming language, 51
Effective resistance, 151
EIA-232 standard, 119
Electrically erasable programmable read only memory (EEPROM), 10
Electric field, 146
Electro-chemical cell, 147, 493–495
Electronic engine controllers, 137–138
Electronic Goldmine, 696
Electronic Industries Alliance (EIA), 119, 193
Embedded control, 8
Embedded microcontrollers, 1

Emitter, 206
Emitter–detector pair modules, 299–300
Equivalent resistance of a capacitor, 158
Evaluation boards, 27
Even parity, 112
Event, 58
Event checkers, 58–62
Event driven programming
 building, 62–63, 69
 definition, 58
Executing instructions, 49, 50
Exponent, of floating point data type, 35

F

Fairchild Semiconductor FDMS2380, 407
Falling edge, 379
Falling-edge active, 108
Farad, 155
Faraday's law, 536
Fast recovery time, 556
FastTrack Schedule, 709
Feedback loop, 234
Field programmable gate arrays (FPGAs), 8–9
Filters
 active, 352–360
 band-pass, 340
 band-reject, 341
 Bessel, 354
 Butterworth, 354
 Chebyshev, 354
 high-pass, 172
 linear phase, 354
 low-pass, 170
 maximally flat, 354
 notch, 341
 order, 340
 passive, 342–343
 quality factor (Q), 341
 Sallen-Key, 355
 voltage controlled voltage source (VCVS), 355
Filter kernel, 363
Finite element analysis (FEA), 676
Finite impulse response (FIR) filter, 363
Finite state automata (FSA), 65
Finite state diagram (FSD), 66
Finite state machine (FSM), 65, 66
 states, 66
 transitions, 66
Fixed size variables, 36–37
Flash EEPROM, 10, 15, 18, 21
Flat response in a pass-band in filtering, 339
Floating
 inputs, 374
 power supply outputs, 149, 489
Floating point number, 35
Flyback, 169, 554
Flyback diode, 555
Flyback regulator, 484
Flying capacitor voltage converters, 482
Foamcore, 676–677
Force-balance accelerometer, 273
Forth interpreter, 51–52
Forth language, 51
Forward bias, of diode, 200
Forward voltage (V_f), of diode, 200
Fourier series for a square wave, 349
Fourier series representation, 162
Fourier theorem, 158
Framing bits, 112
Framing error, 115
Free-running counters, 134
Freescale MC9S12C32 16-bit microcontroller, 21–24, 130
Freescale MC9S12 family, 131
Frequency domain, 161–162
Frequency modulation (FM), 122
Frequency shift keying (FSK), 122
Fuel gauge chip, 506
Full-bridges, 409–411
Full-duplex communications, 120
Fuse, 506
Fused deposition modeling (FDM), 676, 686

G

Gage factor, of strain gage, 303
Gain error
 for an A/D converter, 446
 for a D/A converter, 449
Gain of op-amp, 231
Gantt chart, 708–709, 711
Gate, 212
Gate-source voltage (V_{GS}), 214
Gauge pressure, hydraulic analogy with voltage, 147
Gearhead efficiency, 547
Gear ratios, 547
Golden Rules for op-amps, 233
Graphical user interface (GUI), 58
Grippers, pneumatic, 635
Guard conditions, in state machines, 69

H

Half-bridges, 409–411
Half-duplex communications, 120
Hardware status register, 129
Harvard architecture, 19–21
H-bridges. *See also* Full-bridges
 commercial, 564
 configuration, 410, 561–564
 for higher current applications, 566–567
HCS12C family devices, 23
HCS12 family of controllers, 24
 Freescale MC9S12C32 16-bit microcontroller, 21–22
Heat dissipation, 477
 equivalent thermal circuit analysis, 411–414, 478–479
 thermal design issues, 411–414
Heat sink, 414, 479
 maximum thermal resistance of a heat sink, 414
 rise over ambient (ROA) temperature, 413
Henry, unit of inductance, 159
Hexadecimal (hex), 31–32
High-level languages, 48
High-pass filter, 172
High-side drivers, 408–409
Hold time, 381
Hole, 199
Hybrid compiler/interpreters, 51–53
Hydraulic analogy
 Kirchhoff's current law (KCL), 150
 Kirchhoff's voltage law (KVL), 150
 capacitor, 155
 diode, 199–200
 inductor, 159–160
 N-channel MOSFET, 213–214
 NPN BJT, 206
 resistor, 150
 voltage regulator, 469
Hysteresis
 band, 60
 in comparator, 247
 in software, 60–61
 for MMA1250, 280
 of a sensor, 274–275

I

I^2C (Inter-IC) Bus, 16, 109
Ideal current source, 147
Idle state, 108
IF-THEN-ELSE statements, 67–68
Impedance, 159
Incremental integration, 84, 721–722, 727
Inductive coupled noise
 origins, 523
 reduction, best practices, 524
 reduction at receptor, 524
 reduction at source, 523–524
 reduction of coupling mechanism, 524
Inductive coupling, 511
Inductive isolation, 527
Inductive kickback, 169
 for brushed permanent magnet DC motors, 553–555
 for solenoids, 581
 for stepper motors, 622–623
Inductive proximity sensors, 314
Inductors, 159–161
 equivalent resistance for, 160
 integral and differential forms of inductor behavior, 159
 real, 161
 and time-varying signals, 160–161
 unit measure of inductance, 159
Infineon TLE-5206, 566
Infinite impulse response (IIR) filter, 363
Information hiding, 76
Infrared Data Association (IrDA), 125
In-line assembly, 54
Input capture (IC) function, 136–137
Input high current, 374
Input high voltage, 370–371
Input hysteresis, 372
Input leakage current, 373
Input low current, 373–374
Input low voltage, 370–371
Input/output hardware (I/O), 10
Input/output register(s), 130–131
Input time rise, 379
Insanity, 721
Instruction mnemonics, in assembly language, 48

Instruction set of a CPU, 46
Instrumentation amplifier, 241
Insulators, 198
Int data type, in C, 36
Integers, 35
Integral nonlinearity error (INL)
  for an A/D converter, 447
  for a D/A converter, 450
Integrated circuits (ICs), 198
Integrated development environments (IDEs), 53
Integration plan, for software, 80
Intelligent liquid crystal display (LCD) module, 117
Interface electronics, for sensors, 272
Interface specifications, for software, 81
Internal oscillator circuit, 11
Interpreters, 48–50
Inter-processor communications
  amplitude modulation (AM), 121–122
  asynchronous serial, 111–117
  bit-parallel, 117–118
  bit-serial, 107–108
  falling-edge active, 108
  frequency modulation (FM), 122
  idle state, 108
  I^2C (Inter-IC) Bus, 16, 109
  with IR spectrum, 123–126
  with light, 123–126
  limited bandwidth channels, 120–121
  master and slave, 108–109
  Master In Slave Out (MISO), 109
  Master Out Slave In (MOSI), 109
  media and standards, 106
  Microwire, 109
  modulation techniques, 121–123
  over a radio, 126–127
  phase modulation (PM), 122–123
  quadrature amplitude modulation (QAM), 123
  RF data links, 127
  RF networks, 127
  RF remote controls, 126–127
  rising-edge active, 108
  RS-232 standard, 118–119
  RS-485 standard, 119–120
  Serial Peripheral Interface (SPI), 109
  signaling level standards, 118–120
  slave select line, 109
  SPI modes of operation, 110
  synchronous serial, 108–111
  TTL/CMOS standard, 118
Interrupts, 143–144
Interrupt Service Routine (ISR), 143
Inverting amplifier, 235–236
Inverting comparator, 246
Inverting input, of an op-amp, 231
Inverting summer, 242
IR communications, 123–126
  IrDA standard, 125–126
  IR light emitting diode (LED), 124
  preamble, 124
  pulse-coded waveform, 125
  shift-coded waveform, 125
  space-coded waveform, 125
Isolation technologies
  capacitive, 527
  comparison of technologies, 528
  inductive, 527
  optical, 525–527
Isolation voltage, 525

J

Jameco Electronics, 696
Java compiler, 52
Java language, 51
Java virtual machine (JVM), 52
Johnson noise, 276
Joules, 146

K

Kickback diode, 555–556, 581, 622–623
Kilobits per second (kbps), 111
Kirchhoff's Current Law (KCL), 149–150
Kirchhoff's Voltage Law (KVL), 149–150

L

LaserCAMM, 676
Laser cutting tools, 680–681
Lead zirconate titanate (PZT), 187
Leakage, of charge in a capacitor, 194
Least significant bit (LSB), 112
Least significant digit, 30
Left shift operation, 39
Limited bandwidth channels, 120–121
Linear actuators, pneumatic, 633–634
Linear closed-loop control, 654
Linear drive, in solenoids, 580
Linear phase filter, 354
Linear region
  for BJTs, 207–208
  for MOSFETs, 214
Line filter, 491
Line voltage
  for power supplies, 488, 491
  electrical isolation from, 492, 525
Lithium ion (Li-ion) and lithium polymer (Li-poly) batteries, 501–502
LM324, 260–263, 266
LM339, 402, 404
LM6144, 266–267
LMC6484, 266–267
Load cells, 304–305
Logic chips, 426–427
  expanding inputs using multiplexer, 427–428
  expanding inputs using shift registers, 429–430
  expanding outputs using decoder, 428–429
  expanding outputs using shift registers, 431–432
  SPI (or equivalent) subsystem, using to interface with shift registers, 432
Logic-level MOSFETs, 213
Logic-level outputs, 405
Long integer data type, in C, 36
Loopholes, 703
Lorentz force law, 536
Low-level programming languages, 48
Low-pass filter, 170
Low-side drive, 206, 405–408
LTC6088, 266–267

M

Machine language of a computer, 46
MacOS, 58
Magnetic field sensors, 310–314
Mantissa, 35
Marking state, in serial communications, 111
Masking, of bits, 41
Master In Slave Out (MISO), 109
Master Out Slave In (MOSI), 109
MathCad, 50, 174
Mathematica, 50, 174
Mathematics and number manipulation, of microcontrollers
  Boolean algebra, 40
  data sizes and ranges of values, 36
  data types, 35–36
  fixed size variables, 36–37
  individual bits or groups of bits, 40–42
  modulo arithmetic, 37–38
  number base and counting, 29–33
  representation of negative numbers, 33–34
  shortcut operations, 39
  sizes of common data types, 36
  testing for multiple bits or individual bits, 42
Matlab, 50, 174
Maximally flat filters, 354
Maxim Integrated Products
  MAX3000E, 391
  MAX6816, 284
MC3479 stepper motor driver, 379, 381, 383, 387
MC9S12C32 microcontroller, 26, 373, 385
McMaster-Carr, 697
Measurand, 270
Mechatronics
  curriculum, 4
  defined, 2
  engineer in, 4
  philosophy, 3
Memory map, 14–15
Memory space, 10
Metal oxide semiconductor field effect transistor (MOSFET), 212–215, 519
  N-channel, 406–407
  P-channel, 408
  peak detection, 347
  reading data sheet of, 221–224
  totem-pole output stage, 399
Micro
  meaning of, 7
  instructions to, 10
Microchip PIC12F609 microcontroller, 24–26
Microcomputers, 8
Microcontrollers, 1, 8, 368
  address bus, 14
  architecture of, 10–11
  arithmetic logic unit, 12–13
  4-bit, 12

Microcontrollers, *(continued)*
- 8-bit, 12–13
- 16-bit, 13
- 32-bit, 13
- data bus, 13
- Harvard architecture, 19–21
- major manufacturers and distributors of, 27
- memory locations, 14–15
- number representation in, 11–12
- real world examples, 21–26
- subsystems and peripherals, 15–17
- von Neumann architecture, 17–19

Microfarads, 155
Microprocessors, 8
Microsoft Project, 709
Microwire, 109
Mixed signal device, 432
Mnemonics, 47–48
Modem (MOdulator DE-Modulator), 121
Modulation techniques, 121–123
Module level variables, 82–83
Modules, in software, 75, 80
Modulo arithmetic, 37–38
Monolithic ceramic capacitors, 337, 344, 514
Morphological chart, 704–706, 742–743
Morse code receiver, developing
- AddDash() function, 91
- AddDot() function, 91
- Button module interface specification, 89–90
- CheckMorseEvents() function, 90
- ClearMorseChar() function, 91
- compiled language implementations, 102
- DecodeMorseChar() function, 91
- detailed design, 92–102
- digital signals, interpretation of, 85–86
- InitDisplay() function, 91–92
- initialization sequences, 100–101
- InitMorseElements() function, 90
- integration of subsystems, 103–104
- LCD Display module interface specification, 91–92
- module interface specifications, 89–92
- Morse code, defined, 84–85
- Morse Decode module interface specification, 91
- Morse elements and printable characters, 99–100
- Morse Elements functionality, 93
- Morse Elements module interface specification, 90
- performance specifications, 89
- pseudo-code development, 100, 102
- requirements, 85
- software architecture, 86–89
- StartCalibration() function, 90
- system architecture, 85–86
- TestCalibration function, 93
- unit testing, 102–103
- Waiting for End of Character functionality, 93
- WriteChar() function, 92

Most significant bit (MSB), 112
MPLAB, 26
Multidrop communications, 120
Multimaster system, 109
Multiple channel conversions, of A/D converters, 143
Multiply-accumulate (MAC) operation, 363–364

N

NAND logic chip, 399
National Semiconductor's Microwire, 109
N-channel MOSFETs, 213, 406–143, 409
Negative temperature coefficient (NTC) thermistors, 308
Network diagrams, 711
Newark, 696
Nibble, 36
Nickel-cadmium (NiCd) batteries, 501, 503
Nickel-metal hydride batteries (NiMH), 501, 503
Noise
- coupling channels, 510–511
- dominant paths of, 341
- of a sensor, 275–276
- specifications for the MMA1250, 280

No-load speed, of brushed permanent magnet DC motor, 539
Nonblocking code, 58
Noninverting amplifier, 233–235
Noninverting comparator, 246
Noninverting input, of an op-amp, 231
Nonlinearity error
- for an A/D converter, 446
- for a D/A converter, 450

Nonlinearity for the MMA1250, 280
Nonlinearity of a sensor, 274
Non-volatile memory, 10
Notch filter, 341
NOT operator, 40–41
NPN Darlington transistors, 407
N-P-N sequence, 206
NPN transistors, 206, 406
N-type semiconductor, 199
Number base for a representation, 29–33
Numerals, 30
Nyquist limit, 442
Nyquist Theorem, 441

O

Object-oriented programming (OOP) language, 51
Odd parity, 112
Offset adjustments
- by AC coupling, 336–337
- amplification relative to, 334–335

Offset error
- for A/D converter, 446
- for D/A converter, 449

Ohms (Ω), 150
Ohm's law, 150–151, 154, 345, 511
ON14490, 284
On-chip analog-to-digital converters, 17
On-off closed-loop control, 653–654
On-off keying (OOK), 121
On-off-shift keying (OOSK), 121
Op-amps, 231
Open circuit, 148
Open-circuit voltage, 154
- of a battery, 497

Open collector outputs, 246, 401–403
Open drain outputs, 246, 401–403
Open-loop control, 652–653
Operational amplifiers. *See also* Real op-amp, characteristics
- Absolute Maximum Ratings section of a data sheet, 259
- circuits, 233–244
- commercial, 265–268
- common mode rejection ratio (CMRR), 263
- computation with, 243–244
- definition, 231
- difference, 240–241
- Electrical Characteristics section of a data sheet, 259–263
- gain (G), open-loop, 232
- Golden Rules, 233
- ideal, 232–233
- input bias current specification, 262
- input common mode voltage, 262
- input offset current specification, 262
- input offset voltage specification, 262
- internal current drawn by, 263
- inverting, 235–237
- large signal voltage gain, 263
- maxima, minima, and typical values of a data sheet, 259
- modern, 264
- negative feedback, 232
- noninverting, 233–235
- op-amps, 231–232
- open-loop gain, 232
- output current specification, 263
- packaging, 263–264
- reading data sheet of, 258–264
- saturation in, 244
- summer, 241–242
- trans-resistive, 242–243
- typical application section, 264
- unity gain buffer of, 238–240

Opto-interrupters, 299
Opto-isolators, 525–527
Order of filter, 340
OR operator, 40
Output compare (OC) system, 140–141
- function, 135–136

Output fall time, 383
Output high current, 382
Output high voltage, 381
Output impedance, 175
Output leakage current, 403
Output low current, 382
Output low voltage, 381
Output rise time, 383
Output saturation voltage, 403
Overdrive, of comparator, 265
Overflow condition, 34, 134–135
Oversampling, in a sigma-delta A/D converter, 465

P

Parallax's BASIC Stamp (PBASIC) language, 53
Parallel input/output (I/O) peripherals, 16

Parallelogram load cells, 304
Parallel resistors, 151–153
Parity bit, 112
Pass-band, 339
Pass element, in linear voltage regulators, 474
Passive filters, 342–343
P-channel MOSFETs, 408
Peak and hold technique, 579
Peak detection, 346–347
Pentium microprocessor, 9
Perforated circuit board, 692
Peripheral component interconnect (PCI) buses, 107
Peripheral drivers, 406
Peripherals, microcontroller
analog-to-digital (A/D) converter subsystem, 17, 141–143
automating of A/D converter subsystems, 143
clearing the source of the interrupt, 144
control registers, 16, 129–130
counters, 16
data direction register (DDR), 130
digital input/output (I/O) pins, 16, 130–132
input capture (IC) system, 136–137
input/output register(s), 130–131
interrupts, 143–144
interrupt service routine (ISR), 143
output compare (OC) system, 135–136
polling, 143
pulse width modulation (PWM), 138–141
serial I/O, 16–17
shared function pins, 131–132
timer subsystems, 16, 132–138
Permanent magnet brushed DC motor
back-EMF, 536
bidirectional control of, 561–567
characteristics constants, 536–537
characteristics equations for constant voltage, 538–540
construction, 533–535
electrical model, 535
gearheads, 547–548
generator effect, 536
inductive kickback, 552–560
motor efficiency, 543–546
power characteristics, 541–542
speed control with pulse width modulation, 567–570
Phase modulation (PM), 122–123
Phase shift keying (PSK), 122
Philips' I^2C (Inter-Integrated Circuit) Bus, 109–110
Photo-cells, 300–301
Photo-detectors, 300–301
Photo diodes
definition, 205, 293
interface circuit, 293
response time, 295
semiconductor junction, 205, 293
use, 295
V-I characteristics, 205
Photo-transistors, 211–212, 295–299
interface circuits, 297–299
PIC12F609 microcontroller, 26, 396
PIC16F690, specifications, 406
PIC16C84 microcontroller, 117
PIC machine code, 47
PIC microcontrollers, 46
Picofarads, 155
PID controller, 662
Piezo actuators, 637–639
bimorph, 640
characteristics, 638–639
stacked, 639–640
Piezo buzzers, 642–644
Piezoelectric effect, 637
Piezo motors, 641–642
Pins, 16
Place, in base-N numbers, 30
Platform independent, 52
Pneumatic and hydraulic systems
actuators, 632–635
direct acting solenoid valves, 627–630
grippers, 635
piloted solenoid valves, 630–631
servo valves, 631–632
solenoid valves, 627–631
PNP bipolar transistors, 205–210, 408
P-N-P sequence, 205–206
PNP transistor, 205–210, 408
Poisson's ratio, 303
Poles
of a filter, 340
of a switch, 281
Polling, of bits, 143
Ports, 16
Freescale MC9S12C32 16-bit microcontroller, 23
PIC12F609 microcontroller, 26
Position sensors
capacitive displacement, 321–322
flex, 323
inductive pickups/gear tooth, 319–320
optical encoders, 317–319
potentiometers, 315–317
reflective infrared, 320–321
ultrasonic displacement, 322–323
Positive temperature coefficient (PTC) thermistors, 309–310
Potential of electric field, 146
Potentiometer, 183–184, 315–317
Power, in electrical terms, 146–147
Power drivers, 369, 406
Power MOSFETS, 214–215
Power requirements, 468–469
Power sources, 468–469
Power supplies, 147
linear, 487–490
switching, 490–493
Preamble, 124
Pressure, as hydraulic analogy for voltage, 147
Primary coil, in transformer, 485
Printed circuit board (PCB), 513
Private functions, in software, 81
Program design language (PDL), 77
Programmable logic, 9, 368
Programmable systems-on-chips (PSoCs), 9
Programming languages. *See also* Ada language, assembly language, BASIC programming language, C programming language, C++ programming language, EC++ programing language, Forth language, Java language, PBASIC language, Smalltalk language
assembly language, 47–48
backward compatible, 47
compilers, 50–51
high-level, 48
hybrid compiler/interpreters, 51–53
integrated development environments (IDEs), 53
interpreters, 48–50
low-level, 48
machine, 46–47
modern software development process, 45
selection of, 54–55
Program structures
event checkers, 58–62
event driven programming, developing of, 62–63
event driven programming model, 58
example, 64–65
service routines, 62
state machines, 65–71
Project management, defined, 708
Project planning and management, 699
brainstorming, 704
generating design candidates, 703
communication and documentation, 714–715
design concept evaluation, 707
management tools, 708–714
morphological chart, 704
need for, 700
pitfalls and suggestions, 715
planning and executing project, 700–708
specifications, 707–708
system requirements, 701–703
Propagation delay, 384
Proximity sensors, 314–315
Pseudo-code, 59, 77. *See also* Program design language (PDL)
PS/2 protocol, 9
P-type semiconductor, 199
Public interface, in software, 81
Public functions, in software, 81
Pull-down resistor, 374
Pull-up resistor, 374
for open-collector output, 401–402
for comparator output, 246
Pulse-coded waveform, 125
Pulse width modulation (PWM)
as D/A output, 450–451
microcontroller peripheral, 17, 23, 138–141
for speed control of brushed permanent magnet DC motors, 567–570
for stepper motor microstepping, 612–614
Push-pull (totem pole) output, 245

Q

Quadrature amplitude modulation (QAM), 123
Quality factor (Q), 341
Quantization error, 441
  for an A/D converter, 446
Quickparts.com, 698

R

Radiative coupling, of noise, 511
Rail-splitter, 337
Random access memory (RAM), 11
Rapid prototyping
  circuit boards, 689–692
  concept, 674
  2-D part drawings of components, 682
  of electrical systems, 687–696
  foamcore, 676–677
  fused deposition modeling (FDM), 686
  hot-melt glue, 678–680
  laser cutting to, 680–681
  PCBs, 692–694
  philosophies, 674–675
  schematic capture and circuit simulation tools, 687–689
  selective laser sintering (SLS), 685
  soft tooling, 686–687
  soldering technique, 694–696
  solderless breadboard, 689
  of solid modeling tools, 675–676
  stereolithography (SLA), 684
  suppliers and resources, 696–698
  tab and slot construction, 682
  toys, 682–684
RC oscillator
  as microcontroller clock source, 11
  as a means of measuring capacitance, 291–292
RC servos, 636–637
Reactance, 158
Reactive systems, 65
Reading of data sheet
  bipolar junction transistors (BJT), 219–221
  comparator, 264–265
  digital I/O system, 369–370
  MOSFET, 221–222
  operational amplifiers, 258–264
Read-modify-write operations, 131
Read only memory (ROM), 10
Real op-amp, characteristics. *See also* Operational amplifiers
  input bias current and input offset current, 253–254
  input common mode rejection ratio (CMRR), 257–258
  input common mode voltage range, 256–257
  input impedance, 254–255
  input offset voltage, 256
  noninfinite gain, 251–252
  open-loop gain, variation with frequency, 255–256
  output impedance, 255–256
  output voltage swing, 257
  power supplies, 256
  temperature effects, 258
  voltage sources, 255–256
Recursive filter, 363
Reduced instruction set computers (RISC), 47
Reflective optical sensor, 299
Registers, 12
Regression testing, 80
Reload value, in timer systems, 134
Resettable counters, 134
Resistance, 150–154
Resistive sensor, interface circuits for
  measuring resistance using a current source, 287–288
  in a voltage divider, 286–287
  Wheatstone bridge circuit, 288–290
  Resistive temperature detectors (RTDs), 308
Resistors
  carbon film, 178–180
  choosing for an application, 184
  color code and standard values, 755
  construction basics, 178–179
  metal film, 180–181
  power dissipating, 181–183
  real, model, 178
  shunt, 177
  variable, 183
Resolution
  of A/D and D/A converters, 441, 442–443
  of a sensor, 276
  for MMA1250 accelerometer, 280
Reverse-breakdown voltage
  of a diode, 200
  of a Zener diode, 200, 202–203
Reverse recovery time, 556
RF communications, 126
RF data links, 127
RF networks, 127
RF remote controls, 126–127
Right shift operation, 39
Ripple
  in filters, 339
  with PWM drive of brushed permanent magnet DC motor, 568–570
Rising edge, 379
Rising-edge active, 108
RobotShop, 697
Robot Store, 697
Roll over condition, for timer systems, 134–135
Rotary actuators, pneumatic, 634–635
Rotor, of brushed permanent magnet DC motor, 533
Round-off errors, 35
RRIO CMOS op-amps, 347
RS-232 standard, 118–119
RS-485 standard, 119–120
Run time, 46

S

Sallen-Key filter, 355
Sampling, 439–440
Sampling interval, 440
Saturation
  operational amplifiers, 244
  of bipolar junction transistors, 207
  of Darlington pairs, 211
  of MOSFETs, 214
Saturation region
  of bipolar junction transistors, 207
  of MOSFETs, 214
Schmitt trigger, 247–248, 284
SCI (Serial Communications Interface), 23, 113, 132
Sealed lead acid (SLA) batteries, 501, 505
Secondary coil, in transformer, 485
Seebeck effect, 306
Selective laser sintering (SLS), 685–686
Semiconductor devices
  bipolar junction transistor (BJT), 205–212
  choosing between BJTs and MOSFETs, 216–217
  diodes, 199–205
  metal oxide semiconductor field effect transistor (MOSFET), 212–215
  multitransistor circuits, 217–218
  reading transistor data sheets, 218–225
Semiconductors, defined, 198
Semiconductor temperature sensor, 310
Sensitivity
  for MMA1250 accelerometer, 278
  of sensor, 274
Sensors
  acceleration, 323–326
  as voltage source, 147
  capacitive, 290–292
  design, 271–281
  force, 326–328
  light, 293–301
  magnetic field, 310–314
  microcontrollers and, 270–271
  performance characteristics, 273–281
  placement of, 286
  position, 315–323
  pressure, 328–329
  proximity, 314–315
  purpose of, 270
  resistive, 284–290
  strain, 302–305
  switches as, 281–282
  switch interface circuit, 282–284
  temperature, 306–310
Sequential logic
  counter, 423–424
  D-type flip-flop, 422
  flip-flop types, 421
  J-K flip-flop, 422
  NOR gate, 421
  Reset and Set inputs, 421
  RS latch (also SR latch, S-R flip-flop, R-S flip-flop), 421
  shift register, 425–426
Serial clock (SCK), 109
Serial communications interface (SCI), 132
Serial I/O peripherals, 16
Serial peripheral interface (SPI), 16, 109–110, 132

Serial ports, 118
Series control element, in linear voltage regulators, 474
Series resistors, 151–153
Service routine, 58
Servo accelerometer, 273
Set a bit in a variable, 41
Setup time, 381
Shape memory alloy (SMA) actuators, 644–648
Shear beam load cells, 304
Shift-coded waveform, 125
Shoot-through current, 398, 411, 563
Short circuit, 148
Shortcut operations, in microcontroller math, 39
Shorting, 148
Side effects, in software, 81
Signal conditioning
  amplification, 337–339
  basic operations, 333
  case studies, 347–350
  filtering, 339–343
  offset removal, 333–337
  peak detection, 346–347
  timing, 348–349
  using instrumentation amplifier, 344–346
Signal generators, 147
Signaling level standards, 118–120
Signal level MOSFETS, 214–215
Signal to noise ratio (SNR), 341
Sign bit, 33
Sign convention for current, 374
Sign-magnitude, 33
Silicon (Si), 199
Single-ended signaling standard, 119
Single-ended signals, 344
Single-purpose systems, 9
Sinusoidal oscillations, 158
Sinusoidal voltages, 158
Slave select line, 109
Small outline package (SOP), 409
Smalltalk language, 52
Snubbing, 169, 555–560, 622–623
Soft tooling, 686
Software architecture for a program, 79–80
Software design
  abstraction, 76–77
  decomposition, 75–76
  detailed design, 81
  developing pseudo-code, 82
  information hiding, 76–77
  integrating subsystems, 84
  interface specifications, 80–81
  intra-module organization, 82–83
  model, 78
  performance specifications, 80
  planning, 73–75
  process, 78–84
  program architecture, 79–80
  pseudo-code, 77
  requirements, development of, 78–79
  for a simple machine, 64–65
  unit testing, 84
Solar powered lighting systems, 1
Solderless breadboard, 689
Solder mask, 692
Solenoids
  ampere-turns of, 578
  applications, 582
  C- and D-frame, 575
  construction, 575–577
  driving, 580–581
  force vs. stroke plot, 579
  magnetic permeability, 577
  maximum duty cycle, 578
  maximum on time, 578
  mechanical response time, 582
  performance, 577–580
  pull-type, 575
  push-type, 576
  rotary, 576
  tubular, 575
Source program, 46
Space-coded waveform, 125
Spacing state, in serial communications, 111
Span/Dynamic Range
  for MMA1250 accelerometer, 278
  of sensor, 274
Sparkfun, 697
Speed regulation constant, 539
SPICE (Simulation Program with Integrated Circuit Emphasis), 174, 675
SPI (Serial Peripheral Interface), 16
  modes of operation, 110
SPST (single pole, single throw) switch, 282
Stall current, 540
Stall torque, 540
Start bit, in serial communications, 111
State machine(s), 65–71
  cockroach behavior as, 69–70
  implementation example in C, 763–767
  in software, 67–69
State transition diagram (STD), 66
Stator, in brushed permanent magnet DC motor, 533
Step-down regulator, 482
Stepper motor
  comparison of types, 605–606
  construction, 600–603
  drive sequences, 609–615
  driving a, 608–609
  dynamics of, 615–617
  hybrid, 604–605
  internal wiring, 606–608
  optimization of performance, 620–622
  performance specifications, 617–619
  permanent magnet (PM), 601
  phases, 603
  role of snubbing in performance, 622–623
  stepping sequences, 609–615
  variable reluctance, 603–604
Step-up regulator, 482–483
Stereolithography (SLA), 684
STMicroelectronics
  L293B, 411–413
  L297, 614–615
  L298, 614–615
  VN808CM, 408–409
Stop attenuation, in filters, 340
Stop-band, in filters, 340
Stop bit, in serial communications, 112
Strain, defined, 302
Strain gages, 302
Summer amplifier circuit, 241–242
Superposition principle, 158
Surface-mount components, 696
Switch
  definition, 281
  interface chips, 284
  interface circuit, 282
  interfacing with, 282–284
  poles of, 281
  throws of, 282
  types, 281
Switch bounce, 283
SWITCH:CASE structure, in C programming language, 68–69
Switch debounce, 283
Switching voltage regulators, 491
Switch interface chips, 284
Synchronous communication, 16, 23, 108–111
Synchronous sampling, in filtering, 364–365
Synchronous serial communications, 108–111
Systems engineering, 712–713

T

Temperature measurement example with thermistor, 271–272
Test harness, in software, 81, 84, 727
Texas Instruments
  DVR8402, 566–567
  TXB0104, 391–394
Thermal management, 411–414, 478–479
Thermistors, 271–272, 307–310
Thermocouples, 306
Thermoelectric effect, 306
Thevenin equivalent resistance, 154
Thevenin equivalent voltage, 154
Thevenin impedance, 175
Thevenin's theorem, 154
Three-op-amp instrumentation amplifier, 344
Through-hole components, 696
Throws of a switch, 282
Time domain, 161–162
Timer/counter register, 135
Timers, 16, 132–141
  definition, 132
  Freescale MC9S12C32 16-bit microcontroller, 23
  PIC12F609 microcontroller, 26
Timer subsystems
  basics, 133–134
  electronic engine controllers, 137–138
  input capture (IC) function, 136–137
  output compare (OC) function, 135–136
  overflow condition, 134–135
  roll over condition, 134–135
Torque ripple, 569, 596
Total error
  for A/D converter, 447
  for D/A converter, 450
Totem-pole output, 245

Totem-pole output specifications, 398–401
Tower Hobbies, 697
Toys, as a means of prototyping, 682–684
Transducer, 270
Transfer function
for MMA1250 accelerometer, 277–278
ideal A/D converter, 445
ideal D/A converter, 448
of sensor, 274
Transistor-transistor-logic (TTL), 426
Transition region, in filters, 340
Translator program, 49
Trans-resistive amplifier circuit, 243
Trapezoidal rule, 657
Traveling wave ultrasonic motors, 642
TRIS register, 132
Tri-state outputs, 403–405
Troubleshooting, 718
application of, 721–728
benefits of planning for, 721
identifying the problem and its source, 719–721
recommendations on preparations and procedures, 728–729
Truth table, 417–418
TTL/CMOS standard, 118

U

ULN2003 specifications, 407, 408
Ultrafast recovery time for diodes, 556
Ultrasonic
displacement sensors, 322–323
motors, 642
proximity sensors, 315
Unit testing, of software, 84
Unity gain buffer, 238–239
Universal asynchronous receiver transmitter (UART), 113–115
Universal Serial Bus (USB) protocol, 9, 16
Universal synchronous/asynchronous receiver transmitter (USART), 113
USB (Universal Serial Bus), 9, 16

V

V-I curves
for bipolar junction transistors (BJTs): 207–209
for diodes, 200–201
for light emitting diodes (LEDs), 204
for MOSFETs, 214
for photo-diodes, 205
for photo-transistors, 212, 297
for Zener diodes, 202
Virtual ground, 236–237, 337
Volatile memory, 10
Volt (V), 146
Voltage, 145–147
Voltage controlled voltage source filter (VCVS), 355
Voltage divider, 153
Voltage divider equation, 154
Voltage level translators, 391
Voltage regulator
adjustable, 475
dropout, 472
efficiency of, 472–473
fixed, 475
linear, 473–475
line regulation, 470
load regulation, 470–471
low dropout linear, 475–476
low dropout regulators, 472
output noise, 471
power dissipation, 476–479
power supply rejection ratio (PSRR), 471–472
switching, 485–487
terms and definitions, 470–473
Voltage sources, 147
Von Neumann architecture, 17–19

W

Waterfall model of software generation, 78
WebBench Tools, 486
Wheatstone bridge, 288–290
WiFi (IEEE 802.11a/b/g/n), 127
Windings, in a brushed permanent magnet DC motor, 533
Wiper, of a potentiometer, 183
Wired NOR, 402
Wired OR, 402
Wire wrap, 690–691
Working voltage (WV) of a capacitor, 185

X

XBee OEM RF Module, 389, 390
XOR operator, 40, 42
X-Windows, 58

Z

Zener diode reverse-breakdown voltage, 560
Zero-order hold (ZOH), 441
Zero-overhead loops, in DSPs, 363
Zeros, of a filter, 340
Zero-software-overhead PWM, 140
Ziegler-Nichols approach to tuning the PID gains, 664–665
Zigbee (IEEE 802.15.4), 127, 389–390